An Annotated Catalogue of the
Edward C. Atwater Collection of
American Popular Medicine and Health Reform

An Annotated Catalogue of the Edward C. Atwater Collection of American Popular Medicine and Health Reform

Volume I: A–L

Compiled and Annotated by
Christopher Hoolihan

Ⓡ University of Rochester Press

First published 2001
by the University of Rochester Press

The University of Rochester Press

668 Mt. Hope Avenue, Rochester, NY 14620, USA
and   Boydell & Brewer, Ltd.
P.O. Box 9, Woodbridge, Suffolk 1P12 3DF, UK

ISBN 1–58046–098–4

Library of Congress Cataloging-in-Publication Data

Hoolihan, Christopher.
    An Annotated Catalogue of the Edward C. Atwater
    Collection of American Popular Medicine and
    Health Reform
        p.   cm.
    Includes index.
    Contents: v. 1. A-L.
    ISBN 1-58046-098-4 (alk. paper)
        1. Social medicine—United States—History—
    Catalogs.   2. Medical care—Unitd States—
    History—Catalogs.   I. Atwater, Edward C.
    II. Title.

    RA395.A3 H64 2001
    362.1'09793—dc21   2001048044

British Library Cataloguing-in-Publication Data
A catalogue record of this book is available from the
British Library.

Set in New Baskerville Roman type by:
Straight Creek Bookmakers of Boulder, Colorado
Printed in the United States of America
This publication is printed on acid-free paper

CONTENTS

Illustrations

List of Tables

Foreword

When I first arrived at the Edward G. Miner Library in 1980 I was struck by the attention paid to the historical collections in medicine that resided in the Library. Being a lover of technology, and definitely not a "history buff," I struggled a bit to understand why the printed historical record of healthcare and medicine was afforded such considerable resources. Even George Corner, the University of Rochester School of Medicine's first Library Committee chairman, expressed his doubts in a 1926 report when he wrote, "Opinions differ as to the extent to which it is advisable to put money into books which cannot be considered essential to the modern worker." Corner and the Library Committee concluded, however, that it was their "duty to plan at least a modest collection of historical books sufficient to illustrate teaching in the history of medicine."

Rather than a duty, I now consider it a great privilege to support the acquisition, organization, and cataloging of the Library's collections of historical materials and rare books—collections that rank among the best in the nation. With the help of my friend and mentor Lucretia McClure (Director, Edward G. Miner Library, 1979-1993), I began to understand the importance of the past in approaching the present, and the future. Christopher Hoolihan, the Library's History of Medicine Librarian since 1985, through his spoken and written words and his extraordinary dedication, has created for me a clear image of medical history as a living and breathing force that continues to shape the world of healthcare. The treasures of the Library's rare book and manuscript holdings—the scope, quality, and uniqueness of these collections—have guided thousands of students, researchers, teachers, and clinicians from the University and across the eastern United States and Canada, to an understanding of medical history.

"Because it deals with the vital interests of both individuals and societies—with life and death, and with so much that matters in between—medicine has long had an unusually complex and intimate relationship to social and cultural developments at large. Hence the term 'medical history'... is broad indeed, involving the history of disease and of all attempts—magical or scientific, biologic or social—to promote health and to prevent, cure, or ameliorate illness." (Richard H. Shyrock, *Medicine in America: Historical Essays.* Johns Hopkins Press, Baltimore, 1966. p. xiii) The The Edward C. Atwater Collection of American Popular Medicine and Health Reform so richly described in this volume is a premier example of printed material that illustrates this strong connection between medical and social history. The collection presents an intriguing, colorful, sometimes amusing, but always insightful picture of this era in American history.

We are fortunate indeed that Edward C. Atwater, M.D., decided in 1994 to transfer his collection to the Edward G. Miner Library. The Library and Medical Center are exceedingly grateful that he continues to support the collection both with his intellectual contributions and his bibliographic gifts. He is a friend to the Library and to future generations of medical historians and book collectors. It is important to note that it is the mutual respect and admiration between Edward Atwater and Christopher Hoolihan that has made this partnership such a positive and productive one, and I thank both of them for their efforts, their enthusiasm, and their good will.

Julia Sollenberger
Director, Edward G. Miner Library

Introduction to the Collection

A friend once asked me whether it was worth collecting books known as "popular medicine." The negative implication of his question made me think he probably envisioned an accumulation of advertisements for proprietary remedies and devices, books promoting cure-all regimens, and other types of nostrums and quackery. Such materials are certainly part of the popular medical literature. They are fun to collect because they are often imaginative, colorful, and outrageous. They are also part of the early history of modern advertising, and reflect the remarkable combination of purveyor cupidity and consumer credulity in matters related to health, as strong today as ever. Of such stuff there is a generous sampling in this collection. But it is a small part of the whole.

There is a much larger popular literature, primarily educational in nature, intended to improve the health and well-being of each individual. Though written mostly by those with professional competence, it is directed to a non-professional audience. It discusses human anatomy, physiology, hygiene, sanitation, temperance, diet; how to maintain or regain health, or how to cope with illness, especially when no professional help could be had; reproduction, how to do it, how to limit it; how to deliver and care for a baby; the health and special needs of women; the closeted world of venereal disease; physical fitness, exercise, recreation and travel for health; what to do till the doctor comes; what to do in times of epidemics; home nursing and cooking for invalids; and, of course, how to treat all manner of sickness and injury. These materials make up the major portion of this collection.

The literature of popular medicine does not focus on milestones of scientific discovery such as vaccination, anesthesia, aseptic surgery, germ theory, x-ray, or antibiotics. It is concerned more with what actually happened to the patient. What did the patient understand about himself; how did it affect his life? Most health-related activities and medical care are undertaken without professional intervention. The popular literature gives some idea of what the self-treating layman may have known, when he knew it, and how he found out about it. It usually was not from his doctor. The popular literature also records the major role played by non-mainstream doctors and laymen in reforming medical practice. In a time when the medical profession had satisfactory answers to few of the problems with which it had to deal, it responded with ever more aggressive therapeutic methods. It was generally the popular writers who were strong voices for moderation, who emphasized the importance of preventive medicine and of healthful regimen, and the need for public sex education. In these areas, largely ignored by the regular profession, the popular literature made important contributions to the health of citizens and to the history of medicine.

The lineage of books offering general medical advice to the American layman started in England. In the 18th and early 19th Centuries such books were imported or were American editions, legal or pirated, of English authors such as Nicholas Culpeper, the Pseudo-Aristotle, John Wesley and William Buchan. Culpeper simplified the English pharmacopoeia for the benefit of everyman. The so-called Pseudo-Aristotle was the dominant, if erroneous, sex manual until the 1830s. Wesley's book, *Primitive physic,* provided health advice to his followers who were among the less prosperous members of English society. It appeared in at least sixteen American editions between 1764 and 1859. William Buchan's *Domestic medicine,* most popular of all with sixty-one American editions between 1771 and 1863, was the standard domestic medical text until John Gunn's book was published in 1830.

Among early native American authors such as Horatio Jameson Gates, James Thacher, James Ewell, Thomas Ewell, and John C. Gunn, Gunn was the most enduring. Between

1830 and 1929 Gunn's book, with title varying, appeared in 214 editions, a record unsurpassed by any subsequent publication. Two
books popular later in the 19th and early 20th
centuries were Edward Bliss Foote's *Medical
common sense* (later *Plain home talk*), that appeared in more than forty editions between
1858 and 1909, and Ray Vaughn Pierce's
People's common sense medical adviser, with 100
editions between 1875 and 1935. Foote's
book, as will be mentioned shortly, was important for more than its durability.

It is hard to overestimate the influence of
popular medical literature as an instrument
of reform. Before the Civil War, when effective therapeutic intervention was limited, and
evaluation of therapeutic efficacy dependent
largely on testimonial, and concepts of public health and of hygiene were rudimentary
at best, the medical profession, honoring the
dictum of Philadelphia's Benjamin Rush that
Nature must be driven from the sick-room,
became increasingly aggressive in treatment,
egged on no doubt by the expectations of patients that something be done. In response
to this therapeutic extremism, a host of would-
be reformers appeared. Some within the established profession tried but were not very
successful in the attempt. In Boston, generally more conservative than Philadelphia in
therapeutic methods, Jacob Bigelow published an essay *On self-limited diseases* (1835)
in which he said that some diseases run their
course independent of any therapy. But neither this, nor the attempt to introduce nu-
merical analysis of therapeutic outcome—a
method introduced from Paris, which compared groups of patients similar in all but the
therapy they received—met with much enthusiasm among most mainstream members of
the profession. This was perhaps understand-
able since the profession then had little to offer but advice and strong medicine. Most
Americans seemed to prefer the latter, and
so the responsibility for reform passed to those
outside the Medical Establishment.

One of the earliest of these reformers was
a New Hampshire farmer named Samuel
Thomson. His family had not fared well at the
hands of several successive physicians. Thomson developed his own therapies, using only
medicines derived from indigenous plants,
and proscribed use of all "mineral" medicines

such as calomel, as well as bleeding. The patient still had sweats and puking, but no long-
term mercury poisoning or circulatory depletion. Furthermore, the plants employed grew
abundantly in the countryside, free to all, and
one did not need the services of a physician.
This was the origin of what came to be known
as the botanic medicine movement, which
competed quite effectively with the "regulars"
for over three decades prior to the Civil War.
Thomson's *Narrative* of his own life and *New
guide to health* (1822) went through many editions and are classics in the popular literature.
Even scarcer are the books of some of
Thomson's many imitators, often printed in
rural communities, places like Bennington or
Cornwall, Vt., Milford or Dover, N.H.,
Taunton, Ma., Castile, Cooperstown, Warsaw,
or Westfield, N.Y., Bethania or New Berlin, Pa.,
Bridgeton, N.J., St. Clairsville, Ohio, or
Forsyth, Ga., to list but a few.

The botanic movement became so strong
that it finally brought the regular profession
to its knees. The latter had sought and won
increasingly stringent laws to protect it from
the competition of the Thomsonians, whom
it derogated with the terms "steamers" and
"pukers." Each new restriction, however, was
used by the Thomsonians to strengthen the
claim that they were being oppressed by rules
that further enfranchised a medical aristocracy. In the 1830s and early 1840s every state
abolished its medical licensure laws, allowing
thereafter anyone who wished to do so to practice medicine and be liable only for negli-
gence or malpractice.

Without the strawman of oppression the
Thomsonian movement rapidly faded. Its
banner was taken up by the Eclectics, also
botanics to a degree, but, unlike the botanics,
less anti-professional and not anti-intellectual.
They and their allies would become the major medical reformers of the mid-19th Century. Contemporaries of the Eclectic movement were the more famous and more prosperous adherents of Samuel Hahnemann's
Homoeopathy. The greater success of this
group can probably be attributed to the fact
that it was based on a clear and simple theory:
like cures like. It used tiny doses of medicine
that, even if they did little else, did no harm,
and its physicians were employed by many of
the prominent citizens in a community. They,

not the Eclectics, became the chief competitors of the regular profession for patients. The Eclectics, on the other hand, as their name implies, had neither such a clear focus, nor the popularity. Their contribution was an important one, but of a different kind. They became the framework to which many important health reform activities attached themselves.

Still another group that flourished at mid-century, in many ways similar to the Eclectics, were the hydropathists or Water-Cure doctors. Like the homoeopaths, they had a theory and a therapeutic focus: water. They emphasized a regimen that included vegetarianism, avoidance of alcohol and coffee, regular exercise, sensible clothing, and lots of water, internally and externally. Also, like the homoeopaths, at least part of their success came from the fact they did less harm. In this corps were prolific (and often prolix) writers like Joel Shew, Russell Trall, Orson Squire Fowler and other members of his family, James Caleb Jackson, Martin Holbrook, and Simon Mohler Landis. Another aspect of the water-cure was the popularity of mineral springs and "taking the waters." Some of the vast promotional literature for these institutions will be found in the collection.

The idea of preventive medicine, of health through daily regimen, and with them of learning physiology was not new in the 1830s and 1840s, but it was becoming popular—not just among the Eclectics and hydropathists but among prominent lay writers as well. Sylvester Graham, William Alcott, Catharine Beecher, and Edward Hitchcock all promoted the popular study of anatomy and physiology. In a time when dietary excess, especially of fat and alcohol, was the rule and the ideas of proper diet and regular exercise as preventive medicine were concepts virtually unknown, and of little interest to the mainstream of the medical profession, it was persons like these who stressed the need for reform in daily life.

In the 1880s a major change took place that began to bring the regular profession back into favor. The recognition of several different bacteria and the fact that each was associated with a different disease, gave medicine something to work with. The idea of the germ as a disease-causing agent gradually entered the public consciousness. It is surprising how limited the number of popular books written

on this subject was, considering its importance. A partial explanation may be that much of the education was done by periodicals and newspapers which, by 1895, could be easily distributed by rural free delivery. Furthermore, such materials are ephemeral. There is no question, though, that the idea of "germs" caught the public's fancy, spawning a whole new industry of germicides for every imaginable purpose. Sanitation, hygiene, and personal cleanliness, now of commercial interest, were preached with ever greater urgency.

Another aspect of the preventive medicine effort was the introduction into the public school curriculum of the study of anatomy, physiology and hygiene. The earliest American book of this kind was probably William Mavor's *Catechism of health* (1819), but it was not until the 1830s and later that such books began to come into more general use, starting with George Hayward's *Outlines of human physiology* (1834), William Alcott's *The house I live in* (1834), J.V.C. Smith's *Classbook of anatomy* (1837), and Jane Taylor's *Physiology for children* (1839). It would not be long before such books were focused even more precisely, with separate editions designed for primary, intermediate, and advanced grades. These texts were notable for their consistent advocacy of abstinence from alcohol and for their impressive silence on matters related to reproduction. Not until 1891 did such a book, Eli F. Brown's *Sex and life,* address this subject. It remained for some years an exception and was probably not much used in schools.

Reproduction was, of course, one of the most interesting subjects addressed in the popular literature of medicine. The popular literature on the subject of sex education was especially important because neither the schools nor the medical profession provided much information. Sex was not a publicly recognized subject. The medical profession did not consider it their place to instruct their patients on the matter. They held the view that it was gentle young husbands (themselves quite ignorant unless they had consorted with prostitutes) who would instruct their presumably virgin wives. Nor was it considered a woman's right to understand, much less control, her reproductive capacities.

This viewpoint was clearly demonstrated at the 1899 meeting of the American Medical

Association in Chicago. A prominent Chicago gynecologist, Denslow Lewis, presented a paper on the physiology of sexual intercourse to fellow physicians in the hope that they might become more active and effective teachers of their patients. In the firestorm that ensued, Lewis was severely criticized for discussing such a subject by almost every person that chose to comment. Howard A. Kelly, professor of gynecology at Johns Hopkins Medical School and a leader in his specialty, told the assemblage that he wished such a paper had not been given. The *Journal of the American Medical Association* refused to publish the paper. It did not appear in that journal until eighty-four years later, accompanied by historical commentary. [JAMA July 8, 1983, 250:228.] More than fifty years earlier, in 1845, the popular writer, Dr. Frederick Hollick had written that this very subject "is one of universal importance; it concerns everybody, and all are interested in it, whether they will or not. It occupies the mind of nearly all persons, more perhaps than any other subject, no matter what may be the extent, or deficiency, of their information. The thoughts of the ignorant man are as full of it, as those of his more enlightened neighbor; probably much more so. Keeping persons in ignorance upon this subject therefore, does not prevent them thinking about it . . . We have therefore merely a choice, between giving correct information, and leaving the mind to be filled with error and vain surmise, for with one or the other it will most assuredly be occupied."

In view of the position taken by most of the medical profession, it is not surprising, that sexual education was left to those outside the mainstream medical and educational establishments. It was the reformers and liberals who did the job. Unfortunately, because society closeted the subject, there were inevitably entrepreneurs who distorted and commercialized the forbidden topic and gave the whole genre of sex education materials a bad name. Much of the material, however, was an honest attempt to improve the lot of everyman. Some of the information was incorrect, as we now know, but in its time was consonant with contemporary understanding of physiology. Many publications were blatant promotions, often using scare tactics, for proprietary products and services. Some of it was

the best information available at the time. The works of Frederick Hollick and Edward Bliss Foote stand pre-eminent in this respect.

Most influential in the matter of contraception were the social reformers Robert Dale Owen and John Humphrey Noyes, along with several physicians of the Eclectic school: Frederick Hollick, Simon Landis, Thomas Low Nichols, Harmon Knox Root and Edward Bliss Foote. The only important writer on contraception at mid-century who was a regular physician, Charles Knowlton, was shunned by his professional colleagues after publication of his book *Fruits of philosophy* in 1831. Though others did, Knowlton did not again publish his book. Of one other physician-writer on contraception, James Ashton, little is known. The titles of the two earliest American books on contraception—Owen's *Moral physiology* and Knowlton's *Fruits of philosophy*—signify the importance the authors placed on justification for preventing conception. Only the last fifth of these books deals with method. The idea that sexual partners were entitled to prevent conception was new one to most Americans. Knowlton went to jail for his book, as did many after him who wrote on this subject. Other writers obscured themselves by using pseudonyms: Eugene Becklard, A.M. Mauriceau and P.C. Dunne. Some editions of the Pseudo-Aristotle, though published in the United States, have London imprints, presumably to protect the publishers.

In the 1870s, with the urging of Anthony Comstock, then agent of the New York Society for the Suppression of Vice and, later, special agent for the United States Post Office, Congress hastily passed a law by which it became illegal to send through the mails materials that some persons considered obscene. Mr. Comstock, who was effectively responsible for enforcing the provisions of the law, viewed as obscene any book or pamphlet that discussed human sexual activity, any image that portrayed even the partially naked body, or anything that dealt with the subject of contraception. Dr. Foote and, later, Margaret Sanger were among those prosecuted under this law. New editions of books that had formerly included material on contraception (those of Foote and Cowan, for example) appeared with several blank pages where the offending information had previously been.

It would not be until after the First World War that the subject of sex education began to come out of the closet and was considered an appropriate topic for serious discussion by the medical profession and the public itself.

As late as 1914, when Margaret Sanger first published her explicit pamphlet on *Family limitation*, addressed primarily to working-class women, she, like Knowlton, Hollick, Landis and Foote before her, was indicted and tried on charges of obscenity. This landmark pamphlet continued the association of popular sex education with social reform, even revolt, partly because of its language and also because it was at first sponsored by the I.W.W., a labor organization considered by many to be at least socialistic, if not anarchistic.

The literature of popular medicine was generally looked down on by the medical profession: it was often superficial; it not infrequently gave incorrect information; and some of it was commercial promotion. It also represented another form of competition for physicians, who competed for patients and had to collect their own fees. There was, indeed, a fine line between what was educational or would pass as public service, and what was self-promotion or commercial self-interest. However, the professional literature of the mainstream is not wholly devoid of these latter attributes. Perhaps the fairest thing to say is that popular medicine includes a broader spectrum of offerings, ranging from serious scholarly efforts at educating the public to the most flagrantly fraudulent commercial scams. In the case of sex education, contraception and the prevention of venereal disease, the unwillingness of the regular profession to provide instruction or devices inevitably abandoned the field to others. Being subjects not discussed in public, they lent themselves and those concerned to exploitation. The operators of public "anatomy museums" found in many large cities, not to mention quite a few authors, took advantage of the secrecy surrounding venereal disease and of the American obsession with masturbation, preying on the gullible with extended, useless (and costly) treatments. These con men and those who promoted fanciful or complicated dietary regimens, breathing exercises, cathartic orgies, high colonic irrigations, and other forms of orificial therapy as

being the royal road to health gave the rest of the popularists a bad name. A generous sampling of this proprietary literature will be found in the collection. Appended to the second volume (M-Z) will be an extensive list of patent medicine almanacs.

Americans love to take medicine. Most of us are probably overtreated, by ourselves or by our physicians. This is not a new situation. Reform movements of the 19th century were successful partly because they were able to substitute harmless or even beneficial regimens that allowed this craving to be satisfied more safely. Improvements in the health of Americans in the 19th (as in the 20th) century came not alone from the efforts of the medical profession. They resulted also from educating individual citizens about their anatomy, physiology, hygiene, sexual function, and illnesses. Today we are much better informed about these matters, or at least the information is at hand if we want it. What is lacking today is another kind of learning that we might be better able to evaluate the flood of information to which we are exposed, to distinguish what are conclusions from valid scientific studies and what are testimonials. There are those who still believe, as in the example of jurors who heard a recent case involving illness attributed to a breast implant, that testimonials were sufficient proof of cause and effect relationship and that "the science really didn't matter."

The books, pamphlets and broadsides listed in this bibliography were collected over a period of thirty-five years. Book dealers play an important role in assembling such a collection. They buy libraries, estates, collections, even single volumes. They search attics, cellars, and barns. They save from destruction things no longer wanted by those who have them and find for them those who do want them. They make them useful again. In more recent years, the growth of interest in paper ephemera, promoted by the Ephemera Society, has, by making it desired, attracted dealers to this genre as well, and thus saved a lot of historically useful materials that might otherwise have perished. Good book dealers are crucial agents in the preservation of printed materials.

My early contacts with book dealers go back to college days when Frank Opperman,

Worden Gilboy, and John Weiss each had shops in Rochester in which I spent many pleasant hours browsing. After completing medical training, my collecting interests became somewhat more focused on the history of medicine. In this course I was considerably influenced by two persons. Roger Butterfield, an historian who turned book dealer after his retirement, always had a supply of wonderful items, especially ephemeral ones, and sold them at reasonable prices. Robert Volz, then rare book librarian at the University of Rochester's Rush Rhees Library, persuaded me to write a piece on the early medical authors of upstate New York, and also broadened my vision to include more costly volumes.

Among the many dealers, some of them now gone, who helped especially over the years in building the collection, with books found in their open stock, or their catalogs, or through direct offer are Kinsey Baker of The Book Haven in Lancaster, Pa., George Beebe of St. Albans, Vt., James Brunner of Livonia, N.Y., Douglas and Marlene Calhoun of Geneva, N.Y., Robert Campbell of Ex Libris in Montreal, Henry Chavetz and Sidney Solomon of Pageant Book Store in New York City, Ronald Cozzi of Limited Editions in Buffalo, N.Y., John DeMarco of Lyrical Ballad in Saratoga Springs, N.Y., Scott DeWolfe and Frank Wood of Alfred, Me., Webb Dordick of Somerville, Ma., Richard Foster of Rittenhouse Bookstore in Philadelphia, Bruce Fye of Marshfield, Wisc., Ed Glaser of Sausalito, Calif., Joanna and Ralph Grimes of Old Hickory in Brinklow, Md., Malcolm Kottler of Scientia in Arlington, Ma., Peter Luke of New Baltimore, N.Y., Willis Monie of Cooperstown, N.Y., Sam Morrill of Edward Morrill & Son in Boston, Ian and Helen Morrison of Old Number Six Book Depot in Henniker, N.H., Cheryl Needle of Chelmsford, Ma., Maurice and Helen Poitras of Towson, Md., Stephen and Carol Resnick of Cazenovia, N.Y., Craig Ross of Medina, N.Y., Dan Siegel of M & S Books in Providence, R.I., John Spencer of Riverow in Owego, N.Y., Arnold Silverman of Goodspeed's in Boston, Douglas and Doris Swarthout of Berry Hill in Deansboro, N.Y., and Hans Tanner of Groveland, N.Y.. These and many others made building the collection both possible and fun. To them, and to the many other dealers whose shops I visited and who occasionally provided a book or two I am grateful. Even those whose catalogs had usually been picked over by the time I received them sometimes provided clues of items to search for in the shops or on the internet. I thank them all. I would also like to thank Mme. Alain Brieux, of Paris, for providing some of the information on Dr. Auzoux.

My interest in history dates from high school days. Later, as an undergraduate at the University of Rochester, I majored in history, with professors Arthur May, Dexter Perkins, Glyndon Van Deusen, and John Christopher as my teachers. A sabbatical year spent as a Josiah Macy, Jr. Fellow in Johns Hopkins' Institute of the History of Medicine exposed me to discussions with fellow students and professors, especially Lloyd G. Stevenson, and considerably broadened my historical viewpoint. It was a stimulating year for which I am grateful.

To my colleague, Christopher Hoolihan, history of medicine librarian at the University of Rochester's Edward G. Miner Library, I am greatly indebted. His informative and lucid notations have turned what otherwise was just a big collection of books that would, no doubt, just have collected dust from disuse, into a useful research tool. More than that, it is fun to read what he has provided. I have known few history of medicine librarians who are as knowledgeable in medical history or bibliography. His ability to focus totally on a task at hand is impressive; it has made it possible for him to carry on other professional duties while doing this work.

Most of all, thanks should go to my wife, Ruth, and in the earlier days when they still lived at home, my children Rebecca and Ned. Sharing your husband or father with such an all-consuming passion as book-collecting is not a minor sacrifice. Numerous are the books that Ruth unearthed in shops we visited together. Many, also, are the hours that Ruth sat patiently in the car, reading her book, while I rummaged in second-hand bookshops; many are the extra miles—whole trips sometimes—she went with me.

Edward C. Atwater, M.D.
Rochester, New York
February 1, 2001

Introduction to the Catalog

Over dinner one evening a decade ago at one of the annual meetings of the American Association for the History of Medicine, the conversation turned to an increased interest among historians in books and pamphlets on popular medicine. "Who's collecting this stuff?" someone asked. Though we assumed that the National Library of Medicine must certainly be the nation's most complete repository for this genre of literature, no one could name a medical library that included popular medicine among its active collecting interests. Each of us was proud of the particular collection or collections for which our library is known, and about which we might often speak or write. Few of us, however, had ever turned our attention to the well-worn copies of Buchan, Gunn or Foote in our collections, or given more thought to the tattered school physiologies on our shelves than as candidates for eventual weeding.

I was aware that Edward C. Atwater, M.D. collected American popular medicine. For half a dozen years before this dinner I had been employed by Edward to enter his extensive pamphlet collection (19th-century introductories, valedictories, medical school catalogs, etc.) into his personal database. Each Monday evening I visited his home, and after the ritual glass (or two) of wine, passed bookcases stuffed with seemingly indistinguishable editions of Chase, Hollick, Pierce *et al.* before seating myself at his computer and getting on with the evening's work.

Several times Edward casually proposed transferring the collection from his home to the Miner Library. I wasn't certain how serious he was about this proposal, and was always occupied with something seemingly more important. Edward's patience gradually bore fruit. When he again brought the matter up early in 1994, I had by then realized that not only was popular medicine a subject of increasing interest in the scholarly community, but that the collection cumulating in the Atwater home for more than two decades had become an important repository for the study of this particular genre of medical literature. As I began to examine the collection more closely, it was obvious that the titles it included would not only serve the research interests of medical historians, but also scholars at work in American social history, religious history, gender studies and the history of mass-circulation publishing and advertising.

In the autumn of 1994, the first boxes of books were transferred to the Rare Books & Manuscripts Section of the Edward G. Miner Library. More than 200 related titles were subsequently culled from the Library's 18th and 19th century holdings and added to the Edward C. Atwater Collection of American Popular Medicine and Health Reform. At its inception, the Atwater Collection numbered nearly 2,000 titles. The Miner Library has continued to purchase for the collection, adding another 200 titles between January 1995 and the end of 2000. Our collecting efforts pale in comparison to those of Edward Atwater, however, whose energy and focus have resulted in the addition of nearly 1,800 books, pamphlets, periodicals and pieces of ephemera to the collection over the same period.

The Edward G. Miner Library has been fortunate in having a long line of directors who have supported its historical collections—dating to 1925 when George Washington Corner, M.D. (1889-1981) chaired the Dept. of Anatomy, headed the Library Committee, and over the next fifteen years built the core of our early printed book collection. The Library's two most recent directors have played key roles in the publication of this catalog. Valerie Florance, Ph.D., Director of the Miner Library between 1993 and 1998, embraced the idea of adding the Atwater Collection to the Miner Library's rare book holdings and of publishing a catalog. Julia Sollenberger succeeded Dr. Florance as Library Director in the autumn of 1998, and has since provided unstinting support and encouragement in every aspect of the collection's growth and care, as well as in the publication of this catalog.

No form of acknowledgment to the donor of this wonderful collection would ever be adequate. Acknowledgment must be forthcoming, however, for Edward Atwater's intellectual contribution to this catalog. The biographical essays on James Caleb Jackson, Frederick Hollick, Ralph Glover, S.S. Fitch, Simon M. Landis, the Grindle family, the Jordan-Kahn clans, and others are from Edward's pen. In addition, the many references to the Boston and New York city directories throughout the catalog are the result of Edward's many hours in the libraries of those cities.

Christopher Hoolihan
Head of Rare Books & Manuscripts
Edward G. Miner Library

How to Use This Catalog

Our purpose in publishing this catalog is two-fold: first, to make known to the scholarly world the existence of this important collection; and second, to provide accurate bibliographic descriptions and commentary that will serve the diverse needs of librarians, booksellers, collectors and scholars of many stripes. It is our intention that the catalog be more than the record of one library's holdings, but that it serve as a guide to the literature of American popular medicine and health reform generally.

Although the scope of the Atwater Collection extends to the the Second World War, America's entry into the First World War serves as the chronological limit for inclusion in this catalog. The vast majority of catalog entries have American and Canadian imprints. There are, however, a considerable number of entries for works with British imprints, due to the widespread marketing of such books during the colonial period and early decades of the 19th century, and because of the continued availablity of British imprints in the Dominion of Canada. Non-British foreign imprints have been included when a foreign author was reprinted in the United States, or on those rare occasions when a book by an American author was reprinted in Europe.

The catalog is arranged alphabetically by main entry. This first volume includes entries A-L; and the second volume will contain entries M-Z. In the majority of instances, the main entry takes the form of an author's name, though corporate and title main entries are common. Entries for authors of multiple works are sub-arranged alphabetically by title. Multiple editions of the same title are arranged by date of publication. The two exceptions to this rule are the works of the Pseudo-Aristotle and Edward B. Foote, the complexity of whose publishing history warranted an alternative arrangement explained in the text

Each entry includes a main entry (in bold type); a transcription of the title-page; a detailed physical description; and, in the majority of instances, an annotation. Each entry is numbered. Within the catalog these numbers allow the brief citation of works in the cross-references, in the annotations, and in the subject index. This obviates the need to provide lengthy citations. The entry numbers in the cross-references and in the annotations are each preceded by a number symbol (#).

Subject access to the catalog is provided in subject indices for each volume. The subject index to volume one (A-L) appears at the end of this volume. The choice of subject headings has been limited to providing access to the largest subject concentrations in the collection. Under each subject heading, entries are arranged by date of publication, followed by the author's name and the entry number.

The second volume will contain entries for personal, corporate or title main entries M-Z. This volume will also include a supplement to volume one (i.e., A-L entries added to the collection since the publication of the first volume), a subject index, an index of publishers to both volumes, and an appendix listing almanacs in the Atwater Collection.

C. H.

An Annotated Catalogue of the Edward C. Atwater Collection

A-L

A

A.G. see **GOULD** (#1387).

1. ABBEY, Emery C.
The sexual system and its derangements . . .
Buffalo, N.Y., 1875.
    48 p. : ill. ; 16 cm.

In this promotional publication, Abbey proclaims his life-long devotion to diseases of the generative organs and the invincibility of his remedies for their cure. The first thirty-one pages are devoted to disorders affecting masculine vitality, i.e., masturbation, spermatorrhoea, impotence and venereal diseases. Abbey believes "but one venereal virus exists" with "many complications and anomalies"; and sets his readers' minds at rest by declaring that "three-fourths of the cases of syphilis which occur require no treatment at all" (p.[28]). All these disorders are curable by "vegetable remedies" which can be had of Dr. Abbey for the sum of ten dollars monthly. The remaining sixteen pages discuss menstruation disorders, uterine diseases, sterility, abortion and nymphomania.

Those considering consultation are advised to send Dr. Abbey one dollar with their first letter (no physical examination required). "As it takes time and brings knowledge into question to read letters, consider cases and write replies, no one should expect this service for nothing . . . To think of sitting at the desk all day and working hard gratuitously is preposterous" (p. 47).

A native of Buffalo, Abbey was an 1863 graduate of the Eclectic Medical College of Pennsylvania, according to Polk's 1886 *Medical and surgical directory of the United States.* The AMA's *Directory of deceased American physicians, 1804-1929* lists Abbey as an 1863 graduate of the Philadelphia University of Medicine & Surgery. Both institutions were extinct by 1880. In printed wrapper; wood-engraved portrait of author on title-page.

2. ABBOT, Abiel, 1770-1828.
An address, delivered before the Massachusetts Society for Suppressing Intemperance at their anniversary meeting June 2, 1815, on the objects of their institution . . . Cambridge: Printed by Hilliard and Metcalf, 1815.
    23 p. ; 23 cm.

A 1791 Harvard graduate, Abbot was pastor at the First Church in Beverly, Mass. from 1803 to 1828. Many of his sermons and addresses were published. Abbot died of yellow fever while returning from Cuba, where he had gone to recover impaired health.

Includes the "Third annual report of the Massachusetts Society for Suppressing Intemperance" (p. [17]-23).

3. ABBOTT, Simon B.
The Southern botanic physician: being a treatise on the character, causes, symptoms and treatment of the diseases of men, women and children of all climates, on vegetable or botanic principles, as taught at the reformed medical colleges in the U.S. Containing also many valuable recipes for preparing medicines. The whole preceded by practical rules for the prevention of disease and the preservation of health. Compiled from the best works now published on the reformed practice . . . Charleston: Published for the author, 1844.
    xii (i.e., xiii), [14]-395 p. ; 19 cm.

Botanic medicine flourished in South Carolina as it did in the northern and western states. In his *History of medicine in South Carolina, 1825-1900,* J.I. Waring credits the Thomsonians with the repeal of the state's 1817 licensing act. *The Southern botanic physician* was the only published work of S.B. Abbott, who, according to Waring (p. 115), established a Thomsonian Infirmary at Charleston in 1837.

In his argument for the popularization of medical instruction, Abbott writes, "There is no man so simple that cannot be taught to cultivate grain; and there is no woman who cannot be taught to make it into bread. And shall the means of preserving our health by the culture and preparation of aliment be so intelligible, and yet the means of restoring it when lost, so abstruse, that we must take years to study, to discover and apply them? To suppose this, is to call in question the goodness of the Deity" (p. v). Extolling the naturalness and simplicity of Thomsonian practice, Abbott declares, "A revolution in medical practice is nigh at hand . . . There is no time spent in looking after names, symptoms, theories, causes and indications; the name is out, the cause out, the indication out; and in a few hours . . . the patient is relieved, restored, requires food, recovers strength, sleeps and rises, and returns to the business of life" (p. vii).

The Southern botanic physician is divided into three parts. The first part, "Means of preventing disease, and promoting health and longevity," consists of fifty-five pages of recommendations on food, drink, air, exercise, clothing, cleanliness, sleep, etc. The first sixty-two pages of part two, "Practice of medicine," provide the theoretical background of Thomsonianism (i.e., the unity of disease; the botanic therapeutic armamentarium, etc.), followed by a 232-page catalog of diseases arranged alphabetically (from *abscess* to *yellow fever*) that provides a description of each disease and its treatment. Part three is a "Pharmacy and dispensatory."

4. ABEL, Mary Hinman, b. 1850.
Practical sanitary and economic cooking adapted to persons of moderate and small means . . . The Lomb Prize essay . . . [S.l.] : Published by the American Public Health Association, 1890 . . . Printed by E.R. Andrews, Rochester, N.Y.
    viii, [2], 188, 40 p. ; 19 cm.

"The object of this work," as described in the preface, "is for the information of the housewife, to whose requirements the average cookbook is ill adapted, as well as to bring to her attention healthful and economic methods and receipts." It was awarded the Lomb Prize in 1888, an essay contest sponsored by the American Public Health Association on a given theme. Entrants were instructed to orient the work toward "(1) those of moderate means; (2) those of small means; (3) those who may be called poor" (p. iii). Abel was author of various treatises on domestic economy, including *Success-*

ful family life on the moderate income (1921) and *The science of nutrition* (1891). Several were published as part of the U.S. Dept. of Agriculture's *Farmers' bulletin* series. Mary Hinman Abel was the wife of the pharmacologist John Jacob Abel (1857-1938).

Practical sanitary and economic cooking was issued in several editions. It was thrice issued at Rochester in 1890, including a German-English version. It includes two additional essays by Abel: "Food of children" and "Some points in practical household management" (40 p.). Spine-title: *Prize essay of the American Public Health Association.*

5. ABEL, Mary Hinman, b. 1850.
Practical sanitary and economic cooking adapted to persons of moderate and small means . . . The Lomb Prize essay . . . [S.l.] : Published by the American Public Health Association, 1890 . . . Rochester, N.Y.: Printed by E.R. Andrews, [1900, c1889].
    viii, [2], 188, 40 p. ; 19 cm.

Spine-title: *Prize essay of the American Public Health Association.*

6. ABERNETHY, John, 1764-1831.
Abernethy's family physician; or, ready prescriber in cases of illness and accident, where medical attendance is not desired or cannot be procured. Containing the causes, symptoms and treatment of diseases, the proper remedies in burns, contusions, poisoning, and all varieties of physical casualties. Forming a complete guide to the retention and recovery of health. First American, from the thirtieth London edition. Revised and enlarged by H. Bostwick, M.D. New York: Burgess and Stringer, 1844.
    5, vi-viii, 9-120 p. ; 17 cm.

The present work is the American edition of a 215 page treatise published at London sometime in the 1830s under the title *The new family physician.* A second edition was issued at London in 1841. The "thirtieth London edition" statement on the title-page of the New York edition is entirely fanciful. The book is divided into five sections that provide domestic recipes for the treatment of diseases of the digestive organs, the respiratory system, the blood & circulation, and the nerves. The American editor, Homer Bostwick (#376-379), has appended four pages of cases and prescriptions from his own practice. Issued in printed wrapper; series: *Useful books for the people; no. 3.* A second and somewhat expanded edition of *Abnernethy's family physician* was issued at New York by Stringer &Townsend in 1849.

An eminent London surgeon associated with St. Bartholomew's Hospital for nearly three decades, Abernethy was a gifted teacher and a prolific author. In addition to his well known surgical, anatomical and physiological writings, he authored at least one other treatise for a lay audience. In 1829, a twenty-four page pamphlet entitled *The Abernethian code of health and longevity* was published at London (reprinted New York, 1831). John L. Thornton failed to include this title among Abernethy's published works in his 1953 biography, either because he was unaware of the pamphlet's existence, did not think it worthy of his subject, or doubted its authenticity. Another pamphlet, *The mother's oracle, and nurse's guide,* was published at London with Abernethy's name on the title-page in 1840.

7. ACADÉMIE ROYALE DE MÉDECINE (PARIS).

Rapport sur une pièce d'anatomie clastique du Docteur Auzoux. Commissaires: MM. Adelon, Dubois, Cruveilhier, Breschet, Hte Cloquet, Ribes, et Baffos, rapporteur. Précédé d'une notice sur les travaux anatomiques . . . Se trouve à Paris: Chez M. Auzoux . . . et à Madras, chez le docteur Knox, 1835.

24 p. ; 20 cm.

Title and imprint from wrapper. At head of title: *Académie Royale de Médecine, séance du 10 mai 1831.* The "commissaires" of the Académie royale de médecine charged with evaluating Auzoux's anatomical models reported to their colleagues that "M. Auzoux has succeeded in reproducing everything that pertains to the myology, angiology, neurology and splanchnology; and it is not just the boney structures that are reproduced with such veracity, but if one had not seem them beforehand, they might be taken for the real thing. The thinnest, most delicate parts, as well as the most voluminous, the softest parts, as well as the hard, the super-ficial parts, as well as those most profound, are represented with strict exactitude in their forms, colors, relations and connections" (p. 10-11).

Prior to Auzoux, European artisans had created models of the human anatomy for instructional purposes, usually in wax, but also made of wood or cardboard. None of these artisans created a commercially viable product, however, nor was this necessarily their intention. By 1822, however, Louis Thomas Jerome Auzoux (1797-1878) had perfected a method of creating *papier-maché* models at his workshop in Saint Aubin d'Ecrosville in Normandy. "He succeeded . . . in compounding a paper paste that was flexible, and when hardened, was solid, supple, light, and unbreakable. The Auzoux models were made using either a plaster or lead mold. The model organs produced from these molds started as wet sheets of paper, wheat starch paste, and glue paper laid down in alternating layers in the plaster molds, or as a wet pulp, reinforced with cork, in the lead molds" (Mark Dreyfuss, "The anatomical models of Dr. Auzoux," *Medical heritage,* 1986, 2:60). Vessels made of fine wire wrapped in hemp were attached to the parts, and the models were painted and varnished. By 1858 Auzoux marketed fifty-three anatomical models, and by the First World War 326 models, of which 100 pertained to human anatomy, and the rest to zoology or botany.

In his introductory essay to this report of the Académie, Auzoux outlines nine advantages that his "préparations d'anatomie clastique" bring to the study of anatomy, including the following: they "render the study of anatomy practicable for all classes of society," and they "make possible the realization of a desire long expressed by those occupied with the education of youth, i.e., to see the study of anatomy made part of public instruction" (p. 3). He boasts that his anatomical models are used not only for instruction in hospitals, medical schools and other public institutions, but that "a large number of physicians have acquired them for their own personal benefit, and have found in these preparations a stroke of good fortune in that they make anatomical instruction possible in countries where previously this kind of teaching was deemed impracticable" (p. 5).

As public lectures on human anatomy, physiology, hygiene and sex education proliferated in the United States post-1830, the demand for visual aids to instruction such as Auzoux's models increased. Auzoux himself notes that even before 1835 his models were available in the United States (p. 5). They were an integral part of the lectures of such figures as Frederick Hollick (#1697), who possessed a full-size manikin of the human body from Auzoux's workshop, as well as models of the male and female reproductive organs.

American firms such as Codman & Shurtleff (#696) marketed Auzoux *papier-maché* models from the 1830s. The sixth edition of Chas. Truax, Greene & Co.'s *Price list of physicians' supplies* (Chicago, c1893) devotes seven pages to the Auzoux anatomical models the firm sold, ranging from a "complete model of a male human body, five feet ten inches high; composed of ninety-two parts," to a set of "eight uteri, containing the product of conception at the first, second, third, fourth, eighth, and ninth month," etc. Customers were advised that the "popularity of these models, combined with the delicacy and care necessary in their construction, renders the demand for them more than equal to the supply, and there is often considerable delay in getting an order filled . . . Under ordinary circumstances, the time necessary to fill an order for manikins is from four months to a year" (p. 1035).

8. AN ACCOUNT OF THE ORIGIN . . .
An Account of the origin, symptoms, and cure of the influenza or epidemic catarrh; with some hints respecting common colds and incipient pulmonary consumption. Philadelphia: Henry H. Porter, publisher, 1832.
    4, 80 p. ; 23 cm.

This brief and anonymously written treatise on influenza was apparently published in response to that winter's epidemics in New York, Boston and Philadelphia (p. 30-2). It is just one of the hygienic publications issued by Henry H. Porter (ﬁ. 44-1146, 2053, 2868) in a thirty-month publishing career. A thorough review of Porter's brief but intense career as a publisher of popular literature is provided in T. Horrock's "Promoting good health in the age of reform: the medical publications of Henry H. Porter of Philadelphia, 1829-32," *Canadian bull. of med. hist.,* 1995, 12:259-87.

In a lengthy and inconclusive discussion of influenza's nature and origins, the author determines that it is not contagious, but provides no understanding of its etiology beyond the remark, " . . . its simultaneous occurence in places widely separated from each other; the number of persons attacked by it at one and the same time; its quick extension from one country to another . . . prove . . . very clearly, that it depends for its origin in the first instance, as well as for its subsequent propagation, upon some cause connected with the atmosphere" (p. 22). The greater part of the treatise is devoted to preventive measures and therapeutics. The reader is cautioned to avoid the damp and cold, "high living," and rapid changes in body temperature. Therapeutics is divided into those remedies that can be self-administered, such as warm foot baths, mild laxatives, "mucilaginous fluids" (e.g., barley water), and inhalations; and those treatments that should be administered by a physician, such as bleeding and the prescription of emetics or opiates. Cover-title: *Origin, symptoms and cure of the influenza, or epidemic catarrh.*

ADAM, Graeme Mercer see **SANDOW** (#3081).

ADAMS, Daniel see *MEDICAL AND AGRICULTURAL REGISTER* (#2414).

9. ADAMS, Myron Howell, b. 1846.
A practical guide to homeopathic treatment. Designed and arranged for the use of families, prescribers of limited experience and students of homeopathy . . . Philadelphia: Boericke & Tafel, 1913.
    xvi, 455 p. ; 21 cm.

An 1870 graduate of the Hahnemann Medical College of Philadelphia, Adams was elected to the Homoeopathic Medical Society of the State of New York in 1892. He maintained an office on Rochester's Park Avenue, just around the corner from the Rochester Homeopathic Hospital where he was a consulting physician.

The *Practical guide* is divided into three parts. The first part is a brief discussion of the principles of homeopathy for a general audience. The second part is devoted to the definition, symptoms and treatment of diseases grouped under regions of the body; and the third to a homeopathic materia medica. The Atwater copy is inscribed by the author to the Rochester Homeopathic Training School, the nursing school of the Rochester Homeopathic Hospital.

10. ADAMS, Samuel Hopkins, 1871-1958.
The great American fraud . . . A series of articles on the patent medicine evil, reprinted from Collier's Weekly . . . Also the patent medicine conspiracy against the freedom of the press. [New York]: Copyright 1905 by P.F. Collier & Son; [Chicago: Press of the American Medical Association, 1906].
    95 p. : ill., facsims. ; 21 cm.

Title from printed wrapper. Chicago imprint from colophon on p. 95. Author and journalist, Adams acquired national renown with two series of articles for *Collier's Weekly* on patent medicines and quackery that ran from 7 October 1905 through 5 January 1907. The series had an immense impact: "together with such men as Harvey Wiley and Upton Sinclair he was credited with having inspired the passage by Congress of the Pure Food and Drug Act" (*Dict. Amer. biog.*, suppl. 6:6). The A.M.A. sponsored the publication of the *Collier's* articles as Adam's first book: *The Great American fraud* (1906). Adams went on to write some fifty books and numerous short stories and articles. His fictional work enjoyed great commercial success. Seventeen of his novels and short stories were made into motion pictures, including *It happened one night* starring Claudette Colbert and Clark Gable.

11. ADAMS, Samuel Hopkins, 1871-1958.
The great American fraud . . . Articles on the nostrum evil and quacks, reprinted from Collier's Weekly . . . Fourth edition. [New York]: P.F. Collier & Son, copyright 1905, 1906, 1907.
    168 p., [1] folding plate : ill., facsims. ; 21 cm.

12. ADAMS, Samuel Hopkins, 1871-1958.
The health master . . . Boston and New York: Houghton Mifflin Company, 1913.
    vii, [3], 338, [4] p. ; 19 cm.

The health master is a didactic novel. Through the dialogue of its principal character (Dr. Strong) with his employers (the family of Thomas Clyde), Adams promotes progressive notions of "the relation of the physician and of the citizen to the social and ethical phases of public health." The book is dedicated to George W. Goler (#3012), described by Adams as "a type of the courageous, unselfish, and far-sighted health official, whom the enlightened and progressive city of Rochester, N.Y., hires to keep it well" (p. [v]).

ADAMS, William see **BIRCH DALE MINERAL SPRING WATERS** (#326).

13. ADEE, Daniel.
IMPORTANT TO CONSUMPTIVES! DO NOT DESPAIR! 3,000 CASES CURED. A gentleman having the good fortune to find a remedy for the consumption, whereby he saved the life of his son in the last stages of that dread disease, after having been given up to die by the most eminent physicians, will gladly make known the same to all who are similarly afflicted, and will send them the recipe, which proves a sure cure for coughs, colds, asthma, bronchtis, CONSUMPTION, and all kindred affections of the throat and lungs—free of charge . . . DANIEL ADEE, 176 Fulton Street, New York, N.Y. . . . [187-?].
    Broadside ; 27.5 × 21.5 cm.

14. ADEE, Daniel.
IMPORTANT TO CONSUMPTIVES! DO NOT DESPAIR! 3,000 CASES CURED. A gentleman having the good fortune to find a remedy for the consumption, whereby he saved the life of his son in the last stages of that dread disease, after having been given up to die by the most eminent physicians, will gladly make known the same to all who are similarly afflicted, and will send them the recipe, which proves a sure cure for coughs, colds, asthma, bronchtis, CONSUMPTION, and all kindred affections of the throat and lungs—free of charge . . . DANIEL ADEE, P.O. Box 3531, New York, N.Y. . . . [187-?].
    Broadside ; 27 × 20.5 cm.

ADKIN, Thomas F. see **NEW YORK INSTITUTE OF PHYSICIANS AND SURGEONS** (#2597).

AIKIN, John see **ARMSTRONG** (#152-156) and **GREEN** (#1407).

15. ALBERTUS MAGNUS.
Albertus Magnus: approved and verified, both sympathetic and natural Egyptian secrets, for man and beast; containing to secure men and beasts against evil spirits; to become strong; to stop the blood; when suffering from burns; for the wild fire; for the sweeny disease; for cramps; for worms; for all kinds of fevers; for colic; to heal ruptures of young and aged people; for epilepsy; for scabs; for putrid mouths; for sprains; for sore eyes; for erysipelas; for pestilence; when a child is liver grown; for consumptive lungs; for gravel; for dysentery; for cancer; how to detect a thief; for the gout; for arthritis; for sore breasts; how to recover stolen property; to make a thief own up; for hysterics; to prevent danger of fire from your house; to secure a house so that no fire will ever go out therein; how to quench fire; for toothache; for the itch; for bad hearing; to destroy bed bugs; to make an incombustible oil; to drive away spiders and house flies; to destroy rats and mice; a curious performance to improve common wine, and make the same good in a quick way; how to make wine good and wholesome; to make wine clear in a quick manner; to disern [sic] all diseases by examining the urine; for hydrophobia; and many other approved wonderful performances, hitherto unknown, and now printed for the benefit of mankind, for the first time . . . Translated from the German. Harrisburg, Pa., 1875.
   3 v. in 1 (80; 78; 62 p.) ; 19 cm.

A collection of incantations and prayers to be recited for the removal of illnesses, curses or dangers. Also included are recipes for folk remedies that heal various diseases.This genre of literature was published in Germany as early as the 15th century, and was brought to North America by German immigrants. The so-called "pow-wow book" was common among the de-scendants of the Pennsylvania Germans into the 20th century. ("The word 'pow-wow' seems to have originated with the New England first settlers who adopted it from the Indian title describing the treatment of the sick, etc., by their medicine men," A.M. Aurand, *The 'pow-wow' book*, Harrisburg, Pa.: Aurand Press, 1929, p. 21).

The formulas contained in "pow-wow books" are commonly attributed to famous figures whose reputations as magicians in early modern Europe equalled or exceeded their actual accomplishments. Hence the contents of these books are frequently attributed to Moses and the cabala, to Aristotle, to the medieval scholar Albert Magnus, or to other figures historical and legendary.

16. ALBERTUS MAGNUS.
Albertus Magnus. Being the approved, verified, sympathetic and natural Egyptian secrets white and black art for man and beast. The book of nature and the hidden secrets and mysteries of life unveiled; being the forbidden knowledge of ancient philosophers by that celebrated occult student, philosopher, chemist, naturalist, psychomist, astrologer, alchemist, metallurgist, sorcerer, explanator of the mysteries of wizards and witchcraft; together with recondite views of numerous secret arts and sciences—obscure, plain, practical, etc., etc. New revised, enlarged edition prepared for publication under the editorship of Dr. L.W. de Laurence. Three books which were faithfully translated from the German original and now published in one new revised large volume. The very same being the finest and most complete work of Albertus Magnus, famous magician and prince of philosophers, extant to-day. Chicago, Ill., U.S.A.: de Laurence Scott & Co., 1910.
   [2], iv, [5]-208 p. ; 20 cm.

Edited and published by Lauron William De Laurence (b. 1868), who published over the first four decades of the 20th century numerous texts on hypnotism, telepathy, clairvoyance and other occult topics. De Laurence's edition of *Albertus Magnus* was reissued under his imprint in 1930.
   Spine-title: *Egyptian secrets.*

ALCEDO, Antonio de see BUCHAN
(#500).

17. ALCOHOL AND TOBACCO.
Alcohol and tobacco. Alcohol: its place and
power, by James Miller. The use and abuse
of tobacco, by John Lizars. Philadelphia:
Lindsay & Blakiston, 1867.
   2 pts. in 1 v. (xi, [12]-179, [1] ; [vii]-xi,
[12]-138 p.) ; 18 cm.

Miller's treatise on alcohol and Lizars' on
tobacco had separate publishing histories in
Scotland before their combination in a single
volume by American publishers. The firm of
Lindsay & Blakiston issued at least nine editions
of the combined work at Philadelphia between
1858 and 1881. *Alcohol and tobacco* was also published by the National Temperance Society &
Publication House in New York in 1867, 1870
and 1875. Each treatise was issued separately
by Lindsay & Blakiston in 1859.
   The prominent Scottish surgeon, James
Miller (1812-1864) succeeded Sir Charles Bell
as professor of surgery at Edinburgh in 1842.
He was the author of two standard texts: *The
practice of surgery* and *The principles of surgery,* the
American editions of which were also published
by Lindsay & Blakiston. In later years Miller
became increasingly involved in religious and
social issues, including the temperance movement. *Alcohol: its place and power* was written at
the behest of the Scottish Temperance League.
Miller found no use for alcohol, either as a food
or a medicine. He regarded it simply as a poison affecting every organ and system. On purely
moral grounds, he insisted on the pernicious
effects of even moderate consumption to intellect and judgement.
   Early in his career at Edinburgh, John Lizars
(1787?-1860) was in partnership with the prominent anatomist and surgeon John Bell (1763-
1820), his former tutor. Later he was Robert
Liston's colleague at the Royal Infirmary. Lizars
is perhaps best remembered for his atlas *A system of anatomical plates of the human body,* illustrated with 101 engraved folio plates. Early in
his essay on tobacco, Lizars specifies "nicotina,"
an alkaloid, and "nicotinian," an oil produced
in drying tobacco leaves, as the double agents
of tobacco's deleterious effects on the human
system. He blames tobacco for numerous ul-

cerations of the lips, gums, tongue and throat
he observed among tobacco smokers; and for
broader constitutional effects, especially on the
circulatory, digestive and nervous systems. The
abuse of tobacco has its moral as well as physiological consequences. Lizars attributes much
of the blame for Spain's prolonged decline to
widespread use of the drug imported from her
colonies (pp. 21, 42).

18. ALCOTT, William Andrus, 1798-1859.
The boy's guide to usefulness. Designed to
prepare the way for the "Young man's
guide" . . . Boston: Waite, Peirce, and Company, 1846.
   180 p., [1] leaf of plates ; 16 cm.

This is the third of four editions of *The boy's
guide to usefulness* published between 1844 and
1849. It contains twelve chapters on good manners, good habits, good morals, and good
health for a juvenile male audience. Alcott issues recommendations and admonitions on
such topics as diet, work, play, conversation,
reading (*Robinson Crusoe* is to be avoided), care
of the body and care of the soul. Sexual physiology and behavior are not discussed. Added
lithographic title-page.
   William Andrus Alcott shared his cousin
Bronson Alcott's passion for educational reform. Both were influenced by the educational
theories of Europeans such as Rousseau,
Pestalozzi and Robert Owen, and both devoted
the greater part of their prodigious energies to
innovative educational practice. William
Alcott's educational reforms extended beyond
the usual curriculum to what he termed "physical education," i.e., instructing youth in the
principles of physiology and hygiene. He became nationally renowned as a health reformer
through his lecture tours and the multiple editions of his many published treatises on hygienic topics. It has been asserted that Alcott's
"was the most vigorous voice and prolific pen
in health reform, and his the richest vision of
the means and demands of hygiene" (J.
Whorton, "Christian physiology: William
Alcott's prescription for the millenium," *Bull.
Hist. Med.,* 1975, 49:466). Herbert Thoms
praised Alcott for his "remarkable intuitive perception into the principles of popular education" and for his "extraordinary foresight into

the modern development of medical practice, particularly preventive medicine" ("William Andrus Alcott," *Bull. Soc. Med. Hist. of Chicago,* 1928, 4:[123]).

Alcott became interested in medicine partly because of the vagaries of his own health. He began the study of medicine during and between periods of teaching by preceptorships and self-study. Having taken his medical degree from Yale following a brief course of lectures in 1825-26, Alcott returned to education, only to find his health once more in jeopardy. After a period of convalescence, he took up a rural medical practice. In doubt as to his constitutional ability either to teach or practice medicine, he considered farming, but was once again induced to return to the classroom. After moving to Boston in January, 1832, Alcott became fully committed to the popularization of his ideas on health. It was at this period that he began his prodigious literary output, both as an author of books and the editor of popular serial publications. According to the *Dict. Am. biog.* (1:143), his literary output may be classified: "educational—nineteen volumes; medical, physical education, and health—thirty-one volumes; books for the family and school library—fourteen volumes." With Sylvester Graham, Alcott was one of the founding members (and first president) of the American Physiological Society (1837-1839), and was the first president of the American Vegetarian Society (#95).

19. ALCOTT, William Andrus, 1798-1859.

Familiar letters to young men on various subjects. Designed as a companion to the Young man's guide . . . Buffalo : Geo. H. Derby and Co., 1850.

312 p. ; 19 cm.

Second edition; originally published at Buffalo, N.Y. in 1849. Apart from letter XIV on physiology, the *Familiar letters* are devoted to matters of "self knowledge" and the development of character. In letter XIV, Alcott defines "physiology" and stresses the importance of its study: "Really, my young friends, I know of no one subject, so rarely forced upon your notice, as the world now is, and yet so immensely important to your happiness as this same thing— this obedience to physical law—the laws of Anatomy, Physiology, and Hygiene. Be per-

suaded then, oh be persuaded, to make them, next to the laws contained in the Bible, the man of your counsel and the guide of your youthful steps" (p. 130). Spine-title: *Letters to young men.* Series: *Alcott's new series.*

20. ALCOTT, William Andrus, 1798-1859.

Forty years in the wilderness of pills and powders; or, the cogitations and confessions of an aged physician. Boston: John P. Jewett and Company . . . [et al.], 1859.

x, [2], 384 p., [1] leaf of plates : port. ; 20 cm.

First and only edition. Alcott describes this anonymously published work as "my medical confessions." It is one of two of Alcott's autobiographical works, the other being *Confessions of a schoolmaster* (1839). Alcott recounts his early interest in health and healing, his medical training, the practice of medicine in rural Connecticut, and observations on the limitations and obvious dangers of existing therapeutics. It is indeed a confession of the inadequacy of regular medicine as practiced by Alcott and his professional brethren. The book's message can be summarized in one line, "The best preventive of disease is good health" (p. 29).

"What I most ardently desired was to know the causes of disease," Alcott writes of his early interest in medicine, "and how far they were or were not within human control. Such a science as that of *hygiene*—nay, even the word itself, and the phrase *laws of health*—was at that time wholly unknown in the world in which I moved" (p. 29). Incidents derived from his own medical practice provide the medium for revealing the general ignorance of the medical profession, the dangers of medicines arbitrarily administered, and his gradual realization of the healing powers of nature and the hygienic principles by which disease may be averted altogether.

Regarding the medical wars raging about him, Alcott observes, "Do they not almost, if not quite, prove that when we take medicine, properly so called, or receive active medical treatment, we recover in spite of it? Is there any other rational way of accounting for the almost equal success of all sorts of treatment, —allopathic, botanic, homoeopathic, hydropathic, etc.—when in the hands of good common sense, and conjoined with good nursing and attendance? Is it not that man is made to live, and is tough, so that it is not easy to poison him to death?" (p. 368). He later betrays the indefatigable optimism of the chiliast, "There is a wide difference between the practice of our profession which . . . excludes medicine, and that which . . . includes it. And an entire change from the latter to the former is, perhaps, too great to be expected immediately. Yet, in the progress of society toward a more perfect millenial state of things, must it not come?" (p. 377). Steel-engraved frontis. port. of author.

21. ALCOTT, William Andrus, 1798-1859.
Gift book for young ladies; or familiar letters on their acquaintances, male and female, employments, friendships, &c. . . . Auburn [N.Y.] and Buffalo: Miller, Orton & Mulligan, 1854.
    xv, [1], [13]-307 p., [1] leaf of plates : port. ; 18 cm.

Originally published in 1852, Alcott intended the *Gift book* "as a kind of second volume to the 'Young Woman's Guide'." Frontispiece portrait of Martha Washington.

22. ALCOTT, William Andrus, 1798-1859.
The home-book of life and health; or, the laws and means of physical culture. Adapted to practical use. Embracing laws of digestion, breathing, ventilation, uses of the lungs, circulation and renovation, laws and diseases of the skin, bathing, how to prevent consumption, clothing and temperature, food and cooking, poisons, exercise and rest, the right use of physicians, &c. &c. &c. . . . Boston: Phillipps, Sampson and Company, 1856.
    iv, 500 p. : ill. ; 20 cm.

The home-book of life and health was originally published at Boston in 1853 under the title: *Lectures on life and health* (#34). This 1856 reissue was printed from the same stereotype plates as the 1853 edition, and still has the spine-title of the first edition. Two more issues of *The home-book* were published in 1858 and 1860.

23. ALCOTT, William Andrus, 1798-1859.
The house I live in. Part first: the frame. For the use of families and schools . . . Bos-

ton: Lilly, Wait, Colman, & Holden, 1834.
    xii, [13]-144 p. : ill. ; 15 cm.

First edition. "Our minds are the tenants of bodies so complicated as to be continually liable to derangement and premature decay, yet we seldom know the means of preventing either. Is it not strange that knowledge of such vast importance should have been so long overlooked, and, practically, disregarded?" (p. [v]). Alcott adds, "Man has a body as well as a mind; and such is the connection between them, that it is as much his duty to understand the nature of the one, as of the other. A system of education which overlooks either, is essentially defective" (p. vi).

The house I live in has its origins in a series of essays published under that title in the first volume of *The Juvenile rambler; or, family and school journal* (Boston, 1832), a periodical Alcott edited after moving to Boston late in 1831. The book takes the form of an elaborate analogy likening the human frame to the structure of a house. Each chapter of this edition includes text in large and small type, the former in language suitable to a juvenile or less literate audience. Though the book is devoted to anatomy, it includes occasional references to hygiene. Most chapters conclude with review questions for school use. This slender volume is illustrated with thirty wood engravings—an unusually large number of illustrations in a graphic medium that during the 1830s was just coming into general use in American book illustration.

24. ALCOTT, William Andrus, 1798-1859.
The house I live in; or the human body. For the use of families and schools . . . Third stereotype edition. Boston: George W. Light, 1838.
    264, [4] p. : ill. ; 16 cm.

The third edition was the definitive edition for later issues of this title; its text and format remained the model for successive editions. The anatomy provided in the third edition extends beyond the skeletal and muscular systems described in the first edition to include the anatomy—and, to a lesser extent, the physiology—of the organs of sense, digestion, circulation, and the nerves. For a fuller treatment of physiology, Alcott refers his readers to George Hayward's 1834 *Outlines of human physiology* (#1596).

25. ALCOTT, William Andrus, 1798-1859.
The house I live in; or the human body. For the use of families and schools . . . Fifth stereotype edition. Boston: George W. Light, 1839.
264, [4] p. : ill. ; 16 cm.

26. ALCOTT, William Andrus, 1798-1859.
The house I live in; or the human body. For the use of families and schools . . . Seventh stereotype edition. Boston: George W. Light, 1839.
[2]-264, [4] p., [1] leaf of plates : ill. ; 16 cm.

27. ALCOTT, William Andrus, 1798-1859.
The house I live in; or the human body. For the use of families and schools . . . Eleventh stereotype edition. Boston: Wait, Peirce & Co., 1844.
264 p. : ill. ; 16 cm.

28. ALCOTT, William Andrus, 1798-1859.
The house I live in; or the human body. For the use of families and schools . . . Thirteenth stereotype editon. Boston: Charles H. Peirce, 1847.
264, 22 p. : ill. ; 16 cm.

The 13th appears to be the last numbered edition of *The house I live in*. It was issued again in 1849. Editions after the 13th are described on the title-page as "New edition." The 13th edition is also the first to include twenty-two pages of "Questions to 'The house I live in'" appended to the text. Otherwise, the text is identical to the third edition (#24).

29. ALCOTT, William Andrus, 1798-1859.
The house I live in: or the human body. For the use of families and schools . . . New edition, with questions and explanations. Boston: C.D. Strong, 1854.
264, 22 p. : ill. ; 16 cm.

First issue of the "new edition," which was last published at Boston in 1857. The so-called "new edition" is identical in format, text and illustration to the 13th edition.

30. ALCOTT, William Andrus, 1798-1859.
The house I live in or popular illustrations of the structure and functions of the human body. For the use of families and schools. Edited by Thomas C. Girtin, surgeon. New edition. London: Longmans, Green, and Co., 1872.
xv, [1], 181 p. : ill. ; 16 cm.

The first edition of Thomas Calvert Girtin's edition of Alcott's domestic physiology was published at London in 1837, three years after the appearance of the first American edition (#23). In his preface Girtin writes: "This little work was founded on one published under the same title by Dr. Alcott, a popular American writer. In presenting it to the English public, the Editor proposes to supply a deficiency, which is acknowledged to have been long felt by teachers in schools, and by instructors of youth in general." However, Girtin adds: "In preparing the present edition for the English reader, much was necessary to be done. With a large proportion of what is highly valuable, there were, in the original work, many passages altogether inadmissible, and others that required considerable alteration. Idiomatic forms of expression, peculiar to our transatlantic brethren, a redundancy of words, and frequent inaccuracies in the subject-matter . . . rendered not merely a careful revision, but, for the most part, an entire reconstruction necessary. Besides this, some entirely new articles have been added, and a fuller exposition of many of those in the original given, in order to make the work more complete" (p. [vii]-viii). Girtin's Alcott appeared in ten London editions through 1863 before the publication of this "new edition."

31. ALCOTT, William Andrus, 1798-1859.
The laws of health: or, sequel to "The house I live in" . . . Designed for families and schools. Boston: Published by John P. Jewett and Company . . . [et al.], 1857.
x, 424 p. ; 20 cm.

First edition. "The knowledge of ourselves, physically, until within a comparatively recent period, has been practically overlooked and neglected. But, during the last twenty-five years, a new era has dawned upon our race. Books of anatomy and physiology have been written for families and schools; and the study of the 'house we live in' has become, at least in theory, the order of the day. And yet . . . experience has shown that it is not anatomy . . . nor physiology . . . which is so much needed by the mass of our citizens, as knowledge of our relations to the things around us; or, in other words, HYGIENE" (*Pref.*, p. ix). The text consists of 2,230 numbered "laws" grouped under ten chapter headings.

"Americans, of course, were so mired in their dissipations that the conversion of his generation to the gospel of hygiene would be a heroic labor. But Alcott, while professing humility, clearly saw himself as equal to the task. What else is one to make of his repeated references to himself as a medical missionary, of his claim of having spent forty years in the medical wilderness, or of his account of literally going up on a mountain to have the laws of health revealed to him? Even when not in his prophetic frame of mind, Alcott was ecstatic about physiology. 'In the present blaze of physiological light,' he once exulted, 'we can, in ways and processes almost innumerable, manufacture human health to an extent not formerly dreamed of.' For the dream to be realized, however, physiological knowledge had to be publicized. The science had not yet been productive of human good because it had been 'locked up in the dead languages of the medical man's library.' An interpreter was needed, someone with at least a physician's understanding of physiology and an educator's ability to simplify complex ideas: someone like Alcott" (J. Whorton, "Christian physiology: William Alcott's prescription for the millenium," *Bull. Hist. Med.*, 1975, 49:471).

32. ALCOTT, William Andrus, 1798-1859.
The laws of health, or, sequel to "The house I live in" . . . Boston: Published by John P. Jewett and Company . . . [et al.], 1860.
    x, 424 p. ; 20 cm.

The third and final edition. The second edition was published at Boston in 1859.

34. ALCOTT, William Andrus, 1798-1859.
Lectures on life and health; or, the law and means of physical culture . . . Boston: Phillips, Sampson, and Company, 1853.
    iv, 500 p., [1] leaf of plates : port. ; 20 cm.

First issue. The three later issues of this work (1856 [#22], 1858, 1860) were published under

the title: *The home-book of life and health.*"While travelling in the United States, during the last twelve or fifteen years, as a public lecturer and missionary of health, the question has been asked me a thousand times, What work have you written, not unlike your course of lectures, which may be regarded as containing the substance of your views on health?" (p. iii). *Lectures on life and health* is Alcott's response to this demand. The lectures "are not strictly physiological lectures; but, more properly, lectures on hygiene, or the laws of health. Our popular books in this department,—and so of our public lectures,—though good . . . are for the most part on anatomy and physiology, with a mere 'sprinkling' of hygiene. They are like the trunk and limbs of a most invaluable tree; but they have not borne enough of foliage and fruit. My lectures are the reverse of this. They are chiefly and substantially on the laws of health, with continual appeals to anatomy and physiology as their basis" (p. iii-iv).

35. ALCOTT, Wiliam Andrus, 1798-1859.
The Library of health, and teacher on the human constitution . . . [Vol. 1 (1837)-v. 6 (1842)]. Boston: George W. Light, 1837-42.
    6 v. : ill. ; 18 cm.

Alcott's monthly continues *The Moral Reformer* (#37) and is continued by *The Teacher of health, and the laws of the human constitution* (#47). Printed wrapper for numbers issued in 1837 also enumerated "Vol. III." Atwater set includes vols. 1,2,3 and 5 only.

36. ALCOTT, William Andrus, 1798-1859.
The moral philosophy of courtship and marriage. Designed as a companion to the "Physiology of marriage" . . . Sixth edition. Boston: John P. Jewett & Company; Cleveland, Ohio: H.P.B. Jewett, 1860.
    iv, 308 p. ; 19 cm.

First edition published at Boston in 1857. Alcott's *Physiology of marriage* (#40-45) dealt with the sexual aspects of marriage. The present work is a consideration of the purpose of marriage; the qualifications for marriage; and the duties, virtues, habits, accomplishments and disposition required for a successful union between man and woman.

37. ALCOTT, William Andrus, 1798h-1859.
The Moral reformer, and teacher on the human constitution. [Vol. 1, no. 1 (Jan. 1835)-v. 2, no. 12 (Dec. 1836)]. Boston: Published by Light & Horton, 1835-36.
    2 v. : ill. ; 18 cm.

Issued monthy. Alcott began the *Moral reformer* to fill the need that he felt had been left by cessation of the *Journal of health* (#2053), the only American periodical to that time devoted exclusively to popular hygiene. Alcott had several years experience editing periodicals by the time the first issue of the *Moral reformer* appeared in January 1835. Shortly after moving to Boston in 1831, he had assumed editorship of the *Juvenile rambler* and *Peter Parley's magazine,* in both of which he published his own articles on hygiene.

Alcott both edited this periodical and wrote most of its articles. "The *Moral Reformer* . . . illustrated the range and detail of advice he offered. In a single issue one might find discussions of such topics as 'Cider-Drinking,' 'Uses and Abuses of Light,' 'Education of the Stomach,' 'Featherbeds,' and 'Evening Parties.' In each case, physiological wisdom and moral sensibility combined to show the reader the right way to handle each detail of life" (R.H. Abzug, *Cosmos crumbling,* New York, 1994, p. 170).

Abzug speculates on the reason for the level of detailed attention to "little things" in the works of authors such as Alcott, "To put it plainly, the moral physiologists hoped to create nothing less than a neo-Mosaic law, a system of everyday piety expressed in the language and style of Enlightenment science but resembling the Books of Leviticus and Deuteronomy . . . These reformers, in fact, identified strongly with Moses in the wilderness . . . Moses received the Ten Commandments and other laws—criminal, sexual, dietary, marital—and gave them to his people as a sign and system of regeneration. Exodus, Leviticus, Numbers, and deuteronomy thus recounted a passage from spiritual wilderness to national renewal" (*ibid.,* p. 171). Abzug notes that Alcott's advice "envisioned a comprehensive Christian life designed to permeate the individual's 'every action, word, thought, and feeling, in the performance of the most ordinary duties of life' . . . Revival conversions . . . allowed some to be 'pulled out of the fire' for a time, but he had no faith in the permanence of such changes of heart. Alcott's systematic sacralization of diet, dress, and other aspects of personal life made a conversion of the stuff of daily life" (p. 172).

38. ALCOTT, William Andrus, 1798-1859.
The mother's medical guide in children's diseases . . . Boston: Published by T.R. Marvin, 1842.
    314, [4] p., [1] leaf of plates : port. ; 17 cm.

First edition. Alcott uses his introductory "Address to mothers" to warn his readers that diseases are not easily defined entities for which there are established, normative cures; and to remind them: "The fact is, that we seldom cure disease at all; nor, if we would cure it, would it be safe, at least as a general rule. Disease is the consequence or sequel of violating one or more laws of the human constitution" (p. 16). In place of "dosing and drugging," he encourages the "general diffusion of a knowledge of the *structure, functions,* and *relations,* of the human being—or, in other words, of anatomy, physiology, and hygiene" (p. 18). Mothers should reject "the study of medicine . . . [and] seek to know the laws of health, and the causes of disease, that, by obeying the former, and teaching their offspring to obey them, they might escape the penalty of disobedience in the infliction of the latter." Alcott would see mothers "banish their medicine-chests and closets of medicine, and expend that labor and thought on *prevention,* which was formerly expended on cure."

The text is an alphabetical catalog of eighty childhood disorders. Alcott describes the symptoms and causes of each, but seldom discusses active treatment. In part, this is intended "to show the mother how far, in the management of infantile disease, she can safely go, and when she ought to call for medical advice or aid" (p. 3-4). The author's principal intention, however, is to discourage active medical intervention altogether (i.e., the use of drugs), and to encourage hygienic measures in both the management of disease and its prevention. Frontispiece portrait of author.

39. ALCOTT, William Andrus, 1798-1859.
The mother's medical guide in children's diseases . . . Boston: Published by T.R. Marvin, 1845.
    314 p., [1] leaf of plates : port. ; 16 cm.

Second edition; first issued at Boston in 1842 (#38).

40. ALCOTT, William Andrus, 1798-1859.
The physiology of marriage. By an Old Physician. Fourth thousand. Boston: Published by John P. Jewett & Co.; Cleveland, Ohio: Jewett, Proctor & Worthington; New York: Sheldon, Lamport & Blakeman, 1856.
   vi, [7]-259 p. ; 18 cm.

First edition published at Boston in 1856. "There can be very little doubt that if among the motives that impel our sex to marriage, appetite were not included, celibacy would be vastly more common than it is. In all countries, as luxury advances, celibacy keeps pace with it . . . This appetite, like other appetites, gives us efficiency of character— impelling us forward in some direction or other—just as the heavy mall, by its physical force, impels the wedge which is before it. But the appetites are as blind as the mall itself is, and, unless properly directed by reason and conscience, as likely to go wrong as right" (p. 17-18). Thus it is that Alcott devotes himself in this frequently reprinted work to the social and moral, as well as the physical consideration of the relations of men and women in marriage.
   Alcott provides admonitions on "premature marriage" (i.e., not before age 25 for males and 21 for females); the importance of an early understanding of reproductive physiology; the dangers in courtship of the "prurient and ungovernable passions and appetites" of young males upon the "citadel of female character" ("For though woman is naturally pure . . . still she is easily and largely susceptible of perversion," p. 50); the deleterious effects of "premature sexual indulgences," i.e., fornication and masturbation; "the physical laws of marriage" (e.g., the legitimate frequency of coitus; influence at the time of conception of the physical condition of sexual partners upon their offspring); pregnancy (sexual abstinence; the concerns husbands should have for their wives' physical & mental health); abortion; contraception (coitus interruptus, rhythm method, abstinence); sexual behavior during lactation; venereal disease; and two concluding chapters on the laws of hygiene.

41. ALCOTT, William Andrus, 1798-1859.
The physiology of marriage. By an Old Physician. Tenth thousand. Boston: Published by John P. Jewett . . . [et al.], 1856.
   vi, [7]-259 p. ; 17 cm.

The publisher of Alcott's manual of sexual physiology, John P. Jewett, "spent most of his life in the publishing business, but he found neither fame nor wealth even though he was the first publisher of *Uncle Tom's cabin* . . . He published William Alcott's *The physiology of marriage* in Boston in 1855, but, in the depression of 1857, left publishing to become a vendor of watches and patent medicines and from 1871 to 1882 a partner in the Wakefield Earth Closet Company. He returned to publishing in the mid-1870s, managing Edward Bliss Foote's (#1213-1256) Murray Hill Publishing Company; after Foote's trial and conviction in July 1876 for publishing birth control information, he lost his job" (Janet F. Brodie, *Contraception and abortion in nineteenth century America*, Ithaca: Cornell Univ. Press, 1994, p. 197).

42. ALCOTT, William Andrus, 1798-1859.
The physiology of marriage. By an Old Physician. Fifteenth thousand. Boston: Published by John P. Jewett & Co. . . . [et al.], 1856.
   vi, [7]-259, [1] p. ; 17 cm.

43. ALCOTT, William Andrus, 1798-1859.
The physiology of marriage . . . Sixteenth thousand. Boston: Published by John P. Jewett & Co. . . . [et al.], 1857.
   vi, [7]-259, [1] p. ; 17 cm.

44. ALCOTT, William Andrus, 1798-1859.
The physiology of marriage . . . Twentieth thousand. Boston: John P. Jewett & Company; Cleveland, Ohio: H.P.B. Jewett, 1859.
   vi, [7]-259, [1] p. ; 17 cm.

45. ALCOTT, William Andrus, 1798-1859.
The physiology of marriage . . . Twenty-seventh thousand. Boston: Dinsmoor and Company, 1866.
   vi, [7]-259, [1] p. ; 19 cm.

46. ALCOTT, William Andrus, 1798-1859.
Tea and coffee: their physical, intellectual, & moral effects on the human system . . . A new edition, revised and condensed, by T. Baker . . . Tenth edition. Manchester [Eng.]: John Heywood, [ca. 1876].
   31 p. ; 22 cm.

Alcott regards tea as a narcotic and a dangerous stimulant to the nervous system. He goes so far as to write: "The female who restores strength by tea, the labourer by spirituous liquors, and the Turk by opium, are in precisely the same condition, as far as the stimulation is concerned" (p. 5). Habitual use of tea is the cause of numerous maladies, among them headache, indigestion, and nervous prostration. In like manner Alcott labels coffee a narcotic, and provides a litany of quotations from Beaumont, Willich, Graham, Hitchcock, Dunglison, etc. regarding its pernicious effects on the nervous system. Habitual coffee drinkers are afflicted with the *coffee disease.* "Intestinal motion is more difficult and more painful, muscular motion generally irksome, the extremities chilly, ill-humour is excited, a sort of gnawing comes on, and there is more or less oppression of the head and stomach" (p. 22). Instead of squandering so much time and money on tea and coffee drinking, Alcott asks: "How much might be done in promoting social, intellectual, and moral improvement? How many school, village, and town libraries might be purchased? How many teachers of temperance, physiology, and moral reform might be scattered?" If yet unmoved, "Let them reflect on the health lost to those who use poisons; the number of diseases excited or rendered fatal; and especially the nervous complaints" (p. 26). *Tea and coffee* was originally published at Boston in 1839. The imprint date of this London edition is based on dates found in Baker's footnotes and additions to the text. Thomas Baker (b. 1819), "of the Inner Temple, barrister-at-law," edited several legal compilations, including *The laws relating to public health* (London, 1865).

47. ALCOTT, William Andrus, 1798-1859.
The Teacher of health and the laws of the human constitution . . . [Jan. 1843-Dec. 1843]. Boston: D.S. King & Co., 1843.
    1 v. (384 p.) ; 18 cm.

"The present series of this periodical was commenced nine years ago, and was called the 'Moral Reformer . . . ' [#37]. After two volumes were completed, a second series of six volumes was published, the first part of the title being changed to 'Library of health' [#35]. This volume completes the third series, and is the conclusion of the work" (*Advert.*, p. [3]). Most of the articles in this monthly were written by

Alcott and range in length from a single paragraph to several pages. They are supplemented with equally brief extracts from books, letters, circulars, etc. selected by Alcott.

48. ALCOTT, William Andrus, 1798-1859.
Vegetable diet: as sanctioned by medical men, and by experience in all ages . . . Boston: Marsh, Capen & Lyon, 1838.
    xi, [1], 276 p. ; 19 cm.

First edition. Though not all dietary reform advocates were vegetarians, many of the most prominent banned meat from the human diet. Animal flesh was thought to overstimulate the nerves, tax the digestion, and divert energies generally. More importantly, gluttony, dyspepsia and other culinary ills were seen as a threat to the nation's physical and moral well-being. Advocates of a vegetable diet were certain that through the transformation of the nation's eating habits, its citizenry would be revitalized, the republic transformed, and the millenium inaugurated. Alcott was an early and leading advocate of the vegetarian movement, and a founding member and first president of the American Vegetarian Society (est. 1850).

In the 11 March 1835 issue of the *Boston medical & surgical journal* (12:76-77) and the May 1835 issue of the *American journal of the medical sciences* (16:259-60), Milo North, M.D., of Hartford Ct. (#2641-2644), published eleven questions soliciting from his colleagues their experiences and opinions on the use of an exclusively vegetable diet. North received seventeen responses (including one from Alcott), which he intended to publish in a monograph. Ill-health forced North to transfer this responsibility to Alcott's willing shoulders. The *Vegetable diet* includes the full text of North's correspondence with his medical brethren; Alcott's observations on these letters; additional correspondence solicited by Alcott; an extensive review of the opinions of such figures as George Cheyne (#632, 633), William Cullen, Benjamin Rush (#3054-3056), Sylvester Graham (#1392-1395), Percy Bysshe Shelley (a vegetarian from age 21), and William Lambe (#2176, 2177), the English physician who was the first to scientifically examine the effects of a vegetable diet. Alcott's concluding chapter contains his "anatomical, physiological, medical, political, economical, experimental and moral arguments" for the universal adoption of vegetarianism. Atwater copy lacking p. xi.

49. ALCOTT, William Andrus, 1798-1859.
Vegetable diet as sanctioned by medical men, and by experience in all ages. Including a system of vegetable cookery . . . Second edition, revised and enlarged. New York: Fowlers and Wells, publishers, 1849.
    xi, [12]-312 p. ; 19 cm.

Vegetarianism was one of the many facets of health reform embraced by the publishing firm of Fowlers and Wells. After a lapse of more than a decade since the first Boston edition, the firm added Alcott's *Vegetable diet* to its list, publishing five issues of this second edition between 1849 and 1859. The second edition contains "nearly fifty pages of new and original matter" (*Advert.*, p. [vii]), largely in the form of additions to the litany of supporting medical opinion (e.g., Ticknor (#3523, 3524), Coles (#711-721), Shew (#3180-3206), O.S. Fowler (#1279-1302), etc., and in the addition of "Outlines of a new system of food and cookery," a philosophical cook book providing the categorical arrangement of acceptable foods into classes, divisions and sections, as well as recipes for their preparation. The Atwater copy bears the inscription of Henry Foster, M.D. (#1265), founder of the Clifton Springs Water-Cure near Rochester, N.Y.

50. ALCOTT, William Andrus, 1798-1859.
The young house-keeper, or thoughts on food and cookery . . . Eighth stereotype edition. Boston: Charles H. Peirce, 1848.
    424 p. ; 17 cm.

Originally published at Boston in 1838, *The young house-keeper* was one of Alcott's more successful books, appearing in at least twenty editions through 1851. Having devoted other volumes to the "young wife" (#57-58) and the "young mother" (#52-56.1), Alcott in this book addresses woman's role as "house-keeper." He uses the terms "mother" and "house-keeper" in the same breath, explaining that those women who might object to their equation "have not only been trained to the neglect of household duties, and to the belief that they are quite beneath them . . . but to the erroneous idea that they involve a life of hardship, pain

and drudgery. Whereas the truth is, that there are no labors which taken together are more easy or more healthful; and, if properly conducted, there are few which give more freedom as well as more leisure for recreation and study" (p. 31). In actuality, Alcott devotes only the first five of his forty-eight chapters to the dignity and principles of domestic science. The greater part of the book is given to the principles of nutrition and reformed cooking. Alcott explains the nutrional benefits of grains, legumes, fruits, vegetables and nuts, as well as their preparation and consumption. He also discusses the pitfalls of various foods, particularly in the consumption of animal flesh and excessive reliance on the dairy. "I regret to be obliged to recognize animal food as, in any sense, a primary aliment; for I consider the resort to it as proper only in the case of infants, diseased persons, and the people of those regions and places where better food cannot be obtained" (p. [265]).

In summarizing the principles of diet and cookery, Alcott states that the house-keeper should "observe simplicity" in the quantity and variety of foods served at home; that meals should be served at regular hours and accomodated to the age groups of those at table. He insists that food should not be served hot; and that it should be well-masticated in order to aid taste, nutrition and digestion. Recognizing the volume of work facing the the house-keeper, Alcott concludes his book with a time-effort study of normal household tasks; provides recommendations on how time may be economized; and discusses how the house-keeper can implement the "reformation" of domestic routine.

51. ALCOTT, William Andrus, 1798-1859.
The young man's guide . . . Fourteenth edition. Boston: Perkins and Marvin, 1839.
    360 p., [2] leaves of plates : frontis. ; 16 cm.

"The purpose of the Young Man's Guide, is the formation of such character in our young men as shall render them the worthy and useful and happy members of a great republic. To this end, the author enters largely into the means of improving the *mind*, the *manners* and the *morals*—as well as the proper management of *business*. Something is also said on *amusements*, and *bad habits*" (*Advertisement,* p. [5]). This manual on the formation of character and the conduct of life includes advice on such pertinent health matters as sleep & early rising, drunkenness & gluttony, bathing, the use of tobacco, and the moral and physical consequences of sexual vice. The chapter on marriage considers the institution from a social and moral point of view. Alcott defers his discussion of the physiology of marriage to works written specifically for a married audience (#40-45). *The young man's guide* went through at least twenty-one editions between 1833 and 1851. Added engraved title-page dated 1836.

52. ALCOTT, William Andrus, 1798-1859.
The young mother, or management of children in regard to health . . . Boston: Light & Stearns, 1836.
336 p., [1] leaf of plates ; 18 cm.

First edition. "The 'Young mother' is designed as an every-day manual for those who are desirous of conducting the physical education of the young . . . on such principles as Physiology and Chemistry indicate" (from the publisher's advertisement in the 1839 edition of Alcott's *The house I live in,* #25).
CONTENTS: I. The nursery. II. Temperature. III. Ventilation. IV. The Child's dress. V. Cleanliness. VI. Bathing. VII. Food. VIII. Drinks. IX. Giving medicine. X. Exercise. XI. Amusements. XII. Crying. XIII. Laughing. XIV. Sleep. XV. Early rising. XVI. Hardening the constitution. XVII. Society. XVIII. Employments. XIX. Education of the senses. XX. Abuses.
In his preface to this first edition of *The young mother,* Alcott reveals his reading of Rousseau by asking, "Now if it be the intention of divine Providence . . . that the animal body should be capable of resisting with impunity the impressions of heat, cold, light, air, and the various external influences to which, at birth, it is subjected, it may be properly asked why this primitive state of health cannot be maintained, and diseases, and medicines, and even preventives, wholly avoided. But the reason is obvious. Civilized society has placed the human race in artificial circumstances. Instead of listening to the dictates of reason, making ourselves acquainted with the nature of the human constitution, and studying to preserve it in health and vigor, we yield to the government of ignorance and pre-

sumption" (p. 25-6). Added steel-engraved title-page.

53. ALCOTT, William Andrus, 1798-1859.
The young mother, or management of children in regard to health . . . Second edition. Boston: Light & Stearns, 1836.
336 p. ; 18 cm.

Added engraved title-page.

54. ALCOTT, William Andrus, 1798-1859.
The young mother or management of children in regard to health . . . Sixth stereotype edition. Boston: George W. Light, 1838.
342 p., [1] leaf of plates ; 18 cm.

The sixth edition retains the same twenty chapters and headings as the first edition, with minor additions to chapters one and seven. Added steel-engraved title-page. As in most of his publications, Alcott is never hesitant in quoting or paraphrasing the published work of his fellows in the medical profession when their opinions are in accord with his own. The authority most frequently quoted in the pages of *The young mother* is William Dewees' *A treatise on the physical and medical treatment of children,* which went through eleven editions between 1825 and 1858. Alcott lauds Dewees' *Treatise* as "one of the most valuable works on Physical Education in the English language" (p. 30).

55. ALCOTT, William Andrus, 1798-1859.
The young mother, or management of children in regard to health . . . Tenth stereotype edition. Boston: Waite, Peirce and Company, 1846.
342 p. ; 18 cm.

56. ALCOTT, William Andrus, 1798-1859.
The young mother, or management of children in regard to health . . . Sixteenth stereotype edition. Boston: Charles H. Peirce, 1848.
342 p. ; 16 cm.

56.1. ALCOTT, William Andrus, 1798-1859.
The young mother, or management of children in regard to health . . . Twentieth stereotype edition. New York: James C. Derby;

Boston: Phillips, Sampson & Co.; Cincinnati: H.W. Derby, 1855.
342 p. ; 19 cm.

The twentieth is the final edition of this title, originally published in 1836 (#52).

57. ALCOTT, William Andrus, 1798-1859.
The young wife, or duties of woman in the marriage relation . . . Third stereotype edition. Boston: George W. Light, 1837.
376 p., [2] leaves of plates : frontis. ; 18 cm.

The first edition of *The young* wife was also published at Boston in 1837. This work was one of Alcott's most popular, appearing in twenty editions through 1855. In his opening chapter Alcott provides general remarks on marriage. Each of the next twenty-four chapters is devoted to a particular quality that Alcott thinks essential to a young wife and a successful marriage, e.g., submission, kindness, cheerfulness, confidence, sympathy, delicacy, love of home, purity of character, simplicity, etc. The remaining chapters touch on other wifely concerns and obligations, such as dress, hygiene, attending the sick, intellectual improvement, and moral influence on the husband. Sexual ethics and physiology are not discussed. Includes steel-engraved frontispiece and an added engraved title-page.

58. ALCOTT, William Andrus, 1798-1859.
The young wife, or duties of woman in the marriage relation . . . Ninth stereotype edition. Boston: George W. Light; New York: 126 Fulton Street, 1840.
376 p., [2] leaves of plates : frontis. ; 18 cm.

This ninth edition includes the steel-engraved frontispiece and added engraved title-page of the 1837 first edition.

59. ALCOTT, William Andrus, 1798-1859.
The young woman's book of health . . . Boston: Tappan, Whittemore and Mason, 1850.
311, [1] p. ; 19 cm.

First edition. Alcott's first chapter is a paradigm of the relation expressed or assumed in all his published works, i.e., the mutual dependence of holiness and bodily health. There is no desert asceticism in Alcott's theology, and no place for that contempt of flesh that nurtures the spirit at the body's expense. Alcott's is a Christianity of healthy complexions, firm muscles and sound digestions. The perfected body that is resurrected and accompanies the soul in Paradise is to be paralleled in this world by a body purified on hygienic principles.

In his preliminary remarks, Alcott notes that although few young women are truly healthy, even fewer are "enough inclined to sober thought" (p. 14) to be duly concerned. He warns that without such concern young females may never attain to that "sanctification and perfection of the body" that results in "freedom from absolute disease" (p. 21), i.e., liberation from the numerous physical and nervous ailments that burden most women. Both the problem and the solution lie in the education of young females. In chapter II, Alcott enumerates the result of contemporary "female miseducation," i.e., "a whole generation of young women, trained as a whole, to tenderness, delicacy, nervousness, feebleness of muscle, want of appetite and imperfect digestion." He wonders, "Shall we not be almost driven to the conclusion . . . that if female education is to be conducted, for three hundred years to come, as it has been for a century past, our race must become extinct?" (p. 37).

Having lamented "the present condition of female character and health," Alcott suggests its remedy in chapter III. through "the laws of exercise, respiration, cleanliness, diet, circulation, mind and heart" (p. 39). The remaining chapters describe how neglect of these principles results in such uterine disorders as menstrual abnormalities, hysteria, sterility, uterine displacements, neoplasms, etc. Throughout his discussion of these maladies, Alcott lays greater emphasis on prevention than cure, and hygienic management rather than self-treatment through "injudicious dosing and drugging."

60. ALCOTT, William Andrus, 1798-1859.
The young woman's book of health . . . New York and Auburn [N.Y.]: Miller, Orton & Mulligan, 1855.
311, [9] p. ; 19 cm.

A re-issue of the 1850 Boston edition with new title-page and nine leaves of publisher's advertisements after the text.

61. ALCOTT, William Andrus, 1798-1859.
The young woman's guide to excellence .
. . Thirteenth edition. Boston: Charles H.
Peirce . . . [et al.], 1847.
356 p., [2] leaves of plates ; 16 cm.

First edition: Boston, 1840 (copyrighted
1836; preface dated Nov. 1839). In this work
Alcott devotes his attention to female education, which, in this and his other published
works, he regarded as grossly deficient. By "education" he means the "development, or formation of any part of our character, physical, intellectual, social or moral" (p. 25). Alcott emphasizes that the formation of female character is not an end in itself, but, given women's
influence on husbands and children, has inestimable social import. Most of this book's thirty-
two chapters are devoted to a discussion of the
social and moral duties of young women. Six
chapters, either wholly or in part, discuss health
issues, i.e., ch. xx, exercise, ch. xxi, rest and
sleep; ch. xxv, health and beauty; ch. xxi, neatness and cleanliness; ch. xxvii, dress and orna-
ment; ch. xxix, taking care of the sick. The hygienic training of females was the exclusive focus of a later work entitled *Young woman's book
of health* (#59-60). Includes steel-engraved frontispiece and added title-page.

ALCOTT, William Andrus see also
ARMSTRONG (#156).

62. ALDEN, Walter.
The human eye; its use and abuse: a popular treatise on far, near and impaired sight,
and the methods of preservation by the
proper use of spectacles, and other acknowledged aids of vision . . . Cincinnati:
Published by the author, 1866.
[2], 138, [14] p., [2] leaves of plates :
ill. ; 22 cm.

The author was a Cincinnati optician. Cover-
title: *Preservation of sight.* Atwater copy lacking
pages 133-4.

63. ALDRICH, Auretta Roys, b. 1829.
Life and how to live it . . . Philadelphia:
Drexel Biddle, publisher, 1901.
[4], 186 p., [3] leaves of plates, [10] p.
of plates : ill., port. ; 19 cm.

Aldrich maintains that all life is motion, and
that in its proper functioning this motion is
based on elemental rhythms. Her purpose in
this manual of exercise and breathing is help
her readers discover and restore the "perfect
adjustment of all the life forces," i.e., the natural rhythms that are the basis of health. "Can it
be possible," she asks, "that the material universe is governed by unalterable law and perfect in its operation while man is the plaything
or victim of every chance microbe or germ that
exists? Surely there is a law of energy; of vital
rhythm, that would keep us from such a humili-
ating position" (p. 28).

64. ALLEN, Chilion Brown.
The man wonderful in the house beautiful. An allegory teaching the principles of
physiology and hygiene, and the effects of
stimulants and narcotics. For home reading;
also adapted as a reader for high schools and
as a text-book for grammar, intermediate,
and district schools. By Chilion B. Allen . . .
and Mary A. Allen . . . New York: The Educational Publishing Co., 1886.
[4], 366 p. : ill. ; 19 cm.

The Allens' juvenile physiology was published at least six times between 1883 and 1891,
and appeared in a Dutch translation at
Amsterdam in 1887. The authors are described
as members of the Broome County (N.Y.) Medical Society on the title-page. Chilion B. Allen
was an 1881 graduate of the medical department of the University of the City of New York;
Mary Augusta Allen later published as Mary
Wood-Allen (#3851-3860). Atwater copy inscribed by Chilion B. Brown.

65. ALLEN, Joshua.
Consumption of the lungs, is not a contagious nor a germ disease. Its cause, prevention and cure . . . [Philadelphia, 1901?].
11, [9] p. ; 22 cm.

"In the month of February in the year 1890,
as a result of long continued study and observation of nervous phenomena, as a factor in the
production and cure of diseases, I came to the
conclusion that consumption of the lungs was
of *nervous origin*" (p. 5). At the same time Allen
discovered a cure for consumption, "the force

of which is expended almost entirely on the nervous organization." He dismisses Koch's discovery as "not yet proven," and attributes the presence of bacilli in the sputa of some consumptives to conditions favorable to their development in the patients' diseased state. According to Allen, the lungs are "dependent upon a proper supply of nerve force distributed to them" (p. 9). When this supply is weakened or exhausted, then conditions develop in the lungs symptomatic of consumption. The nine unnumbered pages contain testimonials of consumptives who were cured by Allen. His method of treatment is specificed neither in his text nor in the testimonials. Allen was an 1878 graduate of the Hahnemann Medical College (Philadelphia). Title from printed wrapper.

ALLEN, Mary Augusta see **ALLEN** (#64).

ALLEN, Mary E. see **SAFFORD** (#3066).

66. ALLEN, Monfort B.
The glory of woman or love, marriage and maternity. Containing full information on all the marvelous and complex matters pertaining to women including creative science; bearing, nursing and rearing childen; hereditary descent; hints on courtship and marriage; promoting health and beauty, vigor of mind and body, etc., etc. Together with diseases peculiar to the female sex. Their causes, symptoms and treatment. The whole forming a complete medical guide for women. By Monfort B. Allen, M.D. and Amelia C. McGregor, M.D. . . . [S.l.: s.n., c1896].
    xiv, 17-511 p., [17] leaves of plates, [16] p. of plates : ill. ; 21 cm.

    The treatise of Allen and McGregor on love, marriage, reproduction, sexual conduct, child care, feminine beauty and social proprieties was published under several titles between 1896 and 1902: *The glory of woman or love, marriage and maternity* (1896-1900); *Ladies' home manual of physical culture and beauty* (1896); *The woman beautiful, or maidenhood, marriage and maternity* (1901); and *Woman's guide to health and happiness* (1902).

67. ALLEN, Monfort B.
The glory of woman or love, marriage and maternity. Containing full information on

all the marvelous and complex matters pertaining to women including creative science; bearing, nursing and rearing childen; hereditary descent; hints on courtship and marriage; promoting health and beauty, vigor of mind and body, etc., etc. Together with diseases peculiar to the female sex. Their causes, symptoms and treatment. The whole forming a complete medical guide for women. By Monfort B. Allen, M.D. and Amelia C. McGregor, M.D. . . . Paris, Ontario, Can.: John S. Brown & Sons, [1900, c1896].
    xiv, 17-511 p., [17] leaves of plates, [16] p. of plates : ill. ; 21 cm.

    A presentation leaf is bound before the frontispiece in this Paris, Ontario edition: "Presented to [*Mrs. Hamill*] by ___ [*September*] 19[*00*]."

68. ALLEN, Richard L., 1808-1873.
An analysis of the principal mineral fountains at Saratoga Springs, embracing an account of their history; their chemical and curative properties; together with general directions for their use; also, some remarks upon the natural history, and objects of general interest in the County of Saratoga . . . New York: Ross & Tousey, 1858.
    113, [1] p., [4] leaves of plates : ill. ; 16 cm.

    This work is so extensively expanded and revised that it cannot be regarded simply as the third edition of his 1844 and 1848 treatises (#69, 70). The 1858 *Analysis* provides an invaluable history of the region around Saratoga Springs, chemical analyses of its mineral springs, and an exposition of the principles of balneology and hydrotherapy for the lay reader. Later editions (1859, 1866) were issued under the title *Hand-book of Saratoga, and strangers' guide.* The wood-engraved plates provide views of the High Rock, Congress, Columbian and Empire springs.

69. ALLEN, Richard L., 1808-1873.
A historical, chemical and therapeutical analysis of the principal mineral fountains at Saratoga Springs; together with general directions for their use . . . Saratoga

Springs: B. Huling; New-York: W.H. Graham, 1844.
    iv, [5]-71 p., [1] leaf of plates : ill. ; 15 cm.

First edition. Long known to the native American population, the natural mineral springs at Saratoga were attracting white visitors by the late 1760s. Interest in the springs was continuous in the last decades of the 18th century, and Saratoga had its first hotel by 1791. Throughout the 19th century, Saratoga Springs remained the nation's most fashionable spa and resort. R.L. Allen arrived in the village in 1833, the same year the already popular resort was connected by rail to Albany, an important connection in the state's growing network of railroads and the terminus for Erie Canal and Hudson River traffic. Allen became the professional associate of John Honeywood Steel (1780-1838), whose published analyses of Saratoga's mineral springs he continues and revises: "As there has not been a uniform analysis of the principal mineral fountains here since the one by Dr. Steel [#3331-3336]; some fountains of importance having since been added, and old ones materially altered by additional and complicated tubing since the Doctor's death; it became desirable to give them all a uniform examination" (p. [5]). Cover-title: *Allen's analysis 1844.*

70. ALLEN, Richard L., 1808-1873.
A historical, chemical and therapeutical analysis of the principal mineral fountains at Saratoga Springs; together with general directions for their use . . . Second edition, revised . . . Saratoga Springs: B. Huling, 1848.
    iv, [5]-72 p., [1] leaf of plates : ill. ; 15 cm.

The first edition of Allen's treatise was published at Saratoga Springs in 1844 (#69). Cover-title: *Allen's analysis.* Wood-engraved frontispiece: "A view of the High Rock Spring."

71. ALLGEMEIN VERSTÄNDLICHE BELEHRUNG.
Allgemein verständliche Belehrung über homöopathische oder Hahnemannische Aerzte, Kuren, Arzeneien, Diät, u.s.w. Geschrieben von einem Manne, der alles Homöopathische recht gut kennt, ob er gleich kein Doctor ist. Allentaun [Pa.]: [s.n.], 1834.
    24 p. ; 20 cm.

This brief introduction to homeopathy, "Written by one who knows homeopathy quite well, although himself not a doctor," concludes with the exhortation: "Die Homöopathik ist besser! Wer ihr sich in die Arme wirst, thut wohl!!!" [Homeopathy is better. You do well if you throw yourself into her arms]. Pencilled on the title-page of this anonymous pamphlet are the words: "Dr. Hering [#1625-1634] will know by whom this is written. W."

72. ALSAKER, Rasmus Larssen, b. 1883.
Curing catarrh and colds . . . New York: Frank E. Morrison, publisher, 1917.
    117, [3] p. ; 18 cm.

First edition. In 1917 Frank E. Morrison published eight *Alsaker health books: Conquering consumption; Curing catarrh, coughs and colds; Curing constipation and appendicitis; Dieting, diabetes and Bright's disease; Eating for health and efficiency; Getting rid of rheumatism; How to live on three meals a day* (#73); and *Maintaining health.* Most of these titles appeared in more than one edition. Alsaker wrote half a dozen other popular health treatises, two of which he wrote and published in the mid-1960s.

73. ALSAKER, Rasmus Larssen, b. 1883.
How to live on 3 meals a day . . . New York: Frank E. Morrison, publisher, 1917.
    x, 13-96 p. ; 18 cm.

First edition.

74. AMERICAN AND CONTINENTAL "SANITAS" CO.
Sanitas disinfectants, antiseptics, deodorants and oxidents. This pamphlet is issued for the use of all interested in disinfection and the maintenance of sanitary conditions—whether in cases of sickness or everyday life. It mentions under appropriate headings the proper use of the numerous forms of disinfectants prepared at our factories. These directions, if followed, will assure thorough disinfection at minimum expense and trouble. We trust the contents

will prove of interest not only to the general public, but also to physicians, nurses, veterinary surgeons and sanitarians. New York: The American & Continental "Sanitas" Co., [c1895].
    96 p. : ill. ; 18 cm.

Title on printed wrapper: *How to disinfect: a guide to practical disinfection in everyday life, and during cases of infectious illness. By C.T. Kingzett.* The various *Sanitas* products advertised were available as fluids for atomization or direct application to surfaces, fumigating powders and candles, laundry soaps, ointments, oils, powders, mouth-washes, disinfectant sawdust; tooth powders, jellies etc. The name of Charles Thomas Kingzett (1852-1935), the British chemist, is attached to several of these products, and his name appears on the title-page beneath the trademark and copyright date.

AMERICAN ANTIQUARIAN SOCIETY see **MATHER** (#2839).

75. AMERICAN ANTI-TOBACCO SOCIETY.
Boys cured of the tobacco mania . . . Fitchburg, Mass.: Anti-Tobacco Tract Depository, [185-?].
    [2] leaves : ill. ; 19 cm.

"Strong drink, tobacco, and their affinities are cutting down our youth by thousands, and nothing so meets the emergency as BANDS OF HOPE to be thoroughly organized as soon as convenient" (2nd leaf recto). Each Band of Hope was an association "to encourage the young to abstain from intoxicating liquors, tobacco, and profanity." A pledge form ("HATERS OF TOBACCO") is printed on the verso of the second leaf. The leaflet is signed George Trask (1798-1875). Caption title.

76. AMERICAN ECLECTIC MEDICAL REGISTER.
The American eclectic medical register. For the year commencing Jan. 1st, 1868. Editor, Robert S. Newton, M.D. New York: The Trow & Smith Book Manufacturing Co., 1868.
    1 v. ([1-4], [2], 5-16, 13-176, [28] p., [1] leaf of plates) : ill., port. ; 24 cm.

"The first number of this work contains a list of every state eclectic medical society in this country, so far as they could be obtained, and every county and district auxiliary society, together with the constitutions and by-laws, and the names of their officers and members and the times and places of all their meetings. It also contains a complete list of our standard eclectic medical works as used in our professional schools and libraries, together with a list of our colleges and the names of their faculties." (pref., p. [3]). No more volumes were published after this first volume.
    Robert Safford Newton (1818-1881) graduated from the allopathic Louisville Medical College, but in the mid-1840s transferred his professional allegiance to the "reformed practice." After a brief association with the Memphis Medical Institute, he joined the faculty of the Eclectic Medical Institute in Cincinnati (1851-1862), where from 1851 to 1861 he also edited and published the *Eclectic medical journal.* In 1863 Newton moved to New York to join the faculty of the newly established Eclectic Medical College of the City of New York. He was one of the founding members of the Eclectic Medical Society of the State of New York, and its president in 1866-67.

AMERICAN EDUCATIONAL SERIES see **HITCHCOCK** (#1653).

77. AMERICAN FEDERATION FOR SEX HYGIENE.
    Report of the Special Committee on the Matter and Methods of Sex Education. Presented before the Subsection on Sex Hygiene of the Fifteenth International Congress on Hygiene and Demography held in Washington, D.C. September twenty-third to twenty-eighth nineteen hundred and twelve . . . New York City: Issued by the American Federation for Sex Hygiene, 1912.
    [2], 34 p. ; 23 cm.

Charged with considering the practicability of introducing sex education into school and college curricula, and the "matter and methods" of such instruction," the Special Committee made twenty-three recommendations to the Federation at its December 1912 meeting. The

Committee stresses that the aim of such instruction is "to develop a healthier public sentiment in regard to sex, which will make it possible to discuss with more freedom than is now customary the grave hygienic and moral dangers to the individual and to the community which grow out of the violation of the physical and moral laws governing sex life and the sacred processes of human reproduction" (p. 2). However, the instructional process should aim "to keep sex consciousness and sex emotions at the minimum," to which end "detailed descriptions of the external human anatomy are to be avoided; and . . . descriptions of internal anatomy should be limited to what is necessary to make clear and to impress the hygienic bearing of the facts to be taught" (p. 3). Scientific instruction is to be complemented by ethical instruction, and therefore: "It is of immense consequence that during the adolescent years the pupils' minds be saturated with the great masterpieces in both poetry and prose, which deal with romantic love in its purest forms" (p. 4). The value of sports and physical exercise is also to be stressed as a means of "control of the sex instinct." As girls reach sexual maturity earlier than boys, their instruction should begin "a year and a half earlier than boys . . . and emphasis should be laid upon instruction in regard to the special care of their health at the change of life called puberty" (p. 8). Sex education should not be taught in specially set classes, but should be integrated into courses on "nature study, biology, hygiene, and ethics." Instruction on the reproduction of plants and lower animals can be given in co-educational classes; instruction in mammalian reproduction, however, "should always be given in separate classes, and each class should be taught by a teacher of their own sex" (p. 9). The Special Committee was composed of the biologist Maurice Alpehus Bigelow (#319), the educator Thomas Minard Balliet (1852-1942) and the dermatologist Prince Albert Morrow (1846-1913).

78. AMERICAN GENTLEMAN'S MEDICAL POCKET-BOOK.

The American gentleman's medical pocket-book, and health-adviser: containing a statement of the modes of curing every disease to which he is liable; and directions in case of accidents on the road or at sea.

With a full account of epidemic cholera, of dyspepsia, and of sick-headach; their causes, cure, and prevention: and a popular description of the human teeth; their formation, diseases and treatment. By the author of The Lady's medical pocket-book. Philadelphia: James Kay, Jun. and Brother; Pittsburgh: C.H. Kay and Co., [c1833].

254, [2] p., [2] leaves of plates : ill., port. ; 13 cm.

The anonymous author of this book promises his readers that the principles of medicine are "few, short and easily understood." He provides anecdotal evidence justifying the widespread dissemination of books such as this: "I have known a considerable district in one of our western states, which contained but one doctor, and him I have met on his way to his patients by eight o'clock in the morning so drunk, as to render it necessary to lift him into his gig . . . What becomes of the sick, under such circumstances!" (p. [3]). The book is divided into two parts. Part one contains three chapters: "Of fevers," "Of diseases, commencing in, and chiefly confined to particular parts," and "Of accidents." The "Preface to new edition" on page four states that "part second" (on cholera, dyspepsia, headache and the teeth) is new to this edition. Added engraved title-page (with "John I. Kay & Co." in imprint) and frontispiece portrait of Benjamin Rush. Spine-title: *Gentleman's medical pocket-book.*

An earlier edition had been published anonymously at Philadelphia in 1832 under the title *The Gentleman's medical pocket-book, and health-adviser* (160 p.). This 1832 publication is also described as a "new edition." We have been unable to trace the title of its earliest publication, however. *The American gentleman's medical pocket-book* was re-issued at Philadelphia in 1836 (#79) and again in 1837 (#80).

79. AMERICAN GENTLEMAN'S MEDICAL POCKET-BOOK.

The American gentleman's medical pocket-book and health-adviser: containing a statement of the modes of curing every disease to which he is liable; and directions in case of accidents on the road or at sea. With a full account of epidemic cholera, of dyspepsia, and of sick-headach; their causes, cure, and prevention: and a popular description of the human teeth; their formation, diseases and treatment. By the author of The Lady's medical pocket-book. Philadelphia: James Kay, Jun. and Brother; Pittsburgh: John I. Kay and Co., [1836, c1833].

254, [2] p., [2] leaves of plates : ill., port. ; 13 cm.

The title-page and added engraved title-page are undated. The verso of the title-page bears an 1833 copyright date. At the base of the spine is stamped the date 1836. Spine-title: *Gentleman's medical pocket-book.*

80. AMERICAN GENTLEMAN'S MEDICAL POCKET-BOOK.

The American gentleman's medical pocket-book and health-adviser: containing a statement of the modes of curing every disease to which he is liable; and directions in case of accidents on the road or at sea. With a full account of epidemic cholera, of dyspepsia, and of sick-headach; their causes, cure, and prevention: and a popular description of the human teeth; their formation, diseases and treatment. By the author of The Lady's medical pocket-book. Philadelphia: James Kay, Jun. and Brother; Pittsburgh: John I. Kay and Co., [1837, c1833].

254, [2] p., [2] leaves of plates : ill., port. ; 13 cm.

The title-page and added engraved title-page are undated. The verso of the title-page bears an 1833 copyright date. At the base of the spine is stamped the date 1837. Spine-title: *Gentleman's medical pocket-book.*

81. AMERICAN HEALTH CONVENTION.

A report of the proceedings of the second American Health Convention, held in the tabernacle of the Presbyterian Church in Broadway, N. York, on the afternoon of Wednesday, May 8, 1839. Peported [sic] by R. Sutton, professional short hand reporter . . . Boston; New York: [s.n., 1839].

16 p. ; 22 cm.

The American Health Convention was one of several Grahamite organizations active in Massachusetts and New York from the late 1830s. The second convention drew 182 participants from twelve states and the West Indies. The business portion of the convention is not recorded in the *Report,* which consists largely of the transcription of speakers' remarks. A. Ball, M.D., president of the New York Physiological Society, proposed a resolution that "the efforts of medical men, and the success of medicine, are and will be lost, or worst than lost, until the community in general, and mothers in particular, are more thoroughly and effectually enlightened on the subject of health and disease" (p. 3). Dr. Enoch Mack, of Dover, N.H., reminded the convention that in the Gospel of St. John it is written that "if all our Savior's deeds and words were written, the world itself would not contain the books which those transactions and doctrines would fill," meaning, he supposed, that "the subject of physiological reformation . . . would constitute a large amount of that doctrine which our Savior inculcated in his preaching" (p. 4). Sylvester Graham (#1392-1395) addressed the convention on the vegetable diet. Title and imprint from wrapper.

AMERICAN HEALTH PRIMERS see **BULKLEY** (#516), **BURNETT** (#530), **PACKARD** (#2700), **WHITE** (#3765), **WILSON** (#3831).

82. AMERICAN HYGIENIC AND HYDRO-PATHIC ASSOCIATION OF PHYSICIANS AND SURGEONS.

Constitution of the American Hygienic and Hydropathic Association of Physicians and Surgeons: together with the list of officers, standing committees and members, proceedings of the first and second annual meetings; and the first annual report of the Committee on hygiene . . . New York: Published by Fowlers and Wells, 1851.

36 p. ; 19 cm.

The preamble to the Association's constitution proclaims its adherence to "the doctrine of the *vis medicatrix naturae*" and the belief that "of all the remedial agents which the experience of ages has shown to be requisite to assist nature in her operations, WATER is by far the best, the safest, and most universal in its applications" (p. [3]). Sixteen members were in attendance at the Association's first annual meeting in New York on June 19th, 1850. At its second meeting on May 9, 1851, Mary Gove Nichols (#2623-2625), Rachel Brooks Gleason (#1362-1367), and Lydia N. Fowler (#1277, 1278) were elected "honorary members," the first women to join the Association. The proceedings of both meetings make fascinating reading, as attendees hammer out wording of the constitution or debate membership criteria. Of less interest, perhaps, is the "Report of the Committee on Hygiene" (p. 23-32). Influenced by the work of Chadwick in England and Shattuck in Boston, the report is a review of contemporary public hygiene intended to encourage members' interest in the "sanitary movement." Signed by R.S. Houghton (#1797), the report concludes that public health "transcends the importance of all other sciences, and in its beneficent operation seems to embody the spirit, and to fulfill the intentions, of practical Christianity" (p. 32). In printed wrapper.

AMERICAN HYGIENIC AND HYDRO-PATHIC ASSOCIATION OF PHYSICIANS AND SURGEONS see also HOUGHTON in *BULWER AND FORBES ON THE WATER-TREATMENT* (#521).

83. AMERICAN LADY'S MEDICAL POCKET-BOOK.
The American lady's medical pocket-book and nursery-adviser: containing rules for preserving the health of unmarried females; directions to pregnant and lying-in women; and an account of their diseases; together with instructions for the rearing of children from the hour of their birth; and an account of the diseases of infancy. By the author of The Gentleman's medical pocket-book. Philadelphia: S.C. Hayes; T. Ellwood Zell, publisher, 1860.
296 p. ; 15 cm.

Originally published at Philadelphia in 1833 and again in 1854. This is the third and probably last edition of this treatise. Spine-title: *Lady's medical pocket book.*

84. AMERICAN MAGAZINE DEVOTED TO HOMOEOPATHY & HYDROPATHY.
American magazine devoted to homoeopathy & hydropathy. Containing also popular articles on anatomy, physiology, hygiene and dietetics. [Vol. 1, no. 1 (Oct. 1851)-v. 1, no. 12 (Sept. 1852)]. Cincinnati, 1851-52.
1 v. ; 23 cm.

Only the first issue of this title (October 1851) is in the Atwater Collection. The editors, J.H. Pulte (#2891-2897) and H.P. Gatchell, state that "the most important object of our publication will be the advocacy of a union of homoeopathy and hydropathy, at least in so far as to make the latter part of the former . . . " (p. 4). Issued monthly, the publication changed its title to the *American magazine of homoeopathy* in its second and final volume (Oct. 1852-Dec. 1853).

AN AMERICAN MATRON see *THE MATERNAL PHYSICIAN* (#2388).

85. AMERICAN MEDICAL ASSOCIATION.
"Female-weakness" cures and allied frauds . . . Chicago: The Journal of the American Medical Association, 1914.
72, [4] p. : ill., facsims. ; 22 cm.

An examination of twenty-four fraudulent mail order cures for the treatment of female disorders. The lengthier articles include discussions of the Viavi treatment (#2213-2220, 3645-3652), Bertha C. Day (#918), Cora B. Miller, the "Mitchella compound" of John H. Dye (#1022), and the "Wine of Cardui" manufactured by the Chattanooga Medicine Company. The publication of pamphlets such as this was part of the AMA's effort to educate the public regarding quack healers and ineffective proprietary medicines. Ranging in length from a single gathering to a hundred or more pages, these pamphlets often contain material published earlier in the pages of the *Journal of the American Medical Association*. Many of these articles and pamphlets were later gathered and reprinted in the AMA's three-volume *Nostrums and quackery* (#86). In printed wrapper. Issued with "Wine of Cardui," reprinted from *JAMA*, Dec. 5, 1914 (laid-in).

86. AMERICAN MEDICAL ASSOCIATION.
Nostrums and quackery. Articles on the nostrum evil and quackery reprinted from the Journal of the American Medical Association . . . Chicago: American Medical Association, [1911]-36.
3 v. (509 ; 832 ; xii, [2], 232 p.) : ill. ; 22 cm.

By 1906, the year the Pure Food and Drugs Act was passed by Congress, the *Journal of the American Medical Association* increasingly (if belatedly) gave attention to discussion of fraudulent practices in the manufacture and advertising of proprietary medicines. The articles gave a more professional tone to the exposure of the nostrum industry already taking place in the pages of such journals as *Colliers Weekly* and *The Ladies home journal*. In 1905 the AMA barred advertising for nostrums in the pages of its *Journal*, and in 1906 established a Propaganda Dept., headed by Arthur J. Cramp. In cooperation with the AMA's Council on Pharmacy and Chemistry (est. 1905), the Propaganda Dept. systematically investigated proprietary medicines and published the findings as articles in the *Journal*, and in some instances separately as pamphlets. Many of these articles/pamphlets were gathered together and published in 1911 as the first volume of *Nostrums and quackery*. (A second edition of the first volume, larger than the first by 199 pages, was issued in 1912.) The second volume was published in 1921; and the third volume (*Nostrums and quackery and pseudomedicine*) in 1936, the year after Cramp's retirement from what was by then known as the Bureau of Investigation. *Nostrums and quackery* remained in print through mid-century and admirably served the purpose of informing and protecting the public from "'patent medicine' exploitation and quackery." It remains a thorough and invaluable resource for understanding a complex phenomenon in the history of American popular medicine.

AMERICAN MEDICAL ASSOCIATION. COUNCIL ON HEALTH AND PUBLIC INSTRUCTION. *SEX HYGIENE PAMPHLET* see **HALL** (#1524, 1527-1529).

AN AMERICAN PHYSICIAN see **ROZIER** (#3041), **SMITH** (#3258).

AMERICAN PHYSIOLOGICAL SOCIETY (est. 1837) see **BARTLETT** (#227).

AMERICAN PUBLIC HEALTH ASSOCIATION. *LOMB PRIZE ESSAY* see **ABEL** (#4-5).

87. AMERICAN PUBLISHING AND PATENT SPECIALTY COMPANY.
The elite household guide and advertising register containing treatise on the science of health, diet, dress, personal magnetism and hypnotism, physical culture, palmistry, character reading and many of the leading subjects of the day, compiled from the writings and lectures of our greatest authorities who deal with practical experience and not theory. Also a perfect advertising medium to reach the masses, and to keep the public in touch with the leading styles, designs, fashions, prices, and inventions of the day . . . New York: American Publishing and Patent Specialty Company, [c1903].
[18], [7]-266, [32] p. ; 24 cm.

The *Elite household guide* was a mechanism for bringing advertising directly into the home by uniting it to a subject of obvious domestic importance. The text is bracketed by eighteen unnumbered preliminary pages and thirty-two final pages of advertisements. As all the firms advertised in the Atwater copy have Cincinnati addresses, it must have similarly been marketed in other cities.

88. AMERICAN PURITY ALLIANCE.
Regulation reviewed. Reasons against state sanction of social evil. New York City: Published by the American Purity Alliance, [after 1902].
16 p. ; 14 cm.

An argument against the legal and sanitary regulation of prostitution on the grounds that "vice tolerated in spots, and permitted in theory and practice, will eventually infect the whole social fabric" (p. 5). "As to ordinary disease," the pamphlet argues, "the object of wise treatment, sanitary and medical, is preventive and curative, and has for a result the final stamping out of the disease. The same purpose should be at the heart of all attempts to deal with the social evil, whether directed toward its disease elements or its moral elements" (*ibid.*).

89. AMERICAN SANITARY ASSOCIATION.
American Sanitary Association. (Egyptian Discovery Utilizied) [*oval wood-engraved logo*] "Carbolic Purifying Powder." This powder consists of a number of well-known

sanitary agents, chemically combined with carbolic acid . . . IT IS THE BEST DISINFECTANT KNOWN . . . It will be found valuable for ships, water closets, sinks, drains, damp, mouldy places, good to sprinkle on all kinds of offal . . . It is a sure preventive of small pox, cholera, typhoid, ship fever, & all other contagious diseases . . . It is cheap and perfectly safe to use in any quantity desired. No family should neglect to use it more or less . . . AMERICAN SANITARY ASSOCIATION . . . Boston . . . [ca. 1871, c1870].

[1] leaf printed on both sides ; 32 × 14 cm.

AMERICAN SCHOOL OF HOME ECONOMICS see **ELLIOTT** (#1052).

AMERICAN SCIENCE SERIES see **MARTIN** (#2379-2383).

AMERICAN SOCIAL HYGIENE ASSOCIATION see **CADY** (#549).

90. AMERICAN SOCIETY OF SANITARY AND MORAL PROPHYLAXIS.
Health and the hygiene of sex for college students. By a Member of the American Society of Sanitary and Moral Prophylaxis. [New York: American Society of Sanitary and Moral Prophylaxis, c1911].
32 p. ; 17 cm.

Series: *Educational pamphlet; no. 6.* The author considers four principle topics: "The physiological laws of sex" (most notably, that the "specific and supreme purpose of the sex function is reproduction, and not sensual pleasure"); "Fallacies about the sex function" (e.g., "sexual intercourse is a physical necessity for men and essential to their health'); "The physical consequences of unclean living" (i.e., sexually transmitted diseases); and "The control of the sex impulse" (e.g., "Every young man should build up high ideals of sex relations"; and if all else fails, try "cold sponging of the lower spine and genital organs"). Title from printed wrapper.
Composed of "members of the medical profession and of the laity, including women," the American Society of Sanitary and Moral Prophylaxis saw as its object "to limit the spread of diseases which have their origin in the Social

Evil" (*Educ. pamphlet no. 1,* #91, inside front wrapper). Toward this end, the Society published its *Transactions* (v. 1-3, 1905-10), the *Journal of social diseases* (v, 1-6, 1910-15), and seven titles in its *Educational pamphlet* series. In 1912 the organization changed its names to the Society of Sanitary and Moral Prophylaxis, and in 1916 to the New York Social Hygiene Society.

91. AMERICAN SOCIETY OF SANITARY AND MORAL PROPHYLAXIS.
The young man's problem. By a Member of the American Society of Sanitary and Moral Prophylaxis . . . Special edition printed for the Maryland Society of Social Hygiene. [New York: American Society of Sanitary and Moral Prophylaxis, 190?].
32 p. ; 17 cm.

Series: *Educational pamphlet; no. 1.* The anonymous author considers topics affecting both the physical and moral health of males between sixteen and twenty five years of age, e.g., continence, masturbation, frequenting prostitutes, sexually transmitted diseases, the importance of teaching sexual physiology and hygiene, etc. Title from printed wrapper.

92. AMERICAN SUNDAY-SCHOOL UNION.
Good health: the possibility, duty, and means, of obtaining and keeping it . . . Philadelphia: American Sunday-School Union . . . ; London: Religious Tract Society, [after 1846].
192 p. ; 15 cm.

The American Sunday-School Union's reprint (by agreement) of the undated London edition issued by the Religious Tract Society. The anonymously penned *Good health* is typical of popular manuals on hygiene in its discussion of exercise, nutrition, digestion, pure air, cleanliness and rest. Chapter VIII is entitled "On some mental and moral causes of disease and ill health," subheaded the "reciprocal influence of mind and body." The third edition of Carpenter's *Physiology* (1846) is quoted on p. 37, making this the *annus post quem* the work was published.

93. AMERICAN TEMPERANCE SOCIETY.
Permanent temperance documents of the American Temperance Society. Vol. I. Bos-

ton: Seth Bliss . . . and Perkins, Marvin, and Co. . . . [et al.], 1835 [i.e., 1836].

[6], 58, [63]-104, [111]-165, 198-295, [300]-434, 441-514 [i.e., 455], [455]-560 p. ; 22 cm.

Includes the Society's fourth (1831) through ninth (1836) annual reports. Each of these reports was also issued separately. No other volumes were published.

94. AMERICAN TRACT SOCIETY.
An appeal on the subject of cholera. [New York: Published by the American Tract Society, 183-?].
8 p. ; 13 cm.

"It has pleased God, whose judgements are unsearchable, to permit pestilence to reach our shores. The dark wing of the destroying angel, whose sword has smitten the nations of Asia and Europe, now shadows our own threshold. It is too probable that all precautions will be ineffectual to prevent the ravages of this subtle and mortal foe" (p. 1). On this fatal note, the author speculates that perhaps God would stay the hand of the avenging angel, in spite of the "many and awful . . . national sins" that justify this visitation (p. 3). The only reasonable response is resignation and repentance, for if one has "shown to others how a Christian should live, may he not teach them how a Christian may die?" (p. 8).

The American Tract Society was established at New York in 1825, and became one of the most prolific and successful publishing firms of the 19th century. Caption title. Series statement (at top of page 1): *Occasional* [no.] *4.* Imprint on page 8.

AMERICAN TRACT SOCIETY see also **WILLISON** (#3814).

95. AMERICAN VEGETARIAN SOCIETY.
Proceedings of the eleventh annual meeting of the American Vegetarian Society, held in Philadelphia, Sept. 19, 1860; and the address on the scientific basis of vegetarianism. By R.T. Trall, M.D. New York: Published for the Society, 1860.
44 p. ; 18 cm.

The modest gathering of the American Vegetarian Society's (est. 1850) eleventh annual

meeting adopted nine resolutions, epitomized by the third stating "That the various moral and physiological reforms of the age . . . will always be subject to sudden and disastrous relapses, so long as their advocates and proselytes alike indulge in such dietetic habits as pervert the normal instincts and provoke morbid habits" (p. 7), i.e., as long as their diet includes animal flesh. In all, the resolutions and commentary that precede them reveal a movement that perceived itself as small band of misunderstood zealots. The greater part of this publication is given to Russell Thacher Trall's (#3558-3593) address entitled "The scientific basis of vegetarianism" (p. 10-44). The first third of Trall's address is a rhetorical exercise; the remaining pages address the objections of popular and medical opinion to vegetarianism. In printed wrapper. Atwater copy inscribed by R.T. Trall.

96. AMERICAN WOMAN.
Suggestions for the sick room. Compiled by an American woman. New York: Anson D.F. Randolph & Company, [c1876].
72 p. ; 17 cm.

Written by a professionally trained nurse, the author acknowledges that her book draws not only from personal experience, but from the published works of such authors as Florence Lees and Florence Nightingale. The manual begins with a discussion of the personal appearance, decorum and hygiene of sick room attendants. It then succinctly covers such topics as ventilation, contagion, baths, bedding and diet, as well as those "remedies and appliances" that fall within the nurse's range of practice.

ANALYTIC SERIES OF TEACHERS' AIDS see **CLARK** (#653).

97. ANDERS, James Meschter, 1854-1936.
House-plants as sanitary agents; or, the relation of growing vegetation to health and disease. Comprising also a consideration of the subject of practical floriculture, and of the sanitary influences of forests and plantations . . . Philadelphia: J.B. Lippincott Company, 1887.
334 p. ; 20 cm.

An 1877 medical graduate of the University of Pennsylvania, Anders was on the faculty of the Medico-Chirurgical College of Philadelphia from 1892 to 1916, when it was taken over as a graduate school by the University of Pennsylvania. He remained on its faculty through 1928. Anders was the author of two important textbooks of the period: *A text-book of of the practice of medicine,* which went through fourteen editions between 1897 and 1917; and *A text-book of medical diagnosis,* first issued in 1911.

House-plants as sanitary agents was Anders first book. Based upon "personal experiments extending over a period of eight years," Anders sets forth "in plain terms, the latest light regarding the effects of some of the various physiological functions in plants and flowers upon the atmosphere in general, and the air of dwellings in particular, as well as the application of this knowledge to the laws of health" (p. 6). He discusses how house plants modify atmospheric moisture and temperature; how ozone production purifies the air (of bodily emanations from the ill as well as impurities from without); how plants mitigate the influence of communicable disease (especially chronic respiratory affections); and their beneficial effect on convalescence (both moral/aesthetic and in the promotion of sleep, appetite and digestion). The final two chapters discuss the salutary role of forests.

98. ANDERSON, Thomas A.

The practical monitor, for the preservation of health, and the prevention of diseases . . . First edition. Published and sold by Z. Jayne, of Philadelphia, Monroe County, Tennessee. F.S. Heiskell, printer, Knoxville, T., 1831.

xxii, [2], 253, [1], iv p. ; 21 cm.

"The present work has been written with a three-fold view. 1st—To hold up to the indignant reprobation of the community, the murderous atrocities of knavish imposture. 2d—To exhibit the just pretentions of the Regular Faculty, and 3d, To place in the hands of the people, a book that will honestly instruct them when, and how far they may safely interfere in the treatment of disease, and under what circumstances they should resort to professional advice" (p. [v]). *The practical monitor* "is not intended as a 'family physician,' or as a 'domestic medicine,' as those titles are generally re-

ceived and understood, but it is . . . intended to occupy the place between the physician and the people" (p. xxii). Topics are arranged in no apparent order (a non-alphabetic "index" provides a topical overview in page order). The text concludes with an essay entitled "A treatise, on the causes, character and cure of the summer and autumnal complaints, peculiar to the climate of East Tennessee" (p. 221-253).

99. ANDERSON, William Gilbert, 1860-1947.

Anderson's physical education. Health and strength, grace and symmetry . . . New York: Published by A.D. Dana, c1897.

102, [2] p. : ill. ; 19 cm.

Anderson received his medical degree from Western Reserve in 1883. He became president of the Brooklyn Normal School of Physical Education in 1885, and was associate director of the Yale Gymnasium from 1892 until his appointment as director in 1894. Anderson served as dean of the Chautauqua School of Physical Education each summer between 1885 and 1904; and was one of the organizers of the American Association for the Advancement of Physical Education. He authored several books and a number of other publications on physical training. In printed wrapper.

100. ANDREW, Thomas.

A cyclopedia of domestic medicine and surgery: being an alphabetical account of the various diseases incident to the human frame; with directions for their treatment, and for performing the more simple operations of surgery, also, instructions for administering the various substances used in medicine; for the regulation of diet and regimen; and the management of the diseases of women and children . . . Glasgow; Edinburgh; London: Blackie and Son, MDCCCXLIII. [1843]

iv, 692 p., IX leaves of plates : ill. (some col.) ; 25 cm.

"This work addresses itself especially to mothers and nurses, and to all who are intrusted with the management of a family; and to clergymen, missionaries, and other members of the community who devote no small portion of their time and means to mitigate the sufferings of their poorer brethren. To such parties this book will prove a valuable digest of medical knowledge; and from the attention that has been paid to foreign and tropical climes, the work offers a no less useful source of reference to emigrants and to travellers by sea and land . . ." (p.[iii]).

The number of North American libraries that record holdings for Andrew's *Cyclopedia* (issued six times between 1842 and 1865) is indicative of its popularity among Scottish emigrants to the United States and Canada. The text is illustrated by numerous wood engravings and is supplemented by nine leaves of engravings. Plates V-VIII contain hand-colored illustrations of medicinal plants.

101. ANDREWS, Jane, 1833-1887.
Child's health primer for primary classes. With special reference to the effects of alcoholic drinks, stimulants, and narcotics upon the human system. New York and Chicago: A.S. Barnes & Company, [c1885].
  viii, [9]-124 p. : ill. ; 19 cm.

The first of three volumes in the *Pathfinder physiology* series aimed at instructing students in the primary grades through high school in the principles of physiology and hygiene, including the effects of narcotics and alcohol. The *Child's health primer* was written anonymously by the educator and children's author Jane Andrews. In 1860 Andrews opened a primary school in her native Newburyport where she implemented many innovative ideas over the twenty-five years she maintained the school. She may best be remembered for her often re-issued children's book: *Seven little sisters who live on the round ball that floats in the air* (1st ed., 1861). Spine-title: *Pathfinder primer.*

102. ANGELL, Henry Clay, 1829-1911.
How to take care of our eyes: with advice to parents and teachers in regard to the management of the eyes of children . . . Boston: Roberts Brothers, 1878.
  70, [2] p. : ill. ; 16 cm.

First edition. "The growing prevalence of weak sight in this country, both in children and adults, would seem to make a wider knowledge of the eye, and how to take care of it, of some importance" (p. [4]). Angell provides succinct and clearly written explanations of the physiology of vision, errors of accomodation & refraction, corrective lenses, diseases of the eyes and ocular hygiene. This small manual appeared in five editions through 1891. An 1853 graduate of the Hahnemann Medical College (Philadelphia), Angell was professor of ophthalmology at Boston University School of Medicine for twenty years. He is best remembered as the author of *A treatise on the diseases of the eye,* which went through seven editions between 1870 and 1891.

103. ANTI-CIGARETTE LEAGUE.
Anti-Cigarette League song book. Containing also constitution and by-laws for local anti-cigarette leagues . . . Chicago: Anti-Cigarette League, Incorporated, 1900.
  [8] leaves ; 15 × 9 cm.

Title and imprint from wrapper. "The object of this organization shall be to combat by all legitimate means, the use of tobacco by boys especially in the form of cigarettes" (leaf [5] verso). Membership was limited to "boys under twenty-one years of age," who signed a pledge to refrain from tobacco use "at least until I reach the age of twenty-one." The order of business required much singing, to which end thirteen battle hymns are provided. To the tune of "Yankee Doodle," for instance, youthful voices might intone:

> Why don't they practice what they preach,
> And set a good example?
> As boys who know what's right and wrong
> In us you have a sample:
> We know it's wrong to smoke and chew,
> We never mean to do it;
> We'll let the dirty stuff alone,
> A decent hog won't chew it.

ANTI-TOBACCO TRACT DEPOSITORY see **AMERICAN ANTI-TOBACCO SOCIETY** (#75).

APPERT, Nicolas see **COOPER** (#788).

APPLETON'S HOME BOOKS see **GUERNSEY** (#1442).

APPLETON'S SCIENCE TEXT-BOOKS see
   TRACY (#3551, 3552)

ARISTOTLE, pseud. *Note that the listing of works attributed to the Pseudo-Aristotle in this catalog (#104-147) departs from the normal order of arrangement. Entries are arranged chronologically by date of publication rather than alphabetically by title.*

---

104. ARISTOTLE, pseud.

Aristotle's compleat and experienc'd midwife; In two parts. I. A guide for child-bearing women, in the time of their conception, bearing and suckling their children; with the best means of helping them, both in natural, and unnatural labours: together with suitable remedies for the various indispositions of new-born infants. II. Proper and safe remedies for the curing all those distempers that are incident to the female sex ; and more especially those that are any obstruction to their bearing of children . . . Made English by W__ S__, M.D. The fifth edition. London: Printed, and sold by the booksellers, [172-?]

[4], iv, 156, [4] p. : ill. ; 15 cm. (12mo)

The first edition of the *Compleat and experienc'd midwife* appears to have been the 1700 London edition. It enjoyed a long publishing history as a separately issued title and as part of the many editions of the "complete works" of the Pseudo-Aristotle published during the 18th and 19th centuries. Only isolated passages in these treatises can be traced back to the actual works of Aristotle, who wrote extensively on generation, but nothing on human sexuality, reproduction or parturition exclusively. As Otho Beall noted, the treatises that comprise the Pseudo-Aristotle "more often show a sublime disregard of his [Aristotle's] teachings." Nonetheless, "These random passages provide the slender link between the real Aristotle and the manuals which bear his name" ("Aristotle's Master Piece in America," *William & Mary Q.*, 1963, 3rd ser., 20:210).

As to authorship Beall speculates, "A literary pirate must have used as a starting point Aristotle's remarks on the subject of human reproduction, then have added material from certain spurious and doubtful works of Aristotle, especially the *Problems of Aristotle*" (*ibid.*). The material added was considerable, and far more

extensive than the sections on generation in the *Problemata*. Several names have been preferred as bearing partial or whole responsibility for the various Pseudo-Aristotle texts. This fifth edition of the *Compleat and experienc'd masterpiece* was "made English" [i.e., edited] by "W.S." or William Salmon (1644-1713). Charitably described in the *Dict. nat. biog.* as an "empiric," Salmon "treated all diseases, sold special prescriptions of his own, as well as drugs in general, cast horoscopes, and professed alchemy" (17:698). Salmon was a prolific author, including such domestic medical treatises as *Seplasium. The compleat English physician* (1693), *The family dictionary* (1698) and *The country physician: or, a choice collection of physick: fitted for vulgar use* (1703). Salmon enraged the College of Physicians by translating the *Pharmacopoeia Londinensis* into English, as did another possible compiler, Nicholas Culpeper (1616-1654). Self-trained in the astrological and medical arts, Culpeper was the author of numerous popular treatises on medicine. The likelihood of Culpeper's involvement (either as an editor or as one whose works were gleaned) is evident in a comparison of descriptions of the clitoris from his *Directory for midwives* (1st ed., 1651) and the *Compleat and experienc'd midwife.*

Culpeper

"The clitoris is a sinewy and hard body, full of spongy and black matter within, as the side ligaments of the yard are; in form it represents the yard of a man, and suffers erection and falling, as that doth; that this is what causeth lust in women, and gives delight in copulation, for without this a woman neither desires copulation, nor hath pleasure in it, nor conceives by it."

Pseudo-Aristotle

". . . just above the urinary passage, may be observed the clitoris, which is a sinewy and hard body, full of spongy and black matter within, like the side-ligaments of the yard, representing in form the yard of a man, and suffers erection and falling as that doth; in proportion to the desire a woman hath in copulation: and this also is what gives a woman delight in copulation: for without this a woman hath neither a desire to copulation, and delight in it, nor conceive by it."

Whatever the roles of Salmon, Culpeper and "Dr. Borman" (translator of the *Last legacy*) in the compilation of the Pseudo-Aristotle, the editors obviously drew from a wide selection of published authorities, both ancient and modern. Much work remains to determine the the fields gleaned by the compilers.

105. ARISTOTLE, pseud.

Aristotle's works compleated. In four parts. Containing I. The Compleat master-piece: displaying the secrets of nature in the generation of man. II. His Compleat and experienced midwife: being a guide for childbearing women. III. His Book of problems: wherein is contained divers questions and answers touching the state of man's body. IV. His Last legacy; or, his golden cabinet of secrets opened. London: Printed, and sold by the booksellers of London. M.DCC.XXXIII. [1733]

4 pts. in 1 v. ([1-2], [2], 3-144 p., [1] folding plate ; [4], iv, 156, [4] p. ; [4], 152 p. ; [8], 111 p.) : ill. ; 15 cm. (12mo)

This 1733 London edition is the earliest in the Atwater Collection to include all four titles that traditionally constitute the complete Pseudo-Aristotle. This edition may well have been the first of the Pseudo-Aristotle compilations that were published throughout the 18th and early 19th centuries as *Aristotle's works completed, or The works of Aristotle, the famous philosopher.*

Each of the four parts has a separate title-page, collation and pagination, and had been issued separately before their compilation in the present volume. The title-page to the *Compleat master-piece* is designated the 19th edition, and like the general title-page, is dated 1733. Its typefaces and typesetting closely resemble that of the *Compleat and experienc'd midwife* that follows. The title-page of the *Compleat and experienc'd midwife* is designated the sixth edition, and is undated. Though the pagination is the same as the 5th edition of this title (#104) and its typefaces similar, the type and type ornaments are different, and the pages not quite identical in composition. The 6th edition does include the same woodcut frontispiece as the 5th edition. The title-page of the *Book of problems* is designated the 25th edition, and is undated. The undated imprint of the *Last legacy*

bears the names of A. Bettesworth, C. Hitch, J. Osborne and T. Hodges, all of whom were active in London in the 1720s and 1730s.

These treatises mix ideas on anatomy and reproductive physiology derived from classical, medieval and Renaissance medical sources with gynecological lore derived from folk-sources. Issued separately or together, they were the most popular English-language manual(s) of sexual fact and lore on both sides of the Atlantic for a century and a half. Practical information of varying degrees of accuracy is balanced with frank avowals of conjugal pleasure and pious reminders of the virtues of marriage, monogamy and child-bearing. Notably lacking from the pages of the Pseudo-Aristotle, however, is any reflection of the "scientific midwifery" that emerged in late 17th- and 18th-century London and Edinburgh.

Part I. of the *Compleat master-piece* is devoted to the anatomy of the male and female genitalia. Twice as much text is given to a description of the "secret parts of women" than to the male organ. In part II. on conception, the "seed of the man" is described as the "active principle, or efficient cause of the foetus, the matter of which is arterial blood, and animal spirits" (p. 38), and the ovum as the "passive principle." Impregnation takes place in the womb. The signs of conception are briefly noted, including how to determine whether the fetus be male or female. The author warns women trying to conceive not to engage in intercourse too often, "For satiety glutes the womb, and renders it unfit to do its office." By way of example he points out that "common whores (who often use copulation) have never, or very rarely, any children; for the grass seldom grows in a path that is commonly trodden in" (p. 46). Considerable attention is given to the causes of barrenness and "insufficiency in men." One of the author's recommendations for those of "frigid disposition" and "cold distemper" is a diet of partridges, quails and sparrows: "Being extremely addicted to venery, they work the same effects in those who eat them" (p. 54). There follows discussions of the management of pregnancy, the qualifications of the midwife, and surprisingly brief directions for managing natural and preternatural labors.

Part II. concludes with a discussion of "moles or false conceptions" and teratology. Moles result from "corrupt and barren seed in the man" mixed with menstrual blood in the womb. Monsters might result from God's wrath upon the parents, the marking of the fetus by maternal imagination, coition during menstruation, or copulation with animals. This section includes four crudely designed and executed woodcuts copied from one of the several London editions of *The workes of that famous chirurgeon Ambrose Parey* (i.e., Ambroise Pare, 1510?-1590), "Lib. XXV. Of monsters and prodigies." The English translation of Pare's treatises on generation and teratology was probably just one of the sources upon which the compilers of the Pseudo-Aristotle drew.

Part III. of the *Compleat master-piece* is a treatise on "physiognomy," describing natural signs that indicate the "inclinations and dispositions" of character visible in the form of every part of the body. D'Arcy Power thought this part taken from the physiognomy of Polemon (2nd cent. A.D.), whose treatise in Greek was widely available in Latin and vernacular translations. Other than the subject, the physiognomy in the *Master-piece* bears little resemblance to Polemon. Appended to the *Master-piece* in most editions is the *Family physician* (in some editions, *A treasure of health*). It is a miscellaneous compilation of recipes for the cure of a wide variety of disorders that is sometimes attributed to Hippocrates, with about as much accuracy as assigning the *Master-piece* to Aristotle.

The *Compleat and experienc'd midwife* is divided into two parts: the first entitled "A guide for child bearing women" and the second on "proper and safe remedies for curing all those distempers that are peculiar to the female sex." Part I is very similar to the wording and topic order to the *Compleat master-piece,* except that the anatomical descriptions are more detailed and the chapter on barrenness has been moved further back in the text. Chapters II and III in part II. discuss gynecological disorders, and are unique to the *Experienc'd midwife.*

The *Book of problems* is a medieval compilation of questions and answers to a wide variety of topics in natural history. It was long attributed to Aristotle, and in most bibliographies the *Problemata* are included among his doubtful and spurious works. Comparatively little of the book pertains to reproduction *per se*, apart from half a dozen sections headed "Of carnal copulation," "Of the seed of man or beast," "Of hermaphrodites," etc. Its attribution to Ari-

stotle, the topicality of part of its contents, and a long publishing history seem to have been enough to warrant its inclusion in the complete Pseudo-Aristotle.

The *Last legacy* is a much briefer treatise on the same themes as the *Master-piece* and *Experienc'd midwife.* It has nine chapters: on virginity, the anatomy of the female genitalia (similar in wording and order to the corresponding chapters of the *Master-piece*), copulation, the male genitalia, conception, barrenness, on the pleasures and advantages of marriage, conduct during coition to insure conception, and directions for midwives.

106. ARISTOTLE, pseud.

Aristotle's compleat master piece. In three parts: Displaying the secrets of nature in the generation of man: Regularly digested into chapters and sections, rendering it far more useful and easy than any yet extant. To which is added, A treasure of health; or, the family physician: Being choice and approved remedies for all the several distempers incident to human bodies. The twenty-second edition. [London]: Printed and sold by the booksellers, 1741.

viii, 9-144 p. : ill. ; 15 cm. (12mo)

The 1733 19th edition of *Aristotle's compleat master piece* (#105) is the earliest of five 144 page editions of this title in the Atwater collection. Though their pagination is the same, the typesetting is new in each edition. Each of these 144 page editions is illustrated in the same iconographic tradition, if not always from the same blocks. For example, seven of the nine woodcuts that illustrate this 1741 edition of the *Master-piece* were printed from blocks used in the illustration of the 1733 London edition (#105). The physiognomical head and the palmistry hand are new (though possibly not to the 22nd edition), and were used in the illustration of the 1749 (#107), 1759 (#108) and 1763 (#109) editions. By the time the 1749 edition was issued, the woodcut depicting the boy with one head, four arms, four legs, etc. had been recut. It appears again in the 1759 and 1763 editions. The woodcut block depicting the zodiac man in the 1759 edition differs from the block used in the 1749 edition, and was used again in illustrating the 1763 edition. Even as late as the 1763 29th edition (#109), five of the nine blocks used in its illustration had been used in the 1733 edition. This 144 page edition of *Aristotle's compleat master piece* was issued as late at 1782 (31st ed.), presumably with illustrations in this same iconographic tradition.

107. ARISTOTLE, pseud.

Aristotle's compleat master piece. In three parts: Displaying the secrets of nature in the generation of man: Regularly digested into chapters and sections, rendering it far more useful and easy than any yet extant. To which is added, A treasure of health; or, the family physician: Being choice and approved remedies for all the several distempers incident to human bodies. The twenty-third edition. [London]: Printed and sold by the booksellers, 1749.

    viii, 9-144 p. : ill. ; 15 cm. (12mo)

108. ARISTOTLE, pseud.

Aristotle's compleat master piece. In three parts: Displaying the secrets of nature in the generation of man. Regularly digested into chapters and sections, rendering it far more useful and easy than any yet extant. To which is added, A treasure of health; or, the family physician: Being choice and approved remedies for all the several distempers incident to human bodies. The twenty-seventh edition. [London]: Printed and sold by the booksellers, 1759.

    viii, 9-144 p. : ill. ; 15 cm. (12mo)

109. ARISTOTLE, pseud.

Aristotle's compleat master piece, in three parts: Displaying the secrets of nature in the generation of man: Regularly digested into chapters and sections, rendering it far more useful and easy than any yet extant. To which is added, A treasure of health; or, the family physician: Being choice and approved remedies for all the several distempers incident to human bodies. The twenty-ninth edition. London: Printed and sold by the booksellers, 1763.

    viii, 9-144 p. : ill. ; 15 cm. (12mo)

110. ARISTOTLE, pseud.

[The works of Aristotle, the famous philosopher. London, ca. 1766].

    4 pts. in 1 v. (viii, 9-114 [i.e., 142], [2] p., [1] folding leaf of plates; [4], iv, 156,

Fig. 1. Frontispiece to Aristotle's compleat master piece, *#108.*

[4] p. ; [4], 152; viii, 9-120 p.) : ill. ; 15 cm. (18mo)

This volume does not have the title-page that usually introduces compilations of the Pseudo-Aristotle's complete works. As in many such collections, each of the four titles was published separately at an earlier date before being gathered in a single volume. This edition of the complete Pseudo-Aristotle includes: *Aristotle's compleat master piece . . . To which is added, A treasure of health; or, the family physician. The twenty-seventh edition* ([London]: Printed and sold by the booksellers, 1766); *Aristotle's compleat and experienc'd midwife . . . The eleventh edition* (London: Printed and sold by the booksellers, [175-?]); *Aristotle's book of problems . . . The twenty-sixth edition* (London: Printed for and sold by J.W.

... [et al.], [175-?]); *Aristotle's last legacy*... (London: Printed and sold by the booksellers, 1764).

111. ARISTOTLE, pseud.
The whole of Aristotle's compleat masterpiece, in three parts: displaying the secrets of nature in the generation of man. Regularly digested into chapters and sections, rendering it far more easy than any yet extant. To which is added A treasure of health: or the family physician. Being choice and approved remedies for all the several distempers incident to human bodies. The fifty second edition. [Edinburgh?]: Printed in the year M,DCC,LXXIV. [1774]
[2], x [i.e., ix], [1], 152 p., [1] folding plate : ill. ; 16 cm. (12mo)

Printer's imprint (p. 152): "Edinburgh; printed and sold at the printing-office, in James' Court. Where may be had Culpepper's English physician, and compleat midwifery."

112. ARISTOTLE, pseud.
The works of Aristotle, the famous philosopher. In four parts. I. His Complete master-piece; displaying the secrets of nature in the generation of man.—To which is added, The family physician . . . II. His Experienced midwife; absolutely necessary for surgeons, midwives, nurses, and childbearing women. III. His Book of problems, containing various questions and answers, relative to the state of man's body. IV. His last legacy; unfolding the secrets of nature

in the generation of man. A new edition. London: Printed for the booksellers. MDCCXI. [i.e., 1791-92]
289 p. : ill. ; 14 cm.

Unlike many earlier compilations of the Pseudo-Aristotle, the pagination is consecutive throughout this edition, i.e., all four treatises were printed specifically for this edition and had no prior, independent publishing history. The title-page is dated "MDCCXI," a typesetting error that should probably read "MDCCXCI" or "MDCCXCII." The only treatise of the four to have a distinct title-page in this edition is the *Last legacy*, which is dated "MDCCXCII." Atwater copy lacking pages 15h-44.

113. ARISTOTLE, pseud.
Aristotle's master-piece completed. In two parts. The first containing the secrets of generation in all the parts thereof . . . The second part being a private looking-glass for the female sex . . . New-York, Printed for the Company of Flying Stationers. 1793.
130 p. : ill. ; 13 cm.

The 1793 New York edition represents a different textual tradition from the preceding editions, as evidenced by the following table of chapter headings in part one of the New York edition compared with parts one and two of the London editions:

Table 1. A comparison of headings in the 1733 London and 1793 New York edition of the Pseudo-Aristotle (#104-147).

Text of the 1733, 1741 & 1774 editions:	1793 New York edition:
PART ONE. I. A particular description of the parts or instruments of generation, both in men and women.	I. Of marriage . . .
II. Of the restriction laid upon men in the use of carnal copulation, by the institution of marriage . . .	II. How to get a male or female child
III. Of virginity, what it is, how it may be known . . .	III. The reason why children are like their parents, and that the mother's imagination contributes thereto . . .
PART TWO. I. What conception is; what is pre-requisite thereto . . . II. How a woman should order herself that desires to conceive . . .	IV. A discourse of man's soul, that it is not propagated by the parents . . . V. Of monsters, and monstrous births . . .

III. How the child lieth, and how it groweth up in the womb of the mother after conception.	VI. A discourse of the happy state of matrimony . . .
IV. Of the obstructions of conception; with the cause and cure of barrenness . . .	VII. Of errors in marriage . . .
V. How child-bearing women ought to govern themselves during the time of their pregnancy.	VIII. The opinion of the learned concerning children, conceived and born within seven months . . .
VI. Directions for midwives . . .	IX. Of the green sickness in virgins . . .
VII. What unnatural labour is . . .	X. Virginity . . .
VIII. How child-bed women ought to be ordered after delivery.	XI. Directions and cautions for midwives, and how first a midwife ought to be qualified
IX. Of a mole, or false conception; and also of monsters and monstrous births . . .	XII. Further directions for midwives . . .
	XIII. Of the genitals of women . . . XIV. A description of the woman's fabrick . . . XV. A description of the use and abuse of several parts in women appointed in generation. XVI. Of the organs of generation of man. XVII. A word of advice to both sexes: being several directions respecting copulation.

Much of the text of the 1793 New York edition derives from the same textual tradition as the London editions, though it appears in different order and under different chapter headings. "Of marriage," for instance, the first chapter in the New York edition, is nearly identical in text to pt. I, ch. I, para. 3 of the London editions. The discussion of "barrenness" that appears in ch. IX of the New York edition is similar to pt. II, ch. 4 in the London editions. The description of the female genitalia in chs. XIII and XIV of the New York edition is an expurgated version of pt. I, ch. I, para. 2 in the London editions, etc.

One of the obvious differences between the 1793 New York edition and the London editions is that even when the text is nearly identical, the former lacks many of the more descriptive and erotic passages of the latter, and includes none of the verses (often erotic) that are interspersed throughout the London editions. Finally, this New York edition of the *Master-piece* does not include the discussion of physiognomy that constitutes the third part of the London editions.

The illustration of this New York edition varies as well. It includes the frontispiece of the "maid all hairy and infant born black," and a reduced copy of the depiction of the gravid uterus found in the London editions. The New York edition has six rather than four illustra-

tions in the section on monsters, however. These illustrations may be woodcuts, but given their crude execution, might also be metal relief cuts or perhaps even wood engraving, which in the 1790s was in its experimental stages in New York.

REFERENCES: Not in *Austin*.

114. ARISTOTLE, pseud.
The works of Aristotle, the famous philosopher. In four parts. I. His Complete master-piece: displaying the secrets of nature in the generation of man.—To which is added, The family physician; being approved remedies for the several distempers incident to the human body. II. His Experienced midwife; absolutely necessary for surgeons, midwives, nurses, and child-bearing women. III. His Book of problems, containing various questions and answers, relative to the state of man's body. IV. His Last legacy; unfolding the secrets of nature respecting the generation of man. A new edition. London: Printed for the booksellers. MDCCXCIV [1794, i.e., 1795].
    iv, [5]-224, [4], [225]-312, [243]-289 p. : ill. ; 14 cm. (12mo)

The *Books of problems* and the *Last legacy* have separate title-pages dated 1795.

115. ARISTOTLE, pseud.
The works of Aristotle, the famous philosopher. In four parts. Containing I. His Complete master-piece; displaying the secrets of nature in the generation of man. To which is added, The family physician . . . II. His Experienced midwife; absolutely necessary for surgeons, midwives, nurses, and child-bearing women. III. His Book of problems, containing various questions and answers, relative to the state of man's body. IV. His Last legacy; unfolding the secrets of nature respecting the generation of man. A new edition. Philadelphia: Printed for the booksellers. MDCCXCVIII. [1798].
72, 100, 68, 39 (i.e., 34) p. : ill. ; 17 cm. (12mo)

In this compilation, the *Complete master-piece* does not have a distinct title-page. The text of this edition follows the tradition of the 1733, 1741, 1774 and 1791 editions in the Atwater Collection, differing only in the exclusion of the physiognomical treatise. The chapter on monstrous births is illustrated with relief cut copies of the same four woodcuts in the 1733 London edition. The 1798 Philadelphia edition was apparently not issued with the "maid all hairy and infant born black" frontispiece, and does not include the depiction of the gravid uterus common to other editions of the *Masterpiece* in this textual tradition.

The Experienced midwife has a separate title-page dated 1799, and was issued separately that year (*Austin #73*). The *Book of problems* and the *Last legacy* also have separate title-pages, each dated 1792.

Otho Beall has remarked that the Pseudo-Aristotle "provided the only works on sex and gynecology . . . which were widely available to eighteenth-century Americans . . . presenting . . . gleanings from the gynaecological lore of western civilization, including Greek, Roman, Arabic, medieval, Renaissance, and later traditions. Probably in no area of popular medicine has the influence of folklore been so persistent as in the field of sex and human reproduction." ("Aristotle's Master Piece in America," *William and Mary Q.*, 1963, 3rd ser., 20:208-09).
REFERENCES: *Austin #74*.

116. ARISTOTLE, pseud.
The works of Aristotle, the famous philosopher. In four parts. Containing, His Complete master piece: displaying the secrets of nature, in the generation of man,—To which is added, The family physician: being approved remedies for the several distempers incident to the human body. II. His Experienced midwife: absolutely necessary for surgeons, midwives, nurses, and child-bearing women. III. His Book of problems: containing various questions and answers relative to the state of man's body. IV. His Last legacy: unfolding the secrets of nature respecting the generation of man. The most approved edition. London [i.e., New York?]: Printed for the booksellers, 1801.
342 p. : ill. ; 16 cm. (12mo)

The extraordinarily crude relief cuts which illustrate this 1801 edition (wood engravings, metal cuts?) are identical to those used in the illustration of the 1793 New York edition (#113). It is more than likely, therefore, that the London imprint is false and that this "London" edition was actually printed at New York in 1801. Atwater copy lacking pp. 333-342; the text along the outer margins of pp. 328-332 is missing.

117. ARISTOTLE, pseud.
The works of Aristotle, the famous philosopher. In four parts. Containing I. His Complete master-piece; displaying the secrets of nature in the generation of man. To which is added, The family physician; being approved remedies for the several distempers incident to the human body. II. His Experienced midwife; absolutely necessary for surgeons, midwives, nurses, and child-bearing women. III. His Book of problems, containing various questions, and answers, relative to the state of man's body. IV. His Last legacy; unfolding the secrets of nature respecting the generation of man. A new edition. London: Printed for the booksellers, 1802.
240 p. : ill. ; 18 cm. (12mo)

The text is illustrated with four competently executed wood engravings of monstrous births unrelated to the blocks used in the illustration of any other edition of the Pseudo-Aristotle in the Atwater collection.

118. ARISTOTLE, pseud.

Aristotle's complete master-piece, in two parts: displaying the secrets of nature in the generation of man. Regularly digested into chapters, rendering it far more useful and easy than any yet extant. To which is added A treasure of health: or the family physician . . . The sixth New-England edition. New-England: Printed and sold by all the principal booksellers in the United States, 1811.
    [2], [9]-78 p. : ill. ; 17 cm.

The textual tradition in this sixth New England edition is that to which the 1793 New York edition belongs (#113). The chapter "Of monsters, & monstrous birth" is illustrated with five wood engravings.

119. ARISTOTLE, pseud.

The works of Aristotle. In four parts. Containing, I. His Complete master-piece; displaying the secrets of nature in the generation of man. To which is added, The family physician, being approved remedies for the several distempers incident to the human body. II. His Experienced midwife; absolutely necessary for surgeons, midwives, nurses, and child-bearing women. III. His Book of problems; containing various questions and answers relative to the state of man's body. IV. His Last legacy; unfolding the secrets of nature respecting the generation of man. An enlarged edition, embellished with several fine engravings. London: Published for the booksellers, 1812.
    [2], iv, [5]-360 p. : ill. ; 14 cm. (12mo)

This edition of the complete Pseudo-Aristotle is illustrated with eleven wood engravings. The "embellished . . . fine engravings" referred to on the title-page are doubtless the depiction of the gravid uterus (p. [104]) and twins *in utero* (p. 175) that appear in the *Experienced midwife*. Both were copied from engravings published in William Smellie's obstetrical atlas. Half-title: *A new and enlarged edition of the Works of Aristotle*. At bottom of half-title: *J. Kendrew, printer, Collier-Gate, York.*

120. ARISTOTLE, pseud.

The works of Aristotle, the famous philosopher in four parts. Containing I. His Complete masterpiece; displaying the secrets of

Fig. 2. Frontispiece to The works of Aristotle, *#119.*

nature in the generation of man. To which is added, The family physician . . . II. His Experienced midwife; absolutely necessary for surgeons, midwives, nurses and child bearing women. III. His Book of problems, containing various questions and answers, relative to the state of man's body. IV. His Last legacy; unfolding the secrets of nature respecting the generation of man. A new edition. New England: Printed for the proprietor, February 1813.
    264 p. : ill. ; 18 cm.

The 1813 New-England edition follows the textual tradition of the 1733 *et al.* London editions. According to Otho Beall, "The lavish emphasis on the mysterious female sex suggests the book was designed primarily for masculine readers" ("Aristotle's Master Piece in America," *William & Mary Q.,* 1963, 3rd ser., 20:217). Stephen Nussenbaum echoes this opinion, "*The Works of Aristotle* . . . are essentially an eighteenth century male fantasy about what women are really like" (*Sex, debility and diet in Jacksonian America,* Westport, Ct., 1980, p. 27). By these remarks both authors seem to imply that 18th- and 19th-century readers' interest in the Pseudo-Aristotle was primarily pornographic, which is both undocumented and unlikely. Even in terms of the books erotic undertone, they assume that women were not as interested in reading about sex as men. It seems more likely that the book's popularity lay primarily in its provision of information on reproduction, sexual physiology and sexual behavior, and only secondarily as entertainment. The existence of a textual tradition of the *Master-piece* that was expurgated of most of the erotic passages and all the poetry (e.g., #113, 118) confirms the book's importance as popular instruction. The several copies of the Pseudo-Aristotle in the Atwater Collection inscribed with women's names provides further evidence of the book's appeal to a female audience.

REFERENCES: *Austin* #76.

121. ARISTOTLE, pseud.

Aristotle's master-piece, completed. In two parts. The first containing the secrets of generation, in all the parts thereof: treating of the benefit of marriage, and the prejudice of unequal matches. Signs of insufficiency in men or women. Of the infusion of the soul. Of the likeness of children to parents. Of monstrous births. The cause and cure of the green sickness. A discourse on virginity. Directions and cautions for midwives. Of the organs of generation in women, and the fabric of the womb. The use and action of the genitals. Signs of conception, and whether a male or female; with a word of advice to both sexes in the act of copulation. And the pictures of several monstrous births, &c. The second part being a private looking-glass for the female sex. Treating of various maladies of the womb, and of all other distempers incident to women of all ages, with proper remedies for the cure of each. The whole being more correct, than anything of the kind hitherto published. New-York: Printed for the Company of Flying Stationers, 1816.

132 p. : ill. ; 17 cm.

The 1816 New York edition is of the same abridged textual tradition as the 1793 New York edition (#113). The illustrations, however, represent an entirely new tradition in the iconography of the Pseudo-Aristotle in the United States. The crude execution of the relief cuts in every earlier English and American edition is supplanted here by wood engravings of more skillfull design and execution. Proof copies of the figures that illustrate the chapter on monstrous births exist in the scrapbooks of the American wood engraver Alexander Anderson (1775-1870) at the New York Public Library. The illustrations in this 1816 New York edition (and presumably earlier editions with the same imprint?) were patterned on these proofs, but not printed from the same blocks. Jane Pomeroy has noted that Anderson was working on metal relief cuts for the *Master-piece* as early as 1793 ("Alexander Anderson's life and engravings before 1800," *Proc. of the Amer. Antiquarian Soc.,* 1990, p. 181). It is uncertain whether these designs were original to Anderson, or whether he copied them from an earlier English edition. In the Atwater Collection, this iconographic tradition for the monster figures in the *Master-piece* reappears in a series of London editions published in the 1830s and 1840s (e.g., #129).

REFERENCES: Not in *Austin,* which lists a 126 p. edition of this title published in New York in 1816 (*Austin* #68).

122. ARISTOTLE, pseud.

The works of Aristotle. In four parts. Containing, I. His Complete master-piece; displaying the secrets of nature in the generation of man. To which is added, The family physician, being approved remedies for the several distempers incident to the human body. II. His Experienced midwife; absolutely necessary for surgeons, midwives, nurses, and child-bearing women. III. His Book of problems; containing various questions and answers relative to the state of man's body. IV. His Last legacy; unfolding the secrets of nature respecting the generation of man. Third edition . . . London: Printed for the booksellers, 1819.

[2], iv, [5]-167, 170-360 p. : ill. ; 14 cm.

The half-title (*A new and enlarged edition of the Works of Aristotle*) includes the printer's imprint: *J. Kendrew, printer, Colliergate, York.* This edition of the Pseudo-Aristotle is illustrated from the same wood-engraved blocks used in the illustration of the 1812 London edition (#119), which was also printed by J. Kendrew at "Collier-Gate, York."

123. ARISTOTLE, pseud.

The works of Aristotle, the famous philosopher. In four parts. Containing, I. His Complete master piece: displaying the secrets of nature, in the generation of man. To which is added, The family physician . . . II. His Experienced midwife: absolutely necessary for surgeons, midwives, nurses, and child-bearing women. III. His Book of problems: containing various questions and answers, relative to the state of man's body. IV. His Last legacy: unfolding the secrets of nature, respecting the generation of man. A new and improved edition. London: Printed for Tonson, Fletcher, and Knopton, and sold by all other booksellers, [182-?].
   [3]-6, 253 (i.e., 535) p. : ill. ; 16 cm.

Conjugate leaves Y3,4 and Hh1,4 have watermarks dated 1811. This edition has identical pagination with that issued at London by Miller, Law & Carter between 1820 and 1830. The text of the *Master-piece* is the abridged or expurgated version. The text of the *Experienced midwife* is also a variant from the tradition of the 1733 *et al.* London editions. The *Last legacy* is much abridged from the 1733 *et al.* London editions. Notably missing from this last treatise are the chapters describing the male and female genitalia and advice on copulation.

124. ARISTOTLE, pseud.

The works of Aristotle, the famous philosopher: containing his complete master-piece, displaying the secrets of nature in the generation of man, and the maladies incident to females, with proper remedies for their cure: to which is added, The family physician, being remedies for various distempers incident to the human body; also, his Experienced midwife, containing important information, absolutely necessary for surgeons, midwives, nurses, and all child-bearing women; and, his Last legacy, unfolding the secrets of nature in the generation of man. A new edition, with cuts. Derby:

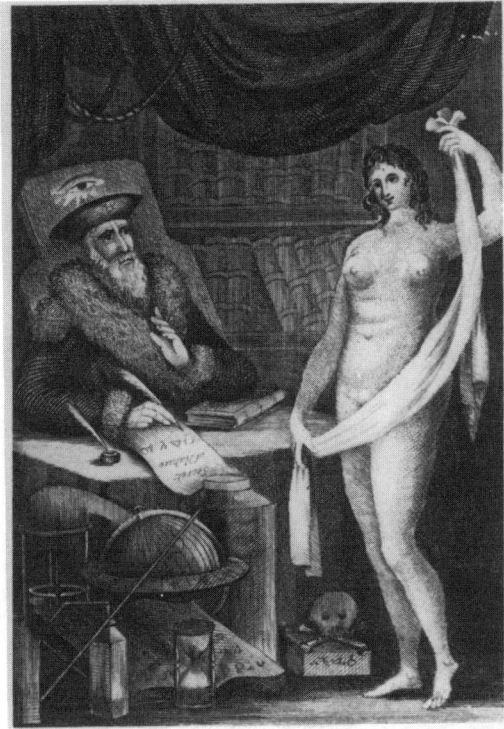

Fig. 3. Frontispiece to The works of Aristotle, *#124.*

Published by Thomas Richardson, [182-?].
   iv, 314 p., [2] leaves of plates : ill. ; 15 cm.

With an engraved frontispiece and an added engraved title-page (also undated). The text is the abridged version. The "cuts" indicated in the editon statement are wood engravings.

125. ARISTOTLE, pseud.

The works of Aristotle, the famous philosopher, in four parts. Containing I. His Complete Master-piece; displaying the secrets of nature in the generation of man. To which is added, The family physician; being approved remedies for the several distempers incident to the human body. II. His Experienced midwife; absolutely necessary for surgeons, midwives, nurses and child bearing women. III. His Book of problems, containing various questions and answers, relative to the state of man's

Fig. 4. Frontispiece to The works of Aristotle, *#126.*

body. IV. His Last legacy; unfolding the
secrets of nature respecting the generation
of man. A new edition. New-England:
Printed for the publishers, 1821.
    286 p. : ill. ; 18 cm.

126. ARISTOTLE, pseud.
The works of Aristotle, the famous philosopher: containing, his Complete master-
piece, displaying the secrets of nature in
the generation of man: to which is added,
The family physician . . . Also, his Experienced midwife, absolutely necessary for
surgeons, midwives, nurses, and childbearing women: and his Last legacy, unfolding
the secrets of nature in the generation of
man. A new and improved edition . . . London: Printed for Miller, Law, and Carter,
and sold by all the booksellers, 1826.
    vi, [11]-216, [2] p., [6] leaves of plates :
ill. ; 15 cm.

The abridged edition. The text of the *Last
legacy* in this edition has been reduced to four
chapters, in comparison with the nine chapters
of the full text in the 1733 London edition
(#105).

127. ARISTOTLE, pseud.
The works, of Aristotle, the famous philosopher, in four parts, containing I. His Complete master-piece; displaying the secrets
of nature in the generation of man. To
which is added, The family physician . . .
II. His Experienced midwife; absolutely
necessary for surgeons, midwives, nurses,
and child bearing women. III. His Book of
problems, containing various questions
and answers, relative to the state of man's
body. IV. His Last legacy; unfolding the
secrets of nature respecting the generation
of man. A new edition, with engravings.
New-England: Printed for the publishers,
1828.
    288 p. : ill. ; 18 cm.

The unexpurgated version of the text, including the third part on physiognomy in the
Master-piece, often omitted from American editions. The *Experienced midwife* in this edition is
unusual in its illustration with a single wood-
engraved illustration ("Explanation of the different parts of the womb, &c."). Atwater copy
lacking pp. 67-70.

128. ARISTOTLE, pseud.
The works of Aristotle, the famous philosopher. In four parts. Containing, I.
His Complete masterpiece, displaying
the secrets of nature in the generation
of man. To which is added, The family
physician . . . II. His Experienced midwife, absolutely necessary for surgeons,
midwives, nurses, and child-bearing
women. III. His Book of problems, containing various questions and answers
relative to the state of man's body. IV.
His Last legacy, unfolding the secrets of
nature in the generation of man. A new
and improved edition. [London]: Miller,
Law, and Carter, [183?].
    iv, [5]-300 p., [1] leaf of plates : ill. ; 16
cm.

129. ARISTOTLE, pseud.
Aristotle's master-piece, completed in two parts. The first containing the secrets of generation, in all the parts thereof . . . The second part being a private looking glass for the female sex . . . London: Published by G. Davis, [183-?].
iii, [4], 142 p., [1] leaf of plates : ill. ; 15 cm.

The abridged version. The earliest 142 p. edition of the *Master-piece* published at London by G. Davis appears to have been the 1830 edition. The frontispiece in this issue is a separately printed, hand-colored wood-engraved plate (labelled "Widow" in this issue; "An amorous widow" in other issues). It is interesting that the iconographic tradition of the wood engravings derives from the same model used for the illustration of the 1816 New York edition (#121). Cover-title (paper over boards): *Aristotle's master-piece. Illustrated.* [wood-engraved vignette] *Complete edition.*

130. ARISTOTLE, pseud.
The works of Aristotle, the famous philosopher. In four parts. Part First. His Complete master-piece, in two books. The first book displays the secrets of nature in the generation of man, and the second treats of the maladies incident to females, with proper remedies for their cure. To which is added, The family physician . . . Part second. His Experienced midwife, in two books. Containing important information, absolutely necessary for surgeons, midwives, nurses, and all child-bearing women. Part third. His Book of problems, containing questions and answers relative to the state of man's body. Part fourth. His Last legacy, unfolding the secrets of nature in the generation of man. A new and improved edition. London: Printed for Cocker, Harris, and Finn, and sold by all booksellers, [183-?].
vi, [7]-141, [3], 156 p : ill. ; 16 cm.

Cocker, Harris & Finn published at least two other editions of the Pseudo-Aristotle circa 1830, one of 342 pages and the other of 322 pages. There are two issues of this edition in the Atwater Collection: one in stamped green linen over boards with spine-title: *The Midwife;*

Fig. 5. Frontispiece to The works of Aristotle, *#130.*

the other quarter linen with printed paper spine label: *Aristotle's works. Complete.*

131. ARISTOTLE, pseud.
The works of Aristotle, the famous philosopher containing, his Complete masterpiece, displaying the secrets of nature in the generation of man: to which is added, The family physician, being approved remedies for the various distempers incident to the human body: also his Experienced midwife, absolutely necessary for surgeons, midwives, nurses, and childbearing women: and his Last legacy, unfolding the secrets of nature in the generation of man. A new and improved edition . . . London: Printed for Miller, Law, and Carter, and sold by all the booksellers, 1830.
viii, [9]-214, [2] p., [6] leaves of plates : ill. ; 14 cm.

132. ARISTOTLE, pseud.
The works of Aristotle, the famous philosopher, in four parts. Containing, 1. His Com-

plete master-piece; displaying the secrets of nature in the generation of man. To which is added, The family physician . . . 2. His Experienced midwife; absolutely necessary for surgeons, midwives, nurses and child bearing women. 3. His Book of problems, containing various questions and answers, relative to the state of man's body. 4. His Last legacy; unfolding the secrets of nature respecting the generation of man. A new edition with engravings. New-England: Printed for the publishers, 1831.
    247 p. : ill. ; 15 cm.

The 1831 New England editions represents the full and unexpurgated text, including part three of the *Master-piece* on physiognomy.
    In the early 1830s a fresh crop of sex manuals began to appear on the American market written by native authors such as William A. Alcott (#18-61) and Sylvester Graham (#1392-1395). The new literature was characterized by a notably more reserved tone than what was found in the numerous editions of the Pseudo-Aristotle—until then, the most available source for information on sex and reproduction.
    "The difference between the sexual hygiene literature of the 1830s and the *Works of Aristotle* is obvious at any level. Where 'Aristotle' waxed lyric about sex, the writers of the 1830s were tense and suspicious. Where 'Aristotle' warned that ill-health would result from sexual abstinence, these writers warned that ill-health would result from sexual overindulgence. Where he assumed that intense sexual desire was altogether natural, they argued that it was inherently pathological. Where his advice for young people in the throes of lust was to get married as soon as possible, theirs was to eliminate the cause of lust by means of careful attention to diet and regimen" (S. Nissenbaum. *Sex, diet and debility in Jacksonian America.* Westport, Ct., 1980, p. 28-9).

133. ARISTOTLE, pseud.
The complete master-piece of Aristotle, the famous philosopher, displaying the secrets of nature in the generation of man. To which is added The family physician; being approved remedies for the several distempers incident to the human body. A new and very superior edition. New York, [184-?].

[2]-291 p., [4] leaves of plates : ill. (some col.) ; 19 cm.

Unabridged text. Three of the four plates are hand-colored wood engravings. This is an oddly cased volume: the first leaf of plates and the final leaf of letterpress are used as pastedowns.

134. ARISTOTLE, pseud.
The complete master-piece of Aristotle, the famous philosopher: displaying the secrets of nature in the generation of man. To which is added The family physician; being approved remedies for the several distempers incident to the human body. A new and very superior edition. New-York: Printed for the publishers, 1842.
    [2], 32, [4], 33-62, [4], 63-66, [4], 67-96 p. : ill. ; 16 cm.

Atwater copy lacking pages 93-96.

135. ARISTOTLE, pseud.
Aristotle's master-piece, completed. In two parts. The first containing the secrets of generation, in all the parts thereof. Treating of the benefit of marriage, and the prejudice of unequal matches. Signs of insufficiency in men and women. Of the infusion of the soul. Of the likeness of children to parents. Of monstrous births. The cause and cure of the green sickness. A discourse of virginity. Directions and cautions for midwives. Of the organs of generation in women, and the fabric of the womb. The use and action of the genitals. Signs of conception, and whether a male or female, with a word of advice to both sexes in the act of copulation . . . The second part being a private looking glass for the female sex. Treating of various maladies of the womb, and of all other distempers incident to women of all ages, with proper remedies for the cure of each . . . London: Published by G. Davis, [184?].
    iii, [4]-142 p., [1] leaf of plates : ill. ; 16 cm.

Cover-title: *Aristotle's master-piece illustrated. Complete edition.* Hand-colored, wood-engraved frontispiece ("An amorous widow").

136. ARISTOTLE, pseud.

Aristotle's master-piece, completed. In two parts. The first containing the secrets of generation, in all the parts thereof. Treating of the benefit of marriage, and the prejudice of unequal matches. Signs of insufficiency in men and women. Of the infusion of the soul. Of the likeness of children to parents. Of monstrous births. The cause and cure of the green sickness. A discourse of virginity. Directions and cautions for midwives. Of the organs of generation in women, and the fabric of the womb. The use and action of the genitals. Signs of conception, and whether a male or female, with a word of advice to both sexes in the act of copulation . . . The second part being a private looking glass for the female sex. Treating of various maladies of the womb, and of all other distempers incident to women of all ages, with proper remedies for the cure of each . . . London: Published by G. Davis, [1845].
   [2], iii, [4]-142 p. : ill. ; 16 cm.

The title-page is undated. The front cover (paper over boards) is dated 1845. The earliest 142 p. edition of the *Master-piece* published at London by G. Davis appears to have been the 1830 edition. The wood-engraved frontispiece was printed from the same block as that in #135. It appears on a leaf contiguous with other leaves of the first gathering, however, and was not colored.

137. ARISTOTLE, pseud.

Aristotle's master piece. Illustrated edition. New York: Published for the trade, 1846.
   vi, [7]-141, [1] p. : ill. ; 16 cm.

The abridged version. Both the wood-engraved frontispiece (seated female nude) and the title-page vignette are hand-colored. Cover-title (printed paper over boards): *Aristotle's master-piece, illustrated. Complete edition.*

138. ARISTOTLE, pseud.

The works of Aristotle, the famous philosopher, in four parts. Containing 1. His Complete master-piece; displaying the secrets of nature in the generation of man. To which is added, The family physician . . . 2. His Experienced midwife; absolutely nec-

essary for surgeons, midwives, nurses and child bearing women. 3. His Book of problems, containing various questions and answers, relative to the state of man's body. 4. His Last legacy; unfolding the secrets of nature respecting the generation of man. A new edition with engravings. New-England: Printed for the publishers, 1846.
   247 p. : ill. ; 14 cm.

Either a re-issue of the sheets of the 1831 New-England edition (#131) with a new title-page, or a stereotype reprint. Issued without the frontispiece of the earlier edition. Spine-label: *Aristotle's complete works. Boston edition. 1846.*

139. ARISTOTLE, pseud.

Aristotle's master piece. Illustrated edition. New York: Published for the trade, 1847.
   [iii]-vi, [7]-141, [1] p., [1] leaf of plates : ill. ; 16 cm.

A re-issue of the 1846 New York edition (#137). Hand-colored wood-engraved frontispiece. Cover-title (printed paper over boards): *Aristotle's master-piece. Illustrated. Complete edition.*

140. ARISTOTLE, pseud.

The works of Aristotle. Including his Masterpiece, and important advice to females. London, 1849.
   [2], 142 p. : ill. ; 17 cm.

The four wood-engravings in the chapter on monstrous births are printed from the same blocks used in the illustration of the 183-? and 1845 London editions having the imprint of G. Davis (#129, 136). The only difference is that the frames around the figures have been cut away in the 1849 edition. This series is patterned on a model that in the Atwater Collection first appears in the illustration of the 1816 New York edition (#121).
   What is curious about the 1849 London edition is that it includes a reversed copy of the wood-engraved frontispiece portrait of Aristotle that appears in the 1816 New York edition, but in neither of the earlier 142 page London editions in the Atwater Collection. It is even more curious that this frontispiece bears the monogram of Alexander Anderson (1775-1870), who was largely responsible for the introduction of wood-engraved book illustration to New York

in the 1790s, and who was the most recognized American wood engraver until his death.

Jane Pomeroy, the leading authority on the work of Alexander Anderson, has seen an earlier version of the wood-engraved portrait of Aristotle with Anderson's monogram in an edition of the *Master-piece* with an 1806 London imprint (*letter to the compiler dated 7/22/96*). She suggests that this London imprint was false. Given the controversial nature of the content of the *Master-piece* and its historically ambivalent imprints, this is a definite possibility—not only for the 1806 London imprint, but perhaps for all the 142 page London imprints (e.g., #135, 136). It is uncertain, however, whether Anderson's designs for the monster figures were original or copied from an earlier English model. If the former, then these London imprints may well have been issued in New York. If the latter, then their re-use in later London editions is more easily explained. This still does not account for the appearance of Anderson's monogram in the frontispiece to this 1849 London edition. At this date in the history of wood-engraved book illustration, it was not uncommon for American publishers to copy the illustrations of books published in England. The reverse only occasionally occurred, however, and it is unlikely that an engraver's monogram would be retained in the cutting of new blocks.

141. ARISTOTLE, pseud.

The works of Aristotle, the famous philosopher: containing his Complete master-piece, and Family physician; his Experienced midwife; his Book of problems; his Remarks on physiognomy; and his Last legacy. Complete in five parts . . . London: Printed for the booksellers, [185-?]

352 p., [1] double plate : ill. ; 12 cm.

Added steel-engraved title-page and frontispiece. The text is the abridged version. However, the treatise on physiognomy, lacking in most abridged editions, has been re-inserted as "Part IV." This edition also includes an addendum entitled "Syphilis, or venereal disease" (p. [347]-352).

142. ARISTOTLE, pseud.

Aristotle's works: containing the Master-piece, Directions for midwives, and counsel and advice to child-bearing women.

With various useful remedies. London: Published for the booksellers, [185-?].

[2]-5, vi-viii, 9-352 p., [6] leaves of plates + frontis. & added title-page : ill. ; 13 cm.

Includes the abridged version of the *Master-piece*, as well as the *Experienced midwife*, the *Problems*, the physiognomy and "The Midwife's vade-mecum" (from the *Last legacy*). An addendum on venereal diseases (p. 317-352) was taken from the late 18th-century work of an unacknowledged English physician, who by his references was the contemporary of Samuel Chapman and Peter Clare.

The pleasingly executed chromolithographic frontispiece and added title-page are modelled on the design of the steel-engraved title and frontispiece of the 185-? London edition (#141). The wood engravings are also identical to those in #141. The chromolithographic plates illustrate the growth of the fetus at different months and the position of the child in the womb at birth. Printer's imprint (p. 352): "John Smith, Tooly Street, London."

143. ARISTOTLE, pseud.

Aristotle's work: containing directions for midwives, and counsel and advice to child-bearing women. With various remedies. London: Printed for the booksellers, 1858.

viii, 9-320 p., [6] leaves of plates + added title-page : ill. ; 13 cm.

This edition is unusual in that it does not include the *Master-piece*. The *Experienced midwife* is thus the first book, in the abridged version (i.e., without the description of the male genitalia, etc.). The *Book of problems* and the physiognomy follow. All that is present of the *Last legacy* is "The midwife's vade-mecum," which constitutes chapter nine of the unabridged edition. The text concludes with "Syphilis or venereal disease" (p. 295-320). Added chromolithographic title-page: *Aristotle.* The chromolithographic plates are identical to those that illustrate #142.

144. ARISTOTLE, pseud.

Aristotle's works: containing directions for midwives, and counsel and advice to child-bearing women. With various useful remedies. London: Printed for the booksellers, 1860.

viii, 9-320 p., [6] leaves of plates : ill. ; 12 cm.

A re-issue of the 1858 edition (#143). The plates are monochrome copies of the chromolithographs that appear in #142 and #143. Half-title: *Aristotle's experienced midwife.*

145. ARISTOTLE, pseud.
The works of Aristotle. Including his Masterpiece, and important advice to females. London, 1883.
[2], 142 p. ; 17 cm.

This 1883 London edition has either been reprinted from the plates of the 1849 London edition (#140), or is a lithographic facsimile. Even the frontispiece reproduces the wood-engraved portrait and ornamental border that adorn the front board of the 1849 edition. Title on frontispiece: *Aristotle's Master-piece.*

146. ARISTOTLE, pseud.
The works of Aristotle the famous philosopher. Containing his Complete masterpiece and Family physician; his Experienced midwife. His Book of problems, and his Remarks on physiognomy. Complete edition, with engravings. London, 1779: John Smith . . . [London: s.n., 188-?].
[2], [7]-516 p., [9] leaves of plates : ill. ; 12 cm.

Illustrated half-title: *Medical knowledge. London: Published by the Booksellers.*

147. ARISTOTLE, pseud.
The works of Aristotle the famous philosopher. Containing his Complete masterpiece and family physician; his Experienced midwife; His Book of problems, and his Remarks on physiognomy. A complete midwife's vade-mecum and some genuine recipes for causing speedy delivery . . . London, [189-?].
[4]-21, 21-518, [14] p., [8] leaves of plates : ill. ; 13 cm.

This edition does not include the *Last legacy.* Following the physiognomy is an addendum entitled "Ways and means of prevention" (p. 517-518 + [2] p.). It provides a brief but thorough description of extant contraceptive devices.
In a lecture delivered circa 1930, Sir D'Arcy Power described the Pseudo-Aristotle as "a

hoary old debauchee acknowledged by no one . . . sold only in those shops devoted to contraceptives." His indignation seems to have stemmed from the availability not only of antiquarian but of recent editions of the Pseudo-Aristotle in London bookshops, leading him to conclude, "There must be a large class of persons in England possessed of prurient minds and so uneducated that the pseudo-science of the middle ages still appeals to them" (*Foundations of medical history,* Baltimore, 1931, p. 147). Illustrated half-title: *Medical knowledge. London: Published by the Booksellers.*

148. ARMITAGE, Robert B.
Private lessons in the cultivation of sex force, the vital power of attraction between the sexes; its control and transmutation for greater strength and higher development. The most advanced teachings on physical and spiritual regeneration. The only system that will perpetually regenerate the whole body. Chicago, Ill.: Advanced Thought Publishing Co.; London, England: L.N. Fowler & Co., [c1913].
202, [6] p. ; 20 cm.

"Womanhood will reach its highest culmination in this twentieth century," Armitage predicts, "and woman will gradually be accorded full recognition of her natural right to stand by the side of the male on a basis of complete equality" (p. 7). Writing at the opening of a century of global warfare and systematized brutality, Armitage was convinced that the "whole race is gradually advancing"; and that "brute force" would be "replaced by gentleness, kindness, civility and altruism, qualities that, in the rudder [sic] ages, were regarded as effeminate. As man progresses he grows more like woman . . . he is becoming feminized . . . she is therefore justified in asking for a position of equality" (p. 8).
Armitage then proceeds to the "proper relation of the sexes " necessary for this new order to be realized. The liberation of woman that Armitages envisages is not paralleled by a liberation of "sex force." Quite the contrary, he regarded the indulgence of sexual appetite as harmful not only to physical and intellectual vigor, but as perverting the true relationship of man and woman. Sensuality wastes "vitality," but also degrades men and women so that their

perception of one another becomes entirely jaded. By continence, i.e., "voluntary and entire absence from sexual indulgence in any way" (p. 87), men and women preserve their "magnetism and vitality," realize their "higher forces," and come to live in harmony and equality.

Men who indulge their sexual passions in wedlock have in reality reduced women to the status of "slaves to their husband's desires" (p. 94). He argues: "The greatest advancement has occurred where the status of women has been raised and her sphere enlarged, where the two sexes acknowledge their inherent equality and recognize in each other something more than mere physical machines . . . new and better impulses arise, resulting in higher thoughts and better actions" (p. 65-6). In order for this to be achieved, "sex force" must be controlled and directed toward a higher end: "But the main purpose of Sex Force is not to propagate the race; it has a higher and nobler purpose; primarily intended to draw the the sexes into closer communion with each other by means of its subtle attraction, this communion is designed to develop the spiritual nature of man, and transform the individual into a higher and nobler being" (p. 76-7). Cover-title: *Sex force— our vital power.*

149. ARMITAGE, Robert B.
Private sex advice to women . . . for young wives and those who expect to be married . . . New York: Sincere Publishing Co., [c1917].
227, [5] p. ; 19 cm.

In his earlier work on *sex force,* Armitage dealt generally with the topic of sexual attraction and its balance in the relation of the sexes. *Private advice to sex women* provides several practical chapters on sexual physiology (i.e., female sexual anatomy, reproduction and pregnancy), but gives considerably more attention to two topics profoundly related in the author's mind: eugenics and birth control. "The world is no longer to be bombarded by an exuberant stream of babies, good, bad, and indifferent in quality, with mankind to look on calmly at the struggle for existence among them. Whether we like it or not, the quantity is steadily diminishing, and the question of quality is beginning to assume a supreme significance" (p. [103]).

Armitage is described as "M.D." on the title-pages of his published works, but does not appear in Polk or the AMA directories. Between 1913 and 1928 he published at least seven titles on sexual ethics, sex education and general hygiene.

150. ARMSTRONG, John, 1709-1779.
The art of preserving health: a poem. In four books . . . The second edition. London: Printed for A. Millar . . . M.DCC.XLV. [1745].
128 p. ; 19 cm.

Physician and poet, Armstrong was a 1732 Edinburgh medical graduate and the author of several medical works, including a successful treatise on the diseases of children (1767) that was translated into Italian and German. His most successful work by far was *The art of preserving health,* a didactic poem in four books first published at London in 1744. An early sign of the poem's success was the publication of an American edition at Philadelphia the following year. The poem is divided into four books: I. Air, II. Diet, III. Exercise, IV. The Passions. "As in all didactic poetry," writes Armstrong's biographer in the *Dict. nat. biog.,* "the practical directions are of little interest; but those who value austere imagination and weighty diction cannot afford to neglect Armstrong's masterpiece" (I:566). Indeed, remarks this biographer, "No writer of the eighteenth century had so masterful a grasp of blank verse as is shown in parts of this poem." The Atwater copy bears the bookplate of John Farquhar Fulton.

151. ARMSTRONG, John, 1709-1779.
The art of preserving health: a poem. London: Printed for A. Millar, M.DCC.LIV. [1754].

[2], 99, [1] p. ; 18 cm.

152. ARMSTRONG, John, 1709-1779.
The art of preserving health . . . To which is prefixed a critical essay on the poem, by J. Aikin, M.D. London: Printed for T. Cadell, Jun. and W. Davies . . . MDCCXCV. [1795]
[4], 150, [2] p., [4] leaves of plates : ill. ; 16 cm.

Includes John Aikin's (1747-1822) introductory essay to *The art of preserving health* which accompanies nearly every edition of the poem published from 1795. Aikin's essay is more a prose summary of the poem than a medical commentary or literary critique. He prefaces his analysis with the observation: "its subject seems on the whole as happily calculated for didactic poetry, as most of those which have been taken for the purpose. To say that it is a peculiarly proper one for a physician to write upon, is saying nothing of consequence to the reader. But the preservation of health is . . . a matter of general importance, and therefore interesting to readers of every class. Then, although its rules, scientifically considered, belong to a particular profession, and require previous studies for their full comprehension, yet in the popular use, they are level to the understanding and experience of every man of reading and reflection" (p. 6). Aikin states: "by associating these notions with images addressed to the imagination, he may convey them in a more agreeable form; and he may advantageously employ the diction of poetry to give to practical rules an energy and conciseness of expression which may forcibly imprint them on the memory" (p. 7).

Aikin's essay on Armstrong is one of a series of biographical and critical essays on British poets written in his retirement following a stroke. A Leyden medical graduate, Aikin enjoyed a modest London practice before early withdrawal into valetudinarianism and literary study.

This London edition is illustrated with four copper-plate engravings designed by Thomas Stothard (1755h-1834), the painter and illustrator of the novels of Fielding, Smollett, Richardson, Swift and Sterne. "These designs made a new departure in book illustration by their variety of invention, their literary sympathy, their spirit and their grace" (*Dict. nat. biog.*, 18:1321). Stothard was elected to the Royal Academy in 1794, the year his illustrations for Armstrong were engraved.

153. ARMSTRONG, John, 1709-1779.
The art of preserving health . . . To which is prefixed a critical essay on the poem, by J. Aikin, M.D. Boston: Printed by Hosea Sprague, 1802.
xxiii, [1], 71 p. ; 13 cm.

The fourth American edition. The first American edition was printed by Benjamin Franklin at Philadelphia in 1745.
REFERENCES: *Austin* #81.

154. ARMSTRONG, John, 1709-1779.
The art of preserving health . . . To which is prefixed a critical essay on the poem, by J. Aikin, M.D. Philadelphia: Printed for Benjamin Johnson . . . [et al.], 1804.
143 p., [1] leaf of plates : frontis. ; 14 cm.

This edition was issued with Matthew Green's *The spleen, and other poems* (#1407).
REFERENCES: *Austin* #82

155. ARMSTRONG, John, 1709-1779.
The art of preserving health . . . To which is prefixed a critical essay on the poem. By J. Aikin, M.D. Walpole, N.H.: Printed for Thomas & Thomas, by Cheever Felch, 1808.
102 p. ; 17 cm.

REFERENCES: *Austin* #84.

156. ARMSTRONG, John, 1709-1779.
The art of preserving health. A poem, in four books . . . With a critical essay by J. Aiken, M.D. [sic] and notes by Dr. Alcott. Boston: George W. Light, 1838.
120, 8 p. ; 18 cm.

Edited by William Andrus Alcott (#18-61), who states in his preface: "With the exception of a few points to which I have adverted in the notes, and two short paragraphs which it was thought proper to omit, the doctrines of the poem on the influence of air, diet, exercise, and the passions, on health, are such as must commend themselves to the friends of temperance in every age and every country. The two passages which I have omitted, related exclusively to the occasional use of wine, which the doctor, with a strange but very common inconsistency, seems inclined to permit" (p. [5]-6).The final 8 pages contain publisher's advertisements for the works of Alcott and Sylvester Graham.

157. ARMSTRONG, Mary Frances, d. 1903.
A haunted house . . . New York: Published for the Hampton Tract Committee by G.P. Putnam's Sons, 1879.
24 p. ; 16 cm.

Armstrong provides a perhaps fictional account proving that the deaths of successive inhabitants of a reputedly haunted house were due to poor drainage and water supply rather than to ghosts. Series: *Hampton tracts for the people. Sanitary series; no. V.* A physician, Armstrong was a member of the editing committee for the *Hampton tracts,* the purpose of which was to "provide strong and condensed statements of the fundamental laws of health, with illustrations of the results of breaking these laws and advice as to the best and easiest way of living in obedience to them" (from printed wrapper).

158. ARMSTRONG, S. R.
Sexual vitality. A key to health and vigor. A compendium of special information gathered from the most authoritative sources, and presented in a form easily understood . . . New York: Published by the Vim Publishing Co., [c1904].
246, [2] p. : ill. ; 19 cm.

Armstrong describes the anatomy of the male and female genitalia and the physical manifestations of sexual maturity. He devotes six chapters to the "lust disease" and its medical, moral and social consequences; and discusses *The Remedy*: "Active exercise is the only remedy for sexual vice and for most of the ills it produces" (p. 137).

159. ARNOLD, John L.
Arnold's lectures on anatomy, physiology and hygiene; and disease, its cause, prevention, and cure. Written in a familiar style; designed for the general reader . . . Illustrated edition. Cincinnati: H.M. Rulison, Queen City Publishing House; Philadelphia: Duane Rulison, Quaker City Publishing House, 1856.
2 pts. in 1 v. (x, 11-196, [4] p. ; xvi, 13-225, [3] p.) : ill. ; 19 cm.

The first part contains eight lectures on the anatomy, physiology and hygiene of the body's different "systems." Part two consists of ten lectures: eight on the diseases of these systems; and two final "lectures" comprising a domestic materia medica. Part two concludes with an appendix (p. 167-212) on miscellaneous hygienic topics. Rulison issued this title again in 1859

and in 1860. Spine-title: *Arnold's medical lectures for the people.*

160. ARNOLD, John L.
Arnold's medical companion for young men; containing the laws of physiology and health, and a history of every disease; its cause, prevention, and cure. Also, a special lecture on organic generation; its philosophy, singularities, and derangements; containing many strange and useful facts on the subject of procreation . . . Cincinnati: H.M. Rulison, Queen City Publishing House; Philadelphia: Duane Rulison, Quaker City Publishing House, 1856.
[6], xvi, 13-187, [188-211], 212-227, [5] p. : ill. ; 19 cm.

This work is simply a re-issue of the sheets of the second part of *Arnold's lectures on anatomy, physiology and hygiene* (#159), with the substitution of "A supplementary lecture on organic generation" in place of the "Appendix." Arnold's lecture on generation includes a detailed description of the male genitalia; a briefer description of the female organs; a theory of generation that places conception in the ovaries; and admonitions on "promiscuous intercourse" and venereal disease. The wood engravings in the "Supplementary lecture" that depict the organs of generation, the female pelvis and fetal presentations all derive from models established in Samuel Bard's *A compendium of the theory and practice of midwifery* (1st ed., New York, 1807), the first textbook on obstetrics by an American author.

161. ARTS REVEALED.
Arts revealed, and universal guide; containing many rare and valuable recipes and directions for the use of families; from the best authorities. Embracing directions for treating diseases—behavior of ladies and gentlemen—embroidery, and other kinds of needle-work—information as to roots and herbs—compounding medicines—how to be prepared for accidents, &c., &c., &c. Several recipes cost from $20 to $50 each, having never before been published. New York: American Family Publication Establishment, 1853.
136 p. ; 21 cm.

This domestic miscellany contains thirteen parts, five of which provide medical advice: Part VIII. The doctor at home; Part IX. Medical qualities of roots and herbs; Part X. Diseases of children; Part XI. Choice medical compounds to be kept on hand; Part XII. Accidents or emergencies.

162. ARTS REVEALED.

Arts revealed, and universal guide; containing many rare and valuable recipes and directions. For the use of families, from the best authorities. Embracing directions for treating diseases—behavior of ladies and gentlemen—embroidery, and other kinds of needlework—information as to roots and herbs—compounding medicines—how to be prepared for accidents, etc. Several recipes cost from $20 to $50 each, having never before been published. New York: American Family Publication Establishment, 1855.

    134, [2] p. ; 21 cm.

163. ARTS REVEALED.

Arts revealed, and universal guide; containing many rare and invaluable receipes and directions for the use of families, from the best authorities. Embracing directions for treating diseases—embroidery and other kinds of needlework—information as to roots and herbs—compounding medicines—how to be prepared for accidents, etc., etc. Several recipes cost from $20 to $50 each, and have never before been published. New-York: H. Dayton; Indianapolis, Ind.: Dayton & Asher, 1858.

    134, [2] p. ; 24 cm.

164. ASHTON, James.

The book of nature; containing information for young people who think of getting married, on the philosophy of procreation and sexual intercourse; showing how to prevent conception and to avoid childbearing. Also, rules for management during labor and child-birth . . . New York: Published by Wallis & Ashton, 1861.

    iv, [5]-64 p., [9] leaves of plates (4 folding) : col. ill. ; 16 cm.

Though issued in at least five editions between 1859 and 1870, this remarkable treatise has received little attention in the historical literature. Among the popular sex manuals of the period, *The book of nature* is unparalleled in the accuracy of the information it provides, and is superior even to many treatises written by and for a professional audience on sexual physiology. Little is known of its author, who describes himself on the title-page simply as "lecturer on sexual physiology," but who from self-reference in the text was probably a physician, and is listed as such in the 1860/61 and 1861/62 New York City directories.

Ashton opens his treatise with a description of the male genitalia, concluding with a discussion of the role of seminal "animalcules" in generation. His consideration of "zoosperme" leads him to a parallel discussion of the ova. Because the anatomical portion of the treatise follows his discussion of reproductive physiology, the description of the female organs of generation begins with the ovaries and concludes with the labia, contrary to the order of most anatomies. Among his contemporaries in the popular literature, Ashton stands among the few who correctly discerned the nature and purpose of menstruation, and who correlated its monthly course to periods of fertility. He knew that each woman is born with a given number of eggs; that an ovum is expelled monthly in menstruation; and that when the supply of eggs has been expended, woman enters her "turn of life." He states that as long as the ovum "remained in the womb, it is . . . liable to be impregnated by the semen from the male, but the moment it is expelled, no impregnation takes place" (p. 14). Based upon the regularity of ovulation and menstruation, Ashton provides guidelines for determining when conception cannot take place. He also knew that tubal ligation rendered a woman barren.

In an era when most of his fellows in the literature of sexual behavior sentimentalized marriage, Ashton makes the observation: "It is useless to deny that the majority of marriages which are apparently based on the sentiment called love, are nothing more than the result of an involuntary obedience to the imperious voice of our sexual organs" (p. 20). He is one of few mid-19th-century authors to discuss orgasm and the differences between male and female orgasms (p. 34). Although Ashton discards much sexual lore, he retains some traditional absurdities, e.g., that the sex of the more arduous partner during coition will determine the gender of the child conceived (p. 28); or that if the father is absent during gestation, the child will physically resemble its mother (p. 30), etc.

Ashton recommends four methods of birth control. The most effective in his opinion is withdrawal of the penis before ejaculation. He dismisses the physical and moral objections that most of his contemporaries had toward *onanism*. A method more likely "to give the wife confidence" is the injection of cold water or a weak astringent solution into the vagina by means of an India-rubber syringe. He insists on the efficacy of this technique if done "instantly" after coition and if "injected with sufficient force and profuseness." He also describes the insertion of a small sponge into the vagina (attached to a silk string) that has been soaked in a spermicide. His final recommendation is India-rubber "coverings for the penis, which are used in Europe to avoid contracting sexual diseases from prostitutes," but which, "must necessarily prevent conception" (p. 41). He notes that this method is not popular, as condoms interfere with "enjoyment of the nuptial act."

The nine wood-engraved plated are hand-colored. The four folding plates depict the male and female genitalia. Four plates provide full-figure depictions of the female in various states of pregnancy. They are more erotic than didactic in nature, and are based, according to Ashton, on "original drawings from anatomical figures in wax in the New York Anatomical Museum [#2590?]" (p. 33). Atwater copy lacking the three foldings plates of the female genitalia.

165. ASHTON, James.

The book of nature; containing information for young people who think of getting married, on the philosophy of procreation and sexual intercourse; showing how to prevent conception and to avoid childbearing. Also, rules for management during labor and child-birth . . . New York: Published by Wallis & Ashton, 1864.

iv, [5]-64 p., [9] leaves of plates (4 folding) : col. ill. ; 16 cm.

166. ASTHMATICS INSTITUTE.

Asthma cured to stay cured. A thesis with reports of cases. P. Harold Hayes, M.D., F. Mason Hayes, M.D. Buffalo, N.Y., 1886.

40 p. ; 14 cm.

"The Asthmatics Institute was opened at Binghamton, N.Y., in March 1883. It was moved to Buffalo, in May, 1885" (p. [1]). It was established by Pliny Harold Hayes, M.D. (1824-1894) and his son Francis Mason Hayes, M.D. (1853-1916). The Hayes developed a line of proprietary remedies that they claimed were directed at the "asthmatic constitution" rather than the "asthmatic paroxysm." Patients could obtain treatment either in person at the Institute or at a distance by correspondence.

Pliny Hayes was born in Bristol, Ontario County, N.Y. on 7 October 1824 (*Nat. cycl. of Amer. biog.*, 12:293). He was raised on his stepfather's farm, and attended the nearby Canandaigua Academy and Lima Seminary before entering Jefferson Medical College, from which he graduated in 1848. Upon graduation he took charge of a small sanitarium, the Greenwood Springs Water-Cure, at Cuba, N.Y., where he treated chronic diseases. Circa 1850 he bought property in nearby Wyoming, N.Y. and built the Wyoming Cottage Water-Cure Institute (#3891), which he operated for several years. In the years that followed, Hayes was associated with several other sanitaria, including R.T. Trall's Hydropathic and Hygienic Institute (#1877). In 1871 he spent six months in study while visiting New York City hospitals, and in 1875 he settled in Binghamton, N.Y., where he opened a medical practice and where he became interested in asthma. In 1885 he moved to Buffalo, and with his son F. Mason Hayes (M.D. 1877, University of the City of New York), conducted an extensive consultative and therapeutic practice (principally by mail) in the field of allergy. While in Binghamton, Hayes married Cornelia Catharine Hall of West Bloomfield, N.Y. Two of their four children became physicians: Francis Mason Hayes and Harold Augustine Hayes (b. 1859 or 60). Pliny Harold Hayes died in Buffalo on 7 April 1894.

167. ASTHMATICS' INSTITUTE.

Asthmatics' Institute. Dr. P. Harold Hayes Associates . . . Consultations at the office, No. 716 Main Street, Buffalo, N.Y. [Buffalo, N.Y.: Asthmatics Institute, 1898].

28 p. ; 14 cm.

Title on wrapper: *Asthma cured to stay cured.* This pamphlet was issued with a form letter dated 10 March 1898 that begins: "Having been informed that you are a sufferer from asthma, we feel that you will be deeply interested in reading the enclosed copy of our thesis, which states clearly the principle upon which our success is curing this deparate disease is based." Included with the "thesis" and letter is an "Examination Paper" or or questionnaire containing 129 questions regarding the suffer's condition so that Hayes and Associates may "render an opinion as to your curability without charge."

168. E.H. ATHERTON AND COMPANY.

EVERETT'S COUGH MIXTURE. Cannot be surpassed, for Coughs, Colds, Bronchitis, and all Throat and Lung troubles where you want a reliable medicine. It is not a cheap mixture of Squills and Tolu, but a scientific combination of Vegetable substances, which, after years of trial, has been pronounced by the people "good" . . . For Sale Wholesale and Retail, by the Manufacturers and Proprietors, E.H. Atherton & Co., Cavendish, Vermont. [ca. 1878].
   Broadside ; 25 × 20 cm.

169. ATLANTA REMEDY COMPANY.

Health at home for women. A plain, concise statement of facts which every woman should know. Contains valuable information as to how to get well at home and how to keep well. Atlanta, Ga.: Published by Atlanta Remedy Co., [191-?].
   144 p. ; 18 cm.

   The Atlanta Remedy Co. was one of several mail-order businesses that provided proprietary medicines to women owned and operated by William M. Griffin, of Ft. Wayne, Indiana. Griffin's mail-order empire included the Dr. J.W. Kidd Company (#2123, 2124), the Bertha C. Day Co. (#918), and Julia D. Godfrey's Woman's Health Institute (#1373)—all operated out of Ft. Wayne. The medical director of the Atlanta Remedy Co. was Lily M. Norrell, an 1894 graduate of the Woman's Medical College of Georgia.

170. ATTLEBORO SANITARIUM.

The Attleboro Sanitarium, Attleboro, Massachusetts. An institute of physiologic therapeutics and rational medicine. "Where tired folks get well rested, where sick folks get well" . . . [Attleboro, Ma.: The Sanitarium, 191?].
   24 p. : ill. ; 23 × 31 cm.

The mission of the Attleboro Sanitarium was "to afford a haven of rest to those who are nervously exhausted by business or social cares or by home worries; to afford the best known methods and appliances for restoring chronic invalids of all classes to renewed health and strength, and most important of all, to instruct those who come under its influence, not only how to get well, but how to live in harmony with nature's laws as to maintain health permanently" (p. 2). Prospective guests are informed: "Experience has shown that much more can be accomplished in the production of health in a well organized and thoroughly equipped institution than can possibly be done by individual effort without proper appliances. Just as the factory has largely taken the place of the individual and private production in other lines, and as the well-equipped university or college has demonstrated its superiority over home education, just so, in health-getting, the well-equipped institution has an immense advantage over the single-handed methods of the home" (p. 4). The "Sanitarium System" boasted "hyrotherapy, electrotherapy, thermotherapy, phototherapy, massage, Swedish gymnastics, diet, rest, and recreation," as well as an "aseptic operating room and ward," and facilities for "bacteriological, chemical, microscopical and other accurate methods for examination and pathological research" (p. 6).
   The Attleboro Sanitarium was established in 1908. It was under the medical direction of Benn Eugene Nicola (b. 1865), a 1910 graduate of the American Medical Missionary College in Chicago (where students spent their first two years at the Battle Creek Sanitarium), and Mary Kate Byington Nicola (b. 1869), an 1897 medical graduate of the University of Michigan. In thick paper wrapper.

171. AUSTIN, George Lowell, 1849-1893.

Dr. Austin's indispensable hand-book and general educator. Useful and practical information pertaining to the household, the trades and the professions, including domestic medicine and surgery; the management of infancy and childhood; housekeeping in all its relations; cookery; beverages; window gardening; choice recipes for the toilet; the art of confectionery; the management of business and the laws and conditions of success; town and country

life; agriculture; horticulture; the veterinary art; the mechanic arts; etc., etc. . . .
Portland, Me.: George Stinson & Company,
1885.

   xii, [13]-719 p. : ill., port. ; 22 cm.

Cover- and spine-title: *Austin's indispensable
handbook and general educator.*

172. AUSTIN, George Lowell, 1849-1893.
Perils of American women or a doctor's talk
with maiden, wife, and mother . . . With a
recommendatory letter from Mrs. Mary A.
Livermore . . . Boston: Lee and Shepard,
publishers, 1885.

   [4], 6, 240 p. : ill. ; 16 cm.

The *Perils of American women* was originally
published at Boston in 1883, and again in 1907
under the title: *A doctor's talk with maiden, wife,
and mother.* Austin devotes the first eighty pages
of his treatise to a description of the anatomy
of the female genitalia, coition, reproduction,
pregnancy and labor. In chapter VIII, "Hygiene
of the marriage bed," he considers the "physiological effects of the venereal act," i.e., the
negative physical and psychological effects of
sexual excess. In the following chapter he discusses contraception, which he regards as ethi-
cally unsound and medically harmful. He reviews current contraceptive methods (i.e., abortifacient drugs, "one of the most common methods for preventing conception," p. 108; vaginal
injections; the *baudruche* or condom; "pessaries,
sponges, tampons and the like"; and early
withdrawl) in order to point out their harmful
effects. A consideration of "the influences,
which, in our present way of living, tend most
decidedly to develop a predisposition to diseases of the female sexual organs" is the topic
of chapter XI, which is followed by related discussions of the "relations of sexual disorders to
the brain and nervous system" (c. XV) and "nervous prostration among women" (c. XVI).
Cover-title: *A doctor's talk with maiden, wife, and
mother.*

173. AUSTIN, George Lowell, 1849-1893.
Water-analysis. A handbook for water drinkers . . . Boston: Lee and Shepard, publishers; New York: Charles T. Dillingham,
[c1882].

   48 p. ; 15 cm.

After outlining the health dangers consequent on the use of untested pond or river water, or the contamination of well water from
"ash-pits, dumb-wells, and cesspools," Austin
states: "The importance of the purity of drinking water cannot be too strongly urged or insisted upon. When there is any reason for sus-
pecting the purity of water, it should at once be
tested, so as to settle the question. Such work is
left altogether too much to local boards of
health throughout the country; and herein lies
a mistake. Every physician ought to know how
to make an analysis of every sample of water,
the healthfulness of which he suspects; nay,
more, every householder ought to at least understand the simpler tests" (p. 12). Spine- and
cover-titles: *Handbook for water-drinkers.*

174. AUSTIN, Harriet N., 1826-1891.
The American costume for women . . .
Dansville, Livingston Co., N.Y.: F. Wilson
Hurd & Co., publishers, 1859.

   [2], 24, [2] p. ; 19 cm.

Austin was one of nine females in the first
graduating class (1851) of the American Hydropathic Institute in New York. After graduation, she returned to central New York State
and opened a hydropathic practice in the village of Owasco. Upon the recommendation of
James Caleb Jackson (#1902-1955), however,
she abandoned private practice, and in July
1852 joined Jackson on the medical staff of the
Glen Haven Water Cure. Thus began a professional and personal association with Jackson
(who legally adopted her as his daughter) that
lasted thirty-nine years. When Jackson sold his
interest in Glen Haven, he and other members
of the staff opened Our Home-on-the-Hillside
water-cure in Dansville, New York, some fifty
miles south of Rochester. Austin became one-
third partner in the new venture (F. Wilson
Hurd & Co.). After the death of Jackson's elder son Giles in 1864, the firm became Austin,
Hurd & Co. In 1868, she became senior partner in the firm of Austin, Jackson & Co., which
owned and managed Our Home-on-the-Hillside and published numerous books, pamphlets
and periodicals on health reform (most of them
authored or edited by Austin and Jackson). Our
Home-on-the-Hillside became one of the best
known and most prosperous water-cures in the
United States, and successfully made the tran-

sition from hydropathic institute to health resort with the decline of hydropathy as a school
of medical thought after the Civil War.

At Glen Haven, Austin had been physician
to the "ladies' department." At Our Home-on-
the-Hillside, however, she attended both male
and female patients. She enjoyed considerable
reputation as a physician at one of America's
leading health resorts, including the esteem
and friendship of Clara Barton, who had been
one of Austin's patients and who subsequently
bought a country home in Dansville.

Austin was a prominent figure in the dress
reform movement. She and Jackson had both
encouraged dress reform at Glen Haven, and
continued to champion the "American costume"
from their influential seat at Dansville. Austin was
a founding member of the National Dress Reform Association, and served as its president in
1858. *The American costume for women* is regarded
as one of the classic statements of the movement's
principles and goals. Austin writes: "Our issue is
with the mode in which *all* women dress . . . and
we assert that their dress is at variance with the
nature of their *physical* constitution, and so for-
bids a true physical development; that it is a positive and powerfully exciting cause of disease, that
it is a main cause of the feebleness of women
and girls . . ." (p. 2). She states that fashion "starts
from the idea that her Creator fashioned her
awkwardly and improperly, that *half* the powers
which He has given her *in common* with man are
of no use, and that inefficiency and disease are
much more desireable than health and vigor"
(p. 3). Austin suggests that men dress in skirts to
discover just how inconvenient they are: "He
would soon become feeble, inert, listless, dependent, sick; so he would, and so she has already
become." She then asserts: "man's dress is nearly
sufficient in itself to account for his superior
strength, power of endurance, and better general health" (p.8). Pages 22-24 contain a description of the "reform dress" for the benefit of those
who wish to adopt it. Running-title: *National Dress-
Reform Association.*

Austin was a frequent contributor to the
pages of *The water-cure journal,* and edited the
popular health reform periodical *The laws of life*
for more than thirty years.

175. AUSTIN, Harriet N., 1826-1891.

The American costume: or, woman's right
to good health . . . Dansville, Livingston

Fig. 6. Harriet N. Austin in American Costume.

Co., N.Y.: F. Wilson Hurd & Co., publishers, 1861.
22, [2] p. ; 19 cm.

First edition. In spite of its similar title, this
is not a reprint of Austin's 1859 *The American
costume* (#174), but an entirely new essay. The
tone of this new work is decidedly less outspoken and challenging than the earlier pamphlet,
which one cannot help but feel was more in
keeping with Austin's true character. This latter version of *The American costume* was re-issued
at Dansville in 1864, 1866, 1867 and 1868.

176. AUSTIN, Harriet N., 1826-1891.

Health dress . . . Dansville, N.Y.: The Sanatorium Publishing Co., publishers, 1886.
20, [2] p. ; 15 cm.

In printed wrapper.

177. AUSTIN, Harriet N., 1826-1891.
How to take baths . . . Dansville, Livingston Co., N.Y.: F. Wilson Hurd & Co., publishers, 1861.
13, [3] p. ; 19 cm.

A description of the various types of baths employed in hydropathic treatment. This pamphlet appears to be the first separate issue of this essay, which was also published under the title "Baths, and how to take them"as an addendum to the various editions of James C. Jackson's *The sexual organism* (1st ed., Boston, 1861) and his *How to treat the sick without medicine* (1st ed., Dansville, 1868, #1925).

178. AUSTIN, Harriet N., 1826-1891.
How to take baths . . . Dansville, N.Y.: Austin, Jackson & Co., publishers, 1875.
13, [3] p. ; 18 cm.

This pamphlet is advertised as no. 3 in a series of *Health tracts* written by Austin and James C. Jackson. It was reissued for the last time in 1885.

AUSTIN, Harriet N., 1826-1891. *How to take baths* see also **JACKSON** (#1925-1932, 1939-1945).

AUSTIN, Harriet N. see also **JACKSON** (#1902, 1903); **OUR HOME ON THE HILLSIDE** (#2675-2679).

AUTHORIZED PHYSIOLOGY SERIES see *HEALTH FOR LITTLE FOLKS* (#1600); **TRACY** (#3553).

AUZOUX, Louis see **ACADEMIE ROYALE DE MEDICINE** (#7), **BEDFORD** (#271), **CODMAN & SHURTLEFF** (#696), **FLINT** (#1199), **HOLLICK** (#1697).

179. AVERY, Stephen Waterman, 1800-1833.
The dyspeptic's monitor; or the nature, causes, and cure of the diseases called dyspepsia, indigestion, liver complaint, hypochondriasis, melancholy, etc. . . . New-York: E. Bliss, 1830.
xi, [1], [ix]-xix, [22]-152 p. ; 18 cm.

Avery attributes Americans' inclination to dyspepsia to an overabundant table: "as a people we eat far too much hearty food; that is, we take in more rich nutriment than we require, and the consequence is, our system becomes overloaded and oppressed, our organs are clogged in the performance of their several functions, the circulating fluids become too thick and stimulating, and the proneness to derangements and diseased action greatly increased. Hence arises a large proportion of the inflammatory and febrile diseases amongst us, and hence it is that copious blood-letting and active medicines are so much more required in America than in most other countries" (p. 70). The most pernicious element in the American diet according to Avery is animal flesh: "Indeed, it is an uncommon thing for great meat eaters ever to attain old age, while it is recorded of all those . . . who have arrived at or surpassed one hundred years, that they invariably used animal food very sparingly if at all" (p. 71). The cure of the three species of indigestion described in the first half of the book is a regimen consisting of a vegetable diet, moderate portions, thorough mastication, avoidance of "spiritous liquors" and tobacco, exercise in fresh air, and attention to the "moral causes" of dyspepsia, i.e, "immoderate emotions of the mind from whatever cause."

180. AVERY, Stephen Waterman, 1800-1833.
The dyspeptic's monitor; or the nature, causes, and cure of the diseases called dyspepsia, indigestion, liver complaint, hypochondriasis, melancholy, etc. . . . Second edition. New-York: E. Bliss, 1830.
xii, [ix]-xx, [21]-164 p. ; 18 cm.

AYER, Amy G. see *FACTS FOR LADIES* (#1113).

181. AYER, Harriet Hubbard, 1849-1903.
Harriet Hubbard Ayer's book. A complete and authentic treatise on the laws of health and beauty. Including many carefully tested formulas hitherto unpublished. Good health, how to preserve it. Good looks, how to obtain them, with full instructions for physical culture, facial, scalp and general massage . . . Springfield, Mass.: The King-Richardson Company, 1902.
543 p. : ill., port. ; 24 cm.

"Recently," Ayer writes in her opening chapter , "a man told me that nothing to him was so hopeless as a woman striving to be better looking than the Lord intended she should be. 'Why,' he said, 'can't a woman be satisfied with Nature?' And I feebly replied that Nature had been extremely skimpy in dealing out personal attributes to some of us, and a Nature that gave a girl crooked teeth, pink eyelashes, freckles, and knock-knees, was not deserving of the undying gratitude of her victim" (p. 39). The tone of Ayer's volume is characterized by the remark: "I believe that good women can be more helpful, and wield a stronger moral influence if they are lovely to look at, graceful as well as gracious, perpetually young and beautiful, than the reverse" (p. 29-30).

The daughter of a prominent Chicago family who married the son of a wealthy iron manufacturer, Ayer's marriage eventually floundered, as did her husband's business. Thrown to her own devices, Ayer became a sales and decorating consultant for a New York furniture and antiques firm. On a buying trip to Paris, she purchased the formula of a skin cream allegedy used by the famed Mme. Recamier. Financed by a New York businessman, Ayer began the Recamier Manufacturing Company. "With shrewd use of advertising pyschology, she announced that it had been used by . . . a famous beauty of Napoleon's day, whose name she placed on the label with her own. Having thus lent her social prestige to the endorsement of a popular product, she hammered home this combination of snob appeal and ballyhoo by extensive newspaper advertising, proving herself a pioneer in modern merchandising techniques" (*Noteable Amer. women*, 1:73). Her success in marketing the cream was throttled by tensions between Ayer and her partner, who eventually convinced her daughter and ex-husband to have Ayer committed to a mental asylum (1893).

Upon her release, Ayer launched a successful lecture tour on conditions in such institutions. In 1896 she began writing a health and beauty column in the *New York World*: "She targeted her mass-circulation columns toward the working woman . . . She joined the Rainy Daisy moderate dress-reform group and was critical of tightly laced garments . . . Within a short time Ayer doubled as a reporter and feature story writer. She gathered her articles and advice, publishing them in *Harriet Hubbard Ayer's book* . . . Ayer helped change women's thinking regarding healthy eating, proper exercise, and dress reform, while advocating a freer lifestyle" (*Amer. nat. biog.*, 1:791). *Harriet Hubbard Ayer's book* was first published in 1899. This second edition was its final publication during the author's lifetime. The numeration of its forty-nine plates is incorporated into the pagination of the text.

182. J. C. AYER & CO.
Ayer's book of emergencies: a manual for handy reference. Alphabetically arranged . . . Lowell, Mass., U.S.A.: Dr. J.C. Ayer and Company, copyright, 1888.
　　36 p. : ill. ; 19 cm.

An alphabetical list of emergencies and how they are managed interspersed with advertisements for *Ayer's Cherry Pectoral*, *Ayer's Sarsaparilla*, *Ayer's Hair Vigor*, etc. In chromolithographed wrapper.

183. J. C. AYER & CO.
If you have a cold or cough, or any disease of the throat or lungs, do not fail to try at once AYER'S CHERRY PECTORAL . . . J.C. Ayer & Co., Lowell, Mass., U.S.A. [ca. 1882].
　　[1] leaf printed on both sides ; 24 × 15 cm.

Advertised on verso: *Ayer's Hair Vigor.*

Fig. 7. "A well ordered medicine chest," from Ayer's book of emergencies, *#182.*

184. J. C. AYER & CO.
IF YOUR BLOOD IS DISORDERED YOU
NEED A PURIFYING MEDICINE. Whatever
ailment you may have, your recovery from
it will be slow, if not impossible, so long as
your blood is impure . . . AYER'S SARSAPA-
RILLA . . . [188-?].
[1] leaf printed on both sides ; 24 × 15
cm.

"This is a concentrated compound of the
most powerful vegetable alteratives, diuretics,
tonics and blood-purifiers ever brought together in any medicine." *Ayer's Sarsaparilla* is
recommended for all diseases caused by "impurity of the blood," e.g., scrofula, erysipelas,
hives, goitre, "female weaknesses and irregularities," venereal diseases, etc.

AYLWORTH, Benjamin H. see **GARDNER**
(#1331).

B

BACON, Francis see CORNARO (#791, 792).

185. BACON, M. W.
Specific medicines. Hanover's Specific for Nervous Debility and Invigorating Pills for impotency. Prepared by Dr. M.W. Bacon . . . [Chicago: M.W. Bacon, ca. 1880].
[8] leaves : ill. ; 13.5 × 7 cm.

Hanover's Specific for Nervous debility is advertised as a cure for spermatorrhoea, which itself is the result of that "secret and pernicious habit . . . frequently called self-abuse, solitary indulgence, masturbation, onanism, or secret pollution." If allowed to progress unchecked, "The inevitable effect of the disease is to shadow all the prospects and hopes of life, rendering existence itself wretched, until almost in despair the unhappy sufferer is tempted to commit suicide" (leaf [3] recto). Bacons' *Invigorating Pills* are certain to cure impotency. Their "active principle, and main ingredient . . . is the extract of a plant found in Mexico, indigenous to one small section of that country" (leaf [7] recto).

186. B.H. BACON COMPANY.
OTTO'S CURE for coughs, consumption, bronchitis, asthma, grippe, and all diseases of the throat and lungs. The Great German Remedy . . . B.H. Bacon Co., Proprietors, Rochester, N.Y. [191-?].
[1] leaf printed on both sides ; 23 × 15.5 cm.

Advertisement on verso: *Celery King.*

BADEN-BADEN REMEDIES see
MONTGOMERY (#2182).

187. BADGER, A. M.
BADGER'S HEALING SALVE, AND PAIN EXTRACTOR. [Rochester, N.Y.: Shephard's Printing Office, ca. 1850].
4 p. ; 20 cm.

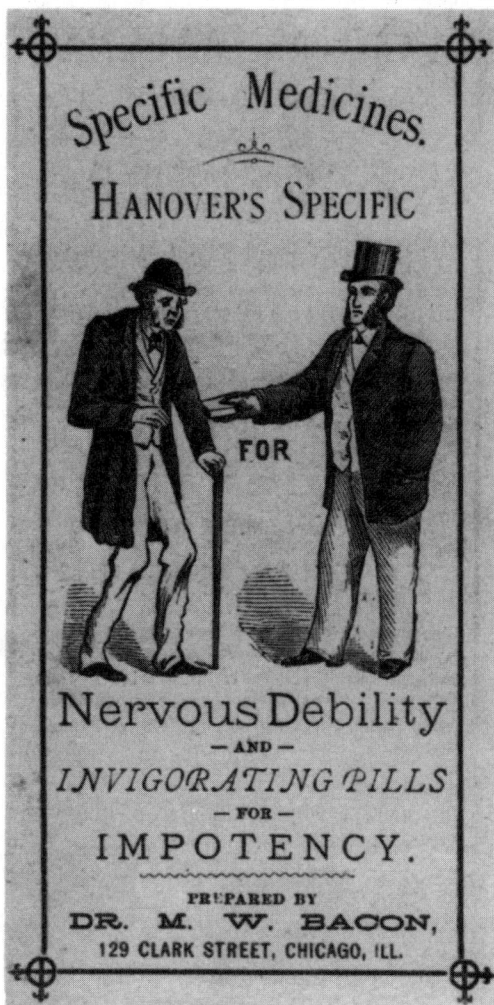

Fig. 8. Front wrapper of M.W. Bacon brochure, #185.

Caption title. Badger claims that his salve will relieve pain "in any flesh wound," rheumatism, "broken and sore breasts," "salt rheum," etc. Pages 3-4 consist of "certificates" or testimonies (dated 1847-1850) from satisfied users of the salve.

BADGER, Paine D. see **THOMSON** (#3482).

BAIKIE, Robert see **FRANCKE** (#1305, 1306).

188. BAILEY, Samuel.
About diphtheria. [Oberlin, Ohio: Samuel Bailey, successor to Hendry & Bailey, ca. 1860].
    4 p. ; 19 cm.

    Caption title. Advertisement for *Dr. Watson's Diphtheria Cure.*

189. BAILEY, William Theodore, 1828-1896.
Richfield Springs illustrated, historical and descriptive . . . Richfield Springs, N.Y.: Mercury office, F.E. Mungor, proprietor, 1886.
    106 p., [12] leaves of plates : ill., map ; 16 cm.

    Located 65 miles west of Albany, Richfield Springs was established in 1791. The healing properties of its springs were long appreciated. In 1820, Horace Manley purchased property surrounding what came to be known as the White Sulphur Spring, and made the waters accessible to the public. The reputation of the springs and the number of visitors increased annually, making Richfield Springs a popular health and summer resort. Bailey's treatise consists of a chemical analysis of the waters; essays on their medicinal properties (by Norman H. Getman and William Baker Crain); descriptions of the major hotels in the village; and a guide to the environs of Canadarago Lake. Cover-title: *Richfield illustrated.*

190. BAINBRIDGE, William Seaman, 1870-1947.
Life's day. Guide-posts and danger-signals in health . . . Second edition. New York: Frederick A. Stokes Company, publishers, [c1909].
    xviii, [2], 308 p. ; 19 cm.

    This treatise on bodily, mental and moral health is based on a series of lectures delivered at Chautauqua by the prominent New York surgeon. Bainbridge approaches his topic via three principle factors: heredity, environment and "function." Part I is concerned with "physiological" and "psychical" heredity, i.e., genetically inherited factors that determine the constitution of the individual and "the race" in matters of health. These factors can be modified, however, by environment, e.g., food, shelter, clothing. Parts II-VI are entitled "Dawn," "Morning," "Midday," "Twilight" and "Night." They discuss "function," i.e., recommendations on hygienic living in each of the six periods of human life, from infancy to old age. Both editions of *Life's day* appeared in 1909.

    Bainbridge was an 1893 graduate of the College of Physicians & Surgeons (New York). At the time *Life's day* was published, he was professor of surgery at the New York Polytechnic Medical School & Hospital, and on the surgical staff of the New York Skin & Cancer Hospital. Bainbridge was the author of numerous articles on obstetrics and cancer surgery. His best known publication was his 1914 monographic entitled *The cancer problem,* which was translated into several languages.

191. BAIRD, William P.
IMPORTANT NOTICE. Dr. Baird having practised in the State of Vermont for nearly fifteen years, has consequently many calls, many patients, and many who are constantly writing to him for advice or treatment . . . DR. BAIRD office at S. Royalton Hotel, S. Royalton, Wednesday, May 18, from 9 to 12 o'clock. Red Lion Inn, Randolph, Wednesday, May 18, from 2 to 5 o'clock . . . [1883?].
    [1] leaf printed on both sides : ill. ; 43 × 19 cm.

    Baird advertises proprietary drugs and electro-magnetic therapy for the treatment of "female complaints," catarrh, eye diseases, cancer, etc. An itinerant physician, Baird could also be consulted by mail at his Boston address.

192. BAKER, Albert Rufus, b. 1858.
Coughs, colds, and catarrh. How to avoid . . . Revised edition. Cleveland, Ohio: The Arthur H. Clark Company, 1904.
    24 p. ; 20 cm.

    "This popular lecture was first delivered before the Cleveland Young Men's Christian Association and was printed in the 'Cleveland Medical Gazette,' June, 1891. Two thousand copies were published separately and were soon exhausted. In 1896 the lecture was again printed in the 'New York Vocalist.' One thousand copies were issued separately and again the edition was soon exhausted" (*Publisher's pref.*, p. [3]).

BAKER, J. see **BUCHAN** (#476).

BAKER, Thomas see **ALCOTT** (#46).

193. S. F. BAKER AND COMPANY.
A chat with the old family doctor about health and disease and how to easily cure many ailments. Keokuk, Iowa: Published by S.F. Baker & Co., [c1901].
   63 p. ; 18 cm.

Promotional literature for patent medicines available from S.F. Baker & Co. Title and imprint from wrapper.

194. S. F. BAKER AND COMPANY.
A chat with the old family doctor about how to relieve at home, without calling him, many ailments, and how best to assist his work. New and revised edition. Keokuk, Iowa: Published by S.F. Baker & Co., [c1904].
   80 p. : map ; 19 cm.

"The publishers of this book felt the need of a family medical adviser which would be unlike all others ever printed in two particulars; that it should not be divided into parts by names of diseases, when the hardest thing in the practice of medicine is to tell what disease the patient has; and that it should not be confined to the conditions which somebody's remedies are designed to cure. In this book, the divisions are according to symptoms . . . The thing that is often left out of this treatise is the name of the disease; but what to do under certain conditions is told in a plain way that anybody can understand" (p. [1]). In printed wrapper.

195. BAKEWELL, Mrs. J.
The mother's practical guide in the early training of her children: containing directions for their physical, intellectual, and moral education . . . New-York: Published by G. Lane & P.P. Sandford, for the Methodist Episcopal Church, 1843.
   224 p. ; 15 cm.

First American edition. Both London editions of *The mother's practical guide* were published in 1836. The American edition was issued again at New York in 1846, 1848 and 1850. In her preface Bakewell insists that the "physical, moral and intellectual training of children ought to be systematically, vigorously, and perseveringly pursued from their earliest infancy" (p. 14). The book's thirteen chapters provide advice to expectant mothers (e.g., conduct during pregnancy, guarding against impressions made by "disagreeable and deformed objects," choice of a "monthly nurse," etc.); infant feeding; hygiene (e.g., clothing, sleep, exercise); "intellectual training"; religious and moral training; punishment & reward as stimuli to good conduct; and advice on "domestic affliction," i.e., coping with the death of a child. Bakewell does not discuss children's diseases *per se* ("none but medical professors can with propriety write on these subjects," p. 12); but does describe the hygienic means for their prevention.

196. BALBIRNIE, John, 1810-1895.
The philosophy of the water-cure: a development of the true principles of health and longevity . . . Bath: Binns and Goodwin; London: Simpkin, Marshall, and Co., 1845.
   xxxix, [1],48, [2], 49-127, [3], [129]-276, [2], [277]-638 (i.e., 368) p. ; 18 cm.

First English edition. The first American edition was published at New York in 1846 (#197).

197. BALBIRNIE, John, 1810-1895.
The philosophy of the water-cure: a development of the true principles of health and longevity . . . Illustrated with the Confessions and observations of Sir Edward Lytton Bulwer. First American from the second London edition. New York: Wilson and Company, 1846.
   144 p., [1] leaf of plates : ill. ; 17 cm.

In his preface Balbirnie insists that "it is acknowledged that it is not physic, or the physician, that cures; but that the functions of the living organism, the unshackled play of its physiological mechanisms (the *vis medicatrix naturae*), are the prime agents in the restoration, as in the conservation, of health. Water, in its varied modes of application, can be made to produce, *demonstratively*, every *salutary* physiological action, and every curative effort of the economy, that the most successful administration of drugs is *said* to produce; *and that, too, in a manner, beyond all compare, more certainly, safely, promptly, and efficaciously*" (p. 6-7). Part I of this work consists of twenty-nine aphorisms establishing the "philosophical principles of health and disease." Part II describes the "physiological action" of water and provides thirty-three "therapeutical canons" for its application. Part III describes the hydropathic armamentarium, i.e., the various modes of its application. Part IV is a treatise on hygiene consisting of 121 recommendations on diet, clothing, air, sleep, etc. Part V attempts to establish the "legitimate province of physic," and concludes with thirty-five roman-numeraled "axioms" that summarize the merits of the water-cure. The book concludes with Edward Lytton Bulwer's (1803-1873) oft published "Confessions and observations of a water patient" (1846). Inspired by the publications of R.T. Claridge (#646-648) and James Wilson (#3829, 3830), Bulwer sought to restore his broken health by following a hydropathic regimen at Wilson's water-cure in Malvern.

Although numerous bibliographic references exist for the first English edition of *The philosophy of the water cure* published at Bath in 1845 (#196), we have been unsuccessful in finding record of the "second London edition." In printed wrapper.

Balbirnie was a Glasgow medical graduate whose thesis (*The speculum applied to the diagnosis and treatment of the organic diseases of the womb*) was published at London by Longmans *et al.* in 1836. He was also author of *A treatise on the organic diseases of the womb* (London, 1836?; 2nd ed., London 1837). His remaining publications, however, were devoted to hydropathy and balneology, including *The water-cure in consumption*, issued in three editions between 1854 and 1856; and *Hydropathic aphorisms* (1856). Neither of these titles appeared in an American edition.

198, BALBIRNIE, John, 1810-1895.
The philosophy of the water-cure: a development of the true principles of health and longevity . . . Illustrated with the Confessions and observations of Sir Edward Lytton Bulwer. First American from the second London edition. New York: Wilson and Company, 1847.
144 p., [1] leaf of plates : ill. ; 16 cm.

Second printing of the first American edition, originally issued in 1846 (#197).

199. BALBIRNIE, John, 1810-1895.
The philosophy of the water-cure: a development of the true principles of health and longevity . . . Illustrated with the Confessions and observations of Sir Edward Lytton Bulwer. First American from the second London edition. New York: Fowlers and Wells, hydropathic publishers, Water-Cure Journal office, 1852.
144 p. : ill. ; 16 cm.

200. BALBIRNIE, John, 1810-1895.
The philosophy of the water-cure: a development of the true principles of health and longevity . . . Illustrated with the Confessions and observations of Sir Edward Lytton Bulwer. First American from the second London edition. New York: Fowler and Wells, 1857.
144 p. ; 18 cm.

201. BALDWIN, David A., 1826-1905.
The family pocket homoeopathist. A concise manual of homoeopathic practice, for families and travelers . . . Rochester [N.Y.]: Erastus Darrow, publisher, 1870.
142, [2] p. ; 12 cm.

First edition, second issue. Baldwin was an 1849 graduate of the University of the City of New York Medical Department who nevertheless practiced medicine as a homeopath in Rochester, New York from 1851 through 1866. It was during this period that this manual was first published in Rochester (1863). Baldwin was licensed to practice medicine in the State of New Jersey in 1881, and died in Englewood. Cover-title: *Baldwin's pocket homoeopathist.*

202. BALDWIN, David A., 1826-1905.
The family pocket homoeopathist. A concise manual of homoeopathic practice, for families and travelers . . . Rochester [N.Y.]: Erastus Darrow, publisher, 1871.
142 p. ; 12 cm.

First edition, third issue. Cover-title: *Baldwin's pocket homoeopathist.*

203. BALDWIN, David A., 1826-1905.

The family pocket homoeopathist, a concise manual of homoeopathic practice, for families and travelers . . . Third edition. Rochester, N.Y.: E. Darrow & Co., publishers, 1894.
    148 p. ; 15 cm.

A second edition of this manual was published at Rochester in 1886 (re-issued 1890). The texts of the second and third editions contain modest revisions of the first edition (1863, 1870, 1871). Cover-title: *Baldwin's pocket homoeopathist.*

204. BALDWIN, Winfred Eugene.

Advanced lessons in human physiology and hygiene for grammar, ungraded, and high schools . . . New York; Cincinnati; Chicago: American Book Company, [c1899].
    412 p., [1] leaf of plates : ill. ; 19 cm.

Baldwin was an 1891 graduate of the Bellevue Hospital Medical College. He wrote three graded textbooks on physiology and hygiene for primary and high school students. All three texts were copyrighted by the Werner School Book Company of Chicago and were issued as part of the *Werner educational series.* The *Essential lessons* (#205, 206) and *Advanced lessons* were also issued by the American Book Company, a major publisher of school physiologies, as part of their *Practical series of school physiologies.*

205. BALDWIN, Winfred Eugene.

Essential lessons in human physiology and hygiene for schools . . . New York; Chicago; Boston: Werner School Book Company, [c1896].
    192 p. : ill. ; 19 cm.

First edition. Series: *The Werner educational series.*

206. BALDWIN, Winfred Eugene.

Essential lessons in human physiology and hygiene for schools . . . Chicago; New York: Werner School Book Company, [c1898].
    200 p., [2] leaves of plates : ill. ; 19 cm.

First published by the Werner School Book Company in 1896 (#205). Series: *Werner educational series.* At head of title: *Authorized edition.*

207. BALDWIN, Winfred Eugene.

Primary lessons in human physiology and hygiene for schools . . . New York; Chicago; Boston: Werner School Book Company, [c1896].
    144 p. : ill. ; 19 cm.

Series: *Werner educational series.*

208. BALKAM, J. D.

Private instructions in practical massage . . . [Boston]: The Author, c1887.
    21, [3] p. ; 19 cm.

Title and imprint from wrapper.

209. BALL, Isaac.

An analytical view of the animal economy. Calculated for the students of medicine, as well as private gentlemen; interspersed with many allegories and moral reflections, drawn from the subject, to awaken the mind to an elevated sense of the Great Author of Nature . . . New-York: Printed for the author, by G.J. Hunt, 1808.
    64, 67-90 p. : ill. ; 18 cm.

The *Analytical view* is one of the earliest examples of what was to become a veritable *genre* in American hygienic literature, i.e., the popular treatise that combined anatomy and physiology with a discussion of their hygienic significance. A physician writing before the emergence of the health reform movement, Ball has no hygienic ideology to propose, and actually is rather diffuse in his selection of topics. The most cogent part of the work is an elementary anatomy of the human body punctuated with frequent exclamations about its wondrous complexity. Ball's purpose is to induce respect for the human fabrick, reverence for its Creator, and interest in the body's maintenance. The last quarter of the book is a discussion of unrelated hygienic topics, e.g., the adaptability of the body to extremes in climate, the dangers of intemperate alcohol consumption, the physiologic pitfalls of old age, etc.

The hand-colored frontispiece of the heart and the allegorical vignette on page seven are wood engravings from the hand of Alexander Anderson, the physician-engraver who effectively established wood engraving as a book illustration medium in the United States.

REFERENCES: *Austin* #101.

210. BALL, Isaac.
An analytical view of the animal economy. Calculated for the students of medicine, as well as private gentlemen: interspersed with many allegories, and moral reflections, drawn from the subject, to awaken the mind to an elevated sense of the Great Author of Nature . . . The second edition, with large additions. New-York: Printed for the author, by Samuel Wood, 1808.
[6], viii, [9]-137, [25] p. : ill. ; 18 cm.

The anatomical section of the second edition is the same as the first but for the addition of several paragraphs to the text and numerous prolix footnotes (few of which touch on anatomy). The major difference between this edition and its predecessor is the addition of a section on the circulatory system, a three-part "Analogical view of plants and animals,"and several "Miscellaneous tracts."
REFERENCES: *Austin* #102.

211. BALL, Isaac.
An analytical view of the animal economy. Calculated for the students of medicine, as well as private gentlemen: interspersed with many allegories, and moral reflections, drawn from the subject, to awaken the mind to an elevated sense of the Great Author of Nature . . . The third edition, with large additions. New-York: Printed for the author, by Samuel Wood, 1808 [i.e., 1809].
[2], [i-ii], [2], [iii]-viii, [9]-137, [1], [141]-164 p. : ill. ; 18 cm.

A re-issue of the second edition (minus the table of contents and list of subscribers) with the addition of an appendix on the senses (pp. [141]-164). The dedicatory leaf ("To Samuel L. Mitchell, M.D."), dated 1 June 1809, is tipped-in after the title-page.
REFERENCES: *Austin* #103.

BALLIET, Thomas Minard see **AMERICAN FEDERATION FOR SEX HYGIENE** (#77).

212. BALLIN, Hans.
Physical training in the schoolroom. A system of bodily movements prepared for the American schools. A manual containing 450 consecutive exercises arranged for daily lessons, graded for primary, grammar and secondary schools; to which is added games and plays adapted for the cultivation of the senses in primary grades . . . [Little Rock, Ark.: Press of Arkansas Democrat Company, c1901].
[4], 183 p. : ill. ; 23 cm. + folding chart (63.5 × 97 cm.)

The "Chart of illustrations for Ballin's 'Physical training in the schoolroom'" is secured by a linen strap to the back pastedown.

213. BALLMER, Daniel.
Eine Sammlung von neuen Recepten und bewährten Curen, für Menschen und Vieh . . . Schellsburg: Gedruckt bey Friedrich Goeb, 1827.
40, [4] p. ; 18 cm.

Goeb also published this title in English at Schellsburg in 1827 under the title: *A collection of new receipts and approved cures for man and beast.* Title on printed wrapper: *Neue Recepte und bewahrte Curen, für Menschen und Vieh.*

214. BALLOU, George D'Estin, b. 1849.
The seven essentials to life and health. A practical primary treatise on hygiene . . . Los Angeles, California: Modern Hygiene Publishing Company, [c1909].
[12], [9]-421 p., [1] leaf of plates : ill., port. ; 20 cm.

BANISTER, E. see **LARMONT** (#2199).

BANKSTON, Lanaier see **REFORM MEDICAL COLLEGE OF GEORGIA** (#2953).

215. BANNING, Edmund Prior, 1810-1892.
A common sense essay . . . on the prevention & auxiliary mechanical cure of chronic, vocal, pulmonary, digestive, nervous, spinal, and female weaknesses: and also on the radical cure of hernia, piles, and prolapsus uteri, by braces, props, girdles, springs, levers and bandages . . . Dr. Banning may be consulted gratuitously for a few days, in this city, at [. . .] by such

as are anxious to improve the form and graceful bearing of their bodies, or to be relieved of any of the above maladies. A lady in attendance. [New York: Pudney & Russell, printers, ca. 1856].

32 p. : ill. ; 11 cm.

"It is my object in this essay, to show why it is that the best *internal* treatment by MEDICINE ALONE, so often fails of curing an extensive class of weaknesses, (among which, spinal, dyspeptic and female weaknesses are prominent,) and how it *may* be, that mechanical support, when physiologically constructed and applied, may of itself, or in conjunction with medicine, cure the same by removing the MECHANICAL CAUSE" (P. 3).

Banning was born in Canfield, Ohio on 3 June 1810, and was raised there by his mother's parents. Sometime after his marriage in 1834, he moved to New York City. Banning claimed to have received his medical degree from the Medical College of Evansville (Ind.) in 1850. His name is listed in the New York City business directories for many (but not all) years between 1850 and 1886. He died on 8 January 1892 in the home of his son, Archibald T. Banning (M.D., 1873, Cincinnati College of Medicine & Surgery), in Mt. Vernon, N.Y.

Banning was the designer and manufacturer of various types of mechanical body supports or braces. His writings are largely devoted to an explanation of the physiological and anatomical principles behind their design, and to the promotion of their sale. Anticipating chiropractic by some years, Banning contended that the body was subject to two physiologies, the vital and the mechanical, and that displacements in either led to various diseases.

A Banning genealogy makes the following claims: "[Banning] was a surgeon of advanced ideas, and founded the field of orthopedics. In 1838, he was the first physician to open the abdominal cavity to correct an intestinal obstruction. In 1839, he invented the first mechanical device for the permanent cure of enteroptosis. He was a writer and lecturer of great power" (Leroy F. Banning, *Banning branches,* Bowie, Md.: Heritage Press, 1994). Of Banning's ten children, at least two were physicians. Title from printed wrapper.

216. BANNING, Edmund Prior, 1810-1892.

Common sense on chronic diseases; or, a rational treatise on the mechanical cause and cure of most chronic affections of the truncal organs. Of both the male and female systems, embracing the author's views on physical education, and the present popular system of artificial life . . . Boston: Waite, Peirce & Company, 1844.

402, [14] p. : ill. ; 20 cm.

First edition. Banning posits "two distinct physiologies, the vital and the mechanical, each having a distinct dominion" (p. 30). The vital physiology relates to the "vital and inorganic functions, as in the case of the heart's perpetual action, the offices of the liver," etc. The mechanical physiology "relates to the physical and corporeal arrangement of man . . . independent of the vital principle" (p. 31). Each physiology has its own class of diseases that are of different origin and require different therapeutic approaches. Banning is unconcerned with the vital physiology, whose disorders fall within the purview of medicine. *Common sense on chronic diseases* addresses mechanical physiology, i.e., the structure and functioning of the body distinct from its vital principle. The book is not a general treatise on hygiene, however. Banning's emphasis is on the proper position of the "truncal organs," and how their displacement results in disease. The latter half of the book is devoted to mechanical appliances that correct these disorders, and includes numerous case histories that testify to the benefits of the corrective, mechanical devices manufactured by Banning.

217. BANNING, Edmund Prior, 1810-1892.

Common sense on chronic diseases; or, a rational treatise on the mechanical cause and cure of most chronic affections of the truncal organs. Of both male and female systems. Embracing the author's views on physical education, and the present popular system of artificial life . . . Third edition. New York: Paine & Burgess, 1846.

viii, [9]-199, [15] p. : ill. ; 18 cm.

Though half the number of pages of the first edition, the third edition of *Common sense in chronic diseases* contains the full text in smaller type. *Common sense on chronic diseases* went through at least twenty-one editions (under several titles) between 1844 and 1852.

BANNING, Edmund Prior, 1810-1892. *Dr. Banning on spine deformities* see **BANNING** (#219).

218. BANNING, Edmund Prior, 1810-1892. The house you live in. A rational treatise on the mechanical cause and nature of chronic diseases. Showing the origin and cause of so much vocal, pulmonary, digestive, nervous, spinal and female weakness . . . 65th edition. [New York: E.P. Banning, Sr., ca. 1877, c1875].
  [2], 64, [2] p. ; 17 cm.

"The writer of this *essay* or *circular* (which was written chiefly for the perusal of unprofessional eyes) desires to call the attention of the medical profession to the fact that he has perfected a system of mechanical supports, which are respectively so in harmony with the forces of the body as to be the most effective *adjuvants* in practice, meeting in a large class of cases, those muscular, osseous, and visceral weaknesses, deformities, and displacements, which medicine cannot successfully meet" (recto of preliminary leaf). Banning discusses various chronic disorders and the mechanical supports that relieve them. The text is illustrated with wood engravings and abounds in testimonials, the latest of which is dated 1877. Title and edition statement from printed wrapper.

219. BANNING, Edmund Prior, 1810-1892. Uterine displacements. Delivered in brief before the New York Academy of Medicine, Section on Obstetrics, May 12, 1866 . . . [New York: Banning & Co., 1867].
  2 pts. in 1 v. (7, [1] ; 12 p.) : ill. ; 23 cm.

Uterine displacements is reprinted from *The Medical and Surgical Reporter*, in which it was published as an eight-page supplement to vol. XVI, no. 25, 22 June 1867. It is issued here with a separately printed piece entitled "Dr. Banning on spine deformities."
  The "fundamental propositions" underlying Banning's etiology and therapeutics are stated in the opening paragraphs of the portion of this work on spinal deformities. First, "the human body is purely mechanical in the forma-

tion and arrangement of all its parts, from the grossest organs to the finest cells" (p. [1]); second, "the viscera are as much under the law of a specific orbit of being and bearing as the bones, and any departure from this will constitute a practical dislocation"; third, as the normal status of any group of organs, muscles, bones, etc. is its maintenance in an erect or "primary ascendant position," any variation from the norm results in a disordered state that is correctible only by restoring its "normal status." Page 12 of "Dr. Banning on spinal deformities" lists fifteen *Banning Surgico-Mechanical Adjuncts* or braces, e.g., No. 1. The abdominal and spinal shoulder brace," No. 4. The pile and prolapsus ani brace," No. 15. The varicocele brace," etc.

220. BANTING, William, 1797-1878. Letter on corpulence, addressed to the public . . . Third edition, with addenda. New-York: Mohun, Ebbs & Hough, 1864.
  22 p. ; 22 cm.

Banting, a retired London undertaker, recounts his unsuccessful struggle against obesity until at age sixty-six he assumed a diet that excluded "bread, butter, milk, sugar, beer, and potatoes," which, he claims, "contain starch and saccharine matter, tending to create fat" (p. 11). Banting's dietary system is based on advice received from William Harvey, M.D. (1806-1876), surgeon to the Royal Dispensary for Diseases of the Ear, to whom Banting had gone because of deafness. Harvey attributed his patient's hardness of hearing to obesity, and put him on a diet restricted to meat, fish and dry toast. "The result of this treatment was a gradual reduction of forty-six pounds in weight, with better health at the end of several weeks than had been enjoyed for the previous twenty years" (*Dict. nat. biog.*, I:1063). Originally published at London in 1863, the pamphlet appeared in at least thirteen American editions through 1880. The third American edition was first published at Philadelphia in 1863, again at Philadelphia in 1864, and in this New York edition. According to the 5th edition of *Morton's medical bibliography*, this regimen "became known as 'Bantingism' or the 'Banting diet," and its publication "is probably the first of the endless stream of bestselling books . . . on how to lose weight" (*Garrison-Morton* #3914).

221. BANTING, William, 1797-1878.
Letter on corpulence, addressed to the public . . . Fourth edition, with prefatory remarks by the author, copious information from correspondents, and confirmatory evidence of the benefit of the dietary system which he recommended to public notice. London: Published by Harrison, 1869.
  xvi, 127 p. ; 22 cm.

To the text of Banting's original *Letter* has been added a lengthy preface by the author, seventy-seven pages of testimonial correspondence, and Felix von Niemeyer's (1820-1871) "The treatment of corpulence by the so-called Banting system" (p. 101-116), originally delivered as a lecture at Stuttgart in 1865.

222. BARBOUR, George M.
Florida for tourists, invalids, and settlers: containing practical information regarding climate, soil, and productions; cities, towns, and people; the culture of the orange and other tropical fruits; farming and gardening; scenery and resorts; sport; routes of travel, etc., etc. . . . New York: D. Appleton and Company, 1882.
  [2], 310, [12] p., [1] folding leaf of plates : ill., map ; 19 cm.

Originally published in 1881.

BARLOW, Samuel Bancroft see **CHEPMELL** (#631).

223. BARNESBY, Norman, b. 1873.
Medical chaos and crime . . . London and New York: Mitchell Kennerley, 1910.
  384 p. ; 21 cm.

"This book is mainly an exposure of the abuses that exist in the medical profession in this country, abuses that not only degrade the practice of medicine, but contribute not a little to the physical and moral deterioration of the American people. It would be futile to attempt to estimate the amont of human suffering caused by the ignorance, incompetence, commercialism, and criminal indifference of those who call themselves disciples of Aesculapius, but the evil at least may be pointed out and denounced, and this I have done" (p. [9]).

Barnesby was an 1898 graduate of Rush Medical College.

224. BARNUM, H. L.
Family receipts, or practical guide for the husbandman and housewife, containing a great variety of valuable recipes, relating to agriculture, gardening, brewery, cookery, dairy, confectionary, diseases, farriery, ingrafting, and the various branches of rural and domestic economy . . . Cincinnati: Published by A.B. Roff; Lincoln & Co., printers, 1831.
  [2], iv, [5]-400 p. : ill. ; 20 cm.

The chapter entitled "Diseases" in Barnum's manual of domestic economy (p. [279]-302) provides receipts for the treatment of about two dozen non-threatening disorders, such as bruises, colds, headaches, etc. The greater part of the chapter is devoted to "general directions for preserving health, and attaining long life" and "rules for nursing sick persons." Barnum is described on the title-page as editor of the *United States agriculturalist and farmers' reporter* (1830-31). He was the author of *The American farrier* (1832); *The farmer's own book* (1832); and school textbooks on arithmetic, geography and spelling. Barnum was also author of *The spy unmasked* (1828), a memoir of Enoch Crosby, who served as the model for J. Fenimore Cooper's fictional character Harvey Birch.

BARNUM, Richard Clarence see *THE PEOPLE'S HOME LIBRARY* (#2768, 2769).

BARRETT, Thomas Squire see **KITCHINER** (#2147, 2148).

225. BARRINGTON, Paul Jones, b. 1829.
Gems of knowledge. Common sense prescriptions and practical information. A systematic treatment in the domestic practice of medicine . . . Chicago: Ottaway & Company, printers, 1882.
  326 p., [1] leaf of plates : port. ; 19 cm.

Second edition. A manual of homeopathic domestic medicine originally published at Chicago in 1881. The book is apparently a collaborative effort. The preface is signed "Mrs. Dr. Barrington" (Mary V. Ogden Barrington?), and the introduction "V.V. Barrington, M.D." Both Mary V. Ogden Barrington (b. 1840) and V.V. Barrington were 1880 graduates of the Chicago Homoeopathic Medical College.

226. BARTLETT, Elisha, 1804-1855.
The "laws of sobriety," and "the temperance reform:" an address delivered before the Young Men's Temperance Society in Lowell, March 8, 1835 . . . Lowell: Printed by Dearborn & Bellows, [1835].
    23, [1] p. ; 22 cm.

An 1826 medical graduate of Brown University, Bartlett spent the succeeding months in Paris attending the lectures of such figures as Cloquet and Cuvier before returning to Massachusetts. Bartlett was a masterful teacher, and over a period of twenty-three years lectured on the faculties of the Berkshire Medical School (1832), Dartmouth College (1839), Transylvania University (1841; 1846); the University of Maryland (1844), Louisville (1849), New York University (1850), the College of Physicians and Surgeons, New York (1852), and the Vermont Medical College each summer between 1843 and 1852. A prolific author, Bartlett made several important contributions to American medical literature, the most significant being *The history, diagnosis, and treatment of the fevers of the United States,* which appeared in four editions between 1842 and 1856.

Less attention has been given by biographers to Bartlett's interest in health reform. Osler acknowledged that "Bartlett was at his best in the occasional address" (*Cycl. Amer. Med. Biog.,* 1912, I:52), of which the present piece is a fine example. A temperate treatise on temperance, Bartlett embraces the cause of moderation in food and drink, but also warns of the dangers that beset such a movement, i.e., too rigid a position, too high expectations, and a spirit of contentiousness that he describes as "an *intemperance* more dreadful that that occasioned by alcohol" (p. 19). The unnumbered verso of the final leaf contains the "Constitution of the Lowell Young Men's Temperance Society" (est. 1833).

227. BARTLETT, Elisha, 1804-1855.
Obedience to the laws of health a moral duty. A lecture, delivered before the American Physiological Society, January 30, 1838

. . . Boston: Published by Julius A. Noble, [1838].
    24 p. ; 19 cm.

"True physiology, by which I mean the whole study of man,—his bodily and mental organization, with their thousand fold relations and mutual dependencies,—constituting the broad and comprehensive STUDY OF HUMANITY is a new science. As a science, and as the chiefest,—as the very end and crowning glory of all the rest,—it is of modern origin. The *laws,* which have impressed upon man's bodily and spiritual nature,—written there in characters as imperishable as if they had been inscribed in starry words, on the arch of the midnight sky,—these laws, impressed upon his whole being, for its government and good, and acting as invariably as those which rule the motions of the planets, are, now, only *beginning* to be studied and known." (p. 21).

James Whorton has described this essay as the best presentation written during the Jacksonian period on the "morality of physiology," i.e., the obligation to live by physiological principles less for the benefit of the individual than for society and succeeding generations. ("Christian physiology: William Alcott's prescription for the Millenium," *Bull. Hist. Med.,* 1975, 49:471, fn.16).The American Physiological Society was established at Boston in 1837. It propagated an understanding of hygiene in line with the teachings of Sylvester Graham (#1392-1395), at the time the leading figure of the American health reform movement. In printed wrapper. Atwater copy inscribed by author.

228. BARTLETT, Ezra A.
Cholera, its history, cause and prevention . . . Albany, N.Y.: H.H. Bender, 1885.
    105 p. ; 16 cm.

Bartlett was an 1879 graduate of the Albany Medical College.

229. BARUCH, Simon, 1840-1921.
A plea for public baths, together with an inexpensive method for their hygienic utilization . . . [New York]: Reprint from the Dietetic Gazette, May, 1891.
    45 p. : ill. ; 20 cm.

Baruch commends the City of New York for its generous funding of relief for the poor, but notes that "while so much has been accomplished for the *relief* of distress and disease, very little is done outside of governmental effort to *prevent* disease and to improve the condition of the poor while they are still in health" (p. 5). Toward this end Baruch recommends the establishment of public baths for "our employees, servants, laborers, trades-people and mechanics." By constructing bathing establishments "in the immediate vicinity of the populous tenement districts," the poor might partake of "the vast hygienic benefits of the bath" at an expense that is trifling compared to the costs incurred by their illness.

Baruch's *Plea* was originally read before the Section of Public Health and Hygiene of the New York Academy of Medicine.

Born in Prussia, Baruch emigrated to the United States in 1855. He received his medical degree in 1862 from the Medical College of Virginia, and immediately enlisted as a surgeon in the Confederate Army. After the war he practiced for thirteen years in South Carolina, where he was also an advocate for mandatory smallpox vaccinaton. In 1881 Baruch moved to New York City, where his career flourished and where he became an influential figure in public health advocacy: "In the 1890s Baruch began the sanitarian activism that resulted, in 1895, in the passage of a New York State law requiring all cities with populations over 50,000 to provide public 'rain baths' (showers)" (*Amer. nat.biog.*, 2:301). Baruch was perhaps the best known American proponent of hydrotherapy at the turn of the 20th century, and the author of *The uses of water in modern medicine* (1892), *The principles and practice of hydrotherapy* (1898), and *An epitome of hydrotherapy for physicians, architects, and nurses* (1920).

230. BARWELL, Louisa Mary, 1800-1885.
Infant treatment: with directions to mothers for self-management before, during, and after pregnancy. Addressed to mothers and nurses . . . First American edition. Revised, enlarged and adapted to habits and climate in the United States, and by a Physician of New York, under the approval and recommendation of Valentine Mott, M.D. New-York: Published by James Mowatt & Co., 1844.
    [4], [ix]-xi, [12]-148 p. ; 19 cm.

Barwell was an author of numerous instructional works for children who has received little or no attention from historians of Victorian children's literature. "After her marriage with John Barwell, wine merchant at Norwich . . . she devoted much attention to the composition of educational works, developing a remarkable gift for the comprehension of child nature, physical and mental. She frequently contributed to the 'Quarterly Journal of Education' from about the year 1831, anticipating some of the modern views and plans of education" (*Dict. nat. biog.*, I:1269). Her many contributions to the juvenile literature were popular on both sides of the Atlantic.

In *Infant treatment* Barwell extends her talents as an educator and author to instructing the mothers of her usual audience. In this American edition, her fifteen chapters on infant hygiene and care are supplemented with a sixteenth entitled "Infantile diseases and treatment in the United States." In the preface the anonymous American editor writes, "We have several able and judicious treatises on infancy by medical men, but they are inevitably deficient on many points, to which a woman alone is competent to do justice" (p. [ix]). Originally published at London in 1836 under the title *Infant treatment under two years of age*, this work appeared in a "third American edition" at New York in 1846 (the 2nd Amer. ed. is unrecorded), an 1847 Philadelphia edition, and three "revised" editions issued at Philadelphia in 1846, 1850 and 1853 under the title *Advice to mothers.* Cover-title: *Mrs. Barwell's infant treatment.*

BASS, J. C. see EXCELSIOR BOTANICAL CO. (#1110)

231. BATCHELDER, Josiah, 1776-1857.
Every seaman his own physician. Adapted to an ample arrangement of ships' medicine chests . . . First edition. Boston: Printed for the compiler, 1817.
    40 p. ; 17 cm.

"While the man sick at home, is surrounded by his friends, and choice medical council, to the one at sea, his medicine chest and book of directions, are his all, and every hope" (p. [3]). In the first of this two-part treatise, Batchelder provides a description of sixty-two medicines that should comprise the ship's medicine chest.

The second part is devoted to medical and surgical topics pertinent to men at sea, e.g., scurvy, venereal diseases (including the woeful advice, "A man who has a gonorrheea [sic], may if careful to wash the parts by syringing perfectly clean, cohabit with a sound woman and not communicate to her the disease"), amputation, how "to unite divided flesh," "jaw out," etc. In plain paper wrapper.

Josiah Batchelder was born on 3 January 1776 in Beverly, Ma. He graduated from Dartmouth College in 1796, received a medical degree from the same institution in 1799, and a second medical degree from Harvard in 1811. He was admitted to the Massachusetts Medical Society in 1810. Batchelder died in Falmouth, MA, on 5 February 1857.

REFERENCES: Not in *Austin*.

232. BATE, John W.
Dr. Bate's true marriage guide, a treatise for the married and marriageable, both male and female, containing information and salutary hints for everyone . . . Chicago, [c1889].
270, [2] p., [3] leaves of plates : ill. ; 18 cm.

An 1876 graduate of the eclectic Bennett Medical College (Chicago), Bate describes himself as a "confidential physician" since 1861, i.e., a specialist in women's diseases and in the genital disorders of both sexes. He was proprietor of Dr. Bate's Institute in Chicago, where he could be consulted in person or by mail, and from where he dispensed advice and proprietary medicines. Bate is listed in *Polk* as an eclectic physician from 1884 through 1898, but not thereafter.

Bate's marriage guide includes discussions of the male and female generative organs; the physiology of generation (he believed conception took place in the ovaries); the importance of sex to bodily and mental health; the causes of sterility (he ascribes the low birth rate of the wealthy not to birth control but to too much and too rich food); obstacles to marriage (e.g., age, consanguinity, hereditary diseases, venereal diseases, malformations of the genitals, etc.); sex disorders (e.g., masturbation, spermatorrhea, etc.); gynecological problems (e.g., menstrual diseases, the diseases of pregnancy, etc.); venereal diseases; and "intermarriage," or

the ill effects of union between persons of disparate physical and emotional temperaments. The only mention of contraception is the chapter on withdrawal (p. 122-24), in which Bate condemns the practice as having an even more deleterious effect on the male system than masturbation. Though later in the book he advertises female syringes for sale, their use is described as hygienic. The Atwater copy is bound in paper.

233. BATHS AND BATHING.
Baths and bathing. New York: D. Appleton and Company, 1897.
[2], 93 p. ; 15 cm.

CONTENTS: I. The physiological action of baths. II. The varieties of baths. III. Bathing localities. IV. The uses of baths. V. A visit to the bath. Series: *Health primers; no. 6.*

234. BATTLE CREEK SANITARIUM.
The Battle Creek Sanitarium. Its buildings and guests. Anniversary souvenir, 1866-1898 . . . [Battle Creek, Mich.: Review & Herald Pub'g Co., September 1898].
[16] leaves : ill. ; 14.5 × 22 cm.

A photographic souvenir album of the Sanitarium.

235. BATTLE CREEK SANITARIUM.
The building of a temple of health. [Battle Creek, Mich., 1902?].
[2], 28 p. : ill. ; 14.5 × 18.5 cm.

On 18 February 1902, the main buildings of the Battle Creek Sanitarium were destroyed by fire. "At the time of the fire," according to R.W. Schwarz, "Dr. Kellogg was returning to Battle Creek from the West Coast. He first learned of the disaster from a newspaper reporter while changing trains in Chicago. When the reporter asked if the sanitarium would be rebuilt, John Harvey, although shocked by the news, replied that of course it would. As soon as he was seated on the Battle Creek train, he . . . spent the remainder of the trip drawing preliminary plans for a new building" (*John Harvey Kellogg, M.D.,* Nashville, 1970, p. 70). By 11 May 1902, the cornerstone had already been laid for the new building. *The building of a temple of health* describes the plans for the new facility, and depicts in photographs the progress of construction.

BATTLE CREEK SANITARIUM see also *SANITARIUM GUESTS' HANDY GUIDE TO BATTLE CREEK* (#3098).

236. BAUNSCHEIDT, Carl, 1809-1873.
Der Baunscheidtismus . . . Auszug aus der neunten sehr bereicherten Auflage, nebst einem Anhange: Das Auge, seine Krankheiten und deren Heilung durch den Baunscheidtismus, zum praktischen Gebrauch für Jedermann . . . Cleveland, O.: John Linden, Baunscheidtist, [1865].
xii, 13-316 p., [1] leaf of plates : ill., port. ; 21 cm.

After his father's death (ca. 1831), it is claimed, Baunscheidt studied medicine briefly in Switzerland, served a year in the military (1832-33), and subsequently became the inventor of several military and agricultural implements. His interest in medicine followed a moment of enlightenment described in the 1859 English translation of this treatise. Suffering from "a tenacious gout which had fallen into the left arm . . . and which at last had assumed the form of a chronic ulcer on the hand," Baunscheidt was one day sitting by the open window when "a swarm of large gnats" flew into the room. The gnats "were humming around the suffering hand . . . and at last settled on it, the hand was thickly covered by the molesting insects; instead of driving them away, he permitted them to remain, although uninvited guests, and perforate the hand all over with their little stings, and after they had done their work and flown away, the back of the hand was speckled all over like the top of a thimble . . . when, in a few seconds afterward, he perceived that the whole of the pain in his hand had passed away through the openings made by the stinging probosces of the gnats . . . Thus, this accidental circumstance served him as a basis whereon to found the invention of a mechanical instrument, which is calculated, at all times, to apply, artificially, the salutary stings of the gnats, to awaken again into life the benumbed limbs. Baunscheidt invented his *Lebenswecker* and his famous 'Oleum Baunscheidtii,' the latter supplying the kind of liquid left in the wounds made by the little piercers of the gnats" (#237, p. xxxiv-xxxv).

In this treatise Baunscheidt outlines his peculiar theories of life and disease, the importance of applying therapeutic processes to the surface of the skin (rather than poisoning the internal organs with medicines), the practical application of the *Lebenswecker* and *Oleum Baunscheidtii,* and the "abducent" principle on which this therapy rests. The *Lebenswecker* consisted of needles released by a spring mechanism in the handle that punctured the skin to a depth determined by the operator. Baunscheidt stresses that the instrument is of limited effectiveness without the oil, which is responsible for the exanthematic irritation that "tends to remove quickly, and efficaciously, all infectious diseases" (#237, p. [41]). A wood engraving of the *Lebenswecker* is printed on page 78.

"Baunscheidtism," the author declares, "desires to deliver the world from the errors and abuses of the old faculty of physic and offers to the people the *Lebenswecker,* instead of the illegible and in most cases steril [sic] receipts, as a true theory of healing, and faithful family doctor, to be easily understood by every body, and always ready at hand" (#237, p. xli). The ninth German edition upon which this Cleveland edition is based was published at Bonn in 1864. *Das Auge* has a separate title-page dated 1865. Frontispiece portrait of the author.

237. BAUNSCHEIDT, Carl, 1809-1873.
Baunscheidtismus . . . First English edition translated from the sixth original edition by John Cheyne and L. Haymann . . . Bonn: By J. Witman and in Endenich, near Bonn, at the office of The Baunscheidtismus, [1859?].
lii, 268 p., [2] leaves of plates : ill., port. ; 22 cm.

Translation of: *Der Baunscheidtismus.* Atwater copy issued with author's *The eye* (Bonn, 1860, #238). The frontispiece portrait of author is dated 1859.

238. BAUNSCHEIDT, Carl, 1809-1873.
The eye, its diseases and their cure by Baunscheidtismus . . . Translated from the original by John Cheyne . . . Bonn: Published by J. Wittmann . . . and at the office of C. Baunscheidt, Endenich, 1860.
[4], 84 p. : diagrs. ; 22 cm.

"Most eye diseases are not local . . . but originate in the condition of the general organism .
. . The scrofula, cancer, scurvy, incipient gout,
congealing of the blood, or want of blood, rheumatic or nervous affections are by far the most
general causes of disease in the eye . . . It is well
known and acknowledged that a disease of the
body attacks the weakest and most susceptible
part . . . None will deny that the eye belongs to
the tenderest and least able organs to resist disease, and therefore are the more exposed to
attack by any complaint lurking in the body . . .
For such evils, of what help will a local treatment, to which physicians confine themselves,
prove? None whatsoever! . . . With our Lebenswecker we seize the root of the evil, and if that
can be completely uprooted it cannot spring
up afresh" (p. 36-37). Translation of: *Das Auge,
seine Krankheiten und deren Heilung durch den
Baunscheidtismus.* Atwater copy issued with the
1859 (?) English translation of the author's
Baunscheidtismus (#237).

BAUNSCHEIDT, Carl see also **LINDEN**
(#2258).

BAYARD, Edward see **JAHR** (#1971).

239. BAZAR BOOK OF HEALTH.
The Bazar book of health. The dwelling,
the nursery, the bedroom, the dining-
room, the parlor, the library, the kitchen,
the sick-room . . . New York: Harper &
Brothers, 1873.
280 p. ; 17 cm.

A manual of domestic sanitation and hygiene treating of such topics as ventilation, air
contamination, home heating, and light, as well
as chapters on child care & general hygiene.
The book's anonymous author claims to be a
physician (p. 6).

240. BEACH, Wooster, 1794-1868.
The American practice abridged, or the
family physician: being the scientific system
of medicine; on vegetable principles, designed for all classes. In nine parts. Part I.—
The means of preventing disease and promoting health. Part II.—General principles
of the reformed practice of medicine and
indications of cure. Part III.—Internal diseases. Part IV.—Surgical diseases. Part V.—

Midwifery. Part VI.—Vegetable materia
medica. Part VII. — Pharmacy and dispensatory, or compounds. Part VIII.—Diet for
the healthy and the sick. Part IX.—Outlines
of anatomy and physiology . . . Ninth edition. New York: Published and for sale by
Messrs. Andrews & Co., 1846.
xlviii, 788 p., [26] leaves of plates : ill.,
port. ; 23 cm.

"Eclectic medicine was the only botanical
movement to become a major force in American medical practice after the Civil War . . . The
professional Thomsonian and other botanical
practitioners who sought to turn Thomsonism
into a medical sect succeeded only in fragment-
ing it into a number of factions . . . One group
emerged out of this melange as a successful and
influential medical sect . . . Its founder was
Wooster Beach . . . Beach became a critic of
heroic therapy and turned to Thomsonians,
Indian doctors, herb doctors, and others for
ideas on medical practice. Because he drew his
ideas from all these sources, he decided to call
himself an eclectic. This was an apt description,
because Beach never did develop any systematic or sectarian dogma and did not limit his
practice to botanical medicines" (William G.
Rothstein, *American physicians in the nineteenth
century,* Baltimore, 1972, p. 218).
A native of Trumbull, Ct., Beach studied
medicine in Huntingdon County, New Jersey
with the German herbalist Jacob Tidd. Upon
Tidd's death in 1825, he studied in the Medical Department of the University of New York,
from which he claimed to have received a medical degree. Whatever his formal qualifications,
Beach was in practice in New York City as a regular physician from 1821. Early on, however, he
took exception to the heroic therapeutics of
his colleagues:
"When I first came to New York," Beach
wrote in the introduction to his *American practice of medicine,* "there were but one or two obscure botanic shops and scarcely a botanic practitioner to be found in the city. I therefore saw
the necessity of a reformation in the science of
medicine, and made every effort to acccomplish
it. I pursued an eclectic or analytical course in
the acquisition of an improved system of practice, and tested the information thus acquired
by clinical practice at the beside of the sick . . .
I conducted a daily and several weekly papers,

periodicals, and tracts, to enlighten the public generally on this subject; and, still further to carry out my views, built an Infirmary, where hundreds resorted for medical aid, which soon gave me excellent opportunities for testing *vegetable remedies,* and thus proving their superiority over the minerals and blood-letting in general use. I passed through the studies, not only of the *botanic* but also of the *allopathic* schools of medicine, rejecting everything which I regarded as erroneous, and adopting all the improvements which I found of importance, thus establishing an improved system of medical and surgical practice" (2nd ed., New York, 1852, I:[v]).

In 1827 Beach openly parted ranks with regular medicine and founded the United States Infirmary, where patients were treated on "reformed" principles and students were trained. In 1829 the infirmary was expanded and renamed the Reformed Medical Academy. Noting the success of the Thomsonians in Ohio and the midwest, Beach was instrumental in founding the Reformed Medical College in Worthington, Ohio (1830). The reputation of the reformed or eclectic movement was enhanced by Beach's "success" in treating cholera patients during the 1832 New York epidemic. The death rate among the afflicted was 50% in most districts of the city. The mortality rate in the 10th ward, where Beach had charge of indigent patients, was 20%. After the closure of the Worthington school, Beach joined several of its faculty in the formation of the Eclectic Medical Institute in Cincinnati (1845), which remained the intellectual flagship of the eclectic movement into the twentieth century. In 1836 Beach gave eclecticism its official voice with the founding of the *Eclectic medical journal* (1836) and the National Eclectic Medical Association (est. 1849; pres. 1855). He moved to Boston in 1852 where he established the Reformed Medical College.

The American practice abridged is a condensation of Beach's three-volume *The American practice of medicine* (#244, 245). This single-volume abridgement of Beach's larger work underwent several title changes in the course of its publication, e.g., the editions preceding *The American practice abridged* were published under the title *The family physician.* These various editions of Beach's text "fell within the accepted genre of advice literature and domestic medicine in the early nineteenth century, combining commonsense observations, uplifting poetry, domestic practice, popular wisdom, and testimonials . . . Like many authors of advice books, he devoted pages to the importance of temperance, exercise, pure air, and proper clothing; the benefits of bathing; and the evils of tight lacing. He also gave rules for preserving health and obtaining longevity" (John Haller, *Medical Protestants,* Carbondale, Ill., 1994, p. 73-74). Haller points out that Beach "was not especially meticulous when it came to recognizing and acknowledging those from whom he had drawn his information and illustrations" (*ibid.,* p. 74). They include Rafinesque, Jacob Bigelow and Elisha Smith (#3253). Alex Berman noted that "one is forced to conclude, after reviewing Beach's work on the plant materia medica, that his contribution was noteworthy mainly for its mediocrity and excessive borrowing" ("A striving for scientific respectability," *Bull. hist. med.,* 1956, 30:13). Spine-title: *The family physician*; frontispiece portrait of the author.

241. BEACH, Wooster, 1794-1868.

The American practice condensed. Or the family physician: being the scientific system of medicine; on vegetable principles, designed for all classes. In nine parts. Part I.—The means of preventing diseases and promoting health. Part II.—General principles of the reformed practice of medicine and indications of cure. Part III.—Internal diseases. Part IV.—Surgical diseases. Part V.—Midwifery. Part VI.—Vegetable materia medica. Part VII.—Pharmacy and dispensatory, or compounds. Part VIII.—Diet for the healthy and the sick. Part IX.—Outlines of anatomy and physiology, with illustrations . . . Twelfth edition. New York: Published and for sale by James M'Alister, 1847.

xlviii, 800 p., [28] leaves of plates : ill., port. ; 23 cm.

The single-volume abridgement of Beach's *The American practice of medicine* underwent several title changes in the course of its publishing history. The first abridged edition appears to have been *Beach's family physician, or the reformed system of medicine,* the earliest recorded issue of which is the 1842 New York edition. Between 1843 and 1846 the abridgement was issued in New York under the title *The family physician*; and again in New York, between 1846 and 1847 under the title *The American practice abridged*. In New York and Cincinnati, it was issued between 1847 and 1879 as *The American practice condensed* (in Boston as *Beach's American practice condensed*); and between 1859 and 1862 in Cincinnati as *Beach's family physician and home guide*. This edition includes an engraved portrait of the author (copied from a daguerrotype) and an engraved frontispiece view of the Place Vendome in Paris.

242. BEACH, Wooster, 1794-1868.
The American practice condensed. Or the family physician. Being the scientific system of medicine; on vegetable principles, designed for all classes. In nine parts. Part I.—The means of preventing diseases and promoting health. Part II.—General principles of the reformed practice of medicine and indications of cure. Part III.—Internal diseases. Part IV.—Surgical diseases. Part V.—Midwifery. Part VI.—Vegetable materia medica. Part VII.—Pharmacy and dispensatory, or compounds. Part VIII.—Diet for the healthy and the sick. Part IX.—Outlines of anatomy and physiology, with illustrations . . . Sixteenth edition, revised with the latest improvements. New York: Published and for sale by James M'Alister, 1850.
    xlviii, 800 p., [26] leaves of plates : ill., port. ; 24 cm.

    "Although allopaths claimed that Wooster Beach and Samuel Thomson were spun from the same heretical cloth, reform medical practice was as unlike Thomsonianism as it was allopathy. To be sure, Beach adhered to a botanic system of cure; however, he did not share Thomson's strong anti-intellectual and class biases. Beach was educated, having reputedly attended the lectures of Philip W. Post, David Hosack, and Valentine Mott at the University of New York. Moreover, Beach insisted on a more catholic view of botanic medicine than the restrictive dogma of Thomson, having been grounded in a liberal foundation emphasizing professionalism and education as its twin pillars of strength . . . Those who discovered the folly of Thomson's system or declared themselves independent of the Thomsonians and enlarged the area of their therapeutical views, as well as the vocabulary of their materia

medica, preferred the more liberal appellation *reformed medical practice*" (John Haller, *Medical Protestants*, Carbondale, Ill., 1994, p. 71-72).

243. BEACH, Wooster, 1794-1868,
The American practice condensed. Or the family physician: being the scientific system of medicine: on vegetable principles, designed for all classes. In nine parts. Part I.—The means of preventing diseases and promoting health. Part II.—General principles of the reformed practice of medicine and indications of cure. Part III.—Internal diseases. Part IV.—Surgical diseases. Part V.—Midwifery. Part VI.—Vegetable materia medica. Part VII.—Pharmacy and dispensatory, or compounds. Part VIII.—Diet for the healthy and the sick. Part IX.—Outlines of anatomy and physiology, with illustrations . . . Twentieth edition, revised with the latest improvements. Cincinnati: Moore, Wilstach, Keys & Co., 1862.
    xlviii, 49-873 p., [5] leaves of plates + frontis. : ill., port. ; 24 cm.

    The forty pages of wood-engraved botanical plates in part six, though separately printed, are included in the pagination of this edition.

Fig. 9. Wooster Beach, from his American practice condensed, *#243.*

244. BEACH, Wooster, 1794-1868.

The American practice of medicine; being a treatise on the character, causes, symptoms, morbid appearances, and treatment of the diseases of men, women, and children, of all climates, on vegetable or botanical principles: as taught at the reformed medical colleges in the United States: containing also a treatise on materia medica and pharmacy, or the various articles prescribed, their description, history, properties, preparations, and uses; with an appendix, on the cholera, etc. . . . The whole preceded by practical rules for the prevention of disease and the preservation of health . . . New York: Betts & Anstice, 1833.

3 v. (19, [3], [19]-679 p., [1] leaf of plates ; 530, 84, [iii]-vi p., 12, [1] leaves of plates ; [2], 279, [1], [iii]-x, [2] p., [95] leaves of plates : ill., port. ; 22 cm.

First edition (the third volume of the Atwater set is from the 1836 2nd New York edition). Beach states in his introduction that this work is "calculated to effect a change, or introduce a reformation in the noble science of medicine" by supplanting "poisonous minerals" and products of the "chemical laboratory" with botanicals in the materia medica. He intends his *major opus* to serve as a textbook for students in reformed medical colleges; to enlighten "physicians of the old school," who are almost entirely ignorant of medical botany; to correct deficiencies in the knowledge of the several grades of so-called "botanical physicians"; and inform the public generally. "We do not mean that every man should become a physician," he writes, "All we plead for is, that men of sense and learning should be so far acquainted with the general principles of medicine as to be in a condition to derive from it some of those advantages with which it is fraught; and at the same time to guard themselves against the destructive influence of ignorance, superstition, and quackery" (p. 11).

The first volume of *The American practice of medicine* is divided into three parts, beginning with a 125 page introductory on the "means of preventing disease, and promoting health and longevity." The second part describes the "general principles of the American practice of medicine." Part three is devoted to a description of "physical diseases" arranged in a classi-

fication that Beach intends to supplant the nosologies of Cullen, Sauvages and Hosack. The second volume contains Beach's surgical treatise, to which is appended a timely treatise on the "Indian or spasmodic cholera," which made its appearance in New York in 1832. The third volume, Beach's materia medica, is devoted largely to medical botany, and is illustrated with wood-engraved plates (printed in green), some of which have been hand-colored. Beach describes and illustrates 141 plants, and includes their localities, qualities and medicinal properties.

245. BEACH, Wooster, 1794-1868.

The American practice of medicine, revised, enlarged and improved: being a practical exposition of pathology, therapeutics, surgery, materia medica, and pharmacy, on reformed principles; embracing the most useful portions of the former work, with corrections, additions, new remedies, and improvements; and exhibiting the results of the author's investigations in medicine in this country, and in a year's tour in Europe . . . New York: Published by the author. For sale by Charles Scribner, 1852.

3 v. ([iii]-lviii, [17]-720 p., 57 leaves of plates ; [4], vii, [1], [5]-584 p., 54 leaves of plates ; [2], i, [1], [5]-604 p., 70 leaves of plates) : ill. (col. plates) ; 25 cm.

Printed beneath imprint: *Second edition.* The first edition of Beach's three-volume *American practice of medicine* was published at New York in 1833 (#244), and the second in 1836. A single-volume abridgement of this work for more general or popular use appeared in 1842, and remained in publication under various titles until 1879.

These three volumes are illustrated with 181 hand-colored lithographic plates and numerous wood engravings in-text. The utility of chromatic illustrations of pathological conditions doubtless came to Beach during his medical tour of Europe in 1848-49. Excepting the botanical plates in volume three, the numerous colored lithographs constitute a kind of poor man's Cruveilhier, whose pathological specimens at the Museum Dupuytren in Paris Beach notes in the introduction (I:xxxiv), and whose atlas he must have known. The surgical volume contains the greatest number of wood engravings, many of which are crudely copied from the surgical texts of such authors as Astley Cooper and Robert Liston. Spine-title: *American reformed practice*. The first volume of the Atwater set is lacking plates XI, F, LX & LXIV.

246. BEACH, Wooster, 1794-1868.

Beach's American practice condensed, or, the family physician: being the scientific system of medicine; on vegetable principles, designed for all classes. In nine parts. Part I.—The means of preventing diseases and promoting health. Part II.—General principles of the reformed practice of medicine and indications of cure. Part III.—Internal diseases. Part IV.—Surgical diseases. Part V.—Midwifery. Part VI.—Vegetable materia medica. Part VII.—Pharmacy and dispensatory, or compounds. Part VIII.—Diet for the healthy and the sick. Part IX.—Outlines of anatomy and physiology, with illusrations [sic] . . . Sixteenth edition, revised with the latest improvements. Boston: Published by Sanborn, Carter & Bazin, 1855.

xlviii, 800 p., [26] leaves of plates : ill. ; 24 cm.

Reissue of the sheets of the New York editions of the *American practice condensed* (#241, 242) with new title-page. Steel-engraved frontispiece portrait of author.

247. BEACH, Wooster, 1794-1868.

Beach's family physician and home guide. For the treatment of the diseases of men, women, and children, on reform principles . . . Forty-fifth thousand, revised and enlarged, with an appendix, from eminent writers, giving the laws of health . . . Cincinnati: Moore, Wilstach, Keys & Co., 1862.

xlviii, 49-873, [1], 873-1000 p., [5] leaves of plates : ill. ; 25 cm.

Identical to the 1862 Cincinnati edition of *The American practice condensed* (#243) with a different title-page, the addition of an appendix entitled "Physiology, and the laws of health" (p. [873]-992), and a glossary (p. 993-1000). The forty pages of wood-engraved plates that illustrate part six ("Vegetable materia medica") are included in the pagination. Spine-title:

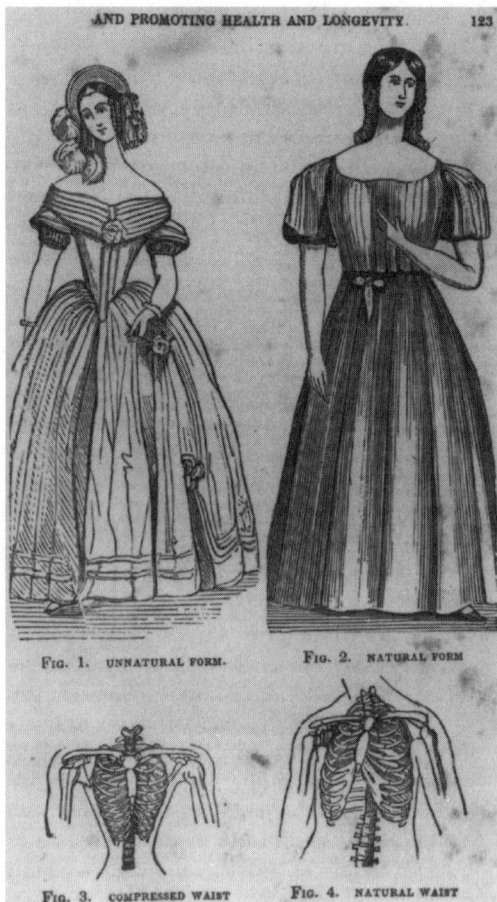

Fig. 10. Incorrect and correct dress for women, from Beach's family physician, #247.

Beach's family physician and home guide to health and happiness.

248. BEACH, Wooster, 1794-1868.

The family physician; or, the reformed system of medicine: on vegetable or botanical principles. Being a compendium of the "American practice." Designed for all classes. In nine parts. Part I.—The means of preventing disease, and promoting health. Part II.—General principles of the reformed practice of medicine, and indications of cure. Part III.—Internal disease. Part IV.—Surgical disease. Part V.—Midwifery. Part VI.—Vegetable materia medica. Part VII.—Pharmacy and dispensatory; or compounds. Part VIII.—Diet for the healthy and for the sick. Part IX.—Outlines

of anatomy and physiology, with illustrations. Appendix. This work embraces the
character, causes, symptoms, and treatment
of the diseases of men, women, and children of all climates . . . Complete in one
volume . . . Fourth edition. New York: Published and for sale by the author; Boston:
Saxton & Peirce, 1844.

xlviii, 782, [4] p., [9] leaves of plates,
[38] p. of plates : ill., port. ; 23 cm.

The Family physician is a condensation of
Beach's three-volume *The American practice of
medicine* (#244). This single-volume abridgement of Beach's larger work underwent several
title changes during the course of its publication (*see annotation*, #241). Engraved frontispiece portrait of the author.

249. BEACH, Wooster, 1794-1868.
The family physician; or, the reformed system of medicine: on vegetable or botanical principles. Being a compendium of the
"American practice." Designed for all
classes. In nine parts. Part I.—The means
of preventing disease, and promoting
health. Part II.—General principles of the
reformed practice of medicine, and indications of cure. Part III.—Internal diseases.
Part IV.—Surgical diseases. Part V.—Midwifery. Part VI.—Vegetable materia medica.
Part VII.—Pharmacy and dispensatory; or
compounds. Part VIII.—Diet for the
healthy and for the sick. Part IX.—Outlines
of anatomy and physiology, with illustrations. Appendix. This work embraces the
character, causes, symptoms, and treatment
of the diseases of men, women, and children of all climates . . . Complete in one
volume . . . Sixth edition. New York: Published and for sale by the author, 1845.

xlviii, 782 p., [9] leaves of plates, [22]
p. of plates + frontis. : ill., port. ; 23 cm.

250. BEACH, Wooster, 1794-1868.
An improved system of midwifery, adapted
to the reformed practice of medicine; illustrated by numerous plates. To which is
annexed, a compendium of the treatment
of female and infantile diseases, with remarks on physiological and moral elevation
. . . New York: Jas. McAlister, 1847.

272 p., 51 leaves of plates : ill., port. ; 29 cm.

First edition (re-issued at New York in 1848,
1850, 1851, 1854 and 1857). Although written
for physicians, medical students and midwives,
and drawing from a wide variety of standard
obstetrical authors (e.g., Denman, Bard,
Blundell, etc.), Beach's midwifery includes
many features common to popular obstetrical
literature of the period, e.g., illustrations of female figures that border on the erotic; descriptions of the organs of generation, sexual physiology and signs of pregnancy in language comprehensible to laymen; a chapter on monsters
that includes wood engravings printed from the
same blocks used to illustrate the 1846 edition
of *Aristotle's master piece* (#137); and a final section on child rearing entitled "The physiological and moral elevation of mankind." This
fourth part includes sections on the physical
training of children (i.e., hygiene); their moral
and religious education; and "miscellaneous
subjects" on sex and reproduction.

251. BEACH, Wooster, 1794-1868.
A treatise on anatomy, physiology, and
health, designed for students, schools, and
popular use . . . New York: Published by
Baker & Scribner, 1848.

viii, [9]-220 p., [28] leaves of plates : ill. ;
24 cm.

The two intaglio and twenty-six wood-engraved plates are hand-colored.

BEACH, Wooster see also **REFORM
MEDICAL COLLEGE** (#2952).

252. BEADLE'S DIME FAMILY PHYSICIAN.
Beadle's dime family physician and manual
for the sick room. With family diseases and
their treatment; hints on nursing and rearing; children's complaints; physiological
facts; rules of health, recipes for preparing well-known curatives, etc., etc. Based
upon the authority of Drs. Warren, Donna
[i.e., Donne], Parker, and others. Expressly
prepared for the Dime family text books.
New York: Beadle and Company, [c1861].

79 p. ; 15 cm.

REFERENCES: Johannsen, A. *The house of
Beadle and Adams.* Norman, 1950, I;375, "Family handbooks," #11.

253. BEALL, Edgar Charles, b. 1853.
The life sexual. A study of the philosophy, physiology, science, art, and hygiene of love . . . New York City: Published by Vim Publishing Company, [c1905].
   [2], vi, [10], 215, [7] p., [28] p. of plates + frontis. : ill., ports. ; 19 cm.

Beall's frame of reference for this discussion of marriage and sexual behavior is *phrenophysics,* the author's term for a discipline that comprehends "the sciences of phrenology, physiognomy, cheirognomy, graphology, temperaments, etc." The "purpose of Nature," according to Beall, "is to maintain the integrity of the race as a whole, rather than to secure the happiness of particular individuals. From this fact it follows that people should be *careful in trusting their natural instincts* in matters pertaining to love and marriage. In order to be truly happy, they must learn to choose their conjugal mates with reference to their own necessities, limitations, etc., as well as with reference to the needs of posterity" (p. 13). With this greater purpose in mind, Beall leads his readers to consider phrenophysical principles in the selection of a mate who is temperamentally and physically compatible ("conjugal affinity"), and in consideration of sexual ethics; sexual hygiene (i.e., the role of mental and physical health in the success of sexual life); and impediments to the sexual life (e.g., "impaired virility," venereal diseases, "sexual perversions"). Beall does not address sexual physiology, conception or contraception in this work.
   Beall is described on the title-page as an M.D. and president of the New York College of Phrenophysics. He was editor of *Vim,* a periodical published in six volumes between 1903 and 1905 that merged with the periodical *Health* (New York, 1845-1910).

254. BEALL, Otho T.
Cotton Mather, first significant figure in American medicine. By Otho T. Beall, Jr. and Richard H. Shryock. Baltimore: The Johns Hopkins University Press, 1954.
   ix, [1], 241 p., [2] leaves of plates : ill., facsim., port. ; 25 cm.

The son of Increase Mather was twelve years old when he entered Harvard College in 1674. Later to become the minister of Boston's Old North Church, Cotton Mather (1663-1728) became an influential figure in the city's religious and political life, laboring to maintain his fellow citizens' sense of their divine mission in New England. His published writings were prodigious in both number and intent, grounding every aspect of man's existence and activity in his relation to God. Less commonly known has been Mather's interest in medical matters—the subject of this study by Beall and Shryock. The authors describe the background of Mather's knowledge of medicine, and its expression in both his non-medical works and in *The angel of Bethseda.* A theologico-medical treatise intended for both physicians and the public, *The Angel* was completed in 1724, though not published until 1972 (#2389). Pages 127-234 contain selections from the manuscript of *The Angel* in the collections of the American Antiquarian Society. Series: *Publications of the Institute of the History of Medicine. 1st series. Monographs; v. 5.* Reprinted from volume 63 of the *Proceedings of the American Antiquarian Society.*

BEAN, J.V. see **ROBB** (#2984).

255. BEARD, George Miller, 1839-1883.
American nervousness. Its causes and consequences. A supplement to Nervous exhaustion (neurasthenia) . . . New York: G.P. Putnam's Sons, 1881.
   [2], xxii, 352, [2] p. ; 20 cm.

"Regulation of 'the passions of the mind' was Galen's sixth non-natural, and one given heavy emphasis by subsequent writers down through the Enlightenment. But coping came to take on new dimensions in the nineteenth century, especially in the United States. From the achievement of independence onward, in fact, Americans found themselves wrestling with ambivalent feelings about their country's open politcal-social system. The precious freedom to get ahead, to capitalize on the main chance, was also responsible for a competetiveness and uncertainty in life that scraped away unceasingly at the individual's mental and emotional reserves. That civilization begets nervousness became a cliche, and its was perhaps unavoidable that as the pace of life quickened, public uneasiness would mount and be translated into medical dogma . . . Nervousness was a self-fulfilling medical prophecy, and if there was a prophet, it has to have been

New York practitioner George Beard. 'Neurasthenia,' or nerve weakness, was the term Beard
applied to what he regarded as the epidemic of
advanced civilization. Its manifestations were
many and protean; his partial list included anxiety, indecisiveness, insomnia, peripheral numb-
ness, 'flying neuralgias' . . . [etc.]. By Beard's
analysis, neurasthenia was a state of exhaustion
brought about by too heavy taxation of the body's
nerve force . . . *American nervousness* (1881), his
chef d'oeuvre, blared through its title his conviction that no nation on earth approached this one
in the degree of its population's nervous exhaustion" (James C. Whorton, *Crusaders for fitness,*
Princeton, NJ, 1982, p. 148-49). Charles
Rosenberg points out that "Within a decade of
Beard's death in 1883, the diagnosis of nervous
exhaustion had become part of the office furniture of most physicians. Few textbooks and systems of medicine failed to discuss it, and in 1893
neurasthenia received its ultimate legitimiza-
tion—the publication of a German *Handbuch der
Neurasthenie*" ("The place of George M. Beard in
nineteenth-century psychiatry," *Bull. hist. med.,*
1962, 36:258). Rosenberg points out that the first
fifty pages of this handbook consist of "a bibliography and history of the new disease" in 5.5 point
type (*ibid.,* fn. 32).

Beard's original description of neurasthenia
appeared in an article published in the *Boston
Medical and Surgical Journal* in 1869 (80:217-21).
The "classic" description of "Beard's disease,"
however, is his *A practical treatise on nervous exhaustion (neurasthenia),* first published at New
York in 1880. *American nervousness* was intended
by Beard as a supplement to the earlier book.
In his preface Beard points out that although
the earlier work was written for the medical profession, the present work "is of a more distinctly
philosophical and popular character" (p. iii-iv).

An 1866 graduate of the College of Physicians
and Surgeons of New York, Beard was a pioneer
in the United States in the study of neurology
and mental illnesses. He established an international reputation with the publication in 1871 of
*A practical treatise on the medical and surgical uses of
electricity,* co-authored with A.D. Rockwell.

256. BEARD, George Miller, 1839-1883.
Hay-fever; or, summer catarrh. Its nature
and treatment . . . New York: Harper &
Brothers, publishers, 1876.
    [4], x, [11]-266, [2] p. ; 19 cm.

257. BEARD, George Miller, 1839-1883.
Our home physician. A new and popular
guide to the art of preserving health and
treating disease; with plain advice for all
the medical and surgical emergencies of
the family . . . New York: E.B. Treat & Co. .
. . [et al.], 1870.
    [2], xxxi, [32]-1065 p. : ill. ; 23 cm.

First edition.

258. BEARD, George Miller, 1839-1883.
A practical treatise on sea-sickness. Its symptoms, nature and treatment . . . Enlarged
edition; fourth thousand. New York: E.B.
Treat, 1881.
    vii, [8]-108, [6] p. ; 19 cm.

"To him science . . . owes the conception of
sea-sickness as a functional neurosis induced
mechanically by concussion, and he, too, introduced to the profession and to the public the
treatment of sea-sickness by bromides" (*Dict.
Amer. biog.,* I:93). Cover-title: *Sea-sickness.*

259. BEARDSLEY, Benjamin F., b. 1841.
A lecture on food and digestion . . . What
to eat and how to eat it. Rochester, N.Y.:
Published by Wilmot Castle & Co., [c1885].
    28 p. ; 14 cm.

"Delivered at Music Hall, Waltham, Mass.,
Oct. 20 '85." A sales pitch for the Arnold Auto-
matic Steam Cooker (manufactured by Wilmot
Castle & Co.) guised as a lecture on nutrition.
Beardsley was an 1865 medical graduate of the
University of Buffalo. In printed wrapper.

260. BEARDSLEY, Benjamin F., b. 1841.
A lecture on food and digestion . . . What
to eat, and how to eat it. Rochester, N.Y.:
Published by Wilmot Castle & Co., [1887,
c1885].
    28 p. ; 17 cm.

"Delivered in Unity Hall, Hartford, Conn.,
Tuesday evening, September 13th, 1887." This
edition includes illustrated advertisements (p.
18-28). On the back of the printed wrapper is a
wood engraving of the Statue of Liberty (dedicated by Pres. Grover Cleveland in Oct. 1886)
holding aloft the Arnold Automatic Steam
Cooker in place of her torch. The caption reads:

"Liberty Enlightening the World (on Steam Cooking)."

261. BEAUMONT, Philip.
A series of medical essays . . . Atlanta, Ga.: Published for the author, 1891.
  252 p. : ill. ; 24 cm.

Beaumont's essays are typical of those found in most 19th-century manuals on sexual physiology and ethics. He includes chapters on reproduction, marriage, masturbation, menstrual disorders, etc. The author identifies himself as an M.D. on the title-page and in his preface. Beaumont does not appear in any edition of *Polk* or the AMA directory, however.

BEAUMONT, William see **GRAHAM** (#1393).

262. BEAUTY.
Beauty: its attainment and preservation. Second edition. (Revised). New York: The Butterick Publishing Company, 1892.
  528 p. ; 19 cm.

"She who has no ambition to be beautiful will never wield the sceptre of womanly power" (p. 15). Being born beautiful is not sufficient, and even those born so will in time pale against their lesser endowed sisters if they do not do what is within their powers to enhance and maintain physical beauty. In giving consideration to the factors that contribute toward this end (e.g., sleep, bathing, diet, etc.), *Beauty* becomes a woman's health manual. Originally published at New York in 1890. Series: *Metropolitan culture series.*

BECK, Samuel T. E. see **NEW YORK MUSEUM OF ANATOMY** (#2598).

BECKLARD, Eugene, pseud. *Onanism and its cure* see **BECKLARD** (#266).

263. BECKLARD, Eugene, pseud.
Physiological mysteries and revelations in love, courtship and marriage; an infallible guide-book for married and single persons, in matters of utmost importance to the human race . . . Among the things duly considered in this work are matters of serious importance to single and young married persons—The causes of, and cures for sterility—The art of beauty and courtship—The danger of solitary practices, and how the habit may be removed—The causes of love and jealousy, with a remedy for eradicating from the system the seeds of a hopeless or an unhappy passion—Offspring, including modes for the propitiation or prevention thereof—Tests for knowing the sexes of unborn children—Intermarriage—Persons who ought and ought not to marry—The most auspicious season for wedlock, &c. &c. Translated from the third Paris edition, by Philip M. Howard, M.D. New-York, [1842].
  x, [11]-192 p., [1] leaf of plates : port. ; 11 cm.

First edition? The imprint date is torn from the title-page of the Atwater copy, but the copyright date on the verso is dated 1842. Engraved frontispiece portrait of author.

The Howard translation of Becklard's manual of sexual physiology was published under the title *Physiological mysteries and revelations* again at New York in 1844 (#264, 266), and with a Philadelphia imprint in 1843 (#265), 1845 (#267) and 1848 (#268). It appeared in Cincinnati in 1855, apparently with the title *Becklard's physiology.* There are no entries for Eugene Becklard in the print catalog of the Bibliotheque nationale, and the sole entry for Becklard in the *British Library general catalogue to 1975* (#270) treats the name as a pseudonym, which it probably is. There is no reason why the American "translator" could not have provided the author's actual name on the title-page; and there certainly would have been no legal recriminations at this period for the unauthorized translation and publication of an American edition of a French author. There may well have been recriminations for the publication of a manual of this nature, however, thus suggesting the need for a pseudonym. *Physiological mysteries* may therefore be the work of an American who was comfortable making reference to prominent French medical figures, who was familiar with the sexual physiologies of such authors as Auguste Debay or Morel de Rubempre, and who thought to protect his identity while profiting from the reputation of the French for esoteric sexual knowledge.

Becklard's manual was certainly written for a male audience, if for no other reason than it lacks the chapters on reproductive anatomy & physiology, female genital disorders and pregnancy that are common to most sex manuals of the period. Even the chapter on marriage is written from a male perspective, focusing more on sexual compatability than emotional compatability, and noting the harmful effects of sexual abstinence on the male system. His chapter on sterility and the use of "stimulants" to promote fecundity is actually a discussion of aphrodisiacs.

Among Becklard's reasons for the use of contraceptive methods was the possibility that "young unmarried females, in a moment of excitement" might "fall into errors, from the consequence of which they might probably recover, were a lawful and *crimeless* mode left open to them for avoiding the ban of the public, and burying their shame in their own bosoms" (p. 35-36). Becklard lists eight contraceptive methods: exercise to "disturb" the embryo (e.g., "dancing about the room, before repose, for a few minutes," or "trotting a horse briskly over a rough road on the following day"); strong cathartics ("even a month after intercourse"); stimulating foods and drinks before and after coition; early withdrawl ("such a system, however, requires too much presence of mind"); bathing in salt water after intercourse; the insertion of a sponge in the vagina before intercourse to absorb "the generating fluid"; "an oiled silk covering worn by the male, and sold by most of the toy shops of this city (Paris)"; and a syringe douche "soon after the act" with "one or two drops of vitriol in the water."

The treatise abounds in traditional sexlore, e.g.: distinguishing a male from a female child *in utero*; red-haired women are more amorous; a woman cannot conceive as a result of rape, etc. Cover-title: *Becklard's physiology.*

264. BECKLARD, Eugene, pseud.

Physiological mysteries and revelations in love, courtship and marriage; an infallible guide-book for married and single persons, in matters of utmost importance to the human race . . . Among the things duly considered in this work are matters of serious importance to single and young married persons—The causes of, and cures for sterility—The art of beauty and courtship—The danger of solitary practices, and how the habit may be removed—The

causes of love and jealousy, with a remedy for eradicating from the system the seeds of a hopeless or an unhappy passion—Offspring, including modes for the propitiation or prevention thereof—Tests for knowing the sexes of unborn children—Intermarriage—Persons who ought and ought not to marry—The most auspicious season for wedlock, &c. &c. Translated from the third Paris edition, by Philip M. Howard, M.D. New-York, 1842 [1844?].
    x, [11]-192 p. ; 11 cm.

Though the title-page is dated 1842, the copyright on its verso is dated 1844.

265. BECKLARD, Eugene, pseud.

Physiological mysteries and revelations in love, courtship and marriage; an infallible guide-book for married and single persons, in matters of utmost importance to the human race . . . Among the things duly considered in this work are matters of serious importance to single and young married persons—The causes of, and cures for sterility—The art of beauty and courtship—The danger of solitary practices, and how the habit may be removed—The causes of love and jealousy, with a remedy for eradicating from the system the seeds of a hopeless or an unhappy passion—Offspring, including modes for propitiation or prevention thereof—Tests for knowing the sexes of unborn children—Intermarriage—Persons who ought and ought not to marry—The most auspicious season for wedlock, &c. &c. Translated from the third Paris edition, by Phillip M. Howard, M.D. Philadelphia, 1843.
    viii, 9-178 p. ; 12 cm.

266. BECKLARD, Eugene, pseud.

Physiological mysteries and revelations in love, courtship and marriage. An infallible guide-book for married and single persons, in matters of the utmost importance to the human race . . . Translated from the third Paris edition, with the revision and additions of the sixth Paris edition, by Philip M. Howard . . . Among the things duly considered in this work are matters of serious importance to single and young married persons—The causes of, and cures for sterility—The art of beauty and courtship—The danger of solitary practices, and how the habit may be removed—The causes of love and jealousy, with a remedy for eradicating from the system the seeds of a hopeless or an unhappy passion—Offspring, including modes for propitiation or prevention thereof—Tests ifor [sic] knowing the sexes of unborn children, and procuring others according to choice—Intermarriage—Persons who ought and ought not to marry—The most auspicious season for wedlock, &c., &c. . . . New York: Holland & Glover, 1844 [i.e., 1845].

x, [11]-256 p., [16] leaves of plates : ill., port. ; 11 cm.

A reissue of the first 192 pages of the 1842 New York edition (#263) with the addition of sixty-four pages of printed matter and sixteen wood-engraved plates. The added matter (p. [193]-256) is a treatise on masturbation, and has a separate title-page: *Onanism and its cure. An infallible text book for the cure of all diseases in the male or female, produced by over indulgence, onanism, or masturbation. Translated from the French of Henriot, Tissot, Deslandes, and Becklard, by James Guierson. New-York: Holland & Glover. 1845.* There is no bibliographic evidence that this treatise was ever issued separately from the larger work. *Onanism and its cure* eschews descriptions of the diseases produced by the habit of masturbation; it is devoted rather to the regimen and remedies necessary for its cure. It does not consist of extracts from the works of Tissot, Deslandes, etc., but is a continuous narrative by a single author. "Becklard" claims to have been awarded "the principal agency in the publication" by the authors cited on the title-page (p. [195]). Six of the wood-engraved plates which illustrate the *Physiological mysteries and revelations* accompany this treatise on masturbation. They represent two series of three figures for a male and a female in the beginning, middle and end stages of chronic masturbation. Frontispiece portrait of author; spine-title: *Becklard's physiology.*

267. BECKLARD, Eugene, pseud.

Physiological mysteries and revelations in love, courtship and marriage; an infallible guide-book for married and single persons, in matters of the utmost importance to the human race . . . Among the things duly considered in this work are matters of serious importance to single and young married persons—The causes of, and cures for sterility—The art of beauty and courtship—The danger of solitary practices, and how the habit may be removed—The causes of love and jealousy, with a remedy for eradicating from the system the seeds of a hopeless or an unhappy passion—Offspring, including modes for propitiation or prevention thereof—Tests for knowing the sexes of unborn children—Intermarriage—Persons who ought and ought not to marry—The most auspicious season for wedlock, &c. &c. Translated from the third Paris edition, by Philip M. Howard, M.D. With a supplement from Canfield's sexual physiology, on coquetry, venereal madness, marriage, etc. Philadelphia: [John B. Perry, c1845].

5, vi-viii, 9-192, 26 p. ; 12 cm.

The American origin of the "Supplement from Canfield's sexual physiology" (p. 179-92) is evident in the mention of Robert Culverwell (#844-848) on p. 181, and the author's reference to lectures he delivered in New York (p. 184). The final 26 pages are a catalog of books published by the firm of John B. Perry, Philadelphia. Cover-title: *Becklard's physiology.*

268. BECKLARD, Eugene, pseud.

Physiological mysteries and revelations in love, courtship and marriage; an infallible guide-book for married and single persons, in matters of the utmost importance to the human race . . . Among the things duly considered in this work are matters of serious importance to single and young married persons—The causes of, and cures for sterility—The art of beauty and courtship—The danger of solitary practices, and how the habit may be removed—The causes of love and jealousy, with a remedy for eradicating from the system the seeds of a hopeless or an unhappy passion—Offspring, including modes for propitiation or prevention thereof—Tests for knowing the sexes of unborn children—Intermarriage—Persons who ought and ought not to marry—The most auspicious season for wedlock, &c. &c. Translated from the third

Paris edition, by Philip M. Howard, M.D. With a supplement from Canfield's sexual physiology, on coquetry, venereal madness, marriage, etc. Philadelphia: [John B. Perry], 1848.

viii, 9-192, 21, [3] p. ; 12 cm.

A reissue of the sheets of the 1845 Philadelphia edition (#267). The final twenty-four pages contain advertisements for the Philadelphia publisher John B. Perry. Cover-title: *Becklard's physiology.*

269. BECKLARD, Eugene, pseud.
Physiological mysteries revealed for the benefit of both sexes. An infallible guidebook for married and single persons, in matters of the utmost importance to the human race . . . Translated from the third Paris edition. New York, 1843.

vi, [7]-192 p. ; 12 cm.

269.1. BECKLARD, Eugene, pseud.
The physiologist: an infallible guide to health and happiness, for both sexes. Containing such information as is highly requisite every adult member of the human family should, for their own welfare, be in possession of; but such, at present, as can be obtained by those only who have the advantage of a medical education; and even these are often deficient in many of the subjects here treated of. Being physiological information concerning love, courtship, marriage, and many other things, which were treated of by the famous philosopher Aristotle . . . Translated from the fourth Paris edition, with corrections and additions, by M. Sherman Wharton, M.D. Second edition—or 10th thousand. Boston, 1844.

[iii]-viii, [9]-108 p. ; 15 cm.

All issues of the Sherman "translation" of Becklard were published at Boston under the title *The physiologist.* The earliest recorded copy of the Sherman translation appears to be this second Boston edition of 1844. Later issues were published at Boston in 1845, 1846 (#270) and 1859. The Boston editions purport to be translations of the fourth Paris edition. Except for two additional chapters (I & XVI), however, the Sherman translation of the fourth Paris

edition is identical word-for-word to the Howard translation of the third Paris edition (#263-268). Title on wrapper: *Physiological mysteries and revelations.*

270. BECKLARD, Eugene, pseud.
The physiologist; or sexual physiology revealed. Being mysteries and revelations in matters of great importance to the married and unmarried of both sexes;—and useful hints to lovers, husbands, and wives. A complete guide to health, happiness, and personal beauty.—Containing such information as can be had by those only who have the advantage of a medical education. With practical remarks on manhood, with the causes of its premature decline, and modes of perfect restoration. Economy and abuse of the generative organs. Effects of excessive indulgence. Love, courtship, and marriage. Its proper season. Directions for choosing a partner. Mysteries of generation. Causes and cure of barrenness. Prevention of offspring. Solitary practices, with their best mode of treatment, &c. &c. . . . Translated from the fourth Paris edition, with corrections and additions, by M. Sherman Wharton, M.D. . . . Boston: Published by the translator, 1846.

x, [11]-107 p. ; 16 cm.

At head of title: "Know thyself." There are two issues of this edition in the Atwater Collection. The linen cased issue has the cover-title: *Becklard's physiology.* The issue in printed wrapper has the cover-title: *Mysteries and revelations in matters of vital importance to the married and unmarried of both sexes . . ."*

271. BEDFORD, Gunning S., 1806-1870.
Reports on the artificial anatomy of Dr. Auzoux . . . New-York: Printed by Charles Vinten, 1840.

iv, [5]-11 p. ; 19 cm.

In printed wrapper. Bedford provides extracts regarding anatomical manikins from the Paris manufactory of Louis Auzoux from half a dozen French sources, including the 1835 report of the Académie Royale de Médecine (#7). "It is unnececessary for me to dilate on the excellencies or public utility of this most extraordinary mechanism," Bedford writes, adding that the study of human anatomy can now "be pursued by all classes of society without the fear of encountering the putrescent atmosphere of the dead." Bedford states: "It shall be my effort to make anatomy tributary to public instruction; and with this in view, I shall devote my leisure hours in explaining the beauties and wonders of our physical organization" (*Pref.*, p. [iii]).

Bedford was one of many in the 1830s and 1840s who enlisted in the crusade to instruct the American public on the structure and functioning of their bodies—providing each citizen knowledge of the laws of nature neccessary for the individual to assume responsibility for bodily health. Bedford received his medical degree from Rutgers College and for two years after continued his medical studies in Europe. On his return in 1833, he was appointed to the faculty of the Charleston Medical College. Following a brief tenure at Albany Medical College (N.Y.), Bedford moved to New York (1836) where he established a flourishing obstetrical practice. In 1840 he was appointed professor of midwifery at the University of New York (1840), where he opened the first obstetrical clinical in the United States. Bedford's two principles books, *Clinical lectures on the diseases of women and children* (1st ed., 1855) and *Principles and practice of midwifery* (1st ed., 1851), went through multiple editions and became standard texts among medical students and practitioners. His translations of the works of Baudelocque and Chailly-Honoré made available to his American colleagues a more thorough knowledge of European obstetrical practice.

272. BEDORTHA, Norman.

Practical medication, or the invalid's guide: with directions for the treatment of disease . . . Albany, N.Y.: Munsell & Rowland, 1860.
    iv, [5]-281, [7] p. ; 19 cm.

Bedortha received his medical degree in 1843 from the Eclectic Medical College of Ohio. He was soon associated with Joel Shew (#3180-3206), who placed him in charge of the newly established Lebanon Springs Water Cure (New York), where he remained until 1851. In June 1849, fifteen hydropathic physicians founded the American Hydropathic Society with Shew as its president, and Bedortha as one of four vice-presidents. Subsequently, Bedortha practiced in Troy, New York, until he established

his own water-cure in Saratoga Springs (1852), opposite the famous Congress Spring. The Saratoga Water Cure burned to the ground on 4 July 1864, when a carelessly thrown firecracker ignited a curtain. It was rebuilt in 1871 by Bedortha with B.T. Bedortha (M.D. 1870, Eclectic Medical College of Pennsylvania), probably his son. Both men are listed in *Butler* as practicing in Saratoga Springs until 1877; neither appear in the first edition of *Polk* (1886). The elder Bedortha was president of the Saratoga Springs Eclectic Medical Society in 1877.

Practical medication is a popular treatise on hydropathy. The first fifty-eight pages are devoted to the usual hygienic topics, i.e., diet, clothing, exercise, etc.. The greater part of the text describes the hydropathic armamentarium and its application to the treatment of a wide variety of disorders. Included among the advertisements at the end of the volume is a wood-engraved view of the Saratoga Water Cure and a brief description. Spine-title: *Invalid's guide.*

273. BEDORTHA, Norman.

Practical medication, or the invalid's guide: with directions for the treatment of disease . . . Second edition. Albany, N.Y.: J. Munsell, 1866.
    iv, [5]-281, [3] p. ; 19 cm.

Bedortha defines disease as "an inability of any organ or organs of the system to perform their normal functions" (p. 60). He subscribes to the "unity of disease" and the *vix medicatrix naturae*, and asserts that in the treatment of disease, water "can be adapted to a greater variety of the forms of disease, than any or all other articles used as medicines" (p. 69).

274. BEECHER, Catharine Esther, 1800-1878.

The American woman's home: or, principles of domestic science; being a guide to the formation and maintenance of economical, healthful, beautiful, and Christian homes. By Catherine E. Beecher and Harriet Beecher Stowe. New-York: J.B. Ford and Company . . . [et al.], 1869.
    [4], xii, [13]-500 p., [2] leaves of frontis. plates : ill. ; 21 cm.

First edition. Catherine Beecher was the eldest child of Lyman Beecher (1775-1863), Pres-

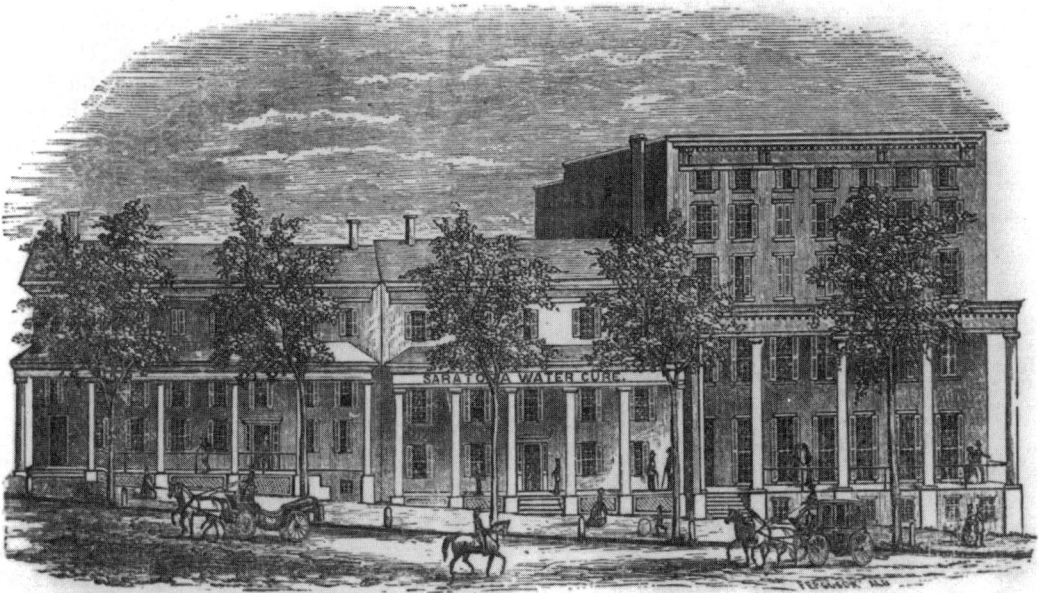

Fig. 11. The Saratoga Water Cure operated by N. Bedortha, from his Practical medication, *#272.*

byterian minister, the most influential evangelical preacher of his day, and an important figure in the temperance movement. Though heir to her father's call to transform the world, Beecher rejected his Calvinism, subscribing instead to a vision of the perfectability of man through education. Beecher's particular concern was the education of females, whom she regarded as the dominant moral force in society. In 1823 she opened (with her sister Mary) a female seminary in Hartford, Ct. where she was able to implement many of her ideas, including the incorporation of calisthenics into the curriculum. Beecher resigned from the Hartford Female Seminary in 1831 due to the stress of her endeavors, and accompanied her father to Cincinnati, where she opened the Western Female Institute (1832-1837). Her lack of success in Cincinnati is attributed by Robert Abzug to local perception of Beecher as a "self-proclaimed missionary for New-England 'civilization' in the rude West" (*Cosmos crumbling,* New York, 1994, p. 202).

During the 1840s she travelled incessantly, giving lectures on the educational needs of the American West. She was instrumental in the formation of the Board of National Popular Education (1847), for which she recruited more than 500 New England teachers to serve in western schools. After parting with the Board, she established the American Women's Educational Association (1852) for the purpose of establishing teacher training schools in the West. Her written output was voluminous. Beecher was the author of treatises on slavery, suffrage, the role of women in society, health reform, physical fitness, home economics, educational reform, religion, and ethics, as well as elementary school texts on grammar, arithmetic and geography.

Among her many writings was an eminently successful manual on home economy: *A treatise on domestic economy* (1st ed., Boston, 1841), and its supplement, *Miss Beecher's domestic receipt book* (1st ed., New York, 1846). "One of the most popular revisions of the *Treatise,* published in 1869 as *The American woman's home,* was prepared in collaboration with her sister Harriet Beecher Stowe. In these books the curious or baffled housewife found practical instructions on cooking, family health, infant care, and children's education, as well as general observations on proper home management" (*Notable American women, 1607-1950,* I:123). Added wood-engraved title-page and frontispiece.

Beecher was an outspoken advocate for women's education and the role of women in society, but within a decidedly domestic and religious context: "The authors of this volume,

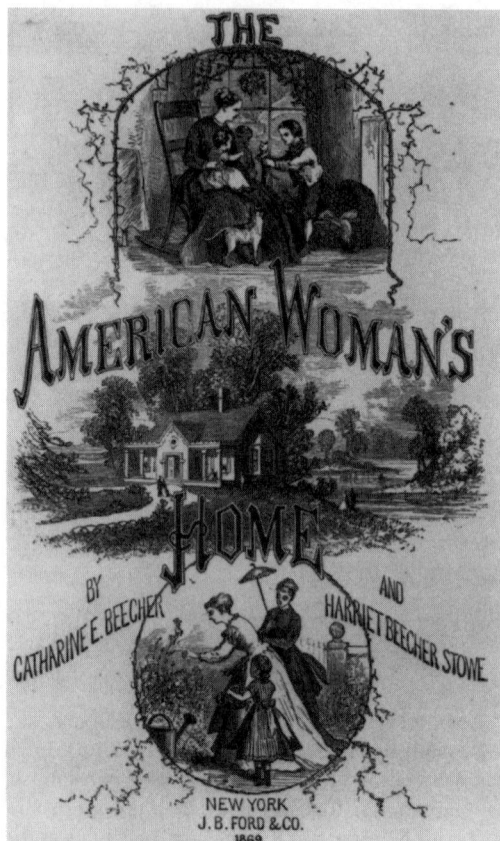

Fig. 12. Added wood-engraved title-page to Beecher's The American woman's home, *#274.*

while they sympathize with every honest effort to relieve the disabilities and sufferings of their sex, are confident that the chief cause of these evils is the fact that the honor and duties of the family state are not duly appreciated, that women are not trained for these duties as men are trained for their trades and professions, and that, as the consequence, family labor is poorly done, poorly paid, and regarded as menial and disgraceful" (p. [13]).

275. BEECHER, Catharine Esther, 1800-1878.

Letters to the people on health and happiness . . . New York: Harper & Brothers, publishers, 1855.

    vi, [7]-192, [1*]-29* p. : ill. ; 17 cm.

    First edition. In the first of her twenty-five letters Beecher expresses anxieties reflective of

a nation in the midst of unpredictable economic, social and religious change: "I have facts to communicate, that will prove that the American people are pursuing a course, in their own habits and practices, which is destroying health and happiness to an extent that is perfectly appalling. Nay more, I think I shall be able to show, that the majority of parents in this nation are systematically educating the rising generation to be feeble, deformed, homely, sickly, and miserable; as much so as if it were their express aim to commit so monstrous a folly" (p. [7]). Like Sylvester Graham and other health reform authors, Beecher harkens to a simpler and more pure era in American life, when children did not have "sallow or pale complexions, or look delicate or partially misformed" (p. 9). To remedy the national decline is the purpose of Beecher's *Letters,* which is divided into five parts: I. Organs of the human body; II. Laws of health; III. Abuses of the bodily organs; IV. Evils resulting from such abuses; V. Remedies for these evils.

    Letter XVII provides a fascinating personal account of Beecher's attempts to regain her own health at the hands of practitioners of various schools. Her eighteenth letter is a statistical survey of the health of American women undertaken during tours of the United States. Letter XXV describes the American Women's Educational Association, whose object was "to aid in securing to American women a liberal education, honorable position, and remunerative employment *in their appropriate profession;* the distinctive profession of women being considered as embracing the training of the human mind, the care of the human body in infancy and sickness, and the conservation of the family state" (p. 189). Appended to the *Letters* is a "Communication from Mrs. Dr. R.B. Gleason" (p. [1*]-16*) on various issues of women's health. Rachel Brooks Gleason (#1362-1367) was co-proprietor of the Elmira Water Cure, Elmira, N.Y. (#1063, 1064). Beecher died in Elmira, where she had gone to live in 1877 with her half-brother Thomas.

276. BEECHER, Catharine Esther, 1800-1878.

Letters to the people on health and happiness . . . New York: Harper & Brothers, publishers, 1856.

    vi, [7]-192, [1*]-29* p. : ill. ; 17 cm.

277. BEECHER, Catharine Esther, 1800-1878.
Miss Beecher's housekeeper and health-keeper: containing five hundred recipes for economical and healthful cooking; also, many directions for securing health and happiness. Approved by physicians of all classes. New York: Harper & Brothers, publishers, 1873.
    482 p. : ill. ; 19 cm.

First edition. In her opening chapter Beecher writes, "This volume embraces, in a concise form, many valuable portions of my other works on Domestic Economy . . . together with other new and interesting matter." For example, the book reprints many chapters from Beecher's 1869 *American woman's home* (#274), and omits others. She continues, "it is designed to be a complete encyclopaedia of all that relates to a woman duties as a housekeeper, wife, mother, and nurse . . . The First Part embraces a large variety of recipes for food that is both healthful and economical . . . The Second Part contains interesting information as to the construction of the body . . . These are so simplfied and illustrated, that by aid of this, both servants and children can be made to understand the *reasons* for the laws of health, as to render that willing and intelligent obedience which can be gained in no other way" (p. [15]-16). *Miss Beecher's housekeeper and healthkeeper* was reissued in 1874, 1876 and 1900.

278. BEECHER, Catharine Esther, 1800-1878.
Physiology and calisthenics. For schools and families . . . New York: Harper & Brothers, publishers, 1856.
    viii, [9]-193, [1], vi, [7]-58 p. : ill. ; 17 cm.

First edition. The first three parts of Beecher's manual are devoted to physiology and hygiene. The fourth part, entitled *Calisthentic exercises,* has a separate title-page and pagination, though the signing of its gatherings is continuous with the rest of the volume. There is no evidence that it was ever issued separately. It describes exercises "designed to promote health, and thus encourage beauty and strength" (*pref.,* p. iv), and is supposedly based on the system of Per Henrik Ling. In the separate intro-

duction to part four, Beecher stresses the importance of exercise by stating that she "has known cures performed in health establishments and elsewhere, by means of these exercises, in connection with a strict enforcement of the laws of health, very much greater than any thing she has ever known effected by *any* method of medical treatment" (p. [iii]). *Physiology and calisthenics* was reissued by Harper & Brothers in 1859, 1860 (#280), 1862, 1867 and 1871 (#281), and in at least one undated reissue (#279).

Although Jan Todd acknowledges Beecher's influence on the development of "purposive exercise" in her history of the physical education of 19th-century American women, she attributes much of that influence to Beecher's "distinguished New England lineage and . . . the sophisticated promotional campaign she organized to sell her books on health and exercise in the mid-1850s" (*Physical culture and the body beautiful,* Macon, Ga.: Mercer Univ. Press, 1998, p. [137]). To the charge of self-promotion, Beecher's practice differed little from that of many other health reformers of the period. Todd does ask, however: "While Beecher is widely regarded by historians as the most significant popularizer of calisthenics for women, little attention has been paid by these same historians as to how, at age fifty-six, Beecher came to publish a work on calisthenics." Todd's earlier chapters chronicle the influence of such figures as Mary Lyon, Mary Titcomb, Sarah Hale and others. In her chapter entitled "Becoming Catharine Beecher," Todd writes: "The highlight of Beecher's career as a physical educator came in the mid-1850s when she embarked on her physical culture campaign. Capitalizing on her status as a celebrity, Beecher organized what appears to have been the first true promotional campaign in the United States for any exercise system. She set up a book tour, created a press packet to get her ideas and the news of her imminent visit into the local papers, and involved local women in both her publicity campaign and her direct marketing scheme to sell books" (*ibid.,* p. 158). Although Todd dismisses Beecher's claim to have been the inventor of systematic calisthenics for women in America, she acknowledges her influence as the synthesizer of others' ideas; for being able to successfully market these ideas; and for laying the groundwork that would enable later advocates of exercise for women (e.g., Dio Lewis, #2233-2243, 2245, 2246) to find a ready market for their own ideas and publications.

279. BEECHER, Catharine Esther, 1800-1878.
Physiology and calisthenics. For schools and families . . . New York: Harper & Brothers, publishers, [c1856].
    viii, [9]-193, [1], vi, [7]-58 p. : ill. ; 17 cm.

280. BEECHER, Catharine Esther, 1800-1878.
Physiology and calisthenics. For schools and families . . . New York: Harper & Brothers, publishers, 1860.
    viii, [9]-193, [1], vi, [7]-58 p. : ill. ; 17 cm.

281. BEECHER, Catharine Esther, 1800-1878.
Physiology and calisthenics. For schools and families . . . New York: Harper & Brothers, publishers, 1871.
    viii, [9]-193, [1], vi, [7]-58 p. : ill. ; 17 cm.

282. BEECHER, Catharine Esther, 1800-1878.
Principles of domestic science; as applied to the duties and pleasures of home. A textbook for the use of young-ladies in schools, seminaries, and colleges. By Catharine E. Beecher and Harriet Beecher Stowe. New-York: J.B. Ford and Company, 1873.
    [2], x, [13]-16, [21]-380 p. : ill. ; 21 cm.

First published at New York by J.B. Ford in 1870 (and again in 1871), the *Principles of domestic science* is an abridged version of Beecher's and Stowe's *The American woman's home* (#274). Ten chapters in the latter title have been eliminated from the *Principles*.

283. BEECHER, Catharine Esther, 1800-1878.
A treatise on domestic economy, for the use of young ladies at home, and at school . . . Revised edition . . . Boston: Thomas H Webb & Co., 1843.
    [3]-383 p. : ill. ; 19 cm.

"The care of a female seminary, for some twelve years, and subsequent extensive travels,

have given such a view of female health, in this Nation, and of the causes which tend to weaken and destroy the constitution of young women, together with the sufferings consequent on want of early domestic knowledge and habits, that, though others may be better qualified to attempt some efficient remedy, the writer has been led to contribute, at least what was in her power, for this end" (*Pref.*, p. [7]). The *Treatise's* thirty-seven chapters cover such diverse topics as the civic, religious and domestic responsibilities of American women; the importance of studying "domestic economy" in order to fully assume these responsibilties; multiple chapters on the laws of health (food, drink, clothing, cleanliness, exercise, rest); "domestic manners"; the virtues required of a housekeeper; the principles of household management; the supervision of servants; infant and child care; nursing the sick; domestic sanitation; sewing; gardening and the care of domestic animals.

"Although she did not use the term, Beecher's writings advocated 'domestic feminism'—that is, expanded power for women within domestic life and ancillary power in the wider society . . . Catherine Beecher was best known for her *Treatise on domestic economy*, first published in 1841 and reprinted annually through 1856. She co-authored a greatly expanded version with Harriet Beecher Stowe, *The American woman's home; or, principles of domestic science* [#274] . . . Beecher's advice to women reflected the contemporary economic changes that were relocating male labor and much of what had formerly been household production outside the household. She advised married women to exercise greater control over domestic life, including family finances, and to value their work as an honorable calling with great significance for the future of American democracy" (*Amer. nat. biog.*, 2:463).

284. BEECHER, Catharine Esther, 1800-1878.
A treatise on domestic economy, for the use of young ladies at home, and at school . . . Revised edition . . . New York: Harper & Brothers, publishers, 1847.
    369, [13] p. : ill. ; 19 cm.

"The number of young women whose health is crushed, ere the first few years of married life are past, would seem incredible to one who has not investigated this subject . . . The writer became early convinced that this evil results mainly from the fact, that young girls, especially in the more wealthy classes, *are not trained for their profession.* In early life, they go through a course of school training which results in great debility of constitution, while, at the same time, their physical and domestic education is almost wholly neglected. Thus they enter on their most arduous and sacred duties so inexperienced and uninformed, and with so little muscular and nervous strength, that probably there is not one chance in ten, that young women of the present day, will pass through the first years of married life without such prostration of health and spirits as makes life a burden to themselves . . . The measure which . . . would tend to remedy this evil, would be to place *domestic economy* on an equality with the other sciences in female schools" (*Preface to the third edition,* p. [5]-6).

285. BEECHER, Catharine Esther, 1800-1878.
A treatise on domestic economy, for the use of young ladies at home, and at school . . . Revised edition . . . New York: Harper & Brothers, publishers, 1849.
    369, [13] p. : ill. ; 20 cm.

286. BEER, Georg Josef, 1763-1821.
The art of preserving the sight unimpaired to an extreme old age; and of re-establishing it and strengthening it when it becomes weak: with instructions how to proceed in accidental cases, which do not require the assistance of professional men, and the mode of treatment proper for the eyes during, and immediately after, the small pox. To which are added, observations on the inconveniences and dangers arising from the use of common spectacles, &c. &c. By an Experienced Oculist. Second edition, considerably enlarged and improved. London: Printed for Henry Colburn, 1815.
    xvi, 247, [1] p., [1] leaf of plates : ill. ; 18 cm. (12mo)

    Beer's manual of ocular hygiene, first published at Vienna in 1800 under the title *Pflege*

gesunder und geschwachter Augen, was translated into French, Italian, Portuguese and English. The English translation appeared in six London editions between 1813 and 1824 (the 1822 5th ed. was reissued in 1828). Beer was a brilliant clinician, an innovative ophthalmic surgeon, a gifted teacher, and a prolific author. In his biographical essay on Beer, Daniel Albert justly remarks that the establishment of ophthalmology as an independent scientific specialty came about largely through Beer's efforts at Vienna early in the nineteenth century (*Men of vision,* Philadelphia: Saunders, 1993, p. [114]).

287. BEGY, Joseph A.
Practical hand-book of toilet preparations and their uses. Also recipes for the household. formulas that have been tested by the author in a long professional career, which will enable the reader to make, at a small price, a class of preparations that are in universal demand, and are positively harmless. The information contained in this book is the result of original investigations, and cannot be obtained in its entirety without a lifetime of research . . . New York: Wm. L. Allison, publisher, [c1889].
    258, 16 p. ; 19 cm.

    A Rochester, N.Y. druggist, Begy provides toiletry recipes that his female readership can make at home or have compounded by a druggist. He includes formulas for the care of the skin, teeth and hair, as well as recipes for toilet waters and colognes, smelling salts, and "miscellaneous formulas or family recipes," i.e., household cleansers, stain removers, polishes, etc. Pages 200-251 contain recipes for domestic medicinals.

BEIGHLE, Nellie Craib see **OWEN** (#2691).

BELDING, Homer see **FARLIN** (#1129).

BELFIELD, William Thomas, 1856-1929.
Sterilization of the unfit a need of the state see **NATIONAL CHRISTIAN LEAGUE FOR THE PROMOTION OF PURITY** (#2570).

BELFIELD, William Thomas see also **LYMAN** (#2302-2306).

288. BELINAYE, Henry.
The sources of health and disease in communities; or, elementary views of "hygiene," illustrating its importance to legislators, heads of families, &c. . . . Boston: Allen and Ticknor, 1833.
vii, [1], 160 p. ; 18 cm.

First and only American edition of a title originally published at London in 1832. The term "hygiene" on the title-page refers to *public hygiene* rather than *private hygiene*. "It is a subject, to which, when once the attention of all educated persons is sufficiently aroused, their own unaided sense and experience will bring continual accession;—a result the more desireable, as it rarely occurs, that persons who have made the science of medicine their immediate study, are called to the performance of those duties which have the most direct influence on the physical improvement of communities" (*Pref.*, p. [vi]). The American publishers describe this work as "the first elementary book on the subject of hygiene [i.e., public health], which the American press has offered" (p.v).

Described on the title-page as "Surgeon Extraordinary to Her Royal Highness the Duchess of Kent," Belinaye was the author of *On the removal of stone from the bladder without the use of cutting instruments* (London, 1825; published again in 1837 under the title *Compendium of lithotripsy*).

BELL, Clark, 1832-1918. *The cigarette, does it contain any ingredient other than tobacco and paper?* see *THE TRUTH ABOUT CIGARETTES* (#3603).

289. BELL, John, 1796-1872.
Dietetical and medical hydrology. A treatise on baths; including cold, sea, warm, hot, vapour, gas, and mud baths; also, on the watery regimen, hydropathy, and pulmonary inhalation; with a description of bathing in ancient and modern times . . . Philadelphia: Barrington and Haswell, 1850.
660 p. : ill. ; 21 cm.

Bell's comma-happy necrologist in the *Medical and surgical reporter* (Philadelphia, 1872, 27:219) regarded this as his "chief work," in which Bell gives "a systematic view of the operations and affects of the different kinds of baths on the animal economy, as well as in its healthy as in its diseased state . . . It must be deemed strange, in medical literature, that though there were essays on the cold bath, on sea-bathing, on the warm bath, and on the vapor bath, and sometimes on two of these in the same volume, yet, until the appearance of the treatise of Dr. Bell, there was no one in the English language in which they were all severally considered, and their resemblances, contrasts, their successive and alternate uses, pointed out."

In his preface Bell writes, "The hygienic and physiological portions,—all that relates to the preservation of health, and the avoidance of disease, by recourse to baths and auxiliary processes,—must be regarded as property common to all. Very different is the case in whatever relates to the therapeutical application of baths, or their employment for the cure of diseases. These portions of the volume are intended by the author to be exclusively appropriated by his medical brethren" (p. 6). Substantial parts of this work were taken word-for-word from his 1831 treatise *On baths and mineral waters* (#291). So much new material has been added, however, that *Dietetical and medical hydrology* can hardly be regarded as a re-issue of the earlier work. Spine-title: *Bell on baths and the watery regimen*. The Atwater copy bears the inscription of James Caleb Jackson (#1902-1955).

John Bell was born in Ireland where he lived until 1810 when his parents emigrated to Virginia. In 1817 he received his medical degree from the University of Pennsylvania. Bell remained in Philadelphia at the encouragement of his mentor, the former Virginian Nathaniel Chapman. In addition to his private practice, Bell lectured on the institutes of medicine and materia medica at the Philadelphia Medical Institute, served on the staff of the City Hospital, and was active in the College of Physicians of Philadelphia (member, 1827), the Philadelphia County Medical Society (pres., 1858), the American Philosophical Society, and the American Medical Association (from its initial meeting at Baltimore in May, 1848).

He was a prolific author. In addition to the monographs listed below, Bell wrote *A practical dictionary of materia medica* (Philadelphia, 1841); co-authored with D. Francis Condie *All the ma-*

terial facts in the history of epidemic cholera (Philadelphia, 1832; 2nd ed., 1832); and authored the
*Report on the importance and economy of sanitary
measures to cities* (New York, 1860) for New York's
Board of Councilmen. Bell co-edited with
Condie the *Journal of health* (#2053), one of the
most seminal and influential contributions to
the the the early literature of American health reform; and edited *The Eclectic journal of medicine,*
published in four volumes at Philadelphia between 1837 and 1840.

Somehow Bell also found time to edit and
make available Americans editions of a surprising number of standard European medical authors, including Averill's *A short treatise on operative surgery* (1823); Armstrong's *Lectures on
morbid anatomy* (1837); Broussais' physiology
(1826); Mueller's physiology (1843); Combe's
The mother's guide for the care of her children (1840);
a folio edition of Rayer's treatise on skin diseases illustrated with forty chromolithographic
plates (1845); and Underwood's text on pediatrics (1841). Certainly his best known contribution to this genre of medical literature, however, are the second through fourth American
editions of William Stokes *Lectures on the theory
and practice of medicine* (2nd ed.: 1840, 1842; 3rd,
1845; 4th, 1848). Bell's contributions to Stoke's
text were so great that he is acknowledged as
co-author of the American editions of this work.

290. BELL, John, 1796-1872.
The mineral and thermal springs of the
United States and Canada . . . Philadelphia:
Parry and McMillan, 1855.
xx, [13]-394 p. ; 16 cm.

"The want of a manual in which travellers
for curiosity and pleasure, and invalids in quest
of health, might learn where to go, how to go,
and what to find, in relation to the mineral and
thermal springs of our country, has been generally felt . . . But, with the exception of a work
by the author, issued twenty-five years ago, no
attempt has hitherto been made to collect and
arrange methodically the numerous and separate and scattered histories and descriptions of
the different mineral and thermal springs of
the United States" (*Pref.,* p. [v]-vi).

In his first three chapters Bell discusses the
properties of mineral waters, their efficacy as
remedial agents, and the regimen to be followed by those taking the waters. He devotes

five chapters to the springs of New York State,
one chapter each to the springs of New England
and Pennsylvania, two chapters to springs in
Virginia, three chapters to those in the other
southern states, three chapters to springs found
between the Mississippi and the Pacific, and a
final chapter on Ontario and Quebec.

291. BELL, John, 1796-1872.
On baths and mineral waters. In two parts.
Part I. A full account of the hygienic and
curative powers of cold, tepid, warm, hot
and vapour baths, and of sea bathing. Part
II. A history of the chemical composition,
and medicinal properties of the chief mineral springs of the United States and of
Europe . . . Philadelphia: Office of the Journal of Health and the Family Library of
Health, &c., Henry H. Porter, proprietor,
1831.
xviii, [2], [17]-532 p. ; 20 cm.

The inspiration for this work, according to
the author's preface, is Giacomo Franceschi's
Igea de' bagne (Lucca, 1820), which Bell discovered on a tour of Italy and translated for publication. Though his translation was rejected by
the publishers, Bell continued to pursue his
interest in balneology in his native land, and
wrote a series of articles of the same title as this
book that were published in the *Philadelphia
journal of the medical and physical sciences* (8:311-
48, 1823; 9:69-97, 1824). These articles, together with copious notes from his travels and
reading, coalesced into the present volume.

Thomas Horrocks points out that although
Bell was intensely interested in public education on hygienic matters and had worked with
the Philadelphia publisher Henry Porter on
several health reform publications, his "work
on baths . . . lacks the urgency of reform and
the moral tone present in the pages of the *Journal of health* [which Bell co-edited] . . . In fact,
the book's relationship to the nascent health
reform movement is marginal at best. It reads
more like a clinical description of the various
kinds of baths and mineral springs and their
effects on the body than an advice manual on
health and moral living. This raises the questions concerning the intended audience; the
book seems to have been directed more toward
physicians as well as those upper-class Americans who could afford the time and expense of

traveling to these places than to middle-class readers" ("Promoting good health in the age of reform: the medical publications of Henry H. Porter of Philadelphia, 1829-32," *Canadian bull. med. hist.*, 1995, 12:270). Much of the matter in this work was incorporated into the text of Bell's 1850 *Dietetical and medical hydrology* (#289).

292. BELL, John, 1796-1872.

On regimen and longevity: comprising materia alimentaria, national dietetic usages and the influence of civilization on health and the duration of life . . . Philadelphia: Haswell & Johnson, 1842.
    [2], xiv, [13]-420 p. ; 20 cm.

Bell describes this work as a *materia alimentaria* on both the title-page and in the preface. His ambition in this work includes but also goes beyond the integration of correct dietary principles into a regimen of personal hygiene. He is interested not only in what kinds of foods benefit individuals, but also in introducing to a more influential audience (i.e., physicians, "the political economist, and the moralist") the exigencies of rapidly increasing populations and the need to provide them "alimentaria" that is both sufficient in quanity and beneficial to health. With the intention of broadening the American concept of nutrition, he describes the "national dietetic usages" of the ancients as well as the moderns (from Europe to S. America to Polynesia). The potential benefits of these varied aliments on the individual human system is not ignored, however. In his final chapters Bell covers ground familiar to readers of such authors as Graham, Alcott or Combe in consideration of vegetable and meat diets, the pernicious effects of alcohol, etc. Spine-title: *Regimen and longevity*.

293. BELL, John, 1796-1872.

Water, as a preservative of health, and a remedy in disease. A treatise on baths; including cold, sea, warm, hot, vapour, gas, and mud baths; also, on hydropathy, and pulmonary inhalation; with a description of bathing in ancient and modern times . . . Second edition. Philadelphia: Lindsay & Blakiston, 1859.
    658 p. : ill. ; 19 cm.

A reissue of Bell's *Dietetical and medical hydrology*, Philadelphia, 1850 (#289) with new title-page and preface. Spine-title: *Bell on baths*.

BELL, John see also **COMBE** (#745-747), *JOURNAL OF HEALTH* (#2053); *PORTER'S HEALTH ALMANAC* (#2868).

294. BELL, Robert, 1845-1926.

Our children, how to keep them well and treat them when they are ill: a guide to mothers . . . Third edition. New York: Frederick A.. Stokes Company, 1890.
    viii, 232 p. ; 19 cm.

Originally published at Glasgow in 1887. The prominent Scottish physician was the author of numerous works on obstetrics and gynecology, and later in his career, on cancer. *Our children* appears to be his only book addressed to a popular audience.

295. BELL, Victor Charles, b. 1867.

Popular essays upon the care of the teeth and mouth . . . [New York]: Published by the author, 1894.
    vi, [7]-103 p. : ill. ; 21 cm.

First edition. The *Popular essays* was issued in eleven editions through 1914, including at least two London editions. The author discusses such topics as oral hygiene, fillings, dentures, extractions, the growth of teeth in children, etc., and includes a chapter on "home remedies" for toothache, sore gums, "foul breath," etc. Bell is described on the title-page as "Director of the Special Prosthetic Department of the New York College of Dentistry." Atwater copy inscribed by author.

296. BELL, Victor Charles, b. 1867.

Popular essays upon the care of the teeth and mouth . . . Fifth edition. New York; London: Victor C. Bell, 1902.
    112, [4] p. : ill. ; 19 cm.

297. BELLEVUE MEDICAL INSTITUTE.

A treatise for men only on marriage and health. (Revised edition). Vital facts concerning diseases of mind and body. Chicago, Ill.: Published by the Bellevue Medical Institute, [1892?].
    64 p. : ill. ; 13 cm.

Promotional literature for the Bellevue Medical Institute in Chicago ("devoted exclusively to the treatment of all chronic, nervous and private diseases of men"), and for propri-etary remedies treating disorders of the male genitalia, e.g., impotence, varicocele, syphilis, the consequences of masturbation, etc. The Bellevue Medical Institute was operated by Buna Newman, an 1866 graduate of the Bellevue Hospital Medical College (a wood en-graving of his alma mater adorns p. [44]). In printed wrapper.

298. BELLEVUE MEDICAL INSTITUTE.
[A treatise for men only on marriage and health. Chicago: Bellevue Medical Insti-tute, c1897].
64 p. : ill. ; 13 cm.

Issued without a title-page, this promotional pamphlet advertising Buna Newman's Bellevue Medical Institute reprints the text of the ear-lier edition published under the title: *A treatise for men only on marriage and health* (#297).

299. BELLEVUE MEDICAL INSTITUTE.
A treatise on marriage and health. (Re-vised edition.) Vital facts concerning dis-eases of mind and body. Chicago, Ill.: Published by the Bellevue Medical Insti-tute, [c1897].
64 p. : ill. ; 13 cm.

300. BELLEVUE PLACE SANITARIUM.
Bellevue Place Sanitarium, Batavia, Illinois. [Chicago: The Frank Company Engravers and Printers, [ca. 1900].
14, [2] p. : ill. ; 13 × 18 cm.

"Bellevue Place Sanitarium was founded in 1867 by Dr. R.J. Patterson, for the treatment of nervous and mental diseases of *women only*" (p. 3). Neurasthenia is the only disorder specifi-cally mentioned in this promotional brochure: "The successful treatment of neurasthenia . . . often depends upon properly nourishing the patient; impaired nutrition being a feature of nearly all these cases. Our experience has been that judicious feeding, supplemented with oil and sea salt rubs, baths and careful nursing, have been the means of curing many a hith-erto intractable nervous trouble" (p. 7). "Sun light, pure air, water and electricity" are also included in the Sanitarium's treatment.
Richard John Patterson (1817-1893) was an 1842 graduate of the Berkshire Medical Col-

lege. Though his name appears on the title-page as president, this brochure may have been is-sued after his death. Frederick Henry Daniels (1855-1928), superintendent, was an 1882 graduate of the Medical School of Maine, but was not licensed to practice in Illinois until 1893. A later date for this brochure is also ar-guable from the status of its Assistant Physician. Helen L. Ryerson is probably the same person whose entry in the 1906 *American medical direc-tory* lists her as a 1903 graduate of the College of Physicians & Surgeons (Chicago). It would have been difficult for her to legally practice in Illinois in view of the state's Medical Practice Act of 1877.

301. BELLOWS, Albert Jones, 1804-1869.
How not to be sick. A sequel to "Philoso-phy of eating" . . . Second edition. New York: Published by Hurd and Houghton, 1869.
364, [4] p. ; 20 cm.

Bellows' "philosophy of eating" might be summarized in the statement that "one set of principles in food enables us to use the muscles . . . another enables us to keep up the animal heat, and another promotes the action of the brain and nerves" (p. 11). *How not to be sick,* however, is less an attempt to provide a practi-cal regimen for the application of these nutri-tional principles than a diffuse collection of his thoughts on diet, hygiene, medical ethics and domestic practice. The lengthiest of these di-gressions is Bellows' *apologia* for homeopathy on pages 203-308. The first edition of *How not to be sick* was published at New York in 1868. The second, third and fourth editions all ap-peared in 1869.

302. BELLOWS, Albert Jones, 1804-1869.
How not to be sick. A sequel to "Philoso-phy of eating" . . . Third edition. New York: Published by Hurd and Houghton, 1869.
364, [4] p. ; 20 cm.

303. BELLOWS, Albert Jones, 1804-1869.
The philosophy of eating . . . Second edi-tion. New York: Published by Hurd and Houghton; Boston: E.P. Dutton and Com-pany, 1868.
[2], 4, [v]-viii, 9-344 p. : ill. ; 19 cm.

"If science in farming is important, as it is proved to be, may not science in eating be more important? The scientific farmer analyzes his soil, and ascertains what it contains; then analyzes his grains and vegetables, and ascertains what elements they require; then analyzes the different manures and composts, and ascertains which contains, in the best combination, the elements to be supplied . . . I propose, upon the same principles, to give an analysis of the human system,—show the elements it contains, and the necessity for their constant supply,—and then to give an analysis of the food which Nature has furnished for the supply of these necessities" (p. 13).

Bellows discusses the principles of nutrition; the major food groups; individual food stuffs; and the role of diet in sickness and in health. His analyses and recommendations offered one of the more scientific approaches to nutrition available to a lay audience in the second half of the 19th century, which explains why *The Philosophy of eating* went through sixteen American editions between its original issue at New York in 1867, through the 16th Boston edition of 1887. A Glasgow edition was published in 1892 based upon the 15th American edition.

304. BELLOWS, Albert Jones, 1804-1869.
The philosophy of eating . . . Eleventh edition, revised and enlarged. Boston: Houghton, Osgood and Company; the Riverside Press, Cambridge, 1880.
[4], 3-4, [v]-viii, 9-426 p. ; 20 cm.

305. BELMONT LITHIA WATER.
Belmont Lithia Water, Spring Number 1. Drink and be cured. Description of spring. Analysis of water. Certificates. Testimonials. R.E. Richardson, proprietor. Talleysville, Va., [ca. 1894].
30, [2] p. ; 8 × 13 cm.

Title from printed wrapper.

BEMIS, E. H. see GLEN FALLS EYE SANITARIUM (#1370).

306. BENDINER AND SCHLESINGER.
A treatise on consumption. Giving a detailed account of a case of incipient, or hasty consumption, sent by the New York Journal to Prof. Hoff of Vienna, to prove to the world that the disease is curable . . . A hand-book giving cause, symptoms, treatment, diet, etc. New York: Published by Bendiner & Schlesinger, Manufacturing Chemists, [c1901].
30, [2] p. : facsims. ; 17 × 15.5 cm.

Promotional literature for *Prof. Hoff's Cure for Consumption,* available from the firm of Bendiner & Schlesinger. "5th edition, 130th thousand" (from printed wrapper).

307. BENNETT, Arnold, 1867-1931.
How to live on 24 hours a day . . . New York: George H. Doran Company, [c1910].
75 p. ; 20 cm.

Novelist, playwright and essayist, Bennett also dispensed advice on marriage, friendship, "mental efficiency" and hygiene in volumes devoted specifically to each.

308. BENNETT, Arnold, 1867-1931.
The human machine . . . New York: George H. Doran Company, [191?].
123 p. ; 20 cm.

309. BENNETT, Sanford, b. 1841.
Exercising in bed. The story of an old body and face made young. The simplest and most effective system of exercise ever devised . . . New York, U.S.A.: Published by the Physical Culture Publishing Co., [c1910].
[18], v, [1], 15-36, vi-xii, [37]-268, [2] p., [1] folding leaf of plates : ill., ports. ; 20 cm.

Second edition. "The movements, or muscular contractions and alternate relaxations, described in this system of Physical Culture are not especially new, but my adaption of those exercises, that they may be effectively and systematically practiced under the most comfortable conditions possible (that is, while lying in bed) is, to the best of my knowledge, a novel idea and a step in advance in the most important of all sciences, the science of health, strength, elasticity of body, and longevity" (*Pref.,* p. [17]). The chapters in this book originally appeared as a series of articles in the *San Francisco Chronicle* in 1906 and 1907. Bennett first published them in a book format at San Francisco in 1907. The Physical Culture Publishing Company was founded in 1898 by Bernarr Macfadden (#2316-2335).

310. BENNETT, Sanford, b. 1841.
Old age its cause and prevention. The story of an old body and face made young . . . New York City: The Physical Culture Publishing Company, [c1912].
[4], xi, [1], 17-394 p. : ill., ports. ; 20 cm.

Bennett's opening chapter begins with a confession common to much of the literature of health reform, i.e., testimony regarding the salvation of the author's health from the brink of irremediable ruin. As did many of his predecessors, Bennett pulled himself from the edge of death by the personal discovery of a regimen that completely altered his life, both physically and emotionally. Like Cornaro several centuries earlier, Bennett's conversion from a life of bad habits took place in middle age: "At fifty I was physically an old man. Many years of a too active business career had resulted in a general break-down. I was then wrinkled, partially bald, cheeks sunken, face drawn and haggard, muscles atrophied, and thirty years of chronic dyspepsia finally resulted in catarrh of the stomach" (p. 18). By attention to "sunlight, pure air, pure water, nourishing food, cleanliness and exercise" Bennett was redeemed, and encourages his elderly readers: "Follow my example and success will be yours. I have been an old man, and now at over 'three score and ten' I am a young man again, and look it. Really, I am a younger man physically than I was when in the best period of my early manhood" (p. 20). Though new chapters have been added, most of the material in *Old age* is taken directly from his earlier work *Exercising in bed* (#309). *Old age* was issued at least four times between 1912 and 1927.

311. BENSON, Luther, b. 1847.
Fifteen years in hell. An autobiography . . . Indianapolis: Tilford & Carlon, 1877.
4, vii, [1], 5-208 p., [1] leaf of plates : port. ; 20 cm.

Benson's contribution to the temperance literature is an autobiographical account of his struggles with alcoholism, commitment to an insane asylum, his religious conversion, fre-

quent backsliding, etc. In his final chapter both author and reader are left in uncertainty regarding his ultimate fate. Lithographic frontispiece portrait of author.

312. BENSON, Reuel Allen, 1877-1956.
A nursery manual. The care and feeding of children in health and disease . . . Philadelphia: Boericke & Tafel, 1908.
[2], 184 p. ; 18 cm.

"This book was originally written for the use of my own patients and nurses . . . It has largely been the outgrowth of lectures delivered in the Flower Hospital Training School for Nurses . . . The book is intended for the use of homoeopathic physicians and homoeopathic families and all those who believe with me that a child who has been properly fed and reared under the homoeopathic regime, is physically better equipped for life than any other" (p. [3]). Benson was a 1903 graduate of the New York Homoeopathic Medical College and Flower Hospital, where he lectured at the time this work was published. Atwater copy inscribed by the author.

313. BERTODY, Francis, 1736?-1800.
L'art de se guérir soi-même, des maladies vénériennes par le moyen de l'eau du sieur . . . Troisième édition. The art of curing one's self of the venereal disease, by means of the water . . . Third edition. Boston: Printed and sold by the author, 1800.
43 p. ; 19 cm.

Text is in French and English on facing pages. Bertody describes in detail the symptoms and prognosis of several venereal disorders, and recommends the internal or external application of one or more of nineteen remedies at each stage of the disease's progress. Pages 37-43 contain "A table of the remedies referred to in the foregoing treatise," i.e., recipes for the "waters," purgatives, injections, decoctions, astringents, cataplasms, mercurial ointment, unguent and liniments specifically recommended in the text. These same medicines were also prepared and sold by Bertody in Boston. Little is known of the author, who on the title-page claims to be a Padua medical graduate.
REFERENCES: Not in *Austin, Blake (NLM 18th Cent.)* or Shipton & Mooney's short-title

Evans. Bertody states that he originally published the details of his therapeutic system in French in 1789 (p. [2]). *Evans* (#31805) cites a 1797 Boston edition; the only edition cited in *Austin* (#201) is the 1804 fourth Boston edition.

BERWICK INSTITUTE see **DR. COLLINS' REMEDY COMPANY** (#727).

314. BESANT, Annie Wood, 1847-1933.
The law of population. Its consequences and its bearing upon human conduct and morals . . . Third ten thousand. Author's American edition from the 35th thousand, English edition. Bound Brook, N.J.: Asa K. Butts, 1886.
    47, [1] p., [1] leaf of plates : port. ; 19 cm.

Married in 1867 to an Anglican clergyman, Besant separated from her husband after the appearance of her anonymously published *On the deity of Jesus of Nazareth* (1873), a profession of her withdrawal from Christianity to theism, and eventually to atheism. Impoverished and on her own, Besant wrote pamphlets for the freethought publisher Charles Scott. In 1874 she met Charles Bradlaugh, who put her on the staff of his *National reformer.* Besant joined the lecture circuit, and soon became notorious for her outspoken social and religious views, her association with Bradlaugh, their trial in 1877 for publication of Knowlton's *Fruits of philosophy* (#2154-2157), etc. Graduating from freethought to socialism, she had a brief affair with Bernard Shaw. In 1889, after reviewing Mme. Blavatsky's *The secret doctrine* for the *National reformer,* she joined the Theosophical Society. She subsequently resigned from the National Secular Society, severed her connections with the Fabians, withdrew from the Malthusian League, and banned further publication of *The law of population.* From 1895 she lived in India, where the theosophical movement's headquarters had been moved in 1878. There Besant adapted her lifestyle and views to Hindu culture, founded the Central Hindu College (precursor of Benares Hindu University), and became involved in the independence movement. In 1916 she launched the Home Rule League, and was elected president of the Indian National Congress in 1918. S. Chandrasekhar draws an interesting sketch of Besant:

"In her life of some eighty-six years she played, with remarkable dedication and apparent sincerity, many roles, some of them contradicting others: she was a devout Christian theologian, an ardent atheist and freethinker, and avowed birth-controller and Neo-Malthusian, a Fabian socialist and feminist, a science teacher and trade unionist, an author, editor, publisher, 'Indian nationalist,' orator, social reformer, and a Theosophist" ("*A dirty, filthy book,*" Berkeley: Univ. of California Pres, 1981, p. 30).

Besant wrote *The law of population* because of her dissatisfaction with Knowlton's *Fruits of philosophy,* which had been published some four decades earlier. According to Bradlaugh, he and Besant had published Knowlton's treatise simply "to test the right of issuing cheap physiological knowledge, merely because that particular pamphlet had just been prosecuted. Having after hard struggle won the right, we dropped the particular book which had been forced on us . . . and issued the same information in a better form" (quoted in: Peter Fryer, *The birth controllers,* New York: Stein & Day, 1966, p. 164-65). The Bradlaugh-Besant trial not only legitimized the publication of information on birth control in England, but influenced the attitude of all classes toward family planning, and probably impacted the declining British birth rate after 1877.

"Shortly after the right of publication of medical information on contraception was clarified in England by legal decision," writes Norman Himes, "the Knowlton pamphlet was replaced by Annie Besant's *Law of population.* This elucidated in even greater detail than the *Fruits of philosophy* methods of controlling conception. Mrs. Besant's tract was first issued in January, 1879 (though possibly late in 1878), to replace the out-of-date Knowlton treatise . . . Though Mrs. Besant's treatment of contraception was thoroughly Malthusian, it was essentially sound both as regards her elucidation of Malthusian principles and her discussion of the tendency of population pressure to cause poverty, misery, low wages, child labor, etc." (*Medical history of contraception,* New York: Gamut Press, 1963, p. 245).

In her discussion of contraceptive technique, Besant addresses the "safe period," *coitus interruptus,* vaginal injections (according to Knowlton's formula), condoms, and the

sponge. "The sponge . . . was Mrs. Besant's favorite check . . . It was only in later editions . . .
that vaginal suppositories were recommended"
(*ibid.,* p. 246). "The influence of the *Law of population* was remarkable," Himes writes. "Up to
1891, that is within twelve years after its first
issue, 175,000 copies were sold in England . . .
It was reprinted in America as well as in Australia . . . It was translated into German, Dutch,
Italian, French and doubtless other languages.
When Mrs. Besant became a theosophist, and
temporarily renounced her Neo-Malthusian
views, the pamphlet was withdrawn" (p. 250).
The first American edition of Besant's treatise
was published by Asa K. Butts at New York in
1878. This 1886 edition is in a printed wrapper
and includes a frontispiece portrait of the author.

315. BESANT, Annie Wood, 1847-1933.
The law of population. Its consequences
and its bearing upon human conduct and
morals . . . Thirty-fourth thousand. New
American edition from the thirty-fifth thousand, English edition. Valley Falls, Kansas:
Fair Play Publishing Co., 1889.
    47, [1] p. ; 20 cm.

316. BESANT, Annie Wood, 1847-1933.
The law of population. Its consequences
and its bearing upon human conduct and
morals . . . San Francisco, May, 1893.
    73 p. ; 19 cm.

Series: *The Readers' library; no. 8.* In printed
wrapper.

317. BESANT, Annie Wood, 1847-1933.
The law of population . . . Its consequences
and its bearing upon human morals . . .
[S.l.: s.n., 1916?].
    3-53, [13] p. ; 18 cm.

Publication date based on quotation from
The Tribune (13 August 1916) printed on back
of wrapper.

BESANT, Annie Wood see also
    KNOWLTON (#2154-2157).

BIEGLER, Augustus P. see **ROCHESTER
LAKE VIEW WATER-CURE
INSTITUTION** (#3014).

318. BIGELOW, Horatio Ripley, 1844-1909.
Social physiology; or, familiar talks on the
mysteries of life . . . Schenectady, N.Y.: Central Publishing House, [c1891].
    429 p., [16] leaves of plates : ill. ; 23 cm.

In this unusually frank work, Bigelow devotes
a chapter each to the anatomy and physiology
of the male and female organs of generation,
and a chapter to conception and gestation. His
fifth chapter, "Facts for married people," discusses such topics as "the immediate effects of
coitus"; "conjugal frauds" (i.e., onanism); and
"sensual artifices" (e.g., "Different positions
from the normal may create desire in a wife
who ordinarily is apathetic"). The sixth and
seventh chapters are devoted to the causes of
male and female sterility, and his eighth chapter to the deleterious physical and moral effects
of masturbation, fornication and adultery.
Chapter IX is a 68-page history of prostitution,
including the modern history of attempts at its
regulation. In the following chapter on marriage, Bigelow returns to a consideration of
sexual ethics and behavior within marriage.
Cover-title: *Mysteries of life.* Bigelow was an 1879
graduate of the Columbian University Medical
Department (D.C.). He was the author of numerous books and pamphlets on gynecology
and electrotherapeutics.

**319. BIGELOW, Maurice Alpheus, 1872-
1955.**
Sex-education. A series of lectures concerning knowledge of sex in its relation to human life . . . New York: The Macmillan
Company, 1916.
    xi, [1], 251 p., [1] leaf of plates : port. ;
19 cm.

First edition. "Many of the lectures printed
in this volume have formed the basis of a series
given at Teachers College, Columbia University,
during the summer sessions of 1914 and 1915,
and during the academic year 1914-15. Others
were addressed to parents, to groups of men, to
women's clubs, and to conferences on sex education" (p. vii). The book is not a manual of
sexual ethics or hygiene, but a series of lectures
addressed to educators and to the public on the
social and ethical importance of sex education. *Sex
education* was reissued in 1918, 1924, and 1929. An
expanded "revised edition" was published in 1936.

Bigelow received his Ph.D. in biology from Harvard in 1901. He joined the faculty of the Teachers College, Columbia Univesity in 1899, and was known for his studies on the early develpment of crustacea. Bigelow was the author of several standard textbooks of biology, as well as popular health manuals such as *Health for every day* (1st ed., 1924) and *Health in home and neighborhood* (1st ed., 1924). Frontispiece portrait of Prince A. Morrow, "chief organizer of the American movement for sex-education," to whose memory the book is dedicated.

BIGELOW, Maurice Alpheus see also
AMERICAN FEDERATION FOR SEX HYGIENE (#77).

320. BIGGLE, Jacob.
Biggle health book. A family monitor and guide to good health and long life. Before the doctor comes . . . Philadelphia: Wilmer Atkinson Company, 1900.
184 p. : ill. ; 14 cm.

"Though neither a doctor nor the son of a doctor, I have always taken great interest in matters relating to the care and preservation of health. I observed the laws of health in my own case as nearly as I could ascertain them, have taken note of the things that have tended to injure the health of my neighbors, have read much, thought much, and studied considerably upon the subject" (*Pref.*, p. [5]). The *Biggle health book* is divided into three sections: hygiene, common diseases and care of the sick. It is number six in the *Biggle farm library* series. An advertisement for the series on p. 184 includes such unhailed classics as the *Biggle berry book* (already in its 2nd edition), the *Biggle cow book* ("all about cows") and the intriguingly titled *Biggle swine book*.

321. BIGGLE, Jacob.
Biggle health book. A family monitor and guide to good health and long life. Before the doctor comes . . . Philadelphia: Wilmer Atkinson Company, 1904.
184 p. : ill. ; 14 cm.

On verso of title-page: Second edition.

322. BILL, Ledyard, 1836-1907.
A winter in Florida; or, observations on the soil, climate, and products of our semi-tropical state; with sketches of the principal towns and cities in Eastern Florida. To which is added a brief historical summary; together with hints to the tourist, invalid, and sportsman . . . New York: Published by Wood & Holbrook, 1869.
222, [2], 3, [1] p., [5] leaves of plates : ill., map ; 19 cm.

First edition. "Within . . . the past few years, Florida has attracted considerable attention as a winter resort for invalids and pleasure-seekers. It is, practically, the only strip of tropical land within our boundaries, and the only state where the invalid can find an equable and mild temperature through the greater portion of the year" (*Pref.*, p. 7). *A winter in Florida* appeared in four editions between 1869 and 1870. Cover-title: *Florida*.

Born in Ledyard, Ct., Bill was a publisher in Louisville at the outbreak of the Civil War, but was forced to leave Kentucky in 1862 because of his Union sympathies. Bill resumed his publishing career in New York, where he remained until 1870, when he moved to Massachusetts. Bill later served as a member of the Massachusetts House (1891) and Senate (1894-96).

323. BILL, Ledyard, 1836-1907.
Minnesota; its character and climate. Likewise sketches of other resorts favorable to invalids; together with copious notes on health; also hints to tourists and emigrants . . . New York: Published by Wood & Holbrook, 1871.
205, [5] p., [1] leaf of plates : frontis. ; 19 cm.

"By general consent," Bill informs us, "Minnesota has enjoyed a superior reputation for climate, soil, and scenery beyond that of any other state in the Union, with, perhaps, a single exception [i.e., Florida]" (p. 6). Bill gives particular attention to Minnesota's reputation as a resort for consumptives. He attributes the salubrious effect of the state's climate on those suffering from pulmonary tuberculosis to "dry continental winds from the interior," which combined with Minnesota's high altitude, provide "an equable and a relatively dry atmosphere" that has "a bracing, tonic effect on the whole man, affording opportunity for unrestrained exercise in the open air, causing good digestion to wait on appetite, and with these the advent of fresh wholesome blood, which is *the* physician to heal the diseased portions of the lungs, and restore healthful action to the inflamed parts" (p.87-88). Cover-title: *Minnesota, California, Florida, Nassau, Fayal, Adirondacks*. Spine-title: *Climates for invalids*.

324. BILLINGS, CLAPP AND COMPANY.
"Bethesda." A traveller's criticism on our health resorts, their scenery, climatic peculiarities and curative influence . . . Boston: Billings, Clapp & Co., publishers, 1885.
[16] leaves : ill. ; 16 cm.

"A large number of persons in the United States are suffering from diseases, the majority of which may be greatly relieved by a change of climate, in combination with the use of some simple and pure tonic" (leaf 2 recto). The tonic is *Nichols' Elixir Peruvian Bark and Protoxide of Iron,* manufactured by Billings, Clapp & Co. Commentary on various American health resorts, mostly in the South and West, are interspersed with commendations of *Nichol's Elixir.* Title on wrapper: *Bethesda or health resorts.*.

325. BILZ, Friedrich Eduard, 1842-1922.
The natural method of healing. A new and complete guide to health . . . Translated from the latest German edition . . . Leipzig; London; Paris: F.E. Bilz; New York: The International News Co., [c1898].
2 v. (iv, [2], 1056 p., 7 paper manikins mounted on trifold sheet, [1] folding table ; [2], 1057-1652, [2], [1653]-2046 p., [2] paper manikins, [6] leaves of plates, [2] folding tables) : ill. ; 24 cm.

First English language edition of Bilz's *Das neue Naturheilverfahren,* a popular medical handbook that in addition to numerous German editions was issued in English (1898, 1900, 1901), Dutch (189-?), Spanish (189-?), French (1900) and Russian (1902) translations. Bilz was not a physician, but the founder of a therapeutic system centered at the Bilz Sanatorium at Dresden-Radebeul. His system is succinctly summarized in the preface: "For with the simple prescription of diet, water, air and light the Natural Healer cures all diseases, of whatever standing, provided they are curable; for this method of treatment requires that there should be sufficient vitality of the body to enable a reaction to follow and the cri-

sis to be got over" (I:[iii]). His therapeutics is largely based on complex system of baths reminiscent of hydropathy, supplemented by diet, exercise and massage. Topics in both volumes are arranged alphabetically (i.e., from ABDOMINAL COMPLAINTS to WRITER'S CRAMP).

326. BIRCH DALE MINERAL SPRING WATERS.
THE BIRCH DALE Mineral Spring Waters. THE BIRCH DALE MINERAL SPRING WATERS have proved, by the scores of cures effected by their use in the last ten years, to be unsurpassed in medical properties by any mineral waters on this continent. They have cured Consumption, Fevers, Cancers, Scrofula, Kidney Complaints . . . These Springs are situated in a beautiful pine grove in the southwest part of the city of Concord, about four miles from the Depot, and only a few rods from the Birch Dale Springs Hotel . . . Wm. Adams, H.K. Stanton, proprietors . . . [187-?].
Broadside : ill. ; 26.5 × 18 cm.

The Birch Dale Mineral Springs were located in Concord, New Hampshire.

327. BIRTH CONTROL REVIEW.
The Birth control review. Dedicated to the principle of intelligent and voluntary motherhood. Volume one. February 1917. Number One. [New York, 1917].
16 p. ; 30 cm.

"The time has come when those who would cast off the bondage of involuntary parenthood must have a voice, one that shall speak their protest and enforce their demands. Too long have they been silent on this most vital of all questions in human existence. The time has come for an organ devoted to the fight for birth control in America. This Review comes into being, therefore, not as our creation, but as the herald of a new freedom. It comes into being to render articulate the aspiration of humanity toward conscious and voluntary motherhood" (p. [3]).
The Birth control review was founded by Margaret Sanger (#3083-3097) and was edited by her until 1928. She regarded the *Review* as her principal organ for the education of the general public on population and birth control issues. The journal remained in publication

through 1939: vol. 1, no. 1 (Feb. 1917)-vol. 17, no. 7 (July 1933); new ser., vol. 1, no. 1 (Aug. 1933)-vol. 4, no. 9 (June 1937); vol. 22, no. 1 (Oct. 1937)-vol. 24, no. 3 (Jan. 1940).

Sanger launched *The Birth control review* with the assistance of Frederick A. Blossom. After his arrival in New York from Cleveland the previous year, Blossom helped organize the New York Birth Control League and worked out financing for the *Review*. He served as general manager of the new journal only briefly: "Sometime in the spring of 1917, Frederick Blossom . . . suddenly quit, taking with him lists of subscribers, account books, funds, and even the office furniture. He so disorganized the operation that no issues of the *Review* appeared between June and December 1917" (David M. Kennedy. *Birth control in America*. New Haven: Yale Univ. Press, 1970, p. 92). Kennedy attributes Blossom's motives to "his ambitious desire to dominate the movement," though his disagreement with Sanger over the United States' entry into the world war was also a factor. Sanger's response to Blossom's action was to go to the district attorney—an act that resulted in her expulsion from the New York Birth Control League and chastisement from her supporters in the ranks of socialists and other radicals for having appealed to the capitalist justice system.

"After its erratic start, the *Review* became the movement's forward thrust. Not only was it the link among adherents, but also the reporter of local, national, and world developments and the special resource for speakers in all parts of the country" (Emily T. Douglas. *Margaret Sanger: pioneer of the future*. Garrett Park, MD: Garrett Park Press, 1975, p. 131). This inaugural issue of *The Birth control review* includes Sanger's lead article "Shall we break this law?"; and an article by Havelock Ellis entitled "Birth control in relation to morality" (p. 6-7).

328. BISHOP, Emily Montague Mulkin, 1858-1916.
Daily ways to health . . . New York: B.W. Huebsch, 1910.
310 p. ; 20 cm.

BISHOP, Louis Faugeres see MORTON (#2513).

329. BLACK, James Rush, b. 1827.
The ten laws of health; or, how diseases are produced and prevented: and family guide to protection against epidemic diseases and other dangerous infections . . . Philadelphia: J.B. Lippincott Company, 1885.
xix, [1], 13-413 p. ; 20 cm.

Third edition (1st ed., Philadephia, 1872). Black divides his treatise into two parts. In the first he reviews the ten laws of health: 1st. Breathing a pure air; 2nd. Adequate and wholesome food and drink; 3rd. Adequate out-door exercise; 4th. Adequate and unconstraining clothing for the body; 5th. The exercise of sexual function for . . . the natural course of reproduction; 6th. A habitation in the climate for which the constitution of the body is adapted; 7th. Pursuits which do not cramp or overstrain any part of the body, or subject it to irritating and poisonous substances; 8th. Personal cleanliness; 9th. Tranquil states of the mind, and adequate rest and sleep; 10th. No intermarriage of near blood relations.

The content of the second part is new to this third edition. In the "Preface to the third and enlarged edition," Black writes of "a new departure to the present volume, opening up a field of knowledge almost unknown to the popular reader, and one that abounds with the keenest anxieties in reference to health, life and death." He is referring to germs and their relation to disease, a knowledge of which "as the sole means of propagating deadly infections, being new to physicians everywhere" also possesses "an importance to everyone's welfare difficulte to exaggerate." The author thinks it "high time the subject should be made familiar to the public," adding: "The immense benefits an easily-comprehended knowledge of this doctrine is capable of conferring should no longer be left to repose on the book-shelves of physicians . . . but should be brought home to the occupants of every household" (p. v-vi). The author was an 1849 graduate of the Medical Dept. of the University of the City of New York.

330. BLACK, John Janvier, 1837-1909.
Eating to live. With some advice to the gouty, the rheumatic, and the diabetic. A book for everybody . . . Philadelphia and London: J.B. Lippincott Company, 1906.
412 p. ; 20 cm.

First edition. Black was an 1862 medical graduate of the University of Pennsylvania. Four years before the appearance of *Eating to live,* Black published his autobiography, in which he laments "the average want of thrift among the plainer people and the utter ignorance, especially of the women, of the value of one food above another, of the kind of food to purchase, looking to its food value, and the proper method of preparing it. Our people need general instruction in all of these matters . . . Physicians should be able to instruct the families they attend as to their food supply and the nutrients required by different families to keep them in strength and good health . . . Practically no thought is taken of this either by families or by physicians. I am sure, as society goes on to the evolution of a higher civilization, it will be the duty and the every-day work of the family physician to take charge of such matters in the families they attend, and thus lead to a more general rational living, and to better, happier, and healthier lives for the individual" (*Forty years in the medical profession, 1858-1898,* Philadelphia: J.B. Lippincott, 1900, p. 216).

331. BLACKSTONE, Thomas.
Accidents and emergencies. What should, and should not, be done before the doctor comes . . . Cincinnati: Cranston & Curts; New York: Hunt & Eaton, 1894.
    122, [6] p. : ill. ; 18 cm.

"An interval of time always elapses between the receipt of an injury, or the arising of an emergency, and the moment when a physician can be called and take actual charge of the case. The length of this interval, of course, varies according to the surrounding circumstances" (p. 9). In an era such as ours when emergency medical help is readily available, the purpose of a work such as this becomes somewhat obscured. For those who lived perhaps hours from medical assistance of any kind, knowing procedures to follow after an accident were essential. Blackstone cautions, "Upon the occurrence of a serious emergency that calls for the speedy presence of a physician, the messenger sent for him should be a person of some intelligence. He should be able to give a reasonably accurate description of the injury to be treated, so the medical man may know what it will be nec-

essary to take with him in order to treat the case properly. This especially applies to accidents far out in the country, where the delay in sending back to the office for necessary articles or appliances might cause great delay and inconvenience, even if nothing worse" (p. 11). The author was probably the Thomas Blackstone who graduated from Bellevue Hospital Medical College in 1873, and is listed in the 1886 edition of *Polk* as practicing in Circleville, Ohio, south of Columbus.

332. BLACKWELL, Antoinette Louisa Brown, 1825-1921.
The sexes throughout nature . . . New York: G.P. Putnam's Sons, 1875.
    240 p. ; 19 cm.

A native of Henrietta, N.Y., near Rochester, Blackwell (nee Brown) grew up in a family and region of the nation much influenced by the evangelism of such figures as Charles G. Finney. After teaching school briefly, Blackwell attended Oberlin College where she completed the literary course in 1847. To the consternation of family, friends and faculty, she then pursued theological studies at Oberlin, completing her studies in 1850, but without being granted a diploma. For the next two years she toured Ohio and the East giving lectures on women's rights, abolition and temperance. Her ambition for a career in the pulpit was rewarded in 1853 when she was ordained a Congregational minister, the first ordained woman in a major denomination in the United States. After only a year, however, Blackwell severed her affiliation with the conservative Congregationalists and became a Unitarian. She worked in the slums of New York and wrote a series of articles on social issues for Horace Greeley's *New York Tribune.* In 1856 she married Samuel C. Blackwell, the brother of Elizabeth Blackwell, the first woman to receive a medical degree in the United States. For the next two decades, Blackwell was out of the public eye as she devoted herself to raising a family. She remained intellectually active, however, giving much of her attention to the social implications of evolutionary theory.

It was during this period, before her return to public life in 1878, that Blackwell wrote and published *The sexes throughout nature.* It is regarded as Blackwell's most interesting book by her biographer in *Notable American women, 1607-1950*: "she emphasized that the male perspective of Darwin and Spencer limited their analysis of 'women's place in nature'," and that in supposing male superiority in evolution they "had simply offered further evidence of the the need for female emancipation: 'Evolution has given and is still giving to woman an increasing complexity of development which cannot find a legitimate field for exercise of all its powers within the household'" (1:160).

In the essay entitled "Sex and work," Blackwell assails E.H. Clarke's (#659-662) thesis that the education of females should be accommodated to a physiological condition more fragile than that of males: "Any healthful school regimen, which will enable a boy to develop his larger muscles and brain . . . must be one which will allow a girl to develop her lesser muscles and brain and her more complex organism, with equal facility and robustness" (p. 153). She argues that as much mental and physical rigor should be expected of girls in their training as from boys: "let every girl exercise mind and body fearlessly . . . She is not a porcelain to be easily broken, nor are the adjustments of her nature so weakly put together that they can be readily disturbed, or so fearfully balanced that the slightest overwork in one direction must mean weakness or disease in some other as penalty" (p. 161).

333. BLACKWELL, Elizabeth, 1821-1910.
The religion of health . . . Edinburgh and Glasgow: John Menzies & Co., 1878.
    22 p. ; 18 cm.

Blackwell's pioneering role as the first female medical graduate in the United States (Geneva Medical College, 1849), her work at the New York York Infirmary for Women and Children, her part in the formation of the U.S. Sanitary Commission, and her establishment of the Woman's Medical College in New York, etc., are too well known to devote much space here. Suffice it is to say that in 1869 she returned permanently to her native England, where she established a successful London practice and resumed her social activism.

By "the religion of health," Blackwell means knowledge of and obedience to the divinely established laws that govern the human body. In this pamphlet, however, Blackwell espouses more than personal responsibility for the maintenance of health. The "physical and moral perfectibility of man" cannot be left entirely in the hands of the individual. The sanitary conditions in which many live are evidence of forces beyond individual control that affect health. "One of the most important problems of the present time," she writes, "is how to embody the sanitary knowledge which we possess in the life of the nation, so that a higher standard of health may be gained by the present and succeeding generations. The solution of this great problem . . . must be sought in the power of legislative action; in the wide-spreading influence of education; and in the strength of social combination" (p. 12).

In an interesting analogy Blackwell writes, "Legislation is the human imitation, or visible representation, of the greatest fact in the universe—law." She continues: "as the divine laws of the human organisation limit its powers, and directs its modes of action, so the human laws which rule a people, determine their modes of thought, and their relations to one another. Legislation, therefore, not only represents the life of the present generation, but is the most powerful educator of the rising generation" (p. 12). Legislation has a decided moral content. "Divine law rewards the good (i.e., the obedient)" and "punishes the bad (i.e., the disobedient)." In like manner, "human law must never temporise with evil . . . In dealing with evils, legislation is bound to investigate the causes of evil and attack them. Herein lies the superiority of legislation or individual effort . . . It is only on this sound basis that wise legislative measures can be framed; only in this way that great questions of national health can judiciously be dealt with" (p. 13). In this pamphlet, at least, Blackwell limits her legislative zeal to "sanitary instruction," "the education of youth in health," "the formation of healthy habits of life," etc. *The religion of health* was never published in the United States.

334. BLAIKIE, William, 1843-1904.
How to get strong and how to stay so . . . New York: Harper & Brothers, publishers, 1884.
    296 p. : ill. ; 17 cm.

"The identification of morality with muscu-
larity was to grow as an article of hygienic faith
through the final third of the century and the
Progesssive years, though the arena would become congested with competing programs of
health building" (James C. Whorton, *Crusaders
for fitness*, Princeton, 1982, p. 282). In 1879
Dudley Allen Sargent established a chair of
physical education at Harvard, and in 1884 began a teacher training program—the turning
point for physical education as a profession in
the United States. Blaikie's *How to get strong and
how to stay so* was one of the key works by which
this movement was given public circulation.

Blaikie was an 1868 Harvard Law School
graduate and a successful lawyer in New York
when this book was first published. Since boyhood he had been an outstanding athlete: "He
was captain of a winning football team at the
Boston Latin School, and in 1866 stroked
Harvard's victorious crew. For ten years he held
the amateur long distance walking record, having covered the 225 miles between Boston and
New York in four and one half days" (*Dict. Amer.
biog.*, I:322). *How to get strong and how to stay so*
was first published in 1879, and appeared in
seven more editions through 1902, including
two London editions.

335. BLAIKIE, William, 1843-1904.
Sound bodies for our boys and girls . . .
New York: Harper & Brothers, 1884.
    ix, [1], 168 p. : ill. ; 18 cm.

Blaikie's manual of physical fitness designed
for school use was first published at New York
in 1883, and appeared in seven more issues
through 1898, including a London edition.

To Blaikie and his contemporaries, the benefits of exercise transcended the merely physical: "For a sensible education of the body causes
the blood-making machinery to make good
blood instead of poor. This good blood is sent
to the brain, and fits that organ to do more and
better work, without risk, than it can do when
fed by a poor article . . . One who is trained in
this way will safely pass the overwork of the brain
and nerves which to-day breaks down so many
useful, but physically untrained men and
women, while they should still be in their prime,
until nervous exhaustion has become a disorder familiar to almost every physician in the
land" (p. [iii]).

336. BLAIR, Henry William, 1834-1920.
The temperance movement: or, the conflict between man and alcohol . . . Boston:
William E. Smyth Company, 1888.
    [4], ix-xxiii, [1], 583 p., [67] leaves of
plates (1 folding) : ill., map, ports. ; 24 cm.

"The plan attempted has been to place clearly
before the mind the nature of alcohol as a poison to the healthy human system; its destructive
effects upon the body and soul of its victim; to
portray its tremendous proportions and malignant influence upon society, nations and races
of men; to discuss the remedies of this great evil
by the exercise of moral suasion and educative
forces, both spiritual and physical, and by the
action of society in the enactment and enforcement of law. This is followed by some account of
the organizations and agencies, religious, secular and political, which are and must be engaged
in the effort to remove the gigantic evil and crime
of alcoholic intemperance from the world" (*Pref.*,
p. ix-x). Except for a folding map of Manhattan
illustrating (in red ink) the concentration of New
York's 10,000 drinking establishments, the plates
are portraits of prominent figures in the temperance movement.

Henry William Blair was a U.S. representative and senator from New Hampshire. He is
described in the *Amer. nat. biog.* as a "rarity in
late nineteenth-century American politics, a
passionate reformer who also managed to win
and hold important public office. In 1876, during his first term in the House, he introduced a
constitutional amendment to bar the manufacture, importation, and sale of liquor after 1900.
During his tenure in Congress he repeatedly
reintroduced this measure, and in 1888 published a massive prohibitionist tome, *The temperance movement; or the conflict between man and
alcohol.* In Congress he was equally persevering
in pressing for a constitutional amendment for
woman suffrage" (2:913).

337. BLAIR REMEDY COMPANY.
Disease its cause, and cure by Dr. Caldwell's
Triplex Method. Rheumatism, kidney trouble,
bladder trouble, stomach trouble, skin diseases,
constipation, catarrh, piles, etc., etc. . . . [Chi-
cago: Blair Remedy Co., c1910].
    40 p. ; 20 cm.

A re-issue of #338 with new title on wrapper.

338. BLAIR REMEDY COMPANY.

GERMS the real cause of rheumatism and its CURE by the Dr. Francis Cook Caldwell Triplex Method . . . [Chicago : Blair Remedy Co., c1908].

    40 p. : ill., ports. ; 21 cm.

    The Triplex Method consists simply in correct diagnosis, treatment, and educating the patient on how to stay well after the cure. The pamphlet maintains that rheumatism is a "microbe disease," and that the "little germ . . . by setting up an acid fermentation" during digestion produces excessive uric acid, which "let loose upon the joints, muscles and tissues" causes rheumatism. The reader is given a simple test for determining the presence of uric acid "by simply pressing your finger on any part of your body, for if there is an excessive amount of this poisonous acid in the blood, a little white spot will remain there for several seconds" (p. 8). The remedy sold over Caldwell's signature is based on "genuine Indian herbs" ("mountain grape root" and "sacred bark"). It not only to neutralizes excessive uric acid, but "also destroys the germs which create the poisonous acid itself" (p. 9).

    Francis Cook Caldwell (1850-1924) was an 1883 graduate of the University of Illinois College of Medicine (Chicago), and a Chicago practitioner. The Blair Remedy Co. was one of several ever-moving firms owned by Thomas R. Bradford for the manufacture and mail order sale of fraudulent medicines. The firm operated in Chicago circa 1906-10. Title from printed wrapper.

339. BLAISDELL, Albert Franklin, 1847-1927.

The child's book of health. In easy lessons for schools . . . Boston: Lee and Shepard, publishers; New York: Charles T. Dillingham, 1886.

    viii, 117 p. : ill. ; 17 cm.

    First edition. At head of title-page: *Physiology for little folks.* Blaisdell received his medical degree from Harvard in 1879 and practiced as a physician and surgeon in Providence. He was the author of numerous works on hygiene and physiology for a juvenile audience, as well as an equal or greater number of school-texts on literature and American history.

Health Hints for the Blackboard.

Fig. 13. "Health hints for the blackboard," from Blaisdell's The child's book of health, *#340.*

340. BLAISDELL, Albert Franklin, 1847-1927.

The child's book of health. In easy lessons for schools . . . Revised edition. Boston: Ginn & Company, publishers, 1901.

    viii, 136 p. : ill. ; 17 cm.

    This revised edition of Blaisdell's physiology for the primary grades includes "four new chapters on the subject of alcoholic drinks and other narcotics." In so doing, it received the *imprimatur* of Mary H. Hunt and the Dept. of Scientific Instruction of the Women's Christian Temperance Union. The WCTU endorsement on the verso of the title-page is common to most juvenile physiologies of this period, and reflects a national concern regarding the exposure of school-age children to alcohol and drugs and an effort to educate the young regarding their perils. The inclusion of such material in school physiologies was required by law in some states. The "revised edition" of *The child's book of health* was first issued in 1893. It appeared in five issues between 1893 and 1905. Most of Blaisdell's many books were published by Ginn & Company of Boston. Founded by Edwin Ginn, Ginn & Company was one of the nation's foremost publishers of school and college textbooks. Series: *Blaisdell's series of physiologies.*

341. BLAISDELL, Albert Franklin, 1847-1927.
How to keep well. A text-book of health for use in the lower grade of schools with special reference to the effects of alcoholic drinks, tobacco and other narcotics on the bodily life . . . Revised edition. Boston: Ginn & Company, publishers, 1893.
vi, 250 p. : ill. ; 18 cm.

How to keep well was first issued at Baltimore in 1885. The first issue of the "revised edition" was published in 1893; it was last published in 1921.

342. BLAISDELL, Albert Franklin, 1847-1927.
How to keep well. A text-book of health for use in the lower grades of schools with special reference to the effects of alcoholic drinks and other narcotics on the bodily life . . . Revised edition. Boston: Ginn & Company, publishers, 1896.
[2], vi, 250, [6] p. : ill. ; 18 cm.

Series: *Blaisdell's revised series of physiologies.*

343. BLAISDELL, Albert Franklin, 1847-1927.
How to keep well. A text-book of physiology and hygiene for the lower grades of schools . . . Revised edition. Boston: Ginn & Company, [c1904].
vi, 265 p. : ill. ; 19 cm.

Series: *Blaisdell's series of physiologies.*

344. BLAISDELL, Albert Franklin, 1847-1927.
Life and health. A text-book on physiology for high schools, academies and normal schools . . . Boston, U.S.A., and London: Ginn & Company, publishers, 1902.
vi, 346 p. : ill. ; 19 cm.

345. BLAISDELL, Albert Franklin, 1847-1927.
Our bodies and how we live. An elementary text-book of physiology and hygiene for use in the common schools, with special reference to the effects of stimulants and narcotics on the human system . . . Boston: Lee and Shepard, publishers; New York: Charles T. Dillingham, 1887.
[2], vi, 285, [3] p. : ill. ; 18 cm.

The first edition of *Our bodies and how we live* was published at Boston in 1885. A revised edition was issued in 1893. Series: *Young folks' physiology.*

346. BLAISDELL, Albert Franklin, 1847-1927.
Our bodies and how we live. An elementary text-book of physiology and hygiene for use in schools with special reference to the effects of alcoholic drinks, tobacco and other narcotics on the bodily life . . . Revised edition. Boston: Ginn & Company, publishers, 1894.
vi, [2], 412, [4] p. : ill. ; 18 cm.

The revised edition (1st issue, 1893) includes new chapters on gymnastics and alcohol consumption. Series: *Blaisdell's revised series of physiologies.*

347. BLAISDELL, Albert Franklin, 1847-1927.
Our bodies and how we live. An elementary text-book of physiology and hygiene for use in schools with special reference to the effects of alcoholic drinks, tobacco and other narcotics on the bodily life . . . Revised edition. Boston, U.S.A.: Ginn & Company, publishers, 1896.
[2], vi, 412, [4] p. : ill. ; 18 cm.

Series: *Blaisdell's revised series of physiologies.*

348. BLAISDELL, Albert Franklin, 1847-1927.
Our bodies and how we live. An elementary text-book of physiology and hygiene for use in schools with special reference to the effects of alcoholic drinks, tobacco and other narcotics on bodily life . . . Revised edition. Boston: Ginn & Company, publishers, 1901.
[2], vi, 412, [4] p. : ill. ; 18 cm.

Series: *Blaisdell's series of physiologies.*

349. BLAISDELL, Albert Franklin, 1847-1927.
Our bodies and how we live. An elementary text-book of physiology and hygiene for use in schools . . . Revised edition. Boston, U.S.A., and London: Ginn & Company, publishers; The Athenaeum Press, 1904.
vii, [1], 352 p., [1] leaf of plates : ill. ; 19 cm.

This 1904 edition is actually a revised edition of the 1893 "revised edition." New material has been added on "important discoveries" concerning "the nature and propagation of bacteria," and their relation to "contagious diseases and the promotion of personal and public health" (p. iii). Series: *Blaisdell's series of physiologies.*

350. BLAISDELL, Albert Franklin, 1847-1927.
Our bodies and how we live. An elementary text-book of physiology and hygiene for use in schools . . . Revised edition. Boston; New York; Chicago; London: Ginn & Company, [c1904].
vii, [1], 352 p., [1] leaf of plates : ill. ; 19 cm.

351. BLAISDELL, Albert Franklin, 1847-1927.
Our bodies and how we live. An elementary text-book of physiology and hygiene for use in schools . . . Revised edition. Boston; New York; Chicago; London: Ginn & Company, [c1910].
vii, [1], 369, [7] p., [1] leaf of plates : ill. ; 19 cm.

"The author of this book has written an additional chapter . . . on tuberculosis, or consumption, and its prevention. This has been done in accordance with the advice and suggestion of those educators and physicians who believe that pupils in our public schools should be taught the simplest facts concerning the cause and prevention of this disease" (*Publisher's note*, p. iv). Series: *Blaisdell's series of physiologies.*

352. BLAKE, John Lauris, 1788-1857.
The farmer's every-day book; or, sketches of social life in the country: with the popular elements of practical and theoretical agriculture, and twelve hundred laconics and apothegms relating to ethics, religion, and general literature; also, five hundred receipts on hygeian, domestic, and rural economy . . . Auburn, N.Y.: Published by Derby, Miller and Company, 1850.
xii, [13]-654, [2] p., [10] leaves of plates : ill. ; 23 cm.

First edition. Blake retired from the Episcopal clergy in 1830 due to ill-health, and devoted the next two decades to literary work. He was the author of nearly fifty books, many of them illustrated textbooks for the elementary grades. Blake wrote two eminently successful family books: the *Family encyclopedia of useful knowledge* (1834), which appeared in ten editions; and *The general biographical dictionary* (1835), which appeared in thirteen editions. *The farmer's every-day book* was one of several agricultural books that Blake wrote, in which he combined useful farm information with encomiums of rural life. In addition to the expected information on soils, manure, seeds, cattle and barns, Blake considers the health aspects of rural life in "Hints on the preservation of health" (p. [97]-104), "Liability of laborers to intemperance" (p. 104-11), "Hints on the benefits of bathing" (p. 141-45), and "Miscellanies in hygeian economy; including receipts, hints, and laconics on the subject of preserving health, and restoring it, under the ordinary circumstances of human life" (p. 524-51). In his chapter "Advice to Western emigrants," Blake advises: "Every family removing into a country that is new . . . should be furnished with a small medicine-chest . . . Not less important is a good Book of Health, or family prescriptions and advice in relation to the preservation of health. By such a precaution, every family is in a measure prepared to take care of themselves; to be their own doctors" (p. 312). *The farmer's every-day book* was reissued in 1851, 1853, 1854, 1856 and 1857.

353. BLAND, Thomas Augustus, b. 1830.
How to get well and how to keep well. A family physician and guide to health . . . Boston: Plymouth Publishing Co., 1894.
202, [2] p., [1] leaf of plates : ill., port. ; 19 cm.

If nothing else, the following quotation from Bland's introduction indicates how rooted eclecticism remained in its early 19th century origins and theories, and how many of its adherents failed to grasp the implications of laboratory science at the century's end: "The medical system presented in this book is based upon the theory that disease is simply a departure from health, and is a unit, presenting a variety of forms and symptoms. The celebrated Dr. Benjamin Rush held to the same view . . . Holding to this doctrine of Dr. Rush, the author believes in general rather than specific treatment. Hence he prescribes substantially the same remedies for diseases of the same class. This not only simplifies the practice of medicine, but is strictly scientific. All true sciences are simple, and can be readily understood by all people of common sense" (p. 11).

Bland last appears in the 7th edition (1902) of *Polk's medical register of the United States and Canada.* At the time this work was published, he was president of the Eclectic Medical Society of the District of Columbia. T.A. Bland was married to the eclectic physician Cora M. Bland.

354. BLECH, Gustavus Maximilian, b. 1870.
A handbook of first aid in accidents, emergencies, poisoning, sunstroke, etc. Second edition . . . Chicago and New York: Bauer & Black, [c1916].
128 p. : ill. ; 17 cm.

Blech was an 1894 graduate of Barnes Medical College in St. Louis, and appears to have spent his entire career in Chicago. The publishers, Bauer & Black, were manufacturers and distributors of medical supplies.

355. BLOOD, C. L.
A century of life, health and happiness. A cyclopedia of medical information for home life and domestic economy. Seventh edition—eightieth thousand. 65,000 copies sold in nine months . . . Boston: Published by the author, 1882.
625, [1] p. : ill. ; 24 cm.

On the final pages of this idiosyncratic treatise on the maintenance of health and treatment of disease, Blood advertises proprietary remedies for the cure of piles, tape worm, liver disorders, etc., as well as his "Suspensory bandage and electric battery combined for the radi-

cal cure of partial or complete impotency, seminal weakness or spermatorrhea."

BLOOMFIELD-MOORE, Clara Sophia see **FARRAR** (#1135).

356. DR. BLOSSER COMPANY.
Testimonials from doctors, ministers and prominent men and women who have used Dr. Blosser's Catarrh Cure. [Atlanta: Dr. Blosser Company, ca. 1905].
32 p. ; 15 cm.

"This remedy in pipe form is put up in tin boxes containing a sufficient amount for one month's treatment . . . We manufacture Dr. Blosser's Catarrh Cure into medical cigarettes for those who prefer it in that form, using carefully selected rice paper" (p. 32). The latest testimonials are dated 1905. "Testimonial booklet no. 3" (p. [1]).

BLUNT, W. S. see **ZANDER INSTITUTE** (#3909).

357. BLY, Douglas. d. 1876.
A new and important invention . . . By frequent dissections, Dr. Bly has succeeded in embodying the principles of the natural leg in an artificial one, and in giving it lateral, or side motion at the ankle, the same as the natural one. By so doing he has produced the most complete and successful invention ever attained in artificial legs. Rochester [N.Y.]: Press of Curtis, Butts & Co., 1862.
30 p. : ill. ; 22 cm.

The pamphlet consists largely of endorsements from physicians and testimonials from laymen regarding the artificial leg designed by the Rochester physician Douglas Bly. The inventor's earliest published promotion of his mechanical limb appears to have been *A description of a new, curious, and important invention* issued at Rochester circa 1859-60. In the 1880s, the manufacture and sale of *Bly's artificial limbs* was assumed by the Rochester firm of George R. Fuller (#1318).

358. BOBO, Walter Thompson, b. 1876.
A guide for goitre sufferers. Containing important suggestions and hygienic "helps"

in the treatment of goitre . . . Battle Creek, Mich., 1916.

30, [2] p., [1] leaf of plates : port. ; 19 cm.

Bobo informs the prospective patient that "I cure hundreds of goitres every year . . . many of them being cases of twenty-five to thirty years' and even fifty years' standing . . . which physician after physician had treated in vain" (p. 7). His therapy for goitre is twofold: a course of unspecified proprietary medicines and "the observance of certain hygienic and dietetic measures designed to improve the general health and vitality of the thyroid gland" (p. 8). The *Guide* addresses the latter. Those desiring the full remedial program need address the author, who was an 1899 graduate of the Marion Sims College of Medicine (St. Louis) and co-proprietor of the Dr. Peebles Institute of Health (Battle Creek, Mich.), manufacturers of such remedies as *Brain Restorative* and *Nerve-Tonic*. In printed wrapper. Frontispiece portrait of the author.

BOBO, Walter Thompson see also **PEEBLES** (#2760).

359. BODENHAMER, William, 1808-1905.
Practical observations on some of the diseases of the rectum, anus, and contiguous textures; giving their nature, seat, causes, symptoms, consequences, and prevention; especially addressed to the non-medical reader . . . Cincinnati: Printed for the publisher, by A.G. Sparhawk, 1847.

iv, 150, [6] p. ; 22 cm.

First edition. Bodenhamer maintains that incidence of the disorders which his treatise addresses has increased, a phenomenon he attributes primarily to "the abuse of purgative medicines," i.e., the idea that it is necessary to remedy "every slight indisposition" by swallowing "some of the numerous and various drastic purgative nostrums which literally fill our country" (p. 9). Bodenhamer provides a description of the symptoms and causes of seventeen anal and rectal disorders, supplemented by published case histories and correspondence from his own patients. His litany of the many possible causes of hemorrhoids reveals the author's grounding in contemporary health reform, and include "high seasoned and stimulating food,

free use of heating wines, such as champaigne, &c., the intemperate use of spiritous drinks, generally, excessive venery . . . [etc.]" (p. 17).

Although Bodenhamer boasts, "I doubt whether any other surgeon, in this or any other country, has ever treated so large a number of this one disease in so short a time, and with such complete success" (p. 12), he refrains from revealing his own method of treatment, other than to suggest its efficacy compared to the medicinal and surgical approaches of his contemporaries (condescendingly described and often dismissed). Chapter XXII contains twenty pages of testimonials from Bodenhamer's patients, all of whom rejoice in the healing of their respective maladies without resort to surgery. On the final page of text, Bodenhamer solicits consultation at his residences in New Orleans and Louisville, or by correspondence ("All persons writing me for medical advice . . . must invariably accompany their letters with at least FIVE DOLLARS" (p. 150).

360. BODENHAMER, William, 1808-1905.
Practical observations on some of the diseases of the rectum, anus, and contiguous textures; giving their nature, seat, causes, symptoms, consequences, and prevention: especially addressed to the non-medical reader . . . Second edition . . . New York: Published for the author, by J.S. Redfield, 1855.

257, [1], 10 p., VIII leaves of plates + frontis. : ill., port. ; 23 cm.

"For omitting the treatment of those diseases in the present work, some may be disposed to censure the author; but he would remark that he is decidedly opposed to '*publishing cures for the multitude.*' Too many works of this class are already in existence, for the good of the community at large. The instruction they need, is how to preserve their health . . . When, however, they do become sick, it is then the especial and exclusive province of the physician to restore them to health" (p. 4). Bodenhamer disparages much popular medical literature: "The dictionaries of *popular* medicine, and gazettes of health slay annually their thousands; not *directly,* by the actual injury of the remedies which they congregate without knowledge or discrimination, but *eventually,* by procrastinating the interference of the regular practitioner till the period of cure is past" (p. 6).

This second edition of the *Practical observations* contains nearly a hundred pages of new matter. Most chapters have been expanded, and new chapters on the anatomy of the rectum and anus, on the physiology of the lower intestine and on constipation have been added. The text is illustrated with seven lithographic plates. Engraved frontispiece portrait of author.

In 1860 Bodenhamer earned his place in the annals of American medical literature with the publication of *A practical treatise on the etiology, pathology, and treatment of the congenital malformations of the rectum and anus,* a work that has been described as "the foundation stone in the surgery of these distressing abnormalities" (*Dict. Amer. med. biog.* New York, 1928, p. 115); and that Rutkow designates "one of the major American works on anal-rectal conditions" (*The history of surgery in the United States, 1775-1900,* San Francisco, 1998, 1:367).

Bodenhamer was an 1839 graduate of the Worthington Medical College of Ohio University. "In 1859, Bodenhamer settled in New York City and began to produce an impressive body of full-length written texts on colorectal diseases . . . Whether Bodenhamer was the country's earliest specialist in rectal diseases . . . is open to historical inquiry. In the 1870s, Bodehamer listed himself as 'professor of the diseases, injuries, and malformations of the rectum, anus, and genitourinary organs.' He never named the medical institution at which he held professorial rank, and he clearly did not limit himself to rectal difficulties" (Ira Rutkow. *American surgery.* Philadelphia, 1998, p. 457). Nevertheless, from modest beginnings as a medical itinerant in the Western and Southern states, Bodenhamer's rose to medical respectability and the patronage of such patients as Cornelius Vanderbilt and Andrew Gregg Curtin.

BOECKMANN, Paul von see *VITALITY* (#3657).

361. BOERICKE AND RUNYON.
The homoeopathic primer. A pocket guide to the popular use of homoepathic remedies. Together with a list of homoeopathic specialties, homoeopathic domestic books and cases, etc., etc. San Francisco: Boericke & Runyon, homoeopathic pharmacy, 1894.
    128 p. : ill. ; 11.5 cm.

In printed wrapper.

362. BOERICKE AND RUNYON.
The homoeopathic primer. A pocket guide to the popular use of homoepathic remedies. Together with a list of homoeopathic specialties, homoeopathic domestic books and cases, etc., etc. New York: Boericke & Runyon, homoeopathic pharmacy, [after 1913].
    144 p. : ill. ; 13 cm.

Sixty pages of medical advice is followed by eighty-four pages of advertisements for homeopathic books, medicines, soaps, dietary supplements, etc. The dating of this piece is based on the publication of one of the books advertised, i.e., Myron Adam's *A practical guide to homoeopathic treatment,* first issued in 1913. In printed wrapper.

363. BOERICKE AND TAFEL.
A descriptive catalogue of homoeopathic works, for domestic and veterinary practice, and also of the standard works of reference, together with price lists of medicine cases . . . [Philadelphia: Boericke & Tafel, ca. 1885].
    34, [14] p. ; 18 cm.

The catalog is prefaced with a fifteen page introduction explaining the principles and effectiveness of homeopathy and homeopathic remedies. The fourteen unnumbered pages contain advertisements for various remedies and food supplements manufactured by Boericke & Tafel, one of the nation's largest manufacturers of homeopathic drugs, and one of the most prolific publishers of homeopathic books: "Among the medical publishing houses, in which Philadelphia excelled, were Boericke & Tafel, which started as a homeopathic pharmacy in 1835, but began to publish homeopathic books in 1870 and by 1935, was publishing about 85 percent of all such books in the United States" (John Tebbell, *A history of book publishing in the United States,* New York: R.R. Bowker, 1975, 2:427). In printed wrapper.

364. BOERICKE AND TAFEL.
Homoeopathic guide for the people: containing descriptive catalogue of domestic works and useful notes and hints for the treatment of the sick, diet. Philadelphia: A.J. Tafel, [ca. 1872].
    36 p. ; 18 cm.

Pages 18-36 contain advertisements for domestic homoeopathic books, medicine chests and medicines available from the firm of Boericke & Tafel. Date of imprint based on publication date of the third American edition of Laurie's *Homoeopathic domestic medicine* (1872), advertised on page 19. In printed wrapper.

365. BOERICKE AND TAFEL.
Medical index, and key to the treatment of common ailments, and list of medicines from the homoeopathic pharmacy of Boericke & Tafel . . . [New York]: Boericke & Tafel, [c1879].
32 p. : ill. ; 16 cm.

366. BOLAND, Mary A.
A handbook of invalid cooking for the use of nurses in training-schools, nurses in private practice and others who care for the sick. Containing explanatory lessons on the properties and value of different kinds of food, and recipes for the making of various dishes . . . New York: The Century Co., 1893.
vii, [1], 323 p. ; 20 cm.

First edition. Boland was an instructor in cooking at the Johns Hopkins Hospital Training-School for Nurses.

367. BOLAND, Mary A.
A handbook of invalid cooking for the use of nurses in training schools, nurses in private practice and others who care for the sick. Containing explanatory lessons on the properties and value of different kinds of food, and recipes for the making of various dishes . . . Revised edition. New York: The Century Co., 1902.
vii, [1], 326 p. ; 19 cm.

The revised edition was reissued in 1900, 1902 and 1906.

368. BOLLER, William George.
Vigorous manhood. A manual of drugless self-treatment for sexual diseases of men. Together with self-treatment by natural means for indigestion, dyspepsia, constipation and piles. Revised, enlarged and rewritten from the author's previous work,

Men's pocket physician . . . New York City: [Published by Vim Publishing Co.], 1903.
80, [4] p. : ill., port. ; 15 cm.

Publisher's name stamped on front board: Vim Publishing Co. The copyright on the verso of the title-page is assigned to the Home Treatment Publishing Co., the name of which has been stamped over on the advertisement at the end of the volume with the name of the Health Publishing Company.

BON VIVANT see **HAYWARD** (#1595).

368.1. DRS. BONAPARTE & REYNOLDS.
Drs. Bonaparte & Reynolds' medical pocket guide. Free to all for 6 cents . . . [Cincinnati, after 1868.]
[16] leaves ; 13 cm.

In this pamphlet Drs. Bonaparte and Reynolds promote their treatment for masturbation, spermatorrhea, "seminal weakness," sexually transmitted diseases, menstruation disorders, sterility and impotence. They advertise eight varieties of "French imported male safes," as well as "The Womb Orifice Veil," a female contraceptive device. Remedies advertised include "Dr. Bonaparte's Combination Aphrodisiac Remedy," "Blood Purifier," "Dr. Bonaparte's Anti-Venereal Lotion" ("a sure preventive against contagion"), and "Dr. Bonaparte's Band d'Amour" ("It will increase the size of the penis at least one-third"). Wood engraving of Dr. Bonaparte's Private Dispensary in Cincinnati on recto of first leaf.

369. BOND, Allen Kerr, 1859-
How can I cure my indigestion? . . . New York: The Contemporary Publishing Co., [c1902].
viii, [9]-180 p. ; 18 cm.

370. BOOK OF HEALTH.
The Book of health; a compendium of domestic medicine, deduced from the experience of the most eminent modern practitioners; entirely divested of technicalities and rendered familiar to the general reader: including the mode of treatment for diseases in general; a plan for the management of infants and children; rules for the preservation of health; and for diet,

exercise, air, and the preparation of food; remedies in cases of accident and suspended animation; rules for preventing contagion; a table of poisons most frequently taken, with the symptoms, and directions how to act when medical aid is not at hand. A domestic materia medica, &c. &c. First American, from the second London edition; revised and conformed to the practice of the United States, with additions by a fellow of the Massachusetts Medical Society . . . Boston: Richardson, Lord and Holbrook, 1830.
    xii, [9]-179 p. ; 23 cm.

As the American editor notes in his preface, the text of this manual of domestic medicine and hygiene quotes abundantly from the works of such standard British medical worthies as John Abernethy, John Armstrong, Henry Clutterbuck and Astley Cooper. This Boston edition is based on the second English edition, which was published at London in 1840.

371. BOOK OF WONDERS.
The Book of wonders, mysteries and disclosures, for males and females. A guide to love and beauty, and the road to happiness, wealth, fame and honor. New York: Dick & Fitzgerald, publishers, [188-?].
    128 p. : ill. ; 15 cm.

An odd compilation of domestic and trade recipes, recommendations for enhancing feminine beauty, card tricks, prescriptions for treating the diseases of both men and animals, advice for lovelorn, illustrated instructions for the construction of animal traps, etc. *The Book of wonders* was first published at New York by Orville Augustus Roorbach (1803-1861) in 1866, who may also have been its compiler. Title and imprint from wrapper.

372. BOOTH, John T., 1842-1922.
Booth's manual of domestic medicine and guide to health and long life: a book for every family. Describing in plain practical language the laws and rules of health and how to regain and maintain it. A treatise on anatomy, physiology and hygiene, including the treatment and cure of diseases of all kinds, plain directions for preparing and administering medicines. Special at-

tention given to the treatment of diseases of women and children. Careful instructions for the nursing and care of the sick, a copious materia medica, etc. etc. . . . Cincinnati: F.M. Dillie & Co., publishers, 1884.
    1024 p. : ill. ; 25 cm.

Booth was an 1879 graduate of the Cincinnati College of Medicine and Surgery. This was the only edition of his domestic medicine.

373. BOSTON (MA.). BOARD OF HEALTH.
Rules for the management of infants and children. Prepared and published under the direction of the Board of Health of the City of Boston. Boston: Rockwell and Churchill, city printers, 1876.
    7 p. ; 23 cm.

This pamphlet for mothers discusses cleanliness, fresh air, clothing, sleep, diet, weaning, etc.

374. BOSTON HYGIENIC INSTITUTE AND SCHOOL OF PREVENTIVE MEDICINE.
Now open. Boston Hygienic Institute and School of Preventive Medicine, 59 Indiana Place, Boston, Mass. Prevention better than cure. [Boston, 1871?]
    [2] leaves ; 23 cm.

Caption title. George Dutton, M.D. (1830-1903), "principal physician" of the Boston Hygienic Institute, was available for consultation by patients, received students, delivered public lectures on physiology, and sold two kinds of proprietary remedies: those directed to the prevention of a list of sixteen diseases; and those directed to their cure. Dutton (#1018) was the principal advocate of *Etiopathy,* "a religious science and a scientific religion." He was the author of half a dozen popular medical treatises, the most successful of which was *Anatomy scientific and popular,* which appeared in three Boston editions between 1886 and 1900.

A BOSTON PHYSICIAN see YEOMAN (#3898).

BOSTON SOCIETY FOR THE DIFFUSION OF USEFUL KNOWLEDGE see HOLMES (#1741).

375. BOSTON TRUE THOMSONIAN.

The Boston true Thomsonian . . . [Vol. 1, no. 1 (Aug. 15, 1840)-v. 3, no. 24 (Aug. 15, 1843)]. Boston: Published by Daniel Lee Hale, 1840-43.
    3 v. ; 24 cm.

The *Boston true Thomsonian* was issued bi-weekly and edited by James Osgood. Each issue is divided between original articles and extracts from Thomsonian and non-Thomsonian periodicals, including a surprising number from the *London Lancet*. Only the third and final volume of this short-lived periodical is in the Atwater collection.

376. BOSTWICK, Homer, 1806-1883.

An inquiry into the cause of natural death, or death from old age. Developing a new and certain method of preventing the consolidation or ossification of the body, and of thus indefinitely prolonging vigorous, elastic, and buoyant health; and of rendering parturition easy and safe . . . New York: Stringer & Townsend, publishers, 1851.
    viii, 152 p. ; 20 cm.

Bostwick was born in Edinburg, Ohio on 25 October 1806. He became a dentist, and in 1828 established himself in New York City. In that same year he published *The family dentist*. In 1830 Bostwick began the study of medicine in the office of Dr. Kearney Rodgers, and later with Dr. Brownlee. He attended lectures at the College of Physician and Surgeons of New York (1830-32), but did not receive a degree. Bostwick practiced in New York thereafter. The author of many books, he was also physician to the City Prison, and founded the New York Surgical Institute. Bostwick died in New York on 14 August 1883. At least one of his eight children (Homer) became a physician.

Bostwick attributes aging's deleterious effects on the body to "calcareous earthy matter" derived from food and drink that virtually hardens the tendons, ligaments, heart, muscles, blood vessels, etc. until the entire body, "once elastic, healthy, active and lively . . . becomes rigid, diseased, feeble" (p. 3). "Life is extinguished by successive gradations," Bostwick observes, "and death is the last term in the succession. " Bostwick maintains that phosphate of lime is "the solid earthy matter which by

gradual accumulation in the body, brings on ossification, rigidity, decrepitude, and death" (p. 10). This matter is introduced into every cell and fibre of the body by food and drink, the most notorious of which are salt, grains, food additives (e.g., chalk, gypsum, bone ash), river and spring water, etc. Bostwick recommends foods low in "calcareous matter," such as fruits, vegetables, fish, fowls and game (but not the flesh of domesticated cattle whose feed contains high levels of dangerous solids). He concludes, "in proportion as individuals, classes, or even nations, subsist upon aliment containing the smallest proportion of earth elements, do they prevent or retard the process of consolidation, maintain a state of health and activity, and prolong their existence" (p. 120). This was the only edition of Bostwick's *Inquiry*.

377. BOSTWICK, Homer, 1806-1883.

A treatise on the nature and treatment of seminal diseases, impotency, and other kindred affections: with practical directions for the management and removal of the cause producing them; together with hints to young men . . . New York: Burgess, Stringer & Co., 1847.
    ix, [1], iii, [4]-251 p. : ill. ; 20 cm.

The first edition of Bostwick's most successful work, which went through twelve editions between 1847 and 1860. In his opening chapter, Bostwick provides his estimation of contemporary manuals of sexual education and disorders: "It is true that many publications affecting to remove popular ignorance on these subjects have been made; and that numerous treatises of greater or less pretention in style and volume, on the abuses of the procreative organs, have been given to the world. But they have almost wholly failed to accomplish any salutary results. With the exception of one or two respectable works . . . these publications have emanated from sources not entitled to regard; and have afforded . . . ample evidence of the unworthy motives and objects of their authors. Many of these productions have evidently been designed to minister to depraved tastes and prurient imaginations, and by the disgust which they have excited in properly constituted minds, have, doubtless, contributed in no small degree to strengthen the prejudices against popular attempts to instruct the young with regard to these subjects . . . Others have been gotten up as clap-trap advertisements for unprincipled pretenders, whose object has been only to alarm the reader by artful exaggerations, and thus terrify him into reliance on their specious promises of relief . . . Such, indeed, has been the general character of these publications, that medical writers of respectability have shrunk from entering the field . . . " (p. 11-12).

The second chapter of Bostwick's *Treatise* provides a description of the male and female organs of generation (illustrated with full-page wood engravings). The rest of the book is devoted to case histories of "seminal diseases," i.e., disorders characterized by weakness of the male organ, taking the form either of impotence or limited sexual vigor symptomized by involuntary seminal emissions. These latter were caused by masturbation (which receives its own chapter and admonitions throughout), sexually transmitted diseases, excessive venery, urethral stricture, etc. Bostwick's treatment combined a regimen of specifics, diet, baths, the insertion of bougies into the urethra, or, in special cases, urethral cauterization. Spine- and half-titles: *Bostwick on seminal diseases.*

378. BOSTWICK, Homer, 1806-1883.
A treatise on the nature and treatment of seminal diseases, impotency, and other kindred affections: with practical directions for the management and removal of the cause producing them; together with hints to young men . . . Eighth edition. New York: Stringer & Townsend, 1855.
2, [2], ix, [1], 251 p. : ill. ; 19 cm.

A reissue of the sheets of the first edition with new title-page and preliminary leaves, and the addition of six wood engravings to the text. It is this preliminary matter that provides the little of what we know of Bostwick. It includes notice of the New York Medical and Surgical Institute (est. 1832), at which Bostwick was "operating surgeon and attending physician," and, of which, in all likelihood, he was proprietor. Bostwick's name appears in the 1850-51 *New York Mercantile Union directory,* and the *New York City business directory* for 1859, 1872 and 1882. "H. Bostwick" is listed as a member of the New York County Medical Society in 1864, but appears in no issue of the *Medical register of the city*

of New York and vicinity after that date; nor is he listed in *Butler* (1877).

379. BOSTWICK, Homer, 1806-1883.
A treatise on the nature and treatment of seminal diseases, impotency, and other kindred affections: with practical directions for the management and removal of the cause producing them; together with hints to young men . . . Twelfth edition. New York: Rogers & Company, 1860.
2, [2], ix, [1], 254 p. : ill. ; 19 cm.

The twelfth edition contains "Additional remarks" (p. 252-54) addressed to females, encouraging them to consult Bostwick for gynecologic disorders. Spine- and half-titles: *Bostwick on seminal diseases and hints before marriage.*

BOSTWICK, Homer, see also **ABERNETHY** (#6).

380. BOTANIC SENTINEL.
Botanic sentinel and literary gazette. [Vol. 1, no. 1 (Aug. 12, 1835)-v. 2, no. 52 (Aug. 17, 1837)]. Philadelphia: John Coates, Jr., 1835-1837.
2 v. ; 36 cm.

A Thomsonian weekly devoted to the promulgation of "the simple truths and plain *cure*ological practice which characterizes our national system of *American Modern Medicine*" (v. 2, no. 32, p. 253). The *Botanic sentinel* was edited by John Coates, Jr., who also maintained a book and stationer's shop on South Street in Philadelphia just above Second. Volumes 3-4 (1837-1839) were published under the title: *Philadelphia botanic sentinel and Thomsonian medical revolutionist.* The Atwater set is lacking volume one, and nos. 5 (21 Sep. 1836), 20 (5 Jan. 1837), 39 (18 May 1837) and 52 (17 Aug. 1837) from volume two.

381. BOTANICO-MEDICAL RECORDER.
The Botanico-medical recorder; or impartial advocate of botanic medicine, and the principles which govern the botanico-medical practice . . . [Vol. 6 (Oct. 7, 1837)-v. 16 (Dec. 1848)]. Columbus, Ohio: Jonathan Phillips, printer, 1837-1848.
11 v. : ill. ; 25 cm.

The Botanico-medical recorder continues the *Thomsonian recorder or impartial advocate of botanic medicine* (#3518) after the fifth volume of the latter title was published at Columbus in 1837. Alva Curtis (#852-857) had taken over ownership and editorship of *The Thomsonian recorder* from Thomas Hersey (#1637, 1638), and continued to edit and publish *The Botanico-medical recorder* until 1855, "when, under the name of *Physio-medical recorder,* he sold it to a colleague and later adversary, William Henry Cook [#775-777] . . . another leading exponent of Thomsonism, who, along with Curtis, established much of the canon of physio-medical thought" (John S. Haller, *Kindly medicine,* Kent, Ohio: Kent State Univ. Press, 1997, p. 38). Of the eleven volumes published, volumes VI (1838), VII (1839) and IX (1841) are in the Atwater Collection.

382. BOTH, Carl.

Questions asked in "The Boston Medical and Surgical Journal," under the heading, "Irregular and quackish advertisements," answered . . . Boston, 1864.

12 p. ; 24 cm.

"In the 'Boston Medical and Surgical Journal' of Nov. 17, 1864, I am accused of irregular and quackish advertisements, of dispensing secret nostrums, of proclaiming alleged superior claims to the confidence of the community, of injuring the honor of the 'Medical Society,' &c." (p. [1]). Rather than defend himself against these accusations, Both goes on the offensive, calling into question the medical knowledge and judgement of his accusers.

382.1. BOURNE, George Melksham.

The home doctor: a guide to health . . . [San Francisco]: San Francisco News Company, 1878.

xx, 505, [1] p., [1] leaf of plates : ill., port. ; 18 cm.

A hydropath in practice at San Francisco since the early 1850s, Bourne makes no claim to formal medical study or to a medical degree. On page vi he states simply, "At the mature age of thirty-six [ca. 1852] I accepted the philosophy and teachings of the Water-Cure as far as then promulgated." The center-piece of Bourne's version of hydropathy was the addition of sweating or the steam-bath to its thera-

peutics—a procedure rejected by most earlier hydropaths. In *The home doctor* Bourne describes the use of his "thermal sudatory system" in the treatment of diphtheria, smallpox, etc., as well as the application of more traditional cold-water baths and "magnetic sleep," or hypnotism.

BOUTON, Eugene see JOHONNOT (#2038, 2039).

383. BOWDITCH, HENRY INGERSOLL, 1808-1892.

Intemperance in the light of cosmic laws . . . From the third annual report of the State Board of Health of Massachusetts. New York: The United States Brewers' Association, 1888.

47 p. ; 23 cm.

Bowditch's report is based on correspondence with American ambassadors and consuls in fifty-two foreign capitals "to learn the nature and character of the stimulants used . . . by the inhabitants of countries to which said correspondents were accredited, and the influence of such indulgence on the health and prosperity of the people" (p. 2). He also sought to know "the relative amount of intoxication in said countries compared with that . . . in the United States." Bowditch determined that "the tendency to inordinate indulgence and intoxication, seems to vary with varying climatic law; with the character of the race, and according to the fashion or taste that may have been modifid by centuries of custom or law; with the character of the stimulant used; and finally, with the cultivation or non-cultivation of the grape" (p. 8). He concluded that the abuse of stimulating drinks diminishes the nearer a peoples lives to the equator. In Europe, he observes, the southernmost regions "are covered by the richest vineyards . . . These wines are drunk freely by all the inhabitants. In some districts the babes drink them as they would their mother's milk. They are used by young and old" (p. 9-10). Above the isothermal line of 50 degrees, however, where the grape will not flourish, "Russia and all the great Scandanavian people, the Anglo-Saxon and Celt . . . drink deeply of more fiery liquor than the men of the south." Indeed, observes Bowditch, "Instead of simple exhilaration such as is generally seen on the shores of the Adriatic and the vine-clad hills of southern Germany and Spain, the dwellers along the

Baltic and the northern seas drink even to narcotism, and lie in beastly intoxication perchance in the very gutters of many a northern city" (p. 10).

The English, Irish and Scots come off worst in Bowditch's estimation of European drinking habits. As the American republic, by reason of immigration, inherited "British tastes for strong liquors," America, therefore, "in her fertility of resources, has increased this love of strong liquor" (p. 14). Bowditch notes two national habits which only increase Americans' culturally acquired tendency to abuse alcohol, i.e., "the evil custom of 'treating' one's acquaintance to a 'drink' at any hour of the day"; and "the support given by the community to open dram shops, where coarser liquors are exposed for sale to all comers" (p. 17).

Bowditch disclaims the temperance cry that "alcohol in any form is *always a poison.*" What does concern him is the nature of the drink and the culture in which it is consumed. "I take the following position, and I fearlessly assert that clinical experience proves . . . that every form of stimulant now in use can be made a blessing, if used temperately and on proper occasions . . . Yet more, I believe that even when used intemperately, light beer, ale, lager-beer, wines, like claret, etc., do vastly less harm than the stronger ardent spirits . . . Man, under the conditioned influence of the tribe of compositions known in America under the various slang terms of 'gin-sling,' 'cock-tail,' 'brandy smashes,' 'mint juleps,' 'eye-opener,' etc., etc., is liable to become at any time a beastly drunkard" (p. 27).

Bowditch makes a reasoned plea for increased cultivation of viniferous grapes in America, and the introduction of what he regards as an inherently temperate wine-drinking culture. Doubtless, his essay was extracted from the *Third annual report of the State Board of Health of Massachusetts* (Boston, 1872, p. 72-129) and published by the brewers association for his remarks on beer: "the Germans are destined to be really the greatest benefactors of this country, by bringing to us . . . their lager-beer. Lager-beer contains less alcohol than any of the native grape wines. This fact, with the other fact that the Germans have not the pernicious habits of our people of treating to sharp liquors at open bars, would, if we chose to adopt their customs, tend to diminish intemperance in this country" (p. 34).

An 1832 Harvard Medical School graduate, Bowditch became the devoted disciple of P.C.A. Louis in Paris during several years' study abroad, an experience which left its mark in his editing two American editions of Louis' works; the publication of his own *The young stethoscopist* (1846); and a pioneering operation for the removal of pleural effusions described in the *American journal of the medical sciences* in 1852. Bowditch is equally renowned for his devotion to the cause of abolition and to the many public health measures instituted through his efforts. "Bowditch's influence in stimulating the public health movement in the country was probably greater than that of any other man of his time," according to his biographer in the *Dict. Amer. biog.* (I:494).

384. BOWKER, Pierpont F.

The Indian vegetable family instructer [sic]: containing the names and descriptions of all the most useful herbs and plants that grow in this country, with their medicinal qualities annexed. Also a treatise on many of the lingering diseases to which mankind are subject, with new and plain arguments respecting the management of the same, with a large list of recipes, which have been carefully selected from Indian prescriptions and from those very persons who were cured by the same after every other remedy had failed. Designed for the use of families in the United States . . . Boston: Published by the author, 1836.

xii, [13]-180 p. ; 16 cm.

"Our fields abound with vegetable medicine, and the fertile meadows, the witness of your labor, produce many, yes, very many, a valuable root . . . Expense is out of the question entirely. Roots and herbs are at the command of every one, and nature's prescriptions are all free gratis. She demands not your money for her services, but like a kind patron and friend invites you to partake of her blessings . . . There is no excuse for you to lay and suffer if you are sick. No; this physician is kind and charitable. The rich and poor may share the blessing alike. Equality and equal rights is the motto" (p. 14). Bowker divides his text into two parts. The first lists and briefly describes 182 plants and the disorders that they remedy. There is no apparent classification, either botanical or nosological. The second part contain recipes for the healing of 157 common disorders. *The Indian vegetable family instructer* was republished at Utica, N.Y. in 1851.

385. BOWMAN AND BLEWETT.
Soap . . . carbolic acid. New-York: C.C. Shelley, steam book and job printer, 1870.
   20 p. ; 22 cm.

Bowman & Blewett were the manufacturers of carbolic acid based soaps for laundry, medicinal, dental, shaving and skin cleansing purposes. The manufacturers advertise it as a "universal disinfectant and antiseptic," a treatment for "many cutaneous diseases," a "preventive in all cases of gangrene," and for eliminating body vermin. Title on wrapper: *Carbolic acid soaps.*

BOWYER, Thomas M. see **GRANVILLE SANITARIUM** (#1404).

BOYLAND, G. Halstead see **BUFFALO LITHIA SPRINGS** (#515).

386. BOYLE, James.
Chymiatrion, for the treatment of all forms of chronic and constitutional disease in men, women, and children. A comprehensive exposition of the views of disease and its treatment, as theoretically and practically recognized at the Eclectic Institute of Medicine and Surgery, Eye and Ear Infirmary, etc. . . . New York: Published by the proprietor, [1849?].
   iv, [5]-64 p. ; 23 cm.

First edition. Promotional literature for the Eclectic Institute of Medicine and Surgery, No. 26 North Moore Street, New York. On page [ii] Boyle warns his patients that "the building in which my office is at present is liable to be sold and pulled down." However, he has left his card with the nearby grocery of Messrs. Acker & Co., "which will inform them where I may be found."

387. BOYNTON, Worcester E. , 1824-ca. 1900.
Orthopaedic surgery. Cases treated and cured . . . [Lawrence, Mass., ca. 1868.]
   [6] leaves in various pagings : ill. ; 19 cm.

Boynton was born on 26 March 1824. He received his early education in the common schools and the at Newbury Seminary (Vt.). The family genealogy claims that he had medical degrees from the University of the City of New York and the Massachusetts Medical College. The catalogs of these schools do not substantiate this, however. Boynton is said to have started practice in East Hopkinton, N.H. in 1851, but the Historical Society in Hopkinton is unable to confirm this. Sometime before 1877, Boynton moved to Lawrence, Ma., where he practiced until his death.

Boynton's *Orthopaedic surgery* is a promotional brochure containing case histories and patient testimonials that attest to the efficacy of his proprietary remedies (a list of which appears on p. 82). The printed wrapper indicates that "these pages are selected from Dr. B.'s work of 300 pages," presumably the *Family physician,* one of several popular and self-promoting treatises authored by Boynton. Each leaf contains small, crudely executed but somehow fascinating wood engravings. In printed wrapper.

388. BOYNTON, Worcester E., 1824-ca. 1900.
Orthopaedic surgery. Cases treated and cured . . . [Lawrence, Mass.], 1869.
   [8] leaves in various pagings, [1] leaf of plates : ill. ; 24 cm.

In printed wrapper.

BRADFORD, Henry Okes see **BRADFORD** (#389).

389. BRADFORD, Robert.
The mother's medical guide: containing a description of the diseases incident to children; with the mode of treatment, as far as can be pursued with safety, independently of a professional attendant. By R. & H.O. Bradford . . . With notes and amendments, by Jerome V.C. Smith, M.D. Boston: Allen & Ticknor, 1833.
   xiv, 110 p. ; 12 cm.

Noting the success of the original edition published at London in 1832, the American editor adds, "In the interior of the United States, in those extensive districts where families are scattered over large tracts of country, remote from settlements, where it must always be difficult to obtain the immediate services of a medical practitioner, a manual of this kind will be of immense value" (p. vi). Within this small volume, the authors describe treatment for sixty-seven common childhood disorders.

The authors are described on the title-page as "members of the Royal College of Surgeons." Jerome Van Crowninshield Smith (1800-1879) received his medical degree from Williams College (i.e., Berkshire Medical College) in 1822, and, among many and varied accomplishments, authored several popular health publications (#3260-3263).

BRADLAUGH, Charles see **BESANT** (#314); **KNOWLTON** (#2155, 2156).

390. BRADLEE, David F.
By the Queen's patent. BUCHAN'S HUNGARIAN BALSAM OF LIFE. The Great English Remedy for the Cure of Colds, Coughs, Croup, Asthma, and CONSUMPTION. Discovered by Dr. Buchan, of London, England, and tested for upwards of seven years in Great Britain and on the Continent of Europe, where it has proved the great and only remedy for the worst forms of Pulmonary Disease. David F. Bradlee . . . Boston, sole agent for the United States and British American provinces . . . [ca. 1845].
15, [1] p. ; 22 cm.

"The principal ingredient of Buchan's Balsam . . . is a Hungarian Gum .. The tree from which this gum is extracted has been known for ages in the northern part of China . . . and is called by the natives "Tschang" . . . It is only within the last three hundred years that this tree has attracted attention in the Western part of Europe . . . It is called in Latin 'Vitae balsamus' . . . It is found in mountainous parts of Hungary, and other countries in the north of Europe and Asia" (p. 8). "Hungarian balsam, *Balsamum Hungaricum,* is a product of *Pinus Pumilio*" (Charles F. Millspaugh, *American medical plants,* New York: Boericke & Tafel, c1887, 2:163-2). The *vitae balsamus* described in this pamphlet may well be a tincture derived from balsamic resins common in the Western pharmacopoeia under such names as "friar's balsam," "Jesuit's drops," "Jerusalem balsam," etc.

This aromatic gum resin was admitted into the *United States pharmacopoeia* as benzoin. It was used as an expectorant, and was a common ingedient in 19th-century cough and cold remedies.

391. BRADY, William, b. 1880.
Personal health. A doctor book for discriminating people . . . Philadelphia and London: W.B. Saunders Company, 1916.
407 p. ; 20 cm.

The author was a 1901 graduate of the University of Buffalo Medical Department, and practiced medicine in Elmira, New York.

392. BRAGG, H. L.
Things worth knowing. Vol. 1. [Washington, Vt.?: H.L. Bragg, 1894?].
12 p. ; 11 cm.

Title from printed wrapper. Promotional literature for *Fragrant Rose Balm,* bottled by H.K. Bragg, of Washington, Vt.

393. BRANDRETH, Benjamin, 1807-1880.
Brandreth's Pills, entirely vegetable and innocent . . . [New York], c1873.
4 p. ; 28 cm.

Brandreth was a native of Leeds, England. "In the mid-eighteenth century, his physician grandfather, William Brandreth, had concocted and sold a Vegetable Universal Pill. Inheriting the formula, Brandreth marketed the pill in 1828 . . . In 1835, sensing a larger pill market in the United States, the family migrated to New York City . . . Upon his arrival, Brandreth began marketing his Vegetable Universal Pills, and, after a slow beginning, sales expanded rapidly. Brandreth briefly attended the New York Eclectic Medical College . . . In 1838 he moved his home and manufacturing operations upriver to Sing Sing . . . Brandreth's pills became one of the best-selling proprietaries in the United States as well as overseas . . . The formula contained sarsaparilla and three powerful vegetable cathartics—aloe, colocynth, and gamboge . . . Americans at this time generally ate a stodgy diet, and constipation was a widespread problem. Brandreth shrewdly asserted that constipation was the source of all ills and acclaimed purgation . . . as the universal cure .

. . A congressional committee in 1849 reported that Brandreth was the nation's largest proprietary advertiser, spending $100,000 a year on advertisements. Between 1862 and 1863, Brandreth's average annual gross income surpassed $600,000" (James Harvey Young in *Amer. nat. biog.*, 3:424). This steel-engraved certificate contains a riverfront view of Brandreth's "manufactory" at Sing Sing (i.e., Ossining), N.Y. Two proprietary stamps attached to the first page.

394. BRANDRETH, Benjamin, 1807-1880.
DR. BENJAMIN BRANDRETH'S CERTIFICATE OF AGENCY. THIS CERTIFICATE is intended as a GUARANTEE to the People of the United States that [. . .] is my Agent, for the sale of BENJ.n BRANDRETH'S VEGETABLE UNIVERSAL PILLS . . . Given at my Principal Office, 241 Broadway in the City of New York, this first day of November One thousand eight hundred and forty. [ms. signature: *B. Brandreth, M.D.*]
  [1] sheet ; 36 × 29.5 cm.

395. BRANDRETH, Benjamin, 1807-1880.
Dr. Benjamin Brandreth's VEGETABLE UNIVERSAL PILLS. Introduced by him into the United States in 1835 . . . [New York, ca. 1843].
  [2] leaves ; 26 cm.

Brandreth is of the opinion "with his grandfather, the late celebrated Dr. William Brandreth, that 'As there is only one principle of life, so there is only ONE principle of DISEASE,' which is an impurity of the blood, which by impeding the circulation, brings on inflammation, and consequent derangement in the organ, or part where such impurity of the blood settles" (leaf 1 recto). Date of imprint based on testimonial letter (leaf 2 verso).

396. BRANDRETH, Benjamin, 1807-1880.
The doctrine of purgation. Curiosities from ancient and modern literature. A collection of quotations on the use of purgatives, from Hippocrates, and other medical writers, covering a period of over two thousand years, proving purgation is the corner-stone of all curatives . . . New York: Baker & Godwin, printers, 1867.
  224 p. ; 26 cm.

First edition. "We can remove disease in two ways," Brandreth writes in his introduction, "by the upper and by the lower passages; by vomiting and by purging . . . For forty years I have directed my attention to the cure of disease on this plan; and facts derived from experience have long since confirmed me in the belief that this is calm Nature's own method of cure, because it assists her in removing impurities by the means and outlets she has so wisely provided for herself" (p. 4). He continues: "The quotations from the writings of medical men embodied in this pamphlet, prove the talent that has been at work upon this theory of purgation for over a period of two thousand years—and in vain. Then what has prevented its complete success? Simply this, in my opinion: Not a single writer has given a medicine which, out of their own hands, would successfully and safely enforce the purgative theory" (p. 5). Fortunately, the public may rely on Brandreth's Pills. Cover-title: *Authorities on purgation.* Atwater copy inscribed by author.

397. BRANDRETH, Benjamin, 1807-1880.
The doctrine of purgation. Curiosities from ancient and modern literature. A collection of quotations on the use of purgatives, from Hippocrates, and other medical writers, covering a period of over two thousand years, proving purgation is the corner-stone of all curatives . . . Third edition. New York: Baker & Godwin, printers, 1873.
  246 p. ; 26 cm.

Cover-title: *Authorities on purgation.*

398. BRANDRETH, Benjamin, 1807-1880.
Theory and practice of purgation. Numerous illustrations of cure by Brandreth's Pills . . . New York: Published by B. Brandreth, M.D., Brandreth House, 1876.
  120 p. ; 23 cm.

"My theory is that purgation is purification. With Brandreth's Pills I can cleanse the inside of the human body as perfectly as water does the outside" (p. 3). In printed wrapper. At head of title: *Confirmed by fifty years' experience.*

399. BRANDS, Orestes M., b. 1843.
An academic physiology and hygiene. Embracing special chapters on foods and

their preparation; water and other beverages; air and ventilation; the removal of waste matters; exercise, rest, and recreation; bathing and clothing; hygiene of the special senses, and the effects of stimulants and narcotics on the human system. By Orestes M. Brands . . . and Henry C. Van Gieson . . . Boston; New York; Chicago: Leach, Shewell, and Sanborn, [c1893].
   xv, [1], 386 p. : ill. ; 19 cm.

400. BRANDS, Orestes M., b. 1843.
Lessons on the human body. An elementary treatise upon physiology, hygiene, and the effects of stimulants and narcotics on the human system . . . Boston and New York: Leach, Shewell, & Sanborn, [c1883].
   xiii, [1], 255 p. : ill. ; 17 cm.

Brands was a school principal in Paterson, N.J. when this first edition of his school physiology was published. It is the earliest of several contibutions that Brands made to juvenile health literature.

401. BRANDS, Orestes M., b. 1843.
Lessons on the human body. An elementary treatise upon physiology, hygiene, and the effects of stimulants and narcotics on the human system . . . Boston; New York; Chicago: Leach, Shewell, & Sanborn, [c1885].
   [2], xii, [2], 240 p. : ill. ; 18 cm.

In this second edition, the lessons that appeared in the section on alcohol in the 1883 edition have been rearranged in the sections on blood and digestion. New lessons have been added on tobacco as well. On cover: *New York edition.*

402. BRASTED, Horace Kimball.
Facts on homoeopathy . . . Lima, N.Y.: A. Tiffany Norton, printer, Recorder office, 1883.
   92 p. ; 23 cm.

An *apologia* for homeopathy intended for a lay audience.

BRATTLEBORO WATER CURE see GRANGER (#1400).

403. BREMER, Ludwig, 1844-1914.
Tobacco, insanity and nervousness . . . St. Louis, Mo.: Published by the Meyer Brothers Druggist, 1892.
   16 p. ; 23 cm.

Formerly physician to the St. Vincent's Institution for the Insane, Bremer observes that "there is an alarming increase of juvenile smokers; and, basing my assertion on the experience, gathered in my private practice and at the St. Vincent's Institution in this city, I will broadly state that THE BOY WHO SMOKES AT SEVEN, WILL DRINK WHISKEY AT FOURTEEN, TAKE TO MORPHINE AT 20 OR 25, AND WIND UP WITH COCAINE AND THE REST OF THE NARCOTICS, AT AGE 30 AND LATER ON" (p. 6). The habitual use of tobacco is not only the cause of a wide variety of physical ailments—those of the nervous system particularly—but often leads to insanity. Bremer believes that the anti-tobacco message could effectively be brought to the young by their teachers, the clergy and physicians—if these individuals were not themselves among the "army of tobacco slaves." He discouragingly concludes, "under existing conditions the habit will not want for new generations of victims, as long as the cigar is looked upon as a symbol of manliness by the young, and the pipe as that of peace and comfort by the adult, and as long as tobacco is praised in word and picture. To what extent the habit is fostered among boys on the theory of forbidden fruit, it is not my province to discuss" (p. 14). This pamphlet was originally presented as a lecture before the St. Louis Medical Society in October, 1891. The author was an 1870 graduate of the St. Louis Medical College and a frequent contributor of articles on diabetes and neurologial-psychiatric topics to the periodical literature. In printed wrapper.

404. BREVITT, Joseph, 1769-1839.
The female medical repository, To which is added, a treatise on the primary diseases of infants: adapted to the use of the female practitioners and intelligent mothers; the technical terms are explained, and an attempt hath been made to reduce these branches of "the healing art," to conciseness and perspicuity . . . Baltimore: Published by Hunter & Robinson . . . J. Robinson, print., 1810.
   252 p. ; 16 cm. (12mo)

"This publication is more particularly offered for the assistance of women likely to become mothers, in regulating aright their conduct, during their pregnancy, labor and after-management; and also to afford them a knowledge, in these eventful periods, of the propriety, or otherwise in the attendants and their measures pursued. It is further offered for the assistance of female practitioners, or of others who have been prematurely, and improperly introduced, into the practice of midwifery, without the opportunity of receiving a scientific knowledge . . . " (p. 6).

Brevitt provides chapters on the anatomies of the female pelvis, the fetal head and the female organs of generation; menstruation and its disorders; diseases of the female genitalia; pregnancy and its disorders; "practical midwifery" (including praeternatural labors, the use of forceps and caesarian section); the puerperium; and the diseases of infants.

REFERENCES: *Austin* #271.

405. BRIANTE, John Goodale.

The old root and herb doctor, or the Indian method of healing. By Dr. John Goodale Briante, for many years with the St. Francis tribe of Indians, at Green Bay; also, for several years with the Pattawattamies and other tribes. Containing directions for preparing and using their most valuable remedies, as used by him, in his extensive practice throughout the Eastern and Middle states. Claremont, N.H.: Granite Book Company, 1870.

viii, [9]-103 p. ; 17 cm.

406. BRIDGER, Adolphus Edward.

Man and his maladies or the way to health. A popular handbook of physiology and domestic medicine in accord with the advance in medical science . . . New York: Harper & Brothers, 1889.

xv, [1], 593 p. ; 19 cm.

Harper & Brothers issued this book the same year that the London edition was published. Bridger, who is described on the title-page as an M.D. and F.R.C.P. (England), was the author of several popular medical treatises published during the 1880s and 1890s. *Man and his maladies* was the only one of these books also to be published in the United States.

407. BRIGGS, James Edwin.

Nervous diseases and magnetic therapeutics . . . New York: Geo. W. Wheat, printer, 1881. 60 p. ; 17 cm.

"That nervous affections precede other physical ills, appears conclusive. The physician should consider this fact in his diagnosis. Whatever medical agency is employed should be selected with reference to its influence on the nervous system . . . What is called *animal magnetism* is a soother of the nerves, and therefore invaluable for that reason. It is no imaginary agency. the product of charlatans, but belongs to a superior science. Its power over the faculties of the body at large, and especially over the brain and nervous system, is immense; and it is therefore capable of application to prevent and remove suffering, and to cure disease, far beyonds the means hitherto pursued by the art of medicine" (p. 21-22).

Briggs states that the the simplest suggestion of the magnetic healer becomes "a spiritual force, acting in harmony with the mind and will of the patient to the end of influencing beneficially the unwholesome states of the body. Every disorder which will admit of a cure will give way to this agency . . . The simple suggestion of the magnetic physician . . . will increase or diminish the action of the heart, change the breathing, affect the functional movements of the stomach, liver, spleen, kidneys, and intestinal canal. The blood will be sensibly modified in character, both chemically and physiologically" (p. 37). After positing magnetism's ability to alter physiological function, Briggs concludes with a long litany of the diseases that it is known to cure. Appended to this treatise are "Magnetism scientifically and successfully applied, by B.L. Cetlinski" (p. 42-51), and "Magnetism, by Giles B. Stebbins" (p. 52-57). The author was an 1875 graduate of the Eclectic Medical College of the City of New York.

408. BRIGGS, Lansing, 1807-1888.

Dr. Briggs' guide to health. A brief account of wonderful cures, and a few personal letters from well known citizens of Auburn . . . Auburn, N.Y.: Published by the Dr. Briggs Medicine Co., 1895.

[8] leaves : ports. ; 23 cm.

The eight unnumbered leaves in this brochure promote Dr. Briggs' proprietary remedies, and largely consist of testimonials under such headings as: "Was told there was no hope for him by physicians, yet 5 packages cured him"; "He prayed for death, but the Kidney cure cured him"; "She had nearly all the diseases peculiar to women, but the Female Regulator, Female Cure and Household Ointment made her a well woman," etc. Lansing Briggs was an 1830 graduate of the Berkshire Medical College.

BRIGHAM, Amariah, 1798-1849. *Influence of mental cultivation in producing dyspepsia* see **HOLBROOK** (#1687).

409. BRIGHAM, Amariah, 1798-1849.
Remarks on the influence of mental cultivation and mental excitement upon health . . . Second edition. Boston: Marsh, Capen & Lyon, 1833.
    xii, [13]-130 p. ; 19 cm.

"The object of this work is to awaken public attention to the importance of making some modification in the method of educating children, which now prevails in this country. It is intended to show the necessity of giving more attention to the health and growth of the body, and less to the cultivation of the mind, especially in early life, than is now given" (p. [vii]). Brigham is also concerned about the condition of young American females, whom he sees as universally "pale, slender, and apparently unhealthy." The reason, he maintains, is that "there is no other country where females generally receive so early and so much intellectual culture, and where so little attention is paid to their physical education" (p. ix). Brigham concludes his preface with a statement soon to be echoed by a multitude of health reformers: "I hope also that my remarks may serve to awaken some attention to the study of human *Anatomy* and *Physiology;* on which all plans of education ought to be founded. The general neglect of these sciences is one of the most extraordinary facts of this kind that this inquiring age presents" (p. x).

Noting the "prevalent eagerness for intellectual improvement in our republic" and the "constant search after new and sure methods by which the education of children may be promoted" (p. 14), Brigham warns "that in cultivating the mental powers of children, we should

be less anxious to ascertain how rapidly and to how great an extent they may be developed, than how much the delicate organ or organs by which the mind acts may be excited without injury to the body or the mind" (p. 16). Brigham maintains that "the brain is the material organ of thought"; that "what ever strongly excites the mind or its organ, whether it be study or intense feeling" alters the physical state of the brain; and that the degree of these changes often results in a diseased state (p. 22). Brigham wonders that "while people are exceedingly fearful of enfeebling and destroying digestion, by exciting and overtasking the stomach, they do not appear to think they may enfeeble or derange the operation of the mind by exciting the brain, by tasking it when it is tender and imperfectly developed, as it is in childhood" (p. 33).

Brigham advises parents "to pause before they attempt to make prodigies of their own children. Though they may not destroy them by the measures they adopt to effect this purpose, yet they will surely enfeeble their bodies, and greatly dispose them to nervous affections" (p. 55). He encourages parents to let children "manifest a true philosophical spirit of inquiry," i.e., to pursue their natural curiosity, even when such activity seems like play. Brigham supports his thesis by citing what he regards as the enlightened educational doctrines current in Europe (e.g., Tissot, Hufeland, Spurzheim, Sinabaldi, Ratier, Londe, Broussais, and, of course, Rousseau).

In Sections V-VIII Brigham extends his study of mental overstimulation to the population as a whole, and adds heart disease and dyspepsia to the list of its resulting disorders. In concluding he encourages his readers to be more moderate in their intellectual endeavors, and to "throw aside their *bitters, blue pills, mustard seed, bran bread,* &c. &c. and seek bodily health and future mental vigor, in judicious exercise of the body, innocent amusements, cheerful company, ordinary diet, and *mental* relaxation" (p. 121).

The *Remarks* was originally published at Hartford in 1832. It appeared in two more American editions (2nd, 1833; 3rd, 1845, #411) and in at least seven editions issued in the United Kingdom, not including an 1874 London reprint based on the 2nd American edition.

Amariah Brigham profoundly influenced Americans' concept of mental illness and the treatment of the mentally ill. He provided national leadership in his direction of the Hartford Retreat and the New York State Lunatic Asylum (Utica). Brigham was the founder and first editor of the *American journal of insanity*, and was one of the founding members of the Association of Medical Superintendents of American Institutions for the Insane.

410. BRIGHAM, Amariah, 1798-1849.
Remarks on the influence of mental cultivation and mental excitement upon health . . . London: J. Cleave ; Manchester: Abel Heywood, 1839.
iv, [5]-48 p. ; 22 cm.

Reprint of the second American edition published at Boston in 1833 (#409). The *Remarks* had a more extensive publishing history in the United Kingdom than in the United States, appearing in at least seven Glasgow, Edinburgh and London editions between the 1836 and 1844.

411. BRIGHAM, Amariah, 1798-1849.
Remarks on the influence of mental cultivation and mental excitement upon health . . . Third edition. Philadelphia: Lea & Blanchard, 1845.
xxviii, [37]-204 p. ; 18 cm.

The third American edition includes the notes added to the Glasgow and London editions by their editors Robert Macnish and James Simpson.

412. BRIGHT, John W., 1791-1880.
The mother's medical guide; a plain, practical treatise on midwifery, and the diseases of women and children. In seven parts. Designed especially for the use of mothers and nurses . . . Louisville, Ky., 1844.
xvi, 792, [2] p., VI (i.e., 5) leaves of plates + frontis. : ill., port. ; 22 cm.

John W. Bright was born near Lexington, Ky. on 29 April 1791. He remained on the family farm until age twenty-one, when he attended a private school, learned Latin, and began to read medicine in the office of Dr. John Bemis. Three years later, Bright opened a medical practice in New Castle, Henry County, Kentucky. In 1820 he was commissioned surgeon's mate in the 88th Regiment of the Kentucky militia. In

1821 he began to attend medical lectures at Transylvania University, from which he is supposed to have received his medical degree the following year. After practicing in New Orleans early in the 1830s, Bright moved to Louisville where he practiced medicine between 1834 and 1859. He subsequently returned to Lexington, where he divided his time between writing and medical practice. Besides his medical work, Bright was intensely interested in religious matters, was a preacher in the Methodist Episcopal church for forty years, and established churches in several previously unchurched communities. He died on 19 July 1880.

"THE MOTHER'S MEDICAL GUIDE comprises a plain, practical treatise on Midwifery, the diseases of pregnancy, and all the diseases peculiar to women in every stage of life. It also treats of those of children of both sexes, from birth to the age of thirteen or fourteen years; and the remedies for the same, in all their various stages. It further contains full directions for nursing women and children. The chapter on gestation—embracing comparative gestation of stock . . . will be found to be a very important one to farmers and stock-raisers" (*Pref.*, p. [iii]). Bright acknowledges his indebtedness to the published works of "Burns, Baudeloque [sic], Denman, Clark, Dewees, Hall, Smellie . . . [*et al.*]" (p. iv). Frontispiece portrait of author. Printer's imprint on verso of title-page: "Printed by A.S. Tilden, Jeffersonville, Ind'a."

413. BRIGHT, John W., 1791-1880.
A plain system of medical practice, adapted to the use of families . . . Louisville, Ky.: Published by Morton & Griswold, for the author, [ca. 1850, c1847].
[4], xv, [1], 1039 p. : ill. ; 24 cm.

Bright's treatise is divided into sections on "Diseases common to men and women," "Diseases of pregnancy," "Midwifery," "Diseases of child-bed," "Children, diseases and management of," an illustrated botanic "materia medica," "Diseases of girls," "Diseases of women," and an appendix that covers topics such as hydropathy, homeopathy, the use of chloroform in midwifery, etc. Some sections are taken verbatim from Bright's 1844 *Mother's medical guide* (#412). For example, "Of the diseases of pregnancy" (p. 341-70) reprints part II, chapters I-XVIII of the earlier work, etc.

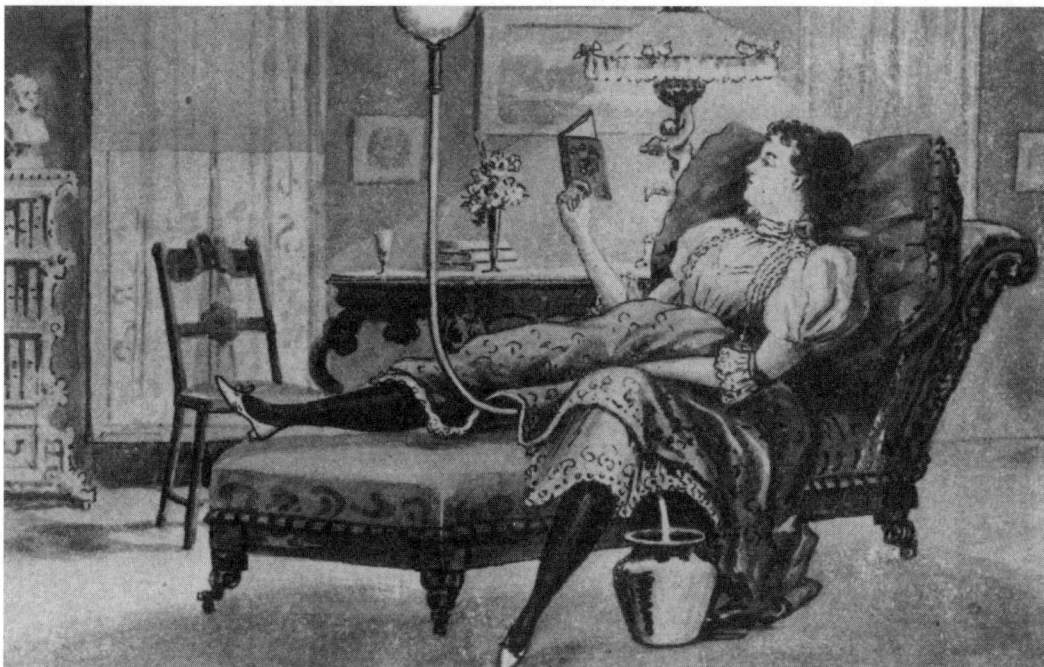

Fig. 14. "Lady in evening costume, taking a vaginal bath" from Brinkman's Illustrated common sense treatment, *#415*

Originally published at Louisville in 1847 (and again in 1848). The publication date of this expanded [third?] edition is based on a quotation taken from the *Dublin Journal* for 1850 on page 1034. The four unnumbered pages of advertisements that precede the title-page include endorsements by Charles Caldwell, Samuel David Gross, John C. Gunn, etc. Spine-title: *Bright's family practice.*

414. BRIGHT'S KIDNEY BEAN COMPANY.

The kidney remedy that brings joy to all in need. This remedy is bright's Kidney Beans for kidney, liver or bladder troubles. [Little Falls, N.Y.: Bright's Kidney Bean Co., after 1896].
    16 p. ; 15 cm.

415. BRINKMAN, A. B.

The illustrated common sense treatment for the cure and prevention of disease without the use of medicine, by use of the Reversible Tube Filter Syringe and Irrigator . . . [Philadelphia: A.B. Brinkman, after 1892].
    48 p. : ill. ; 17 cm.

As did many of his contemporaries, Brinkman maintains that most diseases are the product of auto-intoxication, i.e., the poisoning of the system due to decayed fecal matter retained in the colon. To prevent this condition, Brinkman recommends regularly flushing the colon with his Reversible Tube Filter Syringe, which has the advantage over competing mechanisms of purifying the water injected into the intestines. Brinkman also recommends using the Reversible Tube Filter with his Extension Irrigator for the prevention or treatment of vaginal disorders. He affects horror at the accusation that his Irrigator was simply "an ingeniously invented contrivance for the purpose of defeating nature by preventing conception," but nonetheless quotes a fictitious French opponent's remark that it is "without doubt the greatest and most effective invention of the kind that he had ever saw or heard of" toward that end (p. 27). Brinkman claims to be a medical graduate of "St. Petersburg University," to have lectured in Paris, etc. Title on printed wrapper: *Professor A.B. Brinkman's common sense treatment without the use of medicine. The new American edition. (From the St. Petersburg edition).*

416. BRINTON, Daniel Garrison, 1837-1899.

The laws of health in relation to the human form. By D.G. Brinton . . . and Geo. H. Napheys . . . Springfield, Mass.: W.J. Howland, 1871.

346, [2] p., [5] leaves of plates : ill. ; 19 cm.

"But apart from the fact that in a sense health and good looks are synonymous, the hour is now come when the improvement, the maintenance, if possible the creation of personal beauty, deserve to be recognized as a legitimate and worthy department of medicine . . . Health is the source of beauty, but the stream does not stay forever by its fountain. All the precepts, all the hints, which the diligent study of the healing art gives, all the suggestions proferred by hygiene and physiology, we shall attentively consider and apply." (p. 14-15). The authors define the ideal of feminine beauty in every bodily part, discuss its preservation through hygiene, and examine how deviations from these standards or deformities may be remedied "either by medical or by cosmetic art."

An 1861 graduate of Jefferson Medical College, Brinton served with distinction in the Federal army during the Civil War, following which he practiced medicine in Philadelphia and edited the *Medical and Surgical Reporter* (1867-87). Brinton's considerable reputation rests not on his contributions to medicine, however, but on his pioneer work in the anthropology of native Americans while professor of ethnology and archaeology at the Academy of Natural Sciences (Philadelphia) and professor of American linguistics and archaeology at the University of Pennsylvania. His co-author, George Henry Napheys, was author of several of the most popular treatises on domestic hygiene, sex education and sexual ethics published during the latter half of the 19th century (#2540-2568).

417. BROADBENT, Charles R.

A medical treatise on the causes and curability of consumption, laryngitis, chronic catarrh, and diseases of the air-passages. Combining the treatment by inhalation of medicated vapors. Also, a new and accurate method for the diagnosis of consumption; or, how to detect its signs and symptoms in its various stages. Also, including many chronic and nervous diseases, humors, fits,

Fig. 15. Apparatus for the inhalation of medicated vapors, from Broadbent's Medical treatise, *#417.*

&c. With an appendix on tobacco, showing its injurious effects on both body and mind; phrenology and mesmerism . . . Boston: Damrell & Welch, printers, 1862.

227 p., [1] folding leaf of plates : ill. ; 20 cm.

Broadbent is listed in the Boston business directories from 1858 through 1871. In 1859 and 1860, the Boston Lung Institute is associated with his name.

418. BROADBENT, Charles R.

THE PHYSIOLOGICAL LECTURER, of Providence, R.I. Who has between 1847 and the corresponding month of 1857, delivered eight hundred and seventy-five lectures to large audiences on Physiology, Health, etc. in all the principal cities and large towns of New England, and has in the same time been consulted by fifteen thousand persons upon all diseases to which human flesh is heir. [1857?]

Broadside : port. ; 45.5 cm. × 33 cm.

Broadbent's portrait (beneath the text) dominates this impressive lithographed broadside, printed by the firm of J.H. Bufford.

419. BROCKETT, Linus Pierpont, 1820-1893.

Asiatic cholera: its origin, history, and progress, for over two hundred years, and the devastations it has caused in the East and West; its ravages in Europe and America in 1831-2, in 1848-9, in 1854-55,

and in 1865-66: with a full description of the causes, nature, and character of the disease; its means of propagation, whether by the atmosphere or by contagion; its premonitory and distinctive symptoms; the best known means of preventing its attack, both in communities and individuals; and the most effectual remedies for it according to the celebrated physicians who have treated it; together with simple and plain directions for the care of those who from any cause can not obtain medical aid . . . Hartford: L. Stebbins; Chicago, Ill.: A. Kidder, 1866.
    375 p., [1] leaf of plates : ill. ; 19 cm.

Brockett took his medical degree from Yale University in 1843. Following a brief period interlude of medical practice, he transferred his energies to the publishing trade, and later to historical writing. He published some fifty historical texts, most of them on the American Civil War. Wood-engraved frontispiece: "View of Benares—birthplace of the cholera."

BROMLEY, Pliny M. see **ROCHESTER LAKE VIEW WATER-CURE INSTITUTION** (#3014).

420. BROMO-CHEMICAL COMPANY.
BROMO-CHLORALUM For the Prevention and Treatment of SMALL-POX. In view of the prevalence of small-pox in many cities and towns, we invite your earliest attention to the importance of using Bromo-Chloralum, the odorless and non-poisonous deodorizer and disinfectant which has received the approval of eminent medical men, and has been adopted in Asylums, Hospitals, Prisons, School Rooms and Private Dwellings, as a means of protection against the atmospheric influences which contribute to the spreading of disease . . . [ca. 1880].
    [1] leaf printed on both sides ; 36 × 16.5 cm.

Bromo-Chloralum may be applied in solution to the skin to relieve the "itching, burning surface and inflamed postules" of smallpox; may be used as a gargle for sore throat; may be used to disinfect sick rooms; or as a solution in which to disinfect "chamber utensils, bed pans, etc." The Bromo-Chemical Company was located in New York City.

421. BROOKS, Dr.
The physician's assistant, consisting of a short and comprehensive materia medica, together with a summary view of the whole practice of physic, surgery and midwifery . . . [S.l.]: Published by Benjamin French, 1833.
    viii, [9]-183 p. : ill. ; 21 cm.

According to the publisher's preface, *The physician's assistant* is posthumously published from Dr. Brook's original manuscript, "containing a particular account of his mode of practice in Physic and Midwifery" (p. [iii]). The "celebrated Dr. Brooks, late of Hebron, Me." was an herbalist. The first thirty-two pages describe American plants of medicinal value; most of the remainder of the text discusses a wide variety of diseases and injuries, and their botanical treatment. The work concludes with a section on midwifery, most of which is missing in the Atwater copy (lacking p. 161-76). Indices of plants and diseases precede the text.

422. BROWN, A. R.
Treatise on direct medical administration and renovation through acupuncturation . . . Litchfield, Mich.: Published by Brown & Herrick, 1867.
    261, [3] p. ; 17 cm.

In a therapeutic system suspiciously similar to Baunscheidtism (#236-238), the author describes the administration of medicines by "acupuncturation," i.e., the introduction of liquid medications through skin puncture directly into the blood stream. He promotes an instrument of his own design (the "Renovator") for this purpose, and the proprietary liquid medications (Nos. 1-3) that he insists answer the therapeutic call of most disorders. Brown claims to be a graduate of one of Cincinnati's eclectic medical schools (p. 6, 7). The Atwater copy bears the inscription of Henry Foster (#1265), founder of the water-cure at Clifton Springs, N.Y.

423. BROWN, Bertha Millard, b. 1870.
Good health for girls and boys . . . Boston, U.S.A.: D.C. Heath & Co., publishers, 1906.
    ix, [3], 152 p. : ill. ; 19 cm.

A school text of physiology and hygiene for the lower grades. The author is described on the title-page as "instructor in biology at the State Normal School, Hyannis, Massachusetts." Series: *The Colton series of physiologies.*

BROWN, C. A. see **ROCHESTER CANCER INFIRMARY** (#3013).

424. BROWN, Charles Reynolds, 1862-1950.
Faith and health . . . New York: Thomas Y. Crowell & Company, [c1910].
    [2], vi, 234, [2] p. ; 20 cm.

Examining Jesus' miracles Brown concludes, "Would that each one might know in some more vital way that the help of the ever-present Christ who thus healed men of old is still available for health, for guidance, and for moral recovery" (p. 24). Confident in "the healing power of suggestion," and that the "Saviour of the soul" is also the "Great Physician," Brown examines two "modern faith cures," i.e., Christian Science ("a colossal humbug"); and the Emmanuel Movement (he is concerned about taking medicine out of the hands of physicians and entrusting it to clergy, not to mention the resulting distortion of religion). Brown's final chapter, "The gospel of good health," was originally published as a pamphlet in 1908. Its intent may be summarized by Brown's remark, "If you will stand up, your mind and heart made right with God to the fullest extent you know, and in God's name say, 'Let there be health,' and keep on saying it resolutely, trustfully, hopefully, that very action of your inner life will work wonders. I do not say that no disease can stand before you . . . but I do say that you will set in operation one of the great healing forces of the world" (p. 187).

A Congregationalist clergyman, Brown established a national reputation as a social ethicist, labor arbitrator, educator, author and preacher. In 1911 he was named dean of the Yale Divinity School, which revived under his leadership.

425. BROWN, Eli F.
The eclectic physiology. For use in schools . . . New York; Cincinnati; Chicago: American Book Company, [c1884].
    v, [6]-189, [3] p., IV leaves of plates : ill. ; 19 cm.

Brown was an 1879 graduate of the Medical College of Indiana (Indianapolis).

426. BROWN, Eli F.
The eclectic physiology or guide to health. With special reference to the nature of alcoholic drinks and narcotics and their effects upon the human system . . . New York; Cincinnati; Chicago: American Book Company, [1888?, c1886].
    v, [6]-189, [3] p., IV leaves of plates : ill. ; 19 cm.

Series: *Eclectic series of temperance physiologies.* Cover-title: *The eclectic guide to health.*

427. BROWN, Eli F.
The house I live in, or an elementary physiology for children in the public schools. With special reference to the nature of alcoholic drinks and narcotics, and their effects upon the human system . . . Cincinnati and New York: Van Antwerp, Bragg & Co., [1888, c1887].
    106, [2] p., IV leaves of plates : ill. ; 19 cm.

Author's preface dated December, 1888. Series: *Eclectic educational series.*

STORY OF A BITE OF BREAD.

Fig. 16. "Story of a bite of bread," from Brown's The house I live in, *1888, #427.*

428. BROWN, Eli F.
Sex and life: the physiology and hygiene
of the sexual organization . . . Chicago: F.J.
Schulte & Co., publishers, [c1891].
    vi, 7-142 p. : ill. ; 19 cm.

A manual of sex education intended for
adolescents and their parents.

429. BROWN, John, 1810-1882.
Health: five lay sermons to working people
. . . New York: Robert Carter & Brothers,
1862.
    xii, [13]-90, [6] p. ; 17 cm.

An 1833 Edinburgh medical graduate,
Brown practiced medicine in that city most of
his life. His reputation, though presently diminished, was literary and considerable, numbering such figures as Thackeray and Ruskin
among his friends. He was the author of the
Horae subsecivae, three volumes of essays published between 1859 and 1882, and the immensely popular story *Rab and his friends* (1859).
Health is a collection of five "sermons" delivered
to or written for "weather and work-beaten
understandings." It includes: "The doctor—our
duties to him," "The doctor—his duties to us,"
"Children, and how to guide them," "Health,"
and "Medical odds and ends."

430. BROWN, John A., b. 1810.
The family guide to health, containing a
description of the botanic Thomsonian
system of medicine. With a biographical
sketch of the author . . . Providence: B.T.
Albro, printer, 1837.
    ix, [10]-224 p. ; 13 cm.

"That the physicians in what is termed the
regular practice, sometimes effect apparent
cures, we have no disposition to deny; and a
portion of those cures, where diseases are slight,
and taken in their incipient stages, may be real
and permanent. But, by far the greater portion
of the supposed cures, and especially in cases
of obstinate fevers, we are well satisfied, from
our own observation and the testimony of others, leave the patient the victim of lingering
disease and torment, created by the very remedies employed for the restoration of health,
and worse, even than the original malady, under which he suffered. The Botanic Medical

Practice furnishes safe, efficacious, and *healthy*
substitutes for those poisonous remedial agents.
It furnishes the healing balm from the products of our own soil, to remove disease, to restore the sufferer to sound health, and without
leaving him subject to those enervating and
deleterious effects, produced from mineral
poisons, with which a false science and a heart-
less cupidity have combined to curse suffering
humanity" (p. vi-vii).
    Brown begins his treatise with a lengthy au-
tobiographical sketch in which he describes his
own conversion to the Thomsonian cult. It is
followed by an equally lengthy "Historical account of the Thomsonian system" (p. [35]-90).
The remainder of the treatise is devoted to the
principles and practice of botanic medicine.
Brown is described on the title-page as "proprietor and physician to the R.I. Botanic Infirmary,
Pawtucket, R.I., and President of the R.I.
Botanic Association."

431. BROWN, Lawrason, 1871-1937.
Rules for recovery from pulmonary tuber-
culosis. A layman's handbook of treatment
. . . Second edition, thoroughly revised.
Philadelphia and New York: Lea & Febiger,
[c1916].
    iv, [5]-184 p. ; 18 cm.

Brown provides very basic explanations of
the role of the tubercle bacillus as the causative
agent of pulmonary tuberculosis and how the
disease is acquired. He describes a recuperative regimen that focuses on exercise, recreation and sleep out of doors, as well as diet and
hygienic measures. Brown emphasizes the limited role that the physician plays in recovery,
stressing responsibilities that lies with the tubercular patient: "We cannot make them get
well; all we can do is to give them help. We want
them to think over the whole thing seriously
and earnestly . . . We want them to realize that
the matter is largely in their own hands. We can
watch them for six months and send them to
work; and then, in a week, a day, an hour, they
can upset all we have done . . . Constant care,
day in and day out, must be the watchword for
the next four or five years . . . and if they do that,
we think the vast majority of them will be in excellent shape at the end of that time" (p. 18).
    A tuberculosis specialist, Brown was a 1900
medical graduate of Johns Hopkins University.

"Brown's choice of a medical specialty was determined by a break in his own health. During the third year of his medical course he was found to have developed pulmonary tuberculosis. His faculty advisers sent him to the Adirondack Cottage Sanitarium near Saranac Lake, N.Y., where he became a patient of America's pioneer student of the treatment of tuberculosis, Edward Livingston Trudeau" (*Dict. Amer. biog.*, 11:70). Following graduation from medical school, Brown immediately returned to Trudeau's sanitarium where he became its assistant resident physician. "Promotion followed rapidly, and he soon became the dominating intellectual and professional force in this pioneering sanitarium, whose methods, developed in large measure by Brown himself, set standards in the care of tuberculosis for the rest of country" (*ibid.*). Brown was a pioneer in the roentgenographic diagnosis of tuberculosis, and was one of the founders of the National Tuberculosis Association as well as the American Sanatorium Association. He was author of several books and more than 150 articles. The *Rules* was by far his most successful publication. The first edition was printed privately at Saranac Lake, N.Y. in 1915. A sixth and final edition appeared in 1934.

BROWN, M. W. see *GENEVA GAZETTE* (#1341).

432. BROWN, Oliver Phelps.
The complete herbalist, or the people their own physicians by the use of nature's remedies; showing the great curative properties of all herbs, gums, balsams, barks, flowers and roots; how they should be prepared; when and under what influence selected; at what times gathered; and for what diseases administered. Also, separate treatises on food and drinks; clothing, exercise, the regulation of the passions, life, health, and disease; longevity; medication; air and sunshine; bathing; sleep, etc. Also, symptoms of prevalent diseases; special treatment in special cases; and a new and plain system of hygienic principles . . . Jersey City, N.J.: Published by the author, 1867.
    407, [1] p., [12] leaves of col. plates : ill. ; 19 cm.

Originally published at Jersey City in 1865. Brown's description of medicinal plants, both foreign and domestic, comprises the lengthiest portion of the book, and is illustrated with twelve hand-colored lithographic plates. From the author's pespective, however, the most important section is that which describes his four proprietary herbal remedies: the *Magic Assimilant* (for "fits, indigestion, dyspepsia, liver complaints, mismenstruation, etc."), the *Acacian Balsam* (for consumption and diseases of the lungs), the *Renovating Pill* ("for keeping the bowels in motion, and cleansing the system of all impurities"), and the *Ethereal Ointment* (a "pain eradicator"). All were available by mail order. Brown learned the apothecary business from his father in Jersey City during the 1850s.

433. BROWN, Oliver Phelps.
The complete herbalist; or, the people their own physicians, by the use of nature's remedies; describing the great curative properties found in the herbal kingdom. A new and plain system of hygienic principles. Together with comprehensive essays on sexual philosophy, marriage, divorce, &c. . . . Jersey City, N.J.: Published by the author, [c1872].
    504 p., [1] leaf of plates : ill., port. ; 19 cm.

This edition of *The complete herbalist* has been restructured into three parts. The first consists of a description of medicinal plants, and includes the section on hygiene. Part II is a greatly expanded catalog of diseases. Many more disorders are described, and the author has obviously consulted standard medical texts in its composition. Part III, "The Philosophy of the Sexes," is devoted more to marriage and sexual ethics than reproductive physiology. "Dr. O. Phelps Browns Standard Herbal Remedies" has been relegated to the end of the text, but includes an expanded list of the compounds and fluid extracts advertized in earlier editions. Steel-engraved frontispiece portrait of author.

434. BROWN, Oliver Phelps.
The complete herbalist; or, the people their own physicians, by the use of nature's remedies; describing the great curative properties found in the herbal kingdom. A new and plain system of hygienic principles, together with comprehensive essays on sexual philosophy, marriage, divorce,

&c. . . . Jersey City, N.J.: Published by the author, 1875.
504 p., [1] leaf of plates : ill., port. ; 19 cm.

Steel-engraved frontispiece portrait of author.

435. BROWN, Oliver Phelps.
A treatise on epilepsy or fits . . . A treatise on consumption . . . [London, Eng.], 1871.
46 p., [1] leaf of plates : ill. ; 18 cm.

A brochure promoting Brown's four herbal preparations: the *Restorative Assimilant,* the *Acacian Balsam,* the *Renovating Pill* and the *Herbal Ointment.* Though the names differ somewhat, the description of these compounds is nearly identical to the text of pages 302-23 in the 1867 *Complete herbalist* (#432), where they are also described. The testimonies that follow all bear English addresses and are dated 1865-1867. Brown had apparently transferred operations from Jersey City, N.J. to London from the late 1860s to at least 1871. He provides a Covent Garden address to which correspondence might be sent and where he was available for consultation.

436. BROWN, Ryland Thomas, 1807-1890.
Elements of physiology and hygiene . . . Cincinnati; New York: Van Antwerp, Bragg & Co., [c1872].
vi, 7-286 p. : ill. ; 19 cm.

At the time this work was published, Brown was professor of physiology in the Medical Department of Indiana University. The copyright to the *Elements* was held by the publishing firm of Wilson, Hinkle & Co., who also issued this title under their imprint with an 1872 copyright date. In the 1870s and 1880s, Van Antwerp, Bragg & Co. were the largest publishers of school texts off the East Coast. Spine-title: *Brown's physiology and hygiene.*

437. BROWN, Solyman, 1790-1876.
Dental hygeia, a poem, on the health and preservation of the teeth . . . New-York: Published for the proprietors; Kelley & Fraetas, printers, 1838.
[2], iii, [1], 54 p. ; 19 cm.

Trained at Yale College for the ministry (1813), Brown occupied several pulpits in Con-

necticut before leaving the ministry when his license to preach was not renewed. In 1820 he moved to New York City where for the next twelve years he taught in private schools, wrote (and published) poetry, and became a Swedenborgian minister. In 1832 Brown befriended Eleazar Parmly (1797-1874), with whom he lived and studied dentistry. In 1833 Brown published his best known work, a didactic poem entitled *Dentologia: a poem on the diseases of the teeth and their proper remedies,* with notes by Parmly. Brown and Parmly together established the Society of Surgeon Dentists of the City and State of New York, the world's first dental society (1834); and its succesor: the American Society of Dental Surgeons (1840), the first national dental society.

Dental hygeia, in three cantos, was regarded by some as equal or superior to the *Dentologia* (e.g., B.L. Thorpe, *History of dental surgery,* Ft. Wayne, Ind., 1910, 3:92). Unlike the latter title, it was published only once. In the first canto Brown writes of maternal responsibility for the health of her offspring, and the importance of exercise, food, bathing, clothing, etc. The second canto is "addressed to young ladies" on hygienic topics peculiar to "the fair," including the state and care of their teeth. The final canto is "addresed to young gentlemen" on identical themes, with special admonitions regarding intemperance, tobacco and snuff.

Subsequently, Brown was a frequent contributor to the dental literature (especially the *American journal of dental science,* which he helped found); continued to write poetry; helped establish the Baltimore College of Dental Surgery; was a member of the short-lived Fourieristic commune established in 1844 at Leraysville, Pa.; practiced dentistry in Ithaca, N.Y. and vicinity (1846-50); was involved in dental manufacturing in New York City (1854-60); and between 1862 and 1870 was pastor of a Swedenborgian congregation in Danby, N.Y. Burton Lee Thorpe writes that Brown "took a leading part in the three great events that lifted dentistry from a craft to a profession, i.e., the organization of the first dental college; first dental journal and first dental society, and did a great deal more for dentistry than he has been credited with. He was a useful and ornamental member of the profession and did much to enhance our worth in the estimation of the laity" (3:95).

BROWN, T. A. H. see STEBBINS (#3330).

438. BROWN-SEQUARD, Charles Edouard, 1817-1894.
The "Elixir of life." Dr. Brown-Sequard's own account of his famous alleged remedy for debility and old age, Dr. Variot's experiments, and contemporaneous comments of the profession and the press. To which is prefixed a sketch of Dr. Brown-Sequard's life, with portrait. Edited by Newell Dunbar. Boston: J.G. Cupples Company, publishers, [c1889].
   [4], 119 p., [1] leaf of plates : port. ; 15 cm.

A controversial contribution to the literature of organotherapy by the renowned physiologist. "On June 1, 1889, Brown-Sequard stunned the Société de Biologie in Paris by a brief paper in which he suggested that the testicles contain an active invigorating substance that can be extracted from animals and injected into humans" (M. Aminoff, *Brown-Sequard: a visionary of science,* New York, 1993, p. 163). Brown-Sequard maintained that the debilities of old age result from "a natural series of organic changes and the gradually diminishing action of the spermatic glands" (*Elixir,* p. 22). At age 72, he "prepared a solution in distilled water of testicular blood, seminal fluid, and testicular extract derived from healthy dogs and guinea pigs. After filtering this exotic cocktail, he injected himself with it subcutaneously on some ten occasions over a three week period, and reported . . . a dramatic effect on his various bodily functions. His mental concentration and physical endurance increased, his bowel habits improved, he was able to walk without his customary respites, and the power in the flexor muscles of his forearm became stronger" (Aminoff, p. 164).
   This slender volume contains Dunbar's fifteen-page biographical essay; the text of Brown-Sequard's article in the *Lancet* (1889, 2:105-07) describing his experiments and their results; selections translated from the author's original announcement in the *Comptes rendus hebdomadaires* of the Société de Biologie (1889, v. 41); remarks by other members of the Société in the same publication (including G. Variot's experiments confirming Brown-Sequard's results); and a supplement containing extracts

from both the medical literature and popular press on the "Elixir of Life." Once popular interest was aroused, it did not take long for the nostrum trade to market various rejuvenating extracts and elixirs, with direct or indirect references to Brown-Sequard. Cover-title: *Brown-Sequard's own account of the "Elixir of life."* Wood-engraved frontis. portrait of an elderly Brown-Sequard.

439. BRUSH, Edward Fletcher.
Kumyss . . . [S.l.: s.n., ca. 1901].
   32 p. ; 23 cm.

Originally a fermented drink made from mares' milk by nomads of the Asian steppes, Brush states that since 1876 he has been producing kumyss from cows' milk. He describes its history, physiological action and therapeutic use. Quotations from the medical literature on the the effectiveness of kumyss and a bibliography conclude the pamphlet. In printed wrapper.

440. BRYSON, Charles Lee, b. 1868.
Health and how to get it . . . Racine, Wisconsin: Hamilton-Beach Mfg. Co., [c1912].
   300 p. : ill. ; 20 cm.

Intended as a guide for use with the "New Life" vibrator manufactured by the Hamilton-Beach Manufacturing Company. The several attachments that could be fastened to the shaft of this electronic mechanism provided vibration therapy for 109 conditions and disorders listed alphabetically in the index.

BUCHAN, Alexander Peter see BUCHAN (#486, 487, 503).

441. BUCHAN, William, 1729-1805.
Advice to mothers, on the subject of their own health; and on the means of promoting the health, strength, and beauty of their offspring . . . Philadelphia: Printed and sold by John Bioren, 1804.
   [4], 344 p. ; 20 cm.

First American edition; originally published at London in 1803. This popular treatise on the management of pregnancy, delivery, and the health & rearing of children was issued in four more American editions between 1807 and 1815, again at London in 1811, and was ap-

Fig. 17. "Curing herself of indigestion with the 'New-Life' machine," from Bryson's Health, *#440.*

pended to several American editions of the author's *Domestic medicine*. The *Advice to mothers* was also translated into French (1804) and Italian (1806). The appendix (p. 297-344) consists largely of quotations (with some commentary) from William Cadogan's (1711-1797) *An essay upon nursing, and the management of children* (London, 1748; 1st Am. ed., Boston, 1772).

"I am sure of being listened to with kind attention by the tender and rational mother," Buchan writes in the introduction, "while I am pointing out to her the certain means of preserving her own health, of securing the attachment of the man she holds dear, and of promoting the health, strength, and beauty of her offspring. She will not take alarm at the idea of medical advice, when I tell her that my object is to enable her to do without medicine, and to obtain every desirable end without painful sacrifice" (p. [1]). His remarks reflect the tone of female-oriented medical self-help literature for the rest of the century.

Upon completing his medical studies at Edinburgh, Buchan settled in Yorkshire where in 1759 he was appointed medical officer to the recently opened Foundling Hospital at Ack-

worth. It was there he wrote his M.D. thesis (*De infantum vita conservanda*) and acquired his interest and skill in treating the diseases of women and children displayed in both the *Advice to mothers* and the *Domestic medicine*.

REFERENCES: *Austin* #303.

BUCHAN, William, 1729-1805. *Advice to mothers*, see also **BUCHAN** (#444, 478, 479, 481, 483, 484).

442. BUCHAN, William, 1729-1805.
Buchan's domestic medicine: or the family physician. Designed to render the medical art more generally useful, by showing people what is in their own power; both with respect to the prevention and cure of diseases; chiefly calculated to recommend a proper attention to regimen, and simple medicines . . . Tenth American edition, from the third Edinburgh, enlarged and improved by the author. Hartford: Silas Andrus & Son, 1853.
    xi, [13]-442 p. ; 23 cm.

443. BUCHAN, William, 1729-1805.
Buchan's domestic medicine: or the family physician. Designed to render the medical art more generally useful by showing people what is in their own power; both with respect to the prevention and cure of diseases; chiefly calculated to recommend a proper attention to regimen, and simple medicines . . . Tenth American edition, from the third Edinburgh . . . New York: Derby & Jackson, 1858.
    xii, [13]-442 p. ; 19 cm.

Though considerably smaller in format, this 1858 New York issue is printed from the same plates as the 1853 Hartford edition (#442).

444. BUCHAN, William, 1729-1805.
Dr. Buchan's family medical works: containing the Domestic medicine, enlarged: and the Advice to mothers, on the subject of their own health; and on the means of promoting the health, strength, and beauty of their offspring. [Charleston, S.C.: Printed and sold by John Hoff, at his wholesale and retail bookstore, 1807].
    2 v. in 1 (xxxi, [32]-482; [i-iv], vi, [7]-120 p.) ; 22 cm. (8vo)

The title for this composite volume is taken from the half-title; and the imprint from the separate title-pages for the *Domestic medicine* and *Advice to mothers* that consitute these "family medical works." Though the majority of recorded copies contain both works, there is evidence that each was also issued separately in 1807 (e.g., the Atwater copy of each title was acquired separately).

The title-page to the *Domestic medicine* identifies it as the "The first Charleston edition, enlarged, from the author's last revisal." Hoff's "Advertisement" states: "The present edition of the "*DOMESTIC MEDICINE*," claims no other pre-eminence over former impressions, than that it is a copy of the Author's last revisal . . . The articles which have been added to this . . . whether original or selected, will no doubt enhance its value. The new matter in the body of the work, is marked with a *SECTION* [i.e., a symbol following the chapter title] and that of the notes will readily be distinguished by the letters, A.E. [i.e., American editor?]" (p. [iii]). Considerable additions were made to the text of this Charleston edition. In addition to Buchan's "Observations on diet," they include entirely new chapters on yellow fever (neither Griffits' nor Cathrall's versions) and cow-pox, as well as chapter sub-sections on "aliment," "locked-jaw," and burns.

Atwater copy lacking pp. 371-74. The collective half-title for this 1807 edition in the Atwater collection appears between pages [iv-v] of the *Advice to mothers*. In most copies, it would have been bound immediately before the title-page of the *Domestic medicine*.

REFERENCES: *Austin* #305.

445. BUCHAN, William, 1729-1805.

Domestic medicine; or, the family physician: being an attempt to render the medical art more generally useful, by shewing people what is in their own power both with respect to prevention and cure of diseases. Chiefly calculated to recommend a proper attention to regimen and simple medicines . . . Edinburgh: Printed by Balfour, Auld, and Smellie, M,DCC,LXIX. [1769].

xv, [1], 624 p. ; 21 cm. (8vo)

First edition. No English-language handbook of popular medicine and hygiene enjoyed greater popularity or a longer publishing history than Buchan's *Domestic medicine*, which ap-

peared in at least 70 English, Scottish and Irish editions through 1846, and was published in the United States in more than 60 editions and issues between the first Philadelphia edition of 1771 and the outbreak of the Civil War. Buchan's treatise was also translated into Dutch, French (#501), German, Italian (#499), Russian, Spanish (#500) and Swedish.

A native of Ancrum in Roxburghshire, Buchan went to Edinburgh with the intention of studying divinity, but chose medicine instead. He left Edinburgh after nine years to practice in Yorkshire, and returned to Edinburgh in 1766. Three years later the first edition of the *Domestic medicine* was published. In 1772 Buchan was made a fellow of the Royal College of Physicians of Edinburgh and hoped to succeed the elder Gregory to the chair of the institutes of medicine at the University when the latter died in 1773. Buchan moved to London in 1778 when Gregory's son was appointed to the chair instead. Buchan gained a large practice in London, where he remained until his death.

Buchan was assisted in authorship of the *Domestic medicine* by the Edinburgh printer William Smellie (1740-1795), whose name appears on the title-page of the first edition only (among the printers), and never thereafter. Buchan had entertained the idea of a popular medical treatise years before its actual publication. Early on in the project he turned to Smellie because of the latter's experience in the printing trade and his longstanding interest in botany and medicine. According to Smellie's son, his father not only read the proofs of Buchan's treatise, but entirely rewrote it. C.J. Lawrence has concluded that "it seems likely that Buchan wrote an extensive domestic medicine which Smellie compressed, corrected, and to some extent rewrote" ("William Buchan: medicine laid open," *Med. Hist.*, 1975, 19:22).

Smellie was born in Edinburgh and was apprenticed to a printer at age twelve. "Printing remained his trade for the rest of his life, notwithstanding several attempts to enter other fields. In 1760, with Buchan and others, he was a founding member of a Newtonian Society at Edinburgh. A keen botanist, he lectured at the university, and was unfortunate not to attain the Chair of Natural History. He was an original member of the Society of Antiquaries of Scotland. Popularization was in a sense Smellie's occupation. He compiled, edited and contributed to the first edition of the *Encyclopaedia Britannica*

(1771). He wrote a large and successful *Philosophy of natural history* and translated Buffon's *Natural history*. Smellie had a lifelong interest in medicine . . . A typical polymath of the period, his close friends included David Hume, Lord Monboddo, Lord Kames and Robert Burns" (*ibid.,* p. 21).

The *Domestic medicine* is a product of the intellectual climate of Edinburgh in the latter half of the 18th-century. Knowledge was regarded not merely as the attainment of a learned elite, but as essential for the moral and material progress of all classes of society. Interest in such fields as natural philosophy extended beyond University faculty and students to include an active group of educated laymen. Popular education (e.g., learned societies open to non-academic membership, public lectures, publications) became a common mechanism for disseminating both practical knowledge and reforming ideals. It was in this spirit that Buchan wrote, "If medicine be a rational science, and founded in nature, it will never lose its reputation by being exposed to public view. If it not be able to bear the light, it is high time that it were exploded" (p. [iii]).

Buchan combined in a single volume the traditional manual on hygiene (represented in the 18th century by the earlier works of Cheyne [#632, 633] and Tissot [#3534-3542]), and a treatise on specific diseases and their treatment. "Both these factors resulted in a comprehensiveness unrivalled by most earlier popular works" (Lawrence, p. 27). The *Domestic medicine* is divided into two parts. Part I is devoted to hygiene (i.e., diet, air, exercise, sleep, the consequences of "the passions" on health), etc. Part II is a compendium of disorders that provides descriptions of each disease, as well as their causes, symptoms and treatment (both by regimen and medicaments). Downplaying some of the more vigorous therapeutics of his colleagues, Buchan emphasizes the healing power of nature, a factor that must greatly have contributed to the book's popularity.

Charles Rosenberg speculates that, "the purchasers of Buchan's *Domestic medicine* represented a cross-section of the literate, servant-employing, and self-consciously improving middle-orders— men and women who would think twice before spending hard-earned shillings and pounds on a physician, and who sought a rational guide both to prudent domestic practice and to a moral and health-ensuring regimen. The growth in num-

bers and self-consciousness of this new class may well explain the enormous success of Buchan's guide and the specific shape of the genre it created" ("Medical text and social context," *Bull. hist. med.,* 1983, 57:26).

Rosenberg describes the balance that Buchan sought to maintain between domestic and professional practice: "Buchan was aware that beneath the level of formal academic medicine was an elaborate and regionally-differentiated tradition of folk and domestic medicine . . . Buchan adopted a revealingly ambiguous position in regard to household practice; he never contemplated its entire suppression, but sought instead to limit the boundaries of lay practice and define the physician's relation to it" (*ibid.,* p. 27).

446. BUCHAN, William, 1729-1805.
Domestic medicine: or, a treatise on the prevention and cure of diseases by regimen and simple medicines . . . The second edition, with considerable additions. London: Printed for W. Strahan; T. Cadell . . . and A. Kincaid & W. Creech, and J. Balfour, at Edinburgh. MDCCLXXII. [1772]
xxxvi, 758, [2] p. ; 21 cm. (8vo)

WILLIAM BUCHAN M.D.

Fig. 18. William Buchan.

This second edition provides a new introduction in which Buchan argues for "laying open medicine" to the lay public, as well as wholly new chapters on occupational disorders ("Of artificers") and venereal diseases. The text of nearly every chapter has been expanded and/or revised.

447. BUCHAN, William, 1729-1805.
Domestic medicine; or, the family physician: being an attempt to render the medical art more generally useful, by shewing people what is in their own power both with respect to the prevention and cure of diseases. Chiefly calculated to recommend a proper attention to regimen and simple medicines . . . To which is added, Dr. Cadogan's Dissertation on the gout. Philadelphia: Printed by John Dunlap, for R. Aitken . . . M.DCC.LXXII. [1772]
    vii, [5], 368, iv, 39, [1] p. ; 20 cm. (8vo)

Like the first Philadelphia edition of 1771, this second American edition is based on the text of the first Edinburgh edition (#445). Appended to this edition is the full text of William Cadogan's *Dissertation on the gout*. Although this printing of the *Dissertation* has a separate title-page (Aitken's imprint; no date), collation and paginaton, it does not appear to have been issued apart from Buchan.
    REFERENCES: *Austin* #308.

448. BUCHAN, William, 1729-1805.
Domestic medicine; or, the family physician: being an attempt to render the medical art more generally useful, by shewing people what is in their own power with respect to the prevention and cure of diseases. Chiefly calculated to recommend a proper attention to regimen and simple medicines . . . The second American edition, with considerable additions, by the author. Philadelphia: Printed by Joseph Crukshank, for A. Aitken, at his book-store . . . M.DCC.LXXIV. [1774]
    xxiv, 461 p. ; 20 cm.

This second American edition is a reprint of the 1772 second London edition (#446).
    REFERENCES: *Austin* #310.

449. BUCHAN, William, 1729-1805.
Domestic medicine; or, the family physician: being an attempt to render the medical art more generally useful, by shewing people what is in their own power both with respect to the prevention and cure of diseases. Chiefly calculated to recommend a proper attention to regimen and simple medicines . . . The third American edition, with considerable additions, by the author. Norwich [CT]: Printed by John Trumbull, for Robert Hodge, J.D. M'Dougall and William Green, in Boston, M.DCC.LXXVIII. [1778]
    [2], xiv, [4], [17]-436 p. ; 18 cm. (8vo)

This third American edition is a reprint of the second London edition, minus the dedication to Sir John Pringle.
    REFERENCES: *Austin* #311.

450. BUCHAN, William, 1729-1805.
Domestic medicine: or, a treatise on the prevention and cure of diseases by regimen and simple medicines. With an appendix, containing a dispensatory for the use of private practitioners . . . Carefully corrected from the latest London edition, to which is now added, a complete index. Philadelphia: Printed for Joseph Crukshank, Robert Bell, and James Muir . . . and for Robert Hodge, of New-York. M.DCC.LXXXIV. [1784]
    xxiv, [25]-540 p. ; 20 cm. (8vo)

In the preface Buchan acknowledges the influence of other authors, and takes the opportunity to make an important distinction between the *Domestic medicine* and previously published works on the topic, "Those to whom I have been most obliged were, Ramazzini, Arbuthnot, and Tissot [#3534-3542]; the last of which, in his *Avis au peuple*, comes the nearest to my own views of any author which I have seen. Had the Doctor's plan been as complete as the execution is masterly, we should have had no occasion for any new treatise of this kind soon; but by confining himself to the acute diseases, he has, in my opinion, omitted the most useful part of his subject. People in acute diseases may sometimes be their own physicians, but in the chronic the cure must ever depend chiefly upon the patient's own endeavours. The Doctor has also passed over *Prophylaxis*, or preventive part

of medicine . . . though it is certainly of the greatest importance in such a work" (p. viii-ix).
REFERENCES: *Austin* #312.

451. BUCHAN, William, 1729-1805.
Domestic-medicine: or, a treatise on the prevention and cure of diseases by regimen and simple medicines. With an appendix, containing a dispensatory for the use of private practitioners . . . The eighth edition; corrected and enlarged. To which is now added, a complete and copious index. London: Printed for W. Strahan; T. Cadell . . . and J. Balfour, and W. Creech, at Edinburgh, MDCCLXXXIV. [1784]
xxxvi, 767, [37] p. ; 21 cm. (8vo)

452. BUCHAN, William, 1729-1805.
Domestic medicine: or, a treatise on the prevention and cure of diseases by regimen and simple medicines. With an appendix, containing a dispensatory for the use of private practitioners . . . The ninth edition; to which is now added, an additional chapter on cold bathing, and drinking the mineral waters. London: Printed for A. Strahan; T. Cadell . . . and J. Balfour, and W. Creech, at Edinburgh. MDCCLXXXVI. [1786]
[4], xxxii, 767, [37] ; 21 cm. (8vo)

The ninth London edition appears to be the first that includes the chapter on bathing and mineral waters.

453. BUCHAN, William, 1729-1805.
Domestic medicine: or, a treatise on the prevention and cure of diseases by regimen and simple medicines. With an appendix, containing a dispensatory for the use of private practitioners . . . The eleventh edition, corrected and improved. Hartford: Printed by Nathaniel Patten. M.DCC.LXXXIX. [1789]
xxxii, 780, [36] p. ; 18 cm.

One wonders how the publisher enumerated this Hartford edition the 11th? It is not based on the 11th London edition, which was published the following year (#454). Assuming Austin's *Early American medical imprints* to be complete, this Hartford edition is the seventh edition of Buchan to be published with an American imprint. It includes an addition to

the fifty-four chapters of the previous American edition, i.e., "Chap. LV. Cautions concerning cold bathing, and drinking the mineral waters," which first appeared in the 9th London edition of 1786 (#452).
REFERENCES: *Austin* #313.

454. BUCHAN, William, 1726-1805.
Domestic medicine: or, a treatise on the cure and prevention of diseases by regimen and simple medicines. With an appendix, containing a dispensatory for the use of private practitioners . . . The eleventh edition. London: Printed for A. Strahan; T. Cadell . . . and J. Balfour, and W. Creech, at Edinburgh. MDCCXC. [1790]
xxxvi, 712, [36] p. ; 21 cm (8vo)

Atwater copy lacking pages 533-712.

455. BUCHAN, William, 1729-1805.
Domestic medicine: or, a treatise on the prevention and cure of diseases by regimen and simple medicines. With an appendix, containing a dispensatory for the use of private practitioners . . . The thirteenth edition. Philadelphia: Printed and sold by J. Crukshank, W. Spotswood, W. Young, T. Dobson, M. Carey, and H. & P. Rice, 1793.
xxxvi, 712, [36] p. ; 20 cm. (8vo)

REFERENCES: *Austin* #314.

456. BUCHAN, William, 1729-1805.
Domestic medicine: or, a treatise on the prevention and cure of diseases by regimen and simple medicines. With an appendix, containing a dispensatory for the use of private practitioners . . . The fourteenth edition. Boston: Printed by Joseph Bumstead, for James White . . . and Ebenezer Larkin . . . MDCCXCIII. [1793]
xx, 484, [40] p. ; 21 cm.

REFERENCES: *Austin* #315.

457. BUCHAN, William, 1729-1805.
Domestic medicine: or, a treatise on the prevention and cure of diseases, by regimen and simple medicines. With an appendix, containing a dispensatory for the use of private practitioners . . . Revised and adapted to the diseases and climate of the

United States of America, by Samuel Powel Griffitts . . . Philadelphia: Printed by Thomas Dobson . . . 1795.

    xxxi, [1], 757, [3] p. ; 21 cm. (8vo)

Each of the four editions of Buchan published by Thomas Dobson between 1795 and 1809 (#463, 474, 477) were edited by Samuel Powel Griffitts, who followed Buchan's editorial lead in making most of his additions or comments in the form of footnotes. The exception is the chapter on yellow fever, which is unique to the Dobson-Grifitts editions.

    A native of Philadelphia, Samuel Powel Griffitts (1759-1826) began his medical studies at the College of Philadelphia, and continued his training in Paris, Montpelier, London and Edinburgh (1781-84). Returning to Philadelphia, Griffitts became a member of the American Philosophical Society; founded the Philadelphia Dispensary (1786); was an original member of the College of Physicians; and was appointed professor of materia medica at the University of Pennsylvania (1792-96). He was also one of the original members of the Vaccine Society (1809), a representative to the national pharmacopeia convention of 1820, and an active opponent of slavery. Griffitts' list of publications is modest. He was one of the editors of the *Eclectic repertory,* in which several of his own papers were published.

    REFERENCES: *Austin* #317.

458. BUCHAN, William, 1729-1805.

Domestic medicine: or, a treatise on the prevention and cure of diseases by regimen and simple medicines. With an appendix containing a dispensatory for the use of private practitioners . . . The eighteenth edition. New-London: Printed by Thomas C. Green, and sold at his printing-office and book-store, 1795.

    xxvi, [27]-533, [39] p. ; 20 cm. (8vo)

REFERENCES: *Austin* #316.

459. BUCHAN, William, 1729-1805.

Domestic medicine: or, a treatise on the prevention and cure of diseases, by regimen and simple medicines; with an appendix, containing a dispensatory for the use of private practitioners . . . Adapted to the climate and diseases of America, by Isaac

Cathrall. Philadelphia: Printed by Richard Folwell, 1797.

    512 p. ; 21 cm. (8vo)

"Among the motives which have induced me to republish Dr. Buchan's book on Domestic Medicine, is that of rendering it more extensively useful to the inhabitants of the United States of America, by accommodating it to their diseases" (editor's *Pref.,* p. [3]). Cathrall has made abundant commentary throughout Buchan's text in the form of footnotes (signed 'I.C.'). He also added ch. XXIII on yellow fever, in which he discusses the epidemic at Philadelphia, provides a detailed description of its symptoms and pathology, and describes treatment.

    A native of Philadelphia, Isaac Cathrall (1763-1819) began the study of medicine under John Redman, and continued his medical education in London, Edinburgh and Paris before returning to his native city in 1793. Cathrall distinguished himself by his conduct during the yellow fever epidemics which regularly afflicted Philadelphia in the final decade of the 18th century. In order to understand the aetiology and pathology of the disease, he analyzed the black vomit of yellow fever victims; and in order to determine the fever's contagiousness, injected himself subcutaneously with the blood and vomit of the afflicted. His findings were published in several brief but notable treatises.

    The present work is one of seven issues of Cathrall's edition of the *Domestic medicine* printed by Richard Folwell in 1797, six of which are cited in Robert Austin's *Early American medical imprints* (#s 318-23). These issues differ only in their imprints. The Cathrall edition of Buchan was published again by Folwell at Philadelphia in 1799 (#468) and 1801 (#470).

    REFERENCES: *Austin* #318.

460. BUCHAN, William, 1729-1805.

Domestic medicine: or, a treatise on the prevention and cure of diseases, by regimen and simple medicines: with an appendix, containing a dispensatory for the use of private practitioners . . . Adapted to the climate and diseases of America, by Isaac Cathrall. Philadelphia: Printed by Richard Folwell, for James Carey, 1797.

    512 p. ; 22 cm. (8vo)

Austin's *Early American medical imprints, 1668-1820* lists six issues of the 1797 Philadelphia edition of Buchan with varying imprints, all of which were printed by Richard Folwell (#318-323). Austin does not record this issue printed by Folwell for James Carey, however. The Atwater copy of the Carey issue is lacking the preface and table of contents (signature 'A', p. [3]-8).

461. BUCHAN, William, 1729-1805.
Domestic medicine: or, a treatise on the prevention and cure of diseases, by regimen and simple medicines: with an appendix, containing a dispensatory for the use of private practitioners . . . Adapted to the climate and diseases of America, by Isaac Cathrall. Philadelphia: Printed by Richard Folwell, for Joseph & James Crukshank, 1797.
    512 p. ; 22 cm. (8vo)

REFERENCES: *Austin* #320.

462. BUCHAN, William, 1729-1805.
Domestic medicine: or, a treatise on the prevention and cure of diseases, by regimen and simple medicines: with an appendix, containing a dispensatory for the use of private practitioners . . . Adapted to the climate and diseases of America, by Isaac Cathrall. Philadelphia: Printed by Richard Folwell, for William Young, 1797.
    512 p. ; 21 cm. (8vo)

Not in Austin.

463. BUCHAN, William, 1729-1805.
Domestic medicine; or, a treatise on the prevention and cure of diseases, by regimen and simple medicines. With an appendix, containing a dispensatory for the use of private practitioners . . . Revised and adapted to the diseases and climate of the United States of America, by Samuel Powel Griffitts . . . Second edition. Philadelphia: Printed by Thomas Dobson, 1797.
    xxxi, [1], 757, [3] p. ; 22 cm.

REFERENCES: *Austin* #325.

464. BUCHAN, William, 1729-1805.
Domestic medicine: or, a treatise on the prevention and cure of diseases by regimen and simple medicines. With an appendix, containing a dispensatory for the use of private practitioners . . . The fifteenth edition; to which is added, observations concerning the diet of the common people, recommending a method of living less expensive, and more conducive to health, than the present. London: Printed for A. Strahan, and T. Cadell jun. and W. Davies, (successors to Mr. Cadell) . . . and J. Balfour, and W. Creech, Edinburgh, MDCCXCVII. [1797].
    xl, 746, [38] p. ; 21 cm.

465. BUCHAN, William, 1729-1805.
Domestic medicine: or, a treatise on the prevention and cure of diseases by regimen and simple medicines. With an appendix, containing a dispensatory for the use of private practitioners . . . The twentieth edition, containing all the improvements. Waterford [N.Y.]: Printed by and for James Lyon & Co., 1797.
    xxiv, 475, [21] p. ; 21 cm. (8vo)

Waterford is a village on the Hudson River located ten miles north of Albany.
REFERENCES: *Austin* #326.

467. BUCHAN, William, 1729-1805.
Domestic medicine: or, a treatise on the prevention and cure of diseases, by regimen and simple medicines. With an appendix, containing a dispensatory for the use of private practitioners . . . The first Vermont edition, containing all the improvements. Fairhaven: Printed by and for James Lyon, 1798.
    xxiv, 476, [20] p. ; 21 cm. (8vo)

The first Vermont edition is based on the 11th London edition (#454).
REFERENCES: *Austin* #327.

468. BUCHAN, William, 1729-1805.
Domestic medicine: or, a treatise on the prevention and cure of diseases, by regimen and simple medicines: with an appendix, containing a dispensatory for the use of private practitioners . . . Adapted to the climate and diseases of America, by Isaac Cathrall. Philadelphia: printed by Richard Folwell. 1799.
    512 p. ; 22 cm. (8vo)

A re-issue of the sheets of Folwell's 1797 edition (#459) with the exception of gathering 'A,' in which the type has been entirely reset.

REFERENCES: *Austin* #328.

469. BUCHAN, William, 1729-1805.
Domestic medicine: or, a treatise on the prevention and cure of diseases by regimen and simple medicines. With an appendix, containing a dispensatory for the use of private practitioners. To which are added, Observations on the diet of the common people; recommending a method of living less expensive, and more conducive to health, than the present . . . The seventeenth edition. London: Printed by A. Strahan, printers . . . for A. Strahan; T. Cadell Jun. and W. Davies . . . and J. Balfour, and W. Creech, Edinburgh, 1800.

xl, 746, [38] p. ; 21 cm. (8vo)

470. BUCHAN, William, 1729-1805.
Domestic medicine: or, a treatise on the prevention and cure of diseases, by regimen and simple medicines: with an appendix, containing a dispensatory for the use of private practitioners . . . Adapted to the climate and diseases of America, by Isaac Cathrall. {Second edition.} Philadelphia: Printed by R. Folwell, 1801.

512 p. ; 20 cm. (8vo)

REFERENCES: *Austin* #329.

471. BUCHAN, William, 1729-1805.
Domestic medicine: or, a treatise on the prevention and cure of diseases by regimen and simple medicines. With an appendix, containing a dispensatory for the use of private practitioners. To which are added, Observations on the diet of the common people; recommending a method of living less expensive, and more conducive to health, than the present . . . The seventeenth edition. Halifax [N.C.]: Printed and sold by Abraham Hodge, 1801.

[4], xxii, 445, [33] p. ; 21 cm. (8vo)

In 1797 *Observations concerning the diet of the common people* (p. 376-95) was published at London both separately and as part of the 15th edition of the *Domestic medicine* (#464). This Halifax edition appears to be the first American edition in which the *Observations* is included in the *Domestic medicine*. In his introductory remarks to this essay Buchan writes, "The late distress of the poor has called forth many publications intended for their relief . . . The following observations . . . are intended to recommend a plan of living, which will render the people less dependent on bread and animal food, for their subsistence, & consequently not liable to suffer from a scarcity or dearth of either of these articles in the future" (p. 377). Buchan is interested in more than the economics of feeding the poor. He asserts that too much flesh or bread is harmful to health, and recommends a diet incorporating more rice, corn, and legumes, as well as fruits and "roots" (i.e., potatoes, parsnips, turnips, &c.).

REFERENCES: *Austin* #330

472. BUCHAN, William, 1729-1805.
Domestic medicine: or, a treatise on the prevention and cure of diseases by regimen and simple medicines. With an appendix, containing a dispensatory for the use of private practitioners. The eighteenth edition, to which are now added, some important observations concerning sea-bathing, and the use of mineral waters; with many other additions . . . London: Printed for A. Strahan; T. Cadell and W. Davies . . . and J. Balfour, and W. Creech, Edinburgh. 1803.

xliv, 750, [2] p. ; 21 cm. (8vo)

The new chapter entitled "Of cold bathing, and drinking the mineral waters" (c. lv) is followed by Buchan's "Observations concerning the diet of the common people" (c. lvi). In the "Advertisement to the eighteenth edition," Buchan addresses pirated editions of this work, noting that in order "to evade the law, by acting under the sanction of an old statute for limiting the period of copyright, they have reprinted the *early copies* of my book, published between thirty and forty years ago, which, to say nothing of inaccuracies, did not contain above half the matter inserted in the later editions" (p. ix). Half-title: *Dr. Buchan's domestic medicine.*

473. BUCHAN, William, 1729-1805.
Domestic medicine: or, a valuable treatise on the prevention and cure of diseases by

regimen and simple medicines. With an appendix, containing a new dispensatory, for the use of private practitioners. To which are added, Observations on the diet of the common people; recommending a method of living less expensive, and more conducive to health, than the present . . . Leominster: Printed by Adams & Wilder, for Isaiah Thomas, Jun. Sold by him in Worcester, and by Thomas & Whipple, in Newburyport., March, 1804.

xxiv, [17]-480 (i.e. 484) p. ; 21 cm. (8vo)

REFERENCES: *Austin* #331.

474. BUCHAN, William, 1729-1805.
Domestic medicine; or, a treatise on the prevention and cure of diseases, by regimen and simple medicines. With an appendix, containing a dispensatory for the use of private practitioners . . . Revised and adapted to the diseases and climate of the United States of America, by Samuel Powel Griffitts . . . A new edition. Philadelphia: Printed by A. Bartram, for Thomas Dobson, 1805.

xxxi, [1], 757, [3] p. ; 22 cm.

The third of Dobson's issues of Buchan is a reissue of the sheets of the 1797 2nd edition (#463).

REFERENCES: *Austin* #332.

475. BUCHAN, William, 1729-1805.
Domestic medicine: or, a treatise on the prevention and cure of diseases, by regimen and simple medicines. With an appendix, containing a dispensatory for the use of private practitioners . . . The thirty-sixth edition. Arbroath: Printed by J. Findlay, 1806.

xxxviii, 430 p. ; 17 cm.

Arbroath is a port in Scotland on the North Sea above the Firth of Tay.

476. BUCHAN, William, 1729-1805.
Domestic medicine: or, a treatise on the prevention and cure of diseases by regimen and simple medicines. With an appendix, containing a dispensatory for the use of private practitioners . . . To which is now

first added the following new treatises: sea-bathing—mineral-waters—cow-pox—recovery of drowned persons—ruptures—medical electricity—child-birth—diet. By J. Baker, M.D. A new edition, greatly improved and enlarged . . . London: Printed by R. Butters, 1808.

xxiv, 606, [2], 73, [35] p. ; 22 cm.

Butters first issued this edition of Buchan in 1808 and again in 1809. It was edited by J. Baker, who writes in his preface (dated London, August, 1808): "To render the present posthumous edition more valuable, the editor has carefully revised the whole, and made such amendments and additions, as the late improvements in medical science, and his own extensive practice, have put in his power, endeavouring as much as possible to imitate the perspicuity and simplicity of the author. In this edition, therefore, will be found new treatises on sea-bathing—mineral waters—inoculation from the cow-pox—on the recovery of persons apparently drowned—ruptures—midwifery—electricity" (p. [iii]). These "additions" replace identically titled chapters in earlier editions of Buchan's text, and in some instances employ similar wording. The provision of copper-plate illustration is indicated both on the title-page and in the preface. Illustrations appear neither in this nor the 1809 issue, however.

477. BUCHAN, William, 1729-1805.
Domestic medicine: or, a treatise on the prevention and cure of diseases, by regimen and simple medicines. With an appendix, containing a dispensatory for the use of private practitioners . . . Revised, and adapted to the diseases and climate of the United States of America, by Samuel Powel Griffits . . . Philadelphia: Published by Thomas Dobson; Fry and Kammerer, printers, 1809.

xxxi, [1], 252, 251-757, [3] p. ; 21 cm. (8vo)

The fourth and final issue of Buchan from the press of Thomas Dobson (#457, 463, 474). Dobson's 1809 issue includes a new section on vaccination (p. 241-52) appended to the chapter on smallpox. These new leaves added to the sheets of the 1805 issue account for the discrepancy in pagination between the two issues.

478. BUCHAN, William, 1729-1805.
Domestic medicine: or, a treatise on the prevention and cure of diseases, by regimen and simple medicines. With an appendix, containing a dispensatory for the use of private practitioners. To which are added, Observations on diet; recommending a method of living less expensive, and more conducive to health, than the present. Also, Advice to mothers, on the subject of their own health; and of the means of promoting the health, strength, and beauty of their offspring . . . New, correct edition, enlarged—from the author's last revisal. Boston: Printed for Joseph Bumstead, 1809.
     629, [1] p. ; 22 cm.

Bumstead's three issues of the *Domestic medicine* (see also #479, 483) were based on the augmented text of the 1807 Charleston edition (#444), which added two new chapters, several chapter sub-sections, and footnotes to Buchan's text. Bumstead includes Buchan's *Advice to mothers* in this work, which he also issued separately in 1809 (*Austin* #304). Appended to the *Advice* are extensive quotations from William Cadogan's *An essay upon nursing, and the management of children* (p. [613]-629).
     REFERENCES: *Austin* #334.

479. BUCHAN William, 1729-1805.
Domestic medicine: or, a treatise on the prevention and cure of diseases, by regimen and simple medicines. With an appendix, containing a dispensatory for the use of private practitioners. To which are added, Observations on diet; recommending a method of living less expensive, and more conducive to health, than the present. Also, Advice to mothers, on the subject of their own health; and the means of promoting the health, strength, and beauty of their offspring . . . New, correct edition, enlarged—from the author's last revisal. Boston: Printed for Joseph Bumstead, 1811.
     629, [3] p. ; 23 cm.

The second of three issues of Buchan published by Joseph Bumstead between 1809 (#478) and 1813 (#483).
     REFERENCES: *Austin* #335.

480. BUCHAN, William, 1729-1805.
Domestic medicine; or, a treatise on the prevention and cure of diseases, by regimen and simple medicines. With an appendix, containing a dispensatory for the use of private practitioners. To which are added, Observations on diet; recommending a method of living less expensive, and more conducive to health, than the present . . . Enlarged from the author's last revisal. New-York: Published by Richard Scott, 1812.
     xxvii, [28]-436 p. : 17 cm.

Scott's edition of Buchan is based on the version issued at Boston by Joseph Bumstead between 1809 and 1813 (#478, 479, 483), except that it does not include Buchan's *Advice to mothers.*
     REFERENCES: *Austin* #336.

481. BUCHAN, William, 1729-1805.
Domestic medicine, or, a treatise on the prevention and cure of diseases by regimen and simple medicines. With an appendix, containing a dispensatory for the use of private practitioners. To which are added, Observations on diet; recommending a method of living less expensive, and more conducive to health than the present . . . New-York: Published by George Lindsay, 1812.
     xxxii, [33]-482, vi, [7]-118 p. ; 22 cm.

Lindsay's edition of Buchan is based on the version issued by Joseph Bumstead between 1809 and 1813 (#478, 479, 483). Though not indicated on the title-page, this edition includes the entire text of Buchan's *Advice to mothers.* Although the *Advice to mothers* has a separate title-page and pagination, its collation is continuous with the rest of the text. Appended to the *Advice* are extensive quotations from William Cadogan's *An essay upon nursing, and the management of children.* The Atwater copy apparently was not issued with the half-title present in some copies, i.e., *Dr. Buchan's family medicial works: containing the Domestic medicine enlarged, and the Advice to mothers.* This edition is not listed in Austin's *Early American medical imprints.* The sheets of Lindsay's Buchan were reissued in 1815 with the imprint of Richard Scott (#484).

482. BUCHAN, William, 1729-1805.
Domestic medicine; or, a treatise on the prevention and cure of diseases, by regimen and simple medicines: with an appendix, containing a dispensatory for the use of private practitioners . . . Edinburgh: Printed by J. & C. Muirhead, for W. Somerville, A. Fullarton, J. Blackie & Co., Glasgow, 1813.
xxiv, [25]-656 p. ; 21 cm.

483. BUCHAN, William, 1729-1805.
Domestic medicine: or, a treatise on the prevention and cure of diseases, by regimen and simple medicines. With an appendix, containing a dispensatory for the use of private practitioners. To which are added, Observations on diet; recommending a method of living less expensive, and more conducive to health, than the present. Also, Advice to mothers, on the subject of their own health; and the means of promoting the health, strength, and beauty of their offspring . . . New, correct edition, enlarged—from the author's last revisal. Boston: Printed for Joseph Bumstead, 1813.
629, [3] p. ; 22 cm.

The third and final issue of Buchan published by Bumstead. Atwater copy lacking pp. 345-348.
REFERENCES: *Austin* #337.

484. BUCHAN, William, 1729-1805.
Domestic medicine, or, a treatise on the prevention and cure of diseases by regimen and simple medicines. With an appendix, containing a dispensatory for the use of private practitioners. To which are added, Observations on diet; recommending a mode of living less expensive, and more conducive to health than the present . . . New-York: Published by Richard Scott, 1815.
xxxii, [33]-482, vi, 7-118 p. ; 22 cm.

A reissue of the 1812 New York edition published by George Lindsay (#481) with a new title-page. It includes Buchan's *Advice to mothers*, though this is not indicated on the title-page. The title-page to the *Advice* is dated 1812 and bears Scott's imprint. It does not appear to have been issued separately that year, nor was it included in Scott's 1812 edition of the *Domestic medicine* (#480).
REFERENCES: *Austin* #338.

485. BUCHAN, William, 1729-1805.
Domestic medicine; or, a treatise on the prevention and cure of diseases, by regimen and simple medicines: containing observations on the comparative advantages of vaccine inoculation, with instructions for performing the operation: an essay enabling ruptured persons to manage themselves; and a family herbal . . . To which are added, such useful discoveries in medicine and surgery as have transpired since the demise of the author. London: Printed and published by W. Lewis, 1818.
[2], 511 p., [1] leaf of plates : ill., port. ; 21 cm.

Frontispiece portrait of author.

486. BUCHAN, William, 1729-1805.
Domestic medicine; or, a treatise on the cure and prevention of diseases by regimen and simple medicines. Containing a dispensatory for the use of private practitioners . . . With considerable additions, and various notes, by A.P. Buchan . . . To which is added, A family herbal. A new edition, revised and amended, by John G. Coffin . . . Boston: Published by Phelps and Farnham . . . and Nathaniel S. Simpkins, 1825.
viii, 652, [4] p. ; 23 cm.

"The present edition of the Domestic medicine is printed from the twenty-first London edition, published by the author's son" (*Advert.,* p. [iii]).

487. BUCHAN, William, 1729-1805.
Domestic medicine. A treatise on the prevention and cure of diseases, by regimen and simple medicine . . . A new edition. Improved, by W. Buchan, Jun. With remarks on the properties of food, vaccination, electricity, galvanism, bathing, &c. London: Printed for John Bumpus, 1828.
xxiv (i.e., xii), [25]-528 p. ; 14 cm.

488. BUCHAN, William, 1729-1805.
Domestic medicine, or, a treatise on the prevention and cure of diseases, by regimen and simple medicines. With observations on sea-bathing, and the use of mineral waters. To which is annexed, a dispensatory for the use of private practitioners . . . From the twenty-second English edition, with considerable additions, and notes. Exeter [N.H.]: J. & B. Williams, 1828.
    xvii, [18]-495, [1], xlviii p. ; 22 cm.

This the first of seven editions of the *Domestic medicine* published at Exeter, N.H. by J. & B. Williams (1828, 1832, 1834, 1836, 1839, 1842, 1843), all of which were based on the 22nd English edition issued at London in 1826. Extensive alterations and additions had been made to British editions of the *Domestic medicine* by the time of (and with) the publication of the 22nd edition. This edition opens with "Observations on diet," a treatise usually appended to the *Domestic medicine.* "Of children," which normally is the opening chapter of part I, has been moved to the end of the text. Extensive additions have been made throughout, and, "obsolete or irrelevant matter . . . has been expunged, and its place supplied with information of a more useful and more recent nature" (*Pref.,* p. [iii])..

This 22nd London edition of Buchan was edited by J.S. Forsyth, the author of *The mother's medical pocket book* (1824) *The natural and medical dieteticon* (1824) *The new domestic medical manual* (1824), *The new London medical pocket book* (1825), *A practical treatise on diet* (1827) and *A dictionary of diet* (1833).

489. BUCHAN, William, 1729-1805.
Domestic medicine, or, a treatise on the prevention and cure of diseases, by regimen and simple medicines: with observations on sea-bathing, and the use of the mineral waters. To which is annexed, a dispensatory for the use of private practitioners . . . From the twenty-second English edition, with considerable additions, and notes. Exeter [N.H.]: J. & B. Williams, 1832.
    xvii, [18]-495, [1], xlviii p. ; 21 cm.

490. BUCHAN, William, 1729-1805.
Domestic medicine, or, a treatise on the prevention and cure of diseases, by regimen and simple medicines: with observa-

tions on sea-bathing, and the use of mineral waters. To which is annexed a dispensatory for the use of private practitioners . . . From the twenty-second English edition, with considerable additions, and notes. Exeter [N.H.]: J. & B. Williams, 1834.
    xvii, [18]-495, [1], xlviii p. ; 22 cm.

The third Exeter edition of the *Domestic medicine.*

491. BUCHAN, William, 1729-1805.
Domestic medicine; or, a treatise on the prevention and cure of diseases, by regimen and simple medicines: with observations on sea-bathing, and the use of mineral waters. To which is annexed, a dispensatory for the use of private practitioners . . . From the twenty-second English edition, with considerable additions, and notes. Exeter: J. & B. Williams, 1839.
    xvii, [18]-495, [1], xlviii p. ; 21 cm.

The fifth Exeter edition of the *Domestic medicine.*

492. BUCHAN, William, 1729-1805.
Domestic medicine. A treatise on the prevention and cure of diseases, by regimen and simple medicines . . . A new stereotype edition. London: Published by J. Smith, [184-?]
    [6], [iii]-xiii, [14]-581 p., [12] leaves of plates : ill. ; 16 cm.

Illustrated with twelve wood-engraved anatomical plates described in "A compendium of anatomy" (p. [563]-581).

493. BUCHAN, William, 1729-1805.
Domestic medicine, or, a treatise on the prevention and cure of diseases, by regimen and simple medicines: with observations on sea-bathing, and the use of mineral waters. To which is annexed, a dispensatory for the use of private practitioners . . . From the twenty-second English edition, with considerable additions, and notes. Exeter: J. & B. Williams, 1842.
    xvii, [18]-495, [1], xlviii p. ; 22 cm.

The sixth Exeter edition of the *Domestic medicine.*

494. BUCHAN, William, 1729-1805.
Domestic medicine, or, a treatise on the prevention and cure of diseases, by regimen and simple medicines: with observations on sea-bathing, and the use of mineral waters. To which is annexed, a dispensatory for the use of private practitioners . . . From the twenty-second English edition, with considerable addtions, and notes. Exeter [N.H.]: J. & B. Williams, 1843.
xvii, [18]-495, [1], xlviii p. ; 22 cm.

The last of the seven editions of the *Domestic medicine* published at Exeter, N.H. by J. & B. Williams.

495. BUCHAN, William, 1729-1805.
Domestic medicine, or, a treatise on the prevention and cure of diseases, by regimen and simple medicines: with observations on sea-bathing, and the use of mineral waters. To which is annexed, a dispensatory for the use of private practitioners . . . From the twenty-second English edition, with considerable additions, and notes. Boston: Otis, Broaders, and Company, 1846.
xvii, [18]-495, [1], xlviii p. ; 23 cm.

This Boston edition was printed from the same plates used in the printing of the seven Exeter, N.H. editions (#488-491, 493-494) of the *Domestic Medicine*. These plates were used again in the printing of the 1871 Philadelphia edition (#502).

496. BUCHAN, William, 1729-1805.
Domestic medicine; or, a treatise on the prevention and cure of diseases, by regimen and simple medicines: with an appendix, containing a dispensatory for the use of private practitioners . . . A new and enlarged edition containing new treatises on sea-bathing, mineral waters, vaccine inoculation, diet, &c. &c. London: Milner and Sowerby, [186?].
xxiv, 534 p., xii p. of plates, [1] folding plate (col.) : ill. ; 18 cm.

The firm of Milner & Sowerby were active from the late 1850s through the 1860s. The firm published an edition of Buchan dated 1863 with a Halifax imprint (West Riding, Yorkshire),

where this undated London edition was also printed (printer's imprint on p. 534). Spine-title: *Buchan's domestic medicine—A compendium of anatomy*. The twelve pages of wood-engraved plates are much reduced copies of the Vesalian muscle and skeletal figures. The plates are explained in the "Compendium of anatomy" (p. [521]-534) added to this edition.

497. BUCHAN, William, 1729-1805.
Domestic medicine; or, a treatise on the prevention and cure of diseases, by regimen and simple medicines: with an appendix, containing a dispensatory for the use of private practitioners . . . A new and enlarged edition containing new treatises on sea-bathing, mineral waters, vaccine inoculation, diet, &c. &c. London: Milner and Company, [186?].
xxiv, 534 p., xii p. of plates, [1] folding plate (col.) : ill. ; 18 cm.

Spine-title: *Buchan's domestic medicine—A compendium of anatomy*.

498. BUCHAN, William, 1729-1805.
Every man his own doctor; or, a treatise on the prevention and cure of diseases, by regimen and simple medicines . . . To which is added, a treatise on the materia medica; in which the medicinal qualities of indigenous plants are given and adapted to common practice. With an appendix, containing a complete treatise on the art of farriery; with directions to the purchasers of horses; and practical receipts for the cure of distempers incident to horses, cattle, sheep, and swine—To all of which are added, a choice collection of receipts, useful in every branch of domestic life—making in all a complete family directory. New-Haven: Published by Nathan Whiting, 1816.
viii, [9]-464, iv, [5]-144 p. ; 22 cm.

"In this first edition of *Every man his own doctor*, the treatise on the prevention and cure of diseases, as laid down by Dr. Buchan has been followed without any alteration, except the omission of some general observations, which were designed, principally, for physicians; and some articles which have become obsolete." (*Pref.*, p. [iv]). Among the articles omitted from this edition was the chapter on smallpox. The Whiting edition does include the chapters on yellow fever and cow pox that appeared in the Bumstead editions of the *Domestic medicine* (#478, 479, 483).

Added to this edition are "A concise account of the medicinal qualities, of some of the most common indigenous, and naturalized plants, of New England, and the middle states, extracted from Thacher's Dispensatory, and Barton's Collections," p. [417]-464; and "A Complete treatise on the art of farriery." The latter treatise has a separate title-page, collation & pagination, and was also issued separately by Whiting in 1816. Its attribution to Buchan (see Amer. Antiq. Soc., *Dict. cat. of Amer. books pertaining to the 17th through 19th cent.*, 3:101) is false. The *Complete treatise* includes "The Universal receipt book, or complete family directory" (p. [87]-138), a collection of home remedies.

REFERENCES: *Austin* #339.

499. BUCHAN, William, 1729-1805.

Medicina domestica o sia trattato completo di mezzi semplici per conservarsi in salute impedire e risanare le malattie. Opera utile e adattata all'intelligenza di ciascuno di Guglielmo Buchan . . . Tradotta dall'Inglese e arricchita di molte aggiunte ed annotazioni dal sig. Duplanil . . . Edizione seconda italiana riveduta, ricorretta e notabilmente accresciuta su la settima di Londra e la quarta di Parigi . . . In Padova, MDCCLXXXIX. [1789] Nella Stamperia del Seminario appresso Tommaso Bettinelli.

5 v. ; 20 cm. (8vo)

The first Italian edition of Buchan's *Domestic medicine* was published at Naples in 1781-82. There were at least four Italian editions published before this "edizione seconda italiana," which may refer simply to the second edition published at Padua (1st Padua ed., 1783). It is based on the 7th London edition (1781), and translated from the 4th Paris edition (1785), edited and greatly augmented by Jean-Denis Duplanil (*see* #501).

500. BUCHAN, William, 1729-1805.

Medicina doméstica ó tratado del método de precaver y curar las enfermedades con el regimen y medicinas simples, y un apéndice que contiene la farmacopea necesaria para el uso de un particular. Escrito en Inglés por el doctor Jorge Buchan . . . traducido en Castellano por . . . Antonio de Alcedo . . . Madrid: En la imprenta de Ramon Ruiz, 1798.

xl, 256, 253-268, 273-688 p. ; 20 cm. (4to)

Alcedo's translation of Buchan's *Domestic medicine* was published in 1785, 1786, 1792, 1798 and 1818. Antonio de Alcedo (1736-1812) is perhaps best remembered as author of the *Diccionario geografico-historico de las Indies Occidentales o America* (1st ed., Madrid, 1786; English trans., London, 1812-15). It is not inconceivable that his translations of Buchan were brought to Hispanic America as Edinburgh and London editions were brought to English-speaking America.

501. BUCHAN, William, 1729-1805.

Médecine domestique, ou traité complet des moyens de se conserver en santé, et de guérir les maladies par le régime et les remèdes simples. Ouvrage mis à la portée de tout le monde, par G. Buchan . . . Traduit de l'Anglais par J.D. Duplanil . . . Cinquième édition, revue, corrigée et considérablement augmentée, et spécialement d'un article sur la vaccine; de la nouvelle nomenclature chimique, et de la dénomination des nouveaux poids et mesures . . . A Paris: Chez Moutardier, imprimeur-libraire, An. X—1802.

5 v. : port. ; 20 cm.

According to Dezeimeris' *Dictionnaire historique de la médecine* (Paris, 1834, 1:159), the translation of Buchan's *Domestic medicine* by Jean-Denis Duplanil (1740-1802) was first published at Paris in 1775. This 5th edition appears to be its last. Duplanil states in his preface (p. xxxviii) that the superiority of the *Domestic medicine* over Tissot's *L'avis au peuple* (#3540) rests on the equal importance that Buchan places on hygiene (i.e., *la médecine prophylactique ou préservatif*), as well as the attention he gives to the treatment of disease. Duplanil has augmented Buchan's text with abundant footnotes. This French edition became the basis for later Portuguese, Spanish and Italian (#499) translations of the *Domestic medicine*. A Montpellier medical graduate, Duplanil was the author of *Médecine du voyageur* (3 vols., Paris, 1801), a work intended for *les gens du monde,* addressing the hygienic precautions that need to be taken on long voyages and the treatment of accidents and diseases encountered during travel.

The following survey of American, English, Scottish and Irish editions of the *Domestic medicine* was compiled from the OCLC database, *NUC Pre-1956 Imprints,* Austin's *Early American medical imprints,* the *British Library Catalog to 1975,* and the American Antiquarian Society's *Dictionary catalog of American books pertaining to the 17th through 19th centuries.* The purpose of this list is to demonstrate the remarkable publishing history of Buchan's text, and its influence on both sides of the Atlantic. There are editions that were doubtless published, but remain unrecorded (e.g., the 4th London edition, the 2nd Dublin edition, etc.). There are sixty-seven American editions on this list, forty-one of which are in the Atwater collection. American editions are highlighted in bold type. Editions in Atwater Collection are asterisked.

502. BUCHAN, William, 1729-1805.

New and enlarged edition of Dr. Buchan's domestic medicine; or, a treatise on the prevention and cure of diseases by regimen and simple medicines. With the latest corrections and improvements, and full directions in regard to air, exercise, bathing, clothing, sleep, diet, &c. &c. and the general management of the diseases of women and children. To which is annexed a complete family dispensatory, for the use of private practitioners . . . Fifty-ninth American, from the last London edition. With considerable additions and corrections by an American physician. Philadelphia: Claxton, Remsen & Haffelfinger, 1871.

xvii, [18]-495, [1], xlviii p. ; 24 cm.

Preface signed "J.S.F." (i.e, J.S. Forsyth). Printed from the same plates used for the seven Exeter, N.H. editions (#488-491, 493-494) and the 1846 Boston edition (#495).

Table 2. A check-list of recorded editions of Buchan's *Domestic medicine* (#445-502).

Year	Place of Publication	Enumeration	Notes
1769*	Edinburgh	1st	
1771	**Philadelphia**		
1772*	London	2nd	
1772*	**Philadelphia**		
1772?	**Philadelphia**		
1772	Edinburgh		
1774	London	3rd	
1774	Dublin	3rd	
1774*	**Philadelphia**	**2nd Am. ed.**	
1776	London	5th	
1777	Dublin	6th	
1778*	**Norwich, CT**	**3rd Am. ed.**	
1779	London	6th	
1781	London	7th	
1781	Dublin	7th	
1783	London		
1784*	London	8th ed.	
1784	Dublin	9th	
1784*	**Philadelphia**		
1786*	London	9th	
1788	London	10th	
1789	London	9th	
1789*	**Hartford, CT**	**11th ed.**	

Table 2. *Continued*

Year	Place of Publication	Enumeration	Notes
1790*	London	11th ed.	
1791	London	12th	
1792	Dublin	12th	
1792	London	13th	
1793*	Philadelphia	13th ed.	
1793	Philadelphia	14th ed.	
1793*	Boston	14th ed.	
1794	London	14th	
1795*	Philadelphia		
1795*	New London, CT	18th ed.	
1796	Dublin	9th	
1797*	London	15th	
1797	Dublin	15th	
1797*	Philadelphia		Published by Richard Folwell
1797	Philadelphia		Printed by R. Folwell for R. Campbell
1797*	Philadelphia		Printed by R. Folwell for J. &. J. Crukshank
1797	Philadelphia		Printed by R. Folwell for H. Sweitzer
1797	Philadelphia		Printed by R. Folwell for W. Woodward
1797	Philadelphia		Printed by R. Folwell for H. & P. Rice
1797*	Philadelphia		Printed by R. Folwell for James Carey
1797*	Philadelphia		Printed by R. Folwell for William Young
1797*	Philadelphia	2nd ed.	T. Dobson's 2nd ed. (1st, 1795)
1797*	Waterford, NY	20th ed.	
1798	London	16th	
1798*	Fairhaven, VT	1st Vermont ed.	
1799*	Philadelphia		
1800*	London	17th ed.	
1800	Manchester		
1801*	Philadelphia	2nd ed.	
1801	Halifax, NC	10th ed.	
1801*	Halifax, NC	17th ed.	
1802	London	New ed.	
1802	Edinburgh		
1803*	London	18th ed.	
1803	Manchester		
1804	Manchester		
1804*	Leominster, MA		
1805	London	19th ed.	
1805*	Philadelphia	New ed.	
1806*	Arbroath	36th ed.	
1806	Philadelphia	New ed.	
1807	London	20th ed.	
1807*	Charleston, SC	1st Charleston ed.	
1808	Harrop		
1808	Glasgow		
1808*	London	New ed.	
1809	London	New ed.	Kelly (publ.)
1809	London	New ed.	Butters (publ.)
1809*	Philadelphia		

Table 2. *Continued*

Year	Place of Publication	Enumeration	Notes
1809*	**Boston**	New corr. ed.	
1810	London		
1810	Manchester		
1810	Burslem		
1811*	**Boston**	New corr. ed.	
1812*	**New York**		Lindsay (publ.)
1812*	**New York**		Scott (publ.)
1812	**Philadelphia**		
1812	Edinburgh		
1812	Newcastle		
1812	Leith		
1813*	Edinburgh		
1813	London	21st	
1813	Dublin	21st	
1813*	**Boston**	New corr. ed.	
1813	**Philadelphia**	2nd ed.	
1813	**New York**		
1814	London	New ed.	
1814	Edinburgh	New & enl. ed.	
1815*	**New York**		
1816*	**New Haven, CT**		Title: *Every man his own doctor*
1816	**New York**		Title: *Complete family physician*
1818*	London		
1818	Dunbar	New & enl. ed.	
1820	Edinburgh	New & enl. ed.	
1821	London		
1822	Manchester		
1823	London		
1824?	London		Engr. port. dated 1824
1824	Edinburgh	New & enl. ed.	
1824	London	New ed.	Kelly (publ.)
1824	London		Fischer (publ.)
1824	London	New ed.	Bumpus (publ.)
1825	London	3rd ed.	Title: *The cottage physician and family adviser*
1825*	**Boston**	New ed.	
1826	London	22nd ed.	
1826	London		McGowan (publ.)
1827	London	New ed.	
1828*	London	New ed.	
1828*	**Exeter, NH**	From 22nd Engl. ed.	
1829?	**New York**		
1830?	Bungay		
1830	London		
1830	London	3rd ed.	Title: *The cottage physician and family adviser*
1832*	**Exeter, NH**	From 22nd Engl. ed.	
1834*	**Exeter, NH**	From 22nd Engl. ed.	
1836	**Exeter, NH**	From 22nd Engl. ed.	
1838	**Cincinnati**	Rev. & enl. ed.	
1839	London	New ed.	

Table 2. *Continued*

Year	Place of Publication	Enumeration	Notes
1839*	Exeter, NH	From 22nd Engl. ed.	
1840?	London		Fischer, publisher
1840?*	London	New stereotype ed.	J. Smith, publisher
1840	Hartford, CT	2nd Am. ed.	
1841	Cincinnati	Rev. & enl. ed.	
1842*	Exeter, NH	From 22nd Engl. ed.	
1843	Cincinnati	Rev. & enl. ed.	
1843*	Exeter, NH	From 22nd Engl. ed.	
1844	Derby		
1846	London	New ed.	
1846*	Boston	From 22nd Engl. ed.	
1848	Boston		
1849	Philadelphia	New ed.	
1851	Hartford, CT	10th Am. ed.	
1851	Philadelphia	29th ed.	
1852	Philadelphia	29th ed.	
1852	Cincinnati	Rev. & enl.	
1853*	Hartford, CT		
1854	Philadelphia	29th ed.	
1857	New York	10th Am. ed.	
1858*	New York	10th Am ed.	
1863	Halifax [Yorkshire]	New & enl. ed.	Milner & Sowerby
186?*	London	New & enl. ed.	Milner & Sowerby
186?*	London	New & enl. ed.	Milner & Co.
187-?	London		
1871*	Philadelphia, 1871	59th Amer. ed.	

503. BUCHAN, William, 1729-1805.

Observations concerning the prevention and cure of the venereal disease: Intended to guard the ignorant and unwary against the baneful effects of that insidious malady . . . The third edition, augmented and improved by the author. To which is added, a supplement, containing remarks on some anomalous venereal affections; a pharmacopoeia syphilitica, &c. by Dr. Buchan, Jun. . . . London: Printed for T. Cadell and W. Davies, 1803.

xvi, [17]-300, civ p. ; 23 cm.

Buchan's *Observations* appeared in four editions betwen 1796 and 1808. "There is no disease which exhibits such striking proofs of the advantages of diffusing medical knowledge as this," according to Buchan (p. 66). Among the reasons that induced him to write a treatise on venereal diseases for a lay readership was the need he felt to inform the public that treatment for this disorder can only be trusted to medical professionals. He bitterly notes, "Unfortunately for those who labour under the venereal disease, its treatment has fallen into bad hands. Not only quacks of all descriptions undertake to cure it; but every idle fellow who does not chuse to follow some useful employment, sets up for doctor, assumes some well known name, and advertises an infallible remedy for the venereal disease. The apothecary's man, or even the apothecary's man's man, often passes for an adept in curing this malady. Nor is it uncommon for the fellow who brushed the surgeon's coat, or cleaned his shop, to step into his master's shoes, and sometimes into his chariot, by his pretended skill in curing the lues venerea" (p. 19).

Buchan provides a chapter on the prevention of venereal diseases, describes venereal disorders and the differences among them, and details a course of medical treatment. He is insistent on the efficacy of mercurial preparations for the cure of any venereal disorder, provided the medication is adminstered in time and the patient follows a prescribed regimen. The supplement (civ p.) by Alexander Peter Buchan (1764-1824), the author's son, concludes with a formulary of mercurial preparations for the treatment of venereal diseases. The younger Buchan had published his own brief treatise on syphilis (*Enchiridion syphiliticum*) at London in 1797.

504. BUCHANAN, John.

A practical treatise on the diseases of children . . . Adapted to the use of families and physicians. Philadelphia: John Buchanan, M.D., 1866.

iv, [5]-104 p., [1] leaf of plates : port. ; 24 cm.

A catalog of pediatric diseases including their treatment. An alphabetical index of disorders precedes the text. Lithographic frontispiece portrait of the author. The Atwater copy bears the bookplate of Sarah Adamson Dolley (1829-1909), an 1851 graduate of the eclectic Central Medical College, Rochester, N.Y.

Buchanan was a member of the faculty of the Eclectic Medical College of Pennsylvania, Philadelphia (est. 1850); wrote several treatises on the eclectic practice for professional and lay readers; and edited *The Eclectic medical journal of Pennsylvania*. In the 1870s Buchanan was exposed for "having sold sixty thousand diplomas over a forty-year period at prices ranging from ten to two hundred dollars" (J. Haller, *Medical Protestants,* Carbondale, Ill., 1994, p. 149).

505. BUCHANAN, Joseph Rodes, 1814-1899.

Outlines of lectures on the neurological system of anthropology, as discovered, demonstrated and taught in 1841 and 1842 . . . In four parts: Part I. Phrenology. Part II. Cerebral physiology. Part III. Pathognomy. Part IV. Sarcognomy. Cincinnati: Printed at the office of Buchanan's Journal of Man, 1854.

[4], 16, 384, 16 p., [1] leaf of plates : ill. ; 23 cm.

A native of Frankfort, Ky., Buchanan was an 1842 graduate of the medical department of the University of Louisville. "While still in college he laid the foundations for two new so-called sciences, which he later elaborated, 'psychometry' and 'sarcognomy,' the former demonstrating the excitability of cerebral tissue by the aura of another person, the latter dealing with the sympathetic relations between other parts of the body and the indwelling soul. The trained psychometer, he held, could diagnose any disease at the sight of a patient or even by letter, while the sarcognomist could heal all diseases by making dispersive passes over the body . . . Buchanan lectured upon his alledged discoveries with great success throughout both the North and the South, and established a periodical, *The Journal of man,* in which for many years he promulgated his extraordinary views" (*Dict. Amer. biog.,* 2:216-17).

"Buchanan coined the word *sarcognomy* in 1842 while he was affiliated with the American Medical Institute in Cincinnati. Derived from *sarx* . . . (flesh) and *gnoma* (opinion), the term meant 'a knowledge of the flesh, or recognition of its character and relations.' From a practical standpoint, the new science implied a knowledge of the physiological and psychological powers that belonged to the body in health and in disease, and it implied the understanding of the correlation of soul, brain, and body in that relationship . . . The exercise of every psychic faculty . . . produced a characteristic effect on the body, while the exercise of any portion of the body produced a characteristic effect within the brain and mind, the locality of which could be specified on the brain. In this regard, diseases had mental as well as physical symptomatologies" (John Haller, *Medical Protestants,* Carbondale, Ill.: Southern Illinois U. Press, 1994, p. 112).

Buchanan joined the faculty of the Eclectic Medical Institute of Cincinnati in 1846, and became both the school's and eclecticism's principal spokesman. Forced out of the Institute by internal dissension in 1856, he immediately founded the rival Eclectic College of Medicine. Once again falling into disagreement with his colleagues, Buchanan left the College and Cincinnati in 1863 to run (unsuccessfully) for Congress in Kentucky. In 1867 he joined the faculty of the Eclectic Medical College of New York City, where he remained until establishing the College of Therapeutics in Boston in 1881.

506. BUCHANAN, Joseph Rodes, 1814-1899.
Periodicity: the absolute law of the entire universe long known to control all matter now revealed as the law of all life . . . Third edition. Evanston, Ill.: Kosmos Sanatorium, 1912.
[3]-135, [10] p., [1] folding table ; 17 cm.

Buchanan's last and perhaps oddest book, *Periodicity* is a numerological/occult study of cycles. Summarizing the laws of periodicity in the first chapter Buchanan writes: "all vital operations proceed in a varying course, measured by the number seven. This septimal division I expect to find in the life of every individual from youth to age, in the progress of diseases, in the history of nations, societies, enterprises, and everything that has progress and decline—in short in all life, for all life has its periods of birth, growth, decline and death" (p. 7). Buchanan's "septimal system" can be applied to the study of "vital periodicity," i.e., cyclicism as it applies to health and hygiene, as well as to determining inevitable "evil periods." *Periodicity* appeared in four editions between 1897 and 1913. At head of title: *A scientific secret revealed.*

507. BUCHANAN, Joseph Rodes, 1814-1899.
Therapeutic sarcognomy. The application of sarcognomy, the science of the soul, brain and body, to the therapeutic philosophy and treatment of bodily and mental diseases by means of electricity, nervaura, medicine and haemospasia, with a review of authors on animal magnetism and massage, and presentation of new instruments for electro-therapeutics . . . Boston: J.G. Cupples Co., publishers, 1891.
[4], xiii, [1], 671, [13] p., [3] leaves of plates : ill. ; 25 cm.

Buchanan regarded his neurological system as the logical extension of the theories of Gall. According to John Haller, "The Gallian system failed to explain the laws of sympathy between mind and body, while Buchanan's neurological system demonstrated that every portion of the brain sympathized with and was connected to a corresponding part of the body. The connection

between the brain and the corporeal organs made it possible, he believed, to construct a psychological map of the body corresponding to that of the brain . . . Just as physiognomy interpreted the character of the face, so sacrgnomy interpreted the character of the body, revealing its laws, connections, and sympathies to the physician." (*Medical Protestants,* Carbondale, Ill.: Southern Illinois U. Press, 1994, p. 111).

In spite of its length and complexity, Buchanan intended this book as a mechanism for enlightening the public on his ideas regarding mind, body and healing: "The knowledge of Therapeutic Sarcognomy, when widely diffused and incorporated into popular education, will bring the grand philosophy of Anthropology into familiar contact with daily life . . . It will give the mother and father a power of controlling their offspring in a manner heretofore unknown, by which the development of both soul and body may be gradually carried on toward perfection . . . In the treatment of disease it gives . . . many remedial measures . . . which upon impressible constitutions produces cures so marvellous and so speedy as to excite a stubborn scepticism among those who have been kept in ignorance of the powers of the nervous system" (p. 4). Purely "nervauric" healing could be achieved either by placing hands on the affected part of the body, or some cases, by mental power alone. Buchanan also describes electrotherapy, massage and other therapies.

508. BUCKELEW, Sarah Frances, b. 1835.
Practical work in the school room. Part I. A transcript of the object lessons on the human body given in Primary Department, Grammar School No. 49, New York City. Third edition. New York: A. Lovell & Company, 1885.
ix, [1], 157 p. : ill. ; 19 cm.

509. H. E. BUCKLEN & COMPANY.
Dr. King's guide to health and household instructor. H.E. Bucklen & Co., props. Hamilton, Can.; Chicago, Ill.: [H.E. Bucklen & Co., ca. 1886].
32 p. ; 23 cm.

Promotional literature for *Dr. King's New Discovery, Bucklen's Arnica Salve, Electric Bitters* and other proprietary remedies. Title from chromolithographed wrapper. There are two issues in the Atwater Collecton. Both display on the front wrapper a six-story red brick building that presumably is the headquarters of H.E. Bucklen & Co. On one wrapper a steam locomotive with passenger cars runs along the bank of a river in front of the building; in the second, a fairy-like female figure reads from an oversize volume to a *putto*, who stands to her right mixing ingredients in a mortar with pestle.

Herbert E. Bucklen bought rights to all the formulas of Dr. Z.L. King of Elkhart, Ind., and in 1879 moved to Chicago. With the help of intensive advertising, the *Dr. King's New Discovery* became one of America's best selling proprietary remedies by 1900.

510. H. E. BUCKLEN & COMPANY.
Dr. King's lucky book. Charms, signs, omens, marriage predictions and fortune telling. New discovery for consumption, coughs & colds. [Chicago]: H.E. Bucklen & Co., copyrighted 1904.
     32 p. : ill. ; 23 cm.

Title and imprint from chromolithographed wrapper.

511. H. E. BUCKLEN & COMPANY.
Dr. King's new beauty book [and] prize cook book. Chicago, Ill.; Windsor, Can.: H.E. Bucklen & Co., [ca. 1905].
     31, [1] p. ; 23 cm.

Title from chromolithographed wrapper.

512. H. E. BUCKLEN & COMPANY.
Dr. King's new guide to health, household instructor and family prize cook book. Chicago, Ill.; Windsor, Can.: H.E. Bucklen & Co., [ca. 1906].
     31, [1] p. ; 23 cm.

Title from chromolithographed wrapper.

513. H. E. BUCKLEN & COMPANY.
Dr. King's world's wonder book, health guide and family prize cook book. Chicago, Ill.: H.E. Bucklen & Co., [ca. 1911].
     31, [1] p. ; 23 cm.

Title from chromolithographed wrapper.

514. BUCKTON, Catherine M.
Health in the house. Twenty-five lectures on elementary physiology in its application to the daily wants of man and animals, delivered to the wives and children of working-men in Leeds and Saltaire . . . Tenth edition. Toronto: W.J. Gage & Co., 1880.
     [4], [ix]-xx, 207 p. : ill. ; 18 cm.

Buckton's treatise on anatomy, physiology and hygiene was originally delivered as a series of lectures to the wives and children of English working-class families in 1874 and 1875. They were collected and published in 1875. *Health in the house* appeared in at least fifteen English editions through 1890; a 12th Toronto edition was published ca. 1882. At head of title-page: *Miller & Co.'s educational series*; on front board: *Gage & Co.'s educational series*.

515. BUFFALO LITHIA SPRINGS.
Buffalo Lithia Springs, Mecklenburg County, Virginia. Health primer. Stone in the bladder (in which this water is the only known solvent), Bright's disease of the kidneys, gout, rheumatic gout, rheumatism, disorders of the stomach and nervous system, affections peculiar to women, eczema, acne, and other facial eruptions. Resident physician: G. Halstead Boyland . . . Medical opinions—clinical reports. Thomas F. Goode, proprietor . . . Philadelphia: McCalla & Stavely, printers, 1884.
     20 p. ; 23 cm.

Title and imprint from wrapper. Contains testimonials regarding the efficacy of the Buffalo Lithia Springs.

516. BULKLEY, Lucius Duncan, 1845-1928.
The skin in health and disease . . . Philadelphia: Presley Blakiston, 1880.
     vi, 7-148, [2] p. : ill. ; 15 cm.

Bulkley has divided his text into four chapters: anatomy and physiology of the skin, the care of the skin in health, diseases of the skin, and diet & hygiene in diseases of the skin. *The skin in health and disease* is no. 10 in the *American health primers* series, edited by William Williams Keen (1837-1932). By the time Bulkley's text was issued, twelve volumes in the series had been published. "This series of *American health primers* is prepared to diffuse as widely and cheaply as possible, among all classes, a knowledge of the elementary facts of Preventive Medicine, and the bearings and applications of the latest and best researches in every branch of medical and hygienic science. They are not intended (save incidentally) to assist in curing disease, but to teach people how to take care of themselves, their children, pupils, employees, etc." (from front pastedown).

A pioneer of American dermatology, Bulkley was an 1869 graduate of the College of Physicians and Surgeons (New York). He continued his dermatological studies in Europe under such figures as von Hebra and Hardy. Returning to New York in 1872, Bulkley was active in the professionalization of his speciality, helping to establish local and national dermatological societies, founding the *Archives of dermatology* (1874), lecturing on diseases of the skin at the New York Hospital, and later at the New York Skin and Cancer Hospital (which he founded in 1882). Bulkley was a prolific author, with more than forty books on dermatology and cancer to his name.

BULL, A. T. see *CANADIAN JOURNAL OF HOMEOPATHY* (#554).

517. BULL, Thomas.
Hints to mothers, for the management of health during the period of pregnancy, and in the lying-in-room; with an exposure of popular errors in connexion with those subjects . . . From the third London edition, with additions by an American physician. To which is added, the Ladies perpetual calendar. New York: Wiley & Putnam, 1842.
268 p. ; 16 cm.

First American edition, based on the third London edition of 1841. *Hints to mothers* appeared in sixteen London editions between 1837 and 1865, was issued in the United States by the Wiley firm as late as 1877.

"In the minds of married women, and especially in young females, those feelings of delicacy naturally and commendably exist which prevent a full disclosure of their circumstances, when they find it necessary to consult their medical advisers. To meet this difficulty, and

also to counter-act the ill-advised suggestions of ignorant persons during the period of confinement, is the chief object of the following pages. While it is believed that much of the information contained in this volume is highly important to the comfort, and even to the well-doing of the married female, much of it is, at the same time, of a character upon which she cannot easily obtain satisfaction. She will find no difficulty in *reading* information, for which she would find it insuperably difficult *to ask*" (*Pref.*, p. [3]).

Bull's treatise is divided into seven chapters: I. Of popular errors on the subject of pregnancy; II. Of the mode by which pregnancy may be determined; III. Of the diseases of pregnancy, and hints for their prevention and relief; IV. On the prevention of miscarriage; V. Hints for the lying-in room; VI. Suckling; VII. Hints for the management of health during infancy. The *Ladies perpetual calendar* (for reckoning the time of confinement from the date of conception) is taken from Anton F.A. Desperger's (b. 1789) *Schwangerschaftskalender*.

518. BULL, Thomas.
Hints to mothers, for the management of health during the period of pregnancy, and in the lying-in room; with an exposure of popular errors in connexion with those subjects . . . Second edition, with additions by an American physician. To which is added, the Ladies perpetual calendar. New York: Wiley & Putnam, 1843.
268 p. ; 16 cm.

519. BULL, Thomas.
Hints to mothers, for the management of health during the period of pregnancy, and also in the lying-in room; with an exposure of popular errors in connexion with those subjects . . . From the third London edition, with additions by an American physician. To which is added, the Ladies perpetual calendar. New-York: John Wiley, 1848.
268 p. ; 16 cm.

Wiley continued to issue the third London edition of *Hints to mothers* as late as 1868, even though by this time the sixteenth London edition had appeared.

520. BULL, Thomas.
The maternal management of children, in
health and disease . . . Second edition.
Philadelphia: Lindsay and Blakiston, 1853.
    xii, [13]-424, [8] p. ; 19 cm.

Second American edition. Originally published at London in 1840, the first American edition was issued at Philadelphia in
1849, and was based on the third London
edition (1848). Bull divides his text into two
parts: the management of children in health
(i.e., nursing, diet, bathing, clothing, vaccination, etc.); and the management of children in disease.

BULLOCK, Helen L. see **NATIONAL
WOMAN'S CHRISTIAN TEMPERANCE
UNION** (#2574).

**521. BULWER AND FORBES ON THE
WATER-TREATMENT.**
Bulwer and Forbes on the water-treatment.
A compilation of papers on the subject of
hygiene and rational hydropathy edited,
with additional matter, by Roland S.
Houghton . . . New and revised edition,
stereotyped: with additions and improvements. New York: Fowlers and Wells, publishers, 1851.
    [2], xii, [13]-258 p. ; 19 cm.

CONTENTS: I. Edward Bulwer-Lytton. *Confessions of a water patient*; II. John Forbes. *A review of hydropathy*; III. Erasmus Wilson. *Two chapters on bathing and the water treatment*; IV. Charles
Scudamore. *A medical investigation of the water-
cure treatment*; V. Herbert Mayo. *The cold water
cure: its use and misuse*; VI. Roland S. Houghton.
Observations on hygiene and the water treatment.
    Series: *Fowlers and Wells water cure library.* Half-
title: *Bulwer and Forbes on the water cure with lectures by Houghton.* Issued with: Houghton, R.S.
Three lectures on hygiene and hydropathy. New York,
1851.

BULWER-LYTTON, Edward see **LYTTON**
(#2308).

522. BUNCE, Charlie White-Moon.
The Cowboy "Herbalist." [Louisville, Ky.:
C.W. Bunce, c1911].
    64 p. : ports. ; 24.5 × 10 cm.

Promotional literature for *Charlie White-
Moon Cheyenne Indian Remedies.* Banking on
America's infatuation with the West, Bunce
claims to be a representative of the disappear-
ing cowboy culture: "His was the fearless, daunt-
less, brave yet gentle character which knew no
danger, feared no hardship & blazed the way
for the great farms & cities that now cover the
old time ranges" (p. 4). Bunce claims to have
left the family farm in Kansas at age 17 for the
life of a wrangler. After being shot by a cattle
thief and given up for dead by the doctors,
Bunce came under the care of a "Cheyenne
Indian woman who gave me Root & Herb medicines & within 18 days I was on my feet" (p. 6).
After being unsuccessfully treated by physicians
in Texas for malaria, he went to an "Arrapahoe
Indian woman," who again cured him with roots
and herbs. Bunce claims that in 1888 he was
made a member of the Cheyenne tribe, with
whom he studied medicinal roots and herbs.
In 1898 went into the herbal medicine business
with "193 different roots, herbs, leaves, berries,
barks, twigs & flowers."
    Bunce boasts: "I have had a bitter fight with
so-called 'medical men,' but I have always
whipped all enemies of God's medicines, Roots
& Herbs, because I have always been in the
right, & there is a law, older far than any made
by man, THE LAW OF HUMANITY, which all men
are compelled to respect, & this is the law under which I have always worked, selling these
pure God-given Root & Herb medicines" (p.
7). On pages 24 and 25 Bunce lists twenty-four
proprietary remedies, including *Charlie White-
Moon's Cheyenne Indian Blood Purifier* for syphilis, *Charlie White-Moon's Cheyenne Indian Flatuent
Remedy*, and *Charlie White-Moon's Cheyenne Indian
Leucorrhea Douche.* A full-length, half-tone photograph of Bunce in full cowboy regalia graces
the front wrapper. Atwater copy lacking pages
57-62.

523. BURBANK, Luther, 1849-1926.
The training of the human plant . . . New
York: The Century Co., 1907.
    [8], 99 p. : port. ; 18 cm.

Burbank conducted thousands of experiments in plant breeding and cross-breeding
over half a century that resulted in the development of hundreds of new varieties of fruits,
flowers and vegetables, and that contributed on

an unprecedented level to understanding the principles of heredity by which these were attained. In this brief work Burbank projects "the adapation of the principles of plant culture and development" to the development and progress of the human species. Burbank saw in America at the turn of the 20th century "the grandest opportunity ever presented of developing the finest race the world has ever known out of the vast mingling of races brought here by immigration" (p. 5). On the human plane Burbank was never the determinist. He stresses the importance of "selective environment" on the outcome of this great experiment, i.e., the influence of training, health ("sunshine, good air, nourishing food"), and the prohibition of marriage among the "physically, mentally and morally unfit." Frontispiece portrait of author.

524. DRS. BURBANK & BURNS.

Drs. Burbank & Burns, Indian physicians, would most respectfully inform the citizens of this place and vicinity that they have taken room [*90*] at [*American House, St. Albans*] where (at the earnest solicitation of many) we will spend a few days professionally. All who are sick and in any way debilitated would do well to call early. Drs. B. & B. compound and prepare all their own medicines from the best ingredients to be found in the *Materia Medica.* No poisonous drugs or minerals are used. And the very best remedies are selected from the Allopathic Practice (of which class one of our party is a graduate) . . . We are meeting with the greatest success in curing CANCERS AND CANCEROUS HUMORS . . . M.W. Wilson's Print, late of the Burnt district, leave orders at Merifields Bookstore, St. Albans. [186-?].
      Broadside ; 23.5 × 15.5 cm.

BURDELL, John see **CORNARO** (#796).

525. BURGESS, William Harvey, 1842-1913.

Chronic disease. The natural method of diagnosis and successful treatment . . . (Avondale) Chattanooga, Tenn., [c1907].
      320 p. : port. 19 cm.

Burgess traces the origins of disease to one of five causes: the *retention* of morbid matter,

the *invasion* of the body by microscopic pathogens, *enervation, trauma,* or *poison.* His panacea for the treatment of diseases arising from these causes is the epsom bath. In his *Easy lessons* (#526), Burgess writes: "the warm Epsom sponge bath . . . opens the pores of the skin, energizes the dermal glands, relaxes nerve tension, neutralizes toxin matters in the blood, sedates the nerves and relieves pain, swelling and congestion. How it lowers fever not alone by evaporation, as plain water may do, but by destroying bacteria and toxins, which cause fever" (p. 1). In *Chronic disease* we find an explanation of how the epsom bath operates in relieving disorders caused by retention: "This opens the pores and the absorbed drug attacks the thirty or forty enzymes, ptomaines and leucomaines which have been discovered, robbing them of their carbon and checking the decomposition present and putting it in shape for elimination and energizing the emunctories for the purpose" (p. 197). Equivalent descriptions are provided of the action of epsom salts on diseases caused by invasion, enervation, etc. A list of thirty-three proprietary remedies available from Burgess (p. 302-09) are intended to supplement the epsom bath. Burgess was an 1880 graduate of the Medical College of South Carolina.

526. BURGESS, William Harvey, 1842-1913.

Easy lessons (medical). Third edition . . . East Chattanooga, Tenn., copyrighted, 1908 by W.H. Burgess.
      16 p. ; 19 cm.

      Caption title.

527. BURGESS, William Harvey, 1842-1913.

Little ailments and consequences . . . East Chattanooga, Tenn., [c1912].
      192, [8] p. ; 19 cm.

Burgess restates the theoretical basis of his "New Field" school of medical thought (*see* #525), and concludes with an expanded list of 44 "double sulfide" and "cell salt" proprietary remedies to be used in combination with the epsom bath for the cure of almost any disorder. For example, *No. 38 Bull Nettle Eclectic Combination* cures syphilitic affections; *No. 44 Sugar Cure Tablet* "cures a woman's headache in 10 minutes," etc.

528. BURGESS, William Harvey, 1842-1913.
The new field . . . [Chattanooga, Tenn.,
c1904-05].
    287 p. ; 17 cm.

The "New Field" was the term Burgess applied to his theory of the five causes of disease
(i.e., retention, invasion, enervation, trauma,
poison), "tracing" (i.e., diagnostics) and therapeutic armentarium (epsom salts and proprietary remedies). *The new field* is the earliest of
Burgess' largely redundant publications.

529. BURKE, William.
The Virginia mineral springs, with remarks
on their use, the diseases to which they are
applicable, and in which they are contra-
indicated, accompanied by a map of routes
and distances. A new work. Second edition,
improved and enlarged . . . Richmond, Va.:
Printed and published by Ritchies &
Dunnavant, 1853.
    [4], 376 p., [1] folding leaf of plates :
map ; 20 cm.

"The treatment of ACUTE diseases is the
exclusive province of the physician . . . It is when
disease degenerates into a chronic form, that
the skill of the physician is at fault and frequently powerless, and it is just here that nature, the most beneficent physician, presents
from her generous bosom those pure streamlets
that restore health and vigor, and elasticity to
the afflicted invalid. It is in this condition that
Mineral Springs are sought after, and in which
it becomes desirable to the invalid to know
whither he must direct his course to effect the
object he has in view" (p. 2). The first edition
was published under the title: *The mineral springs
of Virginia* (1851). Burke was also the author of
The mineral springs of Western Virginia (New York,
1842; 2nd ed., New York, 1846).

BURNER, H. R. see **GLEASON** (#1361).

530. BURNETT, Charles Henry, 1842-1902.
Hearing, and how to keep it . . . Philadelphia: Lindsay & Blakiston, 1879.
    viii, 9-152 p. : ill. ; 15 cm.

Burnett discusses the anatomy and physiology of the ear, the physics of sound and hearing, ear diseases and the care of the ear both in
health and disease. Burnett's treatise was the
first of the *American health primers* series edited
by William Williams Keen (1837-1932) for Lindsay & Blakiston. The object of this series was
"to diffuse as widely and as cheaply as possible,
among all classes, a knowledge of the elementary facts of Preventive Medicine, and the bearings and applications of the latest and best researches in every branch of Medical and Hygienic Science" (advert. on front pastedown).
    Burnett was an 1867 medical graduate of the
University of Pennsylvania. Following graduation he spent ten months (1868-69) continuing his medical studies in Europe. Strongly attracted to the study of otology, he returned to
Europe in 1870, "where he worked for over a
year, especially in the laboratories of Helmholtz
and Virchow, and in the clinic of Politzer . . .
With Helmholtz, in particular, he established
most cordial relations, conducting in his laboratory his invaluable series of investigations into
the condition of the membrane of the round
window during the movement of the auditory
ossicles and upon the various effects of changes
in intralabyrinthine pressure. This research . . .
placed him at once among the most eminent
investigators into the physiology of hearing"
(*Dict. of Amer. med. biog.*, New York: D. Appleton
& Co., 1928, p. 174). Burnett returned to Philadelphia in 1872, and devoted himself to the
diseases and surgery of the ear.

531. BURNETT, Francis Eaton.
Human pearls . . . Chicago: R.R. Donnelley
& Sons Co., printers, 1908.
    86 p. : ill. ; 18 cm.

A manual on oral hygiene by an 1898 graduate of Northwestern University Dental School.

**532. BURNETT, James Compton, 1840-
1901.**
Curability of tumours by medicines . . .
London: The Homoeopathic Publishing
Company, 1893.
    xiv, 332 p. ; 17 cm.

First edition. Burnett, "a Scotsman of commanding presence, studied medicine in the
great medical schools of Vienna . . . He returned
to take his Glasgow M.B. in 1872, and while
holding a resident post became so depressed
by the results of medicine, as practised, that he

had serious thoughts of throwing up his profession for something more useful. But he was persuaded to study homoeopathy with the idea of refuting it." Instead, Burnett defected to the homoeopathic camp, though he "could not agree that all medicine was a blank before Hahnemann, and that all medical progress had been interred with his bones" (John Weir, "British homoeopathy during the last hundred years," *Brit. med. j.,* 1932, 2:604).

Burnett was the author of numerous brief treatises on the cure of specific diseases by the prescription of specific homoeopathic medicines. Even in diseases for which surgical intervention was considered the normal treatment (e.g., tonsils, cataracts, tumors), Burnett proposed *medicinal* treatment. Though most of these books were written for his homeopathic brethren, their brevity and colloquial style must have assured them a more general readership. Another feature of Burnett's books is that the London edition was nearly always followed by an American edition. The first American appearance of the *Curability of tumours by medicines* was the 1901 Philadelphia edition, based on the second London edition of 1898.

Burnett admits that even among allopaths "Tumours have occasionally been cured by medicines, records whereof may be found in literature . . . But though such may be the case, the indications for the use of remedies for the withering of tumours are so meagre that few believe in, and still fewer undertake, their medicinal cure" (p. 29). What is required for the successful treatment of neoplasms medicinally is to address treatment homeopathically, i.e., not to treat symptoms, an approach which in most cases is palliative at best, and to avoid the temptation of surgical excision, but to find remedies that go beyond symptoms and even the "symptomatic simillimum" to the "pathologic simillimum," i.e., "zoic remedies" that "have a range of action equal to the disease-processes" (p. 24). Burnett maintains that even malignant neoplasms are curable if treatment is begun early enough.

533. BURNETT, James Compton, 1840-1901.
Delicate, backward, puny, and stunted children: their developmental defects, and physical, mental, and moral peculiarities

considered as ailments amenable to treatment by medicines . . . Philadelphia: Boericke & Tafel, 1896.
    iv, 164 p. ; 17 cm.

Burnett describes the treatment of mental and physical growth disorders in children by "the use of specific homoeopathic, organopathic, and various other constitutional remedies systematically administered, so as to rectify the wrong underlying the said backwardness, to cure the diseased organ or parts, to rouse them medicinally from torpidity, or to cure the diseases of the individual as an entirety, or to get rid of the perverted or other morbid conditions due to hereditary diseases and taints, falls, blows, fits, or other . . . accidents and diseases" (p. 2). Originally published at London in 1895.

534. BURNETT, James Compton, 1840-1901.
Enlarged tonsils cured by medicines . . . Philadelphia: Boericke & Tafel, 1901.
    100 p. ; 17 cm.

"Most medical men have made up their minds that enlarged tonsils can *not* be cured by medicines, but must be cut off, and therefore for most people, professional and lay, there exists no question of enlarged tonsils, and whether they should or should not be removed. But for the past twenty years I have treated my cases of enlarged tonsils by medicines, and have, moreover, succeeded in curing the great bulk of them" (p. [3]-4). The London and Philadelphia editions of *Enlarged tonsils* were both published in 1901.

535. BURNETT, James Compton, 1840-1901.
Gout and its cure . . . Philadelphia: Boericke & Tafel, 1895.
    4, 172 p. ; 17 cm.

536. BURNS, JOHN, 1774-1850.
Popular directions for the treatment of the diseases of women and children . . . New-York: Published by Thomas A. Ronalds, 1811.
    iv, [5]-324 p. ; 18 cm.
    First (and only) American edition (1st ed., London, 1811). The *Popular directions* is divided

into four parts: I. Of pregnancy; II. Of labour, and the child bed state; III. Of the management and diseases of children; and IV. Of diseases of grown-up women.

In his preface Burns writes, "it has been far from my intention to encourage unprofessional readers, to undertake the office of prescribing for diseases, which often baffle the skill of the most experienced practitioner. But there are situations, in which a little medical knowledge, or a general acquaintance with the most frequent complaints, incident to women and children, may be of essential consequence . . . before regular assistance could be procured . . . The reader is only to expect a general view of the subject. In my 'Principles of Midwifery,' I have treated the matter professionally, for the use of medical readers" (p. [iii]-iv). The *Principles of midwifery* (1st ed., London, 1809) was Burns' most important and oft published work.

The tone of the work is exemplified in part 2, chapter 1 ("Of labour"). Rather than detail the mechanisms of parturition, Burns provides his female readers fifteen "general rules of conduct" to observe during labor, "which though very simple, will yet, if attended to, be of great benefit" (p. 70). Similarly in chapter 2 ("Of tedious and preternatural labor"), Burns does not describe complex presentations or manipulation of the fetus, but discusses what the woman might expect and do during difficult deliveries.

Born in Glasgow, Burns became a lecturer on anatomy, obstetrics and surgery in his native city, and was the author of four influential treatises on obstetrics and gynecology. In an age when many Americans studied and/or took medical degrees in the United Kingdom and when obstetrical publishing in the United States was dominated by Scottish and English authors, it is not surprising that all these works were issued in American editions soon after their original publication.

REFERENCES: *Austin* #361.

537. M. S. BURR AND COMPANY.
Copied from the Boston Medical and Surgical Journal, of February 21, 1861. "BICOLORATA BARK." By Edward E. Phelps . . . Having observed the communication of of Edward E. Phelps, M.D., of Dartmouth College . . .

on the use and properties of "Bicolorata Bark," we have procured a limited supply, and have exclusive *control of all the article known to commerce.* It is ground and put up in one pound packages, and we are prepared to supply physicians and the trade generally with the same. M.S. BURR, & CO. . . . Boston . . . [ca. 1861].
Broadside ; 38 × 33 cm.

Includes the text of Phelps article in two columns, as well as notice of the availability of Oxygenated Bitters and other "proprietary medicines, drugs, essential oils, sponges, hair restoratives, toilet soaps, perfumery, hair dyes, &c."

538. BURR AND PERRY.
Dr. Warren's BILIOUS BITTERS! The Great Blood Purifier and Regulator, cures liver complaint, jaundice, biliousness, weakness, debility, colds and fevers, fever and ague, headache, dizziness, eruptions on the skin, humors of the blood, loss of appetite, female complaints, costiveness, piles, and all complaints caused by impure blood, imperfect or obstructed circulation, or a deranged and diseased condition of the stomach, liver, kidneys and bowels . . . John A. Perry, chemist, Boston, proprietor . . . Burr & Perry, general agents . . . Boston, Mass. [ca. 1868].
[1] leaf printed on both sides ; 23.5 × 14.5 cm.

Advertised on verso: *Burr's Patent Nursing Bottle.*

539. BURT, Samuel H.
The universal household assistant or what everyone should know. A cyclopedia of practical information. Complete directions for making and doing over 5,000 things necessary in business, the trades, the shop, the home, the farm, and the kitchen. Recipes, prescriptions, medicines, manufacturing processes, trade secrets, chemical preparations, mechanical appliances, aid to injured, antidotes, business information, every day law, ornaments, home decorations, art work, fancy work, agriculture, fruit culture, stock raising, and hundreds of other useful hints and helps, gathered

from the most reliable sources . . . New York: S.H. Moore & Co., 1885.
    510, [2] p. ; 17 cm.

540. BUTLER, William Frederick.
Ventilation of buildings . . . New York: D. Van Nostrand Publishers, 1873.

    77 p. : plans ; 15 cm.

First edition. "I fear that even persons who build houses for their own occupation, are but little in advance of the speculative builder, as far as any recognition of the absolute necessity of efficient ventilation is concerned. Many hold to such crude devices as open windows and doors . . . It then becomes the duty of scientific men . . . to educate the public up to the recognition of the fact, that ventilation is every whit as important as drainage, to individual houses, and that man can no more live in a foul atmosphere than he can while constantly imbibing poisonous water" (p. 6).

Series: *Van Nostrand's science series; no. 5*. Established in 1848, David Van Nostrand's (d. 1888) firm quickly became the largest publisher of scientific books in the United States. "George W. Plympton, professor of physics and engineering at Brooklyn Polytechnic Institute, edited the famous 'Van Nostrand Science Series,' whose distinctive green-cloth volumes were known everywhere in America. There were 127 titles on that list when it was completed in 1902 . . . It was often said that this series helped to educate a generation of engineers. They were still being sold as a series in the 1920s" (John Tebbel, *A history of book publishing in the United States,* New York: R.R. Bowker, 1975, 2:239). The front pastedown and free endpaper of Butler's book list thirty-two titles in *Van Nostrand's Science Series,* ranging in topic from civil engineering to energy, mining and sanitation. Van Nostrand revised and reissued the *Ventilation of buildings* in 1885.

541. BUTTERICK PUBLISHING COMPANY.
Nursing and nourishment for invalids . . . London and New York: Published by the Butterick Publishing Co., [c1892].
    [2], 32, iv, [2] p. ; 24 cm.

Title and imprint from wrapper. Series: *Metropolitan pamphlet series; v. 5, no. 4.*

BYERS, W. C. see **COPELAND MEDICAL INSTITUTE** (#790).

542. BYRN, Marcus Lafayette, 1826-1903.
The mystery of medicine explained; a family physician, and household companion; prepared for the use of families, plantations, ships, travelers, &c. . . . New York: Published by Dr. M. Lafayette Byrn, 1869.
    iv, [5]-473 p., [5] leaves of plates : ill. ; 23 cm.

First edition. *The mystery of medicine explained* appeared in seven editions between 1869 and 1889. Typical of books of its genre, it is intended "as a guide for preserving health and prolonging life, by giving that kind of information . . . which has long been needed by the masses." "It has been my aim," Byrn writes in his preface, "not only to simplify the laws of health and physical education, but to give such plain descriptions of the various ailments which our bodies are subject to, that everyone may know from the symptoms, each ailment or disease, and be enabled to give the best remedies, where a physician cannot be had, or, in cases of emergency, to know what to do before the physician arrives, so as to alleviate suffering or be the means of saving life" (p. [iii]).

The book's first seventy pages are devoted to hygiene, followed by discussions of the diseases of children, the diseases of females, and midwifery. The latter two thirds of the text describe diseases not peculiar to women or children, arranged in no discernible order. An alphabetical index at the end of the volume, however, provides subject access to the text.

An 1851 medical graduate of the University of the City of New York, Byrn wrote half a dozen popular medical treatises, including two sexual physiologies. A man of diverse ambitions, he was also the author of *The complete practical brewer* (1852), which appeared in six editions, and *The complete practical distiller* (1853), issued in three editions. Not the least of his accomplishments was a very successful series of humorous fiction that Byrn wrote under the pseudonym David Rattlehead.

543. BYRN, Marcus Lafayette, 1826-1903.
The mystery of medicine explained; a family physician, and household companion; prepared for the use of families, planta-

tions, ships, travelers, &c. . . . New York: Published by Dr. M. Lafayette Byrn, 1871.
iv, [5]-474, [6] p., [1] leaf of plates : ill. ; 23 cm.

Second edition. The six unnumbered pages at the end of the volume advertise the *Magneto-Electric Machine*, the *Family Medicine Chest* "and auricles or artificial ears" available from Byrn's Cedar St. office in New York.

544. BYRN, Marcus Lafayette, 1826-1903.

The mystery of medicine explained; a family physician, and household companion; prepared for the use of families, plantations, ships, travelers, &c. . . . New York: Published by M. Lafayette Byrn, 1880.
508, [4] p., [1] leaf of plates : ill. ; 21 cm.

Fifth edition. At head of title: "Sixtieth edition—enlarged and revised."

545. BYRN, Marcus Lafayette, 1826-1903.

[Letter], 1877 June 27, New York, [to] Wm. R. Bland, Wellville, Virginia.
[1] leaf ; 19 cm.

On letterhead stationery, Byrn informs Bland that he has shipped the electrotherapeutic "magnetic machine" that he ordered, and provides instructions for its use. Includes postmarked envelope and a hand-written prescription (dated 25 Sept. 1877).

DR. BYRNE'S NEW YORK MUSEUM OF MEDICINE see **NEW YORK MUSEUM OF MEDICINE AND EUGENICS** (#2599).

C

546. CABELL, James Lawrence, 1813-1889.
Hot Springs, Bath County, Virginia, with some account of their medicinal properties, and an analysis of the waters, with cases of cure of gout, rheumatism, diseases of the liver, paralysis, neuralgia, chronic diarrhoea, enlarged glands, old injuries, deafness, etc., etc., etc. Richmond: Clemmitt & Jones, printers, 1871.
   96 p. ; 20 cm.

A description of the ladies' and gentlemens' baths available at the Hot Springs spa is followed by J.L. Cabell's essay entitled "On the hygienic and therapeutical effects of thermal baths" (p. 14-38). The 60-page appendix that follows is composed entirely of "certificates of cure," i.e., testimonies from persons healed of varying maladies by the local waters.
   Described as resident physician at Hot Springs on the verso of the title-page, Cabell was an 1834 medical graduate of the University of Maryland. He joined the medical faculty of the University of Virginia (Charlottesville) in 1837, serving on its faculty for fifty years. Cabell was active in the American Public Health Association, the National Board of Health and the American Medical Association. In printed wrapper.

547. CABOT, John F.
READ AND CONSIDER. It may sa[v]e many lives and much money. New York, August, 1877 . . .
   Broadside ; 28 × 21.5 cm.

Advertisement for *Phoenix Disinfectant*, effective against the "evil effects of putrefying foecal matter in the privies of large manufactories, where many hands are employed."

548. CABOT, Richard Clarke, 1868-1939.
A layman's handbook of medicine with special reference to social workers . . . Boston and New York: Houghton Mifflin Company, the Riverside Press, Cambridge, 1916.
   [2], xvii, [3], 523, [3] p. : ill. ; 21 cm.

"'A little knowledge is a dangerous thing . . . But since every body and soul in the civilized world has now been thoroughly exposed to this dangerous contagion, I know no way to reduce the risk of disaster except by injecting into all who will submit a larger dose of knowledge in the least irritating form procurable. Gradually I hope an immunity to its dangers may thus be produced" (*Pref.*, p. v). Cabot's book took shape from lectures delivered to Boston social workers in 1915-16. He describes a wide variety of disorders in an easy style obviously directed to an educated, but non-medical audience. Disease entities are grouped by bodily system (i.e., respiratory, circulatory, gastro-intestinal, generative, nervous, etc.), with additional chapters on infectious diseases, occupational diseases and emergencies.
   An 1892 Harvard medical graduate, Cabot served on the staff of the Massachusetts General Hospital from 1898 to 1921, was on the faculty of the Harvard Medical School from 1899 to 1933, and was professor of social ethics at Harvard College from 1920 to 1934. Cabot was a formidable medical author. Books such as *A guide to clinical examination of the blood for diagnostic purposes* (1897), *Physical diagnosis* (1905) and *Differential diagnosis* (1911) went through multiple editions. He is also noted for his integration of the social sciences into medicine, writing several landmarks books on the subject. *A layman's handbook of medicine* was reissued in 1923.

CADOGAN, William, 1711-1797. *A dissertation on gout*, see **BUCHAN** (#447).

CADOGAN, William, 1711-1797. *An essay upon nursing, and the management of children*, see **BUCHAN** (#441).

549. CADY, Bertha Chapman, 1873-1956.
The way life begins. An introduction to sex education. Text and illustrations by Bertha Chapman Cady and Vernon Mosher Cady. With a foreword by William Freeman Snow,

M.D. New York: The American Social Hygiene Association, [c1917].

vii, [1], 78, [2] p. : ill. (most col.) ; 24 cm.

A brief manual of plant, animal and human reproduction for parents and teachers. The nine color plates (tipped into the text) were reproduced from watercolors by Bertha Chapman Cady. The book was printed by Douglas Crawford McMurtrie (1888-1944), type-designer, printer and historian of American printing. Series: *American Social Hygiene Association serial publication, no. 85.*

550. CADY, Edward Everett.
Dentistry by specialists and the gentle art of painlessness . . . [Brooklyn, N.Y., c1910].

47, [1] p. ; 15 cm.

A brochure advertising Cady's dental practice in Brooklyn. The author was an 1885 graduate of the Chicago College of Dental Surgery. In printed wrapper.

CADY, Vernon Mosher see **CADY** (#549).

551. CALDWELL, Charles, 1772-1853.
Thoughts on physical education: being a discourse delivered to a convention of teachers in Lexington, Ky., on the 6th and 7th of Nov. 1833 . . . Boston: Marsh, Capen & Lyon, 1834.

4, [13]-133 p. : ill. ; 19 cm.

Caldwell was one of the more remarkable medical figures of his era. Undoubtedly a gifted man intellectually, Caldwell has also been described as "ambitious, pompous, vain, and possessing a colossal ego" (*Amer. nat. biog.,* 4:199). He thrived on controversy, in which he enthusiastically engaged both in person and in print. His autobiography (1855) is a continuous (and oddly entertaining) narrative of this kind of engagement. Caldwell began the study of medicine in 1791 under a preceptor in his native North Carolina. The following year he enrolled in the Medical Department of the University of Pennsylvania, courageously provided medical succor during the devasting yellow fever epidemic of 1793, took time out from his studies to serve as an army surgeon during the Whiskey Rebellion, and then returned to the

University of Pennsylvania, from which he received his medical degree in 1796. Always hopeful of an appointment to the medical faculty of his *alma mater,* Caldwell remained in Philadelphia until 1819. Although he delivered lectures on geology at the University's newly formed Faculty of Sciences (est. 1815), a medical appointment was never forthcoming. Chagrined, Caldwell accepted the professorship of the institutes of medicine at Transylvania University (Lexington, Ky.), which he transformed into one of the finest medical schools in the nation. After the inevitable falling out in Lexington, Caldwell moved in 1837 to Louisville, where he made the failing Medical Institute of the City of Louisville into a reputable institution. A list of Caldwell's publications occupies eight pages in his autobiography.

Caldwell begins his treatise on physical education by remarking on the importance of education for "the advancement of the people in intelligence and virtue." To the usual division of education into moral and intellectual branches, Caldwell adds a third: physical education. Whereas moral and intellectual education "consists in amending the condition of the brain," physical education "is the improvement of the other parts of the body." He continues, "Physical education is to the other two, what the root, trunk, and branches of the tree are to its leaves, blossoms, and fruit. It is the source and *sine qua non* of their existence" (p. 26). To Caldwell, physical education is "tantamount to an entire system of Hygeiene [sic]. It would embrace every thing, that, by bearing in any way on the human body, might injure or benefit it in its health, vigor, and fitness for action" (p. 29). Caldwell stresses the importance of heredity in establishing healthy constitutions, and reviews the usual components of human physiology and hygiene, including infant care, child rearing and the "physical education of the brain," i.e., mental health.

CALDWELL, Francis Cook see **BLAIR REMEDY CO.** (#337, 338).

552. CALDWELL, John Edwards, 1769-1819.
A tour through part of Virginia, in the summer of 1808. In a series of letters, including an account of Harper's Ferry, the Natural Bridge, the new discovery called Weir's

Cave, Monticello, and the different medicinal springs, hot and cold baths, visited by the author. New-York: Printed for the author; H.C. Southwick, printer, 1809.

31 p. ; 23 cm.

Variously attributed to Samuel Latham Mitchill, Charles Caldwell, Joseph Caldwell, and T. Caldwel (cf. Sabin #9917). John Caldwell's authorship is affirmed by David B. Warden in *A statistical, political, and historical account of the United States of North America*, Edinburgh & Philadelphia, 1819, 2:227. See also W.M.E. Rachal's preface to the 1951 Richmond edition of Caldwell's *Tour.*

553. CAMPBELL, John Bunyan, b. 1820.

Appendix and supplement to the Encyclopedia of nature and full vitapathic practice, with special lessons . . . Cincinnati, Ohio, 1881.

[2], 26, xvi p. ; 19 cm.

The *Encyclopedia of nature containing the full and complete vita-pathic system of medical practice* was first published in Cincinnati in 1878 and again in 1881.

554. CANADIAN JOURNAL OF HOMEOPATHY.

The Canadian journal of homeopathy. [Vol. 1, no. 1 (Jan. 1856)-v. 2, no. 3 (March 1857)?]. [St. Catharines; Hamilton, Ont.], 1856-57.

2 v. ; 23 cm.

"If, in bringing more prominently before the public the principle advocated by Homeopathy, of '*similia similibus curantur*,' a safer and surer method of treating disease is introduced to popular notice, it would alone be sufficient reason for the publication of the 'JOURNAL'" (v. 1, no. 1, p. [1]).

This monthly journal was edited by William A. Greenleaf and A.T. Bull, homeopathic physicians of Hamilton and London, Ontario respectively. The first three issues of the journal were published at St. Catharines until Greenleaf moved his practice to nearby Hamilton. Although the editors appeal to the "public at large" in their opening essay, much of the journal's content was directed to Canadian homeopaths. The interest of the editors in the

activities of homeopathic medical societies and colleges south of the border is apparent throughout.

Volume I in the Atwater Collection is complete (176 p.). Bound into the Atwater copy is the title-page for vol. II (1857), though no issues for this second volume are present. It is uncertain how much of vol. II actually appeared. According to the *Union list of serials in libraries of the United States and Canada* (3rd ed., New York: H.W. Wilson, 1965, 2:911), the New York Academy of Medicine holds the first three issues of vol. II, which may be all that was published.

REFERENCES: Roland, C.G. *An annotated bibliography of Canadian medical periodicals, 1826-1875,* [Hamilton, Ont.]: Hannah Institute, c1979, #58 ("No copy located").

555. CANFIELD, F. G.

PROF. F.G. CANFIELD, the only surviving representative of the GREAT BONE SETTERS OF MANCHESTER, ENGLAND, would respectfully inform the inhabitants of this place, after numerous applications, has consented to make a tour through the United States and Canadas, previous to retiring to private life. The Prof. will call upon you at your residence . . . Canfield's remedies for the following: LUNGS, LIVER, HEART, SPINE SCROFULA AND NERVOUS AFFECTIONS, HIP DISEASES . . . [188-?].

Broadside ; 25.5 × 14.5 cm.

556. CAPP, William Musser, 1842-1924.

The daughter. Her health, education and wedlock. Homely suggestions for mothers and daughters . . . Philadelphia and London: F.A. Davis, publisher, 1891.

viii, 144 p. ; 18 cm.

A manual of advice for mothers to enable them to guide their daughters "through the circle of infancy, girlhood, wifehood, and maternity—the four stages in the round of woman's life" (p. v). He continues, "The aim is to enable the mother to second more intelligently the efforts of the medical adviser when he comes professionally into the family, and to offer some practical considerations affecting woman in her family relation" (p. vi). The author was an 1885 graduate of the Jefferson Medical College.

557. CAPP, William Musser, 1842-1924.
The daughter. Her health, education and wedlock. Homely suggestions for mothers and daughters . . . Philadelphia: The F.A. Davis Company; London: F.J. Rebman, 1893.
viii, 144, [8] p. ; 18 cm.

At head of title: *Sixth thousand.*

558. CAPRON, George, 1802-1882.
New England popular medicine. By George Capron, M.D., assisted by David B. Slack, M.D. Providence: J.F. Moore, printer, 1846.
iv, [9]-608 p. ; 23 cm.

"The Domestic Medicine of Buchan, printed seventy-five years ago, is too old a book to be much of a guide at the present day. The discoveries since made, the improvements introduced, and the alterations which have taken place in the modes of practice, have made Medicine comparatively a new science. In the following work . . . an endeavor has been made to give a complete and comprehensive description of diseases, as well as full and definite directions for their treatment. In the choice of a proper arrangement for the work, many plans presented themselves, but the one selected, as the most convenient for a book of reference, is that of an alphabetical or dictionary form, placing the common name of the disease, medicine, or other subject, first, and the corresponding systematic or professional name after it. The properties and effects of all the most important articles of medicines, both simple and compound, have been faithfully described, and the subjects of food, exercise, air, sleep, and regimen generally, received such consideration as is needful for the prevention and treatment of diseases" (*Pref.*, p. [iii]-iv).

First edition. In the preface to their second edition, the authors note, "Nearly the whole of the first edition, of one thousand copies, was disposed of in the State of Rhode Island; not more than a hundred copies went abroad" (#559, p. [v]). The second edition of the *New England popular medicine* was published in 1848 (#559). Subsequent editions were published at New York in 1849 (#560), 1853 and 1854. In 1856 this work was issued under the title *Popular medicine, or the American family physician* (#561).

Self-educated during a difficult adolescence, Capron began his medical studies at age eighteen in preceptorship with a Providence physician. He attended courses of medical lectures in both Boston and Providence before starting his own practice in 1823. The following year he received his medical degree from Brown University. According to Atkinson, Capron's practice was largely obstetrical (*The physicians and surgeons of the United States*, Philadelphia, 1878, p. 679). He was a frequent contributor to the *Boston medical & surgical journal.*, writing primarily on obstetrical topics. Articles by the Providence physician David B. Slack (1798-1871) are also to be found in the pages of the *BMSJ*. In 1837 he published *Lectures on drunkenness* at Providence; and in 1845 *An essay on the human color*, an anthropological treatise on skin pigmentation.

559. CAPRON, George, 1802-1882.
New England popular medicine: a work in which the principles and practice of medicine are familiarly explained. Designed for the use of families in all parts of the United States. By George Capron, M.D., and David B. Slack, M.D. Boston: Stereotyped by George A. Curtis; A.J. Wright, printer, 1848.
viii, [9]-620, 2 p. ; 23 cm.

Second edition. "This edition, by employing a smaller-sized type . . . has been enlarged by the addition of between forty and fifty pages of new matter, embracing more articles of medicine, and further delineations of diseases. The whole has been revised, corrected, and improved" (*Pref. to the stereotype ed.*, p. vi).

560. CAPRON, George, 1802-1882.
New England popular medicine: a work in which the principles and practice of medicine are familiarly explained. Designed for the use of families in all parts of the United States. By George Capron, M.D. and David B. Slack, M.D. Fifth edition. Boston: Stereotyped by George A. Curtis, A.J. Wright, printer 1849.
viii, [9]-620, [2] p. ; 23 cm.

561. CAPRON, George, 1802-1882.
Popular medicine, or the American family physician; a work in which the principles and practice of medicine are familiarly

explained. Comprising a complete and comprehensive description of diseases, with full and definite directions for their treatment. The properties of many new and important medicines, with their doses and mode of administration, are accurately described. Designed for general use, and adapted to all climates. By George Capron, M.D., and David B. Slack, M.D. . . . New York: Ray & Brother, publishers, 1856.
    vii, [8]-705 p. ; 23 cm.

The first 620 pages were printed from the stereotype plates made for the 1848 edition (#559). The considerable addititions made to the text take the form of a supplement, i.e. fifty-eight pages of new material in a second alphabetical series (p. [625]-687), as well as a formulary and an index to both parts.

562. CARDWELL, David P.
The book of private knowledge and advice, of the highest importance to individuals in the detection and cure of "a certain disease," (venereal.) Which if neglected or improperly treated, produces the most ruinous consequences to the human constitution, with particular, and all necessary directions, to a speedy and safe recovery. By A Physician. New-York: Printed for the publisher, 1833.
    108 p. ; 15 cm.

In most bibliographies this work is attributed to D.P. Cardwell, whose name does not appear on the title-page, but in the copyright on its verso. In the preface he writes, "Our design is to impart to the non-professional reader such information, with regard to the nature, symptoms and treatment of venereal diseases, as shall warn him of the first appearance of the maladies under consideration, enable him to discriminate between them and diseases of a different character, caution him against the dangerous remedies of empirics, and acquaint him with the most rational and improved methods of treatment. We have no wish to render every man his own physician" (p. [3]).
    The author maintains that gonorrhea and syphilis are distinct diseases, in some circles still a matter of debate in the 1830s. He provides a lengthy description of the symptoms of gonorrhea, and an equally detailed description of its

treatment. His preferred remedies are solutions of zinc sulfate or silver nitrate injected through the urinary meatus. His description of symptoms and treatment for women totals four brief paragraphs, indicating that the book was intended for a male audience. Cardwell divides the manifestations of syphilis into two stages: the "primary" and the "secondary." He devotes considerable attention to the appropriateness of mercurials in the treatment of syphilis, and concludes: "All venereal diseases are curable without mercury" (p. 81). Confident that the disease could be arrested early, Cardwell devotes considerably more attention to the treatment of the chancrous stage (recommending mild caustics and poultices) than to its later, constitutional manifestations. His non-mercurial treatment of the secondary stage consisted of a regimen of bloodletting, purgatives, warm baths, vegetable diet, abundant draughts of sarsaparilla and solutions of nitric acid taken orally. Although he rejects mercurials in the symptomatic treatment of primary syphilis, he did employ such remedies as "the blue pill, calomel, and corosive sublimate" when the patient did not respond to non-mercurial treatment. "Errata" slip pasted onto p. 108.

CARLETON, Will see LOGAN (#2272).

563. CARLIN, William, 1780-1870.
Old Doctor Carlin's recipes. Being a complete collection of recipes on every known subject, as selected from the mss. of Old Doctor William Carlin, of Bedford, England, together with additions by the American editor, on various subjects; embracing also a department for the household of most thoroughly tried recipes; a treatise on bees; a treatise on poultry, etc. Being the latest and most reliable collection of recipes for the farm, the household, the sick room, the kitchen. Toledo, O., and Boston, Mass.: The Locke Publishing Co., 1881.
    602 p. ; 20 cm.

CONTENTS: Accidents and poisons; Bee-keeping; Cement, glue and paste; The cook; Disinfectants; For farmers; Flower and gem alphabet; General household; House-plants and birds; Medical and surgical; Poultry; The stable and its occupants; The toilet; Trapping, tanning

and fishing; Varnishing, polishing and paper-
ing; Miscellaneous.

564. CARLISLE, Anthony, Sir, 1768-1840.
An essay on the disorders of old age, and
on the means for prolonging human life
. . . Philadelphia: Published by Edward
Earle . . . W. Myer, New-Brunswick, 1819.
   iv, [5]-74 p. ; 21 cm.

"The age of sixty may, in general, be fixed
upon as the commencement of senility. About
that period it commonly happens, that some
signs of bodily infirmity begin to appear . . .
Long continued professional experience has
taught me to seek for such incipient disorders
in the evidences of the state of the stomach,
and in its dependencies, and from the condition of the blood and its vessels. Over-fulness
of the vessels, contamination of the blood, impaired digestion . . . obstructed bowels, and all
the dangers which result from impediments to
that source of keeping the body pure and
wholesome, are to be reckoned as the leading
causes of many diseases" (p. 13-14).
   The 1819 Philadelphia edition is the only
American issue of Carlisle's *Essay*, which first
appeared at London in 1817, and in a second
edition in 1818. It was one of numerous publications by the English surgeon, ranging in subject matter from medicine to electricity, and
from biology to the fine arts. Carlisle was president of the Royal College of Surgeons (1815),
and was knighted in 1821.
   REFERENCES: *Austin* #416.

CARMICHAEL, John Hosea see
FAULKNER (#1139, 1140).

565. CARPENTER, B.
The sick man's companion, or a short note
on fever, and a new method of cure: together with an enumeration of the symptoms which usually attend a number of diseases and a mode of treatment prescribed
with medicine, the production of our own
country . . . Johnstown [N.Y.]: Printed for
the author, 1812.
   46, [2] p. ; 20 cm.

In the treatise's first thirty-one pages the author provides brief descriptions of forty-five common diseases and their treatment with (mostly)

botanic remedies. Carpenter is particularly fond
of prescribing lobelia and mandrake root. Pages
32-45 contains "Observations upon the medicine
recommended in the foregoing treatise." An index of diseases mentioned in the text appears
on the recto of the final, unnumbered leaf.
   REFERENCES: *Austin* #418.

566. CARPENTER, Edward, 1844-1929.
Love's coming of age. A series of papers
on the relations of the sexes . . . Chicago:
Stockham Publishing Co., 1902.
   [3]-162 p., [1] leaf of plates. : port. ; 17
cm.

Originally published at Manchester, Eng. in
1896, *Love's coming of age* was the most frequently
reprinted work of the English social critic and
visionary, whose way of life is characterized in
the *Dict. nat. biog.* as "a reaction against Victorian convention and respectability" (suppl. 4,
p. 161). The scope of Carpenter's many published works reflect his interests in art, socialism, homosexuality, feminism, etc.; and, more
in the spirit of William Morris than Karl Marx,
his attempt to integrate his ideas into the reformation of English society.
   Love's coming of age consists of eight essays:
*The sex-passion; Man, the ungrown; Woman: the serf;
Woman in Freedom; Marriage: a retrospect; Marriage:
a forecast; The free society; Some remarks on the early
star and sex worships.* The tone of Carpenter's
thought is revealed in the following quotation
from *The sex passion,* "While the glory of sex
pervades and suffuses all Nature; while the flowers are rayed and starred out toward the sun in
the very ecstasy of generation; while the nostrils
of animals dilate, and their forms become instinct, under the passion, with a proud and fiery
beauty; while even the human lover is transformed, and in the great splendors of the mountains and the sky perceives something to which
he had not the key before—yet it is curious that
just here, in Man, we find the magic wand of
Nature suddenly broken, and doubt and conflict
and division entering in, where a kind of uncon-
scious harmony had erst prevailed" (p. 10-11).

567. CARPENTER, Edward, 1844-1929.
Love's coming of age. A series of papers
on the relations of the sexes . . . New York
& London: Mitchell Kennerley, 1911.
   [2], 199 p. ; 20 cm.

The first American edition of *Love's coming of age* appeared at Chicago in 1902 and was published by Alice Stockham (#3358-3374), whose own thoughts on intercourse without emission as a method of birth control are quoted and elaborated by Carpenter (p. 179-80). The present work is the first of three New York editions issued by Kennerly between 1911 and 1927. The now forgotten publishing firm of E. Haldeman-Julius, that published numerous pamphlets on sexuality from its office in Girard, Kansas from the 1920s through the 1940s, issued the final American edition of *Love's coming of age* in 1931.

Its long publishing history on both sides of the Atlantic (the final London edition appeared in 1948) testifies to this book's readership and influence. Many ideas contained in its pages that seem commonplace or mildly unconventional today were radical in 1896 and the years that followed. In *Marriage: a retrospect*, for instance, Carpenter writes: "As long as man is only half-grown, and woman is a serf or a parasite, it can hardly be expected that Marriage should be particularly successful. Two people come together, who know but little of each other . . . who certainly do not understand each other's nature; whose mental interests and occupations are different, whose worldly interests and advantage are also different; to one of whom the subject of sex is probably a sealed book, to the other perhaps a book whose most dismal page has been opened first. The man needs an outlet for his passion; the girl is looking for a 'home' and a proprietor. A glamor of illusion descends upon the two, and drives them into each other's arms. It envelops in a gracious and misty halo all their differences and misapprehensions . . . But at a later hour, and with calmer thought, they begin to realise that it is a life-sentence which has so suavely passed upon them—not reducible (as in the case of ordinary convicts) even to a term of 20 years" (p. 80).

CARPENTER, Russell Lant, 1816-1892. *A lecture on tobacco* see **LIVERMORE** (#2269).

568. CARPENTER, William Benjamin, 1813-1885.
On the use and abuse of alcoholic liquors, in health and disease. Prize essay . . . Boston: Published for the Massachusetts Temperance Society by Wm. Crosby & H.P. Nichols, 1851.
xxviii, 264 p. ; 20 cm.

Fifteen essays were submitted for a prize of one hundred guineas offered in 1849 by a private donor on the topic of the effects of alcohol on the human system. Carpenter's essay was judged the most meritorious by a committee of three London physicians that included John Forbes, Physician to the Queen's Household. In this treatise Carpenter takes the temperance argument beyond the usual moral plane and places it on a scientific footing that it had not often enjoyed. He reaches eight conclusions regarding the deleterious effects of alcohol consumption (summarized in his preface). In the seventh Carpenter quite simply concludes that "it is the duty of the medical practitioner to discourage as much as possible the habitual use of alcoholic liquors, in however 'moderate' a quantity, by all persons in ordinary health" (p. xvi). The "Preface to the Boston edition" was written by John Collins Warren (#3712-3715), president of the Massachusetts Temperance Society. The Boston edition has been augmented with numerous notes.

So well known are Carpenter's contributions to the literature of physiology, mental physiology, microscopy and marine zoology, that it is easy to lose sight of this modest treatise in such an estimable pile. *On the use and abuse of alcoholic liquors* appeared in two London editions (1850, 1851). On this side of the Atlantic, Carpenter's treatise appeared in as many editions as any of his more prominent works. It was published at Philadelphia in 1850, 1853, 1854, 1855, 1856, 1858, 1861 and 1866 (#569); at Boston in 1851 and 1861; and at Hamilton, Ontario in 1852.

569. CARPENTER, William Benjamin, 1813-1885.
On the use and abuse of alcoholic liquors, in health and disease . . . With a preface by D.F. Condie . . . Philadelphia: Henry C. Lea, 1866.
xxiv, 25-178 p. ; 20 cm.

The first American edition of Carpenter's essay on alcohol consumption was published at Philadelphia in 1850. This 1866 Philadelphia issue was the book's final American printing.

In his preface, the Philadelphia physician and health reform advocate David F. Condie emphasizes the most important distinction made in Carpenter's treatise: "All are agreed as to the baneful influence upon health and morals resulting from the excessive use of alcoholic drinks . . . But as long as the opinion prevails, that in moderate quantities the use of these drinks is both proper and salutary, it will scarcely be possible to guard the masses against indulgence in them to excess . . . To test the truth of the opinion referred to, by an examination of the effects produced upon the human frame by the use of alcoholic drinks, whether in moderate or excessive doses, is the object of the present essay. And we know of nothing that has been written upon this important question better calculated to eradicate the prejudices which still exist to intoxicating liquors . . . by showing that their occasional moderate use, so far from promoting the health and vigor of the human frame . . . is on the contrary, under all circumstances, rather injurious than beneficial" (p. vi)

569.1. CARPENTER, William Benjamin, 1813-1885.

The physiology of temperance and total abstinence. Being an examination of the effects of the excessive, moderate, and occasional use of alcoholic liquors on the healthy human system . . . London: Henry G. Bohn, MDCCCLIII. [1853].
    vi, [2], 184 p. ; 18 cm.

First edition. "This little volume may be considered, in some degree, as a new edition of the essay 'On the use and abuse of alcoholic liquors in health and disease' . . . Two large editions of that essay . . . having been disposed of, the author has remodelled it with a view to its more extensive diffusion among the general public; substituting a large amount of matter of more general interest, in the place of purely professional details; and so changing the arrangement of the whole, as to make the sequence of subjects more natural, and the line of argument more obvious" (*Pref.*, p. [iii]). Carpenter's object in this revision remains the same, i.e., to show that "many forms of disease is [sic] engendered by the habitual *moderate* use of these beverages." *The physiology of temperance* does not appear to have been reprinted in the United States, where Carpenter's *On the use and*

abuse of alcoholic liquors (#568, 569) enjoyed great success.

570. CARRINGTON, Thomas Spees.

Fresh air and how to use it . . . New York: The National Association for the Study and Prevention of Tuberculosis, 1912.
    xviii, [2], 15-250 p. : ill. ; 20 cm.

"There is no point at which the campaign against tuberculosis has had a more benficial effect on popular opinion than in the change of attitude toward the value of fresh air both in health and disease. Emphasized first as an agent of cure, the public is now beginning to recognize its value as a mode of prevention . . . In preparing this book, Dr. Carrington has kept constantly in mind the practical difficulties which the modern house dweller must meet in his attempt to avoid the evils of our present methods of construction" (L. Farrand, *Pref.*, p. vii). Carrington discusses the interior ventilation of domestic buildings; the provision of "roof bungalows," sleeping porches, loggias, etc. in the construction of new homes; and the building of tent houses and open air cottages in the country.

571. CARTER, J. E.

The botanic physician, or family medical adviser: being an improved system, founded on correct physiological principles. Comprising a brief view of anatomy, physiology, pathology, hygieine [sic], or art of preserving health: a materia medica, exclusively botanical, containing a description of more than two hundred and thirty of the most valuable vegetable remedies: to which is added a dispensatory, embracing more than two hundred recipes for preparing and administering medicine. The diseases of the United States, with their symptoms, causes, cures, and means of prevention. Likewise, a treatise on the diseases peculiar to women and children. By J.E. Carter. Written by A.H. Mathes. Madisonville, Ten.: Published by B. Parker & Co., 1837.
    688 p. ; 21 cm.

"We are apprised that a very common prejudice prevails, that the great mass of the human family are incapable of administering to their own, or to each others wants in case of sickness. This opinion has been got up, and is still maintained by the artifice of the profession in clothing the whole 'healing art' with the veil of mysticism, and obscuring its sense by learned technicals, which are often applied where they are as unmeaning, as they are uncouth . . . The same earth which yields our food, produces our medicine—all learn how to prepare their food, which requires as much art as the preparation of many of the simple and most valuable remedies do, and the preparation of the latter is as easy learned as that of the former" (p. [7]-8).

The authors preface their catalog of medicinal plants with a discourse on the nature of disease: "The most distinguished medical theorists, of later times, are Cullen, Brown and Rush, of the mineral school, and Thomson, the father of the botanic school of medicine. And it is worthy of remark that the respective theories of all these great men, except Thomson's, failed, or at least proved uncertain and inefficient in their own hands, when reduced to practice" (p. 115). Unlike those who founded medical practice on rational speculation, Thomson based his theory on practice: "his practice was first established, and therefore, not dependent upon his theory . . . He entered the school of experience without having his mind trammeled by the rules of any previous favorite theory" (p. 119). Even so, Carter and Mathes write, "We do not perfectly agree with Dr. Thomson in all things; yet we do believe that he has done more towards establishing a simple, safe and efficient mode of removing the maladies, that afflict the body, and destroy life than any other man. He has, notwithstanding many imperfections . . . opened the fountain of true medical science, from which issues a clear stream, bearing on its bosom a potent balm for the healing of most of the maladies of man" (p. 120). Disease, according to our authors, may be considered "as the *effect* of one general cause, under all its different types or modifications. These different modified forms are the symptoms of disease: they are not properly different diseases, but the different types or forms assumed by disease" (p.123). The botanic materia medica that follows this discussion is arranged "according to their most obvious effects upon the system," i.e., antispasmodics, antiseptics, astringents, diaphoretics, diuretics, expectorants, stimulants, tonics, cathartics and emetics. Atwater copy lacking p. 687-88.

572. CARTWRIGHT, Samuel Adolphus, 1793-1863.

Some account of the Asiatic cholera, cholera asphyxia or pulseless plague; with a sketch of its pathology and treatment and advice, relative to its prevention on plantations, premonitory symptoms and treatment . . . Natchez, Miss.: Printed at "The Natchez" office, 1833.
34, [2] p. ; 19 cm.

"Several of my medical acquaintances, living in remote situations, have requested: First. Some account of the cholera in reference to its history, contagion, &c. Secondly. A succint view of its pathology and treatment as set forth by the best authors, and also my own views. Thirdly. Many of my friends at a distance and of the surrounding neighborhood, particularly planters, have desired to be informed of the means best calculated to prevent cholera among negroes on the plantations—to mitigate its violence if it should occur—the premonitory symptoms, and the treatment necessary . . . Fourthly. The medicines and things necessary to have always at hand, especially on plantations remote from any physician or apothecary" (p. [4]).

A native of Virginia, Cartwright apprenticed to Benjamin Rush and studied medicine at the University of Pennsylvania. He practiced in Natchez from 1822 to 1848, and according to the *Dict. Amer. med. biog.,* "gained respect and admiration among planters around Natchez for his successful and untiring efforts during the 1832 cholera epidemic" (p. 125). Cartwright moved to New Orleans in 1848, where he remained until the outbreak of the Civil War. As a physician in the Confederate army, Cartwright applied himself to sanitary conditions in military camps around Vicksburg and Port Hudson, Miss. Among his contributions to the medical literature were articles maintaining that blacks are medically different from whites, and that slavery could be justified on the basis of these differences. Less controversial was his idea that Southern medicine should be a distinct study since the region differs from the rest of the nation in climate, diseases and their treatment.

572.1. CARTWRIGHT, Sarah, b. 1828.

Magnetism clairvoyantly discerned. Lessons from nature. Inherited characteristics explained. New light on the treatment of diseases. Medicine and how to take it. With treatises on various subjects of general interest . . . Detroit, Mich.: O.S. Gulley, Bornman & Co., 1884.

vii, [1], 269, [3] p. ; 19 cm.

Cartwright was raised in St. Clair, Michigan. After a long illness at age twenty-three, she suddenly found herself open to occult influences. A spiritual force guided her hand in writing out the prescription of a botanic remedy that was the antidote to the debilitating effects of the calomel she had been taking (p. 16). Soon after, she experienced electric shocks "which vibrated through every nerve in my body," and which she claims restored her to health. She immediately realized that "there was outside myself an intelligence which had directed and performed the cure" (p. 17), and that she herself was clairvoyant and capable of clairvoyant healing.

Cartwright explains clairvoyant healing: "This world being a material body with spirit life, and the spiritual intelligence of man being the highest, a power is given to him over all things of a lower grade in the scale of spiritual development. When the inner spirit-sight becomes active, directed by spirit intelligence, the currents are already formed for the transmission of light, sound and sight, through all material substance that is permeated with spiritual life, wherever the mind of man has penetrated. Leaving the magnetism of his thought and presence, the spirit-sight can be trained to follow, being able to examine the internal organs of the physical system, analyze the blood, discern diseased particles, trace diseases to their origin, determine the proper remedies to neutralize their effects, and bless the world with the truths thus revealed" (p. 44).

The clairvoyant healer works through a magnetic force that is not only "the great living principle" but also the "great healing principle": "The magnetic atmosphere of a human body extends for the distance of two or three feet outward more or less, according to the health and strength of the individual . . . The diseases of the physical body are represented in this atmosphere much the same as malaria is represented in the atmosphere of the earth. All accessions to this magnetic force convey to the weaker body greater power to free itself from the dead matter by increasing its vitality and

natural strength" (p. 75). Massage is a great adjunct to magnetic healing by increasing the vitality of the body's surface, making it more susceptible to positive magnetic influences from without, and able to draw disease "to the surface through the excretory actions of the pores" (p. 83).

573. CASPARI, Carl Gottlob, 1798-1828.
Dr. Caspari's homoeopathic domestic physician, edited by F. Hartmann, M.D. . . . Translated from the eighth German edition, and enriched by A treatise on anatomy and physiology, by W.P. Esrey, M.D. With additions and a preface by C. Hering, M.D. Containing also a chapter on Mesmerism and magnetism; directions to enable patients living at a distance from a homoeopathic physician, to describe their symptoms; a tabular index of the medicines and the diseases in which they are used; and a sketch of the biography of Samuel Hahnemann, the founder of homoeopathy . . . Philadelphia: Published by Rademacher & Sheek . . . [et al.], 1852.

iv, 475, 5 p. : ill. ; 20 cm.

A translation and the only American edition of Caspari's *Homoeopathischer Haus- und Reisearzt*, originally published at Leipzig in 1826. The diseases and other topics that comprise the bulk of Caspari's text are arranged alphabetically, from "Abortion" to "Wounds." William P. Esrey's (1818-1854) popular anatomy & physiology occupies nearly forty percent of the volume. It appears to have been printed from the same plates used for the original publication of Esrey's *Treatise* by Rademacher & Sheek in 1851 (#1079).

CASTLE, Frederick Augustus see *WOOD'S HOUSEHOLD PRACTICE* (#3867).

574. CASTORINE CERATE COMPANY.
STAFFORD'S CASTORINE SERATE is a prompt remedy for burns—scalds, bruises, cuts, chapped hands, fever sores, frosted feet, felons, chilblains, galls on horses, sore teats on cows . . . Manufactured by THE CASTORINE CERATE CO., East Bloomfield, Ontario Co. N.Y. . . . [187-?].

Broadside ; 31 × 24 cm.

Stafford's Castorine Hoof Ointment and *Stafford's Cathartic Pills for Horses* are also advertised.

CATECHISM OF HEALTH see **FAUST** (#1144-1146).

CATHRALL, Isaac see **BUCHAN** (#459-462, 468, 470).

575. CATLIN, George, 1796-1872.
The breath of life or mal-respiration, and its effects upon the enjoyments & life of man (manu-graph) by Geo. Catlin . . . London: Trubner & Co., 1862.
    75 p. : ill. ; 22 cm.

Trained in law, Catlin practiced in Luzerne, Pa. until 1823, when he moved to Philadelphia in order to pursue to his real vocation—painting in oils. In 1824 he moved to Washington, D.C., where he specialized in portraits of prominent figures. Catlin was so intrigued by a delegation of Native Americans in the nation's capital that he decided to devote his talents to recording on canvas the lives of an already vanishing people. From 1829 through the 1830s, Catlin spent his summers among the Indians, returning home each winter to paint. His hundreds of portraits of distinguished Native Americans and scenes of life among the peoples he visited were widely exhibited in Europe and the United States. His published works, *Letters and notes on the manners, customs, and condition of the North American Indians* (1841), *Catlin's North American Indian portfolio* (1845), *Life among the Indians* (1867), etc., enjoyed great popularity on both sides of the Atlantic.

Less known is *The breath of life,* originally published at New York in 1861. Catlin begin this brief treatise: "With the reading portion of the world it is generally known that I have devoted the greater part of my life in visiting and recording the looks of, the various races of N. & South America; and during those researches, observing the healthy condition and physical perfection of those people, in their primitive state, as contrasted with the deplorable mortality—the numerous diseases and deformities, in civilized communities, I have been lead to search for, and am able, I believe, to discover, the main causes leading to such different results" (p. [3]). The reason is the habit of civilized peoples to sleep with their mouths open: "The Savage infant . . . breathing the natural and wholesome air . . .

closes its mouth during its sleep; and in all cases of exception the mother rigidly (and cruelly, if necessary) enforces Nature's Law . . . until the habit is fixed for life, of the importance of which she seems to be perfectly well aware. But when we return to civilized life, with all its comforts, its luxuries—its science and its medical skill, our pity is enlisted for the tender germs of humanity, brought forth and caressed in smothered atmospheres which they can only breath with their mouths wide open, and nurtured with too much thoughtlessness to prevent their contracting a habit which is to shorten their days with the croup in infancy, or to turn their brains to idiocy or lunacy, and their spines to curvatures—or in manhood, their sleep to fatigue and the nightmare, and their lungs and their lives to premature decay" (p. 17-18). Both the text and the illustrations of Catlin's "manu-graph" were executed on lithographic stones by the author.

576. CATLIN, George, 1796-1872.
The breath of life or mal-respiration. And its effects upon the enjoyments & life of man . . . New York: John Wiley, 1864.
    76, [2] p. : ill. ; 23 cm.

This edition is typeset and its illustrations reproduced in wood engraving.

CAUSTIC, Christopher see **FESSENDEN** (#1157-1159).

CECIL, Richard. *A friendly visit to the house of mourning* see **WILLISON** (#3812, 3813).

CENTRAL VIAVI COMPANY see **VIAVI COMPANY** (#3652).

577. CENTURY BOOK OF HEALTH.
Century book of health. The maintenance of health, prevention and cure of disease, motherhood, care, feeding and diseases of children, modern home nursing, accidents and emergencies, injurious habits, a complete practical guide based upon the latest medical practice, recent discoveries in science and U.S. Pharmacopoeia revision of 1905. J.H. McCormick, M.D., editor-in-chief . . . Springfield, Mass.: The King-Richardson Company, [c1906].
    [3]-872 p., [5] leaves of plates : ill. ; 24 cm.

First edition. Acknowledging the impact of both laboratory medicine on the public imagination and the health education movement in American schools, John Henry McCormick (b. 1870) writes in his introduction, "At no period of the world's history have the people been so well qualified to take up the study of health as they are now. Physiology and other science subjects, taken up of late in the schools of the country, have given a solid foundation upon which a sound knowledge of the human body and its needs may rest. This fact and the wonderful advances in medical science require that the subject be approached in a very different way from that which satisfied our immediate ancestors. Ten years have made a wonderful change in our knowledge of health and disease. Treatises that were excellent in their day now fail to meet the new conditions" (p. [9]). Whatever changes were appearing in the popular medical literature, it is obvious that works such as this were still addressed largely to women, who remained responsible for infant care, treating childhood diseases, maintaining a sanitary environment in the home, and nursing the sick. As in earlier generations of domestic medical texts, special chapters are provided on "Womanhood and motherhood," home nursing, marriage advice, the female toilette, etc.

578. CENTURY BOOK OF HEALTH.

Century book of health. The maintenance of health, prevention and cure of disease, motherhood, care, feeding and diseases of children, modern home nursing, accidents and emergencies, injurious habits, a complete practical guide based upon the latest medical practice, recent discoveries in science and U.S. Pharmacopoeia revision of 1905. J.H. McCormick, M.D., editor-in-chief . . . Springfield, Mass.: The King-Richardson Company, [c1907].
  [3]-872 p., [5] leaves of plates : ill. ; 24 cm.

579. CENTURY BOOK OF HEALTH.

Century book of health. The maintenance of health, prevention and cure of disease, motherhood, care, feeding and diseases of children, modern home nursing, accidents and emergencies, injurious habits, a complete practical guide based upon the latest medical practice, recent discoveries in science and U.S. Pharmacopoeia revision of 1905. J.H. McCormick, M.D., editor-in-chief . . . Springfield, Mass.; Buffalo, N.Y.; Chicago, Ill.: The King-Richardson Company, [c1909].
  [3]-872 p., [5] leaves of plates : ill. ; 24 cm. + paper manikin

This issue has a paper anatomical manikin attached to the front pastedown.

580. CHABOT, E. Charles.

Physiology of the wedding night; by M. Octavius de St. Ernest: with an introduction, philosophical, hygeinical [sic], and moral, by Morel de Rubempre, M.D., of the Faculty of Paris. Boston, 1844.
  95 p., [2] leaves of plates : ill. ; 12 cm.

Octavius de St. Ernest was the pseudonym of E. Charles Chabot (fl. 1827-1854). The *Physiology of the wedding night* is a translation of *Physiologie de la première nuit des noces,* published at Paris in 1842. This slender volume is not a manual of sexual physiology, but a series of often very clever reflections on prospective brides, grooms, their families, and the pending union. J. Morel de Rubempre, who wrote the introduction, was the author several sex manuals in French, including *Les Secrets de la génération.* Cover-title: *Wedding night.* Two lithographic plates of female nudes precede the text.

CHAMBERLAIN, Annie Lord see HULL AND CHAMBERLAIN (#1839).

CHAMBERLIN, Caroline P. see ELECTRO-PATHIC INSTITUTE (#1047).

CHAMBERLIN, Dr. see *EVERY LADY HER OWN PHYSICIAN* (#1096).

CHAMBERLIN, Otis K. see ELECTRO-PATHIC INSTITUTE (#1047).

581. CHAMBERS, Reuben.

The Thomsonian practice of medicine; containing the names, and a description of the virtues and uses of the medicines belonging to this system of practice; also, directions for giving the proper quantity

of each article for a dose; together with the names and symptoms of the different forms of each disease, and ample directions for curing the same, with those excellent remedies . . . Bethania, Pa.: [R. Chambers, printer], 1842 [i.e., 1843].
449, [3] p. ; 19 cm.

The "Note to subscribers" (unnumbered verso of p. 449) is dated "4th month, 1843." Having purchased a "right" to Thomson's system, Chambers justifies the publication of an additional and seemingly extra-legal text on the botanic practice in his preface: "I have owned a right to this incomparable system of medicine about seven years; and, by attention and application thereto, have obtained so competent a knowledge of it, that I confidently believe that no one will be disappointed, or ever regret purchasing this book . . . But there are many individuals and families who are yet unsupplied with any book of directions for using these medicines, and for the proper management of the sick. This has been, in some cases, owing to a want of correct information respecting the system, and in others, to the PRICE (twenty dollars) at which this work has been sold. This work is, therefore, intended to supply the deficiency, at a trifling cost . . . " (p. [3]). A second issue was published at Bethania in 1843.

CHAMBERS, Thomas K. see NATIONAL TRAINING SCHOOL FOR COOKERY (#2573).

CHAMPION, Aristarchus see CHIPMAN (#635).

CHANNING, Walter, 1786-1876. *Physical education* see WINSLOW (#3837).

582. CHAPIN, Charles Value, 1856-1941.
How to avoid infection . . . Cambridge: Harvard University Press, 1917.
[2], 88 p. ; 17 cm.

Chapin provides a popular explanation of the "germ theory" in its relation to communicable disease. As were all authors in the *Harvard health talks* series, Chapin was a member of the Harvard faculty, which he joined in 1913. James H. Cassedy summarizes Chapin's work in the *Dictionary of American medical biography* (West-

port, Ct., 1984, 1:130): "Leader of the American movement to bring the findings of bacteriology into public health work. Discredited outmoded general sanitary theories and procedures by use of specific measures directed against specific diseases. Organized the first municipal bacteriological laboratory (1888) . . . Helped rejuvenate the specialty of epidemiology in this country. Conducted the first evaluation of state public health work (1913-1916) and established a realistic scale of priorities for both state and municipal health work."

CHARITY ORGANIZATION OF THE CITY OF NEW YORK, see NEW YORK (N.Y.). DEPT. OF HEALTH (#2601).

583. CHARLESTOWN (MA.). HEALTH OFFICE.
City of Charlestown. Vaccination and small pox. Extracts from the laws of the Commonwealth . . . Stephen P. Kelley, Health Officer . . . July 16, 1860.
Broadside ; 25 × 19 cm.

584. CHASE, Alvin Wood, 1817-1885.
Dr. Chase's combination receipt book. Being a combination of three books: 1. The favorite medical receipt book and home doctor. 2. Dr. Chase's receipt book. 3. A practical law and business guide for home and office . . . Enlarged by over one hundred physicians and authors. Detroit, Michigan: Published by the F.B. Dickerson Company, 1917.
[14], 504, [4], iii-x, 505-1042, [6], 1043-1292 p., [36] leaves of plates : ill., ports. ; 23 cm.

At head of title: *Combination edition. The Favorite medical receipt book.* Pt. 1 was compiled by Josephus Goodenough. It includes a section on the diseases of women and children (p. 380-462) written by Helen F. Warner. Part 2 is the "memorial edition" of *Dr. Chase's receipt book.* Part 3, *A Practical law and business guide,* was compiled by the publisher Freeman B. Dickerson. Each part has special title-page.

585. CHASE, Alvin Wood, 1817-1885.
Dr. Chase's family physician, farrier, beekeeper, and second receipt book. Being an entirely new and complete treatise, point-

ing out, in plain and familiar language, diseases of persons, horses, and cattle, upon common-sense principles; giving instructions in relation to butter and cheese manufacturing and manufactories, also full instructions in bee-keeping, and entirely new methods of horse-taming or handling vicious horses, breaking colts, etc.; embracing also a large number of entirely new receipts, in all departments of household affairs, and every branch of mechanical industry . . . Toledo, Ohio: Chase Publishing Company, 1875.

xxiv, 25-644 p. : ill. ; 20 cm.

At head of title: Fortieth thousand. In 1869 ill health forced A.W. Chase to sell his publishing firm and copyrights to the books he was publishing, including his most successful title: *Dr. Chase's recipes; or, information for everybody* (#587-595). After a period of convalescence Chase resumed his publishing career in Toledo, Ohio. As indicated in the title, *Dr. Chase's family physician* was indeed his "second receipt book," and lead the firm's list much as his first book had done. The first edition of *Dr. Chase's family physician* was published at Toledo in 1872. In all it appeared in seven issues between 1872 and 1878. In 1880 a revised edition was published that appeared in four more issues by 1887.

The recipes in *Dr. Chase's family physician* are not divided into separate "departments" as they were in his earlier work. ABORTION, AXLE, OR LUBRICATING GREASE, BEE-KEEPING, BANDAGING, etc. appear in a single alphabetical series throughout the text.

586. CHASE, Alvin Wood, 1817-1885.

Dr. Chase's family physician, farrier, bee-keeper, and second receipt book. Being an entirely new and complete treatise, including the diseases of women and children, pointing out, in plain and familiar language, diseases of persons, horses, and cattle, upon common-sense principles; giving instructions in relation to butter and cheese manufacturing and manufactories, also full instructions in bee-keeping, and entirely new methods of horse-taming or handling vicious horses, breaking colts, etc.; embracing also a large number of entirely new receipts, in all departments of household affairs, and every branch of

mechanical industry . . . Revised and enlarged by the author . . . Toledo, Ohio: Chase Publishing Company, 1884.

xxiv, 25-652 p., [1] leaf of plates : ill., port. ; 20 cm.

At head of title: Seventy-seventh thousand. Frontispiece portrait of author.

587. CHASE, Alvin Wood, 1817-1885.

Dr. Chase's recipes; or, information for everybody: an invaluable collection of about eight hundred practical recipes for merchants, grocers, saloon-keepers, physicians, druggists, tanners, shoemakers, harness makers, painters, jewellers, blacksmiths, tinners, gunsmiths, farriers, barbers, bakers, dyers, renovators, farmers, and families generally. To which has been added a rational treatment of pleurisy, inflammation of the lungs, and other inflammatory diseases, and also for general female debility and irregularities . . . Second Canadian edition. Stereotyped. Carefully revised, and much enlarged, with remarks and full explanations . . . London, Ont.: Published by E.A. Taylor, bookseller and stationer, [1864?].

xxxii, [33]-385 p. ; 17 cm.

Chase attended courses of lectures in the medical department of the University of Michigan, Ann Arbor (1856-58), and claims to be a graduate the Eclectic Medical Institute in Cincinnati. In the preface he writes, "The author, after having carried on the drug and grocery business for a number of years, read medicine, after being thirty-eight years of age, and graduated as a physician to qualify himself for the work he was undertaking; for, having been familiar with some of the recipes, adapted to these branches of trade, more than twenty years, he began in "fifty-six" . . . to publish them in a pamphlet of only a few pages" (p. [v]). The first edition of this pamphlet, the predecessor of *Dr. Chase's recipes*, was published at Ann Arbor under the title: *A guide to wealth! Over one hundred valuable recipes*. According to Chase, it was first published in 1856; the earliest copy recorded in *NUC Pre-1956 imprints* is an 1858 edition of 32 pages. By the end of 1858 a fifth edition was in print, increased in size to 42 pages. A sixth edition of 56 pages was published in 1859 under the title: *Information for everybody.* By 1860 the work had increased to 224 pages, and thereafter was published as *Dr. Chases's recipes; or, information for everybody.* The text was stereotyped with the tenth edition in 1864. The text of the tenth edition remained in print through the early 1870s, though the numeration of each issue changed. This second Canadian edition includes the "Preface to the 10th [American] edition."

The "Medical department" (p. 75-214) constitutes thirty-six percent of the text, making it the lengthiest section in the book. An index at the front of the volume provides alphabetical access to names of diseases, conditions, remedies and procedures (e.g., bleeding). Unlike most domestic encyclopedias, the entries for many diseases in Chase include descriptions of symptoms and prognosis, or hygienic and preventive measures (as well as treatments) equivalent to what may be found in most domestic medicines of the period. Many of the remedies are from the eclectic armamentarium, but the majority from folk practice gathered by the author. Extravagant claims are made for quite a few recipes, e.g., Chase's syrup for "consumption" (p. 119); the "toad ointment" for "sprains, strains, lame-back, rheumatism, caked breasts, caked udders, &c., &c." (p. 130); the "Dutch remedy" (goose oil and urine) for croup (p. 150); and the twenty-two drops, salves, washes, poultices, etc. for sore or inflamed eyes (p. 154-59). No recipes are provided for the treatment of sexually transmitted diseases. Spine-title: *Chase's recipes.* Atwater copy lacking pp. 383-85.

588. CHASE, Alvin Wood, 1817-1885.
Dr. Chase's recipes; or, information for everybody: an invaluable collection of about eight hundred practical recipes for merchants, grocers, saloon-keepers, physicians, druggists, tanners, shoe makers, harness makers, painters, jewelers, blacksmiths, tinners, gunsmiths, farriers, barbers, bakers, dyers, renovators, farmers, and families generally. To which has been added a rational treatment of pleurisy, inflammation of the lungs, and other inflammatory diseases, and also for general female debility and irregularities . . . Twenty-fourth edition . . . Ann Arbor, Mich.: Published by the author, 1865.
    xxxi, [32]-384 p. : ill., port. ; 17 cm.

589. CHASE, Alvin Wood, 1817-1885.
Dr. Chase's recipes; or, information for everybody: an invaluable collection of about eight hundred practical recipes, for merchants, grocers, saloon-keepers, physicians, druggists, tanners, shoe makers, harness makers, painters, jewelers, blacksmiths, tinners, gunsmiths, farriers, barbers, bakers, dyers, renovators, farmers, and families generally, to which have been added a rational treatment of pleurisy, inflammation of the lungs, and other inflammatory diseases, and also for general female debility and irregularities . . . Thirty-second edition . . . Ann Arbor, Michigan: Published by the author, 1866.
    xxxi, [32]-384 p. : ill., port. ; 17 cm.

590. CHASE, Alvin Wood, 1817-1885.
Dr. Chase's recipes; or, information for everybody: an invaluable collection of about eight hundred practical recipes, for merchants, grocers, saloon-keepers, physicians, druggists, tanners, shoe makers, harness makers, painters, jewelers, blacksmiths, tinners, gunsmiths, farriers, barbers, bakers, dyers, renovators, farmers, and families generally, to which have been added a rational treatment of pleurisy, inflammation of the lungs, and other inflammatory diseases, and also for general female debility and irregularities . . . Thirty-ninth edition . . . Ann Arbor, Michigan: Published by the author, 1866.
    xxxi, [32]-384 p. : ill., port. ; 17 cm.

591. CHASE, Alvin Wood, 1817-1885.
Dr. Chase's recipes; or, information for everybody: an invaluable collection of about eight hundred practical recipes, for merchants, grocers, saloon-keepers, physicians, druggists, tanners, shoe makers, harness makers, painters, jewelers, blacksmiths, tinners, gunsmiths, farriers, barbers, bakers, dyers, renovators, farmers, and families generally, to which have been added a rational treatment of pleurisy, inflammation of the lungs, and other inflammatory diseases, and also for general female debility and irregularities . . . Thirty-seventh edition . . . Ann Arbor, Michigan: Published by the author, 1866.
    xxxi, [32]-384 p. : ill., port. ; 17 cm.

592. CHASE, Alvin Wood, 1817-1885.

Dr. Chase's recipes; or, information for everybody: an invaluable collection of about eight hundred practical recipes, for merchants, grocers, saloon-keepers, physicians, druggists, tanners, shoe makers, harness makers, painters, jewelers, blacksmiths, tinners, gunsmiths, farriers, barbers, bakers, dyers, renovators, farmers, and families generally, to which have been added a rational treatment of pleurisy, inflammation of the lungs, and other inflammatory diseases, and also for general female debility and irregularities . . . Fortieth edition . . . Ann Arbor, Michigan: Published by the author, 1866.

[2], xxxii, [33]-384 p., [4] p. of plates : ill., port. ; 17 cm.

The text is preceded by four pages of wood-engraved plates that includes a view of Dr. Chase's Publication Office and Book Bindery in Ann Arbor, "built expressly to enable him to introduce greater facilities to get out his 'Recipes' sufficiently fast to satisfy the increasing demand for the work."

593. CHASE, Alvin Wood, 1817-1885.

Dr. Chase's recipes; or, information for everybody: an invaluable collection of about eight hundred practical recipes, for merchants, grocers, saloon-keepers, physicians, druggists, tanners, shoe makers, harness makers, painters, jewelers, blacksmiths, tinners, gunsmiths, farriers, barbers, bakers, dyers, renovators, farmers, and families generally, to which have been added a rational treatment of pleurisy, inflammation of the lungs, and other inflammatory diseases, and also for general female debility and irregularities . . . Ann Arbor, Michigan: Published by the author, 1867.

[2], xxix, [1], [33]-384 p. : ill. , port.; 17 cm.

On verso of title-page: Forty-third edition.

594. CHASE, Alvin Wood, 1817-1885.

Dr. Chase's recipes; or, information for everybody: an invaluable collection of about eight hundred practical recipes, for mer-

Fig. 19. "Dr. Chase's office," from the 40th ed. of Dr. Chase's recipes, #592.

chants, grocers, saloon-keepers, physicians, druggists, tanners, shoe makers, harness makers, painters, jewelers, blacksmiths, tinners, gunsmiths, farriers, barbers, bakers, dyers, renovators, farmers, and families generally. To which have been added a rational treatment of pleurisy, inflammation of the lungs, and other inflammatory diseases, and also for general female debility and irregularities . . . Ann Arbor, Michigan: Published by the author, 1867.

[2], xxix, [1], [33]-384 p. : ill. , port.; 17 cm.

On verso of title-page: Forty-fourth edition.

595. CHASE, Alvin Wood, 1817-1885.
Dr. Chase's recipes; or, information for everybody: an invaluable collection of about eight hundred practical recipes, for merchants, grocers, saloon-keepers, physicians, druggists, tanners, shoe makers, harness makers, painters, jewelers, blacksmiths, tinners, gunsmiths, farriers, barbers, bakers, dyers, renovators, farmers, and families generally. To which have been added a rational treatment of pleurisy, inflammation of the lungs, and other inflammatory diseases, and also for general female debility and irregularities . . . Ann Arbor, Michigan: Published by the author, 1867.

[2], xxix, [1], [33]-384 p. : ill. , port.; 17 cm.

On verso of title-page: Forty-ninth edition.

596. CHASE, Alvin Wood, 1817-1885.
Dr. Chase's recipes; or, information for everybody: an invaluable collection of about eight hundred practical recipes, for merchants, grocers, saloon-keepers, physicians, druggists, tanners, shoe makers, harness makers, painters, jewelers, blacksmiths, tinners, gunsmiths, farriers, barbers, bakers, dyers, renovators, framers, and families generally, to which have been added a rational treatment of pleurisy, inflammation of the lungs, and other inflammatory diseases, and also for general female debility and irregularities . . . Ann Arbor, Michigan: Published by R.A. Beal, 1870.

[2], xxix, [1], [33]-384 p. : ill., port. ; 18 cm.

On verso of title-page: Fifty-first edition. In *Dr. Chase's family physician* (#585), the author notes that in 1869 ill-health forced him to sell his publishing business and the copyrights to the books he was then publishing (p. iv). Thereafter, the edition of *Dr. Chase's recipes* that had been in print from 1864 through the mid-1870s appeared under the imprint of R.A. Beal.

597. CHASE, Alvin Wood, 1817-1885.
Dr. Chase's recipes; or, information for everybody: an invaluable collection of about eight hundred practical recipes, for merchants, grocers, saloon-keepers, physicians, druggists, tanners, shoe makers, harness makers, painters, jewelers, blacksmiths, tinners, gunsmiths, farriers, barbers, bakers, dyers, renovators, framers, and families generally, to which have been added a rational treatment of pleurisy, inflammation of the lungs, and other inflammatory diseases, and also for general female debility and irregularities . . . Ann Arbor, Michigan: Published by R.A. Beal, 1870.

[2], xxix, [1], [33]-384 p. : ill., port. ; 18 cm.

On verso of title-page: Fifty-fourth edition.

598. CHASE, Alvin Wood, 1817-1885.
Dr. Chase's recipes; or, information for everybody: an invaluable collection of about eight hundred practical recipes, for merchants, grocers, saloon-keepers, physicians, druggists, tanners, shoe makers, harness makers, painters, jewelers, blacksmiths, tinners, gunsmiths, farriers, barbers, bakers, dyers, renovators, framers, and families generally, to which have been added a rational treatment of pleurisy, inflammation of the lungs, and other inflammatory diseases, and also for general female debility and irregularities . . . Ann Arbor, Michigan: Published by R.A. Beal, 1870.

[2], xxix, [1], [33]-384 p. : ill., port. ; 18 cm.

On verso of title-page: Fifty-fifth edition.

599. CHASE, Alvin Wood, 1817-1885.
Dr. Chase's recipes; or, information for everybody: an invaluable collection of about eight hundred practical recipes, for mer-

chants, grocers, saloon-keepers, physicians, druggists, tanners, shoe makers, harness makers, painters, jewelers, blacksmiths, tinners, gunsmiths, farriers, barbers, bakers, dyers, renovators, farmers, and families generally, to which have been added a rational treatment of pleurisy, inflammation of the lungs, and other inflammatory diseases, and also for general female debility and irregularities . . . Ann Arbor, Michigan: Published by R.A. Beal, 1870.

[2], xxix, [1], [33]-384 p. : ill., port. ; 17 cm.

On verso of title-page: Fifty-sixth edition.

600. CHASE, Alvin Wood, 1817-1885.
Dr. Chase's recipes; or, information for everybody; an invaluable collection of about eight hundred practical recipes for merchants, grocers, saloon keepers, physicians, druggists, tanners, shoe makers, harness makers, painters, jewelers, blacksmiths, tinners, gunsmiths, farriers, barbers, bakers, dyers, renovators, farmers, and families generally. To which have been added a rational treatment of pleurisy, inflammation of the lungs, and other inflammatory diseases, and also for general female debility and irregularities . . . New edition . . . Ann Arbor, Michigan: Published by R.A. Beal, 1872.

400 p. : ill., port. ; 18 cm.

On verso of title-page: "Seventy-third edition."

601. CHASE, Alvin Wood, 1817-1885.
Dr. Chase's recipes; or, information for everybody: an invaluable collection of about eight hundred practical recipes for merchants, grocers, saloon-keepers, physicians, druggists, tanners, shoemakers, harness makers, painters, jewelers, blacksmiths, tinners, gunsmiths, farriers, barbers, bakers, dyers, renovators, farmers, and families generally, with a rational treatment of pleurisy, inflammation of the lungs, and other inflammatory diseases, and also for general female debility and irregularities . . . Greatly enlarged and improved by the publisher, who has added appendices to the medical, saloon, farriers', barbers' and

toilet, bakers' and cooking, miscellaneous, and coloring departments, and also several new departments, viz.: "Advice to mothers," "Rules for the preservation of health," "Accidents and emergencies," "Hints upon etiquette and personal manners," "Hints on housekeeping," "Amusements for the young," and "Bee-keeping" . . . Ann Arbor, Mich.: Published by R.A. Beal, 1875.

648 p. : ill., port. ; 21 cm.

On verso of title-page: Sixth new and enlarged edition. The first issue of the "new and enlarged edition" was published by Beal in 1874. "The "Medical Department," the "Appendix to the Medical Department," and the added chapters on first aid, hygiene and infant care comprise 57% of the text in the new edition, compared to 36% of the text that the "Medical Department" constituted in earlier editions. Chase's original "Medical department" remains intact in this edition. The appendix, however, is drawn from entirely new sources, which the publisher points out is based on the "reformed practice" of medicine (*Pref.*, p. [7]). Chase lost the copyright to this title in 1869. It continued to be published by Beal until 1900. From 1875 it was in competition with Chase's two new compilations: *Dr. Chase's family physician* (#585, 586) and *Dr. Chase's third, last and complete receipt book* (#609-611).

602. CHASE, Alvin Wood, 1817-1885.
Dr. Chase's recipes; or, information for everybody: an invaluable collection of about eight hundred practical recipes for merchants, grocers, saloon-keepers, physicians, druggists, tanners, shoemakers, harness makers, painters, jewelers, blacksmiths, tinners, gunsmiths, farriers, barbers, bakers, dyers, renovators, farmers, and families generally, with a rational treatment of pleurisy, inflammation of the lungs, and other inflammatory diseases, and also for general female debility and irregularities . . . Greatly enlarged and improved by the publisher, who has added appendices to the medical, saloon, farriers', barbers' and toilet, bakers' and cooking, miscellaneous, and coloring departments, and also several new departments, viz.: "Advice to mothers," "Rules for the preservation of health," "Accidents and emergencies," "Hints upon

etiquette and personal manners," "Hints on housekeeping," "Amusements for the young," and "Bee-keeping" . . . Ann Arbor, Mich.: Published by R.A. Beal, 1880.

648 p. : ill., port. ; 21 cm.

On verso of title-page: Fifteenth new and enlarged edition. The text of this 15th edition is prefaced by a 32 page catalog of proprietary remedies entitled: *Dr. A.W. Chase's . . . family medicines . . . Ann Arbor, Mich.: Dr. A.W. Chase Family Medicine Company, 1880.*

603. CHASE, Alvin Wood, 1817-1885.
Dr. Chase's recipes; or, information for everybody: an invaluable collection of about eight hundred practical recipes for merchants, grocers, saloon-keepers, physicians, druggists, tanners, shoemakers, harness makers, painters, jewelers, blacksmiths, tinners, gunsmiths, farriers, barbers, bakers, dyers, renovators, farmers, and families generally, with a rational treatment of pleurisy, inflammation of the lungs, and other inflammatory diseases, and also for general female debility and irregularities . . . Greatly enlarged and improved by the publisher, who has added appendices to the medical, saloon, farriers', barbers' and toilet, bakers' and cooking, miscellaneous, and coloring departments, and also several new departments, viz.: "Advice to mothers," "Rules for the preservation of health," "Accidents and emergencies," "Hints upon etiquette and personal manners," "Hints on housekeeping," "Amusements for the young," and "Bee-keeping" . . . Ann Arbor, Mich.: Published by R.A. Beal, 1881.

648 p. : ill., port. ; 21 cm.

On verso of title-page: Nineteenth new and enlarged edition.

604. CHASE, Alvin Wood, 1817-1885.
Dr. Chase's recipes; or, information for everybody: an invaluable collection of about eight hundred practical recipes for merchants, grocers, saloon-keepers, physicians, druggists, tanners, shoemakers, harness makers, painters, jewelers, blacksmiths, tinners, gunsmiths, farriers, barbers, bakers, dyers, renovators, farmers, and families generally, with a rational treatment of

pleurisy, inflammation of the lungs, and other inflammatory diseases, and also for general female debility and irregularities . . . Greatly enlarged and improved by the publisher, who has added appendices to the medical, saloon, farriers', barbers' and toilet, bakers' and cooking, miscellaneous, and coloring departments, and also several new departments, viz.: "Advice to mothers," "Rules for the preservation of health," "Accidents and emergencies," "Hints upon etiquette and personal manners," "Hints on housekeeping," "Amusements for the young," and "Bee-keeping" . . . Ann Arbor, Mich.: Published by R.A. Beal, 1883.

648 p. : ill., port. ; 21 cm.

On verso of title-page: Twenty-first new and enlarged edition.

605. CHASE, Alvin Wood, 1817-1885.
Dr. Chase's recipes; or, information for everybody: an invaluable collection of about eight hundred practical recipes for merchants, grocers, saloon-keepers, physicians, druggists, tanners, shoemakers, harness makers, painters, jewelers, blacksmiths, tinners, gunsmiths, farriers, barbers, bakers, dyers, renovators, farmers, and families generally, with a rational treatment of pleurisy, inflammation of the lungs, and other inflammatory diseases, and also for general female debility and irregularities . . . Greatly enlarged and improved by the publisher, who has added appendices to the medical, saloon, farriers', barbers' and toilet, bakers' and cooking, miscellaneous, and coloring departments, and also several new departments, viz.: "Advice to mothers," "Rules for the preservation of health," "Accidents and emergencies," "Hints upon etiquette and personal manners," "Hints on housekeeping," "Amusements for the young," and "Bee-keeping" . . . Ann Arbor, Mich.: Published by R.A. Beal, 1884.

648 p. : ill., port. ; 21 cm.

On verso of title-page: Twenty-second new and enlarged edition.

606. CHASE, Alvin Wood, 1817-1885.
Dr. Chase's recipes; or, information for everybody: an invaluable collection of about

eight hundred practical recipes for merchants, grocers, saloon-keepers, physicians, druggists, tanners, shoemakers, harness makers, painters, jewelers, blacksmiths, tinners, gunsmiths, farriers, barbers, bakers, dyers, renovators, farmers, and families generally. With a rational treatment of pleurisy, inflammation of the lungs, and other inflammatory diseases, and also for general female debility and irregularities . . . Greatly enlarged and improved by the publisher, who has added appendices to the medical, saloon, farriers', barbers' and toilet, bakers' and cooking, miscellaneous, and coloring departments, and also several new departments, viz.: "Advice to mothers," "Rules for the preservation of health," "Accidents and emergencies," "Hints upon etiquette and personal manners," "Hints on housekkeping," "Amusements for the young," and "Bee-keeping" . . . Ann Arbor, Mich.: Published by R.A. Beal, 1887.

648 p. : ill., port. ; 21 cm.

On verso of title-page: "Twenty second new and enlarged edition." Title-page is a *cancellans,* and is pasted onto stub of the former title-page.

607. CHASE, Alvin Wood, 1817-1885.

Dr. Chase's recipes or information for everybody. An invaluable collection of practical recipes for merchants, grocers, saloon keepers, physicians, druggists, tanners, shoemakers and harness makers, painters, jewelers, blacksmiths, miners, gunsmiths, furriers, barbers, bakers, dyers, renovators, farmers and families generally. To which have been added a rational treatment of pleurisy, inflamation [sic] of the lungs and other inflamatory [sic] diseases, and also for general female weakness and irregularities . . . Revised and enlarged by Wm. Wesley Cook . . . With additions to many departments by the publishers. Chicago: Thompson & Thomas, [c1902].

[3]-601 p. : ill. ; 21 cm.

The editor, William Wesley Cook (1859-1936), was an 1882 graduate of the Physio-Medical Institute in Cincinnati. He is described on the title-page as "Professor of Physiological Medicine in the National Medical University, Chicago."

608. CHASE, Alvin Wood, 1817-1885.

Dr. Chase's recipes; or information for everybody: an invaluable collection of about eight hundred practical recipes for merchants, grocers, saloon-keepers, physicians, druggists, tanners, shoemakers, harness makers, painters, jewelers, blacksmiths, tinners, gunsmiths, farriers, barbers, bakers, dyers, renovators, farmers, and families generally. With a rational treatment of pleurisy, inflammation of the lungs, and other inflammatory diseases, and also for general female debility and irregularities . . . Greatly enlarged and improved by the publisher, who has added appendices to the medical, saloon, farriers', barbers' and toilet, bakers' and cooking, miscellaneous, and coloring departments, and also several new departments, viz.: "Advice to mothers," "Rules for the preservation of health," "Accidents and emergencies," "Hints upon etiquette and personal manners," "Hints on housekeeping," "Amusements for the young," and "Bee-keeping" . . . Ann Arbor, Mich.: Sheehan & Co.; Detroit, Mich.: National Book Co., 1903.

648 p. : ill., port. ; 21 cm.

At head of title: Beal edition.

609. CHASE, Alvin Wood, 1817-1885.

Dr. Chase's third, last and complete receipt book and household physician or practical knowledge for the people, from the life-long observations of the author, embracing the choicest, most valuable and entirely new receipts in every department of medicine, mechanics, and household economy; including a treatise on the diseases of women and children. In fact, the book for the million . . . Detroit, Mich., and Windsor, Ont.: Published by F.B. Dickerson Company, 1898.

ix, [1], xii-xiii, [2], 865, [3] p., [16] leaves of plates : ill., port. ; 23 cm.

At head of title: *Memorial edition.* The medical and health-related sections occupy the first 317 pages of Chase's third collection of receipts. The book is once again divided into sections (e.g., "Medical receipts," "Culinary or cooking department," "Domestic animals," etc.), the format of his first book (#587-595) that was abandoned in his second (#585-586). The contents of this third collection of receipts are entirely different from his first two compilations. Chase held the original copyright to this "third, last and complete receipt book," but did not live to publish the volume. The firm of F.B. Dickerson held the copyright its entire publishing history, appearing in at least eighteen issues between 1887 and 1909.

610. CHASE, Alvin Wood, 1817-1885.
Dr. Chase's third, last and complete receipt book and household physician or practical knowledge for the people, from the lifelong observations of the author, embracing the choicest, most valuable and entirely new receipts in every department of medicine, mechanics, and household economy; including a treatise on the diseases of women and children; in fact, the book for the million . . . Detroit, Mich., and Windsor, Ont.: Published by F.B. Dickerson Company, 1903.
ix, [1], xii-xiii, [2], 865 p., [24] leaves of plates : ill., port. ; 23 cm.

At head of title: *Memorial edition.*

611. CHASE, Alvin Wood, 1817-1885.
Dr. Chase's third, last and complete receipt book and household physician or practical knowledge for the people from the lifelong observations of the author, embracing the choicest, most valuable and entirely new receipts in every department of medicine, mechanics, and household economy; including a treatise on the diseases of women and children, in fact, the book for the million . . . Detroit, Michigan: Published by F.B. Dickerson Co., 1909.
ix, [1], xii-xiii, [2], 865 p., [24] leaves of plates : ill. ; 23 cm.

The final issue of this title, originally published at Detroit in 1887. At head of title: *Memorial edition.* Spine-title: *Dr. Chase's last receipt book and household physician.* Cover-title: *Dr. Chase's last complete work.*

612. CHASE, Alvin Wood, 1817-1885.
Recepte von Dr. Chase, oder Belehrung für Jedermann, eine sehr werthvolle Sammlung von ungefähr 800 praktischen Recepten für Kaufleute, Specereihändler, Salonhalter, Aerzte, Apotheker, Gerber, Schuhmacher, Sattler, Anstreicher, Goldarbeiter, Grobschmiede, Flaschner, Büchsenmacher, Thierärzte, Barbier, Bäcker, Färber, Renovirer, Farmer und Familien im Allgemeinen, eine gründliche Behandlung von Brustsellentzündung, Lungenenzuendung &c., sowie der Krankheiten des weiblichen Geschlechts . . . in's Deutsche übersetzt von Ch.F. Spring . . . und durchgesehen von Ch. Eberbach . . . Electrotypirte Ausgabe. Ann Arbor, Mich.: Verlag des Verfassers, 1865.
x, [11]-282 p. : ill., port. ; 21 cm.

Translation of: *Dr. Chase's recipes; or, information for everybody* (#587-608)

CHASE, Deiadamia Button see **WHITNEY** (#3771).

614. CHAVASSE, Pye Henry, 1810-1879.
Advice to a mother on the management of her children . . . Thirteenth edition. To which is added Counsel to a mother by the same author. Philadelphia: J.B. Lippincott & Co., 1873.
2 pts. in 1 v. ([2], xii, [13]-408 p. ; xii, 13-169 p.) ; 19 cm.

Chavasse's treatise on the hygiene, medical management and rearing of children was the longest published and arguably the most influential manual of its kind ever printed in the United Kingdom. The first edition was published at London in 1839 under the title *Advice to mothers on the management of their offspring,* and appeared in at least twenty-two editions over more than a century. The final London edition was published in 1948. The first American edition was published at New York in 1844 from the third London edition (1843). It was frequently reprinted by publishers in New York, Chicago and Philadelphia over the next half-century. The 18th Philadelphia edition (1898) appears to be its final American publication.
Chavasse's treatise is divided into three parts: infancy, childhood, boyhood and girlhood. With obvious exceptions, the chapters within each part include the identical topics in the same order (e.g., ablution, clothing, diet, exercise, sleep, etc.). It is supplemented by *Counsel*

to a mother: being a continuation and the completion of "Advice to a mother" (with separate title-page and pagination, but continuous collation). Spine-title: *Advice and counsel to a mother.*

According to his obituary in the *Brit. med. journal* (1879, 2:521), Pye Henry Chavasse "at an early age selected the medical profession for his career in life, and became a pupil of his cousin, Mr. Thomas Chavasse, who then carried on an extensive practice in . . . Birmingham. Having completed his pupilage, Mr. Chavasse entered as a student at University College, London; and in 1833, was admitted a member of the College of Surgeons and a Licentiate of Apothecaries' Hall, and shortly afterwards commenced to practice in Birmingham, directing his attention more particularly to the diseases of women and children . . . In 1852, the deceased was elected a Fellow of the Royal College of Surgeons."

615. CHAVASSE, Pye Henry, 1810-1879.
Advice to a mother on the management of her children and on the treatment on the moment of some of their more pressing illnesses and accidents . . . Canadian copyright edition. Toronto and Winnipeg: The T. Eaton Co. Limited, [c1879].
  [4], 328 p. ; 19 cm.

Issued with the author's *Advice to a wife on the management of her own health* (Canadian copyright ed., Toronto, c1879, #619). Spine-title: *Manual for wives and mothers.*

616. CHAVASSE, Pye Henry, 1810-1879.
Advice to a mother on the management of her children and on the treatment on the moment of some of their more pressing illnesses and accidents . . . First American edition. New York: Albert Cogswell, publisher, 1880.
  273 p. ; 19 cm.

Issued with author's *Advice to a wife on the management of her own health* (1st Amer. ed., New York, 1880, #620). Cover-title: *Advice to a wife and mother.*

617. CHAVASSE, Pye Henry, 1810-1879.
Advice to a mother on the management of her children, and on the treatment on the moment of some of their more pressing illnesses and accidents . . . Ninth edition

revised and brought up to date by a leading physician. London, New York, Toronto and Melbourne: Cassell and Company, Limited, 1911.
  iv, 316 p. ; 19 cm.

618. CHAVASSE, Pye Henry, 1810-1879.
Advice to a wife on the management of her own health and on the treatment of some of the complaints incidental to pregnancy, labour, and suckling. With an introductory chapter especially addressed to a young wife . . . First American edition. Chicago: Belford, Clarke & Co., [after 1878]
  2 v. in 1 ([6], 15-264 p. ; 273 p.) ; 20 cm.

The title-page of this Chicago edition is undated; its preface is dated June, 1878. This treatise was originally published as the second part of Chavasse's *The young wife's and mother's book* (London, 1842), the first part of which was the second edition of his *Advice to mothers.* The first American edition (*Advice to wives on the management of themselves* . . .) was published at New York in 1844 from the second London edition (1843). This Chicago edition of *Advice to a wife* is issued with the author's *Advice to a mother on the management of her children.* Spine-title: *Advice to a wife and mother.*

619. CHAVASSE, Pye Henry, 1810-1879.
Advice to a wife on the management of her own health and on the treatment of some of the complaints incidental to pregnancy, labour, and suckling, with an introductory chapter especially addressed to a young wife . . . Canadian copyright edition. Toronto and Winnipeg: The T. Eaton Co. Limited, [c1879].
  [8], 307 p. ; 19 cm.

Issued with author's *Advice to a mother on the management of her children* (Canadian copyright ed., Toronto, c1879, #615). Spine-title: *Manual for wives and mothers*; half-title: *Advice to a wife and Advice to a mother. Two volumes in one.*

620. CHAVASSE, Pye Henry, 1810-1879.
Advice to a wife on the management of her own health and on the treatment of some of the complaints incidental to pregnancy, labour, and suckling. With an introductory chapter especially addressed to a young wife . . . First American edition. New York:

Albert Cogswell, publisher, 1880.
  [4], 9-[10], 15-264 p. ; 19 cm.

Issued with author's *Advice to a mother on the management of her children* (1st Amer. ed., New York, 1880, #616). Cover-title: *Advice to a wife and mother.*

621. CHAVASSE, Pye Henry, 1810-1879.
Counsel to a mother: being a continuation and the completion of "Advice to a mother" . . . Philadelphia: J.B. Lippincott & Co., 1871.
  xii, 13-169 p. ; 17 cm.

"'Advice to a mother' [#614-617] has, by repeated additions, so increased in size that it will be quite out of the question to enlarge it. I still have a store of 'advice'—the fruits of a long and extensive experience—to offer a mother. I have, therefore, been induced to publish a second series, under the title of 'Counsel to a mother,' which I would fain hope will be as acceptable to my readers as the original volume" (*Pref.*, p. vii).

622. CHAVASSE, Pye Henry, 1810-1879.
Physical life of man and woman: or advice to both sexes; being a compendium of the laws whose observance brings health and happiness, and whose infraction, disease and misery. Advice to wife and mother by P. Henry Chavasse . . . and Advice to a maiden, husband, and son, from the most recent French and German works; with notes and additions, by an American medical writer. Cincinnati, O.; Memphis, Tenn.: National Publishing Company; Chicago, Ills.: Jones Brothers & Co., 1873.
  ix, [10]-431 p., [1] leaf of plates ; 20 cm.

This work has three parts. The second and lengthiest is a reprint of Chavasse's *Advice to a wife.* The first ("Advice to a maiden") and third ("Advice to a man") are treatises on marriage, sexual ethics, the laws of hereditary influence, etc. "compiled mainly from the recent works of A. Debay, Van Ammon and other French and German writers" (*Pref.*, p. [10]).

623. CHAVASSE, Pye Henry, 1810-1879.
The physical training of children . . . With a preliminary dissertation, by F.H. Getchell

. . . Philadelphia, Pa. . . . [etc.]: New-World Publishing Company; Burlington, Iowa: R.T. Root, 1872.
  xvi, 17-368 p., [4] leaves of plates : ill. ; 22 cm.

The physical training of children is a reprint of *Advice to a mother.* The New World Publishing Company issued this title every year between 1871 and 1875. Francis Horace Getchell's (1836-1907) "preliminary dissertation" is simply an introduction.

624. CHAVASSE, Pye Henry, 1810-1879.
The physical training of children . . . With a preliminary disseration by F.H. Getchell . . . Philadelphia, Pa. . . . [etc.]: New-World Publishing Company; Burlington, Iowa: R.T. Root, 1874.
  xvi, 17-368 p., [4] leaves of plates : ill. ; 22 cm. + atlas ([16] leaves)

625. CHAVASSE, Pye Henry, 1810-1879.
The physical training of children . . . With a preliminary dissertation by F.H. Getchell . . . Philadelphia: W.S. Fortescue & Co., 1881.
  xvi, 17-368 p., [4] leaves of plates : ill. ; 22 cm.

626. CHAVASSE, Pye Henry, 1810-1879.
What every woman should know. Containing facts of vital importance to every wife, mother and maiden. Adapted from the writings of Pye Henry Chavasse . . . together with an introduction by Sarah Hackett Stevenson, M.D. . . . Chicago and Philadelphia: H.J. Smith & Co., 1892.
  [2], x, 11-606 p., [29] leaves of plates : ill. ; 22 cm.

A reprint of *Advice to a mother* and *Advice to a wife.* This combined reprint was published earlier under the title *Wife and mother* (#627-629), with several new chapters. In the pocket on the back pastedown is an insert entitled: *Plates illustrating the female reproductive organs* ([17] leaves) dated 1892.

627. CHAVASSE, Pye Henry, 1810-1879.
Wife and mother; or, information for every woman. Adapted from the writings of Pye Henry Chavasse . . . Together with an

introduction by Sarah Hackett Stevenson, M.D. . . . Chicago and Philadelphia: H.J. Smith & Co., 1887.
    [2], x, 11-526 p. ; 20 cm. + atlas

A reprint of *Advice to a mother* and *Advice to a wife*. The Atwater copy is lacking the plates illustrating the female reproductive organs that was inserted into the pocket on the back pastedown.

628. CHAVASSE, Pye Henry, 1810-1879.
Wife and mother; or, information for every woman. Adapted from the writings of Pye Henry Chavasse . . . Together with an introduction by Sarah Hackett Stevenson, M.D. . . . Revised and enlarged . . . Thirty-fifth edition. Philadelphia . . . [etc.]: H.J. Smith & Co., 1888.
    [2], x, 11-575 p. ; 20 cm. + atlas

Atwater copy lacking insert in back pocket.

629. CHAVASSE, Pye Henry, 1810-1879.
Wife and mother; or, information for every woman. Adapted from the writings of Pye Henry Chavasse . . . Together with an introduction by Sarah Hackett Stevenson, M.D. . . . Revised and enlarged . . . Fiftieth edition. Philadelphia . . . [etc.]: H.J. Smith & Co., 1889.
    [2], x, 11-575 p. ; 20 cm. + atlas ([17] leaves : ill. ; 18 cm.)

Insert in back pocket: *Plates illustrating the female reproductive organs. H.J. Smith & Co., Philadelphia . . . 1889.* The *Plates* contains thirty wood engravings and explanatory text.

630. F. J. CHENEY AND COMPANY.
Hall's Catarrh Cure. Cures catarrh of the nasal cavity, chronic and ulcerative, catarrh of the eye, ear or throat, stomach, bowels or bladder. It is taken internally, and acts directly upon the blood and mucus surfaces of the sytsem. It is the best blood purifier in the world, and is worth all that is charged for it, for that alone. Only internal cure for catarrh on the market. And we offer $100 for any case of catarrh it will not cure . . . Toledo, Ohio: F.J. Cheney & Co., [ca. 1889].
    7, [1] p. : ill. ; 21 cm.

Issued with several other pieces of printed ephemera advertising *Hall's Catarrh Cure*, including a notarized declaration that "Frank J. Cheney makes oath that he is the senior partner in the firm of F.J. Cheney & Co. . . . and that said firm will pay the sum of one hundred dollars for each and every case of catarrh that cannot be cured by the use of Hall's Catarrh Cure" (dated 6 Dec. 1886).

631. CHEPMELL, Edward Charles.
A domestic homoeopathy, restricted to its legitimate sphere of practice; together with rules for diet and regimen . . . First American edition, with additions and improvements, by Samuel Barlow, M.D. New-York: William Radde, 1849.
    xliii, [1], 224 p. ; 16 cm.

"A *domestic* practice of medicine, in order to be useful, must be confined to its legitimate sphere of action; being called into requisition, either in times of sudden emergency, when no better means are within reach, or in those ordinary cases for which no further knowledge is needed beyond the elementary notions of anatomy and diseases in general, which most intelligent persons possess" (*Pref.*, p. [iii]). Chepmell has assured that his manual is used within these bounds by "leaving out the whole of that class of diseases, which none but qualified persons should undertake" (p. v). His treatise is divided into two parts. The first is discusses hygiene; the second is devoted to "common disorders" (e.g., boils, stye, toothache, colic, etc.), "accidental disorders" (from 'fright' to "sprains and strains"), "diseases of infants and children," and "complaints of women." Originally published at London in 1848.

632. CHEYNE, George, 1673-1743.
An essay of health and long life . . . The fifth edition . . . London: Printed for George Strahan . . . and J. Leake, bookseller at Bath, 1725.
    [4], xx, [24], 232 p. ; 19 cm. (8vo)

"Most men know when they are ill, but very few when they are well. And yet it is most certain, that 'tis easier to *preserve* health than to *recover* it, and to *prevent* diseases than to *cure* them. Towards the first, the means are mostly in our power . . . But towards the latter, the means are perplexed and uncertain; and for the knowledge of them the far greatest part of mankind must apply to others, of whose skill and honesty they are in a great measure ignorant, and the benefit of whose art they can but conditionally and precariously obtain" (p. 1-2).

A successful London practitioner, ill-health and corpulence forced Cheyne to abandon the *bon vivant* tone of his early life and habits in the City, and to adopt a more sombre tone in which he decried of the evils of luxury and extolled the benefits of moderation. Cheyne was the author of several iatromathematical treatises, now of little interest; but also of *The English malady* (London, 1733), his best known work, and an early exposition of nervous diseases and of the hypochondria from which he suffered. He also wrote several popular treatises on health and medicine that his biographer in the *Dict. nat. biog.* characterizes as "preaching temperance to an intemperate generation" (IV:218). They include his *Observations concerning the nature and due method of treating the gout* (1st ed., 1720), *An essay on health and long life* (1st ed., 1724), *An essay on regimen* (1st ed., 1740), and *The natural method of cureing the diseases of the body* (1st ed., 1742).

Cheyne devotes the seven chapters of *An essay of health and long life* to the principal factors influencing health: air, diet, rest, exercise, "evacuations and their obstructions," the "passions," etc. The temper of Cheyne's readership is evident in his remark that "most people think the only remedy for gluttony is drunkenness, or that the cure of a surfeit of meat is a surfeit of wine" (p. 47). Having suffered the consequences of this mode of life, Cheyne here recommends a "simple, low, and almost vegetable diet" (p. 39). Athough he does not specifically forbid the consumption of meat in the pages of the *Essay*, he leaves little to desire when describing the quality of animal flesh available on the London market, and recommends that his readers at least lessen the quantity of meat they consume by half. Cheyne advises his readers to drink water or wine mixed with water, and to avoid the consumption of undiluted wines, malt liquors and spiritous liquours. "Wine is now become as common as water," he warns, "and the better sort scarce ever dilute their food with any other liquor. And we see, by daily experience, that . . . their blood becomes inflamed with gout, stone, and rheumatism, raging fe-

vers, pleurisies, small pox, or measles; their passions are enraged into quarrals [sic], murder, and blasphemy; their juices are dried up; and their solids scorch'd and shrivel'd" (p. 44).

Cheyne recommends that his largely sedentary readership take exercise by "walking, riding a horse back, or in a coach, fencing, dancing, playing at billiards, bowls, or tennis, digging, working at a pump, ringing a dumb bell, &c." (p. 94), so that the "joints are render'd pliable and strong," the "blood continues sweet, and proper for a full circulation," the perspiration is free and the "organs stretched out . . . to their proper extension" (p. 96). He reminds his readers that "The three principal evacuations are, by siege [an obsolete term for excrement], by water and by perspiration" (p. 109); and includes their unobstructed functioning in his concluding remarks: "The Grand Secret, and sole mean of long life, is to keep the blood and juices in a due state of thinness and fluidity, whereby they may be able to make those rounds and circulations through the animal fibres, wherein life and health consist, with the fewest rubs, and least resistance, that may be" (p. 220).

633. CHEYNE, George, 1673-1743.
An essay on health and long life . . . New-York: Printed and published by Edward Gillespy, 1813.
vi, [7]-192 p. ; 17 cm.

Although Cheyne is occasionally quoted in the works of American health reformers pre-1850, this 1813 edition of his *Essay* is the only one of his several popular treatises on health and medicine published in the United States. His books did not enjoy the transatlantic success accorded similar works by Buchan, Cornaro, Tissot, *et al.* The reason may be, perhaps, that his science as well as his rhetoric were perceived as outdated. Nonetheless, there are many elements in his hygienic regimen that agree with the precepts expressed by American health reformers throughout the 19th century.
REFERENCES: *Austin* #455.

CHEYNE, John see **BAUNSCHEIDT** (#237, 238).

634. CHILD, Lydia Maria Francis, 1802-1880.
The family nurse; or companion of the frugal housewife . . . Revised by a member of

the Massachusetts Medical Society . . . Boston: Charles J. Hendee, 1837.
      156 p. ; 19 cm.

In the earlier part of her career, Child was known as a novelist, a writer of children's stories (her *Juvenile miscellany*, 1826-34, was the first American periodical for children), and author of *The frugal housewife* (1829), a manual of domestic economy reprinted more often than any of her other books. She later gained notoriety as an abolitionist and author of such treatises as *An appeal in favor of that class of Americans called Africans* (1833) and *Correspondence between Lydia Maria Child and Gov. Wise and Mrs. Mason of Virginia* (1860). 300,000 copies of the latter pamphlet were circulated. After publication of the *Appeal,* Child's popularity dropped dramatically. Her *Juvenile miscellany* folded, and sales of her other books suffered, including *The family nurse,* which appeared in this one edition. *The family nurse* includes chapters on hygiene, home nursing, children's diseases, a domestic formularly, an herbal, and recipes for poultices, ointments, blisters, etc.

CHILD'S HEALTH PRIMER see **ANDREWS** (#101).

635. CHIPMAN, Samuel, b. 1786.
Report of an examination of poor-houses, jails, &c., in the State of New-York, and in the counties of Berkshire, Massachusetts; Litchfield, Connecticut; and Bennington, Vermont, &c. Addressed to Aristarchus Champion, Esq. of Rochester, N.Y. . . . Albany: Printed by the Executive Committee of the New-York State Temperance Society; Hoffman & White, printers, 1834.
      96 p. ; 19 cm.

Over a period of nine months, Chipman travelled 4,500 miles through New York State visiting the prisons and poor-houses of each county. This fact-finding journey was funded by Aristarchus Champion, Rochester's wealthiest citizen. Its purpose was to determine the number of inmates in each tax-supported institution whose presence was the result of intemperance. In his introductory letter to Champion, Chipman writes: "I had become fully satisfied that in our efforts to advance the cause of temperance, facts must be our principal reliance. I saw one great field yet but partially

explored, where a rich harvest of facts might be gathered" (p. [3]). Chipman's intention was not only to statistically verify the consequences of alcohol consumption (i.e., crime, poverty, etc.), but to assess the cost to taxpayers of maintaining intemperate inmates in publicly supported institutions.

636. CHOATE, Clara Elizabeth, 1851-1916.
Modern science of body; The Christ cure; lectures and miscellanies . . . Boston, Mass.: H.H. Carter & Co., publishers, 1889.
      180, [2] p. ; 20 cm.

"God is a law, a supreme and demonstrable principle," Choate declares in the first of these collected lectures and sermons (p. [9])."Therefore, "God is science, and must be understood." Having established God as a principle evident to mind, Choate concludes that man, like Jesus, can discern the "Divine Law" by which all things operate. "Surely, then, when this is comprehended, health will triumph . . . Above and beyond the so-called interruptions of birth and death, and their attending limitations, man ever expresses the realities of soul, and the truth of immortal science governs mind and body" (p. 12). Assuming as axioms that "Spirit is substance and the only substance," and that "Body is the expression of mind" (p. 15), Choate declares that mind, which is spirit, can control the body, which is a manifestation of the same spirit, and heal the body's ills and deformities. For Jesus "understood better than others . . . that body is manifest mind; that whatever mind establishes as law the body, moulded only by thought, must reflect the same. Never by word or deed did he teach that man must suffer or sin, or that he must die, that matter is power, or that man with truth cannot overcome these dreaded foes to happiness and health" (p. 17-18). The acceptance of sickness and death is based on ignorance. "Man should be taught the truth of his being, he should understand its science, then his body, like himself, would express the wondrous immortality of health and holiness" (p. 20). Health is the consequence and one of the benefits of knowing Divine Law. The resurrection of the body is now, and death and illness are reduced to the needless consequences of ignorance of the truth of science.
      Choate was president of the Choate Metaphysical College. Her obituary appears in the

22 January 1916 issue of the *Boston evening transcript* (9:4), and describes Choate as "one of the oldest Christian Science teachers and one of the first associates of Mrs. Mary Baker Eddy in presenting Christian Science to the world . . . Mrs. Choate was born in Salem in 1851, the daughter of Mr. and Mrs. Warren F. Childs. After she married George Warren Choate of Salem, they came to Boston to live. Mr. Choate was one of the first Christian Science practitioners, and was assisted by his wife, who had become a great friend of Mrs. Eddy. The Choates practiced in Newbury Street for more than twenty years, and before the Christian Science Church was completed their home was a meeting place for many notable followers of the faith. Mrs. Choate was considered the leading Christian Science lecturer . . . and delivered more than seven hundred lectures in different parts of the country . . . She was the oldest contributor to both the Monitor and the Christian Science Journal."

637. CHRISTIAN, Eugene, 1860-1930.
Eat and be well. Eat and get well . . . New York: Alfred A. Knopf, MCMXVI [1916].
[14], 133 p. ; 19 cm.

Christian was the author of more than a dozen popular treatises on diet and health, including his five volume *Encyclopedia of diet* (New York, 1914). A native of McMinnville, Tenn., Christian was in sales and manufacturing until 1900, when he began to teach and practice as a dietician. He was prosecuted by the Medical Society of the County of New York for his venture into diet therapy, which resulted in a decision by the Supreme Court of New York that a food scientist does have the right to diagnose and prescribe diet in the treatment of illness.

638. CHRISTIAN, Eugene, 1860-1930.
How to live 100 years. What to eat according to your age, your occupation, and the time of the year. New York: The Christian Dietetic Society, 1914.
xi, [1], 121, [1] p., [1] leaf of plates : port. ; 16 cm.

The Christian Dietetic Society was one of several ventures founded by the author that included a realty company, a hotel, and vitamin & health food retailing

639. CHRISTIAN, Eugene, 1860-1930.
Little lessons in scientific eating . . . Maywood, N.J.: Published by Corrective Eating Society, [c1916?].
24 pamphlets ; 17 cm.

Also published under the title: *Eugene Christian's course in scientific eating.*

Table 3. Contents of Christian's *Little lessons in scientific eating* (#639)

CONTENTS:	
Lesson I.	The three great laws that govern life—human nutrition—corrective eating
Lesson II.	What food is and its true purpose
Lesson III.	Digestion, assimilation and metabolism explained in plain language
Lesson IV.	The chemistry of the body and the chemistry of food explained and simplified
Lesson V.	How wrong eating causes disease
Lesson VI.	How foods cure, or how foods establish health by removing the causes of disease
Lesson VII.	Scientific eating explained, simplified and made practical—with sample menus
Lesson VIII.	Harmonious combinations of food and digestive inharmonies
Lesson IX.	How to select, combine and proportion your food according to age—with sample menus
Lesson X.	How to select, combine and proportion your food according to your occupation and the season of the year—with sample menus
Lesson XI.	Obesity, its cause and cure—with sample menus
Lesson XII.	Emaciation, its cause and cure—with sample menus
Lesson XIII.	The business man—right and wrong way of living—sample menus

Table 3. *Continued*

CHRISTIAN, Eugene see also
CORRECTIVE EATING SOCIETY
(#803).

640. CHRISTIE, Abel H.
The wonders of galvanism, and its application as a remedial agent . . . New-York: D.C. Morehead, M.D., [c1848].
8 p. : ill. ; 20 cm.

Promotional brochure for Christie's galvanic belts, necklaces, bracelets and fluid. In printed wrapper.

641. CHRYSTAL, Andrew.
How electricity cures. A treatise on various forms of organic and nervous affections, describing a certain, safe and scientific cure. Marshall, Mich.: Prof. Andrew Chrystal, c1903.
48 p. ; 14 × 7 cm.

Chrystal maintains that his patented *Electronic Invigorator* is "the true means of reviving waning manhood and vitality" (p. 7). When the male organs "lie dormant, shrunken, cold and lifeless, they are paralyzed, the blood has ceased to circulate, and they are dead and useless as those of a mummy; but when put under treatment of my Electronic Invigorator and the blood is gradually and carefully brought back

through the dried-up and shruken arteries and blood vessels, they develop in size and strength and capacity every day" (p. 23). Chrystal's *Invigorator* not only revives the dysfunctional male organ, but is a valuable tool in the treatment of rheumatism, catarrh, urogential disorders, headache, locomotor ataxia, neuralgia, and consumption. It even invigorates the brain: "No man with a weak, vacillating will, a brain incapable of understanding and applying the great facts of life can attain to true eminence . . . If you are deficient in electrical energy, your brain, that great storehouse of power, is being cheated of its sustenance . . . My Invigorator gives strength of purpose, depth of understanding, power, energy, ability and resourcefulness" (p. 45-46). The *Electronic Invigorator* was available for $30.00. Title on printed wrapper: *An Electric Invigorator and how it cures.*

642. CHURCHILL, Thomas Furlong.
The new practical family physician; or, improved domestic medical guide: containing a very plain account of the causes, symptoms, and method of curing every disease incident to the human body; with the most safe and rational means of preventing them, by an approved plan of regimen, air, and exercise, adapted to the use of private families. Together with observations on the effects of the passions, and their influence on the pro-

duction and prevention of disease. Including ample instructions to parents and nurses respecting the disorders peculiar to children. To which is added a very extensive collection of medicines and prescriptions, according to the most recent practice of the London and Edinburgh colleges of physicians. With observations on bathing and mineral waters. Rendered so easy and familiar as to be perfectly intelligible to every capacity. By Dr. Buchan. The whole improved and corrected by Thos. Furlong Churchill . . . London: Albion Press; printed for James Cundee, 1817.
  xviii, [6], [19]-622, [10] p., [1] leaf of plates : port. ; 22 cm.

  Second edition (1st ed., London, 1808). Bibliographically, *The new practical family physician* is a confusedly hybrid work. On the one hand, it is so solidly grounded on Buchan's text (#442-502) that it might be described as an injudiciously edited version of the *Domestic medicine*. On the other hand, Churchill has made so many omissions and additions to Buchan's text as to practically constitute a new work. Without comparing the two works, however, it is impossible to distinguish what is Buchan's and what Churchill has added, deleted or simply paraphrased. There can be little doubt that Churchill intended the work to be regarded primarily as his own, as indicated by the inclusion of footnotes that begin: "If the reader will look into Buchan's Domestic Medicine"; or, "The late Dr. Buchan says"; or, "See Buchan, p. 32," to indicate that Buchan's is a distinct work. Frontispiece portrait of Churchill. Binder's title: *Buchan's Medicine.* Atwater copy lacking pages 577-84 (4E1-4) due to the accidental omission of gathering 4E during collation for binding and the repitition of gathering 4F.

643. CICCOLINA, Sophia A., marchesa.
Deep breathing, as a means of promoting the art of song, and of curing weaknesses and affections of the throat and lungs, especially consumption . . . Translated from the German by Edgar S. Werner. New York: M.L. Holbrook & Co., [c1883].
  48, [24] p. : ill. ; 19 cm.

  Translation of: *Die Tief-Einathmung* (1880), which in turn is a translation from the original Dutch title: *De diepe ademhaling, hare verhouding tot de gezondheid en de zangkunst* (1877). Preface by the publisher, Martin Luther Holbrook (#1678-1695). The final [24] p. consists of publisher's advertisements.

CITIZEN OF BOSTON see PERRY (#2781).

644. CIVIALE AGENCY.
Glad tidings of good things. [New York: Civiale Agency, 189?].
  32 p. : ill. ; 17 cm.

  "Thirteenth edition" (verso of front wrapper). A brochure promoting proprietary remedies for the cure of various sexual disorders (spermatorrhoea, impotence, sterility), as well as urological diseases. A staff of consulting physicians is also advertised.

645. CIVIALE AGENCY.
Manhood perfectly restored. The Prof. Jean Civiale soluble urethral crayons, as a quick, painless and certain cure for impotence, lost manhood, spermatorrhoea, losses, weakness and nervous debility. Also for prostatitis and varicocele. Facts for men of all ages. Eleventh edition, enlarged, revised and illustrated. New York: Issued by the Civiale Remedial Agency, 1886.
  48 p. : ill. ; 16 cm.

  Cover-title: *Glad tidings of good things.* In printed wrapper.

646. CLARIDGE, R. T.
Abstract of Hydropathy; or, the cold water cure, as practised by Vincent Priessnitz, at Graefenberg, Silesia, Austria . . . Eighth edition. London: James Madden and Co., MDCCCXLIII [1843].
  72 p. ; 20 cm.

  The first edition of the *Abstract* was also issued in 1843. This pamphlet extracts "as much original matter as could be compressed into so small a space at One Shilling" (p.[iii]) from the larger work (1842), an American edition of which was published in 1843 (#648).

647. CLARIDGE, R. T.
Every man his own doctor. The cold water, tepid water, and friction-cure, as applicable

to every disease to which the human frame is subject and also to the cure of diseases in horses and cattle . . . New York: John Wiley, 1849.
    ix, [10]-212, [2] p. ; 20 cm.

This revised paean to Vincent Priessnitz and his hydropathic establishment at Graefenberg includes quotations on hydropathy from the most recently published literature in the British medical press, a description of the hydropathic armamentarium, and a review of diseases to which the water-cure is both applicable and beneficial. This New York edition was issued the same year as the London edition.

648. CLARIDGE, R. T.
Hydropathy; or, the cold water cure, as practiced by Vincent Priessnitz, at Graefenberg, Silesia, Austria . . . First American, from the third London edition. With notes by Dr. P. Lapham. New-York: Published by the proprietor [P. Lapham], 1843.
    v, [6]-285, [3] p. ; 19 cm.

This American edition of Claridge's treatise of the same title is based on the third London edition (1842), and is one of the earliest works on hydropathy published in the United States. Claridge had been treated at Priessnitz's water cure at Graefenberg, and upon his return home became one of hydropathy's earliest proponents in Britain. The author provides a detailed description of Priessnitz's methods, and quotes liberally from the hydropathic writings of German-speaking authors such as C. Munde (#2521), J. Gross, and others. Claridge wrote two other popular medical treatises: *Abstracts of hydropathy* (1842) and *Every man his own doctor* (1849). Lapham, the American publisher, describes himself in an advertisement at the end of the volume as a Thomsonian physician of twenty years' experience who was available for consultation at his Botanical Medicine Store in the Bowery.
    Claridge was not a physician, but a man of means who could afford to live as an invalid in Venice and Rome, travel on the continent in a private carriage with his family, and spend several months at a German water-cure. In 1837, Capt. Claridge had published an account of his travels in northern Italy, the Balkans and Turkey in *A guide along the Danube*.

649. CLARK, Dr.
Dr. Clark's new illustrated marriage guide. A treatise on the anatomy and physiology of the generative organs of both sexes, and their organic and functional diseases. Being a complete and comprehensive private medical instructor and guide to the married and those contemplating marriage, in language within the comprehension of all. Newark, N.J.: The Union Publishing Company, 1886 [c1873].
    384, 71, [55] p. : ill. ; 13 cm.

Divided into two pagination sequences, the first and longest describes the anatomy of the male and female genitalia, reproduction, pregnancy, parturition, diseases of the male and female genitalia that affect generation (apart from the effects of masturbation or sexually transmitted diseases), diseases of the genitalia resulting from abnormal behavior (e.g., masturbation, excessive venery, nymphomania, etc.), and miscellaneous advice on marriage, etc. This first section was obviously written by a medical professional. The second numbered sequence, which has the running-title "Clark's book of secrets," is a book of recipes for home remedies, stain removers, dyes, alcoholic beverages, fertilizers, etc. Spine-title: *Dr. Clark's medical instructor;* cover-title: *Dr. Clark's marriage guide.* The final [55] p. contain publisher's advertisements.

CLARK, James, Sir see COMBE (#729, 730).

650. CLARK, James Henry, 1814-1869.
Sight and hearing, how preserved, and how lost . . . New York: C. Scribner, 1856.
    xiv, [15]-351 p. : ill. ; 19 cm.

The work is divided into separate parts on sight and on hearing, each of which is devoted to the anatomy, diseases and hygiene of the sensory organ. At head of title: *A popular handbook.* Atwater copy inscribed by the author, who received his medical degree from the College of Physicians and Surgeons (New York) in 1841, and practiced in Newark, N.J. for more than twenty years.

651. CLARK, W. C.
The family medical adviser or the science of medicine simplified. Written to meet the

wants of the people in plain, simple language, especially adapted for family and general use. Embracing a symptom directory, a list of invalid's foods and recipes, homoeopathic therapeutics, a medicine dose list, treatment and antidotes in cases of poisoning, a medical dictionary, and three treatments under each disease . . . Milwaukee, Wis.: [W.C. Clark, c1884].
    237, [11] p. : port. ; 20 cm.

652. CLARK, W. C.
The family medical adviser; or the science of medicine simplified. Written to meet the wants of the people in plain, simple language, especially adapted for family and general use. Embracing a symptom directory, a list of invalid's foods and recipes, homoeopathic therapeutics, a medicine dose list, treatment and antidotes in cases of poisoning, a medical dictionary, and three treatments under each disease . . . Rochester, N.Y.: Polypathic Medical Institute, [c1884].
    237, [11] p. : port. ; 20 cm.

Rochester imprint printed on slip pasted over the original Milwaukee imprint.

653. CLARK, William Arthur.
Physiology: a manual of 1000 questions and answers systematically arranged, containing a full treatment of the physiological effects of alcohol and narcotics, with a complete analytic outline of the subject, and notes on teaching. An aid in teaching and in preparing for examination . . . Lebanon, Ohio: C.K. Hamilton & Co., [c1888].
    [2], viii, [9]-150 p. : ill. ; 20 cm.

Series: *The analytic series of teachers' aids.*

654. C. G. CLARK AND COMPANY.
COE'S Cough Balsam . . . For sale by all druggists . . . C.G. Clark & Co., proprietors, New Haven, Conn. [ca. 1863].
    Broadside ; 24 × 15 cm.

655. N. L. CLARK AND COMPANY.
History of the Peruvian Syrup. With letters from the minister plenipotentiary from Peru to the U. States, and other distinguished gentlemen of this and foreign countries, in proof of its value. New York: N.L. Clark & Co., importing druggists, [c1854].
    35 p. ; 18 cm.

Promotional literature for the *Peruvian Syrup* available from N.L. Clark & Company. Following a brief history of the remedy's origins in Peru are letters testifying to the remedy's efficacy in curing dyspepsia, epilepsy, dropsy, neuralgia, etc.

656. N. L. CLARK AND COMPANY.
A letter to clergymen on the preservation of health, and the use of Peruvian Syrup, or protected solution of protoxide of iron combined, as a medicinal agent . . . Boston: Published by N.L. Clark & Co., [c1859].
    48 p. ; 22 cm.

"Gentleman of the clerical profession, are peculiarly subject to those diseases which result from the digestion becoming impaired by cerebral excitement and bodily inactivity. Not only excessive intellectual labor, but also extreme tension of the moral and emotional sentiments, tends to produce this effect . . . When the brain is overworked, the stomach and digestive organs do not receive their natural nervous stimulus . . . Dyspepsia poisons the water of life at the fountain head" (p. 13). The profession is also liable to the disorder called "clergyman's sorethroat" (pharyngitis). The "Peruvian Syrup" is advertised by N.L. Clark, its manufacturer, as an iron tonic that acts as both a prophylactic and a restorative in cases of "imperfect digestion," clergyman's sore throat, "nervous affections" and general debility. In printed wrapper.

657. CLARKE, Albert, 1840-1911.
St. Albans and vicinity as a summer resort . . . St. Albans [Vt.]: Messenger Job Printing House, 1872.
    40 p., [3] leaves of plates : ill. ; 22 cm.

Twenty-five miles north of Burlington and sixty-five miles south of Montreal near Lake Champlain, St. Albans became a popular resort in the 19th century. In addition to the scenery, hunting and fishing, Clarke boasts of St. Albans' comfortable hotels and of the medical efficacy of its mineral springs. In printed wrapper.

658. CLARKE, Arthur, Sir, 1778-1857.
The mother's medical assistant, containing
instructions for the prevention and treatment of the diseases of infants and children . . . London: Printed for Henry
Colburn and Co., 1820.
   xii, 148, [8] p. ; 18 cm.

First edition. No American edition of this
or any of Clarke's books (professional or popular) were reissued in the United States. Even
so, popular manuals of this sort were often imported for the American and Canadian mar-
kets, or were brought into these countries by
immigrants and travellers. An 1803 St. Andrews
medical graduate, Clarke practiced in Dublin,
where he was physician to the Bank of Ireland
and founder of the Dublin Fever Hospital. He
was elected F.R.C.S. in 1844. His several popular medical treatises include *A practical manual
for the preservation of health and the prevention of
diseases incidental to the middle and advanced stages
of life* (1824), *Lecture on sea-bathing* (1828), and
*A code of instructions for the treatment of sufferers
from railroad and steam-boat accidents* (1849). A
second edition of *The mother's medical assistant*
was published at London in 1822 under the
title: *The young mother's assistant.*

659. CLARKE, Edward Hammond, 1820-1877.
The building of a brain . . . Boston: James
R. Osgood and Company, 1874.
   [2], 153 p. ; 18 cm.

First edition. Following the publication of *Sex
in education* (#661, 662), Clarke was invited to
speak before the 1874 meeting of the National
Education Association in Detroit. *The building of
a brain* is an elaboration of that talk. Clarke discusses the physiological principles by which he
imagines the brain assimilates ideas, i.e., "brain-
building" or education, and the psychological
and pathological consequences of ignoring these
principles. In the training of boys and girls, he
continues the discussion of "their appropriate
and consequently different education" based
upon the physiological similarities and differences that was the theme of *Sex in education.*

660. CLARKE, Edward Hammond, 1820-1877.
The building of a brain . . . Sixth edition.
Boston: Houghton, Mifflin and Company;
Cambridge: The Riverside Press, [c1874].
   153 p. ; 18 cm.

661. CLARKE, Edward Hammond, 1820-1877.
Sex in education; or, a fair chance for the
girls . . . Boston: James R. Osgood and Company, 1873.
   181 p. ; 18 cm.

Second edition. The second edition of this
work was printed within a week of the publication of the first (*Pref.*, p. [8]). In the debate on
the equality of men and women, Clarke inter-
jects, "The real question is not, *Shall* women
learn the alphabet? but *How* they shall learn it?
. . . The principle and duty are not denied. The
method is not so plain" (p. 16). Clarke opposes
those who would educate boys and girls by the
same method. He agrees that women "can master the humanities and mathematics, encoun-
ter the labor of the law and the pulpit, endure
the hardness of physic and the conflicts of politics; but they must do it all in a woman's way"
(p. 18-19). He posits a fundamental difference
in their physiologies that makes it unlikely that
a female will "retain uninjured health and a
future secure from neuralgia, uterine disease,
hysteria, and other derangements of the nervous system, if she follows the same method that
boys are trained in" (p. 18).
   Part of the reason for the "singular pal-
lor of American girls" Clarke attributes to
deficiencies of diet, clothing and exercise.
His greater concern, however, is with the
American system of education and its failure to take into account the peculiarities of
adolescent female physiology. In Part II he
discusses the physiological reasons for the
maldevelopment of American females, focusing on the nutritive, nervous and reproductive systems. It is the last that determines the
difference in how the sexes should be
trained. He insists that the demands that
their developing reproductive systems put on
young girls must be considered in their education. If the nervous, muscular and reproductive systems are forced to develop fully
and coterminously, inevitable harm will ensue. The more educators ignore the physiological conditions of female sexual development, the larger will be the number of
cases of "amenorrhoea, menorrhagia, dys-
menorrhoea, hysteria, anemia, chorea and
the like" (p. 48), not to mention imperfect
development of the sexual organs. Clarke
provides many case histories, among them

Miss A., whose teacher "made her brain and muscles work actively, and diverted blood and force to them when her organization demanded active work, with blood and force for evolution in another region . . . He not only made his pupil's brain manipulate Latin, chemistry, philosophy, geography, grammar, arithmetic, music, French, German, and the whole extraordinary catalogue of an American young lady's school curriculum with acrobatic skill; but he made her do this irrespective of the periodical tides of her organism" (p. 71). The result of ignoring her reproductive development was invalidism.

"The principle or condition peculiar to the female sex," according to Clarke, "is the management of the catamenial function, which, from the age of fourteen to nineteen, includes the building of the reproductive apparatus. This imposes upon women, and especially upon the young woman, a great care, a corresponding duty, and compensating privileges . . . This lends to her development and to all her work a rhythmical or periodical order, which must be recognized and obeyed" (p. 120). Clarke targets two mistakes in American education which ignore this "periodical order": co-education and the wholesale adoption of curricula successful in boys' schools for female academies. "Because the education of boys has met with tolerable success . . . in developing them into men, there are those who would make girls grow into women by the same process" (p.127). He concludes: "if the education of the sexes remains identical, instead of being appropriate and special . . . then the sterilizing influence of such a training, acting with tenfold more force upon the female than upon the male, will go on, and the race will be propagated from its inferior classes" (p. 139). In particular Clarke was concerned with Irish immigration, and betrays his Bostonian prejudices with the caution: "The stream of life that is to flow into the future will be Celtic rather than American: it will come from the collieries, and not from the peerage" (p. 140).

662. CLARKE, Edward Hammond, 1820-1877.
Sex in education; or, a fair chance for the girls . . . Boston: James R. Osgood and Company, 1873.

7, [1], 11-181 p. ; 18 cm.

In this apparently latter issue of the second edition, the table of contents appears on p. [4], whereas it appears on p. 9 in the earlier issue.

CLARKE, Edward Hammond see also **COMFORT** (#748); *SEX AND EDUCATION* (#3144).

663. CLARKE, John Henry, 1852-1931.
Catarrh, colds and grippe, including prevention and cure, with chapters on nasal polypus, hay fever and influenza . . . American edition. Revised by the author from the fourth English edition. Philadelphia: Boericke & Tafel, 1899.
xiv, 122 p. ; 17 cm.

The prominent British homeopath was physician to the London Homoeopathic Hospital, lecturer at its medical school, editor of the periodical *Homoeopathic world*, and the author of books for both professional and lay audiences. His most important contribution to the former was the *Dictionary of practical materia medica* (London, 1900-1902), a revision of the homeopathic materia medica that occupied Clarke for sixteen years.

Several of his popular homeopathic treatises were published in the United States, such as *Catarrh, colds and grippe*, which originally appeared at London in 1888 under the title *Cold-catching, cold-preventing, cold-curing*. This Philadelphia edition is based upon the fourth London edition (1896).

664. CLARKE, John Henry, 1852-1931.
Cold-catching, cold-preventing, cold-curing . . . Third edition. London: James Epps & Co., 1890.
xi, [1], 75, [1] p. ; 17 cm.

Clarke terms the common cold "The British disease *par excellence*." "Foreigners," he goes on to observe, "reserve that name for what they call 'the spleen,' meaning by this, low spirits, melancholy, and tendency to suicide, which they attribute to our foggy climate. Naturally we British are not so keenly alive to the prevalence and gravity of 'the spleen' as our observant neighbours; but there is no denying that we do possess peculiar facilities for catching cold" (p. vi). Clarke's treatise on the common cold appeared in at least six London editions between 1888 and 1903. An American edition was published at Philadelphia in 1899 under the title: *Catarrh, colds and grippe* (#663).

665. CLARKE, John Henry, 1852-1931.
A dictionary of domestic medicine. Giving a description of diseases, directions for their general management and homoeopathic treatment. With a special section on diseases of infants. American edition revised and enlarged by the author. Philadelphia: Boericke & Tafel, 1901.
363 p. ; 19 cm.

An alphabetically arranged dictionary of diseases and their homeopathic treatment. Clarke's *Dictionary* was originally published with a London and New York imprint in 1890. A fourth edition of Clarke's *Dictionary* was published at London by the Homoeopathic Publishing Company in 1949.

666. CLARKE, John Henry, 1852-1931.
Indigestion: its causes and cure . . . Third edition, revised. London: James Epps & Co., [ca. 1890].
119, [1] p. ; 17 cm.

The first edition of *Indigestion* was published at London in 1888.

667. CLARKE, John Henry, 1852-1931.
Indigestion: its causes and cure . . . American edition. Revised and enlarged by the author, from the fifth English edition. Philadelphia: Boericke & Tafel, 1900.
147 p. ; 17 cm.

668. C. V. CLICKENER AND COMPANY.
The family hand-book. Containing many valuable recipes for cooking, dyeing, making perfumery, &c. With directions for the care and treatment of horses and cattle. And much valuable information not to be obtained elsewhere. [N.Y.: Ezra N. Grossman, printer, 1855].
[12] leaves : ill. ; 18 cm.

A brochure advertising *Dalley's Magical Pain Extractor*, which could be applied to the relief of eighty-seven disorders listed on the verso of the second leaf. "C.V. Clickener & Co., whole-sale druggists" are listed as the agents for the *Pain Extractor* and several other advertised remedies. An "Almanac for 1855" appears on the verso of the last leaf.

669. CLIFTON SPRINGS SANITARIUM.
Clifton Springs Sanitarium, Clifton Springs, N.Y. [S.l., between 1896 and 1900].
[10] leaves : ill. ; 9 × 16 cm.

An illustrated brochure describing the sulphur springs, the medical staff, the facilities, grounds, fees, and accessibility by railway. One of the most successful water-cures in New York State, the Clifton Springs Water-Cure was founded in 1850 by Henry Foster, M.D. (1821-1901). After the Civil War, the hydropathic movement declined generally in the United States. Those water-cure establishments which survived, such as Clifton Springs, did so by successfully making the transition from water-cures to hygienic sanitaria, where balneology and hydropathy were combined with regular medical care. Foster changed the name of his establishment to reflect its altered philosophy in 1871. The institution functions today as a community hospital.

670. CLIFTON SPRINGS SANITARIUM.
The Clifton Springs Sanitarium . . . Clifton Springs, New York. [Newark, N.Y.: The Du Bois Press, ca. 1902].
15, [1] p. : ill. ; 23 cm.

Promotional brochure. Date of imprint from a dated analysis of the sulfur waters on p. 13. Title on printed wrapper: *The Sanitarium, Clifton Springs, N.Y.*

671. CLIFTON SPRINGS SANITARIUM.
Clifton Springs Sanitarium [*wood-engraved view of the Sanitarium*] Henry Foster, M.D. Sup't. [Clifton Springs, N.Y.: The Sanitarium, 1885].
[2] leaves : ill. ; 21 cm.

The inside pages contain the Sanitarium's dinner menu for 24 September 1885. The piece is accompanied by an envelope postmarked 25 September 1885, and was presumably sent from a visitor at the Sanitarium to a friend or family member in Philadelphia.

672. CLIFTON SPRINGS WATER-CURE.
Report of the addresses and sermon at the dedication of the Clifton Springs Water-Cure held July 25, 1856. Published by vote of the audience. Rochester, N.Y.: Daily Democrat Steam Book and Job Printing House, 1856.
15, [1] p. ; 22 cm.

The collection of addresses printed here celebrate the recent and extensive expansion of the water-cure's facilties within only six years of construction of the original building. The "Circular" on the verso of p. 15 notes that "The Clifton Springs Water Cure does not depend in its treatment *solely* on the use of water, but adds to the effect thus produced whatever can *safely* be done by medicine; nor does it limit itself to any one school of medical practice, but it employs all *known remedies* . . . " In addition to its springs and baths, the water-cure included a gymnasium and provided compressed air and electro-thermal baths. The medical staff of six physicians included two women: Rachel T. Speakman and Mary Dunbar. Wood-engraved view of the Clifton Springs Water-Cure on back of printed wrapper.

Henry Foster (1821-1901) was founder of the Clifton Springs Water-Cure, 29 miles east of Rochester, New York. Foster began his medical studies under his brother Hubbard in Milan, Ohio, and received his medical degree from the Western Reserve University in 1848. Immediately following graduation he became physician to the New Graefenberg Water-Cure near Utica, New York, and in 1849 moved to Clifton Springs where he opened his own water-cure the following year. The institution survived the decline of the hydropathic movement, and under Foster's direction became more scientifically oriented in its therapeutics as medicine itself became more scientific. His biographer in the *Dict. Amer. med. biog.* (1928) wrote that Foster was "the first in this country to establish a highly reputable sanitarium, as well as to adopt the now universal group practice in building up a large efficient staff of coadjutors, representing all the important medical specialties" (p. 428).

Foster was not the first to operate a health spa at Clifton Springs. In the 24 July 1848 issue of the *Rochester Daily Democrat,* an advertisement signed "M. Parke" describes The Grove, where those travelling between Niagara Falls and Saratoga Springs might break their railway journey to sample the springs for refreshment or medical treatment: "As a medicinal remedy for all the various cutaneous diseases, rheumatic affections, dyspepsia, and in cases of loss of appetite and debility of the general organization, &c., &c., they invite a trial and solicit a comparison with any curative waters in the State." That Foster shared Parke's estimate of the locale is evidenced in his purchase of property in Clifton Springs the following year.

CLIFTON SPRINGS WATER-CURE see also **FOSTER** (#1265).

673. CLOSE, Charles William, 1859-1915.
Sexual law and the philosophy of perfect health . . . Bangor, Me.: Published by C.W. Close, [c1898].
16 p. ; 23 cm.

Close maintains that the body "is composed of innumerable points of atomic Life, each of which is in itself perfect, diseaseless, and deathless," and that they are held together "in bodily form by the magnetic force of the Supreme Power in man, the ideal human *Ego.*" Therefore, "When the entire mentality of the individual is in perfect harmony with the supreme Ego, or perfect Self, the personal bodily expression is one of perfect health; but when from any cause there is a lack of harmony . . . the inharmony or lack of ease expresses itself in the body as disease, the magnetic force of the ideal human has been deflected, and consequently the harmonious arrangement of the atomic substance of the body has been broken in upon" (p. 5).

"Mind cure," or what Close terms "phrenopathic healing," is the ability of the healer to channel "a new supply of healthy vitality to the subconsciousness of his patient with a force which increases the magentic quality of the latter's personality, thus enabling him to subconsciously appropriate an increased amount of healthy vitality" (p. 8). Other than his introductory remarks about "sex magnetism," i.e., the attraction or repulsion of auras, Close fails to coherently integrate "sexual law" into his "philosophy of perfect health." The author received his Ph.D. and "S.S.D." in 1888 from the Spiritual Science University (Chicago). Between 1897 and 1901, he edited and published *The Free Man*, a "New Thought" monthly. In printed wrapper.

674. CLOSE, Charles William, 1859-1915.
The value of esoteric thought and the philosophy of absent healing . . . Bangor, Me.: Published by C.W. Close, [c1901].
16 p. ; 23 cm.

"When a patient sends to a phrenopathic healer for 'absent treatment' he (or she) is . . . receptive to the healer's thought; i.e., there is a desire for health in both conscious and subconscious mentality and this desire makes the organ of the subconscious vitality, the back brain, receptive to the vibrations of a healthy vitality, and this establishes an unbroken circuit between patient and healer. The healer now acts from his own self center and through the instrumentality of the cerebrum sends out a vitalizing thought vibrating through the universal life till its force reaches the atmosphere of the patient where it is absorbed by the sub-conscious mentality of the patient . . . and is received and used by the back brain for the benefit of the patient" (p. 15).

675. CLOUD, J. Albert, 1841-1900.
Homoeopathy: its difficulties and some of the principal errors against it. With hints in dietetics, viewed in relation to the laws of digestion; also, the practice of homoeopathic medicine simplified, for the use of families, and a history of the Cincinnati Homoeopathic Medical Dispensary . . . Cincinnati: V.C. Tidball, printer, 1869.
[4], 82, [10] p. ; 18 cm.

Homoeopathy is divided into four parts. The first is an apologia for homeopathy directed toward the general reader; the second an examination of "the principles of dietetics viewed in relation to the laws of digestion"; the third a brief alphabetical catalog of diseases and their homeopathic treatment; and the fourth a description of the Cincinnati Homoeopathic Hospital (est. 1867), of which Cloud was resident physician. The unnumbered pages before and after the text contain advertisements for a wide variety of Cincinnati merchants. From his reference to studies at the "Penn Hospital," it would appear that the author is the Cloud listed

in the *Directory of deceased American physicians, 1804-1929* (I:296) as an 1866 medical graduate of the Univ. of Pennsylvania. Atwater copy inscribed by author.

676. CLUM, Franklin D., 1853-1925.
Health. The physical life of men and women. Their structure and function and how to supply their wants, direct their powers, avoid their afflictions, and sustain their lives . . . Boston: D. Lothrop Company, [c1887].
400 p. ; 18 cm.

A treatise on physiology and preventive hygiene by an 1875 Yale medical graduate.

677. CHAS. CLUTHE AND SONS.
Cluthe's advice to the ruptured. By Chas. Cluthe & Sons. New York City: Cluthe Rupture Institute, [c1910].
64 p. : ill., ports. ; 20 cm.

Advertisement for the *Cluthe Truss*, which, "in addition to keeping the rupture comfortably and continuously in place, automatically gives a soothing, strengthening and healing massage treatment" (p. 17). Charles Cluthe and his four sons claim to "have treated by mail, and in person at our Institute, over 270,000 ruptures" (p. 37). An illustrated questionnaire/ order blank for the *Cluthe Truss* is laid-in.

678. CHAS. CLUTHE AND SONS.
Cluthe's advice to the ruptured. By Chas. Cluthe & Sons. Bloomfield, N.J.: Cluthe Rupture Institute, [ca. 1913, c1912].
104 p. : ill., ports. ; 20 cm.

Publication date based upon dates of most recent testimonials.

CLUTHE RUPTURE INSTITUTE see **CHAS. CLUTHE AND SONS** (#677, 678).

679. CLYMER, Reuben Swinburne, b. 1878.
The Thomsonian system of medicine. With complete rules for the treatment of disease. Also a short materia medica . . . Allentown, Pa.: Published by the Philosophical Publishing Co., [c1905].
196, [2] p., [1] leaf of plates : ill. ; 20 cm.

"The philosophy and principles of the Thomsonian system is that which is taught by Dr. Thomson himself [#3482-3515] and later formulated by the well-known Dr. Comfort [#749-751]. Every true Herbalist, and Physio-Medicalist subscribes and holds to this philosophy, as it is the only true philosophy and the only one in which no flaw can be found . . . Although this philosophy was first taught nearly a hundred years ago, no change has been found necessary, it is therefore such a philosophy as no other school of medicine or healing has ever been able to give the world. This alone proves that the Physio-Medical system is founded on truth and that sooner or later, all schools must subscribe to it" (*Pref.*, p. 4-5).

Writing of physio-medicalism at the turn of the 19th century, John Haller observes, "Despite pretences, the physios never felt at ease with the expectations of the new age. Having prided themselves in their native roots and herbs and in a commitment to a domestic practice based on a regimen of kindly medicines, they were in no positon— intellectual or financial—to manage the changes that scientific medicine demanded. All but a few physios viewed laboratory science and experimental medicine as foreign and elitist" (*Kindly medicine*, Kent, OH, 1997, p. 150). Clymer made at least one obeisance to modernity. The sole plate is a photograph of "The Magnetic Oxygen-Ozone Producer invented and perfected . . . by Drs. Clymer & Woodhouse," and portrays either Clymer or his partner at the controls. The "greatest appartus for the healing and curing of the sick extant" is described on pp. 192-196, and was one of the feature's of Clymer's Thomsonian Institute in Allentown, Pa. Among the credentials beneath Clymer's name on the title-page is "Diplomate in Osteopathy of the English School of Osteopathy." Clymer was also the author of numerous manuals of Rosicrucian thought, Biblical commentaries, etc., that remained in print through the middle of the century.

680. COALE, William Edward, 1816?-1865.
Hints on health; with familiar instructions for the treatment and preservation of the skin, hair, teeth, eyes, etc. . . . Boston: Phillips, Sampson & Company, 1852.
[2], iv, 207 p. : ill. ; 17 cm.

First edition (3rd ed. , 1857). A native of Baltimore, Coale began his medical studies with Eli Geddings, M.D., of the University of Maryland. He entered the Navy in 1837 as an assistant surgeon. He left the service in 1843 and moved to Boston, where he practiced for nearly twenty years. In 1862 he again offered his services to the U.S. military. Several years later he succumbed to a "hepatic disorder" aggravated by his military duties and service with the U.S. Sanitary Commission. Coale had been an elected officer of the Massachusetts Medical Society and instructor at Harvard Medical School.

COATES, John, Jr. see *BOTANIC SENTINEL* (#380).

681. COATES, Reynell, 1802-1886.
First lines of physiology, being an introduction to the science of life; written in popular language. Designed for the use of common schools, academies, and general readers . . . Sixth edition, revised; with an appendix. Philadelphia: Published by E.H. Butler & Co., 1847.
vi, 11-xii, xiii-xxiv, 25-340 p. : ill. ; 19 cm.

Coates received his medical degree from the Univ. of Pennsylvania in 1823. After a year-long voyage as surgeon on an East Indiaman, he returned to Philadelphia. By the early 1830s, Coates reportedly had given up medical practice to devote himself to writing and lecturing. He contributed several articles to Isaac Hays' *American encyclopedia of practical medicine and surgery* (2 vols., Philadelphia, 1834-36) and published in the medical journals. Coates literary specialty was the popularization of scientific and medical instruction, best represented by his *Popular medicine, or family adviser* (#682), *First lines of natural philosophy* (1846), and the present work. The *First lines of physiology* was his most successful publication. It was first issued at Philadelphia in 1840 under the title: *Physiology for schools.* A seventh and apparently final edition was issued in 1850.

682. COATES, Reynell, 1802-1886.
Popular medicine; or, family adviser; consisting of outlines of anatomy, physiology, and hygiene, with such hints on the practice of physic, surgery, and the diseases of women and children, as may prove useful in families when regular physicians cannot

be procured: being a companion and guide for intelligent principals of manu-factories, plantations, and boarding-schools, heads of families, masters of vessels, missionaries, or travellers; and a useful sketch for young men about commencing the study of medicine . . . Philadelphia: Carey, Lea & Blanchard, 1838.

xx, [17]-614, [2] p. : ill. ; 24 cm.

Among his reasons for extending medical information to the public, Coates considers "that portion of the population . . . too poor to remunerate a physician, and too proud to appeal to his charity" (p. vii). "In the West," he writes, "where the tide of immigration sweeps on so rapidly, that new communities are created every year, and in such numbers that heaven and earth are ransacked to find appellatives for the growing host of towns and villages—a very considerable portion of the population is placed entirely beyond the aid of the profession; and in the South "still greater difficulties exist. On the large estates of the planters, a family composed of thirty, fifty, or even a hundred or more individuals, often remains cut off in a great degree from the rest of the world, and dependent almost exclusively upon the agriculturalist for hygienic, medical, and even surgical treatment" (*ibid.*).

683. COATES, Reynell, 1802-1886.

Syllabus of a course of popular lectures on physiology, with an outline of the principles on which depend the improvement of the faculties of mind and body . . . Philadel-phia: T.K. & P.G. Collins, printers, 1840.

27 p. ; 24 cm.

Coates provides the outlines of twelve lectures intended "to diffuse as widely as possible throughout the community a knowledge of certain general principles connected with the development of the animal frame, with the applicability of these principles to the every-day business of life, the promotion of health and happiness" (p. [3]). In printed wrapper.

684. COBB, Daniel J., b. 1796.

The family adviser; calculated to teach the principles of botany. Compiled with a strict regard to logick. Containing directions for preserving health, and curing diseases, for

the use of families and private individuals . . . Rochester, N.Y.: Printed for the author, by Marshall & Dean, 1828.

131 p. ; 17 cm.

Daniel J. Cobb was born in Connecticut in 1796 and moved to New York State sometime before 1830. According to census records, he lived in Lima (Livingston County), N.Y. in 1830, and in nearby Mt. Morris in 1850. (His name does not appear in the census records for 1840 or 1860.) Between these dates he apparently lived in Castile, N.Y., where he was operating a dispensary in 1837 and where his final book was published in 1846.

On pages 10-39 Cobb describes 113 medicinal plants, and assigns them to one or more of the twenty-six classes (A-Z) of botanic medicines described on pages 14-62. "The classes of medicine are calculated to effect different objects" (p. 44), e.g., A = "Antispticks," B = "Antispasmodicks," C = "Aromaticks," etc. Pages 68-122 contain an alphabetical list of seventy-three common disorders and the class(es) of medicine applicable to their cure. An index (p. 125-31) provides name access to plants, classes, and diseases in the text. *The family adviser* was reprinted at New Bedford, Ma. in 1832.

685. COBB, Daniel J., b. 1796.

The medical botanist, and expositor of diseases and remedies. In two volumes. VOL-UME FIRST—in six parts. Part first—contains a description of medical plants, with their localities, and directions for collecting and curing them for medical purposes. Part second—contains an arrangement of medical plants into classes and other important divisions . . . Part third—contains an invaluable selection of recipes. Part fourth—contains a description of the symptoms of diseases, with references to the different remedies for their removal. Part fifth—contains remarks on dyspepsia, diet, dress, and exercise. Part sixth—contains remarks on the temperaments of the human system, and on phrenology . . . VOLUME SECOND. The mother's guide: contains an abridged system of midwifery, and much useful instruction upon the diseases and conditions peculiar to females . . . Castile, N.Y.: Published by the author, 1846.

iv, [5]-255, [1] p. ; 17 cm.

A greatly expanded version of *The family adviser* (#684) in which Cobb takes into account various elements of the health reform movement, e.g., dress reform, hydropathy, phrenology, etc. The number of medicinal plants described in the first part has increased to 213 from the 113 provided in his 1828 work. The second volume described on the title-page (*The Mother's guide*) was apparently never published.

686. COBB, Daniel J., b. 1796.
The medical instructer: containing the symptoms of diseases, and directions for their removal, by the use of Dr. Daniel J. Cobb's highly celebrated and invaluable family medicines; to which are prefixed, certificates and observations recommendatory of their superiour virtues in the prevention and cure of diseases . . . Warsaw [N.Y.]: J.A. Hadley, printer, 1837.
    60 p. ; 17 cm.

"It is universally admitted, even by those who oppose the botanick practice, that every country contains vegetable remedies for the removal of all curable diseases its inhabitants . . . It is also equally certain to those who have investigated the subject that enough is known of the curative properties of the indigenous plants of America, to justify the exclusion of all minerals, and deadly poisonous vegetables, from the list of medicines; and I look forward with pleasing anticipation to the time when these, and the lancet, shall give place to the more humane and efficacious use of botanick remedies, exclusively" (*Pref.*, p. 4).
    In the "Advertisement," Cobb "gives notice to the publick . . . that he is now prepared to furnish his invaluable family medicines" at his "Dispensary, situated in the town of Castile, Genesee county, N.Y." (p. [5]). The opening pages of *The medical instructer* are devoted to testimonial letters and remarks on diet, dress and hygiene. The greater part of the text (p. 23-58) is an alphabetical catalog of diseases that includes brief description of symptoms and the proprietary remedy Cobb recommends for its cure.

687. COBB, Daniel J., b. 1796.
A supplement to The family adviser: containing some remarks on the diseases of women and children. Also a plain system of midwifery; with directions for preventing, or relieving, most, if not all, the diffi-culties connected therewith; to which are added, some remarks on the importance of regularity and concord . . . Westfield, N.Y.: Hull and Newcomb, printers, 1831.
    95, [1] p. ; 16 cm.

The introduction to Cobb's *Supplement* is an eight-page diatribe against the regular medical profession, their stranglehold on the practice of medicine, their extortionate fees, and the difficulties of females entering the practice of midwifery. Cobb's text begins with a discussion of various aspects of women's health, particularly menstruation, the dangers of bloodletting during pregnancy and tight lacing. Pages 19-80 contain "A plain system of midwifery" (also the title of an unrecorded book Cobb claims to have published, *see*#685). He provides descriptions of the pelvis and the "copulative and generative organs" (quoting from Thomas Denman and others), and a discussion of menstruation and its disorders, conception ("nothing with regard to it has been established"), pregnancy and parturition (ample quotes from Denman on "difficult labours").

688. COCKBURN, Samuel.
Medical reform: being an examination into the nature of the prevailing system of medicine; and an exposition of some of its chief evils; with allopathic revelations. A remedy for the evil . . . First American edition. Philadelphia: Rademacher & Sheek; New York: William Radde, 1857.
    6, 3-180 p. ; 16 cm.

Cockburn's apologia for homeopathy was originally published at London in 1856. A licenciate of the Royal College of Surgeons of Edinburgh, Cockburn later converted to homeopathy. He declares the allopathic school in which he was trained "a thing of the past, a product of the dark ages" that "shuns the light and cannot stand investigation" (p. 11). The American publisher describes this work as "peculiarily adapted to the popular mind, being learned without pedantry, analytical without being abstruse, and scientific without technicalities" (*Pref.*, p. 5).

689. COCROFT, Susanna, b. 1862.
Aids to beauty. [Chicago: Published by the Physical Culture Extension Society, c1907].
    66, [2] p. ; 23 cm.

"Be woman's ambition wealth, fame, the admiration of the many—or of one—health, and the natural grace and beauty that health brings, are her surest capital; and it is the duty and privilege of each to make herself as attractive, magnetic and pleasing to the eye as her natural gifts will admit of development." (p. 5). Title on printed wrapper: *Beauty*. Series: *Know thyself series*.

690. COCROFT, Susanna, b. 1862.
Body manikin and position of vital organs . . . Third edition. Chicago, Ill.: Published by the Headington Publishing Company, [c1914].
[4], 65, [3] p., [2] leaves of plates + paper manikin : ill. ; 19 cm.

"Sanitariums are filled with women being operated upon for displacements of various organs, and they do not realize that incorrect habits of standing, sitting, walking and breathing are largely responsible. The Body Manikin, herewith, shows the relative position of the vital organs. The purpose here is to point out the most frequent causes for the common displacement or dislocation of any organ, and to show the effect upon the health" (p. 6). Series: *Know thyself series*.

691. COCROFT, Susanna, b. 1862.
The circulatory system. [Chicago: Published by the Physical Culture Extension Society, c1905].
37, [1] p. ; 23 cm.

Series: *Know thyself series*.

692. COCROFT, Susanna, b. 1862.
Motherhood. [Chicago: Published by the Physical Culture Extension Society, c1906].
[2], 69, [1] p. : ill. ; 23 cm.

"A realization of the fact that too many women pride themselves upon their ignorance of the laws of generation through a mistaken and unwholesome idea that such knowledge would mean less modesty and delicate refinement, furnishes sufficient incentive for this lecture" (p. 19). Cocroft's brief essay on the anatomy of the female genitalia, pregnancy and uterine disorders is preceded by an essay on the ideal of motherhood in which she stresses

the importance of pre-natal influences on the moral as well as the physical development of the child. Title on printed wrapper: *Motherhood and the generative organs: their purpose and derangements*. Series: *Know thyself series*.

693. COCROFT, Susanna, b. 1862.
The vital organs. Their use and abuse. The stomach, pancreas, liver, intestines, kidneys. [Chicago: Published by the Physical Culture Extension Society, c1906].
45, [1] p. ; 23 cm.

First edition. "The purpose of these lectures is to describe as briefly as may be done comprehensively, the action of the digestive organs upon the foods in converting them into nourishment for use by the body, and to explain the disorders of the digestive tract, which give rise to chronic ailments" (p. 5). Series: *Know thyself series*.

694. COCROFT, Susanna, b. 1862.
The vital organs, their use and abuse . . . First edition 1906, second edition 1911. Chicago, Ill.: Published by the Physical Culture Extension Society, 1911.
119, [9] p. ; 19 cm.

Series: *Know thyself series*.

695. COCROFT, Susanna, b. 1862.
The vital organs, their use and abuse . . . Fifth edition. New York City: Published by the Headington Publishing Company, [c1914].
119, [5] p. ; 18 cm.

The author is described on the title-page as "originator of the Physical Culture Extension Society" and "commandante, United States Training Corps for Women." Series: *Know thyself series*.

696. CODMAN & SHURTLEFF.
Catalogue of manikins, anatomical preparations, models, and maps, for sale, or imported to order, by Codman & Shurtleff, makers and importers of surgical instruments . . . [Boston, ca. 1855-70].
3, [1] p. ; 25 cm.

The firm of Codman & Shurtleff (Boston) was established in 1838 and remained in busi-

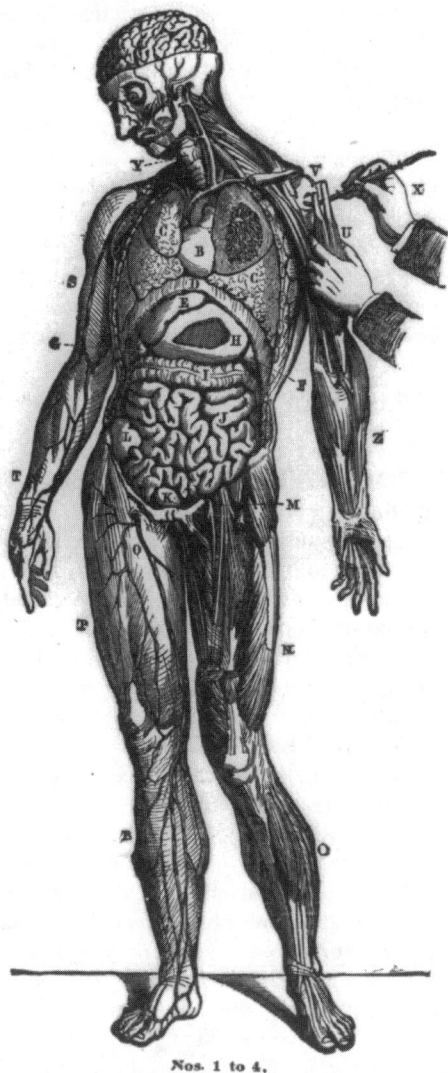

Nos. 1 to 4.

Fig. 20. An Auzoux papier-maché manikin, from the 6th edition (c1893) of Truax, Greene & Co.'s Price list of physicians' supplies.

ness until 1953. The paper, typefaces, references to Auzoux's paper mache manikins (sold out of Albany, N.Y. as early as 1844) and references to the lithographic plates that accompany the anatomical atlases of Jean Marc Bourgery (1797-1849), suggest that the *Catalogue* was issued ca. 1855-1870.

The first two pages describe in some detail the papier-maché manikins and models made by Louis Auzoux (1797-1878) of St. Aubin d'Ecrosville, France. It was manikins such as

these that Frederick Hollick (#1697) used to illustrate his public anatomy and physiology lectures. Hollick had a set of sixteen manikins and models that are described in the introduction to his 1845 *Origin of life* (#1724, p. xv-xvi). The use of these figures was intended to make his lectures more interesting and comprehensible, but also contributed to Hollick's arrest and trial on charges of obscenity. The importance of these manikins to popular lecturers such as Gunning S. Bedford (#271), C.W. Gleason (#1351), the Cutters (#864), Hollick *et al.* is hard to overestimate. In his introduction to *The origin of life* Hollick writes, "One great difficulty was, to illustrate [my] discourses in an efficient manner. A mere verbal description would be of little service, while pictures, and diagrams give but imperfect and inaccurate ideas. I therefore sent to Paris for a set of M. Auzoux's Anatomical Models, expressly adapted for my purpose. These preparations answer better than real dissections, for teaching an audience, inasmuch as the parts retain their natural appearance, and the exhibition of them is unattended by anything repugnant to the sensibilities. Their completeness and accuracy, is astonishing; there is not a single detail that is needed, however minute it may be, but what is there, and so true is the whole to nature, that the eye could scarcely detect it as being a work of art."

697. COFFEE, William Oakley, 1859-1927.

The new system of home treatment for diseases of the eye by the absorption method . . . Des Moines, Iowa: Published by Dr. W.O. Coffee Eye Infirmary, [1900?].

79, [1] p., VIII leaves of plates : ill. (some col.) ; 21 cm.

"This is an age of grand and important discoveries. Anesthetics, antiseptic methods in surgery, anti-toxin, the Roentgen ray, and the various triumphs of science associated with the name of Pasteur and a majestic host of others in both America and Europe are illustrations of the trend in the present age. In the same rank with these eminent men, a high place must be accorded to W.O. COFFEE, M.D., of Des Moines, Iowa, whose wonderful success in the application of the absorption method in treating all diseases of the eye . . . has made his name a household word throughout the United States" (p. [3]).

Coffee assures his readers that not only can he cure cataract without surgical extraction or depression by his absorption method, but that the treatment "can be carried on at home as well under Dr. Coffee's care" (p. 23). W.O. Coffee was an 1881 graduate of the Missouri Medical College. In printed wrapper.

698. COFFIN, Albert Isaiah, ca. 1798-1866.
A botanic guide to health, and the natural pathology of disease . . . Fourth edition. Manchester [Eng.]: Printed for the author by Wm. Irwin, 1846.
[12], xvi, [17]-347 p., [1] leaf of plates : port. ; 18 cm.

Coffin was a native of New York State who emigrated to England (ca. 1837), where he lectured on botanical medicine, formed botanical societies, and published this his principle work. Coffin describes his practice in the opening "Address": "The principles contained in this work are in many respects similar to those introduced in the United States by Samuel Thomson [#3482-3515], and the great success attending the practice in that country, determined him to prepare the Work, which he now submits to the test of public investigation. The Author has much travelled in America, and has associated a great deal with the Indians of that country, as well as with the naturalist, Thompson [sic], and from them has derived much useful knowledge; and all the information thus acquired, both from the writings of the one and the oral instructions of the other, he has carefully adapted to the circumstances of this country" (p. [7]). *A botanic guide to health* remained in print through the mid-1860s. It was translated into Welsh (1844), French (1849, #699) and German (1856), but was never published in the United States. Engraved frontispiece portrait of author. *A Catalogue of printed books in the Wellcome Historical Medical Library* (II:366) supplies Coffin's dates of birth and death: "c. 1798-c. 1862."

699. COFFIN, Albert Isaiah, ca. 1798-1866.
Guide botanique de la santé, ou traité simple des maladies et des herbes qu'il faut employer pour les guérir . . . Traduit sur la quatorzième édition anglaise. A Paris: Chez Charpentier, libraire . . . et chez l'auteur, 1849.
xi, [1], xv, [16]-371 p., [1] leaf of plates : port. ; 18 cm.

Translation of: *A botanic guide to health.* "Dans mon pays (je suis Américain de naissance), j'ai vu les Indiens rouges, je les ai suivis au travers de leurs forêts interminables . . . J'ai été longtemps parmi les Indiens; j'ai habité avec eux dans leurs wigwams, et j'y ai recueilli des connaissances bien plus importantes que ne peuvent en fournir les archives poudreuses de tous les colleges de l'Europe" (p. iv). Lithographed frontispiece portrait of author.

700. COFFIN, Albert Isaiah, ca. 1798-1866.
A treatise on midwifery, and the diseases of women and children; together with a full description of the herbs, roots, and other preparations prescribed as remedies in the work . . . Third thousand. Manchester [Eng.]: Printed at the British Medico-Botanic Press; published in London by W.B. Ford, herbalist, 1849.
[8], vi, [9]-184 p., 4 leaves of plates + frontis. : ill., port. ; 18 cm.

"The practice of midwifery, in every part of God's fair earth, except where it is said science and civilization prevails, is directed by the laws, and under the sole superintendence of nature. In those districts there seldom or ever occurs a death, either with the mother or child; and in many places there is not any attendant . . . and what is better, the mother suffers little or no pain"(p. iv). Coffin wrote this manual for the instruction of women. He was vehemently opposed to the intrusion of men into the management of pregnancy or labor. Coffin's *Treatise* appeared in at least sixteen editions between 1849 and 1878. The four engraved plates illustrating Coffin's text are much reduced (and unacknowledged) copies of engravings in William Smellie's classic obstetrical atlas.

701. COIT, Daniel.
DOCTOR COIT'S FAMILY PILLS. Doctor Daniel Coit of Burlington, late of St. Albans, in the State of Vermont, has obtained a patent, conformable to the laws of the Union, for a composition called

COIT'S FAMILY PILL, which is now recommended to the patronage of the public.
The proprietor of this pill is not insensible
of the opposition generally made to patent
medicine. Having, however for a long time
been impressed with an idea that some
valuable composition might be found,
which would in a great degree prevent disorders, and thereby render a variety of
medicine less necessary, and having for
more than eight years been improving on
the one now offered to the public, he
boldly and unhesitatingly intrudes himself
upon the unprejudiced and candid, in the
recommendation of his Family Pill, as possessing more virtues in the prevention and
cure of the most common disorders incident to mankind, than any other particular composition heretofore made use of.
This pill will be found beneficial by all of
either sex, of whatever age and in every
climate. It is principally composed of vegetable gums and roots, a considerable part
of which are the natural productions of
America, in whose balsamic medicinal
qualities its excellence consists . . . These
pills if timely administered will remove the
causes which commonly produce yellow
fever, bilious fevers, ague and fever, cholic
pains, flatulencies, indigestion, costiveness,
hypochondrical and hysterical complaints,
stranguary, gravel, rheumatism and gout.
They are particularly serviceable in female
disorders . . . [1804].
   Broadside ; 20 × 13 cm.

   Manuscript signature: *D. Coit.*

702. COLBURN, Richard T.
The salt-eating habit. Its effects on the animal organism in health and disease. A contribution toward the study of the rational
food of man . . . Dansville, N.Y. Published
for the author, by Austin, Jackson & Company, 1878.
   29, [7] p. ; 19 cm.

   Originally published in 1875. In printed
wrapper.

703. COLBY, Benjamin.
Colby's guide to health: being an exposition of the principles of the botanic system

of practice, and their mode of application
in the cure of every form of disease. Embracing a concise view of the various theories of ancient and modern practice . . .
Sixth edition, enlarged and revised . . .
Cambridge, Mass., 1879.
   [2], 6, [4], [13]-72, [2], 73-78, [2], 79-
84, [2], 85-88, [2], 89-90, [2], 91-92, [2],
93-96, [4], 97-116, [2], 117-120, [2], 121-
186, [2] p. : ill. ; 18 cm.

704. COLBY, Benjamin.
Colby's guide to health: being an exposition of the principles of the botanic system
of practice, and their mode of application
in the cure of every form of disease. Embracing a concise view of the various theories of ancient and modern practice . . .
Sixth edition, enlarged and revised . . .
Cambridge, Mass., 1880.
   [2], 6, [4], [13]-72, [2], 73-78, [2], 79-
84, [2], 85-88, [2], 89-90, [2], 91-92, [2],
93-96, [4], 97-116, [2], 117-120, [2], 121-
186, [2] p. : ill. ; 18 cm.

705. COLBY, Benjamin.
A guide to health; being an exposition of
the principles of the Thomsonian system
of practice, and their mode of application
in the cure of every form of disease; embracing a concise view of the various theories of ancient and modern practice . . .
Nashua, N.H.: Published by Charles T. Gill,
1844.
   x, [11]-144 p. ; 19 cm.

   "Therefore, if every man was his own physician, the interest of physician and patient would
be identified. Those who make the practice of
medicine a source of gain, will ridicule the idea
of every man being his own physician. So have
priests ridiculed the idea of letting every man
read the bible and judge for himself the important truths therein contained. As well might
the village baker ridicule the idea of the good
housewife making her own bread; alleging that
it required a long course of study to make bread
. . . The preparation and use of medicine to
cure disease, requires no more science than the
preparation of bread" (p. ix).
   One thousand copies of this first edition of
A guide to health were issued (*v.* pref. to 2nd ed.).
Colby's *Guide* is divided into two parts. The

chapters of the first part attack allopathic and homeopathic therapeutics with extracts from the writings of their learned practitioners, and conclude with an emotional appeal to the truths of the Thomsonian system: "We now find the illiterate farmer a doctor—a graduate of the school of nature, with almost universal access for his diploma" (p. 38). The second part is devoted to a practical explanation of Thomsonism, in particular "the indications necessary to be accomplished in the cure of different forms of disease" (i.e., the use of relaxants, stimulants, tonics, etc.) and a listing and discussion of "the articles calculated to answer each of these indications." Chapter IX is an alphabetical catalog of thirty-nine common diseases and their treatment. Cover-title: *Colby's guide to health.*

706. COLBY, Benjamin.

A guide to health, being an exposition of the principles of the Thomsonian system of practice, and their mode of application in the cure of every form of disease; embracing a concise view of the various theories of ancient and modern practice . . . Second edition, enlarged and revised . . . Concord, N.H.: Colby and Collins, 1845.

[2], xii, [13]-72, [2], 73-78, [2], 79-84, [2], 85-88, [2], 89-90, [2], 91-92, [2], 93-96, [4], 97-116, [2], 117-120, [2], 121-181, [1] p. : ill. ; 18 cm.

Part one contains modest additions and revisions. The second part was substantially revised and expanded, however, including added matter on hygiene, an expansion of the catalog of diseases from thirty-three to fifty-six, and the addition of ten full-page wood engravings of plants. The unnumbered verso of p. 181 bears an advertisement for the Concord Thomsonian Botanic Medicine Store, maintained by Benjamin Colby and his partner David O. Collins (b. 1805). After "eight years extensive and successful practice," Drs. Colby & Collins announce that they "will attend upon all who may wish their professional services . . . All calls in the village or vicinity promptly attended to." D.O. Collins apprenticed with a "regular" physician, but became a botanic. He practiced in Manchester, Hopkinton and Contoocook, N.H. The second edition was also issued at at Milford, N.H. in 1845 with the imprint of John Burns (#707).

707. COLBY, Benjamin.

A guide to health, being an exposition of the principles of the Thomsonian system of practice, and their mode of application in the cure of every form of disease; embracing a concise view of the various theories of ancient and modern practice . . . Second edition, enlarged and revised . . . Milford, N.H.: John Burns, 1845.

[2], xii, [13]-72, [2], 73-78, [2], 79-84, [2], 85-88, [2], 89-90, [2], 91-92, [2], 93-96, [4], 97-116, [2], 117-120, [2], 121-181, [1] p. : ill. ; 18 cm.

708. COLBY, Benjamin.

A guide to health, being an exposition of the principles of the Thomsonian system of practice, and their mode of application in the cure of every form of disease; embracing a concise view of the various theories of ancient and modern practice . . . Third [i.e., fourth] edition, enlarged and revised . . . Milford, N.H.: John Burns, 1846 [i.e., 1848].

[2], xii, [13]-72, [2], 73-78, [2], 79-84, [2], 85-88, [2], 89-90, [2], 91-92, [2], 93-96, [4], 97-116, [2], 117-120, [2], 121-181, [1] p. : ill. ; 18 cm.

This fourth edition is designated the third edition on the title-page, which is dated 1846. On the verso of the title-page, however, appears the "Preface to the fourth edition" dated "May, 1848." The sheets were reprinted from the stereotype plates of the second edition (#706, 707). The 1860 *New England business directory* finds Colby in Saco, Me. as a botanic physician.

709. COLCORD AND BABCOCK.

The family medical guide, being a catalogue of family medicines. With directions for their use, with a short account of such poisons as are usually in use in families, and their antidotes: to which is added a catalogue of useful fancy articles, perfumery, brushes, superior soaps, etc., such as are constantly on hand and for sale by the publishers. By Colcord & Babcock, druggists & apothecaries . . . Boston, 1845.

33 (i.e., 32) p. ; 13 cm.

710. COLEMAN, Walter Moore, 1863-1926.

The elements of physiology for schools . . . New York: The Macmillan Company; London: Macmillan & Co., 1915.

xi, [1], 364 p., VIII leaves of plates : ill. ; 20 cm.

Coleman's *Elements* was first published in 1903. This 1915 issue was the last of its twelve printings. Intended for secondary schools, it is the companion to his primary school text, *Physiology for beginners* (1908), which together form *Coleman's physiological series*. Coleman distinguishes the *Elements* from other texts in two respects: "Physiology is the study of the cells and tissues in their related activities. Yet the usual school physiology neglects the functions of the cells and tissues and studies organs almost as isolated things, rather than as components of the bodily structure" (p. v). He also takes issue with the manner in which "temperance" is normally presented in school physiologies: "In this book the purpose is to present the blessings of natural incitants, and to show how, if the incitants are taken advantage of by a healthful mode of living, no craving for poisonous stimulants need ever arise. In other words, the temperance teaching is positive rather than negative" (p. vii).

Coleman was an 1889 graduate of Washington and Lee University. He followed his undergraduate degree with studies at the Univ. of Berlin and the Royal School of Science. Returning to the United States, Coleman pursued a career in secondary education in his native Texas. In 1908 Coleman returned to Berlin and London for two years further research/study. He was a prolific author of school science texts and manuals of popular health.

711. COLES, Larkin Baker, 1803-1856.
The beauties and deformities of tobacco-using; or its ludicrous and its solemn realities . . . Twelfth thousand—revised. Boston: Ticknor, Reed, and Fields, 1855.
144 p. ; 20 cm.

"From a pretty extended examination into the nature of the article, and the prevalence of its use, it is my settled conviction that it is now doing a more deadly work to the physical welfare of the American people than alcoholic liquors. The devastations of alcohol are fearful beyond the power of pen or tongue to tell; but the destructiveness of this dreadful poison to the physical system, though now comparatively unperceived by the popular eye, is more cer-

tain and irresistible" (p. [7]). Issued with the 41st edition of Cole's *Philosophy of health* (Boston, 1855, #714).

712. COLES, Larkin Baker, 1803-1856.
The beauties and deformities of tobacco-using; or its ludicrous and its solemn realities . . . Fourteenth thousand—revised. Boston: Ticknor and Fields, 1855.
144 p. ; 20 cm.

Issued with the 43rd edition of Coles' *Philosophy of health* (Boston, 1857, #715). Most editions of the latter title were issued with Coles' anti-tobacco treatise. The 1851, 1853, 1854 and 1855 issues of *The beauties and deformities* were also issued separately.

713. COLES, Larkin Baker, 1803-1856.
Philosophy of health: natural principles of health and cure; or, health and cure without drugs. Also, the moral bearing sof erroneous appetites . . . Twenty-sixth thousand—revised and enlarged. Boston: Ticknor, Reed, & Fields, 1851.
vi, [7]-260 p. ; 18 cm.

714. COLES, Larkin Baker, 1803-1856.
Philosophy of health: natural principles of health and cure; or, health and cure without drugs. Also, the moral bearings of erroneous appetites . . . Forty-first edition—revised and enlarged. Boston: Ticknor, Reed, & Fields, 1855.
vi, [7]-312 p., XIII leaves of plates + frontis. : ill. ; 20 cm.

"A former treatise on this subject, of smaller size, was written as an experiment, to prove whether the people in general were willing to be informed on the science of right living; and whether they would appreciate truth in its warfare against their much-loved and destructive appetites and habits. In proof of the success of that experiment, it may suffice to say, that the work has, in six years, passed into THIRTY-SEVEN EDITIONS, and the demand for it still continues. It has, therefore, been thought best to revise and enlarge it, so that it may contain much more instruction upon matters of such vital importance in practical life" (*Pref.,* p. [iii]).

The present work is a revised and much expanded version of *The philosophy of health; or health without medicine* (#716-721). Most issues of this expanded edition were issued with the author's *The beauties and deformities of tobacco-using,* including this 41st edition, which was issued with the 1855 edition of author's anti-tobacco treatise (#711).

715. COLES, Larkin Baker, 1803-1856.
Philosophy of health: natural principles of health and cure; or, health and cure without drugs. Also, the moral bearings of erroneous appetites . . . Forty-third edition—revised and enlarged. Boston: Charles A. Cummings, 1857.
  vi, [7]-312 p., XV leaves of plates + frontis. : ill. ; 20 cm.

Issued with Coles' *The beauties and deformities of tobacco-using* (Boston, 1855, #712).

716. COLES, Larkin Baker, 1803-1856.
The philosophy of health; or health without medicine: a treatise on the laws of the human system . . . Second edition. Boston: William D. Ticknor & Company, 1848.
  vi, [7]-119 p. ; 16 cm.

The philosophy of health may have been issued in seven editions in 1848 (its first year of publication), though only four of these are recorded. In 1851, the work was greatly expanded and published under the title: *Philosophy of health: natural principles of health and cure* (#713-715)
  Coles' manual of hygiene focuses principally on diet and the digestive process: "There is no part of the human system which has such controlling over the whole body, as respects health or disease, as the Digestive Organs . . . Nearly all the morbid actions found in the general system are produced from causes first operating on the stomach. Hence, keeping the digestive system in a healthy state, secures . . . a healthy action in every other part of the physical organization" (p. [9]). Even when elements of the hygienic regimen such as exercise and sleep are discussed, they are related to digestion.
  Coles lays out his argument for vegetarianism on nutritional, physiological, and what we would term "psychophysiological" principles. On pages 34-35 he makes the point that "All our nutrition comes primarily from the vegetable kingdom. If we eat flesh, the nourishment which made that flesh came from vegetables. The nutrition from the corn on which the hog is fatted

becomes assimilated into his flesh, and by eating that pork we get the nutrition of the corn, animalized." Coles recommends "taking vegetable nutrition in its original state" for several reasons, one of the more important of which is that "eating animal food . . . increases the proportion of our animality," i.e., "When the nutrition of vegetation comes to us through the flesh of an animal, it has undergone a sort of animalization; and as it passes into our circulation the proportion of the animality in our natures is increased" (p. 40). He explains: "When we increase the proportion of our animal nature, we oppress the intellectual and moral" (p. 41). The result of an animal diet is that our "carnality" overwhelms our "moral powers." Coles states, "We are naturally savage enough in our dispositions, and fleshly enough in our appetites, without taking a course that will increase those qualities."

717. COLES, Larkin Baker, 1803-1856.
The philosophy of health; or, health without medicine: a treatise on the laws of the human system . . . Third edition. Boston: William D. Ticknor & Company, 1848.
  vi, [7]-119 p. ; 16 cm.

718. COLES, Larkin Baker, 1803-1856.
The philosophy of health; or, health without medicine: a treatise on the laws of the human system . . . Seventh edition. Boston: William D. Ticknor & Company, 1848.
  vi, [7]-120 p. ; 16 cm.

719. COLES, Larkin Baker, 1803-1856.
The philosophy of health; or, health without medicine: a treatise on the laws of the human system . . . Twelfth edition. Boston: William D. Ticknor & Company, 1849.
  vi, [7]-120 p. ; 16 cm.

720. COLES, Larkin Baker, 1803-1856.
The philosophy of health; or, health without medicine: a treatise on the laws of the human system . . . Fifteenth edition. Boston: William D. Ticknor & Company, 1849.
  vi, [7]-120 p. ; 16 cm.

721. COLES, Larkin Baker, 1803-1856.
The philosophy of health; or health without medicine: a treatise on the laws of the human system . . . Eighteenth edition. Boston: Published by the author, 1850.

vi, [7]-120 p. ; 16 cm.

Coles has entitled the appendix to *The philosophy of health* "Philosophy of healthy reproduction" (p. [101]-116). In these pages Coles warns that any deviation from the laws of health will result in degenerate offspring: "On the healthy condition of the bodily system depends the vital energy of the germinating principles" (p. 102). "Numerous experiments of learned physiologists" confirm this reproductive law. Coles provides the example of a father of nine who until the birth of his fourth child remained temperate, "but being unfortunate in business, he suddenly became, and continued, addicted to his cups." His first four children were normal and healthy. Of the remaining five, generated during his alcoholic period, "one of these was convicted of robbery and served an apprencticeship in the state prison; another of theft; another of larceny; another became a drunkard; the fifth was an idiot" (p. 104). Similarly, "There is great sympathy between the female mind and her own reproductive system. The offspring, while in its foetal state, receives an imprint from the maternal mind, which . . . can never be eradicated. It there receives a mental and moral mould, the great outlines of which can never be obliterated" (p. 111).

722. COLES, Walter, 1839-1892.
The nurse and mother. A manual for the guidance of monthly nurses and mothers; comprising instructions in regard to pregnancy and preparation for child-birth; with minute directions as to care during confinement, and for the management and feeding of infants . . . St. Louis, Mo.: J.H. Chambers & Co., 1887.
vi, [7]-153 p. ; 22 cm.

Originally published at St. Louis in 1881. An 1860 graduate of the New York University Medical College, Coles was professor of obstetrics at the Beaumont Hospital Medical College (St. Louis) and president of the St. Louis Obstetrical & Gynecological Society.

723. COLLECTION OF AFFIDAVITS AND CERTIFICATES.
A Collection of affidavits and certificates, relative to the wonderful cure of Mrs. Ann Mattingly, which took place in the City of Washington, D.C. on the tenth of March, 1824. City of Washington: James Wilson printer, 1824.
20, 22-41 p. ; 18 cm.

In 1817, Ann Mattingly began to manifest symptoms of breast cancer. She underwent medical treatment ("external applications of hemlock and mercurial ointment") to no avail. Her illness and pain only increased over the years. Anticipating her end, Mrs. Mattingly prepared herself spiritually, made her confession, and on the morning of 10 March 1824 received communion. She was instanteously cured of her malady after receiving the sacrament. This collection of documents testifying to the miraculous healing of Ann Mattingly (1782-1855) was compiled by William Mathews (1770-1854), rector of St. Patrick's Roman Catholic Church in Washington, D.C. It includes the testimony of John Marshall (1755-1835), Chief Justice of the U.S. Supreme Court from 1801 to 1835.

724. COLLETT, John.
First lecture free. Highly amusing and instructive entertainment for ladies and gentlemen. Popular lectures on human anatomy, physiology, and the laws of health . . . At the [*Jackson Hall, Canterbury*] on the evening of [*Decr 23d*] and continue the same for five consecutive evenings, (except Sunday,) . . . before mixed audiences. After which he will give an extra lecture to ladies exclusively . . . and another to gentlemen exclusively . . . The Doctor will suitably illustrate his subjects by beautiful plates, models, manikins, &c. &c. . . . [1859?].
[1] leaf printed on both sides : ill. ; 68 × 18 cm.

The verso has the same bold face heading and wood-engraved anatomical figures as the recto, followed by "Canadian testimonials" of lectures delivered in 1858 and 1859. Canterbury is near Cornwall, Orange County, New York.
"At the annual meeting of the Medical Society of Westchester in 1855, those present passed a resolution expressing confidence in their fellow member, Dr. John Collett of Peekskill, who was engaged in giving a course of public lec-

tures on physiology, and whose personal and professional reputation had been assailed by a rival lecturer. The resolution recommends Dr. Collett to the public as one worthy of confidence. It was strenuously opposed by some members on the ground that a certificate of membership in the Society was sufficient to vindicate the doctor's professional and moral standing" (Laurance C. Redway, *History of the Medical Society of the County of Westchester, 1797-1947*, [White Plains?]: Westchester Co. Med. Soc., 1947, p. 10).

725. COLLINGS, W. A.
Dangers that are before us. How to avoid them . . . [Watertown, N.Y.? after 1905].
    16 p. : ill. ; 16 cm.

Brochure promoting *Capt. Collings' Truss and Compound* for the cure of "ruptures."

726. COLLINS, George T.
The cholera: a familiar treatise on its history, causes, symptoms and treatment, with the most effective remedies, and proper mode of their administration, without the aid of a physician, the whole in language free from medical terms, especially adapted for the use of the public generally. Also containing a history of the epidemics of the Middle Ages . . . New York: First National Manufacturing and Publishing Co.; Cincinnati: J.R. Hawley & Company, 1866.
    [2], vi, [3]-162 p. ; 20 cm.

727. DR. COLLINS' REMEDY COMPANY.
The beacon light on the Dr. Collins' mode of treatment, the great European blood and nerve specialist, for the permanent cure of nervous debility and sexual diseases, including spermatorrhoea, impotence, etc., etc., etc. . . . Sixth edition. Boston, Mass.: Dr. Collins' Remedy Co., "Berwick Institute," [190-?].
    64 p. ; 11 cm.

In addition to spermatorrhea and impotence, *Dr. Collins' Remedy* is promoted as a cure for venereal diseases and a wide variety of urological disorders. The remedy must be used under the supervision of one of the physicians attendant at the Berwick Institute, however, whether the patient apply for treatment in person or by correspondence. In printed wrapper.

728. COLTON, Buel Preston, 1852-1906.
Elementary physiology and hygiene . . . Boston: D.C. Heath & Co., publishers, 1902.
    viii, 317, [3] p. : ill. ; 19 cm.

First edition. A revision and condensation of the author's *Physiology: experimental and descriptive* (1st ed., 1898). Colton was the author of a several school textbooks on zoology and physiology published by D.C. Heath & Co. (est. 1885), one of the nations foremost school text publishers. Colton was professor of natural science at Illinois State Normal University.

COLTON SERIES OF PHYSIOLOGIES see BROWN (#423).

729. COMBE, Andrew, 1797-1847.
The management of infancy, physiological and moral. Intended chiefly for the use of parents . . . Revised and edited by Sir James Clark . . . First American from the tenth London edition. New York: D. Appleton and Company, 1871.
    302 p. ; 19 cm.

An Edinburgh native, Combe began an apprenticeship in surgery in 1812. After receiving his diploma from Surgeon's Hall (1817), Combe immediately left for Paris to continue his medical studies. He came under the influence of Spurzheim, whose lectures on the brain he attended in the French capital. Soon after his return to Edinburgh in 1819, Combe was compelled by illness to convalesce in the south of France and in Italy. In 1823 he returned to Edinburgh both to establish a practice and to complete his studies with a medical degree from the University (1825). Under the influence of Spruzheim and the often controversial ideas of his elder brother George, Combe became an outspoken advocate of phrenology, and in 1827 was elected president of the Phrenological Society. By the early 1830s ill health again forced Combe from full-time medical practice, and even made it necessary for him to relinquish his appointment as physician to the King of Belgium. In spite of his chronic illness (pulmonary tuberculosis), Combe was able to continue his phrenological investigations, complete several successful medical books, and serve as Physician Extraordinary to the Queen in Scotland.

Combe dedicated the first edition (1840) of his *Treatise on the physiological and moral management of infancy* (#744-747) to his friend Sir James Clark (1788-1870), the prominent London physician and author of several books on medical climatology and pulmonary tuberculosis. Clark assumed editorship of his late friend's *Treatise* with the 9th Edinburgh edition (1860), changing the title to its present form. The 10th edition, again revised by Clark, was issued in 1870.

In his introduction Clark writes: "I have altered the order of some of the chapters, with the view of bringing the subjects treated of more consecutively before the reader. I have also ventured to omit some portions, chiefly in the earlier chapters, as less necessary now than during the author's lifetime. I have given some additional information on the causes and extent of infant mortality . . . and, lastly, I have added an appendix, in which will be found some useful matter, not so well fitted for the body of the work" (p. 6). Clark recommends this work not only to parents (Combe's intended audience), but to "young medical practitioners," teachers ("to guide them in conducting education in accordance with the progressive development of the mental faculties"), and to governesses ("A very large proportion of the upper, and even middle classes in this country, have no sooner left the nursery than they are consigned to the care of a governess").

730. COMBE, Andrew, 1797-1847.
The mother's guide for the care of her children; or, the management of infancy. Intended for the use of parents . . . Revised and edited by Sir James Clark . . . New York: Continental Publishing Company, [c1872].
303 p. ; 19 cm.

A reprint of the 10th edition (Edinburgh, 1870) of *The management of infancy, physiological and moral*, Sir James Clark's (1788-1870) revision of Combe's *Treatise on the physiological and moral management of infancy* (#744-747).

731. COMBE, Andrew, 1797-1847.
The physiology of digestion considered with relation to the principles of dietetics . . . New-York: Published by Howe & Bates, 1836.
ix, [10]-310 p. : ill. ; 16 cm.

First American edition; originally published at Edinburgh in 1836. *The physiology of digestion* was intended as a continuation of Combe's *The principles of physiology* (1st ed., Edinburgh, 1834). "It may, at first sight, be doubted whether I have not exceeded proper bounds in thus dedicating a whole volume to the consideration of a single subject; but the more we consider the real complication of the function of digestion,—the extensive influence which it exercises at every period of life over the whole of the bodily organization,—the degree to which its morbid derangements undermines health, happiness, and social usefulness . . . we shall become more and more convinced of the deep practical interest which attaches to a minute acquaintance with the laws by which it is regulated" (*Pref.*, p. iv-v). In a nation that had so recently and enthusiastically embraced the dietary crusade preached by such figures as Graham (#1392-1395), Hitchcock (#1651, 1652) and Alcott (#18-61), it is not surprising that American publishers responded by issuing at least a dozen editions of Combe's treatise between 1836 and 1850.

"In many of our best works," Combe states, "the relation subsisting between the human body on the one hand, and the qualities of the alimentary substances on the other, as the only solid principle on which their proper adaption to each other can be based, is almost altogether lost sight of; so that, while the attention is carefully directed to the consideration of the abstract qualities of the different kinds of aliment, little or no regard is paid to the relation in which they stand to the individual constitution, as modified by age, sex, season, and circumstances, or to the observance of the fundamental laws of digestion" (p. vi). In establishing and clarifying these laws, Combe acknowledges the originality and influence of William Beaumont's very recently published researches (a "work still inaccessible to the British reader"). Combe's treatise is divided into two parts: the first on the physiology of digestion; and the second on dietetics, i.e., the "times of eating," quantity, selection of foods, "conditions to be observed before and after eating," thirst, and "the proper regulation of the bowels."

732. COMBE, Andrew, 1797-1847.
The physiology of digestion considered with relation to the principles of dietetics

... Third American edition. Boston: Marsh, Capen, & Lyon; New-York: Daniel Appleton & Co., 1836.
ix, [10]-310 p. : ill. ; 16 cm.

733. COMBE, Andrew, 1797-1847.
The physiology of digestion considered with relation to the principles of dietetics ... From the third Edinburgh revised and enlarged edition ... New York: William H. Colyer; Boston: Lewis & Sampson, 1844.
[i-ii], [2], [iii]-xii, [25]-287 p. : ill. ; 15 cm.

The third Edinburgh edition was published in 1841.

734. COMBE, Andrew, 1797-1846.
The physiology of digestion considered with relation to the principles of dietetics ... From the third Edinburgh revised and enlarged edition ... New-York: William H. Colyer; Boston: Phillips & Sampson, 1846.
xii, [25]-287 p. : ill. ; 16 cm.

735. COMBE, Andrew, 1797-1847.
The physiology of digestion considered with relation to the principles of dietetics ... Sixth American edition. Rochester [N.Y.]: Published by Alling, Seymour & Co., 1846.
ix, [10]-310 p. : ill. ; 16 cm.

This Rochester edition is a reissue of the 1836 New York edition (#732).

736. COMBE, Andrew, 1797-1847.
The principles of physiology applied to the preservation of health, and to the improvement of physical and mental education ... New-York: Harper & Brothers, 1834.
[4], vii, [8]-291 p. : ill. ; 16 cm.

First American edition; originally published at Edinburgh in 1834. Combe's physiological treatises were enormously successful. Within two years the first three Edinburgh editions sold out, the third alone numbering 3,000 copies. By the year of Combe's death, 28,000 copies of this title had been sold in the United Kingdom (*Dict. nat. biog.* 4:881). The publishing history of Combe's physiology in the United States was equally impressive. The New York firm of

Harper & Bros. stereotyped the first Edinburgh edition and issued it a dozen times between 1834 and 1846 as part of its *Harper family library* [no. 71]. Between 1846 and 1876, Fowler & Wells published the text of the 7th Edinburgh edition in eight issues from their New York office.

Combe's *Principles* was one of the earliest titles in the genre of popular physiology and hygiene available to readers in the United States, and remained a standard text even as the market became glutted with domestic and school physiologies. The *Principles* remained in print for more than forty years, making it the longest published popular physiology on the American market.

737. COMBE, Andrew, 1797-1847.
The principles of physiology applied to the preservation of health, and to the improvement of physical and mental education ... Fourth edition, revised and enlarged. Edinburgh: Maclachlan & Stewart; London: Simpkin, Marshall, Co., 1836.
xii, 438 p. : ill. ; 19 cm.

738. COMBE, Andrew, 1797-1847.
The principles of physiology applied to the preservation of health, and to the improvement of physical and mental education ... New-York: Harper & Brothers, 1836.
[2], vii, [8]-291 p. ; 15 cm.

739. COMBE, Andrew, 1797-1847.
The principles of physiology applied to the preservation of health, and to the improvement of physical and mental education ... From the seventh Edinburgh edition. New-York: Harper & Brothers, 1844.
x, [11]-396, [12] p. : ill. ; 16 cm.

On spine: *The family library; no. 71. Dr. Combe's principles of physiology.*

740. COMBE, Andrew, 1797-'847.
The principles of physiology applied to the preservation of health, and to the improvement of physical and mental education ... To which is added, notes and observations, by O.S. Fowler . . . From the seventh Edinburgh edition, enlarged and improved. New-York: Fowler and Wells, 1848.
xii, [13]-320 p. : ill. ; 23 cm.

In his preface, O.S. Fowler notes that although Harper & Bros. hold the American rights to Combe's physiology, they had done little more with the book than reissue the stereotyped text of the first edition for a dozen years, thus failing to provide the American public with the improved text of successive Edinburgh editions. He then provides a fascinating narrative of his attempts to get the 7th Edinburgh edition (1838) into print, the refusal of Harper Bros. to do so, the hesitation of other publishers to challenge the powerful Harper firm, and his personal decision to undertake its publication himself.

741. COMBE, Andrew, 1797-1847.
The principles of physiology applied to the preservation of health, and to the improvement of physical and mental education . . . To which is added, notes and observations, by O.S. Fowler . . . From the seventh Edinburgh edition, enlarged and improved. New-York: Fowler & Wells, 1849.
xii, [13]-320 p. : ill. ; 23 cm.

742. COMBE, Andrew, 1797-1847.
The principles of physiology applied to the preservation of health, and to the improvement of physical and mental education . . . To which is added, notes and observations, by O.S. Fowler . . . From the seventh Edinburgh edition, enlarged and improved. New York: Fowlers and Wells, publishers, 1854.
xii, [13]-320 p. : ill. ; 23 cm.

743. COMBE, Andrew, 1797-1847.
The principles of physiology, applied to the preservation of health, and to the improvement of physical and mental education . . . To which is added, notes and observations, by O.S. Fowler. From the seventh Edinburgh edition, enlarged and improved. New York: Fowler & Wells Co., publishers, 753 Broadway, [between 1880 and 1887].
viii, [13]-320 p. : ill. ; 23 cm.

Fowler & Wells was located at 753 Broadway between 1880 and 1887.

744. COMBE, Andrew, 1797-1847.
A treatise on the physiological and moral management of infancy. For the use of parents . . . From the fourth Edinburgh edition. New York: William H. Colyer; Boston: Phillips & Sampson, 1845.
xi, [12]-296 p. : ill. ; 16 cm.

The fourth Edinburgh edition on which Colyer's text is based was published in 1844. Spine-title: *Combe's management of infancy.*

745. COMBE, Andrew, 1797-1847.
Treatise on the physiological and moral management of infancy . . . With notes and a supplementary chapter. By John Bell . . . Third edition. Philadelphia: Carey & Hart, 1842.
307 p. ; 19 cm.

The *Treatise* was the last book Combe authored. It was first published at Edinburgh in 1840. Carey & Hart brought out the first and second American editions that same year. In his preface Combe notes two deficiences of the popular manuals then in print, i.e., "Most of those hitherto published, touch briefly upon the general management of early childhood merely as preliminary to an exposition of its diseases," which Combe regarded as potentially dangerous and inappropriate for "any except medical readers." He also faults contemporary manuals on infancy for "presenting their rules and admonitions as so many abstract and individual opinions" without grounding them on "physiological laws or principles." (p. [5]). Combe therefore avoids all descriptions of disease, emphasizing instead the physiological principles upon which every hygienic principle is based.

In his fifteen chapters Combe discusses infant mortality (a sobering prelude to infant hygiene), the "delicacy of the constitution in infancy"; the influence of mothers' health on the health of their children; neonatal care; infant food; criteria for selecting a wetnurse; bathing, exercise & sleep; teething; and the "moral management of infancy." John Bell's (#289-293) supplementary chapter places Combes' recommendations on infant hygiene in the context of American climates and "the greater dangers, both physical and moral, to which children are exposed in this country" (p. 281). Combe's text has also been supplemented with numerous footnotes by Bell. Spine-title: *Combe on infancy.*

746. COMBE, Andrew, 1797-1847.
Treatise on the physiological and moral
management of infancy . . . With notes and
a supplementary chapter. By John Bell . . .
Sixth edition—illustrated. New York; Boston: Fowlers & Wells, 1852.
  [2], 9, 11-15, 17-307 p. : ill. ; 22 cm.

This sixth edition was printed from the same
plates used for the Carey & Hart editions of the
Treatise. Series: *Fowlers and Wells phrenological library.*

747. COMBE, Andrew, 1797-1847.
Treatise on the physiological and moral
management of infancy . . . With notes and
a supplementary chapter. By John Bell . . .
Sixth edition—illustrated . . . New York:
Fowler & Wells, publishers, No. 389 Broadway, [186?].
  307 p. : ill. ; 19 cm.

The dating of this issue of Fowler & Well's
sixth edition is based on the firm's move to 389
Broadway in the mid-1860s.

COMFORT, Anna Manning see **COMFORT**
(#748).

748. COMFORT, George Fisk, 1833-1910.
Woman's education, and woman's health:
chiefly in reply to "Sex in education." By
George F. Comfort . . . and Mrs. Anna Manning Comfort, M.D. Syracuse [N.Y.]: Thos.
W. Durston & Co., publishers, 1874.
  xii, [13]-155 p. ; 18 cm.

A critical response to E.H. Clarke's *Sex in
education* (#661, 662), in which Clarke blamed
contemporary methods of education for the ill-
health of adolescent American females. The
Comforts "hold his work to be utterly wrong in
all its essential features," and in their introduction note that as a result of Clarke's book,
"Some parents have already withdrawn their
daughters from school through fear of the dire-
ful consequences which Dr. C. predicts. Some
young women have decided to shorten their
courses of study through the same fear" (p. viii).
The authors intend "to point out the erroneous conclusions and inferences to which he
would lead his readers," and "to show to what
extent his premises, his conclusions, and his

illustrations are inappropriate and untenable"
(p. ix).
  At the time this work was published George
Fisk Comfort was dean of the College of Fine
Arts at Syracuse University. His wife, Anna Manning Comfort (1845-1931), was an 1865 graduate of the New York Medical College for
Women, the first female medical graduate licensed to practice in Connecticut, and a pioneer suffragist.

749. COMFORT, John William, 1821-1854.
The practice of medicine on Thomsonian
principles, adapted as well to the use of
families as to that of the practitioner. Containing propositions illustrative of the philosophy of Thomsonism, a brief history of
the symptoms, peculiarities, and general
course of disease in its different forms and
varieties; with practical directions for administering a Thomsonian course of medicine, including the various methods of
administering vapour baths and emetics,
and a materia medica, adapted to the work
. . . Philadelphia: Published by A. Comfort,
1843.
  xv, [1], 514, [2] p. ; 22 cm.

Though it appeared rather late in the annals of Thomsonian literature, this first edition
of Comfort's treatise was one of the most successful manuals of the botanic practice, appearing in eight editions through 1888.
  Comfort was born in Logan County, Kentucky, in 1821. Where, or with whom, he studied medicine is unknown. He settled in New
Orleans and was married there, but must have
spent some time in Kosciusko, Mississippi. The
1850 census places Comfort there, and at least
two of his four children were born there (1850,
1852). Comfort died of yellow fever in New
Orleans on 6 August 1854, prior to the birth of
his last child.

750. COMFORT, John William, 1821-1854.
The practice of medicine on Thomsonian
principles, adapted as well to the use of
families as to that of the practitioner. Containing a biographical sketch of Dr.
Thomson, propositions illustrative of the
philosophy of Thomsonianism; a brief history of the symptoms, peculiarities, and
general course of disease in its different

forms and varieties; with practical directions for administering the Thomsonian medicines, including the various methods of adminstering vapour baths, emetics, &c. and a materia medica, adapted to the work . . . A new and revised edition. Philadelphia: For sale by A. Comfort, 1850.

[4], liv, 55-582 p. ; 23 cm.

The 1850 "new and revised edition" is the third edition of Comfort's *Practice*. Atwater copy lacking pp. 579-582.

751. COMFORT, John William, 1821-1854.

Thomsonian practice of midwifery, and treatment of complaints peculiar to women and children . . . Philadelphia: Published by Aaron Comfort, 1845.

viii, 215, [1] p. ; 22 cm.

Comfort's manual covers a range of topics common to obstetrical manuals of the period, i.e., the signs of pregnancy, its disorders, the conduct of labor, puerperal disorders, and diseases peculiar to newborns. He quotes abundantly from the obstetrical treatises of Dewees, Meigs, Robert Lee, etc., whose understanding of the mechanics of labor, the management of difficult presentations, etc., he obviously respects. Comfort's book differs from these treatises in its appeal to a lay audience and in its prescription of Thomsonian remedies for every medical need during pregnancy, delivery and the post-partum care of mother and infant. Lobelia, for example, a staple of the Thomsonian materia medica, is recommended for the relief or treatment of such varied symptoms and disorders as morning sickness, uterine hemorrhage, uterine convulsions, puerperal infection, etc.

The relation of the publisher Aaron Comfort to the author is undetermined. A.C. operated a "Thomsonian medicine store" on Market St. in Philadelphia where he compounded and sold Thomsonian medicines as well as published and sold Thomsonian books.

752. COMINGS, Benjamin N., d. 1899.

Class-book of physiology; for the use of schools and families. Comprising the structure and functions of the organs of man, illustrated by comparative reference to those of inferior animals . . . New-York: D. Appleton & Co., 1853.

270, [6] p., XXIV leaves of plates : ill. ; 19 cm.

First edition. The *Class-book* is arranged on similar principles to Comings' and Comstock's 1851 *Principles of physiology* (#756), and was intended as a lower-priced alternative to the earlier folio-size volume. A second edition of the *Class-book* was published in 1855. It was issued for the final time in 1866. The numbered lithographic plates (I-XXIV) are also included in the pagination of the text. Atwater copy lacking plate XVII. Comings was an 1845 graduate of Castleton Medical College (Vt.).

753. COMINGS, Benjamin N., d. 1899.

Preservation of health, and prevention of disease: including practical suggestions on diet, mental development, exercise, venti-

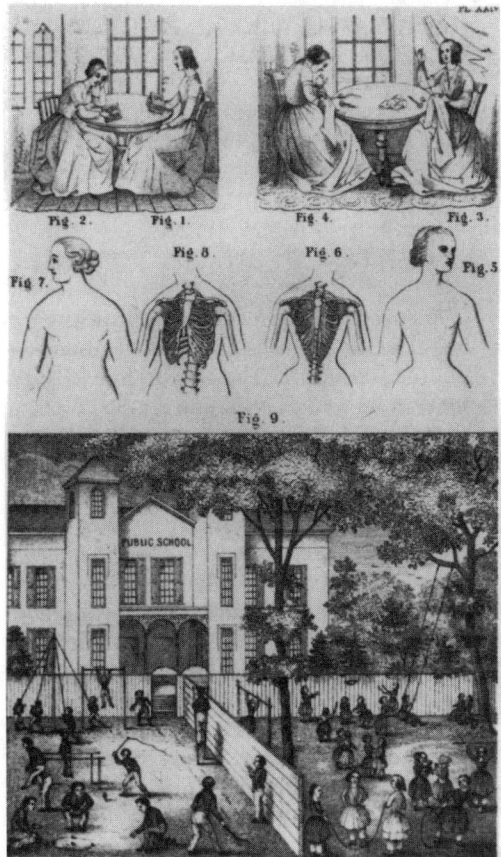

Fig. 21. From Coming's Class-book of physiology *(1853), #752*

lation, bathing, use of medicines, management of the sick, etc. . . . New York: D. Appleton & Co., 1854.
208 p., X leaves of plates : ill. ; 19 cm.

The thirteen chapters of Comings' hygienic treatise may be divided into four parts: diseases of the organs of respiration; diet and digestion; the nervous system and its diseases; and regimen (i.e., exercise, bathing, ventilation). In the final chapter on management of the sick, Comings advises against self-prescription and provides advice on the selection of a physician. He includes the text of the A.M.A. *Code of ethics* in this discussion. The numbered lithographic plates (I-X) are also included in the pagination of the text.

COMINGS, Benjamin N. see also
COMSTOCK (#756).

COMINGS, Isaac Miller see **REFORM
MEDICAL COLLEGE OF GEORGIA**
(#2953).

754. COMSTOCK, Anthony, 1844-1915.
Traps for the young . . . With introduction by J.M. Buckley, D.D. New York: Funk & Wagnalls, publishers, 1883.
xii, [7]-253 p. ; 19 cm.

First edition. Comstock worked as a retail clerk in several states before abandoning business in 1873 to devote himself full-time to the reform of public morals. By 1871 he was at the service of the Y.M.C.A., for which he established a Committee for the Suppression of Vice. In 1873 he founded the New York Society for the Suppression of Vice. His earliest major success came two years later when he forced postal legislation that banned the shipment of obscene literature in the mails. As a special (and unpaid) agent of the Post Office Dept., Comstock effected the arrest of more than 3,600 persons between 1872 and 1915 for selling obscene literature, pictures, contraceptive devices, abortificants and gambling materials through the mails.
"Utterly incorruptible and tirelessly zealous in the pursuit of what he considered his duty, he spent the rest of his years in furious raids upon publishers of obscene and fraudulent literature, quacks, abortionists, gamblers, managers of lotteries, dishonest advertisers, patent-medicine venders, and artists in the nude" (*Dict. Amer. biog.*, 2:330). Comstock attained national fame (and much ridicule) for his efforts to cleanse public morals. His zealous pursuit of the enemies of purity inspired the term "comstockery" ("Excessive opposition to, or censorship of, supposed immorality in art or literature," *OED*, 2nd ed., III:642)—a word coined by Bernard Shaw in a 1905 article published in the *New York Times*.

In addition to a number of pamphlets, Comstock was the author of three books: *Frauds exposed* (1880), *Traps for the young* (1883), and *Gambling outrages* (1887). In his preface to *Traps for the young* Comstock writes, "This book is designed to awaken thought upon the subject of *Evil Reading*, and to expose to the minds of parents, teachers, guardians, and pastors, some of the mightly forces for evil that are to-day exerting a controlling influence over the young" (p. [ix]). The author goes on to discuss the peril to adolescent morals from newspapers, dime novels, advertising and blatantly obscene literature, as well as the unwholesome attractions of gambling, free-love, etc. *Traps for the young* was first published at New York in 1883. A second edition was issued in 1884, followed by the third editions of 1884 (#755) and 1890 (#755.1).

The "Comstock laws" not only restrained the distribution of explicitly pornographic literature in the last quarter of the 19th century, but extended its influence to the dissemination of literature on sex education and contraception, both legitimate and prurient. The eclectic physician E.B. Foote (#1213-1256) was convicted in 1876 for having sent a pamphlet on contraceptive methods through the mails. Authors and publishers (including Foote) became increasingly wary of the kind of contraceptive advice or advertisements they might include (or exclude) in books, pamphlets and journals. Janet Farrell Brodie has concluded that as a direct result of the "Comstock laws," the quality of published advice on contraceptive methods available to American women in the last quarter of the 19th century was inferior to what women were able to read fifty years earlier (*Contraception and abortion in nineteenth-century America*, Ithaca: Cornell Univ. Press, 1994, p. 281).

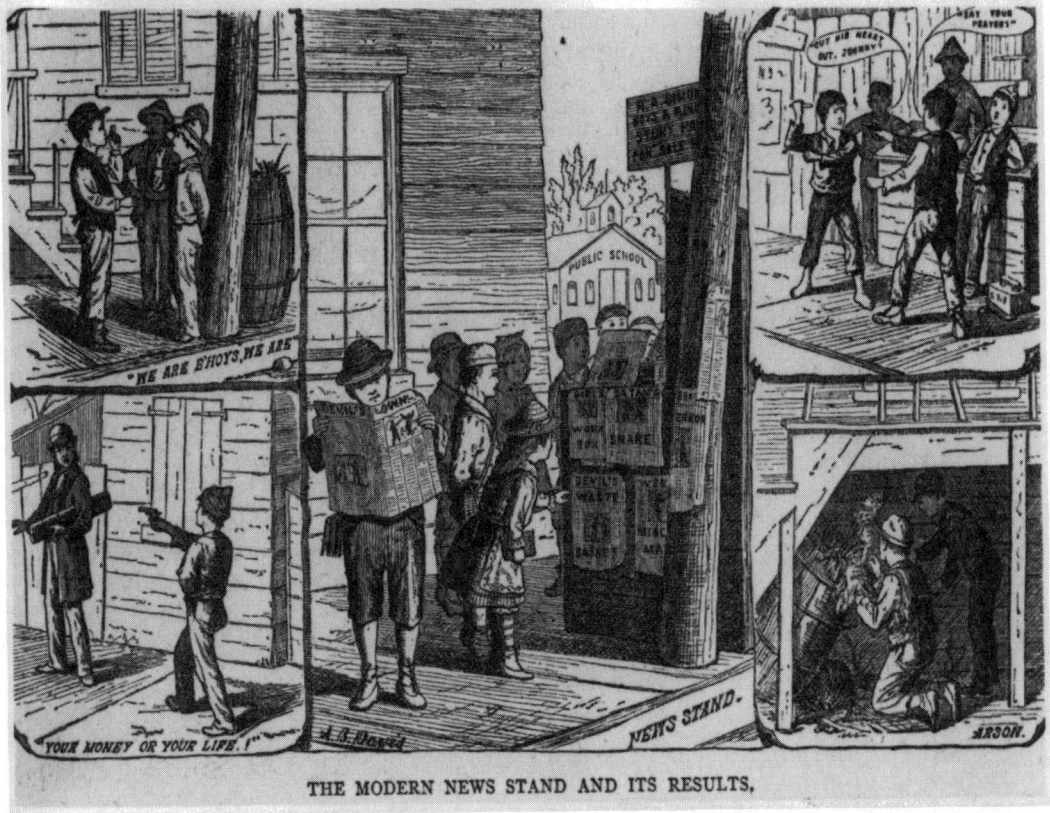

Fig. 22. From Comstock's Traps for the young *(1884), #755.*

755. COMSTOCK, Anthony, 1844-1915.
Traps for the young . . . With introduction by J.M. Buckley, D.D. Third edition. New York: Funk & Wagnalls, publishers, 1884.
xii, [7]-253 p. ; 19 cm.

Both the second and third editions of *Traps for the young* were issued in 1884.

755. 1. COMSTOCK, Anthony, 1844-1915.
Traps for the young . . . With introduction by J.M. Buckley, D.D. Third edition. New York: Funk & Wagnalls, publishers, 1890.
xii, [7]-253 p. ; 19 cm.

756. COMSTOCK, J. C.
Principles of physiology; designed for the use of schools, academies, colleges, and the general reader. Comprising a familiar explanation of the structure and functions of the organs of man, illustrated by comparative reference to those of the inferior animals. Also, an essay on the preservation of health. By J.C. Comstock and B.N. Comings, M.D. New-York: Pratt, Woodford & Co.; Hartford: E.C. Kellogg, 1851.
110 p., XIV leaves of plates : ill. ; 32 cm.

Remarkable in this folio volume are the fourteen hand-colored plates.

757. COMSTOCK, John Lee, 1789-1858.
Outlines of physiology, both comparative and human; in which are described the mechanical, animal, vital, and sensorial organs, and functions; including those of respiration, circulation, digestion, audition and vision, as they exist in the different orders of animals, from the sponge to man. Also, the application of these principles to muscular exercise, and female fashions, and deformities . . . Intended for the use of schools and heads of families . . . New York: Robinson, Pratt & Co., 1836.
vii, [1], [13]-314 p. : ill. ; 19 cm.

Comstock's *Outlines* was one of the earliest popular physiologies available to the American reading public, and one of several popular texts on the natural sciences he authored. Spine-title: *Comstock's physiology.*

758. COMSTOCK, John Lee, 1789-1858.
Outlines of physiology, both comparative and human; in which are described the mechanical, animal, vital, and sensorial organs, and functions; including those of respiration, circulation, digestion, audition and vision, as they exist in the different orders of animals, from the sponge to man. Also, the application of these principles to muscular exercise, and female fashions, and deformities . . . Intended for the use of schools and heads of families. Second edition . . . New York: Robinson, Pratt & Co., 1837.
vii, [8]-310 p. : ill. ; 19 cm.

Spine-title: *Comstock's physiology.*

759. COMSTOCK, W. H.
Dr. Morse's Indian Root pills. W.H. Comstock, proprietor, Morristown, St. Lawrence Co., N.Y. . . . [S.l., ca. 1878].
15, [1] p. ; 14.5 cm.

"It is an established fact, that all diseases spring from one source, namely—impurity of the blood . . . When the various passages become clogged, and do not act in perfect harmony with the different functions of the body, the blood looses its action, becomes thick, corrupted, and diseased . . . How important, then, that we should keep the various passages of the body free and open" (p. 3-4). To attain that end, *Morse's Indian Root Pills* contain four botanicals: "One of the roots . . . is a sudorific, which opens the pores of the skin, and assists nature in throwing out the finer parts of the corruption within. The second is a plant which is an expectorant, that . . . unclogs the passage to the lungs, and thus . . . performs its duty by throwing off the phlegm and other humors from the lungs by copious spitting. The third is a diuretic, which gives ease and double strength to the kidneys; thus encouraged, they draw large amounts of impurity from the blood . . . The fourth is a cathartic, and accompanies the other properties of the Pills while engaged in purifying the

blood; the coarser particles of impurity which cannot pass by the other outlets, are thus taken up and conveyed off in great quantities by the bowels" (p. 4).

COMTE, Achille see **MILNE-EDWARDS** (#2459).

CONDIE, David Francis see **CARPENTER** (#569); **FAUST** (#1144-1146); *JOURNAL OF HEALTH* (#2053); *PORTER'S HEALTH ALMANAC* (#2868).

759.1 CONFIDENTIAL MEDICAL TALKS TO WOMEN.
Confidential medical talks to women. Health of women and children, woman's physical structure, maidenhood—education of women, womanhood and wifehood, motherhood, babyhood, childhood. Diseases of women. General hygiene . . . New York; Akron, Ohio; Chicago: The Werner Company, 1899.
xi, [12]-388 p., [24] leaves of plates : ill. ; 20 cm.

"In the very outset, the book deals boldy with the question of personal beauty, and points out the secret of its development and perpetuation . . . Passing from this matter of personal beauty, the book then deals with the important question of dining . . . Fifty pages of this invaluable book are devoted to a wise resume of practical hints on all the important points connected with this very important department of social life . . . The book next deals with the subject of house decoration . . . The building of a house may tax the skill of architect and builder, but the finishing and decoration that change a mere 'house' into a 'beautiful home,' require taste and skill and tact . . . The last part of the book vies with all the rest in practical importance. 'Man know thyself' is a good motto, but if the maidens and mothers of the world were better acquainted with their physical organism, and the simple secrets of health, they would spare themselves endless and unnecessary suffering, and the future races of men would be caste in a more heroic mold" (p.iii-vi).

There is no author statement on the title-page of the *Confidential medical talks to women,* but the spine-title is followed by the statement:

"By Robert A. Gunn. " However, the text of the *Confidential medical talks* is identical to that of *Facts for ladies*, edited and published by Amy G. Ayer at Chicago in 1890 (#1113). Though a smaller format, the *Confidential medical talks* is printed from the same plates used in the production of *Facts for ladies*. The latter work acknowledges Gunn on its title-page as the author of the chapters on the health of women and children.

760. CONGER, Horace O.

Obstetrics and womanly beauty. A treatise on the physical life of woman, embracing full information on all important matters for both mothers and maidens. A complete guide to health and beauty, with hints on courtship, marriage, hereditary descent, mental conditions, etc., etc. Together with diseases peculiar to the female sex, their cause, symptoms and treatment. By Horace O. Conger, M.D., and Caroline P. Crane, M.D. Chicago, Ill.: Published by American Publishing House, [189-?].

xv, [16]-541 p., [2] leaves of plates : ports. ; 20 cm.

The premise of this volume is that physical beauty is an indication that all else in a woman's life is well. Beauty, therefore, is not only to be desired but is to be actively sought by enhancing what nature has provided and what artifice may attain. "True beauty of face and form are more to be desired than the wealth of a Croesus, because true beauty implies not only a clear complexion and well formed features, but health of body, purity of mind and nobleness of character" (p. 25). There is a fascinating tension in this volume between "modern" ideas on science and women's status that had infiltrated the popular consciousness and very traditional concepts of women's value and role. Hence the authors can in one paragraph make the statement, "the time must come—is even here— when, notwithstanding the galling servitude of ages, all womankind will secure in full measure, liberty and equality with man," and in the following paragraph write that even if women "have written no great poem . . . formed no governments . . . composed no famous opera and written no 'Hamlet,'" they have "performed a work far greater and grander than all this, for at their knees the purest and most god-like men

and women have been trained, and these are the best products of all creation" (p. 21-22).

The volume is divided into six parts: the first on becoming beautiful; the second on criteria for selecting a husband, the physical and mental changes of puberty, and the philosophy of marriage; part three on the sexual life of married women; part four on what to expect during pregnancy and in labor; part five on child care; and the final part on home nursing, with chapters on etiquette and physiognomy. A pocket on the back pastedown indicates the presence of a pamphlet missing in the Atwater copy, that was probably an atlas of reproductive anatomy and physiology.

761. CONGER, HORACE O.

"Vitalogy." Private words to men and women. Sex pamphlet and the beautiful story of life illustrated . . . by Horace O. Conger, M.D. [and] Caroline P. Crane, M.D. [Chicago? ca. 1900].

48 p. : ill. ; 22 cm.

Title from printed wrapper. This pamphlet is comprised of black and white illustrations (and descriptive text) intended as a separately issued atlas to accompany the authors' *Vitalogy*, a "practical family medical work." Lacking any bibliographic record, *Vitalogy* may never have been published. Pages 6-13 of this pamphlet describe and illustrate "local inadaption," described as "one of the commonest causes of barrenness." Local inadaption is an incongruity between the size and/or form of the penis and the placement of the uterus during intercourse.

762. CONGRESS AND EMPIRE SPRING CO.

Congress, Empire, and Columbian spring waters, Saratoga Springs, N.Y. Their medicinal character, and their uses in preventing disease, and preserving health. With a description of the Congress Spring Park. Saratoga Springs, N.Y.: Congress and Empire Spring Co., [ca. 1876].

16 p. : ill. ; 13 cm.

763. CONN, Herbert William, 1859-1917.

An elementary physiology and hygiene for use in schools . . . New York: Silver, Burdett and Company, [c1903].

272 p., [2] leaves of plates : ill. ; 19 cm.

First edition. Among Conn's many writings was a series of three graded school physiologies published as the *Conn series of physiologies* by Silver, Burdett & Company: *Introductory physiology and hygiene* (1904, #766), *An elementary physiology and hygiene* (1903), and *Advanced physiology and hygiene* (1909). This second text in the series appeared in five editions or issues between 1903 and 1913.

Conn received his Ph.D. in biology from Johns Hopkins in 1884, and immediately joined the faculty of Wesleyan University, on which he remained his entire career. He was also director of the Cold Spring Harbor Biological Laboratory (1889-97), was bacteriologist at the Storrs Experimental Station (1890-1906), and director of the laboratories of the Connecticut State Board of Health.

764. CONN, Herbert William, 1859-1917.
An elementary physiology and hygiene for use in schools . . . Enlarged edition. New York; Boston; Chicago: Silver, Burdett and Company, [c1906].
       334 p. : ill. ; 18 cm.

765. CONN, Herbert William, 1859-1917.
An elementary physiology and hygiene for use in upper grammar grades . . . New York: Silver, Burdett and Company, [c1913].
       349 p., [2] leaves of plates : ill. ; 19 cm.

766. CONN, Herbert William, 1859-1917.
Introductory physiology and hygiene for use in primary grades . . . New York: Silver, Burdett and Company, [c1904].
       152 p. : ill. ; 19 cm.

       First edition. An enlarged edition was published in 1906 and reissued in 1908.

767. CONN, Herbert William, 1859-1917.
The story of germ life . . . New York: McClure, Phillips & Co., 1904.
       [2], 199 p. : ill. ; 19 cm.

       First published in 1897. Series: *Library of valuable knowledge.*

768. CONN, Herbert William, 1859-1917.
The story of germ life . . . New York and London: D. Appleton and Company, 1913.
       [2], 199 p. : ill. ; 17 cm.

       Series: *The Library of useful stories.*

769. CONNECTICUT SOCIETY OF SOCIAL HYGIENE.
Igiene del sesso. Opusculo no. 1. Statistica generale. Hartford, Conn.: Pubblicato per cura della Connecticut Society of Social Hygiene, 1911.
       9 p. ; 14 cm.

       A pamphlet intended to educate the Italian immigrant population on the prevelance and dangers of *la peste nera*, i.e., syphilis and gonorrhea.

770. CONNECTICUT TRAINING SCHOOL FOR NURSES.
A hand-book of nursing for family and general use. Published under the direction of the Connecticut Training-School for Nurses, State Hospital, New Haven, Connecticut. Philadelphia: J.B. Lippincott & Co., 1882.
       266 p. ; 19 cm.

       "In all cases of serious illness, whether a trained nurse can be secured or not, there must always be some one person in the family who can be responsible for the patient and to the physician. Two or three different persons taking orders and reporting symptoms will invariably make confusion and mistakes. There must be one head. Many of the following directions for a professional nurse will apply with equal force to the member of the family who stands in that relation to the patient. Want of order and common sense in a sister, wife, or daughter is even more distressing to a sick person than the same qualities in a stranger. Let whoever proposes to assume the care of a seriously sick patient read and follow the directions given to the professional nurse" (p. 7). The *Hand-book* is divided into two parts: the first on medical and surgical nursing; the second on midwifery and wet-nursing. It was originally published at Philadelphia in 1879.

771. CONNECTICUT TRAINING SCHOOL FOR NURSES.
A hand-book of nursing for family and general use. Published under the direction of the Connecticut Training-School for Nurses, State Hospital, New Haven, Connecticut. Philadelphia: J.B. Lippincott & Co., 1885.
       266 p. ; 19 cm.

772. CONNECTICUT TRAINING SCHOOL FOR NURSES.
A hand-book of nursing for family and general use. Published under the direction of the Connecticut Training-School for Nurses, State Hospital, New Haven, Connecticut. Philadelphia: J.B. Lippincott Company, 1892.
   266 p. ; 19 cm.

773. CONWELL, Joseph Alfred, b. 1855.
Practical medical therapy. A book for the family, the physician, and the druggist. A popular treatise on the prevention and cure of diseases, and the use and abuse of medicines, embracing the essential principles of medical practice, in accord with the latest scientific research and the most advanced medical treatment; a setting forth of the correct and intrinsic value of drugs and medicines as curative agents, and telling how to save life, health, time, and money in their use; to which is added a great variety of valuable formulas for medicinal, toilet, and household purposes; receipts for disinfectants, insecticides, hygienic measures, etc., etc., etc. . . . Vineland, N.J.: The Conwell Publishing Co., 1892.
   xiv, [15]-656 p. ; 23 cm.

   Cover-title: *Popular medical knowledge*; spine-title: *Medical therapy.*

COOK, George W. see **EPPS** (#1073).

774. COOK, Marc, 1854-1882.
The wilderness cure . . . New York: William Wood & Company, 1881.
   153 p. ; 19 cm.

   Journalist and poet, Cook published verse under the pseudonym Vandyke Brown. Apart from a posthumously published volume of poetry, *The wilderness cure* was his only book. Cook describes his prolonged sojourn in a remote Adirondack camp seeking relief from pulmonary tuberculosis. He provides advice on finding the right guide to lead the invalid into the wilderness, the selection of a salubrious site for the camp, requisites of the tent in which one must reside, how to equip and stock the camp kitchen, and other matters.

775. COOK, William Henry, 1832-1899.
A handbook of practical medicine, for the use of students, practitioners and families; including a formulary, medical ethics and form of will . . . First edition. Cincinnati: William H. Cook, publisher, 1859.
   x, 11-251 p. ; 18 cm.

   William Henry Cook is one of the principals of John Haller's 1997 *Kindly medicine: physio-medicalism in America, 1836-1911*. According to Haller, Cook "established much of the canon of physio-medico thought," one of several movements that emerged from Thomsonism in the 1830s and 1840s. Physio-medicalism remained grounded in botanic practice, including the exclusion of chemicals and minerals from its materia medica. The "physios" insisted, however, that they were a school of medical thought based on scientific principles. Contrary to Thomson's informal training of practitioners, the physio-medicalists recognized as medical brethren only those educated at physio-medical colleges. "Thirteen physio-medical colleges were established between 1836 and 1911," by Haller's count (p. xiii). Though many physio-medicalists continued "to view laboratory science and experimental medicine as foreign and elitist" (p. 87), most of these schools boasted curricula that included not only the botanic materia medica, but human anatomy, physiology, chemistry, obstetrics and surgery.
   Through his work as an educator, author and spokesman, Cook became one of the movement's most influential figures and a proponent of basing physio-medical theory and training on a more scientific basis. When Cook became dean of the Botanico-Medical College (Cincinnati) in the mid-1850s, he restructured the school's curriculum and undertook a re-examination of the botanic materia medica. In 1859 Cook left the College and formed his own Physio-Medical Institute several blocks away. He won important victories in gaining access for his students to the wards of the Cincinnati Hospital; fought for state licensure of physio-medical school graduates; maintained doctrinal standards and legitimacy by combatting the practices of physio-medical diploma mills; and took a leading role in state and national physio-medical societies.
   The *Handbook* is a general treatise on disease divided by bodily system, with additional

chapters on emergencies and a final chapter on medical ethics.

776. COOK, William Henry, 1832-1899.
A handbook of practical medicine for the use of students, practitioners and families; including a formulary, medical ethics and form of will . . . Fifth edition. Cincinnati: William H. Cook, publisher, 1864.
    x, 11-251 p. ; 19 cm.

777. COOK, William Henry, 1832-1899.
Woman's hand-book of health. A guide for the wife, mother, and nurse . . . With addenda of favorite prescriptions by J. Allen Hubbs, M.D. Ninth edition. Charleroi, Pa.: Published by R. Cook Hubbs, 1902.
    vi, [7]-408 p., [1] leaf of plates : ill., port. ; 20 cm.

    "Though surrounded with conveniences, and provided with comforts, and nurtured with a tenderness never before enjoyed by her sex, the constitutional vigor of woman has never before appeared so feeble. Her maladies seem to be increased, and their perniciousness intensified . . ." (p. iii). And yet, Cook informs us, "It is a fact well established, that both the physical and mental strength of the human race depends very largely upon woman. If a mother is strong and healthy, her children will enjoy many prospects of being robust and vigorous beyond the children of a delicate and sickly mother . . . And it is also well known that the strength and clearness of the mother's mind, will rarely fail to reproduce itself upon the minds of her children" (p. [9]-10). Given the tragic debility of American womanhood, and the tremendous responsibility that nonetheless burdens woman, Cook devotes himself to the crucial issues affecting female reproductive health.
    CONTENT: I. Need of a school for the body. II. Influence and peculiarities of woman. III. Structure of the female organs. IV. Female development. V. Diseases of the organs and functions. VI. Marriage. VII. Period of pregnancy. VIII. Diseases of pregnancy. IX. Child-birth. X. Management after delivery. XI. Management of children. XII. Medical preparations. The first edition was published at Cincinnati in 1866. Cover-title: *Woman's book of health.*

COOK, William Wesley see **CHASE** (#607), **PANCOAST** (#2716, 2730).

778. COOKE, George.
Professional experience in various climates: a complete practical treatise on genital maladies; with pathological observations on the philosophy of reproduction, spermatorrhoea, and their immediate and remote consequences: together with the more successful management of diseases of women and children, as adopted at the present day. Second edition . . . Albany, N.Y.: Published by the author, 1852.
    xviii, 312 p., [1] leaf of plates : port. ; 22 cm.

    On the title-page and in various places throughout the book, the author claims to be a former surgeon in the Royal Navy, an M.D. and LL.D, a fellow of the Royal College of Physicians and Surgeons, a former associate of John Abernethy, Astley Cooper and John Hunter, "Chancellor of the university, and president of the medical department of the College of Ripley [Ohio]," proud owner of a copy of his own bust, the original of which "was placed in the Royal Academy of Fine Arts, Trafalgar Square, London," and a general in the Illinois militia. Cooke states that he emigrated to the United States in 1827 and established the Lock Hospital in Albany, N.Y. in 1830. For twenty-two years he claims to have been available "for the prevention and removal of all diseases requiring professional secrecy, without mercury."
    Cooke's *Professional experience* is divided into four parts: I. "Philosophy of reproduction"; II. "Spermatorrhoea"; III. "Venereal diseases"; and IV. "General hospital practice in Europe, Asia, North and South America." In this last part he devotes three chapters to a description of his medical training and arrival in America, as well as testimonial letters.

779. COOKE, Nicholas Francis, 1829-1885.
Satan in society. By a Physician . . . Cincinnati and New York: C.F. Vent; Chicago: J.S. Goodman & Co., 1871.
    412 p. ; 20 cm.

    A survey of *NUC pre-1956 imprints* and the OCLC database indicates that the anonymously published *Satan in society* appeared in some two dozen editions or issues between 1871 and 1895. "Little matters it to us," Cooke writes in his opening *apologia*, "that we shall doubtless obtain many readers from the singularity of our title and the nature of the topics discussed. Those who shall seek in our pages the gratification of a libidinous curiosity, will be disappointed, but, better still, they will be *scared!* Their terror will prove eminently salutary, for, in describing the evils of sexual excesses and unnatural practices, we point with the finger of authority which they dare not despise, at the deplorable consequences involved" (p. 37).

In his first two chapters Cooke considers the education of adolescent males and females, arguing that "promiscuity of the sexes" in the co-educational system, the atmosphere of boarding schools, dancing, etc. stimulate the growth of abnormal sexual passions. The third and fourth chapters on "male masturbation" and "female masturbation" are a medico-ethical discourse on the "national curse." Chapter VI, "The Physiology of marriage," is less an explanation of reproductive physiology than an essay on sexual ethics, covering such topics as frequency of the "conjugal act"; continence as the only acceptable means of birth control; the moral and physical consequences of "conjugal onanism"; intercourse during pregancy, etc. Chapter VIII provides a "Psycho-physiological comparison of the sexes" in which Cooke argues that the "anatomical peculiarities" of woman "no less than her special and periodical function, and the very tendencies of her character, all go to prove that she has not been created to cope with the exigencies of material life, or to place in subjection the hostile elements of the outer world" (p. 272). The chapter that follows is entitled "What can women do in the world?"

Cooke entered Brown University in 1846, but left before taking his degree. He studied medicine for a time before interrupting his studies for a tour abroad. Upon his return, Cooke attended lectures at the University of Pennsylvania and Jefferson Medical College, but shortly after converted to homoeopathy. In 1855 he moved to Chicago, where in 1859 he joined the faculty of the newly established Hahnemann Medical College. Contemporary Christian mores underly the argument of *Satan in society*. Cooke was deeply involved in the Episcopal Church until his conversion to Roman Catholicism in 1875.

780. COOKE, Nicholas Francis, 1829-1885.
Satan in society. By a Physician . . . Cincinnati and New York: C.F. Vent; Chicago: J.S. Goodman & Co., 1874.
    412 p. ; 20 cm.

781. COOKE, Nicholas Francis, 1829-1885.
Satan in society. By a Physician . . . St. Louis: Edward F. Hovey, 1877.
    412 p. ; 20 cm.

782. COOKE, Nicholas Francis, 1829-1885.
Satan in society: By a Physician . . . Cincinnati and San Francico: Edward F. Hovey, 1881.
    412 p. ; 20 cm.

COOLEY, Anna Maria see **KINNE** (#2141).

783. COOLEY, Arnold James.
A cyclopaedia of six thousand practical receipts, and collateral information in the arts, manufactures, and trades, including medicine, pharmacy, and domestic economy. Designed as a compendious book of reference for the manufacturer, tradesman, amateur, and heads of families . . . New York: D. Appleton & Company . . . [*et al.*], 1846.
    576 p. : ill. ; 24 cm.

Originally published at London in 1841, this first American edition is based on the second London edition (1845). A sixth and apparently final American edition was published at New York in 1891. At head of title: *The book of useful knowledge.*

784. COOLIDGE, Algernon, 1860-1939.
Adenoids and tonsils . . . Cambridge: Harvard University Press, 1916.
    [2], 46 p. : ill. ; 17 cm.

Series: *Harvard health talks.* Coolidge was born 24 January 1860, the son of Algernon Sidney Coolidge (M.D., Harvard, 1853). The younger Coolidge received his undergraduate degree from Harvard (*magna cum laude*) in 1881, and his medical degree from Harvard in 1886. He started practice in Boston in 1888. By 1893 Coolidge was with the Massachusetts General Hospital, and at Harvard Medical School was appointed clinical instructor of laryngology (1893-1906), assistant professor (1906-1911), and professor (1911-1925). In 1918 he was the acting dean of the Graduate School in Medicine; and in 1921 became a trustee of the Massachusetts General Hospital. His principal published work was *Diseases of the nose and throat* (1915). Coolidge died in Boston on 16 August 1939.

785. COOPER, Irving Steiger, 1882-1935.
Way to perfect health . . . Revised American edition. Chicago: The Theosophical Press, 1916.
    viii, 118, [2] p. ; 16 cm.

"The value of our physical body is determined by its *usefulness* to us as spiritual beings. We have taken physical incarnation in order to gain experience and to unfold our inward powers, and not merely to indulge and gratify the appetites of the body through which we come into contact with the world. If that body is responsive to outward experiences and to our inward thoughts, it is serving its purpose, but if it is gross, sluggish and animal-like in its tendencies, it is handicapping our development" (p. 10). In this brief treatise Cooper considers the influence of diet in developing this responsiveness. In Chapter II he provides twenty "indictments against flesh-food." The remainder of the treatise considers the kinds, quantities and preparation of foods that produce a healthful mind-body relationship. First published in 1912, the *Way to perfect health* is no. 2 in the *Manuals of occultism* series. A native Californian, Steiger studied at the Univ. of California, followed his metaphysical bent on a pilgrimage to India, and travelled nationally as a lecturer for the Theosophical Society from 1908 to 1919. He became a priest of the Old Catholic Church in 1918, and in 1919 was made a bishop in the Liberal Catholic Church.

786. COOPER, James W.
The experienced botanist or Indian physician. Being a new system of practice founded on botany. 1. A description of medicinal plants—their properties, &c. and the method of preparing and using them. 2. A treatise on the causes, symptoms and cure of diseases incident to the human frame; with a safe and sovereign mode of treatment. For the use of families and practitioners . . . Ebensburg, Pa.: Printed by

Canan & Scott, 1833.
    xxiv, [25]-301 p. ; 14 cm.

First edition. "Does it reflect honor on that kind Providence, who supplies the wants of all creatures, to suppose that the science of health, in which every child of Adam is so deeply concerned, must necessarily be the exclusive privilege of a few? That it should be locked up in an unknown language, or merged in a mass of learned lumber, requiring an age of study to explore and apply its principles to useful purposes? A reference to the Aborigines of our country is sufficient to refute such doctrine" (p. v). "On the pernicious effects of mercury, by James Hamilton," p. [263]-287.

787. COOPER, James W.
The experienced botanist or Indian physician, being a new system of practice, founded on botany; containing: 1. A description of medicinal plants—their properties, &c. and the method of preparing and using them. 2. A treatise on the causes, symptoms and cure of diseases incident to the human frame; with a safe and sovereign mode of treatment. For the use of families and practitioners . . . Lancaster [Pa.]: Printed for the author and publishers; John Bear, printer, 1840.
    xxi, [22]-303 p. ; 15 cm.

Second edition.

788. COOPER, Thomas, 1759-1839.
A treatise of domestic medicine, intended for families, in which the treatment of common disorders are alphabetically enumerated. To which is added, a practical system of domestic cookery, describing the best, most economical, and most wholesome methods of dressing victuals; intended for the use of families who do not affect magnificence in their style of living . . . Also, the art of preserving all kinds of animal and vegetable substances, for many years, by M. Appert. Reading [Pa.]: Published by George Getz, 1824.
    [8], 128 p., [3] leaves of plates : ill. ; 27 cm.

The "Domestic medicine" comprises the *Treatise's* first twenty-three pages. "The art of preserving all kinds of animal and vegetable substances" (p. [111]-128) is excerpted from a translation of Nicolas Appert's *L'art de conserver.* Atwater copy lacking pp. 125-128.

789. COOPER, William Colby, 1835-1913.
Preventive medicine including a disquisition on therapeutic philosophy . . . Cleves, O.: Published by the author, 1905.
[4], 147, [1] p. ; 18 cm.

Cooper's brief treatise is divided into three parts. In part one the author is concerned with two subjects: "stirpiculture" (or "homoculture") and "sanitation, individual and general." Regarding the former he writes, "Surely the time is not so far distant when popular sentiment will justify a law requiring a certificate of physical fitness from each of the contracting parties before their marriage will be permitted. Under this ideal state of affairs, there would exist in every county a board of medical examiners whose duty it would be to examine all matrimonial candidates, male and female . . . There are uncounted thousands of people who are sufficiently civilized to vote for such an arrangement right now" (p. 3-4). Though Cooper thinks it desirable to imprison members of the criminal class for life (thus making it impossible for them to reproduce), he realizes the impracticability of such a step. He therefore recommends the castration of "incorrigibles" as an alternative. Sanitation ranks second for Cooper in the prevention of disease, and hence receives secondary attention in comparison to eugenics. The second and third parts of *Preventive medicine* are a condemnation of what the author regards as physicians' "drug dependance, and servility to authority," i.e., their excessive use of drugs. The appendix contains thirteen pages of "Cooperisms," including: "There is hope for the physician who will admit he commits malpractice every day of his life"; "The legitimate healer is seldom 'well heeled'"; and "The pocket-book is the clitoris of acquisitiveness." Cooper was an 1867 graduate of the Eclectic Medical Institute of Cincinnati.

CO-OPERATIVE HEALTH ASSOCIATION
see *THE IGNORAMUS* (#1880).

COPELAND, William H. see COPELAND MEDICAL INSTITUTE (#790).

790. COPELAND MEDICAL INSTITUTE.
SYMPTOM BLANK prepared by Drs. Copeland, Cowden & Byers. To those living at a distance . . . [Boston, 1894].
[2] leaves ; 22 cm.

"Drs. Copeland, Cowden and Byers have always given special attention to that large class of sufferers of chronic disease who reside out of Boston and cannot call at the office for personal examination. They have adopted a plan of home treatment by which the same excellent results may be derived at the patient's home as at the office. This plan involves first a symptom blank containing such questions as are necessary to disclose to the physicians . . . the nature of the disease and to enable them to prescribe intelligently" (leaf [1] recto).

The symptom blank is accompanied by a typescript letter (on Copeland Medical Institute letterhead) and a postmarked envelope, both dated 14 February 1894. The letter and envelope are addressed to A.W. Swallow, of Brownsville, Vt., who had requested the enclosed symptom blank.

The name of D.P. Doyle also appears beneath the letterhead on the typescript letter. The 4th ed. of Polk's *Medical & surgical register of the United States* (1896) lists a Daniel P. Doyle (M.D., 1890, Univ. of the City of New York) at 180 Tremont St. in Boston, the address of the Copeland Medical Institute. The 1896 Polk also places William H. Copeland at the 180 Tremont St. address. Copeland is described on the symptom blank as a graduate of the Bellevue Hospital Medical College. Both the 5th (1898) and 7th (1902) editions of Polk provide listings for Wm. H. Copeland (M.D., 1885, Bellevue) in New York, Chicago and Cincinnati, but not in Boston. Cowden is described on the symptom blank as an Ohio Medical College (Cincinnati) graduate. The 4th ed. (1896) of Polk lists him in Boston at 180 Tremont St. as J. Morrow Couden (M.D., 1881, Ohio Med. Coll.); while the 5th ed. (1898) places him in the District of Columbia. Byers, whose name does not appear beneath the letterhead on the 14 Feb. 1894 letter, is described on the symptom blank as a Jefferson Medical College graduate. Polk places a Wm C. Byers (M.D., 1872, Jeff. Med. Coll.) in Buffalo and Cleveland in 1896, but not in Boston. He was in Pittsburgh by 1902.

CORIAT, Isador H. see **WORCESTER** (#3873).

791. CORNARO, Luigi, 1475-1566.
The art of living long. A new and improved English version of the treatise by the celebrated Venetian centenarian Louis Cornaro; with essays by Joseph Addison, Lord Bacon, and Sir William Temple . . . Milwaukee: William F. Butler, 1903.
214, [2] p., [4] leaves of plates : ports. ; 23 cm.

"The sixteenth-century Italian [*i.e., Venetian*] nobleman Luigi Cornaro was probably the best known of all the pre-1800 health writers. His *Discourses on a sober and temperate life* was an autobiographical account of a man whose sensual indulgences had virtually ruined his health before the age of forty. Warned by physicians that he would have to trim his sails or die, Cornaro immediately converted to a program of extreme moderation, allowing himself only twelve ounces of food and fourteen ounces of wine daily. Abstemiousness not only relieved his ailments, it lifted him to a higher level of physical, and mental, happiness than he had ever known before. At the age of eighty-three, still enjoying remarkable vitality, he began the *Discourses* which he finally completed in his ninety-fifth [sic] year—and he lived several years longer. Most of the advice of the work is standard 'non-natural' fare [*the six "non-naturals" included air, food & drink, sleep, exercise, evacuation, repletion and the "passions of the mind"*], but it also includes certain emphases that were to have long-ranging influence on hygienic thought. First, Cornaro believed right living created immunity to disease . . . produced an uncommon degree of physical strength and energy, as well as, and even promoted mental clarity, emotional cheerfulness, and spiritual tranquility . . . Presaging a crucial element of [American] health reform ideology, he denied the fatalism that accepted illness as a part of the divine plan . . . Rather, health and contentment were God's desire, as was an extended life-span . . . All these expectations of temperance went into a 'Cornaro tradition' that colored hygienic literature down to the 1830s" (James C. Whorton, *Crusaders for fitness,* Princeton, N.J., 1982, p. 19-20).
"Cornaro based his dietary rules on the classical principle that the world consists of four

basic elements apportioned to provide perfect balance, and also on Galen's adaption of this principle to human life . . . It was therefore necessary, Cornaro reasoned, that the body 'be preserved and maintained by the very same order' as the world outside it. Since food alone failed to provide all four elements . . . it followed that habitual gluttony would provoke an imbalance in nature's harmonious arrangement. Regulation and moderation, then, were the key to a long and healthy life . . . What preserved his own health, Cornaro insisted throughout *Discourses,* was that he consumed only such food and drink as 'agreed with my constitution,' and that he consumed it only in 'the quantity I knew I could digest' . . . Cornaro's dietary rules were restrictive only in a quantitative way . . . " Stephen Nissenbaum, *Sex, diet, and debility in Jacksonian America,* Westport, Ct., 1980, p. 41-42).
Cornaro's *Trattato de la vita sobria* was originally published at Padua in 1558. Some sense of the popularity and influence of Cornaro's original twenty-seven leaf treatise may be gauged by the 263 entries in the OCLC database of Latin, Italian, French, English, German and Russian editions. This 1903 English translation includes Joseph Addison's essay on hygiene in the introduction (p. 15-21); and "Extracts selected and arranged from Lord Bacon's 'History of life and death' and from Sir William Temple's 'Health and long life'" on p. [115]-155.

792. CORNARO, Luigi, 1475-1566.
The art of living long. A new and improved English version of the treatise by the celebrated Venetian centenarian Louis Cornaro; with essays by Joseph Addison, Lord Bacon, and Sir William Temple . . . Milwaukee: William F. Butler, 1905.
214, [2] p., [4] leaves of plates : ports. ; 23 cm.

793. CORNARO, Luigi, 1475-1566.
Discourses on a sober and temperate life . . . Translated from the Italian original. Philadelphia: Printed by T. Dobson, MDCCXCI. [1791]
xi, [12]-202 p. ; 12 cm. (12mo)

Second American edition. The first American edition was published at Portsmouth in 1788 (*Austin* #544).
REFERENCES: *Austin* #541.

794. CORNARO, Luigi, 1475-1566.
Discourses on a sober and temperate life
. . . Wherein is demonstrated, by his own
example, the method of preserving health
to extreme old age. Translated from the
Italian original. A new edition, corrected.
To which is added, Physic of the golden
age, a fragment. Boston: Printed by Thomas B. Wait, and Sons, for the proprietor,
1814.
    vi, [7]-90 p. ; 14 cm.

Printed label on front wrapper: *Health and
long life. By Cornaro.*
REFERENCES: *Austin* 542.

795. CORNARO, Luigi, 1475-1566.
Discourses on a sober and temperate life
. . . Wherein is demomstrated [sic], by his
own example, the method of preserving
health to extreme old age. Translated from
the Italian original. A new edition, corrected. To which is added, Physic of the
golden age, a fragment. New-Haven: Published by Harlow Porter; printed by S. Converse, 1823.
    vi, [7]-90 p. ; 14 cm.

The typesetting of this New Haven edition
is plainly based on the composition of the 1814
Boston edition (#794). According to the epicure William Kitchiner, "The system of Cornaro
has been oftener quoted, than understood—
most people imagine, it was one of rigid abstinence and comfortless self-denial—but this was
not the case:—his code of longevity consisted
in steadily obeying the suggestions of instinct—
and economising his vitality, and living under
his income of health,—carefully regulating his
temper—and cultivating cheerful habits" (*The
art of invigorating and prolonging life*, Philadelphia, 1823, p. 39, #2145).

796. CORNARO, Luigi, 1475-1566.
The discourses and letters of Louis Cornaro, on a sober and temperate life; with a
biography of the author, by Piero Maroncelli. And notes and an appendix, by John
Burdell. New York: Fowlers and Wells, 1847.
    xii, 228 p. : ill. ; 16 cm.

In his preface, the American editor John
Burdell (1806?-1850) writes, "The former edi-

tions of the works of Cornaro, which have appeared in this country, are mere reprints of the
English translation of 1768 (#798), and contains
but four of his 'Discourses,' as they are termed
by the English editor" (p. vi). Noting its several
shortcomings, he continues, "Wishing to give
the world an edition, which might, in some respects, at least, be superior to its predecessors,
I requested of Signor Piero Maroncelli . . . a
biographical memoir of Cornaro" (p. vii). In
addition, Burdell has added to this edition the
dedication to the first Padua edition and "two
very valuable and interesting letters from
Cornaro himself to the Cardinal Cornaro, and
to Sperone Speroni; the last of which merits
the appellation of a fifth 'Discourse.' These letters have been translated from the original Italian expressly for this edition, and are to be
found in no other" (p. ix). The text is supplemented by twenty pages of Burdell's notes ("afforded by the present advanced state of physiological research"), and by an appendix (p.
[183]-228) consisting of three supplementary
essays by the editor: "Original food of man,"
"Digestion," and "The effects of tight dress."
Atwater copy bound with John Forbes' *The water cure, or hydropathy* (1847, #1259).

797. CORNARO, Luigi, 1475-1566.
Discourses on a sober and temperate life
. . . Wherein is demonstrated by his own
example, the method of preserving health
to extreme old age. Translated from the
Italian original. A new edition corrected.
With an introduction and notes, by
Sylvester Graham. London: M.DCC.LXXIX.
. . . New-York: Printed and published by
Mahlon Day, 1833.
    [2], xiv, [15]-178 p. ; 15 cm.

In his introduction, Graham includes
Cornaro among those who "by greatly improving their habits and diet, without ever coming
up to the most correct regimen, greatly improve
their health, and with a very considerable share
of comfort, live to a great age, suffering always
through life, inconveniences and infirmities
equal to the errors of their regimen" (p. xi).
One wonders why Graham bothered to edit yet
another edition of Cornaro in such grudging
admission of the beneficial effects of Cornaro's
change of diet: " . . . and much as it had suffered from former excesses, and irregularities,

it still had renovating power enough to rise, under the comparatively small remaining weight of oppression, to a considerable degree of health. But had that oppression been entirely thrown off, had no portion of the wine been retained, and had the *qualities* of his food been in all respects more consistent with sound physiological principles, he would not only have recovered a more perfect state of health, and, beyond all question, have lived many years longer, but he would entirely have escaped those yearly depressions of health and strength, in which he almost sank into the grave" (p.xii-xiii). Graham concludes, "Deeply interesting and truly valuable, therefore, as the following work in many respects is, yet it is very questionable whether, if sent forth to the world at the present day, without 'note or comment,'it would not do far more mischief than good to mankind" (p. xiv). Graham's edition of Cornaro's discourses is abundantly annotated, correcting the errant nonagenarian whenever he seems him deviating from "sound physiological principles."

798. CORNARO, Luigi, 1475-1566.
Sure methods of attaining a long and healthful life. Written originally in Italian . . . Translated into English by W. Jones, A.B. Edinburgh: Printed by A. Donaldson, and sold at his shops in London and Edinburgh. MDCCLXVIII. [1768]
    [8], 147 p. ; 13 cm.

In his preface to the 1847 New York edition of Cornaro's *Discourses,* John Burdell notes, "The former editions of the Works of Cornaro, which have appeared in this country, are mere reprints of the English translation of 1768, and contain but four of his 'Discourses,' as they are termed by the English editor" (#796, p. vi).

CORNARO, Luigi see also *IMMORTAL MENTOR* (#1884); **KITCHINER** (#2147); **SINCLAIR** (#3220).

799. CORNELL, William Mason, 1802-1895.
Clergymen's and students' health; or, physical and mental hygiene, the true way to enjoy life . . . Fourth edition. New York: I.K. Funk & Co., 1880.
    xviii, 21-360 p. ; 19 cm.

Clergymen's and students' health is a reissue of Cornell's *How to enjoy life,* originally published at Philadelphia in 1860 (#801). The reason for the change of title with this fourth edition is implied in the following quotation from the introduction: "The will of God as revealed to man is found engraved upon his *works,* as well as in *his written word*: and the laws of the former are as obligatory as the commands and precepts of the latter; and never will Christianity become practically what its Divine author intended, in its fullest beauty, perfection and enjoyment . . . till *those laws which govern the body, in connection with the mind, are correctly understood and obeyed.* We take this high position; and here we verily believe the clergy have been much at fault in not teaching and preaching these laws, and practising in accordance with them. This is the reason why the writer has chosen to illustrate so large a portion of physical and mental hygiene by clerical examples. If they will study the *natural laws* of God more, they will find fewer 'mysterious Providences,' and see more clearly 'his Eternal Power and Godhead' in His works; and the people, understanding that no atonement has been made, or will be, for a violated natural law of God, but that the penalty must be paid, will be careful to live in harmony with the Divine economy" (p. xvii-xviii).

Cornell graduated from Brown University in 1827, and in 1830 was ordained a Congregationalist minister. Failing health forced Cornell to leave the ministry in 1839, whereupon he began the study of medicine. After graduating from the Berkshire Medical College in 1844, Cornell moved to Boston where he practiced medicine and found time to author a surprising number of books on topics as varied as English grammar, epilepsy, Pennsylvania history, biography and theology. He was the author of several manuals of popular medicine and hygiene in addition to the present title. Spine-title: *How to enjoy life.*

800. CORNELL, William Mason, 1802-1895.
Consumption forestalled and prevented . . . Boston: James French, 1846.
    vi, [7]-120, [8] p. ; 17 cm.

In his opening sentence Cornell states that "Pulmonary consumption is a hereditary disease" (p. [9]). This predisposition "is developed by colds, inflammation of the lungs, pleurisy, eruptive fevers, unhealthy localities, intemperance in eating and drinking, suppression of any natural evacuations, constitutional syphilis, insufficient clothing and undue exposure, neglect of exercise, abuse of mercury and other medicines, excessive mental exertion, various kinds of mechanical labor, such as the manufacture of needles, filing of iron, laboring in cotton and other manufactories where much of the dust must be inhaled," etc. (p. 10). To obviate these causes, Cornell specifies a hygienic regimen attending to air, bathing, dress, food & drink, exercise, sleep, etc.

801. CORNELL, William Mason, 1802-1895.
How to enjoy life: or, physical and mental hygiene . . . Philadelphia: James Challen & Son . . . [*et al.*], 1860.
    xviii, 21-360 p. ; 19 cm.

    First edition.

802. CORNELL, William Mason, 1802-1895.
How to enjoy life; or, practical hints on the preservation of health . . . Boston: Published by B.B. Russell, 1873.
    xviii, 21-360 p., [2] leaves of plates : ports. ; 19 cm.

    Second edition. We have been unable to locate a recorded copy of the third edition. The fourth edition (#799) was issued under the title *Clergymen's and students' health,* a sixth and final edition of which was published in 1883. Atwater copy lacking pp. 23-24.

CORNELL, William Mason see also *GOOD HEALTH* (#1378), *JOURNAL OF HEALTH AND MONTHLY MISCELLANY* (#2054); *JOURNAL OF HEALTH AND PRACTICAL EDUCATOR* (#2055).

THE CORNER CUPBOARD see **PHILP** (#2790).

803. CORRECTIVE EATING SOCIETY.
Form letter dated 1917 June 11, New York, with facsimile signature of R.W. Lockwood.
    [1] leaf ; 28 cm. + pamphlet ([12] leaves ; 14 cm.) + postcard

    Letter on Corrective Eating Society stationery offering the recipient Eugene Christian's

Little lessons in scientific eating (#639). Enclosed with this letter are R.W. Lockwood's pamphlet entitled *The crimes we commit against our stomachs* (1917) and a return postcard.

CORRECTIVE EATING SOCIETY see also **CHRISTIAN** (#639).

804. CORT, C. A. von.
The household treasure; or, medical adviser . . . New York: Published by the author, 1875.
    xvi, [17]-233 p., [1] leaf of plates : port. ; 18 cm.

    One does not have to read much of *The household treasure* to realize why this work bears the author's imprint. There is little consistency or coherence to connect one paragraph to another, or one page to the next. In a series of disjointed ideas, von Cort expresses concern that in her age "the sexual organs are more fully developed than those of the intellectual." One reason for this (or is it an instance?) is that little girls are allowed to play with dolls, which develops maternal instincts, which in turn only increases the influence of the genitals over their other faculties. In a disconnected stream of thought, Cort alters between condemning the depravity of the age (morals, medicine, women, etc.), exalting the role of woman and her place in the coming age, and providing hints on the domestic management of health and disease. She advocates vegetarianism and a vegetable materia medica, and believes in the healing power of magnetism and music.
    In some parts, the book is typical of contemporary domestic manuals, providing advice for the treatment of a multitude of common accidents and disorders. The stranger aspects of her domestic hygiene are characterized in her consideration of household pets and the matrimonial bed. Cort warns that "inhabitation with dogs" has "evil effects" on children: "The animal often becomes the most prominent . . . the dogs, being the stronger element, draw the finer forces from the human; and the human partake more fully of the animal" (p. 88). The "animal element" is somehow "taken up into the blood, and carrying it through the whole system, blending with the finer organs of the brain, and as this is supplied with animal force, it creates animal propensities with the human."

(Incorrigible pet owners will be pleased that the human element is similarly "taken up" by Fido, much to his advantage.) In a chapter entitled "Sleeping together" Cort warns, "The habit of two persons sleeping in the same bed is productive of depravity, in which all manner of diseases are transmitted to each other, and often creates a nervous derangement, sometimes producing partial insanity" (p. 188). In the next paragraph she concludes, "*Both* must die, as a general thing, and from the evil effects of sleeping together—there is no remedy." Frontispiece lithographic portrait of author.

COTE, Emma H. see **TESTE** (#3460).

THE COTTAGE PHYSICIAN see **FAULKNER** (#1141, 1142).

COUDON, J. Morrow see **COPELAND MEDICAL INSTITUTE** (#790).

805. COUNCILMAN, William Thomas, 1854-1933.
Disease and its causes . . . New York: Henry Holt and Company; London: Williams and Norgate, [c1913].
    [2], viii, [9]-254 p. : ill. ; 17 cm.

No. 68 in the *Home university library of modern knowledge* series, *Diseases and its causes* is a popular explanation of recent advances in bacteriology and infectious disease. At the time this work was published, the author was professor of pathology at Harvard Medical School. An 1880 medical graduate of the University of Maryland, Councilman "did significant original research on several diseases including diphtheria, epidemic cerebrospinal meningitis, chronic nephritis, and smallpox" (*Dict. Amer. med. biog.*, I:161). He was also the first American to to describe the malarial parasite.

806. COUPLAND, Sidney, 1849-1930.
Personal appearances in health and disease . . . New York: D. Appleton and Company, 1879.
    [2], 96 p. : ill. ; 15 cm.

"Now the object of this little book is to try and explain as briefly as possible how and why variations that are so plain on the surface can be taken as indices of disorder within, to give the

reasons for form-changes which occur within the limits of health, and for those which mark the departure from those boundaries" (p. 6). Series: *Health primers; no. 5*. The New York edition was published the same year as the London edition.

The author "entered University College, London, as a medical student. He qualified in 1871 and then held resident posts at University College Hospital. In 1873 he became pathologist to the Middlesex Hospital and in 1875 was elected assistant physician, becoming full physician at age thirty. While at the Middlesex, Coupland lectured on medicine, practical medicine and pathological anatomy and in 1891 was made dean of the Medical School. He resigned from the active staff in 1898 to take up the appointment of commissioner in lunacy" (*Lives of the fellows of the Royal College of Physicians of London, 1826-1925*. London: The College, 1955, IV: 271-72).

807. COVENTRY, Charles Brodhead, 1801-1875.
TO THE VOLUNTEERS. An Old Soldier's Advice. [between 1861 and 1865].
    Broadside ; 20.5 × 9 cm.

Eleven recommendations to soldiers on the maintenance of health, beginning with the warning: "Remember that in a campaign more men die from sickness than by the bullet." Coventry was an 1825 graduate of the College of Physicians & Surgeons of the Western District (Fairfield, N.Y.). He taught on the faculties of the Berkshire Medical College (1828-31), the Geneva Medical College (1839-46) and the Medical Dept. of the University of Buffalo (1846-52). Active in the Medical Society of the State of New York from 1825, Coventry lobbied with the Society and the State Legislature for the improved treatment of the insane. He was instrumental in founding the State Lunatic Asylum (Utica), served on its Board of Managers, and succeeded Amariah Brigham (#409-411) as director at the latter's death in 1849.

808. COWAN, John.
The science of a new life . . . New York: Cowan & Company, publishers, 1869.
    405 p. : ill. ; 22 cm.

First edition. Cowan's treatise is divided into three parts: the first on courtship, marriage, sexual anatomy and physiology, sexual ethics, contraception, and heredity; the second (entitled "The Consummation") on conception, gestation, the management of pregnancy, and "the management of mother and child after delivery"; and the third ("Wrongs righted") on "foeticide," female genital diseases, male genital diseases (including sexually transmitted diseases), masturbation, sterility, impotence," "woman's rights," and concluding remarks on marriage.

According to Norman Himes, who was unaware of the full publishing history of *The science of a new life,* Cowan's "account of contraceptive technique, though unsatisfactory, must have played some role—no one can estimate just how much—in popularizing contraceptive knowledge" (*Medical history of contraception,* New York, 1963, p. 266). Given that the book remained in print through the First World War, its influence was obviously far greater than Himes knew, even if that advice had not changed over a period of nearly fifty years.

Cowan's attitude toward contraception is revealed in the opening sentence of his chapter on the subject: "One of the sequences to licentiousness is the desire to prevent undesirable results in the wife or betrayed woman" (p. 108). Although he admits valid medical reasons for avoiding conception, and acknowledges that "in the great majority of married lives . . . it is the wife who desires this knowledge, so as to guard her health, aye, her very life, against the unbridled passion of the husband" (p. 109), Cowan feels that contraception nevertheless tends "to the physical and spiritual harm of the individual."

In spite of his reservations, Cowan reviews existing methods of preventing conception: the *congressus reservatus* of the Oneida community, i.e., "sexual embrace, without a full orgasm— that is, stopping short of emission" (which Cowan regards as "of a low, animal nature," resulting in the same debilitating effects as nocturnal emissions, and beyond the will power of most males); *coitus interruptus* (which he regards as "beastly, and not one iota different, in its effects on the mind and body of man, from self-abuse"); condoms ("certainly effectual; but there is no pleasure . . . derived from their use," and they irritate the vagina); "sponge or rubber pads, placed against the mouth of the womb, to prevent the entrance of the sperm"

Fig. 23. John Cowan, frontispiece to his Science of a new life *(1870), #809.*

(which "prevents the very pleasure that is the object of licentious people," and "is not in any sense a reliable method"); observing the menstrual cycle and having intercourse during "the two weeks in which the uterus contains no germ-cell"; relying on the theory that women are infertile during lactation; and the injection of cold water into the vagina immediately following intercourse ("there is really nothing reliable in it"). Cowan concludes this chapter with the remark, "There is but one positively sure method of preventing conception—one within the reach of all, and which has no bad effects afterward, and that is *to refrain from the sexual act*" (p. 112).

809. COWAN, John.
The science of a new life . . . New York: Cowan & Company, publishers, 1870.

405, [3] p., [1] leaf of plates : ill., port. ; 22 cm.

810. COWAN, John.
The science of a new life . . . New York: Cowan & Company, publishers, 1871.
10, 405, [3] p., [1] leaf of plates : ill., port. ; 21 cm.

811. COWAN, John.
The science of a new life . . . New York: Cowan & Company, publishers, 1878.
11, [1], [3]-405, [3] p., [1] leaf of plates : ill., port. ; 22 cm.

812. COWAN, John.
The science of a new life . . . New York:
Fowler & Wells, 753 Broadway, [between
1880-87].
11, [1], [3]-405 p. : ill. ; 22 cm.

The Fowler & Wells Co. was headquartered
at 753 Broadway between 1880 and 1887.

813. COWAN, John.
The science of a new life . . . New York:
Fowler & Wells Co., publishers, 753 Broadway, [between 1880-87].
11, [1], [3]-405 p. : ill. ; 23 cm.

814. COWAN, John.
The science of a new life . . . New York: J.S.
Ogilvie and Company, No. 25 Rose Street,
[1881-82].
11, [1], [3]-405 p. : ill. ; 22 cm.

Many issues of Cowan's *Science of life* bear no
date but the 1869 copyright on the verso of the
title-page. Most of these undated issues were
published by the firm of J.S. Ogilvie, whose
imprints always include a street address. By consulting New York City directories it has been
possible to determine the years during which
Ogilvie was at each address, and hence approxi-
mate dates for the publication of each undated
issue (see Table 4):

Table 4. Addresses of John Cowan (#808-836)

Address	Years at this Address
29 Rose Street	1879-1881
25 Rose Street	1881-1882
31 Rose Street	1883-1886
57 Rose Street	1888-1897
54 Rose Street	1900-1901
57 Rose Street	1901-1925

815. COWAN, John.
The science of a new life . . . New York: J.S.
Ogilvie & Company, 31 Rose Street, [1883-86].
11, [1], [3]-405 p. : ill. ; 23 cm.

816. COWAN, John.
The science of a new life . . . New York: J.S.
Ogilvie, publisher, 57 Rose Street, [1888-
97].
11, [1], [3]-405 p. : ill. ; 23 cm.

817. COWAN, John.
The science of a new life . . . New York; Chicago: J.S. Ogilvie, publisher, 57 Rose Street . . .
79 Wabash Avenue, Chicago, [1888-97].
11, [1], [3]-405 p. : ill. ; 23 cm.

818. COWAN John.
The science of a new life . . . New York:
John B. Alden, publisher, 1889.
11, [1], 405 p. : ill. ; 23 cm.

819. COWAN, John.
The science of a new life . . . Troy, N.Y.:
Nims and Knight, 1889.
11, [1], [3]-405 p. : ill. ; 23 cm.

820. COWAN, John.
The science of a new life . . . Chicago: Alice
B. Stockham & Co., 1891.
[12], [3]-405 p. : ill. ; 23 cm.

Stockham's imprint is printed on slip pasted
over original imprint: *New York: J.S. Ogilvie, 57
Rose Street.*

821. COWAN, John.
The science of a new life . . . New York: J.S.
Ogilvie, publisher, 57 Rose Street, [c1897].
[3]-405, [13] p. : ill. ; 21 cm.

822. COWAN, John.
The science of a new life . . . Chicago: Geo.
W. Ogilvie & Co., publishers, [c1903].
320 p. : ill. ; 18 cm.

At head of title: *A book for all who are married
and those who contemplate marriage.*

823. COWAN, John.
The science of a new life . . . New York: J.S.
Ogilvie Publishing Company, 57 Rose
Street, [1911].
[3]-405, [19] p. : ill. ; 21 cm.

"Printed 1911" (verso of title-page).

824. COWAN, John.
The science of a new life . . . New York: J.S.
Ogilvie Publishing Company, 57 Rose
Street, [1913].
[5]-405, [19] p. : ill. ; 21 cm.

"Printed 1913" (verso of title-page).

825. COWAN, John.
The science of a new life . . . New York: J.S.
Ogilvie Publishing Company, 57 Rose
Street, [1915].
    [4], [9]-405, [3], 9, [9] p. : ill. ; 20 cm.

"Printed 1915" (verso of title-page).

826. COWAN, John.
The science of a new life . . . Toronto:
McClelland, Goodchild & Stewart, publish-
ers, [1915].
    [4], [9]-412, [4], 6, [2] p. : ill. ; 20 cm.

COWDEN, J. Morrow see **COPELAND
MEDICAL INSTITUTE** (#790).

827. COX, Abraham Lidden, 1800-1864.
The pathology and treatment of Asiatic
cholera, so called . . . New York: John Wiley,
1849.
    54 p. ; 17 cm.

First edition. Cox considers the nature of
the disease, its symptoms, treatment, and distinction from such disorders as "epidemic diarrhoea." Cox maintains that cholera is not
contagious, but that "the general predisposing
cause is the air we breathe" (p. 46). Pages 52-
54 contain "Directions for treatment in the
absence of a physician." A second edition was
also issued in 1849.

828. COXE, Edward Jenner, 1801-1862.
Domestic medicine; or, medical vade
mecum: a safe companion and guide for
families, planters, commanders of ships
or steamers, or any one who may require
a true friend in time of need. This companion embraces, the medicines in general use; the diseases and accidents of
usual occurrence; the most useful articles of diet or drink for the sick or convalescent, with the best mode of preparing them; and numerous remarks in reference to bathing, exercise, and other
hygienic measures to preserve health, to
repair and strengthen an enervated
constitution, and to cure disease . . .
[Philadelphia]: Printed for the author,
and for sale by all booksellers throughout the United States, [c1854].
    [4], [xvii]-xxiv, [25]-300 p. ; 19 cm.

The "fourth edition" (verso of title-page).
The first three editions of this work were published under the title *Medical vade mecum*. Coxe's
Domestic medicine is divided into four parts: the
first is a description of medicines listed alphabetically; the second has the running-title "Hygienic maxims"; part III is entitled "The practice of medicine, the arrangement of the diseases being alphabetical"; and the fourth is a
domestic surgery with topics again arranged
alphabetically. An errata-slip is pasted onto title-
page.

CRAIB-BEIGHLE, Helen see **OWEN** (#2691).

CRAIGHEAD, Robert D. see **DR.
STRONG'S SANITARIUM** (#3395).

CRAMP, Arthur Joseph see **AMERICAN
MEDICAL ASSOCIATION** (#86).

CRAMPTON'S HYGIENE SERIES see
TOLMAN (#3543).

CRANE, Caroline P. see **CONGER** (#760,
761).

829. CRAWFORD, Mary Merritt, b. 1884.
Before the doctor comes. A ready refer-
ence book, giving the symptoms of com-
mon diseases, and indicating proper emer-
gency treatment in case of a sudden illness
or accident pending the physician's arrival.
By Mary M. Crawford, M.D., and Thurston
S. Welton, M.D. New York: The Christian
Herald, Louis Klopsch, proprietor, Bible
House, [c1909].
    376 p. : ill. ; 21 cm.

The sixteen leaves of plates are included in
the pagination; the frontispiece is not.

830. CROSS, John B.
ELECTRICAL ROOMS, No. 19 Temple
Place. JOHN B. CROSS, MEDICAL ELEC-
TRICIAN . . . [Boston, 1846].
    [2] leaves ; 27.5 cm.

"I do not assume the title of Doctor,"
Cross informs his readers, "and have no
claim to it . . . But I profess a theoretical and
practical knowledge of the phenomenon of
the Nervous System, the physical and men-

tal structure of man, and to be qualified by extensive and successful experience, to make judicious application of the various forms of Electricity I employ as auxiliary remedial agents, in accordance with pathological science" (leaf 1 recto). Cross states that he is available for consultation in his new Electrical Rooms. "The length, number and frequency of the sittings," he informs potential patients, "must depend upon the peculiarities of each case. Yet, as the experience of the past proves, that if any beneficial results follow the treatment, it has generally been in some degree manifested from the developments of eight or ten sittings, I have as a matter of convenience, concluded to fix the course at ten sittings each and . . . in ordinary cases, I shall establish the regular sittings at a quarter hour each. The charge for each regular sitting, will for the present be one dollar" (leaf 1 verso). Leaf 2 (recto) contains a table listing sixty-six diseases that may be cured by "Electrical Treatment," the number of cases of each disease that Cross has treated, and the average number of sittings required in the treatment of each disease.

831. CROSS, John B.
Electricity, as newly applied in Medical Treatment . . . JOHN B. CROSS, Neurologist & Electrician . . . Boston, Sept. 19th, 1844.
    [2] leaves ; 21.5 cm.

Cross informs his reader that he "employs Magneto Electric and Electro-magnetic machines, a revolving Armature, rotary Magnetic Machine, Voltaic Battery, double Helix, an Electrical Bath, &c." in the treatment of such disorders as "tic doloreux [sic]," "dispepsia [sic]," "epileptic, cataleptic and paralytic affections," scrofula, "amenorrhoea," "St Vitas [sic] dance," "curvatures and other diseases of the spine," etc. The printed text appears on the recto of the first leaf only. Leaf [2] recto contains a hand-written solicitation initialled "J.B.C." The document was folded for mailing; the verso of leaf [2] is addressed and postmarked (date illegible).

832. CROSSMAN, T. J.
A catalog of the names of upwards of two hundred persons, operated on for the cure

of strabismus, or squinting . . . With introductory remarks, notices of the public press, and correspondence of patients . . . Philadelphia, July 22, 1841.
    24 p. ; 17.5 cm.

In extolling his own abilities as an ocular surgeon, Crossman maintains that "my mode of operating is far superior to any other . . . not excepting Liston or Dieffenbach" (p. [3]). In defending himself against the regular medical profession, he is unequivocal in his opinion, stating: "A distinguished professor gets a piece of pine board with a knot-hole through it, and calls it a *stethoscope,* and then bamboozles about the chests of ladies with the gravity of a well known animal . . . and calls it *auscultation*—puts up enough cough lozenges, and in order to more effectually pick the pockets of his patients, unites with a drug store, to which he sends his orders, fills his sick chamber with dish-water prescriptions and rye bread pills . . . and then the vile hypocrite and detestable cormorant has the audacity and impudence to prate about quackery" (p. 4). Crossman names the patients of such prominent Philadelphia physicians as Mutter, Horner, Condie, etc. who came to him to surgically correct their strabismus after the aforementioned had failed. He assures his prospective patients, "The operation for strabismus in skilful hands, lasts but one minute, or less. When compared with other operations, the pain is trifling. No confinement, or bandaging of the eye, or dieting, is required afterwards" (p. 13). In printed wrapper.

833. CROWEN, Mrs. T. J.
The management of the sick room, with rules for diet; cookery for the sick and convalescent; and the treatment of the sudden illnesses and various accidents that require prompt and judicious care. Compiled from the latest medical authorities, by a Lady of New York, under the approval and recommendation of Charles A. Lee, M.D. . . . Third edition. New York: Published by J.K. Wellman, 1845.
    ix, [10]-107 p. ; 19 cm.

Originally published at New York in 1844. Mrs. Crowen was author of *The American system of cookery* and *Every lady's book,* which went through multiple editions from the late 1840s through the 1860s. In printed wrapper. Charles Alfred Lee (#2222) taught on the faculties of half a dozen medical schools, and was a prolific author.

834. CROY, Mae Savell, b. 1886.
Putnam's household handbook . . . New York and London: G.P. Putnam's Sons; the Knickerbocker Press, 1916.
    vii, [1], 327 p. ; 19 cm.

Croy includes chapters on child care, child rearing, home nursing and domestic sanitation.

835. CRYSTAL SPRINGS HOTEL.
Crystal Spring: its discovery, its situation and how to get there; its hotel and sanitarium, with an account of the medical properties of the waters and the advantages the locality presents to the tourist and the invalid. Crystal Springs, Yates County, New York, 1880.
    14, [2] p. : ill. ; 23 cm.

The Crystal Springs Sanitarium, erected on the grounds of the Crystal Springs Hotel, was owned by Dr. Alex. De Borra. The hotel was owned and operated by the firm of Fuller Brothers (Corning, N.Y.). Cover-title: *The Crystal Springs.* In printed wrapper.

CRYSTAL SPRINGS SANITARIUM see **CRYSTAL SPRINGS HOTEL** (#835).

836. CULLER, Joseph Albertus, b. 1858.
The first book of anatomy, physiology and hygiene of the human body for pupils in the lower grades . . . Philadelphia and London: J.B. Lippincott Company, [c1905].
    x, 7-148 p. : ill. ; 19 cm.

Series: *Lippincott's physiologies.* Cover-title: *Lippincott's physiology. First book.* Spine-title: *First book of physiology.*

837. CULLIS, Charles.
Faith cures; or, answers to prayer in the healing of the sick . . . Boston; New York; Philadelphia: Willard Tract Repository, [c1879].
    [3]-109, [3] p. ; 19 cm.

Cases of faith healing compiled by a Boston homeopath.

838. CULPEPER, Nicholas, 1616-1654.
Culpeper's complete herbal, to which is now added, upwards of one hundred additional herbs, with a display of their medicinal and occult qualities; physically applied to the cure of all disorders incident to mankind. To which are now first annexed his English physician enlarged, and key to physic, rules for compounding medicine according to the true system of nature. Forming a complete family dispensatory, and natural system of physic. To which is also added upwards of fifty choice receipts, selected from the author's last legacy to his wife. Embellished with engravings of upwards of four hundred different plants . . . London: Published by Thomas Kelly . . . and sold by most booksellers and vendors of publications in the British Empire, 1824.
    vi, 398, [4] p., 40 leaves of plates : hand-col. ill., port. ; 27 cm.

In so many respects, the English apothecary and botanist Nicholas Culpeper presaged those who led the health reform movement in America during the second quarter of the 19th century. Disdaining the medical establishment, Culpeper revised and simplified the existing materia medica; infuriated the College of Physicians with his translation of the *Pharmacopoeia londinensis*; dismissed polypharmacy, uroscopy, venesection and purging; emphasized the importance of diet and exercise in the maintenance of health; defied the self-serving monopoly of London physicians and apothecaries by establishing himself as an unlicensed practitioner; made available "English herbs for English bodies" (i.e., medicines that were readily available, simple to prepare, easy to administer and affordable); and broadly disseminated his unorthodox ideas in a series of popular books.

The publication of Culpeper's works in North America was meagre compared to the hundreds of editions of his treatises issued in the United Kingdom from the mid-17th through mid-19th centuries. Although the popularity of an author in Britain often spurred American publishers to pirate the same, his only works published in America during the 18th century were the 1708 Boston edition of *The English physician* and his translation of the *Pharmacopoeia londinensis* (Boston, 1720). Perhaps his scientific orientation seemed outdated; or perhaps America was simply glutted with English editions of his works brought to the colonies by immigrants from England, Scotland and Northern Ireland. If they survived generations of handling, Culpeper's manuals of domestic and botanic practice should have appealed to the descendants of these colonists, and should have remained popular in a climate that eagerly embraced the teachings of populist medical botanists such as Samuel Thomson.

At age sixteen Culpeper began his studies at Cambridge with the intention of following his late father into the ministry. He quickly tired of the course of studies, longing instead to study medicine, an option to which his family was opposed. Culpeper left Cambridge without a degree, and beginning in 1634 was apprenticed to a series of London apothecaries. During these years he also read medicine, attended anatomical demonstrations, and came under the influence of the astrologer William Lilly. In 1640 Culpeper terminated his apprenticeship, and set up shop as an unlicensed apothecary just outside London. He participated in the civil war on the parliamentary side, and was wounded in action. In 1646 he spent several months in self-imposed exile in France, returning only after the recovery of the man he had wounded in a duel. He may have spent his Paris sojourn familiarizing himself with the writings of Riolan, which he translated into English. In the few years remaining to him, Culpeper began a period of incredible literary productivity. In 1649 he published his translation of the *Pharmacopoeia londinensis* from Latin into English, incurring the wrath of the College of Physicians. In 1651 his *Directory for midwives* was published, and in 1652 the first edition of *The English physitian.*

Culpeper's herbal remained in print for nearly two hundred years under such titles as *The English physician enlarged, The compleat herbal; or family physician, Culpeper's family physician, Culpeper's complete herbal, Culpeper's English physician,* etc. Until the end of the 18th century, editions of the *Herbal* or *English physician* were not illustrated. Each of the figures on the forty leaves of engraved plates that illustrate this 1824 London edition are hand-colored, a feature that appears to have been common to the many editions of Culpeper issued by the firm of Thomas Kelly between 1819 and 1865.

839. CULPEPER, Nicholas, 1616-1654.
Culpeper's English physician; and complete herbal. To which are now first added, upwards of one hundred additional herbs, with a display of their medicinal and occult properties, physically applied to the cure of all disorders incident to mankind. To which are annexed, rules for compounding medicine according to the true system of nature: forming a complete family dispensatory, and natural system of physic. Beautified and enriched with engravings of upwards of four hundred and fifty different plants, and a set of anatomical figures. Illustrated with notes and observations, critical and explanatory. By E. Sibly . . . London: Printed by Lewis and Roden . . . for the proprietor; and sold at the British Directory office . . . and by Champante and Whitrow, M.DCCC.V. [1805-07].

2 v. in 1 (xvi, 238, 241-400 p., 30 leaves of plates ; [2], 256 p., 13 leaves of plates) : ill., port. ; 26 cm.

Few words would have struck a more familiar note to 19th-century American health reformers than those written in Ebenezer Sibly's introduction to the *English physician.* Sibly writes of those in the "middling sphere" and "lower orders"of society: "The nature of their avocations, and the attentions requisite for business, beget infirmities, which, though easily removed by change of air and simple regimen, are frequently increased by irritating drugs, until the constitution receives a shock too violent to restore. The lower orders of society, however, and particularly the poor, are not exposed to this danger. Their misfortunes arise from an unfeeling inattention and neglect on the part of those who are called to their assistance; but by whom they are frequently left either wholly destitute of advice and of medicines, or are obliged to put up with such as it would be much more prudent to avoid. How extensively advantageous then would medical knowledge prove to men in almost every occupation of life? since it would not only teach them to know and to avoid the dangers peculiar to their respective stations, but would enable them to discern the real enjoyments of life, and be conducive to the true happiness of mankind?" (p. xv-xvi).

The *English physician* or *Herbal* was Culpeper's most successful published work. Originally printed in 1652, it appeared in more than 100 editions, most published after 1700. Those edited by Ebenezer Sibly (first issued in 1789) were published in two volumes. The first contains Culpeper's herbal with considerable commentary and "the addition of upwards of one hundred newly-discovered aromatic and balsamic herbs." Sibly retains the original format of Culpeper's alphabetical catalog that provides a four-part description of most plants: physical characteristics, the places where found, the times best gathered, and their "government and virtues," i.e., what planet governs the plant and to what disorders it may be applied. Sibly's second volume is of his own authorship. Entitled "Culpeper's English Physician, Containing the Medical Part" on p. 1, it is divided into two parts: "A physical and astronomical description of man," and "On diseases in general, their prevention and cure." The title-page to the 2nd volume is dated 1807. Ebenezer Sibly (1751-1800) was a 1792 Aberdeen medical graduate who was as devoted to astrological science as to medicine. He was the author of several works on astrology, including the *Key to physic and the occult science of astrology* (179?) and *The medical mirror* (1794). The *Key to physic* was issued with Sibly's 1798 edition of Culpeper in place of the "Medical Part" that appears in this and other editions of the *Herbal.* Ten editions of Sibly's version of the *English physician* were published between 1789 and 1817.

840. CULPEPER, Nicholas, 1616-1654.
Culpepper's [sic] family physician. The English physician enlarged, containing 300 medicines, made of American herbs. Being an astrologo-physical discourse of the vulgar herbs of this nation, containing a complete method of physic, whereby a man may preserve his body in health, or cure himself, being sick, with such things only as grow in America, they being most fit for American bodies . . . Revised, corrected and enlarged by James Scammon. Exeter [N.H.]: Published by James Scammon; printed by J. & B. Williams, 1824.
    360 p. ; 18 cm.

First edition; reissued at Exeter in 1825. According to Olav Thulesius, the difference between the 1824 Exeter edition of Culpeper and the English editions on which it is based is that Scammon reduced the number of plants described from 361 to 300 "because he was not sure that they actually could be found in America" (*Nicholas Culpeper,* [New York], 1992, p. 182). Culpeper's limited appeal on this side of the Atlantic may have been due to skepticism toward the more occult aspects of his system, i.e., his adherence to the doctrine of signatures and his belief in the influence of the heavens on the human body, on the diseases to which it is subject, and on the plants that heal it.

    J. & B. Williams were also the publishers of seven editions of William Buchan's *Domestic medicine* issued at Exeter between 1828 and 1843 (#488-491, 493-494).

841. CULPEPER, Nicholas, 1616-1654.
A directory for midwives: or, a guide for women, in their conception, bearing, and suckling their children. The first part contains, I. The anatomy of the vessels of generation. II. The formation of the child in the womb. III. What furthers conception. V. A guide for women in conception. VI. Of miscarriages in women. VII. A guide for women in their labour. VIII. A guide for women in their lying-in. IX. Of nursing children. To cure all diseases in women, read the second part of this book . . . Newly corrected from many gross errors. Belfast: Printed by James Magee . . . M,DCC,LXVI. [1766].
    vii, [8]-336 p., [1] folding leaf of plates : ill. ; 18 cm. (12mo)

A directory for midwives appeared in seventeen editions between 1651 and 1777, but was never reprinted in the American colonies. The book is divided into two sections. In this edition the first has the running-title *Culpeper's midwife enlarged,* and consists of nine books on the anatomy of the male and female genitalia, conception, the formation of the fetus, sterility, the signs and management of pregnancy, and remarks on labor, nursing and health. Culpeper gives little attention to the stages of labor, to natural and preternatural labors, or to intervention in difficult labors, for the simple reason that he had little experience of the matter. It is illustrated by a folding plate with six woodcut figures depicting the position of the child in the womb, etc. The second section has the running-title *Of practical physick,* and is devoted to uterine diseases, menstruation disorders, and nursing, with additional consideration of conception. The book is largely ignored by historians of British obstetrics and midwifery, apart from an enthusiastic chapter in Olav Thulesius' 1992 *Nicholas Culpeper.* Its longevity in a publishing history that spanned 126 years is attributable less to the book's utility to physicians, surgeons, or midwives, than in its appeal to women as an elementary guide in matters of reproductive physiology and health. In this respect, the *Directory* falls in the same category as the works of the Pseudo-Aristotle (#104-147), a work that plagiarized Culpeper and enjoyed a success on both sides of the Atlantic that the *Directory* never had.

842. CULPEPER, Nicholas, 1616-1654.
The English physician enlarged, with three hundred and sixty-nine medicines, made of English herbs . . . Being an astrologo-physical discourse of the vulgar herbs of this nation, containing a complete method of physic whereby a man may preserve his body in health, or cure himself, being sick, for three-pence charge, with such things as only grow in England, they being most fit for English bodies . . . London: Printed for the booksellers, 1799.
xii, 348 p. ; 17 cm. (12mo)

843. CULPEPER, Nicholas, 1616-1654.
The English physician enlarged, containing three hundred and sixty-nine receipts for medicines made from herbs . . . Taunton [N.J.]: Printed by Samuel W. Mortimer, 1826.
259, [5] p. ; 18 cm.

844. CULVERWELL, Robert James, 1802-1852.
Diseases of winter. On consumption, coughs, colds, asthma, and other diseases of the chest; their remedial and avertive treatment; addressed in popular language to non-medical readers, with copious observations on the diet and regimen necessary for invalids. Also, an appendix, containing two hundred formulae of the latest and most approved remedies, many valuable domestic receipts, and full directions for the practice of inhalation . . . New York: Published by J.S. Redfield, 1850.
91, [5] p. : ill. ; 23 cm.

Originally published at London in 1834 under the title *On consumption, coughs, colds, asthma, and other diseases of the chest.* This title was first issued in America by Redfield in 1849. In printed wrapper.

845. CULVERWELL, Robert James, 1802-1852.
How to be happy. An admonitory essay, for general and family perusal, on regimen, expediency, and mental government . . . New York: Published by J.S. Redfield, 1849.
94, [2] p. ; 23 cm.

Culverwell argues that there is a direct connection between one's mental state and bodily health, or lack thereof. "It is indeed true, that nine tenths of our sufferings, bodily and mental, outwardly and inwardly, our indigestions, our low spirits, even our gouts and rheumatisms, are of our own creation, and consequently what we might avoid we need not have" (p. 55). Originally published at London in 1847. At head of title: *The laws of life, health, and happiness, rendered clear to the humblest intelligence.* In printed wrapper.

846. CULVERWELL, Robert James, 1802-1852.
Married woman's private medical companion, embracing the treatment of menstruation, or monthly turns, during their stoppage, irregularity, or entire suppression. Pregnancy, and how it may be determined; with the treatment of its various diseases. Discovery to prevent pregnancy; its great importance and necessity where malformation or inability exists to give birth. To prevent miscarriage or abortion. When proper and necessary to effect miscarriage. When attended with entire safety. Causes and mode of cure of barrenness, or sterility, masturbation, its effects, treatment, &c. &c. . . . Newly revised, and notes and letters added by M.B. La Croix, M.D. . . . Sixty-

sixth edition. [Albany, N.Y.: M.B. La Croix, 185-?].

viii, [9]-300, [4] p : ill. ; 15 cm.

This is a composite work, i.e., the text of Culverwell's *On single and married life* (#847) is combined with substantial new material written by M.B. La Croix (doubtless a pseudonym), who advertises himself as medical consultant to those seeking advice on sexual matters or who suffer from sex disorders, sexually transmitted diseases, etc. 46% of the text is from Lacroix' pen: pages 80-117 contain an anatomy of the organs of generation, remarks on coition, reference to an electro-galavanic instrument for the prevention of pregnancy (sold by the author), the physiology of impregnation and the signs of pregnancy; pages 133-48 contain a chapter entitled "On choice in marriage phrenologically considered"; and pages 217-90 include a mixture of testimonial letters and case histories affirming the author's success in the treatment of sex-related disorders (and promoting his books and proprietary remedies).

The wording of the title-page is identical to that of the sex manuals of A.M. Mauriceau (#2398-2407) and J.H. Josselyn (#2050) published in the late 1840s and 1850s. Mauriceau's text is entirely different; the texts of Culverwell-La Croix and Josselyn, however, are nearly identical. Spine-title: *Medical companion.* The printed wrapper bears the statement "Fifty-sixth edition." On the back of the printed wrapper is a wood-engraved view of Dr. La Croix's Dispensary.

847. CULVERWELL, Robert James, 1802-1852.

On single and married life; or the institutes of marriage; its intent, obligations, and physical and constitutional disqualifications. To which is added, a poetical essay entitled Callipaediae: or, the art of having and rearing beautiful and healthy children . . . New York: Charles J.C. Kline & Co., publishers, [c1862].

150 p. ; 15 cm.

This treatise is a collection of twenty-two case histories. Most describe the author's treatment of sexual dysfunction due to physical disorders or to debilitating practices, such as masturbation. Others seem to have been included for the sake of their curious sexual pathologies. The title on the printed wrapper of this New York edition is *The green book.* Culverwell's *Green book* was a different work, however, devoted to sexually transmitted diseases that was published at London as *Medical counsellings, or the green book* and *Porneiopathology, or the green book.*

Pages 126-32 contain advertisements for *Dr. Culverwell's Radical Regenerator* and *Dr. Culverwell's Royal Nervine* available from the publisher Charles J.C. Kline, M.D., who claims to be "a friend and relative of that great and lamented physician." Kline also advertises sexual aids such as *L'Anime,* "the celebrated Paris remedy for impotency"; *L'Echauffeur,* "a French apparatus for increasing the size of the penis"; and "French condoms or male safes."

848. CULVERWELL, Robert James, 1802-1852.

Porneiopathology. A popular treatise on venereal and other diseases of the male and female genital system; with remarks on impotence, onanism, sterility, piles, and gravel, and prescriptions for their treatment . . . New York: J.S. Redfield, 1844.

215 p. : ill. ; 15 cm.

"Some years ago the idea first occurred to me that a popular treatise . . . explaining to the non-medical reader the structure and anatomy of the parts primarily affected by venereal disease, and describing its first as well as its subsequent and aggravated symptoms, and pointing out the safest treatment of it in experienced hands . . . would be of much avail in counteracting the effects of the complaint resulting from mal-treatment or neglect among the young and thoughtless" (*Pref.,* p. [3]). The text is abundantly illustrated with wood engravings.

The 1844 New York edition appears to be the first American edition of *Porneiopathology* (Culverwell's preface is dated 1843). The bibliographic record of Culverwell's several treatises on sexual hygiene, behavior, disorders, etc. is very confused. Part of the reason lies in the possibility that completely different works may have been published under the same title on occasion. Another reason lies in the fact that neither individually nor collectively do the *British Library catalogue,* the *National union catalogue of pre-1956 imprints* or the OCLC database provide anything approaching a complete or clear record of this author's publishing output. The *Index-Catalogue of the Library of the Surgeon-General's Office* (1st ser., III:557) records a 20th edition of *Porneiopathology* published at New York in 1849. Whatever the inadequacies of the bibliographic record, it is unlikely that the edition statement of the 20th edition is accurate.

849. CUMMING, William Henry.
Food for babes; or, artificial human milk, and the manner of preparing it and administering it to young children . . . Toronto: Printed for the author, by the Globe Printing Company, 1867.
viii, [9]-100 p. : ill. ; 15 cm.

First Canadian edition; originally published at New York in 1859. Acknowledging that maternal nursing is alway preferable to artificial food substitutes, Cumming nonetheless points out that, "it cannot be denied that there exists a great and increasing degeneracy on the part of the women of this age and country. The number of those who are able to nourish their children fully, is exceedingly small; and we very much fear that in twenty years, it will be smaller still" (p. vii). He attributes this "increasing degeneracy" to "the evils of female education and training during the period of early womanhood." The "artificial human milk" Cumming recommends is a series of dilutions of cow's milk. Wood-engraved frontispiece of a milking bottle. Atwater copy inscribed to A.L. Fowler (i.e., Abbie L. Ayres, O.S. Fowler's third wife?).

850. CUNARD STEAMSHIP COMPANY.
Springs of health in Great Britain and France. [Compiled by Atlas Advertising Agency, Inc. New York: Published by Gaines Thurman, c1917].
104 p. : ill. ; 23 cm.

"The purpose of this book is to supply the American public with a short and necessarily concentrated description of the principal spas of Great Britain and France, together with such essential information as to the character of the waters and local amusements" (p. 3). Thirteen British and forty-four French health resorts are described and illustrated. A table toward the end of the volume divides resorts by the disorders for which they are recommended (e.g., "affections of the liver," rheumatism, "women's

diseases," etc.). The final four pages describe the ships of the Cunard Line and list its offices in the United States and Canada.

851. CURIE, Paul Francis, 1799-1853.
Domestic homoeopathy . . . With additions and improvements by Gideon Humphrey . . . Philadelphia: Printed by Jesper Harding, 1839.
[2], xxxiv, 250 p. ; 15 cm.

Humphrey's "Preface to the American edition" (p. [vii]-xxx) is an apologia for homeopathy, a doctrine recently transplanted to American shores (the first American edition of Hahnemann's *Organon* had just been published in 1836). In the introduction that follows, Curie states, "The aim of this little volume is to place the public in possession of enlightened hygienic rules, applicable to the various periods of life, and referrible [sic] as well to a state of health as to that of suffering. From the limited number of homoeopathic practitioners, such a book is especially important in the present state of the science" (p. xxxii). Curie was one of the earliest proponents of homeopathy in London and a prolific author.

852. CURTIS, Alva, 1797-1881.
Discussions between several members of the regular medical faculty, and the Thomsonian botanic physicians, on the comparative merits of their respective systems . . . Columbus, Ohio: Printed at the office of the Thomsonian Recorder, by Jonathan Phillips, 1836.
400 p. ; 14 cm.

A collection of polemical articles and correspondence published in various periodicals and newspapers regarding Samuel Thomson's botanic system of practice. Curtis' compilation of these materials was intended for a general as well as a professional reading audience.

853. CURTIS, Alva, 1797-1881.
A fair examinaton and criticism of all the medical systems in vogue . . . Cincinnati: Printed for the proprietor, 1855.
iv, 5-208 p. ; 23 cm.

"Ever since the true science of medicine was shadowed forth, by Dr. Samuel Thomson and other pioneers of reform, a constant crusade has been kept up against it by interested men, in the hope of rendering its doctrines and practices ridiculous and unpopular, and thus preventing that thorough regeneration of this noble science, which would greatly mitigate our sufferings, prolong our lives and multiply our pleasures . . . Many friends of reform, and practitioners and teachers of medicine, have done what they could to develop its principles and illustrate its practice; but no one has yet attempted to furnish a full and safe defence of it, against the attacks of its enemies—especially no one has ventured to branch out from his own fortress of defence, and attack the enemy on the high seas of his own crazy craft . . . " (*Pref.*, p. iii).

It is this *venture* that Curtis undertakes in *A fair examination and criticism.* Following "General denunciations of medicine as science," i.e., admissions of medicine's limitations and impostures culled from the writings of European and American medical authors, Curtis launches into an attack on allopathy (p. [25]-99), eclecticism (p. 100-129), homeopathy (p. 130-52), and chrono-thermalism (p. 153-66). Curtis presents a very favorable portrait of hydropathic principles and practice (p. 167-76); and reviews Thomsonism (p. 177-85) before a concluding exposition of physio-medicalism (p. 186-96), one of several attempts before mid-century to establish Thomson's botanic practice on a more scientific and professional basis. Curtis was perhaps the leading intellectual figure in physio-medicalism, and is one of the principals of John Haller's *Kindly medicine, physio-medicalism in America, 1836-1911* (Kent, Ohio, 1997).

854. CURTIS, Alva, 1797-1881.

Lectures on midwifery and the forms of disease peculiar to women and children. Delivered to the members of the Botanico-Medical School, at Columbus, Ohio . . . Columbus, Ohio: Printed for the author, by Jonathan Phillips, 1837.

398 p. : ill. ; 14 cm.

First edition. Although delivered as a course of lectures to students at the Botanico-Medical School in Columbus, the populist nature of Thomsonism extended the relevance and comprehensibility of the *Lectures on midwifery* to a general audience. In his "Advertisement" to this volume, Curtis states that although he presents in this work the best obstetrical practice of the "authors of the old school" (Dewees in particular), he is nonetheless aware of "how much more favourable to our view those opinions would be, were their supporters acquainted with the superiority of our practice" (p. 9). Acknowledging that the "subject of the following pages is admitted to be very delicate," he insists that the book is "intended to be a confidential friend to married ladies," and that it was "the want of this minute and accurate knowledge," dictated by "numerous and pressing calls upon me for information," that necessitated its publication (p. 10).

Curtis provides a brief description of the female genitalia and their diseases; avoids the topic of conception, noting that "the more [authors] write, the more they exhibit their ignorance of the whole subject" (p. 52); briefly discusses the diseases of pregnancy; describes labor, presentations, etc.; and provides instructions for "putting to bed" and other postpartum considerations. The remaining and greater part of the *Lectures* is devoted to the diseases of women and children. The thirteen wood engravings at the end of the text are copied from the first American textbook of obstetrics, Samuel Bard's *A compendium of the theory and practice of midwifery* (1st ed., New York, 1807). In the second edition of the *Lectures* (#855, p. [3]) Curtis writes: "So highly was this work prized on its first appearance, that the whole edition of 4,000 copies was sold in less than than twelve months after it escaped from the press."

855. CURTIS, Alva, 1797-1881.

Lectures on midwifery and the forms of disease peculiar to women and children, delivered to the members of the Botanico-Medical College of Ohio . . . Second edition, corrected and enlarged . . . Columbus, Ohio: Printed for the author, by Jonathan Phillips, 1841.

255 p., [1] folding leaf of plates : ill. ; 21 cm.

856. CURTIS, Alva, 1797-1881.

Lectures on midwifery and the forms of disease peculiar to women and children, delivered to the members of the Botanico-Medical College of the State of Ohio . . .

Third edition, corrected and enlarged. Cincinnati: Printed, for the author, by C. Nagle, 1846.

xvi, 447 p., [1] folding leaf of plates : ill. ; 22 cm.

"Though explicit and decided in its principles, and rigid in its requirements in practice, yet so clear and conclusive are the facts and arguments given in evidence of their correctness, that the work has secured the approbation and and patronage of all classes of persons, from the mothers and nurses of the land, and the Botanico-Medical Practitioners, for whom it was especially designed, through the long catalogue of Reformers of every description— the rooters, the Beachites, the Eclectics, the Homoeopaths, even in the ranks of the regular faculty" (*Pref.,* p. iv). Obviously satisfied with the book's widespread appeal, Curtis continues: "So complete and explicit are the instructions of this volume, that many female nurses, after reading them, have slipped out of the train of their medical advisers, and taken the whole responsibility themselves." Not the least daunted by its adoption by laymen, Curtis adds, "many persons, male and female, have ventured, on the strength of the fundamental principles laid down in this work, to enter on the general practice, and they have been very successful" (p. v).

857. CURTIS, Alva, 1797-1881.

Synopsis of a course of lectures on medical science, delivered to the students of the Botanico-Medical College of Ohio . . . Cincinnati: Printed by Edwin Shepard, for the author, 1846.

xvi, 464 p. : ill. ; 22 cm.

Curtis' physio-medicalism and its cousin the reformed or eclectic practice represent the professionalization of what had begun two decades earlier as a populist attempt to place medical care in the hands of the afflicted, their friends and families. This tension between an emerging cadre of trained botanic physicians and the Thomsonian movement's roots among lay practitioners (self-taught in the garden, field and home) is evident in Curtis' preface to the *Synopsis.* Curtis writes of a "demonstrative science, whose processes should proceed from established principles" (quoting Samuel Jackson), and of bringing "into practice something

of exactness" (quoting N. Chapman), adding: "Though the principles and facts here set forth, will aid persons of all classes in society, in preserving their health, their time and money, to a great and valuable extent; yet the variety, connexion and beauty of those principles and the vast amount of practical conclusions, conduct and consequences they involve, will soon convince all reflecting persons, that, to be safe and successful healers of the sick . . . they must be thoroughly educated in and imbued by these principles, and devote their whole time to the practice of the art" (p. [ii]-iii).

The actual printing of this work was begun circa 1840, "when, from the pressure of the times it was suspended till last year"(p. [ii]). The gatherings signed 1-8 were apparently intended as a supplement to the *Botanico-medico recorder* (#381), and bear a legend to that effect across the top of the first leaf of each gathering.

CURTIS, Alva see also *THE THOMSONIAN RECORDER* (#3518).

858. CUTLER, Elbridge Gerry, 1846-1929.

The care of the sick room . . . Cambridge: Harvard University Press, 1914.

[2], 54 p. ; 18 cm.

Series: *Harvard health talks.* Cutler was an 1872 graduate of Harvard Medical School and a member of its faculty from 1878.

859. CUTLER, John Wesley, 1850-

Crumbs of comfort. How to cure the sick . . . First edition. Madison, Wis.: Democrat Printing Company, 1899.

413 p., [1] leaf of plates : port. ; 21 cm.

Cutler's text is divided into two parts. The first is an alphabetical catalog of diseases that is unusual among manuals of domestic practice in that it lists the drugs employed in the treatment of each disease, but provides no description of the disease, its symptoms, etiology or prognosis. The second part is an alphabetical catalog of the drugs employed in the first part.

860. CUTTER, Calvin, 1807-1873.

Anatomy and physiology: designed for schools and families . . . Second edition . . . Boston: Benjamin J. Mussey . . . [et al.], 1846.

xiv, [15]-322 p. : ill. ; 19 cm.

"As health requires the observance of the laws inherent to the different organs of the human system, so not only boys, but girls, should acquire a knowledge of the laws of their organization. If sound morality depend upon the inculcation of correct principles in youth, equally so does a sound physical system depend on a correct physical education during the same period of life" (*Pref.*, p. [v]).

Cutter's school texts were among the earliest contributions to the juvenile literature of physiology and hygiene in the United States, and dominated this branch of the school book trade through the Civil War. The first edition of *Anatomy and physiology* was published at Boston in 1845. According to "certificates of commendation" on an 1839 broadside advertising his lectures (#864), Cutter was an 1831 graduate of Dartmouth Medical College who also attended surgical lectures with Valentine Mott in New York and George McLellan at Jefferson Medical College. The Dartmouth and Columbia catalogs list Cutter as an 1832 Dartmouth medical graduate. Following his graduation, Cutter attended lectures at the College of Physicians & Surgeons (1833-34). He subsequently practiced in Nashua and Dover, N.H. During the Civil War, Cutter was a surgeon with the 21st Massachusetts Volunteer Infantry.

861. CUTTER, Calvin, 1807-1873.
Anatomy and physiology: designed for academies and families . . . Fourth stereotype edition . . . Boston: Benjamin B. Mussey . . . [et al.], 1847.
viii, [9]-10, [xi]-xiv, [15]-342, 4 p. : ill. ; 19 cm.

862. CUTTER, Calvin, 1807-1873.
Anatomy and physiology: designed for academies and families . . . Sixth stereotype edition . . . Boston: Benjamin B. Mussey and Co. . . . [et al.], 1847.
viii, [9]-10, [xi]-xiv, [15]-342, 4 p. : ill. ; 19 cm.

863. CUTTER, Calvin, 1807-1873.
Anatomy and physiology: designed for academies and families . . . Fifteenth stereotype edition . . . Boston: Benjamin B. Mussey and Co. . . . [et al.], 1848.
viii, [9]-10, [xi]-xiv, [15]-342, 4, 2, 10 p. : ill. ; 19 cm.

The 15th is the final edition of this work published under the title *Anatomy and physiology*. It was revised and published in 1848 with the title *A treatise on anatomy, physiology, and hygiene* (#873-877).

864. CUTTER, Calvin, 1807-1873.
ANATOMY & PHYSIOLOGY! THE LECTURE OF DRS. CUTTER at the [*Babcock*] Hall, THIS EVENING, [*Tuesday, Dec 23*] at [7] o'clock Will be upon the 250 Bones and 400 Muscles of the Human System. Their form, size, color, and situation, will be given by the use of Human SKELETONS and MANIKINS . . . [S.l., ca. 1839-40].
Broadside : ill. ; 37 × 17 cm.

865. CUTTER, Calvin, 1807-1873.
Common school physiology: designed for schools and families . . . First stereotype edition . . . Boston: Benjamin B. Mussey . . . [et al.], 1846.
xiv, [15]-226 p. : ill. ; 19 cm.

The first stereotype was the only edition of *Common school physiology* published. Its similarity to Cutter's *Anatomy and physiology* (#860-863) in text and illustration probably made it redundant on Mussey's list.

866. CUTTER, Calvin, 1807-1873.
First book on anatomy, physiology, and hygiene, for grammar schools and families . . . Stereotype edition. Boston: Benjamin B. Mussey and Co. . . . [et al.], 1849.
viii, [9]-180, [10] p. : ill. ; 19 cm.

Originally published at Boston in 1847 (and again in 1848) under the title *First book on anatomy and physiology*, this work was reissued with its present title in 1849. It was intended for children in grammar schools. Regarding the comprehensibility of the subject to ten year old minds the author writes, "It will demand no more maturity and thought to understand the reasons for adequate clothing, bathing, the necessity of an erect position in standing and sitting, regularity in taking food, the supply of pure air to the lungs, &c., than to comprehend geographical details or moral truths" (p. [v]). The *First book* remained in print into the early 1880s.

867. CUTTER, Calvin, 1807-1873.
First book on anatomy, physiology, and hygiene, for grammar schools and families . . . Stereotype editon. Boston: Benjamin Mussey and Co. . . . [et al.], 1850.
viii, [9]-180, [12] p. : ill. ; 19 cm.

868. CUTTER, Calvin, 1807-1873.
First book on anatomy, physiology, and hygiene, for grammar schools and families . . . Revised stereotype edition. Boston: Benjamin B. Mussey and Co. . . . [et al.], 1852.
viii, [9]-191 p. : ill. ; 16 cm.

Atwater copy lacking pp. 33-34, 47-48 and 191.

869. CUTTER, Calvin, 1807-1873.
First book on anatomy, physiology, and hygiene, for grammar schools and families . . . Revised stereotype edition. New York: Clark, Austin and Smith; Boston: B.B. Mussey & Co., 1855.
viii, [9]-191, [1] p. : ill. ; 19 cm.

870. CUTTER, Calvin, 1807-1873.
First book on anatomy, physiology, and hygiene, for grammar schools and families . . . Revised stereotype edition. New York: Clark, Austin and Smith; Cincinnati: W.B. Smith & Co.; St. Louis, Mo.: Keith & Woods, 1859.
viii, [9]-191, [1] p. : ill. ; 18 cm.

871. CUTTER, Calvin, 1807-1873.
New analytic anatomy, physiology and hygiene, human and comparative. For colleges, academies and families. With questions . . . Philadelphia: J.B. Lippincott & Co., 1874.
[2], 322, vii, [1], [3]-66 p. : ill. ; 19 cm.

"The solicitation of my publishers, and the request of many teachers, have induced me to review and remodel my school-book . . . My former work was published in 1849, and thoroughly revised in 1852 [#874] . . . In general arrangement, the present treatise is modeled after the former. The aim has been to improve the analysis; to bring the chemistry and histology to the present advanced state of these sciences; to make the anatomy and physiology concise and definite, the hygiene plain and practical; to introduce comparative anatomy; and to furnish illustrating cuts, both apposite and artistic" (*Pref.,* p. 5). Originally published in 1870, the *New analytic anatomy* was reissued in 1871, 1873 and for the final time in 1874. The wood engravings were printed from blocks used to illustrate Joseph Leidy's *An elementary treatise on human anatomy,* published by Lippincott in 1861.

872. CUTTER, Calvin, 1807-1873.
The physiological family physician, designed for families and individuals . . . West Brookfield [Ma.]: Merriam and Cooke, printers, 1845.
xii, [13]-218 p. : ill. ; 18 cm.

In this treatise Cutter grafts domestic medicine onto the genre of popular physiology. As in most physiologies, the volume is arranged by system, to each of which Cutter appends advice on diagnosing and treating its most common disorders. In his consideration of respiration, for example, Cutter describes the anatomy and physiology of the lungs, stresses the importance of pure air and ventilation to their proper functioning, and adds sections on such diseases as colds, consumption, asthma, croup, etc.

873. CUTTER, Calvin, 1807-1873.
A treatise on anatomy, physiology, and hygiene: designed for colleges, academies and families . . . Stereotype edition. Boston: Benjamin B. Mussey and Co. . . . [et al.], 1850.
458, [10], 10, [2] p. : ill. ; 19 cm.

The *Treatise* is a revised and expanded edition of the author's *Anatomy and physiology* (#860-863).

874. CUTTER, Calvin, 1807-1873.
A treatise on anatomy, physiology, and hygiene: designed for colleges, academies, and families . . . Revised stereotype edition. Boston: Benjamin B. Mussey and Co. . . . [et al.], 1852.
466 p. : ill. ; 19 cm.

875. CUTTER, Calvin, 1807-1873.
A treatise on anatomy, physiology, and hygiene: designed for colleges, academies, and families . . . Revised stereotype edition.

Boston: Benjamin B. Mussey and Co. . . . [et al.], 1853.

466, [9], 5 p. : ill. ; 19 cm.

876. CUTTER, Calvin, 1807-1873.
A treatise on anatomy, physiology, and hygiene: designed for colleges, academies, and families . . . Revised stereotype edition. New York: Clark, Austin and Smith; Cincinnati: W.B. Smith & Co.; St. Louis, Mo.: Keith & Woods, 1857.

466, [14] p. : ill. ; 19 cm.

877. CUTTER, Calvin, 1807-1873.
A treatise on anatomy, physiology, and hygiene: designed for colleges, academies, and families . . . Revised stereotype edition. New York: Clark, Austin, and Smith; Cincinnati: W.B. Smith & Co.; St. Louis, Mo.: Keith & Woods, 1859.

466, [14] p. : ill. ; 19 cm.

878. CUTTER, Charles, 1837-1912.
Cutter's guide to Mount Clemens, the great health and pleasure resort of Michigan . . . 8th edition. [S.l.: Charles Cutter], 1900.

68 p. : ill., maps ; 14 × 20 cm.

Cover-title. Beginning in 1873, Charles Cutter (1837-1912) became the publisher of a series of guides to American health resorts that included Hot Springs, Ark., Eureka Springs, Ark., Saratoga Springs, N.Y., the White Mountains of New Hampshire, Mineral Wells, Texas, etc. The guides include descriptions of the local waters, but are largely devoted to the hotels, sanitaria and spas in each locale, as well as to the sights in the surrounding area.

879. CUTTER, Charles, 1837-1912.
Cutter's guide to the Hot Springs of Arkansas . . . 49th edition . . . [S.l.: Charles Cutter], 1906.

64 p. : ill. ; 13 × 20 cm.

Cover-title. Printed on front wrapper: *Compliments of the Superior Bath House*, which is described on p. 21.

880. CUTTER, Charles, 1837-1912.
Cutter's official guide to Hot Springs, Arkansas. America's greatest health and pleasure resort. The health resort with a na-

tional backing. 58th edition. [S.l.]: Published by Chas. Cutter & Son, 1914.

64 p. : ill. ; 14 × 22 cm.

Cover-title. "Edited and compiled by John Milton Cutter" (p. 3). Along the top of front wrapper: *Special "Townsend" edition* (a photograph of the Townsend Hotel appears on back wrapper). Guides to the Hot Springs were the Cutter firm's most successful publication, first appearing in 1875.

881. CUTTER, Eunice Powers, 1819-1893.
Human and comparative anatomy, physiology, and hygiene . . . Revised and stereotyped. New York: Clark, Austin, Maynard & Co.; Cincinnati: W.B. Smith & Co.; St. Louis, Mo.: Keith & Woods, [c1854].

132, 8 p. : ill. ; 19 cm.

At head of title: *School edition.* Eunice P. Cutter was the wife of Calvin Cutter (#860-877). Her physiology is intended for grammar schools.

882. CUTTER, Eunice Powers, 1819-1893.
Human and comparative anatomy, physiology, and hygiene . . . Revised and stereotyped. New York: Clark, Austin, and Smith, 1855.

132, 8 p. : ill. ; 19 cm.

At head of title: *School edition.*

CUTTER, Eunice Powers see also **CUTTER** (#864).

883. CUTTER, John Clarence, 1851-1909.
Beginner's anatomy, physiology, and hygiene, including scientific instruction on the effects of stimulants and narcotics on the growing body . . . Philadelphia: J.B. Lippincott Company, [c1887].

144 p. : ill. ; 18 cm.

The first title in *Cutter's series of physiologies* published by Lippincott. The series also includes Cutter's *Intermediate anatomy, physiology, and hygiene* (a revision of Calvin Cutter's *First book on anatomy, physiology and hygiene*, #866-870) and his *Comprehensive anatomy, physiology, and hygiene*. The author was an 1877 Harvard Medical School graduate.

884. CUTTING, Harvey I.
A message from nature . . . Potsdam, New York: Published by Harvey I. Cutting, [ca. 1905].
43, [1] p. : ill. ; 16 cm.

Pamphlet describing the origins and medicinal benefits of *Adirondack Ozonia Natural Spring Water.* Title on printed wrapper: *Adirondack Ozonia Water.*

885. CYCLOPAEDIA OF TEMPERANCE AND PROHIBITION.
The Cyclopaedia of temperance and prohibition. A reference book of facts, statistics, and general information on all phases of the drink question, the temperance movement and the prohibition agitation. London; New York; Toronto: Funk & Wagnalls, 1891.
vi, 7-671 p. ; 26 cm.

A rather remarkable reference work providing an encyclopedic approach to secular and sectarian temperance movements, prohibition legislation, reviews of alcohol consumption in various nations, biographies of important figures in the temperance movement, and varied topics such as *Bible Wines, Crime, Delerium Tremens, Hasheesh, Inebriate Asylums, Phylloxera,* etc.

D

886. DALTON, John Call, 1825-1889.
A treatise on physiology and hygiene; for schools, families, and colleges . . . New York: Harper & Brothers, publishers; London: Simpson, Low, Son, & Marston, 1868.
xvi, [17]-399, [1], 6 p. : ill. ; 19 cm.

First edition. Dalton's popular manual of physiology was published twelve times in two editions between 1868 and 1890. "The first full-time professor of physiology in an American medical school, [Dalton] introduced Claude Bernard's methods of research and teaching to the United States" (*Dict. Amer. med. biog.*, I:177). While on the medical faculty at Buffalo, Dalton "delivered the first lectures in the United States during which experiments on animals were performed" (*ibid.*); and at the College of Physicians and Surgeons of New York, where he was professor of physiology from 1855 until his death, he established the first permanent laboratory of physiological research in the United States. Dalton was the author of *A treatise on human physiology* (1st ed., Philadelphia, 1859), which in its seven editions became the standard physiological text in most American medical schools.

887. DALTON, John Call, 1825-1889.
A treatise on physiology and hygiene; for schools, families, and colleges . . . New York: Harper & Brothers, publishers; London: Sampson, Low, Son & Marston, 1869.
xvi, [17]-399, [1], 7, [1] p. : ill. ; 19 cm.

888. DALTON, John Call, 1825-1889.
A treatise on physiology and hygiene; for schools, families, and colleges . . . New York: Harper & Brothers, publishers, 1875.
xvi, [17]-424, 8 p. : ill. ; 19 cm.

The eighteen chapters of Dalton's *Treatise* remained unchanged through the five issues published between 1868 and 1874. The 1875 edition was expanded to include a nineteenth chapter on human anatomy. This 424-page ver-

sion was the standard text for the remaining fifteen years of the *Treatise*'s publishing history.

889. DALTON, John Call, 1825-1889.
A treatise on physiology and hygiene for schools, families, and colleges . . . New York: Harper & Brothers, publishers, 1890.
xvi, [17]-424, 8 p. : ill. ; 19 cm.

The final issue of Dalton's popular physiological text, first published in 1868 (#886).

890. DAME, STODDARD AND KENDALL.
"The Old Doctor's Ice Pick." [*wood engraving*] . . . This is the invention of an old doctor who has often felt the need of some simple and efficient instrument for cutting ice for the use of the sick, without the usual annoyance of noise, flying of ice, and the wetting of furniture and clothing . . . Sole selling agents, Dame, Stoddard & Kendall . . . Boston, Mass. [189-?]
Broadside : ill. ; 21.5 × 14 cm.

891. DANELSON, J. Edwin.
Dr. Danelson's counselor with recipes: a practical and trusty guide for the family, and a suggestive hand-book for the physician . . . New York: A.L. Burt, publisher, 1881.
viii, [9]-720 p. : ill. ; 20 cm.

In his preface to this second edition Danelson writes, "The book does not, nor is it intended to, usurp or abridge the office of the physician. There are communities entirely without and beyond medical assistance; others with doctors, it is true, but those lacking in many necessary qualifications, and others still . . . where the medical adviser is fastened in the rut of routine" (p. vi). The *Counselor* is divided into four parts: I. Physiology; II. Hygiene; III. Marriage; IV. Medical Practice.
    Danelson was an 1866 graduate of the Eclectic Medical College of the City of New York. He is listed as resident in Buffalo, N.Y. in the 1886

edition of Polk's *Medical and surgical directory of the United States.*

892. DANELSON, J. Edwin.
Dr. Danelson's counselor with recipes: a practical and trusty guide for the family, and a suggestive hand-book for the physician . . . Revised and enlarged. New York: A.L. Burt, publisher, 1889.
    v, [1], [9]-720, [2] p. : ill. ; 19 cm.

893. DANFORTH AND BRISTOL.
Dr. Brown's receipts or information for the million. An invaluable collection of original and practical receipts, household, family, domestic, agricultural, medicinal and miscellaneous, with directions for preparing all the Thomsonian remedies, and directions for course of treatment; to which is added some original suggestions pertaining to the laws of life and health . . . New York City: Published by Danforth & Bristol, 1874.
    viii, 72 p. ; 15 cm.

Title on printed wrapper: *Dr. Brown's medicinal, family, household, agricultural and miscellaneous receipts.*

DANIELS, Ralph Roy see *THE HYGIENIST* (#1879).

894. DANIELS, W. H.
The temperance reform and its great reformers. An illustrated history, including the temperance "blue laws" of the New England colonies; the liquor laws of the states; a sketch of the various temperance organizations; biographies of Lyman Beecher, John B. Gough, Dr. Jewett, etc., a chapter on the medical and chemical phases of the movement; "the crusades"; the Woman's Christian Temperance Unions; "red ribbon" reform under Dr. Reynolds; the evangelistic temperance work unders Messrs. Moody, Sawyer, etc.; history of the Murphy temperance movement; with thrilling experiences of reformed and converted drunkards; with over twenty portraits of the chief reformers, and characteristic selections from their best writings and addresses . . . New York . . . [etc.]: Nelson & Phillips; Cincinnati . . .

[etc.]: Hitchcock & Walden, 1878.
    xi, [1], [3]-612 p. : ill., ports. ; 21 cm.

DARE, Shirley see **POWER** (#2874).

895. DARRIN, Sidney I.
NINTH EDITION. Cures without medicines. Dr. Darrin, naturaepathic physician for the safe and natural treatment of diseases, weaknesses and infirmities without the use of poisonous drugs, and painful surgery, (late from Paris, London, Liverpool, Dublin and Cork,) has taken offices over 56 State Street . . . Rochester, N.Y. where he is permanently located, for the healing of the sick and afflicted. Examination and consultation free . . . [Rochester, N.Y., 1874].
    [2] leaves ; 30 × 22.5 cm.

Newspaper notices place Darrin in Rochester between 1872 and 1875. An advertisement in the 1873 Rochester city directory states that he "cures without medicines . . . For the safe, sure and natural treatment of diseases, weaknesses and infirmities without the use of poisonous drugs and painful surgery . . . Examinations and consultations free." Darrin received good coverage in the Rochester press, starting with the announcement of his arrival in the city in November 1872. The following May a newspaper noted, "Magnetic healer sails for European vacation," and that he would return in November (six months in Europe, or working in another community?). In November of the following year there is notice of his return from a vacation in New York, and that he would "resume his practice at his old location" downtown. Darrin subsequently moved west. The 1886 and 1890 *Polk* directories place him in San Francisco; in Portland, Oregon in 1893 and 1900, and in Tacoma in 1896.

896. DAVID, William King.
Secrets of wise men, chemists and great physicians illustrated . . . Chicago: Wm. K. David, 1889.
    [8], 125, [3] p. : ill., port. ; 18 cm.

First edition. Among the instructions for making beer, building ice houses, and incubating eggs, David includes a section on medicine (p. 103-122): "Part III.—Medical. Prescriptions of eminent physicians. Arranged and revised by Frank V. Luse, M.D." The formulae are arranged alphabetically by disease, and include remedies for menstruation disorders, sexually transmitted diseases and even nymphomania. Frontispiece portrait of author.

897. DAVID, William King.
Secrets of wise men, chemists and great physicians illustrated. Comprising an unusual collection of money-making, money-saving, and health-giving prescriptions, receipts, formulas, processes and trade secrets secured at considerable expense from a multitude of thinkers and workers in practical affairs . . . Philadelphia: Wm. K. David, publisher, 1902.
[8], 125, [3] p. : ill., port. ; 18 cm.

Originally published at Chicago in 1889 (#896).

898. DAVIDSON, W. F.
I bring Health and Happiness, in lieu of Suffering and Pain. G. Ky. L. THE GREAT KENTUCKY LINIMENT . . . It cures Toothache in one minute. Earache in three minutes. Headache and Sore Throat in five minutes. Pain in Stomach, Bowels, Side, Stiff Neck and Cramps, in from five to ten minutes . . . G. Ky. L. is equally serviceable for MAN and BEAST . . . W.F. DAVIDSON, chemist, proprietor . . . Covington, Ky. [ca. 1850].
[1] leaf printed on both sides ; 27 × 16 cm.

Verso contains advertisements for *Davidson's Two-Forty Condition Powders, Davidson's Balsam of Hoarhound*, etc.

DAVIES, Nathaniel Edward see **YORKE-DAVIES** (#3900).

DAVIESON, David see **EUROPEAN ANATOMICAL, PATHOLOGICAL AND ETHNOLOGICAL MUSEUM** (#1081); **GRAND MUSEUM OF ANATOMY** (#1399).

DAVIESON, Samuel see **GRAND MUSEUM OF ANATOMY** (#1399).

899. DAVIS, Alexander H.
Improved theory and practice of medicine, deduced from forty years successful practice . . . Chicago: Published by the author, 1880.
vi, [7]-383, [1] p. : ill. ; 23 cm.

This explication of the author's eclectic-based theories of physiology, pathology and therapeutics was written for a lay audience. The sheer complexity of the work, however, guaranteed that this self-published monograph would never see a second edition. Not until page 114 does Davis abandon his theoretical idiosyncrasies and launch a description of disease entities and recipes for their cure. Pages [288]-357 contain an alphabetically arranged botanic materia medica, followed by recipes for various specifics, recommendations on diet, etc. The author was an 1851 graduate of the Central Medical College (Rochester, N.Y.).

900. DAVIS, Andrew Jackson, 1826-1910.
The genesis and ethics of conjugal love . . . New York: A.J. Davis & Co., Progressive Publishing House, 1874.
xi, [12]-142, [2] p. ; 17 cm.

First edition. One of the most influential American spiritualists of the 19th century, Davis "to a large extent created Spiritualism's juxtaposition of Neoplatonist/Swedenborgian metaphysics, mesmeric trance, and radical/utopian thought on social issues such as marriage" (*Amer. nat. biog.*, 6:165). Davis' *magnum opus* was his five-volume *Great harmonia* (1st ed., 1850-55). The fourth volume (*The reformer*) was devoted to sex and marriage. According to Davis, it "roused thousands of the best women and men to an investigation of the most delicate, and, at the same time, the most familiar of all human questions," but was also "viciously assailed by popular orthodox minds" and "misrepresented by talented social agitators" (p. [iii]). The present work, therefore, "has become a spiritual necessity, in order to reiterate, reinforce, and more thoroughly to impress upon mankind the harmonial principles and arguments" pertaining to marriage, sexual ethics, etc.
Davis opens his treatise by establishing the "divine origin" of sex: "The sexual principles, the male and the female . . . are revelations of

the essential bi-sexual constitution of Deity. The attributes of male and female complement one another," e.g., "Goodness is feminine; truth is masculine. The first is warm; the second is cold," etc. (p. 14). The "universality of the sexual principles" is manifested in every human couple: "A point of life (female) meet and weds an atom of matter (male); the offspring multiply and replenish the earth with their innocence and beauty. The magnetic bosom of vitality draw the electrical atoms of space; there is meeting and copulation, and conception, and legitimate productiveness" (*ibid.*). Marriage is the "highest and holiest" of affections; but there can be no marriage "where equality of sex and of personal rights are not first intelligently recognized, acknowledged, and solemnly accepted as the immovable basis" (p. 19). The purpose of a couple uniting is not simply to achieve terrestial happiness, but to foster "celestial joy in the hearts of the truly mated" (p. 20). To keep the married state unpolluted from base sexual aggression (on the male's part) and collusion (on the female's part), the female should keep not just a separate bed but a separate apartment in the home where she can "make her toilet unmolested and alone" and where she can be assured of "personal freedom from uncongenial association and familiarity" (p. 21).

Davis portrays the "ideal of marriage" on p. 36: "If you would look upon the nearest approach to the condition called heaven, you must visit a harmonially married pair, dwelling amid the loveliness and sweet beauty of Nature, surrounded with a little family contented and happy . . . all participate in the pathos and poetry of communion with the departed, realize the sublimity of immortality, see the unquenchable fire and fascination of the Harmonial Philosophy burning its exalted grandeur in the world's universal reason . . . [and] here behold a glimpse of that which is for all in the Summer-Land." He contrasts this with "false marriages," which result in physically and morally deformed offspring. "It is my impression," Davis writes, "that idiots, cretins, adult-infants, moral imbeciles, and epileptic criminals, need not, and ought not, ever to come into this world, and they certainly will not in the better time, when marriages become harmonial, and personal habits are attuned to the sacred laws of reproduction" (p. 39). Marriage should be re-

stricted to "competent persons," but even so, divorce should be readily allowed those who were mistaken in their choice of a soul mate. Davis castigates free love, prostitution and adultery as destructive mechanisms of sexual gratification, and denies the frequent portrayal of spiritualists as "promiscious passionists."

Davis is hopeful about the spiritual possibilities for marriage in the American republic: "Spirit unions . . . which are perpetually blissful and beautiful, even amid great fiery trials consequent upon this outer-life, may become more frequent; they may be multiplied: first, by true refinement and spiritualization among the people; second, by mingling true ideas of spiritual love and the divine uses of marriage with the practical education of our children. What is now only conjugal wrongheadedness, idle dreaming, bad longing, and vicious practice . . . may, by frank and exalted methods of education, become the world's delight, triumph, and lasting glory" (p. 77-78). Davis regards the 19th century as the end of the Christian era, and the beginning of the "Harmonial Epoch." He devotes the final pages of his book to a meditation upon the dissolution of the old orders of Church and State and the birth of a new epoch.

901. DAVIS, Andrew Jackson, 1826-1910.
The great harmonia; being a philosophical revelation of the natural, spiritual, and celestial universe . . . Vol. I. Boston: Benjamin B. Mussey & Co., 1850.

456 p., [2] leaves of plates : ill., port. ; 19 cm.

"Between 1850 and 1855 Davis wrote his major work, *The Great Harmonia,* in five volumes. Its 'Harmonial Philosophy' provided an intellectual understanding of the varied phenomena of Spiritualism. Beyond Spiritualist circles, the book influenced radical social, educational, medical, and religious thought" (*Amer. nat. biog.,* 6:165). *The great harmonia* is intended as an explication of "the great general Truths . . . which constitute the Foundation of all particulars, and the Basis of all true reasoning" (p. 6). Davis states in his preface that the content of this first and all the succeeding volumes is the product of spiritual dictation: "When physical conditions and surrounding circumstances are favorable to a full development of the author's sensibilities, he disconnects himself, by a peculiar exercise of the will . . . from the obstructing influences of the material world, and passes into the superior condition. Thus, unaided by any individual, he enters this high and superior state of mental exaltation; and while in this independent condition . . . he receives his spiritual impressions, and rapidly records them with a pencil. In the composition of his works, the author derives no assistance from the reading of books" (p. 6). Davis continues: "The author is impressed to devote his life and interior powers to the promotion of human progression, happiness, and spiritual illumination. To obtain means and principles adequate to the accomplishment of these ends, he is impressed to search . . . the Natural, Spiritual, and Celestial departments of God's Universal Temple, and to reveal and suggest the proprer application of such general truths as man's physical and spritual organization requires in this, his rudimental state of existence" (p. 9).

This first volume (spine-title: *The Great Harmonia. Vol. I. The Physician*) is devoted to the revelation of the laws of man's physical well-being. By health, as one might expect, Jackson does "not merely mean that the stomach, or liver, or lungs, or other members of the structural brotherhood, are free from disease, but . . . that state where the immortal spirit is circulating harmoniously through every organ, tissue, and ramification, of the organism" (p. 43). "Health," he writes, "is a state of complete harmony, because, should anything disturb the harmonious circulation of the Spiritual Principle from the Brain through the diversified ramifications of the system, it is positively certain that discord or disease would be the inevitable consequence" (p. 44).

Davis describes the mode of action of the Divine Mind on living matter: "the Deity acts universally upon matter in seven distinct but converging ways—namely: he acts upon matter first, anatomically; second, physiologically; third, mechanically; fourth, chemically; fifth, electrically; sixth, magnetically; and seventh, spiritually" (p. 47). The relations which subsist between mind and matter, body and spirit, require a "new method of imparting medical or physiological instruction, to begin with the spirit and consider its many, and hitherto hidden and mysterious influences upon the organism over which it so majesticaly presides" (p.

71). Davis dismisses the superficial mode of acquiring medical knowledge that characterizes orthodox medicine's short-sighted accumulation of disease entities, symptoms and empirical remedies. "Disease," he writes, "is a want of equilibrium in the circulation of the spiritual principle through the physical organization. In plainer language, disease is discord; and this discord or derangement must exist *primarily* in the spiritual forces by which the organization is actuated and governed" (p. 103). He posits a single disease entity: "By interior perception, I discover that the hundreds of diseases which physicians have distinguished by as many names, are simply but SYMPTOMS OF ONE DISEASE; and that this ONE DISEASE is caused or created by a constitutional disturbance in the circulation of the spiritual principle" (p. 110).

Enamored of the number seven, Davis distinguishes as many "causes of physical inharmonies" that result in disease, i.e., hereditary predispositions, accidents or injuries, atmospherical changes, situation, occupation, habits and spiritual disturbances. His hygienic and therapeutic philosophy, therefore, is directed toward obviating the influence of these "inharmonies" upon the system. To that end, the role of the physician must be reconsidered: "Physicians are designed to minister to the spiritual principle; they should be clergymen, or clergymen should be physicians" (p. 223). They must realize that the "quiet and tranquillizing influence of moral and spiritual principles are indispensable as prophylactic or therapeutic means" (p. 227). Physicians are wrong to regard their work as the categorization of disease entities and the use of the lancet or calomel in the relief of symptoms. "Physicians must cure disease or discord by producing harmony in the human constitution. Their pursuits should be essentially prophylactical. They must banish consumption, scrofula, erysipelas, rheumatism, contagious diseases, and every species of individual affliction, by commencing at their foundation. They must ascend the pulpit, and teach the inhabitants of the earth concerning hereditary impression; explain how parents influence their children while in the embryotic state; explain how the human mind and body are under the control of surrounding circumstances; how the entire individual can be manufactured, perfect or imperfect; how the spiritual principle acts upon the organization; and they must teach the philosophies of Anatomy, of Physiology, of Chemistry, of Mechanism, of Electricity, of Magnetism, and of Psychology" (p. 229). In the final 150 pages of this first volume, Davis applies "the foregoing principles to the preservation of health, and to the curing or amelioration of those important diseases with which individuals, who have transgressed the laws of Nature, are more or less afflicted" (p. 296).

902. DAVIS, Andrew Jackson, 1826-1910.

The harbinger of health; containing medical prescriptions for the human body and mind . . . Fifth thousand. New York: A.J. Davis & Co., 1862.

428, [2] p. : ill. ; 19 cm.

In his preface Davis asks:

"QUESTION: What is the chief end of the earthly life of man?

"ANSWER: To individualize his spirit, and prepare it for the Summer Land.

"Q. What is the first condition of such individualization and preparation?

"A. Physical health.

"Q. In what does physical health consist?

"A. In symmetry of development, energy of will, harmony of function, and bodily purity.

"Q. How can these results be obtained?

"A. First, by inheriting a sound constitution; second, by obeying the law of temperance in regard to foods and drinks; third, by giving free play and equal exercise to the muscular system; fourth, by exerting the Will-power to keep the passions in subjection; fifth and lastly, by sleeping and working and living in accordance with the requirements of Nature's Laws."

In chapters with titles such as "The Pearly Gates of Science," Davis describes a spiritualist system of paired harmonies and "reciprocal operations" that when properly balanced guarantee bodily and mental health, and when disturbed result in disease. Basing his "philosophy of disease" on the principle "that all disturbances of the physical structure originate in the soul-principle," Davis' therapeutics restores the disturbed soul-principle rather than pouring drugs into a diseased body. His readers are urged to recognize "the *periods* when the Self-Healing Energies (the soul-principle) exert themselves for re-establishment of the right physiological conditions" (p. 23). Davis stresses

clairvoyance in the diagnosis and treatment of disease, i.e., in getting beyond the mere physical manifestations of disease to its seat in the disturbed soul-principle. The complexities and limitations of this occult ability may require the patient to employ both a "describing seer" for diagnosis and a "prescribing seer" for healing.

"As the age of honesty and intelligence expands," he writes, "we find patients and physicians more and more agreeing that medicines, at best, but serve and subserve the inherent energies of the organism; that health is possible only by means of self-restoring and conservative principles which the good Father and Mother transmitted to the organs, muscles, nerves, and blood, of the living temple; and, therefore, that all belief, or pretension, that medicines hold and convey the life-giving energies of health and beauty to man's body, is nothing less than mischievous superstition or intentional imposition" (p. 41). Davis expresses sympathy for the Eclectics in their disregard for "time-honored theories" and their therapeutic independence, which in his mind should extend to occult therapies such as magnetism.

"WILL YOURSELF TO BECOME HEALTHY," Davis advises his readers: "all you are required to practice . . . is the art of concentrating your Will and desires simultaneously on the extremities first; then work upward and inward progressively; and when, in the lapse of ten minutes of steady, deep breathing, you have reached the brain, repeat the process in the ascending scale . . . By this Pneumogastric treatment of yourself, you will receive spiritual strength from the air . . . When, by practice, you can breathe deeply and heroically, and at the same time put your Will upon the restoration of the general system, the art of fixing your mind upon some particularly diseased part will become less and less difficult" (p. 52). By such practice one acquires "psychological power over the destinies of your bodily state," and one's home becomes a "Self-healing Institution."

Originally published at New York in 1861, the 19th edition of *The harbinger of health* was issued at Rochester, N.Y. in 1909 (#905). The "Poughkeepsie Seer" began his spiritualist career at age seventeen as the medium for a local tailor who mesmerized Davis in order to apply his clairvoyant powers to the cure of disease. Davis set off on his own path in mesmerism two years later, eventually becoming a pioneer in

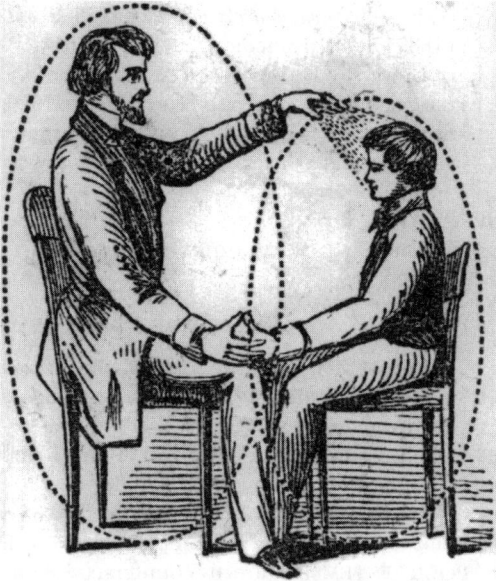

Fig. 24. "Partial blending of magnetic sphere," from Davis' Harbinger of health, #903.

the spiritualist movement in the United States and a sucessful author. According to the *Dict. Amer. biog.,* Davis "gave modern spiritualism much of its phraseology and first formulated its underlying principles" (III:105). His influence on British spiritualists is described in Janet Oppenheim's *The other world* (Cambridge, 1985, p. 101-103).

903. DAVIS, Andrew Jackson, 1826-1910.
The harbinger of health; containing medical prescriptions for the human body and mind . . . Eighth thousand. New York: C.M. Plumb & Co., 1865.
428 p. : ill. ; 20 cm.

904. DAVIS, Andrew Jackson, 1826-1910.
The harbinger of health; containing medical prescriptions for the human body and mind . . . Ninth thousand. Boston: Published by Bela Marsh, 1868.
428, [2] p. : ill. ; 19 cm.

905. DAVIS, Andrew Jackson, 1826-1910.
The harbinger of health; containing medical prescriptions for the human body and mind . . . Nineteenth edition. Rochester, N.Y., U.S.A.: The Austin Publishing Co., 1909.
428 p. : ill. 20 cm.

906. DAVIS, Andrew Jackson, 1826-1910.
Mental disorders; or, diseases of the brain and nerves, developing the origin and philosophy of mania, insanity, and crime, with full directions for their treatment and cure . . . Special edition. New York: American News Company . . . [et al.], 1871.
viii, [9]-487, [5] p., [1] leaf of plates : front. ; 20 cm.

Cover-title: *Mental disorders, causes & treatment.* Spine-title: *Diseases of the brain and nerves.*

907. DAVIS, E. V.
The household guide; or, home remedies and home treatment for all diseases in man or beast. A manual of domestic information for all classes, and a complete encyclopedia of family receipts and celebrated prescriptions . . . Naperville, Ills.: Published by J.L. Nichols, 1891.
279 p. : ill. ; 18 cm.

"The object of this volume is, to instruct every housekeeper and every owner of domestic animals in the use and application of simple domestic remedies. It may be properly called a book of *Self Instruction* in the art of home doctoring . . . The language is simple and technical terms have been carefully omitted, and the book itself makes up a complete series of Home Lessons in Medicine, which can be read and understood by all classes" (*Pref.,* p. [3]).

907.1. DAVIS, Irenaeus P.
Hygiene for girls . . . New York: D. Appleton and Company, 1883.
[2], 210, [4] p. ; 18 cm.

First edition. Davis addresses the moral as much as the physical needs of adolescent females. He considers their nervous temperament and the many forms of "nervousness" that affect females; the effect of "habit and association" on offspring (e.g., "many a child has been born a liar because the parent has practiced a disregard for truth; many a child has been born a sloven because the parent has practiced a disorderly habit; many a child has been born a dunce because the parent has neglected to form habits of study and observation," etc. [p. 58]); the physical manifestations of "sympathy and imagination" (hysteria being one example);

"feminine employments" adapted "to her nature and position" (which exclude the practice of medicine); amusements suitable to the female constitution (although exercise is important, there is hardly a form of it that does not present as many harmful as beneficial effects); the medical consequences of fashion (e.g., tight lacing, neuralgia due to bonnets, the harmful effects of veils on the eyes, etc.); and "hygienic morals," i.e., influences that affect moral integrity. Among these Davis singles out medical advertisements, noting that "the worst and most dangerous of all . . . are those which do not at first appear to be advertisements at all, but profess to be scientific treatises on marriage, the relations of the sexes, and certain forms of immorality. They commonly assume a high moral tone, but their true character is such that no one who is acquainted with them would like to be seen reading them. They appeal to an immodest curiosity, and are directly calculated to rouse unclean desires and toughts" (p. 206). The author was an 1871 graduate of the Bellevue Hospital Medical College (New York).

DAVIS, Irenaeus P. see also **GUERNSEY** (#1442).

908. DAVIS, Nathan Smith, 1858-1920.
Consumption: how to prevent it and how to live with it. Its nature, its causes, its prevention, and the mode of life, climate, exercise, food, clothing necessary for its cure . . . Philadelphia: The F.A. Davis Company; London: F.J. Rebman, 1894.
    viii, 143 p. ; 19 cm.

"I have found it difficult in brief conversations to impress upon consumptives the necessity of rigidly executing certain sanitary rules, whose fulfillment is essential to successful treatment of their disease. This is especially true of patients who live at a distance and are seldom seen. I therefore prepared for my patients a series of hygienic rules, with brief explanations of the effect of their execution. From these rules this small volume has grown" (*Pref.*, p. iii).
    Davis' *Consumption* was first published at Philadelphia in 1891, and was reissued in 1894 and 1902. A second edition was published in 1908. An 1883 graduate of the Chicago Medical College, Davis' practice in Chicago centered around respiratory diseases and diabetes. He

taught at his *alma mater* from 1884 to 1886, and in 1887 joined the medical faculty of Northwestern University. He was vice-president of the U.S. Pharmacopoeia Convention from 1910 to 1920; was prominent in the A.M.A.; and wrote numerous books and articles.

909. DAVIS, Orin.
Welcome words to women . . . Attica, N.Y., [ca. 1883].
    64 p. ; 15 cm.

Promotional literature for Davis' *Wafer pour les Femmes,* a medicament which after being inserted into the vagina, dissolved and was absorbed by the surrounding tissues and blood vessels. Davis recommends its use for a variety of uterine and vaginal disorders. In the treatment of leucorrhea, for example, "The WAFER POUR LES FEMMES directly counteracts inflammation of the uterine tissues. By its anodyne properties it soothes pain and irritation, greatly promotes the return of venous blood, and thus removes venous congestion. The *Wafer* is a persistent stimulant; and hence . . . is permanent in its effects. The *Wafer,* by being absorbed, imparts its definite, specific power, which increases the natural nervous activity of the uterus, and in this respect is not unlike a galvanic force. In these lingering complaints there is a running down, waste and impoverishment of the blood, and a corruption of its fluids . . . When the *Wafer* becomes soluble, and is taken into the blood . . . it supplies an essential reconstituent element which prevents decay and arrests decomposition" (p. 8). The *Wafer pour les Femmes* was only one of the proprietary remedies available from Davis. Publication date estimated from the most recently dated patient testimonials.
    Orin Davis was an 1846 graduate of the Eclectic Medical Institute in Cincinnati. He claims to have established the Dr. Orin Davis' Health Institute at Attica, N.Y. in 1854. He is listed as residing in Attica, N.Y. in the 1st (1886) through 7th (1902) editions of *Polk's medical register and directory of North America.* The 9th (1906) edition of *Polk* places him in "Sawtelle, Los Angeles." Davis' name appears in *Polk* for the last time in the 10th (1908) edition.

DAVIS, William O. see **LIGHTHALL** (#2251).

910. PERRY DAVIS AND SON.
The people's pamphlet. This is the people's pamphlet, read it if you please. It contains the author's account of the origin, name, and use, of PERRY DAVIS'S VEGETABLE PAIN KILLER. [Providence, R.I.]: Perry Davis, [ca. 1848, c1845].
    16 p. : port. ; 21 cm.

At least half the many letters testifying to the efficacy of *Perry Davis' Vegetable Pain Killer* are dated 1848. Full-page wood-engraved portrait of Perry Davis on first page. Atwater copy lacking final leaf (p. 15-16).

911. PERRY DAVIS AND SON.
PERRY DAVIS' VEGETABLE PAIN KILLER . . . TAKEN INTERNALLY, CURES—sudden colds, coughs, fever and ague, asthma and phthisic, dyspepsia, liver complaint, acid stomach, headache, heartburn, indigestion, canker in the mouth and stomach, canker rash, kidney complaints, piles, sea sickness, sick headache, cramp and pain in the stomach, painters' colic, diarrhea, dysentery, summer complaint, cholera morbus, cholera infantum, and cholera. APPLIED EXTERNALLY, IT CURES—scalds, burns, frost bites, chilblains, sprains, bruises, whitlows, felons, boils, ringworms, old sores, rheumatic affections, headache, neuralgia in the face, toothache, pain in the side, pain in the back and loins, neurlagic or rheumatic pains in the joints or limbs, stings of insects, scorpions, centipedes, and the bites of poisonous insects and venomous reptiles . . . [Providence, R.I.]: Perry Davis and Son, [c1860].
    [1] leaf printed on both sides ; 29 × 15.5 cm.

Printed at bottom of verso: PERRY DAVIS & SON, manufacturers and proprietors, Providence, R.I. J.N. HARRIS & CO., Cincinnati, Ohio, proprietors for the Southern and Western States.

912. DAVISON, Alvin, 1868-1915.
Health lessons. Book I. . . . New York; Cincinnati; Chicago: American Book Company, [c1910].
    191 p. : ill. ; 19 cm.

"Scarcely one half of the children of our country continue in school much beyond the fifth grade. It is important, therefore, that so far as possible the knowledge which has most to do with human welfare should be presented in the early years of school life" (*Pref.*, p. 5). Davis is particularly interested in instructing his juvenile audience in the prevention of infectious disease, especially tuberculosis: "After the eleventh year of age, the first cause of death among school children is tuberculosis" (p. 6). The author was professor of biology at Lafayette College.

913. DAVISON, Alvin, 1868-1915.
Health lessons. Book II . . . New York; Cincinnati; Chicago: American Book Company, [c1910].
    288 p. : ill. ; 19 cm.

An expanded edition of Davis' 1909 *The human body and health* (#915). Spine-title: *Davison's health lessons.*

914. DAVISON, Alvin, 1868-1915.
The human body and health. A text-book of essential anatomy, applied physiology, and practical hygiene . . . Advanced. New York; Cincinnati; Chicago: American Book Company, [c1908].
    320 p. : ill., port. ; 19 cm.

915. DAVISON, Alvin, 1868-1915.
The human body and health. An intermediate text-book of essential physiology, applied hygiene, and practical sanitation for schools . . . Intermediate. New York; Cincinnati; Chicago: American Book Company, [c1909].
    223 p. : ill. ; 19 cm.

916. DAWSON, Charles Carroll, b. 1833.
Saratoga: its mineral waters, and their use in preventing and eradicating disease, and as a refreshing beverage . . . New York: Russell Brothers, printers, 1874.
    64 p., [8] leaves of plates : ill., maps ; 16 cm.

917. DAY, Albert, 1812-1894.
Methomania: a treatise on alcoholic poisoning . . . With an appendix, by Horatio R. Storer . . . Boston: James Campbell, 1867.
    70, [2] p. ; 18 cm.

Day considers both the mental and physical consequences of alcoholism. He states that there is no single treatment applicable to all cases of alcoholism, but that any treatment is founded on one prerequisite: "The patient should be made to believe, if possible, that his recovery will probably follow proper treatment. The idea has been so largely held and advanced by medical men that recovery from this disease was impossible . . . that a large amount of skepticism upon this point has pervaded the community. And it is a noticeable fact, that those most likely to adopt this view, and to cling to it with tenacity, are the unfortunate victims of the disease themselves" (p. 35). Day was an 1866 graduate of Harvard Medical School, and superintendent of the Washingtonian Home (Boston) for the treatment of alcoholism. Atwater copy inscribed by Henry Foster, founder of the Clifton Springs Water Cure.

918. DAY, Bertha C.
Diseases of women. Their cause, symptoms and treatment and the medical home guide . . . Fort Wayne, Indiana: Written and published by Dr. Bertha C. Day, [c1909].
　　118, [2] p. : port. ; 19 cm.

Opposite a half-tone photograph of Dr. Bertha C. Day the reader finds three pages of encomiums such as the following: "Due to her very nature and to her vast experience she has great sympathy—great sisterly love—for all women. She knows the burdens they bear and knowing these no one can doubt the advantage she has over other physicians in ministering to the needs of women who are sick and in need of help. Whether a daughter, sister, wife or mother, you can have a sympathetic friend in Dr. Day . . . As a woman her advice is uplifting, as a physician it is most beneficial and certain to bring the best results obtainable. It takes a woman to understand a woman's ills," etc. (p. [5]). In the pages that follow Dr. Day encourages her female reader to begin a correspondence that will enable her to diagnose her illnesses, intitiate a regimen of proprietary medication, and monitor the patient's progress by the return of pre-printed forms.

Bertha C. Day was a 1907 graduate of the Detroit Homoeopathic College who soon after being licensed was hired as a front for the mail-order medicine business operated by William M. Griffin, of Fort Wayne, Indiana. According to a pamphlet published by the A.M.A. in 1914

entitled *"Female weakness" cures and allied frauds* (#85), the letters sent by Day ("printed by the hundred and thousands") were "filled in by girls who have no more medical knowledge than the average school-girl would have." The medicines supplied through the mails were "the cheapest of cheap drugs bought in enormous quantities from the least reputable of drug houses"; and the letters sent to the firm were routinely "sold to other vampires in the same business" (p. 28). Because of the A.M.A.'s exposure of his operation, Griffin renamed his business the Woman's Health Institute and replaced Day with Julia D. Godfrey (#1373). Shortly afterward, Day established her own mail-order business in Hammond, Indiana.

919. DAY, L. Meeker.
The botanic family physician; or the secret of curing all diseases, on improved Hygeian principles, fully disclosed; containing also, formulas, or recipes, for the cure of every disease incidental to human nature: together with a valuable digest on midwifery . . . New-York, 1833.
　　24 p. ; 18 cm.

A botanic formulary issued with Day's *The improved American family physician* (#920).

920. DAY, L. Meeker.
The improved American family physician; or, sick man's guide to health: containing a complete theory of the botanic practice of medicine, on the Thomsonian and Hygeian system, with alterations and improvements. To which is appended, a concise formula for compounding medicines for the cure of every complaint incident to human nature. Also, a complete digest of midwifery, so that the old proverb may be verified, that "Every man be his own physician." New-York, 1833.
　　viii, [9]-120 p. ; 18 cm.

Day was a disciple of both Samuel Thomson (#3482-3515) and of James Morison (1770-1840), the founder of the Hygeian movement (botanic) in Britain. The bulk of *The improved American family physician* is extracted (with acknowledgement) from Thomson's writings, with occasional commentary by Day. Issued with his *The botanic family physician* (#919).

DEAD DOCTOR'S LEGACY see *LEGACY OF A DEAD DOCTOR* (#2225).

DE BORRA, Alex. see **CRYSTAL SPRINGS HOTEL** (#835).

921. DeCAMP, E. J., d. 1910.
Nature's law of health and disease and the ladies' toilet. A concise, up-to-date family medical work and ladies' companion . . . Rochester, N.Y.: George W. Winkelman, publisher, [c1910].
[10], 124, [2], 125-177 p., [9] leaves of plates : ill. ; 19 cm.

DeCamp, an osteopath, divides his treatise into two parts. The first (running-title "Health and Disease") provides a brief description of osteopathy and some remarks on hygiene, but is largely devoted to "concise descriptions of the most common diseases with simple treatment and home remedies, to aid the patient's recovery, until a physician is employed" (p. 16). Part II, "The Ladies' Toilet" (p. [125]-177), is a collection of recipes for skin creams, oils, powders, hair tonics, shampoos, pomades, perfumes, etc. for the enhancement of feminine beauty. The author claims to be a graduate of the New York Institute of Physicians and Surgeons.

DeCOSTA, Benjamin Franklin, 1831-1904.
The White Cross, its origin and progress see **SHEPHERD** (#3173).

922. DeCOURCEY, R.
Man displayed. In four parts. 1st—Being a chemical analysis of the elements which are found to exist in the human frame—thus showing the natural origin of man. 2nd—Also the anatomical and physiological structure o[f] the human frame. 3rd—The powers of the organs of the brain, phrenologically considered. 4th—With the strength and power of expansion of the immortal mind . . . Hamilton [Ont.]: Published by Samuel Morrison, 1857.
313 p. ; 17 cm.

923. DE GROOT ELECTRIC COMPANY.
The Electro-Galvanic Regenerator. No experiments! Applied in a minute! Immediate strength! New York: The De Groot Electric Co., copyrighted 1889.

[1] trifold sheet : ill. ; 15 × 31 cm. (folded to 15 × 8 cm.)

The *Electro-Galvanic Regenerator* advertised in this brochure was a metallic belt "performing cures in the most obstinate cases of lost or failing power, nervous debility, wasting, atrophy, etc., of the parts." An order form (7.5 × 13.5 cm.) is included.

DELAFIELD, Francis see **HAMILTON** (#1542).

924. DE LANEY, B.
A few words on the rational treatment of spermatorrhoea, and its concomitant complaints, by means of Dr. De Laney's newly invented curative instrument. Thirtieth thousand. New-York: Printed for the author, [after 1853].
11, [1] p. ; 18 cm.

The cause of spermatorrhoea, according to the author, is an enlargement and relaxtion of the seminal ducts (due to masturbation or other abuses) that makes seminal retention difficult for many. "To produce a contraction of these seminal vessels, then, is to cure the disease itself." The solution proposed is *Dr. De Laney's Curative Instrument for Spermatorrhoea,* "a small, very simple instrument, to be worn by the patient externally, just above the seminal vessels, acting as a contractor and powerful invigorator" (p. 4.). The instrument may be had for $10.00 by writing to Dr. De Laney, who is also available for medical consultation "personally or by letter."

DE LAURENCE, Lauron William see **ALBERTUS MAGNUS** (#16).

DELAVAN, Edward Cornelius see **MAIR** (#2355).

925. DENISON, Charles, 1845-1909.
Exercise and food for pulmonary invalids . . . Denver: The Chain & Hardy Co., 1895.
71 p. : ill. ; 18 cm.

926. DENISON, Charles, 1845-1909.
Rocky Mountain health resorts. An analytical study of high altitudes in relation to the arrest of chronic pulmonary disease . . .

Boston: Houghton, Osgood and Company; The Riverside Press, Cambridge, 1880.

xii, 192, [18] p., [1] folding leaf of plates : map ; 21 cm.

First edition. Debate regarding the efficacy of "sedative" climates or "stimulating" climates, of warm, moist climates or cool, dry climates, of low altitudes or high altitudes in the treatment of pulmonary tuberculosis was a staple of medical literature the entire 19th century. Denison argues: "As the most favorable climatic qualities, the coolness, diathermancy, and dryness of the air, the amount of sunshine and atmospheric electricity, are increasingly found with increasing elevation and distance from the sea, the localization of the ideal climate we have been seeking is rendered easy in the Rocky Mountain plains and foot-hills, between the altitudes of four to eight thousand feet" (p. 150).

Denison justifies the preparation of this treatise for both a professional and popular audience on two grounds: "First, the growing demand of invalids themselves for a knowledge of the climatic conditions and the benefits which may be reliably anticipated from a change of climate; and, second, the desire of the profession for the results of scientific observation as to the effect of high altitudes upon pulmonary disease in its various stages of progress . . . It is no longer a secret among medical men, that the ordinary therapeutic remedies are of little avail, compared with change of climate, in arresting this disease; and, further, the great majority of the profession are not so located as to make it possible for them to study and personally observe remedial climatic influences . . . Moreover, obviously, in such an undertaking, facts and conclusions should be briefly and concisely stated, and, as far as possible, in terms easily comprehended by the lay reader" (p. vi-vii). Folding lithographic plate: "Climatic Map of the Eastern Slope of the Rocky Mountains"; folding "Chest examination chart" in pocket on front pastedown.

Denison was an 1869 medical graduate of the University of Vermont, and practiced in Hartford, Ct. until suffering a pulmonary hemorrhage. In 1873 he moved to Denver, where he became one of America's leading authorities on pulmonary tuberculosis. Denison wrote numerous articles on the disease, including a comprehensive study on his use of tuberculin in the treatment of 19 cases within six months of Koch's report of the new substance (1891).

927. DENNETT, Roger Herbert, b. 1876.
The healthy baby. The care and feeding of infants in sickness and in health . . . New York: The Macmillan Company, 1912.

xv, [1], 235, [9] p. ; 19 cm.

First edition. Dennett's *Healthy baby* appeared in three editions, the final issue of the third (or "new edition") was published in 1935. A 1902 graduate of Harvard University Medical School, Dennett was on the pediatric faculty of the New York Post-Graduate Medical School and the author of several books on infant care and feeding.

928. DENSMORE, Emmett, 1837-1911.
How nature cures. Comprising a new system of hygiene; also the natural food of man. A statement of the principal arguments against the use of bread, cereals, pulses, potatoes, and all other starch foods . . . London: Swan Sonnenschein & Co.; New York: Stillman & Co., [c1892].

x, [12], 413, [11] p. ; 21 cm.

Densmore prefaces this ill-organized and repetitive treatise on nutrition with the statement: "The doctrine that food and dietetic habits are the chief factors in health and disease is as old as Plato, and received a new and powerful impulse through the life and writings of Louis Cornaro more than three hundred years ago. The doctrine that the use of bread, cereals, pulses and vegetables is not only unwholesome, but is at the very foundation of nervous prostration and modern diseases is, however, sufficiently novel and startling" (*Pref.,* p. [v]).

Early in the treatise he asserts: "A fruit diet, as set forth in the following pages, means the solution of the problems of how to banish disease and intemperance from the race . . . and to give us a food which is at once in accord with our higher instincts and the demands of aesthetics" (p. x). Though advocating a diet that consists largely of nuts and fruits in their natural state, the author is not contrary to the incorporation of flesh and dairy products into the diet. He more or less reasons, "If nuts and sweet fruits be accepted as man's natural foods, the cooking of cereals, flesh, and vegetables common in civilization is but an effort to reduce these various foods to a natural condition" (p. 137). This reduction can be accomplished semi-successfully with meats and dairy products, but not with cereals and vegetables. One must read and read through a burdensome series of extracts from other authors as well as the author's own quasi-scientific ruminations to discover the reasons for his horror of cereals and vegetables. They are objectionable, apparently, because they are difficult to digest, diverting digestive and nervous energies; and because, loaded with potash and soda salts, they introduce calcareous or earthy matter in the body that clogs normal physiological operations, inducing malfunction or disease.

Densmore received his medical degree from New York University Medical College in 1885. He was involved with his brother James in the development of the typewriter manufactured by Remington, which he helped introduce to England. In addition to his business and manufacturing interests, Densmore found time to author *The natural food of man* (1890?); *Consumption in chronic diseases* (1899); and *Sex equality, a solution of the woman problem* (1907). *How nature cures* is dedicated to the second of Densmore's three wives, Helen, who in England and the United States was a "co-worker in the development of the system of hygiene and health which this work aims to unfold." A German translation was published in 1893.

929. DE PUY, Frank A.

The new century home book. A mentor for home life in all its phases; a chronicle of the progress of America and the world; a compendium of the nation's greatest city; and a guide for the great army of home-builders . . . New York: Eaton & Mains, 1900.
    400 p., [8] leaves of plates : ill. ; 19 cm.

"There are fifteen million homes in the United States. There ought not to be one unhappy home. Every one of these millions of of homes can be made happy and kept happy. Right living will do it" (p. 11). De Puy seemingly considers every factor that can possibly contribute to the happiness of the American home at the beginning of the 20th century, a century which promises to surpass even the 19th

in the development of technical marvels ("Wonderful as a dream of enchantment have been the achievements of the nineteenth century," p. 320). After reviewing the necessary behavioral adjuncts to domestic happiness (cheerfulness, considerateness, helpfulness, neatness, etc.), De Puy considers such additions to domestic felicity as "gymnasiums in the home," fresh air and sunlight, the furnishing & ventilation of nurseries, house planning & building, the domestic arts (sewing, crocheting, embroidery), gardening, pets, food & drink, home amusements & entertainment, nursing the sick, first aid, etc. *The new century home book* is a mine of middle-class American values and expectations at the turn of the 20th century.

DE PUY, William Harrison see FOWLER (#1270).

DERBOIS, A. F. see DUNNE (#1015).

930. DERBY, George, 1819-1874.

An inquiry into the influence upon health of anthracite coal when used as a fuel for warming dwelling-houses, with some remarks upon special evaporating apparatus . . . Boston: A. Williams & Company, 1868.
    46 p. ; 18 cm.

First edition. Presented to the Boston Society for Medical Improvement on 10 February 1868. 2nd edition has title: *An inquiry into the influence of anthracite fires upon health.*

931. DESLANDES, Leopold, 1797-1852.

A treatise on the diseases produced by onanism, masturbation, self-pollution, and other excesses . . . Translated from the French, with many additions. Second edition. Boston: Otis, Broaders, & Company, 1839.
    252 p. ; 16 cm.

Deslandes' treatise was originally published at Paris in 1835 as *De l'onanisme et des autres abus vénériens considérés dans leur rapports avec la santé.* The French edition was reprinted in Brussels (1837), and translated into German (1835) and Dutch (1836). *De l'onanisme* was even more successful in the United States, where an English translation was first published at Boston in 1838 under the present title. The title changed to

Manhood: the causes of its premature decline with the publication of the 1842 4th edition.

Deslandes maintains that the genitals influence the entire human system, pointing to the physical, moral and intellectual deficiences of those without them, such as eunuchs. If, Deslandes asks, the genitals have so much influence on human development and physiology in an unexcited state, what is their influence in an excited state, and more to the point, in a state of repeated and excessive excitement, as is the case of those in the habit of masturbation? The answer, of course, is negative, and the influence of such habits is especially pernicious during puberty, the period during which the mind and body undergo extensive development and change. "Here then we have a system of organs forcing their development forward at the expense of the other organs" (p. 34); "This vice . . . does not surely hasten or retard; it deranges: for the derangement of the functions is not generally manifested by irregularities in formation, aspect, and texture, but by material alterations, by diseases." He continues, "now as the susceptibility of the organs varies in individuals, and as in one, the heart, in another the lungs, the stomach, &c. is most liable to be affected, we see why the list of diseases caused by onanism, comprises most of those which afflict the body" (p. 36-37).

The *Treatise* is divided into two parts. In the first Deslandes considers such topics as the age of sexual maturity (i.e., the stage in life when sexual activity is neither precocious nor developmentally harmful); the effects of "venereal excesses" in all stages of life; and the organic and psychiatric pathology of venereal excess (with case histories from his own practice and the medical literature). Part two is devoted to "Rules of preservation and treatment relative to venereal excesses." In addition to monitoring the behavior of children, preventives to chronic masturbation include bloodletting or "the application of leeches or cups around the sexual parts"; cold applications to the genitals, cold hip-baths; and cold injections into the vagina. Female castration is described and considered: "Might not this removal blunt, if it did not deaden, the propensity to solitary enjoyments, and render the other remedies employed more efficient? Although we have but little confidence in this operation, yet, when

we consider that a superficial cauterization of the nymphae and clitoris has cured nymphomania . . . we can conceive that the excision of the internal labia may in some cases present a chance of success" (p. 188). Ovariotomy ("to appease excessive uterine ardor") and male castration are also briefly discussed as last resorts. Finally Deslandes reviews the effects of diet, weather, alcohol consumption, etc. on sexual behavior, and counter-measures to their possible stimulating effects.

932. DESLANDES, Leopold, 1797-1852.
Manhood; the causes of its premature decline, with directions for its perfect restoration; addressed to those suffering from the destructive effects of excessive indulgence, solitary habits, &c. &c. &c. . . . Translated from the French, with many additions, by an American physician. Seventh thousand. Boston: Otis, Broaders, and Company, 1843.
[2], 252 p. ; 16 cm.

The first three American editions of this work were published under title: *A treatise on the diseases produced by onanism, masturbation, self-pollution, and other excesses* (2nd. ed., #931).

933. DESLANDES, Leopold, 1797-1852.
Manhood: the causes of its premature decline, with directions for its perfect restoration; addressed to those suffering from the destructive effects of excessive indulgence, solitary habits, &c. &c. &c. . . . Translated from the French, with many additions, by an American physician. Twelvth [sic] thousand. Boston: Otis, Broaders, and Company, 1843.
[2], 252 p. ; 16 cm.

934. DESLANDES, Leopold, 1797-1852.
Manhood; the causes of its premature decline, with directions for its perfect restoration; addressed to those suffering from the destructive effects of excessive indulgence, solitary habits, &c. &c. &c. . . . Translated from the French, with many additions, by an American physician. Fortieth thousand. Albany [N.Y.]: Broaders & Co., 1850.
252 p. ; 16 cm.

935. DESSAR, Leonard A.
Home treatment for catarrhs and colds. A handy guide for the prevention, care and treatment of catarrhal troubles, cold in the head, sore throat, hay fever, hoarseness, ear affections, etc. Adapted for use in the household, and for vocalists, clergymen, lawyers, actors, lecturers, etc. . . . New York: Home Series Publishing Co.; London: Bailliere, Tindall & Cox, 1894.
[2], vi, [3]-118 p. : ill. ; 19 cm.

Dessar was a New York laryngologist and an 1884 graduate of the Medical College of Indiana (Indianapolis).

936. DE VEAUX, Samuel, 1789-1852.
The travellers' own book, to Saratoga Springs, Niagara Falls and Canada, containing routes, distances, conveyances, expenses, use of mineral waters, baths, description of scenery, etc. A complete guide, for the valetudinarian and for the tourist, seeking for pleasure and amusement . . . Buffalo [N.Y.]: Faxon & Read, 1841.
vii, [3], [13]-135, [1], [139]-173, [1], [177]-258 p., [4] leaves of plates : ill., map ; 15 cm.

First edition. The first ninety-four pages are devoted to a description of Saratoga Springs, its waters and their medicinal use. De Veaux also makes a case for the "superior medical virtues" of the "quality of the atmosphere" in Niagara Falls, where he resided. *The travellers' own book* appeared in five editions between 1841 and 1847. Cover-title: *Niagara Falls and Saratoga. 1841.* Atwater copy lacking folding map.
De Veaux grew up in New York City where his father was a merchant. In 1803 he moved to Canandaigua in central New York, at the time one of the largest settlements in the "west." He simultaneously worked for agents of the Holland Land Company and studied law. In 1807 De Veaux moved further west to the Niagara Frontier, and was appointed clerk to the warehouse at Fort Niagara on Lake Ontario. It is uncertain what his role was during the War of 1812. Even as fighting raged along the Niagara River between Lakes Erie and Ontario, De Veaux courted and married a young woman living on the British side of the river. De Veaux and his bride quickly withdrew to the safety of Leroy, N.Y., eighty miles to the east. With the cessation of hostilities, he returned to the Niagara Frontier, and in 1817 became one of the founding citizens of Manchester (later Niagara Falls, N.Y.). De Veaux resumed business, operating "the first, and for years the only, trading post between Rochester and Detroit" (Donald E. Loker, *A history of DeVeaux School, 1853-1953,* Niagara Falls, N.Y.: Fose Printing Co., 1963, p. 8). Trading, real estate, and banking made De Veaux one of the wealthiest men along the Niagara Frontier. He was appointed a justice in 1821 and held various town posts thereafter. In 1830 he was elected to the State Legislature. Given his pecuniary interest in the region, it is no wonder that De Veaux would attempt in *The travellers' own book* to promote the salubrity of Niagara Falls, an area which was already attracting tourists.

937. D'EWES, B. Elizabeth.
What every woman should know. In trouble certainly; and in health the more, lest the penalty of ignorance and neglect be required of her . . . Philadelphia: Published by 20th Cent. Co., [ca. 1900].
16 p. : ill. ; 15 cm.

Brochure promoting vaginal hygiene with the "20th Century Irrigator." Title on printed wrapper: *Ladies' private hand-book.*

938. DEWEY, Chester, 1784-1867.
An appeal to the friends of temperance, delivered in Pittsfield, Mass. on Sabbath evening, October 7, 1832 . . . Pittsfield: Printed by Phineas Allen and Son, [1832].
24 p. ; 23 cm

The first half of Dewey's address argues against the manufacture and consumption of alcoholic beverages. In the latter half of his address, Dewey states that "the *Cholera* is generated by Intemperance, and feeds on the intemperate and dissolute of like habits" (p. 13). Given cholera's appearance in the United States in the summer of 1832, this must have been an alarming statement to some of his audience, and falsely assuring to others.

A Congregational minister, educator and scientist, Dewey made significant contributions to the fields of geology, mineralogy, zoology and botany. He taught mathematics and the natural sciences at his *alma mater* Williams College from 1810 to 1827. At the time this lecture was delivered, he was principal of the Berkshire Gymnasium in Pittsfield, Ma. (1827-1836). In 1836 Dewey became principal of the Collegiate Institute in Rochester, N.Y.; and in 1851 was appointed professor of chemistry at the newly founded University of Rochester, where he remained until his retirement (1861). Dewey was active in the anti-slavery and temperance movements.

939. DEWEY, Edward Hooker, 1837?-1904.

A new era for woman. Health without drugs. A plain pathway to the "kingdom of health," "without money and without price." The largest possibilities of reaching the natural limit of life assured . . . Introduction by Alice McClellan Birney. Norwich, Conn.: Chas. C. Haskell & Son; London. England: L.N. Fowler & Co., 1903.

xv, [1], 371 p. ; 18 cm.

Dewey was "a regularly trained physician (University of Michigan) who began practice as a military doctor during Sherman's Georgia campaign. His year and a half experience in that capacity taught him that except for relieving pain, there was nothing the physician could do for the patient by the administration of drugs. The lesson was repeated over and over in the civilian practice he began in Meadville, Pennsylvania in 1866, yet he continued to dispense standard remedies, in response to patient demand for medication, rather than hand recovery over entirely to nature. Not until after more than a decade of half-hearted dosing did he find the courage to break with convention" (James C. Whorton, *Crusaders for fitness*, Princeton, NJ, 1982, p. 263). The crux of Dewey's system was to limit the intake of food; specifically, to eliminate breakfast from the daily regimen. In *A new era for woman*, originally published in 1896, Dewey elaborates this theme for a female readership.

940. DEWEY, Edward Hooker, 1837?-1904.

The no-breakfast plan and the fasting-cure . . . Second edition. Passaic, New Jersey:

The Health Culture Co.; London, England: L.N. Fowler & Co., [c1900].

[3]-207 p., [8] leaves of plates : ports. ; 19 cm.

The center of Dewey's new therapeutics was fasting. "If nature conquered disease by slowly using up superfluous matter, Dewey reasoned, disease was probably the gradually developing result of the consumption of excess food. And what common meal was unncessary food taken, if not breakfast! . . . His 'revolutuonary' analysis truly was that uncomplicated, but he dressed his simplicities in a fustian style that allowed them to pass as elegant insights before the eyes of the uncritical . . . Distilled, his philosphy was merely that food in excess of body demand placed a drain on vital energy that weakened all tissues, and especially the blood . . . The benefits of the No Breakfast Plan . . . were typical of ideological hygiene, with nearly unlimited physical improvement transcended by mental and spiritual perfection" (James C. Whorton, *Crusaders for fitness,* Princeton, NJ, 1982, p. 264).

941. DEWEY, Edward Hooker, 1837?-1904.

The true science of living. The new gospel of health. Practical and physiological. Story of an evolution of natural law in the cure of disease. For physicians and laymen. How the sick get well; how the well get sick. Alcoholics freshly considered . . . Introduction by Rev. George F. Pentecost, D.D. Norwich, Conn.: Chas. C. Haskell & Son; London: Gay & Bird, 1899.

323, [5] p. ; 24 cm.

In his introduction Pentecost writes, "Of the two principal matters recommended as the practical outcome of the theory of health developed in this book, the first is fasting, or the abstinence from food until natural hunger calls for it, is the best way to bring about recovery from disease . . . The second is that digestion is best promoted and food so assimilated as to afford the largest amount of nourishment and the greatest quantity of rich blood, by giving the stomach a long rest from the evening till the noon of the next day" (p. 5). The Rev. Mr. Pentecost points out that "rich blood goes not only to preserve the health of the body and prolong life, but also that it makes for righteousness, both in clearing the mind and purifying the body of those humors which make fuel for unholy desires and unrighteous dispositions" (p. 4).

942. DEWEY, Edward Hooker, 1837?-1904.

The true science of living. The new gospel of health. Practical and physiological. Story of an evolution of natural law in the cure of disease for physicians and laymen.How the sick get well; how the well get sick. Alcoholics freshly considered . . . Introduction by George F. Pentecost, D.D. Norwich, Conn.: Charles C. Haskell & Co.; London: L.N. Fowler & Co., 1904.

   323, [5] p. ; 24 cm.

943. DIAMOND, J. Harry.

Lectures delivered before the classes of the Universal School of Mental Science, covering special lessons on hypnotism, Mesmerism, psychology, magnetic healing, suggestive therapeutics, telepathy, clairvoyance, and psychometrical healing, with several of the other and allied sciences and rules for their cultivation, and directions showing how to make use thereof . . . [Rochester, N.Y.], 1900.

   [1], 16, 5, 20, 4, 5, 6 leaves : ill. ; 36 × 20 cm.

These lectures comprise a correspondence course, beginning with an introductory lecture on hypnotism. Diamond distinguishes two principal modes of healing within the scope of "mental science." *Magnetic healing* "is performed by imparting to your patient as much of your personal magnetism or electrical power, as the condition or the nature of the disease may require." Instructions are given on how to accumulate the "nervo-electrical power" that effects healing. *Suggestive therapeutics*, which may be performed with or without inducing a hypnotic state in the patient, removes disease by "radically changing the . . . normal line of thought from bad to good, and restoring natural, healthy thoughts and conditions." Diamond maintains that this alteration from negative to positive (healing) thought is the basis of Christian Science and other systems of mental, metaphysical or faith healing.

944. DICKINS, Job T.

THOMSONIAN BOTANIC MEDICINES, Prepared and Sold by JOB T. DICKINS, M.D. . . . Newburyport, Mass. . . .

   Broadside ; 39.5 × 27.5 cm.

In this broadside (dated May 1848) Dickins advertises more than seventy botanic medicines of his own preparation, as well as books by botanic or eclectic authors, e.g., Morris Mattson (#2394-2397), Alva Curtis (#852-857), Alfred Worthy (#3883), etc. In his concluding paragraph, Dickins notes that during the previous nine months he has treated "between five and six hundred patients in almost every form of disease incident to the climate, with complete success,"employing "the productions of the vegetable kingdom, together with hydropathy; and a strict adherence . . . to Thomsonian principles." Dickins was a graduate of the New England Botanico-Medical College. He is listed as an eclectic physician in the *New England business directory* during the 1860s and 1870s. Printer's imprint at bottom of recto: Huse & Bragdon, Newburyport Advertiser Press.

DICKINSON, Robert Latou see **KNOWLTON** (#2154).

945. DICKSON, Samuel, 1802-1869.

The principles of the chrono-thermal system of medicine, with the fallacies of the Faculty, in a series of lectures . . . Containing also, an introduction and notes by William Turner . . . Second American, from the fifth London edition. New York: H. Long & Brother, 1848.

   xv, [16]-224 p., [1] leaf of plates : port. ; 23 cm.

The first American edition was published in 1845 (1st London ed., 1839). "Fifteen years ago it was my fate . . . to make two most important discoveries in medicine,—namely, the Periodicity of Movement of every organ and atom of all living bodies—and the Intermittency and Unity of all diseases, however named, and by whatever produced. To these I added a third—the Unity of Action of Cause and Cure,—both of which involve Change of Temperature. Such is the groundwork of the Chrono-Thermal System" (p. v). Chrono-thermalism is extensively reviewed in Alva Curtis' 1855 *A fair examination and criticism of all the medical systems in vogue* (#853, p. 153-166). Lithographic frontispiece portrait of Wm. Turner. Atwater copy bound with W. Turner's *Triumphs of 'young physic'* (New York, 1847?).

946. DICKSON, Samuel, 1802-1869.
The principles of the chrono-thermal system of medicine, with the fallacies of the Faculty, in a series of lectures . . . Containing also, an introduction and notes by William Turner . . . Sixteenth edition. New York: J.S. Redfield; S. French, 1854.
    xv, [16]-224 p. ; 21 cm.

947. DICKSON, Samuel Henry, 1798-1872.
Essays on life, sleep, pain, etc. . . . Philadelphia: Blanchard and Lea, 1852.
    301 p. ; 20 cm.

    A literate philosophical consideration of the natural basis of life and vitality. After reviewing the opinions of such authors as Carpenter, Cuvier, Matteucci, Mayo, Richerand, etc., Dickson comes to the conclusion that "life is the consequence of organization," noting, however, that "the special organization adopted is altogether the creation of the principle of vitality, without whose action it could not exist, nor, grow, nor find development" (p. 45). In this spirit, Dickson reflects on the nature of sleep, pain, "intellection," hygiene and death.
    An 1819 medical graduate of the Univ. of Pennsylvania, Dickson was among the founders of the Medical College of South Carolina (1833), and served on the medical faculty of the University of the City of New York (1847-50) and the Jefferson Medical College (1858-72). He was the author of several well-received medical and pathological textbooks, as well as an accurate description of dengue (1839), a disease often confused with yellow fever.

DIDAMA, Henry Darwin see **JOHONNOT**
    (#2038).

948. DIMMOCK, C. E.
DR. C.C. HENRY'S MAGIC NERVE OIL. The great pain killing liniment of the age for the instantaneous relief of pain! It is a permanent cure for rheumatism and neuralgia; a speedy relief for headache, toothache, earache, chilblains, bruises, sprains, flesh cuts, sores, ague in the face, burns, scalds, pain in the side, back or stomach, sore eyes, sore throat, bee stings, dysentery and all summer complaints. C.E. DIM-

MOCK, manufacturer & proprietor, Limington, Me. . . . [186-?].
    Broadside ; 26.5 × 20 cm.

949. DINSMORE, Thomas Hughes, 1855-
First lessons in physiology and hygiene. With scientific instruction concerning the physiological effects of alcoholic stimulants and narcotics on the human body. A text book for the common schools . . . New York; Cincinnati; Chicago: American Book Company, [c1885].
    ix, [10]-163, [5] p., [1] leaf of plates : ill. ; 19 cm.

    The author is described on the title-page as "Professor of physics and chemistry in the State Normal School, Emporia, Kansas."

950. DIRECTIONS FOR DIET.
Directions for diet. In the treatment of the sick according to Hahnemann's method. [New York: Sold by Wm. Radde, ca. 1850].
    4 p. ; 21 cm.

    Caption title.

951. DISEASES OF THE MALE URETHRA.
Diseases of the male urethra, and their treatment by means of the Antrophor . . . New York: The Antrophor, [c1887].
    32 p. : ill. ; 18 cm.

    The *Antrophor* was a nickel-plated tube through which medication could be introduced into the male urethra for the treatment of gonorrhea and other disorders. The pamphlet begins with a description of gonorrhea's symptoms and prognosis. In printed wrapper.

952. DISMUKE, Edward E.
Dismuke's book of formulas and prescriptions. Collected from the most eminent doctors of the United States, for home treatment . . . Waco, Texas: Texas Central Publishing House, [189?].
    28 p. ; 23 cm.

    Title and imprint from wrapper. Dismuke provides sixty-nine formulas for medications that can be prepared in the home. His "Relief for asthma," for example, consists of one ounce each of powdered lobelia, stramonium leaves, nitrate of potash and black tea. "Fluttering or palpitation of the heart" can be controlled by a syrup containing *tincture digitalis* and *elixir Valerian*, etc. Dismuke adds: "Realizing that nearly every family into whose hands this book may fall, has either a horse or a cow, we would consider the work incomplete without a veterinary department" (p. 5). Formulae nos. 48-69 address such disorders as "bloody milk in cows," cough pills for horses, chicken cholera, etc. Dismuke published several similar formularies during the 1890s.

953. DIX, Tandy L., 1829-1902.
The healthy infant, a treatise on the healthy procreation of the human race, embracing the obligations to offspring; the management of the pregnant female; the management of the newly born; the management of the infant; and the infant in sickness . . . Cincinnati: Peter G. Thomson, publisher, 1880.
[2], iv, 134, 6, [4] p. ; 19 cm.

Although the greater part of the book is devoted to fetal development, the management of pregnancy and infant care, Dix is emphatic in his opening chapter about the "duties we owe posterity." "It is manifest," he writes, "that equally with the physical conformation of the parents, their progeny also inherits their moral bias, mental deficiencies, and bodily ailments . . . Thus being held, in some degree, responsible for the impurities of the stream of human life, it then becomes our bounden duty to learn as much as we can of the history of our ancestry, that we may cultivate our inherited virtues and avoid our inherited vices" (p. 7). We must connect to this knowledge adherence to the "natural laws of health," for when "these laws are interfered with and encroached upon, and even positively violated by late hours, frequent parties, abnormal indulgence of the appetite, the midnight dance, and the opera—the human system is radically undermined and rendered utterly unfit for increasing and multiplying its kind . . . The result to the offspring is a patrimony of debility" (p. 8). Dix was an 1853 graduate of the Kentucky School of Medicine, Louisville.

954. DIXON, Edward H., 1808-1880.
A treatise on diseases of the sexual organs: adapted to popular and professional read-

ing, and the exposition of quackery, professional and otherwise . . . New-York: Burgess, Stringer & Co., 1845.
xii, 260, [4] p. : ill. ; 19 cm.

First edition. A pupil of Valentine Mott, Dixon entered medical practice at New York in 1830. He is described in the *Dict. Amer. med. biog.* (1928) as "a man of ability, especially as a genito-urinary surgeon, a line in which he was a pioneer in New York, [who] abandoned scientific medicine to write books and articles for the laity, replete with good sense, but squinting ever in the direction of attracting patients for office and private hospital . . . Though often dimmed by verbosity and extravagance, Dixon's writings constitute a pitiless expose of the follies of the traditional useless and often harmful polypharmacy preceding the advent of rational medicine. Like Florence Nightingale, he found the one ray of hope in hygiene and the 'vis mediatrix naturae,' 'air, warmth, light, food, water, exercise, sleep'" (p. 329).
    Dixon's treatise is divided into two parts, the first on syphilis and its treatment; and the second on gonorrhea and diseases of the male and female genitalia not related to sexually transmitted disease. The work concludes with descriptions and illustrations of several instruments devised by Dixon for the surgical treatment of genito-urinary disorders.

955. DIXON, Edward H., 1808-1880.
A treatise on diseases of the sexual system: adapted to popular and professional reading, and the exposition of quackery . . . Seventh edition. New York: Published for the author, and for sale by Charles H. Ring, 1848.
2, xii, 260, [4] p. : ill. ; 19 cm.

956. DIXON, Edward H., 1808-1880.
A treatise on diseases of the sexual system, adapted to popular instruction . . . Ninth edition, with an appendix. New York: O.A. Roorback, Jr., 1855.
[iii]-xii, 285, [3] p. : ill. ; 19 cm.

957. DIXON, Edward H., 1808-1880.
A treatise on the diseases of the sexual system, adapted to popular instruction . . . Tenth edition, with an appendix. New-York: Robert M. De Witt, publisher, [c1867].
xii, 312 p. : ill. ; 19 cm.

958. DIXON, Edward H., 1808-1880.
Woman, and her diseases, from the cradle to the grave: adapted exclusively to her instruction in the physiology of her system, and all the diseases of her critical periods . . . Fifth edition. New York: Published for the author, and for sale, by Charles H. Ring, 1847.
    2, xii, [5]-309, [3] p. ; 19 cm.

In his introductory chapter Dixon writes, "It is well known to the physician, that very little reliance can be placed upon the best domestic aid our country affords; and that the prescription of the nurse or the patient . . . frequently accompanies, or precedes his own. Would this be so, if the patient possessed a guide for her conduct, in the perilous situations in which she is often placed?" (p. 7). He continues, "If it be conceded, then, that woman should be instructed in the laws of her own existence, a moment's reference to the prejudices of society, may well make us falter on the very threshold of our undertaking. Taught from her earliest existence to look upon man as her sole protector, and to fear the slightest frown of society as the greatest misfortune; she is obliged to forego a thousand methods of strengthening both her mind and body, because they come under the interdictions of a conventionalism as absurd as it is mischevous" (p. 8). Remedying this ignorance is the object of the twenty-two chapters of Dixon's treatise, which discuss such topics as menstruation and its disorders, hysteria, diseases of the reproductive organs, breast cancer, the ills attending pregnancy, etc. *Woman, and her diseases* was first published at New York in 1846. Its popularity and influence are testified by the publication of ten editions over a period of twenty years.

959. DIXON, Edward H., 1808-1880.
Woman and her diseases, from the cradle to the grave, adapted exclusively to her instruction in the physiology of her system, and all the diseases of her critical periods . . . Eighth edition. New-York: Adriance, Sherman & Co., 1851.
    [8], [iii]-xii, [5]-309, [7] p. ; 19 cm.

960. DIXON, Edward H., 1808-1880.
Woman and her diseases, from the cradle to the grave: adapted exclusively to her instruction in the physiology of her system, and all the diseases of her critical periods . . . Tenth edition. New York: Published by A. Ranney, 1855.
    [2], xii, [5]-317, [3] p. ; 19 cm.

"This book was not written for the medical profession, nor is it designed to make woman her own physician, but simply to inform her of those general laws governing the female system, which every woman ought to be acquainted with" (p. [v]).

961. DIXON, Edward H., 1808-1880.
Woman and her diseases, from the cradle to the grave: adapted exclusively to her instruction in the physiology of her system, and all the diseases of her critical periods . . . Tenth edition. Philadelphia: G.G. Evans, 1859.
    xii, [5]-318, [4] p. ; 18 cm.

962. DIXON, Edward H., 1808-1880.
Woman and her diseases, from the cradle to the grave: adapted exclusively to her instruction in the physiology of her system, and all the diseases of her critical periods . . . Tenth edition. Philadelphia: John E. Potter and Company, [c1866].
    xii, [5]-318, [6], 4 p. ; 19 cm.

DIXON, Edward H. see also *THE SCALPEL* (#3110).

963. DIXON, H.
DR. DIXON'S MEDICAL BULLETIN. Issued monthly, at Clyde, Ohio. [Clyde, Ohio: H. Dixon, ca. 1860-75].
    [1] leaf printed on both sides ; 23.5 × 15 cm.

In all likelihood, this is the only issue of this publication. Dixon advertises his ability to cure chronic diseases, and solicits patient consultations through the mail. He also claims to be the "author of several popular medical works," including the *Book of nature,* treating of "PRIVATE MATTERS in which all men and women are deeply interested" (verso, col. 2). No other published works by H. Dixon have been identified.

DOANE, F. Harrison see **DUNNE** (#1015).

DOCTOR FRANK see **PERRY** (#2779, 2780).

964. DODDS, Susanna Way.
The cottage cook book or, health in the household . . . Springfield, Mass. . . . [etc.]: W.C. King & Co., publishers, 1885.
    xi, [1], 602 p. ; 21 cm.

The cottage cook book is a volume on "hygienic cooking" in the tradition of R.T. Trall (#3574, 3580, 3581) and Lucretia Jackson (#1960). In Part I, Dodds discusses the principles of the hygienic dietetics and food preparation. Part II is a collection of recipes on strict hygienic principles; while Part III provides recipes for "the benefit of those who are beginning to try hygienic cookery, but whose families and friends are not thoroughly converted to the system" (p. viii).
    Dodds received her medical degree from the New York Hygeio-Therapeutic College in 1864. She practiced medicine in St. Louis from 1886 to 1909.

DODDS, Susanna Way see also *VITALITY* (#3657).

965. DODGE, Charles Wright, 1863-1934.
[Letter] 1901 December 9, Rochester, N.Y., [to] Marion Craig Potter, Rochester, N.Y.
    [1] leaf ; 27.5 cm.

Typescript (on University of Rochester, Dept. of Biology letterhead). Dodge asks Potter to lecture on the physiology of reproduction to female students in his biology class. He writes: "As the present class is the first in which girls have been regular students I have never before had to take them into consideration when arranging for the lecture. I do not wish to give up the custom nor do I think that the lecture should be given before the entire class, so I come to you to see if you have the time and the inclination to help us in this emergency."
    Dodge came to the University of Rochester in 1890 to organize its biology department. "Trained at the University of Michigan, he came to Rochester on the explicit understanding that no restrictions would be imposed upon teaching the theory of organic evolution. Spoken of as the pioneer scientific experimentalist at the University, Dodge brought with him the only

microscope in the institution and introduced dissection of animals in his teaching . . . Active in the public health movement, Dodge succeeded in producing an antitoxin serum which overcame in 1893 an epidemic of diphtheria in Rochester" (Arthur J. May. *A history of the Univesity of Rochester, 1850-1962.* Rochester, N.Y., 1977, p. 102-03). Marion Craig Potter (1863-1943) received her medical degree in 1884 from the University of Michigan. She practiced medicine with her father for two years following her graduation, and after his death set up her own practice in Rochester, where she specialized in obstetrics, gynecology and pediatrics. Potter was active in the Blackwell Medical Society, one of the first women's medical organizations in the United States, the Women's Medical Society of the State of New York, and the Medical Women's National Association.

966. DODS, John Bovee, 1795-1872.
The philosophy of electrical psychology in a course of twelve lectures . . . Stereotype edition. New York: Fowlers & Wells, 1853.
    252 p., [1] leaf of plates : port. ; 19 cm.

    As a Universalist minister in Provincetown, Mass., Dods "preached gradual spiritual growth and universal salvation, rejecting older notions of sudden conversion, eternal punishment, and the divinity of Christ. He also harbored a growing interest in the phenomena and philosophy of animal magnetism, or mesmerism, which became an American fad during the 1830s and 1840s . . . Dods left the ministry in 1842. His investigation of human spirituality and regeneration led him deeper into mesmermism, according to which an invisible but universally pervasive and empirically demonstrable magnetic fluid operates on the body, mind, and soul, linking them to the higher spiritual powers of the universe and providing a mechanism for physical and emotional healing" (*Amer. nat. biog.*, 6:697). Dods lectured on the science of "electrical pyschology" to audiences throughout the northeastern United States, and published his two principal works: *Six lectures on the philosophy of mesmerism* (1843) and *The philosophy of electrical psychology* (1st ed., 1850).
    In his history of hypnostism, Alan Gauld explains Dods' "electrical psychology" as it relates to therapeutics. In Dods' system, electricity "is the connecting link between mind and

matter, and the agent that mind employs to control the neuro-muscular apparatus. There are two sorts of electricity, positive and negative. Arterial blood contains positive electricity, which it loses in the capillaries, whence it is drawn through the nerves to the brain. Venous blood contains negative electricity . . . When the electricity of the system is thrown out of balance, disease results. The electricity can be thrown out of balance by mental impressions or by physical impressions. Mental impressions can likewise correct the imbalance and cure the disease. The mind will move the electricity to whatever part of the body is thought of or attended to, thereby correcting the imbalance. Curative mental impressions can be self-generated or can be impressed from the outside by a magnetic operator" (*A history of hypnotism*. [Cambridge; New York]: Cambridge Univ. Press, 1992, p. 186).

During the 1850s Dods became increasingly interest in spiritualism. He maintained that spirits used magnetic fluid to contact humanity through mediums. The healing possibilities of this force remained an essential part of his teaching. Frontispiece portrait of the author. Spine-title: *Electrical psychology*.

967. DOHERTY, William Brown, 1847-
You and your doctor. How to prolong life. A practical book on health and the care of it. A fearless expose of all quacks and frauds within and without the medical profession . . . Chicago: Laird & Lee, publishers, [c1900].
260 p., [1] leaf of plates : ill. ; 20 cm.

968. DOLIBER-GOODALE COMPANY.
Advice to mothers on the care and feeding of infants. With useful information for nursing mothers and invalids . . . Twenty-first edition. Boston: Doliber-Goodale Company, 1896.
72 p. ; 17 cm.

Promotional literature for Mellin's Food, "a soluble, dry extract from wheat and malted barley" dissolved in cow's milk and hot water for the nourishment of infants and invalids.

969. DOMESTIC CYCLOPAEDIA OF PRACTICAL INFORMATION.

A Domestic cyclopaedia of practical information. Principally written or revised by the following authorities: Calvert Vaux, architect of the Central Park, and Thomas Wisedell, architect: Locating, building and repairing. Lewis Leeds, sanitary engineer: Warming and ventilation. Col. George E. Waring, of Ogden Farm: Drainage, the garden, the dairy. George Fletcher Babb, architect: Decoration as applied to walls, floors, and furniture. Mrs. Elizabeth S. Miller, author of "In the kitchen," and Giuseppe Rudmani, late cook in the Cooking School, St. Mark's Place, New York, and chef de cuisine, Newport: Cooking and domestic management. Austin Flint, Jr., M.D., professor in Bellevue Medical College: Dietetics and alcoholic beverages. Abraham Jacobi, M.D. professor in the College of Physicians and Surgeons: Diseases and hygiene of children. William T. Lusk, M.D., professor in Bellevue Medical College . . . : General medicine. S.G. Perry, D.D.S.: The teeth. Elwyn Waller, Ph.D., superintendent of laboratory in the Columbia College School of Mines . . . : Domestic chemistry—disinfecting, cleaning, dyeing, etc. Leslie Pell-Clark, veterinary surgeon of Ogden Farm: The horse. Johnson T. Platt, professor in the Yale Law School: Business forms and legal rules. Edited by Todd S. Goodholme . . . New York: Henry Holt and Company, 1878.
iv, 652 p. : ill. ; 26 cm.

Cover- and spine-title: *Goodholme's domestic cyclopaedia.*

970. DOMESTIC MEDICAL PRACTICE.
Domestic medical practice. A household adviser in the treatment of diseases, arranged for family use. Edited by Frank E. Miller . . . [et al.]. Edition 1914. Chicago; New York; London: Domestic Medical Society, [c1913].
xxx, [2], 18, [2], 19-24, [2], 25-122, [2], 123-1463, [3] p., [3] paper manikins, [31] leaves of plates : ill. ; 26 cm.

Most of the twenty-six contributors to this volume were physicians with New York hospital affiliations. The volume is divided into 43 topical "departments," each written by a single author, e.g., Alonzo Benjamin Palmer on "Acute infectious diseases"; L. Emmet Holt (#1743-1751) on "Diseases of the digestive organs of children"; Daniel B. St. John Roosa (#3021-3022) on "The ear and its diseases," etc. A "30th revised edition" of the *Domestic medical practice* was published in 1938.

971. DONNE, Alfred, 1801-1878.
Mothers and infants, nurses and nursing. Translation from the French of a treatise on nursing, weaning, and the general treatment of young children . . . Boston: Phillips, Sampson and Company, 1859.
303 p., VIII leaves of plates : ill. ; 20 cm.

First American edition. Originally published at Paris in 1842 under the title *Conseils aux mères sur la manière d'élever les enfans nouveau-nés.* The final American edition of Donne's treatise was issued in 1860 (#972). Its popularity in France was more enduring; a ninth edition of the *Conseils* was published in 1905.

972. DONNE, Alfred, 1801-1878.
Mothers and infants, nurses and nursing. Translation from the French of a treatise on nursing, weaning, and the general treatment of young children . . . Third edition. Boston: Brown, Taggard & Chase, 1860.
303 p., VIII leaves of plates : ill. ; 20 cm.

The third and final American edition.

973. DOOLITTLE, M. E.
The home physician, a practical medical book designed and prepared especially for family use . . . Household edition, including a brief but complete section on diseases of women and surgery. New Haven, Conn.: Press of the E.B. Sheldon Co., [c1894].
[4], iv, [5]-153 p. ; 17 cm.

974. DOTY, Alvah Hunt, b. 1854.
A manual of instruction in the principles of prompt aid to the injured. Including a chapter on hygiene and the drill regulations for the Hospital Corps, U.S.A. Designed for military and civil use . . . Second edition, revised and enlarged. New York; London: D. Appleton and Company, 1894.

xvi, 304 p., [2] leaves of plates : ill. ; 18 cm.

Originally published at New York in 1889.

975. DOUGLAS, James S., 1801-1878.
Practical homoeopathy, for the people, adapted to the comprehension of the non-professional, and for reference by the young practitioner, including a number of most valuable new remedies and improvements in the treatment of numerous diseases not in general use. Second edition . . . Milwaukee, 1862.
xviii, [2], [23]-85, [1] p. ; 17 cm.

"I do not aim to make accomplished physicians of the public, nor expect that *every* case of disease can be safely treated domestically. But I know, from numerous examples of domestic practice, that a long catalogue of acute and serious diseases, as well as lighter disorders, will be treated much more safely and successfully by families, by following the directions here given, than they are treated by the best drugging physicians" (*Pref.*, p. v) The first edition of *Practical homoeopathy* was published at Milwaukee in 1860; the 15th and apparently final edition was issued at Milwaukee in 1894 (#979).

976. DOUGLAS, James S., 1801-1878.
Practical homoeopathy, for the people, adapted to the comprehension of the non-professional, and for reference by the young practitioner, including a number of most valuable new remedies and improvements in the treatment of numerous diseases not in general use. Sixth edition . . . Milwaukee, 1862.
xviii, [19]-85 p. ; 17 cm.

977. DOUGLAS, James S., 1801-1878.
Practical homoeopathy for the people: adapted to the comprehension of the non-professional, and for reference by the young practitioner, including a number of most valuable new remedies. Tenth edition . . . Milwaukee, Wis.: Douglas & Perrine; Chicago: Halsey Bro.'s, [between 1868 & 1873].
xviii, [19]-132 p. ; 17 cm.

978. DOUGLAS, James S., 1801-1878.
Practical homoeopathy for the people: adapted to the comprehension of the non-professional, and for reference by the young practitioner, including a number of most valuable new remedies. Thirteenth edition . . . Milwaukee, Wis.: Douglas & Sherman; Chicago: Halsey Bros., 1875.
xxiii, [24]-138, [2] p. ; 17 cm.

979. DOUGLAS, James S., 1801-1878.
Practical homoeopathy for the people: adapted to the comprehension of the non-professional, and for reference by the young practitioner. Including a number of most valuable new remedies. Fifteenth edition . . . Milwaukee, Wisconsin: Lewis Sherman, 1894.
xxi, [22]-145 p. ; 17 cm.

980. DOWNING, Beal P.
Doctor Downing's reformed practice and family physician . . . Utica [N.Y.]: Roberts & Sherman, 1851.
iv, [5]-353 p. ; 19 cm.

"When I began to write this work I was a practicing physician, and had no book to look in if I was at a loss. I was compelled to trust entirely to my recollection, and when I thought of a thing I wrote it down for a reference, to save the trouble of thinking it over often, but had no intention of laying it before the public. This is the reason the work is mixed up in the manner it now is. I am now nearly seventy years of age . . . Many that have been acquainted with my practice have advised me to publish it" (*Pref.,* p. iv). Downing was of the opinion that "the Vegetable Kingdom is sufficient to cure all the maladies as fully as to cure hunger . . . I am of the opinion that there is a plant created for every ache that man is subjected to, with the exception of wear and tear of time" (p. [iii]). In spite of the title, the author's botanic practice appears to be based on decades of personal experience and experiment, rather than training in the Thomsonism or "reformed" practice.

Though divided into three parts, there is no discernible order to Downing's text. The work is redundant; and descriptions of medicinal plants are indiscriminantly mingled with descriptions of diseases. An alphabetical index provides access to diseases and plants in the text. Downing's book is nonetheless a fascinating portrait of one physician's therapeutics in the first half of the 19th century, is full of medico-botanical lore, and provides interesting observations, such as the author's opinion on venereal diseases that "When I was young, this complaint was confined to the sea ports, and the country people abhored the name; but of late years, it has become a common complaint, and circumstances bring those that ought to have better luck, to suffer with it many times" (p. 231-32). Downing states that he attended medical lectures at Dartmouth College and spent four years among the western Indians (p. 161). Atwater copy inscribed by the author on the verso of the title-page.

DOYLE, Daniel P. see **COPELAND MEDICAL INSTITUTE** (#790).

981. DRAKE, Daniel, 1785-1852.
"A discourse on intemperance; delivered by appointment, at a public meeting of the Agricultural Society of Hamilton County, Ohio, March, 1st, 1828, and subsequently pronounced, by request, before a popular audience," IN: *The Western journal of the medical and physical sciences,* Cincinnati: E.S. Buxton, No. I, April 1828, p. [11]-34; and No. II, May 1828, p. [65]-91.

In the first part of this article Drake considers the causes of alcohol abuse, discussing in detail "habitual drinking"; "the use of ardent spirits in assemblies of men convened for business or amusement; " gambling ("Vices are gregarious, that is, in the language of agriculture, they go in flocks"); "trades and occupations which, daily, bring numbers of men into close connexion"; "smoking segars" ("tobacco disturbs the nervous system . . . to such a degree, that the stimulus of ardent spirits is . . . necessary to sustain or restore them"); "matrimonial unhappiness" ("that many a weak minded husband has been scolded into drunkenness, is undeniable"); the multiplication of drinking establishments; an increase in the number of distilleries; and the distribution of spirits by contractors to those who labor in "the construction of canals, turnpike roads, and other public works."

In the second part Drake considers "the morbid effects of ardent spirits on the body and mind" (which includes a predisposition to spontaneous combustion); the effects of alcohol abuse on society; and recommendations for the prevention of intemperance.

982. DRAKE, Daniel, 1785-1852.
The Northern lakes a summer residence for invalids of the South . . . Louisville, Ky.: J. Maxwell, Jr., 1842.
    29 p. ; 24 cm.

Following a tour of the Great Lakes in the summer of 1842, Drake relates the advantages of the region's climate for Southern valetudinarians. He recommends travel by ship, noting that "from Buffalo to Chicago and Milwalke [sic], there is a daily line of steam packets, not inferior in size, strength and convenience to any in the United States" (p. 7). The advantages to the invalid of travel on the Great Lakes are the "invigorating atmosphere," "escape from the region of miasms," and stimulation afforded by the voyage "to the faculty of observation." Itineraries and places of historic interest are described. Reprinted from the *Western journal of medicine and surgery,* December, 1842. Title on wrapper: *Northern lakes and southern invalids.*

983. DRAKE, Daniel, 1785-1852.
"Notes of the principal mineral springs of Kentucky and Ohio," IN: *The Western journal of the medical and physical sciences,* Cincinnati, E.S. Buxton, No. III, June 1828, p. 142-167.

984. DRAKE, Daniel, 1785-1852.
A practical treatise on the history, prevention, and treatment of epidemic cholera. Designed both for the profession and the people . . . Cincinnati: Published by Corey and Fairbank, 1832.
    180 p. ; 20 cm.

"The object of the author of this little volume has been, to present the physicians and reading public of the valley of the Mississippi with an authentic digest of the most important facts relative to the causes, symptoms, and treatment of the epidemic which is now impending" (*Pref.,* p. [3]). Drake's *Practical treatise* was written before the epidemic that had appeared in Quebec (June 1832) reached the Ohio River valley. It is based almost entirely on published

sources, a body of literature with which Drake was already familiar. He shared the opinion common to many of his medical contemporaries that epidemic cholera was an aggravated form of the *cholera morbus,* a term applied to acute enteritis and other forms of dysentery. Drake speculated that cholera is non-contagious, and posited an "animalcular hypothesis" regarding its transmission. Still citing the published literature, he describes cholera's symptoms and progress, as well as its morbid appearances on autopsy. In his discussion of treatment at various stages of the disease, Drake recommends bloodletting, emetics, purgatives, injections and other elements of the "heroic" therapy. His discussion of therapeutics concludes with a section entitled "Treatment by the people," "for the management of the disease when it makes a violent onset, and medical aid cannot be immediately had" (p. 162). The book concludes with seven hygienic measures for cholera's prevention.

985. DRAKE, Emma Frances Angell, b. 1849.
Maternity without suffering. These chapters do not promise to free the hour of maternity from all pain, but they do make suggestions which will prevent much suffering . . . Philadelphia, Pa.; London: The Vir Publishing Company; Toronto: Wm. Briggs, [c1902].
    [2], 126 p., [1] leaf of plates : port. ; 17 cm.

Among the causes of painful parturition Drake discusses "the evils of dress," i.e., the corset: "This instrument of torture is answerable for the unnatural figure, which promises suffering in proportion to the deformity wrought" (p. 21). She also cites lack of exercise during pregnancy; "errors in diet"; and state of mind during pregnancy: "If she . . . dwells upon all the old wives' whims she has heard, she will look for and get something similar to these. Nervous and hysterical, she comes to her delivery poorly prepared, and every molehill becomes a mountain to her" (p. 24). Drake disagrees with M.L. Holbrook (#1689-1693) that parturition can be entirely painless for most women (without an anesthetic), but insists that "very much of the suffering can be alleviated by rational methods of living and dressing" (p. 26). The greater part of Drake's book is devoted to hygiene during pregnancy and preparations for delivery. This was the only edition of *Maternity without suffering*.

Drake was an 1882 graduate of the Boston Univesity School of Medicine (homeopathic), which eight years earlier had absorbed the New England Female Medical College. On the title-page she is described as professor of obstetrics at the Denver Homoeopathic Medical School and Hospital, which graduated its first class in 1895. Frontispiece portrait of author.

986. DRAKE, Emma Frances Angell, b. 1849.
What a woman of forty-five ought to know . . . Philadelphia, Pa.; London: The Vir Publishing Company; Toronto: Wm. Briggs, [c1902].
    [8], xiii-xx, 21-211 p., [1] leaf of plates : port. ; 17 cm.

First edition. Drake provides a discussion of the physiological, medical, and psychological aspects of the "climacteric, or change if life, or menopause." "Most women are more or less conscious that there will come a period in their lives when they will pass a change, but what this change portends, they are only dimly conscious . . . A few are well prepared; while many are too well fortified in caution and nervousness, which hinders rather than equips them for the change" (p. 23). At head of title: *Purity and truth*; series: *Self and sex series*. The *Self and sex series* issued by the Vir Publishing Co. in Philadelphia included the popular sex education manuals of Drake, Sylvanus Stall (#3312-3321) and Mary Wood-Allen (#3851-3860).

987. DRAKE, Emma Frances Angell, b. 1849.
What a young wife ought to know . . . Philadelphia, Pa.; London: The Vir Publishing Company; Toronto: Wm. Briggs, [c1902].
    292 p., [1] leaf of plates : port. ; 17 cm.

Originally published at Philadelphia in 1901. The twenty-four chapter's of Drake's treatise cover such topics as women's health (e.g., dress, exercise & fresh air); marriage (e.g., choosing a husband, woman's place in the home, balancing duties in the home with outside activities,); sex (e.g., attitude and behavior in conjugal re-

lations, the importance of heredity, abortion); pregnancy and its disorders; motherhood and fatherhood; infant care and child rearing. At head of title: *Purity and truth*; series: *Self and sex series*.

988. DRAPER, John William, 1811-1882.
A text-book on physiology for the use of schools and colleges. Being an abridgement of the author's larger work on human physiology . . . New York: Harper & Brothers, publishers, 1866.
    iv, [2], 376 p. : ill. ; 19 cm.

A condensation of Draper's *Human physiology* (1st ed., 1856), "sufficiently simple and compendious for . . . general use, and yet representing the state of the science at the present day" (p. [iii]). Born in England, Draper received his undergraduate degree at University College (London) before emigrating to America in 1832. He earned his medical degree from the Univ. of Pennsylvania in 1836, and held several teaching positions thereafter. "In company with Paine, Mott, Bedford, Pattison and Revere he inaugurated the New York University Medical College in 1841, himself occupying the chair of chemistry" (*Cycl. of Amer. med. biog.*, Philadelphia, 1912, I:258).

DRAYTON, Henry Shipman see **SHEW** (#3200).

989. DRESS AND HEALTH.
Dress and health; or, how to be strong. A book for ladies. Revised and enlarged edition . . . Montreal: John Dougall & Son, 1880.
    248 p. : ill. ; 17 cm.

"The term 'Dress Reform' used to suggest only notions of ugliness and impropriety; but within the last few years very different ideas have come to be associated with it. It is now generally known that the ordinary dress of women is exceedingly injurious, and, from its very nature, tends to produce many diseases and deformities . . . Furthermore, it has been discovered that the most important changes required for health may be made without rendering the outside appearances at all conspicuous, and that therefore women may be healthfully clad without suffering martyrdom from transgression of social usages" (*Intro.*, p. [9]).

Dress and health is a compilation of ideas on dress reform extracted from two sources: from the published writings of such authors as Dio Lewis (#2233-2243), M.L. Holbrook (#1678-1695), Harriet Austin (#174-178), Mary Studley, R.T. Trall (#3558-3593), Tullio Verdi (#3642-3644), *et al.*; and from lectures delivered at Boston in the spring of 1874 by Mary J. Safford-Blake, Caroline E. Hastings, Mercy B. Jackson, Arvilla B. Haynes and Abba Goold Woolson that were published in their entirety in 1874 under the title *Dress-reform* (#990). An earlier edition of *Dress and health* (187 p.) had been published at Montreal by John Dougall & Son in 1876. In printed wrapper.

990. DRESS-REFORM.
Dress-reform. A series of lectures delivered in Boston, on dress as it affects the health of women. Edited by Abba Goold Woolson . . . Boston: Roberts Brothers, 1874.
    xviii, [2], 263, [1], 2, [2] p., [3] leaves of plates : ill. ; 18 cm.

As chairman of a committee of the New England Women's Club considering the health implications of women's fashions, Woolson organized a series of five lectures given at Boston in the spring of 1874 by Mary Jane Safford, (1834-1891; M.D. 1869, New York Med. Coll. for Women), Caroline Eliza Hastings (1841-1922; M.D. 1868, New England Female Med. Coll.), Mercy Bisbee Jackson (M.D. 1860, New England Female Med. Coll.), Arvilla Breton Haynes (M.D. 1866, New England Female Med. Coll.), and Abba Louisa Goold Woolson (1838-1921). The purpose of these lectures "was to arouse women to a knowledge of physical laws, and what different garments they should adopt. All, save the last [i.e., Woolson], were written by female physicians of recognized ability and position; and the testimony thus given concerning the injuries inflicted by dress was felt to be authoritative and convincing" (p. [v]).
    The lectures were gathered in the present volume, and are followed by a sixty-nine page appendix (presumably by Woolson) that continues the theme of dress and health. She concludes: "Even in this day of pinching corsets and entangling trains, women are fast learning to respect the nature in themselves; and they will, ere long, forswear bands and burdensome toggery, and roam the meadows and walk the streets, if not kirtled like Diana and her nymphs when equipped for the chase, yet with a dress too simple to absorb their minds, too easy to cripple their movements, too healthful to rob their cheeks of a bloom which should be as fresh and rosy as that of the clover-tops they tread" (p. 252-53). *Dress-reform* followed closely upon the publication of Woolson's *Woman in American society* (Boston, 1873).

991. DRESSLER, Florence.
Feminology. A guide for womankind, giving in detail instructions as to motherhood, maidenhood, and the nursery . . . Revised, fifth edition. Chicago: C.L. Dressler, publishers, 1906.
    702 p., VIII, [11] leaves of plates : ill., port. ; 22 cm.

CONTENTS: I. Marriage; II. Home-building; III. Amative desires; IV. Limitation of offspring; V. Prenatal inheritance; VI. Reproductive physiology; VII. Pregnancy; VIII. Easy labor; IX. Confinement; X. Convalescence and its drawbacks; XI. Abortion, feticide; XII. The mother in relation to the child; XIII. Hygiene of early life; XIV. Normal development from birth to puberty; XV. Parenthood and education; XVI. Diseases of children; XVII. The son; XVIII. The daughter; XIX. Menstruation and its disorders; XX. Diseases peculiar to women; XXI. The menopause; XXII. Diet in health and disease; XXIII. Hints on nursing the sick; XXIV. Some emergencies and accidents; XXV. Care of the person; XXVI. Physical culture; XXVII. Rest; XXVIII. Beauty; XXIX. Social life; XXX. Longevity.
    Feminology appeared in six editions between 1902 and 1911. Its author was a 1900 graduate of the College of Medicine and Surgery (Chicago), a physio-medical school established in 1896. Dressler was professor of gynecology at her *alma mater,* as well as its secretary. Frontispiece portrait of author.

992. DREWRY, George Overend, 1839-1892.
Common-sense management of the stomach . . . New York: Dodd & Mead, publishers, [1875?]
    xvi, 138 p. : ill. ; 17 cm.

Drewry's popular treatise on disorders of the stomach and digestion was first published at London in 1875. He approaches his subject by considering stomach ailments in each of the six periods of life: infancy, childhood, youth, maturity, "the decline of life" and old age.

993. DRYSDALE, Charles Vickery, 1874-1961.
The small family system. Is it injurious or immoral? . . . Revised and enlarged edition with thirteen diagrams of population movements etc., at home and abroad . . . New York: B.W. Huebsch, MCMXVII [1917].
x, 196 p. : diagrs. ; 19 cm.

Second American edition of Drysdale's *apologia* for family planning on "Neo-Malthusian" principles (1st ed., London, 1913; 1st Amer. ed., New York, 1913). "Neo-Malthusianism is an ethical doctrine based on the principal of Malthus that poverty, disease and premature death can only be eliminated by control of reproduction, and on a recognition of the evils inseparable from prolonged abstention from marriage. It therefore advocates early marriage, combined with a selective limitation of offspring to those children whom the parents can give a satisfactory heredity and environment so that they may become desirable members of the community. It further maintains that a universal knowledge of hygienic contraceptive devices among adult men and women would in all probability automatically lead to such a selection through an enlightened self-interest, and thus to the elimination of destitution, and all the more serious social evils, and to the elevation of the race" (p. [ii]).

C.V. Drysdale was the son of Charles Robert Drysdale, M.D. (1829-1907), one of the founders and first president of the Malthusian League (est. 1877), and Alice Vickery, M.D. (1844-1929), who succeeded her husband as president on his death. The League was organized in England in the wake of the Bradlaugh-Besant trial "to agitate for the abolition of all penalties on the public discussion of the Population Question" and "to spread among the people, by all practicable means, a knowledge of the law of population, of its consequences, and of its bearing upon human conduct and morals" (Peter Fryer. *The birth controllers.* New York: Stein & Day, c1965, p. 173).

In addition to an eminent career in electrical engineering, C.V. Drysdale took a leading role in the promulgation of neo-Malthusian doctrine, becoming secretary of the Malthusian League in 1907. According to A.B. Mitchell in the *Dict. nat. biog., 1961-1970,* "He gradually changed the Malthusian League from its earlier emphasis on the relationship between overpopulation and poverty to one of better eugenic selection. His determined and dogged support for Malthusian and neo-Malthusian principles antagonized many of the leading figures of the day and reduced the League's effectiveness in achieving its objectives. Nevertheless it exerted a significant influence upon the family planning movement as a whole and led to Drysdale being made a member of the National Birth Control Council when it was formed in 1930" (p. 311).

DUBOIS, Franklin L. see **GOULD** (#1386).

994. DUFFEY, Eliza Bisbee, d. 1898.
The relations of the sexes . . . New York: Wood & Holbrook, 1876.
320 p. ; 19 cm.

First edition. "What a perplexing tangle do we find, when we take into our hands the meshes which society, and custom, and tradition, and false forms of religion, and ignorance, and human perverseness generally, have woven and knotted togther!" (p. 13). Duffey's anxiety leads to her declamation: "The times seem tending toward disintegration" (p. 14). In this milieu Duffey attempts to establish the familial and social order on firmer foundations through an examination of the relations between the sexes, the significance of marriage, and the threats to these foundations of the "social economy" from advocates of free-love, polygamy, state-regulated prostitution, etc. In discussing conjugal relations Duffey focuses on sexual behavior rather than sexual physiology or hygiene. Similarly, her chapter on "The limitation of offspring" advocates family planning without explaining contraceptive mechanisms.

995. DUFFEY, Eliza Bisbee, d. 1898.
The relations of the sexes . . . New York: M.L. Holbrook, 1885.
320, [4] p. ; 19 cm.

996. DUFFEY, Eliza Bisbee, d. 1898.
What women should know. A woman's book about women. Containing practical information for wives and mothers . . . Philadelphia: J.M. Stoddart & Co. . . . [*et al.*], [c1873].
    320 p. ; 20 cm.

First edition. "The larger class of writers in regard to womanly functions, capabilities and incapabilities, have been men. From Michelet, who regards the normal state of womanhood as one of pretty and useless invalidism, to the unnumbered hosts of English and American authors and essayists, they all set up before us a purely ideal creation, eliminated from their masculine brains, and differing as widely from the reality as—well, say, as man from woman" (p. 17-18).
    CONTENTS: I. Physical and mental differences in the sexes. II. The approach of puberty. III. Puberty. IV. Love in its physical and moral phases. V. When and whom shall women marry. VI. Courtship and engagement. VII. Marriage and the family. VIII. Trials of the young wife. IX. Child-bearing. X-XI. The disorders of pregnancy and their remedies. XII. Delivery. XIII. The sick room. XIV. Can a woman smoke? XV. General diseases of women. XVI. Management of an infant. XVII. The dress of an infant. XVIII. Diseases of infants. XIX. The moral responsibilities of motherhood.

997. DULLES, Charles Winslow, 1850-1921.
Accidents and emergencies. A manual of the treatment of surgical and medical emergencies in the absence of a physician . . . Sixth edition thoroughly revised and enlarged . . . Philadelphia: P. Blakiston's Sons & Co., 1905.
    xii, 9-209 p. : ill. ; 19 cm.

Dulles' handbook of first aid appeared in eight editions between 1904 and 1918. The author was an 1875 medical graduate of the University of Pennsylvania. "After studying in London, Vienna, and Paris during 1876-1877, he settled in West Philadelphia where he remained in practice for over forty years. Dr. Dulles was consulting surgeon to the Rush Hospital, a manager of the University Hospital, and lecturer on the history of medicine at the University of Pennsylvania from 1893 to

1908. He edited . . . 'The Medical and surgical reporter' from 1887 to 1890; he contributed to Keating's 'Encyclopedia on the diseases of children,' 1890, and 'Diseasaes of the bones' to the supplementary volume of Ashhurst's 'International encyclopedia of surgery,' 1896" (*Dict. Amer. med. biog.,* 1928, p. 356).

DUMMIG, Charles see ROSCH (#3029).

DUNBAR, Newell see BROWN-SEQUARD (#438).

998. DUNCAN, Andrew, 1744-1828.
Observations on the distinguishing symptoms of three different species of pulmonary consumption: the catarrhal, the apostematous, and the tuberculous; with some remarks on the remedies and regimen best fitted for the prevention, removal, or alleviation of each species. To which is added, an appendix, on the preparation and use of lactucarium, or lettuce-opium . . . First American, from the second London edition. Philadelphia: Published by Collins & Croft, 1819.
    xix, [5], 155 p. ; 18 cm.

The 2nd Edinburgh edition was published in 1816; first edition, Edinburgh, 1813.
    REFERENCES: *Austin* #705.

999. DUNCAN, Thomas Cation.
The feeding and management of infants and children, and the home treatment of their diseases . . . Chicago: Duncan Brothers, 1880.
    426, [6] p. : ill. ; 20 cm.

First edition. Duncan was an 1866 graduate of the Hahnemann Medical College (Chicago), and the author of several popular homeopathic treatises.

1000. DUNCAN, Thomas Cation.
How to be plump: or talks on physiological feeding . . . Chicago: Duncan Brothers, publishers, 1878.
    60 p. ; 16 cm.

"Why cannot the 'picture of health' be painted in all faces? Why is plumpness associated with health, and leanness with disease?

Why are 'Americans proverbially lean?' These are vital questions, that touch the philanthropic, interest the statesman, and arouse scientific investigation" (*Pref.*, p. v). T.L. Bradford maintains that *How to be plump* appeared in five editions, though this 1878 edition is the only edition recorded (*Homeopathic bibliography*, Philadelphia: Boericke & Tafel, 1892, p. 66).

1001. DUNGLISON, Richard James, 1834-1901.

An elementary physiology with special reference to hygiene, alcohol and narcotics . . . Chicago; New York: The Werner Company, [c1894].
    208 p., [1] leaf of plates : ill. ; 19 cm.

Originally published at Philadelphia in 1885. Series: *The Werner educational series.*

1002. DUNGLISON, Richard James, 1834-1901.

A new school physiology and hygiene. With special reference to the action of alcohol and narcotics . . . Chicago; New York: The Werner Company, [c1894].
    336 p., [2] leaves of plates : ill. ; 19 cm.

Originally published at Philadelphia in 1880. Series: *The Werner educational series.* The son of Robley Dunglison (#1003, 1004), Richard James Dunglison was an 1856 graduate of Jefferson Medical College. He was the author of numerous books and journal articles, and enjoyed a prominent place in Philadelphia medical society.

1003. DUNGLISON, Robley, 1798-1869.

Human health: or the influence of atmosphere and locality; change of air and climate; seasons; food, clothing; bathing and mineral springs; exercise; sleep; corporeal and intellectual pursuits, &c. &c. on healthy man; constituting elements of hygiene . . . A new edition, with many modifications and additions. Philadelphia: Lea & Blanchard, 1844.
    iv, [iii]-viii, [13]-464 p. ; 23 cm.

Originally published in 1835 under the title: *On the influence of atmosphere and locality* (#1004). "The author has subjected the first edition of this work to a complete revision, and has en-

deavoured to notice . . . the recent researches of distinguished hygienists. He has had a two-fold object in view . . . to produce a work which might serve the student as an accompaniment to the observations on hygiene, that are made in lectures on the institutes of medicine—and one which might equally enable the general reader to understand the nature of the action of various influences on human health, and assist him in adopting such means as may tend to its preservation" (p. iv).

1004. DUNGLISON, Robley, 1798-1869.

On the influence of atmosphere and locality; change of air and climate; seasons; food; clothing; bathing; exercise; sleep; corporeal and intellectual pursuits, &c. &c. on human health; constituting elements of hygiene . . . Philadelphia: Carey, Lea & Blanchard, 1835.
    [2], xi, [12]-514 p. ; 22 cm.

First edition (later issued under the title: *Human health, #1003*). Dunglison's treatise was intended as a textbook for use in medical schools (where hygiene or preventive medicine was just being introduced to the curriculum), as well as for the enlightenment of a more general readership. As a treatise on "the influence of physical and moral [i.e., psychological] agents on healthy man, and the means for preserving the healthy play of the functions," Dunglison's is one of the finest publications of the century, displaying his sagacity, industry and wide acquaintance with the medical literature. Born in England, trained in surgery at Edinburgh (1818) and in medicine at Erlangen (1823), Dunglison left England on the invitation of Thomas Jefferson to become professor of medicine at the University of Virginia. He was a prolific author, an influential educator, and one of the more remarkable figures in American medicine of the first half of the 19th century. Atwater copy inscribed by Henry Foster (1821-1891), founder of the Clifton Springs Water-Cure, New York.

1005. DUNN, James B.

The adulteration of liquors, with a description of the poisons used in their manufacture . . . New York: National Temperance Society and Publication House, [ca. 1870].
    24 p. ; 19 cm.

At head of title: *Prize essay.* Dunn was pastor of the Beach Street Presbyterian Church in Boston, and author of numerous tracts on temperance.

1006. DUNN, James B.
Bible wines . . . [New York: Published by the National Temperance Society and Publication House, 186?].
8 p. ; 19 cm.

Examining scriptural references to alcohol consumption, Dunn concludes, ". . . in no single instance can it be proved that it has mentioned intoxicating drinks with approbation; and consequently those who use alcoholic poisons are left without the least sanction from that unerring guide . . . total abstinence is therefore in exact accordance with the letter and spirit of God's Word" (p. 8). Caption title. Series: [*Prize tract*] *no. 9.*

1007. DUNN, James B.
Cholera conductors . . . [New York: Publushed by the National Temperance Society and Publication House, 186?].
4 p. ; 19 cm.

Dunn states, "however writers and theorists may differ about contagion and non contagion, on this point they all agree, that Intemperance predisposes to cholera . . . The *intemperate* are its first victims, and make up ninth tenths of its subjects, and everywhere the cholera has manifested such an affinity for the intemperate that they have been . . . denominated CHOLERA CONDUCTORS" (p. [1]). Caption title. Series: [*Prize tract*] *no. 3.*

1008. DUNN, James B.
The evils of beer legislation . . . [S.l.: s.n., 1870?].
8 p. ; 22 cm.

Caption title.

1009. DUNN, James B.
The Good Samaritan. A sermon . . . New York: National Temperance Society and Publication House, 1871.
29, [3] p. ; 19 cm.

Dunn likens those who produce and market alcoholic beverages to the robbers who stripped and beat the traveller on the road from Jerusalem to Jericho (Luke 10:30-37), the drunkard to the victim of their crime, and those who passed the beaten man on the road to a public indifferent to the availability of alcoholic beverages.

1010. DUNN, James B.
The miracle at Cana . . . [New York: Published by the National Temperance Society and Publication House, 1870?].
4 p. ; 19 cm.

Dunn argues that the miracle at the wedding feast of Cana produced an unfermented wine. Caption title. Series: [*Prize tract*] *no. 94.*

1011. DUNN, James B.
Peter's training, and what came of it . . . [New York: Published by the National Temperance Society and Publication House, 186?].
4 p. ; 19 cm.

Caption title. Series: [*Prize tract*] *no. 8.*

1012. DUNN, James B.
Timothy a teetotaler . . . [New York: Published by the National Temperance Society and Publication House, 186?].
4 p. ; 19 cm.

Expounding on 1 Timothy 5:23 ("No longer drink only water, but use a little wine for the sake of your stomach and your frequent ailments"), Dunn argues, "Timothy was undoubtedly a teetotaler; this Paul knew, and did not reprove him for it; he only asked him to depart from the practice of that principle in a small degree, that his health might be restored" (p. 2). Caption title. Series: [*Prize tract*] *no. 21.*

1013. DUNN, James B.
Why I did not become a brewer; or, what is beer? . . . [New York: National Temperance Society and Publication House, 186?].
4 p. ; 19 cm.

Caption title. Series: [*Prize tract*] *no. 26.*

1014. DUNNE, Brian Boru, 1878-1962?
Cured! The 70 adventures of a dyspeptic . . . Foreword by H.G. Wells . . . Philadel-

phia: The John C. Winston Company, publishers, [c1914].

    240 p., [8] leaves of plates : ill. ; 19 cm.

    First edition. Journalist and author, Brian Boru Dunne was the son of Edmond Dunne, Ulysses S. Grant's appointee as chief justice of the Arizona Territory (1874) until removed from office following a speech to the Arizona legislature advocating the state funding of Catholic schools. The elder Dunne continued his legal career in the eastern states and Canada, during which period he also established Roman Catholic colonies in Florida and Alabama. B.B. Dunne received a classical education at St. Mary's College in N. Carolina followed by several years' travel and study in Belgium, France, Germany and Italy. Dunne memorialized his friendship with George Gissing (1857-1903) in Siena and Rome in a memoir not published until 1999. On his return to the United States, Dunne entered newspaper journalism in a career that began with the *Baltimore Sun* and culminated in his editorship of the *Santa Fe New Mexican,* the oldest newspaper in the southwest.

    In a 1902 letter to Gissing, Dunne states that his health problems were brought on by "the late hours, the irregular and hurried meals, the worry over scoops and the hurry-hurry of the metropolitan newspaperman's life" (quoted in the introduction to Dunne's *With Gissing in Italy,* Athens, Ohio, 1999, p. 26). His attempts to recover health are recorded in *Cured!,* an amusing collection of not-much-exaggerated encounters with the medical profession, in which Dunne tries wet packs, rest cures, milk diets, electric light baths, etc. He described the book to H.G. Wells as "my life's story from the year 1901 to 1911." In 1909 Dunne became a permanent resident of Santa Fe, where he was an activist for statehood, and eventually, one of the city's cultural icons. Some of the chapters in *Cured!* are reprinted from *Collier's.* A second edition was published in 1937. The Atwater copy is inscribed by Dunne to William Groff, his dentist and "the dyspeptic's best friend."

1015. DUNNE, P. C.
The young married lady's private medical guide. Translated from the French of P.C. Dunne and A.F. Derbois . . . With notes, compiled from the public writings and private teachings of those eminent medical men, devoted to a study of the peculiar organs and diseases of females, in the best medical institutions in Europe and America. By F. Harrison Doane . . . Fourth edition. [Boston]: Published for the proprietor, 1854.

    264, 12 p. : ill. ; 19 cm.

    "It is the earnest desire of the authors, as well as the translator and compiler, of this work to extend to every female who has arrived at puberty . . . such information respecting the physiology of her own private organs, their proper and healthy condition, as will enable her to judge of her own diseases" (*Pref.,* p. 12). It is also the intention of this book "to make known to every married lady that means exist, and can be procured by all, by which pregnancy can be effectively controlled at will, with the reasons why it is often necessary to prevent conception in many females, and in very many more to suspend it for a time, in consequence of feeble health or other causes" (p. 15).

    Chapter I describes the female sexual organs; chs. II-X, menstruation and its disorders; ch. XI, conception; chs. XII-XIII, pregnancy and its accompanying diseases; ch. XIV, labor; ch. XV, infant care; ch. XVI, female masturbation and its consequences; chs. XVII-XVIII reasons for the "suspension of conception"; chs. XIX-XXII, the prevention of conception; and ch. XXII, "miscarriage" (i.e., abortion).

    The author-compiler's discussion of the female reproductive system and its disorders is more rhetorical than medical. In the several chapters devoted to the prevention of conception, there is no discussion of the contraceptive methods widely in use at this period. Instead, reference is made to the "discovery of Professors Dunne and Derbois," which is advertized in the appendix with several "remedies especially adapted to cure female complaints and restore health and vigor." Doane offers for sale "preparations for regulating, suspending, or preventing conception," without describing their composition or method of application. Dunne and Derbois were doubtless fictitious personages, but Doane could be contacted at a Boston address for the procurement of his proprietary remedies and for medical advice. In the Boston city directories from 1858 to 1860 Doane is listed under "Physicians, other."

1016. DUNSFORD, Harris F., 1808-1847.
The practical advantages of homoeopathy illustrated by numerous cases . . . Philadelphia: John Penington, 1842.
xii, [13]-201, [3] p. ; 20 cm.

Dunsford's homeopathic *apologia* was originally published at London in 1841.

DUPLANIL, Jean-Denis see BUCHAN (#499, 501).

1017. DURHAM, Joseph Z.
Ladies' physo medical companion, containing the causes and preventatives of a premature decline, with a review of the various changes & derangements of the female constitution; composed of original and selected matter, designed as a guide to both married and unmarried ladies, throughout the various changes of life, with important information concerning the cure of disease . . . First edition . . . Canton [Conn.]: Gotshall and Martin printers, 1850.
v, [6]-288 p. : ill. ; 14 cm.

Durham's *Companion* is divided into three parts: the first on women's health and hygiene; the second on the anatomy and diseases of the female genitalia (including menstrual disorders); and the third on conception, pregnancy, and the management of labor.

1018. DUTTON, George, 1830-1903.
Etiopathy or way of life, being an exposition of ontology, physiology and therapeutics. A religious science and a scientific religion . . . Boston: Cynosure Publishing Company, 1899.
517, [1], 3-121, iii p., [2] leaves of plates : ill. ; 24 cm.

According to the biographical sketch which prefaces this volume, the author was born on 25 March 1830. He received his baccalaureate from Dartmouth in 1855, and between 1848 and 1859 taught school in Vermont, New Hampshire and Massachusetts. During this period, Dutton began the study of medicine under local preceptors, and attended lectures at the National Medical College (Washington, D.C.), from which he received his M.D. in 1861.

From 1861 to 1872, he was both a practicing physician and a school administrator in several New England towns. He spent the Civil War years in Rutland, and from 1869 to 1872 was principal of the West Randolph Academy. While practicing in Rutland, Dutton had "procured a Parisian model of the human system, oil paintings, charts, etc., and prepared himself for giving public lectures on physiology and hygiene . . . Accordingly, for three years, he traveled and lectured in many cities in and towns of Vermont, Massachusetts, New Hampshire, New York and other places. In 1869 he established a school . . . at West Randolph for the purpose of giving instruction in the art of preserving health and in the general principles of medical science; but after continuing it for two years moved to Boston, Mass. (1872), gave some public lectures . . . and opened an office and school of instruction at 60 Essex street" (p. 10-11).

Dutton is listed in the 1871 Boston city directory as principal physician of the Boston Hygienic Institute & School of Preventive Medicine, 59 Indiana Place. While in Boston he organized and became dean of the Vermont Medical College in Rutland (incorporated 1883), a school that initially had a faculty of two, and by 1890 had graduated about forty students (it was listed in the medical directories as active as late as 1900; the first AMA directory [1906] calls it fraudulent). About 1893 Dutton moved to Chicago where he founded and became dean of the American Health University and Dutton Medical College. (The school in Rutland was thereafter known as American Health University No. 2.) At the Chicago institution, he was professor of natural and spiritual science. Dutton died in Chicago on 27 May 1903.

The author propagates "a religious science and a scientific religion" based on the premise that as "human, finite beings we bear the same relationship to the Infinite that a wave of the sea bears to the ocean" (p. 18), i.e., that man is a microcosm, that "one universal law governs in every quarter of the globe; and throughout the spheres"; and that "the law of our being . . . is the Divine Will," that natural law and spiritual law are not "diverse and contradictory, but . . . are ever harmonious, and work in unison," emanating from the "One Supreme and beneficent Mind of the universe" (p. 19). It follows that "Health, strength, longevity and beauty are the result only of glad obedience to the will of the one Supreme Mind" (p. 21); that disease is "always the result of of erroneous unscientific methods of living and acting," and health the result of "correct methods of living."

Etiopathy holds that the "real cause of disease is not microbes, or any material agent or thing, but is always a lack of mental and spiritual unfoldment." Its therapeutics is not based on drugs, but "kind words, loving thoughts, mental illuminaton, rectified conditions of life, both mental and physical." We are all "patients" until "properly instructed in the way of life, or art of living" (p. 28). Immunity from disease is "not secured by poisoning the blood with the virus of any diseased animal, but by *living on a plane of being above disease*" (p. 29). Ideas, in other words, "are safer, surer and better remedies than drugs"; and that to understand and live by the principles of Etiopathy we entrust ourselves to the healing and health maintaining power of "the ever living and universal Spirit" (p. 30). "Etiopathy," Dutton states, "is the missing link that unites Physics and Metaphysics in one harmonious whole. It is that branch of universal science (Ontology) that relates to the recovery of health by the removal of all causes that prevent recovery." He makes the distinction that disease does not destroy life, but "tends to preserve life by calling attention to conditions that need correcting." On the contrary, "It is not disease that kills, but drugs, drug doctors, surgeons, and wrong conditions" (p. 31). Thus, pain is not an evil that we need mask with analgesics, but "a kind messenger that calls . . . attentions to some untoward condition that needs correcting. We must not banish the messenger ever until his message is fully understood" (p. 40).

The section entitled "Technics of medicine" (121 p., 2nd seq.), a medical dictionary, was previously published under the title *Key to the science of medicine* (Boston, 1894).

DUTTON, George see also **BOSTON HYGIENIC INSTITUTE** (#374).

1019. DYE, John H.
Painless childbirth; or healthy mothers and healthy children. A book for women. Containing practical rules how the pains and perils, the difficulties and dangers of childbirth may be effectually avoided; also a practical consideration of the diseases of women, and their common-sense treatment, prescriptions, etc., etc. . . . . Second edition, revised and enlarged. Silver Creek, N.Y.: The Local Printing House, 1882.
    viii, [13]-294 p. ; 19 cm.

Dye maintains that "childbirth is a natural process, when the mother lives in accordance with the laws of health. Natural labor is an easy, short and painless act. Natural labor is never painful. The organic nerves that supply the uterus are never sensitive in a healthy state. Irritation, debility, congestion and inflammation render these nerves sensitive and painful. All pain, difficulty and danger are the consequences of violating natural laws" (p. 51). He continues, "Modern social customs impose upon woman modes of life that impair her constitutional vigor, deform her body, pervert her functions, render her an easy prey to uterine diseases and to prolonged and painful childbirth" (p. 53). To counter the effects of fashion and custom, Dye provides a regimen of proper "hygienic management." Chapter VIII is devoted to anaesthetics in childbirth: "When the patient is healthy and robust, the suffering inconsiderable and the labor promising to be short, I should certainly refuse to give the anaesthetic; but in a weak and feeble woman, suffering intensely . . . I would not hesitate for a moment" (p. 135). Dye also discusses the management of labor and diseases of the female reproductive organs. Cover-title: *Healthy mothers and healthy children.*

Painless childbirth was first published in 1879 at Silver Creek, a village on Lake Erie 28 miles south of Buffalo and a popular summer residence. The 14th and apparently last edition was printed in 1902. Dye, was an 1873 graduate of the Eclectic Medical College of New York. He practiced in Buffalo, and is listed among the medical staff of R.V. Pierce's Invalids & Tourists' Hotel (#1896) at its opening in 1878. Dye's name appears in the 9th edition of *Polk's medical register* (1906), but not in the 1908 10th edition.

1020. DYE, John H.
Painless childbirth; or healthy mothers, and healthy children. A book for all women . . . Eighth edition, revised and enlarged. Buffalo, N.Y.: Published by Baker, Jones & Co., 1889.
    xv, [16]-451, [3] p. ; 20 cm.

1021. DYE, John H.
Painless childbirth; or healthy mothers, and healthy children. A book for all women . . . Ninth edition, revised and enlarged. Buffalo, N.Y.: Published by Baker, Jones & Co., 1890.
xv, [16]-451, [3] p. ; 20 cm.

"Though it may be contrary to what is generally observed, I assert that the terrible agony woman suffers at childbirth is unncessary, and may, in most cases at least, either be greatly modified or entirely overcome. In proof of this assertion scores of women have submitted the plans detailed in the following pages to the test of practical experience with the most gratifying results. They have found that by following a few simple and easy instructions, the duration of labor may be reduced to a few hours or minutes, and that pains may be rendered so slight as to be scarcely worthy of notice" (*Pref.,* p. vi). Cover-title: *Healthy mothers and healthy children.*

1022. DYE, John H.
Talks and testimonials for women. [Buffalo, N.Y.: Dr. J.H. Dye Medical Institute, ca. 1902].
15, [1] p. ; 15 cm.

Pamphlet promoting Dr. Dye's *Mitchella Compound,* a panacea for the disorders and discomforts of women. For example, "When passing through pregnancy put your faith and trust in *Mitchella Compound* and you will not be disappointed. Not only will it relieve or entirely banish those annoying symptoms, morning sickness, backache, nervousness, cramps, swelling of the limbs, etc., but will also make labor natural—that is, easy and painless" (p. 3). *Mitchella Compound* is also recommended for "chronic female weakness," menstruation disorders, "change of life," bladder trouble, etc.

Dye's panacea is singled out in the A.M.A.'s 1912 *Nostrums and quackery* (#86): "But if after a period of drought one went to the woods and raked up a double handful of dried leaves, pieces of bark, and any other debris that happened to be handy, the average man would find

it difficult to distinguish between such rakings and 'Dr.' Dye's Mitchella Compound at $1 a package" (1:180). The remedy was comprised of herbs, roots and bark. "Mitchella Compound is . . . but one more of the innumerable cure-alls on the market in which discarded, unrecognized or useless drugs are pressed into service and invested with miraculous virtues" (1:180-81).

1023. DYER, D.
The eclectic family physician, a scientific system of medicine, on vegetable principles, designed for families. This work embraces the character, symptoms of disease, and treatment for man, woman and child . . . Hallowell [Me.]: Printed for the author, 1855.
252 p. ; 15 cm.

"The character of the work may be known by its title, as the ECLECTIC; which signifies one who does not attach himself to any particular sect, but selects from the opinions and principles or systems, such as he judges to be sound and rational. Many persons desire to know something about the science of medicine, but are not able to read the works that are published, unless they first go to school and study Latin six months or more . . . hence the study of medicine is confined to a few. Another trouble in reading medicine . . . is the reading of books that take a one side view; such as Cold Water Cure, Homoepathic [sic], Alopathic [sic], Thompsonian [sic], &c. . . . But in this you see that one has to buy a large number of medical works, which is attended with much expense. To obviate this difficulty, we need an eclectic work like this" (p. 3-4).

Dyer begins his treatise with a brief discussion of medicinal plants and the elements of hygiene. A catalog of medical and surgical diseases briefly discusses symptoms and treatment, followed by an equally brief section on midwifery. The longest section of the volume is a collection of recipes for liniments, injections, tinctures, plasters, pills, powders, etc., with recommended dosages. An alphabetical index provides unified access to the book's various parts.

E

1024. EASTERN MEDICAL REFORMER.
The Eastern medical reformer: a monthly journal of medical and chirurgical science . . . John B. Hibbard, M.D., editor and proprietor. Vol. I. Rutland, (Vt.) January, 1847. No. 11.
[161]-176 p. ; 21.5 cm.

There is only one issue of *The Eastern medical reformer* in the Atwater Collection, a title so scarce that it is not recorded in Myrl Ebert's "The rise and development of the American medical periodical, 1797-1850" (*Bull. Med. Lib. Assn.,* 1952, 40:243-76), in the first two series of the *Index-catalogue of the library of the Surgeon-General's office,* in the *Union list of serials* (3rd ed., New York: H.W. Wilson, 1965), in the OCLC online union catalog, or in the National Library of Medicine's *SERHOLD* database.

An eclectic medical periodical, the *Eastern medical reformer* began publication with vol. 1, no. 1 in March, 1846. Presumably, the first volume was completed with vol. 1, no. 12 in February, 1847. The editor projects a second volume to be published in March, 1847 (p. 165), but there is no bibliographic record of the journal's continuation beyond the first volume.

Though intended primarily for the eclectic practitioner, the journal was also addressed to "the general reader" (p. 165), who would be informed in its pages of the theory and practice of medicine "on eclectic principles," and who might peruse advertisements such as those for "Dr. J.B. Hibbard's Family Medicines." The January, 1847 issue records the establishment of the Vermont Eclectic Institute at Rutland, Vt. the previous month, noting that it is "the only medical school in the state imbued with the expanded princples of Eclecticism" (p. 162), and dismissing its Vermont competition as the "Castleton or Woodstock Poison Metal Schools" (p. 163). The editor, John B. Hibbard, is listed as professor of pathology, theory and practice of medicine, and obstetrics on the faculty of the new medical school (p. 168).

1025. EASTMAN, Mary F.
The biography of Dio Lewis, A.M., M.D. Prepared at the desire and with the co-operation of Mrs. Dio Lewis . . . New York: Fowler & Wells Co., publishers, 1891.
398 p., [1] leaf of plates : port. ; 19 cm.

Eastman's laudatory biography of Dio Lewis (#2233-2243, 2245, 2246) is still one of the best sources on his life and his pioneering career in physical education and temperance reform. Frontispiece portrait of Dio Lewis. Atwater copy inscribed by Mrs. Dio Lewis (June, 1892).

1026. EATON, Morton Monroe.
Eaton's domestic practice for parents and nurses . . . Cincinnati: M.M. Eaton, Jr., & Co., 1882.
xiv, [15]-703, [1] p. : ill. ; 23 cm.

Eaton was an 1888 graduate of the homeopathic Pulte Medical College. His name last appears in the 10th edition (1908) of *Polk's medical and surgical directory of North America.* Spine-title: *Eaton's illustrated domestic practice for parents and nurses.*

EBERBACH, Ch. see CHASE (#612).

ECLECTIC EDUCATIONAL SERIES see BROWN (#427).

ECLECTIC INSTITUTE OF MEDICINE AND SURGERY, EYE AND EAR INFIRMARY see BOYLE (#386).

1027. ECLECTIC JOURNAL OF MEDICINE.
Eclectic journal of medicine, designed for popular and professional reading. L. Reuben, M.D., & L.C. Dolley, M.D., editors and publishers. Volume IV. From January, 1852, to January, 1853 [i.e., December, 1852]. Rochester [N.Y.]: A. Strong & Co., printers, Democrat office, 1852.
429 (i.e., 529) p. : ill. ; 23 cm.

The *Eclectic journal of medicine* succeeded the *New York eclectic medical and surgical journal*—two of the five titles under which this journal was published between 1849 and 1853 (*see annotation to #1028*). The journal was edited by Levi Reuben (b. 1823) and Lester Clinton Dolley (1825-1872), members of the faculty of the Central Medical College (est. 1849), the eclectic medical school that had tranferred operations from Syrcause to Rochester in April 1850. Although most of the articles were written for practitioners, it was the editors' opinion that the public should be exposed to such literature in order to make it aware of the philosophic basis of eclecticism, its therapeutic validity, and the many errors of the "regular" school. To this end, the volume is also interspersed with articles written specifically for a popular audience.

L.C. Dolley tired of the politics that had plagued the Central Medical College from its beginnings, and remained in Rochester when the College returned to Syracuse in 1853. In June, 1852 he married Sarah Read Adamson, an 1851 graduate of the C.M. College and Rochester's first female practitioner. The Dolleys practiced together in Rochester until L.C.D.'s death in 1872. S.R.A.D. remained in practice at Rochester until her death in 1909.

1028. ECLECTIC MEDICAL AND SURGICAL JOURNAL.

The Eclectic medical & surgical journal. [Vol. 1, no. 1 (July 1849)-v. 2, no. 12 (June 1851)] [Syracuse ; Rochester, N.Y., 1849-51].

2 v. : ill. ; 21 cm.

In his opening remarks to the first issue of this journal, the editor states that "the pages of the JOURNAL will be largely devoted to reading of a general character; such as will interest and impart valuable instruction to readers out of the profession whose leisure or inclination may lead them to the important task of investigating the laws of health, and the *modus operandi* of medicine" (p. 2).

The journal underwent five title changes during the course of its five-year publishing history. It began publication in Randolph, N.Y. in May 1849 (vol. 1, no. 1) as *The New-York eclectic medical and surgical journal* (#2594). The journal resumed publication in July 1849 as the *Eclectic medical & surgical journal* at Syracuse, N.Y.,

where the Central Medical College had been established. By April 1850, the journal was being published in Rochester, where the C.M. College had been relocated after the 1849-50 session. Although vol. 2 was also published as *The Eclectic medical & surgical journal*, the completed second volume was issued with a title-page for the *New York eclectic medical and surgical journal*. This was the title of vol. 3 (1851) as well. With vol. 4 the title changed to the *Eclectic journal of medicine* (#1027). Vol. 5 was published at Syracuse, N.Y. as the *Union journal of medicine* until July, 1853, when the journal ceased publication. These changes of title and place of publication coincide with the fortunes of the Central Medical College with which it was closely associated and whose faculty served as editors. The C.M. College opened its doors in Syracuse for the 1849-50 session, moved to Rochester for the 1850-51 session, and returned to Syracuse in 1853.

Volume 1 of the *Eclectic journal of medicine & surgery* was edited by Stephen H. Potter, who edited the single-issue *New-York eclectic journal of medicine and surgery* while on the faculty of the Eclectic Medical Institute of Randolph, N.Y. Volume 2 was edited by William Warriner Hadley, one of the founding members of the Rochester Eclectic Medical Institute (est. 1848). Both men joined the faculty of the Central Medical College after the decision had been made to fold the Randolph and Rochester institutes into the new medical school at Syracuse in 1849.

Holdings for *The Eclectic medical & surgical journal* in the Atwater collection are incomplete: vol. 1, no. 1 (July 1849)-v. 2, no. 5 (Nov. 1850).

ECLECTIC MEDICAL INSTITUTE OF RANDOLPH see *NEW YORK ECLECTIC MEDICAL AND SURGICAL JOURNAL* (#2594).

ECLECTIC SERIES OF TEMPERANCE PHYSIOLOGIES see **BROWN** (#426).

1029. EDDY, Daniel Clarke, 1823-1896.

The young man's friend; containing admonitions for the erring; counsel for the tempted; encouragement for the desponding; hope for the fallen . . . Boston: Wentworth & Co., 1857.

[4], [9]-260 p., [1] leaf of plates ; 18 cm.

"The object of this book is, to impress upon the minds of young men, such lessons of virtue as will render them useful and successful in life, and by presenting the old Puritan view of sinful pleasures, lead the reader to cultivate the old Puritan integrity" (*Pref.*, p. [3]). Eddy's guide to the formation of character and personal conduct outlines "the four sources of success" (industry, frugality, temperance, honesty); and stresses the distinction between "innocent amusements" and "dangerous amusements" (e.g., the theatre, dancing, social drinking, etc.). A chapter each is devoted to the pitfalls of "wealth and fame" (though the author encourages the acquisition of both), gambling and intemperance. Eddy concludes the book by reminding his readers of the inevitability of God's judgement and the Bible's role as an infallible guide in all life's duties.

The young man's friend was first published in 1849 when Eddy was pastor of the First Baptist Church in Lowell, Ma. It was frequently reissued until its final publication in 1866. According to Eddy's biographer in the *Dict. Amer. biog.* (3:6), *The young man's friend* "sold well, establishing Eddy's reputation in evangelical circles as a sage adviser of the young, and thereafter he gave much of his time to authorship . . . In substance his books are thin and commonplace, but he wrote pleasantly, and for many years he was one of the most popular authors in his denomination." Eddy left the pulpit briefly in 1854 when he was elected to the Massachusetts legislature on the Know-Nothing ticket. "The next election retired him to private life" (*ibid.*)—more specifically, to Boston's Harvard Street Baptist Church, where he was pastor at the time this 1857 issue was printed. Atwater copy lacking plate.

1030. EDDY, Mary Baker, 1821-1910.
Christian healing and The people's idea of God. Sermons delivered at Boston . . . Boston: Published by Allison V. Stewart, 1915.
[8], 20, [2], 14 p. ; 21 cm.

1031. EDDY, Mary Baker, 1821-1910.
Miscellaneous writings. 1883-1896 . . . Twenty-sixth edition. Boston: Published by Joseph Armstrong, 1897.
xv, [1], 471 p., [1] leaf of plates : port. ; 21 cm.

1032. EDDY, Mary Baker, 1821-1910.
No and yes . . . Boston: Published by Allison V. Stewart, 1914.
[8], 46 p. ; 21 cm.

1033. EDDY, Mary Baker, 1821-1910.
Personal contagion. Also What our leader says . . . Boston, Massachusetts: The Christian Science Publishing Society, 1909.
6, [1] p. ; 17 cm.

1034. EDDY, Mary Baker, 1821-1910.
Retrospection and introspection . . . Boston: Published by Allison V. Stewart, 1915.
vi, 95 p. ; 21 cm.

Eddy's autobiographical essay was originally published at Boston in 1891.

1035. EDDY, Mary Baker, 1821-1910.
Science and health with Key to the Scriptures . . . Forty-sixth edition, revised. Boston: Published by the author, 1890.
[4], 590 p. ; 21 cm.

Science and health is the fullest expression of Eddy's metaphysical principles, and is the foundational text of Christian Science. Originally published at Boston in 1875, Eddy made numerous alterations to successive editions until, realizing her literary limitations, she had the text thoroughly revised by James Henry Wiggen (1836-1900), who rewrote "large portions of the book as only a skilful literary editor could, giving clarity to the thought . . . excising irrelevant matter, and reducing vague forms of expression to simple idiomatic language" (*Dict. Amer. biog.*, 3(2):12). Wiggen's revision was issued as the 16th edition (1886).

1036. EDDY, Mary Baker, 1821-1910.
Science and health with Key to the Scriptures . . . Fifty-seventh edition, revised. Boston: Published by W.G. Nixon, 1891.
xii, 651 p. ; 21 cm.

1037. EDDY, Mary Baker, 1821-1910.
Science and health with Key to the Scriptures . . . Eighty-ninth edition. Boston: Published by E.J. Foster Eddy, 1894.
xii, 663, [3] p. ; 21 cm.

"In the year 1866 I discovered the Science of Metaphysical Healing, and named it Christian Science. God had been graciously fitting me, during many years, for the reception of a final revelation of the absolute Principle of Scientific Mind-healing" (p. 1). Eddy makes no mention of the mental healer Phineas Quimby, whose disciple she had been. According to Eddy, the "divine Spirit . . . unfolded to me the demonstrable fact that matter possesses neither sensation nor life; that human experience shows the falsity of all material things"; and that in physical illness, "the only sufferer is mortal mind, since being in God cannot suffer" (p. 2). Healing from both sin and physical illness follow from this realization. In Robert Fuller's words, "Mrs. Eddy followed this line of reasoning to the conclusion that such things as sickness, pain, or evil possess no positive ontological status. They are only delusional appearances created by an erring, mortal mind. The healing ministry in which Christian scientists engage is dedicated to the task of assisting individuals in keeping their minds and mental attitudes centered solely on the higher laws of God's spiritual presence. Mrs. Eddy established churches nationwide . . . to assist sick individuals by teaching them how they might elevate their mental and emotional lives to a higher spiritual level" (*Alternative medicine and American religious life*, New York, 1989, p. 61).

1038. EDDY, Mary Baker, 1821-1910.
Science and health with Key to the Scriptures . . . Boston: Published by Allison V. Stewart, 1910.
xii, 700 p. : port. ; 21 cm.

"The physical healing of Christian Science results now, as in Jesus' time, from the operation of divine Principle, before which sin and disease lose their reality in human consciousness and disappear as naturally and as necessarily as darkness gives place to light and sin to reformation. Now, as then, these mighty works are not supernatural, but supremely natural. They are the sign of Immanuel, or 'God with us,'—a divine influence ever present in human consciousness" (*Pref.,* p. xi).

1039. EDISON, William D.
The people's guide; or, the physiology and hygiene of the organs of generation, and general medical information . . . [S.l.]: Copyright 1891, by William D. Edison.

viii, [9]-149 p., XVI p. of plates : ill. ; 18 cm.

Edison divides his text into four parts. Part I. ("Maiden, Wife and Mother") describes the anatomy of the female genitalia, sexual behavior (e.g., "Is passion undignified?"; "Copulation, shall it be indulged in?"), pregnancy, labor, etc. Part II addresses "the sexual diseases of men" (e.g., sexually transmitted diseases, impotency, masturbation, "neuralgic testicle," phimosis, priapism, spermatorrhoea, etc.). Part III addresses "the sexual diseases of women" (e.g., "barrenness," menstrual disorders, prolapsus uteri, etc.). Part IV contains "miscellaneous formulas" for the domestic treatment of many common diseases.

1040. EDWARDS, Joseph Franics, 1853-1897.
Constipation, plainly treated, and relieved without the use of drugs . . . Second edition, with additions. Philadelphia: Presley Blakiston, 1881.
88 p. ; 17 cm.

Edwards warns against the use of drugs for the stimulation of the bowels, arguing that "The bowels become ultimately so accustomed to the artificial stimulating action of powerful drugs, that they absolutely refuse to move without their aid" (p. 41). In place of the drugs that physicians often prescribe, Edwards recommends changes in diet (i.e., more grains, vegetables, fruits, water, coffee), daily exercise, and such optional palliatives as abdominal manipulation or cold baths.

An 1874 medical graduate of the University of Pennsylvania, Edwards practiced in Philadelphia. He was the founding editor of the *Annals of hygiene* (1884-1897) and the author of several popular treatises on disease and hygiene. Edwards served on both the New Jersey and Pennsylvania boards of health, and was a member of the board of managers of the Trenton Insane Asylum.

1041. EDWARDS, Joseph Francis, 1853-1897.
Dyspepsia. How to avoid it . . . Philadelphia: Presley Blakiston, 1881.
vii, [8]-87 p. ; 17 cm.

To avoid dyspepsia Edwards recommends that his readers eat food that has been cooked slowly, that they masticate thoroughly, avoid satiety, not eat when fatigued or excited, and that they eat at regular times and never between meals or when not hungry.

1042. EDWARDS, Joseph Francis, 1853-1897.

Edwards' catechism of hygiene for use in schools . . . Philadelphia, Pa.: Published by Joseph F. Edwards, M.D., 1893.
   109 p. ; 18 cm.

The *Catechism* was written for elementary school students.

1043. EDWARDS, Joseph Francis, 1853-1897.

Malaria. What it means and how avoided . . . Philadelphia: Presley Blakiston, 1881.
   viii, 9-81 p. ; 17 cm.

Edward's treatise was published the same year (but obviously written prior) to Laveran's description of the malarial parasite in the *Bull. et mém. de la Soc. méd. des Hôp. de Paris*. Edwards states that *mal aria* is not in itself the cause of a specific disease, but is rather the result of unspecified deleterious agent(s) and the medium of its spread. "There is no doubt," he writes, "that many diseases are developed through the agency of impure air, but when we come to *special* diseases, characterized by *special* symptoms, we must look for a *special* cause, and finding it, give it a *special* name" (p. 22). Although unable to attribute this "special name" to any pathogenic entity, Edwards does attribute the resulting *mal aria* to defective sanitation, particularly to defective drains, sewer pipes, water supplies, etc. Effective public health and domestic sanitation are proposed as means of prevention and control.

1044. EDWARDS, Joseph Francis, 1853-1897.

Vaccination: arguments pro and con. With a chapter on the hygiene of smallpox . . . Philadelphia: P. Blakiston, Son & Co., 1882.
   viii, 9-80 p. ; 16 cm.

Edwards examines the vociferous arguments of the anti-vaccination party, against which he aligns the authority of recently published medical literature to argue for the benefits of smallpox vaccination.

1044.1. EGAN, S. T.

LADIES! A Friend in Need, is a Friend Indeed! . . . TURKISH FEMALE REGULATION POWDERS . . . [Boston, ca. 1888].
   Broadside ; 28 × 21 cm.

The *Turkish Female Regulation Powders* are advertised as the "only safe, certain and effectual Emmenagogue known; used today by thousands of American women who declare they speedily restore the Menstrual Secretions or Monthly Sickness where all other remedies fail. If you have been using the many pills that are advertised as a specific in such cases, cast them aside, and send for *The Celebrated Turkish Female Regulation Powders*. They will save you hours of anguish and anxiety, embarassing interviews with physicians, and perfectly restore the Monthly Sickness where all other remedies fail . . . Now you wish something that can be depended upon; so enclose Two Dollars in a letter to Dr. S.T. Egan, and you will receive by return mail a package of CELEBRATED TURKISH FEMALE POWDERS . . . with no marks on the outside."

Emmenagogues are defined in a contemporary source as "agents which stimulate or restore the normal menstrual function of the uterus or cause expulsion of its contents. Among these agents are rue, borax, savin, myrrh, apiol, quinia, and ergot" (J. Thomas, *A complete pronouncing medical dictionary,* Philadelphia: Lippincott, 1889, p. 211). The powder marketed by Egan was doubtless used as an abortifacient. S.T. Egan may be the Boston allopath Sebra Temple Egan listed in the A.M.A.'s *Directory of deceased American physicians, 1804-1929* (1:455). Publication date based on most recently dated testimonial.

1045. EGHIAN, Setrak G., b. 1872.

The mother's nursery guide for the care of the baby in health and in sickness . . . New York and London: G.P. Putnam's Sons, 1907.
   xiii, [1], 263 p., [2] leaves of plates : ill. ; 19 cm.

Eghian (or Eghiaian) was an 1899 medical graduate of the Univ. of Berlin. The 1906 *American medical directory* lists him in practice in Ogden, Utah; the 1912 & 1914 editions place him in Chicago. In the 1916 edition, Eghian is listed on E. 27th Street in New York City. His name last appears in the *American medical directory* in the 1934 edition. His only other published monograph appears to have been his 1899 thesis on diphtheria.

1046. EGLESTON, Nathaniel Hillyer, 1822-1912.
Villages and village life with hints for their improvement . . . New York: Harper & Brothers, publishers, 1878.
[4], viii, [2], [7]-326, 2 p. ; 19 cm.

Villages and village life is an appreciation of country living. The author gives his perceptions of what rural life in New England once was, what it had become (he agonizes over the phenomenon of rural depopulation), and what it could be. In his consideration of "village-improvement," Egleston discusses sanitary matters as well as the social, cultural and aesthetic benefits of rural life. He devotes chapters to the healthful situation of dwellings, to drainage, ventilation and the need of skilled medical care in a rural community. Egleston was the author of several often reprinted books on forestry and agriculture.

1047. ELECTROPATHIC INSTITUTE.
ELECTROPATHIC INSTITUTE. 104 Main Street, Gloversville . . . [Gloversville, N.Y., ca. 1884].
[2] leaves ; 24 cm.

"Some three months ago Dr. Chamberlin located in Gloversville, and announced the important discovery he had made in the polarized action of the electrical forces for the speedy relief and permanent cure of acute and chronic diseases . . . While in Gloversville, Dr. Chamberlin has treated a great number of patients afflicted with various diseases that had long resisted the best medical skill and remedial agencies . . . These cures include affections of the brain, spine, lungs, stomach, liver, bowels, kidneys, bladder, &c., nervous prostration, debility and irritability; also, the crippled and deformed from rheumatism, sciatica and neuralgia—racked with the most excruciating pains—who are now in perfect health, living witnesses of the marvelous curative powers of polarized electricity." Assisting him, "Mrs. C.P. Chamberlin, M.D., graduate of the Woman's Homoeopathic Medical College of New York City, devotes her time to the treatment of ladies" (1st leaf verso).

In addition to treating the afflicted, "Dr. Chamberlin proposes to impart a knowledge of his discoveries in the therapeutic application of Electricity to ladies and gentlemen desiring such information. The lectures and instructions will so combine theory and practice that the student on completion of a course will be fully competent to diagnose, and treat, safely and successfully, even the most delicate and complicated cases" (2nd leaf verso).

The Chamberlins did not remain long in Gloversville. Otis K. Chamberlin, an 1840 Albany Medical College graduate, is listed in the 1886 edition of *Polk's medical and surgical directory of the United States* as practicing in Dayton, Ohio, where he died 20 September 1900. Caroline Parker Chamberlin (b. 1852) was an 1877 graduate of the homeopathic New York Medical College and Hospital for Women. Her only appearance in the medical directories is the 1896 edition of *Polk*, where she is listed as practicing in Chicago.

ELIOT, Ellsworth see **MORTON** (#2513).

1048. ELLET, Elizabeth Fries, 1818-1877.
The practical housekeeper; a cyclopaedia of domestic economy embracing domestic education, the house and its furniture, duties of the mistress, duties of the servant, the store-room and marketing, domestic manipulation, care of children and their food, the table and attendance, the art of cookery, receipts under forty-five heads, family bills of fare, perfumery and the toilet, infusions and cosmetics, pommades, vinegars, soaps, etc., the family medical guide, miscellaneous receipts, etc. . . . New York: Stringer and Townsend, 1857.
599 p. : ill. ; 23 cm.

First edition. Ellet divides her treatise into three parts. The first (p. [13]-152) is titled "Thoughts and maxims on housekeeping," and addresses such topics as furnishings, the duties of servants, the nursery, etc. The second and largest part (p. [153]-534) is a cookbook numbering 1,617 recipes and culinary procedures that includes a chapter on "Cookery for the sick" (p. 517-26). The third part combines chapters on "perfumery and the toilet," a "Family medical guide" (p. [556]-566), and a collection of "miscellaneous receipts." Beginning in 1872, Ellet's manual of home economics was published under the title *The new cyclopaedia of domestic economy* (#1049, 1050).

A native of Sodus Point, N.Y., a village on Lake Ontario east of Rochester, Ellet (nee Lummis) was educated at the Aurora Female Seminary, and attained fluency in French, Italian and German. Beginning in her teens, Ellet contributed poems and stories to various literary magazines. She wrote historical novels, books on contemporary German literature, social history and domestic economy. Ellet's reputation rests on *The women of the American revolution* (1848), *Domestic history of the American revolution* (1850), *Pioneer women of the West* (1852) and *Women artists in all ages and countries* (1859)—books that social historians regard as methodologically ahead of her time for their focus on "ordinary lives rather than extraordinary events" (*Amer. nat. biog.* 7:415).

1049. ELLET, Elizabeth Fries, 1818-1877.

The new cyclopaedia of domestic economy, and practical housekeeper. Adapted to all classes of society, and comprising subjects connected with the interests of every family; such as domestic education, houses, furniture, duties of mistress, duties of domestics, the storeroom, marketing, table and attendance, care and training of children, care of the sick, preparation of food for children and invalids, preservation of health, domestic medicine, the art of cookery, perfumery, the toilet, cosmetics, and five thousand practical receipts and maxims. From the best English, French, German, and American sources . . . Norwich, Conn.: Published by Henry Bill . . . [et al.], 1872.
    [5]-603, [9] p. : ill. ; 24 cm.

Originally published at New York in 1857 as *The practical housekeeper* (#1048), and re-issued under that title in 1867 and 1872. The present edition is the book's first appearance under the title *The new cyclopaedia of domestic economy*, which was issued for the last time in 1873 (#1050). The Atwater copy is lacking at least half the text between pages 213 and 292.

1050. ELLET, Elizabeth Fries, 1818-1877.

The new cyclopaedia of domestic economy, and practical housekeeper. Adapted to all classes of society, and comprising subjects connected with the interests of every family; such as domestic education, houses, furniture, duties of mistress, duties of domestics, the storeroom, marketing, table and attendance, care and training of children, care of the sick, preparation of food for children and invalids, preservation of health, domestic medicine, the art of cookery, perfumery, the toilet, cosmetics, and five thousand practical receipts and maxims. From the best English, French, German, and American sources . . . Norwich, Conn.: Published by the Henry Bill Publishing Company; Ira A. Smith, Milford, Mass., general agent, 1873.
    [5]-603, [9] p., [3] leaves of plates : ill. (some col.) ; 24 cm.

1051. ELLIOT, Sydney Barrington.

Aedoeology. A treatise on generative life. Including pre-natal influence, limitation of offspring, and hygiene of the generative system. A book for every man and woman . . . Boston: Arena Publishing Co., 1893.
    xv, [1], 260 p. ; 20 cm.

Elliot divides *Aedoeology* into three parts. The first consists of five chapters on pre-natal influence, i.e., those physical or mental forces "which modify or distort hereditary tendencies" (p. 3). "As to how the impressions are conveyed to the child, we must confess it a matter of doubt. Many claim that it is through the blood; others that it is through the nerves, and again many claim that it is by means of both . . . Certain it is, that by whatever means the impressions are conveyed, the foetus is undoubtedly influenced through the mother; and whether or not we know the physiology of it, is it not incumbent on us to advise the adoption of measures during gestation that will insure the future well-being of the child?" (p. 51).

In part two Elliot considers two viable methods for the "Limitation of offspring." Chastity is the "ideal procedure," he states, but impracticable: "Most men are born with inordinate sexual passion, and few are endowed with the power of controlling it. Therefore we must have some means more practicable than chastity, without the criminality and danger of abortion, and this we find in the prevention of conception" (p. [148]). The author provides a twenty-one page chapter justifying contraceptive practice, followed by a two-and-a-half page chapter entitled "Methods" in which Elliot writes: "It is much to be regretted that a chapter giving entirely effectual and satifactory . . . methods for preventing conception will have to be omitted, owing to a bill rushed through in the last moments of a recent Congressional session by a few misled, self-styled moralists. This bill makes it unlawful to publish any matter on this subject" (p. 171).

Elliot's third part is entitled "Hygiene and physiology of generative life," in which the author discusses the development of sexual habits in children and adolescents (including a lengthy digression on masturbation), marriage, sexual ethics, etc.

1052. ELLIOTT, Sophronia Maria, b. 1854.
Household hygiene . . . Chicago: American School of Home Economics, 1912.
    iv, [2], 224, [4] p. : ill. ; 20 cm.

In this textbook on domestic hygiene, Elliott explains household drainage, ventilation, heating, lighting, water supply, waste removal, plumbing, pest control, disinfection and cleaning. The author was an instructor in household economics at Simmons College, Boston. On cover: *Text book edition.*

1053. ELLIS, Edward.
What every mother should know . . . Philadelphia: Presley Blakiston, 1881.
    xii, 17-132 p. ; 17 cm.

The present popular work is excerpted from the third edition (1878) of Ellis' *A practical manual of the diseases of children,* published in five London editions between 1869 and 1886 (and as many American editions). "During my residence in New Zealand," Ellis explains in his preface, "the desire for such a small manual has been very frequently expressed by up-country settlers and others, unable readily to procure medical attendance, and also by many mothers, anxious for some book of reference upon . . . general diet for children in health and disease" (p. vi).

1054. ELLIS, Edward Sylvester, 1840-1916.
Health, wealth, and happiness. An invaluable friend to parent and child; to man and woman; to young and old; to rich and poor. HEALTH. How to live to the age of one hundred years; how to be strong, healthy and beautiful; what to do in sickness and health, in accident and sudden emergency; how to save your life and the lives of others, etc., etc., etc. WEALTH. How to become rich; the testimony and rules of those who began poor and became millionaires; how to choose a vocation and how to succeed in it; the road to wealth made as plain as the "road to the mill," etc., etc., etc. HAPPINESS. How to be happy; the only way of attaining happiness in this world and the world to come; the right path or the wrong; golden words of wisdom for the good of all, etc., etc. . . . Philadelphia: Manufacturer's Book Co., [c1892].
    [2], 256, 222, 225 p., [3] leaves of plates : ill. ; 27 cm.

Ellis was superintendent of the Trenton public schools when at age 36 he retired from education to become an incredibly prolific and successful author of dime novels and adventure stories for boys. Later in his career he ventured into non-fiction, including school-texts on arithmetic, history and physiology.

1055. ELLIS, Erastus Ranney, 1832-1914.
Homeopathic family guide and information for the people . . . Detroit, Mich., [c1868].
    144, [8] p. : ill. ; 15 cm.

In justifying his publication of a domestic manual of homeopathy, Ellis writes: "There is no means which the profession can adopt to spread homeopathy so effectually as to bring it right down to the people themselves. Let them try it in their own persons and on their friends, and their vision will surely be delighted with the results. Besides, no physician who has suitable regard for his profession and talents, desires to run to patients every time they have a bit of colic or other of the hundred ailments equally simple" (p. 4). Ellis' text is arranged alphabetically from ABSCESSES to WOUNDS (followed by an appendix: WOMAN, AND SOME OF HER DISEASES). A second edition was published at Detroit in 1882.

Ellis was an 1857 graduate of the Homoeopathic Hospital College of Cleveland. He was professor of surgery at the Detroit Homoeopathic College, and edited the *Michigan journal of homoeopathy* from 1872 to 1875.

ELLIS, Havelock see **KEY** (#2121).

1056. ELLIS, John, 1815-1896.
The avoidable causes of disease, insanity,
and deformity . . . A book for the people as
well as for the profession . . . New York:
Published by Mason Brothers, 1860.
xxi, [22]-348, 48 p. ; 19 cm.

First edition. In this issue, the Mason imprint
appears on a printed slip pasted over original
imprint: New York: Published by the author, 1860.
"The physician, while he confines himself to
the treatment and cure of diseases and deformity, does nothing but plaster over the evils of
humanity. He is at best but a simple scavenger—
useful, it is granted, in a low degree—so long as
he confines himself entirely to the removal of
effects, or symptoms and diseases, which are the
result of causes still operative. Every true lover
of humanity, in the medical profession, has before him a nobler calling, and he neglects the
great duty of his life if he fails to point out to the
community the causes of the ills which he is called
upon to treat" (*Pref.*, p. [v]).
CONTENTS: I. Spiritual or mental causes of
disease. II. Natural causes of disease. III. Use
and abuse of the digestive organs. IV. Violation
of the conditions requisite for physical development and preservation—water, air, sunlight,
exercise. V. Children, their proper and improper management. VI. Our imperfect system
of education among the chief causes of disease.
VII. Fashions and habits of the ladies. VIII.
Neglect of proper amusements, and indulgence
in those which are injurious to physical health
and demoralizing. IX. Improper use of poisons:
narcotics, opium, tobacco. X. Alcohol and fermented drinks. XI. Excessive labor. In "Marriage and its violations" (p. 1-48, 2nd sequence),
Ellis discusses the nature and purpose of marriage, as well as sexual ethics, abortion, adultery, venereal disease, etc. This addendum was
also issued separately in 1860 (#1059).
Ellis was an 1841 graduate of the Berkshire
Medical College in Pittsfield, Mass. Early in his
career he converted to homeopathy, and left
Massachusetts for Michigan. Ellis practiced in
Grand Rapids for two years and Detroit for fifteen years, spending several months out of every twelve in Cleveland teaching at the Western Homoeopathic Medical College. After the
Civil War, he accepted a position on the faculty

of the New York Homoeopathic College, but
gave up the practice of medicine in 1873 to
devote himself to petroleum refining, for which
he had invented a new process.

1057. ELLIS, John, 1815-1896.
The avoidable causes of disease, insanity and
deformity . . . A book for the people as well
as for the profession . . . Fourth edition. New
York: Mason Brothers, 1866.
xxi, [22]-348, 48 p. ; 19 cm.

"Marriage and its violations," p. 1-48 (2nd
sequence).

1058. ELLIS, John, 1815-1896.
Family homoeopathy . . . Thirteenth edition. New York: Boericke & Tafel, 1877.
404, iv p. ; 19 cm.

First published at Cincinnati in 1864, its final appearance seems to be the 1894 Detroit
edition (404 p.).

1059. ELLIS, John, 1815-1896.
Marriage and its violations, licentiousness
and vice . . . New York: Published by the
author, 1860.
48, 4 p. ; 18 cm.

This essay on the the nature and purpose of
marriage was appended to every edition of Ellis'
*The avoidable causes of disease, insanity and defor-
mity* (#1056, 1057). The greater part of this treatise is devoted to the "abuses" of marriage, i.e.,
sexual excess, abortion, adultery and their
medical and moral consequences. The only
contraceptive method mentioned is continence. *Marriage and its violations* was reprinted
in an undated Glasgow edition.

1060. ELLIS, John, 1815-1896.
Personal experience of a physician, with
an appeal to the medical and clerical professions; and an appendix, a review of
"Christ and the temperance question" in
the *Christian Union* . . . Philadelphia:
Hahnemann Publishing House, 1892.
134, [2] p. ; 18 cm.

In his 76th year Ellis wrote this rambling
essay, providing much autobiographical matter,
yet another *apologia* for homeopathy, and an

account of the author's religious opinions. Reflecting on his Baptist upbringing, Ellis recalls his early interest in Unitarianism, his acquaintance with the Millerites, and ultimately, his conversion to the doctrines of Swedenborg. The health of the soul is inseparable from the health of the body in all Ellis' published works, and his doctrinal reflections automatically lead him to expound on such threats to spiritual and bodily health as tobacco, alcohol, and the neglect of hygienic doctrine generally. Many pages are devoted to explaining the "New Dispensation" revealed by Swedenborg, and the need to proselytize in order to usher in an era of renewed spiritual and physical health. In printed wrapper.

1061. ELLSWORTH, Paul.
Direct healing . . . Holyoke, Mass.: Published by the Elizabeth Towne Co., 1914.
[4], 173, [3] p., [1] leaf of plates : port. ; 19 cm.

First edition. Ellsworth was a proponent of the *New Thought* movement, which he addresses thus: "There is a spirit in man, latent usually, which may be quickened and made positive and dominant in the life. When this quickening takes place, the individual ceases to be an insulated unit, at variance with every other unit, and becomes one with that Universal Spirit which permeates and controls all things. Such a man truly becomes a wonder-worker, for he is filled with wisdom and power . . ." (p. 3). *New Thought* will create new citizens for the *New State* and "the elimination of the destructive activities of war, pestilence, [and] unnecessary competition," as well as provide endless "possibilities, with all mankind working together under perfect conditions of training and intellectual education" (p. 8) that will usher in the Kingdom of God here on earth. But first, it is necessary to bring about the "regeneration of the individual." It is prayer, drawing on the wisdom and power of the Universal Spirit, that provides the "spiritual energy" needed for this "self-realization."

In chapter III, Ellsworth discusses "direct healing," which he contrasts to "indirect healing," i.e., the therapeutic method of the various medical schools that addresses symptoms through the administration of drugs. *New Thought* healers, he states, understand that "the only cause of disease is abnormal thought or emotion," and that no matter how skillful the medical regimen, "As long as these corrosive vibrations are turned into the organism it is impossible for permanent health to exist." Rather, when these abnormal thoughts are reversed, "the life power which originally built the body is perfectly capable of rebuilding and setting it to rights. To strive to force it to do so by the use of chemical irritants [i.e., drugs], while at the same time destructive thought is allowed to continue its work, is absurd" (p. 31).

Direct healing was reissued in 1918 and 1919. Elizabeth Towne (b. 1865), the book's publisher, was an author and the publisher of other *New Thought* treatises. In 1982 Ellsworth's book was resuscitated in an edition printed in New Hollywood, Calif., and was reissued the following year in San Bernardino. Frontispiece portrait of author.

1062. ELLWANGER, George Herman, 1848-1906.
Meditations on gout. With a consideration of its cure through the use of wine . . . New York: Dodd, Mead & Company, [c1897].
[2], xvi, [2], 208 p. ; 19 cm.

A native of Rochester, N.Y. and graduate of the University of Rochester, Ellwanger was a partner in the world renowned Mount Hope Nursery (the firm founded by his father George Ellwanger and Patrick Barry), a newspaper publisher, and the author of popular books on nature, gardening, gastronomy, and interior decoration.

In this elegant essay Ellwanger reviews for a lay audience the varying theories of gout's etiology, pathology and management, intending to place in more reasonable perspective "the charges of high-living and intemperance made by vegetarians, Grahamites, and intemperate tea and water devotees" (p. 23). In his consideration of the treatment of gout, Ellwanger points out both the benefits and ill-effects of colchicum, a long-standing botanical remedy for gout, and dutifully reviews the reputed benefits to the gouty temperament of dietary change, abstinence from alcohol, and exercise. Ellwanger choses the middle ground: "To be constantly theorising of what one may eat and what one should leave alone, or considering whether one should grahamise or vegetarianise—measuring this and weighing that—is enough itself to develop gastralgia, and render existence a torture" (p. 111).

In a review of the relation of wine to gout that occupies nearly half the book, he agrees that "some subtle and deleterious property . . . lurks in the fermented fruit of the vine" (p. 124), but argues that this property may not exist in all wines, and that wines harmful to one constitution may actually benefit another. To this end he provides a knowledgeable review of wine varieties that is as entertaining as it is informative. He concludes that rather than eliminate wine from the diet, the patient "…should only drink such species as are best suited to his peculiar constitution, condition, and temperament" (p. 156), quoting Ewart's *Gout and goutiness* (1896): "idiosyncrasy is as apt to be as marked in respect to wine as in respect to articles of diet" (p. 157).

"Clearly," he concludes, "the doctors are at fault; and wine may after all be a good familiar creature, and the sun's best use be 'to warm the grape.' As in angling, 'all winds are hurtful if too hard they blow' … so with wine, all kinds may be harmful if partaken of too generously." He continues, "As wine, with its engaging perfume and exhilarating flavour, is so efficacious in many forms of sickness; as it proves so cheering to the senses and provocative of goodfellowship when partaken of in moderation; as, moreover, the vine grows, and thrives, and ripens its ruby, roseate, and golden fruit on arid soils where no other useful plants could fine a sustenance,— would it not seem a natural consequent that its expressed juices were pre-eminenetly designed by Nature for the health and solace of mankind? And differing so widely as constitutions do, may not a

potent cause of the disease under consideration be the use of certain forms that are inimical to the individual, rather than to the mere use of wine *per se.*" (p. 161).

Who, then, needs a physician in the management of gout? "A conscientious wine-merchant, accordingly, who has the reputation of vineyards of his own at stake, a merchant who is thoroughly familiar with the wines of his district, and who may be absolutely relied upon to send samples not only true to name, but of a *mise irreprochable* so far as the bottling is concerned, becomes the greatest essentiality in a cure of gout by the means of wine" (p. 162). A facsimile of this delightful treatise was published at Rutland, Vt. in 1968.

1063. ELMIRA WATER CURE.
Circular of the Elmira Water Cure … 1882.
Re-opens May 1st. [Elmira, N.Y., 1882].
    [2] leaves ; 20 cm.

The Elmira Water Cure was founded by in 1852 Silas Orsemus Gleason (1811-1899) and his wife Rachel Brooks Gleason (1820-1905) [#1362-1367]. The brochure describes the full range of rooms, amenities and therapeutics provided, including medical consultation with either of the Gleasons.

Fig. 25. The Elmira Water Cure, from R.B. Gleason's Talks to my patients *(#1363).*

1064. ELMIRA WATER CURE.

Turkish and Russian baths. We take this method of informing you, that, having recently fitted up nicely appointed Turkish and Russian bath rooms, they are open to the public . . . Drs. Gleason & Co., Elmira Water Cure, May 1, 1880 . . .
    Broadside ; 21.5 × 14 cm.

1065. ELSBERG, Louis, 1836-1885.

The throat and its functions in swallowing, breathing and the production of voice . . . Lecture delivered in the hall of the Young Men's Christian Association, February 25, 1879, being one of a course of popular-scientific lectures instituted by the New York Academy of Sciences. Second edition—illustrated. Albany, N.Y.: Edgar S. Werner, 1882.
    60, [4] p. : ill. ; 18 cm.

Elsberg was the foremost laryngologist of his generation. An 1857 graduate of Jefferson Medical College, Elsberg studied under Czermak in Vienna (1858), and was appointed professor of laryngology and diseases of the throat at the University of the City of New York in 1861. He was the first in the United States to advocate the diagnostic use of the laryngoscope, was one of the founders (and editor) of the *Archives of Laryngology* (1880), and was one of the founding members of the New York Laryngological Society (1873) and the American Laryngological Society (1878). In printed wrapper.

1066. EMERSON, Charles Wesley, 1837-1908.

Physical culture . . . Tenth edition. Boston: Emerson College Publishing Department, 1905.
    xiii, [1], 40, [4], 41-42, [2], 43-44, [2], [45]-46, [2], 47-48, [2], 49-50, [6], 51-52, [2], 53-54, [2], 55-56, [2], 57-58, [2], 59-60, [4], 61-[62], [2], 63-66, [2], 67-70, [2], 71-72, [4], 73-74, [2], 75-76, [4], 77-78, [2], 79-82, [2], [83]-84, [4], 85-86, [4], 87-88, [4], 89-90, [4], 91-92, [4], 93-94, [4], 95-154 p. : ill. ; port. ; 21 cm.

"The origin of this book is as follows. During the past few years the author has given public lectures upon a system of physical culture arranged by himself and consisting of exercises, many of which he originated, while others were adapted from suggestions received from other systems. By means of this system, together with voice culture, the writer restored himself to health at a time when he had become a confirmed dyspeptic and was a victim of consumption in an incipient stage, and by means of this system he has since developed a most abundant vitality and great muscular power. The system has become part of the curriculum of the Monroe, now the Emerson College of Oratory, where it has been the means of restoring the sick to health, and of harmonious bodily education for the strong" (*Pref.,* p. xiii).

Emerson attended the local schools in his native Pittsfield, Vt. before going on to the study of theology, law and medicine. There is no indication that he took a degree in any of these fields, but he did enter the ministry at an early age. In 1880 Emerson founded the Emerson College of Oratory, from which he resigned in 1902. *Physical culture* appeared in fourteen editions between 1891 and 1924. Emerson's other books, *Evolution of expression* (1888) and *Expressive physical culture* (1900), were devoted to oratory and oratorical gesture.

1067. EMMONS, Ebenezer, 1799-1863.

The Empire Spring: its composition & medical uses; from Prof. E. Emmons' book of analysis of the Empire Spring of Saratoga. Saratoga Springs: Printed at the office of the Saratoga Whig, 1850.
    6, [2] p. ; 16 cm.

Reprinted from Emmons' *The Empire Spring, its composition and medical uses; together with a notice of the mineral waters of Saratoga, and those of other parts of New York* (Albany, 1849).

1068. EMMONS, Samuel Bulfinch.

The vegetable family physician: containing a description of the roots and herbs common to this country, with their medicinal properties and uses; also directions for the treatment of the diseases incident to human nature, by vegetables alone; embracing many valuable Indian recipes . . . Boston: George P. Oakes, 1836.
    [10], [7]-176 p., ; 16 cm.

"From the conviction that the public feeling upon the subject of medicine is decidedly in favor of botanical remedies, the product and growth of our own native soil, rather than the use of the most deadly foreign minerals, a few drops or grains of which have a tendency to destroy life instantaneously . . . the author of this little book has been induced to prepare it for the benefit of the world at large . . . Modern physicians have adopted the use of minerals, which by art are reduced to so small a quantity, that a doctor does not need a pair of saddle-bags larger than two coat pockets, to carry drugs enough to kill or cure all he may have occasion to visit for a considerable time" (*Pref.*, p. [3]).

Emmons has divided his book into four principal parts: "Description of roots and herbs," an annotated alphabetical catalog of 124 medicinal plants; "Remedies for particular diseases," an annotated alphabetical catalog for treating 52 disorders; "Miscellaneous recipes," i.e., directions for making "gargle for sore throat," "Rogers' anti-scrofulous plaster," "acorn coffee," "cancer tea," "bed bug liquid," etc; and a miscellany of brief essays on such topics as "collecting and curing herbs, barks and roots"; the "importance of the steam or vapor bath," excerpts from James Hamilton's essay on mercury, etc. At head of title-page: *Every man his own physician.*

1069. EMMONS, Samuel Bulfinch.

The vegetable family physician: containing a description of the roots and herbs common to this country, with their medicinal properties and uses; also directions for the treatment of the diseases incident to human nature, by vegetables alone; embracing many valuable Indian recipes . . . Boston: Published by Benjamin Adams, 1842.

x, [11]-180 p. ; 16 cm.

Cover-title and at head of title-page: *Every man his own physician.*

1070. ENGLISH, Virgil Primrose, b. 1858.

The doctor's plain talk to young women. Anatomy, physiology and hygiene of the sexual system and the relation of this system to health, beauty and popularity . . . First edition. Cleveland, Ohio: Ohio State Publishing Company, 1902.

ix, [10]-220, [4] p., [1] leaf of plates : ill., port. ; 19 cm.

CONTENTS: I. Feminine beauty and what it signifies. II. Sexuality and its relation to hu-

man beauty and perfection. III. Importance of a correct understanding of nature's sexual laws. IV. Relation of the sexual organs to the rest of the body. V. Female reproductive organs. VI. Male organs of reproduction. VII. How plant reproduction begins in blossoms. VIII. Seeds and eggs and their functions in reproduction. IX. Reproduction in the human body. X. Nature's plans for insuring the most babies from the best parents. XI. Influence of sexuality upon the popularity of men and women. XII. Amativeness and its influence upon character, conduct, personal appearance and popularity. XIII. Masturbation—the philosophy of its evil results. XIV. Signs and symptoms that expose the masturbator's guilt. XV. Sexual excesses—chastity—marriage. XVI. Common sense treatment for sexual wrongs.

"Beauty . . . is a quality of no small significance. And those who have asserted that it is but 'skin deep,' have voiced a mistake. Beauty, on the contrary, extends to the very most inner recesses of the body. Every part and every tissue is characterized by the same degree of beauty and superiority. And it does not stop even there, but it extends to the very soul and permeates even the mind. Whenever the body is a beautiful one, this alone is positive proof that the mind is also beautiful, because the mind builds the body and it always builds a body that corresponds with itself . . . Our thoughts, feelings, emotions and desires, determine the qualities of our bodies" (p. 16). The beautiful woman "is strong and vigorous in every department of body and mind," enjoys the best health, and possesses all the virtues of mind and body that makes her complete and enables her to perform all functions fully, especially those pertaining to marriage and motherhood. English uses the term *sexuality* in a far more general sense than sexual instinct or genital gratification. It is a mode of being, a composite of the physical, mental and emotional qualities that makes one "well sexed," i.e., fully a man or fully a woman. It is *sexuality* in this broad sense that "modifies the entire mind and body. Nothing is more influential in determing the success or failure of a person's life, than the development and health or disease of the sexual system" (p. 26).

This was the only edition of *The doctor's plain talk to young women* (spine-title: *Plain talk to young women*). It was the companion volume to his earlier *The doctor's plain talk to young men.* English was an 1892 graduate of the Cleveland Homoeopathic Medical College. The four unnumbered pages at the end of the volume advertise the author's practice in Cleveland, where he specialized in "nervous disorders and the diseases of women."

1071. ENTRIKIN, Franklin Wayne, 1830-1897.
Woman's monitor . . . Women of America, seek to add to a knowledge of God a knowledge of yourselves, that your lives may be brought to harmonize with his laws. Green Springs, O.: F.W. Entrikin & Co., 1881.
    400, [2] p. ; 20 cm.

"It is not our design to make mothers their own physicians, but to assist them to distinguish between the professional gentleman and the impostor, to cooperate with the family physician in the prevention and cure of disease, and to enable mothers to place in the hands of their daughters, before they leave the maternal roof, a book that will assist to prepare them for the high duties of coming years" (*Pref.*, p. 4). On the following page Entrikin warns, "We frequently see in the hands of mothers small works written by designing men upon some of the diseases of women. So far as we have observed such they have pandered to low tastes and depraved ideas, or puffed some patent nostrum. They usually contain some truth and much error, and are calculated to do more harm than good."

The opening chapter of the *Woman's monitor* is devoted to women's hygiene. Lengthy attention is given to gynecologic diseases (especially uterine and menstrual disorders) before the author enters a discussion of "sexual frauds," by which he means the use of abortifacients and contraceptive devices. In addition to abortion, Entrikin mentions four methods of birth control: the "cundum, named after Dr. Cundum, by whom it was invented"; the vaginal sponge; post-coital use of a syringe; and "conjugal onanism," which he maintains is "most extensively practiced by married persons of the middle and poorer classes" (p. 199). All are contrary to God's law, and each leads to a particular set of negative medical consequences. Entrikin explains the signs and management of pregnancy; and provides extensive coverage of infant care and the diseases of children. In his fi-

nal thirty pages, he returns to the subject of criminal abortion.

Entrikin was an 1873 graduate of the Medical College of Ohio (Cincinnati). According to Polk's *Medical and surgical directory* for 1886, Entrikin was professor of surgery at the Ft. Wayne College of Medicine, physician to the Green Spring Medical and Surgical Sanitarium, and physican to the Findlay Invalids' Home (Findlay, Ohio). The *Woman's monitor* was Entrikin's only book, the first edition of which was published at Cincinnati in 1871.

EPHRATA HYDROPATHIC INSTITUTE
see LANDIS (#2189).

1072. EPPS, John, 1805-1869.
Domestic homoeopathy; or rules for the domestic treatment of the maladies of infants, children, and adults, and for the conduct and the treatment during pregnancy, confinement, and suckling . . . First American edition. Boston: Otis Clapp; New York: William Radde; Philadelphia: J.G. Wesselhoeft, 1843.
    200 p. ; 17 cm.

In his preface to the first edition Epps writes: "The following pages have been compiled because many patients, living at a distance, and in places where no homoeopathist resides, desire to use none but homoeopathic medicines. Having derived benefit from the remedies prescribed by the writer, their wish was not to be obliged to have recourse to the *old* system for the little ailments, which do not require the consultation of the physician, but which require the aid of medicine" (p. [7]).

An 1826 Edinburgh medical graduate, Epps entered practice in London in 1827, where he also began to lecture on phrenology, a subject with which he had become familiar during his medical studies. Dissatisfied with the results of regular medicine, Epps began the study of homeopathy in 1838. Being convinced of its scientific validity, he published in the same year a treatise entitled *What is homoeopathy?* The first edition of *Domestic homoeopathy* was published at London in 1841. It was his most successful published work, and appeared in six London editions through 1863. This first American edition is a reprint of the second London edition (1842). Five American editions were published between 1843 and 1853. In addition to his devotion to phrenology and homeopathy, Epps was active in various radical movements. He was the friend and supporter of Mazzini and Kossuth; the author of numerous monographs on medical, religious and phrenological topics; and the editor of several periodicals.

1073. EPPS, John, 1805-1869.
Domestic homoeopathy: or rules for the domestic treatment of the maladies of infants, children, and adults, and for the conduct and the treatment during pregnancy, confinement, and suckling . . . Fourth American from the fourth London edition. Edited and enlarged by George W. Cook, M.D. Boston: Otis Clapp . . . [et al.], 1849.
   320 p. ; 18 cm.]

Though Cook's name appears on the title-page as editor, the preface to this 4th American edition is signed John Adams Tarbell, 1810-1864 (#3414-3418). Cook's name first appeared on the title-page of the 3rd American edition (1848), his preface to which is included here. In his own preface, Tarbell lists the extensive additions made to this 4th edition (p. [16]), which includes "a condensed account of the "Cold Water Treatment" by himself. Tarbell points out, however, that "Hydropathy, as a system of cure, is not . . . recognized by the adherents of Hahnemann; but, as an auxiliary, merely, to the action of internal remedies." Nonetheless, "its advantages are undeniable."

1074. EPPS, John, 1805-1869.
Domestic homoeopathy; or, rules for the domestic treatment of the maladies of infants, children, and adults, and for the conduct and the treatment during pregnancy, confinement, and suckling . . . Fifth American from the fourth London edition. Edited and enlarged by John A. Tarbell, M.D. Boston: Otis Clapp . . . [et al.], 1853.
   383, [1] p. ; 18 cm.

In this much enlarged 5th edition, Tarbell distinguishes the additions made to Epps text by George W. Cook in the 3rd edition and those made by himself in this 5th edition by inserting "G.W.C." or "J.A.T." after each addition. Tarbell has also added to this edition several chapters of his own composition.

1075. E. H. ERICKSON ARTIFICIAL LIMB COMPANY.
E.H. Erickson Artificial Limb Co. Incorporated, inventors and manufacturers of the E.H. Erickson patent artificial leg. Endorsed and purchased by United States government. Exhibited from May 1 to November 1, 1901, Pan-American Exposition, Buffalo, N.Y. . . . Minneapolis. Minn., [1901].
   [1] leaf : ill. ; 33 × 18 cm.

Three half tone photographs on the verso depict H.C. Pierce, the Company's representative at the Pan American Exposition, wearing the Erickson artificial leg.

1076. ERIE MEDICAL COMPANY.
Perfect manhood and the way to attain it. Buffalo, N.Y.: The Erie Medical Company, [191-?].
   32 p. ; 8 × 13.5 cm.

The Erie Vacuum Treatment is guaranteed to restore and increase "nerve-power and sensation" in any disordered part or limb, but in the male organ particularly. The *Vacuum Strengthening and Developing Appliance* increases blood circulation, and restores "power and strength." It effects "enlargement and strengthening of the muscles," "growth and development of the part," and is applicable to the cure of "sexual weakness"; spermatorrhea; "sterility as connected with impotency"; variocele, urological complaints, and the restoration of "undeveloped, stunted, shrunken, deformed and weak organs" (whether congenitally deformed or as the result of the habit of masturbation). In printed wrapper.

1077. ESDAILE, James, 1808-1859.
Mesmerism in India, and its practical application in surgery and medicine . . . Hartford: Silas Andrus and Son, 1847.
   xxvi, [27]-259 p. ; 20 cm.

First American edition (1st edition: London, 1846). As a hospital surgeon in Hooghly, India, Esdaile's casual interest in Mesmerism increased in seriousness after operating on a Hindu convict whom he had hypnotized. In this work, written for both lay and professional audiences, Esdaile describes the Mesmeric process, the use of Mesmerism as a remedial agent, and its use as an anesthetic. In all, Esdaile claims to have painlessly operated on more than 261 hypnotized patients. On his return to Scotland after twenty years' service in India, Esdaile found the natives markedly less susceptible to hypnosis. His claims for Mesmerism as a surgical anesthetic were doubtless relegated to obscurity by the introduction of chloroform in 1846 and its rapid spread to Europe and even to India by 1848.

1078. ESDAILE, James, 1808-1859.
Mesmerism in India, and its practical application in surgery and medicine . . . Hartford: Silas Andrus and Son, 1850.
xxvi, [27]-259 p. ; 20 cm.

1079. ESREY, William P., 1818-1854.
A treatise on anatomy and physiology . . . Philadelphia: Published by Rademacher & Sheek . . . [et al.], 1851.
195 p. : ill. ; 20 cm.

"The following Treatise was prepared to accompany a work on Domestic Medicine, and without any design of publishing it in a separate form. The Publishers, however, have thought proper to issue it separately, believing that the multiplication of books of the kind, especially when presented in a cheap and condensed form, cannot be other than productive of good results, as a greater number of persons will thereby be induced to read and study them" (*Pref.*, p. [5]). Esrey's *Treatise*, in fact, was published as part of C.G. Caspari's *Homoeopathic domestic physician* (#573), which, though copyrighted in 1851, was not published until 1852. Esrey was an 1844 graduate of the Jefferson Medical College, and a convert to homeopathy.

ESREY, William P. see also **CASPARI** (#573).

EUGENIE, Madame see **LIEB** (#2248).

1080. EUREKA MEDICAL SANITARIUM.
Revised edition of the Magic wand and medical guide. The most wonderful and entertaining book ever published. Full of strange, curious and marvelous disclosures, and practical hints, of use in love, courtship, and marriage. Also, sure methods of

curing and preventing disease. It contains wonderful and mysterious disclosures of the great secrets of ancient and modern days, new revelations in natural and celestial magic, alchemy, transmutation, etc. Edited by R.F. Yungs . . . Brooklyn: Published by the Eureka Medical Sanitarium, 1889.
[2], 166, [2], 167-188, [2], 189-264, [2], 265-310, [2], 311-317, [1], vii p. : ill. ; 14 cm.

The present work is indeed a revised edition of *The magic wand and medical guide* that appears to have originally been published by Dr. J.H. Reeves & Co. in 1866 (#2950). The volume is an odd amalgam of advice on sex disorders, genital diseases, "female weakness," masturbation, menstrual disorders, sexually transmitted diseases, etc. and their cure by remedies made or distributed by the Eureka Medical Sanitarium. The authors also provide advice on selecting a marriage partner, the physiology of sex, recipes for domestic, kitchen and medical use, etc. At the bottom of the title-page: "S. Matherson, M.D., sole proprietor." In printed wrapper.

1081. EUROPEAN ANATOMICAL, PATHOLOGICAL AND ETHNOLOGICAL MUSEUM.
Guide to the European Anatomical, Pathological and Ethnological Museum, at No. 729 Chestnut Street, Philadelphia. Proprietors, Drs. Jordan & Davieson . . . Open daily, for gentlemen only, from 10 A.M. till 10 P.M. [Philadelphia, between 1878 and 1881].
52, 65-66 p. ; 20 cm.

The European Anatomical, Pathological and Ethnological Museum was located at the following Philadelphia addresses between 1872 and 1900: 1872-1877, 807 Chestnut St.; 1878-1881, 729 Chestnut Street; 1883-1900, 708 Chestnut Street. The catalogs issued by the Museum may be dated approximately with this information.
In its first period, the Museum was operated by Philip J. Jordan (#2048), Samuel Davieson and David Davieson. Jordan was a member of the Jordan family that ran similar museums in New York, Boston and San Francisco. Philip J. Jordan received an honorary medical degree

from the Eclectic Medical College of New York in 1872. David Davieson received a degree from the same medical college in 1874. After 1877 Jordan moved to New York City where he was associated with the New York Museum of Anatomy (#2598) from 1879 to 1881. This was the original Jordan family anatomy museum established in 1862 by Henry J. Jordan and Samuel T. Beck. David Davieson moved to St. Louis where he operated the Grand Museum of Anatomy (#1399) with Sydney Davieson.

In 1877 or 1878, the Museum in Philadelphia was taken over by Robert J. LaGrange (#2173-2175), who was soon after joined by Robert J. Jordan (*see Jourdain*, #2051, 2052). LaGrange and Jordan operated the Museum until its demise circa 1900. LaGrange claimed to have received his medical degree from the Eclectic Medical College of New York in 1879. Catalogs of that institution list him as matriculant for the winter sessions of 1877/78 and 1878/79. In the 1880/81 catalog he is denominated M.D. Robert J. Jordan (1830-1909) was born in England, and was probably the son of Henry Jacob Jordan. The younger Jordan had medical credentials from the Royal College of Surgeons (Edinburgh) and the Royal College of Surgeons (London), and in 1881 received a degree from the Eclectic Medical College of New York. Between 1868 and 1881, R.J. Jordan operated the Parisian Gallery of Anatomy and Medical Science in Boston (#2734, 2735) under the name Robert J. Jourdain. LaGrange and Jordan also maintained an office in Philadelphia at 1625 Filbert St. where they were available for "private consultations." LaGrange died at his Philadelphia home on 18 April 1909.

The catalogs of the European Museum are typical of this genre, with emphasis on basic human sexual anatomy and physiology, and on embryological and fetal development. The catalogs list exhibits illustrating the ravages of venereal disease and masturbation designed to persuade the observer that he needed the professional services of one of the doctors. The consultation was often free, but the treatment was usually costly. Special features of the Philadelphia museum included a display of instruments of torture and life-size wax figures of (among others) Venus, Bismarck, Napoleon III on his deathbed (who by means of a mechanical apparatus heaved his chest and moved his eyelids), and a wounded zouave who similarly struggled for breath. In addition to catalogs of the museum, it was the custom to publish books based on lectures given by the physician proprietors (by way of example, see the *Review of the work published by Robt. J. La Grange* [#2174]).

1082. EUROPEAN ANATOMICAL, PATHOLOGICAL AND ETHNOLOGICAL MUSEUM.
Descriptive catalogue of the European Anatomical, Pathological and Ethnological Museum, 708 Chestnut Street, Philadelphia. Open daily, for gentlemen only, from 8 A.M. till 10 P.M. Drs. LaGrange & Jordan, (late Jordan & Davieson) Philadelphia, [188?].
71 p. ; 20 cm.

1083. EUROPEAN ANATOMICAL, PATHOLOGICAL AND ETHNOLOGICAL MUSEUM.
Descriptive catalogue of the European Anatomical, Pathological and Ethnological Museum, 708 Chestnut Street, Philadelphia. Open daily, for gentlemen only, from 8 A.M. till 10 P.M. Drs. LaGrange & Jordan, Philadelphia, [between 1883 and 1900].
71 p. ; 19 cm.

1084. EVANS, Florence, d. 1920.
Purity, birth and sexual problems for mothers . . . Boston, Massachusetts: Purity Publishing Company, [c1908].
120 p. : ill. ; 18 cm.

A mother's guide for training her children on basic sexual morals and physiology.

1085. EVANS, Warren Felt, 1817-1889.
The mental-cure, illustrating the influence of the mind on the body, both in health and disease, and the psychological method of treatment . . . Seventh edition. Boston: Colby and Rich, publishers, 1885.
xviii, [19]-364, [4] p. ; 19 cm.

Evans' *Mental-cure* appeared in nine editions between 1869 and 1887. The author was a Methodist clergyman until 1864, when he withdrew from the Conference of the Methodist Church to join the followers of Swedenborg in the New Church. In 1863 he had become a patient of

Phineas Parkhurst Quimby (1802-1866), the founder of mental healing in America (and the mentor of Mary Baker Eddy). Evans soon became Quimby's disciple, a practitioner of mental healing and the author of numerous books on "mental medicine."

In his preface Evans writes: "We have aimed to illustrate the correspondence of the soul and body, their mutual action and reaction, and to demonstrate the causal relation of disordered mental states to diseased physiological action, and the importance and mode of regulating the intellectual and affectional nature of the invalid under any system of medical treatment. We have also endeavored to demonstrate the value, as remedial agencies, of those subtle forces, both material and spiritual, which the improved science of the age is beginning to recognize, and to explain the laws of our interior being which render the so-called magnetic treatment so efficient in the cure of diseased conditions of the organism, and which bids fair to supplant the current and longer established therapeutic systems" (p. [iii]).

In his biographical essay Robert C. Fuller states that "Evans's healing practices were based squarely on Quimby's system, which combined mesmerizing techniques with instruction designed to counteract faulty ideas and attitudes that Quimby claimed were the true cause of all physical illness. Evans's chief significance is that he progressively developed Quimby's healing system into a distinctive worldview that combined idealistic philosophy, transcendentalism-inspired mysticism, and a pantheism linked with the growing interest in gnostic or occult interpretations of Christianity. In so doing, Evans offered the American reading public a clear alternative to the ideas being developed by one of Quimby's other famous patient-disciples, Mary Baker Eddy [#1030-1038]. Indeed, Evans used the term 'Christian Science' several years before the founding of Eddy's church" (*Amer. nat. biog.*, 7:622).

1086. EVANS, Warren Felt, 1817-1889.
Mental medicine: a theoretical and practical treatise on medical psychology . . . Fourth edition. Boston: Carter & Pettee, publishers, [c1872].
x, 11-216 p. ; 20 cm.

In his preface, Evans writes that this book is "in some degree, supplementary to the previ-

ous volume of the author on the mental aspect of disease and the psychological method of treatment," i.e., his *Mental-cure,* first published in 1869 (7th ed., #1085). The book "contains information everyone needs who has anything to do in the management and care of the sick, and . . . will qualify every person of ordinary intelligence to be his own family physician. It goes forth into the world with the hope that it may contribute some influence towards reviving this divine method of healing the sick, now wholly abandoned by the principal religious and sectarian organizations of the Christian world. May an age of living faith and spiritual power succeed the present reign of materialism and religious impotency . . . It is to be hoped that the world will again witness the spectacle of the doings of men and women who are endued with power from on high, or whose natural faculties and abilities are reinforced and augmented by an influence emanating from Central Life and the omnipresent spiritual realm of existence, intelligence and causation" (p. [iii]-iv). In twenty-four chapters Evans discusses the qualifications of the "psychopathic practitioner"; inducing an "impressionable state" in the patient; the positive and negative influences on the organism of psychic and magnetic forces; the application of "psychic force" to the various regions of the body, etc. *Mental medicine* appeared in fifteen editions between 1872 and 1897.

1087. EVANS, Warren Felt, 1817-1889.
The primitive mind-cure. The nature and power of faith; or, elementary lessons in Christian philosophy and transcendental medicine . . . Fifth edition. Boston: H.H. Carter & Karrick, publishers, [189-?].
x, 215 p. ; 20 cm.

"The philosophy of idealism," Evans writes in his second chapter, ". . . is to be applied to the cure of disease, as it was by Jesus the Christ. All disease, so far as it has a material or bodily expression, must have had a preexistence in us as a fixed mode of thought, that is, as an idea. To expunge from the mind and obliterate from our soul-life the idea of it, is to remove the cause of it, and hence to cure the malady" (p. 10). *The primitive mind-cure* was originally published at Boston in 1884.

1088. EVANS, Warren Felt, 1817-1889.
Soul and body; or the spiritual science of
health and disease . . . Boston: H.H. Carter
and Company, 1876.
    147 p. ; 19 cm.

"In the present earthly stage of our existence
the senses have a preponderating influence
over our judgement and intellectual states.
Hence, in the history of medical opinions (science it can hardly be called), the prevalent theories of disease have been of a purely materialis-
tic character. The various schools of medicine
have come only to the recognition of the effects of disease and of the outward symptoms,
rather than to the perception of disease in its
spiritual causes and reality" (p. 12-13).
    Evans continues, "There is a law of corre-
spondence . . . between the inner man, the living soul, and the external organism . . . When
the soul is restored to health and harmony by
the redemptive agencies of the Gospel, introduced by Christ, the body adjusts itself in accordance with the restored spiritual life, for its
state is always an effect of which the soul is the
cause . . . When the correspondence between
any organ and the answering spiritual principle
is loosened or lost, the vital tone of the part
will be lowered, and its physiological function
disturbed, altered, or suspended" (p. 14-15).

1089. EVANS, William Augustus, 1865-1948.
Some information about the so-called venereal diseases, most of which appeared in
the Chicago Tribune . . . This pamphlet is
being distributed in the hope that it will
decrease the amount of venereal diseases;
that those infected may persist in treatment
until the complete cure is effected; that it
will save some innocent people from infection; that it will promote a saner, fairer
public opinion; and finally that sexual
morality may be promoted. [Chicago:
Dept. of Health], 1913.
    36 p. ; 15 cm.

In printed wrapper.

1090. EVEREST, Thomas Roupell, d. 1855.
A popular view of homoeopathy . . . From
the second London edition. With annotations and a brief survey of the progress and
present state of homoeopathia in Europe.

By A. Gerald Hull, M.D. New York: Published by William Radde, 1842.
    [8], xxiv, [25]-243 p. ; 22 cm.

In the preface to this early contribution to
the apologetic literature of homeopathy, the
American editor, Amos Gerald Hull (1810-
1859) writes, "We tender the following eloquent
pages, from the pen of a learned and able clergyman of the English Church . . . as an explanation of the characteristics of our system, peculiarly adapted to general perusal, and as a
keen and most just rebuke to those of our medical brethren, who daily misrepresent that system among their patients and friends. Mr.
Everest effectively answers the charge circulated
by the classes of Allopathists above indicated,
that there is neither science nor philosophy in
the new system . . . The public should know
that the durable and beneficent cures which
are every where produced, by the real members of our school, are not the work of fortunate conjecture . . . nor of a stumbling routine
of blind empiricism, but that they are results
obtained upon the well defined principles of a
real, and imperishable art" (p. [ii]). The second London edition, upon which this New York
edition is based, was published in 1836. Possibly unknown to Hull, an earlier American edition had been published at Allentown, Pa. in
1835. Spine-title: *Everest on homoeopathy.*

1091. EVERETT, George H.
Health fragments or, steps toward a true
life. Embracing health, digestion, disease,
and the science of the reproductive organs
. . . Part first. By George H. Everett . . . [Part
II. Embracing dress, heredity, child-training, kitchen and dining-room ethics by
Susan Everett . . .]. New York: Published
by the author, 1874.
    viii, 256, [4], 52 p. : ill. ; 24 cm.

First edition. In part one G.H. Everett discusses the laws of hygiene, with special regard
to diet, pure air, and disorders of the respiratory system. In his final chapter, he examines
sexual behavior and the debilities that sometimes result. In part two, S. Everett focuses on
women's dress, child rearing, marriage, and
various aspects of domestic management.
    On p. [240] G. Everett refers to his diploma
from the Univ. of Pennsylvania. He may well be

A WALL STREET MAN HAS ONLY THREE MINUTES FOR DINNER FINDS HE CAN SAVE THIRTY SECONDS BY EMPLOYING AN EATING MACHINE.

Fig. 26. From Everett's Health fragments, *#1091.*

the George Henry Everett (b. 1835) listed in the AMA's *Directory of deceased American physicians, 1804-1929* as having graduated from the Univ. of Pennsylvania School of Medicine in 1869 (I:478). Susan A. Everett graduated from Penn Medical University (Philadelphia) in 1860. The Everetts, man and wife, append the M.D. to their names (p. viii) and describe themselves as "widely known lecturers and health teachers."

1092. EVERETT, George H.
Health fragments or, steps toward a true life. Embracing health, digestion, disease, and the science of the reproductive organs ... Part first. By George H. Everett ... [Part II. Embracing dress, heredity, child-training, kitchen and dining-room ethics by Susan Everett . . .]. Second edition. New York: Charles B. Somerby, 1875.
    viii, 256, [4], 52 p. : ill., port. ; 24 cm.

    Photographic frontispiece portrait of G.H. Everett.

1093. EVERETT, George H.
Health fragments or, steps toward a true life. Embracing health, digestion, disease, and the science of the reproductive organs ... Part first. By George H. Everett ... [Part II. Embracing dress, heredity, child-training, kitchen and dining-room ethics by

Susan Everett . . .]. Third edition. New York: Wood & Holbrook, publishers, 1876.
    viii, 256, [4], 52 p. : ill., port. ; 24 cm.

    Photographic frontispiece portrait of G.H. Everett.

1094. EVERETT, George H.
Health fragments or, steps toward a true life. Embracing health, digestion, disease, and the science of the reproductive organs ... Part first. By George H. Everett ... [Part II. Embracing dress, heredity, child-training, kitchen and dining room ethics by Susan Everett ...]. Sixth edition. New York: Published by the author, 1877.
    viii, 247, [1], 52 p., [1] leaf of plates : ill., port. ; 24 cm.

    Photographic frontispiece portrait of G.H. Everett.

EVERETT, Susan see **EVERETT** (#1091-1094).

1095. EVERY-DAY WONDERS.
Every-day wonders; or, facts in physiology which all should know . . . Lowell: James P. Walker; Boston: Phillips, Sampson & Co., MDCCCLI. [1851]
    vii, [1], 136 p. : ill. ; 16 cm.

    "The object of the writer of this book has been to present a few of the truths of that science, which treats of the structure of the human body, and of the adaption of the external world to it, in such a form, as that they shall be readily apprehended by children and young people" (*Pref.,* p. [iii]). Reprinted from the 1850 London edition.

1096. EVERY LADY HER OWN PHYSICIAN.
Every lady her own physician. By use of Dr. Chamberlin's favorite prescription applied with Chamberlin's syringe . . . A lady is enabled to cure herself of female weakness in her own home without the aid of a physician . . . [Rochester, N.Y., 187-?].
    16 p. : ill. ; 13 cm.

    In printed wrapper.

1097. EVERY MAN HIS OWN DOCTOR.

Every man his own doctor! A medical hand book, containing a brief summary of the symptoms and cure of every disease. Also, a variety of popular prescriptions, and a very full chapter upon the efficacy of plants and herbs; and some remarks on physiology. Hinsdale, N.H.: Hunter & Co., publishers, 1870 [1871 printing].

134, [2] p. ; 17 cm.

Printed wrapper dated 1871.

1098. EVERY MAN HIS OWN DOCTOR.

Every man his own doctor! or, reliable remedies for common complaints. How to be handsome, or, ladies' guide to beauty. Toilet & etiquette. Agents' guide to success. Lover's guide. Art of psychologic fascination: or, soul charming. Showing how to be healthy, wealthy and wise. Together with many other items for pleasure and profit. Newark, N.J.: Union Publishing Company, 1875.

224, [32] p. : ill. ; 12 cm.

In printed wrapper.

1099. EVERY MAN'S DOCTOR.

Every man's doctor: or family guide to health, containing a condensed description of the various causes and symptoms of diseases with an appendix, showing the medical properties of the most valuable roots and herbs. To which is added 150 recipes. New York: J.K. Wellman, 1845.

72 p. ; 16 cm.

An alphabetical catalog of sixty-six disorders, with an appendix of "medical," "miscellaneous" and "domestic cookery" recipes. In printed wrapper. Atwater copy lacking pp. 71-72.

EVERY WOMAN'S LIBRARY see **WALKER** (#3667).

1100. EWELL, James, 1773-1832.

The medical companion: treating, according to the most successful practice, I. The diseases common to warm climates and on ship board. II. Common cases in surgery, as fractures, dislocations, &c. III. The com-plaints peculiar to women and children. With a dispensatory and glossary. To which are added, a brief anatomy of the human body; an essay on hygiene, or the art of preserving health and prolonging life; an American materia medica, instructing country gentlemen in the very important knowledge of the virtues and doses of our medicinal plants; also, a concise and impartial history of the capture of Washington, and the diseases which sprung from that most deplorable disaster. The third edition, greatly improved . . . Philadelphia: Printed for the author, by Anderson & Meehan, 1816.

viii, [vi]-vii, [xi]-xvi, 694 p. ; 22 cm.

A native Virginian, Ewell studied medicine with his uncle James Craik, of Alexandria, Va. (the friend and physician of Washington), and in Baltimore with a Dr. Stevenson. From Baltimore he went to Philadelphia, where he attended the lectures of Rush, Barton and others at the University, without taking a degree. Ewell settled in Lancaster County, Va. circa 1795, where he purchased a plantation and practiced medicine. With the assistance and encouragement of his father's friend Thomas Jefferson, Ewell moved in 1802 to Savannah, Ga., where he was the first to introduce vaccination (again with Jefferson's encouragement), and where in 1807 he published the first edition of *The planter's and mariner's medical companion* (#1105), one of the most popular American manuals of domestic medicine and hygiene published in the first decades of the 19th century.

According to the biographical memoir in the 10th edition of *The medical companion* (#1102, p. xviii), "It was during professional visits to neighboring plantations, that his attention became fixed upon the necessity of affording, through the means of a popular work on medicine, necessary information as a guide for nurses, and heads of families, in the absence of professional aid. A number of planters . . . urged him to furnish them some directions for their guide in case of sickness, in their large families of whites and blacks, in the form of a medical work." In all, Ewell's domestic medicine appeared in eleven editions between 1807 and 1856. According to the *Dict. Amer. biog.*, "With its pleasant mingling of poetical quotations, anecdotes, sentiment, and sound practical counsel, it was a valued possession on isolated plantations of the time" (III:229). Medicine chests could be purchased with many editions, "which were put up under his own direction for the convenience of families, and were eagerly sought for, in every part of the country, especially at the South and West" (#1102, p. xix).

In 1809 Ewell moved to Washington. When the British invaded the city in 1814, Ewell's house, located across from the Capitol, became British headquarters. Ewell remained in the city during the occupation, tending to the sick and wounded among the enemy's forces—solicitude for which he was later criticized. This third edition includes Ewell's detailed account of the British occupation (p. [629]-658). It was with his third edition that Ewell changed the book's title to *The medical companion,* and nearly doubled its content. He begins with a 43 page essay on human anatomy, and follows with 189 pages on "the non-naturals, by which are meant air, food, exercise, sleep, evacuations and passions." His text on domestic medicine and surgery are largely unchanged from the first and second editions, but is followed by a 134 page alphabetical catalog describing 189 American medical plants drawn from the researches of Barton, Chapman, Hosack, Mease, Thacher and others. "There is scarcely a plant that greens the field," he writes, "a flower that gems the pasture, a shrub that tufts the garden, a tree that shades the earth, that does not contain certain medicinal virtues, to remove our pains, and heal our diseases" (p. [461]). In making these substantial additions on anatomy, hygiene and medical botany, Ewell accurately read the expectations and interests of his public, and the direction that popular medical literature would follow the rest of the century.

In 1830, Ewell left Washington to take up residence in New Orleans. His memorialist in the 10th edition relates: "Having sustained repeated and heavy losses, from the unfaithfulness of agents employed in the distribution of his work, Doctor Ewell again turned his attention to the South, and determined to devote his yet unimpaired energies, to the practice of his profession in New Orleans, and while there, to take more especial notice than ever, of the

wants of the South and West in reference to a medical work for the use of families" (p. xx-xxi). Ewell died in the cholera epidemic of 1832 in the home of a friend on Lake Pontchartrain.
REFERENCES: *Austin* #739.

1101. EWELL, James, 1773-1832.
The medical companion, or family physician; treating of the diseases of the United States, with their symptoms, causes, cure, and means of prevention: common cases in surgery, as fractures, dislocations, &c. The management and diseases of women and children. A dispensatory, for preparing family medicines, and a glossary explaining technical terms. To which are added, a brief anatomy and physiology of the human body, showing, on rational principles, the cause and cure of diseases: an essay on hygieine [sic], or the art of preserving health, without the aid of medicine: an American materia medica, pointing out the virtues and doses of our medicinal plants. Also, the nurse's guide ... The ninth edition, revised, enlarged, and very considerably improved. Embracing a treatise on epidemic or malignant cholera. Philadelphia: Carey, Lea and Blanchard, 1836.
xvi, [25]-632 p. ; 23 cm.

1102. EWELL, James, 1773-1832.
The medical companion, or family physician: treating of the diseases of the United States, with their symptoms, causes, cure, and means of prevention: common cases in surgery, as fractures, dislocations, &c. The management and diseases of women and children. A dispensatory for preparing family medicines, and a glossary explaining technical terms. To which are added, a brief anatomy and physiology of the human body, showing, on rational principles, the cause and cure of diseases: an essay on hygieine [sic], or the art of preserving health without the aid of medicine: an American materia medica, pointing out the virtues and doses of our medicinal plants. Also, the nurse's guide ... The tenth edition, revised, enlarged and very considerably improved; embracing a treatise on hydropathy, homaeopathy [sic], and the chronothermal system. Philadelphia: Thomas, Cowperthwait & Co., 1847.

xxvi, [27]-692 p., [2] leaves of plates : ill., port. ; 24 cm.

The content of the tenth edition of Ewell's domestic medicine is largely unchanged from the 1816 third edition (#1100), except for the addition of an anonymously penned biographical memoir (p. [xv]-xxii) and brief reviews of hydropathy (extracted from Shew, #3180-3206), homeopathy (extracted from the American ed. of Croserio), and the chrono-thermal system of Samuel Dickson (#945, 946). Lithographed frontispiece portrait of the author; the other plate contains lithographic copies of the six wood engravings that illustrate Shew's several works on hydropathy

1103. EWELL, James, 1773-1832.
The medical companion, or family physician: treating of the diseases of the United States, with their symptoms, causes, cure, and means of prevention: common cases in surgery, as fractures, dislocations, &c. The management and diseases of women and children. A dispensatory for preparing family medicines, and a glossary explaining technical terms. To which are added, a brief anatomy and physiology of the human body, showing, on rational principles, the cause and cure of diseases: an essay on hygieine [sic], or the art of preserving health without the aid of medicine: an American materia medica, pointing out the virtues and doses of our medicinal plants. Also, the nurse's guide . . . The tenth edition, revised, enlarged, and very considerably improved; embracing a treatise on hydropathy, homaeopathy [sic], and the chronothermal system. Philadelphia: Published by Crissy & Markley, for Jacob A. Geer, 1848.
xxvi, [27]-692 p., [2] leaves of plates : ill., port. ; 24 cm.

1104. EWELL, James, 1773-1832.
The medical companion, or family physician: treating of the diseases of the United States, with their symptoms, cause, cure, and means of prevention: common cases in surgery, as fractures, dislocations, &c.: the management and diseases of women and children. A dispensatory for preparing family medicines, and a glossary ex-plaining technical terms. To which are added, a brief anatomy and physiology of the human body, showing, on rational principles, the cause and cure of diseases: an essay on hygiene, or the art of preserving health without the aid of medicine, also, the nurse's guide . . . Eleventh edition, greatly enlarged and improved, and brought up to the best medical standard of the day. With new treatises upon neuralgia, fevers, female complaints, and other diseases common in the United States. To all which is added, with numerous illustrations, a complete American materia medica, and also an account of the homoeopathic, hydropathic, and chrono-thermal treatment. Philadelphia: Charles Desilver; Chicago: Keen & Lee, 1856.
xxvii, [28]-720 p., [4] leaves of plates : ill. (some col.), port. ; 25 cm.

In addition to the new articles on "intermittent fever, remittent fever, gout, rheumatism, dropsy, neuralgia, and other subjects," the 11th edition is embellished with four hand-colored lithographic plates (including the frontispiece portrait) and wood engravings of plants illustrating the materia medica.

1105. EWELL, James, 1773-1832.
The planter's and mariner's medical companion: treating, according to the most successful practice, I. The diseases common to warm climates and on ship board. II. Common cases in surgery, as fractures, dislocations, &c. &c. III. The complaints peculiar to women and children. To which are subjoined, a dispensatory, shewing how to prepare and administer family medicines, and a glossary, giving an explanation of technical terms . . . Philadelphia: Printed by John Bioren, 1807.
xvi, [4], 328, [2] p. 22 cm.

"On the important subject of domestic medicine, many books have been written, which though excellent in other respects, have greatly failed of usefulness to AMERICANS because they treat of diseases, which existing in very *foreign climates and constitutions,* must widely differ from ours. The book now offered to the public has, therefore, the great advantage of having been written by a native American of long and successful practice in these southern states, and who for years past, has turned much of his attention to the composition of it" (*Pref.*, p., vii).

First edition. *The planter's and mariner's medical companion* went through eleven editions between 1807 and 1856, making it the most popular manual of domestic medicine and hygiene written by an American author through the first three decades of the 19th century. Only Buchan's *Domestic medicine* (#442-502) enjoyed greater popularity during these decades; and it was not until the mid-1830s that *Gunn's domestic medicine* (#1458-1475) challenged its preeminence.

REFERENCES: *Austin* #742.

1106. EWELL, James, 1773-1832.
The planter's and mariner's medical companion: treating, according to the most successful practice, I. The diseases common to warm climates and on ship board. II. Common cases in surgery, as fractures, dislocations, &c. &c. III. The complaints peculiar to women and children. To which are subjoined, a dispensatory, shewing how to prepare and administer family medicines, and a glossary, giving an explanation of technical terms . . . Baltimore: Printed by P. Mauro, 1813.
xvi, [17]-328, 331-381 p. ; 22 cm.

This second edition is largely a reprint of the 1807 first edition (#1105). The greatly expanded third edition (1816) was issued under the title: *The medical companion* (#1100).
REFERENCES: *Austin* #743.

1107. EWELL, Thomas, 1785-1826.
American family physician; detailing important means of preserving health, from infancy to old age: the offices women should perform to each other at births, and the diseases peculiar to the sex: with those of children and adults. With an appendix, containing hints respecting the treatment of domestic animals, and the best means of preserving fish and meat . . . Georgetown, D.C.: Published by James Thomas; James C. Dunn, printer, 1824.
xv, [1], [ix]-xix, [20]-480 p. ; 22 cm.

First edition. "A considerable part of the work is taken from one I published some years

since, entitled 'Letters to ladies concerning themselves and children' [#1109] . . . yet as some parts are deemed exceptionable, they are omitted. Enough, however, it is hoped, is retained to enable women to perform the necessary offices to each other in childbed, without the interference of men-midwives in common cases. The part relating to children is retained in its original state, excepting some additions . . . Some parts were also taken from my Discourses on Chemistry, published in 1806; and much more is taken from the learned, 'Practice of Physic,' by Dr. [Robert] Thomas, which I have freed from its technical phraseology. No one could suppose that a work of this kind could be much more than a compilation, yet it does contain something original in matter and manner; but its chief object is to simplify, to render more intelligible to the common reader, the researches of others; to press more particularly the most simple modes of preserving health and treating diseases" (p. xv-xvi).

Thomas Ewell began the study of medicine with George Graham in his native Dumfries, Va. and with John Weems in Washington, before receiving his medical degree from the Univ. of Pennsylvania (1805). He was a naval surgeon until 1813. During his tenure in Washington, Ewell married the daughter of the Secretary of the Navy and served on a committee that made recommendations for the reorganization of the naval medical service. Ewell wrote several works on medicine and chemistry for a popular audience, and edited the first American edition of David Hume's *Philosophical essays* (1817). He retired early from medical practice to manage his late father-in-law's estates and a gunpowder factory.

1108. EWELL, Thomas, 1785-1826.
American family physician; detailing important means of preserving health, from infancy to old age. The offices women should perform to each other at births, and the diseases peculiar to the sex: also those of children and adults. With an appendix, containing hints respecting the treatment of domestic animals, and the best means of preserving fish and meat . . . Second edition, improved. Georgetown, D.C.: Published by James Thomas, 1826.
503 p. ; 23 cm.

In the preface, Ewell shows no hesitation in evaluating his competition in the literature of domestic medicine. Of Buchan (#442-502) he writes, "His directions for preserving health are exceedingly judicious; those for the practice of physic are very tolerable, at least, for his own country; but his work is replete with false, absurd, exploded doctrines" (p. xxi). Of his elder brother James Ewell's *Medical companion* (#1100-1106) he is even less generous, "it has gone through several editions; not withstanding some glaring defects in its details—great ignorance of authors—much absurdity of doctrine, and an enormity of price" (*ibid.*). Stearn's *American herbal* (#3328) is simply dismissed as "contemptible and disgraceful" (p. xxii), etc.

Ewell describes the *American family physician* as a compilation, incorporating into the text much of his *Letters to ladies* (#1109), parts of his *Plain discourses on the laws or properties of matter* (1806), and extracts from Robert Thomas' *Modern practice of medicine*, "which I have freed from its technical phraseology" (p. xxiii). The book is divided into six "Addresses" and an appendix. The first addresses hygiene; the second, therapeutics (e.g., clysters, cupping, baths, the classification of medicines); the third, the female organs of generation, pregnancy and labor; the fourth, the management of women after labor; and the fifth-sixth, the identification and treatment of common diseases.

1109. EWELL, Thomas, 1785-1826.
Letters to ladies, detailing important information, concerning themselves and infants . . . Philadelphia: Printed by W. Brown, 1817.
xix, [20]-308 p., [1] folding plate : ill. ; 20 cm.

First edition; reissued in 1818 under the title: *The ladies medical companion*. In his preface Ewell writes, "if there be a country in which medical knowledge may be diffused, it is the United States, where professional services are so often dilatory and out of time. If there be a part of the community to whom this knowledge should be confined, it is, assuredly, the female part,—at least concerning themselves and children" (p. xi).

The text is divided into eleven letters, in the first of which Ewell expresses his disapproval of male physicians attending women in pregnancy and childbirth. "No man possessed of a correct and delicate regard for his wife," he writes, "would subject her to any exposure to a doctor, that could be avoided without danger" (p. 27). (On page vii Ewell advocates the establishment of a lying-in hospital in Washington at which women would be formally trained as midwives.) Letters II-XI provide judicious discussions of hygiene; female physiology as it relates to disease; menstruation and its disorders; the anatomy of the pelvis and organs of generation; conception, the signs of pregnancy and its disorders; natural labors; preternatural labors; puerperal disorders; the care of newborns and the diseases of children. Ewell's ample quotation of Bard, Buffon, Burns, Denman, Merriman and Moss provided the lay reader exposure to the best obstetrical thought of the period, and nearly equals in volume his own contribution to the text. The folding engraved plate contains nine figures copied from the wood engravings that illustrate the many editions of Samuel Bard's *A compendium of the theory and practice of midwifery* (1st ed., New York, 1807).

There are two copies of the *Letters to ladies* in the Atwater collection. One is uncut (23 cm.) and contains numerous emendations, deletions and additions in a contemporary hand. Could these corrections and additions be Ewell's as he prepared the second edition? The nature of the marginalia precludes their being some reader's textual commentary.

REFERENCES: *Austin* #745.

1110. EXCELSIOR BOTANICAL COMPANY.
Read this circular carefully and then try the botanical preparations manufactured by Excelsior Botanical Co., at Boonville, N.Y. [Boonville, N.Y.: Willard & Sons' Steam Power Press Print, 1882?].
[1] leaf ; 20 × 38 cm. folded thrice to 20 × 9 cm.

"J.C. Bass, druggist manager."

1111. EXNER, Max Joseph, 1871-
The rational sex life for men . . . New York; London: Association Press, [c1914].
xiii, [1], 95 p. ; 17 cm.

A manual on sexual mores for college-age males by the secretary of the Student Dept. of the Young Men's Christian Association. Exner was the author of half a dozen treatises on popular sex education.

AN EXPERIENCED OCULIST see **BEER** (#286).

AN EXPERIENCED SWIMMER see *THE SCIENCE OF SWIMMING* (#3121).

EXPERIMENTAL KNOWLEDGE see **PERRY** (#2781)

1112. EYRE, James, Sir, 1792-1857.
The stomach and its difficulties . . . Sixth edition. Philadelphia: J.B. Lippincott & Co.; London: J. Churchill & Sons, 1869.
    xiii, [14]-113 p. ; 17 cm.

The first American edition of Eyre's treatise was published at Philadelphia in 1852, the same year the first London edition appeared. His object is to point out how indigestion may be prevented and how it may be removed or alleviated in the dyspepstic. Much of his treatise is concerned with diet and digestion: "The possession of good health can only be realized by the adoption of such rules of diet and regimen as insure a due re-supply of healthy blood to replace that which is daily consumed . . . To renew the quantity of blood daily used up, is the duty of the stomach and other organs of digestion; these prepare our food and drink for conversion into blood . . . Gout, rheumatism, and various other disorders are often produced by the injudicious supplies given to the stomach . . . Irregularities in our mode of living generate lithic and other acids in the stomach; these pass into the circulation of the blood, and lay the foundation of many disorders" (p. 29).

Eyre began his surgical training at St. Bartholomew's Hospital (London), and in 1814 was made a member of the Royal College of Surgeons. He began practice in Hereford. In 1830 Eyre became mayor of that city, and was knighted. Soon afterward, he decided to become a physician, and studied in Paris for a year before receiving his medical degree from Edinburgh in 1834.

F

1113. FACTS FOR LADIES.
Facts for ladies. Beauty. Dining, by Kinsley's. House decoration. Health of women and children by Robert A. Gunn . . . Edited by Amy G. Ayer. Chicago: Amy G. Ayer, publisher, 1890.
    xvi, 13-388 p., [16] leaves of plates : ill. ; 23 cm.

"In the very outset, the book deals boldly with the question of personal beauty, and points out the secret of its development and perpetuation . . . Passing from this matter of personal beauty, the book then deals with the important question of dining . . . Fifty pages of this invaluable book are devoted to a wise resume of practical hints on all the important points connected with this very important department of social life . . . The book next deals with the subject of house decoration . . . The building of a house may tax the skill of architect and builder, but the finishing and decoration that change a mere 'house' into a 'beautiful home,' require taste and skill and tact . . . The last part of the book vies with all the rest in practical importance. 'Man know thyself' is a good motto, but if the maidens and mothers of the world were better acquainted with their physical organism, and the simple secrets of health, they would spare themselves endless and unnecessary suffering, and the future races of men would be caste in a more heroic mold" (p. iii-vi).

Facts for ladies was reprinted at Chicago by the Werner Company in 1899 under the title *Confidential medical talks to women* (#759.1). Ayer's role in the editing and prior publication of this book is not acknowledged in the 1899 reprint; Gunn's name appears only beneath the spine-title.

FAIR, William Cooper. *The people's home stock book* see *THE PEOPLE'S HOME LIBRARY* (#2768, 2769).

1114. FAIRBANKS AND COMPANY.
Circular to the Medical Profession. [Boston: Fairbanks & Co., ca. 1871].
    [2] leaves : ill. ; 23 × 20 cm.

Caption title. Advertisement for the "Fountain Syringe" manufactured by Fairbanks & Co. for use in vaginal douching or enema.

1115. FAIRBANKS AND COMPANY.
The Fountain Syringe. [*wood engraving of syringe & attachments*] SELF-ACTING. NO VALVES! NO PUMPING! NO AIR INJECTED. ENTIRELY NEW! . . . Manufactured and for sale by H. FAIRBANKS & CO., Sole Proprietors . . . Boston, Mass. . . . [ca. 1869].
    [2] leaves : ill. ; 20 cm.

"It is not only a perfect ENEMA-GIVING instrument of itself, but by glancing at the cut it will be perceived that it is a combination of which no other syringe is capable. No. 1 is a sprinkler for a light shower-bath . . . No. 2 is a nasal douche . . . No. 3 is for children and the ear. No. 4 is for the rectum. No. 5, is the vaginal." (leaf 1 verso).

1116. FAIRCHILD, Horace L.
Keep Your Blood Pure, "FOR THE BLOOD IS THE LIFE!" Dr. Dalton's Vegetable Discovery, the World Famed BLOOD MIXTURE, is a guaranteed cure for all blood diseases. It is the most searching blood cleanser ever discovered, and it will free the system from all known blood poisons, be they animal, vegetable or mineral . . . Horace L. Fairchild, M.D., New Brunswick, N.J. . . . sole proprietor. [ca. 1892].
    [1] leaf printed on both sides : port. ; 24.5 × 15 cm.

There are two issues of this advertisement in the Atwater Collection—one with a portrait on the verso of Miss Mary E. Gibbs, an "agent for Dutton's Vegetable Discovery since June 13, 1884"; and the other with a portrait on the verso of Geo. H. Fayerweather, an agent since January, 1892.

1117. FAIRCHILD, Maria Augusta, b. 1834.
How to be well; or, common-sense medical hygiene. A book for the people. Giving

directions for the treatment and cure of acute diseases without the use of drug medicines. Also, hints on general health care . . . New York: S.R. Wells & Company, publishers, 1879.

[4], 180, [8] p. : ill. ; 19 cm.

First edition. Fairchild's introduction to *How to be well* is an autobiographical essay in which she recalls her childhood in the home of her uncle Stephen Fairchild, a homeopath, and her early awareness of a vocation in healing. It is a fascinating if somewhat self-aggrandizing self-portrait, full of the spirit of restless womanhood and societal change. Of her decision to become a physician Fairchild writes, "It was a time when the woman who would become a physician must give up all that life is worth to her—friends, love, and, perhaps, *name*. She must be strong to meet enmity, scorn, ridicule, and a thousand obstacles which society and the medical men of those days were quick to throw in her pathway" (p. 13).

Fairchild graduated from the New York Hygeio-Therapeutic College (hydropathic) in 1860. Early in her career she became associated with perhaps the most prominent figure in the American hydropathic movement, Russell Thacher Trall (#3558-3593). "Availing myself of every opportunity for becoming acquainted with the proper methods of treating invalid women, I was led to extend my medical researches, and thus found myself growing in fitness for my loved profession. I now entered the lecture-field, and have labored there ever since . . . believing that no part of health reform work is so inviting or more useful. To cure people is but to supply part of the demand. The higher is to prevent, by diffusion of hygienic knowledge" (p. 15). Fairchild wrote two books, and was a frequent contributor to the *Water-Cure Journal* (#3720). In the 1870s she established Dr. Fairchild's Hygeian Home and Movement Cure in Quincy, Illinois.

How to be well is a clear explication of Fairchild's medical philosophy. She regarded water and movement as the "two grand remedial agents." Fairchild argues that water purifies the system when when applied hygienically; is the most effective remedial agent in the treatment of many diseases; and safely replaces drugs when used as an emetic, cathartic, diuretic, etc. Movement (i.e., exercise, vibration, massage)

is water's twin in Fairchild's hygienic and therapeutic system.

Fairchild has appended to her text the "Confessions and observations of Sir Edward Lytton Bulwer" (p. 167-180). First published in the *New monthly magazine* (London, 1845), Edward Bulwer Lytton's (#2308) "Confessions of a water-cure patient" was commonly included in American water-cure manuals as testimony to hydropathy's effectivness by an esteemed literary figure.

1118. FAIRCHILD, Maria Augusta, b. 1834.
How to be well; or, common-sense medical hygiene. A book for the people. Giving directions for the treatment and cure of acute diseases without the use of drug medicines. Also, hints on general health care . . . New York: Fowler & Wells, publishers, 1880.

[2], 180, [8] p. : ill. ; 19 cm.

1119. FAIRCHILD, Maria Augusta, b. 1834.
Woman and health. A mother's hygienic hand book and daughter's counselor and guide to the attainment of true womanhood through obedience to the divine laws of woman nature, including specific directions for the treatment and cure of acute and chronic ailments . . . Quincy, Ill.: Published by the author, Dr. Fairchild's Healthery, 1890.

vi, [2], 384 p. ; 24 cm.

Woman and health takes the form of a dialogue between the "Doctor" and two interlocutors, "Viola" and "Julia," on matters pertaining to woman's role in life ("woman nature"), woman's spiritual, mental and physical well-being, and the diseases which arise from a failure to understand and prevent unhealthful emotional and physical influences. Fairchild's Swedenborgian beliefs are evident throughout this work. Atwater copy inscribed by author (1891).

1120. FALCONER, William, 1744-1824.
An essay on the preservation of the health of persons employed in agriculture, and on the cure of the diseases incident to that way of life . . . Bath: Printed and sold by R. Cruttwell; sold also by C. Dilly . . . London, MDCCLXXXIX [1789].

viii, 88 p. ; 20 cm. (8vo)

Falconer took his medical degree at Edinburgh in 1766 and a second medical degree at Leyden in 1767. He moved to Bath in 1770, where he established a successful medical practice. Falconer's writings were highly esteemed by his contemporaries, including his several treatises on the waters at Bath and other hygienic themes. This essay on the health of agricultural workers was originally published in the *Letters and papers* of the Bath & West of England Society for the Encouragement of Agriculture, Arts, Manufactures & Commerce (v. 4, 1783). It was first published separately in this 1789 edition. According to the *Dict. nat. biog.* (6:1030), a third edition appeared at London in 1794.

The *Essay* is addressed not to "persons employed in agriculture," but to those who employ them. In the first half of the treatise, Falconer discusses the advantages to health of agricultural work, as well as the elements that must be considered to ensure continued health, e.g., proper diet, exposure to the elements, exhaustion, miasms, appropriate clothing & footwear, etc. In the second half he discusses "the cure of diseases incident to an agricultural life."

1121. FAMILY DOCTOR.

The Family doctor: a counsellor in sickness, pain and distress, for childhood, manhood, and old age: containing in plain language, free from medical terms, the causes, symptoms, and cure of disease in every form, with important rules for preserving the health, and directions for the sick chamber, and the proper treatment of the sick; the whole drawn from extensive observation and practice, and illustrated with numerous engravings of medicinal plants and herbs . . . Philadelphia: John E. Potter, 1859.

ii, 308 p. : ill. ; 19 cm.

The Family doctor is divided into six parts on hygiene; the sick room; the diseases of children; the diseases of men and women; wounds, accidents, &c.; and the diseases of women.

1122. FAMILY DOCTOR.

The Family doctor, or the home book of health and medicine: a popular treatise on the means of avoiding and curing diseases, and of preserving the health and vigour of the body to the latest period; including an account of the nature and properties of remedies; the treatment of the diseases of women and children, and the management of pregnancy and parturition. By a Physician of Philadelphia. Auburn [N.Y.] and Buffalo: Miller, Orton & Mulligan, 1854.

viii, [9]-630, [10] p. ; 23 cm.

This anonymous treatise is divided into six numbered parts: I. Anatomy and physiology. II. Hygiene, or the means of preserving health. III. Materia medica, or the remedies employed for the cure of diseases. IV. Surgical diseases and accidents. V. Diseases, their symptoms, causes and treatment. VI. Pregnancy and parturition, with the diseases and accidents of those states. The book was obviously written by a trained physician, and provides as sound descriptions of diagnosis, prognosis and treatment (not necessarily self-treatment) as may be expected from a popular manual of the period.

1123. FAMILY HEALTH BOOK.

The Family health book. An up-to-date household medical guide. A manual of domestic medicine and surgery, hygiene, dietetics, and nursing. Dealing in a practical way with the problems relating to the maintenance of health, the prevention and treatment of disease, and the most effective aid in emergencies . . . J. West Roosevelt, M.D., editor. New York and London: D. Appleton and Company, 1913.

xvi, 991 p. : ill. ; 25 cm.

Each of the book's thirteen sections were written by a different author, including chapter III, "Outlines of psychology, or a study of the human mind," by Josiah Royce.

THE FAMILY LIBRARY see UPHAM
(#3636).

1124. FAMILY PHYSICIAN.

The Family physician and the farmer's companion . . . [Syracuse, N.Y.?: Printed for M. Baldwin, 1840?].

24 p. ; 18 cm.

Caption title; this work has neither title-page nor imprint. Though commonly issued with

John Eyre's *The Christian spectator* (Albany, N.Y., 1838) and his *The European stranger in America* (New York, 1839), Eyre was not its author. Citing "the late celebrated Dr. John Redman" (p. 1), the author of *The Family physician* was certainly an American, possibly the M. Baldwin for whom the work is supposed to have been printed. The entry in *NUC Pre-1956 Imprints* (166:357) provides the wording "Printed for M. Baldwin the blind man" in the imprint without citing sources.

The first fifteen pages of this work contains a collection of ninety-eight receipts for the treatment of various human ailments. They are arranged in no particular order, though numbers 48 through 77 provide an alphabetical series ("Apoplexy" through "White Swellings") probably taken from another collection. The rest of the treatise contains forty "Receipts for the care of horses," ten "Receipts for the care of cattle," and four "Receipts for the care of sheep."

1125. FAMILY PHYSICIAN.
The Family physician; or, every man his own doctor, an encyclopedia of medicine, containing knowledge that will promote health, cure disease and prolong life. Describing all diseases, and teaching how to cure them by the simplest medicines. Also, an analysis of everything relating to courtship, marriage, and the production, management, and rearing of healthy families, together with a chapter on the preparation of medicines, giving prescriptions and valuable receipts, full and accurate directions for treating wounds, injuries, poisons, etc. Compiled by leading Canadian medical men . . . Toronto: The Hunter-Rose Company, Limited, [ca. 1900].
viii, 344 p. : ill. ; 20 cm.

1126. FAMILY PHYSICIAN.
The Family physician; or, every man his own doctor, an encyclopedia of medicine, containing knowledge that will promote health, cure disease and prolong life. Describing all diseases, and teaching how to cure them by the simplest medicines. Also, an analysis of everything relating to courtship, marriage, and the production, management, and rearing of healthy families, together with a chapter on the preparation of medicines, giving prescriptions and valu-

able receipts, full and accurate directions for treating wounds, injuries, poisons, etc. Compiled by leading Canadian medical men . . . Toronto: The Hunter-Rose Company, Limited, 1909.
viii, 344 p. : ill. ; 20 cm.

1127. FAMILY RECEIPT BOOK.
The Family receipt book, containing eight hundred valuable receipts in various branches of domestic economy; selected from the works of the most approved writers, ancient and modern; and from the attested communications of scientific friends. Second American edition. Pittsburgh: Published by Randolph Barnes, 1819.
xxxii, [25]-408 p. ; 18 cm.

This work has its origins in England, where it was published at London by John Murray in 1810 as *The New family receipt book, containing seven hundred truly valuable receipts in various branches of domestic economy*. The citation for this title in *A Catalogue of printed books in the Wellcome Historical Library* (IV:226) indicates that it was intended as a companion to Maria Eliza Rundell's (1745-1828) *A new system of domestic cookery*. In this Pittsburgh edition, "Chapter XXVI. Health" (p. 265-295), provides sixty-seven domestic medical recipes. This collection was published the same year by Josiah Spalding in New Haven as: *The New family receipt book . . . A new edition* (#2584).

1128. FANCHER, Jefferson B.
Medical, matrimonial, and scientific expositor; giving the most important information upon every subject relating to man and woman. With character, causes, symptoms, treatment, cure, and prevention of all diseases. Including lessons and advice to lovers and the married. Being a work for the people . . . Containing new and scientific explanations of the sexual functions, their uses and abuses; with qualifications needed for happy marriages; intended for the serious consideration of both married and single. Together with the habits, manners, customs, and condition of the female sex, with their most intimate relation to man, both civilized and savage. Including the metallurgist's guide, with secrets of transmutation . . . New, revised

and enlarged edition. New York: Dr.
Jefferson B. Fancher, publisher, 1867.
xx, 438 p. : ill. ; 19 cm.

Fancher's treatise is divided into four parts,
with an addendum on metallurgy. In Part I.
Fancher describes human anatomy and physiology. In Part II. he explains the causes and
symptoms of disease, without directions for self-
treatment. Human reproduction is included in
Part II. Chapters XXIX-XXXI provide fairly
accurate information on impregnation, gestation and parturition, and are illustrated with
the largest number of wood engravings in the
text. The "home treatment of diseases" is reserved for Part III., which provides recipes for
balsams, bitters, restoratives, gargles, injections,
ointments, plasters, tinctures, electuaries, etc.
Part IV is a treasury of contemporary opinion
on marriage and sexual behavior. Fancher
claims in his preface that a previous edition of
this work sold 100,000 copies. A second printing of this "new, revised and enlarged edition"
appeared in 1868.

FARINHOLT-JONES, May see **JONES**
(#2042).

1129. FARLIN, C. Fred., d. 1896.
THEY ARE COMING! Who are THEY? What
do THEY do? READ AND YOU WILL KNOW!!
C. FRED. FARLIN, A.M., The Lecturer of the
Bi-State Syndicate of Physiologists, Physicians and Surgeons, tells you, upon examination, of the aches, pain and disease that
affects you . . . He Lectures to Ladies and
Gentlemen, Every Evening at 8 o'clock,
upon "House-Building, House-Repairing
and House-Keeping," in relation to that
House Beautiful—the Human Body. His
lectures are illustrated by Charts, Drawings,
a Human Skeleton, Physiological Manikin,
&c. ALL SHOULD HEAR HIS LECTURES . . .
T.C. HARTER, M.D., of the Provident Medical Dispensary, New York City . . . treats with
unparalleled success all Chronic, Nervous,
Skin and Blood Diseases . . . Dr. Harter has
attained the reputation and is known as
one of the BEST diagnosticians in the East
and West . . . after years of experience in
the City Hospital of Baltimore, Md., Pennsylvania Hospital, Johns Hopkins University, and also Toronto, Canada . . . The

Home Office Will Be in Charge of DR. H.W.
STREETER . . . He is a gentleman of profound learning, vast experience and rare
skill . . . 112 East Main St., ROCHESTER,
N.Y. . . . PERFECTED MODERN DEN-
TISTRY! . . . DR. HOMER BELDING, Dentist . . . Lectures and Consultation Free!
. . . [ca. 1888].
Broadside : ill. ; 64 × 23 cm.

This remarkable Rochester, N.Y. broadside
can be dated to the year 1888 by city directory
locations of the principals. It appears to promote a sort of group practice, with Dr. Streeter
the physician, Dr. Belding the dentist, Dr.
Harter the consultant who visits from out of
town, and Dr. Farlin, the physiological lecturer
and "beater" for the other three, and perhaps
for himself as well. The location of Farlin's lectures must have varied, given the blank space
left at the bottom of the sheet.

Farlin received his medical degree in 1883
from the Bellevue Medical College of Massachusetts. Established in 1880, the school's charter
was repealed in 1883. After graduation he came
to Rochester where he practiced medicine for
fourteen years until his premature death. Farlin
specialized in chronic and nervous diseases, as
well as the diseases of women. A heavy-set man,
he died of coronary occlusion on 4 August 1896
while swimming in Lake Ontario. According to
his obituary, Farlin was "a peculiar man, but a
man of varied accomplishments. He had the
reputaton of being a skilled practitioner, was a
confirmed Spiritualist, practiced healing by electricity and preached the Gospel. He was a brilliant conversationalist and fluent talker, ready of
wit and a quick thinker. He had such power of
oratory and possessed such a thorough knowledge of the Scriptures and was of such a peculiarly religious turn of mind that his friends often urged him to forsake the practice of medicine and enter the pulpit" (*Union & Advertiser,* 5
Aug. 1896, 7:3).

Theodore Clarence Harter (1852-ca. 1930)
attended the College of Physicians & Surgeons
of Baltimore (M.D., 1881) and probably did work
at Baltimore's City Hospital. Although Johns
Hopkins University was in existence when Harter
was in Baltimore, its medical school was not, and
the University catalog does not record his name.
Nor has a Provident Dispensary been identified
in New York City (though a clinic by that name

existed in Rochester at the time). There are no entries for Harter in the Rochester directories, making it likely that he was an itinerant physician. Whatever the arrangement described in this broadside, it was short-lived—lasting a year or two at most. After the Rochester venture, Harter settled in the village of Bloomsburg, Pa., where he practiced medicine until his death. H.W. Streeter (1845-1903) was an 1868 graduate of Jefferson Medical College. He practiced in Rochester from 1887 until his death on 29 November 1903. Homer Belding (1833-1903) practiced dentistry in Rochester from 1870.

FARM AND FIRESIDE LIBRARY see **POWER** (#2874).

1130. FARMER, Fannie Merritt, 1857-1915.
Food and cookery for the sick and convalescent... Boston: Little, Brown, and Company, 1904.
xiii, [1], 289, [13] p., [42] leaves of plates : ill. ; 20 cm.

First edition. To "baby-boomers" the name Fannie Farmer brings to mind a chain of candy stores found in nearly every mall in America. To their parents, Fannie Farmer was a name associated with an equally ubiquitous cook book. To their grandparents, however, Fannie Farmer was not just a name on a bookcover, but an actual person—the director of one of the country's best known cooking schools, a regular columnist for the *Woman's home companion,* and a nationally renowned lecturer on culinary matters and invalid cookery.

A paralytic stroke suffered as an adolescent ended Farmer's plans to attend college. "She eventually recovered her health sufficiently to assist in housekeeping, and developed such an interest in cooking that her family urged her to attend the Boston Cooking School. After her graduation from that institution in 1889, she was asked to return as assistant to the director the next year. Upon the death of the director she was elected to that position (1891). She resigned some eleven years later to open (1902) a school of her own, known as Miss Farmer's School of Cookery" (*Dict. Amer. biog.,* 3:276).

Farmer's most successful published work was *The Boston Cooking-School cook book* (Boston, 1896). In 1959, the 11th edition was issued as *The Fannie Farmer cook book,* the 13th edition of

which was published in 1996. In her preface to this first edition of *Food and cookery for the sick and convalescent,* Farmer states that it was "designed to meet the demands made upon me by the numberless classes of trained nurses whom it has been my pleasure and privilege to instruct during my thirteen years of service as a teacher of cookery" (p. [vii]). "It is earnestly hoped," she continues, "that, besides meeting this long felt need, it will do a still broader work in thousands of homes throughout the land."

1131. FARMER, Fannie Merritt, 1857-1915.
Food and cookery for the sick and convalescent . . . Revised, with additions. Boston: Little, Brown, and Company, 1917.
xiii, [1], 305 p., [23] p. of plates : ill. ; 19 cm.

The revised edition of *Food and cookery* was first published in 1912. It contains two additional chapters that provide consideration of the dietary needs of diabetics and those with other "special diseases". The 1917 printing was the final appearance of this title.

1132. FARNSWORTH, Vesta J.
The house we live in or the making of the body. A book for home reading, intended to assist mothers in teaching their children how to care for their bodies, and the evil effects of narcotics and stimulants . . . Oakland, California: Pacific Press Publishing Company, [c1900].
218, [4] p. : ill. ; 22 cm.

1133. FARRAR, Mrs. John, 1791-1870.
The young lady's friend. By A Lady. Boston: American Stationers' Company, 1836.
xi, [1], 432 p. ; 20 cm.

First edition of the the most successful book of the Anglo-American author Eliza Farrar. *The young lady's* friend is a philosophical guide and practical manual for young women of the middle and upper middle classes on household management, family relations, relations with servants & tradespeople, moral & intellectual improvement, entertaining, etiquette & behavior, etc. Several chapters are devoted to domestic health: chap. IV, "Nursing the sick"; chap. V, "Behavior of the sick"; chap. VI, "Dress"; and chap. VII, "Means of preserving health."

"In contrast with other contemporary books in this genre, which stressed the development of moral character and religious principles, Mrs. Farrar emphasized practical details of behavior, counseling women on all aspects of their lives. Although essentially conservative in her concept of woman's sphere, she did believe girls should have a good education and learn 'to find pleasure in intellectual effort.' Despite the criticism of some fastidious readers who found her too explicit in matters of physiology . . . her book became popular in both England and America. It was published anonymously at first, but its success convinced the author that she could acknowledge her writing, and the 1837 edition appeared with her name" (*Notable American women, 1607-1950,* 1:602). Atwater copy lacking pages iii-viii.

1134. FARRAR, Mrs. John, 1791-1870.
The young lady's friend . . . New York: Samuel S. & William Wood, 1838.
  xi, [1], 432 p. ; 19 cm.

1135. FARRAR, Mrs. John, 1791-1870.
The young lady's friend. With introduction. By Mrs. H.O. Ward . . . Philadelphia: Porter & Coates, [c1880].
  375 p. ; 19 cm.

At head of title: *A new edition.* "Mrs. H.O. Ward" was one of the pseudonyms of Clara Sophia Jessup Bloomfield-Moore (1824-1899), the novelist and author on etiquette. In 1873 Bloomfield-Moore brought out a new edition of Mrs. John Farrar's *The young lady's friend,* first published in 1836 (#1133) and a standard work on the intellectual and physical training of young women through the 1850s. In her introducton to this "new edition," Bloomfield-Moore informs her readers that "in a fluctuating society like ours, where the sudden acquisition of wealth brings new families into new positions and surrounds them with new associates, where talents and education carry people into the most refined circles without any previous training in manners, it becomes very necessary to have some means of finding out what belongs to polished life" (p. 6). She summarizes the book's content as "topics that belong to a daughter's education, such as the preservation of health; improvement of time; domestic economy; dress; behavior to parents, rela-

tives, gentlemen, teachers, friends, and to domestics; conduct at public places, female companionship; dinner and evening parties; ceremonious and friendly visits, conversation and mental culture" (p. 8-9). Other than a new chapter on "Character" and replacement of Farrar's chapter on home nursing, Bloomfield-Moore's edition of *The young lady's friend* is faithful to the original. Acknowledgement of Farrar's authorship appears nowhere in this edition, however.

1136. FAUDEL, Henry John, b. 1859.
Horse sense. A school of practical science upon the perplexing problems of to-day, pertaining to life and health, namely: consumption; bacteria in water, soil, air, milk, bread, vinegar and beer; sunshine, the most potent germicide; the up-to-date medical fraternity a menace to life and health; the merits and demerits of the Chicago drainage canal; fresh air and flies; the medical monopoly of ideas; not all so-called quacks are quacks, etc. . . . Independence, Ohio, [c1913].
  v, [1], 266, [2] p. : ill., port. ; 20 cm.

The first edition of this work was an 87 page treatise entitled *Horse sense on consumption* (Cleveland, ca. 1910). In the preface of the 1st edition (reprinted in the 2nd), Faudel launches a one-man crusade against current therapy for pulmonary tuberculosis: "The force of necessity, prompted by the malpractice of the new fresh air cure for consumptives, has made so strong an appeal to my nature that I could resist it no longer, and if there is any virtue in the idea of a superhuman calling for a man to do a certain work, I believe that that call has been assigned to me. I have witnessed the wholesale murder of the helpless and innocent fortunates by the fresh air curists—lacking the courage of taking a stand against them, until I could stand it no longer" (p. 4). Faudel mailed copies of his tract to the anti-tuberculosis societies "of the twenty-five largest cities in this country, and asked them to read the book and write me their opinions as to its merits or demerits" (p. 6). Not one to cease banging his head against the wall, Faudel issued a 2nd edition of his treatise with the inclusion of considerable new matter on the threat that bacteria present to mankind. Faudel's approach to the struggle against bacteria mirrors his therapy for tuberculosis, i.e., warm, dry air kills the offending germ. Throughout the book he rails against the medical profession, medical scientists, public health officials and politicians, who have obviously missed the point in the control of bacteria. Photographic portrait of Faudel on recto of last leaf. Atwater copy inscribed by author.

1137. FAULKNER, Thomas.
The book of nature, a full and explicit explanation of all that can or ought to be known of the structure and uses of the organs of life and generation in man and woman. Intended especially for the married, or those intending to marry, and who conscientiously and honestly desire to inform themselves upon the intent and nature of conjugal pleasures and duties . . . To which is added a complete medical treatise upon all diseases of the generative organs, whether resulting from infection or sexual excesses and abuse . . . Fully illustrating the mysterious process of gestation, from the time of conception to the period of delivery. No such complete panorama of the mysteries of human reproduction has ever before been given to the world. New York: Hurst & Co., publishers, [c1875].
    [2], iv, [5]-183, [3] p. : ill. ; 19 cm.

In addition to his discussion of reproduction and sexual physiology, Faulkner includes "A chapter to young men" (p. 55-67), in which he discusses the moral and physical consequences of fornication; and "A chapter for young women" (p. 67-89) on courtship and marriage. The final chapter, "Ready remedies for common complaints," occupies 20% of the text. There are two copies of this title in the Atwater Collection: one in linen-covered boards and the other in a printed wrapper.

1138. FAULKNER, Thomas.
The book of nature. A full and explicit explanation of all that can or ought to be known by man and woman. Intended especially for the married, or those intend-

ing to marry . . . To which is added a complete medical treatise upon all chronic diseases of the urethra. A complete family doctor. No such complete panorama of the mysteries of human reproduction has ever before been given to the world . . . London [Eng.]: Camden Publishing Co., [ca. 1907].
    223 p. : ill. ; 18 cm.

On printed wrapper: *Original American copyright edition.* Date of imprint based on advertisement for Haydn Brown's *The wife: her book* (London, 1907) on back wrapper.

1139. FAULKNER, Thomas.
The cottage physician. Best know methods of treatment in all diseases, accidents and emergencies of the home, prepared by the ablest physicians in the leading schools of medicine: allopathy, homoeopathy, etc., etc. By Thomas Faulkner . . . [and] J.H. Carmichael . . . Assisted by other able physicians of America and Europe. Complete hand book of medical knowledge for the home . . . Springfield, Mass.; Cincinnati; Sacramento: King, Richardson & Co.; Omaha, Nebraska: Clark Publishing Co., 1892.
    462, 465-642 p., [4] p. of plates, [1] leaf of plates, [1] paper manikin : ill. ; 22 cm.

First published at Springfield, Ma. in 1885. Faulkner is described on the title-page as "President of the Royal Medical Council, London," and Carmichael as a member of the American Institute of Homoeopathy. John Hosea Carmichael (1851-1925) was an 1873 graduate of the Albany Medical College. He is listed in the 4th edition of Polk's *Medical & surgical register of the United States* (1896) as resident in Springfield, Mass., and a member of the Massachusetts Homoeopathic Medical Society and the Western Massachusetts Homoeopathic Medical Society.

1140. FAULKNER, Thomas.
The cottage physician. Best known methods of treatment in all diseases, accidents and emergencies of the home, prepared by the ablest physicians in the leading schools of medicine: allopathy, homoeopathy, etc., etc.

By Thomas Faulkner . . . [and] J.H. Carmichael . . . Assisted by other able physicians and surgeons of America and Europe. Complete hand book of medical knowledge for the home . . . Springfield, Mass.; Cincinnati; Sacramento; Dallas, Texas: King, Richardson & Co., publishers, 1893.
  462, 465-642 p., [4] p. of plates, [1] leaf of plates : ill. ; 22 cm.

1141. FAULKNER, Thomas.
The cottage physician. For individual and family use. Prevention, symptoms and treatment. Best known methods in all diseases, accidents and emergencies of the home. Prepared by the best physicians and surgeons of modern practice. Allopathy, homoeopathy, etc., etc. With introduction by George W. Post . . . Complete hand book of medical knowledge for the home. Springfield, Mass. . . . [etc.]: The King-Richardson Co., 1898.
  [2], 8, [2], 9-295, 295a-295b, 296-462, 465-646 p., [4] p. of plates, [1] leaf of plates : ill. ; 22 cm.

Although the names of Thomas Faulkner and J.H. Carmichael have been removed from the title-page, the text of this edition of *The cottage physician* (first issued in 1885) is almost identical to that of the earlier edition. George W. Post, who wrote the introduction to this edition, was an 1883 graduate of Northwestern University Medical School. He is described on the title-page as professor of the practice of medicine at the College of Physicians and Surgeons, Chicago.

1142. FAULKNER, Thomas.
The cottage physician. For individual and family use. Prevention, symptoms and treatment. Best known methods in all diseases, accidents and emergencies of the home. Prepared by the best physicians and surgeons of modern practice. Allopathy, homoeopathy, etc., etc. With introduction by George W. Post . . . Complete hand book of medical knowledge for the home. Springfield, Mass. . . . [etc.]: The King-Richardson Co., 1900.
  [2], 8, [2], 9-295, 295a-295b, 296-462, 465-646, [2] p., [4] p. of plates, [1] leaf of plates : ill. ; 22 cm.

1143. FAULKNER, Thomas.
The doctor at home; or the unfailing medical instructor, filled with the proven facts of medical science, and family experience, showing in the plainest manner how to retain health, kill disease, lengthen life, and wonderfully increase mental and physical vigor, in man, woman, or child . . . New York: Hurst & Co., publisher, [c1885].
  54, [2], 55-74, [2], 75-94, [2], 95-122, [2], 123-144, [2], 145-186, [2], 187-208, [2], 209-234, [2], 235-238, [2], 239-258, [2], 259-262, [2], 263-298, [2], 299-316, [2], 317-382, [2], 383-414, [2], 415-432, [2], 433-510 p. : ill. ; 23 cm.

At head of title: *Each person his own physician!* The text of the first 339 pages of *The doctor at home* is identical to that of the earliest edition of *The cottage physician* in the Atwater Collection (#1139).

1144. FAUST, Bernhard Christoph, 1755-1842.
The catechism of health; or, plain and simple rules for the preservation of the health and vigour of the constitution from infancy to old age. Philadelphia: Published at the office of the Journal of Health and Journal of Law, 1831.
  [2], x, [7]-195 p. ; 14 cm.

First edition. In some bibliographies and databases, this work is attributed to Henry H. Porter, who was publisher of the journals stated in the imprint. This attribution may be based on the absence of Faust's name on the title-page, or the fact the Porter's *Catechism* is but loosely based on the original. Porter's *Catechism* was extensively edited and/or rewritten by the Philadelphia physician D. Francis Condie, who was closely associated with Porter and John Bell (#289-293) in editing the *Journal of health* (#2053).

Thomas Horrocks points out that Condie reoriented Faust's *Catechism* to a predominantly middle-class, American readership: "Defining middle class life as beneficial to one's health is just one example of how Condie refashioned Faust's message . . . to bring it more in line with the emerging health reform movement of the period." ("Promoting good health in the age of reform," *Can. Bull. Med. Hist.,* 1995, 12:268). In keeping with the theme of personal responsibility inherent to the (closely related) evangelical and health reform movements, Condie placed less emphasis on public health than Faust.

Porter frequently reissued the *Catechism* in its two-year publishing history. In 1831 this first edition of the *Catechism* was published, as well as a 195 page Ladies' edition (#1145), a 202 page Ladies edition, the 5th edition (#1146); and the 7th edition. The 8th and apparently final edition was issued in 1832. The 2nd, 3rd, 4th or 6th editions as yet are unrecorded. Atwater copy issued with printed cloth over boards.

David Francis Condie (1796-1875) was an 1818 medical graduate of the University of Pennsylvania and a prominent figure in Philadelphia medical circles. He is perhaps best remembered for *A practical treatise on the diseases of children,* which went through five editions between 1844 and 1858. He was co-author with John Bell of a timely treatise on cholera (1832), prepared the Philadelphia edition of B. Carpenter's work on temperance (#569), and was the American editor of gynecological and medical works by Fleetwood Churchill and Thomas Watson.

1145. FAUST, Bernhard Christoph, 1755-1842.
The catechism of health; or, plain and simple rules for the preservation of the health and vigour of the constitution from infancy to old age. Ladies' edition. Philadelphia: Literary Rooms . . . ; office of the Journal of Health, Journal of Law, and Family library of health, 1831.
[2], x, [7]-195 p. ; 15 cm.

1146. FAUST, Bernhard Christoph, 1755-1842.
The catechism of health; or, plain and simple rules for the preservation of the health and vigour of the constitution from infancy to old age. For the use of schools. Philadelphia: Literary Rooms . . . Office of the Journal of Health, Journal of Law, and Family library of Health, 1831.
[2], x, [7]-202 p. ; 13 cm.

The statement "Fifth edition" appears beneath the date of imprint on the title-page. The six additional pages of printed matter in this edition consist of endorsements of the *Catechism of health.*

1147. FAUST, Bernhard Christoph, 1755-1842.
A new guide to health. Compiled from the Catechism of Dr. Faust: with additions and improvements, selected from the writings of medical men of eminence. Designed for the use of schools, and private families . . . Newburyport: Published by W. & J. Gilman, printers, booksellers, and stationers, [1810].
iv, [5]-124 p. ; 14 cm.

The first American edition of Faust's catechism was published at Boston in 1795, a work based on the first English translation (London, 1794) of the *Gesundheits-Katechismus* (1792). The next American edition was published at New York in 1798, followed by the 1812 Raleigh edition. The present edition appears to be the fourth American edition. The text is based on the New York edition, which in turn was reprinted from the 1797 Edinburgh edition edited by James Gregory. The New York and Newburyport editions omit some of Gregory's text, while making several additions.

According to Charles Rosenberg, "Faust's manual constitutes what has traditionally been seen as an Enlightenment effort to alter age-old behaviors and thus help the lot of ordinary men and women" ("Catechisms of health," *Bull. Hist. Med.,* 1995, 69:177-8). Organized as a series of questions and answers on such topics as bodily cleanliness, clothing, diet, etc., the work "recalls a classroom setting in which memorization and oral recitation were the primary mode of instruction" (*ibid.,* p. 178). In Rosenberg's words, the catechism "was at once a treatise on domestic medicine, a tract endorsing public health . . . and a tool for classroom use. Perhaps most strikingly, it is a collection of things to do and rules to follow, not a marshalling of arguments justifying those actions . . . Faust's pages . . . show little concern for explicating the mechanisms rationalizing these behavioral admonitions" (*ibid.,* p. 180).

REFERENCES: *Austin* #763.

A FELLOW OF THE COLLEGE OF PHYSICIANS OF PHILADELPHIA see TURNER (#3610, 3611).

FENGER, Christian see **LYMAN** (#2302-2306).

1148. FENN, Artemas Ira, ca. 1828-1879.
The medical and surgical companion. For the use of families, seamen, travellers, miners, &c. Giving a brief description, in plain language, of the diseases of men, women, and children. With the most approved methods of treating them . . . Boston: Published by the author, 1870.
    xii, 384 p., 4 leaves of plates : ill., port. ; 21 cm.

    Fenn is listed in the 1852 Boston city directory as a "clerk." In the 1854 directory he appears as physician; in 1855 and 1856, as an apothecary. In the 1857-60 directories he appears again as a physician; and from 1861 to 1863 as an apothecary. After receiving his medical degree from Harvard in 1863, Fenn practiced in Boston until his death on 19 March 1879. Frontispiece portrait of author.

FENNER, Milton Marion see **M.M. FENNER CO.** (#1149-1153).

1149. M. M. FENNER CO.
Dr. Fenner's cook book. Fredonia, N.Y.: M.M. Fenner Co., [c1907].
    31, [1] p. : port. ; 22 cm.

    Title from printed wrapper. Culinary recipes intermixed with advertisements for the *People's Remedies* available from M.M. Fenner Co. Milton Marion Fenner (d. 1905) was an 1860 graduate of the Eclectic Medical Institute of Cincinnati. He served as an infantryman with the 8th Michigan Volunteers in the opening years of the Civil War, until he was made an assistant surgeon in the U.S. Navy (1863). After his military service, Fenner practiced medicine in Jamestown, N.Y. In 1869 he moved to Fredonia, N.Y. where he began the manufacture of proprietary medicines. Fenner was active in Republican party politics and served two terms in the New York State Assembly. Title from wrapper. Running-title: *Dr. Fenner's Peoples' Remedies cook book.*

1150. M. M. FENNER CO.
Dr. Fenner's cook book. Fredonia, N.Y.: M.M. Fenner Co., [c1912].
    16 p. : ill. ; 22 cm.

1151. M. M. FENNER CO.
Dr. Fenner's People's Remedies, are used all over the world. [Fredonia, N.Y.: M.M. Fenner Co., c1890].
    16 p. ; 18 cm.

    In chromolithographed wrapper.

1152. M. M. FENNER CO.
Dr. Fenner's People's Remedies for the family cook book 1902 . . . Fredonia, N.Y.: M.M. Fenner Co., [c1901?].
    31, [1] p. ; 20 cm.

    Title and imprint from wrapper.

1153. M. M. FENNER CO.
Dr. Fenner's People's Remedies for the family. Copies of letters received from many grateful people . . . Fredonia, N.Y.: M.M. Fenner Co., [c1907].
    31, [1] p. : port. ; 22 cm.

1154. FENTON, Samuel.
The use of brandy and salt, as a remedy for various internal as well as external diseases, inflammation, and local injuries; containing ample directions for making and applying it. Explained by the Rev. Samuel Fenton, M.D., Liverpool, and Wm. Lee, Esq., of La Ferte Imbault, in France. Second edition, revised and corrected. New-York: Carvill & Co., 1842.
    42, [2] p. ; 18 cm.

    "To one quart of the best French brandy add five ounces of salt; cork and shake well together. When mixed, let the salt settle to the bottom, and be particularly careful to use it when clear; the clearer the better" (p. 6). "Salted brandy" may be applied externally for the treatment of headache, eye inflammation, ear ache, rheumatism, gout, etc.; or internally, for use as a mouth wash and gargle, or for the treatment of bowel inflammation, worms, indigestion, etc. This treatise was originally published at London in 1841 under the title: *The excellent properties of salted brandy.* In printed wrapper.

1155. FERNIE, William Thomas, 1830-
Herbal simples approved for modern uses of cure . . . Second edition. Philadelphia: Boericke & Tafel, 1897.
    xxiii, [1], 651, [5] p. ; 19 cm.

First and only American edition. Fernie's *Herbal simples* was first published at Bristol, England in 1895. This Philadelphia edition is based on the second Bristol edition of 1897. A third edition was published at Bristol in 1914. Though the title-page of this American edition bears the Boericke & Tafel imprint, the book was printed and bound in England by John Wright & Co.

1156. FERRIS, C. Sargent.
Vital weakness of man and how to cure it. Man's vital weakness, its causes and its terrible effects. Handsomely illustrated with photographic plates and full information how to cure by a new home treatment. [Cleveland: The Whitworth Bros. Co., printers, 191-?].
[2], 127 p. : ill. ; 19 cm.

"The discovery of the marvelous Vital Life Fluid made by me, has marked a new era in medicine and the treatment of sexual weakness in men" (p. 3). In conjunction with his External Solvent, Ferris claims to be able to cure the full range of disorders of the male sexual organs: impotence, spermatorrhea, the consequences of "self abuse," varicocele, involuntary discharge during intercourse, the effects of conjugal excess, sterility, venereal disease, etc. Title on printed wrapper: *A friend of mankind.*

1157. FESSENDEN, Thomas Green, 1771-1837.
The modern philosopher; or Terrible tractoration! In four cantos, most respectfully addressed to the Royal College of Physicians, London. By Christopher Caustick . . . Second American edition, revised, corrected, and much enlarged by the author. Philadelphia: From the Lorenzo Press of E. Bronson, 1806.
xxxii, 271 p., [2] leaves of plates : ill. ; 21 cm.

First published in London in 1803 under the title: *A poetical petition against tractorising trumpery.* The first American edition was published in New York in 1804 under the title *Terrible tractoration!* (#1158). The Atwater copy is lacking both plates.
REFERENCES: *Austin* #768.

1158. FESSENDEN, Thomas Green, 1771-1837.
Terrible tractoration!! A poetical petition against galvanising trumpery, and the Perkinistic Institution. In four cantos. Most respectfully addressed to the Royal College of Physicians, by Christopher Caustic . . . First American, from the second London edition, revised and corrected by the author, with additional notes. New-York: Printed for Samuel Stansbury, 1804.
xxxv, [1], 190 p., [4] leaves of plates : ill. ; 20 cm.

A native of Walpole, N.H., Fessenden was a 1796 graduate of Dartmouth College. "In May 1801 Fessenden abandoned his law practice at Rutland and sailed to England as agent for a local company to secure English patent rights for a recently invented hydraulic device, which upon further testing proved fraudulent. He spent the next two years and his remaining funds in London in attempts to perfect this device and a new type of grain-mill. In February 1803 he rallied to the defense of another Yankee, Elisha Perkins, whose 'metallic tractors,' after enjoying enormous sale, were being attacked by the reputable medical profession. Fessenden, under the alias 'Christopher Caustic, M.D., LL.D., A.S.S.,' threw together a vigorous Hudibrastic satire, *Terrible Tractoration,* a pretended assault on the tractors, but actually ridiculing the most prominent of the skeptical physicians of England and Scotland. Despite its small merit, the book was surprisingly popular . . . and was several times reprinted" (*Dict. Amer. biog.,* 3:347).

Fessenden's satire was originally published at London in 1803 under the title: *A poetical petition against tractorising trumpery.* The second London edition, upon which this American edition is based, was entitled *Terrible tractoration!!* and was also published in 1803. The four leaves of plates were engraved on metal by Alexander Anderson, M.D. (1775-1870), whose reputation in the annals of book illustration rests on his introduction of wood engraving to American publishing as a viable graphic medium.
REFERENCES: *Austin* #769.

1159. FESSENDEN, Thomas Green, 1771-1837.
Terrible tractoration, and other poems. By Christopher Caustic . . . Third American edition. Boston: Russell, Shattuck & Co. and Tuttle, Weeks and Dennett, 1836.
viii, 264 p., [1] leaf of plates : ill. ; 17 cm.

"In the year 1801 the author . . . was in London . . . In that metropolis, he became acquainted with Mr. Benjamin Douglas Perkins [son of Elisha P., inventor of the tractors], proprietor of a patent right for making and using certain implements, called Metallic Tractors. These were said to cure diseases in all or nearly all cases of topical inflammation, by conducting from the diseased part the surplus of electric fluid which in such cases, causes or accompanies the morbid affection. At the request of that gentleman, the author undertook to make the Tractors the theme of a satirical effusion in Hudibrastic verse. This was originally intended for the corner of a newspaper, but subsequently in the first edition enlarged to a pamphlet of about fifty pages royal octavo. It was published in the summer of 1803, well received, and a second edition called for in less than two months . . . The author never would have written a syllable intended to give Metallic Tractors favorable notoriety, had he not believed in their efficacy. As conductors of what is called animal electricity, and in principles allied to Galvanic stimulants, even their *modus operandi*, he thought, might be in a great measure explained" (*Pref.*, p. iii-iv). The fourth and final edition of *Terrible tractoration!!* was published at Boston in 1837.

1160. FIRST HELP IN ACCIDENTS AND IN SICKNESS.
First help in accidents and in sickness. A guide in the absence, or before the arrival, of medical assistance . . . Boston: Alexander Moore, 1871.
xii, 13-264 p. : ill. ; 19 cm.

". . . arranged and prepared for the press by the editors of Good Health" (p. v).

1161. FISCHER, Louis, 1864-1945.
The health-care of the baby. A handbook for mothers and nurses . . . New York: Funk and Wagnalls Company, 1906.
xi, [1], 144 p. ; 19 cm.

First edition. Fischer was born in Kaschau, Austria-Hungary, and came with his parents to New York early in his childhood. He received his medical degree from New York University Medical College in 1884; travelled to Berlin for post-graduate study; and returned to New York where he specialized in pediatrics. He was author of *The diseases of infancy and childhood* (1st ed., Philadelphia, 1907), a standard pediatric textbook during the first quarter of the 20th century. He also wrote several popular monographs on infant care and feeding. *The health-care of the baby* was issued in eighteen editions between 1906 and 1930.

1162. FISCHER, Louis, 1864-1945.
The health-care of the baby. A handbook for mothers and nurses . . . Seventh revised edition. New York and London: Funk and Wagnalls Company, 1916.
xiii, [1], 150, [4] p., [4] leaves of plates, [4] p. of charts : ill. ; 19 cm.

Issued in dust wrapper for the sixth edition.

1163. FISCHER, Louis, 1864-1945.
The health-care of the baby. A handbook for mothers and nurses . . . Eighth revised edition. New York and London: Funk and Wagnalls Company, 1916.
xiii, [1], 150, [4] p., [4] leaves of plates, [4] p. of charts : ill. ; 19 cm.

1164. FISCHER, Louis, 1864-1945.
The health-care of the baby. A handbook for mothers and nurses . . . Ninth edition. New York and London: Funk and Wagnalls Company, 1917.
xiii, [1], 150, [4] p., [4] leaves of plates, [4] p. of charts : ill. ; 19 cm.

1165. FISCHER-DÜCKELMANN, Anna, 1856-1917.
The wife as the family physician. A practical work of reference for the family in health and sickness, with special attention to diseases of women and children, married life, sexual hygiene, child-birth, care of children, family hygiene, etc. . . . Translated and adapted with the collaboration of a staff of eminent physicians. Milwaukee: International Medical Book Co.; Lon-

Fig. 27. "Sitz bath," from Fischer-Dückelmann's The wife as the family physician, *#1165.*

don: Kegan Paul, Trench, Trubner & Co., 1908.

viii, 911 p., 27 leaves of plates : ill. (some col.), port. ; 25 cm.

The first English-language edition of Fischer-Dückelmann's *Die Frau als Hausärztin,* originally published at Stuttgart in 1901 and subsequently translated into French, Polish, Russian, Spanish, Italian, Dutch, Swedish and Hungarian. A "new and enlarged" edition was published at Milwaukee in 1912; revised German editions remained in print as late as 1967. The author received her medical degree in Zurich and maintained a gynecological practice in Dresden. At head of title: *The golden book of the family.*

1166. FISHER, Irving, 1867-1947.

How to live. Rules for healthful living based on modern science. Authorized and pre-

pared in collaboration with the Hygiene Reference Board of the Life Extension Institute, Inc. by Irving Fisher, Chairman . . . and Eugene Lyman Fisk, M.D., Medical Director of the Institute. Chautauqua, New York: The Chautauqua Press, MCMXVII [1917]. Tenth edition revised.

xix, [1], 345 p. ; 20 cm.

After Fisher received his doctoral degree in mathematics from Yale University in 1891, he immediately joined its faculty in the department of mathematics. His passion was divided between mathematics and economics, however, and in 1895, he transferred to the economics department. Fisher's research and publishing focused on the theory of income, capital and interest. He was a pioneer in the quantification of economics and the application of statistical method to economic problems. His biographer in the *Dict. Amer. biog.* states that Fisher was "in the opinion of many, the leading economic theorist in the United States during the first half of the twentieth century" (Suppl. 4:275).

Later in his career, Fisher became involved in various reform movements, including prohibition and health reform. His devotion to the latter was triggered by a three year leave he was forced take from Yale in 1898 after being diagnosed with pulmonary tuberculosis. Fisher became intensely interested in the work of Horace Fletcher (#1193-1195), and undertook studies measuring the relation between a low protein diet and increases in physical strength and endurance.

As as economist, Fisher "saw as clearly as anyone the significance of health for labor and productivity, and more clearly than most the importance for health of personal (as opposed to environmental) hygiene. The dietetic precepts he had confirmed could thus be given an influential place in the renovation of society . . . His complete design for the healthful society was first given as a report for the National Conservation Commission that had been established by Theodore Roosevelt in 1908 . . . An investigation of human resources and their waste was also included among the commission's duties, and Fisher was appointed to direct that survey. His report, issued the following year, began with a proclamation that perfectly captured the ethos of Progressive physiologic optimism. 'The problem of conserving national resources,' Fisher explained, 'is only one part of the larger problem of conserving national efficiency. The other part relates to the vitality of our population . . . The prevention of disase . . . increases economic productivity.' His elaboration of that theme embraced a host of environmental and personal factors: quarantine, urban sanitation, food and drug legislation, regulation of working conditions, etc. In the realm of personal hygiene, special emphasis was given to the waste caused by 'undue fatigue,' a waste, he felt, 'probably much greater than the waste from serious illness'" (James C. Whorton, *Crusaders for fitness,* Princeton, N.J.: Princeton Univ. Press, 1982, p. 194).

Fisher advocated a national department of health. In lieu of this governmental agency, however, he was able to enlist the support of prominent Americans (e.g., William Howard Taft, William Henry Welch, William C. Gorgas,

etc.) in the establishment of organizations such as the Health and Efficiency League of America, the Commission of One Hundred on National Health, and the Life Extension Institute, to focus and coordinate the effort to improve America's health. He was later to incorporate eugenics into his national health agenda.

Fisher co-authored *How to live* with Eugene Lyman Fisk (1867-1931), an 1888 New York University Medical College graduate, insurance company executive, and Fisher's associate in the Life Extension Institute. A standard text for health courses in high schools and colleges, *How to live* appeared in twenty-one editions between 1915 and 1946. This 10th edition was issued as part of the *Chautauqua home reading series.*

1167. FISHER, Irving, 1867-1947.

How to live. Rules for healthful living based on modern science. Authorized by and prepared in collaboration with the Reference Board of the Life Extension Institute, Inc. By Irving Fisher, Chairman . . . and Eugene Lyman Fisk, M.D., Medical Director of the Institute. Thirteenth edition. New York and London: Funk & Wagnalls Company, 1917.

xix, [1], 345 p., [6] p. of plates + frontis. : diagrs., ports. ; 19 cm.

FISK, Eugene Lyman see FISHER (#1166-1167).

1168. FITCH, Samuel Sheldon, 1801-1876.

Diseases of the chest. A treatise on the uses of the lungs and on the causes and cure of pulmonary consumption. Designed for general as well as professional readers . . . Philadelphia: Hooker and Agnew, 1841.

[2], vii, [8]-175 p. ; 19 cm.

Samuel Sheldon Fitch was born in Plattsburgh, N.Y. on 27 December 1801. His twin brother, Jabez Huntington Fitch, also a physician, died at age twenty-five. The Fitch family had migrated from Norwich, Ct. to Sheldon in northwestern Vermont. Samuel's father, Chauncey, and paternal grandfather, Jabez, were both prominent physicians. The former was a founding member of the Franklin County Medical Society (Vermont), its secretary, a censor, and delegate to the state convention.

Though Samuel S. Fitch claimed to have received his medical degree in Philadelphia in 1827, his name does not appear on the rolls of matriculants either at the University of Pennsylvania or the Jefferson Medical College, the two medical schools extant in Philadelphia at the time. However, Fitch dedicated his first book to Nathaniel Chapman, and proffered thanks to Samuel Jackson—both prominent figures on the medical faculty of the University of Pennsylvania. It seems, therefore, that some professional contact with these men was likely. Fitch practiced in Philadelphia until about 1847 when he moved to New York City. It appears that he started as a surgeon dentist— and is so listed in the Philadelphia directories, as well as on the title-page of his first book : *A system of dental surgery* (Philadelphia, 1829), the first book of its kind in the United States. In 1841 Fitch published *Diseases of the chest,* at which point he seems to have given up dentistry for medicine. After moving to New York, he expanded his book on the diseases of the lung, and over a period of thirty-three years issued the book in thirty editions under three variant titles. Fitch's name also appears in the Boston city directories between 1864 and 1875 under "physicians, other," i.e., not a member of the Massachusetts Medical Society. In addition to his medical practice, Fitch was an entrepreneur. He maintained a publishing business and sold proprietary remedies nationally. S.S. Fitch died in Detroit on 31 December 1871, presumably while on a business trip. He is buried in Sheldon, Vt.

Not only does Fitch claim to have received his medical degree at Philadelphia in 1827, but to have studied in England under Francis Hopkins Ramadge (1793-1867), the author of *Consumption curable* (London, 1834; New York, 1839, #2927). In his preface Fitch writes, "The great body of the medical faculty in all countries at this moment consider pulmonary consumption as a fatal disease; fatal from its commencement, fatal in its progress, and fatal in its termination. So deeply rooted is this prejudice, that the medical practitioner who asserts pulmonary consumption curable, and that he can cure it, at once incurs the hazard of losing, with medical men, all his reputation, however respected or well deserved" (p. vi). Nonetheless, Fitch proclaims pulmonary tuberculosis curable. He attributes the disease to a shrink-

Fig. 28. Samuel Sheldon Fitch, from his Six lectures, *#1181.*

ing of the lungs, caused either by the formation and growth of "cheese-like matter" on their surface, or due to malformations of the chest. Fitch maintains that consumption is cured by "rendering the lungs more voluminous." He acknowledges such standard palliatives for consumptives as exercise, change of climate, etc., but specifically recommends inhalation, i.e., the regular use of an "inhaling apparatus" to inject air into tubercular lungs.

1169. FITCH, Samuel Sheldon, 1801-1876.
Family physician . . . Boston: S.S. Fitch, 1866.
    [2], 76 p. ; 19 cm.

Fitch used the periodically revised *Family physician* as a mechanism to promote his proprietary remedies, publications, etc.. In this 1866 edition he provides his schedule for private consultations in Boston, Greenfield, Ma., Rutland, Vt., New Haven, Buffalo, etc. The itinerant nature of Fitch's practice and business can be gathered from the preface to this edition:

"Since May, 1861, I have spent a considerable time in Europe. Since my return, I have established offices temporarily in Albany, Utica, Buffalo, New York, and Cleveland . . . The medicines I now use are nearly all different from those I used up to 1861. I have made vast improvements in the remedies I usually use, so that I now rarely fail to cure all the cases I treat" (p. 1). The first half of the 1866 *Family physician* consists of brief descriptions of diseases followed by testimonial letters from those who claim to have been cured by Fitch's remedies. The latter half provides similar descriptions of diseases that include simple recipes for their home treatment. In printed wrapper.

1170. FITCH, Samuel Sheldon, 1801-1876.
Family physician . . . New York: Dr. S.S. Fitch, 1868.
    89 p. ; 19 cm.

In printed wrapper.

1171. FITCH, Samuel Sheldon, 1801-1876.
Family physician: teaching how to prevent and cure disease, and prolong life and health to one hundred years . . . New York: Drs. S.S. Fitch & Son, 1875.
    120 p. : ill. ; 17 cm.

The size of this edition of the *Family physician* is swollen by half- or full-page advertisements for his various remedies and therapeutic appartus. In printed wrapper.

1172. FITCH, Samuel Sheldon, 1801-1876.
The influence of intellectual cultivation in improving the social condition and prolonging life. A lecture introductory to a course of lectures, on anatomy and physiology; delivered before the Pennsylvania Library and Literary Institute . . . December 22, 1837 . . . Philadelphia: T.K. & P.G. Collins, printers, 1838.
    24 p. ; 22 cm.

1173. FITCH, Samuel Sheldon, 1801-1876.
Guide to invalids; for persons using the remedies of Samuel Sheldon Fitch . . . With remarks on the cure of consumption, asthma, heart diseases, summer complaints . . . &c. &c. New-York: Calvin M. Fitch, [c1849].
    48 p. ; 18 cm.

Title and imprint from printed wrapper.

1174. FITCH, Samuel Sheldon, 1801-1876.
A popular treatise on the diseases of the heart, palsy, rheumatism, dyspepsia, dysentery, dysentery of children, cholera infantum, cholera and cholera morbus, bilious colic, costiveness, yellow fever, diphtheria, putrid sore throat, with proofs of their curability; and how to treat the diseases of elderly and old people: with medical prescriptions for nine of the above complaints: also rules for preventing sickness and preserving health to one hundred years . . . New York: S.S. Fitch & Co., 1860.
    168, [12] p. : ill. ; 20 cm.

Final [12] pages contain advertisements for Fitch's services, remedies and books.

1175. FITCH, Samuel Sheldon, 1801-1876.
A popular treatise on the diseases of the heart, palsy, rheumatism, dyspepsia, dysentery, dysentery of children, cholera infantum, cholera and cholera morbus, bilious colic, costiveness, yellow fever, diphtheria, putrid sore throat, with proofs of their curability; and how to treat the diseases of elderly and old people: with medical prescriptions for nine of the above complaints: also rules for preventing sickness and preserving health to one hundred years . . . New York: S.S. Fitch & Co., 1861.
    168, [12] p. : ill. ; 18 cm.

Cover-title: *Heart disease, rheumatism, dyspepsia, &c. their cure.* Spine-title: *Fitch on heart disease.*

1176. FITCH, Samuel Sheldon, 1801-1876.
A popular treatise on the diseases of the heart, palsy, rheumatism, dyspepsia, dysentery, dysentery of children, cholera infantum, cholera and cholera morbus, bilious colic, costiveness, yellow fever, diphtheria, putrid sore throat, with proofs of their curability; and how to treat the diseases of elderly and old people: with medical prescriptions for nine of the above complaints: also rules for preventing sickness and preserving health to one hundred years . . . New York: S.S. Fitch, M.D., 1870 [1876 issue].
    168, [10] p. : ill. ; 18 cm.

An advertisement for other of Fitch's publications pasted onto the front pastedown is dated December 1st 1876.

1177. FITCH, Samuel Sheldon, 1801-1876.

A popular treatise upon diseases of the heart, apoplexy, dyspepsia, and other chronic diseases, with proofs of their curability. Also, rules for preventing disease, and preserving health, (especially after forty) to one hundred years . . . New York: S.S. Fitch & Co., 1859.

[8], 112 p., 6 leaves of plates : ill. ; 21 cm.

Fitch provides an idiosyncratic nosology of heart disease, asserts his ability to cure every one of the disorders described, and supports his claims with numerous case histories and testimonial letters. Illustrated with six wood-engraved plates. Cover-title: *Heart disease, apoplexy, dyspesia, &c., their cure.* Atwater copy inscribed by author.

1178. FITCH, Samuel Sheldon, 1801-1876.

Read all the following directions carefully before commencing to take the medicines. M[*iss Mason*] will find marked thus + on the margin the remedies which DR. FITCH prescribes in this case . . . [185-?].

[1] leaf printed on both sides : ill. ; 25 × 19.5 cm.

In the promotional material that supplements his 1860 *A popular treatise on the diseases of the heart* (#1174), Fitch writes: "For the benefit of those who cannot visit the city, he has so arranged his practice and prepared his remedies, that he can treat them successfully and satisfactorily at a distance." He continues, "invalids may write a full statement of their case . . . stating all facts necessary for a full understanding of their symptoms and condition."

The directions printed on this sheet represent Fitch's response to patients who consulted him by mail. A column printed on the left margin of the recto lists thirty-seven medications or apparatus prepared and sold by S.S. Fitch. Miss Mason, whose ailments must have been legion, was shipped twelve items from this list: a shoulder brace, an abdominal supporter, cathartic pills, nervine, pulmonary liniment, pulmonary balsam, cough pills, heart corrector, universal tonic, pain-killer, anti-bilious mixture

and cough curer. The directions describe how Fitch's medications are to be taken, and provide general recommendations regarding bathing, diet, etc.

1179. FITCH, Samuel Sheldon, 1801-1876.

Six discourses on the functions of the lungs; and causes, prevention, and cure of pulmonary consumption, asthma, and diseases of the heart; on the laws of life; and on the mode of preserving male and female health to a hundred years . . . New-York: S.S. Fitch & Co.; London: L.H. Chandler, 1853.

[v]-xix, [3], [19]-368 p., [1] leaf of plates : ill. ; 19 cm.

Earlier editions published under the title: *Six lectures on the use of the lungs* (#1185-1187).

1180. FITCH, Samuel Sheldon, 1801-1876.

Six discourses on the functions of the lungs; and causes, prevention, and cure of pulmonary consumption, asthma, and diseases of the heart; on the laws of life; and on the mode of preserving male and female health to a hundred years . . . New-York: S.S. Fitch & Co.; London: L.H. Chandler, 1855.

[v]-xix, [3], [19]-368, [2] p., [1] leaf of plates : ill. ; 19 cm.

1181. FITCH, Samuel Sheldon, 1801-1876.

Six lectures on the functions of the lungs; and causes, prevention, and cure of pulmonary consumption, asthma, and diseases of the heart; on the laws of life; and on the mode of preserving male and female health to an hundred years. Also a treatise on medicated inhalation . . . Twenty-fourth edition . . . New-York: S.S. Fitch & Co.; London: L.H. Chandler, 1856.

xv, [16]-384, [6] p., [1] leaf of plates : ill., port. ; 19 cm.

Earlier editions published under the title: *Six discourses on the functions of the lungs* (#1179, 1180). Steel-engraved frontispiece portrait of author.

1182. FITCH, Samuel Sheldon, 1801-1876.

Six lectures on the functions of the lungs; and causes, prevention, and cure of pulmonary consumption, asthma, and diseases of the heart; on the laws of life; and on the mode of preserving male and female health to an hundred years. Also a treatise on medicated inhalation . . . Twenty-fourth edition . . . New-York: S.S. Fitch & Co., 1859.

xv, [16]-351, [1], viii, 10 p., [1] leaf of plates : ill., ports. ; 19 cm.

1183. FITCH, Samuel Sheldon, 1801-1876.
Six lectures on the functions of the lungs; and causes, prevention, and cure of pulmonary consumption, asthma, and diseases of the heart; on the laws of life; and on the mode of preserving male and female health to a hundred years . . . Twenty-ninth edition . . . New York: S.S. Fitch & Co., 1861.

xv, [16]-351, [1], viii, 4, [4] p., [1] leaf of plates : ill., port. ; 19 cm.

Steel-engraved frontispiece portrait of author.

1184. FITCH, Samuel Sheldon, 1801-1876.
Six lectures on the functions of the lungs; and causes, prevention, and cure of pulmonary consumption, asthma, and diseases of the heart; on the laws of life; and on the mode of preserving male and female health to a hundred years . . . Forty-first edition . . . New York: King & Chambre, 1865.

xv, [3], [19]-351, [9] p., [1] leaf of plates : ill., port. ; 18 cm.

Spine-title: *Fitch's lectures on consumption.* Engraved frontispiece portrait of author.

1185. FITCH, Samuel Sheldon, 1801-1876.
Six lectures on the uses of the lungs; and causes, prevention, and cure of pulmonary consumption, asthma, and diseases of the heart; on the laws of longevity; and on the mode of preserving male and female health to an hundred years . . . New York: H. Carlisle, 1847.

xi, [12]-324 p. : ill. ; 20 cm.

First edition. The title of the *Six lectures* varied over its publishing history. It was issued under the present title between 1847 and 1852 (#1185-1187); as *Six discourses on*

the functions of the lungs between 1852 and 1855 (#1179, 1180); and as *Six lectures on the functions of the lungs* between 1856 and 1865 (#1181-1184). Though the printed lectures were augmented with successive editions, the basic format and content of the work remained stable throughout. Spine-title: *Consumption cured.*

1186. FITCH, Samuel Sheldon, 1801-1876.
Six lectures on the uses of the lungs; and causes, prevention, and cure of pulmonary consumption, asthma, and diseases of the heart; on the laws of longevity; and on the mode of preserving male and female health to an hundred years . . . New-York: H. Carlisle, 1848.

xi, [12]-324 p. : ill. ; 20 cm.

Spine-title: *Dr. Fitch's lectures on the prevention and cure of consumption.*

1187. FITCH, Samuel Sheldon, 1801-1876.
Six lectures on the uses of the lungs; and causes, prevention, and cure of pulmonary consumption, asthma, and diseases of the heart; on the laws of longevity; and on the mode of preserving male and female health to an hundred years. New edition . . . New York: Calvin M. Fitch, 1852.

xix, [3], [19]-368 p. : ill. ; 19 cm.

Atwater copy inscribed by author (1852).

1188. FITCH, Samuel Sheldon, 1801-1876.
A treatise on health, its aids and hindrances: containing an exposition of the causes and cure of disease, and the laws of life. And noticing the affections of the head, throat, lungs, heart, liver, stomach, bowels, kidneys, bladder, womb, skin, bones, joints, muscles, etc. . . . New York: Pudney and Russell, 1857.

xxxiv, [9]-522, 2 p. ; 23 cm.

1189. S.S. FITCH AND COMPANY.
[Form letter] 1849, New York.

[1] leaf ; 25 cm.

Letter soliciting agents to sell the proprietary medicines, appliances and publications of S.S. Fitch & Co. Atwater copy accompanied by postmarked envelope.

1190. FITZ, George Wells, b. 1860.
Principles of physiology and hygiene . . .
Second edition, revised. New York: Henry
Holt and Company, [1909, c1908].
    xiii, [1], 359 p., IV leaves of plates : ill. ;
19 cm.

This secondary/college level textbook was
first published in 1908. Its author was an 1891
graduate of Harvard Medical School and is described on the title-page of this 2nd edition as
"Sometime Assistant Professor of Physiology
and Hygiene and Medical Visitor, Harvard University." Fitz also edited the 5th editions of H.
Newell Martin's school physiology (#2383).
Though the copyright date only appears on the
verso of the title-page, the "Preface to the second edition" is dated 1909 (p. v). Each plate is
accompanied by a leaf of descriptive letterpress.
Spine-title and title on dust jacket: *Physiology and
hygiene.*

FITZ, George Wells see also **MARTIN**
(#2383).

1191. G. W. FLAVELL AND BROTHER.
SHOULDER BRACES . . . Ladies' and Misses'
Shoulder Braces . . . Men's and Youth's Shoulder Braces . . . G.W. Flavell & Bro., manufacturers . . . Philadelphia. [188-?].
    [1] leaf printed on both sides : ill. ; 23
× 15 cm.

Advertised on verso: *The improved electric truss.*

1192. FLEMING, Lorenzo D., 1808-1867.
Self-pollution, the cause of youthful decay:
showing the dangers and remedy of venereal excess . . . New York: Published by J.K.
Wellman, 1846.
    viii, [9]-70, [2] p. ; 17 cm.

According to Fleming, masturbation is "a
licentiousness . . . which is probably second to
no other in the terribleness of its effects. Few
people are aware of its awful consequences
upon the physical, moral, and mental powers.
Those who have not acquainted themselves with
the subject, can scarcely believe, or even form
an idea of its fearful effects. It may be classed
among the *master evils* of the present day. It probably leads to more disease of body and mind,
and more premature decay, than any other one

evil. If it be asked how this can be, I answer—
because it lays the axe at the root of the tree. It
commences its inroads upon the system while
it is yet tender—at a time when excesses are far
more pernicious to the health and constitution
than at any other period of life" (p. 10-11).
Fleming quotes abundantly from S. Graham
(#1392), Deslandes (#931-934), Alcott (#40-45,
51), Hollick (#1703-1710), O.S. Fowler (#1279),
etc. in describing the pernicious effects of the
solitary vice.
    While preparing for the ministry, Fleming
lost his voice and went to New York for treatment. He subsequently began the study of medicine with his physician. Later treatment by a
homeopath resulted in the return of his voice,
and thereafter Fleming used homeopathic as
well as allopathic remedies. He practiced consecutively in New Bedford, Mass., Canandaigua,
N.Y., and Rochester N.Y. In 1853 he bought the
defunct Rochester Lake View Water Cure
(#3014), which failed and burned the same
year. In the mid-1850s Fleming moved to Ohio,
but returned to Rochester in 1857. He was appointed City Physician in 1862. In 1867, shortly
before his death, Fleming and his physician son
were partners in the Rochester Air Bath Institute.

**FLEMING'S WHOLESALE PATENT
MEDICINE DEPOT** see **FOSGATE**
(#1264).

1193. FLETCHER, Horace, 1849-1919.
Fletcherism. What it is or how I became
young at sixty . . . New York: Frederick A.
Stokes Company, publishers, [c1913].
    xvi, 224 p., [6] leaves of plates : ill., ports. ;
20 cm.

Born in Lawrence, Mass., Fletcher shipped
out on a whaler at age fifteen bound for the
Pacific. Upon his return from the Orient, he
enrolled at Dartmouth, but left after a year.
During the next three decades Fletcher circled
the globe four times, and amassed a personal
fortune as a manufacturer of inks and importer
of merchandise from the Far East. In his for-
ties Fletcher's life took an unexpected turn
when he was refused a life insurance policy
because of his poor physical condition. He immediately set out to change his lifestyle, and
afterward, that of the nation. Fletcher's career

as a health reformer epitomizes the spirit of the Progressive era, and the American businessman's ideal of philanthropic service.

Fletcher based sound health on a dietary regimen summarized by James C. Whorton: "there are several natural powers subject to abuse: appetite, taste, saliva, and mastication. Appetite is nature's announcement of fuel and repair requirements . . . One should eat only when appetite urges, select the food or few foods which seem especially appetizing at that moment, and eat only so long as appetite remains a strong impulse. The second power, taste, is nature's gift of pleasure in nutrition, and one should keep every food item in the mouth until the last trace of taste is extracted . . . Saliva is the first agent of digestion, and would not have been provided unless nature wanted it mixed thoroughly with food before swallowing . . . The final agent, mastication, is nature's mechanism for mixing saliva and extracting taste" (*Crusaders for fitness,* Princeton, N.J.: Princeton Univ. Press, 1982, p. 175-76).

Fletcher believed that his combined program of positive thinking and conscientious eating (with the emphasis on thorough mastication) would result in social as well as individual transformation. The wholesale adoption of *Fletcherism* would introduce an age in which mankind would become immune to disease and live longer, allowing each human being to contribute to the potentially unlimited social and physical evolution of the species.

In the present work, the author reviews the principles of *Fletcherism*, and the experiments and trials to determine its validity made by such figures as Russell Chittenden, Irving Fisher (#1166, 1167), Michael Foster, Walter B. Cannon, John Harvey Kellogg (#2076-2108), etc. Appended to the text is Ernest Van Someren's "Was Luigi Cornaro right?" (p. 197-219), a paper read before the physiological section of the British Medical Association in August 1901 and published in the *Brit. med. journal* (1901, 2:1082-84). A confirmed disciple, Van Someren provides a detailed physiological analysis of mastication on Fletcherite principles. Series: *A.B.C. life series.*

1194. FLETCHER, Horace, 1849-1919.
Menticulture or the A-B-C of true living . . . Chicago: A.C. McClurg & Company, 1895.
145, [3] p. ; 20 cm.

First edition. "By nature a man of action and an optimist, Fletcher was attracted by the supposition that obstacles are imaginary and a person can do whatever he believes he can. The offspring of his analysis of mind power was *Menticulture* (1895). This first book was pedestrian psychology, even by New Thought standards, but in his unabashed offering of banalities disguised as profundities Fletcher completed the first rough draft of his design for health . . . He cheerfully assured readers that clearness of thought, peace of mind, calmness of behavior and other rewards unlimited awaited those who would take the simple step of immediately purging their minds of the two roots of all evil passions: anger and worry . . . Most of the book is given to repetitious exhortations to be happy and to descriptions of the glorious state that would follow 'Mental Emancipation.' It is in these descriptions that Fletcher reveals the already Progressive frame of mind that would endure through his transition from mental to physical hygiene" (James C. Whorton, *Crusaders for fitness,* Princeton, N.J.: Princeton University Press, 1982, p. 170).

1195. FLETCHER, Horace, 1849-1919.
Menticulture or the A-B-C of true living . . . Chicago & New York: Herbert S. Stone & Co., 1899.
280, [4] p. ; 19 cm.

Writing of his vision of the 20th century, Fletcher enthuses: "With a great surplus of means, the matter of attainment of any reasonable hope is not difficult and need not long be delayed . . . It only requires a change in the national point-of-view and a change of the direction of existing energy from wasteful and unprofitable selfishness to profitable co-operative altruism . . . All of the prevailing conditions seem to be favorable to a change from enforced selfishness to co-operative or voluntary altruism, and the nineteen hundredth anniversary of the birth of Christ is a fitting occasion for a Christian nation to re-adjust its manners and its economies on the plan of the Master, as intended by the Declaration of Independence and the Constitution" (p. 151-52). *Menticulture* was first published at Chicago by McClurg & Co. in 1895 (#1194). This "enlarged edition" was first published at Chicago by Stone in 1897. Series: *Menticulture series [no.] 1.*

1196. FLETCHER, Moore Russell.
Our home doctor. Domestic and botanical remedies simplified and explained for family treatment. With a treatise upon suspended animation, the danger of burying alive, and directions for restoration . . . Boston: Wilson Brothers, 1886.
    vii, [8]-332, [2], 70 p., [1] leaf of plates : ill., port. ; 22 cm.

    "The question naturally arises, why the author of this book, after he had been educated in the school known as allopathic, studied medicine, attended lectures in Harvard Medical University in 1834, '35, '36; and also in Bowdoin Medical College, in Brunswick, Me., in 1836, obtained a diploma therefrom, and became a Fellow of the Massachusetts Medical Society, should relinquish that system of practice and adopt another. Our reasons for the change may be thus stated: we learned that minerals and drugs did not nourish, but set up another disease which checked the one then active, destroyed the appetite, prevented sleep, and kept the patient sick for thirty or forty days . . . We also saw that a system known as botanical or vegetable was practised among all nations with safety and success; and every day's experience served to convince us of the value of domestic remedies and treatment" (p. 39).

    Our home doctor appeared in at least four issues between 1883 and 1890. Its author is listed as an 1836 graduate of the Medical College of Maine (Bowdoin) in the 1886 edition of Polk's *Medical & surgical directory of the United States.* Added treatise has title: *One thousand persons buried alive by their best friends.* Wood-engraved frontispiece portrait of author.

1197. FLICK, Lawrence Francis, 1856-1938.
Consumption a curable and preventable disease. What a layman should know about it . . . Philadelphia: David McKay, publisher, 1903.
    295 p. ; 19 cm.

    First edition. Flick was indefatigable in his efforts to establish sanitaria for the treatment of tubercular patients, to provide medical care for patients of all economic and ethnic groups, to educate the the public regarding tuberculosis prevention and treatment, to fund the continued study of the disease, and to establish broadly based organizations of physicians and laymen to coordinate these efforts. He is one of the principal figures in Barbara Bates' *Bargaining for life: a social history of tuberculosis, 1876-1938* (Philadelphia: Univ. of Pennsylvania Press, 1992).

    Though drawn to law, Flick entered the medical profession largely in response to his own adolescent affliction with pulmonary tuberculosis. He received his medical degree from Jefferson Medical College in 1879, but was forced to withdraw from an internship at the Philadelphia General Hospital because of his illness. "Seeking a cure, he went west in 1881 and eventually worked as an orange packer in California. During this period he evolved a regimen of diet, rest, and exercise which he found greatly beneficial; believing that it could be carried out equally well in the East, he returned to Philadelphia in 1883 and resumed his medical practice. Many poor consumptive patients came to him, and this led him to dedicate himself to the conquest of tuberculosis" (*Dict. Amer. biog.,* Suppl. 2:196).

    Against then accepted medical opinion, Flick insisted that tuberculosis is contagious and that isolation in special hospitals is necessary to prevent its spread. In 1892 he established the Pennsylvania Society for the Prevention of Tuberculosis, "a pioneer association of laymen and physicians combined to combat a single disease and a model for many subsequent health organizations" (*ibid.*). Flick attracted the attention of the philanthropists Andrew Carnegie and Henry Phipps, who agreed to finance a tuberculosis sanitarium in Philadelphia. The Henry Phipps Institute for the Study, Prevention and Treatment of Tuberculosis (est. 1903) eventually became part of the Univ. of Pennsylvania. The following year, Flick was influential in founding the National Association for the Study and Prevention of Tuberculosis. Among his many publications are several popular works on pulmonary tuberculosis. *Consumption, a curable and preventable disease* was published in seven editions between 1903 and 1914. At head of title: *The crusade against tuberculosis.*

1198. FLICK, Lawrence Francis, 1856-1938.
Consumption, a curable and preventable disease. What a layman should know about it . . . Fourth edition. Philadelphia: Lawrence F. Flick, M.D., publisher, 1905.
    295 p. ; 19 cm.

At head of title: *The crusade against tuberculosis.*

1199. FLINT, Austin, 1812-1886.

STUDY OF THE HUMAN BODY. The subscriber takes this method of announcing his intention of giving during the coming winter, a course of instruction in Anatomy and Physiology, designed for persons of either sex, who are desirous of acquiring a general knowledge of the structure and mechanism of the Human Body. The course will consist of a series of eight or ten Lectures . . . With reference to this course, arrangements have been made to import one of the SPLENDID PREPARATIONS OF ARTIFICIAL ANATOMY, constructed by Dr. Auzoux, of Paris . . . The course will commence immediately on the arrival of the model, which is expected early in January . . . Buffalo, November, 1841 . . .

Broadside ; 25 × 20 cm.

The *Dictionary of American medical biography* describes Flint as "possibly the most influential American physician of the midnineteenth century" (1:254) because of his "sustained effort to elevate the standards of the teaching and practice of medicine" (*ibid.*) and the influence that his many books and articles on internal medicine had on the American medical profession. An 1833 Harvard medical graduate, Flint practiced in Boston and then in Northampton, Ma. until 1836, when he moved to Buffalo, New York. In 1844 Flint left Buffalo to teach briefly at Rush Medical College in Chicago. He returned to Buffalo in 1845, where he became the founding editor of the *Buffalo medical journal* and one of the founders of the Buffalo Medical College (1847). Flint's peripatetic career as a medical educator took him to Louisville in 1852, back to Buffalo in 1856, to New Orleans in 1858, and to New York City in 1861.

In the promotion of his lecture series in this broadside, Flint emphasizes his pending acquisition of one of the remarkable manikins of Louis Auzoux (1797-1880), whose models were similarly employed in the popular anatomical lectures of Bedford (#271), Hollick (#1697), Gleason (#1351) and others (*see also* Codman & Shurtleff, #696).

1200. H. S. FLINT AND COMPANY.

DR. FLINT'S QUAKER BITTERS. A great medical discovery and remedy, truly one of God's blessings to man. Reader, do not as you scan these lines say to yourself . . . here is another humbug calculated to decoy the weary, and sap their substance. The Quaker Bitters are just what they represented, an invaluable medicine, an antidote for the thousand ills which afflict the human body. No one can remain long unwell . . . after taking a few bottles of Dr. Flint's Quaker Bitters . . . [*wood engraving depicting Quaker offering Bitters to male figure on crutch*] . . . Dr. H.S. Flint & Co., Providence, R.I. . . . [186-?].

[1] leaf printed on both sides : ill. ; 34 × 18 cm.

Wood-engraved view of Dr. H.S. Flint & Co.'s Laboratory printed on verso.

1201. H. S. FLINT AND COMPANY.

QUAKER BITTERS. Dr. Flint's Great Medical Depot, Providence, R.I., proprietors of the Quaker Bitters . . . [ca. 1874].

[2] leaves : ill. ; 26.5 × 17.5 cm.

"These celebrated bitters are composed of choice roots, herbs and barks, among which are gentian, mandrake, sarsaparilla, wild cherry, dandelion, juniper, and other berries,—highly concentrated and so prepared as to retain all their medicinal qualities" (leaf 1 recto). Among the disorders the Bitters "cure, or greatly relieve" are dyspepsia; lassitude; eruptions; "kidney, bladder and urinary derangement;" worms; rheumatism, bronchitis; "all difficult female derangements . . . so prevalent among the American ladies;" "impurities of the blood," etc. Wood-engraved view of Dr. Flint's Great Medical Depot on recto of first leaf above the title. On the inside are wood-engraved illustrations of the "Tank, or Generating Room" and the "Bottling and Packing Department." Printed at bottom of second leaf verso: For sale by Willard Bros., Oxford, N.H.

1202. FLOWER, Alfred Hollis, 1856-1922.

Good health and how to attain it . . . Containing chapters on I. Thought as a therapeutic agent. II. The human hand in disease. III. Music as a medicine. IV. Food as a factor of health. V. Diet for consumption. VI. To preserve health. Boston, Ma. [1897?].

58, [2] p., [1] leaf of plates : port. ; 20 cm.

"The strange power of intuition, or mental receptivity, the possession of which enables one physician to make a brilliant record, while another, equally well versed in books, but lacking this, signally fails, is now coming to be recognized more and more by the most scientific and progressive minds in the medical profession. Its complimentary gift . . . that of inspiring confidence, and, by a subtle mental power, *suggesting health*—is less recognized by scientific men, although it explains to a great degree this mystery . . . exhibited in the fact that the identical remedies which have proved inert in the hands of one doctor, when given by another physician are promptly followed by the desired results. This mental power, which in its extreme manifestation is termed hypnotism, but which is, consciously or unconsciously, exerted to a greater or less extent by every successful practitioner, will be more and more recognized in the future, as the horizon of scientific investigation in the psychic realm broadens. Both these rare gifts are possessed in an extraordinary degree by . . . Dr. Flower" (p. 7-8).

Flower claims to be a graduate of three medical schools (p. 5). According to the 9th edition of *Polk's medical register and directory of North America* (1906), he was an 1889 graduate of the American Health College (Cincinnati), described in *Polk*: "chartered by the State of Ohio in 1876 for teaching and practicing the Religious Spiritual Vitapathic Systems of Health and Life, for Body and Soul" (p. 174). He was in partnership in Boston with his elder brother Richard C. Flower (M.D., 1878, Amer. Health Coll.) from 1890 until 1897. A.H. Flower moved to the District of Columbia, where he is listed in *Polk* by 1898. In printed wrapper. Frontispiece portrait of author.

FLOWER, Sidney Blanchard see *HOME STUDY COURSE IN OSTEOPATHY* (#1757).

1203. FLYNN, William Earl, 1861-
The Flynn system of health culture. Diet and exercise . . . Lincoln, Nebr.: Woodruff Bank Note Co., 1916.
440 p. : ill. ; 20 cm.

According to the "Biographical sketch" prefacing this volume, William Earl Flynn "devotes himself to the spread of the gospel of good health with the earnestness and enthusiasm of a man who is giving the world the message of his own salvation. He began life as a local preacher and evangelist of the Methodist Episcopal Church, but was forced to give up his work on account of ill health . . . His physician told him that his bowels were seriously diseased, and that he could give him no hope of relief. But instead of yielding to his malady, Mr. Flynn entered upon a life and death struggle, during which he developed his system of health culture" (p. 7). Flynn took the message of his "gospel of good health" on the road, conducting "evangelistic health campaigns" in churches "in no less than twenty-two states and two Canadian provinces" (p. 8).

"In order to render more permanent the good accomplished during Mr. Flynn's visits, he has organized clubs and classes for continuing the practice of his system of health culture." We are informed that "these groups of followers have promoted their enthusiasm by calling themselves "Hundred Year Clubs" (*ibid.*) in anticipation of a prolonged span of years. In his first chapter Flynn states: "The foundation stones of this system of health culture are faith, diet, and exercise . . . We use the term faith in a broad sense to cover those more positive forms of mental and spiritual life whose supremacy over man's physical organism is now being fully recognized. We include in it not merely that reverent belief in the sanity and goodness of what God has created, but also the confident assertion of man's power, by his own efforts, to attain a larger measure of life by the systematic observance of the laws regulating his psychical and physical development" (p. 17).

1204. FLYNN, William Earl, 1861-
Short health talks . . . [Lincoln, Nebraska: The Woodruff Press, 191-?].
4 pamphlets (32 p. each) ; 19 cm.

These four "books" contain 106 articles, most of which pertain to diet. In an article entitled "Cooking for health, happiness and success," Flynn places before his reader's imagination the ideal picture of a newly married couple. He is a "soaring, ambitious husband and father who wishes to make a success of life, who has great ambitions for himself and children." Alas, "all these ambitions may reach their goal . . . but never, never, if the wife is not a wise, scientific cook." Indeed, Flynn cautions in an apothegm worthy of *Proverbs,* "A man's ambition may be weighted down with a soggy potato" (Book 4, art. 88, p. 17).

1205. FOIE, F. Marion, Mrs.
[Form letter], 1904 April 10, New York, N.Y., [to] Henry M.W. Eastman, Rosyln, N.Y.
[1] folded leaf ; 20 × 13 cm.

The text begins: "Gentlemen: If you have no Son to perpetuate your 'Name,' inherit your Estates and Fortunes, why not consult Mrs. F.M. Foie." Foie proclaims her ability not only to determine gender *in utero,* but to guarantee a male heir. As evidence of her success she cites the birth of two sons to President Grover Cleveland—by her agency, apparently, and "to his great delight." Included are an addressed, postmarked envelope and Foie's business card, on which she describes her self as "Trained nurse and massage operator," adding by hand: "Teacher of French & Piano."

1206. FOLSOM, James, 1819-1894.
The mariner's medical guide; designed for the use of ships, families, and plantations. Containing the symptoms and treatment of diseases. Also, a list of medicines, their uses, and the mode of administering, when a physician cannot be procured. Selected from standard works. Third edition, revised. Boston: Published by James Folsom, 1864.
189 p. : ill. ; 17 cm.

"After twenty years' experience in compounding and dispensing medicines, and having seen so much neglect in supplying vessels with pure medicines, I have been induced by many ship owners and captains to make it an exclusive branch of business—the furnishing and replenishing of medicine chests. My object in publishing this little work, is to furnish the requisite knowledge for the proper use of these medicines" (p. 3). Born in 1819 in Eastport, now part of Maine, Folsom was engaged in the wholesale and retail drug business in Boston from early manhood. He was the pioneer druggist in Boston's South End, keeping a store at

the corner of Canton and Suffolk streets (now Shawmut Ave.), and afterwards at the corner of Suffolk and Concord streets. In later years he engaged in the business of supplying ship's medicine chests on Eastern Ave. in the North End. Folsom's obituary appears in the *Boston evening transcript* (12 April 1894).

The Atwater copy bears an inscription on the front pastedown indicating that this copy belonged to the bark "John Worcester" out of Boston, May, 1867. It also bears the printed slip of the author and publisher, James Folsom, who is described as "Ship Druggist" with stores in Boston and at "Coenties Slip, corner of Water Street, New York."

1207. FONDA, Sebastian F.
Analysis of Sharon waters, Schoharie County; also of Avon, Richfield, and Bedford mineral waters. With directions for invalids . . . New York: John J. Schroeder, medical bookseller, 1854.
96 p. ; 19 cm.

First edition. Fonda devotes more of his text to a description of "eliminative medicines," the physiology of the skin and the principles of bathing than to the waters at Sharon Springs, Avon and other mineral waters of New York State. The *Analysis* was re-issued in 1857 and 1860 (#1208). A second edition was published at Albany in 1876 (#1209) and 1878.

1208. FONDA, Sebastian F.
Analysis of Sharon Waters, Schoharie County; also of Avon, Richfield, and Bedford mineral waters. With directions for invalids . . . New York: R. Craighead, printer, 1860.
96 p. ; 19 cm.

1209. FONDA, Sebastian F.
The mineral springs of Sharon, Schoharie County, N.Y., comprising an account of the springs, with remarks on the nature and medical applicability of each . . . Second edition: greatly enlarged, with valuable information to invalids. Albany [N.Y.]: Weed, Parsons & Co., printers, 1876.
187 p. ; 18 cm.

This second edition of Fonda's examination of the mineral springs at Sharon, N.Y. has been so extensively revised as to constitute a new work. Cover-title: *Sharon Springs*.

1210. FONERDEN, William Henry.

The institutes of Thomsonism . . . Philadelphia: Printed at the office of the Botanic Sentinel, 1837.

v, [6]-123, [1] p. ; 13 cm.

Fonerden's *Institutes* is somewhat unique in the corpus of Thomsonian literature in that its provides the physiological and pathological principles of the Thomsonian botanic system, rather than its practical application. The book is based on lectures previously delivered by the author. Fonerden was a clergyman and secretary of the Thomsonian Society (Philadelphia) in 1837. He edited the *Southern botanic journal* from 1839 to 1846.

1211. FONSSAGRIVES, Jean Baptiste, 1823-1884.

The mother's work with sick children . . . Translated from the fourth Paris edition. By F.P. Foster . . . New York: G.P. Putnam & Sons, [c1872].

viii, [9]-10, 244 p. ; 19 cm.

Translation of: *Le rôle des mères dans les maladies des enfants* (1st ed., Paris, 1868; 5th ed., Paris, 1883). The translator, Frank Pierce Foster (1841-1911), was an 1862 graduate of the College of Physicians and Surgeons (New York), a prominent medical lexicographer, and editor of the *New York medical journal*. Series: *Putnam's handy book series*.

1212. FONTAINE, A. de, b. 1798.

The book of prudential revelations: or the golden bible of nature and reason, and the confidential doctor at home; expounding to the family circle the laws of human nature and health, and the doctrine, origin, and progression of diseases, and their effectual philanthropic remedies. The prophetic warnings to the transgressors are here recorded, as they resound from the gulf of oblivion and crimes; sufferings and sickness; despair and death; illustrated by the awful disclosures of the mysteries of real life . . . Boston: Published by the author, 1845.

2 pts. in 1 v. (xxii, [23]-507; vi, [7]-53 p.) : ill. ; 22 cm.

"And thousands rescued from disease and pain,
Invoke the willing heaven to bless Fontaine!"

(from the "*Tribute of Gratitude*" that prefaces the *Book of prudential revelations*)

The book of prudential revelations is divided into three parts, the first on sexual behavior, ethics and disorders; the second on anatomy and physiology; and the third on the preservation of health. The first constitutes more than forty percent of the text and is by far the most interesting portion of the book. In part I, sect. I, Fontaine discusses the sexual natures of males and females. The text is laced with passages that must have seemed lascivious to contemporary readers (e.g., the description of adolescent female development on p. 45). Fontaine attributes most female disorders to the emotional and physical consequences of woman's sexual physiology: "Woman has a vital system larger than that of man, and she has also a larger reproductive system; hence her functions are correspondingly more exciting, and capable of greater pleasure than man" (p. 56). This is dangerously complicated by the fact that the "cerebel, or organ of *the will*, is small in woman" (p. 57). Not only are the "pleasures of love . . . more intense, and more essential to her organization," but are "less determinate than in man." As a result, woman is more susceptible to morbid emotional and physical conditions.

Part I, sect. IV is devoted to "female complaints." Again, Fontaine grounds these "complaints" in woman's reproductive physiology, maintaining that it makes her "more tender, delicate, and sensitive than the male, with a more excitable temperament and stronger power of imagination." This emotional and physical unstability has obvious consequences for her "health, long life, and happiness" (p. 115). In fact, Fontaine attributes most female disorders, from nymphomania to suppression of the menses, to woman's "excitable temperament." Part I, sect. V provides a lengthy moral and medical discussion of sexually transmitted diseases—once again, with titillating suggestions of abnormal sexual behavior.

Fontaine recommends proprietary remedies for each disorder that he considers in *The book of prudential revelations*. His *French Philanthropic Remedy,* for example, can be taken to overcome the habit of masturbation, or to kill the worms

Fontaine posits as the primary cause of chronic disease. *Dr. Fontaine's Female Medicine* is recommended for disorders ranging from nymphomania to suppression of the menses. The text is supplemented by *A practical key to the confidential doctor at home* (separate title-page & pagination, but continuous collation). Fontaine provides a catalog of his various proprietary remedies, including the ten applications of the *French Philanthropic Remedy* (no. 5, a contraceptive), and instructions "to be observed by those who request the professional advice, aid, or services of Dr. Fontaine."

The author claims to be of Italian descent (the nephew of Felice Fontana), to have been educated in France, Italy and Germany (completing his medical studies in 1812), to have been exiled from Europe for his liberal political views, and to have arrived in the United States in 1831 or 1832, where he set up a medical practice in New Haven.

Because of the complex publishing history of E.B. Foote's principal work, the following order of entries has been substituted for the normal alphabetical arrangement of this catalog:

1. Medical common sense (#1213-1218)
2. Plain home talk (#1219-1234)
3. Plain home talk [German translations] (#1235-1241)
4. Plain home talk [Finnish translation] (#1242)
5. Dr. Foote's home cyclopedia of popular medical, social and sexual science (#1243)
6. Dr. Foote's new book on health and disease (#1244-1245)
7. Dr. Foote's new plain home talk on love, marriage, and parentage (#1246-1247)
8. Dr. Foote's plain home talk (#1248)
9. Other titles (#1249-1256)

1213. FOOTE, Edward Bliss, 1829-1906.
Medical common sense; applied to the causes, prevention and cure of chronic diseases and unhappiness in marriage . . . Boston: Wentworth, Hewes & Co., 1858.
    iv, [vii]-xviii, 271, [1] p., [1] leaf of plates : ill., port. ; 19 cm.

First edition. A native of Cleveland, Ohio, Foote received little formal education before being apprenticed to the printer of the Cleveland *Herald* at age fifteen. Three years later he left Cleveland for Connecticut, where he become compositor of the New Haven *Journal,* and shortly after, editor of the New Britain *Journal.* Foote's career in journalism culminated in the associate editorship of the Brooklyn *Morning journal.* It was sometime during this period that his interest in medicine emerged, and that he began his preceptorship with a local physician. "I practiced medicine for nearly eight years before receiving my degree," Foote proclaimed in 1872 (*Trans. Eclec. Med. Soc. of the State of New York,* 1871-72, p. 88). Part of this time he practiced in Saratoga Springs, which is described in the early editions of *Medical common sense.* Foote's career as an "unlicensed practitioner" was followed by formal medical training at Penn Medical University (Philadelphia), an eclectic school founded by Joseph S. Longshore in 1853, from which Foote graduated in 1860. By the early 1860s Foote was established in New York, where he practiced medicine, published and operated a mail order business until his retirement.

Medical common sense was one of the most successful manuals of domestic medicine published in the United States during the latter half of the 19th century. Foote claimed that 250,000 copies had been printed between 1858 and 1867. Its much expanded successor, *Plain home talk* (1st ed., 1870, #1219), remained in print into the 20th century. Foote published *Medical common sense* toward the end of eight years' unlicensed medical practice, and two years prior to his graduation from Penn Medical University. He expresses his hope that the ideas contained therein "may prove worthy of Allopathic denunciation" (p. [iii]). The book is divided into two parts. In the first, "Diseases—their causes, prevention and cure," Foote maintains that all disease emanates either from derangements of the brain & nervous system (i.e., from an obstructed flow of "vital electricity") or from "impure blood." He proposes to remedy the disorders resulting from these primary causes by vegetable medicines and a medly of electrotherapies. Much of this first part is devoted to the prevention of such disorders by following common sense hygienic principles.

The second part, "Marriage and sexual philosophy," is devoted to various aspects of the married state and to sexual behavior. In the 1860s Foote had become something of a pioneer in the dissemination of information regarding contraception. Although his reasons justifying contraception and his description of contraceptive methods have received extensive consideration from historians—to the exclusion of the rest of his work—his discussion of contraception occupies only three pages in this first edition of *Medical common sense* (p. 247-50). In this edition he limits himself to expressing concern for the impaired health of females resulting from "a too rapid increase of offspring;" the widespread availability of "quack nostrums, injurious and unreliable recipes;" and the dangers to the health of both males and females from existing birth control methods (e.g., "prevention pills," "caustic injections" into the vagina; "withdrawing" during coition; and abortion). It was not until publication of the 1860 issue of *Medical common sense* (#1216) that Foote expanded the discussion presented here to include his philosophy of contraception and an exposition of contraceptive methods.

In 1872 Foote founded the Murray Hill Publishing Company to manage the printing and distribution of his many publications, as well as other materials distributed through his mail order business (e.g., proprietary remedies, contraceptive devices, syringes, etc.). Foote's devotion to the cause of eclectic medical practice and family planning was motivated as much by his sharp business acumen as by genuine reform ideals. In her study of Foote, Janet F. Brodie notes that: "Foote's career illustrates the conjunction of commercialized birth control with female consumerism in the mid-century United States. He aimed his contraceptive products and information at diverse publics, but he quickly recognized the new and growing importance of women as customers" (*Contraception and abortion in nineteenth-century America*, Ithaca: Cornell Univ. Press, 1994, p. 240). In 1876 he was convicted under the recently enacted "Comstock law" (*see note to* #1224) for distributing a pamphlet through the mails on contraceptive methods (*Words in pearl*). Thenceforth, he desisted from providing advice on birth control in his any of his publications.

1214. FOOTE, Edward Bliss, 1829-1906.

Medical common sense; applied to the causes, prevention and cure of chronic diseases, and unhappiness in marriage . . . Boston: Wentworth, Hewes & Co., 1859.
iv, [vii]-xviii, 271, [1] p., [1] leaf of plates : ill. ; 19 cm.

First edition, second printing. Frontispiece portrait of author.

Fig. 29. Edward Bliss Foote, from his Medical common sense, *#1215.*

1215. FOOTE, Edward Bliss, 1829-1906.

Medical common sense; applied to the causes, prevention and cure of chronic diseases, and unhappiness in marriage . . . Philadelphia: Duane Rulison, 1860.
iv, [vii]-xviii, 271, [1] p., [1] leaf of plates : ill. ; 19 cm.

Steel-engraved frontispiece portrait of author.

1216. FOOTE, Edward Bliss, 1829-1906.

Medical common sense; applied to the causes, prevention and cure of chronic diseases, and unhappiness in marriage . . . Saratoga Springs, N.Y., 1860.
iv, [vii]-xviii, 275 p. : ill., port. ; 19 cm.

This is the final issue of the first edition of *Medical common sense* and the last which includes the chapter on the waters at Saratoga Springs (p. 149-53). The year this issue was published, Foote received his medical degree from Penn

Medical University, and within several years moved his practice to New York City.

This 1860 Saratoga Springs issue of *Medical common sense* differs from the prior two issues in containing three additional pages of matter discussing contraceptive methods (p. 251-54). Foote describes four "*reliable* and *harmless* means which *never* fail in effecting the object," noting that "it is but right and proper that they should be placed in the hands of the married:" 1.) The *membranous envelope,* an "improvement on the ordinary French Male Safe or Cundam," that differs from the usual item in that it "permits the free and unobstructed influx and eflux of individual electricity in the act of coition"; is thinner ("its use does not in the least interfere with the pleasure of the act"); and is stronger. He notes also that the "membraneous envelope is of the greatest utility, because, while it is a sure preventive of conception, it also prevents either party from contracting disease" (p. 251). 2.) The *apex envelope,* "an ingenious contrivance, just large enough to cover the glans penis without enveloping the whole organ." Its advantage is that "while the little sack catches the seminal fluids .. the ring is said to greatly heighten the pleasure of the act on the part of the female." Foote points out its single defect, however: "it does prevent the generation of chemical electricity in the copulative act" (p. 252). 3.) The *electro-magnetic preventive machine,* a device of Foote's own invention," the operative principle of which he fails to describe, but which is offered to couples for twelve dollars. 4.) The *womb veil,* "an India rubber contrivance which the female easily adjusts in the vagina before copulation, and which spreads a thin tissue of rubber before the mouth of the womb so as to prevent the seminal aura from entering" (p. 253). Foote's explicit description of these devices is rather remarkable for the period in which they were published. Frontispiece portrait of author.

1217. FOOTE, Edward Bliss, 1829-1906.
Medical common sense; applied to the causes, prevention and cure of chronic diseases and unhappiness in marriage . . . Revised and enlarged edition. New York: Published by the author, 1866.
xviii, 390 p. : ill., port. ; 19 cm.

The first issue of the "revised and enlarged" edition of *Medical common sense* was published in 1863. It was reissued in 1864, 1865, 1866 and 1867 (#1218). The revised edition contains more than a hundred pages of added material, most of which (eighty-eight pages) provides new chapters (in part one) on urinary tract disorders, diseases of the female organs of generation, sterility (male & female), "seminal weakness," and ruptures & hernias. The second part was enlarged by only nine pages. The section on "Prevention of conception" has nearly been reduced to the proportions of the 1858 edition (#1213). The description of the four contraceptive devices first described in the 1860 edition has been set in smaller type and moved to "Part III. Things useful, and quotations testimonial," in which Foote advertises "mechanical and electrical remedies;" "Preventions," i.e., contraceptive devices (p. 378-80); and proprietary medicines. Frontispiece portrait of author.

1218. FOOTE, Edward Bliss, 1829-1906.
Medical common sense; applied to the causes, prevention and cure of chronic diseases and unhappiness in marriage . . . Revised and enlarged edition. New York: Published by the author, 1867.
xviii, 390 p. : ill., port. ; 19 cm.

Final issue of the "revised and enlarged edition." In his preface to the 1870 edition of *Plain home talk* (#1219), the successor to *Medical common sense,* Foote states that over two hundred and fifty thousand copies of the earlier book had been sold, "a circulation which I venture to affirm has been attained by no other medical work of the same size in this or any other country" (p. v). Frontispiece portrait of author.

1219. FOOTE, Edward Bliss, 1829-1906.
Plain home talk about the human system—the habits of men and women—the causes and prevention of disease—our sexual relations and social natures; embracing Medical common sense applied to causes, prevention, and cure of chronic diseases—the natural relations of men and women to each other—society—love—marriage—parentage—etc., etc. . . . New York: Wells & Coffin; San Francisco: H.H. Bancroft & Co.; London: Johnson, Ferguson & Co., 1870.
xxiv, [25]-912 p., [1] leaf of plates : ill., port. ; 19 cm.

First edition. *Plain home talk* is a greatly expanded version of *Medical common sense* (#1213-1218), containing over five hundred pages of new matter. The original two-part format has been expanded to four, with significant additions to each part: Part I. Disease—its causes, prevention, and cure; Part II. Chronic diseases—their causes and successful treatment; Part III. Plain talk about the sexual organs; the natural relations of the sexes; civilization, society and marriage; Part IV. Suggestions for the improvement of popular marriage, etc. Seventy-three percent of the text (669 p.) is devoted to such topics as marriage, reproduction, sexual physiology, sexual ethics, sex disorders, diseases of the organs of generation, sexually transmitted diseases, etc.

The section entitled "Prevention of conception" (pt. IV, c. vii) differs entirely in content from the same section in *Medical common sense*. Foote limits himself to copious quotations from John Humphrey Noyes' *Male continence* (#2650-2652), adding his own adverse commentary. He justifies the deletion of his earlier description of contraceptive methods by remarking that "First, this work is growing too large; second, the author has already in pamphlet form a work treating fully upon this subject" (p. 876; *see note to* #1224). Not even in the advertisement of "useful articles" (p. 911-12) does Foote mention the four contraceptive devices promoted in *Medical common sense* (*see note to* #1216).

The text of *Plain home talk* remained essentially unchanged through the third revision of 1896. Differences in pagination among issues published between 1870 and 1894 reflect changes in the number of patient testimonials, press reviews and advertisements.

1220. FOOTE, Edward Bliss, 1829-1906.
Plain home talk about the human system—the habits of men and women—the causes and prevention of disease—our sexual relations and social natures; embracing Medical common sense applied to causes, prevention, and cure of chronic diseases—the natural relations of men and women to each other—society—love—marriage—parentage—etc., etc. . . . New York: Wells & Company . . . [*et al.*], 1871.
xxiv, [25]-912 p., [1] leaf of plates : ill., port. ; 19 cm.

1221. FOOTE, Edward Bliss, 1829-1906.
Plain home talk about the human system—the habits of men and women—the causes and prevention of disease—our sexual relations and social natures; embracing Medical common sense applied to causes, prevention, and cure of chronic diseases—the natural relations of men and women to each other—society—love—marriage—parentage—etc., etc. . . . New York: Wells & Company . . . [*et al.*], 1872.
xxiv, [25]-912, [12] p., [1] leaf of plates : ill., port. ; 19 cm.

1222. FOOTE, Edward Bliss, 1829-1906.
Plain home talk about the human system—the habits of men and women—the cause and prevention of disease—our sexual relations and social natures. Embracing Medical common sense applied to causes, prevention, and cure of chronic diseases—the natural relations of men and women to each other—society—love—marriage—parentage—etc., etc. . . . New York: Murray Hill Publishing Company . . . [*et al.*], 1873.
xxiv, [25]-924 p., [1] leaf of plates : ill., port. ; 19 cm.

1223. FOOTE, Edward Bliss, 1829-1906.
Plain home talk about the human system—the habits of men and women—the cause and prevention of disease—our sexual relations and social natures. Embracing Medical common sense applied to causes, prevention, and cure of chronic diseases—the natural relations of men and women to each other—society—love—marriage—parentage—etc., etc. . . . New York: Murray Hill Publishing Company . . . [*et al.*], 1874.
xxiv, [25]-924 p., [1] leaf of plates : ill., port. ; 20 cm.

1224. FOOTE, Edward Bliss, 1829-1906.
Plain home talk about the human system—the habits of men and women—the causes and prevention of disease—our sexual relations and social natures, embracing Medical common sense applied to causes, prevention, and cure of chronic diseases—the natural relations of men and women to each other—society—love—marriage—parentage—etc., etc. . . . New York: Murray

Hill Publishing Company; London, Eng.: Charles Noble; Berlin, Prussia: Burmeister & Stempell, 1877.

xxiv, [25]-936 p., [1] leaf of plates : ill., port. ; 18 cm.

"Largely through the efforts of Anthony Comstock (1844-1915) [#754, 755], secretary of the New York Society for the Suppression of Vice, Congress passed a law in March 1873 closing the mails to every type of 'obscene and indecent matter.' In June 1876 Foote was indicted under this new federal statute, now termed the Comstock law, for distributing contraceptive information through the mails in his pamphlet *Words in Pearl* . . . Foote was found guilty on 11 July 1876 and fined $3,500. A possible ten-year prison sentence was suspended, because the judge 'understood many patients might suffer if a sentence of imprisonment was rendered.' Foote's medical practice was little affected by the trial publicity, although he was publicly denounced by some of his more orthodox colleagues" (Vincent J. Cirillo, "Edward Bliss Foote: pioneer American advocate of birth control," *Bull. med. hist.*, 1973, 47:474-45).

Foote's review of Noyes' *Male continence* in this 1877 edition (and its successors) is preceded by the following notice: "In the 'Centennial year,' when patriotic Americans were celebrating the nation's progress, the author of this work was compelled by the laws of his country to expurgate so much of this essay as in any way related to mechanical devices for the prevention of conception . . . The position taken by the author in earlier editions was practically this: that many people . . . should be provided with means for regulating reproduction, and consistently with this position, the best known means were recommended in a pamphlet entitled 'Words in Pearl' . . . This little tract cost the author a fine of $3,500 . . . for which reason he would beg his readers to excuse him from giving any advice upon this subject until there is a change in our Congressional and state laws" (p. 876). In truth, Foote had deleted his description of "mechanical devices for the prevention of conception" as early as the 1870 first edition of *Plain home talk* (#1219), several years before passage of the "Comstock law."

1225. FOOTE, Edward Bliss, 1829-1906.
Plain home talk about the human system—the habits of men and women—the causes and prevention of disease—our sexual relations and social natures. Embracing Medical common sense applied to causes, prevention, and cure of chronic diseases—the natural relations of men and women to each other—society—love—marriage—parentage—etc., etc. . . . New York: Murray Hill Publishing Company . . . [*et al.*], 1884.

xxiv, [25]-936 p.,[1] leaf of plates : ill., port. ; 19 cm.

1226. FOOTE, Edward Bliss, 1829-1906.
Plain home talk about the human system—the habits of men and women—the causes and prevention of disease—our sexual relations and social natures. Embracing Medical common sense applied to causes, prevention, and cure of chronic diseases—the natural relations of men and women to each other—society—love—marriage—parentage—etc., etc. . . . New York: Murray Hill Publishing Company . . . [*et al.*], 1888.

xxiv, [25]-936 p., VI (i.e., 5) leaves of plates + frontis. : ill., port. ; 19 cm.

1227. FOOTE, Edward Bliss, 1829-1906.
Plain home talk about the human system—the habits of men and women—the causes and prevention of disease—our sexual relations and social natures. Embracing Medical common sense applied to causes, prevention, and cure of chronic diseases—the natural relations of men and women to each other—society—love—marriage—parentage—etc., etc. . . . New York: Murray Hill Publishing Company; London, England: Charles Noble; Berlin, Prussia: Gustav Ney, 1889.

xxiv, [25]-959, [1] p., [6] leaves of plates : ill., port. ; 18 cm.

1228. FOOTE, Edward Bliss, 1829-1906.
Plain home talk about the human system—the habits of men and women—the causes and prevention of disease—our sexual relations and social natures. Embracing Medical common sense applied to causes, prevention, and cure of chronic diseases—the natural relations of men and women

to each other—society—love—marriage—
parentage—etc., etc. . . . New York : Murray
Hill Publishing Company; London: England: Charles Noble; Berlin, Prussia: Gustav
Ney, 1890.

xxiv, [25]-959, [1] p., [6] leaves of plates :
ill., port. ; 18 cm.

Atwater copy lacking plate opposite p. 451.

1229. FOOTE, Edward Bliss, 1829-1906.
Plain home talk about the human system—
the habits of men and women—the causes
and prevention of disease—our sexual relations and social natures. Embracing
Medical common sense applied to causes,
prevention, and cure of chronic diseases—
the natural relations of men and women
to each other—society—love—marriage—
parentage—etc., etc. . . . New York: Murray
Hill Publishing Company; London: England: Charles Noble; Berlin, Prussia: Gustav
Ney, 1891.

xxiv, [25]-959, [1] p., [6] leaves of plates
+ supplementary plate : ill., port. ; 18 cm.

In his section entitled "The causes of
barrenness," Foote discusses *local inadaptation*: "While it is true that some women
are so susceptible to impregnation that
they will conceive if the seminal fluids be
but deposited within the lips of the vagina
. . . there are very many who cannot, unless local adaption is so perfect as to cause
the fluids of the male to be poured directly
into or upon the mouth of the womb" (p.
490). Foote discusses the various "inadap-
tations" (i.e., displacements or diseased
states of the womb) that prevents the semen from entering the womb. These states
are illustrated in figures 127 and 128,
which, beginning with the 1870 edition, are
represented by empty rectangles on pages
492 and 493. The caption to fig. 127 includes the comment that these figures may
be had "by mail . . . without charge, to those
who may be individually or professionally
benefited by their possession." The sheet
bearing the wood-engraved figures that illustrate "local inadaptation" is present in
the Atwater copy of the 1891 edition (*see
fig.* 30).

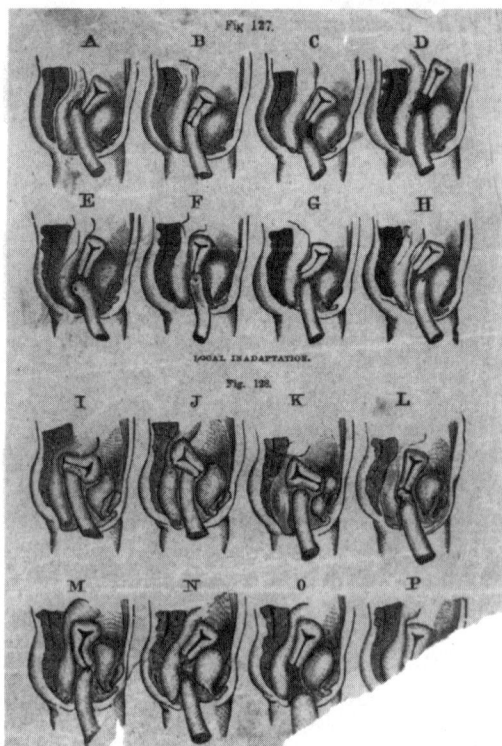

Fig. 30. The separately printed plate illustrating "Local inadaptation," from Foote's
Plain home talk, #1229.

1230. FOOTE, Edward Bliss, 1829-1906.
Plain home talk about the human system—
the habits of men and women—the causes
and prevention of disease—our sexual relations and social natures. Embracing
Medical common sense applied to causes,
prevention, and cure of chronic diseases—
the natural relations of men and women
to each other—society—love—marriage—
parentage—etc., etc. . . . New York: Murray
Hill Publishing Company; London: England: Charles Noble, Berlin, Prussia: Gustav
Ney, 1892.

xxiv, [25]-959, [1] p., plates I-IV, V/VI +
frontis. : ill., port. ; 19 cm.

1231. FOOTE, Edward Bliss, 1829-1906.
Plain home talk about the human system—
the habits of men and women—the causes
and prevention of disease—our sexual relations and social natures. Embracing

Medical common sense applied to causes, prevention, and cure of chronic diseases— the natural relations of men and women to each other—society—love—marriage— parentage—etc., etc. . . . New York: Murray Hill Publishing Company; London: England: Charles Noble, 1894.

xxiv, [25]-959, [1] p., [6] leaves of plates : ill., port. ; 19 cm.

1232. FOOTE, Edward Bliss, 1829-1906.

Plain home talk about the human system— the habits of men and women—the causes and prevention of disease—our sexual relations and social natures. Embracing Medical common sense applied to causes, prevention, and cure of chronic diseases— the natural relations of men and women to each other—society—love—marriage— parentage—etc., etc. . . . Revised in 1896 by Drs. E.B. Foote, Sr. and Jr. . . . Chicago: Thompson & Thomas, [c1896].

xxiv, [25]-973, [3] p., [1] leaf of plates + XVIII p. of plates : ill., port. ; 19 cm.

"After a lapse of a quarter of a century the book appears with considerable new matter . . . and with observations on some of the remarkable advances in the domain of hygiene and medicine. In making this third revision I have associated with me my oldest son" (*Pref.*, p. iv-v). The principal changes to this 1896 revision appear in pt. II, chap. xii—some thirty seven pages in which a discussion of neurasthenia and other "nervous diseases" is added, and the section on syphilis is shortened and updated. The section entitled "Prevention of conception" in previous editions is here entitled "Conjugal prudence" (p. 876-80). The Drs. Foote comment on the 1873 "Comstock law" and the 1896 statute on "*unmailable* things"; ask the question, "what method is most satisfactory, reliable and unobjectionable?"; reply that "until the indignant citizens arouse themselves to effect a repeal of the laws, they will please save themselves postage and the time of the writer by *not asking for what he cannot give them*"; make reference to the works of Noyes (#2649-2652), Kellogg (#2094-2103) and Stockham (#3360-3363) on the topic of continence; and finally refer the reader to E.B.

Foote, Jr.'s *The radical remedy in social science* (#1257) for further discussion.

1233. FOOTE, Edward Bliss, 1829-1906.

Plain home talk about the human system— the habits of men and women—the causes and prevention of disease—our sexual relations and social natures. Embracing Medical common sense applied to causes, prevention, and cure of chronic diseases— the natural relations of men and women to each other—society—love—marriage— parentage—etc., etc. . . . Revised in 1896 by Drs. E.B. Foote, Sr. and Jr. . . . New York: Murray Hill Publishing Company; London, Eng.: L.N. Fowler & Co.; London, W.C., Eng.: L. Underwood, 1899.

xxiv, [25]-959, [17] p., X (i.e., 9) leaves of plates + frontis. : ill., ports. ; 18 cm.

Plates IV-VI actually comprise four pages of plates rather than three leaves. Frontispiece portrait of author.

1234. FOOTE, Edward Bliss, 1829-1906.

Plain home talk about the human system— the habits of men and women—the causes and prevention of disease—our sexual relations and social natures. Embracing Medical common sense applied to causes, prevention, and cure of chronic diseases— the natural relations of men and women to each other—society—love—marriage— parentage—etc., etc. . . . Revised in 1896 by Drs. E.B. Foote, Sr. and Jr. . . . Chicago: Thompson & Thomas, 1902.

xxiv, [25]-973, [3] p., XVIII p. of plates : ill. ; 19 cm.

1235. FOOTE, Edward Bliss, 1829-1906.

Das Menschensystem. Offene Volks-Sprache über das Menschensystem—die Gewohnheiten der Männer und Frauen— die Ursachen und Verhütung der Krankheiten—unsere geschlechtlichen Beziehungen und sociales Leben; ebenso gesunder Menschen-Verstand, erläuternd Ursachen, Verhütung und Heilung chronischer Krankheiten—die natürlichen gegenseitigen Beziehungen der Männer und Frauen—Gesellschaft—Liebe—Ehe— Elternstand—u.s.w., u.s.w. . . . New York

City: Murray Hill Verlags Handlung; Berlin, Prussia: Burmester & Stempell, 1874 [c1870].
xxiv, [25]-844 p., [1] leaf of plates : ill., port. ; 20 cm.

Translation of *Plain home talk*: "Uebersetzt für den Verfasser. Frei nach dem Englischen von A. Loebell" (from verso of title-page). Spine-title: *Offene Volks Sprache und gesunder Menschenverstand.*

1236. FOOTE, Edward Bliss, 1829-1906.
Offene Volks-Sprache über das Menschensystem—die Gewohnheiten der Männer und Frauen—die Ursachen und Verhütung der Krankheiten—unsere geschlechtlichen Beziehungen und sociales Leben; und Gesunder Menschen-Verstand, erläuternd Ursachen, Verhütung und Heilung chronischer Krankheiten—die natürlichen gegenseitigen Beziehungen der Männer und Frauen-Gesellschaft—Liebe—Ehe—Elternstand—u.s.w., u.s.w. . . . New York City: Murray Hill Verlags-Handlung, 1885.
xxiv, [25]-835 p., [3] leaves of plates + frontis. : ill., port. ; 18 cm.

Translation of: *Plain home talk,* originally published under the title *Das Menschensystem* (#1235).

1237. FOOTE, Edward Bliss, 1829-1906.
Offene Volks-Sprache über das Menschensystem—die Gewohnheiten der Männer und Frauen—die Ursachen und Verhütung der Krankheiten—unsere geschlechtlichen Beziehungen und sociales Leben; und Gesunder Menschen-Verstand, erläuternd Ursachen, Verhütung und Heilung chronischer Krankheiten—die natürlichen gegenseitigen Beziehungen der Männer und Frauen-Gesellschaft—Liebe—Ehe—Elternstand—u.s.w., u.s.w. . . . N.Y. City: Murray Hill Verlags-Handlung, 1891.
xxiv, [25]-859, [1] p., [5] leaves of plates + frontis. : ill., port. ; 18 cm.

1238. FOOTE, Edward Bliss, 1829-1906.
Offene Volks-Sprache über das Menschensystem—die Gewohnheiten der Männer

und Frauen—die Ursachen und Verhütung der Krankheiten—unsere geschlechtlichen Beziehungen und sociales Leben; und Gesunder Menschen-Verstand, erläuternd Ursachen, Verhütung und Heilung chronischer Krankheiten—die natürlichen gegenseitigen Beziehungen der Männer und Frauen-Gesellschaft—Liebe—Ehe—Elternstand—u.s.w., u.s.w. . . . N.Y. City: Murray Hill Verlags-Handlung, 1899.
xxiv, [25]-859, [1] p., [5] leaves of plates + frontis. : ill., port. ; 18 cm.

1239. FOOTE, Edward Bliss, 1829-1906.
Offene Volks-Sprache über das Menschensystem—die Gewohnheiten der Männer und Frauen—die Ursachen und Verhütung der Krankheiten—unsere geschlechtlichen Beziehungen und sociales Leben; und Gesunder Menschen-Verstand, erläuternd Ursachen, Verhütung und Heilung chronischer Krankheiten—die natürlichen gegenseitigen Beziehungen der Männer und Frauen-Gesellschaft—Liebe—Ehe—Elternstand—u.s.w., u.s.w. . . . N.Y. City: Murray Hill Verlags-Handlung, 1900.
xxiv, [25]-529, 529/1-529/60, 530-859, [5] p., XVIII p. of plates + frontis. : ill., port. ; 19 cm.

1240. FOOTE, Edward Bliss, 1829-1906.
Offene Volks-Sprache über das Menschensystem—die Gewohnheiten der Männer und Frauen—die Ursachen und Verhütung der Krankheiten—unsere geschlechtlichen Beziehungen und sociales Leben; und Gesunder Menschen-Verstand, erläuternd Ursachen, Verhütung und Heilung chronischer Krankheiten—die natürlichen gegenseitigen Beziehungen der Männer und Frauen-Gesellschaft—Liebe—Ehe—Elternstand—u.s.w., u.s.w. . . . N.Y. City: Murray Hill Verlags-Handlung, 1903.
xxiv, [25]-529, 529/1-529/60, 530-859, [5] p., XVIII p. of plates + frontis. : ill., port. ; 19 cm.

1241. FOOTE, Edward Bliss, 1829-1906.
Offene Volks-Sprache über das Menschensystem—die Gewohnheiten der Männer

und Frauen—die Ursachen und Verhütung der Krankheiten—unsere geschlechtlichen Beziehungen und sociales
Leben; und Gesunder Menschen-Verstand,
erläuternd Ursachen, Verhütung und
Heilung chronischer Krankheiten—die
natürlichen gegenseitigen Beziehungen
der Männer und Frauen-Gesellschaft—
Liebe—Ehe—Elternstand—u.s.w., u.s.w.
... N.Y. City: Murray Hill Verlags-Handlung, 1909.
   xxiv, [25]-529, 529/1-529/60, 530-859,
[5] p., XVIII p. of plates + frontis.: ill., port.; 19
cm.

1242. FOOTE, Edward Bliss, 1829-1906.
Kotilääkäri kirja perheille ja yksityisille
käsittelee terveyttä ja tauteja, aiheita,
kroonillisten tautien ehkäisemista, hoitoa
ja kotiparannusta, lukuisia reseptejä sekä
esitelmiä rakkaudesta, avioliitosta ja
vanhemmuudesta kuin myöskin vakavia,
tarkasteluja, ihmisten yhteiskunnallisista ja
sukupuolisista suhteista, eri luonteen-
laaduista ja niiden sopusuhdasta ynnä eri
avioliittomoudoista, niiden puutteel-
lisuuksista ja parannustavoista ... New York:
Murray Hill Pub. Co., 1909.
   xv, [1], 912 p., XVIII p. of plates, [10]
leaves of plates + frontis. : ill., port. ; 19
cm.

The first and apparently only edition of the
Finnish translation of *Plain home talk*. Atwater
copy lacking pp. 911-12.

1243. FOOTE, Edward Bliss, 1829-1906.
Dr. Foote's home cyclopedia of popular
medical, social and sexual science. Embracing his New book on health and disease,
with recipes, treating of the human system,
hygiene and sanitation—causes, prevention, cure, and home treatment of chronic
diseases, including private words for both
sexes, and 250 practical recipes; also embracing Plain home talk, on love, marriage,
and parentage. A free and earnest discussion of human, social, and sexual relations
in all ages and countries, marriage systems,
defects and their remedies, human temperaments and adaptions . . . New York:
Murray Hill Publishing Co.; London, N.,
England: Camden Publishing Co., 1911.

xiv, [15]-1248 p., [1] leaf of plates +
XVIII p. of plates : ill., port. ; 19 cm.

At head of title: *Twentieth century revised and
enlarged edition.* This is the fourth revision of
Foote's manual of domestic medicine, originally published in 1858 as *Medical common sense*
(#1213) and enlarged in 1870 under the title
Plain home talk (#1219). Foote published the
1900 revision in two versions: in a single volume under the present title (1st ed., 1902); and
in two volumes, the first entitled *Dr. Foote's new
book on health and disease* (#1244, 1245), comprising parts I-II of *Plain home talk*; and the second entitled *Dr. Foote's new plain home talk on love,
marriage, and parentage* (#1246, 1247), comprising parts III-IV.

1244. FOOTE, Edward Bliss, 1829-1906.
Dr. Foote's new book on health and disease, with recipes, including sexology.
Treating of the human system in health and
disease, of hygiene and sanitation—causes,
prevention and home treatment of chronic
diseases, including private words for men
and women, and 250 practical prescriptions . . . Revised and enlarged by the author in 1900. New York: Murray Hill Publishing Company, 1901.
   xiv, 15-805, 1198-1234, 1240-1241, 1245-
1248 p., [2] leaves of plates, XIV p. of color
plates : ill., ports. ; 19 cm.

Foote issued this 1900 fourth revision in two
versions: a single volume entitled *Dr. Foote's home
cyclopedia of popular medical, social and sexual science* (#1243); and in two volumes. The first volume is entitled *Dr. Foote's new book on health and
disease,* and comprises parts I-II of *Plain home
talk.* The second is entitled *Dr. Foote's new plain
home talk on love, marriage, and parentage* (#1246,
1247), and comprises parts III-IV. In the
publisher's preface to the 1900 revision, the first
is described as "the medical part" and the second as "the social part." Both volumes were first
published in 1901.
   In most sections, the text remains identical
to his earlier *Plain home talk.* Several new medical ideas appear in this 1900 revision—due, in
all probability, to the influence of E.B. Foote,
Jr. The germ theory of disease makes its first
appearance in the long publishing history of
Foote's *opus magnum* (p. 26). Foote valiantly tries

to accommodate a theory of disease originally expressed in 1858, his eclectic principles and vegetable materia medica to this newly admitted class of pathogens (p. 323): "It is more than probable that the kindly bacteria wage a war of extermination against those of the baser sort, and under favorable conditions overcome and destroy them. If this philosophy is based on fact, it is self-evident that the true and rational way to get rid of the objectional bacteria is not to take strong and poisonous medicines to kill them, and thereby kill your bacterial friends as well as your bacterial enemies, but to use such remedies as will restore the blood and fortify the nervous system."

Long a believer in the effect of the patient's mental state on the origin and progress of disease, Foote adds that "The time has come when the doubting Thomases of the medical profession must recognize some truth and therapeutic value in Christian science, mental science, mind cure, &c." (p. 323). His discussion of malaria now includes mention of Laveran (p. 31); the section on consumption discusses Koch's bacillus (p. 414), etc. Havelock Ellis is cited in the section on "Sexual perverts and degenerates" (p. 650); and Oscar Wilde is both named and portrayed (in a line drawing) as a classic example of both (p. 653). Cover-title: *Health and disease with recipes, including sexology.*

1245. FOOTE, Edward Bliss, 1829-1906.

Dr. Foote's new book on health and disease, with recipes, including sexology. Treating of the human system in health and disease, of hygiene and sanitation—causes, prevention and home treatment of chronic diseases, including private words for men and women, and 250 practical prescriptions . . . Revised and enlarged by the author in 1900. New York: Murray Hill Publishing Company, 1904.

    xiv, [15]-805, 1198-1238, 1247-1248 p., [1] leaf of plates + XIV p. of plates : ill., port. ; 22 cm.

On cover: *Standard edition. Health and disease with recipes, including sexology.* The first volume of the fourth revision (1900) of Foote's manual of domestic medicine, comprising what had been parts I-II of *Plain home talk.* The companion (or second) volume to

this first volume of the 1904 issue of the "standard edition" was published the same year under the title *Dr. Foote's new plain home talk on love, marriage, and parentage* (#1247).

1246. FOOTE, Edward Bliss, 1829-1906.

Dr. Foote's new plain home talk on love, marriage, and parentage. A fair and earnest discussion of human, social, and marital relations. In all ages and countries—defects in marriage systems and remedies—human temperaments and adaption—based on studies in sexual physiology . . . Revised and enlarged by the author in 1900. Embracing also, Tocology for mothers, by Albert Westland . . . New York: Murray Hill Publishing Company, 1901.

    2 pts. in 1 v. (v, [1], 807-1248 p., [4] p. of plates + frontis. ; [2], 292, [4] p., [12] leaves of plates : ill., port. ; 19 cm.

Foote's plan in the 1900 revision was to publish his expanded and revised work in a single volume (*Dr. Foote's home cyclopedia,* #1243), and in a two-volume version: *Dr. Foote's new book on health and disease* (#1244, 1245), comprising parts I-II of the *Plain home talk*; and the present volume, *Dr. Foote's new plain home talk on love, marriage, and parentage,* comprising parts III-IV. This second volume includes the entire text of Albert Westland's (1853-1915) *Tocology for mothers* (ed. by E.B. Foote, Jr.), a popular guide to marriage, pregnancy and infant care that was also issued separately by Murray Hill in 1901 (#3751). Cover-title: *Plain home talk on love, marriage & parentage.*

1247. FOOTE, Edward Bliss, 1829-1906.

Dr. Foote's new plain home talk on love, marriage, and parentage. A fair and earnest discussion of human, social, and marital relations in all ages and countries—defects in marriage, systems and remedies—human temperaments and adaption—based on studies in sexual physiology . . . Revised and enlarged by the author in 1900. Embracing also, Tocology for mothers, by Albert Westland . . . New York: Murray Hill Publishing Company, 1904.

    2 pts. in 1 v. (v, [1], 807-1223 p., [2] p. of plates; 292, [1223]-1229, 1240-1244 p., [12] leaves of plates : ill., port. ; 22 cm.

On cover: *Standard edition. Plain home talk on love, marriage & parentage.* The second volume of the fourth revision of Foote's manual of domestic medicine (1st ed., 1901).

1248. FOOTE, Edward Bliss, 1829-1906.
Dr. Foote's plain home talk, a cyclopedia of popular medical social [and sexual] science. Embracing his New book on health and disease, with recipes treating of the human system, hygiene and sanitation—causes, prevention, and home treatment of chronic diseases, including private words for both sexes, and 250 practical recipes; a free and earnest discussion of love, marriage and parentage, and human, social, and sexual relations in all ages and countries, marriage systems, defects and their remedies, human temperaments and adaptions. By Edward B. Foote, M.D. Revised by his successor Dr. Joseph Lebenstein . . . New York: Murray Hill Publishing Co; London, N., England: Camden Publishing Co.; Auckland, New Zealand: Wm. Gribble & Co., 1917.
[2], xiv, [15]-615, 668-705, 710-759, 782-806, 852-1197, 1224-1248 p., [1] leaf of plates + 16 p. of plates : ill., port. ; 19 cm.

An expurgated reissue of *Dr. Foote's home cyclopedia of popular medical, social and sexual science* (#1243) with a new title-page and variant title. A publisher's "Notice" preceding the title-page reads: "It has become necessary to eliminate certain parts of this book, owing to new laws, still we have attempted to preserve the great value of the book, and trust that it will be found helpful and valuable to the reader. If there is anything you cannot find in the book and should like to know write to Murray Hill Publishing Co. . . . " Omitted from this issue are: "Private words for men" (p. 616-54); "Impotency" (p. 655-67); "Syphilis" (p. 706-709); "Warranting cures" (p. 760-781); "Introductory chapter" to Part III. (p. 807-9); "The sexual organs" (p. 810-51); and the "Index to Part I. and II." (p. 1198-1223).
At head of title: *Twentieth century revised and enlarged edition.* Cover-title: *Home cyclopedia: popular medical, social & sexual science*; spine-title: *Dr. Foote's plain home talk: a cyclopedia of popular medical and social science.*

1249. FOOTE, Edward Bliss, 1829-1906.
Dr. Foote's hand-book of health-hints and ready recipes. Comprising information of the utmost importance to everybody, concerning their daily habits of eating, drinking, sleeping, dressing, bathing, working, etc. Together with many useful suggestions on the management of various diseases; recipes for relief of common ailments, including some of the private formulae of Dr. Foote and other physicians of high repute, and directions for preparation of delicacies for invalids as pursued in the best hospitals in this country and Europe . . . New York: Murray Hill Publishing Company, 1882.
128, [4] p. ; 17 cm.

First edition. *Dr. Foote's hand-book of health-hints and ready recipes* was issued at least four times between 1882 and 1888. It contains articles gleaned from the pages of *Dr. Foote's health monthly,* a periodical that Foote co-edited with his son Edward Bond Foote (1854-1912), published in 21 volumes between 1876 and the mid-1890s. In printed wrapper.

1250. FOOTE, Edward Bliss, 1829-1906.
Dr. Foote's hand-book of health-hints and ready recipes. Comprising information of the utmost importance to everybody, concerning their daily habits of eating, drinking, sleeping, dressing, bathing, working, etc. Together with many useful suggestions on the management of various diseases; recipes for relief of common ailments, including some of the private formulae of Dr. Foote and other physicians of high repute, and directions for preparation of delicacies for invalids as pursued in the best hospitals in this country and Europe . . . New York: Murray Hill Publishing Company, 1885.
128, [8] p. ; 17 cm.

In printed wrapper.

1251. FOOTE, Edward Bliss, 1829-1906.
Dr. Foote's hand-book of health-hints and ready recipes. Comprising information of the utmost importance to everybody, concerning their daily habits of eating, drinking, sleeping, dressing, bathing, working,

etc. Together with many useful suggestions on the management of various diseases; recipes for relief of common ailments, including some of the private formulae of Dr. Foote and other physicians of high repute, and directions for preparation of delicacies for invalids as pursued in the best hospitals in this country and Europe . . . New York: Murray Hill Publishing Company, 1888.

128, [8] p. ; 17 cm.

In printed wrapper.

1252. FOOTE, Edward Bliss, 1829-1906.
Dr. Foote's replies to the Alphites, giving some cogent reasons for believing that sexual continence is not conducive to health . . . New York: Murray Hill Publishing Company, [1883?].

125, [1] p. ; 17 cm.

Foote's use of the name "Alphite" comes from the *Alpha,* a periodical published by the Moral Education Society of Washington, D.C. (1875-ca. 1888) that advocated sexual continence except for the purpose of reproduction. The editors of the *Alpha* believed that if the sexual behavior of the masses could be altered to refrain from intercourse except for reproductive purposes, then poverty, disease and crime would be drastically reduced or eliminated. While Foote sympathizes with the Society's social goals, he objects to their method of attainment. *Dr. Foote's replies to the Alphites* consists of a dialogue that includes Foote, Celia B. Whitehead, Catherine B. Winslow and other Alphites, originally conducted in the pages of Foote's *Health monthly* beginning in 1881.

Contrary to the position of Whitehead and Winslow, Foote argues (p. 15-16): "We . . . fully believe in animal magnetism; that every living organization generates this force; that there is a sort of individuality in this magnetism; that the interchange of the magnetic forces between persons who are congenial is physically improving and mentally inspiring; that congenial persons of the same sex may benefit each other by social contact, by hand-shaking and agreeable conversation; that the effects of mag-

netic interchange are more markedly exhibited between two congenial persons of the opposite sex; that this interchange may advantageously take place in ordinary social intercourse, but that the most perfect interchange is induced by that relation so strongly demanded by the natural instincts; that the organs or conductors which nature has provided for this interchange are the most perfect of any for the performance of this function"; and that its interruption is unnatural and harmful. He argues that the most effective means to ensure both individual health and social progress is to educate the public regarding methods of birth control, thus enabling them to "interchange magnetic forces" while obviating the harmful social and medical consequences of unlimted offspring. In printed wrapper.

1253. FOOTE, Edward Bliss, 1829-1906.
Flashlights. [New York: Murray Hill Publishing Co., 1902?].

[120] leaves in various pagings : ill. ; 18 cm.

A promotional work consisting of extracts from *Dr. Foote's home cyclopedia of popular medical, social and sexual science* (1st ed., 1902).

1254. FOOTE, Edward Bliss, 1829-1906.
Painless self-cure of congenital phimosis; or other contractions of the prepuce. Dr. E.B. Foote's discovery . . . [New York, c1880].

4 p. ; 20 cm.

Advertisement for *Foote's Prepuce Distender* and *Phimosis Ointment* for the non-surgical cure of this disorder.

1255. FOOTE, Edward Bliss, 1829-1906.
Science in story. Sammy Tubbs, the boy doctor and "Sponsie," the troublesome monkey . . . New York: Murray Hill Publishing Co., 1874-85.

5 v. : ill. ; 15 cm.

CONTENTS: Vol. I. *The boy Tubbs: the bones, cartilages and muscles* (1874). Vol. II. *The student Tubbs: circulation and absorption* (1874). Vol. III.

Fig. 31. Sammy Tubbs receives anatomical instruction, from Foote's Science in story *(v. 2), #1255.*

Practitioner Tubbs: digestive, nutritive, respiratory, and vegetative nervous systems (1874). Vol. IV. *Lecturer Tubbs: brain and nerves, cerebral physiology* (1885). Vol. V. *The gymnast Tubbs* (1882). (Vol. V. in the Atwater set has title-page: *Dr. Foote's sexual physiology for the young*; cover-title: *Sexual physiology for the young*; and spine-title: *Sexual physiology*.)

Murray Hill's standard advertisement for Foote's inventive popular physiology describes the story's characters and purpose: "Sammy Tubbs the boy doctor started in his promising career as the door-boy of a good natured and capable physician, bearing the cognomen of Dr. Samuel Hubbs. He had not been long in his new position before Mrs. Millstone, the wife of a sea-captain and the patient of Dr. H., made the bright colored lad—for Sammy belonged to the oppressed race—a present of a singularly intelligent monkey, to whom was given the euphonious name of Sponsie. Sammy became intensely interested in all he saw and heard in the doctor's office . . . Sammy was bent on self-

improvement, while Sponsie was solely devoted to the pastime of putting everybody and everything into an inextricable muddle. As the reader follows Sammy he finds out, while perusing an amusing and ingenious narrative, all about the bones, cartilages and muscles in the first volume; about the arteries, capillaries, veins, etc. in the second volume; about digestion, nutrition, respiration, and the vegetative nervous system in the third volume, about the brain and nerves in the fourth volume; and all about elimination and reproduction in the fifth volume."

1256. FOOTE, Edward Bliss, 1829-1906.
A step backward. Written . . . in reviewing inconsiderate legislation, concerning articles and things for the prevention of conception. New York: Issued by the author, 1877.
16 p. ; 11 cm.

Falling victim to the "Comstock laws" (#754, 755), Foote was arrested in 1874 and tried for sending his pamphlet describing contraceptive methods and devices through the mails. "To him it was 'purely a medical work' but to Comstock it was an advertisement 'for an infamous article—an incentive to crime to young girls and women.' Foote was found guilty and the judge levied a fine and court costs amounting to five thousand dollars" (Janet Farrell Brodie, *Contraception and abortion in nineteenth century America*, Ithaca: Cornell Univ. Press, 1994, p. 239).

In *A step backward*, Foote puzzles "why articles and things for the *prevention of conception*" should be included under laws that were aimed at preventing the distribution of pornography through the mails. He argues for the social and individual benefits that would result from public education regarding birth control, and lays blame for organized resistance to the dissemination of such information on the Roman Catholic Church in America and "high-church" Protestants. "It is the Romish asceticism of the fourth century sprouting up on the professedly free soil of America," Foote heatedly states, "it is the reabsorption by a liberal Protestantism of some of the Papal poison which was eradicated from a large body of the Christian church by the reformation of Martin Luther. If allowed to penetrate our laws and to saturate our social customs, the realization of the hopes which have inspired the modern reformer, that mankind is to be redeeemed from disease, and in great measure from moral obliquities by studying and disseminating the principles underlying the function of reproduction, will be indefinitely deferred" (p. 10).

FOOTE, Edward Bliss see also **ROOT** (#3024).

1257. FOOTE, Edward Bond, 1854-1912.
The radical remedy in social science; or, borning better babies through regulating reproduction by controlling conception. An earnest essay on pressing problems . . . New-York: Murray Hill Publishing Company, [c1886].
   [2], 148, [10] p. ; 17 cm.

   Edward Bond Foote was the son of Edward , Bliss Foote (#1213-1256). Though his father was an eclectic, the younger Foote received his medical degree in 1876 from the College of Physicians and Surgeons (New York). Nonetheless, Foote was a partner in his father's publishing firm (Murray Hill Publishing Co.) and mail order business; the co-editor of *Dr. Foote's health monthly*; and the editor of later editions of *Plain home talk* and *Dr. Foote's home cyclopedia*. Foote's influence on his father's books is evident in the inclusion of ideas drawn from recent developments in laboratory medicine.
   Foote's "radical remedy" to the evils that beset his era (e.g., low wages, child labor, urban overcrowding, infectious disease, infanticide, infant mortality, etc.) is based on eliminating the ignorance that "leaves man open to the full play of his passional impulses" and the "excessive, reckless, hap-hazard propagation of the race." He writes: "We shall find that in whatever direction we may start out to make a study of social evils, we shall be sure to trace the thread of our investigation to the one great all-pervading factor of ignorance operating through reckless propagation to produce over-population and evil hereditary influences, and a full appreciation of this fact leads us to the conclusion that we must find the remedy in that line of study and education which shall enable and induce all people to regulate reproduction" (p. 14-15).

Foote considers the arguments for and against the three most obvious means of regulating reproduction: complete celibacy, continence (i.e., restraint from sex "except when offspring is desired") and "conjugal prudence" by the use of "preventive checks." His own position, of course, is the promotion of widespread training in the use of contraceptive methods. He notes the courage of Annie Besant during her trial, and that of his father in "stirpicultural propaganda," describing him as "one of the boldest and most persistent preachers of this New Gospel," and crediting him with invention of the "womb veil" or cervical diaphragm (p. 60).
   Foote provides arguments against the moral and physiological objections to contraceptive practices, assails the position of the Roman Catholic Church; considers the proposal of the "Alphites" (#1252); and laments the triumph of Anthony Comstock (#754, 755) in effectively checking the distribution of contraceptive information and devices to those who justly need them. Mindful of his father's conviction under the Comstock laws ten years earlier, Foote confines himself to the medical, moral, social and eugenic arguments supporting the use of "preventive checks," but refrains from their description. Title on printed wrapper: *Borning better babies*.

FOOTE, Edward Bond see also **FOOTE** (#1232-1234, 1244-1247), **WESTLAND** (#3751).

1258. FORBES, Harriet.
Home nursing. Motherhood—care of children. By Harriet Forbes and Harriet Merrill Johnson . . . New York: P.F. Collier & Son, 1905.
   [4], 395 p., [1] leaf of plates : ill. ; 21 cm.

   Home nursing is divided into three parts: I. Home nursing (to "encourage the mother, sister, or wife who is to act the part of nurse *pro tem.* and who needs to know something of the equipment, both material and spiritual, for such an undertaking"); II. Motherhood (i.e., "physiology and hygiene of pregnancy," the management of pregnancy, labor, etc.); and III. Care of children. The authors are described on the title-page as "graduate nurses of the Massachusetts Homeopathic Hospital, Boston, and the Sloane Maternity Hospital, New York; graduates of the Hospital Economics Course, Teachers' College, Columbia University; and Visiting Nurses of Nurses' Settlement and Hartley House." Series: *The household library; v. 3.*

1259. FORBES, John, Sir, 1787-1861.

The water cure, or hydropathy. From the British and foreign medical review . . . Philadelphia: Lindsay and Blakiston, 1847.
    92 p. ; 16 cm.

"In consequence of the modern water-cure having been originated by a non-medical and uneducated man, and having been subsequently . . . adopted and professed by lay practitioners, or by medical men of somewhat equivocal reputation,—and yet more, from the system being held out as a *panacea* or cure for all diseases, with an exclusive scorn of medicinal aid, —the medical profession, as a body, have naturally enough, and not inexcusably, treated it with much contempt . . . and have shown a pretty general determination not to admit it to the catalogue of therapeutic means" (p. [5]). At the risk of offending his medical brethren, Forbes sets out "carefully and calmly to investigate the real merits of the system now so widely established under the name of hydropathy" (p. 7). This was not the first occasion on which Forbes had tread on medically thin ice. In the pages of the *British and foreign medical review,* which he founded and edited between 1836 and 1848, he had published an article on the positive elements of homeopathic practice in January, 1846—much to the consternation of the medical profession. Ten months later, he published the equally unorthodox "Hydropathy, or the cold water cure" (1846, 22:428-58), of which the present work is a reprint.

Forbes enjoyed a remarkable career. He served as a naval surgeon from 1807 to 1816, and then he returned to Scotland to receive his medical degree (Edinburgh, 1817). After practicing with great success in Chichester for many years, he moved to London in 1840 to better manage the *British and foreign medical review,* the leading British medical periodical of its period. Forbes demonstrated great foresight in translating into English Laennec's treatise on auscultation (1821) and Auenbrugger's work on percussion (1824) at a time when these diagnostic techniques were still being ignored or dismissed by the profession. He edited with Alexandeer Tweedie and John Conolly *The Cyclopaedia of practical medicine* (4 v., London, 1833-35), and wrote several other published monographs.

FORBES, John, Sir, 1787-1861. *The water-cure, or hydropathy* see also BULWER AND FORBES (#521).

1260. FORD, J. W.

"The house I live in." A guide to the preservation of health and the attainment of longevity; being a condensed treatise on the importance of physical education, and on the subject of bathing . . . Lowell [Ma.]: L. Huntress, printers, 1838.
    28 p. ; 18 cm.

In this early and little known contribution to the literature of health reform, the author emphasizes the importance of bathing: "The direct and sympathetic relations existing between the skin and the internal organs, render its functions more numerous and conplicated [sic] than those of any other organ in the animal structure. There is, in fact, no part of the human system upon which the whole is so entirely and necessarily dependent, and which consequently requires so strict attention in obedience to the laws of health, as that which forms the natural envelope of the living fabrick" (p. 8). Any number of organic diseases result from the "deranged functions of the skin," and hence "the importance and efficacy of Bathing, not only as a remedial agent, but as a salutary means of preserving health" (p. 14). Ford describes hot baths, cold baths, showers, Turkish and Russian baths, vapor baths, etc. On p. [27] Ford advertises the "Simple and Medicated Vapor Baths . . . administered . . . at his residence" in Lowell for the cure of "gouty and rheumatic affections, palsy, jaundice, scrofula, almost the whole class of cutaneous diseases, and a multitude of other 'ills that the flesh is heir to'."

1261. FORD, Samuel Payne, 1840-1923.

The American cyclopaedia of domestic medicine and household surgery. A reli-

able guide for every family. Containing full descriptions of the various parts of the human body; accounts of the numerous diseases to which man is subject—their causes, symptoms, treatment and prevention—with plain directions how to act in case of accidents and emergencies of every kind; also, full descriptions of the different articles used in medicine, and explanations of medical and scientific terms . . . With an appendix by N.H. Paaren . . . Chicago, Ill.: E.P. Kingsley; Buffalo, N.Y.: J. Hugill; Des Moines, Ia.: W.A. Edwards; Atlanta, Ga.: J.R. Severns, 1882.
    3 v. (v, [6]-420, xx p., [5] leaves of plates ; [2], 421-840 p., [12] leaves of plates ; [2], 841-1152, iv, 1153-1215, [1], vi, [7]-67 p., [22] leaves of plates) : ill., maps ; 26 cm.

    Ford's alphabetically arranged encyclopedia was first issued at Chicago in 1879 in a single volume (1215 p.). It was reissued in three volumes in 1880, 1881, 1882 and 1889. Another single volume issue appeared at Chicago in 1887. "Appendix. Diseases of domestic animals . . . Edited by N.H. Paaren," at end of vol. 3 (vi, [7]-67 p.). Ford was a native of Norwood, Ontario and an 1864 graduate of the University of Buffalo Medical Department.

1262. FOREST, William E.
The new method in health culture. A guide to home treatment of the sick . . . Sixteenth edition, edited, enlarged and illustrated by Albert Turner . . . New York: The Health-Culture Co., publishers, 1902.
    [2], viii, [9]-76, [4], [77]-102, [2], 103-104, [2], 105-106, [2], 107-108, [2], 109-110, [2], 111-140, [2], 141-237, [2], [238]-323, [25] p. : ill. ; 17 cm.

    Forest emphasizes massage, exercise and diet in the maintenance of health. In all likelihood, the edition statement is fanciful.

1263. FOREST, William E.
Physical culture at home . . . New York: Published for the author by Fowler and Wells Co., 1886.
    43 p. : ill. ; 18 cm.

    An illustrated manual of exercises making use of a home gymnasium. In printed wrapper.

FORSYTH, J. S. see **BUCHAN** (#488-491, 493-495).

1264. FOSGATE, B.
FOSGATE'S CELEBRATED ANODYNE CORDIAL, for curing the summer complaints, namely, diarrhoea, and cholera morbus, flatulent and spasmodic cholics, and a CHOLERA PREVENTIVE . . . Sold at FLEMING'S Wholesale Patent Medicine Depot . . . New Orleans, La. [ca. 1866].
    [2] leaves ; 18 cm.

    Advertisement in German on verso of second leaf.

FOSTER, Frank Pierce see **FONSSAGRIVES** (#1211).

1265. FOSTER, Henry, 1821-1901.
An informal address delivered in the Sanitarium Chapel, Clifton Springs, N.Y. . . . In response to a written request from 108 patients, Tuesday evening, October 4, 1892. [Philadelphia?]: Printed at their request, 1892 [Philadelphia: Smith & Salmon, printers].
    26 p., [4] leaves of plates : ill. ; 22 cm.

    "Second edition" (verso of title-page); in printed wrapper. Foster recounts his involvement with the Clifton Springs Sanitarium over half a century—from his establishment of a water-cure at Clifton Springs, N.Y. in 1850 to the commencement of construction of a new building in 1892. In addition to historical reminiscences, Foster reflects on the importance of addressing man's spiritual ills as well as his bodily ailments. In this vein, he addresses the mental healing and Christian Science movements that he found superficially parallel but spiritually alien to his own philosophy.
    Born in Windham County, Vermont, Foster moved with his family in 1835 to a farm near Rochester, N.Y. Soon after, he went to live with his sister and brother-in-law in Ohio, where he attended the Milan Normal School. Foster began the study of medicine with his sister's physician husband, and received his medical degree from the allopathic Cleveland Medical College in 1848. After graduation, Foster became resident physician at the New Graefenberg Water Cure (near Utica, N.Y.) and editor

of *The New Graefenberg Water-Cure reporter* (#2585). Foster felt that he was called by God to launch a new venture. He surveyed several sites before purchasing a tract of land in Ontario County, N.Y. that was blessed with abundant sulphur springs, mineral springs and fresh water. Foster purchased the property and began construction of a water-cure by establishing a joint stock company in which he was a minor partner. In September 1850 he opened the Clifton Springs Water-Cure. His hydropathic establishment flourished, and continued to expand through the end of the century. In 1867 Foster became sole owner of the water-cure, and in 1871 changed its name to the Clifton Springs Sanitarium (#669-672). Like James Caleb Jackson (#1902-1955) in nearby Dansville, Foster was able to prosper long after the decline of hydropathy and the closure of most other water cures by judiciously combining hydrotherapeutics with new developments in medicine, i.e., by transforming his establishment from a traditional water cure into a modern medical facility and health resort. Clifton Springs Sanitarium continued long after Foster's death.

FOSTER, Henry see also **CLIFTON SPRINGS** (#669-672), *GENEVA GAZETTE* (#1341), *NEW GRAEFENBERG WATER-CURE REPORTER* (#2585).

1266. FOTHERGILL, John Milner, 1841-1888.
The diseases of sedentary and advanced life. A work for medical and lay readers . . . New York: D. Appleton and Company, 1887.
     viii, 295, [1] p. : ill. ; 22 cm.

Originally published at London in 1885, *The diseases of sedentary and advanced life* was published at New York the same year (reissued in 1886 & 1887). The American issues of this work were printed in London for Appleton with New York imprints.

An 1865 Edinburgh medical graduate, Fothergill wrote several treatises esteemed by his medical contemporaries, including *The heart and its diseases* (1872), *The practitioner's handbook of treatment* (1876), *Indigestion, biliousness and gout in its protean aspects* (1881), etc. He was also the author of several works on the maintenance

of health written for a lay audience—a somewhat ironic accomplishment given the fact that he was "a man of enormous weight, with a large head and very thick neck, and so continued till he died of diabetes, from which and from gout he had long suffered" (*Dict. nat. biog.,* 7:508).

1267. FOTHERGILL, John Milner, 1841-1888.
The maintenance of health. A medical work for lay readers . . . New York: G.P. Putnam's Sons, 1875.
     [2], 366, [2] p. ; 20 cm.

First American edition (first edition: London, 1874).

E. FOUGERA AND CO. see **ROCHES'S HERBAL EMBROCATION** (#3011).

1268. SETH W. FOWLE AND SONS.
A balsamic ditty. [Boston]: Copyright 1883, by Seth W. Fowle & Sons.
     [8] leaves : ill. ; 14 cm.

An illustrated *ABCD* in verse promoting *Wistar's Balsam of Wild Cherry* and *Peruvian Syrup*, manufactured by the firm of Seth Fowle & Sons. Title and imprint from wrapper.

1269. SETH W. FOWLE AND SONS.
The lay of the lonesome lung. [Boston]: Copyright 1881, by Seth W. Fowle & Sons.
     [8] leaves : ill. ; 9 × 14 cm.

Promotional literature in pictures and verse for *Wistar's Balsam of Wild Cherry*, manufactured by the firm of Seth W. Fowle & Sons. Title and imprint from wrapper.

1270. FOWLER, Charles Henry, 1837-1908.
Home and health and home economics: a cyclopedia of facts and hints for all departments of home life, health, and domestic economy. By C.H. Fowler . . . and W.H. De Puy . . . New York: Phillips & Hunt, 1880.
     352 p. : ill. ; 19 cm.

Fowler and De Puy divide their treatise into three parts: *I. Home* (i.e., marriage, parenting, etiquette, amusements, etc.); *II. Health* (e.g., nutrition, sanitation, bathing, tight-lacing, to-

In London streets he met
A portly man, before he'd been an hour there,
Who said to him, "I'll bet
I know why you so at the Tower stare;
Just read that bill," said he, "you need it, Mister."
It read, "Try BALSAM OF WILD CHERRY, — WISTAR."

Fig. 32. From The lay of the lonesome lung, *#1269.*

bacco, etc.); and *III. Home economics* (i.e., recipes for making soaps, starch, furniture polish, shaving cream, etc.). Eighty per cent of the text is devoted to part two. *Home and health* was first published at Cincinnati in 1879.

Both Fowler and William Harrison De Puy (1821-1901) were Methodist clergymen. Fowler, a native of Burford, Ont., served as a pastor in Chicago for eleven years before he became president of Northwestern Univ. (1872-76). He was made a bishop in 1884, and supervised the organization of missionary activities in the Far East, including the establishment of universities in Beijing and Nanking. A native of Penn Yan, N.Y., southeast of Rochester, De Puy was also editor of *The people's cyclopedia of universal knowledge*—a work first published in two volumes at Cincinnati in 1879 that had expanded to eight volumes by 1905. De Puy edited the *Christian advocate* (New York) for twenty-five years, as well as the *Methodist yearbook* from 1866 to 1889.

1271. FOWLER, Frank C.
Good news for weak, debilitated men. The particulars of a simple and certain means

of self-cure for those suffering from the effects of youthful errors, seminal weakness, spermatorrhea, lost manhood, nervous debility, early decay, loss of memory, impotence, gonorrhea, gleet and premature old age. Showing clearly how any one thus afflicted may speedily, cheaply and permanently restore themselves to health, strength and vigor, by the use of Dr. Rudolphe's Specific Remedy. [Moodus, Ct.]: Copyright 1891 by Prof. F.C. Fowler.
32 p. : ill. ; 15 cm.

Title transcription from printed wrapper.

1272. FOWLER, Frank C.
Good news for weak, debilitated men. The particulars of a simple and certain means of self-cure for those suffering from the effects of imprudence, excesses, wasting weakness, urinary diseases, lost vitality, nervous debility, early decay, loss of memory, despondency, gleet, and premature old age. Showing clearly how any one thus afflicted may speedily, cheaply and permanently restore themselves to health,

strength and vigor, by the use of Dr. Rudolphe's Specific Remedy. [Moodus, Ct.]: Copyright 1891 . . . Revised and recopyrighted in 1894 by Prof. F.C. Fowler. 32 p. : ill. ; 15 cm.

Title transcription from printed wrapper. Atwater copy includes order form laid-in.

1273. FOWLER, Frank C.
Life: how to enjoy it and how to prolong it . . . Twenty-first edition. [S.l.: s.n., c1896]. [8], 210 p., [20] leaves of plates : ill., ports. ; 19 cm.

Fowler divides his treatise into four "chapters": hygiene; physiognomy; "the mating instinct"; and "psychology." Although Fowler maintains that the sole purpose of copulation is to reproduce the species, he devotes nearly a hundred pages to a detailed description of male organs of generation (but not the female), sexual behavior and what he terms "abuses of amativeness." The "abuses" include masturbation and "copulation," i.e., illicit intercourse, excessive venery in the married state and onanism.

1274. FOWLER, Jessie Allen, 1856-1932.
A manual of mental science for teachers and students or childhood: its character and culture . . . London: L.N. Fowler & Co. ; New York: Fowler & Wells Co., [c1897]. xii, [6], 235, [3] p. : ill. ; 19 cm.

By "mental science" Fowler means the study of mind and character revealed by craniology and phrenology. Her textbook begins with a description of the skull and brain; outlines the temperaments; and discusses in detail the faculties (based on traditional phrenological categories), including their cultivation or restraint.

Jessie Allen Fowler was the youngest of three daughters born to Lorenzo Niles Fowler (#1275, 1276) and Lydia Folger Fowler (#1277, 1278). She was a child when her parents left New York and Fowler & Wells, the phrenological institute and publishing house that her father helped establish, to begin careers anew in London. Upon the death of Lydia in 1879, Jessie "took over the work her mother had done before her, studying brain dissection in London's Medical School for Women, and becoming an associate in her father's Fleet Street office," i.e.,

L.N. Fowler & Co. and the Fowler Phrenological Institute (Madeleine B. Stern, *Heads & headlines: the phrenological Fowlers,* Norman: Univ. of Oklahoma Press, 1971, p. 238). In addition to receiving clients and providing phrenological readings at the Institute, Fowler authored or co-authored several books; edited the *Phrenological magazine* (the British equivalent of Fowler & Wells' eminently successful *Phrenological journal*); and assumed a prominent role in the British Women's Temperance Association and the British Phrenological Association.

During the summer of 1896 she joined her father on a visit to New Jersey, where her father was reunited with his octegenarian sister Charlotte. In September, Lorenzo died. Jessie remained in America, joining the staff of the New York office of Fowler & Wells. As heir to the Fowler phrenological dynasty, Jessie assumed leadership of the firm when phrenology was in its decline: "for all her concentrated devotion and ceaseless activity, Jessie Allen Fowler, phrenologist, lecturer, writer, could not stave off the tides that were about to engulf her cherished philosophy. Although Jessie studied experimental psychology, she still believed that 'a phrenologist sees all that a psychologist does, and much more besides.' She appears to have been impervious to, or unimpressed by, the probings of Freud, who was soon to deliver almost single-handed the *coup de grace* to Gall and Spurzheim, Combe and Fowler. Despite the strong intuition with which she was endowed, she did not sense the approaching displacement of phrenoanalysis by psychoanalysis. Yet she could not help sensing in the very air of the twentieth century the ending to so much she had known" (Stern, p. 256). Fowler & Wells and the American Institute of Phrenology split in 1916. Jessie Fowler remained with the latter in a series of increasingly downscale addresses until her death in 1932. With her death, phrenology effectively came to an end in the United States, nearly a century after her uncle and father had begun their efforts to reform and perfect America by the promise of phrenological science.

1275. FOWLER, Lorenzo Niles, 1811-1896.
Lectures on man: being a series of discourses on phrenology and physiology . . . London: L.N. Fowler; New York: Fowler & Wells, 1886. [2], 353 p. ; 19 cm.

Even before leaving Amherst College, Lorenzo Niles Fowler had joined his older brother Orson in the extra-curricular assimilation of the ideas of Gall, Spurzheim and the British phrenologist George Combe. Together they launched a phrenological crusade in America, and soon expanded their mission to the reformation of the many evils that plagued society in the 1830s. The Fowlers toured the nation, delivering lectures and reading heads. In 1836 Orson established an office in Philadelphia and Lorenzo in New York. Their New York headquarters at Clinton Hall on Nassau Street became the center of the phrenological movement in the United States, where they were joined by Samuel R. Wells (#3737) in 1844 to form the publishing firm and phrenological institute of Fowlers & Wells.

Though a less prolific author than his brother, together they co-authored such frequently reprinted staples of phrenological literature as *Phrenology proved, illustrated and applied* (1837), *Illustrated self-instructor in phrenology and physiology* (1840), and *Fowler's practical phrenology* (1847). Lorenzo also wrote *The principles of phrenology and physiology applied to man's social relations* (1842, #1276), and *Marriage: its history and ceremonies* (1847). When Orson left the firm in 1854, it was renamed Fowler & Wells. In 1860, Lorenzo left New York on a lecture tour of Europe with his wife Lydia Folger Fowler (#1277, 1278), the physician, phrenologist and women's rights activist. In 1863 Lorenzo and Lydia again left New York to settle permanently in London, where he established L.N. Fowler & Co., much along the lines of Fowler & Wells in New York. Fowler remained in England until the summer of 1896, when he died on a visit to his sister in New Jersey. *Lectures on man* was originally published at London in 1864.

1276. FOWLER, Lorenzo Niles, 1811-1896.
The principles of phrenology and physiology applied to man's social relations. Together with an analysis of the domestic feelings . . . New York: Published by L.N. & O.S. Fowler; Boston: Saxton & Pierce, 1842.
135, [1] p. : ill. ; 16 cm.

Fowler's *Principles* is divided into two lectures. He begins the first with the statement: "Not only the happiness, but the very existence of man depends on a union of the sexes. This is a law of our nature . . . Not only our own health and happiness require it, but the best interests of posterity demand it" (p. 7). With this in view, Fowler insists that "Phrenology and physiology are the only sciences which make us acquainted with the laws of our organization, and afford that kind of knowledge by means of which we may comply with them to the best advantage" (p. 7-8). He then proceeds to an examination of the "organs of domestic feelings and propensities," i.e., the four phrenological faculties upon which marital happiness and success depend: amativeness, philoprogenitiveness ("the love of offspring"), adhesiveness (those sympathies not founded upon amativeness), and inhabitiveness ("attachment to country, home and residence").

"Lecture II. Hereditary influences" reveals "some of the evils resulting from the perversion of the means put into our hands to continue the race" (p. [67]). These evils always follow ignorance of the principles of phrenology and marriages between persons temperamentally incompatible. The personal and social ills that result from ill-matched marriages can be avoided by prior consideration of the laws of love and generation: "Let the reform be commenced here, upon the principles of phrenology and physiology, and a gradual process of regeneration will be entered upon that will produce the most salutary effects upon the habits, characters, motives and actions of mankind" (p. 85). An appendix includes "Remarks upon tight-lacing" (p. [101]-126). Cover-title: *Phrenology applied to marriage*.

1277. FOWLER, Lydia Folger, 1823-1879.
Familiar lessons on physiology and phrenology, for children and youth . . . New York: Fowlers and Wells, 1850-51.
2 v. in 1 (viii, [9]-95, [5] p. ; xii, [13]-209 p.) : ill. ; 19 cm.

Title transcription from half-title. Each volume has distinct title-page: vol. I: *Familiar lessons on physiology . . . Stereotype edition* (1851); vol. II: *Familiar lessons on phrenology . . . Twelfth thousand* (1850). (In the Atwater copy, vol. II is bound before vol. I.) Spine- & cover-titles: *Physiology and phrenology*. Series: *Fowlers and Wells phrenological library*. The titles that make up this volume were originally published separately by Fowlers and Wells: *Familiar lessons on physiology* in 1847; and *Familiar lessons on phrenology* in 1848. The combined volume was first issued in 1848.

A native of Nantucket, Lydia Folger married Lorenzo Niles Fowler (#1275, 1276) in 1844. She immediately became involved in the activities of Fowlers & Wells, the New York firm in which her husband was partner with his brother Orson Squire Fowler (#1279-1302) and brother-in-law Samuel R. Wells (#3737). In 1847 she launched her career as an itinerant lecturer speaking to women on physiological and hygienic topics. "That year Fowlers & Wells published her *Familiar lessons on physiology* and *Familiar lessons on phrenology* . . . The former gave elementary information on human physiology, along with hints on hygiene, emphasizing exercise, fresh air, a simple diet, and daily bathing; liquor, tobacco, tea, and coffee were all proscribed. The second volume contained rather commonplace admonitions to good behavior couched in phrenological terms" (*Notable American women, 1607-1950,* Cambridge, Ma., 1971, I:654).

In 1849 Fowler enrolled in the Central Medical College, an eclectic school in Syracuse, N.Y. When the college moved seventy-five miles west to Rochester in 1850, Fowler followed, receiving her degree in June, 1850—becoming the second woman in the United States to receive a medical degree. Upon graduation she was made "principal of the female department" at the Central Medical College, i.e., demonstrator of anatomy to female students; and in 1851 became the first female professor in an American medical school with her appointment as professor of midwifery & the diseases of women and children (1851). Fowler's academic career was interrupted briefly when the College dissolved in 1852. Returning to New York, she practiced medicine (at her office on Morton St. and at the Phrenological Depot of Fowlers & Wells on Broadway); lectured at the Metropolitan Medical College (physiopathic) and the New York Hydropathic & Physiological School; visited the mid-western states with her husband on a lecture tour; served as secretary to two national women's rights conventions; and was active in the temperance movement. In 1860 she accompanied her husband on a lecture tour of Europe, during which period she attended lectures in the hospitals and clinics of Paris and London. Upon her return to New York in Au-

gust, 1862, Fowler became professor of midwifery at Russell Trall's Hygeio-Therapeutic College (hydropathic). Her academic career ended, however, when the Fowlers decided to return permanently to England. Though she no longer practiced medicine, Fowler resumed her involvement in temperance activities and published several books, including *Nora, the lost and redeemed* (London, 1863), a temperance novel. Three daughters were born to Lorenzo Niles Fowler and Lydia Folger Fowler: Amelia (b. 1846), Loretta (b. 1850) and Jessie (b. 1856, #1274).

1278. FOWLER, Lydia Folger, 1823-1879.
Familiar lessons on physiology and phrenology for children and youth. New York: Fowlers and Wells, publishers, 1853.

2 v. in 1 (viii, [9]-95, [1] ; xii, [13]-209 p.) : ill. ; 19 cm.

Title transcription from half-title. Each volume has distinct title-page: vol. I. *Familiar lessons on physiology . . . Stereotype edition* (1853); vol. II. *Familiar lessons on phrenology . . . Twelfth thousand* (1853). Spine- and cover-titles: *Physiology and phrenology.*

MRS. LYDIA FOLGER FOWLER.

Fig. 33 . Lydia Folger Fowler, from The Illustrated annuals of phrenology & health almanac, *#1881.*

1279. FOWLER, Orson Squire, 1809-1887.
Amativeness: or evils and remedies of excessive and perverted sexuality. Including warning and advice to the married and single. Being a supplement to "Love and parentage" . . . Thirteenth edition. New York: Fowlers and Wells . . . [*et al.*], 1848.
vi, [7]-72 p. : ill. ; 19 cm.

Although earlier editions of *Amativeness* (1st ed., 1844) were issued separately, from 1848 they appear to have been issued with the author's *Love and parentage* (#1291), to which it is a "supplement" (e.g., this 13th ed. of *Amativeness* was issued in a printed wrapper with the 1848 13th ed. of the parent work.)

"Amativeness" was defined by phrenologists as "reciprocal attachment and love of the sexes," and belongs to Order 1 (affective faculties), Genus 1 (animal propensities), Species 1 ("domestick propensities, or family and social feelings") of the phrenological classification of faculties. In *Amativeness,* Fowler focuses on the depraved exercise of this faculty, i.e., "promiscuous indulgence, matrimonial excess, and self-abuse." He describes the signs of "sensuality," its remedy (e.g., "ABSTINENCE OR DEATH", bathing in cold water, avoiding coffee, tea, tobacco and other stimulants) and prevention.

Orson Squire Fowler was the oldest of the three children born at Cohocton, Steuben County, N.Y. to Horace Fowler's first wife. He was educated in Massachusetts, and graduated from Amherst College in 1834. His biographer in the *Dict. Amer. biog.* notes: "From this training he emerged a characteristic product of the day, with a mass of ill-digested information, many enthusiastic theories, and much reformatory zeal" (3:565). Among the "enthusiastic theories" Fowler embraced as an undergraduate were the doctrines of Gall, Spurzheim and Combe, i.e., the new science of phrenology. "In its simplest form, phrenology postulated that since the brain was the organ of the mind and shaped the skull which neatly cased the brain, there was an observable concomitance between man's mind—his talents, disposition, and character—and the shape of his head. To ascertain the former, one need only examine the latter" (Madeleine B. Stern, *Heads and headlines: the phrenological Fowlers,* Norman: Univ. of Oklahoma Press, 1971, p. x).

Fowler lost no time in launching his career as an apostle of phrenology. In less than a de-cade he had toured the nation, giving lectures and reading heads; began a prolific output of publications; assumed control of the *American phrenological journal;* opened offices in Philadelphia and New York; and established a publishing firm. He was joined in this venture by his brother Lorenzo Niles Fowler (#1275, 1276). In 1842 the Fowler brothers moved their New York phrenologial "cabinet" to Clinton Hall at Beecham and Nassau Streets, which soon became the national headquarters for the phrenological movement in the United States. In 1844 the Fowlers were joined by Samuel Roberts Wells (1820-1875), who had married their sister Charlotte. The firm of Fowlers & Wells was thus established, consisting of a museum of phrenological heads, rooms for analyses, headquarters for the *American phrenological journal,* a mail order business, and a publishing house that became one of the primary disseminators of books on health and social reform in the United States. The importance of Fowlers & Wells has been grossly underestimated in John Tebbel's *A history of book publishing in the United States* (New York: Bowker, 1972, v. 1, p. 363-64).

Phrenology was able to embrace the full panoply of reform (e.g., physiology & hygiene, the water-cure, Mesmerism, temperance, sex education, women's rights, eugenics, child rearing, prison reform, etc.) based on the premise that any of the organs/functions of the brain could be developed (or restrained) if carefully directed. Positive intellectual, artistic, emotional or moral faculties could thus be increased and negative tendencies diminished. If man is thus perfectible, then the progress of the society in which he lives is potentially unlimited. Madeleine Stern accurately assessed the importance of phrenology (and of Fowlers & Wells in particular) to supporting and promoting the many reform movements that swept through America (and Clinton Hall) from the 1830s: "Teaching with dramatic certainty, the improvability of human nature, it opened the way to the perfectibility of the entire race. Thus phrenology could be used as the basis for guidance and the fulcrum of reform" (*ibid.,* p. 33). She adds, "The doctrine was also particularly pertinent to nineteenth-century America. The concept of reform through self-knowledge was first of all, eminently practical; second, it was audaciously idealistic; third, it was good business"

(p. 34).

In 1854, Orson Squire Fowler withdrew from the firm Fowlers & Wells (subsequently known as Fowler & Wells) to devote himself full-time to lecture tours, writing, and the design of his octagonal house outside Wappingers Falls, N.Y. In 1863, the same year his brother Lorenzo emigrated to England, Fowler opened a phrenological office in Boston, unconnected with Fowler & Wells. "His plans were more grandiose than ever," Stern writes, "He would present to the world his 'Complete revised WORKS ON MAN,' including tomes on all the sciences—physiological, phrenological, sexual, religious, and intellectual" (p. 189). His biographer in the *Dict. of Amer. biog.* characterized Fowler's work: "Without special training in philosophy, science, or medicine, he undertook to answer the most difficult questions in these fields. His inordinate conceit, however, saved him from charlatanry . . . The amazing melange of scientific facts, popular superstitions, and personal fancy which came from his fluent pen procured him an immense reputation in his own day. Throughout middle life (1850-1870), he spent most of his time in extensive and lucrative lecture trips in the United States and Canada, charming ignorant audiences equally by his assumption of scientific knowledge and by the extreme sentimentality of his outlook on life" (3:566). Fowler's influence on the American public through his utopian vision of human perfectibility, his adoption of reform movements that aligned with this ideal, and the successful promotion of these movements through Fowlers & Wells' "phrenological depot" was far more important than the idiosyncrasy of his theories, the credulity of his audience, or the profitability of his business.

FOWLER, Orson Squire, 1809-1887.
Amativeness see also *SEXUAL DISEASES* (#3145).

1280. FOWLER, Orson Squire, 1809-1887.
Creative and sexual science: or manhood, womanhood, and their mutual interrelations; love, its laws, power, etc.; selection, or mutual adaption; courtship, married life, and perfect children; their generation, endowment, paternity, maternity, bearing, nursing and rearing; together with puberty, boyhood, girlhood, etc.; sexual impair-

ments restored, male vigor and female health and beauty perpetuated and augmented, etc., as taught by phrenology and physiology . . . Philadelphia, Pa., Chicago, Ill., and St. Louis, Mo.: The National Publishing Company, [c1875].

xxxiv, 35-1065, [9] p., [1] leaf of plates : ill., port. ; 24 cm.

Originally published in 1870 under the title: *Sexual science* (#1300). Madeleine Stern has written that in *Creative and sexual science,* "Orson Fowler distilled his views on sex and sexuality, producing a work which, he claimed, 100,000 women rated 'next to their Bibles.' Copyrighted in 1870—when Freud was fourteen years old—it was an extraordinary volume. Its purpose was to provide sexual education and guidance to the ultimate end of increased sexual enjoyment in marriage and improved offspring. Striking a blow against the evil of Anglo-Saxon civilization that fostered 'female weakliness . . . due chiefly to fashions and education,' Orson frankly championed 'full breasts,' 'a large and well formed pubis,' and '*summum bonum* enjoyment' in coition. At the same time he cautioned against such 'marital errors' as haste or promiscuity, embarassment or non-participation. Sexual ailments, he believed, were caused by 'interrupted love' or 'self-pollution.' His purpose was to convert sexual feelings into sexual enjoyments . . . In fulfilling it, he wrote a book that reads today like a combination of *What every wife/husband should know* and Freud on sexuality. Amazingly frank for its day, it was neither salacious nor pornographic . . . it aspired to teach the married 'how to love *scientifically.*' Like its author, it was well in advance of its time" (*Heads and headlines: the phrenological Fowlers,* Norman: Univ. of Oklahoma Press, 1971, p. 193). Cover-title: *Science of life.* Frontispiece portrait of author.

1281. FOWLER, Orson Squire, 1809-1887.
Creative and sexual science or manhood, womanhood and their mutual inter-relations. Love, its laws, power, etc. Selection, or mutual adaption; courtship, married life, and perfect children. Their generation, endowment, paternity, maternity, bearing, nursing and rearing; together with puberty, boyhood, girlhood, etc., sexual impairments restored, male vigor and female health and beauty perpetuated and

AN INFERIOR MAN AND A SUPERIOR WOMAN.

FIG. 576. — MR. AND MRS. BIBBS.

Fig. 34. Example of an incompatible married couple, from Fowler's Creative and sexual science, *#1281.*

augmented, etc. as taught by phrenology and physiology . . . [S.l.: s.n., 189-?, c1875].

xxxiii, [34]-1052 p., [34] p. of plates : ill., port. ; 23 cm.

Frontispiece portrait of author. The frontispiece and plates are mechanical process plates, most of which reproduce earlier executed wood and metal engravings. The dating of this volume is based on the use of this graphic technology. Cover-title: *Science of life.*

1282. FOWLER, Orson Squire, 1809-1887.
[Education and self-improvement. New York: Fowlers and Wells, 1855.]

3 v. in 1 ([6], xii, [19]-312 ; [2], xiv, [2], [19]-312 ; [ii]-xii, 13-231 p.) : ill. ; 19 cm.

The many editions of *Education and self-improvement* contain three titles that have separate publishing histories in addition to their inclusion here: *Physiology, animal and mental; Self-culture, and perfection of character;* and *Memory and intellectual improvement.* The title transcription for this 1855 edition has been taken from the spine-title. There is no collective title-page as there is in other editions. Each of the three titles included has a separate title-page (dated 1855), pagination and collation.

In the preface to *Memory and intellectual improvement* Fowler writes: "This volume is the third and last of the series on the application of phrenology to 'Education and self improvement.' Volume I. imbodies the importance and means of improving the 'physiology,' including directions for preserving and regaining the health, and the effects of various physical conditions on the mentality. Volume II. shows in what perfection of character consists, and how to attain it . . . Volume III. completes the series by showing how to educate the intellect. It analyzes each of the intellectual faculties, and points out the means of augmenting their efficiency" (p. [v]). Fowler adds, "PERSONAL EFFORT is indispensable to intellectual attainments and greatness. All must be SELF-MADE or not made at all."

1283. FOWLER, Orson Squire, 1809-1887.
Education and self-improvement: comprising Physiology—animal and mental. Self-culture and perfection of character. Memory and intellectual improvement . . . New York: Samuel R. Wells, 1868.
   3 v. in 1 ([2], xii, [4], [19]-312 p. ; xiv, [15]-312 p. ; [2], [v]-xii, 13-231 p.) : ill. ; 20 cm.

Each title has distinct title-page (dated 1868). Samuel Roberts Wells (1820-1875) met the Fowler brothers during a course of lectures delivered in Boston in 1843. In 1844 he married their sister Charlotte, and became partner in the firm of Fowlers & Wells, the single most important publisher of phrenological literature, and an important publisher and distributor of books and journals on popular hygiene, hydropathy, temperance, vegetarianism, etc. When O.S. Fowler left the partnership in 1854, the firm changed its name to Fowler & Wells. After L.N. Fowler emigrated to England in 1863, the firm issued books under the imprint of Samuel R. Wells. The Fowler & Wells imprint was resumed when Charlotte Fowler Wells became head of the firm after her husband's death.

1284. FOWLER, Orson Squire, 1809-1887.
Education and self improvement: comprising Physiology—animal and mental. Self-culture and perfection of character. Memory and intellectual improvement . . . New York: Samuel R. Wells, 1869.
   3 v. in 1 ([6]; xii, [19]-312; xiv, [4], [19]-312; [iii]-xii, 13-231, [7] p.) : ill. ; 20 cm.

1285. FOWLER, Orson Squire, 1809-1887.
The family: in three volumes. Volume I. Matrimony: or, love, selection, courtship,

and married life. Volume II. Parentage: or, a perfect paternity, maternity, sexuality, and infancy. Volume III. Children and home. As expounded by physiology and phrenology . . . New York: O.S. Fowler, publisher, 1859.
   3 v. : ill. ; 19 cm.

Atwater set lacking volumes II-III. O.S. Fowler left Fowlers & Wells in 1854 in part to be able to devote more of himself to writing books that would embrace the whole of human reality within the parameters of phrenological science. He believed that this reality was based on the emotional bonds that men and women establish in marriage, in their sexual relations, in the creation of children, and family life. His writing post-1854 increasingly tends toward this monothematic and all-encompassing vision of life, culminating in 1870 with the publication of the first edition of *Sexual science* (#1300). In his "Preface to the three books" Fowler writes, "Domestic felicity constitutes the great problem of the age. None is more important. None more discussed, and none less understood. Nor is any desideratum as pressing as a thorough, reliable, *scientific* solution of this problem. This, these books attempt" (p.[iii]).
   Matrimony is the only volume of this set present in the Atwater Collection. (It was also issued separately by O.S. Fowler at Boston in 1859.) In the introduction Fowler states: "All human beings are created with certain domestic instincts and faculties. This, Phrenology demonstrates by pointing out a group of cerebral organs, the sole office of whose mental faculties is to carry forward these functions. These predispose to marriage as much as appetite to eating. And this both presupposes and proves the existence of a conjugal entity, which forms as constituent a department of human nature as reason or memory. Of course this department has, must have, its governing laws . . . These laws establish a *science* over this nuptial department. There is, therefore, as much a *matrimonial science* as a mathematical" (p. xx). On the separate title-page for *Matrimony*, we discover that it is divided into three parts: "Part I. Love, its nature, laws, and all-controlling power over human destiny. Part II. Selection: or mutual adaption. Part III. Courtship and married life; their fatal errors, and how to render all marriages happy." Spine-title of first volume: *Fowler on matrimony.*

1286. FOWLER, Orson Squire, 1809-1887.
Fowler on tight lacing. Founded on physiology and phrenology; or the evils inflicted on mind and body by compressing the organs of animal life, and thereby retarding and enfeebling the vital functions. Illustrated with six engravings by J.W. Prentiss . . . "Natural waists or no wives." New-York: O.S. & L.N. Fow[l]er . . . [et al.], 1842.
    16 p., [4] leaves of plates : ill. ; 22 cm.

The four leaves of plates by Josiah W. Prentiss that illustrate Fowler's text are lithographs rather than "engravings."

1287. FOWLER, Orson Squire, 1809-1887.
Hereditary descent: its laws and facts, illustrated and applied to the improvement of mankind; with hints to woman; including directions for forming matrimonial alliances so as to produce, in offspring, whatever physical, mental, or moral qualities are desired; together with preventives of hereditary tendencies . . . New York: O.S. & L.N. Fowler, in Clinton Hall . . . [et al.], 1843.
    270, [2] p. ; 23 cm.

First edition. Fowler spies one dark cloud moving across the otherwise brilliant sky of a scientific and enlightened age: "Though the sun of science has dawned, and is now shining with full effulgence, upon geology, agriculture, chemistry, botany, conchology, natural history, physiology, anthropology, &c., enlightening what was before obscured, dispelling the clouds of ignorance and superstition, improving mechanics and the arts, and shedding on man a flood of happiness, both in their acquisition and application, yet a *sister* science . . . that of PARENTAGE, and the *means of therby improving the race*, remains enshrouded in Egyptian darkness" (*Pref.*, p. [3]). It is time for man to apply the laws of hereditary science by which the farmer has so much improved his stock in order "to produce, in man, personal beauty, physical health and strength, and high intellectual and moral attainments, &c., &c." Fowler enlists his pen in "this cause of God and humanity"; he is confident that his ideas will bear examination "because all his facts *are* facts, and because he has been guided by the lights of Phrenology and Physiology." Indeed, "None but a *Phrenologist*, none but a skilful PRACTICAL Phrenolo-

gist, is at all capable of doing this subject justice," because only he can fully understand both the "physical qualities" of hereditary transmission, *and* the "mental and moral qualities" (p. 4).

Fowler guides his readers among the complex laws governing the transmission of physical, intellectual and moral qualities from parents to their children in the reproductive process. He advises them on the selection of suitable partners; discusses the hereditary characteristics of such racial groups as "the colored race," "the Indian race," Jews, etc.; frightens them with examples of hereditary disease; restores their courage with the positive "moral and intellectual qualities" that are transmissible; and finally, instructs them on the physiological and mental conditions in which parents should conceive and in which women should bear their children. At head of title: *Amer. Phren. Journal, Vol. V, Nos. 9, 10, 11 & 12.* Spine-title: *Fowler on hereditary descent.*

1288. FOWLER, Orson Squire, 1809-1887.
Hereditary descent: its laws and facts applied to human improvement . . . Stereotype edition. New York: Fowlers and Wells, publishers, [c1847].
    6, [vii]-xii, [13]-288 p. : ill. ; 19 cm.

"All the matter of the previous edition has been retained, though considerably condensed . . . Large additions of FACTS have been made to every portion of the work, many new principles, amply illustrated with numerous cases, introduced, many valuable transpositions effected, and the entire work every way improved" (*Preface to the stereotype edition*, p. [3]). Spine-title: *Fowler on hereditary descent.*

1289. FOWLER, Orson Squire, 1809-1887.
Hereditary descent: its laws and facts applied to human improvement . . . Stereotype edition. New York: Fowlers and Wells, 1849.
    6, [vii]-xii, [13]-288 p. : ill. ; 19 cm.

Spine-title: *Fowler on hereditary descent.*

1290. FOWLER, Orson Squire, 1809-1887.
Life: its factors, science, and culture. In two volumes, subdivided into five books. Volume I. Mind, organism, and health by hy-

giene; or phrenology and physiology applied to the laws, preservation, and restoration of health, by self-treatment, without drugs. Including: how organic states manifest and influence mind; the effects of right and wrong eating, breathing, skin and nerve action, etc., upon health, vice, virtue, and happiness; the causes and cures of consumption, brain fever, insanity, neuralgia, meningitis, inflammatory rheumatism, dyspepsia, catarrh, eruptions, bilious and contagious fevers, etc.; together with an analysis of alcoholics and narcotics; twenty-four rules for getting and keeping well; every man his own doctor; nursing vs. medicines . . . Sharon Station, N.Y.: Fowler Publishing House, 1886.

xviii, [2], [33]-357, [3] p., [22] leaves of plates : ill. ; 26 cm.

The second volume indicated on the title-page does not appear to have been published. At head of title: *Study and follow nature.* "In 1880 Fowler rejoined the Fowler firm, now headed by Charlotte Fowler Wells, and moved to Sharon Station, Dutchess County, New York. In 1882, at age seventy-two, he married his third wife, Abbie L. Ayres, by whom he had three children. Late in life Fowler was harassed by debts and charged by some with licentiousness for his pioneering work in sex education. He died at Sharon Station" (*Amer. nat. biog.,* 8:330). Full-page photographic portraits of Fowler and his wife Abbie precede the text. Cover-title: *Life, its science, factors, culture.*

1291. FOWLER, Orson Squire, 1809-1887.
Love and parentage, applied to the improvement of offspring: including important directions and suggestions to lovers and the married concerning the strongest ties and the most sacred and momentous relations of life . . . Thirteenth edition. New York: Fowlers and Wells, Phrenological Cabinet, 1848.

xii, [13]-144 p. : ill. ; 20 cm.

"Early impressions are indelible [sic], are all powerful . . . Yet how few carry this principle back to PARENTAGE, its first, and most eventful application—to the influence, on offspring, of the various conditions of parents at the time the former receive being and constitution from

Fig. 35. O.S. Fowler, frontispiece to Life: its factors, science, and culture, #1290.

the latter. To develop those laws which govern this department of nature, and analyze its facts—to show *what* parental conditions, physical and mental, will stamp the most favorable impress on the primitive organization, health, talents, virtue, &c., of yet uncreated immortals, and what must necessarily entail physical diseases, mental maladies, and vicious predispositions, constitute our subject matter" (*Pref.,* p. vi).

Fowler begins his explanation of the physical basis of hereditary transmission by noting that "All the secretions take part largely, not merely of the general condition of both body and mind, but also of their particular states for the time being" (p. 27). This applies equally to the secretion "employed as the *messenger of life*"—"And the more so, since this secretion *in particular,* is known to be so *intimately* related to the mind, that it *cannot be voided* even in *sleep,* without the concomitant exercise of its corresponding *mental* emotion." Fowler uses this example of the reciprocity between the composition of secretions and state of mind in involuntary seminal discharge to prove that "the existing magnetic conditions of *every element and function of both the mind and body* of parentage . . . are transferred to this secretion, and through it transmitted to that physical and mental constitution of progeny derrived therefrom!" (p. 28).

He continues, "By way of illustrating *how* this transfer of both the permanent and the existing physiology and mentality of parents to this secretion, and through it to the offspring, probably occurs, let us suppose this magnetic *agent* of such transfer to be composed of various ingredients which might be denominated *sub-fluids,* one of which represents and produces anger, another kindness, another love . . . and thus of every other mental faculty, as well as of every bone, muscle, and physical organ of parentage. Now, those children that receive existence and constitution when all these sub-fluids maintain their *usual* relative power and activity in parents, will resemble those parents *in every particular;* but those that receive being and impress when the angry, or kindly, or the intellectual, or any other sub-fluid prevails in parentage . . . will inherit these sub-fluids in their *then existing* predominance to the materials of life" (p. 29). Originally published in 1844, *Love and parentage* was commonly issued with the author's *Amativeness* from 1848. This 13th edition was issued with the 1848 13th edition of *Amativeness* (#1279). In printed wrapper.

1292. FOWLER, Orson Squire, 1809-1887.
Love and parentage, applied to the improvement of offspring. Including important directions and suggestions to lovers and the married concerning the strongest ties and the most momentous relations of life . . . Fortieth edition. [New York]: Published by Samuel R. Wells, 1868.
    xii, [13]-144 p. : ill. ; 19 cm.

In printed wrapper.

1293. FOWLER, Orson Squire, 1809-1887.
Maternity: or the bearing and nursing of children. Including female education and beauty . . . New York: Fowler and Wells, publishers [c1848].
    ix, [10]-221 p. : ill. ; 19 cm.

First edition. In addition to explaining to mothers the importance of the laws of heredi-

tary transmission on their offspring, *Maternity* "teaches mothers what regimen and conditions, in them, will secure the best constituted children; shows how to provide beforehand for a safe and easy delivery; teaches husbands what duties they owe their wives during pregnancy and nursing; gives directions respecting infantile regimen, and the early habits and management of children; and, last but not least, it shows how to prepare girls to bear a far higher order of children, as well as how to rear them after they are borne; that is, it shows how to fit them for the great function of the female, namely, CHILD-BEARING and REARING. It moreover, in doing this, analyzes female beauty. In short, it reflects upon this whole subject the sunlight of Phrenology, Physiology, and Magnetism; and as such, supplies a connecting link between the author's other works on man's social relations" (p. [iii]), i.e., *Matrimony* (#1285), *Hereditary descent* (#1287-1289), and *Love and parentage* (#1291-1292). Spine-title: *Fowler on maternity.*

1294. FOWLER, Orson Squire, 1809-1887.
Maternity: or, the bearing and nursing of children. Including female education and beauty . . . Second thousand. New York: Fowlers and Wells, publishers, 1855.
    ix, [10]-221 p. : ill. ; 19 cm.

Spine-title: *Fowler on maternity.*

1295. FOWLER, Orson Squire, 1809-1887.
Maternity: or the bearing and nursing of children. Including female education and beauty . . . New York: Samuel R. Wells, 1874.
    ix, [10]-221 p. : ill. ; 19 cm.

1296. FOWLER, Orson Squire, 1809-1887.
Physiology, animal and mental: applied to the preservation and restoration of health of body, and power of mind . . . New York: Fowler and Wells, Phrenological Cabinet, 1847.
    xvii, [18]-312 p. : ill. ; 19 cm.

First edition. Having argued that the brain is the organ of mind and that there exists a reciprocal influence of brain/mind and the body, Fowler concludes: "The laws of the mind . . . constitute the highest grade of laws which appertain to our being . . . Hence, since a knowledge of the laws of our being is the most important species of knowledge, and since the laws of mind imbody the highest order of these laws, therefore the study of the LAWS OF MIND—of the physico-mental conditions of happiness and virtue, or the physical conditions as affecting the mentality—constitutes the highest of all human investigations. Since we can control the physical conditions by air, exercise, sleep, diet, and general regimen, and thereby the mentality, and since such mental control is the highest of all human attainments, therefore, to ascertain WHAT states of the physiology will augment moral and intellectual action, and what will occasion mental gloom and wretchedness, or kindle sinful propensity, not only constitutes the highest order of knowledge, but also imports the highest order of power" (p. 36-37).

He then asks, "Yet who understands this subject? What treatise, even on Physiology—that department to which it rightly belongs—even attempts its elucidation? And yet to unfold and enforce this subject, should be the main object of all physiological works" (p. 37). It is certainly the object of Fowler's physiology, which addresses "the momentous questions," i.e., "What physical conditions induce given mental manifestations? into what states shall we throw the body in order thereby to promote particular moral emotions and tendencies, or enhance particular intellectual powers and manifestations?"

Physiology, animal and mental appeared in thirty editions between 1847 and 1884. It was also issued as the first of the three titles that comprise Fowler's *Education and self-improvement* (#1282-1284).

1297. FOWLER, Orson Squire, 1809-1887.
Physiology, animal and mental: applied to the preservation and restoration of health of body, and power of mind . . . Fifth edition. New York: Fowlers and Wells, 1848.
[4], xii, [19]-312 p. : ill. ; 19 cm.

1298. FOWLER, Orson Squire, 1809-1887.
Physiology, animal and mental: applied to the preservation and restoration of health of body, and power of mind . . . Sixth edition. New York: Fowlers and Wells, publishers, 1853.
[6], xii, [19]-312 p. : ill. ; 19 cm.

Series: *Fowlers & Wells phrenological library.*

1299. FOWLER, Orson Squire, 1809-1887.
Private lectures on perfect men, women and children, in happy families; including gender, love, mating, married life, and reproduction, or paternity, maternity, infancy, and puberty; together with male vigor and female health restored, and their ailments self-cured, &c.; as taught by phrenology and natural science . . . New York: E.W. Austin, 1880.
viii, 191 p., [1] leaf of plates : ill. ; 27 cm.

Cover-title: *Perfect "men," "women," and "children."*

1300. FOWLER, Orson Squire, 1809-1887.
Sexual science; including manhood, womanhood, and their mutual interrelations; love its laws, power etc., selection, or mutual adaption; married life made happy; reproduction, and progenal endowment, or paternity, maternity, bearing, nursing, and rearing children; puberty, girlhood, etc.; sexual ailments restored, female beauty perpetuated, etc., etc. as taught by phrenology . . . Philadelphia, Pa.; Cincinnati, Ohio; Chicago, Ill.; St. Louis, Mo.; Atlanta, Ga.: National Publishing Company, [c1870].
xxx, 11-930, [4] p., [1] leaf of plates : ill., port. ; 24 cm.

First edition; the second edition was published under the title: *Creative and sexual science* (#1280, 1281). According to Madeleine Stern, one of the several reasons that moved O.S. Fowler to withdraw from the firm he had founded was Fowlers & Wells' "publishing timidity" in sexual matters (*Heads & headlines: the phrenological Fowlers,* Norman: Univ. of Oklahoma Press, 1971, p. 126). Fowler became increasingly interested in sex, which he thought embraced or touched upon every aspect of human life, not just reproduction. This universal aspect of sexuality Fowler found lacking even in his own publications on reproduction, sexual ethics, marriage, eugenics, etc. In his preface to *Sexual science,* he writes that he now looks back on *Matrimony, Love and parentage, Hereditary descent,* etc. "with marked self-dissatisfaction, because, forsooth, certain subjects, falsely called 'delicate,' must needs be *isolated,* and labelled 'private,' whereas this whole subject, like every other, has its *natural order*, which this separate treatment destroys, and thereby spoils all" (p. viii). It is this "natural order" that he seeks to re-establish in the pages of *Sexual science*.

Fowler continues: "SEXUAL SCIENCE naturally subdivides itself into parts . . . Part I. defines gender, that origin and soul of everything sexual . . . Part II. analyzes LOVE, or the mutual attraction of the sexes, and shows the magic influence its several states wield over human character, conduct, virtues, vices, enjoyments, sufferings, everything human . . . Part III. discusses mating . . . and thereby who can, and who cannot, love whom, and why . . . Part IV. treats of LOVE-MAKING . . . and thereby shows all who may ever love or be loved, how to court *scientifically* . . . Part V. applies to MARRIED LIFE these principles of love-making; shows what always enamours and what alienates, and why . . . many make such miserable affectional shipwreck . . . Part VI. unfolds REPRODUCTION, its laws and conditions as applied to progenal endowments, including those most intimate and sacred inter-relations of husbands and wives . . . Part VII., addressed to the prospective mother, shows her what maternal states, before birth, confer on her unborn robust bodies, sweet tempers, vigorous intellects, and exalted morals . . . Part VIII. expounds the true mode of rearing and governing children . . . including their nursing, health, education, and moral culture . . . Part IX. treats the sexual prostrations and restorations of both sexes; their causes, preventions, and cures, without doctors . . ." (p. ix-x).

The unusual temper of *Sexual science* is revealed in the following quotation from Part VI, in which Fowler discusses the meaning of sexual intercourse: "It is the one and single bond and means of all conjugal union and happiness . . . And most conjugal alienations grow out of its wrong use or non-fulfillment. Matrimonial felicity can no more be maintained without it than noon without the sun . . . Those who do not reciprocate this ultimatum of love, cannot live happily in minor matters . . . Those in concert here will find all minor notes of discord drowned in this key-note of concord . . . therefore those who confer on each other this *summum bonum* are indissolubly bound together by the very strongest bond known to human nature; whilst those who do not or will not confer

and receive this mutual pleasure, cannot possibly love each other, or be happy in other respects. In short, mutual love produces mutual passion. The PERFECT WOMAN, wife, and mother reciprocates it perfectly, even though she does nothing else well; while she is no wife, no woman, who fails here, however excellent in other respects" (p. 638-39). Fowler opposes the use of contraceptives on pages 737-42. Cover-title: *Science of life*. Frontispiece portrait of author.

1301. FOWLER, Orson Squire, 1809-1887.
Sexual science; including manhood, womanhood, and their mutual interrelations; love its laws, power etc., selection, or mutual adaption; married life made happy; reproduction, and progenal endowment, or paternity, maternity, bearing, nursing, and rearing children; puberty, girlhood, etc.; sexual ailments restored, female beauty perpetuated, etc., etc. as taught by phrenology . . . Philadelphia, Pa.; Chicago, Ill.; St. Louis, Mo.,: National Publishing Company, [1873? c1870].

xxx, 11-930, [8] p., [1] leaf of plates : ill., port. ; 24 cm.

The dating of this issue is based on the 1873 publishing date of two titles advertised in the eight unnumbered leaves at the end of the text, i.e., *Behind the scenes in Washington* (New York, c1873), and J.H. Beadle's *The undeveloped West* (1st ed., New York, 1873). Frontispiece portrait of author.

1302. FOWLER, Orson Squire, 1809-1887.
Specimen pages of Sexual science; including manhood, womanhood, and their mutual interrelations; love and its laws, power, etc., selection, or mutual adaption; married life made happy; reproduction, and progenal endowment, or paternity, maternity, bearing, nursing, and rearing children; puberty, girlhood, etc.; sexual ailments restored, female beauty perpetuated, etc., etc. as taught by phrenology . . . Philadelphia, Pa.; Chicago, Ill.; St. Louis, Mo.: National Publishing Company, [ca. 1873, c1870].

1 v. in various pagings : ill. ; 24 cm.

Salesman's sample copy with selected chapters and binding samples (calf and linen).

Bound with specimen copy for John Fleetwood's *Light in the East* (1873).

FOWLER, Orson Squire see also **COMBE** (#740-743).

1303. FOWLER, William Parker, 1849-1927.
Our ears, and how to take care of them . . .
[Rochester, N.Y., n.d.]
    12 p. ; 22 cm.

Fowler was a Rochester homeopath and an 1872 graduate of the Homoeopathic Medical College of New York.

1304. FOWLER AND WELLS COMPANY.
Illustrated descriptive catalogue of new and standard works on phrenology, physiognomy, ethnology, physiology, psychology, health, magnetism, education, etc. New York: Fowler & Wells Co., publishers and booksellers, [between 1884 & 1896].
    [32] leaves : ill., ports. ; 23 cm.

"After leaving college at Amhest, in 1834, O.S. Fowler and his brother L.N. formed a partnership in 1835, and in 1836 located their office in Nassau Street, in what was then the center of business in New York City. In 1845 S.R. Wells became associated with the brothers Fowler. In 1855 O.S. Fowler retired from the firm, and devoted himself principally to public lecturing. In 1860 L.N. Fowler went to Great Britain, where he is now located. The business was then conducted by S.R. Wells until his death, in 1875, when it was continued by Mrs. Wells. On February 29, 1884, the Fowler & Wells Co. was incorporated . . . as a joint stock company, and it stands thus to-day, with Charlotte Fowler Wells as its president . . . Soon after the establishment of the office, the PHRENOLOGICAL JOURNAL was commenced in Philadelphia, and at the end of the third year removed to New York, which publication continues today . . . . The house was the first to publish popular works on health reform for the people, including the publication of the *Water Cure Journal* [#3720] for sixteen years . . . The aim has been to publish nothing on its own account that did not have a practical value to the people. Other houses have supplied a demand; Fowler and Wells created one. The influence exerted by its works has been world-wide, as is marked by a greater knowledge of the laws of health, and a broader, more comprehensive knowledge of human nature and its vast possibilities . . . In connection with its publishing business, the firm has maintained at all times a phrenological cabinet of rare interest and value, free to all visitors, and examination rooms for the delineation of character " (verso of title-leaf).

FOWLERS AND WELLS' WATER CURE LIBRARY see **HORSELL** (#1786-1787); **HOUGHTON** (#1797); **LAMBE** (#2177); **SHEW** (#3180, 3186, 3193, 3196, 3198) **SMITH, J.** (#3265).

FOX-DAVIES, Grace Muriel, tr. see **MUELLER** (#2518).

1305. FRANCKE, Heinrich F., 1805-1848.
Outlines of a new theory of disease, applied to hydropathy, showing that water is the only true remedy. With observations on the errors committed in the practice of hydropathy; notes on the cure of cholera by cold water; and a critique on Priessnitz's mode of treatment. Intended for popular use . . . Translated from the German by Robert Baikie . . . New York: John Wiley, 1849.
    viii, [9]-271 p. ; 20 cm.

A London edition was also published in 1849. Spine-title: *Theory and practice of hydropathy.*

1306. FRANCKE, Heinrich F., 1805-1848.
Outlines of a new theory of disease, applied to hydropathy, showing that water is the only true remedy. With observations on the errors committed in the practice of hydropathy; notes on the cure of cholera by cold water; and a critique of Priessnitz's mode of treatment. Intended for popular use . . . Translated from the German by Robert Baikie . . . New York: John Wiley & Son, 1867.
    [4], viii, [9]-271 p. ; 20 cm.

In his "Preface to the Second Edition," R. Kuczkowski writes: "Finding in my practice the necessity of a work which should explain to patients and others the principles upon which hydropathy is founded, I have procured the republication of this valuable book, regarded as the highest authority in Germany" (p. [i]). This 1867 New York edition is a re-issue of the 1849 Wiley edition, either from the same sheets or printed anew from stereotype plates. Spine-title: *Theory and practice of hydropathy.*

FRANKLIN, Benjamin see *THE IMMORTAL MENTOR* (#1884).

FRANKLIN, H. A. see **HEIDELBERG MEDICAL INSTITUTE** (#1610).

1307. FRANKLIN, Zeus.
Dr. Franklin's illustrated book of life and medical counsellor on the origins, laws, mysteries and wonders of man, structure, use, abuse, redemption and perfection of the sex nature. The naked truth and secrets of nature revealed . . . Laws of hereditary descent and ascent. A new and true social and sexual science of life. How to produce healthy, beautiful, lovely, gifted offspring. Causes and cures of conjugal ills and private diseases. Pains and perils of gestation, labor and lactation, with explicit instructions for self-treatment. Hints to parents, students and physicians. Founded on Holy Writ, the voice and word of God . . . Reading, Penna.: Standard Publishing Company, 1911.
    134, 57, [1] p. : ill. ; 20 cm.

"The love of sex nature is the soul-fountain or foundation of all higher forms and manifestations of love or attributes of the soul, and the cause of all organic life, heavens and hells, joys and sorrows, health and disease; hence the unspeakable importance of its immediate redemption from its present fallen and degraded condition" (*Pref.,* p. 5).

Franklin begins his treatise with a detailed description of the male and female genitalia. In the second part of the *Book of life,* the author justifies its many anatomical illustrations:

These maiden, matron's genital organs revealed,
Could not in a work like this be justly concealed.

Franklin discusses the role of the semen and ova in conception, and provides a lengthy justification of contraception. The only contraceptive method he discussses, however, is withdrawal, which (contrary to the opinion of most of his contemporaries) he

thought causes no physical harm. Franklin carries much sexual lore into the 20th century. Regarding maternal impression he writes: "It is the duty of every woman . . . to govern her passions, to control wisely her emotions, to keep aloof from scenes of a loathsome revolting character, calculated to work powerfully on the mind . . . while carrying in her womb another human being" (p. 99). He also provides advice on how to determine gender at the time of conception: "It is the opinion of many that each testicle secrets its peculiar sperm, and each ovary its peculiar ovum; the right male principle and the left female. This sperm will fructify its corresponding ovum only. In coition the sperm is only ejected from one testicle which is drawn up in the scrotum immediately before the orgasm. To generate either sex at will, it is only necessary to conform to this law by lying on the right side previous to coition and raising for a time the right knee . . . which stimulates and draws up the right testicle in a position to act" (p. 82).

Franklin provides somewhat briefer considerations of pregnancy and its management, labor and care of the newborn. He also discusses sexual hygiene, prostitution, and venereal diseases. The second 57 pages in the numeration of the text contain's the author's vindication of the *Book of life* following his prosecution in Boston on charges of obscenity. The "vindication" takes the form of an epic in nearly forty pages of verse. In printed wrapper.

1308. FRASER, John Frank.
Science of man. A popular treatise on the causes, symptoms and treatment of the common diseases of the generative and urinary functions . . . [New York, ca. 1904].
    95, [1] p. : ill., port. ; 23 cm.

Fraser advertises his ability to surgically treat disorders of the male genitalia such as varicocele, hydrocele, stricture of the urethra, etc.—all of which affect sexual vitality. He also treats the medical consequences of the "solitary vice," neurasthenia (one of the causes of impotence), and sexually transmitted diseases. The author was an 1892 graduate of the Bellevue Hospital Medical College. In printed wrapper.

1309. FREDRICK AND COMPANY.
Dr. Foote's Sanitary Syringe (Patented July 26, 1904). The only scientific vaginal douche. Anatomically correct. Automati-

cally fits. A necessity and luxury in every woman's toilet . . . Fredrick & Co., Toledo, Ohio . . . [ca. 1904].

[1] leaf printed on both sides : ill. ; 24 × 14 cm.

Three line illustrations of the syringe printed on verso.

1310. FREE, John P.
The farmer's guide, and every man his own doctor, a valuable companion, containing over four hundred useful recipes for the cure of the diseases of man and beast, &c. . . . Lancaster, Pa.: Printed at the Daily Express office, 1862.

vii, [8]-85 p. ; 19 cm.

Free provides 435 recipes for every imaginable disorder of man and horse. A handful are incantations or amulets common to "pow-wow" formularies published in German-speaking Pennsylvania during the 19th century.

1311. FRELIGH, Martin, 1813-1889.
The homoeopathic pocket companion; or, a simplified abridgement of the "Homoeopathic practice of medicine." Designed expressly for the use of families and travellers . . . Tenth edition. New York: Charles T. Hurlburt, 1881.

291 p. ; 15 cm.

Eleven editions of this compend of Freligh's *Homoeopathic practice of medicine* (#1312) were published between 1856 and 1890.

1312. FRELIGH, Martin, 1813-1889.
Homoeopathic practice of medicine: embracing the history, diagnosis and treatment of diseases in general, including those peculiar to females; and the management of children. Designed as a text-book for the student, as a concise book of reference for the profession, and simplified and arranged for domestic use . . . New York: Lamport, Blakeman & Law, 1853.

xxx, [31]-576 p. ; 20 cm.

First edition. Fifteen editions of the *Homoeopathic practice of medicine* were published between 1853 and 1884, making Freligh's one of the most frequently published manuals of domestic homeopathy in the second half of the 19th century (Laurie [#2205-2211] appeared in twelve editions between 1843 and 1892, and Pulte [#2891-2896] in thirteen editions between 1850 and 1886). Freligh was an 1858 graduate of the Hahnemann Medical College (Philadelphia).

1313. FREMONT, Augustus.
Health and strength. A system of simple and effective indoor exercises for the proper development of the human body . . . Philadelphia: Physical Culture Society of America, [c1903].

31 p. : ill. ; 16 cm.

1314. FRENCH, Elizabeth J.
A new path in electrical therapeutics: an account of Prof. Elizabeth J. French's great discovery of electrical cranial diagnosis, and the scientific application of ten different currents of electricity to the cure of disease. A complete manual of anatomy and physiology. An historical account of the discoveries in magnetism and electricity, the progress of medical science, and brief sketches of the lives of eminent practitioners, from the earliest ages to the present century; also a thorough system of hygiene; to which are added plain directions for the treatment of disease by Prof. French's system of electrical applications . . . Fourth edition. Philadelphia: J.B. Lippincott & Co., 1877.

xi, [1], 9-259 p., 1 leaf of plates : ill., port. ; 19 cm.

In her opening chapter French states that "medical electricity can be understood and applied as a science, far more exact in its laws and workings than any other remedial agent now in practice," and calls upon "an army of living witnesses to prove the astounding successes which ensue from the application of this force" (p. 10). Aligning herself with Mesmer, Galvani, Franklin and Du Bois-Reymond, French states, "I claim that there is in the human system a sufficient number of phenomena accompanying nervous and muscular action to demonstrate that the said 'life principle' is something analogous to electricity, and that this force properly distributed throughout the system maintains it in health, and when disturbed results in disease. As the electric forces of the human system are simply modifications of those which vitalize nature at large, it follows that any excess or diminution in their quantity, may be supplied by a judicious administration of the electric battery" (p. 19).

French not only proclaims electricity's power to cure, but to give invariably correct diagnoses of disease: "It is many years since I discovered that the human brain is a chart upon which may be found delineated all the organs of the body, and with these, a correct and comprehensible record of the exact condition in which the organs exist" (p. 74). She continues, "I conceived the idea that the brain, as the great nerve-center and focal point to which all sensory and motor nerves report themselves, should be the map on which the organs and their special conditions are represented" (p. 75). French does not describe her method of cranial diagnosis, but assures the reader that "The charts on which the system are laid out with mathematical precision will in due time be given to the world," and that at that time she will no longer be "stigmatized as 'mad'" by the "presumptuous ignorance" of the medical profession or be parodied by "unscrupulous charlatanism" (p. 77). *A new path in electrical therapeutics* appeared in five editions between 1873 and 1886. Frontispiece portrait of author.

FRENCH, Tiberius G. see **TRIAL OF DR. FROST** (#3595).

1315. FRISBEE, John.
Family physician: designed to assist heads of families, travellers and sea-faring people in discerning, distinguishing and curing diseases: with directions for the preparation and use of a numerous collection of the best American remedies, together with a large number of valuable receipts for making plasters, ointments, oils, poultices, decoctions, syrups, or waters made of herbs, the time of gathering all herbs, the way of drying and keeping the herbs all the year, also the way of making and keeping all kinds of useful compounds made of herbs . . . Boston: Published by John Frisbee, 1847.
251 p. ; 19 cm.

The first 63 pages of Frisbee's treatise is a catalog of diseases—their causes, symptoms and treatment. The greater part of the treatise, "American remedies, herbs, plants" (p. [77]-245) is an alphabetically arranged catalog of 174 medicinal plants and herbs.

FROST, Richard K. see **TRIAL OF DR. FROST** (#3595).

1316. FRY, Henry Davidson, 1853-1919.
Maternity . . . New York and Washington: The Neale Publishing Company, 1907.
220 p. ; 19 cm.

"This work is intended for a guide to woman in fulfilling the most sacred duty of her life—maternity. Ignorance and superstition go hand-in-hand to lead her along the road which is beset with many pitfalls. Modern life; the demands of social customs; too early indulgence in the pleasures of society; faulty dress; the lack of fresh air and exercise; carelessness at the menstrual periods; over-development of the mind at the expense of the body—these are some of the pitfalls which hinder the progress of woman in preparing for performing the function of childbearing. The consequences are impaired development and impaired physical strength, sterility, painful menstruation, difficult labors, inability to nurse the offspring, etc. Deterioration is the condition we face to-day—deterioration of mother and offspring" (*Pref.*, p. 11).

Fry was professor of obstetrics and clinical professor of gynecology at the Georgetown University School of Medicine from 1895 to 1917.

1317. FRYER, Jane Eayre, b. 1876.
The Mary Frances first aid book. With ready reference list of ordinary accidents and illnesses, and approved home remedies . . . Illustrations by Jane Allen Boyer . . . Philadelphia: The John C. Winston Company, [c1916].
144 p., [2] leaves of plates (1 folded and laid-in) : ill. ; 22 cm.

Fryer was the author of several books on domestic skills, including a juvenile series that in addition to the present work includes *The Mary Frances cook book* (1912), *The Mary Frances sewing book* (1913), *The Mary Frances gardening book* (1916), *The Mary Frances housekeeper* (1916), and *The Mary Frances knitting and crocheting book* (1918).

1318. GEORGE R. FULLER COMPANY.
From head to foot. [Rochester, N.Y.: George R. Fuller Co., c1907.]
    48 p. : ill. ; 21 × 11 cm.

Manufacturer's catalog advertising abdominal supporters, trusses, shoulder braces, "bow leg and ankle braces," artificial limbs, syringes, "knives for one-armed people," etc. Atwater copy accompanied by printed mailing envelope (postmarked "Rochester, N.Y. 1907").

FULLER BROTHERS see **CRYSTAL SPRINGS HOTEL** (#835).

FURNEAUX, William S. see *MINDER'S ANATOMICAL MANIKIN OF THE HUMAN BODY* (#2460); *WHITTAKER'S ANATOMICAL MODEL* (#3772).

G

1319. GALBRAITH, Anna Mary, 1859-1922.
The four epochs of woman's life. A study in hygiene . . . Philadelphia and London: W.B. Saunders & Company, 1901.
200 p. ; 21 cm.

First edition. Galbraith divides her treatise into four parts: I. Maidenhood (i.e., sexual development, the anatomy & physiology of the female reproductive system, menstrual disorders, etc.); II. Marriage (i.e., sexual ethics and behavior); III. Maternity (i.e., pregnancy, "lying-in," neonatal care); IV. Menopause. *The four epochs of woman's life* was published in three editions: 1st (1901), 2nd (1903, 1911, 1915); 3rd (1917, 1920). A graduate of Vassar College and the Woman's Medical College of Pennsylvania (1884), Galbraith was the author of several treatises directed toward lay women: *Hygiene and physical culture for women* (1895); *Personal hygiene and physical training for women* (1st ed.,1911); and *The family and the new democracy* (1920).

1320. GALBRAITH, Anna Mary, 1859-1922.
The four epochs of woman's life. A study in hygiene . . . Second edition, revised and enlarged. Philadelphia and London: W.B. Saunders & Company, 1915.
[2], 244 p. ; 21 cm.

"The author has added to this edition a section on 'The hygiene of puberty,' one on 'Hemorrhage at the menopause a significant symptom of cancer,' and one on 'The hygiene of the menopause'" (p. 7).

1320.1. GALBREATH, Thomas Crawford, 1876-1916.
Chasing the cure in Colorado. Being some account of the author's experiences in looking for health in the West, with a few observations that should be helpful and encouraging to the tuberculous invalid, who, either from choice or from necessity, remains in his own home to "chase the cure" . . . University Park, Colo." Published by the author, 1909.
[3]-60, [2] p. ; 15 cm.

Originally published in 1907.

1321. GALBREATH, Thomas Crawford, 1876-1916.
T.B. Playing the lone game consumption . . . New York City: Journal of the Outdoor Life Publishing Company, 1915.
[6], 73, [1] p., [4] leaves of plates : ill. ; 18 cm.

Second edition. Galbreath was diagnosed with pulmonary tuberculosis in 1903 while a student at Harvard, and immediately set about "chasing the cure." The present work is an autobiographical account of the author's experiences, including his mistakes, and is intended to shorten the efforts of others in regaining health. Galbreath recommends complete rest, a full diet, fresh air, and complete determination. He discusses the advantages of change of climate (including sanitaria), as well as seeking a cure at home.

1322. GALE, T.
Electricity, or ethereal fire, considered: 1st. Naturally, as the agent of animal and vegetable life: 2d. Astronomically, or as the agent of gravitation and motion: 3d. Medically, or its artificial use in diseases. Comprehending both the theory and practice of medical electricity; and demonstrated to be an infallible cure of fever, inflammation, and many other diseases: constituting the best family physician ever extant . . . Troy [N.Y.]: Printed by Moffit & Lyon, 1802.
285, [3] p. ; 16 cm. (12mo)

In his preface Gale states: "I have shown the probability, if not the certainty, that the sun of our system is an infinite condensation of ethereal fire . . . that all animals and vegetables participate of a degree of the same element; and that the presence of the sun, or the action of his light, induces different states in that diminished degree of ethereal fire which pervades his whole system" (p. 4). Having thus established the cosmic proportions of "the wonderful agency that Heaven hath assigned to the single element of fire," Gale considers "the propriety of improving it artfully, as an excellent antidote of diseases and preservative of health" (p. 5). To this end, he has "unclouded the glory of this inestimable medicine," and in the present work treats the subject "in a very plain, familiar manner; that any man of common ingenuity may perfectly understand the business—may have a clear knowledge of the effect of electric shocks—of the insolation or filling the body with fire—in what manner the shocks must be directed, and repeated, in every particular disease" (p. 6).

A copy of Gale belonged to the family of the young William Andrus Alcott, who many years later regarded it as one of the early influences on his decision to enter medicine: "As I had in those days some leisure for reading, and possessed very few books, I read—and not only read but studied—Dr. Gale's work from beginning to end. It is scarcely too much to say, that I read it till I knew it almost 'by heart'; and my heart assented to it. I believed a new dispensation was at hand to bless the world of mankind . . . " (*Forty years in the wilderness of pills and powders*, #20, p. 10).

REFERENCES: *Austin* #807.

1323. GALLERY OF SCIENTIFIC WONDERS.

Catalogue of Gallery of Scientific Wonders. 203 Main Street. Buffalo, N.Y. Medical offices adjoining. [*Dog-toothed circular device with legend: Doctor Stevens Successor Dr. Linn*] . . . Gallery open from 9 A.M. to 10 P.M. [Buffalo, N.Y., ca. 1911-12].
[3]-74 p. ; 17 cm.

"This is one of the finest collections in the world and the only gallery of its size in the country . . . When we consider the dangers besetting the path of the unwary, and manifold temptations to which the rising generation is exposed, then no intelligent and unprejudiced mind can fail to perceive the advantage derived

. . . from the careful study of a Gallery of Science" (p. [4]). Anatomico-pathological museums such as Buffalo's Gallery of Scientific Wonders (on lower Main St., then a flourishing commercial district close to one of the nation's busiest harbors) operated in many American cities in the late 19th- and early 20th-centuries. Publicly posed as educational and scientific establishments, they exhibited a wide variety of anatomical models, pathological specimens, objects and prints—many pertaining to human reproduction and sexually transmitted diseases, and generally considered prurient by the standards of the age. Many of these galleries maintained "medical offices" specializing in the treatment of venereal diseases and disorders resulting from chronic masturbation.

Until 1911, The Gallery of Scientific Wonders was operated as Dr. Linn's Museum of Anatomy (#2263-2265) by Hugh James Linn, M.D. (1850-1923). After Linn moved to Cleveland in 1911, Edward Thomas Stevens (1859-1919), an 1886 medical graduate of the University of Buffalo, became the museum's proprietor. The 1912 editions of both Polk and the A.M.A. directory give Stevens a Massachusetts Ave. address in Buffalo. In the 1914 editions, both directories list his office address as 203 Main St., i.e., in the Gallery of Scientific Wonders. Title on printed wrapper: *Gallery of Scientific Wonders catalogue.*

1324. GALLERY OF SCIENTIFIC WONDERS.

Catalogue of Gallery of Scientific Wonders. 209 Main Street. Buffalo, N.Y. Medical offices adjoining. [*Dog-toothed circular device with legend: Doctor Stevens Successor Dr. Linn*] . . . Gallery open from 9 A.M. to 10 P.M. [Buffalo, N.Y., ca. 1912].
[3]-74 p. ; 17 cm.

In this edition of the museum's catalog, the "Important Notice" in which Stevens advertises his medical services for "all private, nervous, chronic, blood, and skin diseases" (p. [63]-66) is translated into Italian and Polish (p. 67-74). Italians and Poles were the dominant ethnic groups that emigrated to Buffalo between 1900 and 1930. Some time after Steven's death from pneumonia in 1919, the Gallery of Scientifc Wonders was operated at the same address as Dr. Linn's Museum by William B. Hunt. Title on printed wrapper: *Gallery of Scientific Wonders catalogue.*

1325. GALLUP, Joseph Adams, 1769-1849.

Observations made during a visit to the Clarendon Springs, Vt. in relation to their character and properties, in a part of July and August, 1839. With an analysis of the waters . . . Windsor [Vt.]: Printed by Tracy & Severance, 1840.

14, [2] p. ; 22 cm.

Following his preceptorship in Hartland, Vt., Gallup received a bachelor of medicine degree from Dartmouth College (1798), which also awarded him an M.D. in 1814. In 1820 Gallup joined the faculty of Castleton Medical Academy, and in 1823 joined the medical faculty of the University of Vermont. A year later he returned to medical practice in Woodstock, Vt., where he opened the Clinical School of Medicine in 1827. Gallup was one of the organizing members of the Vermont Academy of Medicine (est. 1813), was known throughout the northeast for his efforts in the reform of medical education, and was the author of *Sketches of epidemic disease in the state of Vermont* (Boston, 1815) and *Outlines of the institutes of medicine* (2 vols., Boston, 1839). In his *Observations* Gallup advises lay readers on the composition and benefits of the waters (as well as accommodations) to be found at the springs. Title on printed wrapper: *Clarendon Springs.*

1326. GARDINER, Charles Fox, 1857-

The care of the consumptive. A consideration of the scientific use of natural therapeutic agencies in the prevention and cure of consumption; together with a chapter on Colorado as a resort for invalids . . . New York and London: G.P. Putnam's Sons, the Knickerbocker Press, 1905.

vii, [1], 182, [2] p., [1] folding table ; 18 cm.

"The purpose of this little book is to give, in as clear and practical a way as posible, the rules that should govern the consumptive in the use of fresh air, sunlight, food, rest, and exercise, so that these natural therapeutic agencies can be applied to the best advantage" (p. v). Gardiner was an advocate of Colorado as a health station for invalids. At the time this work

was published, he had been in practice in Colorado Springs for sixteen years and had published an earlier book entitled *The Colorado Springs region as a health resort* (1898). The author was an 1882 graduate of the Bellevue Hospital Medical College.

1327. GARDNER, Augustus Kinsley, 1821-1876.

Conjugal sins against the laws of life and health and their effects upon the father, mother and child . . . New York: J.S. Redfield, publisher, 1870.

240 p. ; 19 cm.

First edition. "It has been a matter of common observation that the physical status of the women of Christendom has been gradually deteriorating; that their mental energies were uncertain and spasmodic; that they were nervous and irritable; that they were prematurely care-worn, wrinkled and enervated; that they became subject to a host of diseases scarcely ever known to the professional medical men of past times . . ." (p. 13). Gardner attributes this decline to the "refinements of modern life—the listless and enervated condition of the modern woman—the pampered ease which riches and fashion and 'the latest improvements' have brought in their train—the corrupt air of crowded cities—the neglect of healthy occupation—the change from the active housekeeper of our forefathers' pattern, to the vacuity of mind and flabbiness of muscle of the ornamental woman of the present epoch" (p. 17).

CONTENTS: I. The modern woman's physical deterioration. II. Local disease in children and its causes. III. At what age should one marry? IV. Is continence physically injurious? V. Personal pollution. VI. The injurious results of physical excess. VII. Methods used to prevent conception and their consequences. VIII. Infanticide. IX. Conjugal relations during the period of menstruation. X. Conjugal relations between the old. XI. Marriage between old men and young girls. XII. What may be done with health in view, and the fear of God before us.

Much of the work is devoted to the vagaries of sexual behavior that follow the modern woman's mode of living (i.e., masturbation, adultery, sexual excess in the married state, the use of contraceptives, coition during menstruation, etc.), and their injurious consequences to mind, body and offspring. The author enlists lengthy quotations from European medical authors in presenting his arguments.

In his preface to the 1874 revised edition (#1329), Gardner states that 20,000 copies of this first edition of *Conjugal sins* were printed and distributed. The author was an 1844 Harvard medical graduate. His entire career was spent in New York City, where he specialized in gynecology. Gardner was a frequent contributor to the journal literature, and was the editor/translator of the American editions of William Tyler Smith on obstetrics and Scanzoni on the diseases of women.

1328. GARDNER, Augustus Kinsley, 1821-1876.
Conjugal sins against the laws of life and health and their effects upon the father, mother and child ... [Third edition]. New York: J.S. Redfield, publisher, 1870.
240 p. ; 19 cm.

Edition statement from printed wrapper.

1329. GARDNER, Augustus Kinsley, 1821-1876.
Conjugal sins against the laws of life and health and their effects upon the father, mother, and child . . . Twenty-sixth thousand—revised edition. New York: Hurst & Co., publishers, [c1874].
240, 4, [6] p. ; 20 cm.

1330. GARDNER, John, 1804-1880.
Longevity: the means of prolonging life after middle age ... Third edition, revised and enlarged. Boston: William F. Gill and Company, 1875.
191 p. ; 20 cm.

"My purpose in this work, addressed to persons of middle age and in the decline of life, is *not* to supercede the physician in treating their maladies: it *is* to call attention to those peculiarities of the constitution which distinguish age from youth and manhood, to point out those symptoms of deviation from the healthy standard which are usually disregarded, or considered unavoidable incidents of age, but which insensibly glide into fatal diseases if neglected" (*Pref.*, p. 5). *Longevity* was first published at London in 1874. Gardner was also the author of *Household medicine* (London, 1861).

GARDNER, John see also **LIEBIG** (#2249).

1331. GARDNER, Marlin.
The domestic physician, and family assistant, in four parts: Part I.—A short system of anatomy. II.—On materia medica, or a description of medicinal vegetables. III.—On pharmacy, or the preparation of medicines. IV.—On physiology, or the description and treatment of diseases. By Marlin Gardner & Benjamin H. Aylworth, botanic physicians. Cooperstown, N.Y.: Printed for the authors, by H. and E. Phinney, 1836.
iv, [5]-143 p. ; 19 cm.

1332. JOHN H. GARDNER AND SONS.
An illustrated pamphlet relating to Sharon Springs, N.Y. To which is appended a monograph on the properties and uses of the Sharon sulphur water. [New-York: Theo. L. De Vinne & Co. printers, after 1888].
15 p. : ill., plans ; 20 cm.

The Gardner firm's spa at Sharon Springs, Schoharie County, N.Y. included recently constructed bath houses for men and women and the Pavilion Hotel & Cottages. The sulphur waters are recommended for the treatment of rheumatism and other diseases of the joints, as well as "many of the consequences of high living, such as congestion of the liver and kidneys, abdominal plethora, hemorrhoids, etc." (p. 14). Title from printed wrapper.

1333. JOHN H. GARDNER AND SONS.
Introduction of the most approved European methods of using sulphur water at Sharon Springs, N.Y. for treatment of nasal catarrh, pulmonary tuberculosis, and diseases of the respiratory organs. With translations from the notes of French physicians on these methods and the results, as observed in France. New York: Press of Theo. L. DeVinne & Co., 1888.
42 p. : ill., plans ; 20 cm.

Includes selections from Julius Braun's (1821-1878) *On the curative effects of baths and waters* (London, 1875) and the entire text of Arthur C. Hugenschmidt's "Sulphuretted hydrogen inhalations as a method of treatment for pulmonary tuberculosis," published in the *Medical News,* May 7, 1887. Title on printed wrapper: *Most approved European methods of us-ing sulphur water at Sharon Springs, N.Y.*

GARRISON, William H. *A brief for the cigarette* see *THE TRUTH ABOUT CIGARETTES* (#3603).

GASKILL, Silas see **MONROE** (#2480).

1334. GATCHELL, Charles, 1851-1910.
How to feed the sick or, diet in disease. For the profession and the people . . . Second edition, revised and enlarged. Chicago: Gross & Delbridge, 1882.
    150, [10] p. ; 20 cm.

    The first edition was published at Milwaukee in 1880 under the title: *"Doctor, what shall I eat?"* Gatchell was a homeopath, an 1874 graduate of the Pulte Medical College (Cincinnati), and pro-fessor of the theory & practice of medicine at the Hahnemann Medical College in Chicago.

1335. GATCHELL, Charles, 1851-1910.
How to feed the sick. A hand-book of diet in disease . . . Third edition. Chicago: Gross & Delbridge, 1885.
    162, [6] p. ; 20 cm.

GATCHELL, H. P. see *AMERICAN MAGAZINE DEVOTED TO HOMOEOPATHY & HYDROPATHY* (#84).

1336. GATES, Justin, 1797-1869.
MEDICAL REGULATOR. A Diffusible, Elec-trifying, Equalizing Stimulant Pain Killer and Healing Fluid; a universal remedy in all Male and Female Complaints . . . The REGULA-TOR is Prepared and Sold by Dr. Justin Gates . . . Rochester, N.Y. . . . [186-?].
    Broadside ; 36 × 12 cm.

    A botanic panacea, "The Regulator is the only medicine known that will produce perma-nent relief and sound health in all cases and conditions . . . in which a human being ever was or can be placed requiring medical aid. It promotes the life principle and suppresses all deathly effects," curing everything from the common cold to venereal diseases.
    Gates moved into the Genesee country in 1805, making his family pioneers in the settle-ment of western New York. In 1819, still entitled "Mr." and living in the village of Mendon, he married Eliza Peas of nearby Lima. They had at least two sons. In 1834, by then living in Newark (Wayne County), N.Y., Gates attended a Thom-sonian medical convention in Clinton, N.Y., rep-resenting four counties. He was elected vice-presi-dent of the convention, and was appointed to several committees—including the finance com-mittee and the committee that sought to have the State Legislature repeal the recently passed legislation restricting botanical practice. Clearly, Gates had established himself as a leader in state botanico-medical circles. Two years later he at-tended the first annual meeting of the Thom-sonian Medical Society of the State of New York, held at Geddes on 14 June 1836. He was chosen one of five censors at this meeting.
    By 1838 Gates had moved to Rochester, N.Y., then the fastest growing inland city in the United States. He continued to practice medi-cine in Rochester until his death, though not without incurring the hostility of regular prac-titioners. In September 1840, "the steam doc-tor," as the newspaper described him, was ar-raigned for assault (presumably therapeutic) on a Mr. Stratton and held on bail. Four years later he felt it necessary to publish in the local newspaper a defense of his care of a patient who died. That same year he began *The Roches-ter medical truth-teller and monthly journal of health* (#3015), a publication that survived less than two years. Gates maintained a downtown office where he saw patients and sold botanic rem-edies. He also operated a botanic infirmary out of his home. Gates died at age 72 on 3 January 1869. Two sons who moved to California in 1849 were also probably Thomsonian practitioners.

GATES, Justin see also *ROCHESTER MEDICAL TRUTH-TELLER* (#3015).

GATTY, Mrs. Alfred see **MACE** (#2313).

1337. GAYETTY, James C.
Medicated paper, a perfectly pure article for the water closet and a sure prevention and cure of piles . . . *Jas. C. Gayetty*, inven-tor and proprietor . . . N.Y. [c1858].
    Broadside ; 22.5 × 15 cm.

    The advertisement is printed on a sample sheet of Gayetty's medicated toilet paper.

1338. GEM COMPANY.
The Ladies Gem syringe attachment. Attachable to any bulb or fountain syringe. . . . [Kansas City, Mo.: The Gem Co., 1894].
[8] p. on 1 sheet : ill. ; 16 × 36 cm. folded thrice to 16 × 9 cm.

Caption title. "Our vaginal cleanser is the only instrument that properly opens the folds of the vagina, at the same time smoothing them out and holding them away from the discharge pipe of the syringe, by means of the plated frame surrounding it, thus allowing a free and uninterrupted flow of the solution to come in direct contact with the entire lining of the vagina."

1339. GEM COMPANY.
The Star Home Treatment. For women. Kansas City, Mo.: The Gem Company, c1894.
[6] p. on 1 sheet : ill. ; 17 × 28 cm. folded twice to 17 × 9.5 cm.

The "Star Home Treatment" consists of the Star Flexible Pessary ("made of a combination of soft rubber and silk sponge . . . for the purpose of successfully applying, supporting and retaining ointments to those parts, etc."), the Star Applicator for inserting the pessary, and the Star Uterine Ointment ("It is a thorough anticeptic, destroying all germs of disease. It is a deodorizer and disinfectant . . . It quickly reduces inflammatory conditions and heals rapidly, restoring the organs to a strong and healthy state."

1340. GENESEE GINSENG GARDENS.
Ginseng and its culture. Rossburg, New York: The Genesee Ginseng Gardens, [190?].
11, [1] p. ; 15 cm.

The Genesee Ginseng Gardens was a nursery providing plants and seed to those interested in growing and marketing this medicinal root. The Gardens were operated by Perrie Clement Soule (1843-1924), an 1878 graduate of the Cincinnati Medical College who also practiced in Rossburg, N.Y. Title and imprint from wrapper. The price list (p. 11) in the Atwater copy is dated "190[5]."

1341. GENEVA GAZETTE.
GENEVA GAZETTE—Extra. Monday, December 2—Two O'clock P.M. [Geneva, N.Y., 1878].
Broadside ; 35.7 × 22.4 cm.

A special issue of the *Geneva Gazette* publishing the letters of M.W. Brown, of Ithaca, N.Y., and Henry Foster (#1265), of Clifton Springs, N.Y., on the subject of diphtheria, which had recently appeared in Geneva.

GEORGE C. GOODWIN AND CO. see **THOMPSON** (#3473).

1342. GERMAN ELECTRIC BELT AGENCY.
The German electric belts and appliances. Made under foreign patents granted to Prof. Conrad Ziegenfust, and under U.S. Patent No. 357,647 granted to Prof. P.H. Van der Weyde . . . Brooklyn, N.Y.: German Electric Belt Agency, [ca. 1888].
31, [1] p. : ill. ; 12.5 cm.

The several models of German electric belts illustrated and described in this pamphlet are recommended for the treatment of kidney diseases, rheumatism, sciatica, liver complaints, dyspepsia, "female complaints of all kinds," spinal disease, epilepsy, nervous debility, sexual debility, etc. On page four, P.H. Van der Weyde, the patentee, is described as "President of the N.Y. Electrical Society, President of the Electrical section of American Institute, late professor of chemistry in the N.Y. Medical College and Cooper Union," etc. Title on printed wrapper: *The electric era.*

1343. GERNERD, Jeremiah Meitzler Mohr, 1836-1910.
The vaccination imposture: its infliction a crime . . . Williamsport, Pa.: Press of the Gazette and Bulletin, 1906.
52 p. ; 23 cm.

The author maintains that smallpox, like any disease, "is simply manifest as a peculiar set and succession of symptoms that indicate the effort and progress of nature in a self-protective struggle in relation to something pernicious" (p. 6). Eschewing interference by drug therapy, the author states "Nature can be assisted, but

Fig. 36. Front wrapper of German electric belts and appliances, *#1342.*

she wants no interference in her work. Smallpox is best defined as a systematic remedial process, that must have a certain time to do its work, and must 'run its course' . . . Neither doctor nor nurse should 'interfere' at any stage of the defensive process by giving drugs . . . Smallpox is itself the real therapeutic factor. The ejecting process is the disease. If virus gets in, nature must get it out" (p. 6).

Gernerd is puzzled, therefore, by much of the medical profession's advocacy of vaccination: "Medical history is a continuous record of delusive whims, from the earliest times down to the twentieth century, but of all the nonsensical things nothing has yet in the aggregate been more harmful than the practice of puncturing the skin to force filthy animal pus into human veins, and thereby cause a disease to prevent a disease—and a disease, too, that most persons never even contract, and from which 'the vast

majority' recover whether vaccinated or not" (p. 7-8). Gernerd regards "vaccine blood poisoning" as not only unsound medically, but "an outrage upon the personal liberty of the citizens of a 'free' republic" (p. 10).

The author claims to have read medicine "a year or two with the idea of adopting the 'healing art' as a profession," but abandoned the idea when he found "therapeutic processes full of contradictons and uncertainties," and physicians "discordant in opinion and practice." Gernerd's several publications pertain to the history of Lycoming County, Pennsylvania.

1344. GERRARD, Thomas John, 1871-1916.
Marriage and parenthood. The Catholic ideal . . . New York: Joseph F. Wagner, [c1911].
[8], 179 p., [1] leaf of plates : frontis. ; 20 cm.

First edition. The author presents the Roman Catholic position on the institution and purpose of marriage, including mixed marriages, sexual behavior, parenthood, child rearing (including "sexual instruction"), etc. Little more is said regarding contraception and abortion than to note their unacceptability among Catholics. The author was a priest and theologian, who wrote on topics ranging from the present popular manual on marriage to discourses on the work of the French philosopher Henri Bergson and the Orthodox theologian Vladimir Soloviev.

1345. GERRISH, Andrew.
A synopsis on the prevention and cure of disease . . . Boston: Saxton & Peirce, 1841.
viii, [9]-208 p. ; 19 cm.

"Whoever attends to the particulars mentioned in this work will easily perceive, that all the intentions of medicines may be obtained by a long course of diet. Of all subjects diet is the most important. Every indication of cure can be obtained from our food, from exercise, and cleanliness. If mankind would but study the works of creation and providence, disease would in a great measure be banished from the earth. The laws of life and health are easily understood, and man might, in all ordinary cases, become his own physician" (*Pref.,* p. [9]). Gerrish was also an advocate of the medicated vapor bath. He quotes extensively in this work from the correspondence of Charles Whitlaw on the efficacy of vapor baths, which Gerrish provided the Boston public at his Whitlaw Vapor Baths. Atwater copy lacking pp. 203-204.

1346. GERRISH, Frederic Henry, 1845-1920.
Sex-hygiene. A talk to college boys . . . Boston: The Gorham Press; Toronto: The Copp Clark Co., [c1917].
51 p. ; 19 cm.

Series: *Present day problems series.* "In 1911 there was given to Bowdoin College, a memorial of Judge Benjamin Apthorp Gould Fuller, of the class of 1839, a fund the income of which was to be devoted to the instruction of the students in the proper relations of the sexes. As a part of this instruction the following lecture has been given to the freshman class of each succeeding year" (*Pref.*, p. [3]). Gerrish was professor emeritus of surgery at Bowdoin College when this work was written.

1347. GETCHELL, Francis Horace, 1836-1907.
The maternal management of infancy. For the use of parents . . . Philadelphia: J.B. Lippincott & Co., 1868.
vii, [8]-67 p. ; 17 cm.

Getchell devotes chapters to infant food, sleep, clothing bathing, exercise, etc. The author was an 1859 medical graduate of Dartmouth College. After service as a surgeon in the U.S. Army during the Civil War, Getchell settled in Philadelphia where he received a second medical degree from Jefferson Medical College in 1872. He remained in Philadelphia, where he specialized in obstetrics and gynecology, lectured at the Jefferson Medical College, and contributed to the journal literature.

GETCHELL, Francis Horace see also **CHAVASSE** (#623-625).

1348. GETTYSBURG SPRING COMPANY.
The Gettysburg Katalysine Water. Its source. History of its discovery and wonderful cures. Testimony of invalids, physicians, medical journalists and authors.

Whitney Brothers, general agents . . . Philadelphia . . . [S.l.]: Gettysburg Spring Company, 1872 [Philadelphia: Press of Haddock & Son].
64 p. ; 15 cm.

In printed wrapper.

1349. GILMAN, Chandler Robbins, 1802-1865.
Hints to the people on the prevention and early treatment of spasmodic cholera . . . New-York: Charles S. Francis, 1832.
15 p. ; 23 cm.

"The question so often and so anxiously asked 'will cholera cross the Atlantic?' is now answered. This terrible disease has appeared upon our shores and is advancing toward us with fearful rapidity" (p. [5]). Cases of cholera had been reported in Quebec and Montreal early in June 1832. By the end of the month it had travelled down the Hudson to the city of New York. Gilman, who had settled in New York shortly after receiving his medical degree from the University of Pennsylvania, dismisses quarantine as ineffectual. He does recommend, however, that that the public authorities keep the streets clean. This done, "the next object should be to clean the houses of the poor," who he claims "will only be clean upon compulsion" (p. 7). Gilman also insists on the "removal of nuisances," such as slaughterhouses, stables, "cow-houses," "the license enjoyed by so many of keeping pigs in their yards," etc. Gilman advises individuals to attend to personal cleanliness, pursue a diet that is "nutritious but not stimulating," avoid "the intemperate use of ardent spirits, as the most active of the exciting causes of cholera," and avoid "all violent exertion and excessive fatigue," as well the heat of the sun, the night air and "depressing passions." He admonishes his readers that "cholera is similar to all other epidemics, and is most apt to fall upon those, who are most dreading its approach. Strive then to preserve mental tranquility" (p. 11). In the treatment of cholera Gilman recommends a teaspoon of laudanum, "and if no abatement of the symptoms take place in half an hour, another tea-spoonful—Further than this, it would hardly be safe for a non-medical person to go" (p. 13). Hot brandy with cloves or cayenne is also recommended for the relief of the symptoms of cholera, as are hot baths, "fomentations with flannel dipped in hot brandy or spirits of turpentine," mustard plasters, or covering the patient with hot sand.

GIRTIN, Thomas Calvert see **ALCOTT** (#30).

1350. GIVENS, Amos J., 1864-1919.
Nervous exhaustion . . . Stamford, Connecticut, U.S.A., [ca. 1900].
    7, [1] p. : ill. ; 21 cm.

Givens describes the symptoms and treatment of neurasthenia. The verso of page 7 has a view of "Dr. Givens' Sanitarium, Stamford, Connecticut, U.S.A.," also known as Stamford Hall. Givens was an 1886 graduate of the Eclectic Medical Institute (Cincinnati).

1351. GLEASON, Cloye W., 1822-1902.
DR. GLEASON, of Philadelphia, the well-known lecturer on ANATOMY and PHYSIOLOGY, will deliver a chaste, amusing, scientific and instructive LECTURE TO GENTLEMEN ONLY on the ANATOMY, PHYSIOLOGY and PHILOSOPHY of HEALTHY REPRODUCTION, AND THE LAWS OF HEREDITARY DESCENT . . . also the NATURAL LAWS, which govern the development, and maintaining [sic] the health of the reproductive system, and the terrible consequences which result from the violation of these laws, showing that sexual vice is one of the greatest evils of ancient and modern times. [*Two wood-engraved anatomical figures*] Illustrated with the largest collection of models on this subject ever exhibited in the United States, recently imported from Paris, at great expense, consisting of NUMEROUS MANIKINS AND MORE THAN 100 MODELS! showing the development of the HUMAN EMBRYO . . . No boys admitted under 15 . . . [ca. 1850-1860].
    [1] leaf printed both sides : ill. ; 48.5 × 20. 5 cm.

An announcement to the "citizens of Norwich [Chenango Co., New York], that Dr. Gleason . . . will visit this place next week, and deliver his Course of Popular Lectures" (verso).

The two columns of print on the verso largely consist of reviews of Gleason's lectures excerpted from various newspapers printed in the towns of Chemung and Chenango counties. The nature of the manikins and models prominently displayed in this advertisement may be gathered from the *Catalogue of manikins, anatomical preparations, models, and maps, for sale, or imported to order, by Codman & Shurtleff* (#696).

1352. GLEASON, Cloye W., 1822-1902.
Everybody's own physician or how to acquire and preserve health . . . Third thousand. Philadelphia: H.N. McKinney & Co., publishers, [c1873].
    viii, 472 p. : ill. ; 23 cm.

Everybody's own physician is a re-issue of Gleason's *Thirty-eight lectures on how to acquire and preserve health* (#1361) with a new title-page. The author informs us that he "has lectured for nearly thirty years in the principal towns in this country, and is well known, and has acquired a distinguished reputation as a successful *Popular Lecturer,* and a *Physician.*" He describes his book as "a thorough explanation of the human body, its construction, form, and requirements," in which "familiar diseases are explained, their causes and symptoms are given, with the remedies, and the reasons why such remedies are used." He adds, "The causes, symptoms and remedies, are so plainly given, that any one can follow them, and this alone makes the book *invaluable*" (*Pref.,* p. 3). Gleason's final two lectures are devoted to human reproduction and sexual behavior. Lecture XXXVII provides a fairly accurate description of conception, and is unusual in the popular literature for the number, quality and accuracy of its wood-engraved illustration. Gleason discusses the "laws of hereditary descent" in this same chapter, noting the transmission of physical characteristics from parent to offspring, as well as disease and insanity. In lecture XXXVIII, "Abuses of the reproductive organs," Gleason insists that only through the practice of sexual "continence" can human physical development attain perfection, and that "the practice of sensualism, and sexual excesses" not only enervates mind and body (and the minds and bodies of offspring), but actually prevents conception. Gleason was an 1844 medical graduate of the University of Pennsylvania.

How to use the "Inhaling Bottle."

Fig. 37. Self administration of inhaled medicament, from Gleason's Everybody's own physician, *#1352.*

1353. GLEASON, Cloye W., 1822-1902.
Everybody's own physician; or, how to acquire and preserve health . . . Sixth thousand. Philadelphia: H.N. McKinney & Co. . . . [et al.], [c1873].
    viii, 472 p. : ill. ; 23 cm.

1354. GLEASON, Cloye W., 1822-1902.
Everybody's own physician; or, how to acquire and preserve health . . . Toronto: Maclear and Co., publishers, 1874.
    viii, 482, [6] p. : ill. ; 22 cm.

1355. GLEASON, Cloye W., 1822-1902.
Everybody's own physician; or, how to acquire and preserve health . . . Philadelphia: American Publishing Co., 1884..
    viii, 488 p. : ill. ; 23 cm.
    The added matter in this 1884 edition consists of an appendix that contains "Antidotes for poisons, etc." and an index.

1356. GLEASON, Cloye W., 1822-1902.
Everybody's own physician; or, how to acquire and preserve health . . . Philadelphia: American Publishing Co., 1885.
    viii, 488 p. : ill. ; 23 cm.

1357. GLEASON, Cloye W., 1822-1902.
Everybody's own physician; or, how to acquire and preserve health . . . Philadelphia: American Publishing Co., 1886.
    viii, 488 p. : ill. ; 23 cm.

1358. GLEASON, Cloye W., 1822-1902.
Everybody's own physician; or, how to acquire and preserve health . . . Philadelphia, Pa.: American Publishing House, 1894.
    viii, 488 p. : ill. ; 24 cm.

1359. GLEASON, Cloye W., 1822-1902.
Know thyself! No man, however well inclined, can be reasonable upon a subject which he does not study and understand! Dr. Gleason, will delivered his SECOND chaste and appropriate PRIVATE LECTURE TO GENTLEMEN ONLY! On the origin of animal & vegetable life. Splendidly illustrated. With a great variety of costly, curious, wonderful and intensely interesting models, oil paintings, natural preparations, &c. [ca. 1850-60].
    Broadside ; 24 × 12 cm.

1360. GLEASON, Cloye W., 1822-1902.
Seven lectures on the philosophy of life and the art of preserving health . . . Delivered at the Ohio Mechanics' Institute in Cincinnati, and reported by E.P. Stone, phonographer. Columbus: Printed by Scott & Bascom, 1852.
    252, [2] p. : ill. ; 23 cm.

    Gleason devoted nearly thirty years to the lecture circuit, presumably from the time he took his medical degree in 1844. The seven lectures published here reflect not only the content of Gleason's ideas on health and disease, but the spirit in which they were delivered, including audience reaction (e.g., "applause," "great applause," "tumultuous applause," "sensation." "great sensation," "sensation and applause," etc.). These ideas took their final form two decades later in his greatly expanded (and more staid) *Thirty-eight lectures* (#1361). Gleason devotes his seven lectures to diet & digestion; the evils of tobacco ("it blunts the moral sensibilities, and impairs the nervous energies, and weakens the mental faculties," p. 72); the health of the blood and heart (digressing onto the health consequences of excessive political and financial ambition, the pursuit of fashion, etc.); "the pernicious habit of taking medicines"; the health of the lungs, the skin, and nerves; as well as muscular tone (in the discussion of which he contrasts the youth of ancient Greece and Rome with "the thriftless idleness and physical imbecility of the modern Grecian and Italian," p. 196). Gleason's lectures provide a fascinating portrait of the popularization of scientific ideas (as well as pseudo-scientific ideas) at mid-century.

1361. GLEASON, Cloye W., 1822-1902.
Thirty-eight lectures on how to acquire and preserve health . . . By C.W. Gleason, M.D. and H.R. Burner, M.D. [S.l.]: Published by H.R. Burner, M.D., [c1873].
   viii, 472 p. : ill. ; 23 cm.

The first of the thirty-eight lectures is taken nearly verbatim from the lecture published in Gleason's 1852 *Seven lectures* (#1360). Otherwise, this is an entirely new work. The *Thirty-eight lectures* was re-issued in Philadelphia under the title *Everybody's own physician; or, how to acquire and preserve health* (#1352-1358). Gleason's co-author and publisher may have been H. Russell Burner, an 1870 graduate of the Eclectic Medical College of Pennsylvania. Burner's name does not appear on the title-page of *Everybody's own physician*.

1362. GLEASON, Rachel Brooks, 1820-1905.
Talks to my patients; hints on getting well and keeping well . . . New York: Wood & Holbrook, publishers, 1870.
   xii, 13-228 p. : port. ; 19 cm.

First edition. Intended as a guide "in those matters of delicacy with which woman's life is so replete," Gleason addresses such topics as puberty, menstruation & its disorders, the management of pregnancy, confinement & deliv-

ery, the puerperium, infant care, hygiene and menopause.

In 1844, Rachel Brooks married Silas Orsemus Gleason (1811-1899), a recent graduate of Castleton Medical College. Gleason began his career as a travelling lecturer on medical topics, a period during which he abandoned allopathic practice for hydropathy. In 1846 S.O. and R.B. Gleason opened the Greenwood Springs water-cure outside Cuba, New York, where S.O.G. managed the medical department and R.B.G. was matron. One of their patients was James Caleb Jackson (#1902-1955), abolitionist and journalist, whose seriously failing health the Gleasons were able to restore. Jackson underwent a vocational conversion, turning from abolition to health reform. In 1847 Jackson, the Gleasons and Theodosia Gilbert became partners in the Glen Haven Water Cure on Skaneateles Lake in central New York. In November 1849 Rachel Brooks Gleason entered the Central Medical College, an eclectic institution that had recently opened in nearby Syracuse. When the College moved to Rochester the next year, Gleason followed, but received her medical degree from its reincarnation as the Syracuse Medical College in February 1851. Shortly after, the Gleasons sold their share of Glen Haven to Jackson and Gilbert, and moved to Elmira, New York where they opened the Elmira Water Cure (#1063, 1064). The Gleason establishment survived the water-cure movement's decline post-1860 by the Gleasons' judicious combination of hydropathiy and eclecticism in the treatment of patients. The Elmira Water Cure became known as the Gleason Sanitarium from 1893 to 1903, and as the Gleason Health Resort until the Second World War. Rachel Brooks Gleason was the fourth woman in the United States to receive a medical degree. In addition to health reform, she was actively involved in the abolition movement, dress reform, and the advocacy of women's rights. Wood-engraved frontispiece view of the Elmira Water Cure.

1363. GLEASON, Rachel Brooks, 1820-1905.
Talks to my patients; hints on getting well and keeping well . . . New York: Wood & Holbrook, publishers, 1871.
   xii, 13-228, 16-24, [3] p. ; 19 cm.

Second edition. The numbered pages following the text contain advertisements for the
publications of Wood & Holbrook, as well as
the New York Hygienic Institute. Wood-engraved frontispiece view of the Elmira Water
Cure (fig. 25).

**1364. GLEASON, Rachel Brooks, 1820-
1905.**
Talks to my patients; hints on getting well
and keeping well . . . Sixth edition. New
York: M.L. Holbrook, publisher, 1879.
    xii, 13-228 p., [1] leaf of plates : port. ;
19 cm.

Mounted photographic frontispiece portrait
of the author.

**1365. GLEASON, Rachel Brooks, 1820-
1905.**
Talks to my patients; hints on getting well
and keeping well . . . Seventh edition. New
York: M.L. Holbrook, publisher, 1880.
    xii, 13-228 p. : port. ; 19 cm.

Mounted photographic frontispiece portrait
of the author.

**1366. GLEASON, Rachel Brooks, 1820-
1905.**
Talks to my patients; hints on getting well
and keeping well . . . Eighth edition. New
York: M.L. Holbrook, publisher, 1882.
    xii, 13-228 p. ; port. ; 19 cm.

Atwater copy inscribed by author. Mounted
photographic frontispiece portrait of author.

**1367. GLEASON, Rachel Brooks, 1820-
1905.**
Talks to my patients. Hints on getting well
and keeping well. New edition enlarged
with the addition of nineteen "Letters to
ladies" on health, education, society, etc.
. . . New York: M.L. Holbrook Co.; London:
L.N. Fowler & Co., 1895.
    xii, 13-403, [4] p. ; 22 cm.

The nineteen "Letters to ladies" (p. [227]-
403) were originally published in the pages of
M.L. Holbrook's the *Journal of hygiene and herald
of health*. They were edited by Gleason for republication here. Atwater copy inscribed by author.

GLEASON, Rachel Brooks see also
BEECHER (#275, 276); **ELMIRA
WATER CURE** (#1063, 1064).

GLEASON, Silas Orsemus see **ELMIRA
WATER CURE** (#1063, 1064).

**1368. GLEITSMANN, Joseph William,
1840-1914.**
Western North Carolina as a health resort
. . . Read before the American Public
Health Association, November, 1875, at
Baltimore, and reprinted from the Phila-
delphia Medical and Surgical Reporter,
February, 1876. Baltimore: Press of Sher-
wood & Co., 1876.
    8 p. ; 23 cm.

Reprinted from *The Medical and surgical reporter*
(Philadelphia), 19 February 1876, vol. 34, no 8,
pages 141-45. The author recommends the elevated situation and climate of western North
Carolina for the treatment of pulmonary tuberculosis. Title and imprint from printed wrapper.
    Gleitsmann was an 1865 medical graduate
of Wurzburg University. He served as a surgeon
in the German army during the Franco-Prus-
sian War, and in 1871 emigrated to the United
States. He practiced first in Baltimore, and then
moved to Asheville, N.C., where he was "physician in charge" at the Mountain Sanitarium for
Pulmonary Diseases. Gleitsmann's name first
appears in *The Medical register of New York, New
Jersey, and Connecticut* in the 1881-82 edition
(19:xlviii). He remained in New York, where
he specialized in respiratory tract diseases, was
on the faculty at the New York Polyclinic Hospital, served as president of the American
Laryngological Society, and attended the German Hospital and German Dispensary.

1369. GLEIWITZ, George.
The realities of homoeopathy . . . Middlebury, Vt.: A.H. Copeland, publisher, 1859.
    52 p. ; 22 cm.

In this printed lecture Gleiwitz defends homeopathy before a lay audience, and explains
its advantages. In printed wrapper.

1370. GLEN FALLS EYE SANITARIUM.
The eye. The cause of impaired vision and
diseased eyes cured . . . E.H. Bemis, eye

specialist . . . Glen Falls, N.Y.: Glen Falls
Sanitarium, [ca. 1894].
   [14] leaves : ill., ports. ; 22 cm.

An illustrated brochure promoting the Glen
Falls Eye Sanitarium and the therapeutic methods of its founder, E.H. Bemis. Rejecting the surgical treatment of cataract, Bemis claims to be
able to dissolve the "excretions" which cloud the
lens through the use of his *Magnetic Vaporizer* or
by electrotherapy. The former method "suffuses
the eye each time it is used, and through the increased circulation the *excretions* which should
have been thrown off (and are the cause of cataracts) *are thrown off*'; the latter stimulates "into
action the nerve centers and assists in bringing
about a quick reaction . . . making the absorption of cataracts more rapid." Through the application of the *Magnetic Vaporizer*, Bemis' *Antiseptic Drops,* electric stimulation therapy and the
application of magnets to various parts of the
body, Bemis and his medical staff assure the
reader of their ability to cure cataract, detached
retina, glaucoma, amaurosis, etc. The *Vaporizer,*
Bemis' proprietary remedies and various articles
of magnetic clothing are also available through
the mails to those unable to visit the Sanitarium.
E.H. Bemis, who admits to being trained as an
optician, was assisted at Glen Falls Eye Sanitarium
by George W. Little, M.D. (1835-1911), an 1858
graduate of the Albany Medical College.

1371. GLEN SPRINGS.
The Glen Springs, Watkins Glen, N.Y. . . .
Wm. E. Leffingwell, President, John M.
Swan, Medical Director. [Watkins Glen,
N.Y.], 1911.
   32 p. : ill., map ; 15.5 × 20.5 cm.

A publication advertising the Glen Springs
health resort and the mineral waters at Watkins
Glen, N.Y. on the southern end of Seneca Lake
in the Finger Lakes region. In addition to an
analysis of the composition and medicinal properties of the six springs' waters, the brochure
provides a lavish photographic display of the
resort's extensive facilties and grounds. Billing
itself "The American Nauheim," the Glen
Springs hotel-health resort was located on a
panoramic site above the town and lake. In
printed wrapper.
   The medical director of Glen Springs, John
M. Swan, was an 1893 medical graduate of the

University of Pennsylvania. The 11th edition of
Polk's medical register and directory of North America
(1910) provides addresses for Swan in both
Watkins Glen and Philadelphia, indicating they
he may have spent only part of the year at Glen
Springs. The president of Glen Springs, William
Elderkin Leffingwell (1855-1927), was the
younger brother of the public health activist
Albert Leffingwell (#2224) and the nephew of
James Caleb Jackson (#1902-1955). He entered
Cornell University in 1871, but never graduated. Leffingwell became business manager of
Our Home on the Hillside, the Dansville, N.Y.
health resort founded by J.C. Jackson, in 1882—
the year that it burned to the ground. He was
involved in its refinancing, reorganization and
reconstruction. The resort was reopened in
1883 as The Sanatorium. In 1886, Leffingwell
sold his interest in the Dansville resort, and in
1890 established the Glen Springs Corporation
in Watkins Glen. As well as his new sanitarium,
Leffingwell was engaged in banking, salt mining, and Democratic Party politics (he served
as mayor of Dansville and Watkins Glen at different periods). Leffingwell returned to
Dansville in 1917 to assume presidency of the
financially troubled Jackson Health Resort.

1372. GLOVER, Ralph, 1797-1869.
Every mother's book: or the duty of husband and wife physiologically discussed,
being a short treatise on the population
question . . . New York: Published by R.
Glover, 1847.
   [iii]-viii, [9]-107 p., [2] leaves of plates :
ill. ; 15 cm.

Glover was born in Wilbraham, Ma., October 28, 1797. Educated in local academies, he
taught school for a year in Somerville, N.J.,
before returning to central Massachusetts
where, after reading law for two years, he decided on a career in medicine. Glover became
a pupil of Joseph White in Cherry Valley, N.Y.,
attended lecture courses at the College of Physicians & Surgeons of the Western District
(Fairfield, N.Y.) and at the Jefferson Medical
College, where he received his medical degree
in 1826. After practicing for a year and a half
in Virginia, Glover moved to New York City
where he spent the rest of his life, except for a
four-year period (1854-58) during which he
returned to Wilbraham to serve as town clerk

and justice of the peace. Though he had several office sites over the years, they were always on Ann Street, which ran off Broadway near City Hall in lower Manhattan. He died on July 5, 1869.

In 1846 and 1847 Glover edited (with extensive additions) and published three editions of Robert Dale Owen's *Moral physiology* (#2693-2699). Why he became interested in contraception is not apparent. Owen had published six editions of his work in 1831 and another in 1836, after which, except for the English editions, interest in the book seems to have diminished. In the preface to his 1846 edition of *Moral physiology* (#2698), Glover points out that "such a change of public sentiment has taken place, as will render a republication of the work, with such additions and alterations as the discoveries and improvements in this department of physiology have brought to light, acceptable to the author, his friends, and the public." Furthermore, Glover states that Owen's preferred method of contraception (withdrawal) seldom worked and was not popular, especially with the young. He goes on to mention "several years spent in a course of experimental investigation" which led him to develop an instrument that employed galvanic electricity in a manner unspecified. The application of the device must follow ejaculation immediately and the effect would be "instantaneous." Its *modus operandi* is inexplicable, but it is "known to impart a slight momentary impetus to the parts, so that the vivifying influence of the semen is destroyed or expelled" (p. 105-6). This statement is the closest Glover comes to explaining the instrument's action.

A third and considerably expanded edition was published the following year (#2699). In this edition Glover elaborates upon (without clarifying) the theory behind his instrument. He also digresses considerably on the subject of abortion, its evils, and to the currently newsworthy Mme. Restell (the pseudonym of Ann Trow Lohman) who had been arrested in New York City and was about to be tried as an abortionist. She was the sister of Joseph Trow who wrote under the name A.M. Mauriceau (#2398-2407). Glover's point, of course, was that knowledge of contraceptive methods would obviate the need for criminal abortions.

In *Every mother's book*, which appears to have been published in the spring of 1847, Glover replaced Owen's first five chapters (i.e., those that discuss the philosophical, economic and social aspects of contraception) with two of his own, retaining the chapter that deals with the physiology of contraception and the concluding remarks. The jist of Glover's contribution is that sexual activity should be fun and is good for people, but that fear of pregnancy leads to late marriages and an increase in the number of promiscuous relationships, abortions and venereal disease. Those who opposed sex education and the dissemination of information about contraception claimed quite the opposite—that it leads to debauchery and licentiousness. Once again, Glover promotes his contraceptive device, the *Electrogalvania*.

By 1853 Glover had turned his attention to the manufacture and sale of trusses and braces, a business in which he continued thereafter. In that year he published *A treatise on orthopedic surgery and hernia: containing directions for adjusting and applying trusses to every species of rupture, for the purpose of effecting radical cures* (New York, 1853, 175 p.). The business denominated "The New York Truss and Bandage Institute" was located at 12 Ann Street, adjacent to P.T. Barnum's American Museum.

GLOVER, Ralph see also **OWEN** (#2698, 2699).

1373. GODFREY, Julia Day.
Diseases of women. Their causes, symptoms and treatment and the home medical guide. A medical book that should be in every home. Contains information that every woman should know. Fort Wayne, Indiana: Published by Woman's Health Institute, [ca. 1914, c1909].
[2], 121, [5] p. : port. ; 20 cm.

Excepting the preliminary leaves, the content of Godfrey's *Diseases of women* is identical to a monograph of the same title by Bertha C. Day (#918). Julia D. Godfrey, an 1891 graduate of the Hahnemann Medical College & Hospital (Chicago), was hired by William M. Griffin, of Ft. Wayne, as medical director of the Woman's Health Institute to carry on his mail-order medicine business after an A.M.A. investigaton forced closure of his Bertha C. Day Company. Griffin was also proprietor of the Atlanta Remedy Company (#169).

1374. GOFF, C. S.
TO MARRIED MEN AND WOMEN! Also,
those contemplating marriage! [Syracuse,
N.Y., 186-?].
  Broadside ; 17.7 × 11.7 cm.

Goff advertises his *Band d'Amour.* "Some
women from their organization never have any
pleasure in the marriage relations, and the act
of sexual intercourse is a forced operation with
them and affords no charm. On the other hand,
many men from excess of intercourse after marriage, or secret habits before, have partially lost
the virility of the parts, and the erection of the
penis is not strong and full as nature intended it
should be . . . To obviate these difficulties, has
been our study for years, and we are now happy
to inform our patrons, that we have invented a
curious little article which in every case will perform all that is desired. It will increase the size of
the penis, at least one-third . . ."

1375. GOFF, C. S.
SYRACUSE MEDICAL DISPENSARY . . . Syracuse, N.Y. Established in 1850, for the relief of the misguided and imprudent vo-
tary of false pleasure who, finding that he
has imbibed the seeds of a certain loathsome disease, is from an ill-timed sense of
shame deterred from applying to a physician whose knowledge and experience can
alone befriend him in distress . . . [Syra-
cuse, N.Y., 186-?].
  [2] leaves ; 19 cm.

Caption title. Goff advertises "medicated applications" (i.e., suppositories) for the cure of
"seminal weakness" and "excessive irritative debility of the genital organs in both male and
female," as well as tonic pills and "French male
safes."

1376. GOFF, Samuel Bishop, b. 1842.
The 20th century age of reason. A reference work on physiology, physiognomy,
phrenology, psychology, genealogy. Between man and the animal and their creation and characteristics. Man being possessed with the same nature, makes it impossible at all times for the spiritual nature
to overcome the evil propensities. Also
treats on the attributes of the human race
by the unnecessary conditions, making

man different spiritually from that which
his Creator intended him to be. Showing
the necessity of making environments for
the welfare of the present and future generations, thereby home and political
economy will be practiced, causing an
equal distribution of wealth, the same producing a greater amount of happiness . . .
[Camden, N.J.: s.n., c1906].
  316, [4] p. : ill. ; 23 cm.

"As God is perfect in all his works and as man
was created by God, he could not do otherwise
than to imitate his Creator into advancing into a
state of perfection. Unless those attributes are
interfered with by those who possess a spirit of
animal-like nature, thereby a higher development of the human race can only be accomplished by the rejecting of those things which
tend to debase the mind and defile the body.
And as we have advanced in art and science, why
not, in the 20th century, endeavor to use methods to advance the human race, in comparison
to our advancement in the mechanical and business systems of the world?" (p. [9]). The author
is dissatisfied with the whole of the current political, economic, social and physiological order.
He lays much of the blame for this situation on
alcohol consumption and tobacco use, which
reduce men and women to the spiritual level of
animals. In this rambling, repetitive treatise, Goff
condemns the wealthy for holding a disproportionate amount of the nation's wealth; and the
poor, whom he would place in workhouses or
literally tie to the whipping post if they refuse to
become productive members of society. Railing
against the Republican Party that has controlled
government since the Civil War, Goff charges it
with protecting the interests of both the wealthy
and the "vicious and ignorant" among the lower
classes. He calls for the formation of a new political party that would sweep corruption from
government and create an entirely new social
order. Spine- and cover-titles: *The twentieth century age of reason.*

GOLER, George Washington see
**ROCHESTER ATHENAEUM AND
MECHANICS INSTITUTE** (#3012).

1377. GOOD, Peter Peyto, 1789?-1875.
The family flora and materia medica
botanica, containing the botanical analy-

sis, natural history, and chemical and medicinal properties of plants: illustrated by colored engravings of original drawings, copied from nature . . . New York: Published by the author, 1845.
2 v. : col. ill. ; 24 cm.

The complete work was issued serially in 96 parts. The first completed edition was issued at New York by the author in 1845 as *The family flora and materia medica botanica,* and also as *A materia medica botanica* (New York: J.K. Wellman; New Milford, Ct.: Peter P. Good). *The family flora* was reissued in a revised edition at Elizabethtown, N.J. in 1845, 1847 and 1852, and at Cambridge, Mass. in 1854. Atwater set lacking volume 2. The Atwater copy of volume one includes issue numbers 1-66 (i.e., xviii p., [132] leaves of printed text, plates 1-66), and appears to include more text and plates than commonly found in this volume. The lithographic plates are hand-colored. Good published *A materia medica animalia* at Cambridge in 1853.

GOOD, Thomas F. see **BUFFALO LITHIA SPRINGS** (#515).

1378. GOOD HEALTH.
Good health . . . A popular annual on the laws of correct living, as developed by medical science, etc. [Vol. 1, no. 1 (June 1869)-v. 4, no. 6 (Nov. 1872)]. Boston: Alexander Moore, 1869-72.
4 v. ; 24 cm.

Volume 1 in Atwater collection only (title-page dated 1870). Monthly issues for volume 1 have title: *Good health: a journal of physical and mental culture.* The journal was edited by William Mason Cornell (#799-802), who also wrote a considerable part of each issue.

1379. GOODENOUGH, Josephus, 1831-1910.
The favorite medical receipt book and home doctor. Comprising the favorite remedies of over one hundred of the world's best physicians and nurses. Supplied especially for this work. Compiled & edited by Josephus Goodenough, M.D. . . . Detroit, Mich.: Published by F.B. Dickerson, 1908.
[6], 771 p., [35] leaves of plates : ill., ports. ; 23 cm.

"Department II. Diseases of women and children . . . [by] Helen F. Warner, M.D.," p. 380-462. Goodenough was an 1858 graduate of the Eclectic Medical Institute (Cincinnati); Helen Frances Warner (1843-1905) was an 1867 graduate of Vassar College and an 1872 medical graduate of the University of Michigan.

GOODHOLME, Todd S. see **DOMESTIC CYCLOPAEDIA OF PRACTICAL INFORMATION** (#969).

1380. GOODLETT, Adam G., d. 1848.
The family physician, or every man's companion, being a compilation from the most approved medical authors, adapted to the southern and western climates. To which is added an account of herbs, roots and plants, used for medical purposes, with directions how they are to be prepared, so that every man can be his own physician, together with a glossary of medical terms . . . With an appendix, containing a new and successful mode of treating Asiatic cholera . . . Nashville, Tenn.: Printed at Smith and Nesbit's steam press, 1838.
xvi, [17]-792 p. ; 22 cm.

Goodlett claims to have served "for nine years a surgeon in the regular army of the United States, who, after the expiration of the last English and American War, visited the principal hospitals of England, Ireland, Scotland, and France" (p. [xv]). He acknowledges compiling *The family physician* "from the works of Darwin, Cullen, Gunn, Bell, Ewell, and others." The first two hundred pages are devoted to anatomy, physiology and hygiene. "Of diseases" is followed by separate sections on the "Management of the diseases of women" and the "Diseases of children." The volume concludes with an extensive vegetable materia medica (p. [683]-758) and dispensatory.

1381. GOODRICH, Charles Augustus, 1790-1862.
A new family encyclopedia; or, compendium of universal knowledge: comprehending a plain and practical view of those subjects most interesting to persons in the ordinary professions of life . . . Third edition . . . Philadelphia, 1832.
468 p. : ill. ; 19 cm.

Encompassing domestic economy, mechanics, agriculture, etc., Goodrich's encyclopedia includes considerable material pertaining to human health, i.e., anatomy & physiology ("Part I. Man. Sect. I. Natural history, structure, &c.," p. 14-29); nutrition ("Part II. Aliments," p. 35-124); hygiene & domestic medicine "Part III. Preservation of health," p. 124-66). The work is illustrated with beautifully executed wood engravings of plants, animals, etc. in the tradition of Bewick. An 1812 theological graduate of Yale College, Goodrich withdrew from the ministry after a controversial tenure as pastor of the First Congregational Church of Worcester, Ma. "Goodrich then began a second career . . . He returned home to Berlin, Connecticut, in 1820 and established himself as a popular, if pedestrian, writer of (often huge) informational books and texts on religion and travel for juvenile readers" (*Amer. nat. biog.,* 9:257). More interesting than the number of titles he wrote is the duration of their publishing history. Goodrich's *A history of the United States of America* (1822), for example, appeared in more than 150 editions. *A new family encyclopedia* (1831) remained in print until 1869.

1382. GOODRICH, Charles Augustus, 1790-1862.

A new family encyclopedia; or compendium of universal knowledge: comprehending a plain and practical view of those subjects most interesting to persons in the ordinary professions of life . . . Fourth edition . . . [Philadelphia]: Published by T. Belknap, 1833.

468 p. : ill. ; 19 cm.

GORDON, Elizabeth Putnam see also **GREENE** (#1408, 1409).

1383. GOUGE, Henry Albert.

New system of ventilation, which has been thoroughly tested, under the patronage of many distinguished persons, adapted to parlors; dining and sleeping rooms; kitchens and basements; cellars, vaults and water closets . . . &c., &c., &c. . . . Pamphlet edition. New York: Printed by E.S. Dodge & Co., steam printers, 1866.

[2], 46 p. : ill. ; 23 cm.

The author explains and promotes *Gouge's Atmospheric Ventilator,* for which patents were is-

sued 1863-65. "The writer of this is not a physician, but in the course of his professional duties, ventilating kitchens, basements, water-closets, offices, stables, and all sorts of places, he has seen enough to satisfy him that a great deal of disease results from bad air without the cause oftentimes being suspected. The people have yet to learn that pure air is one of the most essential requisites of a healthy existence" (p. 6). In printed wrapper.

1384. GOUGE, Henry Albert.

New system of ventilation which has been thoroughly tested under the patronage of many distinguished persons, being adapted to parlors, dining and sleeping rooms, kitchens, basements, cellars, vaults, water-closets, stables, preserving-rooms, churches, legislative-halls, school and court-rooms, prisons, hospitals, stores, show-windows, hotels, banking-houses, restaurants, coal-mines, powder-magazines, railroad tunnels, factories, pork packing houses, ships, steamboats, etc. A book for the household. Third edition enlarged, with new illustrations . . . New York: D. Van Nostrand, 1870.

[6], [iii]-v, [6]-176 p. : ill. ; 24 cm.

Atwater copy inscribed by the author to Henry Foster (#1265), proprietor of the Clifton Springs Sanitarium.

1385. GOUGH, John Bartholomew, 1817-1886.

Autobiography and personal recollections of John B. Gough, with twenty-six years' experience as a public speaker . . . Springfield, Mass.: Published by Bill, Nichols, & Co.; Chicago, Illinois: Bill & Heron; Philadelphia, Pa.: H.C. Johnson, 1869.

xvii, [18]-552, [2] p., [1] leaf of plates : ill., port. ; 23 cm.

Born in England, Gough emigrated to Oneida County, New York at age twelve in the company of a neighbor's family. Seeking other opportunities, Gough moved to New York City where he trained as a bookbinder. By 1833 he had earned enough to send for his mother and sister from England. In the year that followed, however, he lost his job and mother, and began to drink heavily. Drinking and drifting among east coast cities, Gough eked out a living doing odd jobs at his trade and by comic acting and singing. In 1839 he married and opened a small bookbindery at Newburyport, Ma.. His drinking continued, however, and in 1841 his wife and child died, probably as the consequence of want and exposure. Struggling to reform over the next two years, Gough had gained sufficient mastery over his drinking to launch himself on a temperance lecture tour through New England in 1843. He rapidly developed into the foremost temperance speaker in America, and in 1853 was invited to England by the London Temperance League for the first of his three tours of the United Kingdom. Gough's autobiography was first published in a volume of 168 pages at Boston in 1845; it sold 20,000 copies within a year. In 1869 he greatly expanded the work under the title *Autobiography and personal recollections.*

1386. GOULD, Arthur.

The science of sex regeneration. How to preserve and strengthen and retain the vital powers. A study of the sacred laws that govern the sex forces . . . By A. Gould and Dr. Franklin L. Dubois. Chicago, Ill.: Advanced Thought Publishing Co.; London, England: L.N. Fowler & Co., [c1911].
224 p., [1] leaf of plates : port. ; 20 cm.

At head of title: *Tenth edition contains much new valuable instruction.* Cover-title: *Science of regeneration.*

1387. GOULD, Arthur.

The science of regeneration or sex enlightenment. A study of the sacred laws that govern the sex forces. By A.G. Chicago, Ill.: Advanced Thought Publishing Co.; London, England: L.N. Fowler & Co., 1913.
[2], 161, [5] p., [1] leaf of plates : port. ; 20 cm.

Gould speculates that our uni-sexual species (i.e., one that is divided into distinctly male and female beings) evolved (i.e, degenerated) from an earlier bi-sexual or hermaphroditic state that was self-generating. He cites examples from the plant and animal worlds of self-generating creatures, notes hermaphroditic anomalies among humans, and states that the story of Eve being taken from Adam and the production of children by Greek male deities confirms our an-

cient, hermaphroditic ancestry. Gould states that children should be taught that "sexual intercourse between bi-sexual beings brought about the degeneration of the body, with differentiation into separate sexes: while the maintenance of chastity, with the cultivation of an androgynous mind, at once suggests itself as a means of reversing the degenerative process, and thus prepare the proper conditions for the re-evolvement of bi-sexuality" (p. 46). Gould argues that the pursuit of sexual pleasure is a waste of the body's creative powers, and that by "the retention and reabsorption of the seminal fluids in the body" man draws closer to physical, intellectual and spiritual perfection.

GOVE, Mary Sargeant see NICHOLS
(#2623-2625, 2635).

1388. GRAEFENBERG COMPANY.

The Graefenberg guide to health. (Second edition.) N.Y.: Sibells, printers, 1853.
32 p. ; 13 cm.

An almanac abstracted from *The Graefenberg manual of health* (#1389). In printed wrapper.

1389. GRAEFENBERG COMPANY.

The Graefenberg manual of health . . . Fifth edition. New-York: Published by the Graefenberg Company, 1850.
[2], iii, [4]-292, xviii, [6] p. ; 19 cm.

The "Americo-Graefenberg" system is described as one that transcends the achievements and failures of allopathy, homeopathy, hydropathy, Thomsonism and the chrono-thermal system by combining them all. *The Graefenberg manual of health* is a hygienic and medical guide intended to promote the use of the Graefenberg Company's proprietary medicines for the preservation of health and the cure of disease.

1390. GRAETER, Francis.

Hydriatics: or manual of the water cure, especially as practised by Vincent Priessnitz in Graefenberg. Compiled and translated from the writings of Charles Munde, Dr. Oertel, Dr. Bernhard Hirschel, and other eye-witnesses and practitioners . . . Third edition. New-York: Published by William Radde, 1843.
198, [2] p., [1] leaf of plates : ill. ; 17 cm.

Hydropathy, Graeter writes, "rejects . . . the use of all simple and compound medicinal means save cold water, which she uses in its various applications as the only beverage, in entire cold baths after previous sweating, in half-baths, seat-baths, foot-baths, exciting or cooling fomentations, douche, shower-baths, clysters and other injections, washings and arrosions, for dissolving, dissipating, and carrying off generally and locally situated noxious humors, and preventing their reproduction by improved digestion" (p. 15-16).

In introducing hydropathy to an American audience, Graeter provides a lengthy historical review of the therapeutic use of water before describing the work of Vincent Priessnitz (1790-1851) at Graefenberg. His excerpts from the writings of Munde, Oertel, Hirschel, etc. are numerous, but quoted within the context of his own narrative. Pages 104-64 describe the application of hydropathic therapeutics to sixty-four specific diseases and conditions. *Hydriatics* appeared in four editions between 1842 and 1844.

1391. GRAETER, Francis.

Hydriatics: or manual of the water cure, especially as practised by Vincent Priessnitz in Graefenberg. Compiled and translated from the writings of Charles Munde, Dr. Oertel, Dr. Bernhard Hirschel, and other eye-witnesses and practitioners . . . Fourth edition. New-York: Published by William Radde; London: H. Bailliere, 1844.

198, [2] p., [1] leaf of plates : ill. ; 17 cm.

1392. GRAHAM, Sylvester, 1794-1851.

A lecture to young men, on chastity. Intended also for the serious consideration of parents and guardians . . . Sixth stereotype edition. Boston: George W. Light, 1841.

246 p. ; 16 cm.

"The danger of sexual stimulation had been recognized by Graham and made a part of his lectures by 1833. The following year his conclusions were given wider circulation by the publication of his *Lecture to young men on chastity* . . . No doubt many young men, as well as parents and guardians to whom the book was also directed, were taken back by the content, for it not only substituted a physiological for

the time-honored moral approach to the issue, it also concentrated on new abuses. Chastity literature had previously been occupied with adultery and fornication, and, to a much lesser degree, masturbation. Graham was hardly ready to ignore the former sins, but he demoted those 'social vices' to a position well below the 'solitary vice,' and added to them a vice that had hitherto been a virtue—marital sex. It was standard in pre-Grahamite days to actually prescribe marriage as the cure for masturbatory and libertine tendencies, and even to hint that liberal doses of marital pleasure would do no harm to soul or body. But earlier writers had not been aware of the physiological turmoil caused by sexual excitement. 'The convulsive paroxysms attending venereal indulgence,' an aroused Graham warned, 'are connected with the most intense excitement, and cause the most powerful agitation to the whole system . . . ' Because those nervous paroxysms preceded the sexual climax, orgasm did not have to occur for injury to result . . . A certain amount of contact . . . was necessary for procreation, but that physiological stimulation would become pathological, in Graham's view, if enjoyed more than once a month" (James C. Whorton, *Crusaders for fitness*, Princeton, NJ, 1982, p. 92-94).

Graham's sense of self and mission is revealed in the preface to the first edition of the *Lecture* (Providence, 1834): "When I commenced my public career, as a lecturer on the Science of Human Life, it did not, in any degree, enter into my plan, to treat on this delicate subject: but the continual entreaties, and importunities, and heart-touching . . . appeals which I received from young men, constrained me to dare do that which I was fully convinced ought to be done; and the result has entirely justified my decision and conduct. An incalculable amount of good has already been accomplished by it. Hundreds who have listened to the following lecture, have thereby been saved from the most calamitous evils: and great numbers, among whom are many physicians and clergymen, have urged me to publish it. On this point, I have long hesitated;—not, however, because I doubted the intrinsic propriety of publishing it, but because I doubted whether the world had sufficient virtue to receive it, without attempting to crucify me for the benefaction" (6th ed., p. 12-13). Of the world's response Whorton writes, "the rationale outlined in his

Lecture on chastity clearly affected medical thinking. Many physicians honored his formula for sexual frequency, and other ideas on sexual hygiene. They were as vulnerable as the public to the emotional force of Graham's preface statement that his work 'proved . . . that the Bible doctrine of marriage and sexual continence and purity, is founded on the physiological principles established in the constitutional nature of man'" (*ibid.*, p. 95).

1393. GRAHAM, Sylvester, 1794-1851.
A lecture on epidemic diseases generally, and particularly the spasmodic cholera, delivered in the city of New York, March, 1832, and repeated June, 1832, and in Albany, July 4, 1832, and in New York, June, 1833. With an appendix, containing several testimonials, and a review of Beaumont's Experiments on the gastric juice. New edition, revised and enlarged . . . Boston: Published by David Cambell, 1838.
    viii, [5]-125, [1] p. ; 19 cm.

Second edition; originally published at New York (80 p.) in 1833. "The maturing of what [Graham] was already calling 'the Science of Human Life' was stimulated further by that 'urgent invitation' to New York, for it was during his stay in that city that the dreaded cholera finally appeared in America. Weeks before the first cases were reported, in fact, public fear of the new infection had risen to a height sufficient to inspire Graham to add a lecture on cholera to his repertory. As he well realized, the disease presented an ideal opportunity to promote his Broussaisist scheme of health. The symptoms of profuse diarrhea and abdominal cramps stamped it as a gastrointestinal irritation. All medical advice on prevention gave first place to careful diet for the maintenance of resistance" (James C. Whorton, *Crusaders for fitness*, Princeton, NJ, 1982, p. 45). Whorton explains that according to Graham, "Cholera was a particularly acute manifestation of overstimulation of the stomach . . . but it was basically no different from all other disease and debility. The race had long been deteriorating physically . . . falling from generation to generation as habits of overstimulation continued . . . Sexual excitation, through its effects on the nervous system, could also inflame the stomach, and general inattention to exercise, rest, clothing, the emotions,

and other non-naturals weakened the vital powers and made it easier for inflammation to be established. It was cholera at the moment, but next year it would be something else . . . the real epidemic was not cholera, but overstimulation and improper hygiene" (*ibid.*, p. 46).

1394. GRAHAM, Sylvester, 1794-1851.
Lectures on the science of human life . . . Boston: Marsh, Capen, Lyon and Webb, 1839.
    2 v. (xii, [13]-562, [2] p. ; xii, [13]-660 p.) : ill. ; 20 cm.

First edition. Sylvester Graham was the youngest of seven children from his septuagenarian father's second marriage. After the death of his father and his mother's emotional breakdown, Graham lived with various relations well into adulthood. It was not until his early thirties that Graham settled on a career in the ministry. He soon exchanged the pulpit for the lecture hall, however, when he became general agent for the Pennsylvania Temperance Society in 1830. In this new role Graham discovered his metier. Over the next two years he lectured extensively in Philadelphia and in eastern Pennsylvania and New York, expanding the scope of his interest in what went into the body to include solids as well as liquids. By the mid-1830s, Graham was in demand all along the Atlantic seaboard as a lecturer on human physiology, hygiene, and diet. Nine years of lecturing and publishing on health regimen and diet culminated in his most important and influential published work, the *Lectures on the science of human life.*
    The best study of the work and influence of Sylvester Graham is Stephen Nissenbaum's *Sex, diet, and debility in Jacksonian America.* (Westport, Ct., 1980). Early in his book Nissenbaum points out that "Graham was the first writer to formulate a coherent physiological analysis of the various new anxieties about the human body that had emerged by the 1830s, and to propose a systematic regimen he believed would assuage them. Coming out of both the evangelical ministry and the temperance movement of the late 1820s, he had a direct and significant impact on the development of movements such as vegetarianism, phrenology, and water-cure, in addition to sexual reform. To study Sylvester Graham is to study Victorian physiological theory

and practice in the very act of coming into being—as a complete ideological system governing every aspect of private routine." (p. [ix]).

Nissenbaum traces the sources of Graham's ideas not only to the evangelical ministry, the temperance movement, and the social discontent of the Jacksonian period, but to the writings of such medical authorities as Benjamin Rush, François Broussais and Xavier Bichat. Graham adopted Rush's variation on the 18th-century theme of the balance between stimulation and debility which presumedly exists in the healthy organism. Rush maintained that stimulation, rather than being a counter to debility, was actually its principal source. According to Rush, immoderate diet, the consumption of alcohol, the use of tea or coffee, sexual activity and the passions in general (i.e., fear, grief, despair) overstimulated the system and resulted in a diseased state. All disease, no matter what its symptoms, could be traced to this single cause, i.e., morbid excitement.

From Broussais, Graham adopted the idea that the stomach, with its complex innervation, is the most common site of morbid excitement and hence the originating locus of most disease. For Broussais "gastrointestinal irritation was the effective source of virtually all disease, including cancer, syphilis, tuberculosis, and 'malignant and pestilential fevers'" (Nissenbaum, p. 59). Like Rush, Broussais maintained that there was only one disease, though it might be induced by different forms of morbid excitement. From his reading of Bichat, Graham adopted the theory that life depends upon an indefineable vital principle that resists the disintegrative tendencies of mere physics and chemistry. According to Bichat, the organism lives both externally (i.e., animal life: sensation, intellection, voluntary motion, reproduction) and internally (i.e., organic life: respiration, circulation, secretion, digestion). By the former the organism reacts to its environment; by the latter it assimilates foreign matter. The organic life was by far the more significant for Graham in the maintenance of health, and could be maintained in a healthy state only by the most careful regimen and diet.

The nature and tone of Graham's physiological thinking has been aptly summarized by John C. Whorton, who observed that the science of the *Lectures on the science of human life* "was a Christianized science in which the laws of health were treated as a physical counterpart to the Ten Commandments. And because both sets of rules had been authored by God, and the infinitely wise Deity would never contradict Himself, the realm of physiology had to be always in agreement with the realm of morality . . . Graham simply applied the rule that physiology is congruent with morality to all other areas of of health behavior—drink and diet, exercise and rest, cleanliness and dress, and even (indeed especially) sex" ("Historical development of vegetarianism," *Amer. j. clin. nutr.*, 1994, 59:1105S). The subsequent influence of Graham's ideas on the American popular health movement over the course of the 19th century would be difficult to exaggerate. As much as any author or lecturer, he convinced the public that responsibility for health lay with the individual. He identified the issues, established the assumptions from which popular health reform would be argued, and supplied solutions that were taken up in the work of most of the nation's leading health reform advocates.

GRAHAM, Sylvester, 1794-1851. *Rules and regulations of the Temperance-Boarding House in the City of New York* see **NICHOLSON** (#2636).

GRAHAM, Sylvester, 1794-1851. *The science of human life* see also **KELLOGG** (#2110).

1395. GRAHAM, Sylvester, 1794-1851.
A treatise on bread and bread making . . . [Milwaukee: Lee Foundation for Nutritional Research, between 1990 and 1994]. viii, [9]-131, [1] p. ; 16 cm.

Facsimile of the 1837 Boston edition. Graham deplores what Nissenbaum terms the "commercialization of grain agriculture" during the first quarter of the 19th century, and the negative impact that high-yield, "unnatural" techniques such as manure fertilization had on the end product. Equally at fault for the poor quality of bread available to the American public were millers, who refined the grain to an unnatural state, and commercial bakers, who further diminished the nutritive value of the bread they made and sold by the addition to flour of such substances as alum, zinc sulphate and other substances. "In an implicit concession to the fact that he was addressing a public that no longer had the means to grow its own flour, Graham urged his audience to purchase the best wheat they could find, to grind it themselves with 'a modern patent hand-mill,' and, of course, to bake it in their own ovens" (S. Nissenbaum. *Sex, diet and debility in Jacksonian America.* Westport, Ct.: Greenwood Press, 1980, p. 7). Cover-title: *Graham on bread.*

GRAHAM, Sylvester see also **CORNARO** (#797).

1396. GRAHAM, Thomas John, 1795?-1876.
Modern domestic medicine: a popular treatise, illustrating the character, symptoms, causes, distinction, and correct treatment of all diseases incident to the human frame; embracing all the modern improvements in medicine. To which are added, a domestic materia medica; a copious collection of approved prescriptions adapted to domestic use, &c. &c. The whole intended as a comprehensive medical guide for the use of clergymen, heads of families, and invalids . . . The fifth edition, revised throughout, and considerably enlarged. London: Published for the author by Simpkin & Marshall . . . [*et al.*], 1832.
    xii, 648 p. ; 22 cm.

    Graham received his medical degree from the University of Glasgow in 1828. He was a prolific author of medical treatises for lay audiences. In addition to *Modern domestic medicine* (published at London in twelve editions between 1826 and 1858), he provided its supplement: *On the diseases of females* (1834), and also wrote *Sure methods of improving health* (1827), *A treatise on indigestion* (2nd ed., 1828), *An account of persons remarkable for their health and longevity* (1829), and *The cold water system* (1843), an endorsement of hydropathy. He also edited the 1829 London edition of Hufeland's *The art of prolonging life.* Only the *Sure methods* (#1397) and his treatise on indigestion were reprinted in American editions. Graham was admitted a fellow in the Royal College of Physicians of Edinburgh in 1861.

1397. GRAHAM, Thomas John, 1795?-1876.
Sure methods of improving health, and prolonging life; or, a treatise on the art of living long and comfortably, by regulating the diet and regimen. Embracing all the most approved principles of health and longevity, and exhibiting the remarkable power of proper food, wine, air, exercise, sleep, &c, in the cure of chronic diseases, as well as in the preservation of health, and prolongation of life. To which is added, the art of training for health, rules for reducing corpulence, and maxims of health, for the bilious and nervous, the consumptive, men of letters, and people of fashion. Illustrated by cases. By A Physician. First American edition, with additions. Philadelphia: Carey, Lea & Carey; sold in New York, by G. & C. Carvill; in Boston, by Hilliard, Gray & Co., and Richardson & Lord, 1828.
    viii, [9]-248, [2] p. ; 21 cm.

    The *Sure methods* appeared in at least five London editions between 1827 and 1842. It was one of only two titles among Graham's prolific output of popular medical treatises that was reprinted in the United States.

1398. THE GRAHAM JOURNAL.
The Graham journal of health and longevity. Devoted to the practical illustration of the science of human life, as taught by Sylvester Graham and others. David Cambell, editor . . . [Vol. 1. (Apr. 1837)-v. 3 (Dec. 1839)]. Boston; New York, 1837-39.
    3 v. : ill. ; 23 cm.

    "Founded less than two months after the American Physiological Society by David Cambell, who was at the same time corresponding secretary of the Society, *The Graham Journal* was closely associated with the Society, although not actually sponsored by it. Its circulation among members of the Society was large, sharing honors with Alcott's *Journal of Health* [i.e., *The Library of health*, #35], and it alone carried full reports of the activities of the Physiological Society. It was not intended to present the obstruse physiological principles contained in Alcott's *Journal*, but rather to demonstrate the application of these facts. It became therefore in a great measure a journal of testimonial to the effects of Grahamism . . . Notices of Graham's lectures and books were inserted in prominent positions and Graham contributed many articles. A great number of highly interesting articles on hygiene appeared; bathing, beverages, tobacco, self-abuse, lead colic and tight lacing, being particular favorites . . . The *Journal* did not survive its third year and was 'combined' with Alcott's *Journal of health* in which no trace of it could ever be found." (H.E. Hoff & J.F. Fulton, "The centenary of the first American Physiological Society," *Bull. Inst. Hist. Med.*, 1937, 5:706-7).

The Graham journal was absorbed by William Andrus Alcott's *The Library of health, and teacher on the human constitution* (#35) after its final issue on 14 Dec. 1839. The Atwater set is incomplete and includes the following: v. 1 (1837), nos. 1-38, v. 2 (1838), no. 17; v. 3 (1839), nos. 1-25.

1399. GRAND MUSEUM OF ANATOMY.

Descriptive catalogue of the Grand Museum of Anatomy. Established in St. Louis, 1874, by Drs. S. & D. Davieson, 11 South Broadway, St. Louis, Mo. Open daily, for gentlemen only . . . [St. Louis, 1903].
    48 p. : ill. ; 19 cm.

Title and imprint on wrapper: *Descriptive catalogue of Drs. S. & D. Davieson's Grand Museum of Anatomy . . . St. Louis, Mo. . . . 1903*. In 1874 David Davieson received his degree from the Eclectic Medical College of New York (the College catalog lists a David Davison [sic] from St. Louis). For two years prior to graduation, he was one of the partners in the European Anatomical, Pathological and Ethnological Museum in Philadelphia (#1081-1083). The year of his graduation, Davieson returned to St. Louis where he opened the Grand Museum of Anatomy. From various locations in St. Louis, the Museum moved to 11 S. Broadway in 1898. Sometime in the 1880s, David Davieson was joined by Sydney Davieson, who had obtained his medical degree at Giessen (Germany) in 1873. From the mid-1890s, it appears that Sydney Davieson was sole proprietor of the Museum, which remained in operation into the first decade of the 20th century.

The catalog the Grand Museum of Anatomy reads like those published by similar establishments in New York, Boston or Philadelphia. By 1903 the Museum displayed 1,514 items, an eclectic assortment of stuffed birds, anatomical models, dissected parts, pathological specimens, petrified botanicals, mastodon tusks,

freaks of nature, and such quaint objects as the head of Nat Hoover, said to be the youngest of all American desperados, who was hanged for murder at age sixteen. Interspersed at random were items that reveal the true purpose of the museum, i.e., to provide a (somewhat distorted) view of sex and sex disorders, and to advertise the the the professional services of the Drs. Davieson, whose offices were conveniently located upstairs over the Museum. Drawings, paintings, and models provided detailed illustration of the generative organs of males and females, including the various stages of pregnancy and degrees of fetal development. Syphilitic ulcers on various parts of the body, not to mention the lesions or results of several other venereal diseases, were displayed in lurid and graphic images, as were examples of the pathological consequences of masturbation (which could be reversed with proper treatment).

In addition to the Museum's catalogs, the Daviesons are known to have written at least two books, as proprietors of anatomical museums often did to promote business: *Practical observations on nervous debility and physical exhaustion, to which is added an essay on marriage . . . being a synopsis of lectures delivered at the Museum of Anatomy* (St. Louis, 1871), and *Health book for the people* (St. Louis, 1881). The publication date of the first title suggests that S. Davieson, at least, was active in St. Louis before the establishment of the Grand Museum in 1874.

1400. GRANGER, Isaphine.

[Letter], 1851 April 22, Canandaigua, N.Y., [to] Mrs. Paige, Clifton Springs, N.Y.
    [4] leaves ; 22 cm.

In this remarkable letter, Isaphine Granger, who describes herself as "quite an enthusiast upon the subject of Hydropathy," responds to Mrs. Paige's request for "some particulars concerning the water-cure establishment at Brattleboro," operated by Robert Wesselhoeft. Paige is apparently in residence at the Clifton Springs Water-Cure, operated by Henry Foster (#669-672). "Without wishing to disparage Clifton at all," Granger writes, "which I have no doubt is a very excellent establishment in many respects, no one should hesitate to say Brattleboro is far superior." After "two sojourns" at Wesselhoeft's Brattleboro Water Cure in Vermont, Granger is well acquainted with its situation, facilities and therapeutic regimen, which she describes in detail.

1401. GRANT, Herbert A.
Health and breath culture . . . Lynn, Mass.: Published by the A.F. Coleman Music Co., [c1905].
   [2], 34 p. ; 20 cm.

Writing for his voice students, Grant stresses that "physical culture is a most important aid to vocal culture." In printed wrapper.

1402. GRANVILLE, Augustus Bozzi, 1783-1872.
A catechism of facts, or plain and simple rules respecting the nature, treatment, and prevention of cholera . . . Philadelphia: E.L. Carey & A. Hart, 1832.
   vii, [8]-108 p. ; 15 cm.

Originally published at London in 1831. Granville maintains that the so-called *Asiatic cholera* awaited in England with such forboding is not a new disease entity but another variety of the "usual national cholera" or *cholera morbus*, "with this difference, that its symptoms are more severe, its progress more rapid, and its results more fatal" (p. 15). He maintains that cholera "is not a *contagious* disease, but an *epidemic* one: that it is not a transmitted malady, but one spontaneously evolved in the countries in which it has raged" (p. 33). Preventive measures such a quarantine, therefore, are ineffective. Granville describes cholera's symptoms; its domestic treatment (warm applications, internal medicines, "counter-irritants" applied to the skin); and preventive measures, e.g., moderate diet, warm garments, bodily cleanliness, frequent changes of linen; and avoidance of crowds, poorly ventilated rooms, fatigue, exposure to the wind, etc.
   Of Anglo-Italian descent, Granville (i.e., Bozzi) received his medical degree in 1802 from Pavia, where Spallanzani, Scarpa, Volta and Joseph Frank were his teachers. A patriot and ardent republican whose medical studies were interrupted by imprisonment for giving political speeches and writing lampoons, Bozzi exiled himself from Italy, and successively became second physician to the Turkish fleet, merchant, medical practitioner in Spain, and surgeon in the British navy. Upon the death of his mother

(Rosa Granville, descended from Cornish gentry settled in Italy), Bozzi took her family name. During the greater part of his remarkable career, Granville lived in England but remained involved in Italian politics. He was made F.R.S. (1817); served as physician-accoucheur to the Westminster General Dispensary; edited such journals as the *Medical intelligencer* and the *London medical and physical journal*; wrote books on travel, public health, medicine, obstetrics, health resorts, etc.; was active in the British Medical Association, the Royal Society, etc.; and was active in both medical and governmental politics.

1403. GRANVILLE, Joseph Mortimer, 1833-1900.
Sleep and sleeplessness. By J. Mortimer-Granville. Boston: S.E. Cassino, publisher, 1881.
   vii, [1], 111 p. ; 17 cm.

Originally published at London, England in 1879.

1404. GRANVILLE SANITARIUM.
The Granville Sanitarium, Liberty, Virginia. [S.l., before April 1893].
   [2] leaves ; 20 cm.

"The Trustees of the Granville Sanitarium announce to the medical profession and to the country at large, that the Sanitarium is open for the reception of patients. The medical and surgical management of the Sanitarium will be under the supervision of the physician in charge, Dr. T.M. Bowyer" (leaf 1 recto). Thomas M. Bowyer was an 1853 medical graduate of the University of Pennsylvania. He died at Battle Creek, Mich. in 1900.
   A handwritten note on the verso of the second leaf states that the town of Liberty has recently changed its named to Bedford City, describes the region's natural beauty, provides the costs of daily board and rail fare from New York, and notes that its founder, Mrs. Edith M. Nicholl, is the "daughter of Dean Bradley of Westminster Abbey, London." A library acquisition stamp on the first leaf is dated 18 April 1893.

GREAT EUROPEAN MUSEUM OF ANATOMY see **LA GRANGE** (#2173).

1405. GREEN, George Barrett, 1798-1866.
BIOCOLORATE BARK. Results of Analysis: Sample of "Bicolorate Bark," received from Dr. Geo. B. Green . . . [1860].
[1] leaf printed on both sides ; 34.5 × 24 cm.

The analysis of this tonic bitter is signed "A.A. Hayes, M.D., State Assayer . . . Boston," and is dated 5 Oct. 1860. On the verso it is noted that Bicolorate Bark is "particularly serviceable" in the treatment of "those blood diseases in which there is a dyscratic condition: such as purpura and scurvy," in "anemia in most of its forms," in "constitutional diseases in which there is a marked loss of muscular strength," in an "atonic state of the digestive organs," and "in debility, from acute diseases."

George Barrett Green was the son of Isaac Green (1759-1842) and Ann Barrett (1774-1847). The elder Green was a physician, banker and merchant in Windsor, Vermont. Although his son studied medicine, he never practiced. G.B. Green operated a dry goods store in Windsor, and with his father manufactured and sold *Oxygenated Bitters*. Papers relating to the Green family are retained by the Vermont Historical Society.

1406. GREEN, George Barrett, 1798-1866.
OXYGENATED BITTERS. A sovereign remedy for DYSPEPSIA, in many of its forms, such as Pain in the Stomach, Heartburn, habitual Costiveness, Acid Stomach, Headache, Loss of Appetite, Piles, Night Sweats, and even Consumption (Dyspeptic Phthisis), and Asthma, or Phthisic [sic] attended with derangement of the Stomach (or Dyspeptic Asthma), Difficult Breathing, which often results in imperfect digestion (or Dyspeptic Dyspnoea), is relieved by these Bitters. In short, their use has been proved in the relief of almost all the symptoms that proceed from a debilitated or atonic condition of the Stomach; also in general debility arising from age or from the effects of Fever, particularly Fever and Ague. Females suffering under any uterine derangement arising from weakness, will find the "OXYGENATED BITTERS" an excellent remedy, and not surpassed by any medicine in use . . . GEORGE B. GREEN, Proprietor. Windsor, Vt. . . . 1846.
Broadside ; 87 × 15.5 cm.

This broadside is comprised of two printed sheets pasted together. The better part of the text consists of "Certificates" or testimonials to the efficacy of the Oxygenated Bitters.

1407. GREEN, Matthew, 1696-1737.
The spleen, and other poems . . . With a prefatory essay, by J. Aikin, M.D. Philadelphia: Printed for Benjamin Johnson . . . [*et al.*], 1804.
90 p. ; 14 cm.

In his introductory essay John Aikin (1747-1822) writes, "The longest and most elaborate of Mr. Green's compositions . . . is an epistolary piece entitled THE SPLEEN . . . The writer calls it a motley performance, and apologizes for its want of method: a general subject may, however, be traced through it, which is, the art of attaining a tranquil state of mind, undisturbed by vexatious emotions and gloomy imaginations, and free from that mixture of listlessness and melancholy which has been denominated the spleen. For this purpose, a sort of regimen for the soul is laid down" (p. 7). *The spleen* was published posthumously in 1737, and was much admired by Pope and Gray. The Atwater copy is bound with the 1804 Philadelphia edition of John Armstrong's *The art of preserving health* (#154).

GREENE, Charles Warren see **YORKE-DAVIES** (#3900).

1408. GREENE, Cordelia Agnes, 1831-1905.
The art of keeping well or common sense hygiene for adults and children . . . With a biography by Elizabeth R. Gordon. New York: Dodd, Mead & Company, 1906.
xii, [2], 418 p., [2] leaves of plates : ports. ; 20 cm.

First edition. Born in Lyons, N.Y., a village on the Erie Canal east of Rochester, the precocious Greene earned her county teaching certificate at age sixteen. "More than her intellectual achievement, however, Greene's interest in applied Christianity provides the best clue to understanding her subsequent career. Undergoing religious conversion during one of the waves of religious fervor that periodically swept the region, she transformed her religious impulses into social commitment and the pursuit of a professional career. In 1849, when her father gave up farming to found a water-cure establishment in Castile, New York, Greene became his nurse and assistant. Her later approach to medical therapeutics would always bear the influence of her early exposure to health reform and hydropathy, movements that emphasized a person's ability to manage health through proper diet, exercise, fresh air, and a variety of physical therapy techniques using water" (R. Morantz-Sanchez in: *Amer. nat. biog.*, 9:522).

In the biographical essay included in this work, Gordon notes that after Jabez Greene established his Castile water-cure, his daughter immediately assumed the duties of nurse, at which point "her special adaptations for the medical profession were recognised" (p. 228). In 1853 Greene graduated from the newly established Woman's Medical College of Pennsylvania. Moving to Cleveland to assist at a large sanitarium, Greene continued her medical education and earned a degree from the Cleveland Medical College (1856). She returned to central New York where she spent the next six years as Henry Foster's (#1625) assistant at the Clifton Springs Water Cure (#669-672). After Jabez Greene died in 1864, Greene assumed control of the Castile Sanitarium, of which she was physician and proprietor for the next four decades. Numerous female physicians continued their training at Greene's sanitarium. Cordelia Greene was active in the women's suffrage movement, refusing to pay taxes one year to protest her inability to vote. She was involved in the WCTU, the New York State Medical Society, and the Women's Medical Society of the State of New York.

Underlying Greene's treatise on hygiene (i.e., exercise, diet, dress, etc.) is the importance of "mind over matter." *The art of keeping well* is as much a treatise on mental and spiritual well-being as the maintenance of physical health. For Greene, the latter is impossible without the former: "Control of the physical forces depends upon the ability of the individual to direct at will the vital currents where their good offices are needed. Tumultuous emotions caused by injured feelings, anger, sudden excitement, overwhelming fear, sense of impending danger, excessive weeping, turbulent passion, or hysteria send the living currents strongly to the brain, which for the time is nearly submerged" (p. 9).

The book is plainly directed toward a female readership, who are gently warned that disturbed equilibrium of the nervous system affects the well-being of the entire body. The book is knit together with holistic aphorisms, such as her remark early in the treatise: "The triumph of the soul which directs the brain and body until it moulds both to their noblest work, is for all who will seek it as their highest personal obligation" (p. 12).

1409. GREENE, Cordelia Agnes, 1831-1905. The art of keeping well or common sense hygiene for adults and children . . . With a biography by Elizabeth P. Gordon. New York: Dodd, Mead & Company, 1914.

xii, [2], 418 p., [14] leaves of plates : ill., ports. ; 20 cm.

1410. GREENE, Cordelia Agnes, 1831-1905. Build well. The basis of individual, home, and national elevation. Plain truths relating to the obligations of marriage and parentage . . . Boston: D. Lothrop and Company, [c1885].

ix, [3], 233 p. : ill. ; 19 cm.

"Are the unfortunate results of heredity a necessity?" Greene asks in her opening chapter (p. 1). She continues: "I am sure that many lamentable instances of unfortunate mental and physical deformity, cases of imperfect gift of health and vigor to soul and body, and many individuals with less of power than either parent, may be generally assigned to definite and controllable causes. So far as these harmful influences may be known and avoided, so far it becomes the interest of every parent and citizen to understand them" (p. 2). In her consideration of heredity, Greene discusses the mental, physical and biological influences that produce normal or deformed offspring. These encompass not only mental health, bodily hygiene and sexual physiology, but the proper understanding of love, marriage, gender differences & roles, and the rules of sexual behavior.

Greene writes: "Among the masses of people the sexual desire is looked upon as an uncontrollable impulse . . . but this is only true of its diseased or exaggerated conditions. In men or women where this morbid condition exists, we concede that parentage is not under rational control, which only only be obtained by almost superhuman effort, or by correct ideas and pure habits relative to this function" (p. 80). The consequences of acquiescing to the lower impulses have obvious results: "The bad qualities of parents, deceit, malice, selfishness and the like; their bad habits of stimulation and abuse of creative power,—must of necessity be given to children just in proportion as their conditions of soul and body have been controlled by them" (p. 81). "Every man and woman fairly constituted should attain such conduct of personal life, and such honorable direction of creative power, that no child of theirs could become a defaulter or thief because selfish greed had temporary or permanent possession of the parents' hearts when they gave him being; or become a drunkard or debauchee, because drunkenness, and its ever-consequent sensuality instigated the act which thrust upon him a blighted existence; or receive a murderous spirit by reason of the antagonism or hate which was dominant in his parents at his accidental conception" (p. 82).

1411. GREENE, Cordelia Agnes, 1831-1905.
Build well. The basis of individual, home, and national elevation. Plain truths relating to the obligations of marriage and parentage . . . Revised edition. Boston: Lothrop Publishing Company, [c1885].
   ix, [3], 233 p. : ill. ; 19 cm.

1412. GREENE, Cordelia Agnes, 1831-1905.
The Castile Sanitarium cook book . . . Castile, New York: Fred Norris, printer, 1902.
   101, [9] p., [1] leaf of plates : ill. ; 18 cm.

Greene includes recipes for meat dishes, but notes: "Persons subsisting largely upon meats have nervous excitability and intensity rather than enduring power . . . Those who eat vegetable foods freely such as seeds, grains, edible plants, and fruits are far less excitable and nervous and have much more reserve strength and endurance than those eating much of animal foods" (p. 5). Frontispiece photographic view of the Castile Sanitarium. In thick, pebbled wrapper.

GREENE, Frank A. see **GREENE** (#1413).

1413. GREENE, Jared Alonzo, 1845-1917.
Some health suggestions. [Boston: Drs. F.A. & J.A. Greene, 189-?].
   28 p. ; 20 cm.

Promotional literature for *Dr. Greene's Nervura* and other proprietary remedies available from Drs. F.A. & J.A. Greene. Analysis conducted by government chemists showed *Nervura* "to contain 18 per cent of alcohol, white celery, ginger and other unidentified vegetable material . . . Charges of falsehood and fraud in the therapeutic claims were brought by the government and Frank A. and Jarad [sic] A. Greene were each fined $25" (*Nostrums and quackery*, Chicago: A.M.A., 1921, 2:750). J.A.G. and F.A.G. were the sons of the eclectic physician Reuben Greene (#1415, 1416). J.A. Greene was an 1867 graduate of the Eclectic Medical Institute of Cincinnati; his brother F.A. Greene graduated from the same institution in 1875. Title from printed wrapper.

1414. GREENE, Jared Alonzo, 1845-1917.
What to do in cases of accident, emergencies or poisoning. Also how to cure all nervous diseases . . . New York and Boston, [c1889].
   30 p. : ill. ; 20 cm.

Promotional literature for *Dr. Greene's Nervura* and other proprietary remedies intermixed with advice on first aid and emergencies. In printed wrapper.

1415. GREENE, Reuben, 1817-1900.
The problem of health: how to solve it . . . Boston: Published by B.B. Russell; Philadelphia: Quaker City Publishing House; San Francisco: A.L. Bancroft & Co., 1876.
   4, [3]-10, 13-294, [2] p., [1] leaf of plates : port. ; 19 cm.

In the preface Greene states that his purpose is "to enlighten the people in all that pertains to their general well being; to instruct them in regard to the laws of life, that they may maintain good health; to educate their mental qualities, that they may be intellectually sound; to develop their moral natures, that they may become conversant with the higher duties of life; in short, to do all in our power to solve the great problem of health, and promote human happiness" (p. 3). Greene devotes chapters to the expected hygienic topics (e.g., sleep, exercise, dress), but also to such issues as "why women lose their health and beauty," marriage and "hereditary descent." "Preface to second edition" (p. [3]-4); frontispiece portrait of author.

Reuben Greene received his medical degree in 1850 from the New England Botanico-Medical College, an eclectic instititution organized in 1848 and operated between 1852 and 1859 as as the Worcester Medical College. Greene practiced in Boston for many years. The city business directories from 1857 to 1864 place him at 36 Bromfield Street. In 1860 Greene (hitherto Green) is listed as vice-president of the Indian Medical Institute (est. 1853), which used only "innocent medication" (i.e., no calomel). The Institute provided boarding facilities as well as consultation by mail. After the Civil War it was moved to Temple Place; and from 1875 to the end of the century was operated as the Boston Medical Institute. Greene's sons, Jared Alonzo Greene (#1413, 1414) and Frank A. Greene [Frank E. in *Polk*] joined the staff of the Institute sometime before 1884, and continued the family business after their father's death on 24 February 1900.

Reuben Greene was the author of at least two other books: *Indianopathy, or the science of Indian medicine* (14th ed., Boston, 1858) and *Thoughts for the people* (Boston, 1896). A work entitled *The medical compendium . . . A treatise on diseases and their treatment, designed for the use of invalids and families* is advertised on the final leaf of *The problem of health*. There is no bibliographic record of its publication, however.

1416. GREENE, Reuben, 1817-1900.
Skating Rink, Athol, Friday Evening, Oct. 3. '84. DR. GREENE'S NEW LECTURE, THE NERVOUS SYSTEM Magnificently Illustrated by the Finest Imported Stereopticon! With a Profusion of Superb Dissolving Views . . . The Public are Cordially Invited. ADMISSION FREE. Invalids and the Sick should by no means miss this Lecture. No boys admitted. [1884].
Broadside ; 24 × 15 cm.

The citizens of Athol, Ma. are assured that "These lectures are given in the interests of the new mode of treatment of diseases by harmless

Fig. 38. Reuben Greene, frontispiece to The problem of health, *#1415.*

vegetable remedies." Imprint at bottom of sheet: *Farwell, Pr., Boston.*

1417. LESTER H. GREENE COMPANY.
Household guide and ready reference book . . . Edition of 1895 and 1896. Montpelier, Vermont: The Lester H. Greene Co., [1895]. 32 p. ; 22 cm.

Title and imprint from printed wrapper. Promotional literature for remedies manufactured by the Lester H. Greene Company.

1418. GREENFIELD, William Smith, 1846-1919.
Alcohol its use and abuse . . . New York and London: D. Appleton and Company, 1912. 95 p. ; 15 cm.

Series: *Health primers; no. 2.*

GREENLEAF, William A. see *CANADIAN JOURNAL OF HOMEOPATHY* (#554).

1419. GREER, Joseph H., 1851-1928.
Marriage and fools . . . Chicago, Ills.,
[1900?].
  158, [2] p. : ill. ; 14 cm.

Greer attributes many of the medical and
sexual ills of the age to "nervous debility," which
affects not only parents, but is transmitted to
their offspring as well. "The observation of simpler laws of health . . . and the assistance of some
harmless vegetable remedies which a skillful
and experienced physician has discovered"
promises a cure for this malady. "Such a medicine has been provided by Mother Nature,"
Greer announces, "In the woods and fields are
found the harmless and powerful agencies for
the restoration of lost strength and vigor . . . I
am happy to be able to impart the glad tidings
. . . that I have discovered her simple, perfect
and harmless means of cure" (p. 7). Having
held out hope to his readers for the cure of
"seminal weakness," Greer also discusses venereal diseases, the medical consequences of
masturbation, and various urologic disorders—
all of which fall within the compass of Greer's
skill and practice, which is "limited to special
diseases." In printed wrapper.

1420. GREER, Joseph H., 1851-1928.
A physician in the house for family and
individual consultation. Containing valuable articles on life and its preservation,
the actions of the body in health and disease, the rules of hygiene and proper living, characteristics of food, etc. Also a
complete cyclopaedia of diseases and
their treatment by non-poisonous remedies with descriptions of medicinal
agents and numerous formulas and special articles written in plain language . . .
Chicago: Published by Dr. J.H. Greer,
[c1897].
  [2], xxv-xxviii, 29-816, xvii-xxviii p., [2],
XVI leaves of plates : ill. ; 23 cm.

Greer's domestic medicine was first published in 1897 (767 p.), and last appeared in
1938. Greer was born in Liverpool, England. It
is unclear when he came to the United States.
He received his medical degree in 1875 from
the (eclectic) Bennett Medical College (Chicago). For many years he was on the faculty of
the College of Medicine and Surgery (Chi-

cago), and a specialist in genito-urinary diseases.

1421. GREER, Joseph H., 1851-1928.
Sex science . . . Chicago, Ill.: Published by
the author, [c1911].
  153 p. ; 20 cm.

"Ignorance of that part of man's nature
which belongs to the sexual domain indicates a lack of education, the same as illit-
eracy. The man who is educated in regard to
these things exercises self-control, interest
in the welfare of others, love for humanity
and chivalry to woman. His mental and physical nature is brimming over with life, ambition and manly vigor. The purity of his mind
glows in his face and is a veritable well-spring
of happiness to himself and a benediction
to all who are so fortunate as to come under
its influence. Such men alone should be the
husbands and fathers, the propagators of the
human race and the leaders in all affairs of
city and state. They do not dishonor woman
by purchasing her virtue, but place her
where she rightfully belongs, occupying the
throne of Nature, queen of his home and
ruler of his heart" (p. 12). *Sex science* is largely
devoted to "abnormal sexual desire" and its
consequences (i.e., sexually transmitted diseases), as well as other disorders of the urogenital system.

1422. GREER, Joseph H., 1851-1928.
Talks on nature. Important information for
both sexes. A treatise on the structure,
functions, and passional attractions of men
and women. Anatomy, physiology and hygiene (a true marriage guide) . . . Chicago:
Published by J.H. Greer, [191-?].
  173, [19] p. : ill. ; 18 cm.

  Series: *Greer health club series; no. 1.*

1423. GREER, Joseph H., 1851-1928.
The wholesome woman. A home book of
tokology, hygiene and education for maidens, wives and mothers. A clean and clear
exposition of nature's laws and mysteries
. . . Chicago: The Athenaeum Publishing
House, [c1902].
  vi, [7]-510 p., [16] p. of plates + frontis. : ill.,
port. ; 22 cm.

1424. GREGORY, George.
Facts and important information from distinguished physicians and other sources. Second edition. Boston: D.S. King—Jordan & Co., 1842.
vi, [7]-64 p. ; 13 cm.

"The object of this little work is, to spread information in regard to the dangerous and, oftentimes, fatal consequences of the habit usually denominated Masturbation or Solitary Vice; a practice known to be exceedingly prevalent among all classes of the community, and most desolating in its effects on the body, mind, and moral principle" (*Pref.,* p. iv). The *Facts* was originally published by Gregory in 1841. In printed wrapper.

1425. GREGORY, Holly.
The Rubicon: a medical book on venereal, sensual, sexual, and other delicate or secret maladies of both sexes, with instructions for treatment and cure . . . Fourth edition. [New York]: No. 11 Barclay Street, [c1843].
144 p. : ill. ; 12 cm.

Gregory briefly describes the symptoms of sexually transmitted diseases and other disorders of the genito-urinary tract, including impotence, the consequences of masturbation, and female "barronness." He does not describe treatment, recommending instead that "the patient apply immediately to some physician of known ability." In this regard, he informs his readers that he can "be found at home at all hours of the day or evening," and assures them that "interviews will be characterized by gentleness and kindness, being always specially guarded against asking any questions or making any remarks in any way tending to wound the feelings or self respect of the most sensitive or timid" (p. 52). Gregory's therapeutic armamentarium consists of a variety of proprietary remedies (pills, injections, tonics, medicinal washes, vapor baths). Among his concluding remarks Gregory urges his readers, "Delay not an instant after the first symptoms of the disease to apply for medical assistance; let no false delicacy, no fear of expense deter you; for you need not hesitate to disclose the secret to one whose lot it unfortunately is, every day to wit-

ness the inroads made upon health by sexual indulgence" (p. 126).

Most of the remedies offered for sale by Gregory were priced from fifty cents to two dollars, except his *Female Monthly Pills* for curing suppressed menses, which were sold for $3.00. Under the guise of warning females about possible negative side effects, Gregory actually promotes his pills as an abortifacient: "Females who occasionally indulge in sexual intercourse, but particularly married ladies, should be extremely cautious in their use, on account of the possibility of being pregnant. For should this be the case, the irrepairable [sic] accident of abortion would almost assuredly take place" (p. 122).

The first edition of this work was published in 1839, and at least four editions appeared through 1843. Gregory's background and training are as yet unknown. In an 1846 newspaper advertisement, he claimed to be a regular physician and surgeon . . . of more than forty years experience." His name appears in the New York city directories with the appellation "M.D." or "physician" from 1839 until 1853, with offices at six successive locations.

GRIFFIN, Henry A. see **MORTON** (#2512).

GRIFFIN, William M. see **ATLANTA REMEDY CO.** (#169); **DAY** (#918); **GODFREY** (#1373); **J.W. KIDD CO.** (#2123, 2124).

1426. GRIFFITH, John Price Crozer, 1856-1941.
The care of the baby. A manual for mothers and nurses containing practical directions for the management of infancy and childhood in health and in disease . . . Philadelphia: W.B. Saunders, 1896.
8, 17-392 p., V leaves of plates : ill. ; 20 cm.

The first issue of this first edition was published in 1895. The second edition appeared in 1898, and the seventh and final edition in 1924. Griffith received his medical degree in 1881 from the University of Pennsylvania, where he joined the medical faculty in 1889 and became professor of pediatrics (1913). In 1919 Griffith first published *The diseases of infants and children,* which remained a standard pediatric textbook through the 1940s.

1427. GRIFFITH, John Price Crozer, 1856-1941.
The care of the baby. A manual for mothers and nurses containing practical directions for the management of infancy and childhood in health and in disease . . . Second edition, revised. Philadelphia: W.B. Saunders, 1898.
[4], 11-404 p., V leaves of plates, [1] fold. chart : ill. ; 21 cm.

1428. GRIFFITH, John Price Crozer, 1856-1941.
The care of the baby. A manual for mothers and nurses containing practical directions for the management of infancy and childhood in health and in disease . . . Fifth edition, thoroughly revised. Philadelphia and London: W.B. Saunders Company, 1912.
[2], 7-455 p., IV leaves of plates : ill. ; 21 cm.

1429. GRIFFITH, John Price Crozer, 1856-1941.
The care of the baby. A manual for mothers and nurses. Containing practical directions for the management of infancy and childhood in health and in disease . . . Sixth edition, thoroughly revised. Philadelphia and London: W.B. Saunders Company, 1915.
[5]-463 p. : ill. ; 21 cm.

GRIFFITH, Robert Eglesfeld see **THOMSON** (#3477).

GRIFFITTS, Samuel Powel see **BUCHAN** (#457, 463, 474, 477).

1430. GRINDLE, Henry Dyer, 1826-1902.
The female sexual system, or the ladies' medical guide . . . New York, 1864.
216 p. ; 16 cm.

CONTENTS: I. Elements of generation.—II. Phenomena of pregnancy.—III. Evidences of pregnancy.—IV. Signs of labor.—V. Progress of labor.—VI. After-treatment.—VII. Miscarriages

and premature labors.—VIII. Prevention of pregnancy.—IX. Diseases of women.—X. Diseases of women continued.

In his seventh chapter, Grindle considers the circumstances that justify abortion: when deformities of the pelvis prevent delivery or endanger the life of the woman; in cases where displacement causes the womb to enlarge in a transverse position across the cavity of the pelvis, endangering the life of the woman during pregnancy and making delivery dangerous or impossible; "when the placenta is partially detached from the womb" and hemorrhaging results; and in cases of "excessive and protracted nausea and vomiting during pregnancy" that reduce to the woman to a physical state in which she cannot withstand the demands of labor.

In chapter eight, "Prevention of pregnancy," Grindle points out the medical benefits that contraception affords women who might otherwise resort to abortion, and the benefit that it brings to the "multitudes of families, who have not the means to support and educate a large number of children" (p. 140). He notes that after male ejaculation the semen remains in the vaginal cavity and does not immediately enter the womb. The purpose of contraception, therefore, is "to remove the spermatozoa or vital element of the male semen from the vagina of the female, and . . . to dissolve or destroy them." To this end he recommends the injection of a spermicide into the vagina by means of a "vulcanized gutta-percha syringe manufactured with special reference to the anatomy of the female organs" (p. 162-66). The injection may be repeated "once, twice, or more times." The solution consists of "metallic salts" or "vegetable astringents" dissolved in five or six ounces of water. Grindle promotes his own anti-spermatic compound, "a powder a very little of which added to cold water proves to be an infallible protection against pregnancy" (p. 165). A year's supply of the powder and a syringe are available from the author for fifteen dollars. Spine-title: *Ladies medical guide.*

In reviewing the availability of illicit abortions in New England and the Middle Atlantic states, Janet Farrell Brodie discusses H.D. Grindle in particular: "In the early 1860s the Lying-In Institute at 6 Amity Place in New York City was operated by H.D. and Julia Grindle . . . the Grindles offered rooms for nursing mothers and unwed mothers and their advertising circulars made it clear that abortion facilities were available . . . In 1868, Grindle was indicted for performing an abortion on a woman who subsequently died. The trial ended in his acquittal with only a censure from the judge because the prosecution proved only that the woman died at his institute during childbirth and not during or because of an abortion. In 1872 both Grindles were indicted in New York for abortion, but they were found not guilty because the young woman who charged them with selling her a twenty-dollar bottle of medicine to procure abortion admitted that she had not told the couple she was pregnant" (*Contraception and abortion in nineteenth-century America,* Ithaca, N.Y.: Cornell Univ. Press, 1994, p. 228-29).

Several members of the Grindle family practiced medicine in New York City between 1857 and 1912, including Henry Dyer Grindle, his wife Julia, and their son, John Wesley Grindle. The 1857/58 city directory lists Josiah S. Grindle, Henry's younger brother, as sole proprietor of the firm that manufactured *Dr. Wesley Grindle's Celebrated Magic Compound* and *Grindle's Ancient and Celebrated Japanese Life Pills* (Wesley Grindle was an older brother). The following year H.D. Grindle is listed as a clerk with the firm, and Wesley Grindle (M.D., 1856, U.C.N.Y.) as a physician. J.S. Grindle had become a "broker." By 1864, the year *The female sexual system* was published, Henry Grindle is listed as a physician, though he did not receive his medical degree until 1867 (U.C.N.Y.). After 1881, Grindle practiced with his son John Wesley Grindle at 171 W. 12th Street until his death on 14 Sept. 1902. That same year, the younger Grindle was brought into Police Court following accusations by the County Medical Society that he practiced without proper credentials. He was a graduate of the American University and Eclectic Medical College of Philadelphia, a diploma mill operated by John Buchanan (#504). State law required that an out-of-state degree be endorsed by a local medical school—endorsement that in Grindle's case was forthcoming from the Eclectic Medical College of the City of New York.

1431. GRINDLE, Wesley.
A late important discovery! in medicine. Consumption positively cured . . . Please read, then hand to those it calls for. New York, 1855.

12 p. ; 19 cm.

Title and imprint from printed wrapper. On the first page of this pamphlet Grindle announces "that we have made a discovery which proves to be a *most effectual cure for consumption.* Whether originating in pulmonary tuberculosis, scrofulous diathesis, or cahectic habit of body, it *does* cure this frightful disease." Not that Grindle's earlier remedy for the treatment of consumption was ineffective—but "it lacked one important ingredient, which, after much thought and study, has been supplied" (p. 2). Grindle's *Magic Compound* was available for $3.00 per box at Grindle's Depot on White St., off Broadway, or through the mail postpaid. This pamphlet was reissued in 1856 under the title: *Important announcement: pulmonary consumption positively cured.*

1432. GRISCOM, John Hoskins, 1809-1874.
Animal mechanism and physiology; being a plain and familiar exposition of the structure and functions of the human system. Designed for the use of families and schools . . . New-York: Harper & Brothers, 1839.

[2], xii, [13]-357 p. : ill. ; 16 cm.

Griscom's popular treatise on human anatomy and physiology was first published in this 1839 New York edition, and reissued in 1840 (#1433), 1842 (#1434), 1844, 1846 (#1435), 1848, 1855 and 1858. Series: *The Family library, no 85.*

Griscom's devotion to public health and sanitary reform is commonly attributed to the philanthropic atmosphere of his Quaker upbringing. An 1832 medical graduate of the University of Pennsylvania, Griscom returned to New York in 1833 and was appointed assistant physician at the New York Dispensary. This was the first of the positions Griscom held from which he could survey the sanitary condition of his native city, and the condition of its poor inhabitants particularly. During the course of his career, Griscom was professor of chemistry at the New York College of Pharmacy (1836-38); City Inspector (1842); physician to the New York Hospital (1843-70); and a physician of the New York State Commissioners of Emigration

(1848-51). His contributions to public health reform in New York City are too numerous to outline here. Griscom was a founding member of the New York Academy of Medicine (1846), and an influential member of the American Medical Association, the National Quarantine and Sanitary Conventions, the New York Prison Association, etc. In 1835 Griscom married Henrietta Peale, daughter of the painter Rembrandt Peale.

1433. GRISCOM, John Hoskins, 1809-1874.
Animal mechanism and physiology; being a plain and familiar exposition of the structure and functions of the human system. Designed for the use of families and schools . . . New-York: Harper & Brothers, 1840.
 [2], xii, [13]-357 p. : ill. ; 16 cm.

 Series: *The Family library, no. 85.*

1434. GRISCOM, John Hoskins, 1809-1874.
Animal mechanism and physiology; being a plain and familiar exposition of the structure and functions of the human system. Designed for the use of families and schools . . . New-York: Harper & Brothers, 1842.
 [2], xii, [13]-357 p. : ill. ; 16 cm.

 Series: *The Family library, no. 85.*

1435. GRISCOM, John Hoskins, 1809-1874.
Animal mechanism and physiology; being a plain and familiar exposition of the structure and functions of the human system. Designed for the use of families and schools . . . New-York: Harper & Brothers, 1846.
 [2], xii, [13]-357 p. : ill. ; 16 cm.

 Series: *The Family library, no. 85.*

1436. GRISCOM, John Hoskins, 1809-1874.
The sanitary condition of the laboring population of New York. With suggestions for its improvement. A discourse (with additions) delivered on the 30th December, 1844, at the repository of the American Institute . . . New York: Harper & Brothers, 1845.
 [4], 58 p. ; 22 cm.

This essay was originally presented to the Mayor of New York "with a view to an exposition of the true principles which should regulate the action of public bodies, in matters relating to the health of cities" (*Pref.*). It was offered to the public only after the original manuscript was returned to Griscom as unactionable by the Mayor. In her biographical essay on Griscom, Carolyn Shapiro calls this slender treatise in printed wrappers his most important work: "Modeled on Edwin Chadwick's work on Great Britain, this report correlated the higher morbidity rate among the laboring class with their overcrowded, unventilated tenement living conditions" (*Amer. nat. biog.*, 9:636).

 James C. Whorton writes of this treatise: "The sanitary reform movement of mid-century had been to a significant degree a project of moral, in addition to medical, rescue. Its leaders had expanded the evangelical equation of ungodliness with uncleanliness to introduce factors of crowding and poverty and to calculate the moral and physical damage done by miserable living conditions. New York's John Griscom exemplified the sanitationist spirit, introducing his survey of the sanitary condition of the city's laboring population by asking what was the effect of 'this degraded and filthy manner of life' upon not only health and lifespan of slum residents, but also their 'morals, their self-respect, and appreciation of virtue.' His summation for the implementation of thorough-going public sanitation had as a critical link the argument that a clean population in a decent environment would not be guilty of nearly so much lawlessness, and therefore be less difficult and expensive to govern" (*Crusaders for fitness*, Princeton, N.J.: Princeton Univ. Press, 1982, p. 145). "Unlike [Lemuel] Shattuck, Griscom's efforts bore little direct fruit, but he was a persistent gadfly, irritating municipal officials with his revelations . . . nagging state officials, and, through lectures and publications, constantly seeking to awake the public conscience" (John Duffy, *The sanitarians*, Urbana: Univ. of Illinois Press, 1990, p. 97).

1437. GRISCOM, John Hoskins, 1809-1874.
The uses and abuses of air: showing its influence in sustaining life, and producing disease; with remarks on the ventilation of houses, and the best methods of securing a pure and wholesome atmosphere inside

of dwellings, churches, court-rooms, work-shops, and buildings of all kinds . . . Third edition. New York: Redfield, 1854.
252 p., [12] leaves of plates : ill. ; 20 cm.

The uses and abuses of air appeared in three editions between 1848 and 1854.

GROSS, James E. see **RUDDOCK** (#3044, 3045).

GROSVENOR, Benjamin, 1676-1758. *The mourner; or, the afflicted relieved* see **WILLISON** (#3814).

1438. GROVE HALL UNIVERSALIST CHURCH (BOSTON).
A friend in need is a friend indeed. [Boston: Grove Hall Universalist Church, 1888].
68 p. : ill. ; 23 cm.

Advice on common diseases and their domestic management are intermixed with kitchen recipes and advertisements for baking powder, proprietary medicines, soaps, plumbing & heating contractors, furniture, etc. "The pecuniary object of this publication is to realize a fund for the purchase of an adequate church organ" (p. 4). In printed wrapper.

1439. GROVES, L.
New marriage guide and book of nature. Reproductive physiology and new marriage guide. Being a complete guide, physical, mental and moral in all that appertains to the relation of the sexes. New edition, revised and enlarged, and illustrated with fine engravings of the organs of which it treats, and expressly prepared to be sent through the mails, sealed and under letter postage . . . New York, 1867 [i.e., 1869].
[8], [vii]-viii, [vii]-xii, ix-x, [11]-70, [2], 71-270, [1], 280-414 p. : ill. ; 11 cm.

The title-page is dated 1867; the imprint on the front wrapper, however, reads: *New York: Hilton & Syme, 1869.* The *New marriage guide* is divided into five parts, the first three of which comprise the bulk of the text. The first part includes descriptions of the reproductive organs (illustrated with wood-engraved sagittal sections of the male and female abdomens), a review of the prevalent theories of conception,

the "physiology of love" (i.e., the nature of sexual attraction, behavior, &c.), the "philosophy of marriage" (i.e., selecting a partner with regard to the transmission of hereditary characteristics), and a discussion of the signs and management of pregnancy. The second part is devoted to the "diseases of the reproductive functions," i.e., organic disorders, sexually transmitted diseases, and the consequences of abnormal sexual behavior, including a lengthy discussion of masturbation. The third part covers such "miscellaneous subjects" as the harmful effects of lifelong sexual continence, the proper age for marriage, the "proper time for sexual indulgence," &c. Groves does not discuss contraceptive methods, other than a brief discussion of the causes of abortion that might be interpreted as suggestions "how to" abort spontaneously. In the Atwater copy, pages 183-212 are mistakenly repeated in the collation in place of pages 213-44, which are missing. Cover-title: *Dr. Groves' reproductive physiology and new marriage guide.*

1440. GROVES, L.
New marriage guide and book of nature. Dr. Groves' reproductive physiology and new marriage guide. Being a complete guide, physical, mental and moral, in all that appertains to the relations of the sexes. New edition, revised and enlarged, and illustrated with fine engravings of the organs of which it treats, and expressly prepared to be sent through the mails, sealed and under letter postage. [New York: Published for the trade], [c1872].
x, [2], [11]-18, [2], 19-140, [2], 141-212, [2], 213-266, [279]-302, 307-308, [20] p. ; 13 cm.

The *New marriage guide* was originally copyrighted by Cook & Co. in 1858. The copyright was renewed in this 1872 edition by W.E. Hilton, of New York. Hilton had published at least one earlier edition of this title in 1867 (probably the original issue of #1439). Imprint from printed wrapper.

1441. GUERBER, Helene Adeline, d. 1929.
Yourself and your house wonderful . . . Philadelphia, Pa.: The Uplift Publishing Company, [c1913].
[6], vii, [1], 301 p., [5] leaves of plates : ill. ; 24 cm.

Fig. 39. "Bed clothes should be aired daily," from Guerber's Yourself and your house wonderful, *#1441.*

Originally published in 1902 under the title *Yourself,* Guerber's juvenile physiology was published as *Yourself and your house wonderful* in 1913, 1925 and 1940. Guerber was the prolific author of popular and juvenile books—primarily on mythology and history. *Yourself* was her only venture into juvenile physiology and hygiene.

1442. GUERNSEY, Alfred Hudson, 1824-1902.

Health at home. By A.H. Guernsey and Irenaeus P. Davis . . . New York: D. Appleton and Company, 1884.
155, [5] p. ; 19 cm.

The authors consider such topics as "privies and water closets," ventilation, water supply, lighting, heating, disinfection, and elements of personal hygiene (food & clothing). They also include home remedies for basic family ailments such as stomach aches, nosebleeds, cuts, bruises, etc. Series: *Appletons' home books.*

Guernsey was the author of popular history, geography and biography. His *Harper's pictorial history of the Great Rebellion* (1st ed., 1866), published in two folio volumes, was his greatest success. Irenaeus P. Davis, was an 1871 graduate of the Bellevue Hospital Medical College, and the author of *Hygiene for girls* (#907.1).

1443. GUERNSEY, Alice Margaret, b. 1850.

Skeleton lessons in physiology and hygiene . . . Chicago and Boston: The Interstate Publishing Company, [c1886].
31 p. : ill. ; 18 cm.

Guernsey's *Skeleton lessons* are intended as an introduction to the basic concepts of physiology and hygiene for grammar school students. In her introductory remarks Guernsey writes: "In most high schools . . . physiology has been taught to a greater or lesser extent; but until lately, few superintendents or teachers have introduced it into the lower grades." She points out the importance of reaching younger students by noting that "in many states ninety-five per cent of the children never enter the high school, and fifty per cent do not go beyond the primary school." Thus, "the value to the homes and civilization of the future, of teaching this subject in the lower grades, becomes self-evident" (p. 4).

1444. GUERNSEY, Egbert, 1823-1903.

The gentlemen's hand-book of homoeopathy; especially for travelers, and for domestic practice . . . Boston: Published by Otis Clapp; New York: William Radde, 1855.
xii, [13]-255 p. ; 20 cm.

First edition. "Designing this work especially for gentlemen, it has been our object to make plain those laws of their being which will enable them to ward off disease, and shunning vice and its fearful consequences, harmonize their passions, and make them not alone healthier, but better. Believing much of the disease in the world is the result of improper marriages, we have introduced some important facts upon that subject" (*Pref.,* p. iv). Guernsey's manual is divided into two parts. Part one occupies the first hundred pages, and is devoted to hygiene and to the influence of climate upon health. Part two contains fourteen chapters addressing diseases of the respiratory organs, fevers, cutaneous diseases, affections of the brain and nervous system, etc. A second edition of this work was published in 1857 (#1445).

Guernsey received his medical degree from the University of City of New York in 1846, but within several years of graduation became convinced of homeopathy's therapeutic superiority. By 1850, Guernsey was established in New York City as a homeopathic practitioner. In 1853 he published the first edition of one of the most successful manuals of popular homeopathy of the second half of the 19th century (#1446). Guernsey was a founder of the New York Homoeopathic Medical College (est. 1860), was influential in the establishment of various homeopathic charities and hospitals in the area of New York City, and edited several homeopathic periodicals. Later in his career Guernsey's tolerant attitude toward allopathy antagonized more orthodox homeopaths. He was censured in 1890 by the New York County Homoeopathic Society.

1445. GUERNSEY, Egbert, 1823-1903.
The gentlemen's hand-book of homoeopathy; especially for travelers, and for domestic practice . . . Second edition. New York: William Radde . . . [*et al.*], 1857.
  xii, [13]-255 p. ; 19 cm.

Spine-title: *Hand-book of homoeopathy.*

1446. GUERNSEY, Egbert, 1823-1903.
Homoeopathic domestic practice, containing also chapters on anatomy, physiology, hygiene, and an abridged materia medica . . . New-York: William Radde, 1853.
  xi, [1], 588 p., 5 leaves of plates : ill. ; 19 cm

First edition. "Guernsey's first medical work, *Homoeopathic domestic practice,* appeared in 1853 at a time when there were about fifteen other popular homeopathic manuals . . . Guernsey's was chosen by William Holcombe, an editor of the *North American Journal of Homoeopathy,* as one of the best for its 'superior literary finish.' By 1890 the book had gone through eleven editions and had been translated into four languages" (*Amer. nat. biog.,* 9:697). The final issue of the 11th edition appears to have been published at Philadelphia in 1900.

1447. GUERNSEY, Egbert, 1823-1903.
Homoeopathic domestic practice, containing also chapters on physiology, hygiene, anatomy, and an abridged materia medica . . . Tenth enlarged, revised and improved edition. New York; San Francisco: Boericke & Tafel; Philadelphia: F.E. Boericke, 1874.

xi, [3], 653 p., 5 leaves of plates : ill. ; 21 cm.

1448. GUERNSEY, Henry Newell, 1817-1885.
Plain talks on avoided subjects . . . Philadelphia, Penn'a: F.A. Davis, Att'y, publisher, 1882.
  126 p. ; 16 cm.

First edition. A homeopath, Guernsey was nonetheless an 1844 graduate of the medical department of the University of the City of New York. He published several books on obstetrical and gynecologic topics, the most successful of which was *The application of the principles and practice of homoeopathy to obstetrics* (Philadelphia, 1867). *Plain talks* was first published at Philadelphia in 1882, and subsequently in 1889, 1891, 1899 and 1915. Nearly two thirds of the text addresses the pitfalls that children and adolescents face in order to remain chaste until marriage, and the consequences that result from the "titillation" of sliding down bannisters, the habitual use of vaginal douches, the sexual predations of servants, etc. The remainder of the book is devoted to conduct in marriage.

1449. GUERNSEY, Henry Newell, 1817-1885.
Plain talks on avoided subjects . . . Philadelphia; New York; Chicago: The F.A. Davis Company, publishers, 1899.
  126 p. ; 17 cm.

1450. GULICK, Charlotte Vetter, b. 1865.
Emergencies . . . Boston; New York; Chicago; London: Ginn & Company, [c1909].
  xiv, 173, [5] p. : ill. ; 19 cm.

"A clipping bureau has for nine months furnished the author with newspaper accounts of accidents to children. These have been tabulated and analyzed. Decision as to the classes of accidents which should and which should not receive particular emphasis has been reached by a careful study of these records, supplemented by the lessons of a large range of personal experience" (*Pref.,* p. v). Gulick addresses a juvenile audience on the causes of the accidents to which they are most liable (e.g., handling firearms, playing in the streets, getting on and off moving streetcars, etc.) and how emergencies may be avoided or treated. *Emergencies* is the second of the seven-volume *Gulick hygiene series* edited by Luther H. Gulick (#1451-1454). The other titles in the series were all written by Frances Gulick Jewett (#2012-2015). Charlotte Vetter Gulick was the wife of Luther Halsey Gulick.

1451. GULICK, Luther Halsey, 1865-1918.
The bicycle as a therapeutic agent . . . Reprinted from the Boston Medical and Surgical Journal, January 14, 1904. [S.l.: s.n., 1904].
    18, [2] p. ; 10 × 13 cm.

Gulick advocates the bicycle as "a useful therapeutic agent in those cases in which it is desirable to quicken the general organic functions of circulation, of respiration, of the whole digestive tract, together with ample diversion and out-of-door air" (recto of final leaf). Title from printed wrapper.

1452. GULICK, Luther Halsey, 1865-1918.
The efficient life . . . Edition of Young Men's Christian Association Press. New York: Doubleday Page & Company, 1907.
    xvi, 195 p., [2] leaves of plates : ill. ; 19 cm.

Gulick's treatise examines the "conditions of efficiency" that enable us to "run our machines on the higher levels of quality-efficiency" without depleting our "nerve batteries." Gulick states: "It is not the intention of this book to provide an easy recipe for the development of genius. What it seeks is to enable each man to discover and secure for himself the best attainable conditions for his own daily life. It aims to apply to the various details of that life our present knowledge of physiology and psychology in a common sense and practical way" (p. 11-12). Gulick discusses "states of mind and states of body," exercise, diet and digestion, fatigue, sleep, "the bath—for body and soul," ocular hygiene, etc.
    James C. Whorton uses Gulick as an example of the hygienic ideologists of the Progressive Era: "Hygienists could approach the body as health accountants, evaluating function in terms of deposits of food and rest, and withdrawals of exertion and self-neglect. The quantity, and

quality, of the deposits and withdrawals would determine the degree of physiological wealth, and wise management would yield efficiency of operation—maximum production with minimum waste. And since increased capacity for work improved one's chances for success, physical efficiency actually did translate itself into fiscal efficiency . . . The anticipated product of physiological efficiency was . . . strong and energetic, intellectually agile, morally upright, and dedicated to lifting all fellow creatures to his level . . . Progressive health reformers spoke a *lingua franca* in which the crucial adjectives were natural, harmonious, evolutionary, successful, and efficient" (*Crusaders for fitness*, Princeton: Princeton Univ. Press, 1982, p. 167).
    Gulick received his medical degree from New York University in 1889. His primary interest, however, was the field of physical education. Even before receiving his medical degree, Gulick had been involved in the YMCA, and continued his work with that organization after graduation. "At the outset of Gulick's career the YMCA combined gymnastics and physical exercise with evangelical teachings and religious activities. He led the organization to embrace competitive sports as a means to cultivate and display Christian character. Although he had little experience as an athlete, he hammered out a biological, philosophical and ethical rationale for competetive athletics. Appropriately, his guiding hand lay behind the creation of one of the world's major sports. In 1891 he required a Springfield college graduate class to imagine a new indoor game to supplement tedious gymnastic exercises, and from that assignment came James A. Naismith's invention of basketball" (*Amer. nat. biog.*, 9:720).
    In 1900 Gulick left the Springfield, Mass. YMCA to become principal of the Pratt High School in Brooklyn. He subsequently become president of the American Physical Education Association; director of physical training for the New York City schools; active in the international Olympic movement; one of the founders of the Playground Association of America (1906); director of the children's hygiene division of the Russell Sage Foundation (1908-13); lecturer at the New York School of Pedagogy (1905-09); a founder of the Boy Scouts of America (1910) and the Camp Fire Girls (1912). He was the author of numerous books, many of which were directed toward a popular audience.

1453. GULICK, Luther Halsey, 1865-1918.
The healthful art of dancing . . . Garden City [N.Y.]; New York: Doubleday, Page & Company, 1911.
    xi, [3], 273 p., [64] p. of plates : ill. ; 19 cm.

"While our American nation includes in it representatives from most of the peoples under the sun, we possess less of the folk music, the folk dances, folk lore, folk games, folk festivals of the world than do any of the peoples of which we are made. The reason lies partly in that these folk expressions are social inheritances carried by a community as a whole; when individuals migrate the social customs are lost" (p. 5). Gulick describes efforts to integrate folk dancing into the physical training of New York City school children, and notes its beneficial effects on blood circulation, respiration, digestion and "carriage. "

1454. GULICK, Luther Halsey, 1865-1918.
Physical education by muscular exercise . . . Philadelphia: P. Blakiston's Son & Co., [c1904].
    vii, [1], 67 p. : ill. ; 22 cm.

Originally published in volume VII of *A System of physiologic therapeutics: a practical exposition of the methods, other than drug-giving, useful in the treatment of the sick*, edited by S. Solis-Cohen and published at Philadelphia by Blakiston in eleven volumes between 1901 and 1905.

***GULICK HYGIENE SERIES* see GULICK**
(#1450); **JEWETT** (#2011.1-2015).

1455. GULLY, James Manby, 1808-1883.
The water cure in chronic disease: an exposition of the causes, progress, and terminations of various chronic diseases of the digestive organs, lungs, nerves, limbs, and skin; and of their treatment by water, and other hygienic means . . . New York: Wiley & Putnam, 1846.
    xi, [1], 405 p., [1] leaf of plates : ill. ; 21 cm.

First American edition. Gully's treatise on hydropathy appeared in eleven London editions between 1846 and 1867. "My object in publishing this treatise," Gully states in his preface, "is to afford a truthful and rational exposition of the value of the water treatment in certain chronic diseases. I apprehend that it is wanted, because the works on the subject of the water cure which have hitherto appeared in this country contain . . . much overstatement as to its operation, and are written rather to catch the hopeful invalid, than to enlighten him as to the nature of his disease, or the mode in which the water plan is to relieve it . . . In the First Part of this work the origin, progress, extension, and terminations of Chronic Disease in general, are delineated and explained . . . In the Second Part, this is further developed in the history of individual chronic diseases, the explanation of the pathology and symptoms of each of which is given, and also of the reasons for the water treatment applicable to each . . . Part Third treats, in the first place, of the mode in which the water cure operates in producing its beneficial results . . . in the second place, the details of the water cure are brought forward, the rationale of each process given, and the circumstances which regulate their application stated" (p. [v]-vi).

Gully was a native of Jamaica, where his father owned extensive coffee plantations. He was sent to England to be educated, and studied medicine at both Edinburgh and Paris before receiving his medical degree from the University of Edinburgh in 1829. The following year he set up practice in London. "About 1837 he made the acquaintance of James Wilson, with whom he agreed that the old routine of medication was 'effete and inefficient, if not positively harmful.' This sceptical attitude sent them both searching for a better system. In 1842 Wilson returned from the continent 'filled to the brim' with hydropathy, and convinced his friend of the wonderful power of water treatment both in acute and chronic diseases. They selected Malvern as a locality for the practice of hydropathy, and settled there" (*Dict. nat. biog.*, 8:778). Together Wilson and Gully wrote *The practice of the water cure*, the first American edition of which was published in 1847 (#3830).

1456. GULLY, James Manby, 1808-1883.
The water cure in chronic disease: an exposition of the causes, progress, and terminatons of various chronic diseases of the digestive organs, lungs, nerves, limbs, and skin; and of their treatment by water,

and other hygienic means . . . New York: Wiley & Putnam, 1847.

xi, [1], 405 p., [1] leaf of plates : ill. ; 20 cm.

No other work by a British hydropath was as successful on the American market as Gully's *The water cure in chronic disease*. Following the first American edition of 1846 (#1455), it was reissued by Wiley in in 1847, 1849, 1850 and 1854; and by S.R. Wells in 1873 (#1457) and 1880.

1457. GULLY, James Manby, 1808-1883.
The water-cure in chronic diseases; an exposition of the causes, progress, and terminations of various chronic diseases of the digestive organs, lungs, nerves, limbs, and skin; and of their treatment by water, and other hygienic means . . . New York: Published by Samuel R. Wells, 1873.

xi, [1], 405 p., [1] leaf of plates : ill. ; 19 cm.

GULLY, James Manby see also **WILSON** (#3830).

1458. GUNN, John C., 1795?-1863.
Gunn's domestic medicine. A facsimile of the first edition. With an introduction by Charles E. Rosenberg. Knoxville: The University of Tennessee Press, [c1986].

xxi, [1], 440 p. ; 22 cm.

Facsimile of the first edition (series: *Tennesseana editions*). Gunn was born in Savannah, Ga., which he left circa 1816 to read medicine with an unidentified Virginia physician. As was common among physicians of the period, Gunn's medical training was limited to his preceptorship. In 1827 he moved to Knoxville, Tenn., where he established himself in a prosperous medical practice, and where in 1829 he ran unsuccessfully for a seat in the state senate. After his defeat, Gunn moved his family fifty miles south of Knoxville to Madisonville, Tenn. Thenceforth, Gunn's primary professional interest lay in the publication of successive editions of his eminently successful manual of domestic medicine.

Gunn's domestic medicine, or poor man's friend first appeared at Knoxville in 1830. It became the most successful non-sectarian domestic medical book of its era—gradually supplanting Buchan's *Domestic medicine* (#442-502) in American homes. In considering the reasons for the book's immediate success, Rosenberg writes in his introduction: "The Knoxville practitioner was able to frame his clinical advice in a rhetoric appropriate to its purchaser's expectations, package it in a convenient form, and, most importantly, endorse a style of practice congenial to the habits and presumptions of his Trans-Appalachian contemporaries" (p. vi). His emphasis on medical self-reliance and the medicinal use of locally grown roots and herbs certainly appealed to a readership that was often rural and without access to professional medical assistance. It would also have been of obvious value to those who either could not afford physicians' and pharmacists' fees, or who were imbued with an innate distrust of established authority, medical or otherwise.

Rosenberg points out in his introduction that although Gunn was regularly trained and made frequent references in his book to the opinions of Benjamin Rush and other East Coast medical authorities, he made equally frequent "appeals to the wisdom of the common people, to the beneficence and plenitude of God, to fears of academic learning, [and] to lay suspicion of medical knowledge and motives" (p. vii). Although these ideas had a lineage in popular medical literature dating back to the 17th century, Gunn placed them "in a context of aggressive nationalism" (p. viii). Herbal remedies were accessible in local woods and meadows, obviating the need to import costly drugs from foreign countries or to rely on complex prescriptions written in Latin. Furthermore, knowledge of the availability and use of medicinal plants grown in native soil was available to all.

Rosenberg also notes that unlike Buchan and many other authors of domestic medical literature, "Gunn treated almost every aspect of medicine as appropriate for lay practice. No longer did it seem necessary to define careful boundaries between lay and professional spheres and to place certain areas of practice off-limits, areas into which the nonprofessional could stray only at his or her peril" (p. x). Rosenberg also notes the transgression of class boundaries in Gunn's book. Whereas "Buchan and Tissot . . . presented their works in the guise of advice to members of the upper orders who would be expected to provide aid to their less privileged and 'intelligent' neighbors and tenants ... Gunn was burdened by no such qualms" (p. xviii). His book was intended for the self-reliant common man who was pushing back the frontiers of Jacksonian America.

Gunn's approach to therapeutics was decidedly "heroic." He advocated immediate and aggressive intervention in disease, which included copious bleeding and the use of large doses of drugs for purging and puking. Ironically, the drugs most used for these purposes (e.g., mercury, ipecac, senna) had to be imported. Gunn describes opium as "the monarch of medicinal powers," noting that "Without this valuable and essential medicine, it would be next to impossible for a physician to practice his profession" (p. 401). He regarded Turkish opium as the best variety, but was confident that the white poppy would flourish "in the more southern and western portions of the Union" (p. 402). Gunn's was a domestic form of the therapeutics against which much of the medical reform of the 1830s and 1840s was directed—including the Thomsonians, whose regional and class appeal would most have competed with Gunn's intended market.

Gunn's domestic medicine is divided into six parts: an examination of "the passions" and their effect on health (p. 19-94); hygiene, i.e., sleep, exercise, bathing, food (p. 103-28); the symptoms and treatment of specific diseases (p. 128-88); the diseases of women and children, including pregnancy and childbirth (p. 289-359); medicinal plants (p. 360-408); and therapeutic techniques such as sulphur baths, bleeding, clysters, friction, etc. (p. 409-40).

1459. GUNN, John C., 1795?-1863.
Gunn's domestic medicine, or poor man's friend. Shewing the diseases of men, women and children, and expressly intended for the benefit of families. Containing a description of the medicinal roots and herbs, and how they are to be used in the cure of diseases. Arranged on a new and simple plan ... Second edition. Knoxville, T.: Printed by F.S. Heiskell, 1833.
xviii, [2], 552 p., [2] leaves of plates : ill. ; 20 cm.

1460. GUNN, John C., 1795?-1863.
Gunn's domestic medicine, or poor man's friend. Shewing the diseases of men,

women and children, and expressly intended for the benefit of families. Containing a description of the medicinal roots and herbs, and how they are to be used in the cure of diseases. Arranged on a new and simple plan . . . Second edition. Madisonville, Tenn. : Printed by J.F. Grant, 1834.
xv, [1], 604 p., [1] leaf of plates : ill. ; 20 cm.

The text added to the second edition (p.[529]-574) consists largely of a surgical supplement that includes sections on accidents, wounds, fractures, dislocations and amputation; and a dissertation on the "epidemic cholera" (p.[577]-600) that appeared in the New England and Mid-Atlantic states in 1832. There are two copies of the 1834 Madisonville second edition in the Atwater Collection, one of which includes a hand-colored engraved plate ("A view of the heart," opposite p. [140]). This plate is not present in the other copy. It would appear that the Knoxville and Madisonville editions of Gunn published in the 1830s might be issued with one, two, or no plates at all.

1461. GUNN, John C., 1795?-1863.
Gunn's domestic medicine, or poor man's friend, in the hours of affliction, pain and sickness. This book points out, in plain language, free from doctor's terms, the diseases of men, women, and children, and the latest and most approved means used in their cure, and is expressly written for the use of families in the Western and Southern states. It also contains descriptions of the medicinal roots and herbs of the Western and Southern country, and how they are to be used in the cure of diseases: arranged on a new and simple plan, by which the practice of medicine is reduced to principles of common sense . . . Fourth edition. Madisonville [Tenn.]: Printed at the office of Henderson & Johnston; Edwards & Henderson—printers, 1834.
xv, [1], 604 p. ; 21 cm.

1462. GUNN, John C., 1795?-1863.
Gunn's domestic medicine, or poor man's friend, in the hours of affliction, pain and sickness. This book points out, in plain lan-

guage, free from doctor's terms, the diseases of men, women, and children, and the latest and most approved means used in their cure and is expressly written for the use of families in the Western and Southern states. It also contains descriptions of the medicinal roots and herbs of the Western and Southern country, and how they are to be used in the cure of diseases: arranged on a new and simple plan, by which the practice of medicine is reduced to the principles of common sense . . . Fourth edition. Madisonville [Tenn.]: Printed at the office of Henderson & Johnston; Edwards & Henderson—printers, 1835.

xv, [1], 604 p., [2] leaves of plates : ill. ; 20 cm.

1463. GUNN, John C., 1795?-1863.
Gunn's domestic medicine, or poor man's friend; describing, in plain language, the diseases of men, women, and children, and the latest and most approved means used in their cure, and is intended expressly for the benefit of families in the Western and Southern states. It also contains descriptions of the medical roots and herbs of the Western and Southern country, and how they are to be used in the cure of diseases. Arranged on a new and simple plan, by which the practice of medicine is reduced to principles of common sense . . . Tenth edition. Xenia, Ohio: Printed and published by J.H. Purdy, 1838.

768 p. ; 23 cm.

1464. GUNN, John C., 1795?-1863.
Gunn's domestic medicine, or poor man's friend, in the hours of affliction, pain and sickness. This book points out, in plain language, free from doctors' terms, the diseases of men, women, and children, and the latest and most approved means used in their cure, and is intended expressly for the benefit of families in the Western and Southern states. It also contains descriptions of the medicinal roots and herbs of the Western and Southern country, and how they are to be used in the cure of diseases. Arranged on a new and simple plan, by which the practice of medicine is reduced to principles of common sense . . .

Twelfth edition. Louisville [Ky.]: Published by Charles Pool, 1838.

768 p. ; 22 cm.

The name J.H. Purdy, of Xenia, Ohio, appears on the verso of the title-page of this 12th edition. Purdy's name is associated as printer and/or publisher of a 768 page edition of *Gunn's domestic medicine* issued in several cities between 1837 and 1840.

1465. GUNN, John C., 1795?-1863.
Gunn's domestic medicine, or poor man's friend, in the hours of affliction pain and sickness, this book points out in plain language free from doctor's terms, the diseases of men, women and children, and the latest and most approved means used in their cure, and is expressly written for the benefit of families in the Western and Southern states. It also contains descriptions of medical roots and herbs of the Western and Southern country, and how they are to be used in the cure of diseases; arranged on a new and simple plan, by which the practice of medicine is reduced to principles of common sense . . . Eighth edition. Pumpkintown, Ten: S.M. Johnston, publisher, 1839.

xvi, [17]-635 p. 22 cm.

1466. GUNN, John C., 1795?-1863.
Gunn's domestic medicine, or poor man's friend; describing, in plain language, the diseases of men, women, and children, and the latest and most approved means used in their cure; designed especially for the use of families. It also contains descriptions of the medical roots and herbs of the United States, and how they are to be used in the cure of diseases. Arranged on a new and simple plan, by which the practice of medicine is reduced to the principles of common sense . . . Thirteenth edition. Pittsburgh, Pa.: Published by J. Edwards & J.J. Newman, 1839.

768 p. ; 22 cm.

Printer's imprint (verso of title-page): J.H. Purdy, Xenia, Ohio.

1467. GUNN, John C., 1795?-1863.
Gunn's domestic medicine, or poor man's friend, in the hours of affliction, pain, and

sickness. This book points out, in plain language, free from doctors' terms, the diseases of men, women, and children, and
the latest and most approved means used
in their cure, and is intended expressly for
the benefit of families. It also contains descriptions of the medicinal roots and herbs
of the United States, and how they are to
be used in the cure of diseases. Arranged
on a new and simple plan, by which the
practice of medicine is reduced to the principles of common sense . . . First revised
edition. Philadelphia: Published by G.V.
Raymond, general agent; stereotyped by L.
Johnson, 1839.

893 p., 2 leaves of plates + frontis. : ill.,
port. ; 24 cm.

The major changes in this edition are the
transfer of the chapter on epidemic cholera
from the end of the volume to the section on
specific diseases (p. 216-71), and the addition
of chapters on hysteria and "tight-lacing" to the
section on diseases of women (p. 543-82). Two
lithographed anatomical plates and a frontispiece portrait of author have also been added.

1468. GUNN, John C., 1795?-1863.
Gunn's domestic medicine, or poor man's
friend, in the hours of affliction, pain, and
sickness. This book points out in plain language, free from doctors' terms, the diseases of men, women, and children, and
the latest and most approved means used
in their cure, and is intended expressly for
the benefit of families. It also contains descriptions of the medicinal roots and herbs
of the United States, and how they are to
be used in the cure of diseases. Arranged
on a new and simple plan, by which the
practice of medicine is reduced to principles of common sense . . . Revised edition, enlarged in 1840. [Philadelphia]:
Published by G.V. Raymond, for the proprietors . . . [et al.], 1842.

4, 893 p., 3 leaf of plates + frontis. : ill.,
port. ; 23 cm.

1469. GUNN, John C., 1795?-1863.
Gunn's domestic medicine, or poor man's
friend, in the hours of affliction, pain, and
sickness. This book points out, in plain language, free from doctors' terms, the dis-

eases of men, women, and children, and
the latest and most approved means used
in their cure, and is intended expressly for
the benefit of families. It also contains descriptions of the medicinal roots and herbs
of the United States, and how they are to
be used in the cure of diseases. Arranged
on a new and simple plan, by which the
practice of medicine is reduced to principles of common sense . . . Third revised
edition—improved and enlarged. New
York: Saxton & Miles; Boston: Saxton &
Peirce; Baltimore: John W. Tilyard, 1844.

4, 893 p., 3 leaves of plates + frontis. :
ill., port. ; 23 cm.

1470. GUNN, John C., 1795?-1863.
Gunn's new domestic physician: or, home
book of health. A complete guide for families: giving many valuable suggestions for
avoiding disease and prolonging life, and
pointing out in familiar language the
causes, symptoms, treatment and cure of
the diseases incident to men, women and
children, with the simplest and best remedies; also, describing minutely the properties and uses of hundreds of well-known
medicinal plants . . . With supplementary
treatises on anatomy, physiology, and hygiene, and on nursing the sick, and the
management of the sick room, with hints
on the drainage of premises, the proper
ventilation of dwellings, etc. Cincinnati:
Moore, Wilstach, Keys & Co., 1863.

1129 p., [24] leaves of plates : ill., port. ;
24 cm.

"Purchasers will bear in mind that this is not
a new edition of the old work "*Gunn's domestic
medicine,*" which was published thirty years ago,
but an *entirely new work,* first published in 1857,
and now enlarged and perfected" (*Publisher's
Notice,* p. [4]). The first edition of Gunn's
manual of popular medicine to be entitled
Gunn's new domestic physician was published by
Gunn at Louisville in 1857 (791 p.). That same
year Moore, Wilstach, Keys & Co. entered their
copyright for a book that they issued at Cincinnati in 1858 under the same title (882, 88 p.).
It included the appendix: "Anatomy, physiology, and the laws of health" by Johnson H. Jordan. Moore, Wilstach, Keys & Co. renewed their
copyright in 1859, the year they published an

edition of 1,046 pages; and again in 1863 for this expanded edition. Though the newer work follows the order of topics in *Gunn's domestic medicine,* it is indeed rewritten, as well as expanded in scope. Twenty of its twenty-four plates contain wood engravings of medicinal plants.

1471. GUNN, John C., 1795?-1863.
Gunn's new domestic physician: or, home book of health. A complete guide for families: giving many valuable suggestions for avoiding disease and prolonging life, and pointing out in familiar language the causes, symptoms, treatment and cure of the diseases incident to men, women and children, with the simplest and best remedies; also, describing minutely the properties and uses of hundreds of well-known medicinal plants . . . With supplementary treatises on anatomy, physiology, and hygiene, and on nursing the sick, and the management of the sick room, with hints on the drainage of premises, the proper ventilation of dwellings, etc. Cincinnati: Moore, Wilstach & Baldwin, 1864.
1129 p., [21] leaves of plates : ill., port. ; 24 cm.

"Appendix. Anatomy, physiology, and the laws of health [by Johnson H. Jordan]," p. [961]-1044. Frontispiece portrait of author.

1472. GUNN, John C., 1795?-1863.
Gunn's new family physician: or, home book of health; forming a complete household guide. Giving many valuable suggestions for avoiding disease and prolonging life, with plain directions in cases of emergency, and pointing out in familiar language the causes, symptoms, treatment and cure of diseases incident to men, women and children, with the simplest and best remedies; presenting a manual for nursing the sick, and describing minutely the properties and uses of hundreds of well-known medicinal plants . . . With supplementary treatises on anatomy, physiology, and hygiene, on domestic and sanitary economy, and on physical culture and development. Newly illustrated, and re-stereotyped. Cincinnati: Moore, Wilstach & Baldwin, 1866.

1218, [6] p., [12] leaves of plates : ill., port. ; 25 cm.

At head of title: *Hundredth edition, revised and enlarged.* "Anatomy, physiology, and the laws of health [by Johnson H. Jordan]," p. [991]-1071; "Physical culture and development, by Charles S. Royce," p. [1143]-1178. Frontispiece portrait of author.

1473. GUNN, John C., 1795?-1863.
Gunn's new family physician: or home book of health; forming a complete household guide. Giving many valuable suggestions for avoiding disease and prolonging life, with plain directions in cases of emergency, and pointing out in familiar language the causes, symptoms, treatment and cure of diseases incident to men, women and children, with the simplest and best remedies; presenting a manual for nursing the sick, and describing minutely the properties and uses of hundreds of well-known medicinal plants . . . With supplementary treatises on anatomy, physiology and hygiene, on domestic and sanitary economy, and on physical culture and development. Newly illustrated, and re-stereotyped. Cincinnati: Wilstach, Baldwin & Co., 1887.
1230 p., [12] leaves of plates : ill., port. ; 25 cm.

At head of title: *Two hundredth edition, revised and enlarged.* "Anatomy, physiology, and the laws of health [by Johnson H. Jordan]," p. [991]-1071; "Physical culture and development, by Charles S. Royce," p. [1143]-1178. Frontispiece portrait of author.

1474. GUNN, John C., 1795?-1863.
Gunn's new family physician or home book of health. Forming a complete household guide. Giving many valuable suggestions for avoiding diseases and prolonging life, with plain directions in cases of emergency, and pointing out in familiar language the causes, symptoms, treatment and cure of diseases incident to men, women and children, with the simplest and best remedies; presenting a manual for nursing the sick and describing minutely the properties and uses of hundreds of well-known medicinal

plants . . . With supplementary treatises on anatomy, physiology and hygiene, on domestic and sanitary economy, and on physical culture and development . . . New York; Akron, O.; Chicago: The Saalfield Publishing Co., [c1901].

1005 p., [12] p. of plates, [1] paper manikin : ill. ; 25 cm.

At head of title: *Two hundred and thirtieth edition, revised and enlarged, 1901.* "Anatomy, physiology, and the laws of health [by Johnson H. Jordan]," p. [785]-862; "Physical culture and development, by Charles S. Royce," p. [931]-966.

1475. GUNN, John C., 1795?-1863.
Gunn's newest family physician; or, home-book of health: an improved household guide, for avoiding disease and prolonging life, giving clear directions in cases of drowning, poisoning, wounds and other emergencies, with plain instructions for managing the sick-room and nursing the sick; and pointing out the best methods for securing pure air, pure water, and proper drainage for dwellings and premises, and familiarly indicating the causes and symptoms and the requisite treatment for the cure of the diseases incident to men, women, and children, with the simplest and best remedies; more especially those from the vegetable materia medica. Accurate descriptions are given of the forms, properties and uses of hundreds of well-known medicinal plants. Also, separate treatises on anatomy, physiology, and hygiene, on domestic and sanitary economy, and illustrated lessons for physical recreation in swimming, rowing, riding, jumping, etc. With an appendix, containing some 400 tested practical recipes, and remedies for diseases of domestic animals. By John C. Gunn . . . assisted by Johnson H. Jordan, M.D., and several scientific writers of the highest eminence.. Springfield, Ill.; Philadelphia: Wm. H. Moore & Co., publishers, 1878.

xiv, 15-1190, 1-1¾, 2-40 p., [4] leaves of plates, [16] p. of plates : ill., port. ; 25 cm.

At head of title: *160th revised, or national jubilee edition, 1876.* Frontispiece portrait of author.

1476. GUNN, Robert Alexander, b. 1844.
Everybody's doctor: a new and improved hand-book of hygiene and domestic medicine . . . New York: Nickles Publishing Company, 1883.

xiii, [3], 668 p., [1] leaf of plates : ill., port. ; 24 cm.

Gunn divides his book into three parts: the first on physiology & hygiene; the second on diseases and their treatment; and the third on the "actions and doses of medicines," "mechanical agents" (e.g., blisters, cupping, etc.), and "poisons and their antidotes." Cover-title: *Gunn's new and improved hand-book of domestic medicine.*

Gunn was an 1866 medical graduate of the University of Buffalo. He had been professor of surgery at the United States Medical College in New York, which was established in 1878 but was closed by the Supreme Court of the State of New York in 1882. Gunn is listed in the New York city directories at various addresses from 1890 through 1912. He was elected president of the New York State Eclectic Medical Society (1893); was secretary to the Westside Eclectic Medical Society (New York); was a member of the National Eclectic Medical Society; edited the eclectic periodical *Medical Tribune* early in the 1890s (ceased publication 1895); and is listed in the directories as "surgeon" to St. Elizabeth's Hospital in 1898 and 1900.

1477. GUNN, Robert Alexander, b. 1844.
Everybody's doctor. A new and improved hand-book of hygiene and domestic medicine . . . First edition. New York: Published by the Dennis Manufacturing Company, 1885.

xiii, [14]-692 p., [1] leaf of plates : ill., port. ; 24 cm.

Frontispiece portrait of author.

GUNN, Robert Alexander see also *CONFIDENTIAL MEDICAL TALKS TO WOMEN* (#759.1); *FACTS FOR LADIES* (#1113).

GUTHRIE, Adelaide Wood see **TOLMAN** (#3543).

1478. GUTS MUTHS, Johann Christoph Friedrich, 1759-1839.

Gymnastics for youth: or a practical guide to healthful and amusing exercises for the use of schools. An essay toward the necessary improvement of education, chiefly as it relates to the body; freely translated from the German of C.G. Salzmann . . . Philadelphia: Printed by William Duane, 1802.

xvi, 432 p., [10] leaves of plates (1 folding) : ill. ; 21 cm.

First American edition. A translation of *Gymnastik fur die Jugend,* which was originally published at Schnepfenthal in 1793. Although Christian Gotthilf Salzmann (1744-1811) is described as the author on the title-page, most bibliographies attribute the book to J.C.F. Guts Muths. The English translation was first published at London in 1800. The Philadelphia edition was re-issued in 1803.

Dividing his work in two parts, Guts Muths begins with a Rousseauian meditation on the natural development, health and strength of the naked and noble savage (whether among his ancient Teutonic ancestors or "among people in the state of nature at this very day"), to which he contrasts the training of youth among his contemporaries: "No sooner has the boy entered his sixth year, than the dancing master appears . . . But there is a great difference between learning to dance, and forming the body; between elegance of carriage, and muscular strength; between the timid spirit of the young beau, and the manly mind of rising youth" (p. 13). Through the whole of part one Guts Muths expands his comparison of the physical training of the ancient Germans (or the Indians of Canada, "the Germany of North America") with the effeminate cultivation of modern Europeans, concluding: "no class can impute its physical degeneracy to a gradual

Fig. 40. Engraved plate from Guts Muths' Gymnastics for youth, *#1478.*

decline of the powers of nature, but must seek for the cause in an unnatural education and way of life" (p. 47). The second part systematically describes activities to be employed in the physical education of the young, including leaping, running, "jaculation" (archery, darts, discus), wrestling, climbing, lifting, swimming, etc.

H

1479. HADDEN, Walter J.
The science of human life and eugenics.
The regeneration of the human race. The
privileges and duties of bringing children
into the world, elimination-segregation-
selection, the right to get married, the law
of opposites, the mother the sole arbiter
of her child's fate, how to have perfect children, the effects of environment on children, the science of reproduction in all
animal and plant life, how the human body
can be made immune to disease, physical
and mental training of children and the
development of the brain of a child, the
sexual laws of nature and the effects of their
violation, the Bible on sex hygiene. From
the notes of Walter J. Hadden . . . edited
by Charles H. Robinson . . . Sex-life, love,
marriage, maternity. Life centers, man's
ideal woman, woman's ideal man, choosing a mate, scientific mating, cupid's conquest, the honeymoon, what marriage involves, the reproductive organs, pure sex-
love, womanhood, pre-natal influence, the
gift of motherhood, the successful mother,
the power of the mind, etc., etc., etc. By
Mary Ries Melendy . . . [S.l., c1914.]
   2 pts. in 1 v. ([8], 107, [1] p., VIII leaves
of plates + frontis. ; [15]-596 p.) : ill. ; 24
cm.

At head of title: *The laws of nature revealed.*

1480. HADFIELD, John G.
A plain and simple statement of the origin
and treatment of diseases by the Equalizer
. . . Cincinnati: Bloch & Co., printers, 1873.
   48 p. ; 22 cm.

Hadfield's portable *Equalizer* is a mechanism
consisting of "several air tight receivers, each
adapted to different parts of the body, and a
newly constructed air-pump, by which we extract air from out of the receiver . . . Our large
receiver will enclose the whole body except the
face. In cases where it is not necessary . . . we

have constructed a receiver adapted to every
part of the human body" (p. 10). In using the
large receiver, "the patient sits in a chair, the
instrument is easily adjusted around him, and
made air-tight, the air is exhausted from within
the receiver by the air-pump . . . thus removing
a certain amount of atmospheric pressure
equally from every part of the system . . . causing the blood and nervo-vital fluid to fill the
capillary vessels in every part of the body, thus
removing internal congestions, carrying nutritive fluids equally to every part of the system,
and carrying the waste and dead matter from
every part. This restores an equal amount of
health and vitality to every part of the system"
(p. 12). Hadfield claims that his *Equalizer* is effective in the cure of "paralysis, nervous debility, dyspepsia, neuralgia, weak chests, narrow
chests, asthma, weak lungs, weak backs, weak,
stiff and shrunken limbs, curvatures of the spine
. . . sciatica, rheumatism . . . spasms, epilepsy,
palsy, St. Vitus' dance, derangement of the
mind, prolapsus uteri, obstructed menstruation
. . . impotency, and all kinds of weaknesses of
the generative organs" (p. 11).
   Hadfield advertises a full set of instruments
for $300.00. However, one could purchase a set
of cups "for impotency, seminal weakness, etc."
or "for atrophied, wasted or small breasts" for
$30.00, not including the air pump.

1481. HAGUE, William Grant, 1868-1928.
The eugenic marriage. A personal guide
to the new science of better living and better babies . . . New York: The Review of
Reviews Company, 1914.
   4 v. ([2], xxx, 136 p., [5] leaves of plates;
[2], vii, [1], [137]-329 p., [3] leaves of
plates; [2], v, [330]-494 p., [5] leaves of
plates; [2], v, [1], [495]-656 p., [4] leaves
of plates) : ill., charts ; 18 cm.

In his introduction Hague writes: "The eugenic ideal is a worthy race—a race of men and
women physically and mentally capable of self-
support. The eugenist, therefore, demands that

every child born shall be a worthy child—a child born of healthy, selected parents. No one can successfully assail the ethics of this appeal. It is morally a just contention to strive for a healthy race. It is also an economic necessity" (1:xix). In his assessment of "the task of race-mainte-nance," Hague expected the state to "devise a system of scientific regulation of marriage" to assure that "every applicant for marriage is pos-sessed of the qualities that will ensure healthy, worthy children" (1:xxi). Hague's purpose in these four volumes is not limited, however, to "mating and breeding," but to educate the pub-lic regarding "Every factor that contributes to the well-being and up-lifting of the race, every subject that bespeaks physical or mental regen-eration, that aids moral and social righteous-ness and salvation, and promises a greater so-cial happiness and contentment" (1: xxii). It is not surprising, therefore, that popularizers of eugenics such as Hague played a leading role in the reform movements of the Progressive Era, from sex education, to prohibition (alco-holism being regarded as a cause hereditary defects), to pacifism (war eliminates the best specimens of the race), and birth control.

Hague is concerned not only with common hereditary diseases, but with acquired degen-erative disease as well—especially sexually trans-missible diseases that render otherwise fit men impotent, healthy women sterile, and the un-fortunate children born to infected couples deformed or deficient. The entire age, in Hague's mind, contributes to "race degen-eracy," whether the cause be poverty; its con-trary, "luxurious living, and the life of ease and amusement on the part of the women of wealth," the entry of women into the workforce; neurasthenia; unhygienic living; or not-to-be-trivialized causes such as "the prevalent danc-ing craze." "No sane person," Hague writes, "can regard with complacency the vicious envi-ronment in which the future mothers of the race 'tango' their time, their morals, and their vitality away" (1:xxvii).

Hague begins his treatise with a statement of the "eugenic principle," i.e., "that 'the fit only shall live.' This does not mean that the unfit must die, but that only the fit shall be born" (1:10). He devotes the first four chapters of volume one to the mission and goals of the "sci-ence of eugenics" before he discusses preg-nancy, confinement, labor, child rearing, the sex education of children, marriage, domestic medicine & hygiene, the "patent medicine evil," etc.

Born in Scotland, Hague attended Glasgow University before emigrating to the United States. In 1895 he received his medical degree from the College of Physicians and Surgeons (New York). He remained in New York his en-tire career.

1482. HAGUE, William Grant, 1868-1928.
The eugenic marriage. A personal guide to the new science of better living and bet-ter babies . . . New York: The Review of Reviews Company, 1916.
    4 v. ([2], xxx, 136 p., [5] leaves of plates; [2], vii, [1], [137]-329 p., [3] leaves of plates; [2], v, [330]-494 p., [5] leaves of plates; [2], v, [1], [495]-656 p., [4] leaves of plates) : ill., charts ; 18 cm.

1483. HAGUE, William Grant, 1868-1928.
The eugenic mother and baby. A complete home guide . . . New York City: Published by the Hague Publishing Co., 1913.
    [16], 554 p. ; 28 cm.

    "First edition June, 1913. Second edition December 1913" (verso of title-page).

1484. HALE, Amie M.
The management of children in sickness and in health. A book for mothers . . . Sec-ond edition. Philadelphia: Presley Blaki-ston, 1881.
    viii, 9-110, 8, [2] p. ; 18 cm.

    First edition published in 1880. The author may be mistakenly listed in the AMA's *Directory of deceased American physicians, 1804-1929* (1:634) as "Anna M. Hale," an 1876 graduate of the Northwestern University Women's Medical School. Hale's book is dedicated to Charles W. Earle, professor of pediatrics at Northwestern.

1485. HALE, Sarah Josepha Buell, 1788-1879.
The good housekeeper, or the way to live well and to be well while we live. Containing directions for choosing and preparing food, in regard to health, economy and taste . . . Boston: Weeks, Jordan and Company, 1839.
    xi, [1], 132 p. : ill. ; 19 cm.

First edition. "Foreigners say that our climate is unhealthy; that the Americans have, generally, thin forms, sallow complexions and bad teeth. Is it not most likely that these defects are incurred, in part if not wholly, because the diet and modes of living are unsuitable to the climate, and consequently to the health of the people? Could public attention be drawn to this important subject sufficiently to have a reform in a few points—such as using *animal food* to excess, eating *hot bread,* and swallowing our meals with steam-engine rapidity, the question of climate might more easily be settled" (p. [vii]).

The good housekeeper is the earliest of several books on cookery, manners and domestic economy written by Hale, who, by the time this work was published, was already a successful poet, novelist, and editor of *The Ladies' magazine* (1828-36) and *Godey's lady book* (1837-77). Through her novels, manuals of advice, and editorials in the pages of the journals she edited, Hale became an influential advocate of female education, the entry of women into the professions, dress reform and abolition. (Page 131 is misnumbered 151; page 132 is numbered "132-144.")

1486. HALE, Sarah Josepha Buell, 1788-1879.

The good housekeeper, or the way to live well, and to be well while we live. Containing directions for choosing and preparing food, in regard to health, economy, and taste . . . Sixth edition. Boston: Otis, Broaders, and Company, 1841.

x, [11]-144 p. : ill. ; 18 cm.

1487. HALL, Alexander Wilford, 1819-1902.

Dr. A. Wilford Hall's hygienic treatment, for the cure of disease, preservation of health and the promotion of longevity without medicine . . . New York: Sold only by A. Wilford Hall, [after 1887].

48 p. : ill. ; 15 cm.

Hall maintains that disease arises from "excrementitious impurities from the intestine," i.e., from a "decayed residuum" of putrified food that breeds disease-causing germs absorbed by the body via intestinal circulation. He regards the colon as the "chief reservoir" of this "antagonistic matter," and maintains that

it must frequently be "cleansed of its deleterious contents" (p. 15). Shunning cathartics, Hall's treatment consists of injecting four quarts of warm water into the rectum every second or third day (p. 21). He modestly describes his method of colonic hygiene as "the greatest discovery of all time" (p. 7) and a "most startling revolution which must command the respect of the thinking world" (p. 17). Hall was a Christian evangelist, particularly known for his attacks on Universalism and the theory of evolution.

1488. HALL, Alfred G.

The mother's own book. And practical guide to health; being a collection of necessary and useful information. Designed for females only. Derived from various sources during a period of twenty years practice; in trhee [sic] parts. 1st. Menstruation, &c. &c. 2d. Conception and gestation. 3d. Parturition and abortion. With practical advice; uniting the benefits of the old and new school practice, in the treatment of female weakness, &c. &c. &c. . . . Rochester, N.Y., 1843.

xvi, [17]-144 p. : ill. ; 18 cm.

1489. HALL, Granville Stanley, 1844-1924.

Youth. Its education, regimen, and hygiene . . . New York: D. Appleton and Company, 1911.

x, 379 p. ; 20 cm.

Youth is condensed from the author's two-volume *Adolescence,* a landmark in American adolescent psychology that was first published in 1904. In *Youth,* Hall selected and epitomized "the practical and especially the pedagogical conclusions of my large volumes . . . in such form that they may be available at minimum cost to parents, teachers, reading circles, normal schools, and college classes" (*Pref.,* p. [v]). It was first issued in 1906.

Hall was recipient of the first doctorate in psychology awarded in America (Harvard, 1878); studied under Wilhelm Wundt in Berlin (1878); lectured on psychology and pedagogy at Johns Hopkins (1882), where he established the first psychological laboratory in America; founded the *American journal of psychology* (1887); was made Clark University's first president (1888); served as the first president

of the American Psychological Association
(1892); became an early proponent of Freud-
ian ideas in America, and in 1909 brought
Freud and Jung to Clark University. Hall's biographer in the *Amer. nat. biog.* (9:858) has summarized his importance "primarily as a pro-
moter of academic scientific psychology in the
1880s and early 1890s, and of developmental
psychology, school reform, and popular psychological interest in the Progressive Era."

1490. HALL, Jeanette Winter.
The new century primer of hygiene for
fourth year pupils . . . New York; Cincinnati; Chicago: American Book Company,
[c1901].
　　154, [6] p. : ill. ; 19 cm.

　　Jeanette Winter Hall was the wife and collaborator of Winfield Scott Hall (#1520-1433).
Series: *New century series of anatomy, physiology,
and hygiene.*

HALL, Jeannette Winter see also **HALL**
(#1530, 1532, 1533).

1491. HALL, William Whitty, 1810-1876.
Bronchitis and kindred diseases . . . Eighth
edition. New York: Redfield, 1854.
　　382, ii p. ; 19 cm.

　　A native of Paris, Ky., Hall was an 1836 medical graduate of Transylvania University. That
same year he was ordained into the Presbyterian ministry. "The following year he was a missionary in Texas, and for a number of years he
preached occasionally, but he gradually abandoned preaching for medicine. He practiced
in New Orleans and Cincinnati, and in 1851
removed to New York City, where he established
a consultation practice, 'strictly confined to
chronic ailments of the throat and lungs' (*Dict.
Amer. biog.*, IV(2):147). Hall was the author of
numerous books on hygiene and domestic
medicine, and particularly on the diseases of
the respiratory tract. In 1854 he began publication of *Hall's journal of health,* which he edited until his death. *Hall's journal* continued to
be published until 1894 (v. 41).
　　Bronchitis and kindred diseases was first published in 1852, and was last issued in 1870. It is
a rambling treatise in which Hall not only de-

scribes the lung's various ailments (consumption particularly), but discusses hygienic measures to prevent their onset. He laces his text
with extracts from well-known medical authors,
case histories from his own practice, and anecdotes from the illnesses of various persons.
Atwater copy inscribed by Henry Foster
(#1265), proprietor of the Clifton Springs Water Cure.

1492. HALL, William Witty, 1810-1876.
Consumption . . . New York: Redfield, 1857.
　　276 p. ; 19 cm.

　　First edition. Hall regards pulmonary consumption as the product of "an imperfect nutrition and impure blood, arising in all cases
from an imperfect digestion and the breathing of an impure atmosphere" (p. 266). Its cure,
therefore, lies in the restoration of "perfect
digestion" and exposure to "pure air." "Substantial food, well digested" restores the blood and
increases strength for exercise. Muscular activity out of doors purifies the blood by the action
of "pure air" in the lungs. To these ends, Hall
recommends that his patients "spend six or
eight hours roaming through the woods on
foot, in fishing, hunting, berrying, or in a horseback ride of equal time . . . then mark the vigor
with which he can dispatch a meal of plain meat
and bread" (p. 270). Spine-title: *Handbook of
consumption.*

1493. HALL, William Whitty, 1810-1876.
Dyspepsia, and its kindred diseases . . .
Detroit: Belford Bros., publishers, 1877.
　　vi, [7]-272 p. ; 19 cm.

　　In defining dyspepsia and noting its consequences, Hall writes: "whenever a man has any
uncomfortableness about him at a regular time,
after eating, he is dyspeptic. It is indigestion in
its first beginnings, and if then attacked it can be
promptly and effectually cured, without a particle of medicine, but simply by a judicious regulation of what a person eats and drinks. If no
efficienct attention is paid to these beginning
symptoms, they multiply in number and violence
indefinitely; seldom proving fatal of themselves,
but gradually undermining the constitution,
making it an easy pray to some acute malady" (p.
26-27). Published posthumously, *Dyspepsia* was

also issued in 1877 at Toronto (Belford) and at New York (Worthington).

1494. HALL, William Whitty, 1810-1876.
Dyspepsia and its kindred diseases . . . New York: Worthington Co., 1888.
vi, [7]-272 p. ; 19 cm.

Second and final edition.

1495. HALL, William Whitty, 1810-1876.
Fun better than physic; or, everybody's life-preserver . . . Springfield, Mass.: D.E. Fisk and Company, [c1871].
333, [5] p., [1] leaf of plates : port. ; 20 cm.

First edition. The content of this book is indicated by its running-title, i.e., *Dr. Hall's Maxims*. This collection of apothegms on bodily hygiene, mental health and moral rectitude (arranged in no particular order) was re-issued at Toronto in 1875 (#1516), and at Chicago in 1882 (#1496), 1884 (#1497) and 1889 (#1497.1). Steel-engraved frontispiece portrait of author. Cover-title: *Everybody's life-preserver.*

1496. HALL, William Whitty, 1810-1876.
Fun better than physic; or, everybody's life preserver . . . Chicago, Ill.: Rand, McNally & Co., publishers, 1882.
333, [3] p. ; 19 cm.

1497. HALL, William Whitty, 1810-1876.
Fun better than physic; or, everybody's life-preserver . . . Chicago, Ill.: Rand, McNally & Co., publishers, 1884.
333, [3] p. ; 19 cm.

1497. 1. HALL, William Whitty, 1810-1876.
Fun better than physic; or, everybody's life-preserver . . . Chicago, Ill.: Rand, McNally & Co., publishers, 1889.
333, [3] p. ; 19 cm.

1498. HALL, William Whitty, 1810-1876.
The guide-board to health, peace and competence; or, the road to happy old age . . . Springfield, Mass.: D.E. Fisk and Company; Philadelphia: Crittenden & McKinney, [c1869].

[3]-752, [2] p., [1] leaf of plates : ill., port. ; 23 cm.

A collection of articles on various hygienic topics ranging in length from one to twenty pages, probably culled from the pages of *Hall's journal of health*. Steel-engraved frontispiece portrait of author.

1499. HALL, William Whitty, 1810-1876.
The guide-board to health, peace and competence; or, the road to happy old age . . . Springfield, Mass.: D.E. Fisk and Company; Philadelphia: McKinney & Martin; San Francisco, Cal.: Pacific Publishing Co., [c1869].
[3]-752 p., [1] leaf of plates : port. ; 23 cm.

Frontispiece portrait of author.

1500. HALL, William Whitty, 1810-1876.
The guide-board to health, peace and competence; or, the road to happy old age . . . Springfield, Mass.: D.E. Fisk and Company; Philadelphia: McKinney & Martin; St. Louis, Mo.: F.A. Hutchinson & Co., [c1869].
[3]-752, [2] p., [1] leaf of plates : port. ; 23 cm.

Frontispiece portrait of author.

1501. HALL, William Whitty, 1810-1876.
The guide-board to health, peace and competence; or, the road to happy old age . . . Springfield, Mass.: D.E. Fisk and Company; Philadelphia: H.N. McKinney & Co.; St. Louis, Mo.: F.A. Hutchinson & Co., [c1869].
[3]-752, [2] p., [1] leaf of plates : port. ; 23 cm.

Frontispiece portrait of author.

1502. HALL, William Whitty, 1810-1876.
The guide-board to health, peace and competence; or, the road to happy old age . . . Springfield, Mass.: D.E. Fisk and Company, [c1872].
752, [2] p., [33] leaves of plates + frontis. : ill., port. ; 23 cm.

Fig. 41. From Hall's The guide-board, *#1502.*

1503. HALL, William Whitty, 1810-1876.
The guide-board to health, peace and competence, or, the road to happy old age . . .
Springfield, Mass.: D.E. Fisk and Company;
Philadelphia: H.N. McKinney & Co.,
[c1872].
    752, [2] p., [33] leaves of plates + frontis. :
ill., port. ; 23 cm.

1504. HALL, William Whitty, 1810-1876.
Health and disease; a book for the people
. . . New-York: Published by H.B. Price,
1859.
    298 p. ; 19 cm.

    First edition. "Three fourths of all our ailments
occur, or are kept in continuance, by preventing
the daily food which is eaten, from passing out of
the body, after its substance has been extracted by
the living machinery, for the purpose of renovation
and growth" (p. 6-7). "If the body is in good health
. . . there is a proper proportion between the amount
received and the amount passed out from the system. The very moment that such proportion is altered, disease begins in all constitutions, and under
all circumstances, nay, every additional hour of its
continuance, that disease becomes more aggra-

vated, more destructive, more difficult of arrest, and
more certain of disaster and of death" (p. 8). *Health
and disease* appeared in five editions between 1859
and 1866. It was re-issued in 1870 under the title:
Health and disease as affected by constipation.

1505. HALL, William Whitty, 1810-1876.
Health and disease; a book for the people
. . . Second edition. New-York: Published
by H.B. Price, 1859.
    298, viii, 4 p. ; 19 cm.

    The added pages in this second edition ad-
vertise *Hall's journal of health* and other of the
author's publications.

1506. HALL, William Whitty, 1810-1876.
Health and disease; a book for the people
. . . Third edition. New-York: Published by
W.W. Hall . . . and by Trubner & Co. . . .
London, 1860.
    298, ii, viii, 4 p. ; 19 cm.

1507. HALL, William Whitty, 1810-1876.
Health and disease . . . Fifth edition revised,
with additions. New-York: W.J. Widdleton,
publisher, 1866.

[2], 298, iv, [311]-314, [399]-400 p. ; 19 cm.

The fifth edition was first issued in 1864. The final page sequences are advertisements to those works of Hall listed opposite the title-page in the form of their indices.

1508. HALL, William Whitty, 1810-1876.
Health at home, or Hall's family doctor: showing how to invigorate and preserve health, prolong life, cure diseases, understand the physical conditions of maternity, and the proper management of infants, and discussing the entire physical well-being of man, with a very large collection of the latest and most valuable medical prescriptions . . . Hartford, Conn., and Cincinnati, Ohio: S.M. Betts & Company, 1872.
[2], 781, [13] p., [5] leaves of plates : ill., port. ; 23 cm.

First edition. Steel-engraved frontispiece portrait of author.

1509. HALL, William Whitty, 1810-1876.
Health at home, or Hall's family doctor: showing how to invigorate and preserve health, prolong life, cure diseases, understand the physical conditions of maternity, and the proper management of infants, and discussing the entire physical well-being of man, with a very large collection of the latest and most valuable medical prescriptions . . . Hartford, Conn., and Cincinnati, Ohio: S.M. Betts & Company, 1873.
[2], 799, [9] p. : ill., port. ; 23 cm.

1510. HALL, William Whitty, 1810-1876.
Health at home, or Hall's family doctor: showing how to invigorate and preserve health, prolong life, cure diseases, understand the physical conditions of maternity, and the proper management of infants, and discussing the entire physical well-being of man, with a very large collection of the latest and most valuable medical prescriptions . . . Hartford, Conn., Cincinnati, Ohio, and Chicago, Ills. : Jas. Betts & Company, 1873.

[2], 846 p., [5] leaves of plates : ill. ; 23 cm.

1511. HALL, William Whitty, 1810-1876.
Health at home, or Hall's family doctor: showing how to invigorate and preserve health, prolong life, cure diseases, understand the physical conditions of maternity, and the proper management of infants, and discussing the entire physical well-being of man, with a very large collection of the latest and most valuable medical prescriptions . . . Hartford, Conn., Cincinnati, Ohio, and Chicago, Ills.: S.M. Betts & Company, 1873.
[2], 798, [14] p., [4] leaves of plates + frontis. : ill., port. ; 23 cm.

1512. HALL, William Whitty, 1810-1876.
Health at home, or Hall's family doctor: showing how to invigorate and preserve health, prolong life, cure diseases, understand the physical conditions of maternity, and the proper management of infants, and discussing the entire physical well-being of man, with a very large collection of the latest and most valuable medical prescriptions . . . Albany, N.Y.: Russell Publishing Company, [after 1875].
iv, 891 p., [4] leaves of plates + frontis. : ill., port. ; 23 cm.

In the "Preface to the 18th edition" (p. [1]) reference is made to an expanded chapter on diphtheria (p. 811-18) that refers to the outbreak at New York during the years 1873-1875.

1513. HALL, William Whitty, 1810-1876.
Health at home, or Hall's family doctor: showing how to invigorate and preserve health, prolong life, cure diseases, understand the physical conditions of maternity, and the proper management of infants, and discussing the entire physical well-being of man, with a very large collection of the latest and most valuable medical prescriptions . . . Hartford, Conn.: James Betts & Co., 1884.
iv, 891 p., [5] leaves of plates : ill., port. ; 24 cm.

1514. HALL, William Whitty, 1810-1876.
Health by good living . . . Eighth thousand.
New York: Published by Hurd and Houghton, 1870.
vi, 277 p. ; 19 cm.

By "good living" Hall means "eating the best
food prepared in the best manner" at regular
times time and in moderation. The neglect or
abuse of a proper dietetic regimen is the primary cause of the majority of the ailments of
body and mind.

1515. HALL, William Whitty, 1810-1876.
Health by good living . . . Twentieth thousand.
Philadelphia: H.N. McKinney & Co., 1873.
vi, 285 p. ; 19 cm.

1516. HALL, William Whitty, 1810-1876.
How to live long; or, fun better than physic . . .
Toronto: Belford Bros., publishers, 1875.
293 p. ; 19 cm.

1517. HALL, William Whitty, 1810-1876.
How to live long; or, health maxims, physical, mental, and moral . . . New York: Published by Hurd and Houghton; Cambridge
[Ma.]: The Riverside Press, 1875.
iv, [5]-316 p. ; 19 cm.

1,408 health maxims "in the fewest possible
words, to compel the reading of them and so
impress them on the mind . . . that they cannot
be forgotten in a life-time" (*Pref.,* p. [iii]). Hall
notes that they may be "taken up and laid down
at a moment's notice, on steamship, tramway,
packet, or rail-car . . . because in this age of
restlessness and hurry, the care of the health
. . . is considered one of the things which can
be dispensed with . . . It is hoped that some
who would not spend the time to hear a lecture or read a book may be enticed to peruse a
paragraph now and then in reference to the
care of the body" (p. iv).

1518. HALL, William Whitty, 1810-1876.
Sleep; or, the hygiene of the night . . . New
York: Published by Hurd and Houghton,
1870.
[4], iv, [5]-352, iv p. ; 19 cm.

Hall addresses the role of sleep in bodily and
mental health, as well as the factors that mitigate against it, e.g., city air, poor ventilation,
pernicious sexual habits, "sleeping with consumptives," etc.

1519. HALL, William Whitty, 1810-1876.
Throat-ail, bronchitis, consumption: their
causes, symptoms and cure, in language
adopted to common readers . . . Sixth edition. New York: Published and sold by J.S.
Redfield, 1851.
[2], 87, [1] p. ; 22 cm.

Cover-title: *Clergyman's sore throat, asthma,
bronchitis, consumption.*

1520. HALL, Winfield Scott, b. 1861.
The biology, physiology and sociology of
reproduction. Also sexual hygiene with
special reference to the male . . . Second
edition. Chicago: Herbert A. Ray, publisher, 1907.
[2], 138 p. : ill. ; 19 cm.

Hall was professor of physiology at Northwestern University Medical School when the
first edition of this book was published. Intended as a manual of reproductive biology,
sexual physiology and ethics for college males,
The biology, physiology and sociology of reproduction
appeared in eighteen editions between 1906
and 1919. It is divided into five chapters: I. Reproduction from the standpoint of biology; II.
Adolescence in the male (physical & psychical
changes); III. Anatomy and physiology of the
male genital organs; IV. Sexual hygiene of the
adolescent male (discussing "illicit intercourse,"
sexually transmitted diseases, masturbation,
etc.); and a fifth and final chapter on hygiene
in general (i.e., diet, bathing, exercise, sleep,
etc.). This was the earliest and most successful
of Hall's popular sexual physiologies, and
launched his career as a reputable authority on
adolescent sex education. Cover-title: *Reproduction and sexual hygiene.*
An 1888 graduate of Northwestern University Medical School, Hall also pursued studies
in medicine and physiology at Leipzig. He appears to have been a member of the faculty at
Northwestern most of his career. Hall's name
appears in the first edition (1906) of the
A.M.A.'s *American medical directory,* where he is
listed for the last time in the sixth edition
(1918).

1521. HALL, Winfield Scott, b. 1861.
The biology, physiology and sociology of reproduction. Also sexual hygiene with special reference to the male . . . Sixth edition 1908. Chicago: Wynnewood Publishing Co., [c1907].
[14], [9]-149, [3] p., [1] leaf of plates : ill., port. ; 19 cm.

Cover-title: *Reproduction and sexual hygiene.* Frontispiece portrait of author.

1522. HALL, Winfield Scott, b. 1861.
The biology, physiology and sociology of reproduction. Also sexual hygiene with special reference to the male . . . Twelfth edition. Chicago: Wynnewood Publishing Co., 1911.
[16], 11-149, [5] p., [1] leaf of plates : ill., port. ; 19 cm.

Frontispiece portrait of author. Cover-title: *Reproduction and sexual hygiene.*

1523. HALL, Winfield Scott, b. 1861.
The biology, physiology and sociology of reproduction. Also sexual hygiene with special reference to the male . . . Fourteenth edtion . . . thoroughly revised. Chicago: Wynnewood Publishing Co., 1913.
[6], 149, [3] p., [1] leaf of plates : ill., port. ; 19 cm.

Frontispiece portrait of author. Cover-title: *Reproduction and sexual hygiene.*

1524. HALL, Winfield Scott, b. 1861.
Daughter, mother and father. A story for girls . . . Sex hygiene pamphlet number four, issued by the Council on Health and Public Instruction of the American Medical Association. Chicago: American Medical Association Press, [c1913].
55 p. ; 19 cm.

Title on wrapper: *Life problems. A story for girls.*

1525. HALL, Winfield Scott, b. 1861.
Elementary anatomy, physiology and hygiene for higher grammar grades . . . New York; Cincinnati; Chicago: American Book Company, [c1900].
273, [5] p. : ill. ; 19 cm.

Series: *New century series of anatomy, physiology and hygiene; 2*; series: *New century physiologies.* Spine- and cover-titles: *Elementary physiology.*

1526. HALL, Winfield Scott, b. 1861.
Elementary anatomy, physiology and hygiene for higher grammar grades . . . New York; Cincinnati; Chicago: American Book Company, [c1911].
310 p. : ill. ; 19 cm.

Series: *New century series of anatomy, physiology and hygiene, no. 2.* Cover- and spine-titles: *Elementary physiology.*

1527. HALL, Winfield Scott, b. 1861.
Father and daughter. A story for girls . . . Sex hygiene pamphlet three, issued by the Council on Health and Public Instruction of the American Medical Association. Chicago: American Medical Association Press, [c1913].
46 p. ; 19 cm.

Cover-title: *The doctor's daughter. A story for girls.* In printed wrapper.

1528. HALL, Winfield Scott, b. 1861.
Father and son. Chums. John and his father study some life problems . . . Sex hygiene pamphlet two, issued by the Council on Health and Public Instruction of the American Medical Association. Chicago: American Medical Association Press, [c1913].
47 p. ; 19 cm.

Cover-title: *Chums. A story for boys.* In printed wrapper.

1529. HALL, Winfield Scott, b. 1861.
Father and son. John's vacation. What John saw in the country . . . Sex hygiene pamphlet one, issued by the Council on Health and Public Instruction of the American Medical Association. Chicago: American Medical Association Press, [c1913].
48 p. ; 19 cm.

Cover-title: *John's vacation. A story for boys.* In printed wrapper

1530. HALL, Winfield Scott, b. 1861.
A manual of sex hygiene setting forth in plain and simple terms the instruction which parents and teachers should impart to children regarding the great truths of life, including the origin of life, the care of the body, personal habits and social relationships—showing distinctly how these delicate facts may be successfully unfolded to young minds . . . Assisted by Jeannette Winter Hall. Chicago: The Howard-Severance Company, 1913.
[2], 100 p. ; 23 cm.

1531. HALL, Winfield Scott, b. 1861.
Sex training in the home. Plain talks on sex life covering all periods and relationships from childhood to old age . . . Chicago: W.E. Richardson Company, 1914.
109, [3] p. ; 19 cm.

1532. HALL, Winfield Scott, b. 1861.
Sexual knowledge in plain and simple language; sexology or knowledge of self and sex for both male and female; especially for the instruction of youths and maidens, young wives and young husbands, all fathers and all mothers, school-teachers and nurses, and all others who feel a need of proper and reliable information on sex hygiene, sex problems, and the best way and the best time to impart sexual knowledge to boys and girls about to enter into manhood and womanhood . . . Assisted by Jeanette Winter Hall . . . Vice problems discussed by specialists . . . Philadelphia, Pa.: Published and manufactured by the International Bible House, [c1913].
[4], 320 p. : port. ; 19 cm.

At head of title: *Sexual education for sex problems. Sex hygiene by highest authority.*

1533. HALL, Winfield Scott, b. 1861.
Sexual knowledge. The knowledge of self and sex in simple language: for the instruction of young people, young wives and young husbands, fathers and mothers, teachers and nurses, and all who feel a need of reliable information on the best way and the best time to impart sexual knowledge to boys and girls . . . Assisted by Jeannette Winter Hall . . . Philadelphia:

The John C. Winston Company, publishers, [c1916].
[3]-320 p., [4] p. of plates : ill., port. ; 19 cm.

1534. HALLOCK, Frederick.
The secret medical counsellor; or an essay on the physiology, functions, and diseases of the organs of generation . . . Boston: Published by the proprietor, 1869 [1872 printing].
iv, [5]-92, 95-123, [126]-138, [141]-142, 51-52 p., 14 leaves of plates : ill. ; 16 cm.

The author claims to be physician, surgeon, and proprietor of the Hallock Medical Institute (143 Court Street, Boston), a proprietary clinic specializing in the treatment of "private diseases." *The secret medical counsellor* begins with a description of the male and female organs of generation (illustrated with three wood-engraved plates); discusses the psychology of sexual attraction & "intention of marriage"; describes conception, pregnancy, contraception, menstruation disorders, and masturbation; and concludes with a comparatively lengthy consideration of sexually transmitted diseases. The chapter "On the use of conception-preventers" (p. [58]-59) describes Hallock's "Gastroeges," a shield made of "gum-elastic" inserted at the mouth of the womb. This chapter is accompanied by a wood-engraved plate that shows a saggital section of the female organs and describes the placement of the shield and use of the wire apparatus by which it is inserted. Hallock notes that this shield "may remain two or three weeks, or it may be removed after each intercourse."

In the treatment of the "the diseases of the generative organs" Hallock used herbal remedies of his own preparation. Though available for sale to the public, he does not advise self-medication, "especially [for] those who have private troubles, as there are so many complications and new symptoms . . . that none but a skilled physician can conduct the disease through its various stages" (p. 131). His "Catalogue of remedies" (p. 137-38) lists proprietary medicines for the treatment of menstruation disorders, "seminal weakness" (i.e., the consequences of masturbation), and the venereal diseases. The title-page is dated 1869; the printed wrapper is dated 1872. Six of the four-

Plate 1. This Engraving represents the appearance of the features where the practice of self-pollution has caused general debility.

Plate 2. This Engraving represents the appearance of the features in the last stage of debility from self-pollution.

Fig. 42. From Hallock's The secret medical counsellor, *#1534.*

teen leaves of wood-engraved plates are included in the pagination of the text.

Hallock is first listed in the Boston city directory (1867) as a principal of the Hallock Association. The Hallock name is associated with this establishment through its several name changes until 1917, e.g., the Hallock Medical Institute (early 1870s); Hallock & Co. (early 1880s); the Hallock Medical Institute in 1896; The Old Dr. Hallock Medical Institute (1900-1905); and even a Hallock Museum of Anatomy from the late 1890s through 1905.

1535. HALSTED, Hatfield.
Exposition of motorpathy: a new system of curing disease, by statuminating, vitalizing motion . . . Rochester, N.Y.: Press of Curtis & Butts, Daily Union office, 1853.
vi, [7]-171 p. ; 20 cm.

Halsted had been active in Rochester as a "magnetic physician" and manufacturer of electric pills, galvanized plasters and the like since the late 1830s. In 1844 he bought a three and a half story stone building in the Bull's Head section of Rochester (west of downtown) that he converted into the Halsted Medical Institute. In the 23 April 1846 issue of the *Rochester Daily Advertiser* he announced that "hydropathic treatment or water-cure" was newly available at the Institute. Halsted built a large addition to the Institute in 1851, and renamed the complex Halsted Hall. In this same year he added "motorpathy" to its list of therapies. Though Halsted claimed that the motorpathic regimen was original to himself, it was similar to the "movement-cure" advocated by such contemporaries as Dio Lewis (#2233-2243), C.F. Taylor (#3420, 3421) or G.H. Taylor (#3423-3433).

Halsted's exercises focused specifically on the treatment of gynecological disorders. In 1853 he he published this first edition of the *Exposition*, in which he goes into great detail regarding the symptoms and prognosis of each disease (with case histories), but omits any description of the exercises employed in their cure. In addition to electro-magnetic, hydropathic and motorpathic treatments, Halsted

emphasized "diet . . . pure and fresh air, ventilated sleeping 'apartments,' vigorous general
body exercise and, of course, change of scene.
The importance of psychological factors was
underscored: health was impossible without a
normal state of mind . . . This good sense, however, was topped by a murky and shaky theory.
The body was described as a tabernacle, as
empty without motion as was thought without
action. A 'substantia prima' from a 'higher
source' was transmuted into 'electron animal
magnetic power,' a vita-motive fluid or ether
produced by the brain and emitted to every
corner by the medullary fibres or origins of
nerves . . . Since perfect motion and perfect
health produced harmony and balance, any
distortion of this homeostasis led to disease and
death; the balance could be restored by using
motion as therapy . . . Halsted's book covered a
wide range of sound advice. He included emphasis on personal hygiene, some awareness of
what a century later would be hailed as psychosomatic interactions in the genesis of chronic
disease, a glimpse of the importance of natural
resistance in recovery, and frank acceptance of
the adjuvant value of other types of therapy"
(Edward C. Atwater, "Rochester and the water-
cure, 1844-1854," *Rochester history*, 1970,
32(4):12-13).

In 1854 Halsted left Rochester, and in the
following year bought the Round Hill Water-
Cure, Northampton, Ma. (#3039). Halsted successfully operated Round Hill institute until
1870, when he sold it and presumably retired.

1536. HALSTED, Hatfield.
Magnetic Battery, consisting of Electric
Pills, Galvanized Plaster, Magnetic Aether
No. 1, for consumption, Magnetic Aether
No. 2, for nervous diseases, Magnetic
Aether No. 3, for liver complaints, &c.
[Rochester, N.Y.: H. Halsted & Co., physi-
cians and chemists, 1842].
    [1] leaf printed on both sides : ill. ; 54
× 32 cm.

Few diseases, Halsted claims, can withstand
the combined effect of his "Magnetic Battery,"
consisting of Electric Pills, Galvanized Plaster
and Magnetic Aethers. His Electric Pills, for
example, "are so chemically combined, as to
emit while the process of their digestion is going on, a kind of gas which passes through the

system very much like electricity, giving various
organs a shock or charge, thereby accelerating
their action . . . They operate as the 'refiner's
fire,' completely purifying the system, by removing from it every deleterious substance," etc.
(recto, first column). Numerous testimonial
letters are included (dated 1840-1842).

1537. HALSTED, Hatfield.
A treatise on the Magnetic Battery, consisting of the Electric Pills, Magnetic Aether
or fluid of restoration, and Galvanized Plaster; their application and effects in removing consumption, nervous, bilious &
chronic diseases. Third edition, revised
and enlarged . . . Rochester, N.Y.: Published
by H. Halsted & Co., physicians & chymists,
authors & proprietors; from the power
press of E. Shephard, 1846.
    [iii]-vi, [7]-47, [3] p. : ill. ; 20 cm.

In an 1842 advertisement for his Magnetic
Battery (#1536), Halsted makes mention of "A
book of 48 pages [that] accompanies each package" of his three-fold "magnetic" remedy. The
pamphlet to which he refers is doubtless the
first edition of the present work, which lays
down the principles of his electro-magnetic
pathology; describes each of the illnesses to
which his pills, aethers and plasters may be applied; and outlines the situation, facilities and
therapeutics of the Halsted Medical Institute.

HALSTED, Hatfield see also **ROUND HILL
WATER CURE** (#3039).

1538. HALSTED, Oliver.
A full and accurate account of the new
method of curing dyspepsia . . . With some
observations on diseases of the digestive
organs. New-York: O. Halsted, 1830.
    xii, [13]-155 p. ; 20 cm.

While suffering for years from chronic dyspepsia, Halsted unexpectedly found relief from
his malady "whenever I rode for a succession
of days and nights in the mail-stage." He noticed that the action of the carriage day and
night relaxed his abdominal muscles, and that
"the action was communicated to my stomach
and bowels" (p. vii-viii). Halsted subsequently
devised a regimen of "exercise" that replicated
the action that had relieved his own symptoms.

Halsted believes that the apparent increase in dyspepsia among Americans is attributable to the fact that "many affections, which were formerly known under different names, and treated accordingly, as the spleen, vapours, indigestion, low spirits, and nervous diseases, are now generally comprehended under the sweeping term, dyspepsia" (p. 14). He describes the physiology of digestion, the symptoms of dyspepsia, and the three stages of its progress. Halsted attributes the occurence of dyspepsia to errors in "the quality, the quantity, and the time and manner of taking food" (p. 52) and to want of exercise—both of which leave the abdominal muscles in a more or less permanent state of contraction. His "mode of treatment" aimed first at relaxing these muscles by breathing exercises, by the application of "warm fomentations" to the abdomen, and by the repeated application of "flannel cloths wrung out in a mixture of equal parts of hot vinegar and water." Only then could the restoration of the stomach be attempted by manual stimulation of its nerves. Halsted's treatise concludes with a regimen of diet and exercise for maintaining health restored.

1539. HALSTED, Oliver.
A full and accurate account of the new method of curing dyspepsia . . . with some observations on diseases of the digestive organs. Second edition, with plates and explanatory notes. New-York: O. Halsted, 1831.
    xii, [13]-155, [9] p., IV leaves of plates : ill. ; 19 cm.

The second edition is identical to the first but for the addition of four leaves of engraved plates (three of which depict the method of "communicating the impulse to the stomach") and four leaves of explanatory letterpress.

1540. HAMILTON, Alexander, 1739-1802.
The family female physician: or, a treatise on the management of female complaints, and of children in early infancy . . . First Worcester edition. Printed at Worcester, Massachusetts, by Isaiah Thomas . . . [et al.], MDCCXCIII. [1793]
    vi, 7-351 p. ; 20 cm. (8vo)

Hamilton's treatise is divided into four principal sections: "The observations in the First Part of the Management of Female Complaints, relate to all the diseases which occur in the unimpregnated state, and include also the changes in consequence of pregnancy. In the Second part, the treatment of the complaints during child bearing is detailed; and in the Third Part, directions are given for the management of lying-in women . . . The first chapter on the Management of Children comprehends . . . those rules for their treatment which experience has proved to be the most effectual means for preventing diseases. In the other chapters, the complaints which occur most commonly during the period of nursing are described, and the mode of cure directed" (*Pref.,* p. v).

A popularization of *A treatise of midwifery* (London, 1781), Hamilton's manual of domestic gynecology and infant care was first published at Edinburgh in 1792 under the title: *A treatise on the management of female complaints, and of children in early infancy.* The first American edition (New York, 1792) and later British and American editions were also issued under this title. Hamilton's popular treatise appeared in five American editions between 1792 and 1818.

Hamilton was professor of midwifery at the University of Edinburgh from 1780 to 1800. He succeeded Thomas Young (1726?-1783), who had established Edinburgh as one of the foremost centers of obstetrical training in Europe. The impact of Young and Hamilton on American obstetrical practice was probably considerable given the number of North Americans who attended their lectures over a period of more than four decades. Hamilton's influence was also felt in the publication of his several obstetrical treatises, most notably the *Outlines of the theory and practice of midwifery* (1st ed., Edinburgh, 1784; 1st Amer. ed., Philadelphia, 1790).
    REFERENCES: *Austin* #860.

1541. HAMILTON, Frank Hastings, 1813-1886.
Discourse on the importance of a general diffusion of a knowledge of anatomy, physiology and hygiene. Delivered at the Auburn Female Seminary, May 30th, 1838 . . . Auburn [N.Y.]: Oliphant & Skinner, printers, 1838.
    20 p. ; 22 cm.

In stressing to his adolescent female audience the importance of the knowledge of the structure of the human body and its operations, Hamilton launches into a defense of the regular medical profession against the incursions of homeopaths, Thomsonians ("Thompson [sic], that American prodigy"), phrenologists, "root doctors" ("those miserable mountebanks") and the "ultra dietetic systems of Graham [#1392-1395] and Mussey [#2527-2531]." Title on printed wrapper: *Hamilton's discourse.* Atwater copy inscribed by author.

Hamilton was an 1835 medical graduate of the University of Pennsylvania. He taught anatomy and surgery at the College and Physicians of the Western District (Fairfield, N.Y.) and Geneva Medical College before opening a surgical practice in Buffalo (1843). In 1846 he joined Austin Flint, Sr. and others on the faculty of the Medical Department of the University of Buffalo, which became one of the nation's premier medical schools at mid-century. In 1858 he moved to Brooklyn; served as a volunteer regimental surgeon during the Civil War; and afterward was appointed professor of surgery at the Bellevue Hospital Medical College, where he remained until retirement in 1875. Hamilton was a prolific author and the president of various regional and national medical societies.

1542. HAMILTON, Frank Hastings, 1813-1886.

The essentials of physiology and hygiene. A text-book to accompany White's Physiological Manikin. Prepared under the supervision of Frank H. Hamilton . . . Illustrating the effects of alcohol and narcotics by colored plates, and by a series of microscopical sections, prepared under the authority of the State of New York by Francis Delafield . . . and T. Mitchell Prudden . . . New York: James T. White & Co., publishers, 1886.

47 p. : ill. ; 19 cm.

Hamilton, Francis Delafield (1841-1915) and Theophil Mitchell Prudden (1849-1924) prepared this manual as a classroom guide for use with *White's physiological manikin* (#3767). Published at New York by James T. White & Co. in 1886, *White's physiological manikin* consists of two hinged wooden frames (37" × 26" each) containing heavy cardboard backing on which are mounted superimposed chromolitho-

graphic plates providing successive interior views of the human body. Hanging on the classroom wall, the life-size manikin "is intended to make plain, even to youngest pupils, the operations and functions of the human body, without recourse to the shambles [i.e., the slaughterhouse] or dissecting room. It will greatly assist the teacher by placing before the pupils an *object lesson,* from which they learn *in spite of themselves* the position and relation of the different parts of the body, and the cause and means of prevention of the derangements it is subject to" (p. [3]).

1543. HAMILTON, Riley Leonidas.

The discoveries and unparalleled experience of Prof. R. Leonidas Hamilton, M.D., with regard to the nature and treatment of diseases of the liver, lungs, blood. And other chronic diseases; containing, also, a biographical sketch of his life, (from Harper's Magazine,) with his common sense theory of diseases and the evidence of his wonderful cures . . . [New York, c1870].

112 p., [1] leaf of plates : port. ; 14 cm.

Hamilton's treatise begins with the narration of a "conversion" experience common to health reform evangelism. He claims to have discovered his medical vocation during the course of a critical illness. Realizing the medical profession's ineptitude in the management of his case, Hamilton determined to survive in spite of the "the boasted remedies . . . used in vain," and turned to the healing power of remedies "from the adjacent fields and forests" (p. 8). He fully recovered, of course, and vowed to devote himself thereafter to medicine. Once saved, Hamilton was determined to save others.

Claiming to be a medical school graduate, Hamilton settled in New York where he devoted himself to treating the "diseases of the lungs, liver and blood." He "planned an immense medical business, and secured the most competent and eminent physicians, chemists, and pharmaceutists in the Union" for the preparation of his proprietary remedies and consultation with patients. "It is now an admitted fact that for many years past Dr. Hamilton has made more examinations, and prescribed for more sick people, than any other man in the world" (p. 10).

The *Discoveries* explains Hamilton's theory of disease (i.e., the effect of North America's varied climate on English and Scots-Irish constitutions), discusses the diseases in which he specializes, promotes his line of vegetable remedies, and provides abundant testimonial correspondence regarding their efficacy. In printed wrapper. Frontispiece portrait of author.

1544. HAMILTON, Weston G.

A compend of domestic medicine and household remedies with the treatment of diseases of adult and infant. What to do and how to do it; with a great many valuable formulas and recipes; many of them alone are worth more than five times the price of the book . . . Greensboro, N.C.: Phillips & Stout Bros., printers, 1887 [1888 issue].
   250, [2] p. ; 18 cm.

Errata slip pasted opposite title-page with imprint of "W.C. Phillips, book and job printer" dated 1888.

1545. HAMLIN, Cyrus, 1811-1900.

Cholera and its treatment . . . Boston: Published by Henry Hoyt, [1866?].
   15 p. ; 11 cm.

A graduate of Bowdoin College and the Bangor Theological Seminary, Hamlin was chosen by the American Board of Commissioners for Foreign Missions to open a Protestant school in Turkey. Arriving in Istanbul in February 1839, Hamlin opened the school the following year and was its director until his resignation in 1860. In 1861 he returned to Turkey to open the privately funded Robert College, of which he was president until 1877. In 1878 Hamlin published *Among the Turks,* an account of his thirty-five years in Turkey. He was appointed president of Middlebury College (Vermont) in 1881.
   Having observed outbreaks of cholera during his many years in Turkey, Hamlin writes, "I think there is no disease which may be avoided with so much certainty as the cholera." He notes that "I have personally investigated at least a hundred cases, not less than three fourths could be traced directly to improper diet, or to intoxicating drinks, or to both united" (p. 4-5).

Hamlin recommends the treatment of cholera by the administration anti-diarrhetics (consisting of equal parts of laudanum and rhubarb) and the application of mustard poultices. In printed wrapper. Date of publication based on the publication of this title in Cincinnati (also 15 p.) in 1866.

1546. HAMMOND, Charles D.

Medical informaton for the million: or the true guide to health, on eclectic and reformed principles; to which is added a practical essay on sexual diseases . . . The second edition, enlarged and improved. New York: William Holdredge, 1850.
   xxix, [30]-524, [4] p. : ill. ; 19 cm.

There are two title-pages. The first bears a wood-engraved vignette with motto; the second is headed "Medical Reform," and is nearly identical in transcription to the first. Both are designated second edition and have the imprint of W. Holdredge. Hammond's domestic medicine discusses the usual variety of illnesses, with special emphasis on sexual disorders or disorders caused by abnormal sexual practice—masturbation in particular. The tone of this work is evident in Hammond's discussion of chlorosis in young females, which he states "arises, in most cases, from stifling or suppressing the calls of nature at this vernal season . . . when the primary command of God, 'Increase and multiply,' is most sensibly impressed upon the whole human fabric. Every tube and vessel appertaining to the genital system being now filled with their appropriate stimulus, the procreative fluid, it excites in the female a powerful, if not involuntary excitability or irritation of the parts, which strongly solicits the means of discharging their burthen by venereal embraces. These . . . being often necessarily denied, the prolific essences seize upon the stomach and viscera; dam up and vitiate the catamenia; choke and clog the perspiratory vessels, whereby the venous, arterial, and nervous fluids become stagnant, and the leucophlegmatic or white, flabby dropsical appearance pervades the whole body like one continued tumor, and quickly consigns the unhappy patient to the arms of death. In this manner, I am sorry to say, are thousands of the most delicate and lovely girls plunged into eternity . . . How much, then, does it become the duty of parents and guardians

. . . to suffer them to marry with the men they love, as it will effect the most rational and most natural cure, by removing the causes of the complaint altogether" (p. 155-56). If matrimony is not convenient, however, a mixture of briony, pennyroyal, mugwort and yarrow "unclogs the genital tubes, purges and cools the uterus and vagina . . . promotes the menstrual discharge," etc.

1547. HAMMOND, Charles D.

Medical information for the million: or, the true guide to health, on eclectic and reformed principles; to which is added a practical essay on sexual diseases . . . The third edition, enlarged and improved. New York: William Holdredge, [c1850].

xxix, [30]-524, [4] p. : ill. ; 19 cm.

Charles D. Hammond is listed as a physician in the New York directories from 1849 through 1875, and in Butler's 1877 medical directory. He moved frequently, and had at least eight different addresses during this period. An eclectic, Hammond was, for a brief period, a frequent contributor to the *American medical & surgical journal* (Syracuse, N.Y.) and was prominent in the politics of the eclectic school. He was also at one time connected with the New York Anatomical Gallery (#2590), which operated from 1849 to 1860. In April 1851 he was involved in a lawsuit because of that affiliation.

Medical information for the million appeared in three editions in 1851, each with a printing of 3,000 copies (*Amer. med. & surg. j.*, 1851, 1:121). Hamilton published several other works: *A new theory of life and death—of health and disease* (1850), in which he claimed "to have been the first to promulgate the idea of the existence in all animate matter of an electrical, vital, or nervous circulation, which is anterior and superior in power to the blood circulation." He also published *What is the modus operandi of medicines on the human system* (1851); *The fallacy of cauterization exposed; or, practical observations for young men* (1859); and *Practical observations on the nature and treatment of seminal diseases* (in a 10th ed. by 1866).

In 1851 Hammond served with Stephen H. Potter and two others on the surgery committee of the New York State Eclectic Medical Society. He and Potter were joint delegates to the National Convention that same year. It was probably as a result of his association with Potter, who was dean of the eclectic Syracuse Medical College, that Hammond was asked to be professor of anatomy and pathology at the College for the fall term of 1851. Though announcements of his appointment were made in the *American medical & surgical journal,* Hammond's name never appears on the list of faculty and subsequently drops from sight in eclectic organizations and publications. Hammond continued to practice in New York City at least until the mid-1870s.

1548. HAMMOND, William Alexander, 1828-1900.

The physics and physiology of spiritualism . . . New York: D. Appleton & Company, 1871.

86 p. ; 19 cm.

An 1848 medical graduate of the University of the City of New York, Hammond became an assistant surgeon with the U.S. Army in 1849, serving for ten years at various frontier posts where he conducted physiological and botanical investigations. He re-enlisted in 1861, and by the following year had risen to the post of Surgeon-General. A poor relationship with Secretary of War Edwin M. Stanton, however, resulted in Hammond's court martial and dismissal from the army. He subsequently established himself in New York where "within a short time he became a leader in the practice and teaching of neurology, a specialty then in its infancy" (*Dict. Amer. biog.*, 4:210). Among Hammond's many published works was *A treatise on the diseases of the nervous system* (1st ed., New York, 1871), the first American treatise on neurology and the most successful 19th-century American text on the subject.

In *The physics and physiology of spiritualism,* Hammond reduces all spiritualistic phenomena either to explicable physical causes or to the credulity of receptive individuals—including clairvoyant and mental healing.

1549. HAMMOND, William Alexander, 1828-1900.

Sleep and its derangements . . . Philadelphia: J.B. Lippincott & Co., 1878.

viii, [9]-318 p. ; 19 cm.

Sleep and its derangements was originally published by Lippincott in 1869. It is an enlarged version of Hammond's *On wakefulness,* published by Lippincott in 1866 (93 p.). This work in turn was based on a two-part article published by Hammond in the *New York Medical Journal* entitled "On sleep and insomnia" (1865, I(2):[89]-101, I(3):182-204). Though originating as an article in a medical periodical, the book is addressed to the public as well as the profession.

HAMPTON TRACT COMMITTEE see **ARMSTRONG** (#157).

1550. HANAFORD, J. H., b. 1819.
Mother and child: a guide to the young mother in promoting the health of her children . . . Boston: D. Lothrop and Company, 1878.
    256 p. ; 19 cm.

    First edition.

1551. HANCHETT, Henry Granger, 1853-1918.
Sexual health: a companion to "Modern domestic medicine" . . . Carefully revised by A.H. Laidlaw . . . Third edition. Philadelphia: The Hahnemann Publishing House, 1891.
    86, [2] p. ; 19 cm.

    The first edition of *Sexual health* was published in 1887 as a separately published companion to *The elements of modern domestic medicine* (New York, 1887). In his preface to *Sexual health* Hanchett writes: "A work on domestic medicine to be of any service must be at hand when wanted, and many persons are not willing that such information as these pages contain, should be within easy reach of boys and girls" (p. [3]).
    The author divides *Sexual health* into four chapters: I. "Sexual health of the male"; II. "Sexual health of the female"; III. "Marriage"; and IV. "The medicines and their indications," i.e., homeopathic medications for the treatment of menstruation disorders, irritation of the genitalia, etc. The author stresses the role that parents must play in the sexual training of their offspring: "A warning that danger to health and morals lie before the young, is to no purpose unless the nature of that danger and the path which leads to it be pointed out.

Young ladies do not know that by exposing their persons in evening dress and allowing intimacies and even receiving caresses from young gentlemen, they often awaken passions in the latter which send them to the brothels for gratification. Mothers often do not know that the long foreskin nature frequently gives their boys is a source of more or less constant irritation to their sexual organs, and consequent excitation to the animal passions, from which circumcision offers the only escape. They do not know that the well-dressed decent appearing 'young lady' whom they pass in the street and who is the picture of decorum while they are within earshot, will shamelessly ask their sons to attend her to her chamber, when she chances to meet them alone" (p. 4-5). The author was an 1884 graduate of the New York Homoeopathic Medical College.

1552. HANCHETT, Henry Granger, 1853-1918.
Sexual health: a companion to "Modern domestic medicine" . . . Carefully revised by A.H. Laidlaw . . . Third edition. Philadelphia: Boericke & Tafel, 1897.
    86, [2] p. ; 19 cm.

1553. HANCOCK, Harrie Irving, 1868-1922.
The physical culture life. A guide for all who seek the simple laws of abounding health . . . New York and London: G.P. Putnam's Sons; The Knickerbocker Press, 1905.
    [2], xiv, [2], 229 p., [22] leaves of plates : ill. ; 19 cm.

    "The elements of physical culture? In the main they are air, food, exercise, and water; and the last might be included with food were not the value of proper bathing to be stated. Clothing and housing are mere incidentals, while sleep is purely supplemental to exercise" (p. 3-4). Chemist, journalist, and the author of numerous series of boys' stories, Hancock was also an advocate of the "physical-culture movement" and the author of such works as *Japanese physical training* (1903), *Physical training for women* (1904), *Physical training for business men* (1917), etc.

1554. HANCORN, James Richard, 1808-1861.

Medical guide for mothers, in pregnancy, accouchement, suckling, weaning, etc. and in most of the important d[i]seases of children . . . From the second London edition. New York: Saxton and Miles; Philadelphia: G.B. Zeiber and Co., 1844.

viii, [13]-117, [3] p. ; 19 cm.

Hancorn's treatise was first published at London in 1842. The second edition was issued the same year. The 1844 New York edition was the only American appearance of this title. In printed wrapper. Hancorn's obituary appears in *The Lancet*, 1861, 2:437.

1555. HAND, William M.

The house surgeon and physician; designed to assist heads of families, travellers, and sea-faring people, in discerning, distinguishing, and curing diseases. With concise directions for the preparation and use of a numerous collection of the best American remedies: together with many of the most approved, from the shop of the apothecary. All in plain English. By a Physician and Surgeon. Hartford: Printed by Peter B. Gleason and Co., 1818.

xxiv, [25]-200 p. ; 17 cm.

Fig. 43. Bottle feeding, from Beach's family physician *(#247).*

First edition. "Physicians and surgeons cannot be present in every place; nor can they *alone* do every thing, which should be done for those to whom they are called. The sick and wounded must depend much on nurses and attendants; and almost every individual thing which is done for the sick, is influenced by the notions or prejudices of the attendants. How important, then, that the means of information relating to the healing art, be extended to every one who may suffer, or who can watch . . . In the following work, I have attempted, in the plainest language, to inform the reader what he should do, when he is a witness to pain and sickness, and no one present better informed than himself" (*Pref.,* p. [iii]).

Hand has divided his treatise into three principal sections: "Surgery" (i.e., the management of wounds, burns, animal bites, dislocations, etc.); "Practice" (i.e., the symptoms, causes and treatment of common diseases); and "American remedies" (i.e., an alphabetical catalog of 158 medicinal plants native to the United States). The concluding essay, "Marasmus" (p. [189]-199), is taken from James Hamilton's *Observations on the utility and administration of purgative medicines.*

REFERENCES: *Austin* #873.

1556. HAND, William M.

The house surgeon and physician; designed to assist heads of families, travellers, and sea-faring people, in discerning, distinguishing, and curing diseases; with concise directions for the preparation and use of a numerous collection of the best American remedies: together with many of the most approved, from the shop of the apothecary. All in plain English . . . Second edition, revised and enlarged. New-Haven: Printed by S. Converse, for Silas Andrus, Hartford, 1820.

xii, 288 p. ; 18 cm.

In this second edition the catalog of medicinal plants ("American remedies," p. [177]-252) has been expanded from 158 to 207 plant names. The essay on "Marasmus" (p. 128-38) is from James Hamilton's *Observations on the utility and administration of purgative medicines*; "Of water" (p. [281]-288) was extracted from Anthony Todd Thomson's *London dispensatory.*

REFERENCES: *Austin* #874.

1557. HAND, William M.

The house surgeon and physician: designed to assist heads of families, travellers, and sea-faring people, in discerning, distinguishing, and curing diseases; with concise directions for the preparation and use of a numerous collection of the best American remedies: together with many of the most approved, from the shop of the apothecary. All in plain English . . . With an introduction, by J.L. Comstock, M.D. Third edition, revised and improved. Hartford: Published by S. Andrus and Son, 1847.
    xxiv, [25]-256 p. ; 19 cm.

But for the addition of Comstock's introduction and the reorganization of the indices, the third edition is a reprint of the second. Andrus & Son re-issued this third Hartford edition in 1849 and 1853. Spine-title: *Surgeon and physician.*

1558. HANNEY, P. M.

How to gain health and long life . . . Chicago: The Hazel Pure Food Company, 1899.
    [2], xi, [1], 114, [2] p. ; 18 cm.

"The present writer, after years of study and experimentation, convinced himself of the *absolute worth of Pure Foods*. Forthwith he made it his aim in life to put upon the market . . . Pure Foods. Nearly fifteen years ago the author went to the Bureau of Agriculture in Washington, demonstrated to this bureau the inestimable value of Pure Foods in promoting health, long life, and happiness, and showed the Bureau that its duty was to see to it that all foods upon the market were prepared scientifically and were absolutely pure . . . Then in order to convince his fellow-beings, and through their suffrage force the government to establish Pure Food Laws, the author organized the Hazel Pure Food Company . . . built mills . . . began the manufacture of Pure Foods, especially of cereals, and placed them upon the market" (p. xiii-ix). The author discusses nutrition, food adulteration, exercise, pure air, pure water and sleep.

1559. HARD, M. K.

Woman's medical guide: being a complete review of the peculiarities of the female constitution and the derangements to which it is subject, with a description of simple yet certain means for their cure. Also, a treatise on the management and diseases of children, with a description of a number of valuable medicinal plants and compounds for domestic use, and a glossary . . . Mt. Vernon [Ohio]: Printed by W.H. Cochran, 1848.
    318 [i.e., 319] p. ; 15 cm.

"A condition essential to the perfect enjoyment of the hymeneal privileges, is a state of mental and physical purity. To cultivate, then, virtuous affections, and to preserve those healthy conditions on which the consummation of sexual attachment depends, constitute objects of inconceivable importance to the moralist and physician, and cannot be overlooked or neglected, without danger of disastrous consequences to social peace and physical well-being" (p. 10).

Following a detailed description of the female genitalia, Hard returns to this theme with a consideration of the "marriageable age" for females, warning that "By an early association with the man of her choice . . . the amorous, lovely, fascinating girl escapes a long train of disease, that in afflictive forms are incident to the necessary concealment and suppression of excitements unindulged; propensities which, in no other circumstances, can be innocently gratified" (p. 22). The "debility and decay" that Hard believes is characteristic of American women he attributes to the inherent delicacy of their "construction," to the demands made on the female system by the "performance of all the duties peculiar to her sphere," and to the "false refinements of modern civilization"—particularly fashions in dress that "confine and obstruct the vital organs so that they cannot perform their offices in a healthy manner" (p. 25).

Hard devotes his second chapter to the disorders of menstruation and their treatment (with a vegetable materia medica, injections, baths, etc.). In chapter III he discusses conception, gestation and the signs of pregnancy, as well as the diseases of pregnancy and its successful management. Sexual physiology and contraception are not mentioned. Chapter IV is devoted to the symptoms and treatment of uterine and vaginal disorders; chapter V to "natural" and "unnatural" labors; chapter VI to puerperal disorders; and chapter VII to the "management and diseases of children." Hard's eighth chapter, in which he contrasts the botanic from the allopathic practice, is one of more interesting in his book. Chapter IX, "Medicinal plants, and compounds," provides formulae for stimulants, tonics, astringents, emetics, cathartics, etc. Throughout his treatise Hard quotes an interesting mix of authors such as Alva Curtis (#852-857) on obstetrics, Marc Colombat on gynecology, John Eberle on pediatrics, Morris Mattson (#2394-2397) on the botanic materia medica, etc.

1560. HARDING, Arthur Robert, b. 1871.
Ginseng and other medicinal plants. A book of valuable information for growers as well as collectors of medicinal roots, barks, leaves, etc. . . . (Revised edition.). Columbus, Ohio: Published by A.R. Harding, [c1908].
367, [1] p., [3] folding leaves of plates : ill. ; port. ; 18 cm.

Harding's book is a guide to the cultivation of medicinal cash crops rather than a manual for the domestic use of medicinal plants. It was reprinted as recently as 1972. All of Harding's other published works pertain to hunting, trapping and fur farming.

1561. HARLAND, Marion, 1830-1922.
Eve's daughters or common sense for maid, wife, and mother . . . New York: Charles Scribner's Sons, 1885.
[4], ix, [10]-454 p. ; 19 cm.

Marion Harland is the pseudonym of Mary Virginia Hawes Terhune, novelist and author on domestic topics. Harland's first novel, *Alone* (1854), was an immediate bestseller and launched her into life-long national prominence. Two years later she married Edward Payson Terhune, by whom she had six children. Harland "scrupulously fulfilled the obligations of domesticity in her own home and all the functions of a minister's wife and community leader outside of it. But she never sacrified her writing career for other needs" (*Amer. nat. biog.*, 21:452). In 1871 Harland published a cookbook, *Common sense in the household*, which also became a national bestseller and went through numerous editions. It "was followed by two

dozen additional cookbooks and domestic advice manuals, all of them urging women to regard homemaking not a drudgery but as an honorable profession" (*ibid.*). First published in 1882, "*Eve's's daughters,* aimed at middle-class wives and mothers, details Terhune's ideas about a woman's need for education and an intellectual life, for robust physical exercise and good nutrition, and for some kind of job training" (*ibid.*). Harland's ideas on the education of young females accord with those of Edward H. Clarke (#659-662), to whom she frequently refers or quotes, as she does George M. Beard (#255) on another area of concern, i.e., "American worry."

HARLAND, Marion see also *TALKS UPON PRACTICAL SUBJECTS* (#3413).

1562. HARPER WHISKEY.
Speaking of hygiene. Hygienically, DR. BARNES' TOOTHPICK is the acme of cleanliness and effectiveness. Hygienically, HARPER WHISKEY is the perfection of purity and maturity . . . [188-?].
Broadside ; 16.5 × 12.5 cm.

"Used in moderation, HARPER is a hygienic aid to the digestive organs, stimulating to the secretion of the gastric juices and causing a healthful animation of all the faculties." This small broadside is accompanied by a printed envelope (4.5 × 7 cm.) that once contained *Dr. Barnes' Hygienic Toothpick,* and a tin replica of a Harper Whiskey bottle (5.5 cm. × 1.5 cm.).

1563. J. N. HARRIS AND COMPANY.
"IT SAVED MY LIFE." Words often expressed by invalids after the use of ALLEN'S LUNG BALSAM . . . As an expectorant, it has many rivals but no equal. CONSUMPTION, COLDS, OR COUGH . . . ALLEN'S LUNG BALSAM . . . It acts on the kidneys! It acts on the liver! Which makes it more than a cough medicine . . . It is harmless to the most delicate child. It contains no opium in any form. J.N. HARRIS & CO., proprietors, Cincinnati, Ohio. [ca. 1871].
Broadside ; 18.5 × 8.5 cm.

J. N. HARRIS AND COMPANY see also **P. DAVIS AND SON** (#911).

1564. HARRISON, Frank.
Series of eight popular and literary drawings illustrating the teeth in health and disease . . . New York: Claudius Ash, Sons & Co., c1908.
[8] leaves : ill. ; 28 cm.

Title and imprint from front board. The author was dental surgeon to the Sheffield Royal Hospital, and lecturer in dental surgery at the University of Sheffield.

1565. HARRISON, J. S.
DR. HARRISON'S Peristaltic Lozenges! A Positve Remedy FOR COSTIVENESS AND DYSPEPSIA . . . J.S. HARRISON, Originator and Proprietor . . . Boston . . . [ca. 1857].
[1] leaf printed on both sides ; 36 × 17 cm.

Printed on the verso are advertisements and testimonials for *Iceland Balsam* and other proprietary remedies available from J.S. Harrison, Boston. Date of imprint based on date of the latest testimonial.

1566. HART, Edith Violet.
The baby's physical culture guide . . . Embodying practical information for baby's development; new and improved methods for physical culture exercises. Chicago; New York: Rand McNally & Company, [c1913].
68, [2] p. : ill. ; 17 cm.

The purpose of the infant exercise program outlined in this volume is to "strengthen Baby's body, expand his lungs, and keep him in condition so that germ attacks or self-poisoning are impossible under ordinary conditions" (p. 19). Twenty-four photographs of the author's six month old son exercising various muscle groups illustrate the text.

1567. HARTELIUS, Truls Johan, 1818-1896.
Home gymnastics for the preservation and restoration of health in children and young and old people of both sexes. With a short method of acquiring the art of swimming . . . Translated and adapted from the Swedish original by special permission of the author by C. Lofving . . . Philadelphia: J.B. Lippincott Company, 1891.
x, 94 p. : ill. ; 18 cm.

Translation of: *Hemgymnastik.* Cornelia Lofving's translation was first published at London circa 1881. An 1858 Lund medical graduate, Hartelius became director in 1865 of Stockholm's Central-Institutet, the gymnastic training institute established by Per Henrik Ling (1776-1839).

1568. HARTELIUS, Truls Johan, 1818-1896.
Swedish movements or medical gymnastics . . . Translated by A.B. Olsen, M.D. with introduction and notes by J.H. Kellogg . . . Battle Creek, Mich.: Published by the Modern Medicine Pub. Co., 1896.
ix, [1], 3-162 p., XXI p. of plates + frontis. : ill., port. ; 23 cm.

In 1870, Hartelius published his *Lärobok i sjukgymnastik,* of which the present work is the translation. It is divided into two parts. In the first, Hartelius describes the positions and movements for each exercise; in the second, he describes their application to the treatment of specific diseases. During his first trip to Europe in 1883, John Harvey Kellogg "visited a number of European 'water cures' and also made a side trip to Scandinavia. In Stockholm he investigated the exercise machinery and gymnastic program begun by Dr. Gustaf Zander and T.J. Hartelius, pioneers in exercise therapy. Both men spent hours explaining their ideas and techniques to him" (Richard W. Schwarz, *John Harvey Kellogg, M.D.,* Nashville: Southern Publ. Assn., 1970, p. 34).

HARTER, M. G. see **DR. HARTER MEDICINE CO.** (#1569, 1570).

HARTER, Theodore Clarence see **FARLIN** (#1129).

1569. DR. HARTER MEDICINE COMPANY.
Dr. Harter's Iron Tonic. [*Five portraits through which runs the text*:] President Harrison and Family . . . [ca. 1889].
[1] leaf printed on both sides : ports. ; 31 × 18 cm.

Lithograph. Portraying the first family on the recto, the verso advertises: "DR. HARTER'S IRON TONIC. Acts like a charm on the digestive organs. In a low, weak, tired and tremulous state of the system it is revivifying and peculiarly happy in its results. The *Iron* restores richness and color to the blood . . ." The testimonials printed on the verso are dated 1885-1889. The Dr. Harter Medicine Co. was located in St. Louis.

1570. DR. HARTER MEDICINE COMPANY.
Dr. Harter's valuable information . . . St. Louis, Mo.: The Dr. Harter Medicine Co., [ca. 1894].
        64 p. : ill. ; 10 × 15 cm.

Promotional literature for *Dr. Harter's Family Medicines*. Title from chromolithographed wrapper.

1571. HARTLEY, W. H.
Pollypathic physicians. South American cure! A medicine which performs more cures in indigestion, dyspepsia, catarrh either in the throat or head, rheumatism, liver complaints and kidney diseases, than any other medicine on the face of the earth. Over 30,000 bottles sold in Rochester during our recent visit and since . . . Chronic diseases a specialty. Prof. W.H. Hartley. [n.d.]
        [1] leaf printed on both sides ; 17 × 11 cm.

On verso: *Preserve this. Strange Bible facts . . .*

1572. HARTMAN, Samuel Brubaker, 1830-1918.
The confidential physician or the theory and practice of medicine simplified . . . Principle divisions: lecture—the ills of life—monograph on spermatorrhoea—sexual misfortunes. Female complaints and other diseases. Revised edition. Columbus, O.: Published by Dr. S.B. Hartman & Co., [c1886].
        [8], 112 p. : ill., port. ; 24 cm.

Hartman's manual of medical advice is actually a vehicle for the promotion of a panacea of his own manufacture: "Fully believing, as the author has for many years, that nature was the best doctor any patient could have, that she would cure him if permitted to use her own methods and that all that was required in most

diseases was to simply aid nature in her efforts, he has for years applied himself studiously and with the best means at his command, to form a compound which would if administered judiciously and in time, encourage the vital powers and assist the *vix medicatrix naturae.* So confident is he of having accomplished this great object that he now lays before the public and the medical profession this invaluable result of the labors of his life, PE-RU-NA" (p. 3).

1573. HARTMAN, Samuel Brubaker, 1830-1918.
The confidential physician: or the theory and practice of medicine simplified . . . Revised edition. Columbus, Ohio: Published by S.B. Hartman, M.D., the Hartman Sanitarium, [c1899].
        [8], 174, [6] p. : ill. ; 23 cm.

An 1857 graduate of the Jefferson Medical College, Hartman began a lucrative, twenty-year career as an itinerant physician in 1870 before settling in Columbus, Ohio. This edition of *The confidential physician* includes an illustration of the sprawling complex that Hartman built in Columbus, consisting of the manufactory for *Pe-ru-na* (his most successful proprietary remedy), a hotel, a sanitarium and an administration building. Hartman claims to have discovered the original recipe for *Pe-ru-na* among the formulae of Wooster Beach (#240-251). He continually improved its formula, "adapting it equally well to diseases of all the membranes of the body, the serous membrane as well as the mucuous membrane, and . . . particularly . . . in the treatment of kidney and pelvic diseases" (p. [1]). Thanks to effective advertising, *Pe-ru-na* was by 1900 the best-selling proprietary remedy in the United States. The fact that it was nearly 30% alcohol doubtless contributed to its popularity: "The 'Peruna jag,' a slight intoxication from drinking the nostrum, became so common a phenomenon, especially in areas dry by Prohibition laws, that druggists had trouble keeping it in stock" (*Amer. nat. biog.,* 10:259). *Pe-ru-na* came under attack by Samuel Hopkins Adams (#10-12) during the A.M.A.'s anti-nostrum campaigns, was banned from Indian reservations by the federal government in 1905, and was eventually limited to sale only by licensed liquor agents. Nonetheless, Hartman remained the wealthiest man in Columbus.

1574. HARTMAN, Samuel Brubaker, 1830-1918.
Lectures on chronic catarrh . . . [Columbus, Ohio, after 1896].
[4], 98 p. : ill. ; 23 cm.

"Catarrh is essentially the same everywhere. The remedy that will cure catarrh in one situation will cure it in all situations. It does not require one remedy to cure catarrh of the head and another to cure catarrh of the pelvis, and still another to cure catarrh of the stomach, and another to cure catarrh of the kidneys. A remedy that will cure catarrh anywhere will cure catarrh everywhere. If there is such a remedy in existence that remedy is Pe-ru-na" (p. 2). In promoting his remedy, Hartman could accurately claim, "In every city and hamlet of the land the phrase that Pe-ru-na cures catarrh wherever located has become almost an axiom." Title from front wrapper.

1575. HARTMAN, Samuel Brubaker, 1830-1918.
Our latest. A catarrh book up to date. Columbus, O.: The Peruna Drug M'f'g Company, 1896.
16 p. : ill. ; 18 cm.

1576. HARTMAN, Samuel Brubaker, 1830-1918.
A short treatise on pelvic surgery . . . Columbus, Ohio: [Hartman Sanitarium, c1899].
48, [2] p. : ill., map ; 23 cm.

Purportedly written for the medical profession, Hartman's explanation of "pelvic philosophy" is largely an attempt to foist "fantasy surgery" on a public having too much faith and too little knowledge to distinguish science from pseudo-science, and medicine from medical imposture. Hartman's lead essay on "nerve waste," the resulting irritations, and the surgical interventions required is supplemented by a "Synopsis of treatment" by David Rittenhouse Summy (1853-1923; M.D. Jefferson Med. Coll., 1883), Superintendent of the Hartman Sanitarium. The brochure is illustrated with full-page, half-tone photographs of operating rooms and patient care areas. Title on wrapper: *Pelvic surgery.*

HARTMANN, Franz see **CASPARI** (#573).

HARTSHORN, Edward see **DR. E. HARTSHORN & SONS** (#1577).

1577. DR. E. HARTSHORN & SONS.
Fail not to read the following statement about DR. HARTSHORN'S JAUNDICE AND DYSPEPTIC BITTERS known everywhere as "KEY TO HEALTH." By Dr. E. Hartshorn . . . PURELY VEGETABLE! Dr. Hartshorn's Bitters always cure Sallowness of Skin, Dulness [sic] or Dizziness of Head, Sick Headache, Bad Taste of Mouth, Weakness of Limbs, Weariness, Bad Sleep, Poor Appetite, and all Bilious Complaints . . . These Bitters contain Gentian, Anise, Coriander, Orange, Serpentaria, Galangal, Lacender, Thoroughwart, Prickly Ash, Senna, Rhubarb, Extract White Walnut, Juniper, Cassia, Dandelion, &c., and so combined and prepared as to extract the mild alternative, tonic and laxative qualities of these vegetable remedies, in proportion to every person in every situation . . . DR. E. HARTSHORN & SONS, 71 Blackstone St., Boston. [189-?].
Broadside ; 33 × 24.5 cm.

Edward Hartshorn (1818-1906) received his medical degree from Harvard in 1840. He is listed in the 1893 Boston business directory at 71 Blackstone Street.

1578. HARTSHORNE, Henry, 1823-1897.
The family adviser and guide to the medicine chest. A concise hand-book of domestic medicine . . . Philadelphia: John Wyeth and Brother, 1869.
iv, 5-104, 32 p. ; 17 cm.

Second edition (1st ed., 1868). The final 32 pages consist of advertisements for "medicine chests, medical saddle bags, and medicine pocket cases" manufactured by John Wyeth & Brother, as well as medicines, flavorings, fruit syrups, mineral waters, hair tonics, "appliances for the sick" (e.g., bed pans, syringes, atomizers, etc.) available from the Wyeth brothers. Cover-title: *Guide to the medicine chest.*

1579. HARTSHORNE, Henry, 1823-1897.
The practical household physician. A cyclopedia of family medicine, surgery, nursing and hygiene for daily use in the pres-

ervation of health and care of the sick and injured. Containing a plain description of the parts of the human body and their uses; chapters on "our homes," climate, food and drink, use of intoxicants and narcotics; special chapters giving important information for assisting the skillful efforts of the doctor, and for the treatment of accidents and diseases. Arranged for ready reference to enable one to do instantly what can and ought to be done in emergencies to relieve suffering or save life . . . Philadelphia, Pa.; Chicago, Illinois: John C. Winston & Co., 1889.

xxxi, [32]-971 p., [9] leaves of plates : ill. ; 25 cm.

An 1845 medical graduate of the University of Pennsylvania, Hartshorne held teaching positions at various Philadelphia institutions, including the Women's Medical College of Pennsylvania (1869-76). His extensive list of publications includes several manuals of domestic medicine and hygiene, e.g., *The family adviser and guide to the medical chest* (1st ed., 1868), *A household manual of medicine* (1886), and *Our homes* (1880), a guide to domestic sanitation. Cover- and spine-titles: *Household manual of medicine, surgery, nursing and hygiene.*

HARTSHORNE, Henry see also *THE STANDARD BOOK OF RECIPES* (#3322).

1580. HARVARD, C. S.
New England Cough Syrup, the reputation of which has now become established as the most safe and efficacious remedy ever discovered for influenza, cough, colds, asthma, whooping-cough, spitting of blood, and all affections of the lungs . . . Prepared by C.S. Harvard. Wholesale agent, Thomas Hollis . . . Boston. [184-?].
Broadside ; 21.5 × 11 cm.

HARVARD HEALTH TALKS see **CHAPIN** (#582); **CUTLER** (#858).

1581. HASKELL, Charles Courtney.
Perfect health. How to get it and how to keep it. By one who has it. True scientific living. Norwich, Conn.: Charles C. Haskell; London, England: L.N. Fowler & Co., 1902.
209, [1] p. ; 18 cm.

Charles C. Haskell was the publisher of Edward Hooker Dewey, whose *True science of living* (#941, 942) he agreed to publish after the recovery his own health and from the testimony of other persons whose health had been restored by following Dewey's dietary regimen. Haskell summarizes Dewey's "new gospel of health": "First. To abstain absolutely from the the early morning meal. Second. Never to eat except with natural hunger. Third. To masticate every mouthful of food as long as there is any taste in the food. Fourth. To abstain from all drink with meals" (p. 28). Haskell adds a spiritual component to this regimen, concluding that true "holiness (wholeness)" also "manifests itself in the outer material world" in the form of perfect health.

HASTINGS, Caroline Eliza see *DRESS-REFORM* (#990).

1582. HATFIELD, Marcus Patten, 1849-1909.
The physiology and hygiene of the house in which we live . . . New York: Chautauqua Press, C.L.S.C. Department, 1887.
283, [5] p., [2] leaves of plates : ill. ; 21 cm.

Hatfield was an 1872 graduate of the Chicago Medical College and professor of pediatrics in the Medical Department of Northwestern University.

HATHORN, E. H. see **HATHORN SPRING WATER** (1583#).

1583. HATHORN SPRING WATER.
The Hathorn Spring. What its rivals say of it and what it is. A reply to the statements recently put forth by Congress and Empire Spring Co. Saratoga Springs, New York, July, 1884.
47, [1] p. ; 14 cm.

"The Congress and Empire Spring Company of Saratoga has recently issued, and extensively circulated, a pamphlet characterized by gross misstatements relative to the Hathorn Water. This has been done with the intention of injuring the reputation of that water in the interest of its own commodities" (p. [3]).
E.H. Hathorn defends the Hathorn Spring from his rival's charge that Hathorn waters con-

tain excessive levels of iron. In turn, Hathorn charges the Congress and Empire Spring Co. with treating their water with salt, not bottling directly from the spring, and offering the public water contaminated by "the overflow of the public sewers." Hathorn also claims that waters from the Congress Spring contain higher levels of iron than his own.

1584. HAVENS, F. C.

The possibility of living 200 years. Compiled from the best authorities . . . San Francisco, Cal.: The Two Hundred Company, 1896.
    vii, [1], 216 p. ; 17 cm.

A compilation on proper diet as the guarantee of health and longevity extracted from the writings of Joseph Addison, Luigi Cornaro (#791-798), William Whitty Hall (#1491-1519), Herbert Spencer, etc., to which are added Haven's reflections on the Ralston Health Club (#2912-2926) and the relation of diet to health and aging.

HAYDEN, William R. see NEW YORK PHARMACEUTICAL CO. (#2602-2604).

1585. HAYES, Albert Hamilton, 1840-1911.

A medical treatise on nervous affections. The result of an extensive experience in the treatment of nervous disorders . . . Boston: Published by the Peabody Medical Institute, [c1870].
    vi, [7]-134, [3] p., [1] leaf of plates : port. ; 20 cm.

Hayes was director of the Peabody Medical Institute (Boston), which specialized in the treatment of mental and nervous diseases with proprietary remedies. Hayes addresses five forms of "mental or nervous disease," which he assures his reader may be cured at the Institute or with remedies ordered therefrom: oppression, confusion, delusion, excitement and "diminution" (i.e., "absence of mental energy"). Though the means of cure employed by Hayes and his colleagues at the Peabody Medical Institute are never specified, abundant correspondence from the formerly afflicted testifies to the efficacy of what is several times referred to as simply the "Discovery." Steel-engraved frontispiece portrait of the author.

1586. HAYES, Albert Hamilton, 1840-1911.

Physiology of woman, and her diseases; or, woman, treated of physiologically, pathologically and esthetically . . . Boston: Published by the Peabody Medical Institute, [187?, c1869].
    ix, [1], xv-xxiv, [25]-345, [5] p., [1] leaf of plates : port. ; 20 cm.

"I should say that the majority of women (happily for them) are not very much troubled with sexual feeling of any kind. What men are habitually, women are only exceptionally. It is too true, I admit, that there are some few women who have sexual desires so strong, that they surpass those of men, and shock public feeling by their exhibition. I admit, of course, the existence of sexual excitement, terminating even in nymphomania . . . But with these sad exceptions, there can be no doubt that sexual feeling in the female, is, in a majority of cases, in abeyance, and that it requires positive and considerable excitement to be roused at all; and, even if roused . . . is very moderate, compared with that of the male" (p. 225-26).

Hayes intended the *Physiology of woman* to be the companion volume to his sexual physiology for males, the *Science of life* (#1587-1591). He provides (his presumably female audience) social, moral and medical perspectives on puberty, marriage, sexual physiology, conception, pregnancy, hygiene and the common diseases of women. Hayes devotes a brief chapter to the prevention of conception, in which he condones contraception without discussing methods.

The *Physiology of woman* appeared in at least two issues (with an 1869 copyright date), and was published as well under the title: *Sexual physiology of woman* (#1592). One issue of these two issues of the *Physiology of woman* includes eight chromolithographs depicting female figures of a decidedly erotic nature. These plates do not appear in the issue in the Atwater Collection, which does, however, contain a steel-engraved frontispiece portrait of Hayes not present in the other issue. It is possible that the Atwater copy is a later issue from which Hayes deleted the plates in order to avoid problems similar to those faced by E.B. Foote (#1213-1256) about this same time as a result of the Comstock laws. The Atwater copy was issued without pages x-xiv ("Opinions of the Press"), also indicating a later issue. Half-title: *Sexual physiology of woman.*

1587. HAYES, Albert Hamilton, 1840-1911.
The science of life; or self-preservation. A medical treatise on nervous and physical debility, spermatorrhoea, impotence, and sterility, with practical observations on the treatment of diseases of the generative organs . . . [Boston]: Published by the author, [c1868].
[4], 5, [1], [9]-35, [44]-52, 55-254, 265-278, [2] p. : ill. ; 15 cm.

First edition. *The science of life* was written for the edification of a male audience on sexual physiology and on the social and medical consequences of deviation from normal sexual behavior. In his opening chapter, Hayes describes the female organs of generation and their role in conception and pregnancy. He follows this introduction to reproductive physiology with several chapters on the venereal diseases. Hayes discusses at length the social and moral purposes of marriage, and the role of sexual relations within that context. The greater part of the book, however, is devoted to a discussion of spermatorrhoea and impotence, disorders which he attributes principally to masturbation, a form of behavior described as having "immolated more victims than war, famine, and pestilence put together" (p. 83). A revised edition of *The science of life* was published in 1881 by the Peabody Medical Institute under the names of George H. Jones (#2040) and William H. Parker (#2742-2745).

1588. HAYES, Albert Hamilton, 1840-1911.
The science of life; or self-preservation. A medical treatise on nervous and physical debility, spermatorrhoea, impotence, and sterility, with practical observations on the treatment of diseases of the generative organs . . . Boston: Published by the Peabody Medical Institute, 1871.
[6], 35, [44]-52, 55-254, 265-278, 11 p., [1] leaf of plates : ill., port. ; 15 cm.

Includes a "Preface to the Tenth Edition" (p. [5-8]). The final 11 pages advertise Hayes' books.

1589. HAYES, Albert Hamilton, 1840-1911.
The science of life; or self-preservation. A medical treatise on nervous and physical debility, spermatorrhoea, impotence, and sterility, with practical observations on the treatment of diseases of the generative organs . . . Boston: Published by the Peabody Medical Institute, 1874.
[2], 35, 44-52, 55-254, 265-278, 3-7, 6-13 p. : ill. ; 14 cm.

1590. HAYES, Albert Hamilton, 1840-1911.
The science of life; or self-preservation. A medical treatise on nervous and physical debility, spermatorrhoea, impotence, and sterility, with practical observations on the treatment of diseases of the generative organs . . . Boston: Published by the Peabody Medical Institute, [1876?].
[4], 35, [44]-52, 55-254, 265-276, 3-7, [1], 11-13, [7] p. : ill. ; 15 cm.

Publication date based on references to the recent centennial and a medal presented to Hayes by the National Medical Association (1 Jan. 1876). Atwater copy lacking title-page (supplied in photocopy facsimile).

1591. HAYES, Albert Hamilton, 1840-1911.
The science of life; or self-preservation. A medical treatise on nervous and physical debility, spermatorrhoea, impotence, and sterility, with practical observations on the treatment of diseases of the generative organs . . . Boston: Published by the Peabody Medical Institute, [after 1876].
[8], 35, [44]-52, 55-254, 265-276, 3-7, [1], 11-13, [7] p. : ill. ; 15 cm.

Preliminary leaves include text of the commendation (dated 1 January 1876) that accompanied the medal presented to Hayes by the National Medical Association

1592. HAYES, Albert Hamilton, 1840-1911.
Sexual physiology of woman, and her diseases; or, woman, treated of physiologically, pathologically, and esthetically. With eight elegant illustrative pictures of female beauty . . . Boston: Published by the Peabody Medical Institute, [c1869].

xxiv, 25-345, [3] p., [8] leaves of plates : col. ill. ; 20 cm.

This work was also issued (c1869) under the title: *Physiology of woman* (#1586). The "eight elegant illustrative pictures of female beauty" described on the title-page are chromolithographic plates depicting female nudes, five of which are plainly erotic in their appeal. The plates were eliminated in one of the two issues of the *Physiology of woman*.

HAYES, Albert Hamilton see also **PEABODY MEDICAL INSTITUTE** (#2758).

HAYES, Francis Mason see **ASTHMATICS' INSTITUTE** (#166, 167).

1593. HAYES, Harriet E.
The home nurse and nursery. A practical treatise on the management of the sickroom; its ventilation, temperature and cleanliness; the care of the patient; observations and administering medicines; with chapters on accidents and emergencies; domestic medicines; complete cookery for the sick room; and special department on the management and care of infants . . . New York; Portland, Me.: Union Publishing House, 1888.
  363 p. ; 20 cm.

HAYES, Pliny Harold see **ASTHMATICS' INSTITUTE** (#166, 167), **HYDROPATHIC AND HYGIENIC INSTITUTE** (#1877), **WYOMING COTTAGE WATER-CURE INSTITUTE** (#3891).

1594. HAYES, Thomas, d. 1787.
A serious address, on the dangerous consequences of neglecting common coughs and colds, with ample directions for the prevention and cure of consumptions. To which are added observations on the hooping cough and asthma . . . Walpole, N.H.: Printed for W. Fessenden and G.W. Nichols, 1808.
  ix, [10]-136 p. ; 19 cm.

"Two very sensible writers, Tissot [#3534-3542] and Buchan [#442-502], have addressed

themselves to the public in general, on the subject of medicine; and many excellent directions they have given respecting the management of colds, and other diseases; but on this subject, they have not entered so fully as the disorders require . . . Hence this little tract may not, perhaps, be without its use, as a companion to their celebrated works" (*Pref.*, p. iv).
  Second American edition. The first American edition of Hayes' treatise was published at Boston in 1796. It was originally published at London in 1783, and appeared in four editions through 1786. A third American edition was issued at Philadelphia in 1813. Hayes was a member of the Corporation of Surgeons (London).
  REFERENCES: *Austin* #891.

HAYMANN, L. see **BAUNSCHEIDT** (#237).

HAYNES, Arvilla Breton see *DRESS-REFORM* (#990).

1595. HAYWARD, Abraham, 1801-1884.
The art of dining and of attaining high health; with a few hints on suppers. To which is added anecdotes of dining, connected with distinguished individuals. By a Bon Vivant. New York: Robert M. De Witt, publisher, [c1874].
  288 p. ; 16 cm.

First American edition. The English jurist and essayist originally published his remarks on "aristology" (from the Greek *to ariston*, i.e., meal) in the *Quarterly review* in 1835 and 1836. They extend beyond gastronomy to include observations on "sociability and health," digestion, exercise, etc.

1596. HAYWARD, George, 1791-1863.
Outlines of human physiology; designed for the use of the higher classes in common schools . . . Boston: Marsh, Capen & Lyon, 1834.
  6, [13]-217 p. : ill. ; 19 cm.

First edition. Hayward's text was the earliest juvenile physiology published in the United States. This milestone in the early literature of health reform and popular health education appeared in four Boston editions between 1834 and 1842. Hayward received his medical degree from the Univ. of Pennsylvania in 1812. In 1834 he joined the faculty of Harvard Medical School. Hayward is credited with having performed the first major surgical operation in the United States during which the patient was anesthetized. On 7 November 1846, he amputated the leg of a patient under ether.

1597. HAZARD, Thomas Robinson, 1797-1886,
A family medical instructor. Civil and religious persecution in the State of New York . . . Boston: Colby & Rich, 1876.
        125, [3] p. ; 15 cm.

First edition. With a fortune amassed from textile manufacturing, Hazard retired in his mid-forties to devote himself to various reform issues, including economic reform, education, the treatment of the insane, abolition, etc. Following the death of his wife and two daughters, Hazard became increasingly interested in spiritualism. In this work, Hazard takes issue with legislation passed in the State of New York (1874) requiring practitioners who were not medical school graduates to obtain certification from the State, county or district medical society. Hazard writes: "I see . . . that the M.D.s of New York have commenced broadening their opportunity for mischief in their death-dealing profession by taking the initiatory step to stop by fine and imprisonment a mediumistic doctress [Mrs. Holmes] from healing the sick after the fashion and order prescribed, practiced and *commanded* by Jesus Christ, of imparting health and vitality through 'the laying on of hands'" (p. 4). Throughout this treatise, Hazard contrasts "the criminal practices of the *faculty*" (and the collusion of the legal and clerical professions) with the successes of clairvoyant and spiritualistic healers..

1598. HAZARD, Thomas Robinson, 1797-1886.
A family medical instructor. Civil and religious persecution in the State of New York . . . Fifteenth thousand . . . Boston: Colby & Rich, 1876.
        125, [3] p. ; 15 cm.

1599. HEALD, George Henry, 1861-1934.
Colds. Their cause, prevention, and cure . . . Washington, D.C.: Review and Herald Publishing Assn., 1907.
        62 p. : ill. ; 19 cm.

Originally published in 1905. Heald was editor of *Life and health*, which between 1885 and June 1904 was published as the *Pacific health journal*, and which became *Health and temperance* in June 1915. He was an 1893 graduate of the Cooper Medical College (San Francisco).

1600. HEALTH FOR LITTLE FOLKS.
Health for little folks. New York; Cincinnati; Chicago: American Book Company, [c1890].
        iv, [2], 121 p. : ill. ; 18 cm.

Health for little folks was the first of three titles in the *Authorized physiology series,* and was intended for students in the primary grades.

1601. HEALTH-LIFT COMPANY.
The Health-Lift reduced to a science. Cumulative Exercise. A thorough gymnastic system in ten minutes once a day. Health restored and muscular strength developed by equalizing and invigorating the circulation. The result of twenty years' practical and theoretical study and experiment. The only scientific system of physical training . . . The reactionary lifter. New York: The Health-Lift Company, 1873.
        24 p. : ill. ; 14 cm.

Heavy weight lifting entered the American gymnasium in the 1850s. In her book *Physical culture and the body beautiful* (Macon, Ga.: Mercer Univ. Press, 1998, p. 185), Jan Todd attributes much of this interest to the "strong men" of travelling circuses and to the strength sports brought to America by Scottish and German immigrants.

According to this promotional brochure, the *Reactionary Lifter* is more than an apparatus for muscular development. Through regular exercise on this machine, "The strength of the whole body is augmented and equalized, the weak parts are built up, disease is expelled, and the individual becomes uniformly strong, and consequently healthy . . . All the voluntary and respiratory muscles are brought into harmonious play, expanding the chest, augmenting the breathing capacity, aerating the blood, equalizing the circulation, warming the extremities, and thus vitalizing every part" (p. 4). By increasing blood circulation, exercise on the *Reactionary Lifter* removes "internal congestions wherever located" and accelerates the nutrition of every organ, especially the heart, lungs, liver, kidneys and uterus. In addition, the Health-Lift system corrects "deformities, contractions, and distortions of the chest, limbs and spine." It benefits do not stop with bodily health, however: "All the elements of a perfect manhood are increased, including not only intellectual vigor, but moral power and social purity." (p. 4-5).

1602. HEALTH-LIFT COMPANY.

The Health-lift reduced to a science. Cumulative Exercise. A thorough gymnastic system in ten minutes once a day. Health restored and muscular strength developed by equalizing and invigorating the circulation. The result of twenty years' practical and theoretical study and experiment. The only scientific system of physical training . . . The Reactionary Lifter. New-York: The Health-Lift Comapny, 1876.

48 p. : ill. ; 18 cm.

"It is claimed that a system of Physical Culture has been developed and perfected, in which thousands . . . find—'That fullness of life, that vigorous tone, and that elastic cheerfulness which makes the mere fact of existence a luxury:' 'A regular system of exercise' which 'has brought to many a jaded, weary, worn-down human being the elastic spirit, the simple, eager appetite, the sound sleep of a little child:' 'A hygienic force' which makes 'perfect spiritual religion' possible, by making 'perfect physical religion' practicable . . . An exercise, finally, that is 'adapted to our civilization,' in that it is scientifically accurate, occupies the minimum of time and produces a maximum of results" (p. 4). In printed wrapper.

1603. HEALTH-LIFT COMPANY.

Manual for the use of the Reactionary Lifter in the system of Cumulative Exercise commonly known as the Health Lift or lift-

ing cure. Eleventh edition. New York: The Health-Lift Company, 1876.

22, [4] p. : ill. ;18 cm.

In printed wrapper. Cover-title: *The Reactionary Lifter.*

HEALTH-LIFT COMPANY see also **REILLY** (#2956, 2957).

1604. HEALTH: OR HOW TO LIVE.

Health: or how to live. Number one . . . Battle Creek, Mich.: Steam Press of the Seventh-Day Adventist Publishing Association, 1865.

6 pts. in 1 v. (iv, [5]-64; 64; 64; 64; 80; 64 p.) ; 16 cm.

In his introduction, James White (d. 1881), the husband of Ellen White and co-founder of the Seventh-Day Adventist church, states that the purpose of this "series of pamphlets" is not to cure the sick, but "to draw from personal experience, from the word of God, and from the writings of able and experienced health reformers, facts for the common people, which we ardently hope may teach them how to preserve vital force, live healthily, save doctor's bills, and be better qualified to bear with cheerfulness the ills of mortal life" (p. [iii]). On pages 12-18 of the first "pamphlet," James White describes the visit he and his wife made in September 1864 to Our Home on the Hillside (#2675-2679), the Dansville, N.Y. health resort operated by James Caleb Jackson (#1902-1955) that served as the model for the water-cure opened by the Whites two years later in Battle Creek, Michigan.

James C. Jackson is among the "able and experienced health reformers" most often quoted in this "collection" of pamphlets (they were never issued separately). Other authors whose works are published or abstracted in its pages include John Gunn (#1458-1475), the hydropathic physician Eliza De La Vergne, Our Home's Harriet N. Austin (#174-178) and Horace Mann.

Health was also the vehicle for publication of the six chapters of Ellen White's "Disease and its causes." In this inaugural venture into the literature of health reform, White validates ideas current among many health reformers of the period, and enthusiastically adopts the doctrines of hydropathy, vegetarianism and dress reform. White later claimed that "After I had written my six articles for *How to live,* I then searched the various works on hygiene and was surprised to find them so nearly in harmony with what the Lord had revealed to me. And to show this harmony . . . I determined to publish *How to live,* in which I largely extracted from the works referred to" (quoted in: Ronald Numbers. *Prophetess of health.* Knoxville: Univ. of Tennessee Press, 1992, p. 95). Divinely inspired or not, Numbers notes the "striking similarity" between White's ideas and those expressed in works such as L.B. Coles' *Philosophy of health* (#713-721). Spine-title: *How to live.*

HEALTH PRIMERS: BATHS AND BATHING (#233); COUPLAND (#806); GREENFIELD (#1418); THE HEART AND ITS FUNCTION (#1607); THE HOUSE AND ITS SURROUNDINGS (#1798) RALFE (#2908).

1605. HEALTH REFORMER.
Pork; or the dangers of pork-eating exposed. [Battle Creek, Mich.: Health Reformer, 187-?].
   16, 4 p. : ill. ; 17 cm.

"In the case of no other animal is so large a portion of the dead carcass utilized as food. Pork seems to be considered such a delicacy that not a particle should be wasted. The fat and lean portions are eaten fresh, or carefully preserved by salting or smoking, or both. The tail is roasted, the snout, ears, and feet, are pickled and eaten as souse; the intestines and lungs are eaten as tripe or made into sausages; black pudding is made of the blood; the liver, spleen, and kidneys, are also prized; the pancreas and other glands are considered great delicacies; while even the skin is made into jelly. In fact, nothing is left of the beast but his bristles, which the shoemaker claims. Surely, it must be quite an important matter, and one well-deserving attention, if it can be shown that an animal which is thus literally devoured, and that in such immense quantities, is not only unfit for food, but one of the prime causes of many loathesome and painful maladies" (p. 2). Caption title. Above the title appears the series statement: *Health tract; no 7.*

HEALTH SCIENCE LEAFLETS see INTERNATONAL HEALTH AND TEMPERANCE ASSOCIATION (#1895)

HEALTH TRACT see HEALTH REFORMER (#1605).

1606. HEAP'S PATENT EARTH CLOSET COMPANY.
Healthy homes. How to have them . . . Muskegon, Mich., U.S.A.: Published by Heap's Patent Earth Closet Co., [c1886].
   35, [1] p. : ill. ; 15 cm.

Healthy homes provides advice on domestic sanitation generally (e.g., ventilation, heating, disinfection), as well as explaining the sanitary advantages of Heap's patented *Dry Earth Closet.* Title and imprint from printed wrapper.

HEARD, Franklin Fiske see STORER (#3381).

1607. THE HEART AND ITS FUNCTION.
The Heart and its function . . . New York: D. Appleton and Company, 1881.
   [2], 95 p. : ill. ; 15 cm.

   Series: *Health primers; no. 8.*

1608. HEATH, A. S.
Consumption, diseases of the heart, cancer and chronic diseases, their prevention and cure by Old Dr. Heath's celebrated Japanese and East India medicines . . . With a full account of their discovery in his travels and their curative influence in consumption, coughs, colds, bronchitis, asthma, nervous debility, scrofula, dyspepsia, liver complaint, fever and ague, intermittent, remittent, bilious, typhoid and typhus fevers and chronic diseases. To which is added a large number of letters from patients cured in every stage of disease, now in the full and perfect enjoyment of health . . . Tenth edition. New York City: Drs. Heath, 1860.
   vi, 135, [11] p. : ill. ; 19 cm.

1609. HEDGES, Samuel Parker, 1841-1924.
Home treatment for children with lessons on moral training. Containing various homoeopathic remedies for diseases, acci-

dents, etc., to which infants and children are subject, with full directions to nurses for the care of the sick, diet, etc., etc. . . . Chicago: Blakely, Marsh & Co., printers, 1881.

174 p. ; 20 cm.

Hedges was an 1867 graduate of the Hahnemann Medical College & Hospital. He was later on the faculty of his *alma mater,* as well as that of the Chicago Homoeopathic College. At the time of this book's publication, Hedges was physician-in-chief to the Chicago Nursery & Half Orphan Asylum. He was president of the Illinois Homoeopathic Medical Association in 1883.

1610. HEIDELBERG MEDICAL INSTITUTE.
The private medical adviser. Being a medical treatise on the most prevalent ills, having special reference to the diseases of men. Revised and enlarged edition. St. Paul, Minn.: Published by the Heidelberg Medical Institute, [190-?].

[8], 242, [1] p., [1] leaf of plates, [4] leaves of forms : ill. ; 17 cm.

The Heidelberg Medical Institute in St. Paul, Minn., under the direction of H.A. Franklin, specialized in the treatment of sexual disorders and sexually transmitted diseases. Testimonies abound regarding the efficacy of the "Electro-Medical Treatment" available at the Institute or through the mail. Four tear-out forms inserted at the end of the text allow readers to describe their symptoms by post. A section of this *Symptom Blank* is reserved for the symptoms of "blood poison" (i.e., syphilis, etc.). In printed wrapper.

1611. DR. HEIDEMANN'S MUSEUM.
Catalogue to Dr. Heidemann's Museum. Greatest collection of geological, ethnological and anatomical objects and curiosities. Albany, N.Y.: Weed, Parsons & Co., [188-?].

30 p. : ill. ; 17 cm.

In printed wrapper.

1612. DR. HEIDEMANN'S MUSEUM.
Catalogue of Dr. Heidemann's Museum. Greatest collection of geological and eth-nological and anatomical objects and curiosities. Toledo, Ohio: The Toledo Bee Company, 1888.

30 p. : ill. ; 17 cm.

In printed wrapper.

1613. HEINEN, Henry.
Gesundheits-Schatzkammer, oder kurze, deutliche und richtige Anweisung zur Erhaltung der Gesundheit und Abwendung mancher Krankheiten; so wie auch gute und sichere Mittel zur Wiederherstellung der vorlornen Gesundheit . . . Lancaster [Pa.]: Gedruckt für der Verfasser, von Johann Bär, 1831.

118 p. ; 18 cm.

First edition. The second edition was published by Lippincott at Philadelphia in 1840.

1614. HEINER, Robert Graham, b. 1877.
Physiology, first aid and naval hygiene. A text book for the Department of Naval Hygiene and Physiology at the U.S. Naval Academy, Annapolis, Maryland . . . Annapolis, Md.: U.S. Naval Institute, 1916.

139 p. ; 19 cm.

Heiner was a 1900 medical graduate of the Univ. of Virginia (Charlottesville). In 1916 he was a Navy lieutenant stationed at Annapolis, Md., where he instructed first year midshipmen in physiology and hygiene. In his preface Heiner writes, "A knowledge of the rudiments of hygiene, physiology and first-aid is necessary to every naval officer. Sooner or later each one of them is likely to find himself in charge of a small detachment of men at some isolated station where there is no doctor, and it will devolve upon him to make the necessary arrangements for the preservation of the health of his men, and see that their efficiency is not undermined by sickness" (p. [5]). In the 1920s, Heiner was stationed in Washington, where he served as a medical inspector with the Surgeon-General's Office. In the late 1930s, Capt. Heiner was medical director of the Norfolk Navy Yard. He retired to Annapolis in 1939 or 1940.

1615. HELME, W. C.
The American farrier, and family medical companion; containing the symptoms,

causes and treatment of most diseases incident to man and beast. Together with recipes for compounding medicines for the same, and a list of roots, herbs, plants and barks, with their medical qualities, a brief dictionary of medical terms, and a separate epitome of poisons and their anpidotes [sic], arranged each under its appropriate heading . . . Cleveland: Steam press of Smead and Cowles, 1852.

    97, [3] p. ; 20 cm.

"When the light began to dawn upon the Dark Ages, and the mental obscurity . . . began to give way to the light of science, and men began to assert the God-given right to reason for themselves, then it was that tens of thousands who had previously subsisted upon the superstitious credulity of the multitude, became necessitated to seek another avenue through which they might continue to live upon their prejudices. Thus, vast hordes of rejected family confessors were suddenly transmorphosed into family physicians; thus we are informed of the origin of the Latin terms which still continue to mystify all departments of the medical sciences, and through the means of these mystic cognomens, the medical fraternity have been enabled to extort from the too credulous public exorbitant prices for the very roots, plants, barks and herbs that grow in native profusion in our own fields, gardens and forests" (*Pref.*, p. [3]).

He classifies the diseases of men under such categories as "bilious," "inflammatory," "spasmodic," etc., and describes their symptoms and treatment. For most diseases Helme provides both the "alopathic" and the "reformed botanic" treatments by way of contrasting the excesses of the former with the mildness of the latter. Pages 5-32 contain the veterinary portion of Helme's treatise.

1616. HEMPEL, Charles Julius, 1811-1879.
The homoeopathic domestic physician . . . New-York: Wm. Radde; London: H. Bailliere, 1846.
    [2], 4, iv, [iii]-x, [13]-184 p. ; 15 cm.

"Every homoeopathic physician who loves his art, and feels an interest in its success, should say something popular on its subject to his friends and the public generally; for this is an excellent and a very sure means of enlisting common sense in behalf of the homoeopathic healing art. The reasonableness of the fundamental principle of homoeopathy is so evident to the "*poor in spirit,*" our mode of prescribing is so much superior to the common method, and so much more elegant and pleasant, that the common sense of the public may be appealed to, to sit in judgement between us and our opponents . . . Who would not rather give his child a few pellets of hepar sulphuris, spongia, bichromate of potash, etc., to have cured it of croup, than to have it bled, purged, crammed full of emetics, and to have its skin blistered by vesicatories?" (*Pref.*, p. [iii]-iv).

A native of Solingen in the Ruhr valley, Hempel deferred military service long enough to study at the Collège de France and the Université de Paris before emigrating to New York in 1835. Supporting himself with translating, Hempel quickly mastered English and became familiar with many noted figures in science, literature, social reform and Swedenborgian thought. In 1845 he received his medical degree from the University of the City of New York. Even before his graduation, however, Hempel was inclined toward homeopathy, and in 1846 he published this first edition of *The homoeopathic domestic medicine.* "From 1846 to 1870 Hempel translated twenty-five major works [including the homeopathic texts of Hahnemann, Jahr, Baehr, Gollman, Mure & Tessier] in addition to his busy practice, teaching, and writing of journal articles, books, and nonmedical works. His translations were often enriched with additions from other authors and from his own experience, earning him the title of the 'father of homeopathic literature' in America" (*Amer. nat. biog.*, 10:554). Hempel's most important book was *A new and comprehensive system of materia medica and therapeutics* (Boston, 1859). Like much of his published work, it provoked controversy among American homeopaths who were chagrined as much by Hempel's Swedenborgism as by his deviations from strict Hahnemannism. Hempel moved to Michigan in 1861.

1617. HEMPEL, Charles Julius, 1811-1879.
The homoeopathic domestic physician . . . Second edition, revised and considerably enlarged by the author. New-York: William Radde; London: H. Bailliere, 1850.
    xviii, [19]-216 p. ; 16 cm.

Hempel's text is arranged alphabetically by bodily organ or system, under which are ordered the maladies that affect the organ and system, and their homeopathic treatment.

1618. HEMPEL, Charles Julius, 1811-1879.
Organon of specific homoeopathy; or, an inductive exposition of the principles of the homoeopathic healing art, addressed to physicians and intelligent laymen . . . Philadelphia: Published by Rademacher & Sheek ; New York: William Radde; St. Louis: J.G. Wesselhoeft, 1854.
216 p. ; 23 cm.

Hempel's exposition of homeopathy is directed more to the disciples of Hahnemann than to the enemies of homeopathy or those who were undecided about its merits. In his *Organon,* Hempel takes issue with those in the homoeopathic camp who regard it as "the work and property of Hahnemann" and who thought it outrageous "to alter an iota of its tenets without his consent." Hempel, who continually faced opposition from the homeopathic establishment, argues that homeopathy is a science, and as such is subject to observation, experimentation and change. He challenges those who regarded Hahnemann's doctrines "as infallible testimony . . . a sacred record . . . set down as principles which admit of no improvement, and [which], Minerva-like, started out of the brain of their discoverer in all the fulness of their truth and glory" (p. 10).

HEMPEL, Charles Julius see also **JAHR** (#1970).

1619. HENDERSON, Mary Foote, 1842-1931.
The aristocracy of health. A study of physical culture, our favorite poisons, and a national and international league for the advancement of physical culture . . . Washington, D.C.: The Colton Publishing Company, 1904.
xii, 772 p. : ill. ; 20 cm.

"My study of physical culture began with the problem of tobacco, the so-called solace of mankind; and this led to a study of alcohol, opium, tea, coffee, and our favorite poisons

generally. I was anxious to know how they differ in upsetting physiological law and order. The study of these agents for artificial happiness led to a realizing sense of their connection with the almost universal lack of sound health and happiness on the part of mankind . . . Problems of diet and other questions connected with physical culture naturally followed, as well as the means of relief from what is chiefly instrumental in robbing mankind of its birthright—health, happiness, and success in life" (*Pref.,* p. vi-vii). *The aristocracy of health* was reissued in 1906. Atwater copy inscribed by author.

In the late 1860s, Mary Newton Foote, of New York, was married to John Brooks Henderson (1826-1913), a U.S. Senator from Missouri and a prominent figure in state and national politics. After his defeat in the 1868 election, Henderson "returned to his law practice and settled in St. Louis with his new bride . . . He strongly supported his wife's endeavors on behalf of woman suffrage in the years that followed, as well as her establishment of the St. Louis School of Design in 1876 and the St. Louis Women's Exchange in 1879. They traveled widely and acquired an extensive art collection" (*Amer. nat. biog.,* 10:569-70). The Hendersons moved to Washington, D.C. in 1889, where "they acquired extensive real estate holdings along 16th Street and developed much of the area along what came to be known as 'Embassy Row'" (*ibid.*). M.F. Henderson was the author of several books. Her most successful was *Practical cooking and dinner giving* (1876).

1620. HENDERSON, Mary Foote, 1842-1931.
Diet for the sick. A treatise on the values of foods, their application to special conditions of health and disease, and on the best methods of their preparation . . . New York: Harper & Brothers, 1885.
ix, [1], 234, [4] p. : ill. ; 19 cm.

1621. JOHN F. HENRY & CO.
Henry's household companion. New York: John F. Henry & Co., [c1886].
[16] leaves ; 18 cm.

Title from chromolithographed wrapper. Consists of advertisements for proprietary remedies available from John F. Henry & Co.

1622. HENRY, JOHNSON AND LORD.
Rev. N.H. Downs' Vegetable Balsamic Elixir for the speedy cure of coughs, colds, consumption, whooping cough, croup, asthma, and all diseases of the throat and lungs . . . [Burlington, Vt., ca. 1879].
[8] leaves ; 16 cm.

The first five leaves describe "Downs' Elixir," with testimonials regarding its efficacy. The final three leaves advertise other proprietary remedies available from Henry, Johnson & Lord.

1623. HENRY, JOHNSON AND LORD.
A true story. A tale of other days. Also other facts of practical value. Burlington, Vt.: Henry, Johnson & Lord, [c1882].
33 p. ; 23 cm.

Promotional literature for *N.H. Downs' Vegetable Balsamic Elixir.* Title and imprint from wrapper.

1624. HERB, Ferdinand, b. 1866.
The care-feeding of the baby. A handbook for mothers, midwives and nurses . . . Superior, Wisconsin: Published by the R. & S. Publ. Co., [c1907].
xv, [1], 267 p. : ill. ; 23 cm.

Herb was an 1890 Munich medical graduate, who is listed in the 9th edition of *Polk's medical register and directory of North America* (1906) as practicing in Superior, Wisconsin.

1625. HERING, Constantine, 1800-1880.
C. Hering's domestic physician. Fourth American edition, revised, with additions from the author's manuscript of the sixth German edition. The part relating to the diseases of females and children, by Walter Williamson, M.D. Philadelphia: Published by C.L. Rademacher, 1848.
xxi, [1], 415 p. ; 18 cm.

Hering became convinced of homeopathy's validity while writing a paper on the subject as a medical student at Leipzig. In 1826 he received his medical degree from Wurzburg, and in the early 1830s was in Surinam as part of a scientific expedition sent by the King of Saxony. In 1833 Hering made the decision to emigrate to the United States. From the time of his ar-

rival in Philadelphia, he became one of the leading figures in American homeopathy. He organized the world's first homeopathic medical school in Allentown, Pa. (1836-42), where he also wrote and published the first edition of the present work under the title: *The homoeopathist, or domestic physician* (1835). Hering's was the first domestic manual of homeopathy, which "became the standard homeopathic self-care manual during the westward migrations" (*Amer. nat. biog.*, 10:645). It was eventually translated into nine languages.

Hering was one of the founders of the American Institute of Homoeopathy (1844), the first national medical organization in the United States; as well as a founder of the Homoeopathic Medical College of Pennsylvania. In addition to editing several homeopathic journals, Hering was a prolific author. He wrote six books, and, according to the *Amer. nat. biog.*, 325 journal articles.

1626. HERING, Constantine, 1800-1880.
C. Hering's domestic physician. Revised, with additions from the author's manuscript of the seventh German edition. Containing also a tabular index of the medicines and the diseases in which they are used. Fifth American edition. Philadelphia: Published by Rademacher & Sheek; Cincinnati: J.F. Desilver; New Orleans: A. Leon, 1851.
[4], xxii, 515 p. ; 19 cm.

1627. HERING, Constantine, 1800-1880.
The homoeopathic domestic physician . . . Revised and corrected from the author's last edition, with additional matter, and directions for the administration and repetition of the doses . . . London: John Walker, 1856.
xix, [1], 519 p. ; 20 cm

First London edition.

1628. HERING, Constantine, 1800-1880.
The homoeopathic domestic physician . . . Sixth American edition. by the author himself thoroughly revised and reformed from the original, and augmented by numerous additions from the forthcoming eleventh German edition. Philadephia: Ig. Kohler . . . [*et al.*], 1858.
xxxvi, 393 p. ; 23 cm.

1629. HERING, Constantine, 1800-1880.
The homoeopathic domestic physician . . .
The only authorized English edition, by the
author himself thoroughly revised and reformed from the original, [a]nd augmented by numerous additions from the
forthcoming eleventh German edition.
New York: William Radde; Philadelphia:
F.E. Boericke, 1866.
    xxxvi, 393 p. ; 23 cm.

"Since the last edition of this work in the
English language, had been sold out, about four
years ago, the author had refused to allow another one to be made, either by the former
publishers or others, and declined a great many
offers made to him by those who wished to publish a new edition. The author found his book,
instead of being improved with every new edition . . . had by others, trusted with the revision, been altered in direct opposition to his
own views; it had been, as he calls it, more and
more spoiled . . ." (*Pref.*, p. xv).

1630. HERING, Constantine, 1800-1880.
The homoeopathic domestic physician . . .
The only authorized English edition, by the
author himself thoroughly revised and reformed from the original, [a]nd augmented by numerous additions from the
forthcoming eleventh German edition.
New York: William Radde; Philadelphia:
F.E. Boericke, 1867.
    xxxvi, 393 p. ; 23 cm.

1631. HERING, Constantine, 1800-1880.
The homoeopathic domestic physician . . .
The only authorized English edition, by the
author himself thoroughly revised and re-
formed from the latest German edition.
New York: Boericke & Tafel; Philadelphia:
F.E. Boericke, 1875.
    xxxvi, 399, [1], 3 p. ; 23 cm.

1632. HERING, Constantine, 1800-1880.
The homoeopathic domestic physician . . .
Seventh American edition. Philadelphia:
F.E. Boericke, 1883.
    458 p., [1] leaf of plates : port. ; 24 cm.

This edition was edited by Claude R. Norton,
who writes in his preface: "Not long before the
death of the lamented author of this work, the

correction of the last English edition, now for
some years out of print, was undertaken by his
daughters . . . They compared it, page by page,
with the latest German edition (the fourteenth), which had but a short time previously
been subject to a thorough revision at Dr.
Hering's hands. Some material was eliminated,
and considerable matter was added from the
German" (p. [3]). Frontispiece portrait of the
author. Cover- and spine-titles: *Hering's domestic
physician*.

1633. HERING, Constantine, 1800-1880.
Homöopathischer Hausarzt. Für die
deutschen Bürger der Vereinigten Staaten
nach den besten vaterländischen Werken
und eignen Erfahrungen bearbeitet . . .
Allentaun an der Lecha: Zu haben bei
Jakob Behlert, 1837.
    [iii]-viii, 352 p. ; 19 cm.

First German-language edition. Although
the English-language edition was published first
(*The homoeopathist, or domestic physician*, 1835),
the succeeding German language editions became the standard texts on which the later
English editions were revised. The *Homöopathischer Hausarzt* was published in numerous editions in Germany (23. Aufl., Stuttgart, 1912).
Atwater copy lacking pp. 347-48.

1634. HERING, Constantine, 1800-1880.
The homoeopathist, or domestic physician
. . . Second American, with additions, from
the fourth German edition. Philadelphia,
1844.
    xii, [9]-217 p. ; 22 cm.

The first edition of Hering's manual of domestic homeopathy was published at Allentown,
Pa. in 1835. Printed on title-page beneath imprint of this second edition: "Sold by Jacob and
J.N. Bauersachs . . . In English and German,
with boxes of medicine accompanying each
book . . ." Bradford's *Homoeopathic bibliography
of the United States* (1892) adds: "The box accompanying this book is 7 × 3½ × 2 inches. It is of
mahogany, the vials are only an inch in length
and were filled with infinitesimal pills. The lid
of the box contains a list of medicines, with
numbers corresponding with numbers in the
book. These cases were put up under the personal supervision of Dr. Hering" (p. 145). The

third edition of this work (1845) was published under the title: *Dr. Hering's domestic physician.*

HERING, Constantine see also **CASPARI** (#573).

1635. HERRICK, Christine Terhune, 1859-1944.
Cradle and nursery . . . New York: Harper & Brothers, 1889.
    viii, 298, [6] p. ; 17 cm.

Herrick was the author or co-author of twenty-three books on domestic economy, cookery and hygiene published between 1888 and 1928, as well as the editor-in-chief of two major compilations on domestic science. Her first publication was an article in the premier issue of *Good housekeeping* (1885); and was followed three years later by her first book, *Housekeeping made easy.* "Her books conceived of the home as the chief arena for women's work. To her a woman was her husband's business partner, whose job it was to ensure that the home was pleasant and well-ordered . . . Herrick and others in the domestic science movement sought to legitimize and draw increased recognition to the function of the housewife, by likening her training to that required for jobs primarily held by men . . . The simple housewife would become the economist, scientist, and administrator of her household" (*Amer. nat. biog.,* 10:660-61).

1636. HERRING, Daniel.
[Letter], 1873 October 6, Highland Station, Kansas, [to] T.W. Marsden, New Orleans.
    [1] leaf ; 31.5 × 19.5 cm.

Herring writes: "While driving cattle in Texas I was troubled with a very bad cough," and requests of T.W. Marsden, the New Orleans based manufacturer of proprietary medicines (#2367-2371), two bottles of *Marsden's Pectoral Balm.*

1637. HERSEY, Thomas.
The midwife's practical directory; or, woman's confidential friend: comprising, extensive remarks on the various casualties, and forms of disease, preceding, attending and following the period of gestation. With an appendix. The whole designed for the

special use of the botanic friends in the United States . . . Second edition, enlarged and improved . . . Baltimore: Published by the author, 1836.
    xxxii, [33]-336, [20] p., [1] folding plate : ill. ; 15 cm.

Trained as an allopath, Hersey abandoned allopathy for the botanic or reformed practice of medicine. *The midwife's practical directory* is his principal published work (1st ed., Columbus, Ohio, 1834). Its content differs little from other obstetrical and midwifery texts of the first half of the 19th-century, other than in its recommendation of Thomsonian remedies in the treatment of gynecologic disorders.

Hersey opens his treatise with social and moral reflections on sexual maturity. He moves into the scientific portion of his text with a discussion of the menses, which he regarded as a "secretion *sui-generis*" (i.e., not blood), the purpose of which he declined to speculate. Several chapters are devoted to a description of the female organs of generation. There is no corresponding description of the male organs nor coition *per se.* Hersey's theory of conception was derived from contemporary authors such as William Dewees. He denies uterine conception, maintaining that absorbent vessels near the mouth of the vagina carry the "animalcules" in the male semen to the ovaries: "Having arrived at this place of rendevous, the traveller [i..e., sperm] is provided with lodging and nutrition, by an envelope of mucilaginous fluids which the vesicle [i.e., unfertilized egg] contains" (p. 85). The fertilized ovum then descends the Fallopian tubes to the uterus, where it attaches itself.

The greater part of the text is devoted to a discussion of the signs and disorders of pregnancy; the management of natural and preternatural presentations; the puerperium; and the care of newborn infants. Chapter xxiii is interesting for its argument in favor of female "accoucheurs" and the necessity of institutions for their instruction. The appendix is a formulary of botanic remedies of specific benefit to the disorders of menstruation, pregnancy and the puerperium. The ten wood-engraved illustrations which follow the text are copied from Samuel Bard's *Compendium of the theory and practice midwifery* (1st ed., 1807). Spine-title: *Woman's friend.*

In a letter published in *The Thomsonian botanic watchman* (vol. 1, no. 4, April 1, 1834, p. 52) to John Thomson, the journal's editor and the son of Samuel Thomson, Hersey provides some autobiographical data: "I have been more than forty years engaged in the regular practice of medicine. I was a surgeon during the last war [1812] in the army of the United States. I was by an election, surgeon (*extraordinary*) to the Petersburgh Volunteers, and Major Stodard's two companies of Artillery. I was one of the founders of the Western Medical Society of Pennsylvania, and also a member of the Medical Society of the State of Ohio." Hersey adds, "My practice has been extensive—my experience and opportunity for observation has seldom been exceeded: but I venture to pledge myself upon all I hold sacred and valuable in the profession, that in my estimation, the discoveries made by your honored father, have a decided preference and stand unrivalled by all that bears the stamp of ancient or modern skill." Hersey was editor of the early volumes of the *Thomsonian recorder* (#3518).

1638. HERSEY, Thomas, d. 1834.
The midwife's practical directory; or, woman's confidential friend: comprising, extensive remarks on the various casualties and forms of disease, preceding, attending, and following the period of gestation . . . Philadelphia: Republished by L.D. Stone, M.D., 1861.
[2], [xvii]-xx, [33]-336, [20] p. : ill. ; 15 cm.

Originally published at Columbus, Ohio in 1834, *The midwife's practical directory* was issued in a second edition at Baltimore two years later (#1637). The 1861 Philadelphia edition, edited and published by L.D. Stone, is a re-issue of the original sheets of the 1836 Baltimore edition for which a new title-page and preface were printed.

HERSEY, Thomas see also *THE THOMSONIAN RECORDER* (#3518).

1639. HESS, Charles D.
The anatomy and physiology of the skin. Etiology and treatment . . . Rochester, N.Y.: Prepared and published by the Youthful Tint Mfg. Co., [c189-?].
32 p. ; 13 cm.

Popular discussion of common skin diseases and their treatment with *Mellocuti Remedies*. Title on printed wrapper: *Practical handbook on dermatology, containing a treatise on diseases of the skin, scalp and hair.*

1640. HEWES, Henry Fox, 1867-1926.
Anatomy physiology and hygiene for high schools . . . New York; Cincinnati; Chicago: American Book Company, [c1900].
320 p. : ill. ; 19 cm.

First edition (2nd ed., 1911). Hewes was an 1895 Harvard Medical School graduate, and at the time this book was published, instructor in physiology and clinical chemistry at his *alma mater*. Series: *New century series of anatomy, physiology and hygiene; no. 1.*

1641. HEWETT, Dr.
FURTHER DIRECTIONS. Sometimes Dr. H. recommends four, or more, kinds of external medicines, to be applied each twice a day, or more. Begin early in the morning and finish in the evening . . . [n.d.]
[2] leaves ; 33 cm.

The FURTHER DIRECTIONS consists of three columns of 8-point type printed on the recto of the first leaf. "These directions are only intended to throw light upon what is stated on the other side," i.e., a form in 12 point type with many blanks allowing Hewett to specify what body part is to be bathed or rubbed with liniment, which proprietary remedy is to be used, the frequency of applicaton, etc. In the Atwater copy, Hewett has specified that the patient is to bath his or her leg up to the hip, apply his *Red Nerve* liniment "on the whole hip and small of the back," rub in the "relaxing oil or ointment," take the tonic mixture three times daily "in a little Malaga or Port wine," and take the "nerve alterative" in the evening. On the recto of the first leaf, it is stated: "Dr. Hewitt will here mention that his medicines are very expensive, much more so than is used by other practitioners, which enables him to perform cures which others fail to do. His baths are composed from eleven to twenty-four different articles—vegetable, mineral and animal."

1642. HEYDEN, A. V.
Secrets exposed; or, the sick man's friend, and the sufferers' index. Containing recipes for many valuable compounds never before published. Designed wholly for the use and benefit of families, or an index to simple yet certain remedies for the cure of some of the most common ills that await us . . . [Coldwater, Mich.: Bowen, Dunham & Moore, printers, 1870].
14, [2] p. ; 15 cm.

A collection of fifty-three recipes for "some of the most common ills of the people." The title-page is undated. The imprint and date of publication are taken from the printed wrapper.

HIBBARD, John B. see *EASTERN MEDICAL REFORMER* (#1024).

1643. HIGGINS, Charles Michael, b. 1854.
The crime against the school child. Compulsory vaccination illegal and criminal and non-enforceable upon the people. How to legally defeat this medical evil which now kills more children than smallpox . . . [S.l.: Anti-Vaccination League of America], 1915.
[6], 3-64 p. ; 23 cm.

Higgin's anti-vaccination treatise was motivated by the passage of the Loyster-Tallett Law, passed by the New York State Legislature in 1915, making vaccination compulsory for all school children. In printed wrapper.

1644. HIGGINS, Charles Michael, b. 1854.
Vaccination and lockjaw. The assassins of the blood . . . [S.l.: Anti-Vaccination League of America, 1916?].
31, [1] p. ; 23 cm.

"This public letter exposing the libels of the New York Tribune and demanding their retraction proves that compulsory vaccination is worse than smallpox and shows how frequently it kills little school children by lockjaw and other infections and that it is actually ten times more dangerous and fatal than natural smallpox, and it also shows how these shocking facts are de-

nied and concealed by vaccinists and medical authorities interested in such concealment" (p. 1). Title from wrapper.

1645. HIGGINSON, Thomas Wentworth, 1823-1911.
Out-door papers . . . Boston: Lee and Shepard, publishers; New York: Charles T. Dillingham, 1879.
[4], 370 p. ; 18 cm.

An 1847 graduate of Harvard Divinity School, Higginson was a life-long advocate of such issues as abolition, temperance, the rights of labor and women's suffrage from his earliest years in the pulpit. His political and social activism was channeled into several attempts at elected office, journalism, involvement in various organizations, etc. During the Civil War, Higginson commanded the First South Carolina Volunteers, a regiment comprised entirely of freed slaves. Higginson was regarded as a prominent man of letters by his contemporaries, but today is remembered primarily for his encouragement to women writers, and to Emily Dickinson in particular.

Out-door papers was first published at Boston in 1863, and reissued six times between 1871 and 1894. It is a collection of essays, several of which pertain to exercise, diet and hygiene, e.g., "Saints, and their bodies," "A letter to a dyspeptic," "Gymnastics," "The health of our girls," etc. In "The health of our girls," Higginson correlates the poor health of American adolescent females to their education, i.e., to long hours of study and neglect of physical training. By way of contrast he muses: "The sedentary philosopher, turning from his demonstration of the hopeless inferiority of woman, finds with dismay that his Irish or negro hand-maiden can lift a heavy coal-hod more easily than he" (p. 202).

1646. HILL, Benjamin L., 1813-1871.
An epitome of the homoeopathic healing art, containing the new discoveries and improvements to the present time; designed for the use of families, for travelers on their journey, and as a pocket companion for the physician . . . Cleveland, Ohio: John Hall; Chicago: Halsey & King, 1860.
x, [11]-160, iv p. ; 14 cm.

Hill began his career as an eclectic. He was on the faculty of the Reformed Medical School of Cincinnati when it opened its doors in 1842, and joined the faculty of the Eclectic Medical Institute (Cincinnati) when the former folded in 1845. Hill and his colleague J.R. Buchanan (#505-507) were the catalysts who convinced the board of the Institute to add a chair of homeopathy to its faculty. Shortly thereafter, Hill was made professor of physiology at the (eclectic) Central Medical College in Syracuse, N.Y. Sometime in the early or mid-1850s, Hill joined the faculty of the Western Homoeopathic Medical College in Cleveland. His most important published work was the *Lectures on the American eclectic system of surgery* (Cincinnati, 1850), the first non-allopathic surgical text published in the United States.

The *Epitome* was first published at Cleveland in 1859. This 1860 issue appears to be the 1859 edition (which has an undated title-page) with the date '1860' stamped at the bottom of the title-page. This small manual of domestic homeopathy had a successfully publishing history; a nominally 18th edition was published at Detroit in 1881.

1647. HILL, Benjamin L., 1813-1871.

An epitome of the homoeopathic healing art, containing the new discoveries and improvements to the present time; designed for the use of families, for travelers; and as a pocket companion for the physician. Revised edition . . . Detroit, Michigan: Published at Dr. Lodge's Homoeopathic Pharmacy, 1866.

xii, [13]-142 p. ; 13 cm.

"Dr. Lodge is responsible for the additions made in brackets" (p. ix).

1648. HILL, David Stanley.

Dental physiology and oral hygiene . . . First edition . . . Effingham, Ill.: The LeCrone Press, 1917.

167 p. : ill. ; 20 cm.

The author was a 1907 graduate of the Ohio College of Dental Surgery (Cincinnati).

HILL, Janet Mckenzie see *THE STANDARD BOOK OF RECIPES* (#3322).

1649. HILLMAN, John.

Every family their own physician . . . Auburn [N.Y.]: Printed by Henry Oliphant, 1852.

vi, [7]-128 p. ; 16 cm.

"After I had passed through a legal course of medical studies, I became determined to avail myself of the advantages of attending the practice of physic and surgery in the New York Hospital. I consequently went and purchased a ticket which admitted me to all the sick at the time when the physicians and surgeons went their rounds; and I purchased another ticket which admitted me into the Theater, where all the capital operations of surgery were performed, and special lectures given on all of the most important diseases of the Hospital . . . I likewise in the winter attended a full course of lectures, delivered by the professors of the New York State Medical Society. These advantages gave me a view of the general outlines of the medical profession, and much of its practical application, and by my practice and reading, and by my medically and surgically educating two sons who attended the medical colleges of Philadelphia, Fairfield and Geneva, I think I have a very general knowledge of the profession, and much understanding in regard to the different theories and practices which have arisen of late, all of which . . . appear to be a mere jargon; no agreement any where . . . no two physicians agreeing either in theory or practice . . . Ask twenty physicians separately, in one critical case, what is the disease and what is the remedy, and no two will agree" (p. [iii]-iv).

Hillman, a practitioner in Clyde, N.Y. (a village on the Erie Canal between Rochester and Syracuse), cuts through the jargon of contending schools to provide his readers simple diagnoses and remedies for the management of a variety of ailments. His therapeutics is based upon a regimen of tonics and stimulants, and is opposed to the use of bleeding and purging that "reduce" patients to a "low state" in order to break a fever. The author relies greatly on his "Ague Compound," which was compounded of peppermint, cinnamon, sulphate of quinine, capsicum and ginger (dissolved in water).

HIMES, Norman Edwin see **KNOWLTON** (#2154).

HIRES, Charles Elmer see *HIRES'*
IMPROVED ROOT BEER (#1650).

1650. HIRES' IMPROVED ROOT BEER.
Hires' Improved Root Beer! HIRES' IM-
PROVED ROOT BEER now a favorite beverage for ladies and young persons, to whom
it gives freshness and embonpoint. It has
solved the problem of medicine by imparting strength and pure blood, which soon
gives a person a clear and healthy complexion . . . [ca. 1881].
     [1] leaf printed on both sides : ill. ; 24
× 15 cm.

     Made from such roots as "pipsissewa, sarsa-
parilla, spikenard, dandelion, &c., &c.," this soft
drink "makes the weak strong, and gives the
feeble nerve. It increases muscular force and
doubles the 'staying' power. It imparts a hearty
appetite and insures perfect digestion. It gives
sleep—sound and refreshing—in due season
. . . It strengthens feeble women and makes
puny children strong. It restores the tranquility of nerves shattered by excesses . . . It cleanses
the system of poisonous humors . . . It enriches
the blood," etc. (recto, col. 1). It also refreshes
"farmers and laboring men" in the summer
heat, is healthful for nursing mothers ("It will
give mother and child strength"), and is "recommended and prescribed by some of our best
physicians."
     After leaving the family farm at age sixteen,
Charles Elmer Hires (1851-1937) went to Philadelphia, where he clerked at a pharmacy and
took evening classes at the Jefferson Medical
College and the Philadelphia College of Pharmacy. By age eighteen, he was proprietor of his
own pharmacy. "Hires first tasted the drink that
would bring him both fame and fortune while
staying at a New Jersey boarding house during
his honeymoon [1875]. Served to him by his
landlady . . . the beverage was a mixture made
from sassafras bark and herbs. After conversations with several chemist friends and experimentation with sarsaparilla root and other ingredients, he concocted what became known
as root beer . . . The Philadelphia centennial
Exposition of 1876 offered Hires the opportunity to sell his root beer to the millions of visitors attending the fair. Their response was extremely positive and led Hires to begin mar-

keting and packaging his root beer" (*Amer. nat.
biog.*, 10:852). Hires initially sold his mixture
in bulk, but began bottling the soft drink in
1893. Hires remained chairman of the immensely successful Charles Elmer Hires Company until his death.

1651. HITCHCOCK, Edward, 1793-1864.
Dyspepsy forestalled & resisted: or lectures
on diet, regimen, & employment; delivered
to the students of Amherst College; spring
term, 1830 . . . Amherst: Published by J.S.
& C. Adams and Co.; New York: Jonathan
Leavitt; Boston: Pierce and Williams, 1830.
     viii, [9]-360 p. ; 19 cm.

     "Hitchcock, a well-known clergyman and
geologist—professor of chemistry at Amherst
College in 1830, and later president of the college—was an early temperance advocate. His
books originated as a series of lectures he delivered to the Amherst student body. Hitchcock
himself was a long-time 'dyspeptic'—the term
at this time was a catch-all for a whole series of
chronic 'nervous' complaints that in one way
or another managed to affect the digestive apparatus—and he composed the lectures in order to stem what he called 'the premature prostration and early decay of students and professional men in our country" (Stephen Nissenbaum, *Sex, diet, and debility in Jacksonian America*,
Westport, Ct.: Greenwood Press, 1980, p. 44).
Hitchcock condemned the use of alcohol, tobacco, coffee and tea; and suggested that animal foods were best abstained from. "As for the
rest of Hitchcock's dietary rules, they are all
concerned either with the conditions under
which meals should be eaten ('slowly . . . with
the mind free and the feelings cheerful,' and
never 'while much fatigued') or with the allow-
able quantity of food . . . Gluttony, Hitchcock
insisted, was as harmful as drunkenness; and
the effects on the intellectual and moral faculties produced by 'excess in eating' were similar
to those produced by 'excess in drinking'"
(Nissenbaum, p. 45).
     Dyspepsy forestalled is divided into four parts:
diet; regimen (i.e., exercise, air, clothing, cleanliness, sleep, "evacuations," etc.); employment
(i.e., "influence of different employments upon
health); and dyspepsy. James C. Whorton cites
Hitchcock's book as an example of the "fusion

of science and religion" in pre-Darwinian American thought, i.e., "the ascendance of rationalism in the new religion." Whorton writes: "William Ellery Channing, the leader of American Unitarianism, was so bold as to declare that were Christianity not a 'rational religion . . . I should be ashamed to profess it.' Not all voiced their feelings so bluntly, perhaps, but there was by the 1830s a general consensus that religious truth was rational, and that it necessarily agreed with any other truth discovered by reason, meaning specifically scientific truth" (*Crusaders for fitness,* Princeton: Princeton Univ. Press, 1982, p. 32).

1652. HITCHCOCK, Edward, 1793-1864.
Dyspepsy forestalled and resisted: or lectures on diet, regimen, and employment; delivered to the students of Amherst College, spring term, 1830 . . . Second edition. Corrected and enlarged by the addition of an address delivered before the Mechanical Association in Andover Theological Institution, Sept. 21, 1830; and an appendix of notes. Amherst: Printed and published by J.S. & C. Adams . . . [*et al.*], 1831.
    xii, [13]-452, 4 p. ; 20 cm.

"The present age is characterised by most assiduous efforts, thoroughly, and in the best manner, to discipline all the powers of the mind" (p. 343). Hitchcock complains, however, "except in a few instances, who, among the numerous founders, patrons, guardians, and instructors of our public literary institutions, ever thought of making any regular provision for keeping in play the delicate machinery on which the immortal principle so essentially depends for successful operation? The whole business of physical education has been left . . . almost without time, encouragement, or direction." If the student, however, is "untaught in the fundamental principles of physical education, he suffers the ambition of scholarship, or a natural inclination to gluttony and inaction, to control him; it will be wonderful if bodily health and vigor are not crushed by the load that is laid upon them; and the mind, ever after, do[es] not partake in all its movements, of the feebleness and inefficiency of its material tenement" (p. 344).

In his preface to this much expanded second edition, Hitchcock acknowledges his indebtedness to ideas culled from the *Journal of health* (#2053) and the works of James Johnson (#2023-2028), Philip, Paris (#2733), George Cheyne (#632, 633), Faust (#1144-1147), Beddoes, Lambe (#2176, 2177), Sinclair (#3220, 3221) and Hufeland (#1833-1836); as well as several hygienic treatises in French and even Konstantinos Karatheodore's the *Hygieina parangelmata pros chresin tou Hellenikou laou* published in modern Greek at Paris in 1829.

1653. HITCHCOCK, Edward, 1793-1864.
Elementary anatomy and physiology, for colleges, academies, and other schools. By Edward Hitchcock . . . and Edward Hitchcock, Jr. . . . Revised edition. New York: Ivison, Phinney & Company; Chicago: S.C. Griggs & Company, 1863.
    vi, [5]-443 p. : ill. ; 20 cm.

"At the earnest solicitation of my son, my name stands first on the title-page. But justice requires me to state that most of the body of the work has been prepared by him" (*Pref.,* p.iv). The Hitchcocks' school textbook was first published at New York in 1860. The revised edition was issued in 1861, and reissued six times between 1863 and 1880 (#1654). Series: *American educational series.*

1654. HITCHCOCK, Edward, 1793-1864.
Elementary anatomy and physiology, for colleges, academies, and other schools. By Edward Hitchcock . . . and Edward Hitchcock, Jr. . . . Revised edition. New York and Chicago: Ivison, Blakeman, Taylor & Co., publishers, 1880.
    vi, [5]-443 p. : ill. ; 20 cm.

1655. HITCHCOCK, Edward, 1828-1911.
The Department of Physical Education and Hygiene in Amherst College . . . Boston: Rand, Avery & Co., printers to the Commonwealth, 1879.
    10 p. ; 22 cm.

In printed wrapper. Offprint from the *Tenth annual report of the State Board of Health of Massachusetts, 1879* (Boston, 1879). Hitchcock was the son of Edward Hitchcock (1793-1864), President of Amherst College and author of *Dyspepsia forestalled and resisted* (#1651, 1652). "After completing his medical course at Harvard in 1853, [E.H., Jr.] taught natural sciences and elocution at Williston Seminary until 1860, when . . . he went to England to become the private pupil of Sir Richard Owen of the British Museum. On his return to America in 1861, he was unexpectedly called to the head of a recently organized 'Department of Hygiene and Physical Education' at his alma mater, a position which he held for half a century." (*Dict. Am. biog.*, 5:71) The Department was founded on the premise that "without the support of well-developed bodily powers and functions, the mental faculties could not reach their full development" (p. 4). Its purpose is described in the College's catalogue for 1861-62: " . . . to secure healthful daily exercise and recreation to all the students; to instruct them in the use of the vocal organs, movements of the body, and manners, as connected with oratory; and to teach them, both theoretically and practically, the laws of health. This daily physical training is part of the regular college course."

HITCHCOCK, Edward, Jr. see also **HITCHCOCK** (#1653, 1654).

1656. HITZ, Gertrude.
The importance of knowledge concerning the sexual nature. A suggestive essay . . . [Washington, D.C.]: Published by the Washington Society for Moral Education, 1884.
    32 p. ; 18 cm.

In her consideration of sexual mores, Hitz decries "any woman or man [who] allows the physical relations of marriage to degenerate into legalized prostitution," insisting that "the holy powers of sex" are appropriate only for the purpose of procreation. The habit of masturbation in many children, according to the author, is seated in their parents' indulgence in sexual intercourse even after conception. "Do you realize that the average child is the mere accident of animal passion, and that, during his helpless pre-natal life, he has been disturbed and debased by the irrational sexual indulgence of his parents? In consequence of this unnecessary sexual excitement during pregnancy, it is natural to suppose, as the nervous impressions of the mother are felt by the child, that many children are born with oversensitive sexual natures" (p. 6). Hitz also condemns spanking, which sends "an unnatural amount of blood to a portion of the body, which is near enough to the sexual organs to make the irritation unwise" (p. 6). The author recommends that parents begin sex education of their children from an early age, using examples of reproduction from the plant and animal worlds. "Sexuality, when accepted by what is sweetest and best in childhood, will be ennobled and sanctified in the deeper and more personal appreciation which comes with youth. Under these conditions, the innocence of the child will not be lost at the dawn of manhood and womanhood" (p. 12). In printed wrapper.

HOAR, Samuel see **HOPKINS** (#1784).

HOBURG, William A. see **HOME FORMULARY CO.** (#1756).

1657. HODGES, E. L.
The home doctor book. A functional review of the human body, its care and treatment in disease . . . Victoria, B.C.: Victoria Printing & Publishing Co., 1917.
    410 p. : ill. ; 23 cm.

Hodges' *Home doctor book* is a manual of domestic osteopathy. "The older schools tell us that typhoid fever is caused by a germ called bacillus typhosis; that consumption is caused by bacillus tuberculosis; that diphtheria is caused by the klebs-loeffler bacillus, etc. Now it has been demonstrated that disease germs cannot live and propagate their species in a healthy tissue, and that a freely circulating blood is the best germicide in the world . . . This being true, it is easy to see that the cause of disease is not the germ, but the cause is that which allowed the tissues of the body to become so debilitated that germs can thrive in them . . . Whether the trouble is acute, chronic, or infectious, the remedy is the same: a free and normal distribution of nerve force and a free and unobstructed circulation . . ." (p. 24). The proper circulation of nerve force and blood can be restored by manipulation and massage. These procedures are described in Hodges' text, which is illustrated with 416 half-tone photographs depicting the manipulation of bones, joints and muscles in the treatment of disease.

1658. HOFF, Charles A.

Highways and byways to health. Volume II. Byways: being a guide to physical vigor and purity in the sexual relations, with a complete description of the reproductive organs of both man and woman, the physiology of generation, the laws of procreation, penalties of over-indulgence, etc., with chapters on love, courtship and marriage, including choosing a partner, the wedding, the honeymoon, conception, pregnancy, growth of a new life, confinement, childbirth, rearing children, with the treatment of infantile complaints, diseases peculiar to each sex, venereal diseases, etc. . . . Illustrated. Philadelphia and St. Louis: Planet Book Houses, 1890.
    xx, [21]-364 p. : ill. ; 20 cm.

The second of a two-volume set: Vol. I. *Highways: being a practical household compendium of the healing art*; Vol. II. *Byways: being a guide to physical vigor and purity in the marital relations.* Half-title: *Byways to health.* Originally published in 1887.

1659. HOFF, Charles A.

Highways and byways to health. Volume II. Byways: being a guide to physical vigor and purity in the marital relations, with a complete description of the reproductive organs of both man and woman, the physiology of generation, the laws of procreation, penalties of over-indulgence, etc., with chapters on love, courtship and marriage, including choosing a partner, the wedding, the honeymoon, conception, pregnancy, growth of a new life, confinement, childbirth, rearing children, with the treatment of infantile complaints, diseases peculiar to each sex, venereal diseases, etc. . . . Illustrated. Philadelphia, Pa.: Planet Publishing Company, 1893.
    xx, [21]-364 p. : ill. ; 20 cm.

1660. HOFFMAN, Anna C.

A young mother's tokology . . . Chicago: M.A. Donohue & Co., [c1902].
    148 p. ; 22 cm.

At head of title: *A twentieth century book for mothers and nurses.* The author is described on the title-page as "chief instructor" at the International Nurses Training School.

1661. HOFFMANN, Adolphe, Mrs., b. 1856.

Before marriage. A mother's parting counsel to her son on the eve of his marriage . . . Philadelphia, Pa.: The Vir Publishing Company, [c1908].
    36, [4] p., [1] leaf of plates : port. ; 17 cm.

"Undoubtedly our sons are more exposed to temptations and allurements than our daughters. The world tacitly assumes . . . that every young man who has reached and passed a certain age has learned what 'life' is, in their sense of the word, while in the case of every young girl . . . it takes her maiden purity for granted. For this reason a young man . . . has much greater difficulty than a girl in keeping his morals intact" (p. 7). Hoffmann recounts the moral upbringing of her son Bernard, who is about to marry; and provides a dialogue in which mother and son consider the purposes and responsibilities of marriage, sex, etc. Hoffmann was also the author of *The social duty of our daughters* (1908). Frontispiece portrait of author.

1662. HOGAN, Louise Eleanor Shimer, 1855-1929.

How to feed children. A manual for mothers, nurses, and physicians . . . Fifth edition. Philadelphia: J.B. Lippincott Company, 1900.
    236 p., [5] leaves of plates : ill. ; 19 cm.

Hogan was an expert on dietetics. In 1899, with the cooperation of the Secretary of War, she initiated a program of classes for military personnel on the selection, care and preparation of food. Her most successful published work was *How to feed children,* which appeared in eleven editions between 1896 and 1926. Series: *Practical lessons in nursing.*

1663. HOHMAN, Johann Georg.
Albertus Magnus, oder der lange verborgene Schatz und Haus-Freund, und getreuer und Christlicher Unterricht für Jedermann. Enthaltend wunderbare und erprobte Mittel und Künste für Gebrechen der Menschen und am Vieh. Aus den arabischen Schriften des weisen Alchemisten Omar Arey, Emir Chemir Tschasmir, ins Deutsche übersetzt und mit vielen andern Künsten vermehrt. Pennsylvanien: Gedruckt für den Käufer, 1839.
      100 p. ; 17 cm.

The term "powwow" refers either to a medicine man of the North American Indians, or to the ceremony in which medicine men chanted incantations for the healing of disease. The term was incorporated into English usage by the early New Englanders, and was adopted by the Pennsylvania Germans to describe an occult therapeutic practice that included the recitation of Bible verses, charms and exorcisms to remove disease, as well as the employment of touching, talismans and recipes to be applied externally to diseased body parts. Powwow practitioners were common among the German communities of Pennsylvania throughout the 19th and well into the 20th centuries.
      Hohman's compilation, published under varying titles, became the standard *Brauche, Braucherei* or "pow-wow" book among the Pennsylvania Germans after its first publication at Reading in 1820 (#1672). Don Yoder has traced its sources from the *Romanusbüchlein,* a collection of charms first published in Silesia in 1788; from Albertus Magnus' book of secrets; and from several American works ("Hohman and Romanus: origins and diffusion of the Pennsylvania German powwow manual," In: *American folk medicine,* Berkeley, Univ. of California Press, 1976, p. 238).
      Yoder has described Hohman as "one of the most influential and yet most elusive figures in Pennsylvania German history. He appeared in Pennsylvania in 1802, when he landed at Philadelphia on October 12 on a Hamburg vessel with his wife and son Philip . . . Soon . . . Hohman's name began to appear on broadsides of both an occult and literary nature. For the rest of his life, until 1846, when he disappeared from the scene, we have a stream of print—

books, pamphlets, chapbooks, and broadsides—issuing from his pen. These include the first printed American version of the *Himmelsbrief,* or 'Letter from heaven'; an extensive book on household medicine (1818) [#1667]; the most complete edition of New Testament apocrypha published for the Pennsylvania Germans (1819); the powwow book (1820); and various ballads and spiritual songs, some of which he himself composed or amended. During the time of his broadside peddling, he seems to have lived near Reading, in Rosenthal (Rose Valley) in Alscace Township, where he presumably farmed on a small plot and eked out a meager existence with his publications" (*ibid.,* p. 236).

1664. HOHMAN, Johann Georg.
Albertus Magnus, oder der lange verborgene Schatz und Haus-Freund, und getreuer und Christlicher Unterricht für Jedermann. Enthaltend wunderbare und erprobte Mittel und Künste für Gebrechen der Menschen und am Vieh. Aus den arabischen Schriften des weisen Alchemisten Omar Arey, Emir Chemir Tschasmir, ins Deutsche übersetzt und mit vielen andern Künsten vermehrt. Pennsylvanien: Gedruckt für den Käufer, [ca. 1840].
      95 p. ; 16 cm.

1665. HOHMAN, Johann Georg.
Hohmann's lang verborgener Freund, enthaltend wunderbare und erprobte Heil-Mittel und Künste für Menschen und Vieh . . . Harrisburg, Pa.: Gedruckt bei Theo. F. Scheffer, [186-?].
      110 p. ; 13 cm.

Cover-title: *Kunst Buch.*

1666. HOHMAN, Johann Georg.
Hohmann's lang verborgener Freund, enthaltend wunderbare und erprobte Heil-Mittel und Künste für Menschen und Vieh . . . Harrisburg, Pa.: Gedruckt bei Theo. F. Scheffer, [188-?].
      110 p. ; 13 cm.

Possibly printed from the same plates as Scheffer's 186? edition (#1665), but more likely a lithographic facsimile.

1667. HOHMAN, Johann Georg.
Die Land- und Haus-Apotheke, oder
getreuer und gründlicher Unterricht für
den Bauer und Stadtmann, enthaltend die
allerbesten Mittel, sowohl fur die Menschen als für das Vieh besonders für die
Pferde. Nebst einem groszen Anhang
von der Aechten Farberey, und Turkisch-
Roth, Blau, Satin-Roth, Patent-Grun und
viele andere Farben mehr zu farben.
Erste americanische Auflage . . . Reading: Gedruckt von Carl A. Bruckman,
1818.
    [12], 169, [7] p. ; 17 cm.

"The book contains about one hundred and
thirty recipes for man, mostly medical. but including many miscellaneous ones for preserving meats and other food, for cleaning spots
on goods, for making inks, for fishing, and for
making wines, etc. These are followed by over
one hundred medical recipes for the horse,
about twenty-eight for domestic cattle, three for
swine, and nine for sheep . . . As for the contents of the book itself, practically all of the
medical recipes for man included in the first
forty pages are taken bodily from Bartgis' *Neuer
Erfahrner Americanische Haus-und Stallarzt* . . .
printed at Frederickstown in 1794 . . . Curiously
enough, in view of Hohman's preference for
the occult . . . in editing the Frederickstown
material he has cut out some of the 'sympathie"
remedies and has here included mainly the
sounder material" (T.R. Brendle & C.W. Unger,
Folk medicine of the Pennsylvania Germans, New
York: A.M. Kelley, 1970, p. 239-40).
    REFERENCES: *Austin* #921.

1668. HOHMAN, Johann Georg.
Der lang verborgene Freund, oder getreuer und Christlicher Unterricht für
Jedermann, enthaltend, wunderbare und
probmässige Mittel und Künste Sowohl für
die Menschen als das Vieh. Mit vielen
Zeugen bewiesen in diesem Buch, und
wovon das Mehrste noch wenig bekannt
ist, und zum allerersten Mal in America im
Jahr 1820 im Druck erschienen . . . Und
nun auf Begehren zum zweytenmal gedruckt. Ephrata [Pa.]: Gedruckt bey J.B.
[i.e., Joseph Bauman], 1828.
    94 p. ; 16 cm.

Second edition; originally published at
Reading, Pa. in 1820 (#1672).
    REFERENCES: *First century of German lan-
guage printing in the United States of America*,
#2919.

1669. HOHMAN, Johann Georg.
Der lang verborgene Freund, oder getreuer und Christlicher Unterricht für
Jedermann, enthaltend wunderbare und
probmässige Mittel und Künste, sowohl für
die Menschen als das Vieh. Mit vielen
Zeugen bewiesen in diesem Buch, und
wovon das Mehrste noch wenig bekannt
ist, und zum allererstenmal in America, im
Jahr 1820 im Druck erschienen . . . Und
nun auf Begehren zum drittenmal gedruckt. Harrisburg, Pa.: [s.n.], 1840.
    84 p. ; 16 cm.

1670. HOHMAN, Johann Georg.
Der lang verborgene Freund, oder getreuer und Christlicher Unterricht für
Jedermann; enthaltend wunderbare und
probmässige Mittel und Künste, sowohl für
die Menschen als das Vieh. Mit vielen
Zeugen bewiesen in diesem Buch, und
wovon das Mehrste noch wenig bekannt
ist, und zum allererstenmal in Amerika, im
Jahr 1820 im Druck erschienen . . . Und
nun auf Begehren zum drittenmal gedruckt. Harrisburg, Pa.: [s.n.], 1843.
    84 p. ; 17 cm.

1671. HOHMAN, Johann Georg.
Der lang verborgene Freund, oder getreuer und Christlicher Unterricht für
Jedermann, enthaltend, wunderbare und
probmässige Mittel und Künste, sowohl für
die Menschen, als das Vieh. Mit vielen
Zeugen bewiesen in diesem Buch, und
wovon das Mehrste noch wenig bekannt
ist, und zum allerersten Mal in Amerika im
Jahr 1820 im Druck erschienen . . . Und
nun auf Begehren zum sechstenmal
gedruckt. Harrisburg, Pa., 1846.
    71 p. ; 18 cm.

1672. HOHMAN, Johann Georg.
Der lange verborgene Freund, oder:
getreuer und Christlicher Unterricht für
Jedermann, enthaltend: wunderbare und

probmässige Mittel und Künste, sowohl für die Menschen als das Vieh. Mit vielen Zeugen bewiesen in diesem Buch, und wovon das Mehrste noch wenig bekannt ist, und zum allerersten Mal in America im Druck erschient . . . Reading [Pa.]: Gedruckt [bey Carl Augustus Bruckman?] fur den Verfasser, 1820.

100 p. ; 18 cm. (12mo)

First edition.
REFERENCES: *Austin, #922; First century of German language printing in the United States of America, #2462.*

1673. HOHMAN, Johann Georg.
Der lange verborgene Schatz und Haus-Freund, oder getreuer und Christlicher Unterricht [sic] für Jedermann. Enthaltend wunderbare und erprobte Mittel und Künste, für Gebrechen der Menschen und am Vieh. Aus dem arabischen Schriften, des weisen Algemisten Omar Arey, Emir Chemir Tschasmir, ins Deutsche übersetzt und noch mit vielen andern Kunsten vermehrt, welches zum Erstenmale in Amerika im Druck erscheint . . . Skippacks-ville, Pa.: Gedruckt bei A. Puwelle, 1837.

123 p. ; 16 cm.

1674. HOHMAN, Johann Georg.
Der lange verborgene Schatz und Haus-Freund, oder getreuer und christlicher Unterricht für Jedermann. Enthaltend wunderbare und erprobte Mittel und Künste, für Gebrechen der Menschen und am Vieh. Aus dem arabischen Schriften des weisen Alchymisten Omar Arey, Emir Chemir Tschasmir, ins Deutsche übersetzt und mit noch vielen andern Kunsten vermehrt. Zweite vermehrte und verbesser-te amerikan. Auflage . . . Gedruckt in Pennsylvanien, A.D. 1847.

127 p. ; 17 cm.

1675. HOHMAN, Johann Georg.
The long lost friend. Or faithful & Christian instructions containing wonderous and well-tried arts & remedies, for man as well as animals. With many proofs of their virtue and efficacy in healing diseases, &c. The greater part of which was never published until they appeared in print for the

first time in the U.S. in the year 1820. Literally translated from the German work of John George Hohman . . . Harrisburg, Pa.: [s.n.], 1850.

72 p. ; 17 cm.

"Where is the doctor who has ever banished the panting or palpitation of the heart and hideboundedness? Where is the doctor who has ever banished a wheal? Where is the doctor who ever banished the mother-fits? Where is the doctor who can cure mortification when it once seizes a member of the body? All these cures, and a great many more mysterious things are contained in this book; and its author could take an oath at any time upon the fact of his having successfully applied many of the pre-scriptions on its pages . . . If men were not al-lowed to use sympathetic words, nor the name of the Most High, it would certainly not have been revealed to them; and what is more, the Lord would not help where they are made use of" (*Pref.,* p. 3-4). Hohman adds, "whatever can-not be cured by sympathetic words can much less be cured by any doctor's craft or cunning."

1676. HOHMAN, Johann Georg.
The long lost friend. A collection of mys-terious & invaluable arts & remedies, for man as well as animals. With many proofs of their virtue and efficacy in healing dis-eases, &c., the greater part of which was never published until they appeared in print for the first time in the U.S. in the year 1820 . . . Harrisburg, Pa.: T.F. Scheffer, printer, 1856.

72 p. ; 16 cm.

1677. HOHMAN, Johann Georg.
The long lost friend. Containing mysteri-ous and invaluable arts and remedies for man as well as animals. With many proofs of their virtue and efficacy in healing dis-eases, &c. . . . Lancaster, Pa.: Printed for the purchaser, 1877.

94, [2] p. ; 13 cm.

1678. HOLBROOK, Martin Luther, 1831-1902.
Chastity, its physical, intellectual and moral advantages . . . New York: M.L. Holbrook & Co.; London: L.N. Fowler & Co., [c1894].

[2], 104, [22] p. ; 19 cm.

First edtion. Holbrook states that "the sexual instinct has its foundation deep down in nature," that "it cannot, in its right relation, be a low or base passion," and that the "chaste relation of the sexes" is the source of much good. The unchaste relations of the sexes, however, "have been a curse from the beginning. From unchastity we have our weaklings, our incapable people, our half made and good-for-nothings, our criminals, liars, thieves, our sensualists and our gluttons, our sexual diseases and our sensuality" (p. 73). He asks, "If a perverted sexual instinct is the source of so much evil, can we not in some way alter it and make it pure?" (p. 74). In chapter VIII entitled "THE CURE," Holbrook outlines a regimen consisting of "physical culture," which increases the body's strength while diminishing "the morbid craving for unnatural and unreasonable indulgence of the passional nature" (p. 83). It focuses on diet, cold bathing and "moral training." Issued in softcover with twenty-two unnumbered pages of advertisements. This work was later issued with an expanded appendix under the title: *Physical, intellectual and moral advantages of chastity* (#1694).

1679. HOLBROOK, Martin Luther, 1831-1902.
Eating for strength: a book comprising: 1.—The science of eating. 2.—Receipts for wholesome cookery. 3.—Receipts for wholesome drinks. 4.—Answers to ever recurring questions . . . New York: Wood & Holbrook, 1875.
    157 p. ; 19 cm.

First edition. *Eating for strength* is divided into two parts. The first, occupying a third of the text, explains the functions of foods, their classification, nutritional content, and the physiology of digestion. The second part consists of recipes.
    A native of Mantua, Ohio, Holbrook was a graduate of the Ohio Agricultural College (now Ohio State U.), and for several years (1859-61) edited the *Ohio farmer*. In 1862-63 he trained under Dio Lewis (#2233-2243), and graduated from his Normal School of Physical Culture. Holbrook received his medical degree in 1864 from the Hygeio-Therapeutic College (New York), the hydropathic medical school established by R.T. Trall (#3558-3593) in 1857. In

the year he graduated, Holbrook assumed the long-term editorship of *The Herald of health and journal of physical culture* (which became the *Journal of hygiene* in 1893). Holbrook's influence on the public's conception of personal hygiene and health reform extended beyond authorship of the ten books that he wrote for a general audience. As did the Fowler family, Holbrook regarded the business of publishing as a mechanism for raising public consciousness regarding health issues. Initially in partnership (Wood & Holbrook) and later under his own name, M.L. Holbrook was an important publisher of popular medical and hygienic literature from the early 1870s through the 1890s.

1680. HOLBROOK, Martin Luther, 1831-1902.
Eating for strength: a book comprising: 1.—The science of eating. 2.—Receipts for wholesome cookery. 3.—Receipts for wholesome drinks. 4.—Answers to ever recurring questions . . . Seventh edition. New York: M.L. Holbrook & Co., 1883.
    157, [5] p. ; 19 cm.

1681. HOLBROOK, Martin Luther, 1831-1902.
Eating for strength; or, food and diet in their relation to health and work, together with several hundred recipes for wholesome foods and drinks . . . New York: M.L. Holbrook & Co., [c1888].
    viii, 9-236, [2] p. ; 19 cm.

In his preface to this edition, Holbrook writes, "at no time have so many been engaged in laborious researches on the nature of that which we eat and its relations to health and work. It would almost seem as if the time had nearly arrived when mankind would eat to live, would feed themselves so as to nourish their bodies most perfectly and render themselves capable of the most labor, and least liable to disease . . . There is no doubt that man may double his capacity for work and for enjoyment by improving his dietetic habits. Many have already done this, and multitudes more are only waiting for the knowledge which will help them to do it" (p. [iii]). This 1888 edition has been so greatly revised as to constitute almost an entirely new work. Holbrook's analysis of the principles of dietetics, the nutritional content of foods, the relation of food to energy production, health, etc. has been expanded to occupy more than two thirds of the text, and incorporates the findings of recently published studies on nutrition.

The boundary between science and pseudoscience in Holbrook's writings is indicated in his discussion of apples and grapes—fruits which he regarded as two of the most important foods humans can ingest. Presaging the dietary fads of the 20th century, Holbrook recommends the "grape cure" for the treatment of digestive system disorders, "congestion of the brain," liver disorders, suppressed menstruation, lung diseases (including consumption), "chronic affections of the urinary system" (including calculi), freckles, scurvy, etc. Cover-title: *Food and work.*

1682. HOLBROOK, Martin Luther, 1831-1902.
Eating for strength; or, food and diet in their relation to health and work, together with several hundred recipes for wholesome foods and drinks . . . New York: Fowler & Wells Co., publishers; London: L.N. Fowler, [ca. 1896, c1888].
    viii, 9-246, [2] p. ; 19 cm.

The date of imprint has been estimated from the publication dates of the books advertised on the last two unnumbered pages. Cover-title: *Food and work.*

1683. HOLBROOK, Martin Luther, 1831-1902.
Homo-culture; or, the improvement of offspring through wiser generation . . . A new edition of "Stirpiculture," enlarged and revised. New York: Fowler & Wells Co.; Melbourne; London: L.N. Fowler & Co., [after 1899, c1897].
    238 p. ; 19 cm.

The first 192 pages of *Homo-culture* is identical to Holbrook's earlier *Stirpiculture* (#1695). The pages added to this edition under the heading "Notes" contain various short essays by Holbrook, e.g., "War and parentage," "cases of prenatal influences," "Luxury and parentage," etc. An edition of *Homo-culture* appeared under Holbrook's imprint in 1899. Reference to "the

war with Spain on account of Cuba" on page 202 of this undated Fowler & Wells edition would suggest a publication date no earlier than 1899, and possibly later.

1684. HOLBROOK, Martin Luther, 1831-1902.
How to strengthen the memory; or, natural and scientific methods of never forgetting . . . New York: M.L. Holbrook & Co., [c1886].
    152 p. ; 19 cm.

Cover-title: *Never forgetting.*

1685. HOLBROOK, Martin Luther, 1831-1902.
How to strengthen the memory; or, natural and scientific methods of never forgetting . . . New York: Fowler & Wells Co., publishers; London: L.N. Fowler & Co., [c1886].
    [3]-152, ix p. ; 19 cm.

1686. HOLBROOK, Martin Luther, 1831-1902.
Hygiene of the brain and nerves and the cure of nervousness. With twenty-eight original letters from leading thinkers and writers concerning their physical and intellectual habits . . . New York: M.L. Holbrook & Company, 1878.
    [2], vi, [7]-279 p. : ill. ; 19 cm.

In his consideration of the many forms of "nervous exhaustion," Holbrook considers the causes, i.e., genetic predisposition, defective nutrition, mental strain, the use of stimulants, insufficient sleep, "indulgence in vice and passion," etc. He observes that "Many do not recognize these forms of mental obliquity as diseases, but as moral defects, and treat them by censure. They should be recognized as physical diseases, and treated by hygiene" (p. 50). In the following chapter he narrows his focus: "One thing we know, however, as the blood, rich or poor, pure or impure, that supplies the nerves, is, so will the nervous power be" (p. 58). He continues: "Seeing that the nerves must be supplied with pure blood in proper quantity to enable them to do their duty, can we wonder if neglect of the common rules of health shall cause a feeling of illness, an unstrung state of the system, and misery and wretchedness? The nerves get poisoned with impure blood, starved with thin blood. The blood may be poisoned by bile, by alcohol, by bad food, by tobacco, tea, coffee, opium, henbane, hops, chloral, by breathing polluted air, neglect of the skin. One thing follows—nervous exhaustion" (p. 59). The sufferer from nervousness must adopt a hygienic regimen that counters these effects on the blood and the nervous system.

Part II ("physical and intellectual habits of distinguished men and women") contains the responses of twenty-eight individuals to Holbrook's querie for information regarding their personal regimen for avoiding nervous exhaustion. They include T.L. Nichols (#2626-2633), J.R. Buchanan (#505-507), Wm. Lloyd Garrison, A. Bronson Alcott, Silas O. Gleason, Dio Lewis (#2233-2243), Wm. Cullen Bryant, etc.

1687. HOLBROOK, Martin Luther, 1831-1902.
Liver complaint, nervous dyspepsia, and headache: their causes, prevention, and cure . . . Third edition. New York: M.L. Holbrook & Co., 1878.
    iv (i.e., vi), [7]-141 p. ; 19 cm.

Holbrook reviews the anatomy and physiology of the liver in his first several chapters before considering his principal topic, i.e., "derangements of the liver," more commonly called "torpid liver." He discusses the symptoms of torpid liver (e.g., costiveness, loss of appetite, headache, depression, etc.), and the diseases that may result, particularly those of the nerves and heart. In curing derangements of the liver Holbrook recommends changes in diet, air, exercise, clothing, alcohol consumption, etc. Holbrook's discussion of the liver is followed by a reprint of the chapter entitled "Influence of mental cultivation in producing dyspepsia" (p. [95]-108) from Amariah Brigham's *Remarks on the influence of mental cultivation and excitement upon the health* (#409-411). The book concludes with Holbrook's essay on headache. *Liver complaint* was originally published at New York in 1876. This third edition was reissued in 1894.

1688. HOLBROOK, Martin Luther, 1831-1902.
Marriage and parentage and the sanitary and physiological laws for the production of children of finer health and greater ability. By A Physician and Sanitarian . . . New York: M.L. Holbrook & Co., 1882.
    viii, [3]-185 p. ; 19 cm.

"If the average standard of ability of the race in intellect, in morals, and in physical power were rasied one degree during each century, the results could hardly be estimated. We should, in a comparatively short time, get rid of our thieves, our robbers, our drunkards, the licentious, the feeble and insane, and we should have so many more able men and women than now—far abler than any the world has yet produced—that life, which, with all its drawbacks, is well worth living, would be still grander and better . . . It may, in part, be accomplished by education, by moral culture, and by the gradual processes of evolution, but these agents are too slow, and do not reach all the conditions of life . . . It seems to me the race might be greatly improved by wiser and more sanitary marriages, and by more physiological parentage . . . There are many who will object to the application of the laws of breeding, even so modified as to be applicable to human beings without being in any way unfavorable to the marriage relation as it now exists, but the duty we owe to the future should inspire us to overcome any such objection, and especially when the harvest which may be reaped is so great. Indeed it is not in the least necessary to disturb or destroy marriage and parentage, but only to improve and make them more in accordance with scientific knowledge" (*Pref.,* p. [vii]-viii).

1689. HOLBROOK, Martin Luther, 1831-1902.
Parturition without pain; a code of directions for escaping from the primal curse . . . New York: Wood & Holbrook, 1871.
    113, [7] p. ; 18 cm.

CONTENT: I. Healthfulness of childbearing. II. Dangers of preventions [i.e., induced abortion]. III. Opinions on painless parturition. IV. Preparations for motherhood. V. Exercise and occupation during gestation. VI. The bath;

especially the sitz bath. VII. Painless parturition from fruit diet—food generally. VIII. The mind during pregnancy—"longings"—"mother's marks." IX. The ailments of pregnancy and their treatment. X. Anaesthetics during labor—female physicians. Appendix.

First edition. Holbrook quotes William Potts Dewees to the effect that "pain in childbirth is a morbid symptom," adding that: "it is a per-version of nature caused by modes of living not consistent with the most healthy condition of the system; and that such a regimen as should insure such a completely healthy condition might be counted on with certainty to do away with such pain" (p. 22). The regimen to which Holbrook refers is the *fruit diet system*: "In pro-portion as a woman subsists during pregnancy upon aliment which is free from earthy and bony matter, will she avoid pain and danger in delivery; hence the more ripe fruit . . . and the less kinds of other food . . . is consumed, the less will be danger and sufferings of childbirth" (p.50). In chapter X on anesthetics, Holbrook states his opinion that chloroform might be administered when "the pain is likely to do more harm to the nervous system than the an-aesthetic could possibly do" (p. [91]); but reas-serts that "the fruit diet and the accompanying regimen recommended in this book will be found in most cases to do away with the neces-sity of any anaesthetic, by the prevention of pain" (p. 96).

1690. HOLBROOK, Martin Luther, 1831-1902.
Parturition without pain; a code of direc-tions for escaping from the primal curse . . . Third edition. Enlarged. New York: Wood & Holbrook, 1872.
147, [9] p. ; 18 cm.

Includes an endorsement of the "Health-Lift" (#1601-1603) for pregnant women (p. [115]-127); and Clemence S. Lozier's essay en-titled "Care of children" (p. [129]-147).

1691. HOLBROOK, Martin Luther, 1831-1902.
Parturition without pain; a code of direc-tions for escaping from the primal curse . . . Fourth edition. Enlarged. New York: Wood & Holbrook, 1873.
147, [9] p. ; 18 cm.

1692. HOLBROOK, Martin Luther, 1831-1902.
Parturition without pain; a code of direc-tions for escaping from the primal curse . . . Fifth edition. Enlarged. New York: Wood & Holbrook, 1875.
147, [9] p. ; 18 cm.

The fifth edition was first issued in 1874.

1693. HOLBROOK, Martin Luther, 1831-1902.
Parturition without pain; a code of direc-tions for escaping from the primal curse . . . Fourteenth edition, enlarged. New York: M.L. Holbrook, publisher, 1881.
159, [9] p. ; 18 cm.

The fourteenth edition was re-issued in 1882.

HOLBROOK, Martin Luther, 1831-1902.
Parturition without pain see also
NAPHEYS (#2547, 2548).

1694. HOLBROOK, Martin Luther, 1831-1902.
Physical intellectual and moral advantages of chastity . . . New York: M.L. Holbrook & Co.; London: L.N. Fowler & Co., [c1894].
120, [20] p. ; 21 cm.

The second edition of Holbrook's 1894 *Chas-tity* (#1678), with an expanded appendix. Cover-title: *Advantages of chastity*.

1695. HOLBROOK, Martin Luther, 1831-1902.
Stirpiculture; or, the improvement of off-spring through wiser generation . . . New York: M.L. Holbrook & Co.; London: L.N. Fowler & Co., 1897.
192 p. ; 19 cm.

"During all ages since man came to himself, there have been enlightened ones seeking to improve the race. The methods proposed have been various . . . Some have believed that edu-cation and environment were all-sufficient; oth-ers that abstinence from intoxicating drinks would suffice. A very considerable number have held the idea that by prenatal culture alone the mother can mould her unborn child into any desired form. The disciples of Darwin . . . have held that natural and sexual selection have been the chief factors employed by nature to bring about race improvement. No doubt all these factors have been more or less effectual, but the time has come for man to take special interest in his own evolution, to study and apply . . . all the factors that will promote in any way race improvement . . . We are not yet able to do it perfectly, our knowledge is too deficient, lack of interest is too universal, but we can make a beginning; thoughtfulness may be given to suitable marriages; improved environment may be secured; better hygienic conditions taken advantage of; food may be improved; the knowledge we have gained in improving animals and plants . . . may aid us; air, exercise, water, employment, social conditions, wealth and poverty, prenatal conditions, all have an influence on offspring, and man should be able, to some extent, to make them all tell to the advantage of future generations" (p. [3]-4). This work was later published under the title: *Homo-culture* (#1683).

HOLBROOK, Martin Luther see also *VITALITY* (#3657).

HOLCOMBE, William Henry see **SHIPMAN** (#3209).

1696. HOLLAND, Josiah Gilbert, 1819-1881.
Titcomb's letters to young people, single and married. Timothy Titcomb, Esquire. New York: Scribner, Armstrong & Co., successor to Charles Scribner & Co., 1872.
    xii, [13]-235 p. ; 17 cm.

Holland graduated from Berkshire Medical College in 1844 and established a medical practice in Springfield, Ma. that enjoyed little success. He dabbled in journalism and publishing during this period before abandoning medicine altogether in 1848. Venturing south, Holland taught school in Richmond, Va. and in Vicksburg, Miss., but by 1850 was back in Springfield where he found his true *metier* as joint editor of the *Springfield Republican*. From the mid-1850s Holland began publishing the volumes of poetry, history, novels and essays that made him one of the most widely read American authors of his generation. In 1870 he became

editor of the newly established *Scribner's Monthly* (later *The Century Magazine*).

    First published by Charles Scribner in 1858, *Titcomb's letters to young people* was one of Holland's most successful books. It had reached its 50th edition by 1873, and remained in print as late as 1911. Holland provides his young readers reflections and advice on society, manners, social duties, religion, marriage, family life and even hygiene. "Holland not only conformed to the taste of his generation but he met its moral and spiritual needs, and it is a tribute to his usefulness that half a million volumes of such unsensational works as his were sold." (*Dict. Amer. biog.*, 5:148). At head of title: *Brightwood edition*. Spine-title: *Titcomb's letters*.

HOLLENDER, Marc Hale see **LEWIS** (#2232).

1697. HOLLICK, Frederick, b. 1818.
The diseases of woman, their causes and cure familiarly explained; with practical hints for their prevention, and for the preservation of female health . . . New-York: Stringer & Townsend, 1849.
    xviii, [19]-294, [2] p., V (i.e., 4) leaves of plates : ill. ; 18 cm.

    Born in Manchester, England, Hollick was an early disciple of Robert Owen. After passing examinations in 1838, he became one of Owen's first six "social missionaries," each of whom was assigned to work in a major British city. For the next four years he organized Owenite groups, participated in debates, and gave lectures in Glasgow. He published two pamphlets in 1840: *Murder made moral* (about India) and *What is Christianity? And have the persons calling themselves Christians any right to interfere with the free expression of opinions by other parties?* In the early 1840s, Hollick emigrated to the United States. He received his medical degree in 1846 from the (eclectic) Physio-Medical Institute in Cincinnati, but settled in New York City where he began a career as a lecturer and author on topics related to human health and physiology.

    One of the engaging features of his lectures was his use of a life-size, *papier-maché* manikin imported from France, which could be dismantled organ-by-organ to demonstrate anatomical relationships. Many decades later he wrote, "during a visit to France, I became ac-

Fig. 44. Hollick with an Auzoux manikin, from The origin of life, *#1724.*

quainted with Dr. Auzou [sic], and saw his wonderful models of the human body, made of *papier-maché*, full-sized, and formed and colored to life—so exact, in fact, that it might often be difficult to distinguish the model from the real body. Here, then, I found just what was needed; and I at once purchased a complete set suitable for my purposes . . . and also separate organs of the male and female generative system, with a complete series showing the development of the new being in the womb at every stage" (*The origin of life and process of reproduction,* 1878, #1729, p. vii). A portrait of Hollick with his manikin appears as the frontispiece to the 1845 edition of his *The origin of life* (*see fig. 44*).

Hollick's anatomical lectures embraced physiology, hygiene, and such topics as human reproduction, sexual physiology, and disorders of the reproductive system. Public discussion of reproductive physiology was not widespread during the 1840s, and Hollick was apprehensive about how his lectures might be received. "In the spring of 1844," he later wrote, "I lectured in New York City for three months without intermission, being fully attended most of the time. I never failed of receiving the applause of a single audience, and never heard a word

of censure, or condemnation, from any individual. Many of the most celebrated men in the city attended, and interested themselves early, to induce others to do the same. These lectures, of course, were for gentlemen alone, I however delivered a course on the female structure, female diseases, etc. to ladies only, which, contrary to all expectation, were excellently well attended. Many ladies of note honored me with their presence, and some even gave public testimony afterwards, in behalf of myself and my endeavors" (*Origin of life,* #1729, p. xvii-xviii).

Hollick soon extended the circuit of his lectures to Boston, Hartford, Philadelphia, Baltimore and Washington; and on at least one occasion took a speaking tour by boat down the Ohio and Mississippi rivers. His lectures were extensively advertised in the local press of each community, and, in return, received favorable editorial comment. Many of these comments from the press were presented as testimonials in the back of Hollick's books. (A substantial number were compared with the original text in the newspaper cited and are accurately transcribed.) Hollick enjoyed great success in Philadelphia, where he lectured on at least twenty-six separate occasions (and to as many as 432 women at a single lecture) between 1845 and 1850. After one series of lectures at Philadelphia in 1845, Hollick was presented with a portable writing desk and an engraved gold pen by grateful members of his female following. The gift was accompanied by the words: "The women of this generation have reason to rejoice that, by your efforts, a new and extensive field of information has been opened to them . . . of immense importance to themselves and their posterity, hitherto concealed within professional closures." One audience in Washington included the aged John Quincy Adams and several members of Congress. In spite of the success of his lecture tours, Hollick was tried in the years 1845 and 1846 in Philadelphia on charges of obscenity, ostensibly for his use on stage of the *papier-maché* manikin and for his book *The origin of life* (#1724). The first trial ended in a hung jury; the second in acquittal. Hollick gave up lecturing after 1870, and restricted himself to medical practice, both in person and through the mails.

Though well-known in his own time as a lecturer, Hollick is best remembered for the eleven books he wrote, nine of which were published between 1845 and 1852. Between 1856 and 1895,

Hollick had four different New York offices—all on Broadway between No. 3 and No. 599. He resided, however, on Staten Island. A son, Arthur Hollick, was a scientist and civic leader, and later paleobotanist at the New York Botanical Garden.

Historians, librarians and booksellers have tended to look askance at Hollick's voluminous publishing output. Booksellers relegate his books to the plain-brown-wrapper section of their shops, and historians have dismissed them as self-promoting and primarily proprietary because they advertise products and services. It does not appear that Hollick was so viewed by his contemporaries—if one believes but a part of the testimonials given him. One of his western agents (quoted in *The origin of life*, 1892) wrote that Hollick's books had become "over a large part of the country, household books, so that not a house, cabin, nor miner's camp can be found without them for hundreds of miles. There are few men more extensively known than you are, or more appreciated." Copies of Hollick's books have been found in or traced to the library of more than one "proper" old family.

Fig. 45. The Paris establishment of Louis Auzoux as it appeared in 1999 after closing.

1698. HOLLICK, Frederick, b. 1818.
The diseases of woman, their causes and cure familiarly explained; with practical hints for their prevention, and for the preservation of female health . . . New-York: Stringer & Townsend, 1850.
xviii, [19]-294, [2] p., V (i.e., 4) leaves of plates : ill. ; 18 cm.

"Experience early taught me that the greater part of the female diseases which came under my notice, were caused . . . by the ignorance of the sufferers respecting the constitution and relations of their own systems. This ignorance not only produces their diseases, but also conserves them, because it prevents the adoption of proper means for their removal, and leaves the sufferers liable to constant imposition from uninformed or designing pretenders . . . I feel fully assured that the greatest skill in the world could never cure so many physical evils, as a little timely knowledge would prevent! Therefore, both as a means of teaching females how to avoid disease, and to instruct them in simple means of treatment, and also to prevent imposition, a certain amount of knowledge of themselves is indispensable" (*Pref.*, p. [xii]-xiv). Hollick explains the origin of this work, "In the spring of 1844, I commenced in New York city a course of public lectures to females, on female diseases, illustrated by anatomical models." The present book follows the course of these lectures: "My audiences repeatedly asked me . . . to write such a one, so that they could study it at home" (p. xv). He also remarks that the "growing spirit of enquiry among females respecting themselves, makes such a work as the present absolutely necessary . . . Several books on this subject have lately been issued of a highly objectionable character . . . Some others I have seen, not to be objected to on the same grounds, being written by respectable scientific men . . . nevertheless, fail in giving *popular information,* because they are not simple in their language nor familiar in their explanations" (p. xvii).

Hollick assures his readers: "I have *explained every female derangement yet heard of or seen!*" (p. xvii). Following an anatomical introduction, Hollick addresses uterine displacements, hernia, neoplasms, menstrual disorders, and hysteria, as well as the problems that afflict females in puberty and during "change of life." Hollick devotes a chapter to new technologies, e.g., the vaginal speculum ("because so many females are unacquainted with it") and therapeutic use of the galvanic battery.

The first edition of *The diseases of woman* was published at New York by Burgess, Stringer & Co. in 1847. Stringer left the firm soon after to form Stringer & Townsend, and issued what appears to be the second edition in 1849. This 1850 (3rd?) edition is printed from the same plates as the previous edition. The book is illustrated with four wood-engraved plates and several wood engravings in-text.

1699. HOLLICK, Frederick, b. 1818.
The diseases of woman, their causes and cure familiarly explained; with practical hints for their prevention, and for the preservation of female health . . . New-York: T.W. Strong; Boston: G.W. Cottrell, [ca. 1851-53, c1849].
xviii, [19]-266, [2], 267-294 p., V (i.e, 4) leaves of plates : ill. ; 19 cm.

This edition issued by T.W. Strong bears a copyright date of 1849, and was printed from the same plates used for Stringer & Townsend's 1849 and 1850 editions (#1698). The four wood-engraved plates in this edition are hand-colored, unlike the identical but monochromatic plates of the previous editions.

The firm of T.W. Strong was active circa 1849-1870, and was the primary publisher Hollick's books during the 1850s. "In the second half of the nineteenth century, for a number of earnest, hard-working, upwardly mobile young men the publishing and distributing of literature on reproductive control became a career . . . Thomas W. Strong of New York City . . . found both considerable pecuniary success and social respectability. He was a publisher, engraver, printer, stationer, and dealer in miscellaneous goods (including contraceptives) . . . Strong began his career as a clerk, and in the 1840s, in his twenties, made enough of a profit selling valentines . . . to buy a printing and lithographic press. He began to specialize in inexpensive paperbound literature, particularly advice guides on sexuality, marriage, and reproductive control, and he opened a store to sell it along with other cheap books, stationery, engravings, and 'notions,' which included condoms, douching preparations, cures for

venereal disease, aphrodisiacs, and possibly contraceptive pessaries" (Janet F. Brodie, *Contraception and abortion in nineteenth century America,* Ithaca: Cornell University Press, 1994, p. 197-98).

1700. HOLLICK, Frederick, b. 1818.
The diseases of woman, their causes and cure familiarly explained; with practical hints for their prevention and for the preservation of female health . . . Forty-ninth edition, much improved. New York: T.W. Strong; Boston: G.W. Cottrell & Co., [ca. 1853, c1847].
xviii, 19-417, 2-7, [7] p., [3] leaves of plates : ill. ; 15 cm.

The first 266 pages of the "49th" edition were printed from the same plates as the previous editions. The three colored plates are new (the wood engravings printed as plates in previous editions are now printed in-text); and several new illustrations have been included. More than 150 pages of new material has been added since the first edition, including sections on headache (p. 272-313); "On solitary vices, and other abuses" (p. 314-349); extensive extracts from Hollick's *Marriage guide* (p. 351-398); book reviews of *The diseases of women* (the latest of which is dated Aug., 1853); testimonials; and advertisements for Hollick's other publications.

1701. HOLLICK, Frederick, b. 1818.
The diseases of woman, their causes and cure familiarly explained; with practical hints for their prevention, and for the preservation of female health . . . 53rd edition, much enlarged and improved. New York: T.W. Strong, [c1855].
xv, [16]-467, [1] p., [3] leaves of plates : ill. ; 15 cm.

This "53rd" edition contains additional matter, including Hollick's *Facts for the feeble! or professional notes of curious medical consultations; relating to the various peculiarities, disabilities, and forms of decay of the sexual system* (p. [375]-451), which has a separate title-page but pagination continuous with the rest of the text.

1702. HOLLICK, Frederick, b. 1818.
Diseases of woman, their causes and cure familiarly explained; with practical hints for

their prevention, and for the preservation of female health . . . 63d edition, much enlarged and improved. New York: The American News Company, [ca. 1862, c1849].

xiv, 15-466 p., [3] leaves of plates : ill. ; 15 cm.

Pages 432-45 promote *Dr. Hollick's Aphrodisiac Remedy*, which, "when judiciously employed, invariably increases and maintains sexual power, or restores it when lost" (p. 434). Included are letters (dated 1861-62) purportedly from Hollick's agents in Aden, the Hindostan, the Andes and Leipzig, who were responsible for gathering the secret ingredients that comprise his proprietary aphrodisiac.

1703. HOLLICK, Frederick, b. 1818.

The male generative organs in health and disease, from infancy to old age. Being a complete practical treatise on the anatomy and physiology of the male system; with a description of the causes, symptoms, and treatment, of all the infirmities and diseases to which it is liable. Adapted for every man's own private use, and including an introductory account of all the new discoveries concerning the physiology of the female system and the process of reproduction . . . New York: Published by Nafis & Cornish; St. Louis, Mo.: Nafis, Cornish & Co., 1850.

xi, [12]-432 p., [1] leaf of plates : ill. ; 15 cm.

Fig. 46. "Galvanic process for affection of the lungs," from the Diseases of women, *#1702.*

Hollick's sixth book and the fourth volume in the series dealing with some aspect of human reproductive physiology is addressed primarily to men. It does not consider what are generally thought of as venereal diseases, though fungal and tuberculous infections are discussed. Emphasis is on various physiological dysfunctions, anatomical variations and disruptions, and, of course, normal function. There is a whole chapter on the influence of the brain on function of the generative organs. There are chapters on spermatorrhea (more commonly known today by the vernacular term "wet dream"), on masturbation (a sixty-nine page chapter reviews the opinions of authorities from Hippocrates to Deslandes [#931-934]), and on impotence.

1704. HOLLICK, Frederick, b. 1818.
The male generative organs in health and disease, from infancy to old age. Being a complete practical treatise on the anatomy and physiology of the male system; with a description of the causes, symptoms, and treatment, of all the infirmities and diseases to which it is liable. Adapted for every man's own private use, and including an introductory account of all the new discoveries concerning the physiology of the female system and the process of reproduction . . . New York: T.W. Strong; Boston: G.W. Cottrell, 1851.
    xi, [12]-432 p., [3] leaves of plates : ill. ; 15 cm.

1705. HOLLICK, Frederick, b. 1818.
The male generative organs in health and disease, from infancy to old age. Being a complete practical treatise on the anatomy and physiology of the male system; with a description of the causes, symptoms, and treatment, of all the infirmities and diseases to which it is liable. Adapted for every man's own private use, and including an introductory account of all the new discoveries concerning the physiology of the female system and the process of reproduction . . . New York: Published by Nafis & Cornish; St. Louis, Mo.: Nafis, Cornish & Co., 1851.
    xi, [12]-416 p., [1] leaf of plates : ill. ; 16 cm.

The first edition of *The male generative organs* was published at New York by T.W. Strong in

1849. Nafis & Cornish issued this title at New York in 1850 and in 1851 from the same stereotype plates.

1706. HOLLICK, Frederick, b. 1818.
The male generative organs in health and disease, from infancy to old age. Being a complete practical treatise on the anatomy and physiology of the male system; with a description of the causes, symptoms, and treatment, of all the infirmities and diseases to which it is liable. Adapted for every man's own private use, and including an introductory account of all the new discoveries concerning the physiology of the female system and the process of reproduction . . . New York: T.W. Strong; Boston: G.W. Cottrell, 1853.
    xi, [12]-429, [3] p., [3] leaves of plates : ill. ; 15 cm.

"There are certain difficulties connected with the reproductive system that are very important, as affecting human health and happines, but which are scarcely ever made the subjects of study by medical men at all, at least in this country . . . In the old world there are men of the greatest eminence who devote their sole attention to those matters . . . In this country I am not aware of any one, besides myself, that has embraced this peculiar line of practice, and I have found the greatest want of information prevailing, even amongst medical men, respecting the means of relief that are really at our command" (*Pref.*, p. iv-v).

1707. HOLLICK, Frederick, b. 1818.
The male generative organs in health and disease, from infancy to old age. Being a complete practical treatise on the anatomy and physiology of the male system; with a description of the causes, symptoms, and treatment of all the infirmities and diseases to which it is liable. Adapted for every man's own private use . . . Seventh-sixth [sic] edition, improved. New York: T.W. Strong; Boston: G.W. Cottrell, [1852-54].
    xi, [12]-429, [3] p., [3] leaves of plates : ill. ; 15 cm.

The Atwater copy is lacking the frontispiece and the plate labelled "Penis cut through vertically."

1708. HOLLICK, Frederick, b. 1818.
The male generative organs in health and
disease, from infancy to old age. Being a
complete practical treatise on the anatomy
and physiology of the male system. With a
description of the causes, symptoms, and
treatment of all the infirmities and diseases
to which it is liable. Adapted for every
man's own private use . . . 120th edition,
much enlarged and improved. New York:
T.W. Strong, [ca. 1855].
xi, [12]-461, [7] p., [3] leaves of plates :
ill. ; 15 cm.

The table of contents does not correspond to
the actual arrangement of the text in this "120th"
edition, but belongs rather to the earlier editions.
The 120th edition includes a substantial amount of
new matter, including Hollick's *Facts for the feeble!*,
which has its own title-page (copyrighted 1855), but
pagination continuous with the rest of the text (p.
[367]-398). There are two issues of the 120th edition in the Atwater Collection (#1709). Though
printed from the same stereotype plates, they differ
in the typefaces of their title-pages and in the blocks
from which the color wood-engraved plates were
printed.

1709. HOLLICK, Frederick, b. 1818.
The male generative organs in health and
disease, from infancy to old age. Being a
complete practical treatise on the anatomy
and physiology of the male system. With a
description of the causes, symptoms, and
treatment of all the infirmities and diseases
to which it is liable. Adapted for every
man's own private use . . . 120th edition,
much enlarged and improved. New York:
T.W. Strong, [ca. 1855].
xi, [12]-461, [7] p., [3] leaves of plates :
ill. ; 15 cm.

There are two issues of the "120th" edition
in the Atwater Collection (#1708). Though
printed from the same stereotype plates, they
differ in the typefaces of their title-pages and
in the blocks from which the color wood-engraved plates were printed. In this issue, plate
IX (p. [68]) is printed upsidedown.

1710. HOLLICK, Frederick, b. 1818.
The male generative organs in health and
disease; from infancy to old age; being a

complete practical treatise on the anatomy
and physiology of the male system; with a
description of the causes, symptoms, and
treatment of all the infirmities and diseases
to which it is liable. Adapted for every
man's own private use . . . Two hundredth
edition, much enlarged and improved.
New York: Excelsior Publishing House,
[c1872].
xi, [12]-432, [2], 433-466 p., [3] leaves
of plates : ill. ; 16 cm.

1711. HOLLICK, Frederick, b. 1818.
The marriage guide, or natural history of
generation; a private instructor for married
persons and those about to marry, both male
and female; in everything concerning the
physiology and relations of the sexual system, and the production or prevention of
off spring—including all the new discoveries, never before given in the English language . . . New York: T.W. Strong; Boston:
G.W. Cottrell & Co., [c1850].
4, [v]-xvii, [18]-428, [4] p., [3] leaves of
plates : ill. ; 16 cm.

"Married persons, and those who are thinking of marriage, always need information upon
a variety of topics both interesting and important to them, but which they have no means of
gaining. It is not always agreeable to ask medical men in regard to such matters, particularly
for females; and, besides, many medical men
are not disposed to answer questions of the
kind, and others are not even competent to do
so, because they are not familiar with the subjects inquired about" (*Pref.*, p. [v]).
First edition. *The marriage guide* describes the
anatomy and physiology of the male and female
generative organs; menstruation; ova, semen,
copulation, impregnation, and growth of the
fetus; extra-uterine conceptions, false conceptions and other growths; twins and superfoeta-
tion; causes of the differences in sex; hermaphrodites; coitus during pregnancy; influence of
the mind on the generative organs; impotence
and sterility; the proper age to marry and start
sexual activity; the prevention of conception;
the signs of pregnancy; erotomania and satyri-
asis. In his chapter on aphrodisiacs, cannabis is
mentioned as one of the more effective agents.
Hollick's first discussion of contraception
appears in these pages. Hollick writes: "This

is a subject which many persons may think not necessary to be treated upon, but there are peculiar reasons why it ought not to be passed over in silence. It has been, of late years, so much talked of, and so many unscientific works have been published, pretending to give information about it, that every one is familiar with the idea . . . Even if such information was likely to be productive of great evil, as some imagine, it is now impossible to prevent its dissemination, and it is, therefore, useless to avoid the topic" (p. 332). In his review of contraceptive methods, Hollick discusses withdrawal; douches; condoms, and having "association" only during the last part of the woman's menstrual cycle, when she was thought to be infertile. Hollick seems to favor the last.

1712. HOLLICK, Frederick, b. 1818.
The marriage guide, or natural history of generation; a private instructor for married persons and those about to marry, both male and female; in everything concerning the physiology and relations of the sexual system, and the production or prevention of off spring—including all the new discoveries, never before given in the English language . . . New York: T.W. Strong; Boston: G.W. Cottrell & Co., [ca. 1853].
4, [v]-xvii, [18]-428, [4] p., [3] leaves of plates : ill. ; 16 cm.

A re-issue of the first edition (#1711) distinguished by an advertisement for *The people's medical journal and home doctor* on the verso of the final leaf stating: "First number on July 1st, 1853."

1713. HOLLICK, Frederick, b. 1818.
The marriage guide, or natural history of generation; a private instructor for married persons and those about to marry, both male and female; in everything concerning the physiology and relations of the sexual system, and the production or prevention of off spring—including all the new discoveries, never before given in the English language . . . New York: T.W. Strong; Boston: G.W. Cottrell , 1853.
xviii, [19]-430, [2] p., [3] leaves of plates : ill. ; 16 cm.

1714. HOLLICK, Frederick, b. 1818.
The marriage guide, or natural history of generation; a private instructor for married persons and those about to marry, both male and female, in everything concerning the physiology and relations of the sexual system and the production or prevention of offspring; including all the new discoveries, never before given in the English language . . . 196th edition, much enlarged and improved. New York: T.W. Strong, [c1855].
xx, 21-467, [1] p., [3] leaves of col. plates : ill. ; 15 cm.

1715. HOLLICK, Frederick, b. 1818.
The marriage guide, or natural history of generation; a private instructor for married persons and those about to marry, both male and female; in everything concerning the physiology and relations of the sexual system, and the production or prevention of offspring; including all new discoveries, never before given in the English language . . . 200th edition, much enlarged and improved, and brought down to the present day. New York: T.W. Strong, [c1860].
467, [1] p., [3] leaves of plates : ill. ; 15 cm.

1716. HOLLICK, Frederick, b. 1818.
The marriage guide, or natural history of generation; a private instructor for married persons and those about to marry, both male and female; in everything concerning the physiology and relations of the sexual system, and the production or regulation of offspring; including all new discoveries, never before given in the English language . . . 300th edition!! . . . New York: The American News Co., publishers and agents, 1875.
515 p., [4] leaves of plates : ill. ; 15 cm.

1717. HOLLICK, Frederick, b. 1818.
The marriage guide, or physiological and hygienic instructor, for the married, or those intending to marry, both male and female. Including everything relating to the philosophy of generation, and the mutual relations of man and woman . . . 500th edition revised, with new matter and

illustrations. New York: Excelsior Publishing House, [c1883].

xvii, [1], [13]-108, 108½-[blank], 109-408 p., [13] leaves of plates : ill. ; 15 cm.

Perhaps the most interesting alteration to this "500th" edition are the amendments made to the chapter on contraception. Hollick has shortened the chapter from nine pages (in the previous editions) to five. In his discussion of uterine injections, he deleted his opinion that this is the "most reliable means"; omits any description of the chemical composition of injections; and stresses the negative effects of this method. No reference is made at all to the use of condoms in this edition. Without doubt, Hollick edited this chapter in reaction to the Comstock law (#754, 755). He must certainly have been aware of the prosecution of his fellow New Yorker Edward Bliss Foote (#1213-1256) for disseminating "obscene" literature through the mails.

1718. HOLLICK, Frederick, b. 1818.
The matron's manual of midwifery, and the diseases of women during pregnancy and in childbed, being a familiar and practical treatise, more especially intended for the instruction of females themselves, but adapted also for popular use among students and practitioners of medicine . . . New-York: Published by T.W. Strong, 1849.
xii, 468 p., [1] leaf of plates : ill. ; 16 cm.

The first edition of *The matron's manual* was published at New York by T.W. Strong in 1848. "A short time ago I published a popular treatise on *The Diseases of Woman* (#1697), in the non pregnant state, and in that work I announced my intention of shortly publishing a similar one on *Pregnancy and its diseases*. This book is the fulfilment of that promise" (*Pref.*, p. [iii]). Hollick claims that this is "the first *popular*, and yet strictly *scientific* book on midwifery ever published." He expresses a sentiment common to popular literature of the period when he writes: "Females have been kept in shameful ignorance, of everything connected with their own systems . . . That ignorance has led to untold evils, which can never be corrected till they have become more enlightened respecting themselves" (*ibid.*) Hollick's qualifications to undertake the task of enlightenment rests

on his prior "lectures and books, and in my daily communication with them as patients." "I am therefore, aware," he adds, "both of their great lack of proper information, and of their strong desire for it, and I flatter myself I also know, from experience and careful observation, the best mode of imparting it to them" (*ibid.*).

Hollick's treatise is divided into ten sections: I. Position and uses of the female organs. II. Signs of pregnancy, and the means of detecting it; its duration, and the period at which the foetus can live. III. The form, size, and position of the foetus, and its appendages, at full term. IV. The mechanism of delivery in all the different presentations and positions of the foetus. V. The physiology of spontaneous delivery, or childbirth, and the manner of conducting a natural labor. VI. Protracted and difficult labors. VII. Accidents during labor which may compromise mother's life. VIII. Operations with the hand and with instruments. IX. The diseases of pregnancy. X. The diseases of women in childbed, after lying in. APPENDIX (containing an essay entitled "Use of chloroform in midwifery").

1719. HOLLICK, Frederick, b. 1818.
The matron's manual of midwifery, and the diseases of women during pregnancy and in childbed, being a familiar and practical treatise, more especially intended for the instruction of females themselves, but adapted also for popular use among students and practitioners of medicine . . . New York: Published by T.W. Strong, 1853.
xii, 466 p., [3] leaves of plates : ill. ; 15 cm.

1720. HOLLICK, Frederick, b. 1818.
The matron's manual of midwifery, and the diseases of women during pregnancy and in child bed. Being a familiar and practical treatise, more especially intended for the instruction of females themselves, but adapted also for popular use among students and practitioners of medicine. Forty-seventh edition—much improved. New York: Published by T.W. Strong, [after 1853].
xii, 466, [454]-455 p., [3] leaves of plates : ill. ; 15 cm.

The "Notice of books" (p. 463-64) includes a review dated 1853.

1721. HOLLICK, Frederick, b. 1818.
The matron's manual of midwifery, and the diseases of women during pregnancy and in child bed. Being a familiar and practical treatise, more especially intended for the instruction of females themselves, but adapted also for popular use among students and practitioners of medicine. Forty-seventh edition—much improved. New York: Published by T.W. Strong, [after 1853].
    xii, 469 (i.e., 468) p., [3] leaves of plates : ill. ; 15 cm.

    The "Notice of books" (p. 461-62) includes a review dated 1853.

1722. HOLLICK, Frederick, b. 1818.
The nerves and the nervous. A practical treatise on the anatomy and physiology of the nervous system, with the nature and causes of all kinds of nervous diseases; showing how they may often be prevented, and how they should be treated. Including also, an explanation of the new practice of neuropathy; or, the nerve cure. Intended for popular instruction and use . . . New York: The American News Company, [c1873].
    432 p. : ill. ; 16 cm.

    Hollick's ninth and last medical book, except for the later revised, composite versions of his books. *The nerves and the nervous* is completely different from the earlier *Neuropathy* (#1723). The latter title was an enthusiastic exposition of galvanism and its role in physiology and treatment. Here, galvanism is hardly mentioned and the book appears to be almost a repudiation or correction of Hollick's earlier views. In the preface he writes, "it has been unavoidable that so many old and revered opinions, and beliefs, should be totally dissented from. The progress of modern science makes these opinions and beliefs, now, untenable" (p. 6). Although he continues to argue that electricity and the nervous impulse are probably the same thing (p. 53-54), he no longer discusses galvanism as a therapeutic modality.
    The book is a considered study of what was then known of the anatomy, physiology and diseases of the human nervous system—but "popular, so that all can understand it, but at

the same time . . . strictly *accurate* and *scientific!*" (p. 6). Again, Hollick reaffirms his conviction that popular ignorance of physiology is the cause of much sickness: "Human beings are able to live rationally, so as to avoid suffering and disease, just in proportion as they understand themselves, and their relations to the material world in which they exist."

1723. HOLLICK, Frederick, b. 1818.
Neuropathy; or, the true principles of the art of healing the sick. Being an explanation of the action of galvanism, electricity, and magnetism, in the cure of disease; and a comparison between their powers, and those of drugs, or medicines, of all kinds, with a view to determine their relative value, and proper uses . . . Philadelphia: Printed for the author by King and Baird, 1847.
    xi, [1], 193, [3] p. ; 15 cm.

    "The use of *galvanism, electricity, magnetism,* and other neuropathic agents in the cure of disease, is of recent date, and the philosophy of their action is but little understood. It has, therefore, been fully treated upon in the present work, so that an estimate may be formed of their value, and a comparison made between them and the agents most usually relied upon. It is my belief, confirmed by extensive experience, that these mysterious powers, when rightly used, are the most efficient, and the safest agents we can employ in the cure of disease, and that they will eventually supersede the use of drugs to a great extent, if not altogether" (*Pref.*, p. x-xi).
    Hollick states: "There is no doubt but it [i.e., electricity] is the principle of vitality, or life itself, and that nothing comes into existence, but by its means" (p. 52). He bases his own theory of electrotherapeutics on a line of argumentation that equates electricity with the *nervous fluid* that "makes the heart beat, the stomach digest, the muscles contract," etc. This *nervous fluid* (or "human galvanism") is produced by the action of the body's "natural battery," i.e., "the brain and spinal marrow, and the nervous cords connected with them" (p. 53). After describing the distribution of the nerves and the transmission of *nervous power,* Hollick posits a theory of disease based upon a maldistribution of this power: "The human machine is supplied with a cer-

tain amount of power to keep it going, and if too much of that power be absorbed in any one part, the other parts must of necessity either stop or be weakened in their action" (p. 62).

Having established that "the action of every organ in the human body is caused by nervous powers only," Hollick posits that "All disease . . . no matter where it may be situated, must originate in the nerves, the sole source of all action." As a consequence, "All curative treatment . . . must act upon the nervous system, directly or indirectly." The "most certain and efficient" agents toward that end are mesmerism, "the neuropathic power belonging to one living being . . . made to act upon another living being"; and the application of an electrical current to the affected part, e.g., by means of a galvanic battery (p. 68). Hollick devotes subsequent chapters to mesmerism (or "natural neuropathy") and to magnetism, electricity and galvanism (or "artificial neuropathy").

This was the only edition of *Neuropathy*. The imprint on the title-page of the Atwater copy differs from that of the printed wrapper: *Philadelphia: National Publishing Company, 1847.*

1724. HOLLICK, Frederick, b. 1818.

The origin of life: a popular treatise on the philosophy and physiology of reproduction, in plants and animals, including the details of human generation, with a full description of the male and female organs . . . New York: Published by Nafis & Cornish; St. Louis, Mo.: Nafis, Cornish & Co.; Philadelphia: John B. Perry, [c1845].

[4], xxiii, [1], [5]-233, [1], 57 p., 9 leaves of plates + frontis. : ill., port. ; 13 cm.

This was Hollick's first book, published the year after he started giving public lectures on anatomy and physiology. *The origin of life* is the first popular book to address the subject of reproductive physiology in a scholarly way, using the best information then available and citing sources. Before Hollick, the works of the Pseudo-Aristotle (#104-147) and Becklard (#263-270) were the sex education manuals that were most available to Americans. They were, however, grossly inaccurate and superficial in their treatment of every aspect of sexual physiology. The books of Robert Dale Owen (#2692-2699) and Charles Knowlton (#2154-2157) were pioneering on the subject of contraception, but

focused on the philosophical justification of family limitation and methods of prevention. In the *Origin of life,* Hollick attempts to bring correct, detailed information on sexual physiology to the masses, and discusses areas of reproductive physiology that often were not even addressed in the professional literature.

In the preface, Hollick is at pains to justify his venture, and apprehensive lest he be criticized for being prurient. "The subject," he writes, "is one of universal importance; it concerns everybody, and all are interested in it, whether they will or not. It occupies the mind of nearly all persons, more perhaps than any other subject, no matter what may be the extent, or deficiency, of their information. The thoughts of the ignorant man are as full of it, as those of his more enlightened neighbor; probably much more so. Keeping persons in ignorance upon this subject therefore, does not prevent them thinking about it . . . We have therefore merely a choice, between giving correct information, and leaving the mind to be filled with error and vain surmise, for with one or the other it will most assuredly be occupied" (p. iv-v).

Hollick describes the anatomy and functioning of the male and female sexual systems; impregnation; fetal development; "the sexual feeling"; and masturbation. There are fifty-seven pages of addenda citing the forensic literature on various subjects, from hermaphroditism to determining virginity, etc. The lithographic frontispiece portrait depicts Hollick beside a three-dimensional anatomical manikin of the kind he used in his lectures (*see fig. 44*). "It is taken from a daguerreotype by the celebrated 'Plumbe,' taken at Washington City" (p. xxii). The ten plates which illustrate the text are hand-colored and rather crudely executed lithographs.

1725. HOLLICK, Frederick, b. 1818.

The origin of life: a popular treatise on the philosophy and physiology of reproduction, in plants and animals, including the details of human generation, with a full description of the male and female organs . . . Tenth edition . . . New York: Published by Nafis & Cornish; St. Louis, Mo.: Nafis, Cornish & Co., [ca. 1847].

li, [52]-274, 2, 253-274, 57, [1] p., 10 leaves of plates : ill. ; 13 cm.

The imprint date is based upon the presence of an advertisement for Hollick's *Neuropathy*, which was published in 1847. Hand-colored lithographic plates. In this copy, the final two leaves of signature 21 and the whole of signature 22 (p. 253-264) have been mistakenly bound in place of signature 24 in the second pagination sequence (meaning that p. 3-14 are lacking).

1726. HOLLICK, Frederick, b. 1818.
The origin of life: a popular treatise on the philosophy and physiology of reproduction, in plants and animals, including the details of human generation, with a full description of the male and female organs . . . Twentieth edition . . . New-York: Published by Nafis & Cornish; St. Louis, Mo.: Nafis, Cornish & Co., [ca. 1847].
[2], li, [52]-150, [2], 151-158, [2], 159-274, 57, [1], iii-xvii, [1] p. : ill. ; 13 cm.

Pages v-vii (2nd sequence) contain newspaper reviews dated 1847. Although the title-page states that this edition is "illustrated by fine colored engravings on stone," the color lithographic plates which illustrate some of the earlier editions (e.g., the 10th ed., #1725) have all been replaced by uncolored wood engravings printed on leaves conjugate with the those of the text.

1727. HOLLICK, Frederick, b. 1818.
The origin of life: a popular treatise on the philosophy and physiology of reproduction, in plants and animals, including the details of human generation with a full description of the male and female organs . . . 99th edition . . . New York: Published by Nafis & Cornish; St. Louis, Mo.: Van Dien & McDonald, [ca. 1848].
li, [52]-274, 57, [1], iii-xvii, [1] p., [2] leaves of plates : ill. ; 14 cm.

The imprint date based upon the presence of an advertisement for *The diseases of woman*, which was first published in 1848. In spite of the wording on the title-page, the lithographic illustration of the earlier editions has been supplanted by wood-engraved illustration.

1728. HOLLICK, Frederick, b. 1818.
The origin of life: a popular treatise on the philosophy and physiology of reproduction, in plants and animals, including the

details of human generation with a full description of the male and female organs . . . 99th edition . . . New-York: T.W. Strong; Boston: G.W. Cottrell & Co., [ca. 1849].
[2], li, [52]-274, 57, [1], iii-xvii, [1] p. : ill. ; 15 cm.

The imprint date is based upon the presence of an advertisement for *The male generative organs*, which was first published in 1849.

1729. HOLLICK, Frederick, b. 1818.
The origin of life and process of reproduction in plants and animals, with the anatomy and physiology of the human generative system, male and female, and the causes, prevention and cure of the special diseases to which it is liable. A plain, practical treatise, for popular use . . . New York: The American News Company, [c1878].
xxxvii, [3], 15-924 p., XL leaves of plates : ill. ; 25 cm.

Hollick's preface is an auto-encomium that provides interesting autobiographical detail. He reviews his many publications, and concludes: "The number of the books, convenient for those who wished for information on one special matter only, was, nevertheless, not so well for those who wished the information contained in all of them, but who did not want to buy several separate works. And this led to the issuing of the present volume, which contains the matter of *all the separate works, and much more besides!* thus giving an opportunity for any one to possess in a single volume, and in a compact form, the whole series, with all the new information in addition" (p. xiii). He adds, "It will be understood, therefore, that there is nothing in the single volumes that is not in this one, and that it contains also an amount of *new matter* fully equal to a *new volume*. The old matter has been revised and corrected, and all brought down to the latest date . . . Among the new matter will be found a full and plain account of the new discoveries, opinions, and investigations relating to the *origin of life, spontaneous generation,* and *evolution.*"
Hollick's date of death is as yet unknown. Assuming that he wrote the preface to this compilation, then he was still alive and active as late as 1878.

1730. HOLLICK, Frederick, b. 1818.
The origin of life and process of reproduction in plants and animals, with the anatomy and physiology of the human generative system, male and female, and the causes, prevention and cure of the special diseases to which it is liable. A plain, practical treatise, for popular use . . . New edition, with additions. Philadelphia: David McKay, publisher, [c1902].
[2], xxxviii, [13]-932 p., XLIV leaves of plates : ill. ; 25 cm.

A re-issue of the 1878 edition (#1729) containing an "Addendum. Illustrations of the evolution; and the pedigree of man" (p. [913]-919, and four additional plates). Cover-title: *Dr. Hollick's complete works.*

1731. HOLLICK, Frederick, b. 1818.
Outlines of anatomy & physiology, illustrated by a new dissected plate of the human organization, and by separate views. Designed either to convey a general knowledge of these subjects in itself, or as a key for explaining larger and more complete works . . . Philadelphia: T.B. Peterson, [c1846].
[16], 40, [4] p., [1] paper manikin + frontis. : ill., port. ; 28 cm.

The *Outlines* is Hollick's third book—one of only two published outside New York. It is dedicated to those ladies and gentlemen who supported him during his recent trial on charges of obscenity, or as Hollick put it, "for attempting to destroy the existing monopoly of scientific knowledge." One of the more distinctive features in the *Outlines* is the "dissected plate," designed by Hollick himself. This paper manikin consists of four chromolithographic flaps (printed on both sides) attached to a base plate that together provide successive views into the internal structure of the thoracic and abdominal cavities. It is one of the earliest such devices illustrating an American medical text. Neither the musculoskeletal nor the urogenital systems are considered in the *Outlines,* though they are extensively covered in Hollick's other books. The 1846 edition was also issued under the imprint of the National Publishing Company (#1732). The steel-engraved frontispiece portrait of Hollick was copied from a daguerreotype.

1732. HOLLICK, Frederick, b. 1818.
Outlines of anatomy & physiology, illustrated by a new dissected plate of the human organization, and by separate views. Designed either to convey a general knowledge of these subjects in itself, or as a key for explaining larger and more complete works . . . Philadelphia: National Publishing Company, 1846 [*imprint on front board*: Philadelphia: Published for the author, by King & Baird, 1847].
[16], 40, [4] p., [1] paper manikin + frontis. : ill., port. ; 29 cm.

The frontispiece portrait in this 1847 re-issue is a "specimen of the new art of Plumbeotyping, or transferring daguerreotypes to paper" (legend beneath Hollick's portrait).

1733. HOLLICK, Frederick, b. 1818.
A popular treatise on venereal diseases, in all their forms. Embracing their history, and probable origin; their consequences, both to individuals and to society; and the best modes of treating them. Adapted for general use . . . New-York: T.W. Strong, [c1852].
vii, [8]-315, [1], 87, [1], ix p., 6 leaves of plates : ill. ; 15 cm.

Following Ricord, Hollick divides venereal affections into non-virulent and virulent categories. The non-virulent, such as gonorrhea (which Hollick terms *blennorrhagia*), are confined to the genitalia. They may be transmitted through sexual contact, but often arise through ordinary causes; have no constitutional effects; and are not hereditary. Virulent venereal diseases, such as syphilis, are caused by "a peculiar poison, or virus;" are always contagious (normally through sexual contact); often result in systemic disorders; and are hereditary. Non-virulent and virulent venereal diseases are entirely different, Hollick insists, "and are only associated . . . because they affect the same organs, and because they are both popularly connected together as being more or less the penalty of licentiousness" (p. 145).
Hollick describes the causes, symptoms, prognosis and treatment of both categories of venereal affections. Copaiva is the most important weapon in Hollick's arsenal for the treatment of non-virulent venereal disorders. In the

treatment of syphilis in its secondary stage he
relies on mercurial preparations. Hollick regarded mercury as ineffective in tertiary syphilis, however, and describes the preparation of
iodide of potassium as a remedy. Hollick also
devotes attention to the prevention of sexually
transmitted diseases. "The act of association, in
all suspicious cases," he warns, "should not be
prolonged but should be completed as soon as
possible; and the man should urinate after" (p.
57). He recommends cleansing the genitalia
after intercourse in dubious circumstances (p.
163), and also recommends the use of condoms
as a preventive measure (p. 58). The appendix
contains extracts from *The male generative organs
in health and disease* (#1703-1710) and other
supplementary material.

The "List of plates" (p. 13) enumerates VIII
plates to be included in this work, though no
issue appears to have all eight plates. There are
two issues in the Atwater Collection of this edition published by T.W. Strong and copyrighted
in 1852 (*see also* #1734). Except for the title-
pages, they are printed from the same stereotyped plates (even the advertisements at the end
of the text are identical). The present issue is
illustrated with color lithographic plates; and
the other with color wood-engraved plates. This
issue with lithographs is probably the earlier,
as the wood-engravings reappear in the 1862
(#1735) and 1881 (#1737) editions. The 1862
and 1881 editions were also issued with six
plates, though "VIII" plates continue to be enumerated on the "List of plates." This copy re-
peats pages [13]-24, but is lacking pages 149-
152, 173-176, 185-188, 221-224, 245-248 and
plates 2-6.

1734. HOLLICK, Frederick, b. 1818.
A popular treatise on venereal diseases, in
all their forms. Embracing their history,
and probable origin; their consequences,
both to individuals and to society; and the
best modes of treating them. Adapted for
general use . . . New York: T.W. Strong,
[c1852].
   vii, [8]-315, [1], 87, [1], ix p., 6 leaves
of plates : ill. ; 15 cm.

This issue is identical to #1733 except for its
title-page and wood-engraved plates. This issue
is probably the latter, as the lithographic plates
do not seem to have been re-issued, while the

wood-engraved plates appear in the 1862
(#1735) and 1881 (#1737) editions.

1735. HOLLICK, Frederick, b. 1818.
A popular treatise on venereal diseases, in
all their forms. Embracing their history and
probable origin; their consequences, both
to individuals and to society; and the best
modes of treating them. Adapted for general use . . . New York: T.W. Strong, 98
Nassau-Street, [ca. 1862].
   vii, [8]-413 p., 6 leaves of plates : ill. ; 15
cm.

The date of imprint is estimated on the basis of a letter dated May 15, 1862 in the description of Hollick's *Aphrodisiac Remedy* (p. 391).
The firm of T.W. Strong was located at 98
Nassau Street from 1849 through the early
1860s. The address probably makes this issue
earlier than entry #1736, which differs only in
the typeface of its title-page and the address in
the imprint (599 Broadway).

1736. HOLLICK, Frederick, b. 1818.
A popular treatise on venereal diseases, in
all their forms, embracing their history and
probable origin; their consequences, both
to individuals and to society; and the best
modes of treating them. Adapted for general use . . . New York: Published by T.W.
Strong, 599 Broadway, [186?].
   vii, [8]-413 p., 6 leaves of plates : ill. ; 15
cm.

1737. HOLLICK, Frederick, b. 1818.
A popular treatise on venereal diseases, in
all their forms, embracing their history and
probable origin; their consequences, both
to individuals and to society; and the best
modes of treating them. Adapted for general and private use . . . Fiftieth edition,
with much new matter . . . New York: The
New York Publishing Company, [c1881].
   vii, [8]-407 p., 6 leaves of plates : ill. ; 15 cm.

1738. HOLLIS, Thomas.
A companion to the medicine chest, with
plain rules for taking the medicines, in the
cure of diseases . . . To which are added, rules
for restoring suspended animation, from
drowning . . . Boston: J. Howe, printer, 1834.
   7, [1] p. ; 17 cm.

This pamphlet contains directions for the use of specific drugs included in the medicine chest ("for ships or families") that was sold by Thomas Hollis, "druggist and apothecary" in Boston. Hollis' chests contained emetics, physical bilious pills, jalap & calomel, rhubarb, picra, castor oil, camphor, opodeldoc, burgundy pitch, sulphur, cream of tartar, basilicon, Turner's cerate, Turlington's balsam of life, balsam copaiva, balsam fir, laudanum, adhesive plaster, essence of peppermint, elixir paregoric, epsom salts, senna, tincture of rhubarb, white vitriol and magnesia.

1739. HOLLIS, Thomas.

TO THE PUBLIC. The unparalleled success attending the use of "Dr. Ward's Vegetable Asthmatic Pills," together with urgent and repeated requests from many individuals, induced the proprietor to put them up in some convenient form with directions . . . DR. WARD'S VEGETABLE ASTHMATIC PILLS, One of the most valuable medicines ever discovered, for Coughs, Colds, Asthmatic Complaints, Consumption, Whooping Cough, &c. Prepared only, by the sole Proprietor THOMAS HOLLIS, BOSTON, MASS. . . . [183?].
Three broadsides printed on [1] leaf ; 30.5 × 60.5 cm.

In all likelihood, the three advertisements printed on this sheet represent three distinct broadsides as they came off the press, prior to the sheet being cut. The text of the first two broadsides is identical, though their ornamental type frames and typefaces differ. The third broadside, also within an ornamental frame, advertises *Ointment for the Itch*. All three bear the printers imprint "D. Hooten, printer" beneath their frames.

HOLLIS, Thomas see also **HARVARD** (#1580).

1740. HOLLOPETER, William Clarence, 1855-1927.

Hay-fever its prevention and cure . . . New York and London: Funk & Wagnalls Company, 1917.
   xv, [16]-347, [5] p., [3] leaves of plates : ill. ; 19 cm.

Originally published in 1898 under the title: *Hay-fever and its successful treatment.* A fourth edition was published in 1922. A Philadelphia pediatrician, Hollopeter was an 1877 medical graduate of the University of Pennsylvania.

1741. HOLMES, Oliver Wendell, 1809-1894.

Homoeopathy, and its kindred delusions; two lectures delivered before the Boston Society for the Diffusion of Useful Knowledge . . . Boston: William D. Ticknor, 1842.
   v, [3], 72 p. ; 19 cm.

First edition. "The classic and most effective attack upon homeopathy came at the hands of Oliver Wendell Holmes, physician, poet, and medical educator. He first presented his entertaining and enlightening classic . . . in 1842 as a set of two lectures before the Boston Society for the Diffusion of Useful Knowledge. Later that year it was published and widely disseminated" (Martin Kaufman, *Homeopathy in America,* Baltimore: Johns Hopkins, 1971, p. 35). Kaufman continues: "The real problem in evaluating the merits of homeopathy and other 'delusions,' according to Holmes, was that the vast majority of patients recover under *any* system. He asserted that 90 per cent of the cases commonly seen by a physician 'would recover, sooner or later, with more or less difficulty, provided nothing were done to interfere seriously with the efforts of nature' . . . Thus, Holmes said, even the so-called 'scientific' claims of reputable physicians can be seriously questioned. The remedy may have had nothing at all to do with the so-called 'cure'" (p. 40).
   REFERENCES: *Currier & Tilton,* p. 32.

1742. HOLMES, Oliver Wendell, 1809-1894.

The human wheel, its spokes and felloes, an article by Prof. Oliver Wendell Holmes, in the Atlantic Monthly, May, 1863 . . . Philadelphia: Republished by permission of the author, and publishers, by B. Frank. Palmer, surgeon-artist, [1863].
   15, [1] p. : ill. ; 23 cm.

"It is not two years," Holmes writes in the midst of war, "since the sight of a person who had lost one of his lower limbs was an infrequent occurrence. Now, alas! there are few of us who have not a cripple among our friends, if not in our own families . . . War unmakes legs, and human skill must supply their places as it best may" (p. 8). In a lengthy section of the article, Holmes describes the mechanics and advantages of the "Palmer leg," an artificial limb designed by Benjamin Franklin Palmer. In this clever piece of self-promotion, Palmer advises the public that the *Palmer Leg* is available at his establishments in Boston and Philadelphia. In printed wrapper.

REFERENCES: *Currier & Tilton*, p. 117.

1743. HOLT, Luther Emmett, 1855-1924.
The care and feeding of children. A catechism for the use of mothers and children's nurses . . . New York: D. Appleton and Company, 1895.
   66, [6] p. ; 17 cm.

First edition, 1895 issue. *The care and feeding of children* emanated from Holt's *Catechism for nurses* (1893), a pamphlet prepared for the instruction of nursery-maids during their four months training at the Babies Hospital in New York City (est. by the McNutt sisters in 1887, but directed by Holt from 1888). "The nurses studied the catechism and when they graduated took it with them. There followed requests for copies from their employers and the supply was soon exhausted. It was apparent there was a popular demand for something of the kind. The catechism was expanded, and in 1894 *The care and feeding of children* made its appearance . . . Probably no one would have been more surprised than Dr. Holt or his publishers had they been told that they were launching a volume that was to go through seventy-five printings, and which was to be translated into Spanish, Russian and Chinese . . . In time it was expanded to some two hundred pages and underwent twelve revisions during its author's life . . . The book became the mainstay of many a worried mother and exerted no inconsiderable influence on the profession as well, for the practitioners had to keep abreast of the pediatric knowledge which the mothers possessed" (R.L. Duffus, *L. Emmett Holt*, New York: Appleton-Century, 1940, p. 116-17). The 15th edition of *The care and feeding of children* was published in 1937, edited by L. Emmett Holt, Jr.

A native of Webster, N.Y. (a suburb of Rochester on Lake Ontario), Holt received his under-graduate degree from the Univ. of Rochester (1875) and began his medical studies at the Univ. of Buffalo. Within a year, however, Holt moved to New York, where he resumed his studies at the College of Physicians & Surgeons, and where he came under the influence of the orthopedist Virgil Pendleton Gibney. Holt received his medical degree in 1880 and began an internship at Bellevue Hospital, where he worked in the laboratory of William Henry Welch. "Holt acquired two basic building blocks for a highly successful career: under Gibney, his work with children inspired him to later specialize in pediatrics; and under Welch, he learned the value of scientific research in medical advances" (*Amer. nat. biog.*, 11:103).

Holt remained in New York his entire career. His association with the Babies Hospital, the first of its kind in the nation, lasted until his death. In 1890 he became chair of the diseases of children at the New York Polyclinic; and in 1901 chair of pediatrics at the College of Physicians & Surgeons. Holt was one of the founding members of the American Pediatric Society (1888); advised John D. Rockefeller on the establishment of the Rockefeller Foundation for Medical Research; and founded two major periodicals: the *Archives of pediatrics* (1884) and the *American journal of diseases of children* (1911). In 1897 Holt published the first edition of *The diseases of infancy and childhood*, which quickly became the standard pediatric text in the English language, running through twelve editions. Holt was also one of the "prime movers" in the formation of the Association for the Prevention of Infant Mortality (1909), and was influential in the regulation and improvement of New York's milk supply. As a clinician, teacher, author and public health reformer, Holt was one of the most influential physicians of his generation. "Osler alone in the United States exerted a comparable influence," according to Holt's biographer in the *Dict. Amer. biog.* (5:184). "More than any other single person in this country, he laid the groundwork for the 'age of the child'" (*Pediatric profiles*, ed. B.S. Veeder, St. Louis: Mosby, 1957, p. 60).

1744. HOLT, Luther Emmett, 1855-1924.
The care and feeding of children. A catechism for the use of mothers and children's nurses . . . Second edition, revised and enlarged. New York: D. Appleton and Company, 1900.
   [4], 104, [8] p. ; 17 cm.

"Three years of daily use of the catechism as a manual for nursery maids have shown the need of a fuller treatment of several subjects, notably infant feeding, than was given in the first edition . . . The chapter upon feeding has therefore been entirely rewritten and much new matter introduced . . . Many other points relating to clothing, growth, and training, have been touched upon for the first time, so that the size of the book has been increased by over one half" (*Pref.*, p. 1).

1745. HOLT, Luther Emmett, 1855-1924.

The care and feeding of children. A catechism for the use of mothers and children's nurses . . . Second edition, revised and enlarged. New York: D. Appleton and Company, 1901.

[4], 104, [8] p. ; 17 cm.

1746. HOLT, Luther Emmett, 1855-1924.

The care and feeding of children. A catechism for the use of mothers and children's nurses . . . Third edition, revised and enlarged. New York: D. Appleton and Company, 1906.

149, [19] p. : ill. ; 17 cm.

In this much expanded third edition "the section upon infant feeding has . . . been almost entirely rewritten and much new material introduced, especially that relating to the selection and care of cow's milk . . . An effort has been made in the present volume to state more fully the principles underlying infant feeding, a knowledge of which is indispensable to the nurse or mother who would cope successfully with this difficult problem . . . Many other points relating to growth, training, toys, etc. have been touched upon for the first time" (*Pref.*, p. 5).

According to Richard Meckel, "pediatricians increasingly focused their attention on one specific task of infant and child management . . . That task was infant feeding, and . . . it became the central focus of the specialty in the late nineteenth and early twentieth centuries and provided pediatricians with an area in which they could demonstrate their specific scientific expertise. Indeed, for Thomas Rotch, who along with Jacobi and Holt constituted a very visible triumvirate in turn-of-the-century American pediatrics, perfecting and monitoring infant

feeding was the single most important activity of the new specialty . . . To justify the preeminence they gave to proper infant feeding, pediatricians argued that it was the one form of prophylaxis that addressed all the life-threatening diseases faced by infants and young children . . . Moreover, they contended, the improperly fed infant and young child failed to develop, was constitutionally weak and therefore was much less able to resist and surive such deadly infectious diseases as whooping cough, scarlet fever, and bronchopneumonia" (*Save the babies*, Baltimore: Johns Hopkins, 1990, p. 48-49).

1747. HOLT, Luther Emmett, 1855-1924.

The care and feeding of children. A catechism for the use of mothers and children's nurses . . . Fourth edition, revised and enlarged. New York and London: D. Appleton and Company, 1907.

192, [8] p. ; 17 cm.

"In the present revision . . . the pages upon the diet of children from three to ten years have been entirely rewritten and the subject much expanded; while other questions relating to disturbances of digestion at this period are now considered for the first time. Weight charts have been added . . . The chapters upon infant feeding have been enlarged and, it is hoped, simplified for the average reader" (*Pref.*, p. 5).

"When discussing bottle feeding, Holt . . . emphasized the scientific bases of good artificial feeding. He outlined the known differences between human and cow's milk and used these analyses to rationalize the formulas he proposed. In 1894 he recommended diluting cow's milk with water and adding top-milk and sugar. By the fourth edition (1906), Holt . . . provided readers with a simplified form of the percentage method . . . As his formulas changed in the various editions of his book, so too did the scientific arguments he used to support them" (Rima Apple, *Mothers and medicine*, Madison: Univ. of Wisconsin Press, 1987, p. 110-11).

1748. HOLT, Luther Emmett, 1855-1924.

The care and feeding of children. A catechism for the use of mothers and children's nurses . . . Fifth edition, revised and enlarged. New York and London: D. Appleton and Company, 1909.

190, [8], [191]-195 p. ; 17 cm.

1749. HOLT, Luther Emmett, 1855-1924.
The care and feeding of children. A catechism for the use of mothers and children's nurses . . . Sixth edition, revised and
enlarged. New York and London: D.
Appleton and Company, 1912.
212 p. ; 17 cm.

1750. HOLT, Luther Emmett, 1855-1924.
The care and feeding of children. A catechism for the use of mothers and children's nurses . . . Seventh edition, revised
and enlarged. New York: D. Appleton &
Co., 1914.
218 p. ; 17 cm.

1751. HOLT, Luther Emmett, 1855-1924.
The care and feeding of children. A catechism for the use of mothers and children's nurses . . . Eighth edition, revised
and enlarged. New York and London: D.
Appleton and Company, 1916.
215 p. ; 17 cm.

1752. HOME AND HEALTH.
Home and health. A household manual
containing two thousand recipes and helpful suggestions on the building and care
of the home in harmony with sanitary laws;
the preservation of health by clean, consistent living; and the home treatment of
the more simple ailments and diseases, by
the use of natural, rational remedies, instead of drugs. Prepared and edited by a
competent committee of home-makers
and physicians. Mountain View, California:
Pacific Press Publishing Co., 1907.
xi, [12]-589 p. : ill. ; 22 cm.

Home and health is divided into six sections:
I. The home (e.g., location, sewerage, water
supply, etc.); II. General housekeeping (e.g.,
"cleanliness and order," sleeping-rooms, pests,
etc.); III. Diet; IV. Care of the body (e.g., ventilation, exercise, bathing, "social purity"); V. The
care and training of children; VI. The home
treatment of disease.

**1753. HOME AND HEALTH AND
COMPENDIUM OF USEFUL
KNOWLEDGE.**
Home and health and compendium of
useful knowledge: a cyclopedia of facts and

hints for all departments of home life,
health, and domestic economy, and hand
book of general information. London,
Ont.: The Advertiser Printing and Publishing Co., 1882.
346 p. : ill. ; 17 cm.

Cover-title: *London Advertiser premium. Home
& health: and compendium of useful knowledge.*

**1754. HOME AND HEALTH AND
COMPENDIUM OF USEFUL
KNOWLEDGE.**
Home and health and compendium of
useful knowledge: a cyclopedia of facts and
hints for all departments of home life,
health, and domestic economy, and hand
book of general information. London,
Ont.: The Advertiser Printing and Publishing Co., 1883.
346 p. : ill. ; 17 cm.

Cover-title: *London Advertiser premium. Home
& health: and compendium of useful knowledge.*

**1755. HOME BOOK OF HEALTH AND
MEDICINE.**
The home book of health and medicine: a
popular treatise on the means of avoiding
and curing diseases, and of preserving the
health and vigour of the body to the latest
period; including an account of the nature
and properties of remedies; the treatment
of the diseases of women and children, and
the management of pregnancy and parturition. By a Physician of Philadelphia. Philadelphia: Published by Desilver, Thomas &
Co., 1836.
viii, [9]-630 p. ; 24 cm.

This book is sometimes attributed to William Edmonds Horner (1793-1853). An 1814
medical graduate of the University of Pennsylvania, Horner joined its faculty in 1819 as adjunct professor of anatomy and was appointed
dean in 1822. Horner held that position for
nearly three decades. He is best remembered
for his *Treatise on special and general anatomy*
(1826), which replaced Wistar's as the standard
anatomical text in the United States, and as the
author of *A treatise on pathological anatomy*
(1829), the first pathology textbook published
in the United States. This anonymously pub-

lished domestic medicine was first issued at Philadelphia in 1834, and was reprinted there in 1835, 1836 and 1842.

1756. HOME FORMULARY COMPANY.
The Home formulary. The latest and most valuable toilet and miscellaneous formulas for home use. By William A. Hoburg . . . Cincinnati, Ohio: The Home Formulary Co., [c1900].
30, [2] p. ; 23 cm.

The components of each of the 170 recipes are provided, but not their quantities: "The quantity to be used of each medicine is kept in a separate book on our prescription case" at participating pharmacies. Title and imprint from wrapper.

THE HOME-MAKING SERIES see **KINNE** (#2141).

1757. HOME STUDY COURSE IN OSTEOPATHY.
Home study course in osteopathy, massage and manual therapeutics. Fourth edition. New York: Published by Sydney Flower, [c1902].
130, [6] p. : ill. ; 23 cm.

The publisher, Sydney Blanchard Flower (b. 1867), was the author of several books on hypnotism and "somnopathy," as well as a manual entitled *The mail-order business* (1902). The other publications on Flower's list all concern "new thought" and hypnotism.

1758. HOOD, William Fletcher.
The woman's friend: or Hood's wonderful book of secrets. Containing many most valuable remedies for curing diseases, instructions to mothers; employment to young men, &c., &c. New York, 1864.
16 p. ; 21 cm.

A collection of miscellaneous medical and domestic recipes. In his paragraph on venereal diseases, Hood advises those afflicted "to immediately apply to a good physician . . . Dr. [M.B.] La Croix, of Albany, N.Y., is well known to be very successful in treating this disease" (see Culverwell, #846). In printed wrapper.

1759. C. I. HOOD COMPANY.
Hood's combined cook books. [Lowell, Ma.: C.I. Hood Co., circa 1897].
4 pts. in 1 v. (31, [1]; 16; 31, [1]; 31, [1] p.) ; 19 cm.

Includes *Hood's cook book* numbers 1-3, and the *High-Street cook book*, which were published separately between 1881 and 1895. In each, the recipes are interspersed with descriptions of proprietary medicines available from C.I. Hood Co. Title from wrapper. Publication date estimated from the publication of the first edition of *Hood's practical cook's book* (1897), advertised inside the front wrapper.

1760. C. I. HOOD COMPANY.
Hood's new medical book devoted to the prevention of sickness, promotion of health and diminution of human suffering, by explaining causes and symptoms, and telling how to cure diseases of men, women and children. Revised, 1903. Lowell, Mass.: Published by C.I. Hood & Co., [1903?].
160 p. : ill. ; 14 cm.

An alphabetically arranged dictionary of common diseases that includes brief descriptions of each disorder and the remedy manufactured by C.I. Hood & Co. that cures it. *Hood's Sarsaparilla* is the most frequently recommended medicine from the Hood armamentarium. Sarsaparilla was introduced into Europe in the 16th century for the treatment of syphilis. It has no known physiologic action, but was nonetheless a favorite with manufacturers of patent remedies in the 19th and early 20th centuries. *Hood's Sarsaparilla* contained between sixteen and eighteen per cent alcohol. In printed wrapper.

1761. C.I. HOOD COMPANY.
Hood's standard family medicines and other proprietary articles of established merit. Used with perfect satisfaction in thousands of homes. Valued above all others for the cures they have made, the suffering relieved and the comfort given . . . Lowell, Mass.: Published by C.I. Hood Co., 1907.
16 p. : ill., ports. ; 22.5 × 11.5 cm.

A catalog of twenty remedies prepared by C.I. Hood Company. Cover-title: *Standard family medicines.*

1762. HOOKER, Worthington, 1806-1867.
The child's book of nature. For the use of families and schools. Intended to aid mothers and teachers in training children in the observation of nature. In three parts . . . New York: Harper & Brothers, publishers, 1860-61.
    3 pts. in 1 v. (xii, [13]-120 p. ; vi, [7]-170, [2] p. ; viii, [9]-179 p.) : ill. ; 17 cm.

    CONTENTS: Part I. Plants; Part II. Animals; Part III. Air, water, heat, light, &c. The second and third parts have separate title-pages dated 1861.
    Hooker received his medical degree from Harvard in 1829, and practiced in Norwich, Ct. for more than twenty years before his appointment as professor of the theory & practice of medicine at Yale in 1852. Concerned about the level popular scientific education, Hooker wrote a series of books intended to explain the natural sciences to lay and school-age audiences. They include: *Human physiology* (1854, #1768); *First book on physiology* (1855); *The children's book of nature* (1857); *The child's book of common things* (1858); *Geography for primary schools* (1859); *Natural history* (1860); *First book in chemistry* (1862); and the three-volume *Science for the school and family* (1863-65).

1763. HOOKER, Worthington, 1806-1867.
The child's book of nature. Three parts in one. Part I. Plants. Part II. Animals. Part III. Air, water, heat, light, &c. . . . New York: Published by Harper & Brothers, 1883.
    3 pts. in 1 v. (xii, [13]-120 p.; vi, [7]-166; viii, [9]-179 p.) : ill. ; 17 cm.

    The three parts of *The child's book of nature* were issued as a composite volume or separately throughout the 1860s, 1870s and 1880s.

1764. HOOKER, Worthington, 1806-1867.
Homoeopathy: an examination of its doctrines and evidences . . . New York: Charles Scribner, 1851.
    [2], xiv, [9]-146, [2] p. ; 19 cm.

    First edition. Hooker's essay won the *Fiske Fund Prize* of the Rhode Island Medical Society in 1851, and was one of the more successful attacks on homeopathy from the allopathic

camp published during the 19th century. Series number incorrectly given on the title-page as *No. XIII.*

1765. HOOKER, Wortington, 1806-1867.
Homoeopathy: an examination of its doctrines and evidences . . . Second edition. New York: Charles Scribner, 1852.
    xiv, [9]-146, [157]-162 p. ; 19 cm.

    Fiske Fund Prize Dissertation of the Rhode Island Medical Society, no. XIV.

1766. HOOKER, Worthington, 1806-1867.
Hooker's new physiology, designed as a text-book for institutions of learning . . . Revised by J.A. Sewall . . . New York: Sheldon and Company, 1875.
    xiv, 376, [4] p. : ill. ; 19 cm.

    Earlier editions published under title: *Human physiology* (#1768-1775). Revised edition edited by Joseph Addison Sewall (1830-1917).

1767. HOOKER, Worthington, 1806-1867.
Hooker's new physiology, designed as a text-book for institutions of learning . . . Revised by J.A. Sewall . . . With a chapter on alcohol and narcotics . . . New York & Chicago: Sheldon & Company, [c1874-1884].
    xiv, 401 p. : ill. ; 19 cm.

1768. HOOKER, Worthington, 1806-1867.
Human physiology: designed for colleges and the higher classes in schools, and for general reading . . . New York: Published by Farmer, Brace, & Co., 1854.
    xi, [12]-389 p. : ill. ; 20 cm.

    First edition. Hooker's motivation in undertaking this first of his many popularizations of science is revealed in the preface to *Human physiology*: "We live in the midst of a material world, animate and inanimate, and have daily converse, so to speak, with material forms of every variety, presenting phenomena of the highest interest and of endless diversity. And yet, through almost all the period of childhood . . . this book of nature is in the school-room very nearly a sealed book . . . He is drilled through spelling, reading, grammar, &c., but he is left in total ignorance of the beautiful flowers, and the majestic trees outside of the school-room" (p. viii). Hooker continues, "The whole order of education must be reversed. Instead of beginning the child's education with learning to spell and read, the object should be to make him an observer of nature. Things and not words . . . should from the first, constitute the substantial part of instruction. The child should be made, at home, in the school, and everywhere, a naturalist in the largest sense of that word" (p. ix).

1769. HOOKER, Worthington, 1806-1867.
Human physiology: designed for colleges and the higher classes in schools, and for general reading . . . New York: Published by Farmer, Brace, & Co., 1856.
xi, [12]-454 p. : ill. ; 20 cm.

The author has added a chapter on hygiene (p. 390-410) and an appendix ("Directions to teachers for the use of this book, and questions") to his original text.

1770. HOOKER, Worthington, 1806-1867.
Human physiology: designed for colleges and the higher classes in schools, and for general reading . . . New York: Pratt, Oakley & Company, 1857.
xi, [12]-454, 2 p. : ill. ; 20 cm.

1771. HOOKER, Worthington, 1806-1867.
Human physiology: designed for colleges and the higher classes in schools, and for general reading . . . New York: Pratt, Oakley & Company, 1859.
xi, [12]-454, 2 p. : ill. ; 20 cm.

1772. HOOKER, Worthington, 1806-1867.
Human physiology: designed for colleges and the higher classes in schools, and for general reading . . . New York: Sheldon & Company, publishers, 1865.
xi, [12]-454 p. : ill. ; 20 cm.

1773. HOOKER, Worthington, 1806-1867.
Human physiology: designed for colleges and the higher classes in schools, and for general reading . . . New York: Sheldon & Company, publishers, 1868.

xi, [12]-454, [2] p. : ill. ; 20 cm.

1774. HOOKER, Worthington, 1806-1867.
Human physiology: designed for colleges and the higher classes in schools, and for general reading . . . New York: Sheldon & Company, publishers, 1869.
xi, [12]-454, 2 p. : ill. ; 20 cm.

1775. HOOKER, Worthington, 1806-1867.
Human physiology: designed for colleges and the higher classes in schools, and for general reading . . . New York: Sheldon & Company, publishers, 1871.
xi, [12]-454, 2 p. : ill. ; 20 cm.

1776. HOOKER, Worthington, 1806-1867.
Lessons from the history of medical delusions . . . New York: Baker & Scribner, 1850.
v, [1], [3]-105 p. ; 19 cm.

Hooker was disturbed by the lack of respect shown to the regular medical profession by the American public and by the rise of alternative therapeutic systems. This is one of several books Hooker wrote with the intention of exposing the false premises upon which the public condemned allopathic medicine while embracing the supposed merits of its pseudo-scientific alternatives. Hooker hopes that "If we ponder well the lesson which can be gathered from the consequent delusions which have prevailed in the profession, and among the people, we shall be able to see to what errors we are liable in this busy and restless era, and how we may avoid them" (p. 4). At head of title: *No. XIII. 1850. Fiske Fund Prize Dissertation of the Rhode Island Medical Society.*

1777. HOOKER, Worthington, 1806-1867.
Lessons from the history of medical delusions . . . New York: Baker & Scribner, 1850.
v, [1], [3]-105, [3] p. ; 23 cm.

At head of title: *No. XIII. 1850. Fiske Fund Prize Dissertation of the Rhode Island Medical Society.* This issue is printed from the same plates as the previous entry (#1776), but is taller and adds to the text three unnumbered pages, including "Notices of the press" on the final leaf. There are two copies in the Atwater Collection: one bound in stamped dark green linen (spine-title: *Medical Delusions*); the other in a pale green wrapper, onto the cover of which is pasted a printed label.

1778. HOOKER, Worthington, 1806-1867.
Physician and patient, or, a practical view of mutual duties, relations and interests of the medical profession and the community . . . New York: Baker and Scribner, 1849.
[iii]-xxiv, [25]-453, [1] p. ; 19 cm.

Hooker has several purposes in placing this book before the public. One is to explain the "the medical errors which are common to man everywhere and in every condition," and thus expose "the *modus operandi* by which the genius of imposture has produced from them the fantastic and ever-changing shapes of empiricism" (*Pref.*, p. [vii]). He progresses from an examination of quackery's premises to their specific manifestations—providing chapters on "Thompsonism," homeopathy, natural bone-setters, etc. "Another object," Hooker notes in his preface, "will be to present the claims of the medical profession to the respect and the confidence of the community" (p. viii). "In the practice of medicine," he writes, "there are some points upon which there should be a common understanding between the physician and the friends and attendants of the sick. From the want of such an understanding the purposes and plans of the practitioner are often interfered with, and sometimes are effectually thwarted" (*Pref.*, p.ix-x). In the process of establishing this "common understanding," Hooker produced "the only comprehensive monograph about medical ethics written by an American physician during the nineteenth century" (*Amer. nat. biog.*, 11:139).

1779. HOPE, George H.
Till the doctor comes, and how to help him . . . Revised, with additions by a New York Physician. New York: G.P. Putnam & Sons, 1871.
99 p. : ill. ; 19 cm.

First American edition. Hope's first aid manual was originally published at London in 1870. The "Preface to the American edition" is signed *J.H.E.* Series: *Putnam's handy book series.*

1780. HOPE, George H.
Till the doctor comes, and how to help him . . . Revised, with additions by a New York

Physician. New York: G.P. Putnam & Sons, 1871.
99 p. : ill. ; 19 cm.

Edition statement on printed wrapper: "*From the fifth London edition.* Revised, with additions, by a New York Physician. A complete manual of directions in cases of accidents, indispensable to every household."

1781. HOPE, George H.
Till the doctor comes, and how to help him . . . Revised, with additions by a New York Physician. New York: G.P. Putnam's Sons, 1874.
99 p. : ill. ; 19 cm.

The publisher's advertisements on the endpapers indicates that this is volume IV of *Putnam's handy book series.*

1782. HOPE, George H.
Till the doctor comes and how to help him . . . Revised, and in large part rewritten by Mary M. Kydd, M.D. New York and London: G.P. Putnam's Sons, the Knickerbocker Press, 1901.
xi, [1], 153 p. ; 19 cm.

"In order . . . to add to the comprehensiveness of the work, I have included new material in the shape of chapters on hygiene, diseases of children, and obstetrics; giving special attention in enlarging the text to diseases more prevalent in this country than in England" (*Pref.*, p. iv). Mary Elizabeth Mitchell Kydd (d. 1919) was an 1890 graduate of the Women's Medical College of New York.

1783. HOPKINS, Jane Ellice, 1836-1904.
The power of womanhood or mothers and sons. A book for parents, and those in loco parentis . . . New York: E.P. Dutton & Company, 1899.
viii, 232 p. ; 20 cm.

Hopkins was a social reformer, author, lecturer, and advocate throughout Britain for the moral and legal protection of women and children. In this work she insists "that it is to the woman that we must look for the solution of the deepest moral problems of humanity, and that the key of those problems lies in the hands of the mothers of our race" (p. 2). "We women," she continues, "have to do with the fountain of sweet waters, clear as crystal, that flow from the throne of God; not with the sewer that flows from the foul imaginations and actions of men" (p. 10). It is thus the responsibility of women to ensure the moral safety of their sons, "Because the evil is so rife, the dangers so great and manifold, the temptations so strong and subtle, that your influence must be united to that of the boy's father if you want to safeguard him." As a woman, "there are potent lines of influence open to you . . . from which a man, by the very fact that he is a man, is necessarily debarred" (p. 13).

Having devoted much of her work to the rescue of "friendless girls," Hopkins is insistent on eradicating the attitude that allows a sexual double-standard: "if chastity is a law for women—and no man would deny that—it is a law for every woman without exception; and if it is a law for every woman, it follows necessarily that it must be for every man, unless we are going to indulge in the moral turpitude of accepting a pariah class of women made up of other women's daughters and other women's sisters . . . set apart for the vices of men" (p. 31). "Who so well as a mother," Hopkins asks, "can teach the sacredness of the body as the temple of the Eternal? Who else can implant in her son that habitual reverence for womanhood which to man is 'as fountains of sweet water in the bitter sea' of life? Who, like a mother, as he grows to years of sense and observation, and the curiosity is kindled . . . can so answer the cry and so teach as to make the mysteries of life and truth to be for ever associated for him with all the sacred associations of home and his own mother, and not with the talk of the groom or the dirty-minded schoolboy?" (p. 53). *The power of womanhood* was first published at London as well as in New York in 1899.

1784. HOPKINS, Mark, 1802-1887.
Address to the people of Massachusetts, on the present condition and claims of the temperance reformation. By Mark Hopkins, D.D., Samuel B. Woodward, M.D., Hon. Samuel Hoar. Published by the Massachusetts Temperance Union. Boston: Temperance Standard Press, 1846.
20 p. ; 22 cm.

The authors outline seven effects of intoxicating drink: "1st. That it shortens human life . . . 2d. We find that the use of alcohol as a beverage is a more fruitful source of crime than any other . . . 3d. Alcohol stands highest on the list of those causes by which reason is dethroned, and the various forms of insanity are produced . . . 4th. Aside from the enormous price directly paid for intoxicating drinks . . . they produce the larger proportion of the pauperism with which the country is burdened . . . 5th. The temptations connected with these drinks form a chief ground of anxiety of parents respecting their children . . . 6th. The use of drinks is the cause of most of the corruption in voting . . . 7th. It is also a prominent obstacle in the way of the progress of the religion of Christ among men" (p. 4-6). Title on printed wrapper: *Address to the people of Massachusetts in relation to the temperance reformation.*

Mark Hopkins received his medical degreee from Berkshire Medical College in 1829, and in 1830 joined the faculty of Williams College as professor of moral philosophy. He became president of Williams in 1836. Samuel Bayard Woodward (1787-1850) was a physician and pioneer in the institutionalization of the insane. Samuel Hoar (1788-1856) was a prominent Massachusetts lawyer and Congressman.

1785. HOPKINSON, Mrs. C. A.
Hints for the nursery or the young mother's guide . . . Boston: Little, Brown and Company, 1863.
iv, [5]-169 p. ; 18 cm.

In Part I. "Body and soul," Hopkinson discusses the health and hygiene of the newborn and infant. In Part II., "Soul and body," she discusses discipline, moral instruction and religious training.

HORNER, William Edmonds see *HOME BOOK OF HEALTH AND MEDICINE* (#1755).

1786. HORSELL, William, 1808-1863.
Hydropathy for the people: with plain observations on drugs, diet, water, air, and exercise . . . With notes and observations, by R.T. Trall, M.D. Stereotype edition. New York: Fowlers & Wells, publishers, 1850.
viii, [9]-250 p. ; 19 cm.

First American edition; originally published at London in 1845 under the title: *The board of health & longevity: or hydropathy for the people*. The New York edition was reissued in 1851 and 1855. Horsell was an English clergyman, who having adopted first the temperance regimen and later hydropathy, seeks in these pages "to lead others into the same paths, knowing them to be safe and happy" (p. 13). In his "American preface," Trall (#3558-3593) writes: "If the people can be thoroughly indoctrinated in the general principles of Hydropathy, they will not err much, certainly not fatally, in their home-application of the Water-Cure appliances to the common diseases of the day. If they can go a step further, and make themselves acquainted with the laws of life and health, they will well-nigh emancipate themselves from all need of doctors of any sort" (p. [iii]). Series: *Fowlers and Wells' water cure library*.

1787. HORSELL, William, 1808-1863.
Hydropathy for the people: with plain observations on drugs, diet, water, air, and exercise . . . With notes and observations, by R.T. Trall, M.D. Stereotype edition. New York: Fowlers & Wells, publishers, [ca. 1851].
      viii, [9]-250 p. ; 19 cm.

      Series: *Fowlers and Wells' water-cure library*. Spine-title: *The Water-cure library. Seven volumes. Vol. 3*. Issued in vol. 3 with John Smith's *Curiosities of common water* (New York, ca. 1851).

HOSACK, David, 1769-1835. *Address delivered at the first anniversary of the New York City Temperance Society* see **NEW YORK CITY TEMPERANCE SOCIETY** (#2592).

1788. HOSACK, David, 1769-1835.
Dr. Hosack's address, delivered before the New-York City Temperance Society, May 11, 1830. With other important documents. Stereotype edition. New-York: Published by John P. Haven, 1830.
      24 p. ; 21 cm.

      "I verily believe spiritous liquors to be altogether *unnecessary* and positively *injurious* to persons in *health*" (p. 2). Hosack qualifies his opinion with the remark: "While Temperance

Societies then prohibit the use of spiritous drinks, in those who enjoy health and vigor of constitution, it is no less their duty to except the medicinal use of them as prescribed by the physician." Hosack also addresses tobacco use and the intemperate consumption of food in this address.

HOSACK, David see also **THOMAS** (#3469).

1789. HOSPITAL FOR SICK CHILDREN (TORONTO).
Invalid cookery for the use of the trained nurse and all others who have to cook and serve food for invalids. Individual recipes. Also a chapter on the feeding of infants with full instructions for every mother. Issued by the Alumnae of The Hospital for Sick Children . . . Toronto: [The Hospital for Sick Children], 1907.
      69 p. : ill. ; 18 cm.

      "This book has been compiled by Mrs. George Macbeth, the chief dietician of the Hospital for Sick Children, Toronto . . . The chapter on infant feeding is by Dr. Alan Canfield" (p. 6).

1790. HOUGH, Lewis Sylvester, 1819-1903.
The science of man applied to epidemics: their cause, cure, and prevention . . . Boston: Published by Bela Marsh, 1849.
      290 p., [3] leaves of plates : ill. ; 19 cm.

      Hough concludes his fifty page introductory to the physiology of nervous conduction, circulation of the blood, respiration and digestion with an analogy likening the organs that comprise these systems to a "vital republic, all the parts of which, by the general texture of the tissues, are bound together each to all and all to each, in one community of interest and sympathy" (p. 51). This theme pervades his book, morally as well as physiologically. Hough then considers the hygienic factors that insure the maintenance of these vital functions, as well as the disorders that result from their disruption— particulary by epidemic disease. Through his idiosyncratic exposition of human physiology, pathology and hygiene, Hough calls the nation to personal and collective reform. In his concluding chapter, Hough provides an unselfconscious glimpse into the near-ecstatic mind of religious, moral and social reform that gripped the nation at mid-century. "Sons of Columbia!" he exhorts, "we are slaves. The golden sun rises, shines, and sets, and looks on a nation of slaves . . . Let us then nullify the soul of perdition, and its external expression will die of itself . . . Here, noble Sons of Thunder, is the place for us to begin. *Self-freedom* is the thing for which to strike. Let every one free *himself,* and then we are *all* free, a glorious nation of true patriots. Let every one govern well his own universe, and the great work is done. Sons of Columbia! arise and act . . . We are, with all our imperfections, the chosen nation for reforming earth, and introducing heaven here" (p. 258-59).

1791. HOUGH, Theodore, 1865-1924.

The human mechanism. Its physiology and hygiene and the sanitation of its surroundings. By Theodore Hough . . . and William T. Sedgwick . . . Boston: Ginn & Company, [c1906].
  ix, [1], 564 p., [12] p. of plates : ill. ; 20 cm.

First edition. "In the present text-book, anatomy has been reduced to its lowest terms and microscopic anatomy or histology touched upon only so far as seemed absolutely necessary. Space has thus been gained for more physiology and, especially, for more hygiene than is usual, and also for the elements of sanitation—a new and comparatively easy subject, but one of very first importance in all wholesome modern living. That point of view which regards the human body as a living mechanism is to-day not only the sure foundation of physiology, hygiene, and sanitation, but is also surprisingly helpful in the solution of many questions concerned with intellectual and moral behavior . . . We believe with Mathew Arnold that 'conduct is three fourths of life,' and that this is no less true of the physical than of the moral and intellectual life. We therefore make no apology for fixing upon conduct as the keynote of this work, and *the right conduct of the physical life* as the principal aim and end of all elementary teaching of physiology, hygiene, and sanitation" (*Pref.,* p. iii-iv). Theodore Hough was professor of physiology at the University of Virginia. His co-author, William Thompson Segdwick (1855-1921), was professor of biology at the Massachusets Institute of Technology.

1792. HOUGHTON, J. Skillin.

Extracts from the annual report of J.S. Houghton, M.D., Acting Surgeon of the Howard Relief Association, Philadelphia, Pa. [Philadelphia, 187-?].
  9-16 p. ; 14 cm.

Caption title. Houghton describes himself in his various pamphlets as "Acting Surgeon" of the Howard Association of Philadelphia, the Howard Relief Association, or the Howard Sanitary Aid Association. In so doing, he illegitimately associates his business in the mind of the public with the various Howard Associations that emerged in many American cities from the 1840s through the 1870s to coordinate relief efforts in the wake of such devastating epidemics as yellow fever.

The purpose of Houghton's "Howard Association" is "to treat sexual diseases with all the skill that the resources of Modern Science will permit" (p. 9). Specializing in the treatment of both sexual disorders and venereal diseases, Houghton solicits his readers "to send to the Howard Association a history of their misfortunes, their sufferings, and their present condition . . . with the fullest confidence in the honor, skill, integrity, and humane feelings of the physicians employed to administer its affairs" (p. 14). In return they will receive "the best of modern remedies," and "will not be injured by the unwise use of dangerous and powerful drugs" (p. 15).

1793. HOUGHTON, J. Skillin.

HOWARD SANITARY AID ASSOCIATION. OBSTACLES TO MARRIAGE. Young men are often afflicted with diseases which unfit them for enjoying social life, and especially unfit them to become husbands . . . [Philadelphia, 187-?].
  Broadside ; 29.5 × 12 cm.

Houghton advertises his ability to treat the consequences of "self-abuse," sexual excess, "seminal weakness" and venereal disease with specific remedies of botanic origin.

1794. HOUGHTON, J. Skillin.

Marriage and celibacy. An essay of warning and instruction for young men . . . [Philadelphia, 187-?].
  16 p. ; 14 cm.

To those who men who have not been celibate before marriage, Houghton offers relief to the disorders or diseases that they may have thoughtlessly acquired. Caption title. At head of page 1: HOWARD ASSOCIATION, PHILADELPHIA, PA.

1795. HOUGHTON, J. Skillin.
New treatment of sexual diseases . . . [Philadelphia, 187-?].
    8 p. ; 14 cm.

"Our purpose now is, solely to show how the effects of long-continued sexual abuses may be removed,—how the fearful exaltation of the sexual faculty over all other faculties, may be reduced,—how the injuries to the sexual organs may be repaired,—and how the victim of inordinate desires may be relieved of this terrible mastery of passion, and restored to a calm equilibrium of his senses, a sound and vigorous condition of body and mind" (p. 2). Among the disorders and diseases Houghton claims to be able to cure is "the sexual fever" or "sexual monomania," which may be "wonderfully controlled by appropriate remedies" (p. 5). Caption title.

1796. HOUGHTON, J. Skillin.
Report on spermatorrhoea, or seminal weakness, impotence, the vice of onanism, masturbation, or self-abuse, and other diseases of the sexual organs . . . [Philadelphia, 187-?].
    16 p. ; 14 cm.

Houghton describes the consequences and treatment of masturbation, seminal emissions and venereal diseases. Caption title. At head of title: *Howard Association, Philadelphia, Pa.*.

1797. HOUGHTON, Roland S.
Three lectures on hygiene and hydropathy . . . To which are prefixed the constitution and list of officers of the American Hygienic and Hydropathic Association of Physicians and Surgeons. New York: Fowlers and Wells, 1851.
    x, [11]-132, [4] p. ; cm.

Houghton's book comprise three lectures: "Address delivered at the first annual meeting of 'The American Hygienic and Hydropathic

Association of Physicians and Surgeons,' at Hope Chapel, New York, on the 19th of June, 1850" (p. [15]-40); "Hygiene the true moral of the cholera, a lecture delivered before the Mercantile Library Association, at Clinton Hall, New York, on . . . December 6th, 1849" (p. [41]-83); and "Hydropathy rational, and not empirical, a lecture delivered before the Mercantile Library Association, Clinton Hall, New York, on . . . December 13th, 1849" (p. 84-132). Clinton Hall on Nassau St. in New York was the seat of the Fowler brothers' "phrenological cabinet" and headquarters for the publishing firm of Fowlers & Wells. Series: *Fowlers and Wells water cure library*. Although the Atwater copy was issued with *Bulwer and Forbes on the water-treatment; new and revised edition* (1851, #521), the *Three lectures* was also issued separately.

HOUGHTON, Roland S. see also **BULWER AND FORBES** (#521).

1798. THE HOUSE AND ITS SURROUNDINGS.
The House and its surroundings. New York: D. Appleton & Company, 1879.
    [2], 96 p. ; 15 cm.

Series: *Health primers; no. 3*. A manual of domestic sanitation addressing such factors as the situation and construction of houses, their drainage, water supply, ventilation, etc. Originally published at London under the editorship of J. Langdon-Down (1828-1898), the nine volumes of the *Health primers* series was published in the United States by the firm of D. Appleton & Co. between 1879 and 1882.

THE HOUSEHOLD LIBRARY see **FORBES** (#1258).

1799. HOUSEHOLD TREASURE.
The Household treasure: Containing several hundred valuable receipts for cooking well at moderate expense, making dyes, coloring, cleaning, & cementing. This book also points out in plain language, free from doctors' terms the diseases of men, women, and children, and the latest and most approved means used for their cure. To which is added a description of the medicinal roots and herbs, how they can be used in the cure of diseases . . . Phila-

delphia: Published by Barclay & Co., [c1872].

[2], [19]-112 p. ; 24 cm.

The first publication of *The Household trea-sure* may have been the 1864 edition published by Barclay & Co. in fifty pages. It was re-issued several times in the late 1860s, and in 1872 was expanded to more than double the size of the earlier edition. The preface is signed "P.S." In printed wrapper.

1800. HOUSEKEEPER'S COMPANION.
The Housekeeper's companion. A practi-cal receipt book and household physician, with much other valuable information. The whole forming a complete hand-book of reference for housewives and mothers. Chicago, Ill.: Mercantile Pub. and Adv. Co., [c1883].

425 p. ; 23 cm.

1801. HOW TO BE YOUR OWN DOCTOR.
How to be your own doctor. New York: F.M. Lupton, publisher, [c1887].

[411]-446 p. ; 19 cm.

The 11 February 1888 issue of the *Leisure hour library*, vol. 2, no. 177. Running title: *The Home Physician.*

1802. HOW TO CONTROL AND REGU-LATE THE MATERNAL FACULTY.
How to control and regulate the maternal faculty. [S.l.: s.n., ca. 1850-1870].

4 p. ; 21 cm.

"Women have the right to govern and regu-late the power of conception. They have the right to decide when they will have children and when not" (p. [1]). In the following para-graph the anonymous author recommends an effectual method of preventing conception: "Take a very fine, soft sponge . . . Cut it large enough, when moistened and slightly squeezed, to fill the upper part of the vagina and cover the mouth of the womb . . . The sponge should have a piece of strong but soft silk taste or nar-row ribbon about a foot long, firmly tied to it. It may be kept in a basin of water, soft and ready for use. After connection, withdraw the sponge. This means is effectual in nearly

every case, but when used alone, it is not *al-ways* certain. It can be made *certain,* by com-bining it with the injection of cold water by the syringe" (p. [1]-2).

On page 3 the author states that when the clitoris is not "erected" during sex, women "have little pleasure, and are weakened and injured in health." Clitoral stimulation is briefly described. On page 4 the author states "The begetting of children should never be done in the darkness of night"; and concludes "The penis should rise and set with the sun." Cap-tion title.

1803. HOWARD, Horton, d. 1833.
A brief exposition of the principal circum-stances connected with the publication of the "Improved system of botanic medi-cine;" illustrative of the motives and prin-ciples of the author, in placing that work before the public . . . Columbus, Ohio, 1832.

20 p. ; 22 cm.

Caption title; above the title is a banner-title that reads: THE ECLECTIC, AND MEDICAL BOTA-NIST; [EXTRA], dated November 15, 1832. In this supplement to the first issue of *The Eclectic, and medical botanist* (published at Columbus between 15 Nov. 1832 and 16 Dec. 1833), Howard defends the recent publication of his *Improved system of botanic medicine* (#1811). "The first serious defection from Thomsonism oc-curred when Horton Howard . . . a Columbus printer and publisher and an early agent of Thomson in Ohio and the Middle West, authored his own two-volume *Improved system of botanic medicine* . . . in 1832. Howard formed a dissident group called the 'improved botanics' . . . Howard did not stray far from Thomson, which suggests that he was less a schismatic than a disaffected colleague turned rival. Drawing on the successful marketing techniques of his mentor, Howard employed agents and sub-agents to sell his own medicines, books, and 'improved rights' to practice . . . Thomson, fear-ing the success of his former pupil, directed his own agents to threaten lawsuits for patent in-fringement" (John Haller, *Kindly medicine,* Kent, Ohio, 1997, p. 27).

In the "Brief exposition," Howard relates the stormy history of his business relationship with Samuel Thomson (#3482-3515), as well as the reason for the publication of his own botanic text: "Yet, however enamoured I might have been with Dr. Thomson's system, I had, very early, seen and deplored the imperfections of his book, and the narrow limits of his materia medica; and was not satisfied that the knowledge of botanic medicine should remain in so imperfect a state. I therefore . . . called the friends of this system together and advised the formation of a constitution, and societies, for its improvement . . . [I] took much pains to collect a knowledge of every improvement, and every additional article of value . . . all of which I hoped would enable me, at some future period, to present to the world a better system of medicine than had hitherto previously been offered to its acceptance" (p. 7).

1804. HOWARD, Horton, d. 1833.
Dr. Howard's private medical companion and complete midwife's guide: intended for married females and heads of families; containing very important information concerning conception, with rules for its prevention and control. Together with other matters most invaluable to those expecting to become mothers, for the first time now made public. Philadelphia: Published by Duane Rulison, 1861.
    15, [17] p. : ill. ; 23 cm.

In the preface the publisher writes that this brief work was intended as "a companion to DR. HOWARD'S SYSTEM OF DOMESTIC MEDICINE," published by Rulison at Philadelphia in 1861 (#1808). Its contents are not extracted from the larger work, but consist of brief descriptions of the "organs of generation," pregnancy, and infant care (p. [5]-15). Contrary to the description of the pamphlet's contents on the title-page, there is no discussion of contraception in its pages. The work is primarily a vehicle for the seventeen wood-engraved illustrations that Rulison thought better "bound up separately" (see the *Pref.* to the 1859 ed. of *Howard's domestic medicine,* #1806, p. viii). They are copied from the wood engravings that originally illustrated the five editions of Samuel Bard's *Compendium on the theory and practice of midwifery,* published at New York between 1807 and 1819. In printed wrapper.

1805. HOWARD, Horton, d. 1833.
Howard's domestic medicine, a complete guide to the preservation of health and treatment of disease. Comprising the structure and functions of the human body; the cultivation of health; influence of air, sunlight, exercise, etc.; a full description of all the various diseases, causes of each, symptoms of each, with plain, practical directions as to the best remedies and how to administer them; a full description of botanic medicines, how to gather and prepare them for use; full directions for nursing; best articles of food, how prepared, etc., etc., etc. Covering, in fact, the whole range of HOME MEDICAL ADVISER . . . New York City: Union Publishing House, [c1879].
    2 pts. in 1 v. (18, 23-605 ; 175 p.) : ill. ; 23 cm.

This appears to be the final edition of this title. The *Supplement to Howard's Domestic medicine, being a practical treatise on midwifery and the diseases peculiar to women,* has a separate title-page and pagination. Cover- and spine-titles: *Domestic medicine.*

1806. HOWARD, Horton, d. 1833.
Howard's domestic medicine: being a revised edition of Horton Howard's anatomy and physiology, and midwifery, diseases of women and children, practice of medicine and materia medica, founded upon correct physiological principles, and especially adapted to family use; containing all the important discoveries and improvements down to the present time. Forming a complete family medical guide. New enlarged edition . . . Three volumes in one. Cincinnati: H.M. Rulison; Philadelphia: D. Rulison, 1859.
    xii, 13-989 p. : ill. ; 23 cm.

Howard's manual of domestic medicine was originally published in two volumes at Columbus, Ohio in 1832 under the title *An improved system of botanic medicine* (#1811) Posthumous editions retained that title through 1856, even as the work underwent extensive revision and expansion. Between 1856 and 1858 the Rulisons published three issues of their "new revised edition" of Howard's manual under the title *An improved system of domestic medicine* (#1815, 1816). With this 1859 edition, the title changed to *Howard's domestic medicine.*

Howard's domestic medicine represents a complete revision to the first "volume" of Howard's earlier work. The publisher H.M. Rulison states in his preface, "The department of anatomy and physiology . . . has almost entirely been re-written, and nearly one hundred pages of new matter added . . . Relating to physiology, various modifications have been made and recent discoveries embodied" (p. viii). Minor changes were made to the sections on materia medica and the practice of medicine. The wood engravings that illustrated the section on midwifery have been deleted from this edition, but were still available to the reader as a separate publication (#1804). Spine-title: *Howard's domestic medicine for the people.*

1807. HOWARD, Horton, d. 1833.

Howard's domestic medicine: being a revised edition of Horton Howard's anatomy and physiology, and midwifery, diseases of women and children, practice of medicine and materia medica, founded upon correct physiological principles, and especially adapted to family use; containing all the important discoveries and improvements down to the present time. Forming a complete family medical guide. New enlarged edition . . . Three volumes in one. Cincinnati: H.M. Rulison; Philadelphia: Duane Rulison; Geneva, New York: J. Whitley, Jr.; St. Louis: C. Drew & Co., 1860.

xii, 13-989 p. : ill. ; 24 cm.

1808. HOWARD, Horton, d. 1833.

Howard's domestic medicine: being a revised edition of Horton Howard's anatomy and physiology, and midwifery, diseases of women and children, practice of medicine and materia medica, founded upon correct physiological principles, and especially adapted to family use; containing all the important discoveries and improvements down to the present time. Forming a complete family medical guide. New enlarged edition . . . Three volumes in one. Philadelphia: Published by Duane Rulison, 1861.

xii, 13-989 p. : ill. ; 24 cm.

1809. HOWARD, Horton, d. 1833.

Howard's domestic medicine: being a new and revised edition of Horton Howard's anatomy and physiology, and midwifery, diseases of women and children, practice of medicine and materia medica. Founded upon correct physiological principles, and especially adapted to family use; containing all the important discoveries and improvements down to the present time. Forming a complete family medical guide. New enlarged edition . . . Three volumes in one. Cincinnati: J.R. Hawley, 1863.

xii, 13-992 p. : ill. ; 24 cm.

This edition contains an appendix on diphtheria (p. 990-92).

1810. HOWARD, Horton, d. 1833.

Howard's domestic medicine: being a revised edition of Horton Howard's anatomy and physiology, and midwifery, diseases of women and children, practice of medicine and materia medica, founded upon correct physiological principles, and especially adapted to family use; containing all the important discoveries and improvements down to the present time. Forming a complete family medical guide. New enlarged edition . . . Three volumes in one. Philadelphia: Published by Duane Rulison, 1865.

xiii, 13-992 p. : ill. ; 24 cm.

1811. HOWARD, Horton, d. 1833.

An improved system of botanic medicine, founded upon correct physiological principles; embracing a concise view of anatomy and physiology; together with an illustration of the new theory of medicine . . . In two volumes . . . Columbus: Printed by the author; Charles Scott, printer, 1832.

2 v. : ill. ; 21 cm.

First edition. "Howard, a former agent for Samuel Thomson [#3482-3515], had become involved in a heated controversy with his employer, and had proceeded to set up a Botanic faction of his own. The author's section on plant materia medica took up 85 pages in the second volume . . . What Howard had done was to augment the seventy plant drugs comprising

Samuel Thomson's materia medica with forty-two more, making a total of 112 plants" (Alex Berman, "A striving for scientific respectability," *Bull. hist. med.,* 1956, 30:11). Howard's purpose in writing a treatise that appeared to be in direct competition with the published work of his former employer was to correct the "imperfections of Dr. Thomson's book, and the circumscribed limits of his materia medica" (p. I:vi). Publication of the second volume, lacking in the Atwater set, was delayed by an injunction filed by Thomson in Ohio against Howard.

1812. HOWARD, Horton, d. 1833.

An improved system of botanic medicine, founded upon correct physiological principles; embracing a concise view of anatomy and physiology; together with an illustration of the new system of medicine. To which is added, a treatise on female complaints, midwifery, and the diseases of children . . . In three volumes . . . Second edition; revised and corrected. Columbus, Ohio: Published by the author; Scott & Wright, printers, 1833.
    3 v. : ill. ; 21 cm.

    Atwater set lacking vol. II.

1813. HOWARD, Horton, d. 1833.

An improved system of botanic medicine; founded upon correct physiological principles; embracing a concise view of anatomy and physiology; together with an illustration of the new theory of medicine. To which is added, a treatise on female complaints, midwifery, and the diseases of children . . . In three volumes . . . Third edition revised and corrected. Columbus, Ohio: Published by the proprietors; Horton J. Howard, printer, Saint Clairsville, O., 1836.
    3 v. in 1 (vii, [8]-180 p.; 465, [3] p.; v, [6]-215 p.) : ill. ; 20 cm.

    The three volumes of the *Improved system* are combined into one for the first time in this third edition. This would remain the book's format during the next forty years of its publishing history. The second volume of the Atwater copy is lacking pp. 97-108.

1814. HOWARD, Horton, d. 1833.

An improved system of botanic medicine; founded upon correct physiological principles; embracing a concise view of anatomy and physiology; together with an illustration of the new theory of medicine. To which is added, a treatise on female complaints, midwifery, and the diseases of children . . . In three volumes . . . Fourth edition. Cincinnati: Published by J. Kost, 1850.
    3 v. in 1 (viii, 9-80 p. ; 382 p., [45] leaves of plates; v, [6]-215 p.) : ill. ; 23 cm.

    The eclectic physician and publisher John Kost used the same wood-engraved blocks of botanical specimens for the illustration of this edition of Howard's *Improved* system that he used in his own manuals of domestic medicine (#2161-2163). Spine-title: *Howard's botanic medicine.*

1815. HOWARD, Horton, d. 1833.

An improved system of domestic medicine: founded upon correct physiological principles; comprising a complete treatise on anatomy and physiology; the practice of medicine, with a copious materia medica, and an extensive treatise on midwifery . . . New revised edition. . . . Cincinnati: H.M. Rulison, Queen City Publishing House; Philadelphia: Duane Rulison, Quaker City Publishing House, 1856-57.
    3 v. in 1 (xxviii, 15-201, [1] p.; v, [6]-223, [1] p. ; 541, [1] p.) : ill. ; 23 cm.

    Each volume has separate title-page and pagination. Vol. II has title: *A treatise on midwifery, and the diseases of women and children* (title-page dated 1857); vol. III has title: *Howard's materia medica* (title-page dated 1857).

1816. HOWARD, Horton, d. 1833.

An improved system of domestic medicine: founded upon correct physiological principles; comprising a complete treatise on anatomy and physiology; the practice of medicine, with a copious materia medica, and an extensive treatise on midwifery . . . New revised edition. . . . Cincinnati: H.M. Rulison, Queen City Publishing House; Philadelphia: Duane Rulison, Quaker City Publishing House, 1858.

3 v. in 1 (xxviii, 15-201, [1] p.; v, [6]-223, [1] p. ; 541, [1] p.) : ill. ; 23 cm.

The last publication of *An improved system* under this title. An extensively revised edition was published by Rulison at Philadelphia in 1859 under the title *Howard's domestic medicine* (#1806). Spine-title: *Howard's domestic medicine for the people.*

1817. HOWARD, Horton, d. 1833.

A treatise on midwifery, and the diseases of women and children; adapted to the use of heads of families, and females particularly . . . To which is added, a brief collection of medical plants, illustrated with plates, describing such remedies as are recommended in this work. Cincinnati: Published by J. Kost, 1852.
    v, [6], 223 p. : ill. ; 23 cm.

Until this 1852 publication, Howard's treatise on midwifery had been issued as volume three of his *Improved system of botanic medicine,* beginning with the 1833 second edition (#1812). The text of the *Treatise* appears to have been stereotyped with 1836 third edition (#1813), and was used again in the 1850 fourth edition (#1814) issued at Cincinnati by the reformed physician and publisher John Kost. The only substantial difference between Kost's two issues of Howard's midwifery is the addition to the *Treatise* of an illustrated appendix entitled "Description of medicines recommended in the work" (p. 172-82). Even the signatures of the gatherings in the *Treatise* retain the "Vol. III" designation of Kost's 1850 *Improved system.* The plates are modelled on wood engravings that originally appeared in Samuel Bard's *Compendium of the theory and practice of midwifery* (1st ed., 1807), the first obstetrical treatise by an American author.

1818. HOWARD, Joseph Theophilus (d. 1910).

Gynecology; or treatise on midwifery and physical ailments of women and children; containing an explanation of the phenomena of reproduction; with remarks on sterility; how to care for and raise infants; plural births; chloroform, &c., in confinement; hygiene, &c., &c. . . . Washington City: W.H. & O.H. Morrison, 1871.
    276 p. ; 19 cm.

Howard was an 1859 medical graduate of Georgetown University.

1819. HOWARD, Leland Ossian, 1857-1950.

The house fly. Disease carrier. An account of its dangerous activities and the means of destroying it . . . New York: Frederick A. Stokes Company, publishers, [c1911]
    xix, [1], 312 p., [17] leaves of plates : ill. ; 21 cm.

First edition. "It is only within the last twelve years that the dangerous character of the common house fly has been known; and only within the last two years have the people at large begun to wake up to this danger and to inquire concerning the means by which this fly can be kept down" (*Introduction,* p. xv). Howard states that his book "is not intended to be a scientific monograph." In five chapters he discusses the life history and habits of the housefly, its natural enemies, the housefly as disease carrier, preventive measures, and other species of flies that threaten domestic health.

After receiving his undergraduate degree from Cornell University (1877), Howard joined the U.S. Dept. of Agriculture as an entomologist (1878). He remained with the U.S.D.A. his entire career, and became its chief entomologist in 1894. Howard directed the U.S.D.A.'s earliest efforts in "economic entomology," i.e., the control of insects such as the gypsy moth and boll weevil that threatened agriculture, and advocated the biological control of pests. Howard was also a pioneer in increasing public awareness of insects as disease vectors.

1820. HOWARD, Leland Ossian, 1857-1950.

The house fly. Disease carrier. An account of its dangerous activities and of the means of destroying it . . . Third edition. New York: Frederick A. Stokes Company, publishers, [c1911].
    xix, [1], 312 p., [17] leaves of plates : ill. ; 21 cm.

1821. HOWARD, Leland Ossian, 1857-1950.

Mosquitoes. How they live; how they carry disease; how they are classified; how they may be destroyed . . . New York: McClure, Phillips & Co., 1901.
    xv, [3], 241 p. : ill. ; 21 cm.

HOWARD, Philip M. see **BECKLARD**
(#263-268).

1822. HOWARD, William Lee, b. 1860.
Breathe and be well . . . New York: Edward
J. Clode, publisher, [c1916].
vi, [2], 150, [2] p. ; 19 cm.

1823. HOWARD, William Lee, b. 1860.
Facts for the married . . . New York: Edward J. Clode, publisher, [c1912].
xiv, 161 p. ; 19 cm.

1824. HOWARD, William Lee, b. 1860.
How to rest. Food for tired nerves and
weary bodies . . . New York: Edward J.
Clode, [c1917].
xii, [4], 170 p. ; 19 cm.

1825. HOWARD, William Lee, b. 1860.
Plain facts on sex hygiene . . . New York:
Edward J. Clode, publisher, [c1910].
[4], iv, 171 p. ; 19 cm.

1826. HOWARD, William Lee, b. 1860.
Sex problems solved (those of worry and
work) . . . New York: Edward J. Clode,
[c1915].
x, 204 p. ; 19 cm.

1827. HOWE, Chauncey B., b. 1826.
Dr. C.B. Howe's family receipt book, for
every family in North America it contains
valuable receipts, &c., &c. Dr. C.B. Howe,
proprietor, Seneca Falls, N.Y. Auburn, N.Y.:
W.J. Moses' Publishing House, [ca. 1871].
32 p. ; 13 cm.

A collection of testimonial letters endorsing
*Howe's Concentrated Syrup, Howe's Never-Failing
Ague Cure and Tonic Bitters,* etc. Printed on back
of printed wrapper: For sale by S.A. Newman
. . . Rochester, N.Y.

1828. HOWE, John Moffat, 1806-1885.
Consumption curable. Information respecting the practice of F.H. Ramadge, M.D. . . .
Containing an account of several cases in
relation to this practice in which it has been
beneficial in this country, with other cor-
roborative testimony . . . New York: Published
by the author, [ca. 1872, c1853].
[2], viii, 98 p. ; 19 cm.

The purpose of Howe's essay is to promote
his Inhaling Tube for the treatment of pulmonary tuberculosis: "That while no physician can
cure a consumptive, yet the patient can cure
himself if he is properly instructed, and commences before the lungs are too much wasted
away, and that the inhalation of common air
through a tube prepared for the purpose, is
worth more than all other means, if it be faithfully pursued daily, for a few months . . . thus
exercising, airing and enlarging the capacity of
the lungs, and more perfectly vitalizing the
blood, quickening its circulation, and aiding
digestion—thus sitting in one's chair at home
enjoying all the benefits of vigorous exercise
and air in the lungs where it is so much wanted,
with the fatigue, annoyances and expense of
going from the comforts of home and friends"
(p. [1]). Howe, whose office was in Passaic, N.J.,
claims to have forty years' experience in treating diseases of the lungs. He chose to give his
essay the same title as Francis Hopkin Ramadge's *Consumption curable,* a work first published at London in 1834 and in New York in
1839 (#2927). Howe claims to have known
Ramadge in London, and to have adopted the
latter's "innovation in practice," i.e., the treatment of consumption by "exercise and air instead of drugs and medicines" (p. 5). In printed
wrapper.

1829. HOWE, Joseph William, 1843-1890.
Winter homes for invalids. An account of the
various localities in Europe and America,
suitable for consumptives and other invalids
during the winter months, with special reference to the climatic variations at each
place, and their influence on disease . . . New
York: G.P. Putnam's Sons, 1875.
x, 205 p. ; 19 cm.

"Every profession suffers from one-sided
crude deductions," Howe observes early in this
treatise. "New '*cures*' are advertised daily. The
vegetable-cure, the flesh-cure, milk-cure, water-
cure, an army of cures comes in from every
quarter. Each one has its earnest adherents,
with doubtless some foundation for their faith.
But they go too far in offering to cure every
other sick person on the globe with their special remedies. We know that a diet-vegetarian
assists physical growth . . . but to breakfast, dine,
and sup on vegetables is not conducive to health

or manly vigor . . . The 'milk-cure' has succeeded in many instances . . . but it does not
follow that we are to convert our patients into
miniature walking dairies whenever their internal economy is out of order. There is solace to
be had from the 'grape-cure,' but grapes do not
answer for the cure of every disease. Because
water is known to be beneficial in health and
in disease, it is regarded as a panacea for all the
ills of the flesh" (p. 2-3). In like manner, Howe
cautions his readers regarding the therapeutic
benefits of a change of climate. He discusses
the diseases that are indeed mitigated or cured
by a change of climate (e.g., consumption,
asthma, rheumatism, nervous prostration), and
considers the places where these diseases benefit from the virtues of the local climate.

A native of New Brunswick, Canada, Howe
was an 1866 graduate of the Medical Department of the University of the City of New York.
After graduation he interned "on the surgical
side" at Bellevue Hospital, and in 1868 was appointed clinical professor of surgery at the
University Medical School, a position he held
until 1882. This is the only edition published
of *Winter homes for invalids*. His other popular
treatise, *Emergencies and how to treat them* (1871),
went through several editions, as did several of
his books written for the profession.

HOWE, Julia Ward see *SEX AND
EDUCATION* (#3144).

1830. HOWE, Winfred Lewis, b. 1874.
A treatise on the care of the expectant
mother during pregnancy and childbirth
and the care of children from birth until
puberty . . . Philadelphia: F.A. Davis Company, publishers, 1911.
    viii, 63 p. ; 17 cm.

Howe was a 1900 graduate of Tufts College
Medical School.

1831. HUBBARD, Elbert, 1856-1915.
Health in the making. Being a preachment
on the subject of health, hygiene and right
living . . . New York City: Published by the
Bauer Chemical Co., [c1913].
    32 p. ; 18 cm.

"Neurasthenia catches the great, the extraordinary, the leaders of men. Just about now we

live in times of readjustment. The whole world
is being made over. It is a pivotal point in history, and the strain on the inventors, the inno-
vators, the men of initiative, the builders, and
the creators, was never so severe as it is today
. . . This accounts for the great number of break-
downs which occur in men who bear big burdens, and on whom devolves the responsibility
of thinking things out. We seem to lack the
necessary modicum of phosphorus at a critical
time" (p. 5). To remedy this lack, Hubbard endorses *Sanatogen*, "a scientific blend of pure al-
bumin and phosphorus" manufactured from
cow's milk. Though not printed by the Roycroft
Printing Shop, *Health in the making* resembles
work from Hubbard's press in East Aurora, N.Y.

HUBBARD, Elbert, 1856-1915. *How to keep
well* see **MORAS** (#2491).

1832. J. HUBBARD AND COMPANY.
Hubbard's Disinfectant Deodorizer and
Germicide . . . Boston: [J. Hubbard & Co.,
ca. 1890].
    [1] leaf ; 25 × 51 cm. folded twice to 25
× 17 cm.

A wood engraving on the first leaf depicts
the atomizer by which *Hubbard's Vegetable Disinfectant and Germicide* could be used for "purifying the atmosphere" of any interior space; to
spray disinfectant directly into the throat or
nostrils "for the relief and cure of catarrh, hay
fever, colds, la grippe . . . [etc.]; or onto any
part of the body for healing "cuts, wounds,
sores, ulcers," and even for the cure of neuralgia and headache. The greater part of the document contains testimonials, the most recent of
which are dated 1890.

HUBBS, J. Allen see **COOK** (#777).

**1833. HUFELAND, Christoph Wilhelm,
1762-1836.**
The art of prolonging life . . . Translated
from the German. New York: Raphael J.
Thiry, 1879.
    [2], vii, [8]-106 p. ; 17 cm.

Hufeland's treatise on the hygienic principles and practices necessary to prevent disease and prolong life was originally published
at Jena in 1797 under the the title: *Die Kunst*

das menschlichen Leben zu verlangern. With the third German edition (1805), the title changed to *Makrobiotik*—Hufeland's own term for this branch of knowledge. In Europe, the *Makrobiotik* assumed a place in the literature of popular hygiene equivalent in authority to the manuals of Cornaro (#791-798) and Tissot (#3534-3542), and was translated into most European languages. In the English-speaking world, Hufeland's text was esteemed nearly as much as the works of Buchan (#442-502), Wesley (#3746-3747) and Sinclair (#3220-3221). The first English translation was published in 1797 at London in two volumes. The first American edition appeared at Philadelphia in 1829.

A 1783 Goettingen medical graduate, Hufeland practiced in the Weimar of Goethe and Schiller before being made professor of medicine at Jena. In 1800 Hufeland was appointed professor of medicine at Berlin and director of its Charite hospital. The *Makrobiotik* was the single most successful title of Hufeland's remarkable literary output.

1834. HUFELAND, Christoph Wilhelm, 1762-1836.
The art of prolonging life . . . Edited by Erasmus Wilson . . . From the last London edition. Philadelphia: Lindsay & Blakiston, 1880.
   xii, [13]-298 p. ; 19 cm.

1835. HUFELAND, Christoph Wilhelm, 1762-1836.
Hufeland's art of prolonging life. Edited by Erasmus Wilson, F.R.S. Boston: Ticknor, Reed and Fields, 1854.
   xvi, 328 p. ; 19 cm.

The first American printing of the edition prepared by Sir Erasmus Wilson (1809-1884).

1836. HUFELAND, Christoph Wilhelm, 1762-1836.
Die Kunst das menschliche Leben zu verlangern . . . Zweyte vermehrte Auflage . . . Jena: in der akademischen Buchhandlung, 1798.
   2 pts. in 1 v. (xx, 186; [2], 256, [iii]-xxii p.) ; 20 cm. (8vo)

Originally published at Jena in 1797. The third edition (1805) was issued under the title:

Makrobiotik. Given the numerous German editions in print through 1860 (8. Aufl.), it is likely that this work was to be found in the homes of Pennsylvania Germans and other German-speaking immigrants throughout the United States in the first half of the 19th century. A German-language edition was never published in North America, however.

1837. HUFF, Gershom.
Electro-physiology. A scientific, popular, and practical treatise on the prevention, causes, and cure of disease; or, electricity as a curative agent, supported by theory and fact . . . Second edition . . . New-York: D. Appleton and Company, 1853.
   [4], xv, [1], [19]-385 p. : ill. ; 20 cm.

The first edition of *Electro-physiology* was also published at New York in 1853. Its author has a distinctly nationalistic view of American purity, in matters of health as well as manners; and like Cato, looks back to a less tainted age: "With the stuff and tinsels of the Old World, we import its follies and its vices; the simplicity of republican manners is becoming daily less visible amongst us; our habits of life, our equipages, our houses, seek rather to vie with those of Europe than to look back, for an example, to the stern simplicity which distinguished the sages of the Revolution. American in our political feelings, in the freedom of our civil and religious institutions, we are, too often, not ashamed to become the servile imitators of the Eastern hemisphere in our fashionable and domestic follies . . . To these causes may be attributed in no trifling degree those diseases which, having their origin in a deranged state of the functions of the nervous system, are rapidly increased amongst us; the painful heritage of many among the present generation, the sure inheritance of the future" (p. iv).

On p. 129 Huff states that it is "among children descended from those in whom wealth and luxury have produced habitual indolence, or from those of the opposite extreme, to whom poverty has denied the necessaries of animal existence, that we find the most feeble nervous systems," and consequently, a deficiency of the "nervous force or power" that is necessary for corporeal functions to continue normally. Huff identifies this nervous force or fluid with electricity. He terms it the *vital force* that "appears in living animal tissue, not only as a cause of growth or development, but also as a power acting in opposition to those external agencies which tend to alter the form, structure, and composition of the tissue in which this vital energy resides" (p. 191). He posits "a ceaseless antagonism between the forces of decay and the vital energy." When the former is dominant, disease or even death result. Health, on the other hand, "is that state of the body in which these opposing forces are properly balanced" (p. 192).

Vitality is a nervous or electric power. The object of hygiene is to maintain an optimal level of *vital force,* and that of medicine to restore this force to the body in its diseased or decayed state: "As all the organic functions are performed through the influence of this subtle and powerful agent exciting the nerves, and through them the muscular system, we are thus enabled to regulate those functions at our pleasure. Is the muscular power of the chest diminished . . . and the necessary expansion of the lungs thereby prevented? By electricity we can restore the powers of the former, and thus facilitate the progress of the latter to a healthy expansion. Are the functions of the heart deranged, and the sufferer thereby exposed to sudden death? The electric force will restore its muscular action, the current of blood will again flow in its wonted channel, the electrical relations of the system are changed, until strength is obtained and the disease cured," etc. (p. 217). *Electropathy* ("the animating and sustaining power of human organization) is not dependent on mechanical electrotherapeutic intervention, but upon a regimen that, as conceived by Huff, differs little in its essentials from the recommendations regarding diet, air, rest, etc. proposed by other hygienic advocates and reformers.

1838. HUGHES, Henry.
A treatise on hydrophobia, (taken from the manuscript of a late eminent physician,) to which is appended an infallible remedy, both as a preventative and in confirmed cases . . . New-York: George Dearborn & Co., 1837.
    vi, [7]-30 p. ; 20 cm.

The *Treatise* was originally published at Montreal in 1837, where Hughes, a member of

"H.M. First Royal Regt." (from title-page), was apparently stationed and where he discovered the manuscript on which this pamphlet is based. The method of curing hydrophia (a symptom of rabies commonly used as its synonym) described in this essay consists of the administration of two distinct remedies: the first, a compound of native cinnabar and factitious cinnabar, followed by a compound of tartrate of antimony and nitrate of potash. In printed wrapper.

HULL, Amos Gerald see EVEREST (#1090); LAURIE (#2205-2207).

HULL, Phoebe C. see HULL AND CHAMBERLAIN (#1839).

1839. HULL AND CHAMBERLAIN.
MAGNETIC AND ELECTRIC POWDERS. Something entirely new. *Great Nervine and Regulator.* A complete and reliable family medicine. PURELY VEGETABLE. For the cure of all diseases that can be cured by Medicine, Magnetism or Electricity . . . New York: Hull & Chamberlain, [c1873].
    [2] leaves : ill. ; 23 cm.

The proprietors, Phoebe C. Hull and Annie Lord Chamberlain, advertise magnetic and electric powders for the cure of all chronic and acute disease (including "female diseases" and "general debility"); "magnetized paper" (a sheet of which was placed "over the region of pain or weakness"); and "magnetic and electric uterine wafers, for the cure of female weakness, painful menstruation, prolapsis, inflammation and ulceration of the womb." Hull describes herself as a "magnetic physician"; Chamberlain operated the firm's branch office in Chicago.

1840. THE HUMAN MACHINE.
The Human machine. Its care and repair or how to develop the body, preserve the health, meet emergencies, nurse the sick and treat disease. By twelve authors. Edited by W.E. McVey . . . Topeka, Kansas: Herbert S. Reed, 1905.
    848 p., IX leaves of plates : ill. ; 24 cm.

All but two of the twelve contributors to this volume were physicians. Nine were on the faculty of the Kansas Medical College (Topeka). One was female. The editor, William E. McVey (b. 1864), was an 1888 graduate of the Kansas City Medical College, professor of otorhinolaryngology at the Kansas Medical College, editor of the *Kansas medical journal* and secretary of the Kansas Medical Society.

HUMPHREY, Gideon see **CURIE** (#851).

1841. HUMPHREYS, Frederick, 1816-1900.
Humphreys' homeopathic mentor or family adviser in the use of specific homeopathic medicine ... New York: Humphreys' Specific Homeopathic Medicine Co., 1873.
vi, [3]-352 p. ; 23 cm.

By way of introduction, Humphreys explains the nature of vegetable and animal life, the human organism in its healthy and diseased states, and the principles by which medicines act (contrasting the homeopathic and the allopathic practice). Having laid this groundwork, he devotes the rest of his text to the causes and symptoms of common diseases, and their treatment by proprietary homeopathic remedies available from Humphreys' Specific Homeopathic Medicine Company. Pages 301-31 provide a key to the thirty-five specifics recommended in the text. Each specific has its particular application, e.g., *No. 1* for "fevers, congestions, inflammations, heat, pain," *No. 2* for "verminous affections," *No. 3* for diseases of infants, etc. The final pages advertise the specifics, family medicine chests, "pocket traveling cases," etc. A single-leaf advertisement for *Humphreys' Witch Hazel* is bound-in before the title-page.
Humphreys' homeopathic mentor was first published in 1872. A "revised edition" of the *Mentor* was in print as late as 1943. Humphreys was an 1850 graduate of Hahnemann Medical College. He was on the faculty of the Homoeopathic Medical College of Pennsylvania (Philadelphia), was active in homeopathic medical societies, and wrote several legitimate texts before venturing into the mail order book distribution and patent medicine business at Auburn, N.Y. circa 1854.

1842. HUMPHREYS, Frederick, 1816-1900.
Humphreys' homeopathic mentor or family adviser in the use of specific homeopathic medicine ... New York: Humphreys' Specific Homeopathic Medicine Co., 1876.

vi, [3]-360 p. ; 23 cm.

1843. HUMPHREYS, Frederick, 1816-1900.
Humphreys' homeopathic mentor or family adviser in the use of specific homeopathic medicine ... Revised and enlarged edition ... New York: Humphreys' Homeopathic Medicine Co., 1894.
[2], iv, 523, [25] p., [1] leaf of plates : ill., port. ; 19 cm.

Cover-title: *Humphreys' mentor.*

1844. HUMPHREYS, Frederick, 1816-1900.
Humphreys' homeopathic mentor or family adviser in the use of specific homeopathic medicine ... Revised and enlarged edition ... New York: Humphreys' Homeopathic Medicine Co., 1895.
[2], iv, 523, [25] p., [1] leaf of plates : ill., port. ; 19 cm.

Cover-title: *Humphreys' mentor.*

1845. HUMPHREYS, Fredreick, 1816-1900.
Humphreys' homeopathic mentor or family adviser in the use of specific homeopathic medicine ... Revised and enlarged edition ... New York: Humphreys' Homeopathic Medicine Co., 1899.
[2], iv, 525, [23] p., [1] leaf of plates : port. ; 19 cm.

Cover-title: *Humphreys' mentor.*

1846. HUMPHREYS, Fredreick, 1816-1900.
Humphreys' homeopathic mentor or family adviser in the use of specific homeopathic medicine ... Revised and enlarged edition ... New York: Humphreys' Homeopathic Medicine Co., 1913.
[2], iv, 524, [22] p., [1] leaf of plates : port. ; 19 cm.

Cover-title: *Humphreys' mentor.*

1847. HUMPHREYS, Fredreick, 1816-1900.
Humphreys' manual of specific homeopathy, for the administration of medicine and cure of disease ... New York: Humphreys' Homeopathic Medicine Company, 1889.
[2], 144 p. ; 12 cm.

At head of title: *Revised edition*. In chromo-lithographed wrapper.

1848. HUMPHREYS, Frederick, 1816-1900.
Manual of specific homoeopathy, for the administration of medicine and cure of disease . . . Auburn [N.Y.]: Miller, Orton & Mulligan, stereotypers and printers, 1855.
71, [3] p. ; 11 cm.

First edition. The *Manual* was the medium through which Humphreys first made his ho-meopathic specifics known to the American public. In the annotation to the *Manual* in his *Homoeopathic bibliography of the United States* (1892), T.L. Bradford indicates the astounding scope of Humphreys' mail order business and the importance of this booklet to its success: "The first editions were from 10,000 to 20,000 annually. For ten years the issue has been from 500,000 to 2,500,000 annually. The present rate is 3,000,000 per annum; of this book, 15,000,000 copies have been distributed in the English, German, French, Spanish, and Portuguese languages, of which about 12,000,000 have been distributed in the United States" (p. 167).

1849. HUMPHREYS, Frederick, 1816-1900.
Manual of specific homeopathy, for the administration of medicine and cure of disease . . . [New York]: John A. Gray, printer, 1860.
108 p. ; 11 cm.

"A very large portion of medical practice, especially that which relates to the most impor-tant stage of disease, its beginning, is, and must ever remain in the hands of the people. How important then that they should not only pos-sess correct notions in regard to disease, but have in their hands a few simple, not danger-ous, and yet efficacious elements, which they may apply in the first hours of illness" (p. 3-4). Humphreys not only provides his readers these beneficial homeopathic specifics, but a manual without "long and intricate directions . . . which only puzzle and perplex but rarely explain," consisting of "a concise outline of each particu-lar disease or affection . . . and simple direc-tions how to give the Specific in each particu-lar case" (p. 5). Humphreys' describes *Specific Homoeopathy* as "the discovery of a remedy for each particular disease" (p. 7), and adds: "These

Specifics differ from the usual Allopathic or common medicine in that they cure diseases without producing a cathartic, emetic, or other depletive effect . . . They differ from the usual Homoeopathic medicines in their mode of preparation and in the extent and certainty of their action" (p. 8).

1850. HUMPHREYS, Frederick, 1816-1900.
Manual of specific homeopathy, for the administration of medicine and cure of disease . . . Twentieth edition. New York: Humphreys' Specific Homeopathic Med. Co., [ca. 1872, c1869].
128 p. ; 11 cm.

An advertisement for "Dr. Humphrey's new book," i.e., *Humphreys' homeopathic mentor* (1st ed., 1872), is pasted onto the verso of the front wrapper.

1851. HUMPHREYS, Frederick, 1816-1900.
Manual of specific homoeopathy, for the administration of medicine and cure of disease . . . Five hundredth edition. New York: Humphreys' Specific Homoeopathic Med. Co., [c1869].
144 p. ; 11 cm.

In printed wrapper.

1852. HUMPHREYS, Frederick, 1816-1900.
Manual of specific homeopathy for the administration of medicine and cure of disease . . . One thousandth edition . . . New York: Humphreys' Specific Homeo-pathic Med. Co., [c1869].
144 p. ; 11 cm.

Title on printed wrapper: *Humphreys' homeo-pathic manual.*

1853. HUNT, Ezra Mundy, 1830-1894.
The patients' and physicians' aid; or, how to preserve health; what to do in sudden attacks, or until the doctor comes; and how best to profit by his directions when given . . . New York: C.M. Saxton, Barker & Co.; San Francisco: H.H. Bancroft & Co., 1860.
5, [vi]-xi, [12]-365, [7] p. : ill. ; 19 cm.

"There is a certain degree of information bearing upon human health and disease, which it is right should be the common stock of both physician and patient." Hunt cautions, however, "One who tries to be his own blacksmith, carpenter, and tailor, is very apt to drive some wrong nails and make some miserable outfits, and still more he who flatters himself that he can be his own doctor" (*Pref.*, p. [3]).

Hunt provides advice on hygiene, nursing the sick, first aid, etc. An 1852 graduate of the College of Physicians & Surgeons (New York), Hunt served briefly on the faculty of the Vermont Medical College before returning to his native Metuchen, N.J. to resume private practice. During the decade before the Civil War, Hunt became increasingly concerned about the sanitary conditions of his fellow New Jerseyans, and assumed an active role in health and sanitary reform. These efforts intensified after service as a surgeon with the 29th New Jersey Infantry (1862-63). During the last quarter of the century, Hunt was the leading figure in public health reform in his native state.

1854. HUNT, Mary Hannah (Hanchett), 1830-1906.
Physiology and health. Number two. For intermediate classes. Studies of the human body and of the effects of alcoholic drinks and narcotics upon life and health. Philadelphia: E.H. Butler & Co., [c1889].
vi, 7-134 p. : ill. ; 18 cm.

"Most of our States and Territories now by law require that physiology and hygiene be taught in the public schools, with special reference to the effects of alcoholic drinks and other narcotics on the human system . . . The present work is designed to teach the essential laws of health, and to comply fully with the most stringent provisions of the enactments requiring this subject taught . . . The endeavor has been to make the physiological instruction clear and sufficient—the Temperance teachings thorough, and as radical as the whole truth now revealed by modern scientific investigation. The work throughout has been more or less prepared and wholly supervised by Mrs. Mary H. Hunt, the Superintendent of Scientific Instruction of the National Woman's Christian Temperance Union" (*Pref.*, p. iii-iv).

Hunt became president of the Woman's Christian Temperance Union in 1878, and in 1880 director of its newly formed Department of Scientific Temperance Instruction (STI). "In 1881 the WCTU resolved to seek compulsory state temperance education laws. Commanding a vast army of local WCTU 'superintendents,' Hunt organized sophisticated petition and letter drives to pressure legislators. She also delivered hundreds of addresses across the country . . . She spent much of 1886 in Washington, D.C., winning a congressional measure that required STI in federal territories. Progress was somewhat slower in the South . . . But by 1901 every state mandated some form of the temperance instruction advocated by the WCTU" (*Amer. nat. biog.*, 11:498).

At head of title: *The Union series*. There were three volumes in the *Union series of text-books on physiology and health*: no. 1 for primary classes, no. 2 for intermediate classes, and no. 3 for secondary classes.

HUNT, Mary Hannah (Hanchett) see also **WOMAN'S CHRISTIAN TEMPERANCE UNION** (#3846, 3847).

1855. HUNTER, Robert.
On the curability and cure of consumption, catarrh, bronchitis and asthma . . . Chicago: Jeffery Printing Co., 1888.
32 p. ; 19 cm.

In this pamphlet Hunter endeavors "to make plain *first* the perfect curability of consumption, *second* the murderous quackery of pretending to treat lung diseases through the stomach, and *lastly*, the better results which attend their scientific treatment by medicated air and vapor" (p. [2]). He advertises the availability of his inhalation treatment for lung diseases at 103 State St. in Chicago. An 1846 medical graduate of the University of the City of New York, Hunter shared the State St. address with Edwin William Hunter (1855-1929), probably his son, an 1877 graduate of Rush Medical College. Title and imprint from printed wrapper.

1856. HUTCHINGS, Thomas Gibbons.
The medical pilot; or, new system: being a family medical companion, and compendium of medicine on a totally new plan, and by which all diseases can be treated successfully without minerals, or any poisons whatever; especially the following complaints: consumption, liver complaint,

dyspepsia, dysentery, diarrhoea, rheumatism, diseases of the heart, fevers, bronchitis, dropsy, eruptive diseases, and all their
concomitants; containing also a treatise on
the diseases incident to the female sex. The
whole being adapted in simple and familiar language, suitable to every capacity,
forming a new era in medical practice . . .
New-York: Smithson's Steam Printing Offices, 1855.
     [16], 295, [5] p. : ill. ; 20 cm.

"Sacred History assures us that the 'Life is
the Blood,' and thanks to Harvey, we know that
it circulates through the system by means of
veins, and arteries; that in its periodical revolu-
tions it is subject to important changes for the
well being of life; these changes, whether for
good or evil, depend upon both internal and
external influences, and where such are inimical to well being, then medicine comes in to
the rescue, not indeed by administering poisons . . . but by those simple herbs given by a
wise Providence to be the true restorer of in-
fringed nature. It is this *system* which is here
offered to the world; it is founded on reason,
truth, and Scripture, and now from labor, and
research so judiciously adapted to every form
of disease, that while it restores health it leaves
the constitution in a renovated condition" (p.
6).
     In his consideration of the maladies listed
on the title-page, Hutchings never describes the
botanical elements that constitute his "new system," or their mode of action. He never fails,
however, to note their success in effecting a
cure. Preparations such as *Dr. Hutchings' Inhaling Balm, Dr. Hutchings' Family Vegetable Pills,
Doctor Hutchings' Nervous Antidote*, etc. are offered to the public on the final [5] pages of his
treatise.

1857. HUTCHINGS, Thomas Gibbons.
The medical pilot; or, new system: being a
family medical companion, and compendium of medicine on a totally new plan,
and by which all diseases can be treated
successfully without minerals, or any poisons whatever; especially the following
complaints: consumption, liver complaint,
dyspepsia, dysentery, diarrhoea, rheumatism, diseases of the heart, fevers, bronchitis, dropsy, eruptive diseases, and all their

concomitants; containing also a treatise on
the diseases incident to the female sex. The
whole being adapted in simple and familiar language, suitable to every capacity,
forming a new era in medical practice . . .
New-York: Smithson's Steam Printing Offices, 1855.
     [16], 300 p. : ill. ; 20 cm.

1858. HUTCHINSON, Woods, 1862-1930.
The conquest of consumption . . . Boston
and New York: Houghton Mifflin Company; Cambridge: The Riverside Press,
[1910].
     [8], 138 p., [4] leaves of plates : ill. ; 19
cm.

     At age twelve Hutchinson emigrated with his
family from Yorkshire to Iowa. After graduating from the medical department of the Univ.
of Michigan in 1884, he spent two years studying abroad before returning to Des Moines.
Over the next two decades Hutchinson taught
physiology and/or pathology at the Des Moines
Medical College, the State Univ. of Iowa and
the University of Buffalo. Leaving Western New
York in 1898, Hutchinson spent a year teaching at the London Medical Graduate College
and Univ. of London. In 1905 Hutchinson retired from medical practice, settled in New York
City, and devoted himself to writing. "Realizing that a great number of Americans possessed
little knowledge about the scientific principles
of health and hygiene, he aimed to make such
subjects more intelligible to the general public
by discussing them in layman's terms. His articles and syndicated columns became standard
fare in such popular periodicals as *American
Magazine, Contemporary Review, Cosmopolitan,
Harper's Monthly, McClure's Magazine, Saturday
Evening Post, Success, Woman's Home Companion*,
and the *New York Times*. His favorite topic was
preventive medicine; on this topic he authored
a number of textbooks and handbooks . . ."
(*Amer. nat. biog.*, 11:600).
     Described by one reviewer as "a sort of hy-
gienic John the Baptist," Hutchinson appeals
in *The conquest of consumption* to "the new gospel of bacteriology" and its "message of hope"
in the battle against pulmonary tuberculosis.
He discusses what happens to the progress of
the bacillus in the human body and the importance of fresh air, sunlight, food, "intelligent

idleness" (rest) and the "open-air cure" in effecting a cure. He also considers the monetary cost of treatment to the consumptive, addresses climatological questions, and provides "specifications for the open-air treatment at home."

1859. HUTCHINSON, Woods, 1862-1930.
Exercise and health . . . Special edition published by the Health League of the Young Mens Christian Association . . . New York: Outing Publishing Company, 1911.
    156 p. ; 18 cm.

Series: *Outing handbooks; no. 1.*

1860. HUTCHINSON, Woods, 1862-1930.
Preventable diseases . . . Boston and New York: Houghton Mifflin Company; Cambridge [Ma.]: The Riverside Press, 1909.
    vi, [2], 442 p. ; 19 cm.

Hutchinson considers measures that may be taken to prevent the common cold, tuberculosis, typhoid, malaria, cancer, etc. In his final chapter ("Mental influence in disease), Hutchison writes: "That the mind does exert an influence over the body, and a powerful one, in both health and disease, is obvious. But what we are apt to forget is that the whole history of the progress of medicine has been a record of diminishing resort to this power as a means of cure. The measure of our success and of our control over disease has been, and is yet, in exact proportion to the extent to which we can relegate this resource to the background and avoid resorting to it" (p. 412-13).

1861. HUTCHISON, Joseph Chrisman, 1827-1887.
The laws of health. Physiology, hygiene, stimulants, narcotics. For educational institutions and general readers . . . New York: Clark & Maynard, publishers, [c1884].
    7, [1], [11]-223 p., [2] leaves of plates : ill. ; 18 cm.

First edition. Hutchison received his medical degree from the Univ. of Pennsylvania in 1848. From 1853 he lived in Brooklyn, where he became involved in the medical, sanitary and educational affairs of the city. The greater part of Hutchison's publications were given to his successful series of juvenile textbooks on physiology and hygiene.

1862. HUTCHISON, Joseph Chrisman, 1827-1887.
Our wonderful bodies and how to take care of them. First book—for primary grades. New York: Maynard, Merrill & Co., publishers, [c1894].
    127 p. : ill. ; 17 cm.

First edition. Series: *Hutchison's physiological series.*

1863. HUTCHISON, Joseph Chrisman, 1827-1887.
Our wonderful bodies and how to take care of them. First book—for primary grades. New York: Maynard, Merrill, & Co., publishers, 1904.
    127 p. : ill. ; 17 cm.

Series: *Hutchison's physiological series.*

1864. HUTCHISON, Joseph Chrisman, 1827-1887.
Our wonderful bodies and how to take care of them. First book—for primary grades. New York: Maynard, Merrill & Co., publishers, 1905.
    127 p. : ill. ; 17 cm.

Series: *Hutchison's physiological series.*

1865. HUTCHISON, Joseph Chrisman, 1827-1887.
A treatise on physiology and hygiene for educational institutions and general readers . . . New York: Clark & Maynard, publishers, 1876.
    [2], 23, [4], [4], [24]-40, [4], 41-52, [2], 53-63, [2], 64-79, [2], 80-100, [2], 101-122, [4], 123-147, [4], 148-176, [2], 177-226, [2], 227-270 p., [2] leaves of plates : ill. ; 20 cm.

Hutchison's *Treatise* was his most successful book, and for thirty years one of the most popular manuals of juvenile physiology on the market and in the classroom. It was issued at least twenty-five times between 1870 and 1910.

1866. HUTCHISON, Joseph Chrisman, 1827-1887.
A treatise on physiology and hygiene for educational institutions and general read-

ers . . . New York: Clark & Maynard, publishers, 1878.

[2], 23, [5], 25-40, [4], 41-52, [4], 53-63, [4], 64-79, [4], 80-100, [2], 101-122, [4], 123-147, [4], 148-176, [4], 177-226, [4], 227-235, [2], 236-270 p., [2] leaves of plates : ill. ; 20 cm.

1867. HUTCHISON, Joseph Chrisman, 1827-1887.
A treatise on physiology and hygiene for education institutions and general readers . . . New York: Clark & Maynard, publishers, 1879.

[2], 23, [5], 25-40, [4], 41-52, [4], 53-63, [4], 64-79, [4], 80-100, [2], 101-122, [4], 123-147, [4], 148-176, [4], 177-226, [4], 227-235, [2], 236-270 p., [2] leaves of plates : ill. ; 20 cm.

1868. HUTCHISON, Joseph Chrisman, 1822-1887.
A treatise on physiology and hygiene for educational institutions and general readers . . . New York: Clark & Maynard, publishers, 1887.

[1-2], [2], [3]-319 p., [3] leaves of plates : ill. ; 19 cm.

1869. HUTCHISON, Joseph Chrisman, 1822-1887.
A treatise on physiology and hygiene for educational institutions and general readers . . . New York: Effingham Maynard & Co., successors to Clark & Maynard, publishers, 1890.

[2], 47, [2], [48]-103a, [1], [104]-322 p., [4] leaves of plates : ill. ; 19 cm.

1870. HUTCHISON, Joseph Chrisman, 1827-1887.
A treatise on physiology and hygiene for educational institutions and general readers . . . New York: Maynard, Merrill & Co., 1895.

371 p., [4] leaves of plates : ill. ; 19 cm.

At head of title: New edition, 1895.

1871. HUTCHISON, Joseph Chrisman, 1827-1887.
A treatise on physiology and hygiene for educational institutions and general read-

ers . . . New York: Maynard, Merrill & Co., 1896.

371 p., [4] leaves of plates : ill. ; 19 cm.

At head of title: New edition, 1895.

1872. HUXLEY, Thomas Henry, 1825-1895.
The elements of physiology and hygiene; a text-book for educational institutions, by Thos. H. Huxley . . . and Wm. Jay Youmans . . . Revised edition . . . New York: D. Appleton and Company, 1886.

10, [4], [11]-485, [3], 4 p. : ill. ; 19 cm.

There were two competing editions of Huxley's physiological textbook on the American market during the last quarter of the 19th century. Macmillan & Co., who published the first edition at London in 1866 (#1873), issued Huxley's *Lessons in elementary physiology* on both sides of the Atlantic until 1930. D. Appleton & Co. in New York published its first edition in 1868. It was edited by the American William Jay Youmans (1838-1901), who studied under Huxley in London after completing his medical degree at the University of the City of New York in 1865. "Youmans recast Huxley's *Lessons on elementary physiology* (1866) into a format more suitable for the American market. The *Elements of physiology and hygiene* (1868), containing the addition of seven chapters on hygiene written by Youmans, even won the praise of Huxley's son Leonard, who usually detested American publishers for pirating English works . . . In a second edition of Huxley's work, published in 1873, Youmans more fully reworked his mentor's text. Youmans explained that Huxley's compressed and concise writing style had made the *Elements* too difficult for the average reader, so he had simplified the prose . . . and incorporated more illustrations. The revised edition had been called forth by a growing demand to treat the subject of hygiene within the context of general education" (*Amer. nat. biog.*, 24:146).

1873. HUXLEY, Thomas Henry, 1825-1895.
Lessons in elementary physiology . . . London: Macmillan and Co., 1866.

xix, [1], 319 p. : ill. ; 15 cm.

First edition. Series: *Macmillan's school class books.*

1874. HUYLAR, E. P.

[*At head of p. [1]: wood engraving, 75 × 100 mm, depicting Deborah Noble and her daughter Abigail*]. The above cut represents Mrs. Deborah Noble, as I saw her on my visit to her home, in June, 1868 . ˙. [New York, ca. 1879].

15 p. : ill. ; 21 cm.

Huylar claims to have been introduced to MOTHER NOBLE'S HEALING SYRUP by English friends resident in New York City, where he himself has been in practice for twenty years. Huylar tested the syrup on some of his patients, and discovered that it was indeed "a positve cure for all diseases arising from Impurity of the Blood" (p. [1]), i.e., for *all* diseases, as "all sickness, pains, and diseases of every name are caused by stagnant humors in the blood." Realizing the import of this discovery, Huylar relates, he sailed to England, where he found Deborah Noble at her home outside London. Kindly disposed to her American visitor, she relates to him the discovery of her syrup, and its derivation from a plant accidentally found in a field near her home. "I went to England, prepared to pay the old lady $10,000 for the receipt of this valuable blood-cleanser," Huylar states, but to his surprise, she gave it to him for nothing. "On my passage home I resolved that, in justice to Mother Noble and the world, I should use every means to make this valuable remedy extensively known in the United States, and I planned to build a laboratory on the plan of Mother Noble's in England" (p. 3). At his facility in Williamsburgh, Long Island, Huylar's staff "wash, pick, assort, and prepare the [unspecified] herb for the purpose of distillation" (p. 4). Large bottles may be had for one dollar each, small bottles for fifty cents.

1875. HUYLAR, E. P.

Remarks on kidney diseases, gravel, stone, dropsy, rheumatism, neuralgia, &c., &c. Presenting distinctive and peculiar views concerning the nature and treatment of the same . . . New York: G.A. Whitehorne, book & job printer, 1870.

16 p. ; 23 cm.

"After years of days and nights spent in research and experiments I am prepared to announce that I have made distinctive and peculiar discoveries concerning the nature and treatment of kidney diseases by which I can with unerring accuracy determine the cause of the disease and provide remedies for its removal" (p. [3]). Huylar claims to have discovered a "combination of chemical and vegetable agents" that dissolve urinary calculi, a remedy that cures diabetes by "checking the accumulation of sugar in the blood," and remedies for the cure of Bright's Disease, incontinence, edema, etc. In printed wrapper. Atwater copy lacking pp. 15-16.

1876. HYATT, Thaddeus Pomeroy, 1864-1953.

The teeth and their care . . . Brooklyn-New York: King Press, 1906.

[12], 9-12, [2], 13-24, [2], 25-26, [2], 27-28, [2], 29-34, [2], 35-38, [2], 39-43 p. : ill. ; 16 cm.

The author was an 1890 graduate of the New York College of Dentistry.

1877. HYDROPATHIC AND HYGIENIC INSTITUTE.

HYDROPATHIC AND HYGIENIC [*wood-engraved vignette of the Water Cure and Hydropathic Medical College*] INSTITUTE, R.T. TRALL, M.D., Proprietor, 15 Laight Street, New York . . . [186-?].

[1] leaf printed on both sides ; 24 × 15 cm.

The recto of this sheet advertises Trall's (#3558-3593) New York water-cure, "the oldest and most extensive *City Water-Cure* in the United States," where the public might consult the medical staff (Trall, P.H. Hayes and Anne Inman), receive hydropathic treatment, enroll in the School Department ("for the education of physiological teachers, health-reform lecturers and hygeopathic physicians"), attend popular lectures, participate in calisthenic programs, and procure wholesome food products ("wheaten grits, hominy, oatmeal, farina, crackers, Graham bread, &c."). The verso contains an advertisement for Bockee, Innis & Co., wholesale druggists, Chicago.

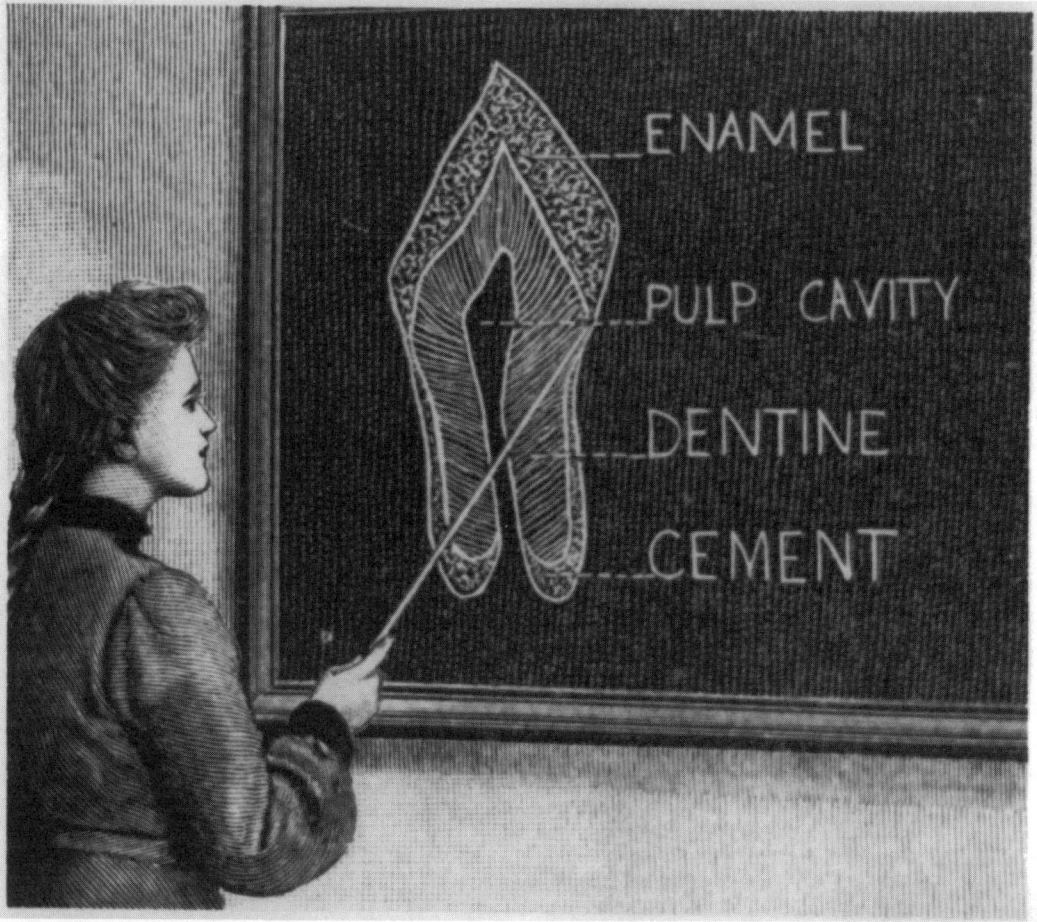

Fig. 47. From A.F. Blaisdell's How to keep well *(#343).*

1878. HYGIENIC TEACHER AND WATER-CURE JOURNAL.

Hygienic teacher and water-cure journal. Devoted to physiology, hydropathy, and the laws of life. [Vol. 34, no. 1 (July 1862)-v. 34, no. 6 (Dec. 1862); whole nos. 198-203]. New York: Fowler & Wells, 1862.

    1 v. : ill. ; 30 cm.

The *Hygienic teacher,* a monthly published in only six issues, continued the *Water-cure journal* (#3720), and was itself continued by the *Herald of Health* in January 1863.

1879. HYGIENIST.

The Hygienist. By Dr. R.R. Daniels . . . [Vol. 1 (1911)-v. 14 (1924)]. Denver, Colorado, 1911-1924.

    14 v. ; 20 cm.

In his opening article in the first issue, Ralph Roy Daniels (b. 1880) explains that "THE HYGIENIST will be devoted to the science of health . . . The cure of disease and the maintenance of health through the proper care of the body, discussed in a concise practical way will be the editor's aim" (1911, 1:3). Daniels adds that he "is not a devotee of any particular fad, does not

HYDROPATHIC AND HYGIENIC

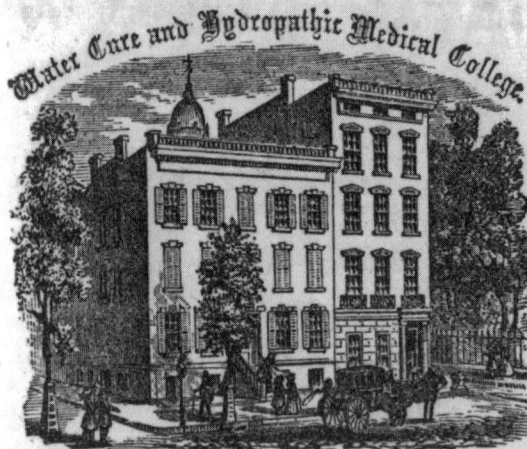

INSTITUTE,

R. T. TRALL, M. D., Proprietor,

15 LAIGHT STREET, NEW YORK.

Fig. 48. R.T. Trall's Hydropathic and Hygienic Institute (New York), #1877.

ride a hobby and has no prejudices either for or against any person or system, except a prejudice toward the truth." The content appears to be largely or wholely from Daniel's pen. Published monthly, *The Hygienist* was continued by *Dr. Daniel's monthly health letter*. Atwater holdings: vol. 1 (1911)-vol. 9 (1919).

I

1880. THE IGNORAMUS.
The Ignoramus . . . [Volume II]. Copyrighted, March, 1912, by the Co-operative
Health Association . . . Denver, Colorado..
[2], 171-330 p., [1] leaf of plates : port. ; 22
cm.

The first volume of *The Ignoramus* is bibliographically unrecorded. The majority of the
contributions to this second volume were written by the editor, Robert F. Maul, and by Leon
Patrick, M.D. The articles traverse popular hygiene, "higher consciousness" and astrology.
Frontispiece portrait of R.F. Maul.

**1881. ILLUSTRATED ANNUALS OF
PHRENOLOGY.**
The Illustrated annuals of phrenology and
health almanacs. New series, from 1874 to
1883 inclusive. Combined in one volume.
New York: Fowler & Wells, 1884.
10 pts. in 1 v. ([2], 3-64; 3-64; 3-64; 40;
40; 32; 32; 32; 32; 32, 6 p.) : ill. ; 24 cm.

Spine-title: *Combined annuals. New series.*

**1882. ILLUSTRATED HYDROPATHIC
REVIEW.**
The Illustrated hydropathic review. [Vol. 1
(1853-1855)]. New York: Fowlers and Wells,
publishers, 1853-55.
1 v. (viii, 760 p.) : ill. ; 20 cm.

Edited by Russell Thacher Trall (#3558-
3593), this quarterly was absorbed after its first
volume into the *Water-cure journal* (#3720). The
final number only of *The Illustrated hydropathic
review* (p. [577]-760) is held in the Atwater Collection. Though such figures as Levi Reuben
(#1027), Joel Shew (#3180-3206) and George
H. Taylor (#3422-3433) contributed articles to
this number, about a quarter of its signed articles were written by Trall, who was also probably responsible for much of its anonymous
content. Almost the only piece of interest
among so much sensible hygienic rumination

is an anonymous article entitled "The last medical pow-wow" (p. 697-704) that chronicles the
4 May 1854 banquet at the annual meeting of
the American Medical Association in St. Louis.
The menu that is reproduced occupies two and
a half pages (including a long list of wines, liquors and cigars provided attendees) and elic-
its the comment: "Did ever a set of men sit down
to a feast where alcoholic liquor pervaded every seasoning and every gravy, without becoming intoxicated? Did ever rum prevail at the
festive board without the imbibers becoming
rowdyish? Did ever a set of gluttons sit down
and gormandize [sic] till the crammed stomach could hold no more, and the morbid and
sensual appetite was stuffed to its utmost capacity of endurance, without acting very much like
'wine-bibbers and rioters, eaters of flesh,' afterwards?" (p. 701).

**1883. AN ILLUSTRATION OF THE
PRESENT PERNICIOUS MODE OF
FASHIONABLE PRACTICE OF
MEDICINE.**
An Illustration of the present pernicious
mode of fashionable practice of medicine.
With plates. Accompanied with a dialogue,
between an apothecary and a physician, on
the subject of the statute regulating the
practice of physic and surgery. After which
is added a few brief anecdotes. Albany
[N.Y.], 1828.
36 p. : ill. ; 17 cm.

"There is a certain class of persons, who
make pretentions to the healing art; but, notwithstanding several hundred years' experience
of their predecessors, are hardly able to cure a
single disease." (p. [3]). They are "persons who
attend and study poisons three years, and cut
and mangle human bodies six months," and
then are "raised on law legs, and authorised to
practice poison to the exclusion of all others"
(p. 8). The anonymous author of this pamphlet
attacks the use of mercurial preparations, phle-
botomy and purgatives by this "privileged or-

510 A Catalog of the Atwater Collection of American Popular Medicine and Health Reform

der of men." He singles out Benjamin Rush as the best known proponent of this school, and libels his conduct during the 1793 yellow fever epidemic at Philadelphia: "When the reader casts his eye on the wretched, half-deserted city, when he sees Rush's sister, his pupils, and perhaps twenty apothecaries apprentices, besides; all making packets of mercury; and when he sees the swift poison committed to the hands of old women and negroes, he will not be surprised at the fatal consequences: instead of astonishment at the vast increase of the bills of mortality, he will find ample occasion for thanksgiving, that a single man was left alive" (p. 14). The occasion for this pamphlet is revealed in "A private conversation between a physician and an apothecary" (p. 19-26), an imagined exchange between two members of the medical establishment who share their relief at the passage of a law in 1827 that excluded root and herb "doctors" from the practice of medicine in New York State. The piece concludes with a pair of poems. The opening stanza of "Modern practice" reads:

> *Much horrid torture every day,*
> *Amongst our neighbors we survey,*
> *If done by Indians it would kill,*
> *If by learned doctors it is skill.*

1884. IMMORTAL MENTOR.
The Immortal mentor: or man's unerring guide to a healthy, wealthy, & happy life. In three parts. By Lewis Cornaro, Dr. Franklin, and Dr. Scott . . . Mill-Hill, near Trenton: Published by Daniel Fenton; Printed by Brown and Merritt, Philadelphia, 1810.
[4], 323, [9] p. ; 17 cm.

The Immortal mentor includes selections translated from Luigi Cornaro's *Discorsi della vita sobria* (#791-798); an appendix containing "Golden rules of health, selected from Hippocrates, Plutarch, and several other eminent physicians and philosophers"; two essays by Benjamin Franklin (1706-1790): "The way to wealth" and "Advice to a young tradesman"; and two essays by Thomas Scott (1747-1821): *A sure guide to happiness* and *On social love.* Franklin's *Advice to a young tradesman* was originally published at Boston in 1762. *The way to wealth* enjoyed even greater popularity, and was also pub-

lished under the title: *Father Abraham's speech to a great number of people.* According to the *Short-title Evans,* it was first printed under the latter title at Boston in 1758.

Scott's essays do not seem to have been published apart from the present compilation, which was edited by Mason Locke Weems (1759-1825). Weems studied medicine at London and Edinburgh, and may have served as a surgeon on a British man-of-war before the Revolution. In 1784 he returned to England to take orders, subsequently serving as a priest in Episcopal parishes in his native Arundel County, Maryland. In 1792 Weems became an agent for the Philadelphia bookseller Mathew Carey. He may best be remembered for his biography of George Washington (1800), the 5th edition of which (1806) first contains the story of Washington and the cherry tree. He was the author of several moralizing books, including *The drunkard's looking-glass,* one of the first American temperance tracts.

Robert Austin lists five editions of *The immortal mentor* in his *Early American medical imprints*: 1796, 1802, 1810, 1815 (Cincinnati) and 1815 (Carlisle, Pa.). The Atwater copy is lacking pages 5-8.

REFERENCES: *Austin #1014.*

1885. IMPERIAL HEALTH BRACE COMPANY.
Imperial health belt . . . An abdominal supporter and health belt combined. Appearance! Health! Comfort! Recommended for large abdomens, kidney, bladder and stomach troubles, lumbago, obesity, weak spines and backs, etc. . . . [New Haven, Ct.]: The Imperial Health Brace Co., [c1915].
7, [1] p. : ill. ; 23 cm.

1886. IMPORTANT FACTS.
IMPORTANT FACTS. A lecture to ladies by a lady. The new era of woman. [S.l.: s.n., 188-?].
[1] leaf printed on both sides ; 23 × 15 cm.

Promotional literature for a vegetable based remedy for the cure of uterine disorders. As there is no name or address printed on either side of the sheet, it may have been handed out at lectures (indicated in the last paragraph on the verso) or accompanied another printed piece.

THE IMPROVED AMERICAN FAMILY
PHYSICIAN see **DAY** (#920).

1887. IMRAY, Keith, d. 1855.
A popular cyclopedia of modern domestic
medicine, comprising every recent improvement in medical knowledge; with a
plain account of the medicines in common
use . . . First American edition. To which
are prefixed by the editor, popular treatises
upon anatomy, physiology, surgery, dietetics, and the management of the sick. Compiled from the works of distinguished physicians and surgeons. Designed for general
use. New York: Gates, Stedman and Company, 1849.
    iv, [5]-855, [1], 6, [2] p., [1] leaf of plates :
ill. ; 24 cm.

    Originally published at London in 1842
(under the title *The Cyclopedia of popular medicine*), Imray's domestic medicine appeared
in only one British edition, but was re-issued
in the United States in 1850 (#1888) and for
the final time in 1866. The American editor
added lengthy chapters to Imray's book on
anatomy, physiology, hygiene, etc., omitted
other matter, and "in very many instances"
changed "the structure of the language . . .
sometimes merely for the sake of brevity and
perspicuity."

1888. IMRAY, Keith, d. 1855.
A popular cyclopedia of modern domestic
medicine, comprising every recent improvement in medical knowledge; with a
plain account of the medicines in common
use . . . American edition. To which are
prefixed by the editor, popular treatises
upon anatomy, physiology, surgery, dietetics, and the management of the sick. Compiled from the works of distinguished physicians and surgeons. Designed for general
use. New-York: Gates, Stedman and Company, 1850.
    iv, [5]-859, [7] p., [1] leaf of plates : ill. ; 24
cm.

    Second American edition. The text of the
1850 edition is identical to that of the first
American edition (#1887) but for the addition of a "Vocabulary" or glossary (p. [845]-
846).

1889. IN SICKNESS AND IN HEALTH.
In sickness and in health. A manual of domestic medicine and surgery, hygiene, dietetics, and nursing, dealing in a practical
way with the problems relating to the maintenance of health, the prevention and
treatment of disease, and the most effective aid in emergencies. By George Waldo
Crary, M.D., Frederic S. Lee, Ph.D., Josiah
Royce, Ph.D. . . . [*et al.*] and J. West
Roosevelt, M.D. editor. New York: D.
Appleton and Company, 1896.
    xvi, 991 p., IV leaves of plates : ill. ; 25
cm.

    CONTENTS: The anatomy of the human
body, by George Waldo Crary; Physiology: the
vital processes in health, by Frederic S. Lee;
Outlines of psychology; or, a study of the human mind, by Josiah Royce; Physical training,
by Joseph Hamblen Sears; Hygiene, by Samuel
Treat Armstrong; Surgical injuries and surgical diseases, by Alexander B. Johnson; Diseases
in general, by J. West Roosevelt and William P.
Northrup; Diseases of digestive organs, heart,
and lungs, by Frank W. Jackson; Diseases of the
kidneys and urinary derangements, by J. West
Roosevelt; Diseases of women and midwifery,
by Samuel Waldron Lambert; Nervous and
mental diseases, by Frederick Peterson; Medicines and treatment, by Henry A. Griffin; Nursing the sick, by Anna Caroline Maxwell.

**1890. INDIANAPOLIS SANITARY
    ASSOCIATION.**
Sketch of the Woman's Sanitary Association
of Indianapolis, Indiana. Prepared by
Hester M. McClung, Secretary. [Indianapo-
lis]: Published by the Indianapolis Sanitary
Association, [1900?].
    40 p. ; 18 cm.

    Established in 1893, the Woman's Sanitary
Association of Indianpolis promoted "general
sanitation by increasing public interest in the
prevention of disease, and by aiding the city
government in the enforcement of its sanitary
ordinances" (p. 8). McLung explains the
Association's workings over a period of seven
years, and its assumed role as an adjunct to the
Board of Health. She describes the assignment
of "visitors" to each of the city's wards, the formation of auxilary societies and standing com-

mittees, the publication of articles in the city's newspapers, etc. So successful were the Association's efforts, that in her preface McClung states: "With the city's street-sweeping, garbage disposal, and sewer systems verging upon perfection, with the milk, anti-sputum, and other beneficent ordinances in very effective working order, and with housekeepers trained to co-operate individually with the various administrative departments, there remains little to do than the avoidance of retrogressive steps" (p. 5). Title on printed wrapper: *One phase of citizenship. Indianapolis Sanitary Association, 1893-1900.*

1891. INGERSOLL, Andrew J.
In health . . . Corning, N.Y.: Published and sold by the author, 1877.
190 p. ; 20 cm.

"Sexual life," Ingersoll states in his preface, "is universally believed to be the life of the sexual organs alone; with this idea, all thoughts and feelings in regard to it congest the organs, and result in lust, and the conclusion follows that sexual life is animal, and must perish with the body. In the following pages, I present an entirely different view of sexual life. I believe it to be the life of the body—the life which brought us into existence; and although it now holds a despised place, I know that Christ is able to redeem it, and beget in us Divine love and reverence for it" (p. 4)

Ingersoll thinks "with all Christians that sexual lust is sinful, and unless we look to Christ to redeem it, we shall be lost. If lust is sinful does it not need a Saviour?" In this seeming ambivalence toward sex, he nonetheless believes that the result of this attitude "is to destroy sexual life," and that this, in turn, results in disease (p. 19). If such is the case, then man must seek "sexual redemption" for the health of the body and the soul. By faith, Ingersoll believes that we can be restored to the condition of Adam before the Fall, when "everything was pure; [and] sexual life was holy." Ingersoll states that "the 'curative power of the system' of which physiologists speak, is sexual life . . . God is all in all; therefore, through the commitment of this life to Him, He becomes the power in us to heal all diseases." (p. 32) If the suppression of sex destroys bodily health, then its consecration to Christ assures health. He

advises a young man diagnosed with "congestion of the brain" that "if he would return thanks to Christ for every sexual sensation, he would regain his health" (p. 24). Later he writes, "When I tell the sick they need to seek the restoration of sexual feeling for the healing of any organ of the body, they fear it will be sinful, and it would be so unless they sought Christ in redemption. The Giver of life has power to be its keeper" (p. 32-33). "Sexual life," then, is more than the power of procreation: "It is the power which assimilates food in the stomach and bowels, causes respiration in the lungs; circulates the blood through the heart, arteries, and veins, and exchanges it into the solids, gives motion to the muscles, and life to the nerves, and controls all the actions of humanity" (p. 34).

Commenting on Ingersoll in her 1903 book *The lover's world,* Alice B. Stockham wrote: "Dr. A.J. Ingersoll teaches that *all* diseases are the result of a wrong attitude of mind toward the creative powers. He says: 'Disease originates in unregenerated sexual life.' He declares that we should not seek to suppress passion, but should desire Christ to redeem it. He more fully explains that we consecrate our intellects, our strength, our being, to the Christ, but that we must make a special consecration of passion. To do this, all condemnation and debasement of the functions and expression of sex must be removed from the mind" (#3365, p. 72).

1892. INGERSOLL, Andrew J.
In health . . . Corning, N.Y.: Published and sold by the author, 1878.
190 p. ; 19 cm.

Second edition.

1893. INGERSOLL, Andrew J.
In health . . . Third edition—enlarged. Corning, N.Y.: Published and sold by the author, 1884.
219 p. ; 19 cm.

A fourth edition of *In health* was published by Lee & Shepard at Boston in 1892.

1894. INGRAHAM, Charles Wilson, 1871-1900.
Don'ts for consumptives, or, the scientific management of pulmonary tuberculosis.

How the pulmonary invalid may make and maintain a modern sanatorium of his home, with additional chapters descriptive of how every consumptive person may apply the forces of nature to assist and hasten recovery. And, also, how the effects of heredity may be best overcome . . . Binghamton [N.Y.]: The Call, 1896.

xviii, [19]-218 p., [1] folding chart ; 19 cm.

Ingraham was an 1892 graduate of the medical department of the University of the City of New York. The cause of his early death is not given in Ingraham's obituary in *JAMA* (1900, 34:187).

INMAN, Anne see HYDROPATHIC AND HYGIENIC INSTITUTE (#1877).

1895. INTERNATIONAL HEALTH AND TEMPERANCE ASSOCIATION.
Pure air. [Battle Creek, Mich.: Health Publishing Co., 189-?].

4 p. ; 17 cm.

"Modern science has revealed the fact that many diseases are directly caused by taking into the system small organisms known as germs. Though microscopic in size, they are most potent agents for mischief on account of their immense numbers, and the marvelous rapidity with which they multiply. Bad air is dangerous partly on account of the presence of these germs, and partly from the presence of foul and poisonous gases" (p. [1]-2). Readers are warned of some of the sources of "impure air,: i.e., "vaults and cesspools," drains, the kitchen sink, "hencoops, piggeries, barnyards, and cowpens," cellars, "unkempt pantries," "carpets piled with dust," "damp, unventilated bedrooms," etc. "But the most constant and dangerous source of air impurities is human beings. Each breath thrown off from the lungs contains impurities sufficient to render three cubic feet . . . of air, unfit to breathe again" (p. 3). Title from wood-engraved vignette at head of page one. Series: *Health science leaflets; no. 3.*

1896. INVALIDS AND TOURISTS' HOTEL.
The Turkish and Russian baths of the Invalids & Tourists' Hotel. Buffalo, N.Y.: In-

Fig. 49. Front wrapper of The Turkish and Russian baths of the Invalids & Tourists' Hotel, *#1896.*

valids & Tourists' Hotel, [c1878].
16 p. : ill. ; 20 cm.

The Invalids and Tourists' Hotel was opened in 1878 by Ray Vaughn Pierce (1840-1914), an eclectic physician whose business acumen made him one of Buffalo's wealthiest citizens in the last quarter of the 19th century. In addition to the hotel, Pierce was proprietor of the World's Dispensary, the headquarters of his mail-order proprietary remedy business; the author of *The people's common sense medical adviser,* which saw 100 editions between 1875 and 1935 (#2802-2843); and a publisher in the mould of E.B. Foote (#1213-1256). Although figures like Hollick (#1697-1737), Foote and Pierce were regarded with suspicion or outright hostility by the regular medical community, Pierce was embraced by his fellow citizens and is distinguished by having been elected to the New York State Senate in 1877 and to the U.S. House of Representatives in 1879.

Pierce began construction of the Invalids & Tourists' Hotel in 1877 on Prospect Avenue on Buffalo's West Side. A master of self-promotion, Pierce celebrated its opening in the summer of 1878 by inviting to the inaugural ceremony local dignitaries, prominent businessmen and 200 members of the press. The hotel-sanitarium accommodated 250 guests under the care of a mixed staff of eclectic and regular physicians. The Hotel boasted state of the art heating and ventilation, the most advanced therapeutic technology (including the baths herein described), and Buffalo's first elevator and telephone. The Hotel was completely destroyed by fire on 16 February 1881. It was succeeded in 1883 by the Invalids' Hotel and Surgical Institute on Main Street, which became part of Pierce's newly incorporated World's Dispensary Medical Association (#3875-3878).

Fig. 50. The Invalids & Tourists' Hotel after the February 1881 fire.

INVALIDS' HOTEL AND SURGICAL INSTITUTE see **WORLD'S DISPENSARY MEDICAL ASSOCIATION** (#3875-3878).

1897. IRVING HOMEOPATHIC INSTITUTE.
The homeopathic treatment of spermatorrhoea, decline of manhood and kidney and bladder diseases . . . New York: Issued by the Irving Homeopathic Institute, [c1892]. 32 p. : ill. ; 16 cm.

In the opening pages of this pamphlet, the anonymous author contrasts "the crude and harsh measures, the nasty potions and strong, irritant methods of the allopaths in Sexual and Generative Diseases, with the advanced and enlightened treatment of scientific homeopathy" (p. 4). He explains to the reader the principles that underly the homeopathic treatment of disease; and relates the causes and ineluctable progress of spermatorrhoea, the third and final stage of which always results in "debility, insanity [and] death." The reader is continually reminded of masturbation's pernicious role in undermining the health of the male system, a habit that results not only in myriad sexual disorders, but in diseases as diverse and destructive as consumption, epilepsy and "softening of the brain." Bogus statistics compare the efficacy of the homeopathic treatment of sexual disorders with that of the allopaths; and finally, a description of Treatments Nos. 1-4 available from the Irving Homeopathic Institute is proffered. Title on printed wrapper: *Homeopathy.*

1898. IRVING HOMEOPATHIC INSTITUTE.
If you desire treatment, please answer the following questions and forward same to us at once, accompanied by a remittance to cover the treatment you need. Thus, if you need NO. 1—TREATMENT FOR IMPOTENCY, send $10.00; or if NO. 2—TREATMENT FOR SPERMATORRHOEA, send $10.00; or if NO. 3—TREATMENT FOR PROSTATIC or BLADDER DISEASE, send $5.00. Each TREATMENT is complete in itself, and covers all the necessary remedies, with full and explicit instructions . . . [189-?].
[1] leaf printed on both sides ; 23 × 20.5 cm.

Patients are queried regarding age, height, weight, marital status, number of children ("Any deformed or idiotic?"), etc.; and to answer such questions as: "Have you lost your sexual power?"; "Are the parts well developed or small and cold?"; "Is there any oozing under excitement?" The Irving Homeopathic Institute was located at 86 Fifth Ave., New York.

1899. IRWIN, John Arthur, 1853-1912.

Hydrotherapy at Saratoga. A treatise on natural mineral waters . . . New York: Cassell Publishing Company, [c1892].

xv, [1], 37. [5], 39-270 p., [2] leaves of plates : ill., port. ; 19 cm.

Irwin laments the "imprudences" of the American city dweller, "his hurrying, restless, nerve-straining life, constant high pressure, too many bracers [i.e., alcoholic drinks], irregular meals, eating too much and chewing too little; but always ready to sacrifice the requirements of nature on the insatiable altars of business or pleasure" (p. 2). A sojourn at Saratoga Springs is described as the ideal means of restoring lost physical and emotional balances. Irwin describes the therapeutic principles of hydropathy and balneology, and the particular benefits of the mineral waters at Saratoga. The author received his medical degree from Dublin University (Ireland), and held several hospital appointments in England before emigrating to New York.

1900. IVES, John.

Electricity as a medicine, and its mode of application . . . New York: Galvano-Faradic M'f'g Co., 1887.

123 p. : ill. ; 20 cm.

Ives explains the principles of electricity, the varieties of electric current, the scientific basis of electrotherapeutics, the instruments used therapeutically, and the diseases which benefit from electrotherapy (which include opium poisoning, alcoholism and the "effects of tobacco"). Ives' treatise appears to have first been published at New York in 1879.

J

1901. JACKSON, C. M.

The memorabilia. A compilation of practical receipts for families, and valuable information for the farmer, and those seeking health . . . [Philadelphia: C.M. Jackson, ca. 1852].

24 p. ; 21 cm.

Descriptions of common diseases, recipes for domestic remedies, kitchen recipes, and advice on farm management are mingled with advertisements for *Dr. Hoofland's German Bitters*. In printed wrapper.

1902. JACKSON, James Caleb, 1811–1895.

The absurdity of sickness and the needlessness of dying therefrom . . . Reprinted from THE LECTURER of July, 1882. Dansville, N.Y.: Published at the office of The Laws of Life, 1884.

[6] leaves ; 26 cm.

James Caleb Jackson was born in Manlius, N.Y. on 28 March 1811. As a young man he was active in abolitionist activities, as agent and later secretary of the American Anti-Slavery Society (1838–40), assistant editor of the newspaper *National antislavery standard* (1840–41), co-founder of the *Madison County abolitionist* (Cazenovia, N.Y., 1841–42), editor of the *Liberty press* (Utica, N.Y., 1842–44), and owner/editor of the *Albany patriot* (1844–46). He was briefly involved in politics as a sponsor of the Liberty League, a fourth-party outgrowth of the Liberty Party, later assimilated into the Republican Party.

Sometime in the mid-1840s Jackson became a patient of Silas Orsemus Gleason, M.D. (1811–1899), and in 1847 they opened a water-cure establishment on Skaneateles Lake with Gleason and his wife Rachel Brooks Gleason (#1362–1367) caring for the patients and Jackson in charge of the business. When the Gleasons moved to Elmira, N.Y., Jackson and his wife took over care of the patients at the Glen Haven Water Cure. During this period

Jackson attended lectures at the Central Medical College (eclectic) in Syracuse, from which he received his medical degree in 1851.

In 1858, Jackson, his wife Lucretia (#1860) and their elder son Giles, together with F. Wilson Hurd and Harriet N. Austin (#174–178) moved to Dansville, N.Y., where they opened a water-cure establishment that was to become known as Our Home on the Hillside (#2675–2679). Though not himself an owner, the elder Jackson was clearly the dominant figure from 1858 until his retirement in 1883. At the Dansville water-cure, Jackson stressed a regimen that included vegetarianism, temperance, water bathing, exercise, outdoor activity, reformed dress for women, and no medicine. After the main building was destroyed by fire in 1882, Jackson and his (by then) adopted daughter Harriet N. Austin retired to Massachusetts. He died on 11 July 1895 at Dansville.

In addition to the five books he wrote between 1861 and 1872, Jackson published at least forty health-related pamphlets and was a major contributor to the house organ, *The Laws of life*. Among the many famous persons who came for treatment at Our Home on the Hillside was Mary Ellen White, founder of the Seventh Day Adventist movement. As a patient she got many ideas for breakfast food reform and took them with her to Battle Creek, Michigan, where they took more lasting form under the Kellogg label.

In *The absurdity of sickness and the needlessness of dying therefrom,* Jackson looks forward to a period "when mankind shall reach such a degree of rational development as to understand that human life has its laws, and that human health is but the legitimate outcome of the operation of these laws" (leaf 1 recto). If we become sick, it is the result of neglecting these laws. "The day will come," Jackson writes, "when men who live in such a way as to become sick, will lose credit and character thereby, because it will be seen that such persons deliberately defy the laws of life and health" (leaf 2 recto). Illness is a choice, whether the cause be

Fig. 51. James Caleb Jackson in 1862, age 51.

gluttony, alcohol, tobacco, etc. If Jackson is unsympathetic with those who bring sickness upon themselves, he has equally little regard for the therapeutics of regular medicine: "The curative force is in the organism and not in any substance under heaven lying outside of it, and the day will come when remedies for disease will be found to lie within the range of substances which are serviceable to the maintenance of health. Persons will not be doctored to be *cured*; they will be treated according to the laws of life and health, that they may *get well*" (leaf 3 recto). In printed wrapper.

1903. JACKSON, James Caleb, 1811–1895.
American womanhood: its peculiarities and necessities . . . Dansville, Livingston Co., N.Y.: Austin, Jackson & Co., publishers; New York: Oakley, Mason & Co., [c1870].

159, [1] p., [1] leaf of plates : frontis. ; 19 cm.

First edition. Jackson opens his first chapter with the observation: "The American woman is a new or original type of womanhood. In no other country, nor in any other age, has the like of her ever been seen" (p. 7). She differs from the woman of ancient Greece or Rome, and from her European contemporaries in three ways: "1. In the style of her physical organization, which *physiologically* demands for her the largest bodily freedom. 2. In the organized combination of her intellectual faculties, which demand for her . . . accomplished education, and full use of her powers for her personal benefit. 3. In the quality and measure of her spiritual endowments" (p. 10). "Much as American men have learned to prize Liberty for themselves," Jackson writes, "and have come to see and feel how vitally necessary to a well-developed manhood she is, in politics, in religion, in business, and in government, they have, as yet, failed to understand how much woman, in these regards, needs her presence and her power" (p. 8).

Jackson considers the physiological organization and needs of women in order to fulfill their role in what American "liberty" has opened for them, as the following chapter headings indicate: I. A peculiar type; II. Physical organization; III. Unhealthy foods; IV. Unhealthy drinks; V. Unhealthy dress; VI. Constrained locomotion. VII. The useful and beautiful in dress; VIII. Life in-doors; IX. Marriage; or, woman who can and do make good wives and good mothers; X. Non-maternity; or, woman who can and do make good wives, but do not make good mothers; XI. Women who do not make good wives, but do make good mothers; XII. Women who, as society goes, can neither make good wives nor mothers ("a class of women who, in all their elements of character, are to me more interesting than any class which America produces" (p. 105); XIII. Competency of this class of women; XIV. Their business capacities; XV. The ballot.

One wonders whether Jackson had Harriet N. Austin in mind when he wrote chapters XII through XIV. She came to the water cure at Skaneateles per Jackson's recommendation, and was legally adopted by him. The dedication of *American womanhood* to Jackson's wife

Lucretia is considerably more formal and less intimate than his dedication of *How to treat the sick* to Harriet Austin (*see annotation to* #1925). Jackson's disgruntled former business partner, W.E. Leffingwell, maintained that Austin was Jackson's mistress. On the back of a photograph of Austin in Leffingwell's possession (*see fig. 6*) he wrote: "The faithful mistress of nearly forty years. In her old age robbed; and died of grief." Steel-engraved frontispiece view of Our Home on the Hillside.

1904. JACKSON, James Caleb, 1811–1895.
American womanhood: its peculiarities and necessities . . . Second edition. Dansville, Livingston Co., N.Y.: Austin, Jackson & Co., publishers; New York: Mason, Baker & Pratt, [c1870].
159, [1] p., [1] leaf of plates : frontis. ; 19 cm.

Steel-engraved frontispiece view of Our Home on the Hillside.

1905. JACKSON, James Caleb, 1811–1895.
American womanhood: its peculiarities and necessities . . . Third edition. Dansville, Livingston Co., N.Y.: Austin, Jackson & Co., publishers; New York: Baker, Pratt & Co., [ca. 1872–83, c1870].
159, [1] p., [1] leaf of plates : frontis. ; 19 cm.

An advertisement for Our Home Hygienic Institute in the advertisements at the back of the volume places its date of publication between the years 1872 and 1883. Steel-engraved frontispiece view of Our Home on the Hillside.

1906. JACKSON, James Caleb, 1811–1895.
Christ as a physician . . . Dansville, N.Y.: Our Home Pub. Dep't, publishers, 1882.
[2], 23, [7] p. ; 19 cm.

In essence, Jackson argues that in order for the body of a Christian to be a temple of the Holy Spirit, it must be a healthy body, and that for the soul to dwell in an unhealthy body is a manifestation of ignorance or transgression of God's laws, and an impediment to the spiritual life. In Jackson's view, that life is nearly impossible in a diseased body: "Now the forms of ailment which a human body can take on, and which, where

any one or more of them exists, can and do make it a poor instrument for the manifestation of the graces of the Holy Spirit, are many. Numerous as they are, so numerous may be the spiritual hindrances" (p. 9). Jackson carries his theme of "Christ as physician" beyond obedience to the physical laws of bodily health (which are really God's natural law, after all), to consider Jesus Christ as the source of a "Spiritual Force" that projected into an invalid heals all bodily ailments: "I am a firm believer in and advocate for those higher forces which the Savior of mankind holds at disposal for the cure of all human ills. These psychical instrumentalities are immense in potency, and no less remarkable in their great adaption . . . The conditions of their application and their effectiveness being complied with, they work wonders; under the inspiration and guidance of the Holy Spirit to whom Jesus has committed them for use, whoever has any curable disease can be made recipient of them" (p.18). In printed wrapper.

1907. JACKSON, James Caleb, 1811–1895.
Clergymen; what they owe to themselves, to their wives, and to society . . . Dansville, Livingston County, N.Y.: Austin, Jackson & Co., publishers, 1868 [1874 printing].
16 p. ; 19 cm.

The title-page is dated 1868, and the printed wrapper 1874. "Now good friends, health is the product of obedience to certain definite or prescribed rules or laws. It is the summing up of accepted and acknowledged conditions of life. He who knows what these conditions are, and cheerfully accepts them, and places himself within their natural operation, will be healthy; while he who fails to do this will be out of health. A clergyman will then be unfit to represent the grand idea underlying his profession, to the degree that he is lacking in health, or is sick" (p. 5).

1908. JACKSON, James Caleb, 1811–1895.
Consumption: how to prevent it, and how to cure it . . . Boston: B. Leverett Emerson; Dansville, Livingston County, N.Y.: Austin, Jackson & Co., [c1862].
vii, [1], 400, [10] p. ; 21 cm.

First edition. In his opening chapter, Jackson makes a passionate summation of his life's work: "I declare my mission to my fellow not to be simply to teach them how to rid themselves of their diseases . . . There is no comfort to me in the contemplation of being able *simply to cure* persons of dyspepsia, liver-complaint, rheumatism, general debility, scrofula, or consumption, and then let them pass into the world ignorant and undisciplined, unrestrained and untrained, only to be resnared. Never! But my mission is to teach my readers how . . . they may not only *rid* themselves of their diseases, but *remain* free from disease, and thus be *cured,* not only in themselves, but in all their generations; to break up forever, between them and their posterity, connection with disease; to set in motion through them such redeemable impulses and influences as shall draw down the blessing of Heaven in the shape of healthy bodies, vigorous minds, and pure hearts, and thus, to their latest descendants, help to create, train, and educate men and women with whom the Spirit of Eternal Wisdom may find it easy to hold intercourse; to impress on them the folly, the shame, and the crime of being sick, so that they shall feel as mortified at being sick as at having committed theft; to make them feel, that whatever their professions of religion, their standing in the church, or their high social position, they cannot 'grow in *grace*' unless they are in possession of health; that sickness is practical selfishness, and that surely and unmistakeably, sooner or later, it shuts out those divine visitations which our Father is so ready to give, and which we all so much need" (p. 8).

Jackson maintains that "pulmonary consumption is dependent largely for its prevalence upon scrofulous conditions of the system," a condition prevalent among Americans. He believes that it is hereditary; and that it can result from "excessive sexual indulgence" (to which he devotes an entire chapter), unhealthy foods (i.e., animal flesh), impure air & water, alcohol consumption (which overcomes the "vital forces" that protect the body from disease), improper dress (particuarly women's dress), etc. Influenza, bronchitis, asthma and other lung diseases frequently produce consumption. Jackson describes the three stages of consumption (citing P.C.A. Louis, Sir James Clark, and other standard authors); and maintains that it is curable in its first and even second stages through exposure to pure air, exercise, a nonstimulating diet, and various hydropathic treat-

ments. That Jackson was unable to cure the tuberculosis that killed his elder son Giles was a lasting burden to him.

1909. JACKSON, James Caleb, 1811–1895.
"The curse" lifted, or maternity made easy . . . Dansville, N.Y.: Austin, Jackson & Co., publishers, 1868 [1874 printing].
    24 p. ; 19 cm.

The title-page is dated 1868, and the printed wrapper 1874. Jackson addresses the proper management of pregnancy to insure the health of the mother, ease of delivery and well-being of the child. He considers such factors as physical labor, diet, dress, air, exercise, sleep, drinking water, and bathing.

1910. JACKSON, James Caleb, 1811–1895.
Dancing: its evils and its benefits . . . Dansville, N.Y.: Austin, Jackson & Co., publishers, 1868 [1873 printing].
    23, [1] p. ; 20 cm.

The title-page is dated 1868, and the printed wrapper 1873. Regarding dancing as "generally conducted in this country," Jackson states: "I have no hesitation in that the evil far overbalances the good that comes from it, so that it is indefensible, and should not be sustained by Christians" (p. 3). He considers the "physiological evils" of dancing, i.e., women dress unhealthfully, constricting the circulation of blood; "halls where people congregate for dancing, are badly ventilated"; dancing causes physical exhaustion in persons "not habituated to this exercise," expending their "vital force"; and dancing parties encourage "undietetic indulgences." Jackson also considers the moral evils resulting from dancing: "Now dancing . . . produces a perverse state of mind." He who dances acquires "under the excitement which it begets, loose habits of thought and, hence, unsubstantial methods of reflection"; his course "becomes fraught with ruinous dissipation, and if followed, spoils his mental energy, depraves the integrity of his moral nature, benumbs his heart, and so concludes the whole matter by *debauching* him" (p. 7). Jackson continues: " I am frank to say that, were it dependent upon my decision whether to have dancing altogether given up, or to have it practised as it is usually done in this country, I should not hesitate for a moment to decide

for its abolishment. But it is not difficult to redeem it from everything improper which now
attaches to it; and, therefore, it is not necessary
to abolish it" (p. 12). He then goes on to describe "recreative dancing" as practiced at Our
Home on the Hillside.

1911. JACKSON, James Caleb, 1811–1895.
Dancing: its evils and benefits . . . Dansville,
N.Y.: Austin, Jackson & Co., publishers,
1881.
   [2], 23, [3] p. ; 19 cm.

In printed wrapper.

1912. JACKSON, James Caleb, 1811–1895.
Dyspepsia and its treatment . . . Dansville,
Livingston Co., N.Y.: F. Wilson Hurd & Co.,
publishers, 1859.
   16 p. ; 18 cm.

"Dyspepsia is generally understood and described as a disease of the stomach *simply*. This
is an error. It gives unmistakeable signs of its
presence in remote portions of the body . . .
Thus every day the physician meets with dyspepsia in the form of sore eyes, catarrh, deafness, partial blindness, ministers' sore-throat,
palpitation of the heart, asthma, dry cough,
quinsy, spasms of the stomach, spinal irritation,
piles, seminal emissions, sick headache, hysteria, hypochondria, gout, rheumatism, epilepsy,
chorea, tonic spasms of the muscles, tic-
doloreux, or neuralgia of the arms, hands, or
heels; or great irritation of the kidneys or bladder, severe constipation, loss of appetite, coldness of hands and feet, etc., etc." (p. 2). In
printed wrapper.

1913. JACKSON, James Caleb, 1811–1895.
Dyspepsia and its treatment . . . Dansville,
N.Y.: Austin, Jackson & Co., publishers,
1876.
   16 p. ; 19 cm.

The title and imprint are taken from the
printed wrapper.

1914. JACKSON, James Caleb, 1811–1895.
Female diseases and the caustic-burners . . .
Dansville, Livingston Co., N.Y.: F. Wilson
Hurd & Co., publishers, 1859.
   17, [3] p. ; 19 cm.

Jackson denounces the use of surgical cau-
tery in the treatment of uterine diseases. He
ranks the "caustic-burners" in the regular medical profession with "blood-letters" and "poison-
givers."

1915. JACKSON, James Caleb, 1811–1895.
Flesh as food for man . . . Dansville, N.Y.:
Austin, Jackson & Co., publishers, 1875.
   18, [2] p. ; 19 cm.

The title and imprint are taken from the
printed wrapper. In this vegetarian tract, Jackson argues that the harm that results from eating meat extends beyond its more obvious dietary and physiological effects: "Meat in the
United States is *the staple* of our food . . . In the
majority of families it is eaten three times a day,
and from the oldest to the babe tied in a chair,
the members eat it. It has its adjuncts or correspondents; these are spices, such as pepper,
black and cayenne; mustard, horseradish, common salt, butter, tea, coffee, and chocolate. Of
vegetables and fruits which are *edible,* there are
aside from potatoes a minimum quantity. Add
to this list *fermented* bread, and you have the
framework of a dietary; but enlarge it, or diminish it as you will, on no consideration is *meat*
to be dispensed with. Now, when in addition to
this universal and habitual use of meat is taken
into account that it excites the nervous system,
increases the heart's action, pushes the digestive and assimilative organs to undue effort; in
fine, that its presence in the stomach and as
pabulum to the blood rouses the whole vital
machinery to exalted extraordinary exhibition,
causing more power to be spent than occasion
warrants, how far does one's imagination need
to wander beyond the limits of fact to take on
the impression that whenever the hour of *reaction* comes, and depression takes the place of
the previous exaltation, the subject will find
within him a clamor for strong drink" (p. 13–
14).

1916. JACKSON, James Caleb, 1811–1895.
The four drunkards . . . Dansville, N.Y.: Austin, Jackson & Co., 1868 [1872 printing].
   10, [6] p. ; 19 cm.

The title-page is dated 1868, and the printed
wrapper 1872. A meditation on the fatal consequences of alcohol consumption.

1917. JACKSON, James Caleb, 1811–1895.
The gluttony plague: or, how persons kill themselves by eating . . . Dansville, N.Y.: Austin, Jackson & Co., printers, 1881.
25, [3] p. ; 19 cm.

Jackson maintains that Americans "work too hard, sleep insufficiently, dress unphysiologically . . . know little or nothing of recreation, except in way of excess; understand very poorly the intimate relation existing between body and mind, and how the expenditure of mental force debilitates the body" (p. 4). The author confines himself in these pages, however, to gluttony, by which he means "eating food which is unhealthy in itself," "eating too much," and "eating at improper times." Originally published at Dansville, N.Y. in 1868. In printed wrapper.

1918. JACKSON, James Caleb, 1811–1895.
Hints on the reproductive organs: their diseases, causes, and cure on hydropathic principles . . . New York; Boston; Philadelphia: Fowler and Wells, publishers, [c1852].
iv, [5]-48, [4] p. ; 19 cm.

Jackson devotes his first two chapters to a consideration of the dignity and responsibilities of the medical and clerical professions, and a proclamation of their failure to provide moral, medical and physiological guidance to the people they serve. Chapter III is devoted to involuntary "noctural emissions," which Jackson attributes to "1. Excessive sexual intercourse. 2. The solitary vice, or self-indulgence. 3. Gluttonous eating and sedentary life. 4. Hereditary transmission. 5. Drugs" (p. 14). Chapter IV discusses "self-pollution," noting that masturbation's harm consists in its "prostrating influence" over the "vital energies," and the debilitating effect it has on the intellectual and moral faculties. Chapter V distinguishes the hierarchical roles of male and female, a position which Jackson was to alter in his later writings. The sixth and final chapter is devoted to "woman, and her diseases," i.e., uterine disorders and their hydropathic treatment. In printed wrapper.

1919. JACKSON, James Caleb, 1811–1895.
Hints on the reproductive organs: their diseases, causes, and cure by hydropathic principles . . . New York; Boston; Philadelphia: Fowler and Wells, publishers, 1857.
iv, [5]-48 p. ; 20 cm.

"Mail edition" (from printed wrapper).

JACKSON, James Caleb, 1811–1895. *Hints on the reproductive organs see also SEXUAL DISEASES* (#3145).

1920. JACKSON, James Caleb, 1811–1895.
How to beget and rear beautiful children . . . Dansville, N.Y. : Published by Austin, Jackson & Co., 1868.
16, [4] p. ; 18 cm,

Jackson discusses in some detail eight suggestions "to improve the breeds of men:" "(1) That of the two parents, *the mother* should be of *better* blood. (2) That neither should possess marked and glaring deficiencies. (3) That both should concentrate in their persons as many desirable qualities as possible. (4) That offspring should not result from the cohabitation of parties *who do not love each other,* though they are married. (5) That loving each other, children should not be begotten when the desire for cohabitation is not *mutual.* (6) That children should not be gotten when either parent has had for some time *in exercise* a special quality of character not desirable to transmit to offspring. (7) That no begetting should be had when either of the parties are fatigued, have just eaten heartily, or while undergoing very prolonged abstinence, or when ill or sick. (8) That cohabitation with the design to have a child should never take place in the darkness, or by any other than solar light" (p. 2–3). At head of title: *Private circular.*

1921. JACKSON, James Caleb, 1811–1895.
How to cure drunkards . . . Dansville, N.Y.: Austin, Jackson & Co., printers, 1868 [1874 printing]..
20, [4] p. ; 19 cm.

The title-page is dated 1868, the printed wrapper 1874. Jackson maintains that one of the principal causes of alcoholism is an overstimulating diet, particularly one laden with meat. Jackson speculates that such a diet overtaxes the digestive and circulatory systems. This level of excitement is followed by a period of physiological depression, which in turn stimulates a compensating appetite for strong drink. In combatting drunkenness, it is not enough "simply to fight alcohol, but to fight all his subordinates—to carry the war, not only against distilled, fermented and brewed liquors used by our people as beverages . . . but also against all those adjunctive substances whose direct and marked effect, when eaten or drunken, is to create an appetite for these liquors. The best way to cure the drunkard is to see that he never becomes one, and the best way to do this is to keep him from having an appetite or desire or feeling or need of stimulants. To do this, take care that he eats nutrient and unstimulating foods . . . and drinks no tea or coffee, chews and smokes no tobacco, and takes no narcotic drugs, nor stimulating medicines when sick . . . Drunkenness is a *disease,* always secondary, and easily managed under right circumstances. It never originates in the use alone of alcoholic liquors, nor can it be kept alive by their use alone." (p. 9–10).

1922. JACKSON, James Caleb, 1811–1895.
How to get well, and how to keep well . . . Dansville, N.Y.: Austin, Jackson & Co., publishers, 1869 [1876 printing].
    21, [3] p. ; 19 cm.

The title-page is dated 1869, the printed wrapper 1876. "During the time in which we have been in practice there have come under our direct personal care . . . over sixteen thousand persons . . . They came to us from all parts of the United States, from the British provinces [i.e., Canada], and the West Indies; and most of them had tried other methods of treatment, with no good effect . . . Of this great number of invalids, not over fifty have ever been treated during the entire course of their sickness without medicine. All of them had at some time taken it in one form or another . . . I think that the greater part of them had spanned the circle of the drug medicating theory. They had tried Allopathy, Homeopathy, Thompsonianism [sic], Eclecticism, Chrono-Thermalism and the Indian practice. Large numbers of them had also tried . . . patent medicines; many of them had tried Galvanism, Electricity, Electropathy, Clairvoyance, and Spiritual medication, and still were sick. Of this whole number, I do not now call to mind *a single person who had been required*

to rely upon his Vital Force as manifested in and through the laws of his organism, for his restoration to health. In every single instance . . . these persons had been dealt with by their physicians in such a way as to leave upon their minds the impression that the Curative Force did not reside in their bodies, but outside of them, and was to be introduced by medical administration" (p. 12–13).

1923. JACKSON, James Caleb, 1811–1895.
How to get well, and how to keep well . . . Dansville, N.Y.: Austin, Jackson & Co., publishers, 1880.
    21, [3] p. ; 18 cm.

In printed wrapper.

1924. JACKSON, James Caleb, 1811–1895.
How to nurse the sick . . . Dansville, N.Y.: Austin, Jackson & Co., publishers, 1878.
    [2], 19, [3] p. ; 19 cm.

"Many more persons die from a lack of good nursing, than is generally supposed, a large majority of those who are taken sick not being at the outset smitten with diseases which in themselves are incurable. There is no cause, therefore, for their deaths but bad management . . . They do not die, either because of want of vital power to live, or by reason of the destructive nature of the diseases with which they are afflicted; but simply because their physicians and nurses do not understand the laws upon which life and health depend, and unwittingly, therefore, become agents in their destruction" (p. 4).

1925. JACKSON, James Caleb, 1811–1895.
How to treat the sick without medicine . . . Dansville, Livingston Co., N.Y.: Austin, Jackson & Co., publishers; New York: Oakley & Mason, 1868.
    537, [5] p., [2] leaves of plates : port. ; 20 cm.

First edition. This was Jackson's third and best-known book. It was issued in twelve American editions between 1868 and 1889. Its final publication appears to have been the 1889 London edition. In it Jackson displays his characteristic prolixity and usual self-image of medical underdog, a stance commonly taken by 19th-century health reformers. Though he praises primarily the work of Russell Thacher Trall (#3558–3593), he also cites more conventional medical authorities such as D. Francis Condie and Thomas Watson.

Jackson describes a materia medica consisting of air, food, water, sunlight, dress, exercise, sleep, rest, social influences, and mental & moral forces—detailing almost every one of these in separate chapters. "In my entire practice," he writes, "I have never given a dose of medicine; not so much as I should have administered had I taken a homeopathic pellet of the seventh-millionth dilution, and dissolving it in Lake Superior, given my patients of its waters" (p. 25). He speaks of his method of treatment as "psycho-hygienic," by which he means "treatment according to the laws of life and health." This approach, at other times known as the water-cure, homeopathy, the Emmanuel movement, psychosomatic medicine, and holistic medicine, becomes more fashionable periodically, alternating with the more aggressive and invasive therapeutic intervention embodied in Benjamin Rush's belief that Nature must be driven from the sick-room.

In this book Jackson addresses many diseases separately, describing his method of treatment in each situation. For example, in the case of epileptics (75 of whom he treated before 1868, claiming to have cured 75%), Jackson advised hot, wet fomentations with woolen cloths applied to the lower dorsal spine and to the neck, and the application of cold, ice or snow cloths to the coccyx for periods of fifteen or twenty minutes. He also advised a cold bath, rubbing the spine with hot and cold cloths alternately, or exposing the patient to sunlight. It is interesting that he never accepted an epileptic patient unless accompanied by an attendant. In cases of lead colic, Jackson claimed success in treating six of seven cases. He treated one patient with two sweating-baths a week, each bath followed by a "plunge," vigorous wiping, and one-and-a-half hours rest in bed. This regimen continued six months. In dealing with scarlet fever, he believed in the efficacy of the wet sheet bath to hasten and maintain the eruption, which in turn, he thought, hastened recovery. He also offered Boerhaave's much-quoted advice: "Keep the head cool, the feet warm and the bowels

Fig. 52. James, Caleb Jackson, frontispiece to How to treat the sick, *#1928.*

open"—to which Jackson added, keep "the circulation to the surface" and "the mind quiet. Keep away all food."

The book is dedicated to Harriet N. Austin (#174–178), "my beloved daughter and friend . . . because to you more than any other person I am indebted for the health and strength whereby I have been able to write it, and also for the sympathy which, under my arduous professional labors, was necessary to make the task of writing a labor of love." Appended to the text is an essay by Dr. Austin entitled "Baths, and how to take them" (p. 523–37), in which she describes half-baths, the plunge, the dripping sheet, the pail-douche, the pack, local baths, fomentations, sweating, emetics, injections (i.e., enemas), the wet-cap, bandages, the wet-jacket, and the throat bandage. The steel-engraved frontispiece portrait of author faces an engraved view of Our Home on the Hillside.

1926. JACKSON, James Caleb, 1811–1895.
How to treat the sick without medicine . . .
Dansville, Livingston Co., N.Y.: Austin, Jackson & Co., publishers; New York: Oakley, Mason & Co., 1869.
537, [5] p., [2] leaves of plates : port. ; 19 cm.

1927. JACKSON, James Caleb, 1811–1895.
How to treat the sick without medicine . . .
Dansville, Livingston Co., N.Y.: Austin, Jackson & Co., publishers; N.Y.: Oakley, Mason & Co., 1871.
537, [5] p., [2] leaves of plates : port. ; 19 cm.

1928. JACKSON, James Caleb, 1811–1895.
How to treat the sick without medicine . . .
Dansville, Livingston Co., N.Y.: Austin, Jackson & Co., publishers; N.Y.: Mason, Baker & Pratt, 1873.
537, [5] p., [2] leaves of plates : port. ; 19 cm.

1929. JACKSON, James Caleb, 1811–1895.
How to treat the sick without medicine . . .
Dansville, Livingston Co., N.Y.: Austin, Jackson & Co., publishers; N.Y. Mason, Baker & Pratt, 1874.
537, [5] p., [2] leaves of plates : port. ; 19 cm.

1930. JACKSON, James Caleb, 1811–1895.
How to treat the sick without medicine . . .
Dansville, Livingston Co., N.Y.: Austin, Jackson & Co., publishers; N.Y.: Baker, Pratt & Co., 1877.
537, [5] p., [2] leaves of plates : port. ; 19 cm.

1931. JACKSON, James Caleb, 1811–1895.
How to treat the sick without medicine . . .
Tenth edition. Dansville, Livingston Co., N.Y.: Austin, Jackson & Co., publishers; New York: Baker, Pratt & Co., 1881.
537, [7] p., [2] leaves of plates : port. 19 cm.

1932. JACKSON, James Caleb, 1811–1895.
How to treat the sick without medicine . . .
Twelfth edition. Dansville, Livingston Co., N.Y.: Our Home Publishing Department, publishers; New York: The Baker & Taylor Co., 1887.
537, [7] p., [2] leaves of plates : port. ; 19 cm.

1933. JACKSON, James Caleb, 1811–1895.
Hygiene and the gospel ministry . . . Dansville, Livingston County, N.Y.: F. Wilson Hurd & Co., publishers, 1859.
17, [3] p. ; 18 cm.

In a pamphlet addressed to the clergy, Jackson cautions that their effectiveness in the ministry is hampered by their "sedentariness," which can only result in "prostration of muscular force" and impaired "digestion, circulation, secretion, and excretion." Jackson writes, "though it be admitted that you are good *theo*logists, I can not admit that you are good *physio*logists, and Physiology *necessarily* as a *substratum* for Theology. Nature *always paves the way* for Grace. Humanity precedes Christianity, and those are always the poorest Christians—the most impefect representatives of Christ's life—who . . . are the sickest in body" (p. 4). He goes on to criticize "your dietetic habits," particularly the consumption of tea and coffee: "They can serve no purpose but to create a flash of excitement, as factitious as it is superficial, ending surely in making you ill-tempered, and at *last stupid*" (p. 6). Tobacco is also stigmatized as "a great obstacle to your successful ministrations." Those who habitually use tobacco "are either in that state where the effect of its use is to exalt their VITAL FORCES, and so are lifted *above* their natural level, or in that state where its effect is powerfully to depress the system, and so they are doubting, despondent, or in despair" (p. 8). Jackson next addresses the subject of "minister's sore-throat"; and concludes with four pages devoted to the "physical condition of the women of the United States," the peril of fashion to the well-being of the female constitution.

1934. JACKSON, James Caleb, 1811–1895.
In memoriam. [S.l.: s.n., 1891.]
14 p. : port. ; 22 cm.

Title from wrapper. Jackson's eulogy of Harriet N. Austin (#174–178) following her death in 1891. A laid-in printed slip states: "The

Fig. 53. Our Home on the Hillside ca. 1882, frontispiece to The sexual organism, *#1945.*

Memorial herein presented is a copy of a letter written by Dr. James C. Jackson to Mrs. Estella McClean, and was read at the annual re-union of the Austin family, held at Ensenore, Owasco Lake, June 27, 1891."

1935. JACKSON, James Caleb, 1811–1895.
Our philosophy of treating the sick . . . Reprinted from The Lecturer of Sept., 1881. Dansville, N.Y.: Published at the office of The Laws of Life, 1884.
   [73]-84 p. ; 26 cm.

Jackson explains the theoretical basis for the regimen that invalids undergo at Our Home on the Hillside: "People come here with a thorough misconception of what we are trying to do and how we are trying to do it. They have heard of Our Home as a water-cure, and really suppose that we consider water to possess great virtues, or what may be called curative properties. They hear that we pay special attention to food, and they imagine that we think there is great curative efficacy in food. They know that we say a good deal about sunlight, and they think, therefore, that we have great confidence

in the curative influence of the sun's rays. They learn that we recommend early retiring and regularity in getting up, and they conclude that we regard this as having in itself curative value. When they find out that our philosophy of restoring the sick does not permit us to contemplate curative force as existing in any substance, influence, or agency lying outside of the patient, they are thrown entirely off their balance, and do not know what to think" (p. [73]). Includes Jackson's "The danger from alcoholic drinks . . . For young men" (p. 81–84). In printed wrapper.

1936. JACKSON, James Caleb, 1811–1895.
Piles, and their treatment . . . Dansville, N.Y.: Austin, Jackson & Co., publishers, 1868 [1876 printing].
   14, [2] p. ; 19 cm.

   The title-page dated 1868; the printed wrapper 1876. Jackson attributes piles to "sedentary habits" and "the use of concentrated, highly seasoned, and stimulating food" (p. 5). "Of course, to sit still . . . day after day, to eat stimulating food, concentrated in its nature—thus leaving but very little factitious matter to pass through the alimentary canal, helping to create fecal bulk, while nervous energy is drawn away from ths stomach, liver and bowels by intellectual task-work—is to so derange the relation between the functional exercise of the bowels and the nervous energy upon which such activity depends, as at length to leave the bowels decidedly deficient in vitalization" (p. 6). "Congestions of the blood-vessels," "morbid or tumorous growths" and enlargement of the veins in the rectum inevitably result. Jackson corrected this condition by the use of injections, sitz- or half-baths, and a diet of fruits and grains.

1937. JACKSON, James Caleb, 1811–1895.
Piles, and their treatment . . . Dansville, N.Y.: Austin, Jackson & Co., publishers, 1882.
    14, [2] p. ; 19 cm.

    In printed wrapper.

1938. JACKSON, James Caleb, 1811–1895.
Scrofula; its nature and treatment . . . Dansville, Livingston Co., N.Y.: F. Wilson Hurd & Co., publishers, 1859.
    15, [1] p. ; 18 cm.

    An archaic term for tuberculosis of the cervical lymph nodes, Jackson attributes its prevalence among Americans to their consumption of the flesh of pigs carrying the disease. He maintains that it is curable by following a strict hydropathic regimen.

1939. JACKSON, James Caleb, 1811–1895.
The sexual organism, and its healthful management . . . Boston: B. Leverett Emerson, 1862.
    iv, [2], [5]-279, [1], 11, [1], 11 p., [1] leaf of plates : port. ; 19 cm.

    "Out of a great variety of causes operating to invert the Divine relations, and to pervert the Divine arrangements, thus destroying health and enthroning sickness, the misuse or abuse of the sexual organism and its functions stands preeminent" (p. 6). To enlighten the public on these dangers, Jackson divides *The sexual organism* into two parts. In the first, he addresses pre-natal influences and other matters pertaining to the management of pregnancy; child rearing & child hygiene; masturbation and its medical consequences; venereal

diseases and their treatment (not by mercurials and other poisons, but by hygienic means); and "diseases of the system arising from over-action or abuse of the reproductive organs."

The second part of the book is devoted to women's health: menstruation and its disorders, uterine diseases, marriage, sex , etc. Jackson believed that sexual intercourse should take place only "at such periods or times as may be followed by the female becoming pregnant: Nature never intending that a man should lose his semen for any other purpose than that of propagating his species; or that woman should ever exercise her sexual faculty so as to reach her highest paroxysm of feeling, unless with a view to conception" (p. 256). Jackson does allow "partial consummation" between married sexual partners, however: "The man and his wife can be brought into sexual embrace without reaching the point of orgasmic action" (p. 258). Jackson warns: "Physicians know full well that a class of diseases arises from undue sexual gratification, such as congestion of the brain, pulmonary diseases; dyspepsia; liver complaint; irritation of the kidneys . . . [etc.]."

The second eleven-page sequence reprints Harriet N. Austin's (#177, 178) "Baths, and how to take them"; the third sequence contains advertisements for Jackson's other publications, *Davidson's Patent Syringe*, etc. Engraved frontispiece portrait of author. In the Atwater copy, the label of Dansville bookseller M.W. Simmons & Co. is glued to the front pastedown.

1940. JACKSON, James Caleb, 1811–1895.
The sexual organism, and its healthful management . . . Boston: B. Leverett Emerson; Dansville, N.Y.: M.W. Simmons & Co., 1864.
    [2], iv, [5]-279, [1], 11, [1], 11, 4 p., [1] leaf of plates : port. ; 20 cm.

    "Baths and how to take them, by Miss Harriet N. Austin, M.D.," (11 p., 2nd sequence).

1941. JACKSON, James Caleb, 1811–1895.
The sexual organism, and its healthful management . . . Boston: B. Leverett Emerson, 1865.
    iv, [5]-279, [1], 11, [5] p., [1] leaf of plates : port. ; 20 cm.

    "Baths and how to take them, by Miss Harriet N. Austin, M.D.," (11 p., 2nd sequence).

1942. JACKSON, James Caleb, 1811–1895.
The sexual organism, and its healthful management . . . Boston: B. Leverett Emerson; Dansville, Livingston County, N.Y.: Austin, Jackson & Co., 1869.
    iv, [5]-279, [1], 11, [7] p., [1] leaf of plates : port. ; 20 cm.

    "Baths and how to take them, by Miss Harriet N. Austin, M.D.," (11 p., 2nd sequence).

1943. JACKSON, James Caleb, 1811–1895.
The sexual organism, and its healthful management . . . Boston: B. Leverett Emerson; Dansville, Livingston Co., N.Y.: Austin, Jackson & Co., 1872.
    iv, [5]-279, [1], 11 p. ; 20 cm.

    "Baths and how to take them, by Miss Harriet N. Austin, M.D.," (11 p., 2nd sequence).

1944. JACKSON, James Caleb, 1811–1895.
The sexual organism, and its healthful management . . . New York: B.L. Emerson & Co.; Dansville, Livingston Co., N.Y.: Austin, Jackson & Co., 1879.
    iv, [5]-279, [1], 11, [9] p. ; 20 cm.

    "Baths and how to take them, by Miss Harriet N. Austin, M.D.," (11 p., 2nd sequence).

1945. JACKSON, James Caleb, 1811–1895.
The sexual organism, and its healthful management . . . New York: B.L. Emerson & Co.; Dansville, Livingston Co., N.Y.: Our Home Pub. Dep't., 1882.
    iv, [5]-279, [1], 11, [5] p., [1] leaf of plates : ill. ; 20 cm.

    The wood-engraved frontispiece depicts "The Dansville Sanitarium" (*see fig. 53*). "Baths and how to take them. By Miss Harriet N. Austin, M.D.," (11 p., 2nd sequence).

1946. JACKSON, James Caleb, 1811–1895.
Shall our girls live or die? . . . Dansville, N.Y.: Austin, Jackson & Co., publishers, 1878.
    [2], 18, [4] p. ; 19 cm.

    Jackson is so pessimistic about the declining physical vitality of the Anglo-Saxon race in America that he predicts it will "become prac-

tically extinct in New England unless something is done to revive and restore it" (p. 4). The key to its revival is to ensure the health of its women, about whose "*constitutional degeneracy*" Jackson is equally despondent, and to which he attributes high infant mortality rates as well as the steady transmission of "weak constitutions." Sounding a theme that would be echoed into the next century, Jackson maintains that "there is a law in Nature, whereby human beings can grow better from generation to generation. Not knowing what the law is, our people have regulated their relations as fancy, caprice, or haphazard might dictate. They have married and begotten and brought forth children without reference to the fact whether these should be able to live to an adult age, and if so, whether the woman-portion of such progeny should have constitutional stamina to give birth to healthy children" (p. 9–10). He laments that inferior races, such as the Germans, the Irish, the Swedes and other recent immigrants, surpass the older Anglo-Saxon population in possessing "healthy blood," which they transmit to their offspring. To remedy the declining state of American womanhood, Jackson insists that girls be raised on the same principles as boys, allowing them free use and development of their physiological, sensory and intellectual powers. Otherwise, "the Anglo-Saxon American woman in the Republic will become extinct at no distant day, and . . . with their destruction, our love for *personal* liberty will be greatly weakened, and the type of our civilization essentially changed" (p. 17–18). In printed wrapper.

1947. JACKSON, James Caleb, 1811–1895.
Shall our girls live or die? . . . Dansville, N.Y.: Sanatorium Publishing Company, 1886.
    [2], 18, [4] p. ; 19 cm.

    In printed wrapper.

1948. JACKSON, James Caleb, 1811–1895.
Speech of James C. Jackson, M.D. delivered in Liberty Hall, March 28th, 1881 . . . the day he was seventy years old. Dansville, N.Y.: Austin, Jackson & Co., publishers, 1881.
    [35]-60 p., [1] leaf of plates : ill. ; 23 cm.

    An autobiographical essay, in which Jackson traces his ancestry, early unhygienic habits, the resulting illnesses, disappointment with the regular medical profession; his discovery of and "conversion" to the laws of health; and his work as a physician and health reformer. The therapeutic system which resulted from this personal transformation is described as a "psycho-hygienic philosophy of treatment" (p. 58). Jackson concludes by congratulating himself on still being in the fullness of his intellectual and bodily powers at age seventy, which he attributes to conformity to Divinely instituted hygienic laws. Caption title. Offprint from *The Lecturer,* v. 3, no. 3 (May 1881). Includes steel-engraved view of Our Home Hygienic Institute (with guard sheet).

1949. JACKSON, James Caleb, 1811–1895.
Spermatorrhea and its treatment . . . Dansville, Livingston Co., N.Y.: F. Wilson Hurd & Co., publishers, 1860.
     16 p. ; 19 cm.

Title and imprint from printed wrapper. Jackson defines spermatorrhea (or "seminal emissions") as "an involuntary flow of semen, or sperm, or seed, from the sexual organs of the male," adding, "The flow generally takes place independently of those natural causes of excitement, which . . . produce secretion in the seminal vessels" (p. 2). Its causes are "1. Functional derangement of a number of organs, or some special organ. 2. Excessive lawful or vicious sexual excitement. 3. Drug medication. 4. Masturbation" (p. 3–4). "Brain-workers" are especially susceptible to this disorder: "They wear themselves out in demands on the sexual propensity much more and much sooner than the *toilers.* Muscular exertion . . . predisposes to *repose,* and that to sleep, while *brain*-work is usually likely to seek relief in *social,* which easily creeps into *sexual* enjoyment" (p. 8). The cure lies in "abstinence from *exciting* causes" (including "brain-work"); increased "muscular exertion," which "sedates the nervous system, and so keeps irritation from being transmitted from the brain or stomach to the genitals" (p. 16); adherence to an "unstimulating diet"; and baths.

1950. JACKSON, James Caleb, 1811–1895.
Spiritual mindedness: a discourse delivered in Liberty Hall . . . and published by request of all the clergymen, Christian patients, and guests at Our Home. Dansville,

N.Y.: Published at the office of The Laws of Life, 1884.
     [59]-70 p. ; 26 cm.

Jackson states that the Christian "must abandon all physical habits which give the body undisciplined passional energy." "I do not wonder," he writes, "that professing Christians make no greater advances in spiritual knowledge. How can they while they are victims to appetitive indulgence? Paul says that to *grow* in grace one must *crucify* the flesh with its affections and lusts. Christians of our day are trying to make advances in the divine life by gratifying the body . . . Great simplicity of physical life is necessary to growth in the divine life. A coarse, gross, overfed, stimulated, half narcotized, drug-medicated body cannot be the temple, the dwelling place, of the Divine Spirit" (p. 63). Reprinted from *The Lecturer.* In printed wrapper.

1951. JACKSON, James Caleb, 1811–1895.
Student life; or, how to work the brain without overworking the body . . . Dansville, Livingston Co., N.Y.: F. Wilson Hurd & Co., publishers, 1860.
     18, [2] p. ; 18 cm.

Jackson opens *Student life* by stating that "of all types of the human race, the Anglo-Saxon possesses the largest degree of intellect: and of *this* type of the race there is none that surpasses . . . the people of the United States" (p. 3). The difference between the Anglo-Saxon in America and in England is that "the American brain is disproportionate in size to the body, while the brain of the Englishman is much better related to his physical structure." While the Englishman may enjoy a better overall balance between mind and body, the mind of the American "possesses more varied capabilities, and, other things being equal, is his superior." Consequently, "We are a nation of thinkers." However, "For their health our people think too much." (p. 4). Jackson continues: "We use up in abstract thought large measures of nervous energy, and therefore deprive the general structure of the nutriment to which it is entitled; and while therefore we are a *large-headed* people, we are coming to be a *small-bodied* people" (p. 5). To remedy this imbalance Jackson recommends the incorporation of rest and play into Americans' daily routine: "There is nothing in

all our educational, social, or domestic arrangements which is so much needed by all classes of
persons as opportunities for Recreation—the
means of *recreating* ourselves against the wear
to which our daily life subjects us" (p. 7). Diet,
sleep and bathing are less extensively discussed.

1952. JACKSON, James Caleb, 1811–1895.
Student-life: or, how to work the brain without overworking the body . . . Dansville,
N.Y.: Austin, Jackson & Co., publishers,
1868 [1876 printing].
    18, [2] p. ; 18 cm.

    The title-page is dated 1868; the printed
wrapper 1876.

1953. JACKSON, James Caleb, 1811–1895.
To the young men of the United States . . .
Dansville, Livingston Co., N.Y.: F. Wilson
Hurd & Co., publishers, 1859.
    16 p. ; 19 cm.

    Jackson forsees the day "when to slight God's
laws, as they are inscribed on *the bodies* of men,
will be considered an immorality; when full-
boned, round-formed, large-muscled children,
whose feet are of the right shape, whose legs
are set on the body well; whose bodies, legs,
and arms are relatively right, and surmounting
all, heads of the right shape—children beautiful as Adonis, shall be in the household, crowns
of rejoicing, and a GLORY, while the feeble, and
the pimping, the pining and the sickly shall be
a grief and a mourning—for their parents shall
be ashamed of their progeny. At *this* day, deformity, disease, defect of build, and want of
growth are laid at the door of the Creator, and
society utters its *amen*. But in the day coming
these shall be laid at the door of the parents,
and the Almighty shall be justified" (p. 3–4).
    In order to hasten "the day coming," Jack-
son advises young men to attend more to their
health and bodily habits that they may produce
stronger offspring and avoid the shame of diseased and deformed children. "You live too fast,
you live wretchedly. You live ignorantly" (p. 5).
Jackson therefore advises his young readers to
avoid the use of coffee, tea and tobacco; provides detailed recommendations on diet; points
out the importance of exercise; and warns
against the use of drugs: "If you get sick, get
well by proper means . . . If you must die, do so

rationally. I should prefer to die rationally than
to die *scientifically*" (p. 12). Title and imprint
from printed wrapper.

1954. JACKSON, James Caleb, 1811–1895.
The training of children; or, how to have
them healthy, handsome and happy . . .
Dansville, N.Y.: Austin, Jackson & Co., publishers, 1872.
    107, [5] p., [1] leaf of plates : frontis. ;
19 cm.

    First edition. "Children," writes Jackson,
"need training just as much as colts. Like them,
they are animals, though something more—
having . . . souls inside of them" (p. 6). Furthermore, "physical health is not a constitutional
endowment in American children. They are
born sickly" (p. 9); and this "because their parents are not physically fit. While animal and
vegetable propagation have improved, human
propagation has not." Jackson states that pregnant women should not over-exert themselves
physically, should get outdoors, be exposed to
sunlight, eat nutritious but unstimulating foods,
and live amid pleasant and agreeable surroundings. Women who cannot nurse should not bear
children. In the last chapter marriage is discussed and Jackson offers the opinion that many
men and women should not marry who are
forced to do so by convention. Engraved frontispiece view of Our Home on the Hillside.

1955. JACKSON, James Caleb, 1811–1895.
The training of children; or, how to have
them healthy, handsome and happy . . .
Dansville, Livingston Co., N.Y.: Our Home
Publishing Department, 1883.
    107, [13] p., [1] leaf of plates : ill. ; 19
cm.

    Wood-engraved frontispiece view of Our
Home on the Hillside.

JACKSON, James Caleb see also **OUR
HOME ON THE HILLSIDE** (#2675–
2679).

**1956. JACKSON, James Hathaway, 1841–
1928.**
Papers on alcohol . . . [Dansville, N.Y.]:
Austin, Jackson & Co., publishers, 1874.
    18, [2] p. ; 18 cm.

In this temperance tract, Jackson pursues a scientific path in describing the ill effects of alcohol consumption. He avoids the rhetorical flourishes that made his father's speeches and pamphlets so popular. The younger Jackson's arguments obviously draw upon his medical school laboratory training, whereas his father appealed to the moral and social sensibilities of his audience. The title and imprint are taken from the printed wrapper.

James Hathaway Jackson, the younger son of James Caleb and Lucretia Edgerton Brewster Jackson, was born in Peterboro, N.Y. in 1841. After spending part of the years 1857–58 in Nebraska with his semi-invalided brother Giles, Jackson returned to the family home in Skaneateles, N.Y. to discover that his father had sold his interest in the Glen Haven Water Cure and planned to open another water cure in Dansville. In 1861 Jackson graduated from Eastman's Commercial College in Rochester, and subsequently served as cashier, bookkeeper and eventually business manager of Our Home on the Hillside. On 13 September 1864, he married Katharine Johnson (#1958, 1959), his father's stenographer. In 1876 Jackson received his medical degree from the Bellevue Hospital Medical College and returned to Dansville both to serve on the medical staff and act as business manager. Jackson assumed increasing responsibillity for the guidance and operation of Our Home as his father's health declined in the late 1870s. After the principal building of Our Home was destroyed by fire in 1882, Jackson planned the financing and construction of its replacement, a fireproof structure that still stands. He was "physician in charge" of "The Sanatorium" (as it was renamed) when the new building was opened in 1883, and served as chief of staff and medical director of the institution succesively known as The Sanatorium, the Jackson Sanatorium and the Jackson Health Resort until his retirement in 1917.

1957. JACKSON, James Hathaway, 1841–1928.

The use and abuse of animal flesh as an article of diet . . . Reprinted from The Lecturer of May, 1880. Dansville, N.Y.: Published at the office of The Laws of Life, 1883.
[33]-43, [5] p. ; 26 cm.

"A claim is made in behalf of animal flesh as an article of diet, that it contains some strength giving properties not found in other foods, particularly of the vegetable class." Jackson sets out to prove that meat is "in no particular essential to life," and that "it is in its constitution absolutely unhealthful and injurious in its influence on the performance of functional life in humans" (p. [33]). Title and imprint from printed wrapper.

JACKSON, James Hathaway see also **JACKSON HEALTH RESORT** (#1962, 1963); **OUR HOME ON THE HILLSIDE** (#2677–2679).

1958. JACKSON, Kate Johnson, 1841–1921.

About babies . . . [Dansville, N.Y., c1878].
[4], 40, [2] p. ; 14 cm.

Jackson addresses the usual health issues pertaining to infant care, and dress in particular. In her concluding remarks to *About babies,* Jackson echoes her father-in-law's millenarian fanstasy: "We believe that when the 'good time coming' comes—that when man by intelligence and obedience gets into harmony with God, and so with nature—then higher and sweeter will be his song than any bird's, more tender his attentions, more joyous his out-goings, and in-comings, and in them more perfectly will he keep time to the tune of the universe" (p. 35). "To help on the good time," Jackson encourages her readers, "it is for all conscientious parents to set about preparing themselves to train up a new and noble race of fathers and mothers." Title from printed wrapper.

In 1864, James H. Jackson (#1956, 1957), the younger son of James Caleb Jackson (#1902–1955), married Katharine Johnson, who had been James Caleb Jackson's stenographer since 1862. From the beginning of their marriage she was engaged in the operation of Our Home on the Hillside, serving as its "overseeing matron" until 1873, when she and her husband went to New York to pursue medical studies. In 1877, Kate Jackson received her medical degree from the Woman's Medical College of the New York Infirmary, headed by Emily Blackwell (the sister of Elizabeth). Upon returning to Dansville, she joined the medical staff at Our Home on the Hillside. When its new building was opened in 1883 (rechristened "The Sanatorium"), James H. and Kate Jackson became its managing physicians. Kate and James H. Jackson had one son, James Arthur Jackson (1868–1922, M.D., 1895), who succeeded his father in the operation of what by then was known as the Jackson Sanatorium.

1959. JACKSON, Kate Johnson, 1841–1921.
Contamination and poisoning from meat . . . [Dansville, N.Y.: The Sanatorium Publishing Co., 1888?].
  11 p. ; 16 cm.

From the beginning, vegetarianism was one of the dietary dogmas at the Jackson family's sanitorium in Dansville, N.Y. Kate Jackson echoes prevailing sentiment when she writes: "It has been repeatedly demonstrated that fresh meat of the best quality contains waste excrementitious substances like urea, uric acid, etc. . . . The presence of these waste poisonous properties, even in small proportion, constitutes a grave objection to meat as a staple article of food" (p. 2). Title from printed wrapper.

JACKSON, Kate Johnson see also OUR HOME ON THE HILLSIDE (#2677–2679).

1960. JACKSON, Lucretia Edgerton Brewster, d. 1890.
The health reformer's cook book: or how to prepare food from grains, fruits, and vegetables . . . Dansville, N.Y.: Austin, Jackson & Co., publishers, 1874.
  70 p. ; 19 cm.

First edition. In 1830, Lucretia Edgerton Brewster, of Mexico, N.Y., married James Caleb Jackson (#1902–1955). She managed the kitchen and dining room first at the Glen Haven Water-Cure in Skaneateles, N.Y., and later at Our Home on the Hillside. In the words of William D. Conklin, she "bore the heat and brunt of the earlier years in the homemaking departments" (*The Jackson Health Resort*, Dansville, N.Y., 1971, p. 13). James C. and Lucretia Jackson had three children: Giles Elderkin Jackson (1835–1864), Mary Jackson (1839–1844), and James Hathaway Jackson (#1956, 1957).

JACKSON, Mercy Bisbee see *DRESS-REFORM* (#990).

1961. JACKSON, Robert Montgomery Smith, 1815–1865.
The mountain . . . Philadelphia: J.B. Lippincott & Co., 1860.
  xii, 3–632 p. ; 19 cm.

Jackson was one of the original "back to nature" enthusiasts. "Can the present tendency to agglomerate in swarms," he asks, "be designated by any such agreeable appellations as love of society, association for mutual refinement and exaltation, or Christian compact for the advancement and more perfect development of the social instincts of the soul?" (p.vii). The question is entirely rhetorical. He castigates the reigning "commercial despotism" that produces "overcrowded cities, at the expense of the well-being and normal life of the whole race," installing in its place the "perverted reign of the elements of anarchy and death." From the mountain top, Jackson calls upon his deluded fellows to realize "the significance of the country life," its "gravity and grandeur." "Might not a friendly voice from the woods be heard in the hum and shock, or possibly reach the ear of some haggard sufferer, writhing in the folds and meshes of the artificial life . . . and inspire him with hope that the blue hills and green fields, the cool sequestered forests, the lonely haunts by the mountain springs in the stillness of the evening or dewy freshness of the morning, might have life, health, and joy for him?" (p. viii).

In writing this volume, Jackson's "real desire has been to get *something* of the natural science of that piece of the venerable spheroid (the earth) called the Alleghany [sic] Mountain, made more generally known to men, also to try to introduce some of its metaphysical elements into the recorded soul of the world; but, above all, to assert its sanitary claims or powers to produce health or happiness" (p. x). In Book I of this unusual treatise, Jackson discusses the geology, hydrography, flora and climate of the Allegheny mountains. Book II is devoted to the "doctorial uses of the mountain," i.e., the benefits of its climate and waters to health. In Book III Jacksons ruminates on health and longevity in more general terms.

In concluding this lengthy paean to The Mountain, Jackson writes: "Thus, from all history, all science of man, from the depths of existence, in all his relations to the world, is heard the still small voice, the pleading prayer of nature, come to *me*—come back to the golden orient of your primal union with *me*—come back to the charmed realm of original life fountains—come back from the wallows of sin—come back from the hot-beds of the artificial life—come back to the little child's simplicity and holiness, for of such . . . is the kingdom of heaven," of health, the paradise of rejuvenescence of the body, and to such only shall be added the perennial raptures of the normal or sound soul" (p. 626). Jackson was the proprietor of a sanitarium at the magnesia springs near Cresson in southwestern Pennsylvania.

1962. JACKSON HEALTH RESORT.
Fiftieth anniversary tidings from the Jackson Health Resort, Dansville, N.Y. 1858–1908. Dansville, N.Y.: F.A. Owen Publishing Co., [1908].
  88 p. : ill., ports. ; 20 cm.

Established as a water-cure in 1858 by James Caleb Jackson (#1902–1955) under the name Our Home on the Hillside, the establishment's name was changed to The Sanatorium in 1883, to the Jackson Sanatorium in 1890, and to the Jackson Health Resort in 1903. Title and imprint from printed wrapper.

1963. JACKSON HEALTH RESORT.
The Jackson Health Resort. An institution distinctive in methods and character . . . Chief of staff, James H. Jackson . . . [Dansville, N.Y., ca. 1905].
  32 p. : ill. ; 15 × 23 cm.

Promotional brochure illustrated with numerous half-tone photographs of the buildings, grounds and surrounding countryside. The date of publication is surmised from a photograph of the 1905 graduating class of the Training School for Nurses, established at the Jackson Health Resort in 1902.

JACKSON HEALTH RESORT see also
OUR HOME ON THE HILLSIDE
(#2675–2679).

1964. JACOBI, Abraham, 1830–1919.
Infant diet . . . Revised, enlarged and adapted to popular use by Mary Putnam Jacobi . . . New York: G.P. Putnam's Sons, 1882.
  viii, [9]-119 p. ; 18 cm.

Series: *Putnam's handy-book series of things worth knowing, no. 15. .*

JACOBI, Mary Putnam see **JACOBI**
(#1964).

1965. JACOBY, M.
A lecture on the treatment and preservation of the hair delivered . . . at the Masonic Temple, Boston, Friday evening, February 23, 1849 . . . Boston: Printed by William Chadwick, 1849.
  15 p. ; 22 cm.

In original printed wrapper. A broadside advertising Jacoby's "Berliner Haar Wasche" and "Chinese wash" is pasted-in.

1966. JACQUES, Daniel Harrison, 1825–1877.
The philosophy of human beauty; or, hints toward physical perfection: showing how to acquire and retain bodily symmetry, health, and vigor, secure long life, and avoid the infirmities and deformities of old age . . . New-York: Wood & Holbrook, publishers, 1871.
  xviii, [19]-244 p. : ill. ; 19 cm.

"Beauty . . . whether in plants and animals or in men and women, is the grand external sign of goodness of organization and integrity of function; and the highest possible beauty can indicate nothing less than perfection in these particulars . . . Physical goodness (or health) and beauty will always be found to bear a strict relation to each other, the latter being everywhere the sign or symbol of the former. A lack of beauty in any member or system of the body indicates a lack of goodness or health in that member or system" (p. 32). Jacques guides his readers to an understanding of how beauty and health may actually be developed. In discussing the "laws of human configuration," he argues that the soul is the essence of the person, and the body is its "external expression and instrument." In turn, "The soul forms, changes, and controls the body through the instrumentality of a nervo-vital fluid, which forms the connecting link between mind and matter." This nervo-vital fluid may be applied to the alteration or development of any part of the body: "The vital fluid, or creative life-spirit, may be thrown upon any organ or part by the exercise of that organ or part, by a simple act of the mind directing the attention intently upon it." (p. 63).

Not surprisingly, therefore, we can improve our outward physical form by the application of inner influences, i.e., through "mental culture." Citing the evidence of phrenology, i.e., the shape of the head is altered as various intellectual or emotional faculties are developed, then "it must be evident that whatever has power to change the shape of the head and the permanent expression of the face may be capable of modifying, in the same degree . . . the contours of the body. The cultivation and continual activity of the intellectual faculties have a tendency to diminish the action of the motive and vital systems, and while they impart expression and refinement to the features, render the body more delicate and, and within the limits of physical health, more beautiful" (p. 91). Thus, physical beauty and health come to be equated with moral goodness: "a life of obedience to the laws of our spriritual being, will promote in the same degree our physical beauty and well being . . . Goodness of heart and purity of life co-operate with an expanded chest, wholesome air, copious breathing, and out-of-door exercise, in imparting to the fair cheek, the coveted rosette tinge . . . Whatever, then, is favorable to goodness, happiness, and ease is, in the same degree, favorable to health and beauty" (p. 102). Not only are religion and cultivation of the arts conducive to the perfection of physical form, but even democracy plays a developmental role: "In this country, the lines which separate the different classes are less clearly defined than in Europe, and individuals and families are constantly rising from lower to higher social grades; acquiring, at the same time . . . the physical traits of the rank which they assume" (p. 114). Jacques' final chapters are devoted to more practical means of improving health and appearance, such as exercise, diet, sleep, etc. Spine-title: *Physical perfection.*

1967. JAEGER, Gustav, 1832–1917.
Selections from essays on health-culture and the Sanitary Woolen System . . . (Trans-

lated from the German). New York: Dr. Jaeger's Sanitary Woolen System Co., 1886. viii, 199, [1] p. : ill. ; 17 cm.

First edition. Although Jaeger studied medicine, he took his degree in comparative zoology from Vienna in 1856, where he became director of the aquarium and zoo. From 1867 through 1884, Jaeger lectured at the agricultural college in Hohenheim and the veterinary college in Stuttgart. In 1884 he gave up his academic status with the state in order to devote himself fully to the propagation of his ideas on "health-culture."

The *Selections* are culled from Jaeger's various books, i.e., *Die menschliche Arbeitskraft* (Munich, 1878), *Seuchenfestigkeit und Constitutionskraft* (Leipig, 1878), *Die Normalkleidung als Gesundheitsschutz* (Stuttgart 1880), and *Die Entdeckung der Seele* (Leipzig, 1880). Jaeger maintains that an excess of fat and water in the tissues weakens the body's ability to resist disease. Jaeger's regimen requires "hardening" the body by the elimination of excess fat and water. Proper clothing is essential to this process of elimination. Jaeger's Sanitary Woolen System introduces garments made entirely of sheep's wool (vegetable fibers such as cotton are to be avoided) that keep the skin uniformly warm, promote cutaneous evaporation of excess water, gently rub the skin "in order to maintain a constant and ample blood supply," and assist in the "self-cleansing process of the skin" (p. 12–13). Woolen garments also protect the body from the cutaneous absorption of germs. In addition, Jaeger also maintains that woolen clothing has a curative power: "A familiar mode of treatment in the removal of solid or fluid morbific deposits from the body, is that which doctors term counter-irritation, wherein it is endeavored . . . to bring out the disease through the skin. The Sanitary Woolen Clothing does this most effectually" (p. 87). In the explication of his "health-culture" regimen, Jaeger also stresses low-fat diets, Turkish baths, physical exercise, the ventilation of rooms, etc. to control excessive levels of body fat and water. After enjoying considerable popularity in Europe and England, the proprietors of Dr. Jaeger's Sanitary Woolen System established a New York office in 1886.

1968. JAEGER, Gustav, 1832–1917.
Selections from essays on health-culture and the Sanitary Woolen System . . . (Trans-

lated from the German). Second edition, revised and enlarged. New York: Published by Dr. Jaeger's Sanitary Woolen System Co., 1891.

viii, 216 p. : ill. ; 17 cm.

1969. DR. JAEGER'S SANITARY WOOLEN SYSTEM CO.

Illustrated catalogue of the Dr. Jaeger's Sanitary Woolen System Co. . . . [New York: Dr. Jaeger's Sanitary Woolen System, c1892].

68 p. : ill. ; 22 cm.

The *Illustrated catalogue* explains the principles of the Sanitary Woolen System (*see* #1967, 1968), and describes the garments available for purchase: underclothes, stockings, shirts, pajamas, skirts, mufflers, jackets, suits, blankets, etc. On printed wrapper: *Illustrated catalogue and price list. Twelfth edition, 1892–93.* Fabric samples mounted on p. 19. With insert ([2] leaves) listing the location of outlets in the United States.

1970. JAHR, Gottlieb Heinrich Georg, 1800–1875.

Jahr's new manual of the homoeopathic materia medica, with Possart's additions. Arranged with reference to well authenticated observations at the sick bed, and accompanied by an alphabetical repertory, to facilitate and secure the selection of a suitable remedy in any given case. Fifth edition, revised and enlarged by the author, and translated and edited by Charles J. Hempel . . . Symptomatology and repertory. New-York: William Radde . . . [*et al.*], 1859.

xvi, 924 p. ; 21 cm.

A translation of Jahr's *Handbuch der Haupt-Arzneien für die richtige Wahl der homöopathischen Heilmittel* (1835) and A. Possart's *Charakteristik der homöopathischen Arzeien* (1851–53). *Jahr's new manual* was intended by Hempel (#1616–1618) as a "convenient book to the physician and lay-practitioner." Part I consists of Jahr's description of 148 of the "most frequently used homoeopathic remedies and their principal pathogenetic and curative symptoms" (p. 1–434), with Hempel's translation of Possart's additions (p. [435]-554). Part II is a "concise repertory to Jahr's and Possart's manuals," re-

lating disease entities to the remedies described in Part I.

1971. JAHR, Gottlieb Heinrich Georg, 1800–1875.

Short elementary treatise upon homoeopathia, and the manner of its practice; with some of the most important effects of ten of the principal homoeopathic remedies, for the use of all honest men who desire to convince themselves by experiment of the truth of the doctrine . . . Second French edition, corrected and enlarged. Translated by Edward Bayard, M.D. New York: Wm. Radde, 1845.

v, [6]-90, [6] p. ; 15 cm.

Edward Bayard (1806–1889) was general secretary of the American Institute of Homoeopathy (est. 1844) between 1845 and 1848, edited the *North American journal of homoeopathy* between 1860 and 1862, and wrote several pamphlets in defense of homeopathy.

1972. JAMES, Bushrod Washington, 1836–1903.

American resorts; with notes upon their climate . . . With a translation from the German by Mr. S. Kauffmann of those chapters of "Die Klimate der Erde," written by Dr. A. Woeikof . . . that relate to North and South America and the islands and oceans contiguous thereto. Intended for invalids and those who desire to preserve good health in a suitable climate. Philadelphia and London: F.A. Davis, 1889.

285 p. : map ; 25 cm.

"Change of climate" was an important concept in the prevention or treatment of many medical disorders throughout the 19th century, from nervous exhaustion to pulmonary tuberculosis. A "change of climate" might take the form of prolonged foreign travel or a visit to a spa. During the course of the 19th century, medical climatology became an increasingly scientific discipline. Most of the extensive literature on this topic was generated by European physicians. Notable American authors, however, include John Bell (#289–293), Robley Dunglison (#1003, 1004), A.N. Bell, and Simon Baruch (#229). Health resorts or sanataria flourished in the United States during the 19th century, particularly during the heydey of hydropathy, during which the nation's landscape was dotted with water-cures of varying repute.

In *American resorts* James provides chapters on sea-side, fresh-water and mountain resorts, as well as winter and summer resorts. Excerpts from *Die Klimate der Erde* of Aleksandr Ivanovich Voeikov (1842–1916), published at Jena in 1887, comprise more than a third of this volume (p. 170–270). Folding map in pocket on back pastedown: *New official railroad, state, and territorial map of the United States of America, Canada and Mexico printed for "American resorts."* Atwater copy inscribed by author (25 Dec. 1899).

Bushrod W. James was an 1857 graduate of the Homoeopathic Medical College of Pennsylvania. He was a prominent figure in American homeopathy (including presidencies of the Pennsylvania Homoeopathic Medical Society in 1873 and the American Institute of Homoeopathy in 1883); and the author of a popular treatise on cholera, as well as several volumes of poetry and fiction.

1973. JAMES, Henry.
Anti-dyspeptic pills, an approved remedy for dyspepsia or indigestion, habitual costiveness, and piles . . . [New-York, c1822].
    8 p. ; 17 cm.

Dyspepsia is described as "one of the most frequent and formidable diseases of our country" (p. [3]), which if left untreated, proceeds from being a simple disorder of the stomach to a chronic disease ensuring a "miserable existence" and a "premature grave." James dismisses the dietary regimens normally prescribed for this disorder, and promotes pills of his own preparation. His "anti-dyspeptic" pills are also recommended for "the distressing acidity and nausea of pregnant women" and for the treatment of piles. James maintained a pharmacy at 32 Pearl Street in New York.

1974. JAMES, John.
The American household book of medicine, or every one's guide in sickness: containing directions on the diseases of men, women, and children; on bathing, diet, exercise, and nursing the sick; on climate, mineral waters, &c., &c.; written in plain language, and adapted to popular use and

ready reference by means of a complete index of symptoms and a natural classification of subjects . . . Cincinnati: R.W. Carroll & Co., publishers, 1879.
    [ii]-xv, [16]-780 p. : ill. ; 23 cm.

Originally published at Cincinnati in 1866.

1975. JAMESON, Horatio Gates, 1778–1855.
The American domestick medicine; or, medical admonisher: containing, some account of anatomy, the senses, diseases, casualties; a dispensatory, and glossary. In which, the observations, and remedies, are adapted to the diseases, &c. of the United States. Designed for the use of families . . . Baltimore: Published by F. Lucas, Jun.; J. Robinson, printer, 1817.
    xii, [17]-675 (i.e., 657), [3] p., [1] leaf of plates : frontis. ; 22 cm.

First edition. Jameson's is one of the earliest manuals of domestic medicine by an American author that compares favorably with Buchan's *Domestic medicine* (#442–502), a text that dominated the American market through the first quarter of the 19th century. Jameson began the study of medicine with his father in 1795, and received his medical degree from the University of Maryland in 1813. When denied an appointment to the medical faculty of his *alma mater*, Jameson founded the rival Washington Medical College (est. 1827). In 1835 he became president of the medical department of Cincinnati College, joining Daniel Drake, Samuel D. Gross and J.N. McDowell on its faculty. His most important publications appeared in the journal literature, and include descriptions of his "extirpation of the upper jaw after ligation of the carotid artery, which he performed for the first time in 1820 . . . and the first removal in America of uterine scirrhus [1824]. After Dorsey and Post, he was the third surgeon to ligate successfully the external iliac artery [1822]" (*Dict. Amer. biog.*, 5:601).
    REFERENCES: *Austin* #1045.

1976. JAMESON, Horatio Gates, 1778–1855.
The American domestick medicine; or, medical admonisher: containing some account of anatomy, the senses, diseases, ca-

sualties; a dispensatory, and glossary. In which the observations, and remedies, are adapted to the diseases, &c. of the United States. Designed for the use of families. Second edition, with additions and improvements . . . Baltimore: Published by the author; John D. Toy, printer, 1818.

xvi, [17]-574, [2] p., [1] leaf of plates : port. ; 22 cm.

REFERENCES: *Austin* #1046.

1977. JAMISON, Alcinous Burton, b. 1851.
The anus and rectum: their physiology, anatomy, and pathology. Together with a description of rectal and anal diseases, and their diagnosis and treatment . . . Sixth edition, revised and enlarged by the author. New York: Published by the author, [ca. 1900, c1897].

64 p. : ill. ; 22 cm.

After devoting the first thirty-six pages to the anatomy and diseases of the rectum, Jamison describes on pages 37–42 proprietary medicines and equipment for both the treatment and prevention of rectal disorders. These include salves, lotions, compounds, laxatives, etc., and appliances for home treatment such as the rectal syringe, salve applicator, hot sitz-bath, rectal douche, etc. Pages 43–64 contain testimonials from grateful patients, the latest dated 1900. Title on wrapper: *How to become strong: health obtained, ill-health prevented.*

1978. JAMISON, Alcinous Burton, b. 1851.
Intestinal ills. Chronic constipation, indigestion, autogenetic poisons, diarrhea, piles, etc. Also auto-infection, auto-intoxication, anemia, emaciation, etc. due to proctitis and colitis . . . New York: The Knickerbocker Press, 1901.

xvi, 244 p. ; 20 cm.

First edition. Jamison maintains that proctitis is "chiefly responsible for chronic constipation, chronic diarrhea, auto-infection; and hence for mal-assimilation, mal-nutrition, anemia; and for a thousand and one reflex functional derangements of the system as well." He continues, "The inflamed surface of the intestinal canal (proctitis) inhibits the passage of feces. Absorbent glands begin to act on the retained sewage, and the whole system becomes more or less infected with poisonous bacteria" (p. xv). This "putrid fecal mass of solid and liquid contents . . . is one of the most common and serious pathogenic (disease-producing) and pyogenic (pus-producing) sources, which, by auto-infection, afflict man from infancy to old age. Here—in the dilated and obstructed sewer—the ptomain and leucomain class of poisons . . . find another and last chance to be taken up by the absorbing cells of the mucuous membrane and returned to the blood" (p. 25). Temporary relief with cathartics and laxatives only aggravates the condition, as these medications are themselves absorbed into the blood stream and increase levels of auto-intoxication.

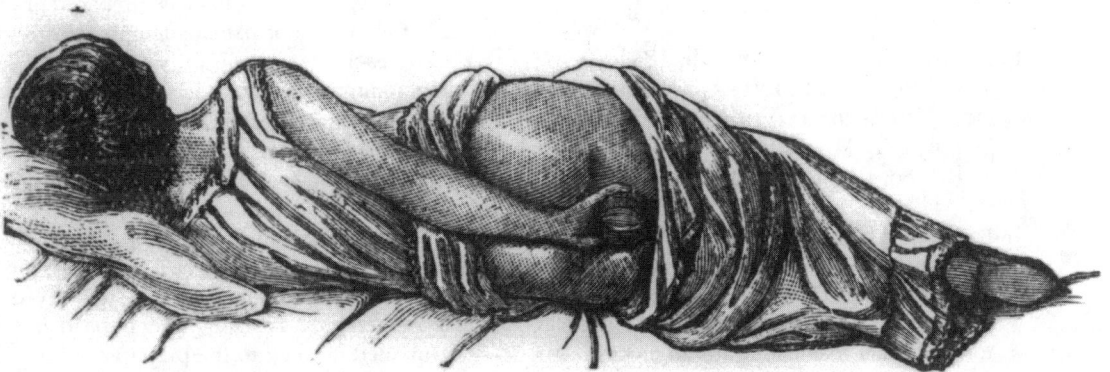

Fig. 54. "Clysma in lateral position," a possibly unique illustration of enema, a commonly recommended procedure for colonic health and for preventing autointoxication. From Fischer-Duckelmann, #1165.

To relieve abnormal conditions of the intestines and to maintain them in a healthy state, Jamison recommends a regimen *per anum* and *per orem*, i.e., the application of warm-water enemas twice daily and the abundant consumption of spring water. Jamison was an 1878 graduate of the Medical College of Fort Wayne (Indiana). He maintained a practice on W. 45th St. in Manhattan, where he specialized in rectal diseases. *Intestinal ills* appears to have been issued for the final time in 1919.

1979. JAMISON, Alcinous Burton, b. 1851.
Intestinal ills. Chronic constipation, indigestion, autogenetic poisons, diarrhea, piles, etc. Also auto-infection, auto-intoxication, anemia, emaciation, etc. due to proctitis and colitis. New York City: Published by the author, 1913.
    xvi, 277 p. ; 21 cm.

    Second edition.

1980. JAMISON, Alcinous Burton, b. 1851.
Intestinal ills. Chronic constipation, indigestion, autogenetic poisons, diarrhea, piles, etc. Also auto-infection, auto-intoxication, anemia, emaciation, etc. due to proctitis and colitis. New York City: Published by Chas. A. Tyrrell, M.D., 1917.
    xv, [1], 277, [5] p. ; 21 cm.

    The five unnumbered pages contain advertisements for remedies and products available from Charles Alfred Tyrrell (1846–1918), author of *The royal road to health* (#3617–3627) and proprietor of the Tyrrell Hygienic Institute in New York.

1981. JANES, Lewis George, b. 1844.
The Butler Health-Lift. Its reasons and its facts. Fifth revised edition. New York: Lewis G. Janes . . . M.L. Holbrook, 1875.
    [6], 88, [2] p. : ill. ; 14 cm.

    In the 1860s, David P. Butler opened in Boston a gymnasium that featured machines for lifting heavy weights. By the 1871, the Butler Health Lift Company had expanded from Boston to New York, where it maintained five branches under the supervision of Lewis G. Janes.
    "Experience and a thorough trial have convinced us that a short series, at brief intervals, of slow but strong efforts in lifting a weight placed directly under the centre of the body, graduated to the strength of the individual, and skilfully poised and adjusted by means of springs, is the best form of any curative or healthful exercise occupying an equal length of time" (p. [5]). Janes maintains that regulated physical exercise increases the circulation of the blood, and hence vitalizes all the body's organs, assists in the removal of wastes, and removes local congestions that lead to disease. "What physicians call the '*vis medicatrix naturae*,' which is their chief reliance in the administration of medicine, is the *reserved vital force of the body* stored away, as it were, in the healthy organs which, when some appropriate stimulus is applied by the physician, is not only aroused to action, but is distributed from the healthy to the unhealthy organs" (p. 8–9). The Butler Health Lift stimulates the muscles ("the most extensive set of organs in the body") to release this vital force throughout the body: "In such a thorough and complete tension of almost every muscular fibre of the body, the '*vis medicatrix*' is appealed to by a stimulus in quantity and quality such as no medicine can apply." Title on printed wrapper: *Reasons and facts*.

1982. JANNEY, Oliver Edward, 1856–1930.
The medical adviser; or how to treat the sick and the injured . . . Baltimore: Published by the Maryland Homoeopathic Pharmacy Co., 1900.
    73 p. ; 16 cm.

    Janney's brief manual was "prepared primarily for the use of patients temporarily out of easy reach of a physician, who desire to use Homoeopathic remedies." An 1881 medical graduate of the University of Maryland, Janney took a second degree from the Hahnemann Medical College (Philadelphia) the following year. He was professor of the practice of medicine at the Southern Homoeopathic Medical College (Baltimore) until he retired from medicine to devote himself to the social reform movements of his era. A Quaker by birth, Janney remained devoted to the principles of the Society of Friends in his involvement in the temperance movement, women's rights, improving race relations, etc.

1983. JARVIS, Edward, 1803–1884.
Elements of physiology. For schools . . . New
York and Chicago: A.S. Barnes & Company,
[c1865].
168 p. : ill. ; 17 cm.

The *Elements of physiology* is a re-issue of Jarvis'
Primary physiology, first published at Philadelphia
in 1848 (#1988). It was published by Alfred
Smith Barnes (1817–1888), whose publishing
house A.S. Barnes & Co. (est. 1838) "may have
been the first to establish itself as primarily an
educational publisher" (John Tebbel, *A history
of book publishing in the United States,* New York:
R.R. Bowker, 1972, 1:294).

An 1830 Harvard medical graduate, Jarvis
practiced in Concord and Northfield, Ma. before moving to Louisville, Ky. in 1837. He returned to Massachusetts in 1842, where he focused his medical interests on the treatment of
the insane. According to Gerald Grob, "Jarvis'
success in psychiatric practice gave him the time
to explore a variety of social and medical problems . . . As a Unitarian who was trained in Scottish Common Sense philosophy, committed to
a Baconian . . . interpretation of science, and
influenced by rising doubts about the efficacy
of traditional therapeutics, he became part of
a group that sought to synthesize medicine,
morality, and social activism . . . His faith in the
existence of a lawful and orderly universe led him
to emphasize the importance of knowledge and
education. In 1847 he published *Practical physiology,* a text on human physiology and health for
families and students. In his eyes, disease was the
product of the willful violation of the natural laws
that governed human behavior and thus was indissolubly linked to filth, immorality, ignorance,
and improper living conditions" (*Amer. nat. biog.,*
11:878). Jarvis was an authoritative and influential figure in the effort to compile accurate statistical data for census and social policy making
purposes in the United States.

1984. JARVIS, Edward, 1803–1884.
Lecture on the necessity of the study of physiology, delivered before the American Institute
of Instruction, at Hartford, August 22, 1845 . . .
Boston: William D. Ticknor & Co., 1845.
55, [1] p. ; 19 cm.

Jarvis explains to educators the "laws of
Physical Life" (and the consequences of ignor-

ing them) in order to impress upon his audience the importance of physiological instruction in the school curriculum. Title on printed
wrapper: *Jarvis on popular physiology.*

1985. JARVIS, Edward, 1803–1884.
Physiology and laws of health. For the use
of schools, academies, and colleges . . . New
York and Chicago: A.S. Barnes & Company,
1872.
427 p. : ill. ; 19 cm.

"In sustaining the body with food, and drink,
and air—in defending it with clothing and shelter—in the use of our muscles and brain—in
applying the body and the mind to whatever
purpose, we use some or all of our organs; and
the health and strength, or pain and weakness,
the good or the evil consequences, must be in
accordance with the wisdom and faithfulness
with which we govern ourselves in these matters. Therefore the knowledge of the Physiological Laws, and of their requirements, becomes
of practical importance in every moment, and
in all the circumstances of our being. This book,
in a somewhat different form, was formerly
published under the name of *Practical physiology* [#1986, 1987]. But as the main purpose of
the work was to teach the Laws of Health, and
as the whole of it has been revised, and much
of it re-written, the title is changed to PHYSIOL-
OGY AND LAWS OF HEALTH, which better describes the work" (p. 4).

Jarvis' *Physiology and the laws of health* was first
published at New York in 1866. This 1872 issue
appears to be its third printing. The book was
part of A.S. Barnes' *National series of standard
school-books,* a series which covered every department of elementary and advanced education.
The firm began publishing titles in this series
before the Civil War, and successfully marketed
it through the end of the century.

1986. JARVIS, Edward, 1803–1884.
Practical physiology; for the use of schools
and families . . . Philadelphia: Thomas,
Cowperthwait & Co. . . . [*et al.*], 1848.
396 p. : ill. ; 19 cm.

This first edition of *Practical physiology* was
issued in 1847, 1848 and 1851 (#1987) before
the appearance of of the revised edition in 1852
and 1859. In 1866 the work was once again re-

vised and published under the title *Physiology and laws of health* (#1985). Issued with 32 p. publisher's catalog.

1987. JARVIS, Edward, 1803–1884.
Practical physiology; for the use of schools and families . . . Philadelphia: Thomas, Cowperthwait & Co. . . . [*et al.*], 1851.
    396 p. : ill. ; 19 cm.

1988. JARVIS, Edward, 1803–1884.
Primary physiology, for schools . . . Philadelphia: Thomas, Cowperthwait & Co. . . . [*et al.*], 1848.
    168 p. : ill. ; 19 cm.

    First edition.

1989. JARVIS, Edward, 1803–1884.
Primary physiology, for schools . . . New York: A.S. Barnes & Co., 1866.
    168 p. : ill. ; 17 cm.

1990. JARVIS, Edward, 1803–1884.
Primary physiology, for schools . . . New York: A.S. Barnes & Co., 1867.
    168 p. : ill. ; 17 cm.

1991. JEFFERIS, Benjamin Grant, 1851– 1929.
The household guide or domestic cyclopedia. A practical family physician. Home remedies and home treatment on all diseases. An instructor on nursing, housekeeping and home adornments. By Prof. B.G. Jefferis . . . and J.L. Nichols . . . Also a complete cook book by Mrs. J.L. Nichols. Twenty-second edition. Rochester, N.Y.: Vick Publishing Co., [c1905].
    [2], 9–[10], xi–xviii, [19]–458 p., [1] leaf of plates : ill. ; 19 cm.

    Cover-title: *Family receipts.*

1992. JEFFERIS, Benjamin Grant, 1851– 1929.
The household guide or domestic cyclopedia. Home remedies for man and beast; a complete receipt book. Home nursing and home treatment; insect extermination; Prof. Henkel's illustrations of the effects of alcohol & cigarettes; care of children; how to cook for the sick, etc. By Prof. B.G.

Jefferis . . . and J.L. Nichols . . . Also, a complete cookbook, by Mrs. J.L. Nichols. Thirteenth edition. Naperville, Ill.: Published by J.L. Nichols, 1894.
    [2], 512 p. : ill. ; 18 cm.

"The object of this volume is, to instruct every housekeeper and every owner of domestic animals in the use and application of simple domestic remedies. It may be properly called a book of *Self Instruction* in the art of home doctoring" (*Publisher's pref.*, p. [4]). The earliest recorded copy of this title in *NUC Pre-1956 imprints* is the second edition of 1891. Jefferis was an 1887 graduate of the Kentucky School of Medicine (Louisville). Canadian by birth, he spent most of his life in Chicago. Cover-title: *Home remedies for man and beast.*

1993. JEFFERIS, Benjamin Grant, 1851– 1929.
Search lights on health or light on dark corners. How to love, how to court, how to marry and how to live. Directions to married and unmarried. By Prof. B.G. Jefferis . . . and J.L. Nichols . . . Second edition. Naperville, Ills.: Published by J.L. Nichols, 1894.
    [2], ii, 393 p. : ill. ; 18 cm.

Jefferis' and Nichols' *Search lights* provides advice on a startling array of topics pertaining to male-female relations, including character formation, "personal purity," courtship, etiquette, personal grooming & dress, "hints on matrimony," sexual ethics, the prevention of conception, infant care & rearing, "home remedies and home treatment," the "secret habit" & "secret diseases."

In her discussion of the impact of Anthony Comstock's (#754, 755) crusade against immoral literature—a crusade that extended to the publication of advice on contraception— Janet F. Brodie remarks: "The general quality of contraceptive advice literature in the last decades of the [19th] century was inferior to that available earlier. The popular and immense work by B.G. Jefferis and J.L. Nichols, *Search lights on health* . . . first published in 1894 and reissued regularly until 1921, is a good example of the consequences of the Comstock laws. The advice on 'prevention of conception' actually was a confused, inaccurate jumble of warnings about the medical and legal dangers of tampering 'with nature's laws,' interspersed with rationales explaining why women needed preventives and discussions of withdrawal, the rhythm method, abortion, and breastfeeding ... The work was decidedly inferior in quality of advice, in organization, and in tone to even the ephemeral pamphlets of fifty years earlier. Jefferis and Nichols themselves escaped prosecution by hedging their stance on contraception with many warnings against it and by publishing simultaneously in out-of-the-way places: Parkersburg, West Virginia; Naperville, Illinois; and Ontario, Canada" (*Contraception and abortion in nineteenth-century America,* Ithaca: Cornell Univ. Press, 1994, p. 283). Cover-title: *Safe counsel.* Atwater copy lacking pages 109–10.

1994. JEFFERIS, Benjamin Grant, 1851–1929.
Search lights on health. Light on dark corners. A complete sexual science and a guide to purity and physical manhood. Advice to maiden, wife, and mother. Love, courtship and marriage. By Prof. B.G. Jefferis ... and J.L. Nichols ... Sixth edition. Toronto, Ontario: Published by J.L. Nichols & Co., 1895.
[4], 432 p. : ill. ; 18 cm.

Cover-title: *Safe counsel.* Spine-title: *Search lights or light on dark corners.*

1995. JEFFERIS, Benjamin Grant, 1851–1929.
Search lights on health. Light on dark corners. A complete sexual science and a guide to purity and physical manhood. Advice to maiden, wife, and mother. Love, courtship and marriage. By Prof. B.G. Jefferis ... and J.L. Nichols ... Eighteenth edition. Halifax, N.S.: W.F. Currie, general agent, [c1894].
[4], 508 p. : ill. ; 18 cm.

Cover-title: *Safe counsel.* Spine-title: *Search lights or light on dark corners.*

1996. JEFFERIS, Benjamin Grant, 1851–1929.
Search lights on health. Light on dark corners. A complete sexual science and a guide to purity and physical manhood.

Advice to maiden, wife and mother. Love, courtship and marriage. By Prof. B.G. Jefferis ... and J.L. Nichols ... Eighteenth edition. Naperville, Ill.; Toronto, Ont.: J.L. Nichols & Co., 1897.
[4], 515, [1] p. : ill. ; 18 cm.

Cover-title: *Safe counsel.* Spine-title: *Search lights or light on dark corners.*

1997. JEFFERIS, Benjamin Grant, 1851–1929.
Search lights on health. Light on dark corners. A complete sexual science and a guide to purity and physical manhood. Advice to maiden, wife, and mother. Love, courtship, and marriage. By Prof. B.G. Jefferis ... and J.L. Nichols ... Atlanta, Ga.; Toronto, Ont.; Naperville, Ill.: J.L. Nichols & Co., [c1904].
487 p., [1] leaf of plates : ill. ; 19 cm.

Cover-title: *Safe counsel.* Spine-title: *Search lights or light on dark corners.*

1998. JEFFERIS, Benjamin Grant, 1851–1929.
Search lights on health. Light on dark corners. A complete sexual science and a guide to purity and physical manhood. Advice to maiden, wife, and mother. Love, courtship, and marriage. By Prof. B.G. Jefferis ... and J.L. Nichols ... Naperville, Ill.; Atlanta, Ga.: J.L. Nichols & Co., [c1904].
[2], 487, [1] p. : ill. ; 18 cm.

Cover-title: *Safe counsel.* Spine-title: *Search lights or light on dark corners.*

1999. JEFFERIS, Benjamin Grant, 1851–1929.
Search lights on health. Light on dark corners. A complete sexual science and a guide to purity and physical manhood. Advice to maiden, wife, and mother. Love, courtship, and marriage. By Prof. B.G. Jefferis ... and J.L. Nichols ... Naperville, Ill.: Published by J.L. Nichols & Co., 1910.
487 p., [1] leaf of plates : ill. ; 18 cm.

Cover-title: *Safe counsel.* Spine-title: *Search lights or light on dark corners.*

2000. JEFFERSON, R.
The family doctor: a dictionary of domestic medicine and surgery, especially adapted for family use . . . Fifteenth edition. Philadelphia: George Gebbie, 1869.
[4], 750 p. : ill. ; 18 cm.

An alphabetically arranged dictionary of disease entities, injuries, medicinal substances (plant & animal), anatomical parts, etc. The text is abundantly illustrated with wood engravings, especially of medicinal plants. This work first appears to have been published at London (2 v.) in 1858–59 under the title: *The Family doctor: being a complete encyclopedia of domestic medicine and household surgery . . . By a Dispensary Surgeon.*
There are two issues of this 1869 Philadelphia edition. The issue in the Atwater Collection has the name "R. Jefferson, M.D., F.R.C.S., London" on the title-page. Gebbie also issued *The Family doctor* in 1869 without Jefferson's name on the title-page. In its place is the statement of responsibility: "By a London Dispensary Surgeon." No other American edition than this 15th is recorded. A final London edition (also 750 p.) was published in 1889 as part of *Routledge's popular library.*

2001. JEGI, John I., 1866–1904.
Practical lessons in human physiology, personal hygiene and public health for schools . . . New York: The Macmillan Company; London: Macmillan & Co., 1903.
[2], xvii, [1], 343, [5] p. : ill. ; 19 cm.

The author was professor of physiology and psychology at the State Normal School, Milwaukee.

JELLIFFE, Smith Ely see *STANDARD FAMILY PHYSICIAN* (#3323).

2002. JENKINS, Oliver Peebles, 1850–1935.
Advanced lessons in human physiology. A treatise of the human body, including an account of its structure, its functions, and the laws of health . . . Indianapolis, Ind.: Indiana School Book Company, [c1896].
319 p., [2] leaves of plates : ill. ; 19 cm.

Originally published at Indianpolis in 1891. Jenkins was a high school administrator until 1876, when he joined the faculty of his *alma mater,* Moores Hill College, as professor of natural history. From 1883 to 1886 he taught at the Indiana State Normal School; and from 1886 to 1891 was professor of biology at DePauw University. In 1891 Jenkins was made professor of physiology and histology at Stanford University, where he was also co-director of the Stanford Marine Laboratory. A physiologist and ichthyologist, Jenkins was noted for his work on the nervous system of invertebrates. Series: *Indiana state series.*

2003. JENKINS, Oliver Peebles, 1850–1935.
Primary lessons in human physiology . . . Indianapolis, Ind.: Indiana School Book Company, [c1896].
209 p., [3] leaves of plates : ill. ; 18 cm.

Originally published at Indianapolis in 1891. Series: *Indiana state series.* The Indiana School Book Company was acquired by the text-book publishing giant the American Book Company about the time Jenkins' physiologies were published.

2004. JENNINGS, Isaac, 1778–1874.
Medical reform: a treatise on man's physical being and disorders, embracing an outline of a theory of human life, and a theory of disease—its nature, cause, and remedy . . . Oberlin: Fitch & Jennings, 1847.
xxiv, [25]-375, [3] p. ; 18 cm.

"An 1812 graduate of Yale's medical school . . . Jennings came to health reform rather late in life. He practiced conventional medicine for some years, but steadily lost faith in heroic treatments and began substituting bread pills for calomel in many cases. The experiment convinced him of the worthlessness of drugs, and by the 1830s he had discarded even placebos. He also soon found confirmation of his new style of practice in both health reform and the Parisian therapeutic philosophy then entering the country. Jenning's articulate discussions of the questions of nature versus art and self-limited disease show he was better versed than other health reformers in contemporary medicine, but he advanced beyond the Paris line of therapeutic skepticism to the position of complete therapeutic nihilism. Nontherapy was for him a positive action, an action he defined as orthopathy" (James C. Whorton, *Crusaders for fitness,* Princeton, N.J., 1982, p. 135).

Jennings was born 7 November 1778 in Fairfield, Ct. After studying medicine, he first practiced in Derby, Ct. In 1813 he married Anne Beach in Trumbull, Ct. They had seven children, five of whom lived to adulthood. In 1839 Jennings moved to Oberlin, Ohio. According to a family genealogist: "After some years of practice, his views on medicine underwent a change and he retired from the practice of his profession. The last twenty-five years of his life he devoted principally to writing and promulgating his views . . . Dr. Jennings was kind, genial and respected by all. He had great patience and charity, never showing any significant irritation or bitterness or fanaticism, as do some reformers whose opinions are not accepted at once" (William Henry Jennings, *A genealogical history of the Jennings families in England and America,* Columbus, Ohio, 1899, 2:427).

2005. JENNINGS, Isaac, 1778–1874.
The philosophy of human life; with especial design to develop the true idea of disease; its nature, immediate occasion, and general remedy . . . Cleveland, Ohio: Jewett, Proctor & Worthington; Boston: John P. Jewett and Company, 1852.
    viii, [9]-276, [2] p. ; 19 cm.

Jennings distinguishes *orthopathy* from what he terms *heteropathy* on p. 215: "many schools, teachers and practitioners still inculcate, theoretically and practically, the importance of taking disease by storm, when practicable, and giving it no quarter till it is exterminated . . . Orthopathists discard this doctrine of armed intervention as opposed to the law of animal economy." "Disease," he wryly observes, "is *not* so much under the control of blood-letting, or other disturbing means, as physicians generally suppose it is. Or rather the animal economy manages her forces so adroitly, that very generally she succeeds in overcoming both the disease and the Doctor; and then, most mischievously, the Doctor gets credit for curing the disease" (p. 221). In his consideration of disease *per se,* Jennings intends to show "by deductions from the Science of Physiology" that "what is called disease is nothing more or less than impaired health, feeble vitality; that the recov-

ery from this state is effected, when effected at all, by a restorative principle, identical with life itself, susceptible of aid only from proper attention to air, diet, motion and rest, affections of the mind, regulation of the temperature, &c., with occasional aid from what may justly be denominated surgical operations and appliances; and that medicine has no adaptation nor tendency to 'help nature' in her restorative work" (p. vi).

2006. JENNINGS, Isaac, 1778–1874.
The tree of life; or, human degeneracy: its nature and remedy, as based on the elevated principle of orthopathy . . . New York: Miller, Wood, & Co., publishers, 1867.
    xvi, 9–279 p. ; 19 cm.

"The orthopathic analysis, developed first at Boston, then Oberlin, where Jennings spent the rest of his life, was performed within the bounds of standard Grahamite pathology. Drugs were stimulants, and thus wasted vitality and undermined the body's efforts to cure itself. Jennings took infinitely more space than that to present his philosophy, but in the end it was still nothing more than an extension of health reform from the dining room to the sick room. It was, however, carried out with an evangelistic flamboyance that exceeded anything else in its class. Orthopathy was presented with camp meeting fervor, held up as the only salvation from the satanic plot to ensnare humankind in stimulation . . . Jennings had an uneasiness about sexuality that was extreme even by Victorian measure; it was the 'Devil's stronghold, the very citadel of his empire on earth.' Thus orthopathy was nothing if it could not cure the illness of sexual passion . . . It was probably not the elimination of sexual pleasure nor even the communistic organization of his utopia, which prevented Jenning's ideas from capturing the public fancy. Orthopathy was handicapped most of all by not really being 'medicine.' It offered no drugs, promised no quick recoveries, left the patient almost entirely on his own. The several other antiheroic systems of treatment being promoted by midcentury fit much more smoothly with the popular conceptions of what medicine should do . . . " (James C. Whorton, *Crusaders for fitness,* Princeton, N.J., 1982, p. 135–36). Whorton cites hydropathy as an example of a competing "antiheroic system" more com-

patible with popular conceptions of health reform theory and therapeutic expectations.

Jennings believed in the perfectability of
man, i.e., in the possibility of rescuing mankind
from its present degeneracy. His vision of how
this would be accomplished extended beyond
obeisance to correct physiological principles.
Jennings combined health reform with spiritual
and social reform to restore man to the being
that God had made him. The tone of Jennings
writings may be gathered from the quotation
that follows: "There is no difficulty in collecting any amount of means for building splendid church edifices, colleges, ladies boarding
halls, and private Christian palaces, which
strengthen rather than weaken Satan's kingdom. But let an attempt be made to inaugu-
rate a psychological, physiological, or ethical
reform that would strike directly at the foundation of Satan's kingdom, and but few are found
ready to engage in it . . . But 'truth is mighty
and will prevail.' It is gradually working its way
into popular favor . . . in breaking up old foundations, and in the laying of new and better
ones; and it will eventually destroy the empire
of darkness, and establish the kingdom of righteousness and peace" (p. xii-xiii).

**2007. JENNINGS, Samuel Kennedy, 1771–
1854.**
The married lady's companion, or poor
man's friend. In four parts. I. An address
to the married lady, who is the mother of
daughters. II. An address to the newly married lady. III. Some important hints to the
midwife. IV. An essay on the management
and common diseases of children. To
which will be added a short note on fever
. . . Second edition, revised, corrected and
enlarged by the author . . . New-York: Published by Lorenzo Dow; J.C. Totten, printer,
1808.
    304 p. ; 13 cm.

"Part first" Jennings devotes to the physical
and moral upbringing of daughters. To that end
he establishes as a *fundamental proposition*:
"Health, morality, and religion, are mutually
and essentially dependent on each other. For
as sound health cannot be continued without
good morals, so neither can sound morals be
preserved without religious sentiment." In
other words, the virtues must be exercised as

well as the body. "Part second" is addressed to
the recent bride. She is advised on the "proper
conduct of the wife toward her husband, as well
as the signs of conception, the management of
pregnancy, and miscarriage. "Part third. Hints
for the midwife" describes "natural" and "preternatural labours," and the various accidents
attendant upon them. Jennings drew upon the
works of Thomas Denman for his discussions
of pregnancy and labor. "Part fourth," a manual
of domestic pediatrics, is the book's lengthiest
section, occupying 133 pages. The first edition
was published at Richmond, Va. circa 1808.

Jennings "studied medicine with his father,
Dr. Jacob Jennings; in 1818 he received an M.D.
(hon.) from the University of Maryland. He was
ordained a minister in the Methodist Episcopal Church, and in 1817 moved to Baltimore,
where he was president of Ashbury College,
1817–18; president of the Medical Society of
Baltimore, 1823–24; a founder of Washington
Medical College, Baltimore, in 1827 and professor in it of materia medica, 1827–9; professor of obstetrics, 1839–42; also professor of
anatomy in the Maryland Academy of Fine Arts,
1838–43. He lived in Tuscaloosa, Alabama, from
1845 to 1853" (Kelly & Burrage, *Dict. Amer. med.
biog.,* New York, 1928, p. 661).

REFERENCES: *Austin* #1055.

2008. JENNINGS, Walter Barry, 1873–1929.
The hygiene of pregnancy . . . New York:
Medical Review of Reviews, [ca. 1905].
    [10], 48, [2] p. : ill. ; 19 cm.

Jennings was an 1898 graduate of the New
York University Medical College.

2009. JENNINGS' SANITARY DEPOT.
JENNINGS' PATENT COMBINED VALVE
WATER CLOSET AND TRAP . . . The want
of a really good and efficient water
closet, not liable to get out of order, and
free from offence has long been felt—
The JENNINGS' WATER CLOSET . . .
SEWER GAS INTO THE HOUSE IS EN-
TIRELY AVOIDED . . . BEST CLOSET EVER
PRODUCED. The ATLANTIC MONTHLY
for Oct. 1875, illustrates these water clos-
ets and sets forth their advantages . . .
JENNINGS' SANITARY DEPOT . . . NEW
YORK. [ca. 1875–79].
    Broadside : ill. ; 27 × 19.5 cm.

JEWELL, Wilson see **PHILADELPHIA. BOARD OF HEALTH** (#2786).

2010. JEWETT, Charles, 1807–1879.
The temperance cause: past, present and future . . . Second edition. Hartford: Press of Case, Lockwood and Company, 1865.
vi, [7]-80 p. ; 22 cm.

Jewett's essay on the temperance movement, originally published at Chicago in 1864, examines the reasons for the "gradual decline and final death of most of those forms of organization and the general discontinuance of those modes of actions which God so abundantly blessed twenty-five years ago" (p. [7]). The lamented "modes of action" that he felt were so influential between 1825 and 1855 were the societies, committees, public assemblies, exhortations from the pulpit, and publications whose "aggregate influence revolutionized the public opinion, social customs and even the laws of the New England States" (p. [7]). Jewett urges a "temperance revival" spearheaded by the "Christian ministry" whose influence over the educated laity would gradually permeate every element of society (p. 71). In printed wrapper. Atwater copy lacking pages 79–80.
"In 1826 he [Jewett] issued for private circulation an address in verse to the town authorities of Lisbon, Conn., his place of residence, setting forth the iniquity of granting liquor licenses. A little later he attended a course of medical lectures at Pittsfield, Mass., and in 1829 began the practice of medicine in East Greenwich, R.I. . . . In 1837 . . . he gave up the practice of medicine to become agent for the Rhode Island State Temperance Society. His lectures were especially valuable and forcible at that time, since his medical training enabled him to treat the drink question in its scientific and physiological aspects . . . In 1840 he accepted the position of agent of the Massachusetts Temperance Union" (*Cyclopaedia of temperance and prohibition*, New York: Funk & Wagnalls, 1891, p. 262–63).

2011. JEWETT, Charles, 1839–1910.
Manual of childbed nursing with notes on infant feeding . . . Fifth edition revised and enlarged. New York: E.B. Treat & Company, 1902.
[2], 84 p. ; 19 cm.

"This manual was originally prepared for the Training School for Nurses at the Long Island College Hospital. It was subsequently rewritten and adapted to general use . . . The author ventures to hope that the book in its present form may be found of service not only to professional nurses but to mothers as well, and to all interested in obstetrics" (p. [1]). Originally published as a treatise of twenty-three pages in 1889 under the title *Manual of rules for child-bed nursing*, Jewett's revised manual appeared under its present title in 1891 (40 p.).
A native of Bath, Maine, Jewett received his medical degree from the College of Physicians & Surgeons (New York) in 1871. He settled and practiced in Brooklyn. In 1880 Jewett was appointed professor of obstetrics at the Long Island Hospital College, where he remained his entire career. He was a member of the American Gynecological Society (pres., 1900), the British Gynaecological Society, and was one of the founders of the International Congress of Obstetricians & Gynecologists. Jewett published more than forty papers and was the author of several standard textbooks of the period, including *Essentials of obstetrics* (1897), *The practice of obstetrics* (1899) and *Syllabus of gynecology* 1900).

2011.1. JEWETT, Frances Gulick, b. 1854.
The body and its defenses . . . Boston; New York; Chicago; London: Ginn and Company, [c1910].
viii, 342 p. : ill. ; 19 cm.

On the verso of this work's title-page, *The Gulick hygiene series* is described as consisting of a numbered five-book series and an unnumbered two-book series. (There was at least one additional title.) *The body and its defenses* is the second of the two-book series (*see also* #2013).
The complete series includes: I. *Good health* (1906, 1927); II. *Emergencies* [by C.V. Gulick] (1909, 1928); III. *Town and city* (1906, 1926); IV. *The body at work* (1909); V. *Control of the body* (1908, 1927); *The body and its defenses* (1910, 1927); *Health and safety* (a reissue of *Good health*, 1916, 1929); and *Physiology, hygiene and sanitation* (1916, 1927). The series was edited by Luther Halsey Gulick (#1451–1454).
Frances Gulick Jewett was born at Ponape in the Micronesian islands, where her parents were missionaries. She was the author of all but

one of the titles that constitute the *Gulick hygiene series* edited by her brother Luther Halsey Gulick and published by Ginn & Company. This firm was one of the foremost American publishers of elementary through college level textbooks. "By 1890," according to John Tebbell, "Ginn & Co. was the sixth largest educational publisher in the United States" (*A history of book publishing in the United States,* New York: R.R. Bowker, 1975, 2:411). The firm's proprietor, Edwin Ginn, was invited to join the conglomerate that constituted the ubiquitous American Book Company in 1890. Ginn declined, however, "preferring to be the independent entrepreneur he had always been" (*ibid.*).

2011.2. JEWETT, Frances Gulick, b. 1854.
The body at work . . . Boston; New York; Chicago; London: Ginn & Company, c1909.
    xvi, 247 p. : ill. ; 19 cm.

    Series: *The Gulick hygiene series, book 4.*

2012. JEWETT, Frances Gulick, b. 1854.
Control of body and mind . . . Boston; New York; Chicago; London: Ginn & Company, [c1908].
    xi, [1], 269, [5] p. : ill. ; 19 cm.

    Series: *The Gulick hygiene series, book 5.*

2013. JEWETT, Frances Gulick, b. 1854.
Good health . . . Boston; New York; Chicago; London: Ginn & Company, [c1906].
    [4], viii, 174, [6] p. : ill. ; 19 cm.

    The Gulick hygiene series was issued in a numbered five-book series and an unnumbered two-book series. *Good health* was the first title in the five-book series as well as the first of the two-book series (see #2011.1). It was re-issued in 1916 under the title: *Health and* safety (#2013.1).
    In his introduction to this title, Luther Halsey Gulick explains the series' educational intent: "The present volume gives detailed instruction in matters of personal health: what to do in caring for eyes, ears, teeth, finger nails, hair, etc.; why we keep clean; how to get pure air into a room and impure out of it; why this is needed, as proved by experiment, etc. In each case the child himself is made to demonstrate

the need. This method of instruction is indeed a dominant characteristic of the series as a whole. Each book has been prepared with the conviction that children are influenced by facts which result in definite courses of reasoning. Assure a child that unwashed people, crowded into unclean rooms, breathing impure air, and drinking impure water are more likely to be ill than people in clean rooms, breathing pure air, and drinking pure water, and he may or may not believe you; but explain to him the nature of those microbes which endanger life through water, air, and food; show by actual facts how the death rate has been raised and lowered; demonstrate by individual example the laws of contagion, and we shall convince the child by the same facts that have convinced his elders" (p. iv-v).

2013.1. JEWETT, Frances Gulick, b. 1854.
Health and safety . . . Boston . . . [*et al.*]: Ginn and Company, c1916.
    viii, 197 p. : ill. ; 19 cm.

    Originally published in 1906 under the title: *Good health* (#2013).

2014. JEWETT, Frances Gulick, b. 1854.
Physiology, hygiene and sanitation . . . Boston; New York; Chicago; London: Ginn and Company, [c1916].
    xiv, 367 p. : ill. ; 19 cm.

    Series: *The Gulick hygiene series* [unnumbered].

2015. JEWETT, Frances Gulick, b. 1854.
Town and city . . . Boston; New York; Chicago; London: Ginn & Company, [c1906].
    [2], viii, 278 p. : ill. ; 19 cm.

    Series: *The Gulick hygiene series, book 3.*

2016. JEWETT, Moses.
Jewett's family physician. The iatroleptic practice of medicine, or the curing of diseases principally by external application and friction . . . Columbus, Ohio: Published by the author, 1838.
    [12], [v]-xvii, [18]-534 p. ; 24 cm.

    Jewett's family physician is divided into four parts. The first describes the liniments, salves, ointments and washes that Jewett both prepared and applied in the treatment of disease. His therapeutic system hangs on his theory of absorption. Jewett maintains that there are two sets of absorbing vessels in the body. The internal absorbent vessels arise from the "alimentary canal," particularly the intestines. This is the mode in which food and other substances are introduced into the body. The second set of absorbing vessels are to be found on "every part of the body—from the whole of its external surface" (p. x). It is through cutaneous absorption that Jewett proposes to introduce "the most powerful therapeutic agents" into the body. He assures his readers that "our remedies, composed of stimulants, tonics, and diaphoretics, externally applied, have produced the most sudden relief to the oppressed organs; the general circulation is promoted; the collected mass of impurities is thrown off by perspiration; the patient is relieved without the occurrence of a tedious state of convalescence" (p. xiv), and without the dangers that often accompany the internal absorption of medicines. In Part II Jewett provides nearly 100 pages of testimonials to the efficacy of his preparations. Part III is devoted to the anatomy and physiology of the human body. Part IV describes chemical, physiological and medical topics that Jewett could did not conveniently fit into the preceding sections.

JEWETT, Paul see **NEW-ENGLAND FARRIER** (#2582).

JOHNSON, Alice A. see *THE STANDARD BOOK OF RECIPES* (#3322).

2017. JOHNSON, Edward, Sir, 1785–1862.
The domestic practice of hydropathy . . . New York; London: John Wiley, 1849.
xlvii, [48]-467 p. : ill. ; 20 cm.

First American edition; originally published at London in 1849. Wiley reissued this title in 1852. In 1854 and again in 1856, it was published under the imprint of New York firm of Fowlers & Wells. Johnson's manual of domestic hydropathy "contains, first, a very minutely detailed description of the various hydropathic processes, and directions as to the proper manner of performing them; with an enumeration of the several kinds of baths in use—their com-

parative powers, their individual effects, their temperature, the manner and times of taking them; observations regarding diet generally, clothing generally, sleep generally, and exercise generally . . . Secondly, it contains general observations on the hydropathic treatment; its mode of action on the living system; with remarks on the nature of general and local disease. Thirdly, it contains a detailed description of the symptoms by which each disease is recognized, with its appropriate treatment; and particular directions as to diet, exercise, clothing, &c." (p. [v]). Johnson states that one of his objects in writing this manual was "to bring the benefits of hydropathy, as much as possible, within the reach of the poor," as well as to "another large class of persons, to whom the advantages of the water treatment have been hitherto a dead letter, on account of their inability to leave their homes, by reason of the pressing claims of business" (*ibid.*). Johnson feels that a manual of this sort "is also rendered doubly necessary, by the difficulty which patients experience of getting any medical man in their neighborhood to consent to treat them on the hydropathic system, although those patients have no faith in any other" (p. vi).

2018. JOHNSON, Edward, Sir, 1785–1862.
Results of hydropathy; or, constipation not a disease of the bowels: indigestion not a disease of the stomach; with an exposition of the true nature amd causes of these ailments, explaining the reason why they are so certainly cured by the hydropathic treatment. To this are added cases cured at Stanstead Bury House, with observations on the treatment generally . . . New York: Wiley and Putnam, 1846.
vii, [1], 181, [3] p. ; 20 cm.

First American edition; originally published at London in 1846. In his opening chapter Johnson poses several important questions regarding the hydropathic system: "What are those particular diseases to which this particular treatment is most applicable? and what, in these cases, is its *amount* of curative power? Is it greater or less than that of drugs? and what advantages has this mode of cure over that of the old method, in those diseases which are capable of cure by *either*?" (p. 3). Several pages later he states: "From my own observation and experience, I believe the hydropathic treatment . . . to be, in its nature, essentially tonic and alterative—*permanently* tonic, because it produces its tonic effects by filling the system with abundance of new and healthy blood—by strengthening the nervous system. and muscular fibre of the heart, and by constringing the capillary blood-vessels, and thereby strengthening the whole circulating system. These are effects which, when once produced, must be permanent, and not temporary—like the tonic effects usually produced by drug-tonics . . . And it is also *alterative,* because it exercises a remarkable influence in promoting and restoring all the secretions, especially those of skin and bowels" (p. 5–6)—and hence its efficacy in the treatment of constipation and indigestion. A student of Priessnitz, Johnson operated a water-cure at Stanstead Bury House, near Ware, Herts.

2019. JOHNSON, Edward, Sir, 1785–1862.
Results of hydropathy; or, constipation not a disease of the bowels: indigestion not a disease of the stomach; with an exposition of the true nature and causes of these ailments, explaining the reason why they are so certainly cured by the hydropathic treatment. To this are added cases cured at Stanstead Bury House, with observations on the treatment generally . . . New York; London: John Wiley, 1849.
    [2], vii, [1], 181 p. ; 20 cm.

    The Atwater copy is inscribed by Silas Orsemus Gleason and his wife Rachel Brooks Gleason (#1362–1367), proprietors of the Elmira Water Cure (#1063, 1064). This title made its final appearance under the imprint of Fowlers & Wells in 1854.

JOHNSON, Harriet Merrill see **FORBES** (#1258).

2020. JOHNSON, Isaac D., 1827–1911.
A guide to homoeopathic practice; designed for the use of families and private individuals . . . Philadelphia: F.E. Boericke, Hahnemann Publishing House, 1879.
    xv, [16]-494 p. ; 23 cm.

    First edition. Johnson's manual of homeopathic domestic practice was published in the United States for more than four decades. It

appeared in ten American editions between 1879 and 1921, including a French translation published at New York in 1884. The author was an 1852 graduate of the Homoeopathic Medical College of Philadelphia.

2021. JOHNSON, Isaac D., 1827–1911.
A guide to homoeopathic practice; designed for the use of families and private individuals . . . Philadelphia: F.E. Boericke, Hahnemann Publishing House, 1880.
    xv, [16]-494 p. ; 23 cm.

2022. JOHNSON, Isaac D., 1827–1911.
A guide to homoeopathic practice; designed for the use of families and private individuals . . . Philadelphia: F.E. Boericke, Hahnemann Publishing House, 1885.
    xv, [16]-494 p. ; 23 cm.

2023. JOHNSON, James, 1777–1845.
Change of air, or the philosophy of travelling; being autumnal excursions through France, Switzerland, Italy, Germany, and Belgium. With observations and reflections on the moral, physical, and medicinal influence of travelling-exercise, change of scene, foreign skies, and voluntary expatriation. To which is prefixed, wear and tear of modern Babylon . . . New York: Samuel Wood and Sons, 1831.
    viii, 326, [2] p. ; 24 cm.

    First and only American edition of a work that appeared in four London editions between 1831 and 1837. A "change of air" was frequently recommended during the 19th century by physicians whose aristocratic and upper middle class patients suffered from any number of disorders commonly labelled diathetic, marasmic, consumptive, or wasting; and who could afford the luxury of months or even years regaining health and strength on a Continental tour. Johnson considers the salutary effects of travel, including separation from the ordinary business of life; the salubrity of specific climates on specific physical and emotional disorders; the benefits that result from the exercise required by carriage travel, horseback riding or sightseeing; the psychological effects of varied and beautiful scenery; as well as the dangers and diseases encountered during travel and in various places. Johnson's *Change of air* gives particular attention to Italy, the climate of which was frequently recommended for invalids.

Johnson was one of the earliest to discourage consumptives from settling in Rome, the climate of which he regarded as fatal to the phthistic.

At age fifteen Johnson was apprenticed to a surgeon-apothecary in Antrim Co., Northern Ireland. After passing examinations at Surgeons' Hall (London) in 1798, he commenced a career as a naval surgeon, serving in both peace and war, and in waters as diverse as the Gulf of St. Lawrence and the Bay of Bengal. At the end of the Napoleonic wars, Johnson was placed on half-pay. He settled in Portsmouth, and then in London, where he began his remarkable literary output, including editorship of the *Medico-chirurgical review*. In 1821 he received his medical degree from St. Andrews, and was admitted a licentiate of the College of Physicians (London). In addition to the half dozen popular medical books Johnson wrote, he was the author of several works for the profession, most notably *The influence of tropical climates, more especially the climate of India, on European constitutions* (1813), which is based upon his experiences in the East.

2024. JOHNSON, James, 1777–1845.
The economy of health; or, the stream of human life, from the cradle to the grave. With reflections, moral, physical, and philosophical, on the septennial phases of human existence . . . New-York: Harper & Brothers, 1837.
   xii, [13]-283, [1] p. ; 16 cm.

First American edition. Johnson remarks on the existing hygienic literature: "The various 'arts of prolonging life,' and the ponderous 'codes of health and longevity,' though read by many, have been remembered by few—and practised by still fewer" (p. [iii]). The reason lies in the failure of these works to consider how man's physical being differs at different periods of his life. In *The economy of health* Johnson divides human life into seven-year periods or *septenniads,* and considers the differing factors required to maintain health during each period. Originally published at London in 1836, *The Economy of health* appeared in three American editions, and remained in print until 1858.

2025. JOHNSON, James, 1777–1845.
The economy of health; or, the stream of human life, from the cradle to the grave. With reflections, moral, physical, and philosophical, on the septennial phases of human existence . . . Third edition. New-York: Harper & Brothers, 1840.
   xii, [13]-283 p. ; 16 cm.

This third New York edition was re-issued by Harper & Brothers in 1854 and 1858.

2026. JOHNSON, James, 1777–1845.
An essay on indigestion; or morbid sensibility of the stomach and bowels, as the proximate cause or characteristic condition of dyspepsy, nervous irritability, mental despondency, hypochondriasis, and many other ailments of body and mind. To which are added, observations on the diseases & regimen of invalids, on their return from hot and unhealthy climates . . . Third Philadelphia, from the sixth London edition, much enlarged. Philadelphia: Nathan Kite, 1831.
   xii, [13]-194 p. ; 19 cm.

First published at London in 1826 under the title: *An essay on the morbid sensibility of the stomach and bowels* (1st American ed., 1828, #2027). With the sixth London edition (1829), the title changed to *An essay on indigestion.*

2027. JOHNSON, James, 1777–1845.
An essay on morbid sensibility of the stomach and bowels as the proximate cause or characteristic condition of indigestion, nervous irritability, mental despondency, hypochondriasis, &c. &c. To which are prefixed, observations on the diseases and regimen of invalids, on their return from hot and unhealthy climates . . . Philadelphia: Published by Thomas Kite, 1828.
   viii, [8]-155 p. ; 17 cm.

First American edition. "In this Essay I have endeavoured to investigate the operation of moral causes on the digestive organs, more minutely than has generally been done; and to trace, with more care, the re-action of these organs on the mental faculties. The amount of suffering which is inflicted on the body through the agency of the mind, is only equalled by the retributive misery reflected on the mind through the medium of the body. The play of affinities and reciprocity of sympathies between the intellectual and material portions of our nature, have not been sufficiently attended to in the investigation and management of diseases" (p. iv). Johnson continues, "above all, I have endeavoured to demonstrate the true principles on which the plan of diet and regimen should be constructed, not only in indigestion, but in a host of mental and corporeal discomforts which are little suspected of having their origin in the stomach" (p. v). This essay was originally published at London in 1826. It was issued both separately and as part of the fourth edition of Johnson's *The influence of tropical climates on European constitutions* (London, 1827, p. [553]-680). With the sixth London edition (3rd American ed., #2026), the work was expanded and the title changed to *An essay on indigestion.*

2028. JOHNSON, James, 1777–1845.
The influence of civic life, sedentary habits, and intellectual refinement, on human health, and human happiness; including an estimate of the balance of enjoyment and suffering in the different gradations of society . . . First American from the London copy. Philadelphia: Printed for Thomas Hope; published by Mathew Carey and Son, Thomas Dobson, and Moses Thomas, 1820.
viii, [9]-110 p. ; 23 cm.

First American edition (2nd American ed., 1820; 1st London ed., 1818). Johnson divides the body into three systems: *organic* (circulation, respiration & digestion), *animal* (voluntary muscular acitivity) and *sentient* (brain & nerves); and emphasizes their influence on one another, e.g., the effect of a toxin ingested into the stomach on the *animal* and *sentient* systems; the influence of a "sudden gust of passion" on the *organic* and *animal* systems, etc. In each brief chapter, Johnson considers the influence of "civic life, sedentary habits and intellectual refinement" on the three systems, and considers such factors as food, drink, air, exercise, sleep, etc. on their health or derangement. He frequently turns to a theme common to all his writings on health, i.e., the central role that the digestive organs play in the health of the whole and its parts.
REFERENCES: *Austin* #1069.

2029. JOHNSON & JOHNSON.
Johnson's first aid manual. Suggestions for prompt aid to the injured in accidents and

emergencies . . . Edited by Fred B. Kilmer. New Brunswick, N.J.: Published by Johnson & Johnson, 1901.
119 p. : ill. ; 22 cm.

First edition. Fred[erick] Barnett Kilmer (1851–1934) was a pharmaceutical chemist who became director of laboratories at Johnson & Johnson in 1889.

2030. JOHNSON & JOHNSON.
Johnson's first aid manual. Suggestions for prompt aid to the injured in accidents and emergencies . . . Edited by Fred B. Kilmer. Third edition. New Brunswick, N.J.: Published by Johnson & Johnson, 1903.
119 p. : ill. ; 22 cm.

2031. JOHNSON & JOHNSON.
Johnson's first aid manual. Suggestions for prompt aid to the injured in accidents and emergencies . . . Edited by Fred B. Kilmer . . . Fourth edition revised. New Brunswick, N.J.: Published by Johnson & Johnson, 1909.
127 p. : ill. ; 22 cm.

2032. JOHNSON & JOHNSON.
Johnson's first aid manual. Suggestions for prompt aid to the injured in accidents and emergencies . . . Edited by Fred B. Kilmer . . . Fifth edition revised. New Brunswick, N.J.: Published by Johnson & Johnson, 1912.
143, [1] p. : ill. ; 22 cm.

2033. JOHNSON & JOHNSON.
Johnson's first aid manual. Suggestions for prompt aid to the injured in accidents and emergencies . . . Edited by Fred B. Kilmer . . . Sixth edition revised. New Brunswick, N.J.: Published by Johnson & Johnson, 1914.
143, [1] p. : ill. ; 22 cm.

2034. JOHNSON & JOHNSON.
Johnson's first aid manual. Suggestions for prompt aid to the injured in accidents and emergencies . . . Edited by Fred B. Kilmer . . . Seventh edition revised. New Brunswick, N.J.: Published by Johnson & Johnson, [c1916].
143, [1] p. : ill. ; 22 cm.

2035. JOHNSTON, Bertha, b. 1864.

Eat and grow fat . . . A handy and efficient guide to the most approved methods of restoring flesh, including menus, potent and palatable. New York: The Sherwood Company, [c1917].

[2], 106, [2] p. ; 19 cm.

2036. JOHNSTON, James Finlay Weir, 1796–1855.

The chemistry of common life . . . Twelfth edition. New York: D. Appleton and Company, 1865.

2 v. (vii, [1], [3]-291, 4 p. ; 381, [1], 2 p.) : ill. ; 19 cm.

When Durham University was founded in 1833, the readership in chemistry and mineralogy was bestowed on Johnston, an appointment he retained until his death. As an author, Johnston is known as the popularizer of recent scientific discoveries and their practical application to agricultural, manufacturing, physiology & hygiene, etc. In his introduction Johnston writes that *The chemistry of life* (1st edition, Edinburgh, 1853–55) "treats . . . of THE AIR WE BREATHE and THE WATER WE DRINK, in their relations to human life and health—THE SOIL WE CULTIVATE and THE PLANT WE REAR, as the sources from which the chief sustenance of all life is obtained—THE BREAD WE EAT and THE BEEF WE COOK, as the representatives of the two grand divisions of human food—THE BEVERAGES WE INFUSE . . ." etc. Though the book was also intended for a general readership, it enjoyed its greatest success as a school text on the chemistry of human foodstuffs. It was published for the final time in 1906. The Atwater set bears the inscription of Henry Foster (#1265), proprietor of the Clifton Springs Water-Cure.

2037. JOHNSTON, William.

The good Samaritan; or, sick man's friend: containing the botanic medical practice necessary for the removal of all curable forms of disease, in strict accordance with the soundest principles of philosophy and common sense: intended as a pocket companion, for Thomsonians, and all others who would wish to prevent, or cure their own diseases . . . Philadelphia: Stereotyped by L. Johnson, 1841.

287, [1] p. ; 14 cm.

"In presenting the following pages to his numerous friends and patrons, the compiler wishes it to be distinctly understood, that it is not his intention that this book should supersede the works of the venerable father and founder of the Thomsonian System of Medical Practice; but, on the contrary, that it may be the means of opening the people's eyes to the excellency, utility, and superiority of *that system* over every other yet discovered" (p. 3). Nonetheless, Johnston's book could be had for one dollar, while Thomson's came with a patent that sold for twenty dollars. Johnston, a resident of Chester County, Pa., places the initials T.B.P. (Thomsonian Botanic Practitioner) after his name.

CONTENTS: Introduction; I. On medicine; II. Remarks on blood-letting by Professor Terry and Dr. Lobstein, showing the inconsistency and barbarity of such an inhuman practice; III. The value of steam in the removal of disease; chiefly extracted from the second and third volumes of the *Thomsonian Recorder* [#3518]; IV. The system on which Dr. Thomson's mode of administering medicine is founded; V. A description of the most common forms of disease, with the mode of treatment according to the true Thomsonian principles; VI. The materia medica and Thomsonian's pharmacopoeia; VII. Containing an abstract of the common medical treatment of disease, compiled from the works of the most celebrated men in the profession, both in Europe and America, by a regular physician. The Appendix includes a description of the mineral & vegetable poisons used in regular practice, e.g., mercury, digitalis purpurea, etc.; "Cased cured," i.e., Johnston's therapeutic successes; "Midwifery," extracted from the *Botanic sentinel*; a glossary, index and a list of references in Lancaster, Chester, Delaware, Philadelphia and Hartford counties.

2038. JOHONNOT, James, 1823–1888.

How we live: or, the human body, and how to take care of it. An elementary course in anatomy, physiology, and hygiene. By James Johonnot and Eugene Bouton . . . Revised and approved by Henry D. Didama . . . New York: D. Appleton and Company, 1886.

178, [2] p. : ill. ; 18 cm.

Johonnot was the author of numerous juvenile texts on zoology, geography, physiology,

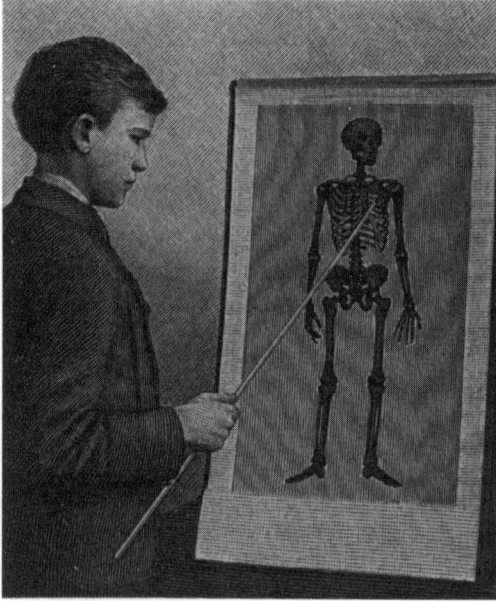

Fig. 55. From Blaisdell's How to keep well, *#343.*

hygiene and history. He collaborated with Eugene Bouton (b. 1850) on both *How we live* (1st ed., 1884) and its successor, *Lessons in hygiene* (#2039). Series: *Authorized physiology series, no. 2.*

2039. JOHONNOT, James, 1823–1888.
Lessons in hygiene or the human body, and how to take care of it. The elements of anatomy, physiology, and hygiene, for intermediate grades. Being an edition of "How we live," revised to comply with the legislation requiring temperance instruction in schools. By James Johonnot and Eugene Bouton . . . New York; Cincinnati; Chicago: American Book Company, [c1889].
    213, [11] p. : ill. ; 19 cm.

2040. JONES, George Howard, 1843–1910.
The science of life; or, self-preservation. A medical treatise on nervous and physical debility, spermatorrhoea, impotence, and sterility, (the gold medal prize essay,) with practical observations on the treatment of diseases of the generative organs . . . Latest edition. Boston: Published by authority of the Peabody Medical Institute, [c1881].

[iii]-xxxvi, 335 p., [2] leaves of plates (1 folding) : facsim., ports. ; 17 cm.

Jones' text in *The science of life* is identical to that first published by Albert H. Hayes in 1868 under the same title (#1587–1591). The differences between the two texts is Jones' inclusion of an "introductory lecture," and the transposition of the chapter on venereal diseases from the beginning of the volume to the end. Jones was an 1864 graduate of Harvard Medical School. Since 1872 he had been chief consulting physician and surgeon at the Peabody Medical Institute (#2758) established in Boston by A.H. Hayes. Later editions of this work bear the name of William H. Parker on the title-page (#2742–2745).

JONES, Gordon W. see **MATHER** (#2389).

JONES, Henry Webster see **LYMAN** (#2302–2306).

2041. JONES, Jesse H.
Scientific marriage, a treatise founded upon the discoveries and teachings of William Byrd Powell, M.D. . . . New York: Murray Hill Publishing Company, [c1887].
    53, [7] p. : ill. ; 17 cm.

"True" or "scientific marriage," according to Jones, requires a "harmony of the temperaments" among partners. He divides temperaments into *vital* (bilious, sanguine) and *non-vital* (encephalic, lymphatic), and maintains that compatible marriages and healthy offspring result from the union of vital and non-vital temperaments, but never from the union of identical temperaments. Jones derived these categories from the doctrine of William Byrd Powell (1799–1866), the eclectic physician, phrenologist and "cerebral physiologist" who earlier taught that marriage between persons with identical temperaments is incestuous, and results in physically and/or mentally defective children. "Prof. Wm. Byrd Powell discovered the supreme secret of human life on this earth," according to Jones," and this discovery in our view was the most important that has been made in modern times. He discovered the secret by which it is practicable so to cleanse the vitality itself of the race that disease will cease from the earth, and all heinous wickedness from among the children of men" (p. [21]).

2042. JONES, May Farinholt, b. 1868.
Keep-well stories for little folks . . . Illustrated by Pauline Wright . . . Philadelphia & London: J.B. Lippincott Company, [c1916].
viii, 140 p. : col. ill. ; 20 cm.

First edition. A charmingly illustrated book for children in the lower grades, upon whom the author hopes to impress "the truths of hygienic living and habits" through the "Land of Story Books." The spector of tuberculosis haunts several chapters, e.g.: "You have seen some plants that you were told not to handle or taste because they were poisonous. Well, these little tuberculosis plants that I am telling you about are more poisonous than the plants you can see. If they get on cups from which you drink, and into your milk or any other food, they may get into your bodies. If you think, I am sure that you will remember some of your friends who have consumption" (p. 38). The third and final edition of *Keep-well stories for little folks* was published in 1930.

JONES, W. see **CORNARO** (#798).

JORDAN, Henry Jacob see **JORDAN, Louis J.** (#2045); **JOURDAIN** (#2051); **DR. KAHN'S MUSEUM OF ANATOMY, SCIENCE AND ART** (#2066); **NEW YORK MUSEUM OF ANATOMY** (#2598).

JORDAN, Johnson H. see **GUNN** (#1471–1475).

2043. JORDAN, Louis J.
Man's mission on earth! A treatise on nervous debility and physical exhaustion, being a synopsis of lectures delivered at the Museum of Anatomy, Science and Art . . . New York . . . New York, [188?, c1871].
x, [11]-181, [3] p. ; 15 cm.

Man's mission on earth is a title that was given to several volumes of collected lectures delivered at anatomy museums in Boston, New York and Philadelphia operated by members of the Jordan-Kahn clans and their partners. However, the volumes published under this title by L.J. Jordan, R.J. Jourdain (#2051, 2052) and R.J. Kahn (#2062–2065) differ in content.

At the same time, the content of L.J. Jordan's *Man's mission on earth* is identical to that of *Manhood; or secrets revealed* (#2173), authored by LaGrange and Jordan: I. Anatomy and physiology of the generative organs; II. General debility; III. Spermatorrhoea, impotency, sterility; IV. Treatment of spermatorrhoea, etc. V. Marriage: its obligations and excesses; VI. Diseases of the generative organs; VII. Special diseases; VIII. Self-diagnosis; IX. Notes from our case book, Cases of complaints, To patients and invalid readers.

2044. JORDAN, Louis J.
Man's mission on earth! A treatise on nervous and muscular debility and physical exhaustion, with practical information on diseases arising from indiscretion, addressed to youth, manhood and old age. Being a synopsis of lectures delivered at Kahn's Anthropological Museum and Gallery of Illustrations . . . [New York, ca. 1890–93, c1871].
x, [11]-181 p. ; 15 cm.

2045. JORDAN, Louis J.
The philosophy of marriage, being four important lectures on the functions and disorders of the nervous system and reproductive organs. Illustrated with cases . . . San Francisco, Cal., [c1865].
vi, 99, [3] p. ; 18 cm.

Louis J. Jordan was the principal or proprietor of the Pacific Museum of Anatomy, located in the Eureka Theater on Geary Street, opposite San Francisco's Union Square. A wood engraving on the final [unnumbered] leaf of this volume provides an interior view of the establishment. Jordan claims to have a medical degree from Edinburgh, to be a member of the Royal College of Surgeons of London (the certificate "hangs in my Surgery"), and to have twenty years' experience in the medical profession. It seems likely that Louis J. Jordan was the son of Henry Jacob Jordan and probably a brother of Robert J. Jordan and Philip J. Jordan (#2048). Though the R.C.S. qualification has not been verified for Louis J. Jordan, it has been for Robert J. Jordan, making Louis' claim more plausible. In 1872, Louis J. Jordan received an honorary medical degree from the Eclectic Medical College of New York. All the

Jordans seem to have emigrated from England prior to 1865. Louis did not long remain in San Francisco. He appears in the New York City directories in 1876 and thereafter, apparently replacing the aging Henry Jacob Jordan in the family enterprises there.

Contrary to the statement on the title-page, *The philosophy of marriage* consists of eight (not four) lectures delivered at the Pacific Museum of Anatomy: I. Male organs of generation; II. Female organs of generation; III. Functions of the generative organs. IV. Philosophy of marriage. V. Abnormal conditions of the generative organs. VI. Spermatorrhoea. VII. False delicacy. VIII. Special diseases (i.e., venereal disorders). IX. Self-diagnosis (five case reports illustrating the effects of masturbation), Instructions to invalids. The statistical data on marriage are almost entirely from England, suggesting that this book (supposedly in its 48th edition, *v.* p. 3) was probably written there.

2046. JORDAN, Louis J.
The philosophy of marriage. Being four important lectures on the functions and disorders of the nervous system and reproductive organs. Illustrated with cases . . . San Francisco, Cal., [c1665, i.e., 1865].
   2 pts. in 1 v. (vi, [7]-125, [3] p.; 59, [3] p.) : ill. ; 15 cm.

This edition of *The philosophy of marriage* is issued with the *Hand-book & descriptive catalogue of the Pacific Museum of Anatomy and Natural Science* (59 p.), of which Jordan was proprietor. The catalog lists 993 exhibits intended "to promote the knowledge and morality of the general public—to act as a beacon to the young, who, ignorant of the dangers attending upon many of their thoughtless actions, are too prone, in their youthful impetuosity, to cultivate tastes and habits injurious and prejudicial" (p. 4). Jordan displayed anatomical and embryological specimens (including human monstrosities), as well as botanical and zoological preparations. He restricted access to the Pathological Room to "medical gentlemen and students only." "In this porton of the Museum," he states, "will be found natural preparations and models in wax, illustrating almost every ill which the flesh is heir to" (p. 50). Most of

these exhibits reveal the pathological effects of masturbation and sexually transmitted diseases in men and women. Having suitably shaken his visitors, Jordan provides the hours that he his available at his home-office on the catalog's final leaf ("for the privacy and convenience of those desirous of consulting him"). A wood-engraved view of the interior of the Pacific Museum of Anatomy and Natural Science serves as the frontispiece to *The philosophy of marriage* (*see fig. 56*).

2047. JORDAN, Louis J.
The philosophy of marriage. Being important lectures on the functions and disorders of the nervous system and reproductive organs. Illustrated with cases . . . San Francisco, Cal.: [Dr. Jordan & Co., c1893]. 188, [4] p. ; 14 cm.

The text is identical to that of the 1865 edition (#2045, 2046), with the addition of a *Handbook and catalogue of Dr. Jordan's Museum of Anatomy and Natural Science* (p. [125]-188), detailing the 993 artifacts on display (a wood engraving of the building's exterior appears on the back of the printed wrapper). Whether the museum had operated continuously since 1865 under local management, or whether Jordan returned to San Francisco after the New York Museum of Anatomy (#2598) was closed down in 1888 is undetermined.

JORDAN, Louis J. see also **EUROPEAN ANATOMICAL, PATHOLOGICAL AND ETHNOLOGICAL MUSEUM** (#1081–1083); **KAHN, Louis J.** (#2060); **KAHN, Robert J.** (#2062–2065); **LAGRANGE** (#2173–2175); **DR. KAHN'S MUSEUM OF ANATOMY, SCIENCE AND ART** (#2066); **NEW YORK MUSEUM OF ANATOMY** (#2598).

2048. JORDAN, Philip J.
The philosophy of marriage. Its duties and obligations. Eight important lectures on the functions and disorders of the nervous system, reproductive organs and special diseases, as delivered at the Parisian Scientific Institution, and Museum of Anatomy . . . Boston, [c1871].

Fig. 56. Interior view of L.J. Jordan's Pacific Museum, from The philosophy of marriage, *#2046.*

xi, [1], 178, [2] p. : ill. ; 14 cm.

The ubiquitous Jordan family may be characterized by their operation of museums of dubious educational intent, their interest in the treatment of sexual diseases and dysfunctions, their remarkable mobility, and a willingness to share manuscripts with one another. P.J. Jordan's *The philosophy of marriage* is an example of the last mentioned attribute; its text is identical to the book of the same title attributed to Louis J. Jordan (#2045–2047). P.J. Jordan received an honorary degree from the Eclectic Medical College of New York in 1872. Both before and after receiving this degree, Jordan was involved in the operation of the family's museums and medical practices in Philadelphia, New York and Boston. At the time this work was published, P.J. Jordan was "principal" of the Parisian Scientific Institution and Museum of Anatomy at 595 Washington St., Boston, and was available for consultation at his private rooms, 19 Joy St., Boston. In printed wrapper.

JORDAN, Philip J. see also **EUROPEAN ANATOMICAL, PATHOLOGICAL AND ETHNOLOGICAL MUSEUM** (#1081–1083); **JORDAN, Louis J.** (#2045); **DR. KAHN'S MUSEUM OF ANATOMY, SCIENCE AND ART** (#2066); **NEW YORK MUSEUM OF ANATOMY** (#2598)

JORDAN, Robert Jacob see **EUROPEAN ANATOMICAL, PATHOLOGICAL AND ETHNOLOGICAL MUSEUM** (#1081–1083); **JORDAN, Louis J.** (#2045); **JOURDAIN** (#2051, 2052); **DR. KAHN'S MUSEUM OF ANATOMY, SCIENCE AND ART** (#2066); **LA GRANGE** (#2173–2175); **NEW YORK MUSEUM OF ANATOMY** (#2598); **PARISIAN GALLERY OF ANATOMY** (#2734, 2735).

DR. JORDAN'S MUSEUM OF ANATOMY AND NATURAL SCIENCE see **JORDAN** (#2047).

2049. JOSLIN, Benjamin Franklin, 1796–1861.
Homoeopathic treatment of epidemic cholera . . . Third edition, with additions. New York: Published by William Radde . . . [*et al.*], 1854.
    252 p. ; 20 cm.

Originally published in 1849, Joslin's treatise "has been used as a guide in the treatment of cholera, not only by the profession, but by many intelligent laymen. In this rapid disease, for which our system affords the only effectual remedies, many families in localities where no homoeopathic physician can soon be procured, will find it necessary at last [least?] to commence the treatment; and some members of the profession who had been previously allopathic will be induced to prescribe homoeopathically for cholera . . . The plain rules for prevention, preliminary treatment and nursing, may either be consulted directly by families, or employed by their physician as a safe and convenient basis for his instructions" (p [7]).

Joslin points out that according to statistics compiled by the Board of Health in New York City, nearly 54% of those admitted to allopathic hospitals died during the cholera epidemic of 1849. He cites statistics compiled by the local homeopathic society during the same epidemic indicating a 15% mortality rate among patients treated by homeopathic physicians (p. 81–83).

2050. JOSSELYN, Joseph Hollis, b. 1820?
Dr. Josselyn's private medical companion, embracing the treatment of menstruation, or monthly turns, during their stoppage, irregularity, or entire suppression. Pregnancy, and how it may be determined; with the treatment of its various diseases. Discovery to prevent pregnancy: its great importance and necessity where malformation or inability exists to give birth. To prevent miscarriage or abortion. When proper and necessary to effect miscarriage. When attended with entire safety. Causes and mode of cure of barrenness, or sterility. Masturbation, its effects, treatment, &c. &c. . . . First edition. Boston, 1853.
    vi, [9]-300, [2] p. : ill. ; 16 cm.

This is a composite work, i.e., the text of R.J. Culverwell's *On single and married life* (#847) is combined with substantial new material written by J.H. Josselyn, who advertises himself as a medical consultant to those seeking advice on sexual matters or who suffer from sex disorders, sexually transmitted diseases, etc. 46% of the text is from Josselyn's pen: pages 80–117 contain an anatomy of the organs of generation, remarks on coition, the physiology of impregnation and the signs of pregnancy; pages 133–48 contain a chapter entitled "On choice in marriage phrenologically considered"; and pages 217–98 include a mixture of testimonial letters and case histories affirming the author's success in the treatment of sex-related disorders. Title on printed wrapper: *The Private medical companion . . . Embracing the causes and treatment of all complaints peculiar to the generative system.*

The wording of the title-page is identical to that of the sex manuals of A.M. Mauriceau (#2398–2407) published in the late 1840s and 1850s. Mauriceau's text is entirely different, however. Josselyn advertises himself in this work as consulting physician at the Howard Botanic Infirmary in Boston. He is listed in the Boston city directories for 1850 and 1852, however, as a "broker." In the 1853 directory, the year his *Private medical companion* was published, Josselyn is listed as a physician. He does not appear in the 1854 or 1855 Boston directories.

In 1866 Josselyn published *"Man know thyself" A treatise upon sexual and other diseases* with a San Francisco imprint. Joseph H. Josselyn first appears in the 1861/62 San Francisco directory as physician at the Electropathic Institute, 645 Washington Street; and is listed for the last time in the city directory in 1904.

2051. JOURDAIN, Robert J.
Man's mission on earth: being a series of lectures at Dr. Jourdain's Parisian Gallery of Anatomy, addressed to those laboring under the baneful effects of self-abuse, excesses, or infection. Also, a familiar explanation of the venereal disease, showing the danger arising from neglect or improper treatment in disorders of the generative system . . . Boston: Rockwell & Churchill, printers, 1871.
    152 p. ; 15 cm.

Dr. Jourdain, before and after his thirteen years of practice in Boston, was Robert Jacob

Jordan (1835–1909), a member of the Jordan family whose members included Henry Jacob Jordan, Louis J. Jordan (#2043–2047) and Philip M. Jordan (#2048), who operated anatomy museums in New York, Philadelphia and San Francisco, as well as in Boston. The francophonic twist to Jordan's (*Jourdain's*) name must have seemed appropriate for a "Parisian Gallery."

Records of the New Synagogue (London, Eng.) list a Robert Jacob Jordan born 27 November 1835. The name also appears in London medical directories from 1859 through 1863, but not thereafter. R.J. Jordan is listed as licentiate of the Royal College of Physicians (Edinburgh, 1859) and a member of the Royal College of Surgeons (London, 1859)—supporting the credentials to which he lays claim in the books he published in America. In 1863, however, Robert Jordan was "deprived" of these licenses, first by the R.C.S. and then by the R.C.P.E. At the 3 Nov. 1863 meeting of the latter body, Jordan was expelled "for conduct unbecoming the character of a physician, in publishing, or causing to be published, an indecent work, entitled 'The Illustrated and Descriptive Catalogue of the Subjects contained in the London Anatomical Museum; to which is annexed the Guide to Masculine Vigour, by a Physician'" (*The Lancet*, 1863, 2:572). Jordan's name was expunged from the *Medical register.*

Soon after this event, the Jordan family emigrated to the United States. Robert established himself in Boston, with home and office at 51 Hancock Street between 1867 and 1870, and at 61 Hancock Street between 1871 and 1881. His Parisian Gallery of Anatomy and Medical Science (#2734, 2735) was located at 307 Washington Street. In 1877–78 Jordan attended lectures at the Eclectic Medical College of New York, along with Louis J. Jordan (#2043–2047) and Robert J. LaGrange (#2173–2175). R.J. Jordan probably received his degree from the E.M.C. in 1881—circa the period *Dr. Jourdain* removed to New York and thence to Philadelphia, where he and LaGrange took over operation of the European Anatomical, Pathological and Ethnological Museum (#1081–1083) from Philip J. Jordan (#2048) and his partners. The Philadelphia museum continued operation into the early 1900s. R.J. Jordan died at his home in Philadelphia on 18 April 1909.

Man's mission on earth is comprised of eight lectures given by Jourdain at the Parisian Gallery of Anatomy. The author addresses consumption, dyspepsia, diabetes and the voice, as well as masturbation and venereal diseases. Additional matter includes a chapter on self-examination, several pages of letters from patients testifying to the doctor's successful treatment of the habit of self-abuse, and instructions on how to consult or contact the doctor. The book is copyrighted 1867. The "Preface to the *nineteenth edition*" may not be a fanciful claim (as was often done), given the distribution of this work through Jourdain's museum and other mechanisms. In printed wrapper. The title *Man's mission on earth* was also common to books published by Louis J. Jordan (#2043, 2044) and Robert J. Kahn (#2062–2065). Although the works of all three authors treat of the same subjects and occasionally paraphrase one another (e.g., the chapters on self-diagnosis), their content differs.

2052. JOURDAIN, Robert J.

Man's mission on earth: being a series of lectures delivered at Dr. Jourdain's Parisian Gallery of Anatomy. Addressed to those laboring under the baneful effects of self-abuse, excesses or infection; also, a familiar explanation of the venereal disease showing the danger arising from neglect or improper treatment in disorders of the generative system . . . Buffalo [N.Y.]: Warren, Johnson & Co., printers, 1872.

152 p. ; 15 cm.

At head of printed wrapper: *Twentieth edition.* On pages 151–52 Jourdain explains how patients may contact him regarding the "permanent cure of the various diseases of the generative organs . . . etc.," stipulating that "The communication must be accompanied with the usual consulting fee of five dollars." The back of the wrapper advertises the Parisian Gallery of Anatomy in Boston, and provides a wood-engraved view of its interior.

JOURDAIN, Robert J. see also **PARISIAN GALLERY OF ANATOMY AND MEDICAL SCIENCE** (#2734, 2735).

2053. JOURNAL OF HEALTH (PHILADELPHIA).

The Journal of health. Conducted by an Association of Physicians . . . [Vol. 1, no. 1

(Sept. 9, 1829)-v. 4 (Aug. 1833)]. Philadelphia: 1830–32.

3 v. : ill. ; 23 cm.

The bi-weekly *Journal of health* was intended to present to the public "plain precepts, in easy style and familiar language, for the regulation of all the physical agents necessary to health, and to point out under what circumstances of excess and misapplication they become injurious and fatal" ("Prospectus," v. 1, p. 1). The *Journal* is notable as the earliest serial publication in the United States that catered to the public's growing concern with personal health, domestic sanitation, temperance, etc. It is also is an example of an emerging phenomenon in American publishing of the 1830s: the mass-circulation periodical. Published by Henry H. Porter, the *Journal of health* was edited by an "Association of Physicians," i.e., the Philadelphia physicians John Bell (#289–293) and David Francis Condie, who were responsible for most of its anonymously contributed content.

According to Thomas Horrocks, "Bell and Condie used the journal to pursue two goals. The first concerned the authority of the medical profession in the management of disease. The editors were unrelenting in their attacks on Thomsonians, botanical doctors, homeopaths, animal magnetists and purveyors of various nostrums . . . The editors' second goal related to the broader reform movement of the period. Bell and Condie believed, as did many other health reformers of the time, that their message of individual responsibility concerning one's health was particularly relevant to their society" ("Promoting good health in the age of reform," *Canadian Bull. Med Hist.,* 1995, 12:263). S. Nissenbaum states that in 1831 the *Journal* became the first American publication to "come out unequivocally against the use of fermented as well as distilled beverages," though this had not been the editors' original position (Stephen Nissenbaum, *Sex, diet, and debility in Jacksonian America,* Westport, Ct., 1980, p. 78). This concern certainly helped place the temperance movement within the context of popular health reform generally.

The *Journal of Health* was perhaps the most important venture in the brief, cash-strapped publishing career of Henry H. Porter. According to Horrocks, Porter "published five journals, six books, and an almanac (and planned to pub-

lish two more books, two more journals, and 'The Family Library of Health' series) in a brief career spanning just under 30 months" (*ibid.,* p. 260).

Volumes 1–2 only are in the Atwater Collection. The first volume of the Atwater set carries the wording "4th improved edition" beneath the imprint, dated 1830. The fourth volume, published in 1833, was issued under the title *The Journal of health and recreation.*

2054. JOURNAL OF HEALTH AND MONTHLY MISCELLANY.

The Journal of health and monthly miscellany . . . [Vol. 1, no. 1 (Jan. 1846)-v. 1, no. 12 (Dec. 1846)]. Boston: Published by the editor, 1846.

1 v. : ill. ; 22 cm.

The first issue of this popular health monthly (Jan. 1846) appeared under the title *The Monthly miscellany and journal of health.* The title was inverted to the *Journal of health and monthly miscellany* with the second issue, but remained under this title only the first year. With the second volume (1847), the title was again changed to the *Practical educator and journal of health;* and with the third volume (1848), the title was inverted to the *Journal of health and practical educator* (#2055).

The *Journal* was edited by William Cornell (#799–802), who writes in the preface: "The proper system of education must commence . . . in the nursery—be carried through the family, the common school, the academy and the college, and into the active and busy scenes of life, and the laws of health incorporated into that system *as a branch of education.* But before this can be accomplished, the science of human life and healthful existence must be divested of technical terms and brought down to the capacity of every instructor of youth . . . To aid in accomplishing these things . . . is the object of this journal" (p. iv). Most of the articles, reviews, etc. in the *Journal* were written by Cornell, who also includes contributions by Walter Channing, John Collins Warren, and others—largely extracted from previously published works.

2055. JOURNAL OF HEALTH AND PRACTICAL EDUCATOR.

The Journal of health and practical educator . . . [Vol. 3, no. 1 (Jan. 1848)-v. 3, no.

12 (Dec. 1848)]. Boston: Published monthly, by Charles Rice, 1848.

1 v. ; 23 cm.

The Journal of health and monthly miscellany (#2054) changed title to *The practical educator and journal of health* with its second volume (1847). The latter title became *The Journal of health and practical educator* the following year. Edited by William M. Cornell (#799–802). Volume III of this title in the Atwater Collection includes nos. 3 (Mar.), 5 (May) and 8 (Aug.) only.

2056. JUST, Adolf.

Return to nature! The true natural method of healing and living and the true salvation of the soul. Paradise regained . . . The care of the body—water, human curative power, light, air, earth, food, fruit culture. Authorized translation from the fourth enlarged German edition by Benedict Lust, naturopath . . . Volume I. New York: Published by the translator, B. Lust, [c1903].

xvii, [1], 309, [1], xxvi p. : ill., ports. ; 22 cm.

"When we look at nature with an open, unprejudiced mind, and are not blinded by the teachings of science, we must arrive at the clear conclusion that man has become sick and miserable only because he no longer heeds the VOICES OF NATURE, and has thus everywhere transgressed the laws of nature, and lost his way . . . *In all cases, and in all diseases,* therefore, man can recover and again become happy only by a *true return to nature*" (p. 5). Just maintains that man has separated himself from nature by his intelligence; cites medical science as the classic example of this perverse phenomenon; and declares, "science is the cunning serpent in paradise which deceived man from the start, led him astray, and gave him false instruction" (p. 6). The only way by which man "can be cured of his dis-

eases with certainty and can again secure entire happiness is to abjure science and all things scientific" (p. 7). He must learn to listen once again to the "voices of nature."

The "nature cure system" advocated by Just "is the same for *all* diseases and *all* cases, even as the origin of diseases has but *one* cause, an unnatural mode of life" (p. 12). Disease is caused by the fermentation of foreign matter in the body, thus poisoning the blood and the entire system. Naturopathic therapy heals or prevents disease by a diet that limits the intake of food, and limits foods to those which will not be retained as foreign matter. Substances which increase internal body heat (and hence fermentation), e.g., alcohol, tobacco, coffee, etc., are to be avoided. The key element of the "nature cure system" is the "natural bath," i.e., bathing in cold water. The natural bath lowers internal heat, cleanses the pores and apertures, and promotes the secretion of foreign matter. The naturopathic armamentarium also includes light-and-air baths (which necessitates going about naked whenever possible); massage and friction; "earth power" (i.e., walking barefoot, sleeping on the ground, earth compresses, sand baths, etc.); and nutrition (especially fruits & nuts). The intention of the "nature cure system" was not simply to restore bodily health, but to restore the mind and the soul to their pre-corrupted (i.e., natural) state.

First American edition. *Return to nature!* is a translation of *Kehrt zur Natur zeruck,* the fourth edition of which was published in 1900. Adolf Just was a student of Sebastien Kneipp and the proprietor of a naturopathic resort in the Harz Mountains. His American disciple and translator, Benedict Lust, was a naturopath, publisher and the founder of *Jungborn,* a naturopathic sanitarium in Butler, N.J.

THE JUVENILE LIBRARY BY MRS. TUTHILL AND OTHERS see **MARTINEAU** (#2385).

K

2057. KAHLER, Peter.
Dress and care of the feet . . . [New York]:
G.P. Putnam's Sons, printers, 1883.
    39, [1] p., [1] leaf of plates : port. ; 15
cm.
    Kahler briefly notes the common diseases
of the foot (e.g., ingrown toe nails, corns, minor deformities), and promotes his "system of
boots and shoes" designed to maintain the
healthy foot in a normal state or to correct
deformities. This slender volume was published
at least five times between 1883 and 1915. A
self-described "surgeon chiropodist," Kahler
was the author of *A modern book on advanced surgical chiropody* (New York, 1915). He claims to
have been in practice and business since 1868.
At head of title: *Important to health and comfort!*
Frontispiece portrait of author.

2058. KAHLER, Peter.
Dress and care of the feet . . . New York,
1894.
    47 p., [1] leaf of plates : port. ; 15 cm.

    At head of title: *Important to health and com-
fort!* Frontispiece portrait of author.

2059. KAHLER, Peter.
Dress and care of the feet . . . New York,
1897.
    48 p., [1] leaf of plates : ill., port.

    At head of title: *Important to health and com-
fort!* Frontispiece portrait of author.

2060. KAHN, Louis J.
Nervous exhaustion: its cause and cure. A
series of lectures on debility and disease,
with practical information on marriage, its
obligations and impediments. Illustrated
with cases. To which is added pictures from
real life, or photographic life-studies, addressed to the young, the old, the grave,
the gay . . . [New York: The New York Printing Company, c1870].
    iv, [5]-198 p. ; 16 cm.

Kahn's treatise on sex disorders, venereal
diseases, "female complaints," etc. is comprised
of lectures delivered by the author at "Dr.
Kahn's New York Museum," one of the several
popular anatomy museums operated by the
Jordan family and their partners in Boston, New
York, Philadelphia and San Francisco from 1865
through the second decade of the 20th century.
    Louis J. Kahn and Louis J. Jordan (#2043-
2047) may actually have been one and the same
person (*see annotation to* #2066). At the same
time, the author states that for many years he
was "head of the greatest Anatomical Institute
in the world—Kahn's Museum, Tichborne St.,
Haymarket, London" (p. 195), and that "one
of the firm (Dr. G.D. Kahn) continues to be
consulted in London" (p. 198)—adding some
plausibility to his separate identity. *Nervous exhaustion* served as the model on which Robert
J. Kahn's *Man's mission on earth* (#2062-2065)
was based.

2061. KAHN, Louis J.
Nervous exhaustion. Its cause and its cure.
Comprising a series of eight lectures on
debility and disease, as delivered nightly
at Dr. Kahn's Museum of Anatomy, with
practical information on marriage, its obligations and impediments. Illustrated with
cases. Including pictures from real life, or
photographic life studies, addressed to the
young, the old, the grave, the gay . . . New
York: No. 51 East Tenth Street, [c1876].
    iv, [5]-196 p. ; 14 cm.

    Title on printed wrapper: *Nervous exhaustion
and special diseases: their causes and cure.* A wood
engraving that depicts the exterior of Dr. Kahn's
Museum is printed on the back of the wrapper.

KAHN, Louis J. see also **KAHN, ROBERT J.**
(#2062-2065); **DR. KAHN'S MUSEUM
OF ANATOMY, SCIENCE AND ART**
(#2066); **NEW YORK MUSEUM OF
ANATOMY** (#2598).

2062. KAHN, ROBERT JOHNSTONE, 1863-1927.
Man's mission on earth. A short treatise on diseases of the genito-urinary organs and accompanying nervous diseases, with a chapter on syphilis. By R.J. Kahn, M.D. and Louis J. Jordan, M.D. . . . Forty-second edition, revised and enlarged, to which is added a glossary of medical terms. New York: Isaac Goldmann Co., printers, 1910.
xiii, [14]-203 p. : ill. ; 18 cm.

Robert Johnstone Kahn was born in Philadelphia on 11 June 1863. According to his death certificate, his father was R. J. Johnston [sic], born in Ireland, and his mother Susanna M. Moss, of Philadelphia. His death certificate also indicates that he moved to New York circa 1880, though he is not listed in the city directories until 1894, from which time he is associated with Louis J. Jordan (#2043-2047) and Louis J. Kahn (#2061, 2061). He graduated from the Albany Medical College (1902) and is listed in the medical directories as practicing in New York thereafter at the same address as L.J. Jordan and L.J. Kahn. It seems likely (though not documented) that the elder Kahn may have adopted him or that the younger man may have assumed the Kahn name for business purposes. R.J. Kahn died at his apartment at 2 East 9th Street on 28 March 1927.

Man's mission on earth is also a title common to books published by Louis J. Jordan (#2043, 2044) and Robert J. Jourdain (#2051, 2052). Though similarities exist among these publications, their content differs. Although the name of Louis J. Jordan appears on the title-page of this and other editions (though not *all* editions) as co-author, the content of this Kahn-Jordan edition differs entirely from Jordan's earlier monograph of the same title—apart from several chapter headings. The book most similar in content to Robert J. Kahn's *Man's mission on earth* is Louis J. Kahn's *Nervous exhaustion* (#2060, 2061). In fact, R.J. Kahn has included in his version of *Man's mission on earth* the "Preface to the first English edition" (p. viii-ix), signed "L.J. Kahn, M.D." and dated London, October 1860. The latter work was apparently intended to update and continue the textual tradition of his mentor's book. Chapters I ("Anatomy and physiology of the urinary and and generative organs"), VI ("Self-pollution—

its effects, etc."), IX ("Venereal disease") and the unnumbered chapter entitled "Pictures from real life" follow closely the text of *Nervous exhaustion*. The authorship of these several works is only confused by the attribution of *Nervous exhaustion* to R.J. Kahn and L.J. Jordan on the title-page of this edition of *Man's mission on earth*. (It has been suggested that L.J. Jordan and L.J. Kahn were the same person, *see* #2066). The original copyright date for this Kahn-Jordan version of *Man's mission on earth* is 1902.

2063. KAHN, ROBERT JOHNSTONE, 1863-1927.
Man's mission on earth. A short treatise on diseases of the genito-urinary organs and accompanying nervous diseases, with a chapter on syphilis. By Robt. J. Kahn . . . Forty-fourth edition, revised and enlarged, to which is added a glossary of medical terms. New York: Isaac Goldmann Co., printers, 1911.
xiii, [14]-200 p. : ill. ; 19 cm.

This issue of the 44th edition of *Man's mission on earth* is identical to #2064 except for the absence of Louis J. Jordan's name on the title-page.

2064. KAHN, ROBERT JOHNSTONE, 1863-1927.
Man's mission on earth. A short treatise on diseases of the genito-urinary organs in health and disease, with a chapter on syphilis. By R.J. Kahn, M.D. and Louis J. Jordan, M.D. . . . Forty-fourth edition, revised and enlarged, to which is added a glossary of medical terms. New York: Isaac Goldmann Co., printers, 1911.
xiii, [14]-203 p. : ill. ; 18 cm.

Softcover.

2065. KAHN, ROBERT JOHNSTONE, 1863-1927.
Man's mission on earth. A contribution to the science of eugenics. A short treatise on the genito-urinary organs of the male in health and disease, with a chapter on syphilis. By R.J. Kahn . . . Forty-eighth edition, revised and enlarged to which is added a glossary of medical

terms. New York: Isaac Goldmann Co., printers, 1915.

    xiii, [14]-221 p. : ill. ; 19 cm.

KAHN, ROBERT JOHNSTONE see also DR. KAHN'S MUSEUM OF ANATOMY AND MEDICAL SCIENCE (#2065.1).

2065.1. DR. KAHN'S MUSEUM OF ANATOMY AND MEDICAL SCIENCE.

Hand book and descriptive catalogue of Dr. Kahn's Museum of Anatomy and Medical Science. On exhibition at No. 312 Bowery above Houston Street—New York City. Open day and evening from 9 A.M. . . . [New York, ca. 1914].

    32 p. ; 18 cm.

Kahn's museum was located at 312 Bowery St. from 1905 to 1914. In the Atwater copy, the Bowery St. address has been lined out and beneath it stamped: "Now at 30 Cooper Square," the address to which the museum moved in 1915 (*see annotation to* #2066). The "Notice" printed inside the back wrapper is signed "Dr. R.J. Kahn," who states that he is available for consultaton at his "residence and consulting rooms" at 21 Fifth Avenue, and that his practice "is confined exclusively to the treatment of men," i.e., the treatment of sexually transmitted diseases, sexual dysfunction, etc. Title on wrapper: *Descriptive catalogue of Dr. Kahn's Museum of Anatomy and Medical Science.*

2066. DR. KAHN'S MUSEUM OF ANATOMY, SCIENCE AND ART.

DR. KAHN'S MUSEUM OF ANATOMY, SCIENCE AND ART. 294 Bowery, Above Houston Street, NEW YORK CITY. Gotham's greatest attraction! . . . Open Every Day and Evening from 9 A.M. . . . The models in this Institution are directly from Nature and illustrate the Human Anatomy in such a faithful manner that admission can be granted ONLY TO MEN. [New York, between 1898 and 1905].

    [2] leaves ; 21 cm.

Opened in 1862 by Henry Jacob Jordan and his brother-in-law Samuel T.E. Beck, the museum was first known as the Parisian Cabinet of Wonders, and was located at 563 Broadway. It was moved the following year to 618 Broadway

and renamed the New York Museum of Anatomy (#2598). Beck dropped out after 1870, and about 1878 H.J. Jordan was succeeded by Philip J. Jordan (#2048), probably a son, who continued the operation until 1881. In the meantime, two other anatomy museums had been opened—one by Louis J. Kahn (#2060, 2061) at 688 Broadway in 1871; and the other by Louis J. Jordan (#2043-2047) at 1146 Broadway in 1879. In 1884, Louis J. Kahn and Louis J. Jordan (they may have been the same person) combined these two museums as Kahn's Museum, which operated successfully at 713 Broadway from 1882 to 1886, and at 708 Broadway from 1887 to 1888. After the police raid of 1888, the museum was out of operation for four years. It re-opened in 1893 under the management of Robert J. Kahn (#2062-2065) and Louis J. Jordan ("Drs. Kahn and Jordan, proprietors," leaf 2 recto) at 252 Bowery through 1897. From 1898 until 1905 Dr. Kahn's Museum was located at 294 Bowery; at 312 Bowery from 1905 to 1914; and at 30 Cooper Square from 1915 until 1920.

Because of the long association of the Jordan family with anatomy museums in New York, and because its members were involved in similar activities in Boston, Philadelphia and San Francisco, this family is of particular interest. There were four Jordans, all born in England, probably a father and three sons. Henry Jacob Jordan, the patriarch, claimed a medical degree from St. Bartholomew's Hospital, London in 1827 (not confirmed). In 1854 he received, in the Queen's name, a patent for the invention of "an improved medicine for the cure of venereal affections" which he denominated "Treisemar," a remedy which might be administered as a lozenge or a tea. Its elements were powdered henbane, compound tragacanth powder, nitrate of potash, tartrate of potash, bicarbonate of soda and sugar.

Of the three presumed sons (Louis J., Robert J. and Philip J. Jordan), only Robert J. Jordan has a confirmed record of professional qualification, although Louis J. Jordan claimed similar credentials. Robert Jacob Jordan is listed as a physician in the London directories from 1859 through 1863, and was L.R.C.P. of Edinburgh (1859) and M.R.C.S. of England (1859). He was "deprived" of these credentials in 1863. It appears that all four Jordans emigrated to the United States sometime after this event. The three younger men later received medical degrees from the Eclectic Medical College of New York—Louis and Philip in 1872 (both honorary), and Robert in 1881 (apparently earned in course).

After coming to the United States, each of the Jordans went to a different city: Henry to New York; Louis to San Francisco; Robert to Boston; and Philip to Philadelphia. In San Francisco, Louis established the Pacific Anatomical Museum and Gallery of Nature History and Medical Science, but returned to New York about 1875. By 1879, Louis J. Jordan was proprietor of an anatomy museum at 1146 Broadway, which he moved to 489 6th Ave. the following year. At Boston in 1866, Robert J. Jordan, using the name Robert J. Jourdain (#2051, 2052), opened the Parisian Gallery of Anatomy and Medical Science (#2734, 2735). Jourdain's museum operated in Boston until 1881, when he moved to Philadelphia to assume charge (with R.J. LaGrange, #2173-2175) of the European Anatomical, Pathological and Ethnological Museum (#1081-1083). The Philadelphia museum had been established in 1872 by Philip J. Jordan and David Davieson. Philip returned to New York in 1879 and took over responsibility for the museum that Henry J. Jordan had established there. Davieson moved to St. Louis, were he became one of the proprietors of the Grand Museum of Anatomy (#1399). Between 1879 and 1881, members of the Jordan family operated three different museums in the city of New York. The most famous and long-lived of these was Kahn's Museum—a presence on Broadway, and later the Bowery, for over thirty-five years.

It is unclear whether Louis J. Jordan and Louis J. Kahn were two different persons or one and the same—a dual identity that was perhaps assumed to elude the authorities. The only documentation found for Louis J. Kahn is his listing in the New York directories. There had been a Dr. Kahn's Anatomical Museum on Coventry St. in London during the 1850s. The association (if any) between its owner, Joseph Kahn, and Louis J. Kahn is unclear. The Coventry St. museum and Kahn were the subject of some rather scathing correspondence in *The Lancet* (1856, 1:28). In July 1857 Joseph Kahn was tried and convicted of medical fraud (*The Lancet*, 1857, 2:150-53). Kahn's museum nonetheless remained in business until March, 1873. A paragraph in the *The Lancet* (1873, 1:354-55)

notes that "Dr. Kahn's Anatomical Museum will be conspicuous by its absence among the institutions of the metropolis" after having been raided and closed by the police on a warrant "granted on the application of the solicitor to the Society for the Supression of Vice." Forty cases of models ("indecent, disgusting, and demoralising") were seized.

None of this necessarily explains the appearance of the Kahn name in New York. There is evidence, nevertheless, for a Kahn-Jordan identity in the fact that they shared the same first names and middle initials, and often the same residential address. From 1871 to 1875, Louis J. Kahn and Henry J. Jordan are listed at the same address; and from 1876 to 1894, Louis J. Kahn and Louis J. Jordan share the same address. Another indication that L.J. Kahn and L.J. Jordan are the same person is that in some editions of Robert J. Kahn's *Man's mission on earth* (#2062, 2064), a work based on L.J. Kahn's *Nervous exhaustion* (#2060, 2061), authorship is attributed to R.J. Kahn and L.J. Jordan—not L.J. Kahn. None of this explains the relationship of R.J. Kahn with L.J. Kahn and Jordan. R.J. Kahn lived at the same address as Louis J. Jordan from 1895 to 1920, and was Jordan's successor in the business. The younger Kahn's death certificate, however, gives his father's name as Robert J. Johnston, making it likely that Kahn was a name assumed for business; or possibly an adoptive name.

Popular proprietary anatomy museums were established in many American cities during the period between the Civil War and the First World War. They differed from the anatomical museums found in medical schools in that the latter were established for professional instruction, whereas the former were located in retail commercial districts and were open to the public for a small fee. According to Michael Sappol, similar establishments had appeared in Europe somewhat earlier, with their roots in the cabinets of scientific curiosities maintained by scholars and the aristocracy.

The early public anatomy museums in New York, such as the one started by Wooster Beach in the 1840s, and later, the New York Anatomical Gallery maintained by Horace B. Tolles (#2590), were part of the reform movement in medicine with the intention of teaching every person about human anatomy and physiology in order to improve health and prevent disease. Though, in the broader sense, anatomical museums remained instructional through out the period of their existence, they evolved from primarily educational institutions to establishments promoting the professional services and wares of their proprietors. The tone of these museums also became more vulgar and explicitly sexual. Instead of focusing on the fundamentals of human anatomy and physiology, including that of the reproductive system, they soon became collections of congenital oddities, freaks of nature, and pathological specimens displaying the consequences of venereal diseases and masturbation (the last two being the therapeutic specialities of most museum physician-proprietors). It is not surprising that city directories of the period list these establishments as "places of amusement."

As anatomy museums became more luridly sexual in orientation, they drew the attention and incurred the displeasure of the authorities, including such figures as Anthony Comstock (#754, 755). Much of what we know of their history comes from newspaper reports of the several legal actions taken against them. Perhaps the most spectacular run-in with the authorities came on Monday, 9 January 1888, when the police, accompanied by Comstock, raided the four anatomy museums then operating in New York. The newspapers reported the event in detail (v. the *New York Times,* 10 Jan. 1888; *New York Tribune,* 10-12 Jan. 1888; *New York Herald,* 10 Jan. 1888; *New York World,* 10 Jan. 1888; *New York Sun,* 10 Jan & 22 Jan. 1888): "Revolting shows closed," headlined the *Times;* "Six truckloads of alleged indecent exhibits seized by the police," reported the *Tribune;* "Four museums turned inside out by Comstock and the police," said the *Herald.* From these reports (and those in the *Sun* and the *World*) dervives the following story. A week before the raids, Inspector Williams "took a trip down the Bowery and casually dropped in all the places that had any of the gaudy signs in front of them, that told the wayfarer that there were human monsters of all kinds within, that could be inspected for the small sum of ten cents. In the ordinary museums, the Inspector could not find much of interest, but when he entered No. 138 [the Egyptian Musee] he was startled to see on every hand the wax figures of both men and women without any clothes on. The figures were grouped so as to show the anatomy of the human body." Continuing his way down the Bowery, Williams found exhibits at No. 81 [the European Museum] that "brought the blush of modesty to his cheek." Another inspector, Steers, "had his sense of decency all torn to pieces" by a visit to Dr. Kahn's Museum on Broadway. Inspector Williams hastened to police headquarters and summoned Anthony Comstock. They studied Section No. 317 of the Penal Code, stating that a person "who sells, lends, or gives away, or shows, or offers to do so, or has in his possession, or advertises any obscene, lewd, lascivious, filthy, indecent or disgusting print, figure, or image is guilty of a misdemeanor and upon conviction shall be sentenced to not less than ten days' nor more than a year's imprisonment or $50-$1000 or both." It was difficult to tell where the nude wax figures came under the statute, but the "immoral exhibitions must be stopped."

Inspector Williams, with affidavits, went before Justice Patterson who issued warrants against all the so-called anatomical museums in the city not directly under the charge of recognized medical colleges. The warrants were served the same afternoon at the three establishments in the Bowery and at Dr. Kahn's Museum, 708 Broadway. That evening, a total of thirteen men, known in their line as "professors" and "doctors," along with "sundry box-window money takers," not to mention a bartender, "rested from their labors in cells at the station houses" of three different precincts. Only Dr. Jordan, "the reputed owner of Kahn's Museum, was not arrested because there is no proof that such a person is connected with the place." During the raids, Comstock "found little that deserved to be spared confiscation and chalked the majority of them, dictated an inventory to his secretary, and sent for a truck with which to remove the stuff to Police Headquarters." Ultimately, five or six van loads were removed. Awaiting the decision of the court, "the figures were laid out quite tenderly . . . [as if] in a first class and well-disinfected dead house."

Kahn's Museum was called "the most pretentious and respectable of the four." It had operated for twenty-six years and never more than a block or so from its previous location. Some members of the community thought that the officials had overstepped their authority. At a meeting of the Medico-Legal Society the day after the raids, resolutions were presented that were strongly critical of Comstock and the police action. The museums, it was said, had been "summarily invaded," arrests made and contents seized on the ground of obscenity. Members of the Society felt that the anatomical collections were a useful form of public instruction. "We agree with Prof. Agassiz that the time has come when scientific truth must cease to be the property of the few, when it must be woven into the common life of the world" (*New York Times*, 12 Jan. 1888).

2067. KARELL, Philipp Jakob, 1806-1886.
The milk-cure . . . A lecture, delivered in the medical society in St. Petersburg . . . Translated from the Russian by the Naturopath and Herald of Health magazine. Butler, N.J.: Published by Benedict Lust, [191-?].
    20, [12] p. ; 24 cm.

    Citing refererences to the therapeutic use of milk in the writings of Graeco-Roman, Byzantine, Arabic and modern European medical authors, Karell, an imperial court physician, states: "according to my own experiences . . . the milk cure surpasses all the other means of treatment which are known to me, in all cases of dropsy, in asthma . . . in obstinate neuralgias . . . also in liver complaints . . . [and] in sufferings of nutrition in general" (p. 5). Far from advocating the milk-cure as a panacea, Karell states: "I could hardly answer the question satisfactorily as to which substance of the quite composite liquid is effective . . . Neither could I give a proper account as to whether the milk-cure has a sudorific or a diuretic, a dissolvent or restorative effect" (p. 16). Nonetheless, he expresses the opinion that milk is not only a mild and easily assimilated nutrient, but that it allows the body to eliminate poisonous elements. "A final word about milk-cure. By Dr. Baden" (p. 19-20), written circa 1900, supports Karell's claims for milk with recent medical opinion. The twenty unnumbered pages at the end of the essay advertise the publications, herbal remedies and and health resorts of the naturopathic entrepreneur Benedict Lust.
    Karell was born in Tallin (Estonia) and received his medical degree in 1832 from the university at Dorpat, i.e., Tartu, Estonia. In 1849 he was appointed physician to Tsar Nicholas I, and continued as courrt physician to Tsar Alexander II. Karell was a contributor to the medical literature, founded the Russian Red Cross (1867), and was esteemed in both the Russian and German imperial courts. His lecture on the milk-cure was originally published in the *St. Petersburger medizinische Zeitschrift* (1865), and later in the *Archives générales de médecine* (1866) and as a pamphlet (*On the milk cure!*) published at Philadelphia in 1870. In printed wrapper.

KAUFFMAN, E. V. see **SYCAMORE MINERAL SPRINGS** (#3410).

2068. KAUFMANN, Carl Ernst.
Kaufmann on disease. Its causes and home cure . . . Boston: A.P. Ordway & Co., publishers, [ca. 1882, c1881].
    96 p., [3] leaves of plates : ill. ; 21 cm.

    "What is the greatest cause of sickness? This is a question I am often asked. And now, my kind reader, I will tell you. Poor, thin, vile, vitiated blood" (p. [2]). This condition of the blood is removed by the use of *Sulphur Bitters,* the discovery of C.E. Kaufmann manufactured by A.P. Ordway & Co. in Boston. Even syphilis is claimed to be cured by the *Sulphur Bitters* when compounded with spirits of nitre, cubeb and copaiba. Imprint from printed wrapper.

2069. KAUFMANN, Joseph Herbert.
Care of the mouth and teeth. A primer of oral hygiene . . . New York: Rebman Company, [c1916].
    vi, [2], 69 p., [6] leaves of plates : ill. ; 24 cm.

    The author was a 1912 graduate of the New York College of Dentistry.

2070. KEATING, John Marie, 1852-1893.
Maternity, infancy, childhood. Hygiene of pregnancy; nursing and weaning of infants; the care of children in health and disease. Adapted especially to the use of mothers or those intrusted with the bringing up of infants and children, and training schools for nurses, as an aid to the teaching of the nursing of women and children . . . Second edition. Philadelphia; London: J.B. Lippincott Company, 1890.
    221, [7] p. ; 19 cm.

Maternity was originally published at Philadelphia in 1887. This second edition was issued at least five times between 1888 and 1895. A third and final edition appeared in 1906. Keating divides his book into three parts. The first is devoted to the management of pregnancy; the second to infant care and feeding; and the third to childhood diseases. The author received his medical degree from the University of Pennsylvania in 1873. He was on the staff of the Blockley Hospital and lectured there on the diseases of women and children. "Mothers and children, how to make them healthy and happy, was the chief life-work and pen-work of the genial John Keating, especially in editing the *Archives of pediatrics* and *The International clinics,* and in working as president of the American Pediatric Society" (*Dict. Amer. med. biog.,* New York, 1928, p. 686). Keating was also the editor of *Clinical gynaecology* (1895) and the five-volume *Cyclopaedia of the diseases of children* (1889-99). Series: *Practical lessons in nursing.*

KEITH, Jennie I. see **KEITH** (#2071).

2071. KEITH, Melville Cox, 1835-1903.
Appendicitis. This little book is written to stamp out fear—man's greatest enemy . . . This is the only book in print which gives a reliable, comprehensive explanation and treatment for this trouble. Second edition. Revised, enlarged, and illustrated by Jennie I. Keith, M.D. Bellville, Ohio, 1915.
[8], 50 p., [1] leaf of plates : ill., port. ; 20 cm.

Regarding the cause of appendicitis, Keith subscribes to the concept of autointoxication current at the time: "The basic reason for all cases . . . comes from the fact that the ascending colon has been crowded full of material and as the feces form in the ascending and transverse colons, these colons being full and perhaps clogged by excessive dryness in the rectum . . . exude or send through the walls of the intestines a quantitty of the thinnest fecal matter . . . It is at this point . . . that the pain, or ache, or uneasy feeling is felt by those persons who are about to have appendicitis" (p. 2-3). The treatment of appendicitis consists in removing the causes of its inflammation, not the appendix: "In plain English, the colons are filled with old material which should be taken out

and, as Nature cannot get it out, therefore, the first remedy would be to remove all the old fecal matter from the intestines" (23). Nature may be assisted in this matter—not by the use of a cathartic, which "ruins the bowels by making them smaller and shrunken," and only irritates them further—but by enemas (a solution of boneset, catnip & lobelia) and herbal emetics ("The regular physician knows no more about an emetic than a hog knows about an eclipse of the sun," p. 36). In printed wrapper.

Jennie I. Keith was an 1890 graduate of the Physio-Medical College of Indiana. In the introduction she writes of the author: "I am a graduate of the same school of medicine, and was active in practice with him for twenty years . . . Having been wife as well as co-worker, the privilege and responsibility of publishing and circulating the Doctor's writings naturally gravitated to my lot."

2072. KEITH, Melville Cox, 1835-1903.
Keith's domestic practice and botanic handbook. A practical treatise on the conditions of the human body called disease and the proper observance of the laws to prevent those conditions. The methods to pursue to overcome disease, or all pathological conditions of the human body, with illustrations, formulas and facts of the most practical value to all who are interested in the welfare of the human race . . . Bellville, Ohio: Published by the author, 1901.
886 p., [20] leaves of plates + frontis. : ill. ; 25 cm.

"PROTOPLASMY," Keith informs his readers, "is the exposition of the laws of atoms, based on the fact that there are atoms, and that these atoms obey the law, and this law is known as the Vital Force, which force is directly from God and is synonymous with the Spirit and Life Force." Each atom contains within it this *Force,* and "When this atom, with life dwelling in it has appropriate surroundings, then we may expect to see the indwelling force act in the most harmonious manner for the benefit of all parts concerned" (p. [37]). The force fufills its role perfectly, whether building up new organic structures or maintaining them in a healthy state once extant.

Keith devotes a considerable number of pages to an idiosyncratic embryology based upon his idea of atoms, the vital force and the law of God. He indignantly singles out the states of Missouri, Kentucky and Texas for their violation of divine law in the practice of cross-breeding mares and jackasses to produce mules. That this violates God's atomic laws is evidenced in the fact that the mule is sterile. Their iniquity does not go unpunished, for "there are no three states and no separate state that has as much crime and lawlessness in it as these mule raising and 'diverse breeding' areas" (p. 39).

All the body's physiological activity is a result of the presence of this force in cells, corpuscles, etc. This activity continues normally unless the atom or cell finds itself in a toxic environment. In this diseased state, the living atoms struggle to overcome the external threat. The role of medicine is to "cleanse the surroundings of the atom and give the force an opportunity to dwell there and rid the atom of its impurities" (p. 44). By way of example Keith asks, "Have you a cancer? See to it that you change these wearily laden corpuscles from their burden of compost and you shall be cured. Why? Because this dreaded cancer is nothing but a great bunch of excrementitious matter that has not yet had any chance to get out of the body by these carriers, the red blood corpuscles, because they have been stopped or shut up" (p. 52).

We become diseased when we violate the God-given laws of health, i.e., of maintaining the vital force in an undisturbed state of purity, a condition impossible to those who fill their bodies with "miserable Irish potatoes, pork, meat, hog fat, oysters, tobacco, wines, beers and alcohols." Health is also impossible to those who have lost it and then seek its restoration from the poisonous medications dispensed by allopathic and homoepathic physicians. "Why so? Because these poisons, all of these minerals, drugs, and many others, kill or drive off the living power, the Vital Force and thus they are worse off than they would have been if they had never taken them" (p. 75). The greater part of the volume is given to Keith's hygienic regimen for the maintenance of health, and to a lengthy catalog of diseases, their symptoms, prognosis and treatment by physio-medical (i.e., botanic) remedies. The author was an 1861 graduate of the Physio-Medical College (Cincinnati) founded by Alva Curtis.

2073. KELLEY, Elbridge Gerry, 1813?-1881.
A popular treatise on the human teeth and dental surgery, being a practical guide for the early management of the health and teeth of children; the preservation of the adult teeth; causes of their diseases; and means of cure: with brief observations on artificial teeth . . . Boston: James Munroe and Company; Newburyport: A.A. Call, 1843.
iv, [5]-196 p. ; 19 cm.

Kelley launches this first edition of his *Popular treatise* with the remark: "And if it should chance to be so universally read and its precepts followed as to obviate the necessity of dentists! for one, we would cheerfully relinquish a profession, three-fourths of whose members dishonor and degrade it, while the services of the remainder are either unknown or unappreciated by the public generally" (*Pref.*, p. iv).
Kelley devotes chapters to dentition (and the alleviation of painful dentition); brushing the teeth (including a discussion of tooth brushes, tooth powders & tooth picks); the importance of tartar removal; diseases of the teeth; dental surgery (i.e., "filling carious teeth" & tooth extraction); gum diseases (and their relation to health generally); and "artificial teeth." The second and final edition of the *Popular treatise* was published at Boston in 1846. The author was an 1838 graduate of the Jefferson Medical College. Spine-title: *Kelley on the human teeth.*

KELLEY, John Clawson see *PHILOSOPH-ICAL MEDICAL JOURNAL* (#2788, 2789).

2074. KELLOGG, Ella Ervilla, 1853-1920.
The good health birthday book. A health thought for each day . . . Battle Creek, Michigan: Good Health Publishing Company, 1907.
[48] leaves : ill. ; 17.5 × 26 cm.

A calendar of health maxims by famous persons interspersd with full-page photographs of places and activities at the Battle Creek Sanitarium.

2075. KELLOGG, Ella Ervilla, 1853-1920.
Science in the kitchen. A scientific treatise on food substances and their dietetic properties, together with a practical explanation of the principles of healthful cookery, and a large number of original, palatable and wholesome recipes . . . Library of health, vol. 5. Battle Creek, Mich.: Published by

Modern Medicine Publishing Company . . . for the Library of Health, [c1892].

[2], 508 p., [40] leaves of plates : ill. ; 24 cm.

First edition. "A little less than ten years ago the Sanitarium at Battle Creek, Mich., established an experimental kitchen and a school of cookery under the supervision of Mrs. E.E. Kellogg . . . During this time, Mrs. Kellogg has had constant oversight of the cuisine of both the Sanitarium and the Sanitarium Hospital, preparing bills of fare for the general and diet tables, and supplying constantly new methods and original recipes to meet the changing and growing demands of an institution numbering always from 500 to 700 inmates. These large opportunities for observation, research, and experience have gradually developed a system of cookery, the leading features of which are so much in advance of the methods heretofore in use that it may justly be styled *A New System of Cookery*" (*Pref.,* p. 3).

In 1879 John Harvey Kellogg (#2076-2108) married Ella Eaton, an Alfred University (N.Y.) graduate who first visited Battle Creek during the summer of 1876. "Ella Kellogg's interests and activities complemented her husband's career in many ways. For approximately forty-three years she served on the editorial staff of *Good health.* During that period most of the monthly issues of the journal contained at least one article from her pen . . . Her journal articles reflected two of her major interests: the care and training of children and food preparation. Mrs. Kellogg also authored three books. Her *Talks with girls* contained advice concerning the emotional and physical problems of the teen-age girl. Her principal book, *Science in the kitchen,* consisted of a record of her experimental work in dietetics and cookery and included a number of recipes for vegetarian foods. In *Studies in character building,* Ella Kellogg summarized her philosophy of child training" (Richard Schwarz, *John Harvey Kellogg,* Nashville: Southern Publishing Assn., 1970, p.155). Although John Harvey and Ella Kellogg had no children, they raised more than forty foster children, several of whom they adopted.

2076. KELLOGG, John Harvey, 1852-1943.

The art of massage: its physiological effects and therapeutic applications . . . Battle Creek, Mich.: Modern Medicine Publishing Co., [c1895].

xvi, 9-282 p., XLV plates [i.e., 17 leaves of plates; 30 p. of plates], 2 tables (1 folding) : ill. ; 23 cm.

"Twenty-two years ago, when the writer first made a study of massage and passive movements as a therapeutic means, even the term 'massage' was unfamiliar to the medical profession of this country, as it still is to the majority of the laity. After obtaining such a knowledge of massage as was at that time possible at home, I visited Sweden, Germany, France, and other European countries, with the purpose of becoming acquainted with the methods employed by the most expert manipulators abroad; and from the knowledge thus obtained, as well as gleaned from the various treatises and papers which have appeared upon the subject within the last twenty years, added to my personal experience in the constant employment of ten to twenty masseurs and masseuses, in the treatment of patients suffering from a great variety of chronic ailments, I have endeavored to summarize and condense in this little work the facts which are essential to a scientific knowledge of the art and science of massage and its rational employment" (*Pref.,* p. iii). Kellogg discusses the physiological effects of massage, its therapeutic effects, and directions for its application. The text is accompanied by photographic depictions of procedures. Opposed to the use of drugs, Kellogg was an enthusiastic proponent of alternative therapies, such as heliotherapy, exercise therapy, electrotherapy, diet therapy, balneology, and massage. *The art of massage* appeared in multiple editions, and was last published in 1937.

John Harvey Kellogg was the son of John Preston Kellogg, a native of Massachusetts who settled on the Michigan frontier early in the 1830s. A prosperous farmer and broom manufacturer, the elder Kellogg was also devoted to such causes as the abolitionist and temperance movements. Initially a Baptist and then a Congregationalist, Kellogg *père* adhered to the prouncements of the prophetess Ellen White from 1852, and moved his family to Battle Creek, headquarters of the burgeoning Seventh Day Adventist movement. From age twelve to sixteen, Kellogg *fils* apprenticed on the staff of the Review & Herald Publishing Co., the printing arm of Seventh Day Adventism. It was in this environment that Kellogg first became interested in health reform. He not only set type for Ellen White's articles on health, but was exposed to the writings of such figures as Sylvester Graham (#1392-1395), Russell Thacher Trall (#3558-3593) and James Caleb Jackson (#1902-1955). In 1872 Kellogg was among a group of Adventist youth sent to New Jersey to study medicine at Trall's Hygieo-Therapeutic College. Dissatisfied with the lack of scientific training at Trall's institution, Kellogg spent the next term studying medicine at the University of Michigan, before he transferred to Bellevue Hospital Medical School, from which he received his diploma in 1875.

Kellogg's name is practically synonymous with the Battle Creek Sanitarium. It was founded in 1866 as the Western Health Reform Institute at the urging of Ellen White (#1604, 3763, 3764), who patterned it on Our-Home-on-the-Hillside in Dansville, N.Y. (#2675-2679), where in 1863 she had been the guest and patient of James Caleb Jackson. John Preston Kellogg was the largest stockholder of the new health resort. From the beginning, the Institute was intended to be "a center for instruction in healthful living and religious principles as well as a place to care for the sick" (Richard W. Schwarz, *John Harvey Kellogg, M.D.*, Nashville: Southern Publishing Assn., 1970, p. 59). In October 1876, James and Ellen White offered its directorship to John Harvey Kellogg, only recently graduated from medical school. The following year Kellogg renamed the Institute the Battle Creek Sanitarium, and within several years became the dominant figure in its operation. Kellogg transformed the failing Institute into a national symbol of modern science and healthful living. Under Kellogg's leadership, the Sanitarium constantly expanded to accommodate the housing, care and education of visitors, to provide facilities for surgical operations and laboratory research, to train Adventist medical missionaries, and to maintain plants for its publishing and health food ventures. "Although Kellogg did not offer an original health regimen, the atmosphere in which he presented it was. The sanitarium became something of a combination of nineteenth-century European health spa and a twentieth-century Mayo Clinic. A patient could enjoy hotel comfort while a group of highly trained medical specialists devoted their energies first to the

scientific diagnosis of his ills, and then to the correction of them through natural means" (*ibid.*, p. 65).

Although an evangelist of health reform in the spirit of Graham, Alcott, Jackson and others, Kellogg brought to the cause the desire to establish what he termed "biologic living" on principles acceptable to contemporary science and medicine. Kellogg still regarded moral principles and physiological principles as based on the same God-given laws, but sought to ground his views on nutrition, physiology, therapeutics, etc. on modern laboratory and clinical science.

The scope of Kellogg's interests and activities is astonishing. In addition to his directorship of the Sanitarium's daily operations, he was a prolific author, edited several journals, served on the board of numerous Adventist organizations (including Battle Creek College, the American Medical Missionary College, the American Health & Temperance Association), lectured on health issues nationally, oversaw the development of various health food products (e.g., granola, flaked cereals, grain-based coffee substitutes, peanut butter, soy milk), invented or supervised the development of exercise and surgical equipment, served on the Michigan State Board of Health, and somehow managed to become one of the most respected gastrointestinal surgeons of his generation.

In 1907 Kellogg was expelled from the Seventh Day Adventists, who in 1903 had moved their headquarters to Washington, D.C. For years his imperious manner had antagonized many within the Adventist hierarchy. More important, however, was the change in Kellogg's religious views. As the century drew to a close, he became uncertain in matters of dogma, began to question the statements of Ellen White, and in his book *The living temple* (1903) expressed theological views that were interpreted as pantheistic. Kellogg also became increasingly impatient with the Adventist clergy and their leaders, whom he regarded as less educated and therefore unable to fully appreciate his work. Additionally, he saw them as indifferent to the principles of health reform, and hence as obstacles to the complete realization of his goals. From its regard, the Adventist hierarchy thought Kellogg was trying to secularize the Battle Creek Sanitarium, and saw the spirit of the place as increasingly humanitarian and less in line with the church's primary mission. None-theless, Kellogg retained control of the Sanitarium, which continued to prosper until the Great Depression. The Battle Creek Sanitarium was placed in receivership in 1933, and in 1937 was reorganized with Kellogg as little more than its figurehead director.

In the late 1870s Kellogg became interested in the production of health foods—initially to enhance the Sanitarium's menu and later as an income generating venture. These products were manufactured and marketed nationally by a subsidiary of the Battle Creek Sanitarium managed by Kellogg's younger brother. By the 1890s, the Sanitarium Food Co. produced forty-two food products. In collaboration with his brother Will Keith Kellogg, J.H. Kellogg continually expanded the business, which was known by 1908 as the Kellogg Food Company. The brothers eventually split, however—filing lawsuits against one another over product rights and commercial use of the Kellogg name.

In the second decade of the 20th century, Kellogg became increasingly involved in the eugenics movement. Kellogg maintained that acquired characteristics could be inherited. The natural corollary of this assumption was that biologic living would produce human specimens capable of transmitting healthful constitutions to their offspring; while to those who defied the laws of health only feeble and unhealthy children would be born—offspring whose existence was guaranteed to perpetuate crime, poverty and insanity. Kellogg's view on this matter was contrary to the opinion of most geneticists, but in agreement with the opinion of most eugenic and health reform activitists. Kellogg founded the Race Betterment Foundation and sponsored several Race Betterment Conferences at the Sanitarium.

2077. KELLOGG, John Harvey, 1852-1943.
The art of massage. Its physiological and therapeutic applications . . . Battle Creek, Mich.: Modern Medicine Publishing Co., [1904, c1895].
xvi, 9-288 p., [19] leaves of plates, [36] p. of plates, [2] leaves of tables (1 folding) : ill. ; 23 cm.

The "Preface to fourth edition" (p. vi) is dated August, 1904.

Fig. 57. J.H. Kellogg, frontispiece to his Plain facts, *#2095.*

2078. KELLOGG, John Harvey, 1852-1943.
Colon hygiene. Comprising new and important facts concerning the physiology of the colon and an account of practical and successful methods of combating intestinal inactivity and toxemia . . . Eighteenth thousand. Battle Creek, Michigan: Good Health Publishing Co., 1916.
[2], 9-415 p., [28] leaves of plates : ill. ; 19 cm.

"In the treatment of every chronic disease, and most acute maladies," Kellogg writes in his preface, "the colon must be reckoned with. That the average colon, in civilized communities, is in a desparately depraved and dangerous condition, can no longer be doubted. The colon must either be removed or reformed" (p. 10).

"Almost all aspects of Kellogg's biologic living appear in one form or another in the teachings of earlier health reformers and in the writings of Ellen White [#1604, 3763, 3764]. In the area which he termed 'colon hygiene,' however, the doctor developed a distinctive tenet which he actively promoted during his later years.

During the first decade of the twentieth century he became fascinated with the ideas being advocated at the Pasteur Institute in Paris by the famous émigré Russian chemist Elie Metchnikoff. Metchnikoff believed that the body absorbed toxic waste products from decomposing food residues in the lower intestine. Labeling the process 'autointoxication,' he declared that the body was literally poisoning itself . . . For several years Metchnikoff's theory won wide acceptance in medical circles. In England the famous surgeon Sir Arbuthnot Lane, an old friend of Kellogg's, developed a special operation in which he endeavored to cure autointoxication by removing a large section of the offending colon. Kellogg, too, first tried surgery as a remedy, but he soon became convinced that a reformed diet and the cultivation of 'natural' bowel habits provided the problem's real solution" (Richard W. Schwarz, *John Harvey Kellogg, M.D.*, Nashville: Southern Publishing Assn., 1970, p. 54).

"For Kellogg, the autointoxication theory provided enough ammunition to support three book-length attacks on meat eating. In *Colon hygiene, Autointoxication,* and *The itinerary of a breakfast,* he elaborated time and again on how the common diet contained so much protein from flesh components that it encouraged the growth and activity of proteolytic bacteria in the colon. As the microbes operated on undigested flesh food, the body would be 'flooded with the most horrble and loathsome poisons,' producing headaches, depression, skin problems, chronic fatigue, damage to the liver, kidneys, and blood vessels, and other injuries . . . Anyone who read to the end of Kellogg's baleful list must have been ready to agree that 'the marvel is not that human life is so short and so full of miseries . . . but that civilized human beings are able to live at all'" (John C. Whorton, "Historical development of vegetarianism," *Amer. j. clin. nutr.,* 1994:1106S-1107S).

Colon hygiene was first published in 1915. One of Kellogg's more successful publications, the book was published eight times between 1915 and 1931.

2079. KELLOGG, John Harvey, 1852-1943.
Colon hygiene. Comprising new and important facts concerning the physiology of the colon and an account of practical and successful methods of combating intestinal inactivity and toxemia . . . Twenty-eighth thousand. Battle Creek, Michigan: Good Health Publishing Co., 1917.

[4], 9-417 p., [29] leaves of plates : ill. ; 19 cm.

2080. KELLOGG, John Harvey, 1852-1943.
Dr. Kellogg's lectures on practical health topics. Volume II. The monster malady: cancer the coming plague of the race. Battle Creek, Michigan: Good Health Publishing Co., 1913.

[8], 17-139 p., [13] leaves of plates : ill., port. ; 19 cm.

"Statistics show that in the United States mortality from cancer has increased 800 per cent in sixty-five years" (p. 18). Kellogg notes that though scientists are unclear as to the cause(s) of cancer, "the most recent and most fruitful and scientific researches in relation to cancer point to at least one common predisposing cause in both plants and animals, viz., poisons resulting from the decomposition of animal substances" (p. 28). From this Kellogg concludes: "There is evidently a close relation between the high protein diet (flesh-eating), of our modern civilization and the rapid increase of cancer." He states that cancers are rare among herbivores, more common among carnivores, and nearly epidemic among urban-dwelling humans. "These facts emphasize the observation frequently made that cancer is a disease of civilization, a result of the conditions of domestic and city life, and that through these conditions it is rapidly extending its ravages" (p. 41). Among the predisposing causes of cancer Kellogg discusses the meat diet, alcohol consumption, caffeine and coal tar. Regarding its cure Kellogg writes: "All scientific authorities are agreed that when the method of curing cancer is found it will not be through discovery of an antidote or a germicide, but through the provision of means by which the natural powers of the body are enabled to deal with and destroy the cancer cell" (p. 126). In this vein he notes: "that the human body is capable of resisting and finally overcoming cancerous disease is shown by numerous cases of spontaneous recovery" (p. 132). Until the time when science identifies and unlocks these natural forces of resistance, Kellogg recommends an "antitoxic diet" as the best means of prevention, i.e., "a diet which discourages the development of putrefactive poisons in the intestine" (p. 135) generated by "flesh foods, tea, coffee, alcohol, tobacco, vinegar, mustard, pepper, peppersauce and other condiments" (p. 136). Frontispiece portrait of author.

2081. KELLOGG, John Harvey, 1852-1943.
First book in physiology and hygiene . . . New and revised edition. New York ; Cincinnati ; Chicago: American Book Company, [c1888].
vi, 174, [12] p., [1] leaf of plates : ill. ; 18 cm.

The first edition of the first of Kellogg's two school physiologies was published in 1888 (167 p.). In 1894 his *Second book in physiology and hygiene* (#2107) for the upper grades appeared.

2082. KELLOGG, John Harvey, 1852-1943.
The home hand-book of domestic hygiene and rational medicine . . . Twentieth thousand.—Revised and enlarged. Battle Creek, Michigan: Health Publishing Company, 1889.
[4], xxviii, 25-1624 p., plates I-XXII, A-E + paper manikin : ill. ; 26 cm.

Kellogg writes in the preface: "It is unnecessary to argue the importance of a work of the character which this volume is intended to possess, since the demand for popular works relating to the preservation of health and the treatment of disease has increased so greatly, particularly within the last few years, that books of no other class are in such constant and general demand, if we except the indispensable family Bible" (p. i).
Originally published in 1880, *The home handbook* underwent only modest changes over the course of its twenty-five year publishing history. Kellogg devotes the first third of the book, more than 500 pages, to the anatomy, physiology and hygiene of the human body. The next 200 pages describe what Kellogg terms "rational remedies" for the treatment of disease, by which he means, "medical treatment as may be safely trusted in the hands of unprofessional persons" (p. 581). These are divided into "hygienic agents" (i.e., water, air, light, heat, electricity, exercise, massage, food & mental influences) and "medicinal agents." The former receive far greater attention than the latter, whose description is limited to sixty-seven pages.

The remainder of this considerable volume is devoted to diseases and their treatment. Kellogg warns his readers that any effort at treatment should "act in harmony with nature," that is, "never to hinder the efforts of nature by officious interference. It is a much safer error in the treatment of the sick to do too little than do too much" (p. 822). During the course of disease, "Nature is at work, endeavoring to free herself from obstruction, to remove obnoxious elements from the system, or in some way to remove existing causes of derangement and to restore harmony to the vital processes." Restoring balance to deranged functions is the cardinal principle of Kellogg's therapeutics, for "nature works blindly, she is not intelligent, and often destroys herself in the effort of self-preservation, by too great intensity of action. Hence, when the morbid action is becoming too intense, it should be checked by the employment of well-known means for lessening vital action . . . When the vital action is sluggish or is of too little intensity for the accomplishment of the object desired . . . such remedies should be applied as will increase or stimulate vital activity, for which purpose heat, electricity, and water properly employed, are among the very best of agents" (p. 822).

2083. KELLOGG, John Harvey, 1852-1943.
The household manual of domestic hygiene, foods and drinks, common diseases, accidents and emergencies, and useful hints and recipes. With many other interesting topics. Battle Creek, Mich.: Published at the office of the Health Reformer, 1875.
3 v. in 1 (iv, [5]-124; 60; 62, [4] p.) : ill. ; 17 cm.

There are two issues of this title in the Atwater Collection. The copy represented by this entry is issued with Russell Thacher Trall's *The health and diseases of woman* (Battle Creek, 1873) and Trall's *The hygienic system* (Battle Creek, 1872).

2084. KELLOGG, John Harvey, 1852-1943.
The household manual of domestic hygiene, foods and drinks, common diseases, accidents and emergencies, and useful hints and recipes. With many other interesting topics. Battle Creek, Mich.: Pub-

lished at the office of the Health Reformer, 1875.

　　3 v. in 1 (iv, [5]-124; 60; vi, [7]-83, [5] p.) : ill. ; 17 cm.

There are two issues of this title in the Atwater Collection. The copy represented by this entry is issued with Russell Thacher Trall's *The health and diseases of woman* (Battle Creek, 1873) and Trall's *An essay on tobacco-using* (Battle Creek, 1872).

2085. KELLOGG, John Harvey, 1852-1943.
A household manual of hygiene, food and diet, common diseases, accidents and emergencies, and useful hints and recipes . . . Battle Creek, Mich.: Published at the office of the Health Reformer, 1877.
　　vi, 7-172, [4] p. : ill. ; 18 cm.

An extensively edited and expanded edition of Kellogg's *The household manual of domestic hygiene* (#2083, 2084). Reissued in 1878.

2086. KELLOGG, John Harvey, 1852-1943.
The household monitor of health . . . Battle Creek, Mich.: Good Health Publishing Company, 1891.
　　xv, [16]-408 p. : ill. ; 20 cm.

"The evident relation of health to morals," Kellogg writes in his preface, "is one of the most powerful arguments for the necessity of works of this kind devoted to popular medical instruction" (p. iv). Reflecting, perhaps, his increasingly secular world-view and growing theological uncertainty, Kellogg continues: "It is high time that those who are seeking to reform the world, should begin to preach the gospel of health. Instead of sending missionaries to the Kaffirs, Hottentots, Kalmucks, and Fiji Islanders, let us send a few messengers bearing the glad tidings of good health to the great 'unwashed,' the badly fed, the poorly slept, the generally neglected, and the physically depraved multitudes of our great cities" (p. vi).
　　The household monitor of health incorporates some of the text of Kellogg's 1877 *A household manual of hygiene* (#2085), but has been so extensively expanded, revised and updated in matters of science as to constitute an entirely new work. Many of the same topics are discussed in the same order as in his earlier manuals of

domestic hygiene, with the addition of new sections on testing for adulterated foods, temperance and a concluding fifty-page chapter entitled "Medical frauds."

2087. KELLOGG, John Harvey, 1852-1943.
Ladies' guide in health and disease. Girlhood, maidenhood, wifehood, motherhood . . . Battle Creek, Mich.: Good Health Publishing Co., 1890.
　　[4], xx, [21]-672 p., plates I-IV, VII-VIII, X-XI, F-H, J-K, M, O : ill. ; 23 cm. + atlas ([2] p., plates V-VI, IX, XII-XVI, A-E, L, N ; 19.5 cm.)

The *Ladies' guide* provides detailed advice to women under six general headings: THE ANATOMY AND PHYSIOLOGY OF REPRODUCTION; THE LITTLE GIRL (the education of girls; their moral culture; clothing, exercise, rest, diet; and concluding remarks on "vicious habits"); THE YOUNG LADY (puberty; the education of young women; "mental equality of the sexes"; the pernicious effects of novel reading and dancing; the danger from drugs, stimulants and narcotics; exercise; "the question of woman's dress"; personal beauty; marriage; and "the social evil" or prostitution); THE WIFE ("the dignity of wifehood"; criminal abortion; menopause); THE MOTHER (eugenics; the management of pregnancy; childbirth and its complications); and THE DISEASES OF WOMEN. The appendix includes chapters on children's diseases, alternative therapies (i.e., balneology, electrotherapy, "postural treatment," massage) and medical and dietetic recipes.
　　One of the most successful of Kellogg's published works, the *Ladies guide in health and disease* was issued at least five times in Des Moines between 1883 and 1890, and a dozen times in Battle Creek between 1888 and 1905. Though the number of plates increased with successive printings, the text of the *Ladies guide* remained the same. The atlas (*Colored plates for Ladies guide*) is in a paper wrapper, and is inserted in a pocket on the back pastedown. The plates selected for the atlas are those pertaining to female reproductive anatomy and embryology.

2088. KELLOGG, John Harvey, 1852-1943.
Ladies' guide in health and disease. Girlhood, maidenhood, wifehood, motherhood . . . Battle Creek, Mich.: Published by the Modern Medicine Publishing Co., 1896.

The Sanitarium at Battle Creek, Mich.

Fig. 58. Wood-engraved view from Kellogg's Plain facts, *1885, #2095.*

[2], xx, [21]-690, [16] p., 2-plate paper manikin, plates I-IV, VII-VIII, X-XI, XIX/ XX-XXI : ill. ; 22 cm. + atlas ([2] p., V-VI, IX, XII-XVIII, A-E, L, N)

Includes a two-leaf paper manikin or "dissected plate." Attached to the second plate of the manikin are layered flaps providing successive interior views of the thoracic cavity. A second illustrated supplement (in wrapper) was issued with some or all editions of the *Ladies guide* from the mid-1890s: *Illustrations of natural grace and symmetry unspoiled by the deforming influences of fashionable dress.* The Atwater copy of this 1896 edition is lacking the atlas and supplement that was inserted in the pocket on the back pastedown.

2089. KELLOGG, John Harvey, 1852-1943.
Ladies' guide in health and disease. Girlhood, maidenhood, wifehood, motherhood . . . Battle Creek, Mich.; Chicago, Ill.; New York City: Modern Medicine Publishing Co., 1904.

[2], xx, 21-690, [4] p., 2-plate paper manikin, plates I-IV, VI-VIII, IX-X, XVII/ XVIII-XIX/XX, XXI, F-H, J-K, M, O : ill. ; 23 cm. + atlas ([2] p., plates A-E, L, N, V-VI, IX, XII-XVI) + pamphlet ([10] leaves)

The pocket on the back pastedown contains both pamphlets that were issued with the *Ladies' guide*: *Colored plates for ladies guide,* and *Illustrations of natural grace and symmetry unspoiled by the deforming influences of fashionable dress.*

2090. KELLOGG, John Harvey, 1852-1943.
Man, the masterpiece, or, plain truths plainly told, about boyhood, youth and manhood . . . Battle Creek, Mich.; Chicago, Ill.; New York City: Published by the Modern Medicine Publishing Co., 1894.

xx, 21-604 p., plates I-XXV, 2-plate paper manikin + frontis. : ill., port. ; 23 cm. + atlas (24 p., plates A-E)

"The sole aim of this work has been to inspire the boys and young men of the rising generation with a higher regard for those bodies which the Almighty 'created in his own image,' and pronounced 'very good'; to encourage a greater love and respect for purity in thought and act; to help those whose aspirations are upward, by exposing the snares and evil enticements by which unwary youth are led astray; and thus to aid in the development of a higher, purer, and nobler type of manhood" (*Pref.*, p. vi).

Man, the masterpiece includes chapters on human anatomy and physiology; reproductive anatomy and physiology; the moral and physiological upbringing of boys; physical culture; religion and morality; courtship and marriage; "How to make life a success"; the central importance of diet to physical and mental well-being; the pernicious effects of alcohol and tobacco; "germs"; "what to wear for health"; bathing; "sexual sins and their consequences"; diseases of the male genitalia (though not sexually transmitted diseases); and a review of "common ailments" and their domestic treatment.

Man, the masterpiece was published at least twice in Des Moines (1886, 1889) before its publication at Battle Creek in 1890. It includes the same "Dissected plate of the human body" that illustrates the *Ladies' guide in health and disease* (#2088), and a supplement illustrating the male sexual organs and the manifestations of venereal disease. The supplement is inserted in a pocket on the back pastedown. The Atwater copy is lacking this insert.

2091. KELLOGG, John Harvey, 1852-1943.
Man, the masterpiece, or, plain truths plainly told, about boyhood, youth, and manhood . . . Battle Creek, Mich.; Chicago, Ill.; New York City: Published by Modern Medicine Pub. Co., 1900.
xx, 21-604 p., plates I-XXV, 2-plate paper manikin + frontis. : ill. ; 22 cm. + atlas (24 p., plates A-E)

Atwater copy lacking the illustrated supplement inserted in pocket on back pastedown.

2092. KELLOGG, John Harvey, 1852-1943.
Neurasthenia or nervous exhaustion. With chapters on Christian Science and hypnotism, habits and the "blues" . . . Twelfth thousand. Battle Creek, Michigan: Good Health Publishing Co., 1916.
339 p., [24] leaves of plates : ill. ; 19 cm.

"From the achievement of independence onward . . . Americans found themselves wrestling with ambivalent feelings about their country's open political-social system. The precious freedom to get ahead, to capitalize on the main chance, was also responsible for a competetiveness and uncertainty in life that scraped away unceasingly at the individual's mental and emotional reserves. That civilization begets nervousness became a cliché, and it was perhaps unavoidable that as the pace of life quickened, public uneasiness would mount and be translated into medical dogma. Physicians, after all, not only shared the feeling that modern life was nerve-wracking; they also repeatedly encountered patients whose symptoms had no definite pathology but could plausibly be related to the nervous system. Nervousness was a self-fufulling medical prophecy, and if there was a prophet, it has to have been New York practitioner George Beard. 'Neurasthenia,' or nerve weakness, wass the term Beard applied to what he regarded as the epidemic of advanced civilization" (James C. Whorton, *Crusaders for fitness,* Princeton, N.J.: Princeton Univ. Press, 1982, p. 148).

In his opening chapter in *Neurasthenia* Kellogg writes: "The writer happened to be pursuing post-graduate studies under Doctor Beard and acting as his assistant in the department of nervous disorders at Demilt Dispensary when the doctor was writing his early treatises on neurasthenia, and had an opportunity to become thoroughly familiar with his views of treatment, one of the characteristic features of which was the almost absolute disuse of drugs, at the time a very heretical position" (p. 15). Beard regarded neurasthenia as a distinct disease, an opinion that Kellogg did not share: "Neurasthenia is simply a state of exhaustion of the vital resources, the result of neglecting to confrom to the great biologic laws which have control over the functions of the mind and body" (p. 21). Kellogg continues: "Neurasthenia is only one of the consequences of our wide digression from the path of physical rectitude. The cure of neurasthenia consists in returning to Nature and in the cultivation of the simple life" (p. 22), i.e., Kellogg's program of "biologic living." Not surprisingly, Kellogg's devotion to the concept of autointoxication led him posit "toxic neurasthenia," a depletion of the mind and body resulting from poisons in the tissues and blood from the consumption of meat, tea, coffee, alcohol, etc. It could be treated by a dietary regimen that suppressed intestinal putrefaction, increased elimination of poisons through bowel movements, and "improved tissue building." Adjuvant therapies include light baths, cold applications, the cold-air bath, correct posture and exercise.

Following his discussion of "religious neurasthenia," Kellogg provides one of his more interesting chapters, "Christian Science, Emmanuel Movement, hypnotism," in which he examines the principles of Christian Science, and dismisses Mrs. Eddy's system as "a veritable quagmire of irrational thought," stating that "Christian Science is neither scientific nor Christian." Kellogg notes the similarities between Christian Science and the Emmanuel Movement (e.g., disease has no reality, it is the result of morbid mental states, and may be cured mentally, etc.). Kellogg notes their principal difference: "The Emmanuel Movement differs from Eddyism in that it claims to be strictly in harmony with modern science, while Mrs. Eddy utterly repudiates all that is commonly recognized as science" (p. 162). He then moves from Worcester's (#3873) reliance on suggestion to a discussion of hypnotism or mental healing, which he also dismissed. "But physical maladies must have physical remedies," Kellogg insists, "A patient whose insomnia is due to autointoxication arising from constipation must be relieved of the constipation before permanent improvement can be secured, no matter how much temporary relief may be secured by suggestion or hypnotism" (p. 176).

Kellogg then returns to the perennial theme of his gospel of good health: "Functional nervous disorders in general are not due to bad mental states, but to bad habits. The use of tobacco and of tea and coffee, the flesh-eating habit, hasty eating, sedentary life and neglect of the bowels, are responsible for a thousand cases of nervous diseases where pessimism is responsible for one" (p. 177). Originally pub-

lished in 1914 as one of the titles in *Dr. Kellogg's lectures on practical health topics* series. The second edition was published in 1915, of which this 1916 issue is the 12th thousand.

2093. KELLOGG, John Harvey, 1852-1943.
The new method in diabetes. The practical treatment of diabetes as conducted at the Battle Creek Sanitarium, adapted to home use, based upon the treatment of more than eleven hundred cases . . . Battle Creek, Michigan: Good Health Publishing Co., 1917.
177 p., [11] leaves of plates : ill. ; 19 cm.

First edition. Kellogg's diet therapy for the treatment of diabetes aimed at reducing excess sugar in the blood and increasing the patient's tolerance for carbohydrates. Included in a pocket on the back pastedown is a folding *Metabolism Chart* to assist the self-treating diabetic in determining his carbohydrate balance. In addition to dietary planning in the control of diabetes, Kellogg discusses therapeutic mechanisms that assist the diabetic, e.g., exercise (in muscular activity "the amount of sugar consumed may be increased from three to ten times the rate of consumption during a state of complete rest," p. 87), hydrothrapy (cold baths aid the body in burning up sugar; hot baths aid in combatting acidosis), etc.

2094. KELLOGG, John Harvey, 1852-1943.
Plain facts for old and young . . . Burlington, Iowa: Published by Segner & Condit, 1881.
xvi, 17-512, [4] p., [1] leaf of plates : port. ; 22 cm.

Kellogg's treatise on the sexual life of humankind was originally published at Battle Creek in 1877 under the title: *Plain facts about sexual life.* Kellogg devotes his first hundred pages to the physiology of reproduction in plants, animals and humans. The content provides less information than equivalent works by Hollick (#1697-1737), Foote (#1213-1256) or Pierce (#2798, 2802-2843, 2847). In spite of his promise to accurately detail sexual physiology, Kellogg is obviously squeamish with the topic, as reflected in his brief and inadequate description of male and female genitalia and the absence of any illustration in the volume. In fact,

Kellogg's treatise is less a manual of sexual science than a book-length condemnation of the transgression of sexual laws by one faithful to the traditions of 19th-century health reform and evangelical Christianity.

In concluding the section on reproduction, Kellogg discusses sexual maturation in the female, proceeds to menstruation and its disorders, and considers at length the physical, moral and social implications of hereditary transmission. In the section entitled "The sexual relations," Kellogg limits women's sexual activity to the years between puberty and age forty-five, at which time "all functional activity should cease. If this law is disregarded, disease, premature decay, possibly local degenerations, will be sure to result. Nature cannot be abused with impunity" (p. 123). He discusses "mutual adaption" (for eugenic purposes as well as the couple's compatability), courtship, etc.; and lists eleven reasons that disqualify a person for marriage, e.g., hereditary disease, congenital deformity, criminal tendencies, disparity of age, different race, or "moral character [that] will not bear the closest scrutiny" (p. 169).

In "Chastity" he reviews the factors that lead to sexual impurity, which may as easily be induced by a stimulating diet (and the consumption of coffee, tea and alcohol) as by pornography or habituation to loose company. Even women's fashions can be harmful in this regard: "Fashion requires a woman to compress her waist with bands or corsets. In consequence, the circulation of the blood toward the heart is obstructed. The venous blood is crowded back into the delicate organs of generation" and stimulation results (p. 192). Kellogg begins "Marital excesses" with the statement: "It seems to be a generally prevalent opinion that the marriage ceremony removes all restraint from the exercise of the sexual functions," and characterizes such behavior as "legalized prostitution" (p. 216). He continues: "The sad results of excessive indulgences are seen on every hand" (p. 225). Frequent sexual activity debilitates the body and results in a multitude of disorders, including premature decay. Excessive intercourse also effects the offspring that result from such couplings. Consider, for example, what happens to the fetus whose parents indulge their sensuality during pregnancy: "The delicate brain, which is being molded . . . receives its cast largely from those mental and nervous sensations and actions of the mother which are the most intense. One of the most certain effects of sexual indulgence at this time is to develop abnormally the sexual instinct of the child" (p. 242).

The practice of contraception only encourages sexual excess: "To even mention all of these would be too great a breach of propriety, even in this plain-spoken work; but accurate description is unnecessary, since those who need this warning are perfectly familiar with all the foul accessories of evil thus employed" (p. 250). Kellogg provides a six-page quotation from A.K. Gardner's *Conjugal sins* (#1327-1329) to bolster his argument and outrage; and labels the Oneida Community a "gigantic brothel" in his condemnation of "the scheming sensualist" J.H. Noyes (#2649-2652).

Kellogg's review of sexual immorality continues in the section entitled "The social evil," his euphemism for prostitution. The propensity of humans to indulge in this behavior he attributes to "libidinous blood" (i.e., "the transmission of sensual propensities" through the marital excesses of parents); gluttony ("the exciting influence upon the genital organs of such articles as pepper, mustard, ginger, spices, truffles, wine, and all alcoholic drinks, is well known," p. 292); fashion ("The temptation of dress, fine clothing, costly jewelry . . . is in many cases too powerful for the weakened virtue of poor seamstresses," p. 294); inadequate moral instruction early in life; "sentimental literature"; poverty; ignorance ("intellectual culture is antagonistic to sensuality," p. 300); and disease (e.g., nymphomania). The "Solitary vice" (i.e., masturbation) warrants 104 pages on the nature of the practice, its origins and consequences. Kellogg concludes the volume with "A chapter for boys" and "A chapter for girls."

2095. KELLOGG, John Harvey, 1852-1943.
Plain facts for old and young . . . Burlington, Iowa: Published by I.F. Segner, 1884.
    xvi, 17-512 p., [1] leaf of plates : port. ; 22 cm.

2096. KELLOGG, John Harvey, 1852-1943.
Plain facts for old and young: embracing the natural history and hygiene of organic life . . . New edition, revised and enlarged. Burlington, Iowa: I.F. Segner, 1886 [c1885].
    xx, 21-644 p. ; 23 cm.

In this "new edition," the order of some of the major divisions or section headings have been rearranged, and their text expanded or abbreviated. Most of the newly added sections ("A chapter for young men," "A chapter for young women," etc.) incorporate material that had appeared in other sections of the book in the previous addition. Several sections that are entirely new have been added to the book, e.g., on women's diseases and general hygiene.

2097. KELLOGG, John Harvey, 1852-1943.
Plain facts for old and young: embracing the natural history and hygiene of organic life . . . New edition, revised and enlarged. Burlington, Iowa: I.F. Segner, 1887.
    xx, 21-644 p. ; 23 cm.

2098. KELLOGG, John Harvey, 1852-1943.
Plain facts for old and young: embracing the natural history and hygiene of organic life . . . New edition, revised and enlarged. Burlington, Iowa: I.F. Segner & Co., 1888.
    xx, 21-644 p. ; 23 cm.

2099. KELLOGG, John Harvey, 1852-1943.
Plain facts for old and young: embracing the natural history and hygiene of organic life . . . New edition, revised and enlarged. Burlington, Iowa: I.F. Segner & Co., 1889.
    xx, 21-644 p. ; 23 cm.

2100. KELLOGG, John Harvey, 1852-1943.
Plain facts for old and young: embracing the natural history and hygiene of organic life . . . New edition, revised and enlarged. Burlington, Iowa: I.F. Segner & Co., 1895.
    xx, 21-720 p. ; 23 cm.

In 1894, the "new edition, revised and enlarged" was itself revised and enlarged. The text remains essentially unaltered, however, from that of the prior two editions. The most extensive change is the addition of thirty-eight pages of "Practical suggestions" on diet, health, etc. at the end of the volume.

2101. KELLOGG, John Harvey, 1852-1943.
Plain facts for old and young or the science of human life from infancy to old age. A cyclopedia of special knowledge for all classes on the hygiene of sex. Comprising the anatomy and physiology of reproduc-

tion in man, illustrating the wonderfully interesting sexual phenomena presented by plants and lower animal forms, and discussing in a comprehensive and practical way, all important questions relating to the functions characteristic of sex in health and disease; with an introductory chapter on general anatomy and physiology, and a concluding chapter on obstetrics and the care and feeding of infants . . . 20th century edition . . . Battle Creek, Mich.: Health Library Association, [c1903].
    [2], xxii, 23-798 p. : ill. ; 23 cm. + insert.

The sixty-four pages of plates are included in the pagination of the text. Atwater copy lacking insert in front pocket.

2102. KELLOGG, John Harvey, 1852-1943.
Plain facts for old and young or the science of human life from infancy to old age. An illustrated cyclopedia of special knowledge for all classes on the hygiene of sex . . . Battle Creek, Mich: Good Health Publishing Company, 1910.
    xix, [20]-788, [2] p. : ill. ; 23 cm.

Cover-title: *Plain facts for both sexes.*

2103. KELLOGG, John Harvey, 1852-1943.
Plain facts . . . Battle Creek, Michigan: Good Health Publishing Co., 1917.
    2 v. ([3]-433 p., [11] leaves of plates; [3], 434-683, [2], 684-932 p., [19] leaves of plates) : ill. ; 19 cm.

This two-volume, small octavo edition of *Plain facts* represents Kellogg's principal work on human sexuality in its final manifestation. The work is prefixed by sixty-five newly written pages on human physiology. The section on human reproduction includes new text on cell division, a lengthier and more accurate description of ovulation, a more scientific consideration of the transmission of inherited characteristics, and illustrations of the female genitalia, the development of the human embryo, etc. His chapters on sexual purity, prostitution, masturbation, and criminal abortion remain largely unchanged, in both tone and content. The principal difference between this and earlier editions of *Plain facts* are the addition of three chapters to volume two: "Heredity and eugenics," "How men and women differ" (citing the works of Havelock Ellis, Galton, Key, Broca, etc.), and "Race degeneracy and race improvement." The first and third of these added chapters reflect Kellogg's growing interest in the science of genetics in the second decade of the 20th century as he attempted to give his long held views on eugenics a more scientific basis. Kellogg's ultimate purpose in this area of study was "to create an aristocracy of health, to form a selected group of human beings possessed of superior characteristics of mind and body" (2:914).

2104. KELLOGG, John Harvey, 1852-1943.
Rational hydrotherapy. A manual of the physiological and therapeutic effects of hydriatic procedures, and the technique of their application in the treatment of disease . . . Philadelphia: The F.A. Davis Company, publishers, 1901.
    xxxi, [1], 21-1193 p., [104] leaves of plates : ill. ; 24 cm.

    "Modern scientific research has placed upon a sure foundation the great truth dimly recognized by the earliest physicians . . . that healing power is not possessed by physicians nor by remedies, but that the curative process is simply a manifestation of the forces which dwell within the body and which are normally manifested in creating and maintaining the organism, in other words, that the body heals itself. Water, applied externally or internally, and at such temperatures as may be required, is a natural agent more capable than any other of co-operating with the healing powers of the body in resisting the onset and development of pathogenic processes" (*Pref.*, p. viii).
    This is the second issue of the first edition of *Rational hydrotherapy,* first published at Philadelphia in 1900. The work began as a series of lectures delivered to Adventist medical students beginning in 1890. Though the published work was intended for medical students, nurses and the profession, it is included here because the therapies described were all in use at the Battle Creek Sanitarium and derive, in part, from the hydropathic armamentarium of the previous century (i.e., procedures with which Kellogg was familiar even before his attendance at Russell Trall's Hygieo-Therapeutic College). Interestingly, Trall is nowhere acknowledged in Kellogg's preface, probably because of his small

regard for the training he received at Trall's College in 1872, and also to disassociate the therapies described here from a similar but outdated school of hydrotherapy.
    Kellogg describes the book's arrangement: "First, a short historical sketch and a brief resume of physical, anatomical, and physiological facts which are especially related to the subject; second, a study of the physiological effects of thermic applications; third, a description of the technique of all useful hydriatic procedures; and fourth, a section on hydriatic prescription making, in which is presented a brief summary of the indications presented by the diseases most commonly encountered in practice and of the hydric measures required to meet the same . . . The number of procedures described in this work is about two hundred . . . Of those procedures which are the outgrowth of his own experience, the author has mentioned only such as have acquired a recognized value by extended use. Of these, the most worthy of mention are the electric-light bath, the percussion douche, cold mitten friction, cold towel rub, alternate spinal sponging, a number of forms of hot and cold compress and the hot and cold pack, and the simultaneous hot and cold douche" (p. vi).
    The second edition of *Rational hydrotherapy* was published by F.A. Davis in Philadelphia in 1902 (#2105), 1903 (#2106) and 1904. A third edition appeared in 1906; and the fourth edition in 1910. The fourth edition was reissued in both Philadelphia and Battle Creek in 1918. An undesignated 1,265 p. edition was published in Philadelphia and Battle Creek in 1923; and reissued for the final time at Battle Creek in 1928.

2105. KELLOGG, John Harvey, 1852-1943.
Rational hydrotherapy. A manual of the physiological and therapeutic effects of hydriatic procedures, and the technique of their application in the treatment of disease . . . Second edition. Philadelphia: F.A. Davis Company, publishers, 1902.
    xxxi, [1], 21-1193 p., [104] leaves of plates : ill. ; 24 cm.

    In the "Preface to the second edition" Kellogg writes," "The interest in therapeutic agents of a non-medicinal character, including heat, light, electricity, exercise, massage, and other physical means, as well as water, is rapidly gaining ground; and, if one may judge by the frequency with which articles and treatises on these subjects are now appearing from the medical press, it may be said that an era of physiological medicine has begun. If this be true, the beginning of the twentieth century will mark the most important epoch in the history of medicine" (p. x). Kellogg could not have anticipated the eclipsing effect that the development of drugs such as diphtheria antitoxins, salversan, insulin, sulfa drugs, etc. would have on the attitude of the medical profession and the public toward "physiological medicine" by the 1930s.

2106. KELLOGG, John Harvey, 1852-1943.
Rational hydrotherapy. A manual of the physiological and therapeutic effects of hydriatic procedures, and the technique of their application in the treatment of disease . . . Second edition. Philadelphia: F.A. Davis Company, publishers, 1903.
      xxxi, [1], 21-1193 p., [104] leaves of plates : ill. ; 24 cm.

2107. KELLOGG, John Harvey, 1852-1943.
Second book in physiology and hygiene . . . New York; Cincinnati; Chicago: American Book Company, 1894.
      vi, [7]-291 p., IV leaves of plates : ill. ; 18 cm.

The second of Kellogg's two school physiology texts. In 1888 he published his *First book in physiology and hygiene* (#2081) for the elementary grades.

2108. KELLOGG, John Harvey, 1852-1943.
The stomach: its disorders, and how to cure them . . . Battle Creek, Mich. . . . [etc.]: Modern Medicine Pub. Co., 1896.
      342, [2], 343-368, [12] p. : ill. ; 20 cm.

"In the study of the practical part of the work, it will be noticed that the measures of treatment recommended are of a remarkably simple character. This may lead to the impression that they are lacking in efficiency; hence it may be proper to say in their defense, that many years of experience in the treatment of a large number of patients of this class, in connection with the Battle Creek (Mich.) Sanitarium, have fully established the author in the belief that in the proper regulation of the diet, in the varied uses of water as described in scientific hydrotherapy, in the employment of electricity, Swedish movements, Swedish gymnastics, massage, and exercise, together with general control of the habits of life, we have means of far greater value than all others for combatting those causes which give rise to indigestion, and in aiding nature in her efforts for recovery from the varied morbid conditions resulting from the numerous disorders commonly included under the general term, 'dyspepsia'" (*Pref.*, p. 4). *The stomach* was also published in 1896 at Burlington, Iowa by I.F. Segner and at London, England by the International Tract Society.

KELLOGG, John Harvey see also
HARTELIUS (#1568).

2109. KELLOGG, Merritt Gardner, 1832-1922.
The hygienic family physician: a complete guide for the preservation of health, and the treatment of the sick without medicine . . . Battle Creek, Mich.: Published at the office of the Health Reformer, 1873.
      vii, [1], [5]-380, 16 p. ; 18 cm.

The eldest son of John Preston Kellogg and the half-brother of John Harvey Kellogg (#2076-2108), Merritt Gardner Kellogg graduated in 1867 from the Hygieo-Therapeutic College, the hydropathic medical school founded and operated by R.T. Trall (#3558-3593). Until 1872 Kellogg lived in California, at which time he was called back to Battle Creek by his family to help manage the troubled Western Health Reform Institute established in 1866 and operated by the Seventh Day Adventists. Almost immediately Merritt Kellogg decided that he needed a second course of lectures at Trall's medical college in Florence Heights, N.J., and took with him his half-brother J.H. Kellogg and several other young Adventists. Merritt returned to California before J.H. Kellogg took over management of the Health Reform Institute in 1876.
      The hygienic family physician is divided into four parts. In the first Kellogg addresses "health and hygienic agents" such as air, water, food, clothing, exercise, etc. In the second he dis-

cusses the nature of disease and its treatment, stating that disease is a remedial effort on the part of the body to remove poisons, "and that, consequently, the only safe and successful way to treat disease is to supply such conditions as will enable the diseased organs to be successful in their efforts" (p. 113). Drug therapy is pointless, since these supposedly remedial agents "mitigate or change the symptoms without removing the cause which occasioned them." In part III Kellogg examines the principles of the hydropathic practice and provides detailed descriptions of water's numerous therapeutic applications. Part IV addresses diseases specifically, noting their symptoms, prognosis and treatment. The sixteeen pages appended to the volume promote the health publications of the Adventist press, its monthly journal *The Health reformer,* and the Health Institute in Battle Creek, Michigan.

2110. KELLOGG, Merritt Gardner, 1832-1922.

The hygienic family physician. A complete guide for the preservation of health, and the treatment of the sick without medicine; comprising also a valuable pamphlet on diet. Battle Creek, Mich.: Published at the office of the Health Reformer, 1874.
    5 pts. in 1 v. (iv, 32 p. ; 48 p. ; 64 p. ; viii, [9]-156 p. ; 3-194, 8 p.) ; 18 cm.

"Owing to the unexpected rapidity with which the first bound edition [#2109] . . . has been sold . . . the Publishers have been induced to prepare this volume as a provisional substitute for it, a reprint of the first edition being at present inexpedient. As will be observed, this work consists of five pamphlets. The first four contain essentially the same reading matter as the bound volume of the Hygienic Family Physician, each of the four parts of that work having been published separately in pamphlet form. The remaining pamphlet comprises three of the most important and interesting of Dr. Graham's celebrated 'Lectures on the science of human life,' and is chiefly devoted to the subject of diet" (p. [ii]).
    The content of this edition corresponds to the four parts of the 1873 edition by the inclusion of Kellogg's *Good health* (no title-page); *The nature and cause of disease, and the so-called "action" of drugs* (no title-page); *The bath* (title-page

dated 1873); and *The treatment of disease by hygienic agencies* (title-page dated 1873). Also included is an abridged edition of Sylvester Graham's *Lectures on the science of human life* (title-page dated 1872).

2111. KENDALL, Burney James, 1845-1922.

The parent's guide in sex problems from 5 to 75 . . . Geneva, Ill.: Published by B.J. Kendall, M.D., [c1911].
    283 p., [1] leaf of plates : port. ; 17 cm.

Kendall's book might have been titled *The prospective parent's guide in sex problems.* He shares his concern with likely parents about the hereditary implications of their future sexual activity: "Some physicians who have written upon this subject have taken the extreme view that unlimited sexual intercourse is essential to the health of man. This theory is not only contrary to experience and the laws of nature, but is a most dangerous idea for many reasons. In the first place, if it was practiced generally, there never would be another child well-born. No man can indulge to excess . . . and become the father of children that are well-born . . . In such a case conception might take place at any time, and would rarely, if ever, be when the man is in a fit condition to procreate children who would inherit all that is best in human nature" (p. 14). In order to guard against the temptation to indiscriminate intercourse, Kendall recommends that couples sleep in separate beds and not dress or undress in one another's presence.
    Following a discussion of sexual etiquette, Kendall devotes considerable attention to the role that parents' physical and emotional condition plays in hereditary transmission: "All children should be superior to their parents . . . This will apply to their physique, or bodily structure, mental capacity, moral character, religious inclination, musical culture, art, science, ease of expression or eloquence in public address. Each generation should be better born than the preceding one and it will be if the laws of heredity are studied and proper pre-natal culture practiced by prospective parents" (p. 108). Kendall advocates that prospective parents improve their own moral, intellectual, social and artistic qualities both before and after conception in order to ensure that their children possess these same qualities in equal or superior measure at birth. In later chapters Kendall ad-

dresses parents on the training of infants and children, the sexual dangers that await their adolescents, and other threats to the physical and moral development of the young.

An 1868 medical graduate of the University of Vermont (Burlington), Kendall authored several manuals of domestic and veterinary medicine. His most successful published work, *A treatise on the horse and his diseases* (1879) was last issued in 1949. He was also founder of the Dr. B.J. Kendall Co., manufacturers of proprietary remedies and grooming aides (#2112). In 1894 Kendall moved from Vermont to Illinois.

2112. DR. B. J. KENDALL CO.

The doctor at home. Illustrated. Treating the diseases of man and the horse. A practical hand-book for the non-professional. Written in plain, simple language so as to be easily understood by the common people. It treats all of the common diseases of the human family, and gives a large number of excellent receipts and favorite prescriptions for various diseases . . . Also, an essay on hygiene, a table of doses, with proportional doses for children; and, in Part Second, the diseases of the horse are plainly treated, with the plainest and best treatment that can be given under most circumstances; and a table of doses for the horse, a table showing the age of the horse, a large number of excellent receipts for the horse, and a large amount of other valuable information . . . Enosburgh Falls, Vt.: Printed at steam printing house of Dr. B.J. Kendall Co., 1884.
    96 p. : ill. ; 19 cm.

In printed wrapper. Includes separately printed advertisement for *Glyceroil for the Hair* and numerous promotions for Burney James Kendall's (#2111) proprietary remedies throughout the text.

2113. KENNEDY, Donald, 1812-1889.

Donald Kennedy's medicines. Including Kennedy on diseases of the skin. Third edition. Edited and published by Donald Kennedy & Co. Roxbury, Mass.: Copyright, Donald Kennedy & Co., 1909.
    76 p. ; 22 cm.

Kennedy's Medical Discovery, a compound of roots and herbs in an alcohol solution, is de-scribed as "a medicine for humors—in use since 1846—to open the channels of the body and remove obstructions in the liver, bowels, kidneys and glands" (p. [5]). The manufacturers profess their ignorance as to "how, why, or by what means its does its work"—"satisfy yourself with the explanation you prefer" (p. 6). Its efficacy is sufficient justification in the treatment of some forty-four ailments, thirty of them skin diseases or diseases with skin manifestations.

2114. KENNEDY, Donald, 1812-1889.

Kennedy on diseases of the skin. Roxbury, Mass.: Published by Donald Kennedy, [after 1858].
    48 p. ; 22 cm.

Title and imprint from wrapper (publication date from p. 13). Kennedy recommends his *Medical Discovery* and other proprietary remedies in the treatment of skin diseases, rheumatism, "female diseases," etc. Nearly half this pamphlet is devoted to testimonials.

2115. KENNEDY, Donald, 1812-1889.

Kennedy on diseases of the skin. Roxbury, Mass.: Published by Donald Kennedy, [after 1860].
    64 p. ; 22 cm.

Title and imprint from wrapper (publication date from p. 17).

2116. KENNEDY, Donald, 1812-1889.

Kennedy on diseases of the skin. Second edition. Roxbury, Mass.: Published by Donald Kennedy, [ca. 1878, c1871].
    128 p., 3 leaves of plates : ill. ; 22 cm.

Title and imprint from wrapper. A four-leaf advertisement for Kennedy's *Prairie Weed* (bound between pages 16 & 17) includes testimonials dated 1878.

KENNEDY, James Donaldson see
KENNEDY AND KERGAN (#2117).

2117. KENNEDY AND KERGAN.

Boyhood, manhood, fatherhood. The three stages of life. [Detroit: Kennedy & Kergan, ca. 1900].
    [2], 32, [2] p. : ill. ; 16 cm.

A brochure promoting the *New Method Treatment* of Drs. Kennedy & Kergan, 148 Shelby St., Detroit. This unspecified remedy is guaranteed to cure all disorders of the male genital system, e.g., spermatorrhoea, varicocele, sexually transmitted diseases, the many consequences of masturbation etc. Drs. Kennedy & Kergan attempt to convince their readers that the structure few laymen would recognize as the epididymis is actually the beginning of varicocele (p. 14); that they can supply an *Adjustable Developer* to lengthen "small parts or organs to a normal state" (p. 12); or that in cases of syphilis, "The virus or poison is neutralized in the blood and eliminated from the system" by use of their *New Method Treatment* (p. 9).

The 4th edition of Polk's *Medical and surgical register of the United States* locates James Donaldson Kennedy (1865-1927) and John Depew Kergan (d. 1904) at 148 Shelby St. in Detroit. Kennedy was an 1888 medical graduate of the University of Western Ontario. In 1913 Kennedy's license to practice medicine in Michigan was revoked: "This action followed the flight of Dr. Kennedy from the state while awaiting sentence on conviction of distributing obscene literature" (*Nostrums and quackery*, #86, 2:433). Kergan is described in Polk as an eclectic, an 1860 graduate of the Royal College of Surgeons (Dublin), a "licentiate of Canada," an "ex-professor" of physiology at the Detroit Homoeopathic College, the "ex-secretary" of the State Homoeopathic Medical Institute, "and an "ex-member" of the Board of Censors of the Eclectic Medical and Surgical Society of Michigan.

2118. KEPPEL, William H.

The ladies' beauty book and the old home doctor. Containing over 1300 toilet receipts and hints to assist in improving one's personal appearance as well as over 2300 doctors' prescriptions, medical formulas and suggestions for relief from the many ailments to which the human flesh is heir . . . Two complete volumes bound in one. Copyright edition. Tiffin, Ohio: The Keppel Publishing Company, [191-?].
    392 p. ; 23 cm.

KERGAN, John Depew see **KENNEDY AND KERGAN** (#2117).

2119. KERLEY, Charles Gilmore, 1863-1945.

Short talks with young mothers on the management of infants and young children . . . New York & London: G.P. Putnam's Sons, the Knickerbocker Press, 1904.
    xiii, [1], 262, [2] p. : ill. ; 19 cm.

Kerley was an 1888 medical graduate of the University of the City of New York. He was professor of pediatrics at the New York Polyclinic Medical School & Hospital, and attending physician at the New York Infant Asylum and the Babies' Hospital. Kerley was an advocate for better school health and the instruction of students on public health issues. In 1908 he was elected president of the American Pediatric Society. Kerley published the *Treatment of the diseases of children* in 1907 and *The practice of pediatrics* in 1914. His *Short talks with young mothers* was first published in 1901; the eighth and final edition appeared in 1926.

2120. KESTEVEN, William Bedford, 1812-1891.

Home doctoring. A guide to domestic medicine and surgery . . . London and New York: Frederick Warne and Co., [1889].
    [4], 156 p. : ill. ; 19 cm.

Kesteven received his medical education at St. Bartholomew's Hospital and University College (London) before entering the practice of medicine in 1838. He was made a Fellow of the Royal College of Surgeons in 1854. In 1856 Kesteven published his *Manual of the domestic practice of medicine,* which appeared in at least four English editions through 1870. *Home doctoring* (1889) appears to be the final manifestation of this work.

2121. KEY, Ellen Karolina Sofia, 1849-1926.

Love and marriage . . . Translated from the Swedish by Arthur G. Chater. With a critical and biographical introduction by Havelock Ellis. New York and London: G.P. Putnam's Sons, [c1911].
    [2], xvi, [2], 399 p. ; 20 cm.

First edition (12th printing); translation of: *Kärleken och äktenskapet.* A disciple of Rousseau and Darwin, the Swedish social reformer argues that monogamous marriage was simply a stage in the social and moral evolution of mankind, but that it has now become a constraint to its further development. Those who adhere to traditional Christian morality have failed to realize that the health of the race henceforth depends not on monogamous relationships, but "on the capacity and willingness of its women to bear and foster children fit to live, and on their husbands' capacity and willingness to protect the national existence" (p. 9). Key writes: "The new conception of morality grows out of the hope of the gradual ascent of the race towards greater perfection. Those forms of sexual life which best serve this progress must therefore become the standards of the new morality" (p. 16). "Each fresh couple, whatever form they may choose for their cohabitation, must themselves prove its moral claim" (p. 17), and must realize that their relationship is "subject to the same law as every other creative force" as they climb toward a "higher humanity." The new morality "no longer accepts commandments from the mountains of Sinai or Galilee" (p. 52); rather, it "will reject what hinders and select what assists *its struggle for the strengthening of its position as humanity and its elevation to superhumanity*" (p. 53). From this perspective, Key addresses the issues of love, motherhood, divorce and marriage in the chapters that follow.

2122. KIBLINGER, Elliott, b. 1873.
Medical facts you should know . . . Moreauville, Louisiana: Published by the author, [c1916].
    150 p. : ill. ; 17 cm.

    Kiblinger covers much ground in this slender volume: personal hygiene, home nursing, the diseases of women and children, a home medical formulary, veterinary medicine, etc. He devotes a four-page chapter to the use of enema for treatment of autointoxication ("the injection of from one to one and one-half gallons of water into the bowels" daily); and in the following chapter describes the domestic treatment of alcoholism by the administration of strychnine tablets ("a powerful tonic"), ergot ("for its effect on the circulation" and the prevention of "wet brain"), a morphine solution ("to keep the patient's nervous system quiet"), and "hyoscamous" (which "in a few days . . . obliterates that constant craving for whiskey"). The author

was a physician in Moreauville, Avoyelles Parish, in central Louisiana.

KIDD, James William see **J.W. KIDD COMPANY** (#2123, 2124).

2123. J. W. KIDD COMPANY.
The ills of humanity. By Dr. James W. Kidd. Fort Wayne, Ind., U.S.A.: The J.W. Kidd Company, [190-?].
    104 p. : ill. ; 19 cm.

    Although under the direction of James William Kidd, the J.W. Kidd Company was one of several proprietary drug companies owned by W.M. Griffin, whose activites "gave Fort Wayne an unenviable reputation as the home of some of the most impudent pieces of mail-order quackery in the world" (*Nostrums and quackery,* #86, 2:298). The Griffin-Kidd partnership was dissolved in 1914 when they were charged with obtaining money fraudulently through the mail. It is noted in *Nostrums and quackery* that: "After two or three years of experience in the theatrical business J.W. Kidd seems to have returned from Muscatine, Iowa, to his old stamping ground at Fort Wayne. Now (1920) Kidd has a medical mail-order concern of his own operated under the name of 'Dr. James W. Kidd'" (*ibid.,* p. 300). In printed wrapper.

2124. J. W. KIDD COMPANY.
The ills of humanity. Their cause and cure . . . Fort Wayne, Indiana: The J.W. Kidd Company, [c1908].
    192 p. : ill. ; 20 cm.

    Prospective patients are solicited to send a description of their symptoms through the mail to James W. Kidd for diagnosis and recommendation of treatment: "It is no longer necessary for anyone suffering with a chronic disease to accept the services of an incompetent physician or inferior medicines of any kind . . . Dr. Kidd has ample facilities for sending the very best medical treatment right to the home of the person afflicted" (p. 9).

2125. KIDDER, E., Mrs.
MRS. E. KIDDER'S CHOLERA, DYSENTERY AND DIARRHOEA CORDIAL. Cliftondale, Mass. . . . [ca. 1849].
    [2] leaves ; 34 cm.

"Where this all-powerful antidote is at hand, Cholera is no longer seriously to be feared, or looked upon with terror, as this Cordial will most assuredly cure the disease in the course of a very few hours, if taken at the commencement. It has been before the public for more than seventeen years, and was the first article made known to the public as an IMMEDIATE AND PERFECT CURE OF THE CHOLERA. It has been thoroughly tested in every country and every climate, and its effect has everywhere proved the same; SURE TO CURE even where the disease has advanced to its last stages" (leaf 1 recto).

2126. KIDDER, Jerome.
The vital resources contributing to capacity, health, and longevity. Including also the great law of the evolution and progression of the human species, by the more comprehensive inheritance resulting from duality of parentage . . . New York: Published by the author, 1878.
181 p. : ill. ; 19 cm.

In his study of the laws of hereditary transmission, Kidder asserts that not only the physical characteristics of parents may be transmitted, but also "the capacities acquired by education and experience [may] . . . be added to the inheritance of descendants" (p. 106).

KILMER, Fred Barnett see **JOHNSON & JOHNSON** (#2029-2034).

KILMER, S. Andral see **DR. KILMER AND COMPANY** (#2128-2131).

2127. KILMER, Theron Wendell, 1872-1946.
The practical care of the baby . . . Philadelphia: F.A. Davis Company, publishers, 1904.
xiv, 158 p. : ill. ; 20 cm.

Kilmer's manual of infant care and domestic pediatrics was issued at least five times between 1903 and 1920. An 1895 graduate of the College of Physicians & Surgeons (New York), Kilmer was professor of pediatrics at the New York Polyclinic Medical School & Hospital.

2128. DR. KILMER AND COMPANY.
GUIDE TO HEALTH. Circulated for the benefit of suffering humanity [*wood-en-graved portrait of Kilmer, 18.2 × 13 cm.*] The Physician at Work. S. ANDRAL KILMER, M.D., BINGHAMTON, N.Y. The Invalid's Benefactor . . . [Binghamton, N.Y.: Dr. Kilmer & Co., 1890].
[8] leaves : ill., port. ; 29 × 22 cm.

The *Guide to health* promotes *Dr. Kilmer's Swamp-Root* ("the wonderful specific for rheumatism, gravel, diabetes, Bright's disease, catarrh of the bladder, and all urinary difficulties") and other herbal remedies. S. Andral Kilmer's (1840-1924) consulting rooms, dispensary and laboratory in Binghamton, N.Y. are also described. Kilmer's medicines were successfully marketed in both the United States and Great Britain.

2129. DR. KILMER AND COMPANY.
Invalid's guide to health. Circulated for the benefit of suffering humanity . . . Binghamton, N.Y.: Dr. Kilmer & Co., [ca. 1895].
[8] leaves : ill., port. ; 28 × 22 cm.

Promotional literature for *Dr. Kilmer's Swamp-Root* and other remedies manufactured by Dr. Kilmer & Co.

2130. DR. KILMER AND COMPANY.
Our guide to health. [Binghamton, N.Y.]: Dr. Kilmer & Co., [c1897-98].
[8] leaves : ill., ports. ; 28 × 21 cm.

Advertisements for *Dr. Kilmer's Swamp-Root* and other remedies available from Dr. Kilmer & Co. The pamphlet is accompanied by a sheet promoting *Swamp-Root* in French, Spanish, Polish and German; and with a "Certificate of Purity," signed by Jonas M. Kilmer and dated 26 April 1898 testifying that the remedy is "purely vegetable and does not contain any calomel, mercury, creosote, morphine, opium, strychnine, cocaine, nitrate potash (salt-peter), bromide potassium, narcotic alkaloid or any other harmful drug."

2131. DR. KILMER AND COMPANY.
TESTIMONIAL SHEET. (Supplement to "Guide to health.") *Every testimonial* published by Dr. Kilmer & Co. is *absolutely true.* The most searching inquiry is invited. The original letters will be sent on application. FOR ALL KIDNEY, LIVER & BLADDER

TROUBLES USE Dr. Kilmer's SWAMP-ROOT, the best in the world . . . [ca. 1890].

[1] leaf printed on both sides : music, port. ; 29 × 22 cm.

Printed on the verso are music and lyrics for *Musica-medicus,* a piece that sings the praises of *Dr. Kilmer's Swamp-Root* with refrains such as: "Now Doctor Kilmer is a man believing in progression, and that is why we find him at the head of his profession." S. Andral Kilmer was an 1875 graduate of the Bennett College of Eclectic Medicine & Surgery (Chicago). His activities merit a lengthy article in the A.M.A.'s *Nostrums and quackery* (#86, 2:674-80).

2132. KING, Dan, 1791-1864.

Quackery unmasked: or a consideration of the most prominent empirical schemes of the present time, with an enumeration of some of the causes which contribute to their support . . . Boston: Printed by David Clapp, 1858.

334 p. ; 20 cm.

First edition. An 1814 Yale medical graduate, King practiced medicine in Rhode Island until 1848, and in Massachusetts until his retirement in 1859. Although unsuccessful as a businessman (cloth manufacturing), King served in the Rhode Island General Assembly and was an outspoken advocate of the suffrage and abolition movements. In 1852 he was awarded an honorary M.D. by the Berkshire Medical Institution. "*Quackery unmasked . . .* while in the main an indictment of homeopathy, was also an eloquent plea for higher standards of medical education" (*Dict. Amer. biog.* 5:386). Although King devotes half the book to homeopathy, *Quackery unmasked* addresses the gamut of alternative therapies available to Americans at mid-century, i.e., hydropathy, Thomsonism, Indian medicine, eclecticism, chrono-thermalism and natural bone-setters. He attributes the proliferation of these schools to newspaper advertising ("the American press is under no legal or moral restraint, and is ever ready for money, to aid impostors in deceiving and defrauding the public," p. 251), "female influence" ("the husband rarely meddles with medical matters in his own family," p. 263), discord within the medical profession, and clerical influence ("instead of leading men to the

fountain of living waters . . . they direct them to the wet sheet or shower bath of Priessnitz," p. 275). A "new edition" of *Quackery unmasked* was published at New York by Wood (1858?).

2133. KING, Dan, 1791-1864.

Tobacco; what it is, and what it does . . . New York: S.S. & W. Wood, publishers, 1861.

viii, [9]-171 p. ; 15 cm.

King argues that tobacco induces or assists the progress of numerous diseases by poisoning the blood. It also diminishes appetite, weakens digestion, debilitates the body's recuperative powers, impairs the intellect and memory, encourages alcohol consumption and leads to other vices—most notably, *individualism*: "Tobacco . . . tends to dissolve the domestic and social ties . . . The life of a confirmed smoker is an abstraction—he is wholly wrapped up in himself—whilst the brain is overpowered with tobacco his existence seems like a blissful dream, and . . . he has little desire for social intercourse, unless, prompted by motives of self interest; his attachment to his pipe or cigar is of the strongest kind, his domestic and social propensities become more or less smothered in that oblivious element, and the attention and affection that he should manifest toward a wife, a child, friend or society is bestowed upon this poisonous weed" (p. 105).

2134. KING, Franklin Hiram, 1848-1911.

Ventilation for dwellings, rural schools and stables . . . Madison, Wis.: Published by the author, 1908.

[4], 128, [4] p. : ill. ; 20 cm.

"In the preparation of this brief treatise the aim has been to reach parents, teachers and school officers of rural and other elementary schools, and the owners and caretakers of all classes of live stock, and lay before them the foundation facts and principles underlying the growing and imperative demand for a more nearly adequate supply of pure air than is being continuously maintained in the vast majority of homes, offices and stables today" (*Pref.*)

Both before and after his two years study of the natural sciences at Cornell University (1877-78), King taught science in the secondary schools of rural Wisconsin. In 1888 he was appointed to the chair of agricultural physics at the Wisconsin College of Agriculture, the first such academic position in the United States. "In this field, his contributions to agriculture were valuable and varied. His most important contributions to farm life were the construction of the round silo, a new method of barn ventilation, and his studies of soil solution" (*Dict. Amer. biog.*, 5:389). From 1901 to 1904, King was chief of the division of soil management for the United States Bureau of Soils. His several other published books and many journal articles were devoted to agricultural topics.

2135. KING, John, 1813-1893.

The American family physician; or, domestic guide to health. For the use of physicians, families, plantations, ships, travelers, etc. . . . Cincinnati: Longley Brothers, publishers, 1858.

viii, 9-798, [2] p., [3] leaves of plates : ill., port. ; 24 cm.

King was an 1838 graduate of the Reformed Medical Academy (New York), the eclectic medical school founded by Wooster Beach (#240-251) in 1829. He practiced in New Bedford, Mass. until 1846, when he migrated west where the Thomsonians and their reformed allies enjoyed greater success. King practiced in Kentucky until 1848, when he moved to Cincinnati. In response to the formation of the A.M.A. in 1847, King called for a congress of "reformed practitioners" that resulted in the founding of the National Eclectic Medical Association. In 1849 he joined the faculty of the Memphis Medical Institute, but returned to Cincinnati in 1851 as professor of obstetrics and the diseases of women & children at the Eclectic Medical Institute. Except for a brief tenure at the rival Cincinnati College of Eclectic Medicine, King remained on the E.M.I. faculty the remainder of his career. John King was one of the leading intellectual figures of the eclectic school. He wrote several books that served as standard texts among students and practitioners of eclecticism. His most enduring contribution to medical literature was *The American dispensatory* (1852), a monumental work on the botanic materia medica that became the standard reference work of eclectic pharmacology, appearing in eighteen editions before the end of the century.

King devotes the first 160 pages of *The American family physician* to matters of hygiene and the preservation of health. The remainder of the work is divided into four parts: Part. I. Symptoms, causes, and treatment of diseases; Part II. Surgical diseases; Part III. Materia medica and pharmacy; and Part IV. Hydropathic appliances (written at King's request by "W.S. Bush, a successful hydropathic practitioner"). Originally published at Cincinnati in 1857, King's manual of domestic hygiene and medicine appeared in at least fifteen editions by 1892. Cover-title: *Guide to health*. Lithographic frontispiece portrait of author.

2136. KING. John, 1813-1893.

The American family physician; or, domestic guide to health. Prepared expressly for the use of families, in language adapted to the understanding of the people . . . Indianapolis, Ind.: A.D. Streight, publisher, 1864.

2 pts. in 1 v. (viii, 9-794 p., [16] leaves of plates + frontis. ; 333, [5] p.) : ill., port. ; 24 cm.

Fig. 59. John King, frontispiece to The American family physician, *#2135.*

The *New American family physician* (spine-title) is divided into two parts. (The 1861 Indianapolis edition appears to be the the the first in which this two-part division appears.) Division I. corresponds to the contents of the first edition (#2135), and was printed from the same plates (with the addition of a new preface and index). In his preface King writes: "For the purpose of imparting to the people much valuable information relative to health and disease . . . this work is more especially prepared . . . No less than *three hundred and seventy forms of disease,* including diseases of women, diseases of children, chronic diseases, as well as those of a surgical nature, are accurately described, and the most successful methods of treatment made known; nearly *four hundred and fifty simple medicines* are described, together with their virtues and medicinal uses; and the recipes for some *two hundred and fifty valuable and successful medicinal compounds and preparations* are given" (*Pref.,* p. vii).

Division II. ("Household Department") is entirely new to this edition. In his introductory remarks to this division King writes: "From the fact that so many persons reside where it is difficult to immediately procure some of the remedies recommended in the first Division of this work, especially those living in country places, who are frequently many miles from a physician or drug-store . . . I have deemed it best to prepare a second Division, from which any *person*—one who can barely read—will be enabled to treat most diseases promptly and successfully" (p. 1). In this second division, "I have not selected my medicines from the *Materia Medica* of the profession alone, but have recommended many of those excellent and efficacious articles which were so successfully employed in cases of disease by the old female nurses of past years, and by the good mothers of our early childhood. These articles may mostly be found either in the kitchen, the pantry, the garden, the fields, along the roadsides, or in the woods, throughout our country; they are simple, safe, and harmless, yet will prove of immense advantage in breaking up disease, if applied in season." Lithographic frontispiece portait of author.

2137. KING, John, 1813-1893.
The American family physician; or, domestic guide to health. Prepared expressly for the use of families, in language adapted to the understanding of the people . . . Indianapolis, Ind.: Streight & Douglass, publishers, 1874.
2 pts. in 1 v. (viii, 9-794 p., [16] leaves of plates + frontis. ; 338, [2] p.) : ill. ; port. ; 24 cm.

Spine-title: *King's new American family physician.*

2138. KING, John, 1813-1893.
Women: their diseases and their treatment . . . Cincinnati: Longley Brothers, publishers, 1858.
vi, 7-366, [2] p. : ill. ; 23 cm.

Unlike his 1855 *American eclectic obstetrics,* the present work was intended to meet to the demands of both the profession and the public: "In the composition of the work, several objects have been held in view, viz: to give a thorough account of the history, symptoms, &c., of the various diseases referred to, in as condensed a manner as possible, so as to retain it within certain limits, and place it at a price within the reach of everybody; to state fully the treatment pursued in these affections by the most successful practitioners . . . and so to arrange the work that while it may prove useful to the practitioner, its pages may also be consulted with advantage by every intelligent female in the land" (p. v).

KING, John see also *THE PHILOSOPHICAL MEDICAL JOURNAL* (#2789).

2139. KINGET, Theodore R.
MANHATTAN MEDICAL INSTITUTE. Evidences of Dr. Kinget's Medical Success . . . [New York, ca. 1879].
Broadside ; 28 × 22 cm.

Includes the text of a notarized "affadavit" (dated 1879) and "testimonies . . . written by patients without invitation or solicitation" attesting to the efficacy of Kinget's proprietary medicines in the treatment of sexual disorders, dyspepsia, asthma, etc. The text of a "medical certificate" (dated 1874) certifying Kinget's membership in the Eclectic Medical Society of the City of New York is appended.

2140. KINGET, Theodore R.
Medical good sense treats of the laws of life and health, and causes, prevention and cure of acute and chronic diseases, and the physiology of the human system. Embracing sexual philosophy, and private reading for married people. It reveals the deepest mysteries of human nature in the generation and development of life, sexual relationship of man and woman, marriage, parentage and offspring, viewed from a secular standpoint . . . Second edition, revised and enlarged. New York: Published by the author, 1886.
    v, [6]-432, [2] p. : ill. ; 20 cm.

Medical good sense is as much a platform for the expression of Kinget's many unorthodox views as it is a treatise on physiology, hygiene and medicine. In its pages Kinget propounds two of the "discoveries" of William Byrd Powell, of Covington, Ky.: the "index of longevity" and the doctrine of temperaments that predetermines marital compatability. In adopting Powell's "vital index," Kinget claims that by measuring the depth of the base of the brain he can determine a patient's "vital tenacity," i.e., his or her ability to recover from illness. The author devotes a chapter to planetary influences on human physiology, maintaining that the same electro-magnetic forces that keep the earth on its axis or in orbit around the sun maintain the human body's equilibrium. In another chapter, he analogizes from the harmony of the body's physiological processes to the workings of a more just and equitable society.

At least half the book is devoted to reproduction, sexual physiology, contraception, sexual ethics, and social issues connected with the relations of men and women. Kinget advocated a healthy and active sex life, but shared with most of his contemporaries the fear that sexual intemperance resulted in dire physical and mental consequences for parents and their offspring. Kinget maintains much traditional sexual lore, e.g., his explanation of how couples can beget male or female children at will. His views on other topics, e.g., contraception, are more "advanced." Kinget describes E.B. Foote's

(#1213-1256) conviction in 1876 for sending contraceptive advice through the mails as an injustice (p. 367) both to "a highly respectable physician of this city" and to the women who are deprived of such knowledge. He adheres to accepted wisdom in the several pages he devotes to the ill effects of withdrawal as a contraceptive method. In the section titled "Means to prevent conception" (p. 372-73), however, he suggests the existence of a medically safe and effective chemical method for preventing conception. Without actually describing the technique, Kinget probably refers to a chemically treated sponge.

Medical good sense is full of surprises. Kinget writes of free love: "To introduce into society, principles and practices at one stage of its progress, that are only applicable at another, is not only very unwise, but quite impracticable . . . In this light we consider the present introduction of what is denominated 'Free-Love.' We cannot doubt that good would flow from its practical application in an advanced communistic system of society which we hope for, but until this happy time does come, we fear that a vast amount of evil would follow from the practice of it" (p. 398). His views on marriage are equally unorthodox for the period. Kinget refers to marriage as "too often the tomb of love" (p. 382), and attributes the existing laws regarding marriage to "an irrational state of society." He writes that at present, "Marriage is more necessary, because women are dependent. Make them independent . . . and the case would be changed" (p. 383). The author advocates private contracts in place of the existing marriage laws, and calls for the liberalization of divorce laws.

KINGZETT, Charles Thomas see
AMERICAN AND CONTINENTAL "SANITAS" CO. (#74).

2141. KINNE, Helen, 1861-1917.
Food and health. An elementary textbook of home making. By Helen Kinne . . . and Anna M. Cooley . . . New York: The Macmillan Company, 1916.
    [2], vi, 312 p. : ill. ; 19 cm.

First edition. Series: *The home-making series.* Kinne and Anna Maria Cooley (b. 1874) both taught "household arts education" at the Teachers College of Columbia University. In a brief but productive collaboration, they co-authored *Shelter and clothing* (1913), *Foods and household management* (1914), *Clothing and health* (1916), and *The home and the family* (1917). In their preface to *Food and health*, the authors write: "This volume, like its companion, *Clothing and health,* is intended for use in the elementary schools in those sections of the country where the home life is of the type described . . . This volume treats largely of food problems, including something of raising food and of selling it, in addition to the preparation of food at school and at home. Such topics as the water supply, disposal of waste, and other sanitary matters are woven in with the lessons on nutrition and cookery" (p. iii).

2142. KINSLEY, Charles.
The circle of useful knowledge; for the use of farmers, mechanics, merchants, manufacturers, surveyors, housekeepers, professional men, etc., etc., etc. . . . Clinton, Iowa: Published by the author, 1885.
   xx, [21]-255, [1] p., [2] folding tables ; 20 cm.

   "The 'Circle of Useful Knowledge' is a system of useful information, and contains hundreds of valuable receipts in the various departments of human effort . . . They tell how to manage a farm, how to cook, all about wines and vinegar, how to fish and tan, how drugs and chemicals are composed, how to be your own doctor and nurse—in short, everything connected with everyday life" (*Pref.*, p. [iii]). "Druggist's, perfumer's, and medical department," p. 72-100. Kinsley's compilation of domestic, farm and trade recipes was issued at least seven times between 1874 and 1893.

KIRK, Alice Gitchell. *The people's home recipe book* see *THE PEOPLE'S HOME LIBRARY* (#2768).

2143. KIRK, Eleanor, 1831-1908.
The woman's way to health and beauty. How to be well, keep well, and look well . . . Brooklyn, N.Y.: Bureau of Criticism and Revision, [c1891].
   39, [11] p. ; 17 cm.

   In printed wrapper.

KIRKPATRICK, James see **TISSOT** (#3534-3538).

2144. KISTLER, Wilson Peter, d. 1912.
Practical medical and surgical family guide in emergencies. A manual explaining the treatment of diseases, accidental injuries and cases of poisoning which demand prompt action in the absence of the physician. Hints and helps on health. Home nursing and remedies. Care of children. How to cook for the sick, etc. Also a complete pronouncing vocabulary of medical terms. Designed for families, students, teachers, and practitioners of medicine . . . [Allentown, Pa.: Berkemeyer, Bechtel & Co. book and job printing, c1894.]
   xiv, [2], 339, [iii]-ix, [2] p., [1] leaf of plates : port. ; 22 cm.

   Kistler was an 1867 graduate of the Bellevue Hospital Medical College.

2145. KITCHINER, William, 1775?-1827.
The art of invigorating and prolonging life, by food, clothes, air, exercise, wine, sleep, &c. and peptic precepts, pointing out agreeable and effectual methods to prevent and relieve indigestion, and to regulate and strengthen the action of the stomach and bowels . . . To which is added, the pleasure of making a will . . . By the author of "The Cook's oracle," &c. &c. &c. From the third London edition. Philadelphia: H.C. Carey & I. Lea, 1823.
   [6], 281 p. ; 18 cm.

   The son of a wealthy London coal merchant, Kitchiner obtained his medical degree in Glasgow. Returning to London with a handsome inheritance, he never took up the practice of medicine, but devoted himself to an interest in the sciences and to entertaining a select circle of learned friends. "Though always an epicure, he was regular and even abstemious in his habits. Convinced that the health depends to a great extent on the proper preparation of food, he experimented in cookery in his own house . . . He soon attained to a considerable culinary skill. His lunches, to which only a few were admitted, were far famed. His dinners were conducted with much ceremony" (*Dict. nat. biog.* 11:231). Kitchiner's culinary avocation resulted in his best known work: *Apicius redivivus, or, the cook's oracle,* which went through at least twenty editions between 1817 and 1855 (1st Amer. ed., Boston, 1822).

The art of invigorating and prolonging life was first published at London in 1821. It went through seven London editions in as many years. This Philadelphia edition is probably the first American edition (*see also* #2146). Both 1823 American editions were based on the third London edition of 1822. It is not surprising that the works of an English epicure would be published in the United States. There were no copyright laws protecting foreign authors, and American publishers found that it was almost always to their advantage to reprint books that had been successful on the English market. In fact, it was often less expensive for publishers to do so than to buy the rights to works by American authors. Only 30% of the books published in the United States during the 1820s were written by American authors. The firm of Carey & Lea, who published this edition, were assiduous in reprinting books that had been successful abroad, and maintained an agent in London for that purpose. The Harper firm in New York republished this work in 1831 (#2147), 1833, 1847 and 1855 (#2148) under the title: *Directions for invigorating and prolonging life,* based upon the sixth London edition (1828).

In this work Kitchiner presents a hygienic program intended "to animate and strengthen enfeebled constitutions—prevent gout—reduce corpulency—cure nervous and chronic weakness—hypochondriac and bilious disorders, &c.—to increase the enjoyment and prolong the duration of feeble life—for which medicine, unassisted by diet and regimen—affords but very trifling and temporary help" (p. 4). He begins with a consideration of the stages of life (which he divides into three), the physiological demands of each, and how we must adjust our manner of living at each stage, particularly after age forty. Given his culinary interests, it is not surprising that the work emphasizes diet and digestion. The author devotes a lengthy chapter to sleep; brief chapters to clothes, "fire" (i.e, bodily warmth), air and exercise; and a chapter on wine that begins with its proper aging and storage, and procedes to what wines assist or impede digestion, its benefits to the aged, etc.

2146. KITCHINER, William, 1775?-1827.
The art of invigorating and prolonging life, by food, clothes, air, exercise, wine, sleep, &c. and peptic precepts, pointing out agreeable and effectual methods to prevent and relieve indigestion, and to regulate and strengthen the action of the stomach and bowels . . . To which is added the pleasure of making a will . . . By the author of "The Cook's oracle," &c. &c. &c. From the third London edition. Lexington, Ky.: Printed by W.W. Worsley, 1823.

[4], 117 p. ; 19 cm.

Although based on the text of the third London edition (1822), this edition is probably a reprint of the 1823 Philadelphia edition (#2145) minus the index.

2147. KITCHINER, William, 1775?-1827.
Directions for invigorating and prolonging life; or, the invalid's oracle. Containing precepts, pointing out agreeable and effectual methods to prevent and relieve indigestion, and to regulate and strengthen the action of the stomach and bowels . . . From the sixth London edition, revised and improved by T.S. Barrett . . . New York: Published by J. & J. Harper . . . [*et al.*], 1831.

4, [5], iii, 252 p. ; 16 cm.

At head of title: *Harper's stereotype edition.* The sixth London edition was published in 1828 under the title: *The art of invigorating and prolonging life.* Includes "Extracts from the writings of Lewis Cornaro . . . on health and long life" (p. 223-43). Spine-title: *Invalid's oracle.* The American editor, Thomas Squire Barrett, is described on the title-page as a "licentiate in medicine and surgery, Fellow of the New-York Medical & Philosophical Society, &c., &c."

2148. KITCHINER, William, 1775?-1827.
Directions for invigorating and prolonging life; or, the invalid's oracle. Containing peptic precepts, pointing out agreeable and effectual methods to prevent and relieve indigestion, and to regulate and strengthen the action of the stomach and bowels . . . From the sixth London edition; revised and improved, by T.S. Barrett . . . New York: Harper & Brothers, publishers, 1855.

252 p. ; 16 cm.

2149. KITCHINER, William, 1775?-1827.
The economy of the eyes: precepts for the improvement and preservation of the sight. Plain rules which will enable all to judge exactly when, and what spectacles are best calculated for their eyes; observations on opera glasses and theatres, and an account of the pancratic magnifier, for double stars, and day telescopes . . . Boston: Wells and Lilly, 1824.
viii, 224 p., [2] leaves of plates : ill. ; 15 cm.

First American edition; originally published at London earlier the same year. The present work, the fruit of the author's lifelong interest in optics, is a popular yet thorough treatise on eyeglasses and ocular hygiene. It includes a discussion of "opera glasses" for the theatre or gallery. A sequel, *The economy of the eyes. Part II.* (London, 1825), is devoted to telescopes. There does not appear to have been an American edition of this second part.

KLOSS, Moritz, 1818-1881. *The dumb-bell instructor for parlor gymnasts* see **LEWIS** (#2238).

2150. KNIGHT, Archibald Patterson, 1849-1935.
Introductory physiology and hygiene. A series of lessons in four parts designed for use in the first four forms of the public schools. Parts I and II contain rules of hygiene for pupils in forms I and II . . . Parts III and IV contain rules of hygiene and elementary physiology for pupils in formers III and IV . . . Toronto: The Copp, Clark Company, Limited, [c1905].
xiv, 198 p. : ill. ; 19 cm.

Knight received his undergraduate and master's degrees from Queen's University, Kingston, Ont., and his medical degree from Victoria University (1889). In 1892 he was appointed professor of biology and physiology at Queen's University, a position that he held until retirement (1919). Knight was the author of several texts on chemistry, physiology and hygiene for primary and secondary school students.

2151. KNIGHT, Archibald Patterson, 1849-1935.
The Ontario public school hygiene . . . Authorized by the Minister of Education for Ontario for use in forms IV and V of the public schools. Toronto: The Copp, Clark Company, Limited, 1910.
viii, 248 p. : ill. ; 20 cm.

2152. KNOPF, Sigard Adolphus, b. 1857.
Tuberculosis a preventable and curable disease. Modern methods for the solution of the tuberculosis problem . . . New York: Moffat, Yard and Company, 1909.
xxxii, 3-394 p., [1] leaf of plates : ill., port. ; 20 cm.

First edition. "The book is intended to be helpful, first, to the patient afflicted with a tuberculous disease, but not with a view of replacing the physician . . . it will aid the sufferer by giving him insight into his affliction as will convince him of the curability of the disease in the earlier stages and the great possibility of improving his condition in the latter stages, providing he places himself under the careful guidance of a physician in his own hygienically arranged home, in a health resort, or in a special institution (sanatorium or hospital)" (*Pref.*, p. ix). Knopf adds, "By reading the following pages, it is hoped that the layman may learn that a sober, proper, and regular mode of living is all that is necessary to overcome a hereditary predisposition or an acquired tendency to the disease, and learn also what he is to do and what not to do if he wishes never to fall a victim to tuberculosis" (p. xiii).
Knopf was an 1888 graduate of the Bellevue Hospital Medical College. He was professor of phthisio-therapy at the New York Post-Graduate Medical School from 1908 to 1922, and attending physician at the Riverside Sanatorium for Consumptives (New York). A prolific author, Knopf also wrote numerous books and pamphlets on tuberculosis and on birth control for popular audiences. His *Tuberculosis as a diseases of the masses* (1901), by way of example, was translated into some thirty languages. Atwater copy inscribed by author.

2153. KNOW THYSELF OR NATURE'S SECRETS REVEALED.
Know thyself or nature's secrets revealed. A word at the right time to the boy, girl, young man, young woman, husband, wife, father and mother, also timely counsel, help and instruction for every member of

every home. Including important hints on social purity, heredity, physical manhood and womanhood by noted specialists. Introduced by Bishop Samuel Fallows . . . Medical department by W.J. Truitt . . . Marietta, Ohio: The S.A. Mullikin Company, exclusive publishers, 1911.

602 p., [16] leaves of plates : ill. (some col.) ; 20 cm.

The first half of *Know thyself* is devoted to a rather sentimental consideration of the importance of women in domestic life; the attainment of feminine beauty; love and courtship; marriage; sexual ethics; child rearing; "diseases peculiar to men," etc. The second half (the "Medical Department") was written by William John Truitt (1867-1918), an 1889 graduate of the Hahnemann Medical College & Hospital (Chicago), and the author or co-author of several popular treatises on marriage and sexual ethics. Later editions of *Know thyself* work were published under the title *Nature's secrets revealed* (#2576-2578).

KNOW THYSELF SERIES see **COCROFT** (#689-695).

2154. KNOWLTON, Charles, 1800-1850.
Fruits of philosophy or the private companion of adult people . . . Edited with an introductory notice by Norman E. Himes . . . With medical emendations by Robert Latou Davidson . . . Mount Vernon [N.Y.]: Peter Pauper Press, MCMXXXVII [1937].
[2], xv, [3], 107, [2] p. ; 24 cm.

A native of central Massachusetts and graduate of Dartmouth College Medical School (1824) and Berkshire Medical College (1827), Knowlton's fame rests principally on this single work published anonymously at New York in 1832. "Knowlton may perhaps be considered the American founder of contraceptive medicine. At all events, this is the first treatise on contraceptive technique by an American physician" (Norman E. Himes, *Medical history of contraception*, Baltimore, 1936, p. 462-64, p. vi). Knowlton's treatise was reprinted at Boston in 1833 and appeared in at least eleven more American editions through 1895. "Knowlton's birth control manual sold well, and he was prosecuted three times under the state common law obscenity statute for selling it. First he was fined fifty dollars in Taunton; then a prosecution in Lowell led to three months

hard labor in the East Cambridge jail; and finally, a charge in Greenfield was dropped after two juries failed to reach a verdict" (*Amer. nat. biog.*, 12:830). The *Fruits of philosophy* had an unusually long publishing history in Britain as well. Himes records 17 British editions (there were at least four more) between 1833 and the 1889.

Knowlton's *Fruits of philosophy* is divided into five chapters that discuss the social and individual needs for population control, the physiology of generation, methods of preventing conception (the shortest chapter), the signs of pregnancy, and "Remarks on the reproductive instinct." The long publishing history of this treatise is attributable to the fact that its discussion of contraceptive technique was more complete and accurate than what was provided in most publications on this topic, particularly after passage of the Comstock law (#754, 755).

In his chapter on contraceptive technique, Knowlton recommended douching two to three times within five minutes of coitus with solutions of alum, sulphate of zinc, saleratus, vinegar or liquid chloride of soda. He states that "the fecundating property of the semen depends on the seminal animalcules" and that "it seems to me more than probable that this property may be easily destroyed by any liquid, which by its chemical or other properties, can produce any considerable change in the semen, when brought in contact with it." (1937 ed., p. 61). No copy of the 1832 first edition is recorded in any published bibliography or bibliographic database.This 1937 reprint was issued in an edition of 450 copies.

2155. KNOWLTON, Charles, 1800-1850.
Fruits of philosophy. An essay on the population question . . . Second new edition, with notes. Fortieth thousand. London: Freethought Publishing Company, [1877?].
vi, [7]-56 p. ; 19 cm.

Preface by Charles Bradlaugh (1833-1891) and Annie Wood Besant (1847-1933), who were prosecuted in London for their publication of Knowlton's *Fruits* in 1877 (*see annotation to* Besant, #314).

2156. KNOWLTON, Charles, 1880-1850.
Fruits of philosophy. A treatise on the population question. By Charles Bradlaugh and Mrs. Annie Besant. [Chicago: G.S. Baldwin, 188-?].
20, [4] p. ; 22 cm.

The Baldwin edition of Knowlton's *Fruits of philosophy* includes a "Publisher's preface" and "Preface to the second new edition" by Charles Bradlaugh and Annie Besant, as well as a "Preface, by one of the former publishers." In printed wrapper.

2157. KNOWLTON, Charles, 1880-1850.

Fruits of philosophy. An essay on the population question . . . New edition—with notes. London: Minerva Publishing Company, 1892.
   [2]-49 p. ; 19 cm.

The insides of the printed wrapper contain advertisements for syringes, vaginal pessaries and "Malthus sheaths" (i.e., condoms) available from E. Lambert & Son, London.

KNOWLTON, Charles see also *LEGACY OF A DEAD DOCTOR* (#2225).

2158. KOCH, G. H.

Der Leibarzt, oder Doktor und Apotheker in jedem Haus. Populäres Familienbuch der bewährtesten und besten Hausarzneimittel gegen Krankheiten des Menschen nebst Belehrung zur Heilung und Verhütung derselben, Anweisung zur Krankenpflege und Diät, Erklärung der medicinischen Ausdrücke &c. Nach den Erfahrungen und Theorien von Dr. Hufeland und anderer berühmter Aerzte Europa's und Amerika's practisch zusammengestellt und dem Volke gewidmet . . . 2te verbesserte und vermehrte Auflage. Cincinnati, O.: Verlag von Max Weil & Co., 1860.
   viii, 647, [1] p. ; 22 cm.

Originally published at Cincinnati in 1843 as *Familien-Heilkunde, oder, ärztlicher Rathgeber für Familien.* This second edition was first published at Cincinnati in 1859.

2159. KOHLER MANUFACTURING COMPANY.

Laws of etiquette. By a Society Lady. Baltimore, Md.: Published by Kohler Manufacturing Co., [189-?].
   [4] leaves ; 13.5 cm.

Title and imprint from wrapper. Promotional literature for *Kohler's Antidote, Kohler's*

Nerve Comfort, etc. The text consists of ninety-nine social "dos and don'ts," and has been attributed to the American lyricist George Morley Vickers (b. 1841).

2160. KOLA IMPORTING COMPANY.

Testimony of 1000 living witnesses to HIMALYA, the kola compound. Nature's specific for the cure of asthma, hay fever, etc. New York; Cincinnati: The Kola Importing Co., [ca. 1895].
   48 p. : ill. ; 21 cm.

"The most important Medical Discovery ever made since the dawn of civilization, is, without doubt, the recent discovery of the wonderful kola plant, found in the Soudan and on the Congo river, in the Heart of Africa. This strange botanical product in its Compound (Himalya) has a peculiar action upon the Nervous System of Man, through which it allays and removes the irritability which is the foundation and cause of spasmodic diseases. Through this singular property it is a Sure and Unfailing cure for asthma at every stage and in every form" (p. [35]). Cover-title: *1,000 living witnesses testify to Himalya.* In printed wrapper.

Arthur Cramp reported that chemists at the North Dakota Agricultural Experiment Station found Himalya to be "a weak hydro-alcoholic solution of potassium iodide, flavored with peppermint and licorice and colored with caramel" (*Nostrums and quackery,* #86, 2:596). Though the preparation contained little if any kola, it was about nine percent alcohol.

2161. KOST, John, 1819-1904.

Domestic medicine. A treatise on the practice of medicine, adapted to the reformed system, comprising a materia medica . . . Cincinnati: Published by H. Burnard, 1851.
   624 p., [1] leaf of plates : ill., port. ; 24 cm.

Originally published at Mount Vernon, Ohio under the title: *The practice of medicine, according to the plan most approved by the reformed or botanic colleges of the U.S.* (#2163). As in the first edition, Kost divides his treatise into two parts: a domestic medicine *per se* and a materia medica. This second edition differs from the first in that the entries on fevers and inflammation have been separated from the alphabetical catalog

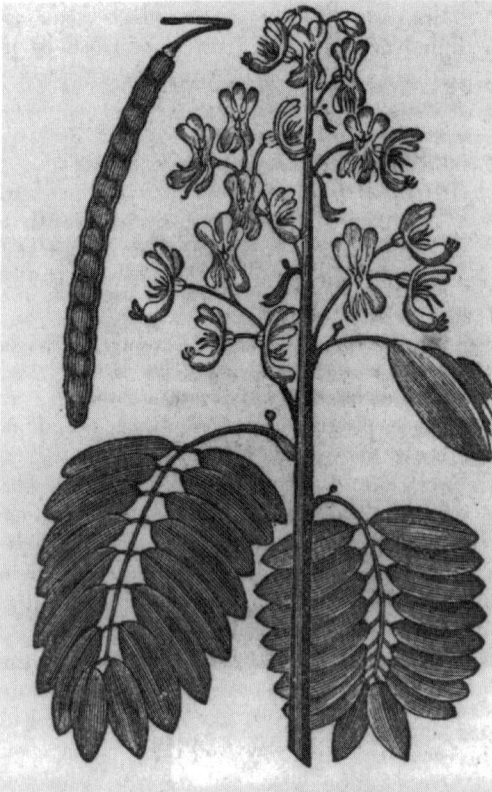

Fig. 60. "American senna," from Kost's Domestic medicine, *#2162.*

of diseases and now preface the text. The number of wood engravings of medicinal plants has been increased to seventy-five from the forty-two that illustrated the first edition. Spine-title: *Kost's domestic medicine.* Frontispiece portrait of the author.

Although his origins and training are obscure, John Kost was a prominent figure in "reformed" medical circles of the 1840s and 1850s—a period when the Thomsonian movement from which the reformed practice emerged had fragmented into competing schools. Although Kost's manual of domestic medicine was frequently published (1847 [#2163], 1851, 1857, 1859 [#2162], 1868, 1872), his authorship of *The elements of materia medica* (Cincinnati, 1849) reflects the transition of Thomsonian medicine from a grass-roots movement of botanically based self-treatment to a system requiring formally trained practitioners and a scientifically based therapeutics.

Kost taught on the faculty of several reformed or eclectic medical schools, most notably the Botanico-Medical College and American Medical College in Cincinnati. He was co-editor of the *Botanico-medical reformer* (2 v., Mt. Vernon, Ohio, 1844-46) and the first several volumes of *The Southern medical reformer and review* (10 v., Macon, Ga., 1852-60). Kost also ventured into publishing in the early 1850s. The 1850 Cincinnati edition of Horton Howard's *An improved system of botanic medicine* (#1814) and the 1852 edition of Howard's *A treatise on midwifery* (#1817) both bear the imprint of J. Kost.

2162. KOST, John, 1819-1904.

Domestic medicine. A treatise on the practice of medicine, adapted to the reformed system, comprising a materia medica . . . Cincinnati: J.W. Sewell, publisher, 1859.

624 p., [1] leaf of plates : ill., port. ; 23 cm.

Spine-title: *Kost's domestic medicine.*

2163. KOST, John, 1819-1904.

The practice of medicine, according to the plan most approved by the reformed or botanic colleges of the U.S. Embracing a treatise on materia medica and pharmacy . . . Designed principally for families . . . Mt. Vernon [Ohio]: Published by the author; printed by E.J. Ellis, 1847.

xv, [16]-508, [4] p. : ill. ; 22 cm.

First edition. Later editions published under the title: *Domestic medicine* (#2161, 2162). Kost divides his treatise into two parts. "The part treating on practice is arranged in alphabetical order [i.e., from APOPLEXY to WOUNDS], as it was supposed to be more simple, and better adapted to the use of private individuals, than any system of classification could possibly be, In this part, all the diseases of common occurrence in this country are embraced,—their character and symptoms fully pointed out, and the most appropriate treatment given, in a style as *brief,* and yet as *comprehensive* as possible" (p. v-vi). A botanic materia medica comprises the second part, arranged under effects, i.e., emetics, cathartics, expectorants, diuretics, etc.

2164. KRAFT, Frank.

Sex of offspring. A modern discovery of a primeval law . . . Cleveland, Ohio: B. Barsuette, publisher, [c1908].
112, [2] p. ; 20 cm.

Kraft was a Cleveland homeopath and devoted Darwinian. In *Sex of offspring*, his only published book, Kraft posits a double potentiality for determing the sex of offspring based upon man's remote origins in an aquatic environment. Kraft revives the long-popular starvation theory of sex determination, i.e., in times of plenty more females are born, while in times of dearth more males are born because of their ability to adapt to physical hardship and to forage. Kraft gives this theory an evolutionary twist in suggesting that the female reproductive system still responds to a primeval cycle of feast and famine based upon conditions that prevailed when water covered the earth: "The tides of those days were . . . monstrous and all enveloping, compared with which our tides are but feeble and impotent imitations—a merely theoretical or ideal time of starvation or plenty instead of an actual" (p. 42). When the tides were in, starvation ensued; when the tides were out, and as our evolutionary forebears became increasingly land-oriented, plenty ensued. According to Kraft: "We are then provided by inheritance, with a physical organization which will produce a being in times of starvation best calculated to contend with starvation, and a being in times of plenty best calculated to exist under the latter conditions" (p. 46).

Kraft maintains that ovulation and menstruation also respond to a cycle established when primeval tides and the organisms that lived in them were governed by lunar influences. Quoting several embryologists regarding the hermaphroditic nature of the unfertilized ovum, Kraft explains: "The moon's movement in going from upper to lower transit every twelve hours of the day, marks off a natural cycle of time . . . During half of the twelve hour cycle the ovum is in a season presenting only its male phase or potentiality; and then it changes to the other season and presents during the other half of the twelve hours its female phase or potentiality" (p. 79). He describes how "the ovum or egg's life moves

synchronously with the tide water and presents during rising water the condition or potentiality which if fertilized then, results in male offspring, and during falling water the condition or potentiality which if fertilized at such a time, results in female offspring" (p. 80). In his final chapter Kraft explains how conception can be timed to coincide with these inherited cycles in order to determine the sex of offspring.

2165. KRAUSKOPF, Joseph, 1858-1923.

The great white plague. A discourse . . . at Temple Keneseth Israel . . . Philadelphia. Series XXII. No. 16 February 21st, 1909. [Philadelphia: Temple Keneseth Israel, 1909].
10 p. ; 23 cm.

Born in Prussia, Krauskopf came to the United States alone at age fourteen. "On the opening of the Hebrew Union College in Cincinnati in 1876, he entered it and graduated as rabbi in its first class, 1883. He began his rabbinical career . . . in Kansas City. On Oct. 22, 1887, he accepted a call to Congregation Keneseth Israel, Philadelphia, to which he ministered for the rest of his life. Under him the congregation grew from 150 to 1,500 families . . . Krauskopf was from the beginning the energetic, practical, public rabbi. He at once made Sunday services a regular feature, and his topical addresses, which attracted large audiences, were often quoted in the local press and were regularly printed and widely disseminated for thirty-six years" (*Dict. Amer. biog.*, 5:500-501). The titles of the twenty-two series of *The Keneseth Israel Sunday discourses* published to Feb. 1909 (most of which were delivered by Krauskopf) are listed on both sides of the paper wrapper. The topics extend beyond Judaism to reflect Rabbi Krauskopf's intense interest in the social, political and literary issues of his age. *The great white plague* appears to be the only lecture among these series specifically oriented toward health. Krauskopf uses the International Tuberculosis Exhibition being held in Philadelphia as a stage from which to launch into an appeal for hygienic housing for the poor and for more sanitary conditions of milk production—remediable causes of the spread of tuberculosis.

2166. KUEPPER, Wilhemine H.
Nature cure (formerly called water cure)
or home treatment without medicine . . .
Philadelphia; Chicago; Toronto: The John
C. Winston Company, 1905.
    254 p., [12] leaves of plates : ill. ; 19 cm.

    "In olden times most people were strong and
healthy, and the weak and sickly were despised.
They were strong because they lived close to
Nature and did not take any medicine . . . Man
has strayed away from Nature, and the punishment is disease. Let us go back to Nature, and
in order to do this let us take for our example
the large animals of the lower order. They are
out-of-doors in all kinds of weather, summer
and winter. As few of us can be out-of-doors all
the time, let us do the next best thing—sleep
with open windows, and take out-of-door exercise as much as possible; walk to our business,
instead of riding in closed cars filled with bad
air, and ventilate well our offices, workshops,
schoolrooms, etc." (p. 10). Kuepper continues:
"This book . . . will teach you to become your
own physician, as everybody naturally ought to
be. And after you are well it will teach you how
to live in order never to grow sick again, for
sickness is punishment for a wrong way of living. We ought to be well and die of old age *without disease*" (p. 12).

2167. KUHNE, Louis.
Neo-naturopathy. The new science of heal-
ing or the doctrine of the unity of diseases
. . . Special authorized American edition.
Butler, N.J.: Published by Benedict Lust,
N.D., M.D., copyright 1917.
    291, [21] p. : ill., port. ; 25 cm.

    Translation of *Die neue Heilwissenschaft* (1st ed.,
1890). In his preface, the American translator
and publisher, Benedict Lust, states that Kuhne's
treatise is the "cornerstone of Naturopathy," a
school of alternative medical thought that
emerged in Germany toward the end of the 19th
century. Lust contends that in rightly discover-
ing the cause and cure of all diseases, Kuhne has
inaugurated a new epoch in drugless medicine:
"Living just prior to the age of serums and vac-
cines, he rightly recognized that all disease was
the result of a CONTAGIUM VIVUM or a MA-
TERIES MORBI, a fermentation of poisonous
matters that were not eliminated from the sys-

tem, and he devoted his energies to the
eliminaton of such poison by the most simple
and natural methods" (p. 4). "Official medicine,"
Lust explains, "seeks to remove disease in an organism by treating the symptoms with a pill or a
potion, and for the most part ignores the simple
agents of cleanliness within and without, pure
air, simple food, sunshine, water, exercise, active
or passive, that strike at the cause of the given
ailment." Kuhne's therapeutics relies heavily on
baths. He uses the term "neo-naturopathy," however, to distinguish his system from the earlier
(and similar) hydropathic school of such figures
as Priessnitz and Rausse.

2168. KUHNE, Louis.
The new science of healing; or the doctrine
of the unity of disease forming the basis of
a uniform method of cure, without drugs
and without operations. An instructor and
advisor for the healthy and the sick . . .
Twelfth authorized English edition translated from the thoroughly revised 50th
German edition by Kenneth Romanes . . .
Leipsic: Published by Louis Kuhne, 1901.
    x, 460, [4] p., [1] leaf of plates : ill., port. ; 23
cm.

    Translation of *Die neue Heilwissenschaft*. Fron-
tispiece portrait of author.

**2169. KURZGEFASSTES ARZNEY-
BÜCHLEIN.**
Kurzgefasstes Arzney-Büchlein, für Menschen
und Vieh, darinnen CXXVIII. auserlesene
Recepten, nebst einer prognostischen Tafel.
Wien gedruckt, und in Ephrata [Pa.] zum
vierten Mal nachgedruckt [bey Solomon
Mayer], Anno 1792.
    24 p. ; 15 cm. (12mo)

    "This may be an edition of the work pub-
lished anonymously in Dresden in 1687 under
title: *Kurz verfasstes Arznei-Buchelin*, attributed to
a certain Samuel Mueller in Holzmann,
Deutsches Anonymen-Lexikon" (Austin, #1101).
    REFERENCES: *Austin, #1102; First century of
German printing in the United States of America, #843.*

**2170. KURZGEFASSTES WEIBER
BÜCHLEIN.**
Kurzgefasstes Weiber-Büchlein. Enthält
Aristoteli und A. Magni Hebammen-Kunst

mit den darzu gehörigen Recepten. Somerset [Pa.]: Gedruckt [bey Friedrich Goeb] für Jacob Sala, 1813.

    53, [1] p. ; 16 cm.

This handbook of Pennsylvania German folk medicine provides information on pregnancy and delivery (with receipts for related disorders) culled from the Pseudo-Aristotle (#104-147), Culpeper (#838-843) and the *De secretis mulierum* attributed to Albertus Magnus. It was originally published in 1796 and in numerous editions thereafter. In their description of the 1796 edition, Brendle & Unger write: "Quite a little bit of the material included seems to come from a little work . . . attributed to Albert Magnus . . . Many editions, or adaptions of it, appeared in Germany under various titles such as, *Albertus Magnus daraus Man alle Heimlichkeit des weiblicher Geschlechts erkennen kan*; *Die Heimlichkeiten Albertus Magnus Hebammen und Kindbaren Frawen Dienlich*; *Von der Geheimnuessen deren Weiber*, etc." (*Folk medicine of the Pennsylvania Germans*, New York: A.M. Kelley, 1970, p. 234-35).

    REFERENCES: *Austin* #1110; *First century of German language printing in the United States of America*, #2006.

KYDD, Mary Mitchell see **HOPE** (#1782).

L

2171. LACHAPELLE, Séverin, 1851-1913.
La santé pour tous ou notions élémentaires de physiologie et d'hygiène à l'usage des familles. Suivies du petit guide de la mère auprès de son enfant malade . . . Montréal: Compagnie d'Imprimerie canadienne, 1880.
  [4], iv, 316 p. ; 18 cm.

First edition. *La santé pour tous* is divided into four parts: I. Hygiène du corps ou hygiène physique (i.e., the physiology and hygiene of digestion, respiration, the organs of sensation, etc.); II. Hygiène morale ou hygiène de l'âme (i.e., a guide to spiritual and moral well-being in which the author cites the Greek tragedies, St. Augustine, Amariah Brigham, etc.); III. Hygiène selon l'âge de l'homme (including occupational health); and the "Petit guide de la mère."

The son of a physician, Lachapelle was born in Saint-Remi, Lower Canada (i.e., Quebec). After completing classical studies at the Petit Séminaire de Montréal (1868), Lachapelle enlisted in the Papal Zouaves, serving in Italy for two years. On his return to Canada in 1870, he enrolled in the École de Médecine et de Chirurgie (Montréal), from which he received a medical degree in 1874. Lachapelle practiced in Montreal, and became a frequent contributor to French-language medical and scientific journals published in Canada. In 1879 he joined the medical faculty of the Université Laval. Lachapelle's career-long interest in public sanitation and in educating the public on its benefits involved him in various federal and provincial initiatives. In addition, he edited the *Journal d'hygiène populaire* (1884) and the short-lived *La mère et l'enfant* (1890-91). Lachapelle wrote several popular treatises: *La santé pour tous* (1880), *Manuel d'hygiène à l'usage des écoles et des familles* (1888) and *Femme et nurse* (1901). In 1892 he was elected conservative MP for Hochelaga, and in 1907 was instrumental in founding the Hôpital Sainte-Justine for sick children.

LA CROIX, M. B. see **CULVERWELL** (#846).

2172. LADIES INDISPENSABLE ASSISTANT.
Ladies' indispensable assistant. Being a companion for the sister, mother, and wife. [Co]ntaining more information for the price than any other wor[k] upon the subject. Here are the very best directions for the behavior and etiquette of ladies and gentlemen, ladies' toilette table, directions for managing canary birds; also, safe directions for the management of children; instructions for ladies under various circumstances; a great variety of valuable recipes, forming a complete system of family medicine, thus enabling each person to become his or her own physician: to which is added one of the best systems of cookery ever published; many of these recipes are entirely new and should be in possession of every person in the land. New-York: Published at 128 Nassau-Street, 1852.
  [10], [7]-72, [7]-48, [121]-138 p. ; 22 cm.

The *ladies indispensable assistant* appeared in at least five New York editions between 1850 and 1860. The "Family Physician" occupies the first fifty-two pages of text, offering advice on the household cure of many common diseases. Deafness, for example, may be cured by taking "ants' eggs and onion juice; mix, and drop into the ear"; and cancer by boiling down "the inner bark of white and red oak, to the consistency of molasses; apply as a plaster, shifting it once a week." It is followed by a description of the medical properties of sixty plants arranged alphabetically. Kitchen recipes and advice on invalid cookery occupy forty-eight pages. The state of contemporary manners may be derived from the chapter entitled "Etiquette for ladies and gentlemen," in which it is advised that at table "the knife should never be put in the mouth," and that one should "carefully abstain from every act or observation that may cause disgust, such as spitting, blowing the nose, gulping, rinsing the mouth, &c." (p. 125). Spine-title: *Family manual.*

LADY OF NEW YORK see **CROWEN** (#833).

2173. LAGRANGE, Robert J.
Manhood; or, secrets revealed. A practical medical work on nervous debility and physical exhaustion. To which is added an essay on marriage, with important chapters on disorders of the reproductive organs. Being a synopsis of lectures delivered at the Museum of Anatomy, 708 Chestnut St., Philadelphia. By Drs. LaGrange & Jordan (late Jordan & Davieson) . . . Philadephia, 1885.
    164, [2] p. ; 14 cm.

CONTENTS: I. Anatomy and physiology of the generative organs; II. General debility; III. Spermatorrhoea, impotency, sterility; IV. Treatment of spermatorrhoea, etc. V. Marriage: its obligations and excesses; VI. Diseases of the generative organs; VII. Special diseases; VIII. Self-diagnosis; IX. Notes from our case book, Cases of complaints, To patients and invalid readers.
    The content of this volume is identical to L.J. Jordan's *Man's misson on earth,* copyrighted in 1871 (#2043, 2044). Both titles are compilations of lectures delivered at the museums operated by varying combinations of Jordans, Kahns and their partners in Philadelphia and New York. In 1877 or 1878, Robert J. LaGrange and Robert J. Jordan assumed ownership of the European Anatomical, Pathological and Ethnological Museum in Philadelphia (#1081-1083) from Philip J. Jordan (#2048) and the Daviesons. In the 1880s, this establishment became known as the Great European Museum of Anatomy.
    Robert J. La Grange is listed in the 7th edition of *Polk's medical and surgical register of the United States* (1902) as an 1879 graduate of the Eclectic Medical College of the City of New York, the same institution from which his partner Robert J. Jordan received a medical degree (*see annotation to* #2051). LaGrange's name does not appear in the 9th edition (1906).

2174. LAGRANGE, Robert J.
Review of the works published by Dr. Robt. J. La Grange. [S.l.: s.n., ca. 1880].
    36 p. ; 15 cm.

In this pamphlet LaGrange reviews the contents of three of his published works: *What to eat, drink and avoid* (1879), *Social evils of the present day* (1879) and *Premature decay* (1879). On page 33 he also claims success as an orthopedic surgeon, and on p. 36 advertises *A treatise on orthopaedic surgery and the removal of deformities*—a work that does not appear to have been published. For those who desire personal consultations, LaGrange provides his readers office addresses in both Philadelphia and New York. In printed wrapper.

2175. LAGRANGE, Robert J.
Secrets revealed. A course of lectures . . . Philadelphia: No. 1625 Filbert Street, 1884.
    145 p. ; 15 cm.

Four of these lectures (on consumption, venereal diseases, diabetes and the voice) are identical to those given by Dr. Jourdain in Boston a decade or so earlier—published in *Man's mission on earth* (#2051, 2052). No matter what the primary topic of these lectures, the organs of generation play a role as "the very mainspring of the corporeal frame." Masturbation, "the detestable vice," is frequently woven into the discourse as well. Cover-title: *La Grange's lectures on nervous diseases, etc.*

LAGRANGE, Robert J. see also **EUROPEAN ANATOMICAL, PATHOLOGICAL AND ETHNOLOGICAL MUSEUM** (#1081), **JOURDAIN** (#2051, 2052); **DR. KAHN'S MUSEUM OF ANATOMY** (#2066).

LAIDLAW, Alexander H. see **HANCHETT** (#1551, 1552).

LAIR, Pierre Aimé, 1769-1853. *On the combustion of the human body* see **TROTTER** (#3597).

2176. LAMBE, William, 1765-1847.
Water and vegetable diet in consumption, scrofula, cancer, asthma, and other chronic diseases. In which the advantages of pure soft water over that which is hard are particularly considered; together with a great variety of facts and arguments showing the superiority of the farinacea and fruits to animal food in the preservation of health

. . . With notes and additions, by Joel Shew, M.D. New York: Fowlers and Wells; London: John Chapman, 1850.

viii, [9]-258, [6] p. ; 19 cm.

First American edition. "The first scientific effort to demonstrate that a vegetable diet was not necessarily debilitating was made early in the nineteenth century by an English physician named William Lambe. Lambe had long been suffering from a variety of chronic diseases when he decided, in 1806, to investigate what would happen to his health if he temporarily abandoned the use of animal food. The results of the experiment were so encouraging he decided to continue it on a permanent basis . . . After trying the same diet on several of his patients, Lambe published the results of his experiment as *Additional reports on the effects of a peculiar regimen, in the cases of cancer, scrofula, consumption, asthma and other chronic diseases* in 1815. This book contained . . . an empirical demonstration that vegetable diet provided enough energy and nutriment to sustain human life, and that far from being debilitating, it was in fact wholly beneficial in cases of certain chronic complaints. Nowhere in the *Additional reports* did Lambe recommend a vegetable diet to persons in good health, or claim that animal food had any ill-effects on such people. The book was not, in other words, a tract for vegetarianism. Nevertheless, by purporting to confute the single most convincing physiological argument against vegetarianism, it provided a firm position for those who might care to produce such a tract" (Stephen Nissenbaum, *Sex, diet and debility in Jacksonian America,* Westport, Ct.: Greenwood Press, 1980, p. 45-46). Receiving his medical degree in 1802, Lambe was admitted a fellow of the Royal College of Physicians of London in 1804.

In the "Preface to the American edition," Joel Shew writes: "Vegetarianism has hitherto been presented to Americans as a means of *preventing* rather than of *curing* disease. This work, then, brings the matter up in a new aspect." One of America's most notable hydropathists, Shew adds, "The perusal of this work will lead many, doubtless, to the conclusion, that if such striking benefits as are here described are to be gained by attention to diet alone, how much greater may we not look for in combining the *water-treatment* with the course pointed out" (p.

iv). Fowlers & Wells published five editions of Lambe's treatise between 1850 and 1855. Copies doubtless accompanied members of the Vegetarian Settlement Company and the Vegetarian Kansas Emigration Company, who in 1855, with encouragement from Fowlers & Wells and their associates, sought to establish vegetarian colonies in remote Kansas (*see* Madeleine B. Stern, *Heads & headlines,* Norman: Univ. of Oklahoma Press, 1971, p. 173-77).

2177. LAMBE, William, 1765-1847.
Water and vegetable diet in consumption, scrofula, cancer, asthma, and other chronic diseases. In which the advantages of pure soft water over that which is hard are particularly considered; together with a great variety of facts and arguments showing the superiority of the farinacea and fruits to animal food in the preservation of health . . . With notes and additions, by Joel Shew, M.D. New York; Boston: Fowlers and Wells; [ca. 1851].

viii, [9]-258 p. ; 19 cm.

Series: *Fowlers and Wells' water-cure library.* Spine-title: *The Water-cure library. Seven volumes. Vol. 5.* Issued in vol. 5 with Joel Shew's *Tobacco: its history, nature and effects on the body and mind* (New York, ca. 1851).

2178. LAMBERT, Alexander, 1861-1939.
Hope for the victims of narcotics . . . [S.l.: The Success Co., c1909].

16 p. ; 16 cm.

Title from wrapper. "The treatment of those addicted to narcotics has heretofore been a more or less rapid withdrawal and then a deprivation of the drug, trusting to the deprivation to cause gradually a cessation of the craving for the narcotic." Lambert notes the frequent failure of this method, and proposes a cure that "begins at the other end," i.e, that obliterates the craving for the narcotic by the administration of "a mixture of belladonna, xanthoxylum (prickly ash), and hyoscyamus, with a proper amount of active catharsis to stimulate the action of the liver and produce rapid and thorough elimination of the narcotic" (p. 3). While the narcotic is being purged from the system, the patient is still given enough of the drug at intervals to prevent symptoms of withdrawal.

Once the system is purged of the narcotic, the patient presumably has no further craving. Lambert devotes the rest of the pamphlet to an examination of various narcotics and their effects. An 1888 graduate of the College of Physicians & Surgeons (New York), Lambert was professor of clinical medicine at Cornell University Medical College (1898-1931) and attending physician at Bellevue Hospital (1894-1933). He was president of the American Medical Association in 1919.

LAMBERT, James. *The Countryman's treasure* see **NEW-ENGLAND FARRIER** (#2582).

LAMBERT, Richard Jay see **LOWRY** (#2293, 2294).

2179. LAMBERT, Thomas Scott, 1819-1897. Human anatomy, physiology and hygiene . . . Tenth edition. Hartford: Brockett, Hutchinson & Co. . . . [et al.], 1855.
    4, [13]-459 p., [8] leaves of plates : ill. ; 19 cm.

A revision of the author's *Popular anatomy and physiology* (#2183).

2180. LAMBERT, Thomas Scott, 1819-1897. Hygienic physiology . . . Portland, Me.: Sanborn and Carter; New-York: Leavitt and Company, 1851.
    143 p., [3] leaves of plates : ill. ; 19 cm.

Spine-title: *First book on physiology.* Written for elementary school students, the *Hygienic physiology* was the first in a series of three school physiologies written by Lambert (*see also* #2183, 2184). It is issued with his *Pictorial anatomy* (#2182). A 20-page advertisement is also present that describes *Lambert's pictorial anatomy and physiology,* a set of twenty-five chromolithographic plates (each two by three feet) that Sanborn & Carter intended to publish in three sets beginning in 1851. The National Library of Medicine holds a portfolio containing the first set (the six plates described here), which may be all that were published.

2181. LAMBERT, Thomas Scott, 1819-1897. Hygienic physiology . . . Portland, Me.: Sanborn and Carter; New-York: Leavitt and Company, 1852.
    143 p., [3] leaves of plates : ill. ; 19 cm.

Spine-title: *Lambert's first book.* Isssued with the author's 1851 *Pictorial anatomy* (#2182).

2182. LAMBERT, Thomas Scott, 1819-1897. Pictorial anatomy . . . Portland, Me.: Sanborn and Carter, 1851.
    20 p. : ill. ; 19 cm.

The *Pictorial anatomy* contains thirty-eight wood engravings depicting the bones of the human body. The copies in the Atwater collection were issued with the 1851 and 1852 issues of Lambert's *Hygienic physiology* (#2180, 2181).

2183. LAMBERT, Thomas Scott, 1819-1897. Popular anatomy and physiology. Adapted to the use of students and general readers . . . Auburn [N.Y.]: Alden, Beardsley & Co., 1852.
    2, [14], 408, [8] p., 5 leaves of plates : ill. ; 19 cm.

Popular anatomy and physiology was first published in 1849, and again in 1851 and 1852. In the publisher's advertisements that precede the text, it is referred to as the "Third Book" in Lambert's series of physiologies, intended for "higher classes of scholars . . . and as a first book for medical students."

2184. LAMBERT, Thomas Scott, 1819-1897. Practical anatomy, physiology, and pathology. Hygiene and therapeutics . . . Portland [Me.]: Sanborn & Carter; New York: Leavitt & Co., 1851.
    7, [1], [11]-257, [1] p., 5 leaves of plates : ill. ; 19 cm.

This abridgement of the author's *Popular anatomy and physiology* was written for a more diverse readership than his earlier text. It was first published in 1850, and in the publisher's advertisements is referred to as the "Second Book" in Lambert's series of school physiologies.

2185. LAMBERT, Thomas Scott, 1819-1897. Systematic human physiology, anatomy, and hygiene: being an analysis and synthesis of the human system, with practical conclusions . . . Second edition. New York: William Wood & Co., 1873.
    iv, [5]-420 p., [15] p. of plates + frontis. : ill. ; 19 cm.

The first edition was published at New York in 1865. Spine-title: *Human physiology, anatomy and hygiene.*

2186. LAMBERT, Thomas Scott, 1819-1897.
They are not dead. Restoration by the "heat method," of those drowned, or otherwise suffocated . . . New York: Wright & Schondelmeier, publishers, 1879.
    [1-4], ii, [5]-120, xlviii, A-B, [26] p., 8 pages of plates : ill., port. ; 20 cm.

A full-figure photographic portrait of Lambert is mounted on board bound among the advertisements.

2187. LAMONT, Gregorie.
Special announcement! To the WEAK AND FEEBLE and the SICK AND AFFLICTED. Chronic Diseases cured in a few days without Pain! DR. GREGORIE LAMONT, The Famous Practical Physician for Chronic Diseases, from Europe, Member of the Reformed Medical College, N.Y. . . . Dr. Lamont is the only physician in this country gifted with THE POWER OF TELLING DISEASES AT SIGHT. By looking into the eye, without the patient saying a word to him, he can tell them how they are affected in every particular, and prescribe for the immediate relief and permanent cure of their complaints. He can be consulted for a short time FREE OF CHARGE, AT His Rooms No. 4, Cumming's Block, Purchase St., New Bedford . . . [1856?].
    Broadside ; 36.5 × 24 cm.

Lamont states: "I have practiced my profession for the last 23 years, in most of the cities of Europe; I have been with the Indian tribes West as far as Independence, south of the Missouri, and at risk of life and limb, I have gained a thorough knowledge of their secrets in the healing art, possessed by no other living white man this side of the head waters of the Upper Missouri"; and assures his readers that he uses "nothing but but roots, herbs, gums, and balsams, leaves and barks" in his remedies.

2188. LAMSON, Armenouhie Tashjian, b. 1886.
My birth. The autobiography of an unborn infant . . . New York: The Macmillan Company, 1916.

xii, [2], 140 p., [18] leaves of plates : ill. ; 19 cm.

First edition. Second edition published in 1926 under the title: *How I came to be: the autobiography of an unborn infant.*

LANDER, Meta see **LAWRENCE** (#2221).

LANDES, Simon M. see **LANDIS** (#2189-2195).

LANDIS, Emma see **LANDIS** (#2192).

2189. LANDIS, Simon Mohler, 1829-1902.
The American improved family physician, or home doctor. A practical explanation of pathology, therapeutics, surgery, midwifery and pharmacy, on reformed principles. By Doctor Simon M. Landes . . . Lancaster [Pa.]: Independent Whig, printer, 1853.
    [2], 343, [3], 18 p., [1] leaf of plates : ill. ; 23 cm.

The Landis (or Landes) family emigrated from Switzerland in the 18th century and settled near Lancaster, Pa. in what became known as Landis Valley. They were Anabaptists, and had emigrated in search of religious freedom. Landis began his medical training at the Ephrata Water-Cure (Ephrata, Pa.) around 1850 under the preceptorship of S.M. Eby, M.D. He claimed to have graduated from the Eclectic Medical College of Pennsylvania in 1853, although the school did not receive its charter until 1856. By 1856 Landis was administering electro-chemical vapor baths at the Philadelphia Model Water-Cure, and organized the Philadelphia Private Hydropathic College, where students were assured "rapid progress in obtaining a practical knowledge of hygeiotherapeutics, anatomy, physiology and the fallacies of drug medication." The school was co-educational. In 1858 he moved to Madison, Wisconsin to operate the Lake Side Water-Cure, but soon returned to Philadelphia.
    Billing himself as a Reform Hygienic Physician & Surgeon, Landis operated a Hygienic Establishment at 13 North 11th Street in Philadelphia, where he gave private lectures on physiology. During the 1860s he also wrote several books and pamphlets, and went on the road

with his lectures (#2191). Landis engaged in other entrepreneurial activities. In 1867 he advertised in the *Water-cure journal* for agents to market *Dr. Landis' Celebrated Patent Compound Electro-Magnetic Hot and Cold Air Bath,* as well as "a most complete water-cure, electrical and Turkish vapor bath establishment in a portable box resembling a wardrobe." He also marketed a *Patent Compound Male and Female Magnitude Syringe and Organic Bath.* Without such accoutrements one might consider oneself left behind in the age of "hygienic improvement."

In the mid-1860s Landis organized the First Progressive Christian Church in Philadelphia, from the pulpit of which he espoused liberal social ideas and more open discussion of sexual matters. In 1870 he was tried, convicted and jailed in Philadelphia County Prison on a charge of obscenity in connection with his 1853 book entitled *Secrets of generation*—though this may have simply been a means to quiet his preaching. The indictment claimed that the book was "so lewd and filthy, and obscene that it is unfit to be spread upon the records of this court . . . and hence to read in the presence of a public audience." The defense claimed it was a scientific work and compared it to Horatio R. Storer's *Why not?* (#3383). Landis was in jail from 2 January 1870 until 18 May, when he received a pardon from Gov. John W. Geary.

Some evidence suggests that he went to New York at this time, where he remained about five years. By the time the second editon of *Secrets of generation* was published in 1885 (#2194), Landis had moved to Detroit. The book was considerably augmented, with the addition of a supplement on self-abuse, a discussion of contraception by "natural means," criticisms of the ideas of Dr. Beecher and Mrs. Woodhull on free love, and a polemic that includes eighteen reasons why he should have a new trial. From 1891 until his death, Landis practiced in Boston with his second wife, Lillian E. Landis, M.D. He died on 25 December 1902.

The author divides his treatise into five parts: I. Special pathology & therapeutics, or theory and practice; II. Surgery (which includes "the external application of lotions, liniments, poultices, plasters, &c." as well as hydropathic procedures); III. Midwifery; IV. Dietetics (vegetarian); and V. Pharmacy (arranged alphabetically from BALSAMS to WASHES). On the verso of p. 343 is a description of the "Eclectic or Hydro-

Fig. 61. S.M. Landis, from his broadside entitled To improve the human race, *#2195.*

pathic Institute . . . on the Harrisbuirg and Philadelphia turnpike, Ephrata, Lancaster county, Pa.," of which Landis was "physician and proprietor." A wood-engraved view of the "Ephrata Hydropathic Institute" is used as the frontispiece.

2190. LANDIS, Simon Mohler, 1829-1902.
Key to love! or Dr. S.M. Landis' celebrated private lecture on psychological fascinations. Secrets worth knowing! Fill the ever-blessed fount of the soul with the spirit of love! Learn to charm those you love! Worship the creator who ordained you to be loveable and loving, and seek his laws whereby you can attain your own ends . . . Philadelphia: Published by the request of thousands, 1865.
24 p. ; 18.5 cm.

"A true physiology teaches that every organ, gland, fibre, and tissue of the human body is sound and fully developed, and each of these performs its own office or function, and no more and no less. This is the state in which man is made in God's own image, because it signifies perfection. Under these circumstances, 'Love' holds its true and powerful position in the great Nervous Centre, the Thinking Brain, which may be called the great 'Human Battery,' from and to which all telegrams of the ethereal telegraph are dispatched through no other agency, essence, or influence, but concentrated and powerful thought. 'Faith,' cannot sustain itself, unless fed by a sound knowledge of the machinery which is used to send 'Love Despatches' from Nervous Centre to Nervous Centre, or Thinking Brain to Thinking Brain, or Battery to Battery; whilst Physiology lays the laws down in wonderful simplicity, and mathematical precision, whereby this 'Human Battery,' and its wires, are kept in working order" (p. 8). In printed wrapper.

2191. LANDIS, Simon Mohler, 1829-1902.
PHYSIOLOGY, PHRENOLOGY & HYGIENE! The Rev. S.M. Landis, M.D., the Celebrated Hygienic Physician and Popular Lecturer, of Philadelphia, will deliver three of his most practical, scientific and simplified LECTURES, to Ladies and Gentlemen, on Human Anatomy and Physiology; including Physiology of the Brain, called Phrenology, Health and Disease; setting forth practically, the causes which destroy health, and the way to preserve and restore it upon physiological, rational, common-sense and Christian principles; moreover, how to prolong life, increase happiness, &c., &c., AT THE TOWN HALL IN Mechanicsburg, Pa., on Thursday, Friday and Saturday evenings, Jan. 28th, 29th & 30th, '64, at 7 1-2 o'clock. Each lecture will be illustrated with some of the largest and best OIL PAINTINGS in the United States; and will be closed by PUBLIC PHRENOLOGICAL EXAMINATIONS . . .
Broadside ; 59 × 32.5 cm.

2192. LANDIS, Simon Mohler, 1829-1902.
Sectarian bigotry exposed without malice or glory. A series of twenty-five original, scientific lectures, entitled War on Christian heathens! . . . Reported, 1882, by Miss Emma Landis. Detroit, Michigan: Published by the Landis Publishing Company, at Dr. Landis' Medical Office, 1885.
33 p. ; 24 cm.

A collection of sermons delivered by Landis from the pulpit of his Scientific Church in which he mixes doctrinal and hygienic issues, e.g.: "The will of our heavenly Father is, to exercise the members, organs, glands, faculties and propensities of our bodies, each and every one aright." Needless to say, he "who is ignorant of what these bodily members and their functons are, cannot expect to obtain the blessings of heaven" (p. 27). Landis must have outraged not a few of his contemporaries with such ideas as only "false Christ worshippers" maintain that Jesus died for our sins. He died, according to Landis, "for principle; to show future generations that, to imitate or follow him, they must forego all pleasures if necessary, and suffer persecution for righteousness sake, for principle and justice" (p. 12)—a role which Landis obviously relished for himself. In printed wrapper.

2193. LANDIS, Simon Mohler, 1829-1902.
Sense and nonsense in relation to all topics concerning human affairs . . . Philadelphia, Pa.: Published by the First Progressive Christian Church of Philadelphia, Pa.; for sale . . . by Dr. Simon M. Landis, agent, at his office in the Church, 1867.
vi, 7-622 p. ; 19 cm.

CONTENT: Part I. Sense and nonsense in relation to body-and-soul doctrines. Part II. Sense and nonsense in relation to popular civilization and normal naturalization. Part III. Sense and nonsense in relation to church and state. Part IV. Sense and nonsense in relation to the explanation and inspiration of the "Holy Bible." Part V. Sense and nonsense in relation to diseases, drugs, medicines, doctors, preachers, and lawyers. Part VI. Sense and nonsense miscellany.

2194. LANDIS, Simon Mohler, 1829-1902.
A strictly private book on marriage; entitled, Secrets of generation! With supplement . . . This book must not lie about the house, but adults, male and female should read it. Detroit, Michigan, 1885.
254, [2] p. ; 11 cm.

In his introductory discussion of reproductive physiology, Landis explains that in truly natural intercourse the hymen is never ruptured, and that to lacerate this membrane is the result of "ungraceful and unphysiological intercourse," the consequence of which is to "ruin certain corresponding delicate organic structures in the womb, egg-chamber, breast, and brain" that results in "various weaknesses in the head, breast, womb and its appendages" (p. 25). In addition, "the vagina and vulva . . . in which man's conductor must work, will be flabby and somewhat lifeless, consequently less pleasure is experienced and more feeble children will come forth from such reckless lacerations" (p. 26). Landis insists that the unbroken hymen adapts to the "husband's conductor" during intercourse, and that "this membrane will grow with, in proportion and in harmony with the child and womb—and will render birth natural and painless" (p. 27).

Landis provides advice on courtship; when and how often couples should engage in sexual intercourse (never during pregnancy); the management of pregnancy; the "seasons and periods to generate healthy and talented babies," i.e., the times when "you should endeavor to generate; and, of course, prevent it at all other seasons either by entirely refraining from sexual intercourse, or by using the most natural and healthy agencies" (p. 54); advice on how to conduct intercourse that is practically unique to the literature of the period (p. 57-63); and "how to avoid too many children of one sex" (argued from the assumption that eggs from the right ovary produce males, from the left ovary females). The "supplement" (p. [83]-254) discusses masturbation and its consequences; provides more advice on selecting a mate and on courtship (from a eugenic standpoint); discusses parenthood, free love, etc.; and concludes with ten pages advertising Landis' "new discoveries and inventions in the healing science."

2195. LANDIS, Simon Mohler, 1829-1902.
To improve the Human Race has been my lively aim for 36 years, in opposition to "Scribes and Pharisees, Hypocrites," and I am still in dead earnest . . . READ! READ!! READ!!! The famous, original scientific publications by [*wood-engraved portrait of Landis*] SIMON M. LANDIS, M.D., D.D. Physiologist, Reform Practicing Physician, Author and Founder of the Scientific Church, of New York City, now residing at 124 Miami Avenue, Detroit, Mich. THE LANDIS PUBLISHING COMPANY has just issued

(Jan. 20, 1885) Dr. Landis' 25 Lectures . . . entitled: War On Christian Heathens . . . Also, his strictly private book on marriage, entitled "Secrets of Generation" . . . Broadside : port.; 36 × 13.5 cm.

2196. LANGLEY, J. O.
BUY ME AND I'LL DO YOU GOOD. The Great Spring and Summer Medicine. [*Wood-engraved portrait*] . . . DR. LANGLEY'S ROOT AND HERB BITTERS, Composed of Sarsaparilla, Wild Cherry, Yellow Dock, Prickley Ash, Thoroughwort, Rhubarb, Mandrake and Dandelion. THE BEST AND CHEAPEST MEDICINE THE WORLD EVER SAW. These Bitters will Cure Liver Complaint. These Bitters will Cure Jaundice in all forms. These Bitters will Cure Bilious Diseases . . . [etc.] The Ladies all like them, for they are sure to do them good. Let everybody use them! for they will Cleanse, Purify, Heal, Strengthen, Build up, and keep in order THE HOUSE YOU LIVE IN! . . . J.O. LANGLEY, Proprietor . . . J.E. Farwell & Co., Printers . . . Boston. [186-?].
Broadside ; 51 × 32 cm.

2197. LANKESTER, Edwin, 1814-1874.
Cholera: what is it? and how to prevent it . . . London; New York: George Routledge and Sons, 1866.
iv, [5]-93, [3] p. ; 16 cm.

Cholera appeared in the south of England in July 1865, and again the following summer. Though confined largely to the east end of London in 1866, Lankester warns against public carelessness or indifference: "a single case imported from the infected district may be the means of a similar explosion in any other part of London" (p. 13). Though he cannot identify the morbific agent, Lankester is aware that cholera is spread by the contamination of drinking water by improperly drained repositories of human effluvia. "The practical lesson to be derived from this knowledge is the fact that no community is safe from cholera that is dependent for its supply of drinking-water on rivers receiving sewage, or on wells that may communicate with drains, cesspools, or water-closets" (p. 60). In printed wrapper.

LAPHAM, Pardon see **CLARIDGE** (#648).

2198. LAPLACE, B.
Laplace's Standard Preparations. INDIAN
TURNIP COMPOUND PECTORAL BALM . . .
Laplace's Santonine Worm Lozenges . . .
FRENCH TROPICAL FEVER & AGUE MIX-
TURE . . . LAPLACE'S PURE SUBNITRATE OF
BISMUTH LOZENGES . . . Laplace's Phenic
Lotion . . . Highly Concentrated Compound
Fluid Extract of Sarsaparilla . . . LAPLACE'S
CONCENTRATED EXTRACT OF JAMAICA
GINGER . . . Laplace's Peruvian Hair Restor-
ative . . . All the above for sale by B.
LAPLACE, Pharmacist . . . New Orleans . . .
[186-?].
    [1] leaf printed on both sides ; 30.5 ×
22 cm.

    The same text in French is printed on the
verso: *Spécialités de la Pharmacie Laplace* . . . [etc.].

2199. LARMONT, Martin.
Medical adviser and marriage guide. Representing all the diseases of the genital organs of the male and female, the most complete and practical work on the physiological mysteries and revelations of the male
and female systems, with the latest experiments and discoveries in reproduction, it
illustrates anatomically and fully with
plates, everything pertaining to the male
& female genital systems; with a full description of the causes, symptoms and most certain mode of cure, of all the infirmities and
diseases to which they are liable from the
secrets habits of youth, and excesses of
mature age, as involuntary loss of semen
and impotency, innocent or unforeseen
affections, and those resulting from contraction, as syphilis, (primary and constitutional,) gonorrhoea, or blennorhagia,
(clap,) gleet, strictures, etc. etc. With numerous certificates of the most unparalleled cures ever performed . . . New York:
Published by the author, & Dr. E. Banister,
1861 [c1859].
    xvii, [18]-410 p. : ill. ; 19 cm.

    At head of title: *Fiftieth edition. Paris, London,
New York.* The first edition of the *Medical adviser
and marriage guide* was published at New York
in 1854. The twenty-eight wood-engraved plates,

though printed separately, are included in the
pagination of the text.
    Larmont divides his book into two parts. He
devotes the first eighty pages of Part I to the
anatomy of the male and female genitalia, the
physiology of reproduction and gestation. In
his discussion of the prevention of conception
(p. 87-93) Larmont dismisses the sponge recommended by other authors of the period as
ineffectual; describes douching as "sometimes"
effective; and regards condoms as effective but
for the fact that "they are troublesome, liable
to be torn in coition, [and] prevent the full
sexual enjoyment" (p. 89). He continues: "as
persons including physicians from all parts of
the country are applying to me every week by
letter or in person, for a sure preventive for
their wives, or some of their female patients . . .
I have after years of study, research and experiment, attained the great object which has been
so unavailingly sought for heretofore, that is,
for a man and his wife to fully enjoy sexual intercourse according to the laws of nature, and
at the same time not beget offspring, or if they
wish any, to limit the number according to their
wishes or circumstances" (p. 89). Larmont does
not describe his contraceptive device but to
state "It is used by the wife without inconvenience, and so secret that it cannot be known
by the husband, and will last until the change
in the wife's life, at about her fortieth year . . .
or in other words for a lifetime." The device
was available for purchase. The next sixty pages
of Part I describe a variety of urogenital diseases,
but focus on gonorrhea and syphilis.
    In Part II Larmont devotes nearly 200 pages
to the deleterious effects on mental and physical health of seminal loss through masturbation and spermatorrhea: "The main cause of
the genital organs exerting such influence over
the whole system, is their intimate connection
with the nerves. As there is no process carried
on in the body requiring so much action of the
nerves, as the evacuation of the semen; that is
the reason it is so exhaustive of the vital energy
. . . And if this nervous energy is exhausted by
emissions, its abstraction necessarily weakens
the power of the whole system . . . Thus it is the
stomach cannot digest the food, the heart cannot, for lack of power, propel the blood, nor
the brain think" (p. 193-94). Impotency, oph-
thalmia, sterility and melancholy are just some
of the disorders resulting from the "unnatural

and excessive loss of semen." In his lengthy discussion of spermatorrhea, Larmont quotes the standard medical authorities, and none more copiously than François Lallemand (1790-1854), whose *Des pertes séminales involontaires* was published in three volumes at Paris between 1836 and 1842. The first complete American edition of Lallemand's treatise was published at Philadelphia in 1848 under the title *A practical treatise on the causes, symptoms, and treatment of spermatorrhea* (5th ed., 1866).

Larmont is listed as a physician in the New York city directories from 1850/51 until 1869/70. He does not appear in Butler's 1877 medical directory nor in the first edition of *Polk* (1886). In the "Professional Notice" printed on the verso of the title-page, he describes E. Banister as his associate in practice, noting that their offices at 647 Broadway are "specially appropriated, so that patients never meet each other." Thus, "Persons in need of medical aid can apply to us with the undoubted assurance of a cure, in the most speedy and convenient manner, with perfect privacy." Those suffering from venereal and other disorders "at a distance, can be cured as secretly and as perfectly through the mail . . . as by a personal visit." Consultations cost five dollars.

2200. LARMONT, Martin.
Medical adviser, marriage guide, and physician for all. Profusely illustrating the most important diseases which afflict mankind, to which is added the history, symptoms and treatment, together with recipes for the cure of acute and chronic diseases peculiar to the genito-urinary organs and pelvic viscera, as well as many diseases of the general system . . . Eighty-first edition. [S.l.]: Sold by news dealers generally, [ca. 1870, c1868].
xi, [12]-373, [374-437], 438-460 p. : ill. ; 18 cm.

At head of title: *Paris, London and New York.* Date of imprint based on Larmont's change of address effective May, 1870 (p. v). This "81st" edition is probably the last appearance of Larmont's treatise.

2201. LATSON, William Richard Cunningham, 1866-1911.
The attainment of efficiency. Rational methods of developing health and per-

sonal power . . . New York; Passaic, N.J.: The Health Culture Co.; London, England: L.N. Fowler & Co., 1911.
[iii]-viii, 92, [12] p. ; 19 cm.

"Life is a struggle," Latson informs his readers, "a struggle in which many are vanquished and few survive. Only those survive who fit most perfectly to their environment. And for the sake of the few who are fitted to survive nature provides millions, most of whom must fail" (p. [30]). Latson defines "efficiency" as "the possession of such powers as will make the possessor strong, healthy, active and skillful at whatever task he may be employed" (p. vi). The possession of these powers derives not from bodily health alone, but more importantly, from the possession of a vigorous mind. In the attainment of efficiency, therefore, Latson considers factors such as diet, exercise, rest, etc. secondarily. He is primarily concerned with mental health and habits: "Man is his mind and nothing else. His body is purely the creature of his mind" (p. [9]). Even ill health is a product of mind, which "creates the body, preserves the body, destroys the body" (p. 13). The conditions that produce disease are the result of mental states. Thus the conquest of disease is the conquest of one's own mental states—resulting in the perfert harmony of mind and body that allows one to develop one's full powers for the struggle of life: "And what is the test by which is determined the fitness of the individual to survive? . . . How shall we know whether or not a given man or woman is destined to go up and bear a line of conquerers; or whether he or she is destined to propagate those who will swell the number of the unfit? . . . Reduced to the simplest terms, it may be said that the conqueror is he who conquers himself; she who conquers herself" (p. 31).

The attainment of efficiency was first published in 1910; the 8th and apparently final edition was published in 1915. A 1904 graduate of the Eclectic Medical College of the City of New York, Latson was the author of several works on popular medicine and hygiene, and edited the journal *Health culture* (v. 1, 1894), which continued publication until 1964.

2202. LATSON, William Richard Cunningham, 1866-1911.
Common disorders with rational methods of treatment, including diet, exercise, mas-

sotherapy, baths, etc. New York: Health-Culture Co., [c1904].

[2], 6, [4], [7]-325, [21] p. : ill., port. ; 19 cm.

Latson was one of the many proponents of the doctrine of autointoxication, i.e., that disease results from the body's inability to evacuate toxic matter. "In the strict sense," he writes, "there is but one disease—one disease, with many and varied symptoms. That is the effort of the organism to throw off accumulated waste matter" (p. 5). The hygienic system outlined in this volume is intended to be both prophylactic and therapeutic with regard to the elimination from the body of those elements whose retention results in disease. Frontispiece portrait of author.

2203. LATZ, Peter J.

Manual of health for women. Plain advice in sickness and health . . . Chicago: J.S. Hyland & Co., publishers, 1906.

326 p., [8] leaves of plates : ill. ; 25 cm.

"The increasing prevalence of female diseases is appalling," Latz writes in his preface, "Female troubles are so frequent especially among the so-called better classes that one rarely finds a healthy woman." He attributes this unhealthy state of an entire class of women to "an unnatural mode of living, i.e., lack of exercise, habitual constipation, highly spiced food, unsuitable dress, gonorrhoeal infection, abortion, imprudent conduct during the menses or in childbed, and the like" (p. 7). Latz divides his treatise into three parts. In the first he consider's "nature's curative agencies," a therapeutic system he terms *naturotherapy*, i.e., pure air, balneology, herbal remedies, sand baths, the movement cure (to be distinguished from gymnastics), massage, electrotherapy, and nutrition. In the second part he discusses the causes, symptoms and treatment of "female diseases." Part III, "The wife and mother," is devoted to marriage, pregnancy, and the care of infants and children. It includes a discussion of "voluntary sterility" (contraception) which Latz brands as "unnatural, immoral, damnable, and injurious to health" (p. 181). Latz was an 1896 graduate of the homeopathic Dunham Medical College and Hospital (Chicago).

2204. LAUFFER, Charles Alpheus, b. 1875.

Resuscitation from electric shock, traumatic shock, drowning, asphyxiation from any cause by means of artificial respiration by the prone pressure (Schaefer) method. With anatomical details of the method, and complete directions for self-instruction . . . First edition, fourth thousand. New York: John Wiley & Sons; London: Chapman & Hall, 1913.

[2], iv, [2], 47 p. : ill. ; 17 cm.

Lauffer received his medical degree from the University of Pennsylvania in 1905. He became resident physician at the Chester Hospital (Chester, Pa.), and later medical director of the Relief Dept. for Westinghouse Electric in East Pittsburgh, Pa.

2205. LAURIE, Joseph, d. 1865.

Homoeopathic domestic medicine . . . Edited with additions by A. Gerald Hull, M.D. New York: William Radde, 1843.

xii, 230 p. ; 16 cm.

Laurie's *Homoeopathic domestic medicine* was first published at London in 1842. (The 30th London edition appeared in 1910.) This first American edition was issued in 1843 by the New York publisher William Radde, who began his career early in the 1830s with editions of the classic German authors. Radde soon found his niche in the American market, however, and became one of the most prolific publishers of homeopathic literature in the United States. Between 1843 and 1853, Radde issued at least six editions of A. Gerald Hull's (1810-1859) version of Laurie's text. Not to be confused with the New York physician Amos Gerald Hull (1775-1835), A. Gerald Hull was one of New York's earliest advocates of homoepathy. He edited American editions of the works of Everest (#1090), Laurie and Jahr; and was editor of the *American journal of homoeopathia* (1835) and *The Homoeopathic examiner* (1840-43). Hull's edition of Laurie was supplanted on the American market by the version edited R.J. McClatchey in 1871 (#2208-2211).

2206. LAURIE, Joseph, d. 1865.

Homoeopathic domestic medicine . . . Arranged as a practical work for students. Containing a glossary of medical terms. Fifth American edition, enlarged and im-

proved, by A. Gerald Hull, M.D. New-York: William Radde . . . [*et al.*], 1849.
  xx, 568 p. ; 20 cm.

This fifth American edition is based on the third London edition (1846).

2207. LAURIE, Joseph, d. 1865.
Homeopathic domestic medicine . . . Arranged as a practical work for students. Containing a glossary of medical terms. Fifth American edition, enlarged and improved, by A. Gerald Hull, M.D. New-York: William Radde . . . [*et al.*], 1850.
  xx, 568 p. ; 19 cm.

2208. LAURIE, Joseph, d. 1865.
The homoeopathic domestic medicine . . . Edited and revised, with numerous important additions, and the introduction of new remedies, and a repertory, by Robert J. McClatchey . . . First American, from the twenty-first English edition. New York; San Francisco: Boericke and Tafel; Philadelphia: F.E. Boericke, 1871.
  xi, [1], 1034 p. ; 24 cm.

This 1871 "First American" edition of *The homoeopathic domestic medicine* is based on the 21st London edition of 1868. It was the first publication of Laurie's text in the United States since the 6th and final edition of Hull's version published by Radde in 1853. No where in the text does the editor, Robert John McClatchey (1836-1883), make mention of the earlier edition. In his "Preface to the American edition," however, McClatchey writes: "The publication of the American edition of *Laurie's Homoeopathic Domestic Medicine* was commenced, about two years ago, in the "Homoeopathic Sun" [1868-69], a family homoeopathic magazine, published in New York . . . As the work appeared with the monthly parts of the magazine, it attracted great and deserved attention, and was pronounced the best and most comprehensive treatise on homoeopathic domestic practice that had been issued in this country. Messrs. Boericke & Tafel, the successors of Mr. Radde, determined to discontinue the publication of the "Homoeopathic Sun," but were induced, by the repeated demands for the "Domestic," to go on with that work; and, subsequently placed it in my hands for completion" (p. vii). The 15th and final

edition of McClatchey's version of Laurie appeared in 1907.

McClatchey was an 1856 graduate of the Homoeopathic Medical College of Pennsylvania. He was a member of the faculty of his *alma mater* (prior to 1871), edited the *Hahnemannian monthly* between 1868 and 1878, and was general secretary of the American Institute of Homoeopathy (1872-1879).

2209. LAURIE, Joseph, d. 1865.
The homoeopathic domestic medicine . . . Edited and revised, with numerous important additions, and the introduction of the new remedies, and a repertory, by Robert J. McClatchey . . . Third American edition. New York; San Francisco: Boericke and Tafel; Philadelphia: F.E. Boericke, 1872.
  xi, [1], 1034 p. ; 24 cm.

2210. LAURIE, Joseph, d. 1865.
The homoeopathic domestic medicine . . . Edited and revised, with numerous important additions, and the introduction of the new remedies, a repertory, and a glossary, by Robert J. McClatchey . . . Eighth American edition. New York; San Francisco: Boericke and Tafel; Philadelphia: F.E. Boericke, 1877.
  xi, [1], 1044 p. ; 24 cm.

2211. LAURIE, Joseph, d. 1865.
The homoeopathic domestic medicine . . . Edited and revised, with numerous important additions, and the introduction of the new remedies, a repertory, and a glossary, by Robert J. McClatchey . . . Twelfth American edition. Philadelphia: Hahnemann Publishing House, 1892.
  viii, [2], ix-xi, [1], 1044 p. ; 24 cm.

2212. LAURIE, Joseph, d. 1865.
The parent's guide: containing the diseases of infancy and childhood and their homoeopathic treatment. To which is added a treatise on the method of rearing children from their earliest infancy; comprising the central branches of moral and physical education . . . Edited, with additions, by Walter Williamson . . . Philadelphia: Published by Rademacher & Sheek . . . [*et al.*], 1854.
  458 p. ; 19 cm.

First and only American edition of a work originally published at London in 1849. The American editor, Walter Williamson (1811-1870), was on the faculty of the Homoeopathic Medical College of Pennsylvania, and was the author of *A short domestic treatise on the homoeopathic treatment of the diseases of females and children* (Philadelphia, 1848).

2213. LAW, Hartland.

Viavi hygiene for women, men and children. By Hartland Law, M.D. [and] Herbert E. Law, F.C.S. Manchester, N.H.: New Hampshire and Vermont Viavi Company; Chicago, Ill.; Washington, D.C.: Eastern Viavi Company; San Francisco, Cal.: The Viavi Company, 1905.
[2], 610 p., [1] leaf of plates : ports. ; 20 cm.

"It is a fact familiar to all persons informed upon the subject," write the Law brothers, "that at least nine-tenths of all women are afflicted to a greater or lesser degree with some disease of the generative organs" (p. 28). "It is lamentable," they continue, "that women as a rule do not fully appreciate the evidences and effects of disease. Not knowing why they suffer, they do not seek means of relief" (p. 29). The *Viavi* treatment of Hartland Law and his brother Herbert Edward Law (b. 1864) was intended to bring suffering womanhood this relief. The phenomenal success of the Viavi regimen, which consisted in little more than basic feminine hygiene, was made possible by capitalizing on women's negative attitude toward their own reproductive systems. According to an article originally published in *J.A.M.A.*, "It is a well known fact that women seem to have the singular and rather unhealthy idea that the sexual organs should be ignored as something 'low,' 'vulgar' or 'indecent.' Most of them do not keep these portions of the anatomy . . . clean . . . Most women suffer more or less from their reproductive organs, and a very considerable amount of this discomfort or suffering is due to lack of common sense and cleanliness" (American Medical Association, *Nostrums and quackery*, 2nd ed., Chicago, 1912, v. 1, p. 239-40).

After reviewing the phenomena of blood circulation and absorption and how they provide the mechanism for the systemic spread of disease, as well as the therapeutic activity of *Viavi*, the authors review the numerous disorders to which women especially are subject. Finally, in chapter LXXIII, the forms of *Viavi* are addressed, e.g., the *Viavi Capsule* placed in the vagina: "its specific action is felt upon the generative tract, but its action is by no means confined to this one part of the body, as the entire system feels largely its curative action" (p. [490]). The contents of the capsule are absorbed through the vaginal membranes. The *Viavi Cerate*, the action of which also depends on absorption, was a lotion "applied [externally] to the diseased tissues and organs and over the nerve centers controlling the supply of blood to these organs" (p. 493). The *Viavi Liquid* was administered as a mouth spray that was carried into the lungs, enabling the *Viavi* globules "to act upon the inflamed tissue and membrane of the bronchi," where it is also taken up into the blood and is "conveyed to the nerve centers that control the blood supply in diseased membranes" (p. 500). There were also the *Viavi Rectal Suppository*, *Viavi Tablettes*, the *Viavi Laxative*, the *Viavi Tonic* and *Viavi Royal* (another tonic). An analysis of *Viavi*, i.e., the curative entity in all the Law brothers' preparations, determined that the *Viavi* remedies were composed of hydrastis and cocoa butter (Americnan Medical Association, *Nostrums and quackery*, 2nd ed., Chicago, 1912, v. 1, p. 243).

The *Viavi* empire spread rapidly across the nation. The Laws licensed agents to establish branch offices in dozens of cities in the United States (and Great Britain). One of the features of the system was that the physical examination of the patient was never necessary, even in the local offices: "The proprietors of the Viavi 'treatment' not only maintain that their agents are competent to suggest the proper treatment without examinaton of the patient, and that the omnipotent wisdom of the officials in the home office . . . can give 'competent advice' by mail, but they refer in terms of greatest horror to physician, gynecologist and surgeon, intimating that more harm than good always results from obtaining professional advice from licensed physicians" (*Nostrums and quackery*, 2nd ed., Chicago, 1912, v. 1, p. 245).

2214. LAW, Hartland.

Viavi hygiene for women, men and children. By Hartland Law, M.D. [and] Herbert E. Law, F.C.S. Boston, Mass.: New

Hampshire and Vermont Viavi Company; Chicago, Ill.; Washington, D.C.: Eastern Viavi Company; San Francisco, Cal.: The Viavi Company, 1906.
[2]-610 p., [1] leaf of plates : ports. ; 20 cm.

2215. LAW, Hartland.
Viavi hygiene. By Hartland Law, M.D. [and] Herbert E. Law, F.C.S. . . . Chicago, Ill.; Washington, D.C.: Eastern Viavi Company; San Francisco, Cal.: The Viavi Company, 1908.
[2]-416 p. ; 19 cm.

The title-page includes a "partial list of cities where there are Viavi offices." Listed alphabetically are ninety-three cities in the United States, three cities in Canada, two cities in Great Britain; and one each in Ireland, France, Australia and South Africa. The *Viavi* empire enabled the Law brothers to translate profits from their preparations into one of the most formidable real estate and financial enterprises in turn-of-the-century San Francisco.

2216. LAW, Hartland.
Viavi hygiene for women, men and children. By Hartland Law, M.D. [and] Herbert E. Law, F.C.S. Chicago, Ill.; Washington, D.C.: Eastern Viavi Company; San Francisco, Cal.: The Viavi Company, 1908.
[3]-610 p. : ill. ; 20 cm.

Laid into the Atwater copy is a hand-written receipt (dated 16 June 1909) for nine Viavi remedies from Ella M. Burt, a Viavi representative in Barre, Mass., to one of her customers.

2217. LAW, Hartland.
Viavi hygiene. Explaining the natural principles upon which the Viavi system of treatment for men, women and children is based. By Hartland Law, M.D. [and] Herbert E. Law, F.C.S. (Revised edition). Chicago, Ill.; Washington, D.C.: Eastern Viavi Company; San Francisco, California: The Viavi Company, Inc., 1909.
[3]-414 p. ; 19 cm.

2218. LAW, Hartland.
Viavi hygiene. Explaining the natural principles upon which the Viavi system of treat-

ment for men, women and children is based. By Hartland Law, M.D. [and] Herbert E. Law, F.C.S. (Revised edition). New York, N.Y.: New York Viavi Co.; San Francisco, California: Published by The Viavi Company, Inc., 1911.
[3]-414 p. ; 19 cm.

2219. LAW, Hartland.
Viavi hygiene. Explaining the natural principles upon which the Viavi system of treatment for men, women and children is based. By Hartland Law, M.D. [and] Herbert E. Law, F.C.S. (Revised edition). Chicago, Ill.; Washington, D.C.: Eastern Viavi Company; San Francisco, California: Published by The Viavi Company, Inc., 1913.
[3]-414 p. ; 19 cm.

2220. LAW, Hartland.
Viavi hygiene. Explaining the natural principles upon which the Viavi system of treatment for men, women and children is based. (Revised edition). Chicago, Ill.; Washington, D.C.: Eastern Viavi Company; San Francisco, California: Published by The Viavi Company, Inc., 1915.
[3]-416 p. ; 19 cm.

LAW, Hartland see also **VIAVI COMPANY** (#3646-3652).

LAW, Herbert Edward see **LAW** (#2213-2220).

2221. LAWRENCE, Margaret Oliver Woods, 1813-1901.
The tobacco problem. By Meta Lander . . . Fourth edition. Boston: De Wolfe, Fiske & Company, [c1885].
viii, 279 p. ; 19 cm.

Meta Lander was the pseudonym of Margaret Woods Lawrence, the poet and author of inspirational biography, whose books enjoyed considerable popularity in the second half of the 19th century. In this anecdotal and often redundant attack on "the ugly brown idol," Lawrence considers the "financial view" (i.e., the monetary burden tobacco consumption places on the individual and society); the "physical and intellectual view: (e.g., the effect of nicotine on circulation and respiration, tobacco's influence on mental processes, its connection to increased alcohol consumption, "tobacco-heredity" or the likelihood of nervous system disorders among smokers' offspring, etc.); "to-bacco-benefits" (i..e., the erroneous notions that tobacco protects against malaria, aids digestion, quiets the nerves, etc.); the "social and aesthetic view" (including a section on "female devotees"); and the "moral and spiritual view" (the tobacco habit deadens the conscience, weakens the will, over-rides reason and man's better sentiments, etc.). The first edition of *The tobacco problem* was published in 1885. An undated sixth edition was issued during the 1890s.

LEBENSTEIN, Joseph see **FOOTE** (#1248).

2222. LEE, Charles Alfred, 1801-1872.
Human physiology, for the use of elementary schools . . . Fourth edition. New York: Turner, Hughes, & Hayden ; Raleigh, N.C.: Turner & Hughes, 1843.
x, [11]-336 p. : ill. ; 19 cm.

An 1825 graduate of the Berkshire Medical College, Lee practiced medicine in his native Salisbury, Ct. for two years before moving to the city of New York. "After 1850 Dr. Lee devoted himself chiefly to teaching various branches of medicine in different medical colleges, among which may be named the University of the City of New York; Geneva Medical College; University of Buffalo . . . Vermont Medical College . . . Maine Medical School . . . Berkshire Medical Institution; Starling Medical College, Columbus, Ohio" (*Dict. Amer. med. biog.*, 1928, p. 730). He often retained two or more faculty appointments simultaneously. Lee retired from the Buffalo Medical College in 1871. He was a prolific author and editor. Among his popular works were the *Elements of geology* (1st ed., 1839) and *Human physiology,* which went through eleven editions between 1838 and 1854. Lee wrote several books for the profession, numerous articles for the periodical literature, and edited the American editions of such standard British medical authors as Paris, Pereira, Guy, etc.

LEE, Charles Alfred see also **CROWEN** (#833).

LEE, Henry see *SOUTHERN BOTANICO-MEDICAL JOURNAL* (#3293).

LEE, William see **FENTON** (#1154).

2223. LEE MANUFACTURING CO.
The Improved Vacuum Cupper and Developer. Restores strength by simply starting up new circulation of the blood which strengthens and cures IMPOTENCY, LOST MANHOOD, SEMINAL WEAKNESS, NERVOUS DEBILITY, PREMATURE DECAY, NIGHTLY EMISSION3 [sic], SPERMATORRHOEA . . . [Grand Crossing, Ill., 190?].
[2] leaves ; 12 cm.

For sexual debilities, the manufacturers recommend use of the vacuum cupper on the afflicted organ in combination with a course of *Lee Sexual Tonic Tablets.*

2224. LEFFINGWELL, Albert, 1845-1917.
American meat and its influence on the public health . . . New York: Theo. E. Schulte; London: G. Bell & Sons, 1910.
vii, [5], 208 p. ; 19 cm.

Leffingwell condemns lax enforcement of the second enactment of the 1906 Pure Food and Drugs Act, which pertains to the inspection of meat, including the slaughter of diseased animals for public consumption. The author was an 1874 graduate of the Long Island College Hospital. Circa 1883-1888, he served on the medical staff of The Sanatorium, the Dansville, N.Y. health resort founded by his maternal uncle James Caleb Jackson (#1902-1955). Leffingwell travelled widely, published an account of his tour of Japan (1892), was one of the founders of the American Society for the Regulation of Vivisection, was elected president of the American Humane Association (1904-5), served as American consul in Warsaw (1905-6), and was the author of several books on animal experimentation, vivisection and animal rights.

LEFFINGWELL, William Elderkin see **GLEN SPRINGS** (#1371).

2225. LEGACY OF A DEAD DOCTOR.
Legacy of a dead doctor; or advice to the married. Bang All, N.Y., 1841.
vi, [7]-75 p. ; 15 cm.

In the "Preface by the Publisher," this anonymously published work is attributed to "a humane and eminent Physician, whose heart has ceased to beat, and whose head is pressed by the clods of the valley" (p. iv). It is actually a paraphrased and/or plagiarized abridgement of the *Fruits of philosophy* by Charles Knowlton (#2154-2157), who died in 1850. Knowlton cautioned his readers (#2154, p. 2-3) about plagiarized versions of his work, specifically Canfield's *Sexual physiology* and Dubois' *Marriage physiologically discussed*. The *Legacy of a dead doctor* (or *Dead doctor's legacy* as it is twice named in the text) may be added to this list. In any event, it would be difficult to imagine a more suitable imprint for a manual of sex education.

The author follows Knowlton's description of the female genitalia, but omits the male, due to the greater role he imagined the female organs play in the process of generation. He again follows Knowlton in agreeing with Dewees' theory of conception, i.e., a belief in "absorbent vessels leading directly from the inner surface of the labia externa and the vagina to the ovaries" that "absorb the semen and convey it to the ovaries," resulting in fertilization of the ovum (p. 32). Conception takes place when the fertilized ovum descends the Fallopian tubes and attaches itself to the uterus.

The whole of chapter three is devoted to contraception. The author reviews such methods as withdrawal, cautioning his readers about the importance of total withdrawal due to the presence of semen absorbing vessels in the area of the external genitalia; the "baudrauche" (i.e., condom), a method that "is by no means calculated to come into general use" (p. 55); and a moistened sponge that cleanses semen from the lining of the vaginal lining as it is withdrawn following coitus. The method most enthusiastically recommend by the author, however, is plainly Knowlton's: "It consists in syringing the vagina immediately after connexion, with a solution of sulfate of zinc, of alum, pearlash, or any salt that acts chemically on the semen and at the same time produces no unfavorable effects on the female" (p. 56).

The author omits Knowlton's chapter on the signs of pregnancy; and concludes with a plagiarization of most of the chapter on "reproductive instinct," with the addition (acknowledged) of remarks on masturbation from the 1832 New York edition of Tissot's *A treatise of the diseases produced by onanism* (#3542).

2226. LEISERING, Friedrich.
Die Tugenden des Wassers für Gesunde und Kranke, worin auseinander gesetzt wird, wie ein vernünftiger Gebrauch des Wassers zum Trinken, Baden, und in Dämpfen die Gesundheit stärkt, vor vielen Krankheiten bewahret, und in Krankheitsanfällen der Mensch sich meistens selbst helfen kann; gestützt auf viele einstimmige Zeugnisse wohlerfahrner Aerzte im In- und Auslande, nebst andern Nachrichten in Beziehung auf Wassergebrauch . . . Philadelphia, im Monat December, 1842.
    46, [2] p. ; 17 cm.

Leisering's is the earliest hydropathic treatise in the Atwater Collection. An English translation was also published at Philadephia in 1842 under the title: *The virtues of water for the healthy and sick*. The author claims to have been a hydropath since the early 1820s in both Germany and the United States. In plain paper wrapper.

2227. LENT, Edward Burcham, b. 1869.
Being done good. An amusing account of a rheumatic's experiences with doctors and specialists who promised to do him good . . . Eleventh thousand. Brooklyn-New York: The Brooklyn Eagle Press, [c1904].
    [2], 345 p. : ill. ; 19 cm.

A New York journalist, Lent was invalided by rheumatism four years before the publication of this book. The author satirically chronicles his treatment at the hands of twenty-five "osteopaths, hydrotherapists, regulars, homoeopaths, Turkish-baths, surgeons, magnetic healers and clairvoyants"—to little avail. Even so, "They have done their best and the author forgives them" (p. 24). Lent's litany of remedial woes includes a lengthy chapter on Christian Science.

2228. LENT, Edward Burcham, b. 1869.
Being done good. Comments on the advance made by medical science during the past 5,500 years in the treatment of rheumatism . . . Second edition. Brooklyn-New York: The Brooklyn Eagle Press, 1904.
    345 p. : ill. ; 20 cm.

2229. LEONARD, Charles Henri, 1850-1925.
A popular treatise on the hair: its growth, care, diseases and their treatment. Designed for the use of the general public ... Fifth thousand. Detroit: Post and Tribune Job Printing Company, publishers, [c1879].
[2], 316 p. : ill. ; 20 cm.

An 1874 graduate of the University of Wooster Medical Department (Cleveland), Leonard was professor of medical and surgical gynecology at the Detroit College of Medicine & Surgery, the compiler of several successful handbooks for physicians on drug dosage and anatomy, and editor of *Leonard's illustrated medical journal* (1880-1906). *The hair* appears to have been his only book for a lay audience. Leonard devotes the first five chapters to the "chemistry, anatomy and physiology of the hair," and provides a single chapter on its care and hygiene before discussing the diseases of the hair and scalp. His final two chapters cover the history of hair styles and the beard.

LEVI, Reuben see *ECLECTIC JOURNAL OF MEDICINE* (#1027).

2230. LEVIS, S. Virginia.
Nursing. A practical treatise giving the fullest directions for the care of the sick in all the simple as well as the more serious ailments ... Philadelphia: The Penn Publishing Company, 1906.
214, 10 p. ; 15 cm.

Second edition (1st ed., Philadelphia, 1901). "The value of the trained nurse can scarcely be overestimated ... Yet, notwithstanding her advent, circumstances frequently demand that the care of the sick one shall devolve upon a member of the family. To such, therefore, instructions which are designed to equip them for better attendance in times of illness, must obviously be welcome" (p. 5-6). Levis discusses the sickroom and its appointment; the importance of understanding the germ theory in the domestic care of the ill; a great many children's diseases ("summer diarrhoea," mumps, measles, chicken pox, etc.); "preparations for confinement"; and diet for the sick and convalescent.

2231. LEWES, George Henry, 1817-1878.
The physiology of common life . . . New York: D. Appleton and Company, 1860.
2 v. (xii, 368 p.; 410 p.) : ill. ; 19 cm.

First American edition (1st ed., 2 vols., Edinburgh, 1859-60). "His education was desultory," begins Lewes' lengthy entry in the *Dictionary of national biography* (11:1043-46). Nonetheless, his range of intellectual interests and literary productivity were remarkable. "For a time he walked the hospitals, but gave up the profession from his dislike to witnessing physical pain" (*ibid.*, p. 1043). Lewes also gave serious consideration to a career in the theatre (in his actor father's footsteps)—and did appear in several stage productions. He found his *metier* as a writer early in life. By his mid-twenties, Lewes already enjoyed a reputation as an authoritative critic on contemporary continental drama. He also found time to write (and see produced) one play, and to author two novels. Lewes was intensely interested in philosophy. He wrote articles on French, German and Italian philosophers for the periodical literature, and in 1845-46 published his multi-volume *Biographical history of philosophy*, which received critical acclaim. Leaving his wife and family in 1854, Lewes lived in Germany, where he completed his *Life of Goethe*—long the standard critical biography of Goethe in the English language.
In the late 1850s Lewes combined his philosophical and physiological interests, turning his attention to the study of physiology, zoology and marine biology. *The physiology of common life* was one of the products of these studies. In his first volume, Lewes discusses the physiology of alimentation & digestion, the blood & circulation, and respiration (concluding with a Comtean assault on the idea of the "vital principle"). The second volume is devoted to his then emerging interest in the implications for psychology (sensation & thought) of the structures of the brain and nervous system. Though based upon scientific research, the work was intended as a popular exposition on these subjects.

2232. LEWIS, Denslow, 1856-1913.
Pioneer publication in American medical education reform. The gynecologic consideration of the sexual act . . . Chicago 1900. And an appendix with an account of Denslow Lewis, pioneer advocate of pub-

lic sexual education and venereal prophylaxis by Marc H. Hollender, M.D. Weston,
Massachusetts: M & S Press, 1970.
    16, [2], 49, [3], 11 p. : ports. ; 23 cm.

In addition to Marc Hale Hollender's (b.
1916) historical introduction, this slender volume includes facsimiles of Lewis' *The gynecologic
consideration of the sexual act* (Chicago, 1900) and
"The advocacy of publicity regarding venereal
prophylaxis: a personal experience," which
Lewis published in the *Pacific medical journal* in
1906 (49:411-21).
    "In 1899 Denslow Lewis read a paper . . . at
the meeting of the American Medical Association in Columbus, Ohio. This paper dealt . . .
with five subjects: 1.) the importance of the
sexual instinct; 2.) the description of the sexual
act; 3.) the sexual education of girls and young
women; 4.) the rights of women; 5.) the nature
of sexual responses in women. The paper was
turned down for publication in the *Journal of
the American Medical Association*" (Hollender's
introduction, p. 7). In response to the rejec-
tion of this paper, Lewis published *The gynecologic consideration of the sexual act* at Chicago in
1900. "The pamphlet contained his original
paper, discussions of it and the correspondence
pertaining to its publication as an article in the
Journal" (*ibid.*, p. 13). In "The advocacy of publicity regarding venereal prophylaxis," Lewis
describes the reaction to his distribution of the
pamphlet at the A.M.A. meeting in Atlantic City
in 1900, and of the medical profession's obligation to speak candidly about sexuality, sexually transmitted diseases, etc. Lewis was an 1878
graduate of the Medical Department of the
University of Michigan, and a prominent member of the Chicago medical community. He was
active in state and national medical societies,
and a frequent contributor to medical periodicals on obstetrical, gynecological and public
health topics.

2233. LEWIS, Dio, 1823-1886.
    Chastity; or, our secret sins . . . New York
City: Clarke Brothers, [c1874].
    320 p., [1] leaf of plates : port. ; 19 cm.

Lewis' treatise on "sex-morality" was first
published at New York in 1871. The author
takes his reader on a rambling discourse
through such topics as the harm caused young

men by women's provocative dress (e.g., "low
necks," "short sleeves," "padded busts," etc.);
the ills of early marriage; sexual behavior during marriage (i.e., the consequences of "unbridled venery"); the reprehensibility of contraception ("The various devices of the French
voluptuaries are familiar to the public. They are,
without exception, unsatisfactory and mischievous"); hereditary influences (i.e., race improvement; the effects of alcoholism, syphilis,
etc. on offspring); masturbation; the perils to
children of "vicious servants" ("put your children to bed yourself") and obscene literature
(including an 1873 letter to the author from
Anthony Comstock, #754, 755); advice to young
women and young men on the merits of celibacy; and a concluding chapter on practical
hygiene.
    Born Dioclesian Lewis near Auburn, N.Y.,
Lewis grew up in the *Burned-Over District* of central and western New York State, "America's
most fertile ground for revivalism and reform
during the Second Great Awakening (1800-
1830)." There he "absorbed revivalism's lesson
of individual improvement through self-disci-
pline and applied it to social problems created
or exacerbated by urbanization and industrial-
ization" (*Amer. nat. biog.*, 13:565). Lewis began
his medical studies with a physician named
Briggs, a preceptor in his native Auburn, and
attended lectures at Harvard Medical School
(1845-46). He never completed his studies,
though he did receive an honorary degree in
1851 from the Homoeopathic Hospital College
of Cleveland. While practicing medicine (without a degree) in Port Byron, N.Y., Lewis was
converted to homeopathy. In 1849, he moved
to Buffalo, where he began publication of the
Homoeopathist (1850-51) and organized his first
gymnastic classes for women. Lewis became
convinced of the efficacy of physical training
and exercise therapy during the successful treatment of his wife for pulmonary tuberculosis.
He devised an exercise program that required
little or no equipment, and that could be
adapted by persons of either sex, and any age
or physical condition. In 1860 Lewis moved to
Boston, where in 1861 he established the Boston Normal Institute for Physical Education, the
first training school in the United States for
physical educators. He also began publication
of *Lewis's gymnastic monthly and journal of physical culture* (1861-62), and published his most

important book, *The new gymnastics* (#2238), in 1862. In 1864 Lewis opened Dr. Lewis's School for Young Ladies in Lexington, Mass. (#2245, 2246), where he was able to fully implement his ideas on physical education and pedagogy. From the late 1860s Lewis turned his attention to temperance reform. The effectiveness of his message on the lecture circuit during the winter of 1873-74 in western New York and southern Ohio resulted in the Women's Temperance Crusade, which gave birth to the Woman's Christian Temperance Union. Lewis moved to New York in 1881, and retired from all but literary activity.

2234. LEWIS, Dio, 1823-1886.
Chastity; or, our secret sins . . . New York: Fowler & Wells, publishers, 1894.
    320 p. ; 19 cm.

Chastity was issued at least eight times between 1871 and 1894.

2235. LEWIS, Dio, 1823-1886.
Five-minute chats with young women and certain other parties . . . New York: Harper & Brothers, publishers, 1874.
    426 p. ; 19 cm.

"This volume is made up of paragraphs, and is mostly devoted to the subject of health" (p. [7]). Lewis' "paragraphs" address marriage, the status of women, tobacco, "physical culture," longevity, diet, etc. Cover-title: *Chats with young women.*

2236. LEWIS, Dio, 1823-1886.
Five-minute chats with young women, and certain other parties . . . New York: Fowler & Wells, publishers, 1892.
    [4], [9]-426 p. ; 19 cm.

Atwater copy inscribed by Mrs. Dio Lewis, December, 1897.

2237. LEWIS, Dio, 1823-1886.
In a nutshell. Suggestions to American college students . . . New York: Clarke Brothers, 1883.
    209 p. : ill. ; 17 cm.

Lewis offers advice on hygienic issues he thinks of importance to college age males.

Fig. 62. Dumbell exercises from The new gymnastics, *#2238.*

Pages [179]-209 contain a "Descriptive catalogue of the works of Dio Lewis, published exclusively by Clarke Brothers . . . New York, 1883."

2238. LEWIS, Dio, 1823-1886.
The new gymnastics for men, women, and children. With a translation of Prof. Kloss's Dumb-bell instructor and Prof. Schreber's Pangymnastikon . . . Boston: Ticknor and Fields, 1862.
    274 p. : ill. ; 20 cm.

First edition. The most successful of Lewis' published works, *The new gymnastics* appeared in twenty-five editions through 1891. Moritz Kloss' (1818-1881) *The dumb-bell instructor for parlor gymnasts* (p. [117]-164) and Daniel Gottlieb Moritz Schreber's (1808-1861) *The Pangymnastikon* (p. [165]-260) have separate title-pages, but their collation and pagination are continuous with the rest of the text.

"Lewis had . . . become convinced that the strength built by gymnastic work was not the true measure of muscular health. The gymnasts and the weight lifters were not gaining freedom from disease or insuring longer life with their bigger muscles, for their movements were too limited and strenuous, and not integrated to give balanced exertion . . . Lewis began formulating his system about 1854, while in the process of curing his wife of tuberculosis with walking, wood-sawing, and other conventional exercises. His goal was to find exercises that would have the excitement of free play but could be combined to develop whole body flexibility, coordination, agility and grace of movement.

Exceptional muscular strength was not an object; mental and moral improvement were" (James C. Whorton, *Crusaders for fitness,* Princeton, N.J., 1982, p. 276).

"Lewis' energetic prose," writes Jan Todd of *The new gymnastics,* "marched the reader through 115 pages of descriptions of exercises that, taken together, he claimed, as *his* system of New Gymnastics. There were thirty bean-bag exrecises, fifty-four exercises involving the use of a partner and six-inch wooden ring, sixty-seven 'wand' exercises using four-foot long hollow rods inside which several pounds of lead shot could be placed to increase resistance, thirty-four dumbell drills, and twenty-two Indian club exercises. The final series of exercises that Lewis claimed as part of his system in 1862 were thirty-eight free-hand, calisthenic movements" (*Physical culture and the body beautiful,* Macon, Ga.: Mercer University Press, 1998, p. 241-42). Todd later writes: "Lewis's most significant contribution to the history of women's exercise, however, was broadly philosophical and difficult to quantify. It was an attitudinal contribution and resulted in both an emerging physicality and a growing independence among American women" (p. 272). She adds: "In a number of ways, Lewis changed forever the basic dialogue concerning women's bodies. Rather than encouraging women to view their physical ideal as small, frail, slightly ill, and painfully weak, Lewis championed health, vigor and substance . . . These were new, important concepts for American women, concepts that freed women to realize that, within limits, they were actually in charge of their physical destiny" (p. 274).

In 1861 Lewis incorporated at Boston the Normal Institute for Physical Education, which within six months graduated the first class of physical education teachers in America. Whorton notes that half the members of the first graduating class were women (p. 278).

LEWIS, Dio, 1823-1886. The new gymnastics see also **WARREN** (#3708, 3709).

2239. LEWIS, Dio, 1823-1886.
Our digestion; or, my jolly friend's secret . . . Philadelphia and Boston: Geo. Maclean; Cincinnati and Chicago: E. Hannaford & Co.; New York: Maclean, Gibson & Co., 1872.

407 p., [5] leaves of plates : ill., port. ; 19 cm.

Originally published in 1871, *Our digestion* discusses mastication, nutrition, digestion, abuses of the digestive system, etc. in a series of brief articles similar in nature and format to some of Lewis' other books, such as *Five minute chats, In a nutshell,* etc. Steel-engraved frontispiece portrait of author.

2240. LEWIS, Dio, 1823-1886.
Our girls . . . New York: Harper & Brothers, publishers, 1871.
vii, [8]-388, [8] p. ; 19 cm.

First edition. *Our girls* is one of Lewis' more interesting books, or perhaps, one of his less fatuous literary efforts. It was issued at least half a dozen times between 1871 and 1885. The table of contents speaks for the volume:

Girls' boots and shoes; How girls should walk ["The corset is a deadly enemy to fine walking"]; The language of dress ["Many a modest woman appears at a party with her arms nude, and so much of her chest exposed that you can see nearly half of the mammal gland"]; Description of dress; Outrages upon the body [subtitled: Fashionable sufferings; Woman tortures her body; Stocking supporters ["the pressure of the garter . . . is very injurious," producing "weakness of the legs, a lack of circulation, and, therefore, coldness of the feet"]; Large *vs.* small women ["the average large-sized woman is . . . superior intellectually and otherwise, to the small-sized one"]; Idleness among girls; Idlness is fashionable [women should have an occupation, so that they may marry from choice, and not as a "contract for board, women must not be compelled to choose between marriage and starvation"]; Work is for the poor; Employments for women [e.g., bank clerks, brokers, dentists, lawyers, physicians, publishers, etc.]; False tests of gentility [a plea for "woman's rights"]; A short sermon about matrimony [actually about women's dress]; Piano music [and music in general, including remarks on Italian opera: "God alone knows the number of pure souls that have been ruined by the insidious poison of the opera"]; Study of French ["French is studied, in most cases, for the same reason that the piano is, it is fashionable"—besides, "Our language is as superior to the French as is our civilization"]; Dancing [square dancing is "graceful, chaste, and healthful," but dancing during which a woman is in "contact with her partner is not a modest one"]; The theatre [not only is the ventilation bad, but it "it keeps you up till midnight" and "the general tone is morbid, not to say impure"]; Sympathy between the stomach and the soul ["if the digestive functions be insane, pure and noble moral impulses are no longer possible"]; About the treatment of diseases; Sunshine and health; A word about baths [bathe daily]; Home gymnasium ["The effeminacy of our civilized life, with the employment of machinery for the hard work, necessitates a resort to artificial physical exercise"]; What you should eat; What you should drink; Additional health thoughts; Amusements for girls [e.g., croquet, skating, dancing, walking, "base-ball"]; True education for girls ["The graduate of a Woman's Seminary should . . . be as much improved in body as in mind"]; Heroic women.

2241. LEWIS, Dio, 1823-1886.
Talks about people's stomachs . . . Boston: Fields, Osgood and Company, 1870.
vii, [8]-320 p. ; 19 cm.

2242. LEWIS, Dio, 1823-1886.
Weak lungs, and how to make them strong. Or diseases of the organs of the chest, with their home treatment by the movement cure . . . Boston: Ticknor and Fields, 1863.
vii; [8]-360 p. : ill. ; 20 cm.

First edition. Early in their marriage, Lewis' wife (*nee* Helen Cecilia Clarke) was diagnosed with pulmonary tuberculosis. Regarding drug therapy as useless, Lewis devised a system of exercises for the treatment of her malady. *Weak lungs* is a guide for those with consumption (or those who may be disposed to the disease) in the prevention or treatment of the disease. Lewis discusses the causes of consumption (e.g., impure air, dust, atmospheric moisture, etc.); the much exaggerated influence of climate in the production or amelioration of consumption; the considerable role that diet, alcohol, tobacco and dress (i.e., tight-lacing) play in producing the disease; and most importantly, exercise as the most effective means of treatment. Lewis endorses modes of exercise that were often recommended by medical authors (e.g., horseback riding, walking, hunting, gar-

dening, etc.). The most important part of this work, however, is the description of his own ten-week course of home exercise, illustrated in a series of 105 wood engravings. *Weak lungs* was issued frequently between 1863 and 1881 (the bibliographic record is certainly not complete). It was reissued in 1885 and 1891 under the title *Weak lungs and the remedy.*

2243. LEWIS, Dio, 1823-1886.
Weak lungs, and how to make them strong. Or diseases of the organs of the chest, with their home treatment by the movement cure . . . Boston: Ticknor and Fields, 1864.
vii, [8]-360 p. : ill. ; 20 cm.

"Fourth edition" (on verso of title-page). The Atwater copy bears the bookplates of Lawrason Brown (#431) and the Trudeau Sanatorium library.

LEWIS, Dio see also **EASTMAN** (#1025); **DR. DIO LEWIS'S SCHOOL FOR YOUNG LADIES** (#2245, 2246).

LEWIS, Helen Cecelia Clarke see **EASTMAN** (#1025).

LEWIS, Margaret Wiseham see **BUCKELEW** (#508).

2244. LEWIS, Thomas.
Rules and regulations for the employment of injections in various diseases; with quotations from distinguished medical authors. Printed to accompany Lewis' Improved Portable Syringes, or Domestic Injecting Apparatus. Boston, 1857.
54 p. : ill. ; 16 cm.

A manual for the domestic use and care of the *Improved Portable Syringe.* Lewis provides recipes for rectal and vaginal "injections" for the treatment of a variety of diseases. Vaginal douching as a contraceptive technique is not described, though one of the apparatus' four tubes is specifically for vaginal use. The text is illustrated with two wood engravings. On printed wrapper: *Eleventh edition.* Thomas Lewis first appears in the Boston directories in 1856 as a seller of syringes at 166 Washington Street. Three years later the business became the firm of Lewis & Richardson, at 13 Water Street.

2245. DR. DIO LEWIS'S SCHOOL FOR YOUNG LADIES.

Second anniversary of Dr. Dio Lewis's School for Young Ladies, Lexington, Mass. Tuesday and Wednesday, June 5 and 6, 1866 . . . [Lexington, Ma., 1866].

12 p. ; 20 cm.

One of Lewis' objections to the "old gymnastics" was that its strenuousness excluded women. He was also convinced that the object of physical training was not simply muscular strength or size: "Lewis's *idee fixe* was that physical education was the indispensable foundation of all other education, and that the American school system could never succeed at its task until it gave physical education as central a place as intellectual and character training" (James C. Whorton, *Crusaders for fitness,* Princeton, N.J., 1982, p. 276).

In 1864 Lewis opened his school for the education of young women, "a school that he hoped would set off a new revolution, a winning of physical independence for women . . . There Lewis intentionally accepted 'delicate' girls of average age seventeen, girls who could not 'go upstairs without symptoms,' and put them on a schedule heavy with exercise and study . . . Each day began with thirty to ninety minutes of gymnastics, and the rest of the day was ordered to give fresh air and sun, plain (but not vegetarian meals), nonrestrictive dress, pleasant social life, and regular hours . . . The school burned in 1867 and was not resurrected, but for that brief period it was a shining demonstration of the potential for female vitality" (*ibid.,* p. 278-9). In *Our girls* (#2240) Lewis describes the school at Lexington as "the truest exponent of education for girls, which has been seen in our country" (p. 352).

2246. DR. DIO LEWIS'S SCHOOL FOR YOUNG LADIES.

Third anniversary of Dr. Dio Lewis's School for Young Ladies, Lexington., Mass., Tuesday and Wednesday, June 4 and 5, 1867 . . . [Lexington, Ma., 1867].

12 p. ; 18 cm.

This third anniversary celebration of Lewis' female seminary was to be its last. "The interesting and successful experiment in a many-sided education for girls which was inaugurated at Lexington came to an untimely end through the burning of the beautiful building in which it was carried on. On the morning of September 7th, 1867, three years after the establishment of the school, the imposing structure, containing one hundred and ten rooms, including a spacious ball-room used for lectures, receptions, and gymnastics, as well as for dancing, was discovered to be on fire, and in a few hours was burned to the ground with almost all its contents" (Eastman, #1025, p. 117). The academic year had been scheduled to resume on September 20. Lewis was able to quickly relocate the school in a summer hotel five miles from Boston, but the school lasted only another year. He had insured the building for two thirds its value, and was unable to finance its reconstruction. Title from printed wrapper.

2247. LIBRARY OF HEALTH.

Library of health. Complete guide to prevention and cure of disease. Containing practical information on anatomy, physiology and preventive medicine; curative medicine, first aid measures, diagnosis, nursing, sexology, simple home remedies, care of the teeth, occupational diseases, garden plant remedies, alcohol and narcotics, treatment by fifteen schools of medicine, beauty culture, physical culture, the science of breathing and the dictionary of drugs. Twenty books—one volume. By B. Frank Scholl . . . John Forsyth Little . . . Frank E. Miller . . . Assisted by a large and competent staff of practitioners, lecturers and teachers representing the foremost universities of the world. Philadelphia, Pa.: American Health Society, [c1916].

[4], xvi, 33-1438, [1441]-1774, [2] p., [61] leaves of plates + [3] chromolithographed paper manikins : ill. ; 26 cm.

Benjamin Franklin Scholl (1860-1933), John Forsyth Little (1880-1918) and Frank Ebenezer Miller (1859-1932) claim to have been assisted by twenty five other contributors in the compilation of this massive volume. Its content is similar to the *Medicology* published at New York in 1906 (#2976), and in many places throughout the text is identical to the latter title in wording and illustration. The *Library of health* also bears some similarity to the *Domestic medical practice* edited by Frank E. Miller and published at Chicago in 1914 (#970).

The final (unnumbered) leaf is a certificate entitling the purchaser of the *Library of health* to membership in the American Health Society, which includes "free, prompt and competent advice for a period of two years" and a free physical examination. Nearly identical certificates for membership and entitlements appear in the 1906 *Medicology* for the University Medical Society (Scholl's name appears at the bottom of the certificate) and in Miller's 1914 *Domestic medical practice* (the Domestic Medical Society).

LIBRARY OF HEALTH, AND TEACHER ON THE HUMAN CONSTITUTION see **ALCOTT** (#35).

LIBRARY OF USEFUL STORIES see **CONN** (#768).

LIBRARY OF VALUABLE KNOWLEDGE see **CONN** (#767).

2248. LIEB, Herman A.
Health, strength and beauty. By Herman A. Lieb, M.D. and Madame Eugenie . . .

Fig. 63. From Blaisdell's How to keep well, *#343.*

Chicago: Manufactured by American Publishing House, [c1902].
    [4], [3]-404 p. : ill., ports. ; 22 cm.

Lieb addresses women's health, child rearing and marriage. He gives special emphasis to "physical culture" and "water applications" in the maintenance of health. "Part III. Beauty," was contributed by Madame Eugenie, "a widely known dermatologist, and a model of health and physique." "Beauty is a woman's special domain," Madame E. informs her readers, "It is to her what strength and courage is to a man. It is the lever by which she moves all to her purpose" (p. 301-2). Toward the attainment of this end, she provides more than two hundred recipes for skin cleansers, creams, "complexion powder," depilatories, perfumes, etc.

2249. LIEBIG, Justus Freiherr von, 1803-1873.
Familiar letters on chemistry, and its relation to commerce, physiology and agriculture . . . Edited by John Gardner . . . New York: D. Appleton & Co.; Philadelphia: George S. Appleton, 1843.
    180 p. ; 16 cm.

Translation of *Chemische Briefe* (1843). In the early 1840s, Liebig began to apply his understanding of quantitative chemical methodologies and his knowledge of the chemical properties of organic compounds to the study of animal physiology. In the *Familiar letters* he provides a popular digest of this work and its significance for "the practical arts of life," i.e., "the bearing of chemistry upon physiology, medicine, and agriculture—which may be said to be only just begun" (p. 6). Although renowned throughout the English-speaking world, Liebig's work received mixed reviews, even by health reformers seeking to establish physiology and hygiene on scientific principles. Liebig posited "a theory of metabolism maintaining that all body heat is produced by the oxidation of fats and carbohydrates (the respiratory foods), and the energy for all muscular movement by the oxidation of protein (plastic food). An extension of the latter position was that an individual's protein consumption should be proprotional to the amount of muscular work he had to do . . . This aspect of Liebig's thought was most unwelcome to vegetarians. Since the typical fleshless diet contained significantly less protein than the standard, vegetarians were theoretically incapable of prolonged exertion" (James C. Whorton, *Crusaders for fitness*, Princeton, N.J.: Princeton University Press, 1982, p. 226).

Liebig entrusted the English edition of his letters to John Gardner, M.D. (1804-1880), one of the founders of the Royal College of Chemistry. The London edition of the *Familiar letters* was published in 1843, and was followed that same year by several American editions in New York and Philadelphia. Spine-title: *Liebig's chemical letters*. The Atwater copy bears the signature of Henry Foster (#1265).

2250. LIFE EXTENSION INSTITUTE.
What to eat. How to use the science of modern dietetics for more efficient living. Published by the Review of Reviews Company under the auspices of the Life Extension Institute of New York. [New York], Copyright, 1917, by the Review of Reviews Company.
48 p., [2] leaves of plates, [8] p. of plates : ill. ; 20 cm.

The Life Extension Institute was established by Irving Fisher (#1166, 1167) as a non-profit foundation for the education of the public on health matters and the provision of low-cost heath examinations.

2251. LIGHTHALL, J. I., b. 1856.
The Indian household medicine guide. By J.I. Lighthall, the Great Indian Medicine Man, and Dr. W.O. Davis, a graduate of the Eclectic Medical Institute, Cincinnati, Ohio . . . Peoria, Illinois, 1882.
iv, [5]-167 p. ; 18 cm.

First edition (2nd ed., 1883). In the "Sketch of J.I. Lighthall's life" (p. [5]-8) we read that he left his native Illinois at age eleven and migrated to the Indian Territory, where he spent the next thirteen years observing Indian doctors in Kansas, Wyoming and Minnesota select plants and prepare them for medicines. In his preface, Lighthall writes that his intention is "to teach the humble and poor, the farmer and mechanic, the merchant and his clerk, that God, in his infinite wisdom, has created and grown an herb with medicinal properties to

prove a balm to every ailment that the human organization is heir to" (p. [iii]). He describes the medical properties of seventy-three plants, followed by recipes for various maladies. Lighthall's collaborator, William O. Davis (b. 1852), was an 1879 graduate of the Eclectic Medical Institute (Cincinnati).

2252. LIGHTHILL, August P.
A popular treatise on deafness: its causes and prevention. By Drs.Lighthill. Edited by E. Bunford Lighthill . . . New York: Carleton, publisher, 1862.
133, [3], 8 p. : ill. ; 19 cm.

The authors discuss the anatomy, physiology and diseases of the ear, as well as preventive measures against deafness and a "review of some of the popular remedies for deafness." August P. Lighthill was an 1851 graduate of the University of Berlin and an 1877 graduate of the Eclectic Medical College of the City of New York. He appears in the Boston directories under the category "Physicians, other" intermittently from 1863 to 1868, and with E.B. Lighthill in 1885. Edward Bunford Lighthill was an 1882 graduate of the Eclectic Medical College of the City of New York. The book includes an eight page publisher's catalog preceded by an advertisement for hearing trumpets manufactured by Otto & Reynders. Atwater copy inscribed by Henry Foster (#1265).

2253. LIGHTHILL, August P.
A popular treatise on deafness: its causes and prevention. By Drs. Lighthill. Edited by E. Bunford Lighthill, M.D. Fourth edition . . . New York: Carleton, publisher, 1862.
133, [1], 8 p. : ill. ; 19 cm.

2254. LIGHTHILL, C. B.
Lewis County Democrat—Extra. DR. C.B. LIGHTHILL will be at HOWELL'S HOTEL, LOWVILLE, Tuesday, Wednesday and Thursday, November 22, 23 and 24, and at BAGG'S HOTEL, UTICA, Friday & Saturday, Nov. 25, and 26, 1870, and will visit no other places in this vicinity. He can be consulted, as usual, on CATARRH, DEAFNESS, BLINDNESS, and all diseases of the EYE, EAR, THROAT, nose and lungs . . .

[1] leaf printed on both sides ; 30 × 11 cm.

The verso contains testimonials from residents of Lowville and Utica, N.Y. regarding Dr. Lighthill's skills. The first testimonial describes Lighthill as "a regular educated German Physician."

LIGHTHILL, Edward Bunford see **LIGHTHILL** (#2252-2253).

2255. LIGNOSKI, R. B.
LIGNOSKI'S ELECTRIC LINIMENT, FOR MAN OR BEAST. Try one bottle and you will never be without another. The principal ingredient in this celebrated medicine is an oil procured from the Indians of the Rocky Mountains. It alone possesses wonderful curative properties, and being chemically combined, as in the *Electric Linament,* with some of the most powerful, yet soothing and healing, remedial agents known to the mind, produces the most perfect and efficacious remedy ever known to the world! . . . Prepared and sold by R.B. Lignoski, druggist and chemist, Monroe, La. . . . New Orleans: "Fire-Fly" print, [186-?].
[1] leaf printed on both sides ; 29.5 × 23.5 cm.

On the verso one reads that the liniment is "For sale at Flemings Patent Medicine Depot . . . New Orleans," where one may also find *McLane's Celebrated Liver Pills and Vermifuge,* the *Chinese Flea Powder,* and Lignoski's *Ninth Wonder, or Oil Extract of Zoepotent.*

2256. LINCOLN, David Francis, 1841-1916.
Hygienic physiology. A text-book for the use of grammar and common schools . . . Boston, U.S.A.: Ginn & Company, publisher, 1891.
[2], v, [1], 206 p. : ill. ; 19 cm.

An 1864 Harvard medical graduate, Lincoln served as a naval surgeon during the Civil War. He lived in Massachusetts until 1881, when he relinquished his practice and moved to Geneva, N.Y., where he lectured at Hobart College and wrote extensively on school and public hygiene.

2257. LIND, George Dallas, b. 1847.
Man. Embracing his origin, antiquity, primitive condition, races, languages, religions, superstitions, customs, peculiarities, civilization, nature and constitution, physical structure, the care and preservation of the body, the mental and moral faculties, etc., etc. . . . Chicago: T.S. Denison, publishers, 1884.
xii, 13-750 p. : ill., ports. ; 23 cm.

Professor of natural sciences at the Central Normal College, Danville, Indiana, Lind's anthropological treatise is divided into three parts: *Book I. Primeval man* (i.e., national mythologies regarding man's origins; the evolutionary theories of Lamarck, Darwin, *et al.*); *Book II. Man in history* (i.e., the languages, customs and religions of the various races of men; the "progress of civilization," etc.); and *Book III. Nature and constitution of man.* This third book is in turn divided into three parts: the first on anatomy & physiology; the second on hygiene; and the third on the mental faculties. Lind was an 1883 graduate of the Central College of Physicians and Surgeons (Indianapolis), and the author of books on education and the popularization of natural science.

LIND, George Dallas see also **AN OLD PRACTITIONER** (#2656).

2258. LINDEN, John.
Manual of the exanthematic method of cure also known as Baunscheidtism. With an appendix on "the eye" and "the ear," their diseases and treatment by means of the exanthematic method of cure. For the practical use of everyone . . . Prepared with special reference to our climatic relations and the diseases peculiar to America. Thoroughly revised and enlarged. Published also in the German language.Twenty-fourth edition. [Cleveland: Publishing house of the Evangelical Association], 1911 [c1878].
xvi, 31, 31a-31b-32, 32a-32b, 33-433, 15, [6] p., [1] leaf of plates : ill., port. ; 20 cm.

Linden maintains that most diseases are the result of corruption of the blood through the body's absorption of impure fluids. This common origin of a multitude of diseases admits their common treatment—which derives from a naturally reactive process of nature: "If these deleterious substances are retained in the organism, fevers will arise; and as the vital forces attempt to remove them from the blood, they will be deposited in various parts of the body, which results in fermentation and fever and consequent pain . . . If the vital forces succeed, by means of the accelerative circulation of the blood excited by the fever, to restore the activity of the skin, and to open its pores once more, then profuse perspiration will follow . . . and thus, in diseases of a milder type, nature will itself effect a cure" (p. iii). An eruption of the skin (exanthema) accompanies this process, signalling the "restored activity of the skin."

In more serious diseases, however, it may not be possible to rely on the restorative powers of nature and the "salutary exanthema." To replicate the process, therefore, Linden placed upon the affected part his *Resuscitator*, "a small instrument . . . in which are fastened 30 galvanized gilt and very finely pointed needles," so arranged that when the operator springs the mechanism in which they are enclosed, "The punctures of the needles produce artificial pores," which will assist "the excretion of the detained morbid and pathogynetic [sic] matter" (p. iv). To assure that these tiny punctures do not heal immediately so that expulsion of morbid matter is assured, "the operated parts are anointed with an irritative oil, called Oleum Baunscheidtii," which induces an artificial exanthema.

Both the *Resuscitator* and the irritating oil were available from Linden, who is at pains to point out that although Baunscheidt (#236-238) may have improved the mechanism, he cannot lay claim to its discovery. About half the volume is devoted to testimonial letters and advertisements. Spine-title: *The exanthematic cure.*

2259. LINDLAHR, Anna.
The nature cure book and ABC of natural dietetics. By Mrs. Anna Lindlahr and Henry Lindlahr, M.D. Fourth edition. Chicago: Published by the Nature Cure Publishing Co., [c1917].
    xii, [2], 469 p. ; 20 cm.

Lindlahr's vegetarian cookbook appeared in fourteen editions between 1915 and 1918. With the 15th edition (1922), the title changed to

Lindlahr's vegetarian cook book. The authors were proprietors of the Lindlahr Nature Cure Institutes, which encompassed the Chicago Home for Nature Cure, the Elmhurst Health Resort, the Lindlahr College of Nature Cure and Osteopathy, and the Nature Cure Publishing Company. All were located in Chicago and are advertised in an eight-page supplement at the end of the text. Henry Lindlahr (1862-1924) was a 1904 graduate of the National Medical University (Chicago) and the author of several books on naturopathy. Series: *Nature cure series, v. 2.*

LINDLAHR, Henry see **LINDLAHR** (#2259).

2261. LINDLEY, E. Marguerite.
Health in the home. A practical work on the promotion and preservation of health with illustrated prescriptions of Swedish gymnastic exercise for home and club practice . . . New York: Published by the author, 1896.
    xii, 414, [4] p. : ill. ; 21 cm.

2262. LINES, Leverett H.
Thirty years of female life. A treatise on the diseases of females, incident to this period, with their causes, symptoms and treatment; including the theory of conception, and the symptoms of pregnancy . . . For popular use . . . New York: Printed by W.E. Hilton, 1862.
    336 p. : ill. ; 19 cm.

There is hardly an author on obstetrics and gynecology available in the English language at mid-century whom Lines does not quote in this popular treatise on female genital disorders and pregnancy. Most frequently cited are the works of Dewees, Meigs, Churchill, Scanzoni, Gooch and West.

LINN, Hugh James see **DR. LINN'S MUSEUM OF ANATOMY** (#2263-2265).

2263. DR. LINN'S MUSEUM OF ANATOMY.
Catalogue of Dr. Linn's Museum of Anatomy. 345 Main Street . . . Buffalo, N.Y. Private office on first floor . . . Museum open from 9 A.M. to 10 P.M. Buffalo, N.Y.: G.M. Hausauer, printer and publisher, 1901.
    61, [15] p. : ill. ; 17 cm.

In his preface, H.J. Linn, proprietor of Dr. Linn's Museum of Anatomy, states that "it is only possible to a very restricted number of learned men to devote all their time to the studies of natural sciences; but, on the other side, it is out of doubt that every man who pretends having received an education should at least possess the rudiment notions, for their knowledge is of incontestible utility in practical life. Nothing, therefore, will generally be more useful than a popular acquirement of knowledge of the human body, its organs and functions" (p. [3]). Such was the purpose of the anatomy museums established in the 1840s by such figures as Wooster Beach and Horace B. Tolles (#2590)—intended as part of the public's education on the functions of the human body and the maintenance of health. After the Civil War, however, a new class of museums opened in American cities whose *raison d'etre* was nominally educational, but were essentially places of amusement that exhibited specimens of congenital oddities, venereal pathology, female anatomy, and the like. The character of these establishments had become more explicitly sexual, were often limited to "gentlemen only," and served as inducements for the medical services (i.e., the treatment sexual diseases and disorders) of their physician-proprietors.

Aware of the reputation that establishments such as his own had earned, Linn writes, "it is inexplicable that the governments which should employ all means to strengthen the citizens' health and to favor their physical development, do not encourage the access and frequentation of anatomical-pathological museums" (*ibid.*). Although his museum contains numerous pathological specimens of syphilitic and gonorrheal conditions, such exhibits have an educational purpose: "We daily see unfortunate ones whose health is destroyed by syphilis . . . If such a patient had the occasion to study in an anatomico-pathological museum, the ravages that inveterated and neglected syphilis may produce in the human body, he would indeed have immediately surmounted his false shame, and looked for help before the illness had assumed a bad appearance" (p. [4]). Unsuspecting visitors might also recognize their disordered state in exhibits such "No. 391. Herpes, a skin disease, sometimes caused by having sexual intercourse with a woman who has leocorrhoea," "No. 397. Penis of a masturba-

tor," etc. Following the catalog of 1,125 items, Linn advertises his medical services "for all private, nervous, chronic, blood and skin diseases." The final [8] pages illustrate some of the wax, plaster and papier mâché figures of A.F. Brouillard, who was associated with Dr. Linn's Museum. A halftone photograph of the Museum appears on the final unnumbered page. Title on printed wrapper: *Dr. Linn's Museum of Anatomy.*

The 7th edition of *Polk's Medical Register and Directory* (1902) provides the address of Hugh James Linn (1850-1923), the only Linn practicing in Buffalo at the turn of the century, as 345 Main St., the same address as Dr. Linn's Museum of Anatomy. Hugh James Linn was an 1878 medical graduate of the University of Pennsylvania. According to the AMA's *Directory of deceased American physicians, 1804-1929* (I:934), Linn practiced in Buffalo, N.Y. from 1878 until 1911 when he moved to Cleveland. Linn practiced medicine in San Francisco from 1913, until he moved to San Jose in 1920. The 1917 Univ. of Pennsylvania catalog lists him as professor of serum therapy at the College of Physicians & Surgeons (San Francisco) and on the staff of St. Winifred's and San Francisco Hospitals. He died in Oakland, Calif. on 8 February 1923.

After Linn's departure for Cleveland in 1911, his museum appears to have been taken over by Edward T. Stevens, M.D. and renamed the Gallery of Scientific Wonders (#1323). After Steven's death in 1919, the museum continued for a time as Dr. Linn's Museum under the management of William B. Hunt.

2264. DR. LINN'S MUSEUM OF ANATOMY.

Catalogue of Dr. Linn's Museum of Anatomy. 345 Main Street . . . Buffalo, N.Y. Private office on first floor . . . Museum open from 9 A.M. to 10 P.M. Buffalo, N.Y.: G.M. Hausauer, printer and publisher, 1902.

61, [15] p. : ill. ; 17 cm.

Title on printed wrapper: *Dr. Linn's Museum of Anatomy.*

2265. DR. LINN'S MUSEUM OF ANATOMY.

Catalogue of Dr. Linn's Museum of Anatomy. 345 Main Street . . . Buffalo, N.Y.

Private office on first floor . . . Museum open from 9 A.M. to 10 P.M. Buffalo, N.Y.: G.M. Hausauer, printer and publisher, 1903.
61, [15] p. : ill. ; 17 cm.

DR. LINN'S MUSEUM OF ANATOMY see also **GALLERY OF SCIENTIFIC WONDERS** (#1323).

2266. LIQUID OZONE COMPANY.
How Liquozone kills germs. Liquozone is the only way to kill germs in the body without killing the tissues too. Chicago: The Liquid Ozone Co., 1904].
48 p. ; 14 cm.

This germicide's "great value to humanity lies in the fact that it is the only way known to kill germs in the body without killing the tissues, too. Liquozone does this without the slightest harm to any animal cell . . . The reason is that germs are vegetables, and Liquozone . . . is deadly to vegetal matter . . . To an animal, Liquozone is exhilarating, vitalizing, purifying; and those effects are immediate . . . but all vegetable matter instantly perishes wherever Liquozone goes. And the fact that germs are vegetables has made it possible for Liquozone to solve the greatest problem that medical men ever met . . . Liquozone carries a germcide into the stomach, into the bowels and into the blood. It goes wherever the blood goes. And no touch of impurity, no germ of disease, can exist in the presence of Liquozone" (p. 3-4). Title from printed wrapper. The postmarked envelope in which the Atwater copy was mailed is dated 26 April 1904.

LISPENARD, C. W., & LISPENARD, W. C. see **REYNOLDS** (#2959, 2960).

2267. LISTON, Robert.
Directions for the use of patients being treated for female weaknesses . . . Albany General Infirmary . . . [Albany, N.Y., 187-?]
6, [2] p. : ill. ; 23 cm.

Directions explaining the use of proprietary remedies specified for the disorders described in this pamphlet.

2268. LISTON, Robert.
READ IT. Volume VII. Rochester, Sept. 25, 26, 27, 28, 1872. No. 11.
[2] leaves ; 30 cm.

Banner title. Advertisement for the medical services of "Dr. Liston, oculist, aurist, and general surgeon . . . Albany Eye & Ear Infirmary . . . Albany, N.Y." Includes a schedule of his tour of eight cities in New York State from August through October 1872. Presumably a different *Read it* was printed for each city.

LITTLE, John Forsyth see *LIBRARY OF HEALTH* (#2247).

2269. LIVERMORE, Abiel Abbot, 1811-1892.
Anti-tobacco . . . With A lecture on tobacco by Rev. Russell Lant Carpenter and On the use of tobacco by G.F. Witter, M.D. Boston: Roberts Brothers, 1888.
[2], 117 p. ; 16 cm.

After completing his studies at Cambridge Divinity School, Livermore was ordained a clergyman in the Congregational Church (1836). He was pastor of Unitarian parishes in Keene, N.H., Cincinnati and Yonkers before being made president of the Meadville Theological School (Pa.). A prolific author, Livermore is perhaps best known for his commentaries on the Gospels and Acts. Though conservative by temperament, he was theologically liberal, and involved himself in such issues as abolition, the education of women and temperance.

Anti-tobacco was originally published at Boston in 1883. It is based on an address given by Livermore before the Meadville Temperance Union (29 Jan. 1882), and is supplemented by the addition of Russell Lant Carpenter's (1816-1892) "A lecture on tobacco" (p. [35]-70), and G.F. Witter's 1881 report to the Wisconsin Board of Health entitled "Tobacco and its effects" (p. [71]-114).

LIZARS, John, 1787?-1860. *The use and abuse of tobacco* see *ALCOHOL AND TOBACCO* (#17).

LLOYD, Seth Louis see **SAND SPRINGS WATER** (#3080).

2270. LOBSTEIN, Jean Frederic Daniel, 1777-1840.
On the several principal diseases in this country, with the physician's advice, including a collection of over two hundred do-

mestic remedies, a work useful to families
. . . New-York: Sold, and to be had of the
author, of Mr. Ch. C. Christman . . . and in
the apothecary store of Drs. Vere & Haynes,
1839.

iv, [5]-143, [1] p. ; 23 cm.

Born in Strasbourg, the author was the son
of the eminent physician Johann Friedrich
Lobstein (1736-1784). After a period of obstetrical study and practice in his native city,
Lobstein *fils* ventured to Paris where he studied under such figures as Leroy and Baudelocque, published several obstetrical treatises,
and served as a military physician during the
Napoleonic period. In 1815 Lobstein returned
to Strasbourg, where in spite of continuing his
life as a physician and author, "jedoch in Folge
von ungeordneten Leben und Verschwendung
bankerott, fluchtete nach Amerika" [as the result of a life of disorder and profligacy, he went
bankrupt and fled to America] (*Biographisches
Lexikon der hervorragenden Arzte*, Berlin, 1931,
3:813). Lobstein settled in New York.

A naturalized citizen of the United States,
Lobstein expresses his happiness in the preface at being "an inhabitant of a country in which
the liberty of speaking, writing and publishing,
is equally permitted to the poor and the rich,"
and where "every one has a right to profess his
opinions and his principles," including physicians whose medical opinions might not conform to medical orthodoxy. Lobstein continues: "Many respectable families, whose physician I have had the honor to be, as well in Philadelphia as in New-York, have often requested
me to publish a work in which I would communicate my treatment of the different diseases,
on account of their having received great benefit from such treatment, and chiefly on account of my not having employed the remedies
so common in this country . . . Many of my patients know by experience that many of their
acquaintances have fallen victims to the treatment of bleeding and *mercury*, and that the convalescence of others was very much retarded
by the extravagant means above mentioned" (p.
6).

Lobstein confines himself in this work to the
principal diseases and the provision of some two
hundred domestic remedies. The treatise takes
the form of dialogues with imagined patients,
each suffering from a different disorder. A pa-

tient suffering with a peristent cough, for example, complains of his physician: "First he *bled*
me, and the following day I was *cupped*; then I
had to take *calomel* and jalap; then salts, then I
was blistered. Notwithstanding all this, I became, every day worse" (p. 44-45). Lobstein diagnoses him to be in an early stage of pulmonary tuberculosis, provides the recipe for a
cough suppressant, and recommends a light
diet. Several days later he writes a prescription
for twelve powders to be had of the apothecary
Vere on Hudson Street [whose name appears
in the imprint]. The powders, of course, have
"an almost miraculous effect"; and Lobstein
completes the cure with a final prescription of
pills.

2271. LOCKWOOD, John L.
Your attention is respectfully called to the Supe-
riority *of the highly effervescent* Clysmic Min-
eral Water *as a table water and a dilutent . . .*
The water is an *indispensable adjuvant* in the
treatment of the following diseases, and
should be used freely: Bright's Disease,
Diabetes, Inflammation of the Bladder or
Kidneys, Catarrh of the Bladder . . . USED
AND RECOMMENDED BY THE MEDICAL
PROFESSION GENERALLY . . . [New York,
ca. 1884].

[2] leaves ; 28 cm.

"John L. Lockwood, proprietor and sole
manager . . . New York City" (leaf 2 verso).

LOCKWOOD, R.W. see **CORRECTIVE
EATING SOCIETY** (#803).

LOEBELL, A. see **FOOTE** (#1235-1241).

LOFVING, Concordia see **HARTELIUS**
(#1567).

**2272. LOGAN, Mary Simmerson
Cunningham, 1838-1923.**
The home manual. Everybody's guide in
social, domestic, and business life. A treasury of useful information for the million
. . . Embracing etiquette, hygiene, household economy, beauty, methods of money-
making, care of children, nursing of invalids, outdoor sports, indoor games, fancy
work, home decoration, business, civil service, history, geography, physiology, writing

for the press, teaching, Italian art, etc., etc. Prepared by Mrs. John A. Logan, assisted by Prof. William Mathews, Catherine Owen and Will Carleton . . . Philadelphia, Pa. . . . [etc.]: H.J. Smith & Co., [c1889].
    [2], vi, [2], 507 p., [13] leaves of plates : ill., ports. ; 26 cm.

The home manual appeared with an 1889 publication or copyright date under the imprint of four different Philadelphia publishers, as well as publishers in Detroit, New York, Boston, Kansas City and Beaver Falls, Pa. Logan was assisted in the compilation of this work by William Mathews (1818-1909), Catherine Owen, the pseudonym of Helen Alice Mathews Nitsch (d. 1889), and Will Carleton (1845-1912).
    Logan was the wife of Union general and U.S. Senator John Alexander Logan (1826-1886), whose memory she perpetuated in *Reminiscences of a soldier's wife* (1913), a book described as "readable, adulatory, and unreliable." After her husband's death she edited *The Home magazine* and was on the syndicated staff of the Hearst News Service.

LOMAX, L. W. see **NEW JERSEY HOMEOPATHIC PHARMACY** (#2588).

LONDON ANATOMICAL MUSEUM see **JOURDAIN** (#2051).

2273. LONDONDERRY LITHIA SPRING WATER.
Have you read this? It is for you personally! Knowing that you will be very likely to require some medicinal spring water, we have decided to call your attention to the famous Londonderry Lithia, which is today the most popular water in the world. It has been brought before the public by the physicians and has by its wonderful control over rheumatism, gout, gravel, acid dyspepsia, Bright's and all kidney diseases won its way into the confidence of all the people . . . [Nashua, N.H.: Londonderry Lithia Spring Water, ca. 1887].
    [2] leaves : ill. ; 27 cm.

2274. LONG, John Ignatius Theodore, 1852-1918.
Lines from a doctor to his son—or—knowledge *vs* ignorance. [S.l.]: Published by the author, 1910.

216 p., [1] leaf of plates : port. ; 19 cm.

"Youth is not only a period of enjoyment and growth, but a period of susceptibility and giddiness; frivolity and waywardness—indiscretion—a period when the tide of appetites and passions runs dangerously high, and without some restraining power; guiding and controlling influence—proper and adequate instruction—you will be impelled, not only into gross absurdities and inconsistencies, but ruinous habits and damning errors" (p. 7). Long provides his son advice on digestion (advocating Fletcherism); neurasthenia; the evils of tobacco; intemperance ("'Tis said that Bacchus has drowned more people than Neptune," p. 76); preventing consumption; the physical consequences of excess in venery; autointoxication; marriage, etc. Long was an 1876 graduate of the College of Physicians and Surgeons of Baltimore. Frontispiece portrait of the author.

2275. LONG, Solomon Levy, b. 1864.
Rights legal and ethical of practitioners of drugless systems of healing . . . Oklahoma City: [s.n.], 1912.
    220 p. ; 23 cm.

General counsel for the National Association of Suggestive Therapeutics, Long attacks state laws restricting the practice of medicine to individuals graduated from a licensed medical college as an infringement of the rights of "drugless healers" guaranteed them by the fourteenth amendment. Arguing that federal law overides state law, he insists that in view of this amendment, "the practice of any method of drugless healing is legitimate, legal and lawful"; and that although "the state . . . has the right to regulate any legitimate calling," and "may prohibit any calling, profession or avocation which is a menace to public health and welfare," it "has no power to impose unnecessary restrictions which may amount to prohibition," i.e., "the state has no power to prohibit indirectly what it may not prohibit in express terms." Therefore, "neither the state, nor a court, may sit as a judge and declare one school of healing orthodox and worthy of support, and another heterodox and unworthy" (p. 47). The state's power is limited to requiring "a reasonable degree of proficiency in those branches of learning or science which the citizen essays to put in operation." Long provides an historical review of the medical practice acts of the various states and to the history of the various schools of drugless healing.

LONGRIDGE, Benjamin see **MONNETT** (#2479).

2276. LONGSHORE-POTTS, Anna Mary, 1829-1912.
Discourses to women on medical subjects . . . English edition. Twenty-second thousand. London: Published by the author, 1887.
xiv, [15]-351 p., [1] leaf of plates : ill., port. ; 19 cm.

Longshore-Potts was the sister-in-law of Hannah E. Myers Longshore (1819-1901), with whom she was enrolled as a student at the Woman's Medical College of Pennsylvania, from which she graduated in its first class (1851). The *Discourses* is dedicated to her brother Joseph Skelton Longshore (1808-1879), with whom she began the study of medicine, and who was one of the founders of the Woman's Medical College as well as its professor of obstetrics and the diseases of women and children.

Upon graduation she began to practice medicine in Philadelphia, where, she writes in the preface, "I resolved to devote a few months out of every year to lecturing upon medical subjects"; adding "my first effort took place in the city of Philadelphia, where I gave several consecutive courses of lectures to women only" (p. viii). Her commitment to the lecture circuit quickly increased, and Longshore-Potts spoke on women's health and physical improvement throughout the United States and Canada, as well as in New Zealand, Australia and the United Kingdom. "Throughout all these countries and colonies there have been earnest entreaties from the large number of listeners, and from hundreds of grateful patients who sought relief at the hands of one of their own sex, for a book to read, from which could be gained a similar course of instruction to that to which they had listened in the public and private lectures, and to this general call I have yielded" (p. ix).

Longshore-Potts divides her book into two parts: "Gynecology," and "Reproduction." In the first she describes the anatomy of the female pelvis and organs of generation; the functions of these parts; menstrual disorders (four chapters); uterine diseases; rectal diseases and urinary tract disorders. In the second part she discusses conception; fertility and sterility; fetal development; pregnancy; contraception (she regards all methods in use as ineffective except abstinence); and parturition.

Pages 347-51 describe the "Italy of America," i.e., the author's sanitarium outside San Diego, "a 'resort' for invalids who desire medical treatment, combined with superior climatic conditions, or for persons who wish to maintain their health, and desire the advantages of hygienic food and other sanitary influences, with the educational opportunities this institution affords" (p. 350). Frontispiece portrait of the author.

2277. LONGSHORE-POTTS, Anna Mary, 1829-1912.
Discourses to women on medical subjects . . . Twelfth English edition. America: Published by the author, National City, San Diego Co., California, U.S.A.; London: published by the author, 1897.
xviii, [19]-355 p., [2] leaves of plates : ill., port. ; 19 cm.

In this twelfth edition, Longshore-Potts lauds medicine for the recent fruits of laboratory science, e.g., the discovery of the microscopic agents that cause tuberculosis, diphtheria, cholera, etc. She predicts that the days of "absolute good health" are nearly arrived, and that "a bountiful harvest of good health, morals, and happiness" is to come. The substance of her lectures has not changed as the result of these developments, however. A description of Dr. Anna M. Longshore-Potts Medical Institute and Sanitarium appears on pages [349]-355.

2278. LONGSHORE-POTTS, Anna Mary, 1829-1912.
The logic of a lifetime . . . Alameda, California: Published by the author, 1911.
[10], [7]-303, [1] p., [1] leaf of plates : port. ; 19 cm.

The sixty-three articles contained in this volume "are all of a religio-moral and metaphysical character, non-sectarian and of spiritual purport" (*Pref.*). They reflect her Quaker background, and however "metaphysical," are never far removed from her fundamental faith in the ineluctable progress of mankind on the planes of moral and material well-being. For example, "The rolling, gushing waters of rational opinions are bearing on the surface and in the depths below, a new unfettered trail of thought which will electrify the world . . . uprooting the forests of error and establishing more generous and humane conceptions of righteousness, justice, reason and truth; the world will then move on to victory like unto the succession of stated seasons," etc. (p. 165-66). Frontispiece portrait of the author.

2279. LONGSHORE-POTTS, Anna Mary, 1829-1912.
Love, courtship and marriage . . . Published by the authoress . . . London, England, and The Sanitarium, National City, San Diego Co., California, U.S.A., [c1887].
    [8], 114, [4] p., [1] leaf of plates : port. ; 18 cm.

Softcover. Frontispiece portrait of the author.

LOOMIS, I. N. see REFORM MEDICAL COLLEGE OF GEORGIA (#2953).

2280. LOOMIS, Justin Rudolph, 1810-1898.
Elements of the anatomy, physiology, and hygiene of the human system . . . New edition, thoroughly revised and greatly improved in illustrations. New York: Sheldon, Lamport & Blakeman . . . [*et al.*], 1855.
    vi, [7]-210, [213]-221 p., 7 [i.e., 8] leaves of plates : ill. ; 19 cm.

Originally published at New York in 1853, this 221 page edition of Loomis' physiological school textbook was issued at least six times through 1870. The revised edition was first published in 1871.

2281. LOOMIS, Justin Rudolph, 1810-1898.
Elements of the anatomy, physiology, and hygiene of the human system . . . New edition, thoroughly revised and greatly improved in illustrations. New York: Sheldon, Blakeman & Co. . . . [*et al.*], 1857.
    vi, [7]-210, [213]-221 p., 7 [i.e., 8] leaves of plates : ill. ; 19 cm.

2282. LOOMIS, Justin Rudolph, 1810-1898.
Elements of the anatomy, physiology, and hygiene of the human system . . . New York: Sheldon & Company, publishers, 1867.
    vi, [7]-210, [213]-221 p. : ill. ; 19 cm.

Issued without the eight lithographic plates common to earlier editions.

2283. LOOMIS, Justin Rudolph, 1810-1898.
Elements of the anatomy, physiology, and hygiene of the human system . . . Revised edition. New York: Sheldon and Company, 1872.
    254 p. : ill. ; 19 cm.

The revised edition of Loomis' *Elements* was first published at New York in 1871. This 1872 edition appears to be the final publication of Loomis' physiological textbook.

2284. LORAND, Arnold.
Old age deferred. The causes of old age and its postponement by hygienic and therapeutic measures . . . Third edition. Translated, with additions, by the author from the third German edition. Philadelphia: F.A. Davis Company, publishers, 1912.
    xii, 480 p., [1] leaf of plates : port. ; 24 cm.

A translation of the third edition (1910?) of *Das Altern, seine Ursachen und seine Behandlung durch hygienische und therapeutische Massnahmen* (1st ed., Leipzig, 1910).The first American edition of *Old age deferred* was published at Philadelphia in 1910; a 6th edition appeared in 1926.

Lorand describes old age as "a chronic disease due to degeneration of the glands with internal secretions (hereinafter referred to as the ductless glands), of the thyroid, the sexual glands, and the adrenals in particular. In this work we will show that this degeneration is amenable to treatment" (*Pref.*, p. iii-iv). Lorand attributes the degeneration of the ductless glands not simply to the accumulation of years, but to such influences as "disease; nervous affections; alterations of the mind . . . chronic infections; numerous pregnancies, etc., or by faulty hygienics: excess in food, alcohol, sexual pleasures, etc." (p. 64). Diseased endocrine glands in turn produce a long list of disorders in every bodily system, and promote premature aging. He emphasizes that "the degeneration of certain glands with internal secretions . . . will produce a condition of auto-intoxication, as poisonous products will not be destroyed in the proper manner, and also not eliminated from the body" (p. 127). This can be remedied "by a careful hygiene of these different ductless glands"; i.e., "The preventive treatment of old age is in no less degree possible than that of any other disease. To prevent old age rationally, we must avoid all those harmful agencies which may be deleterious to the glands with internal secretions, as it is the degeneraton of these glands that brings it about" (p. 115). These agencies include immoderate sexual activity, diet, and the use of stimulants (tobacco, alcohol, coffee, etc.). Lorand devotes two thirds of the book to the description of a hygienic regimen that assures the health of the ductless glands and their unimpeded functioning in eliminating poisons from the body. He discusses the use of hormonal extracts ("animal extracts") in the treatment of old age, citing the work of Brown-Sequard (#438) on p. 125ff., and more thoroughly in chapter LVI. Lorand was also the author of *Health and longevity through rational diet* (1st Amer. ed., 1912).

LOUGEST, Charles A. see **LOUGEST AND COMPANY** (#2285, 2285.1).

2285. LOUGEST AND COMPANY.
Read, mark and learn, or guide to marriage, a medical treatise on consumption and nervous debility, embracing physiological admonitions to the married and single of both sexes, also indicating the means by which the human race may be made happy, healthy and vigorous. By Lougest & Co., European Medical Depot . . . [Boston, c1893].
    64 p. ; 11 cm.

Lougest & Co. specializes in the treatment of the constitutional effects of masturbation, spermatorrhea and sexually transmitted diseases. Patients are encouraged to apply in person to Dr. Lougest at his Columbus Ave. office, or to forward a description of their symptoms through the mail. Treatment relies on remedial agencies provided by Lougest & Co. Charles A. Lougest (d. 1907) appears in the first edition of *Polk* (1886); he is last listed in the 9th edition (1906). In printed wrapper.

2285.1. LOUGEST AND COMPANY.
Words of warning, or guide to marriage, a medical treatise on consumption and nervous debility, embracing physiological admonitions to the married and single of both sexes, and the true cause of unproductive unions, also indicating the means by which the human race may be made happy, healthy and vigorous. By Lougest & Co., European Medical Depot . . . [Boston, n.d.].
    64 p. ; 12 cm.

2286. LOVE, Isaac Newton, 1848-1903.
Painful menstruation in virgins . . . [St. Louis, Mo.: Mellier Drug Company, 189-?].
    [10] leaves ; 15 cm.

Promotional brochure for "Ponca Compound, a uterine alterative" and other remedies manufactured by the Mellier Drug Company.

LOVE, Mrs. S. G. *An essay on animal magnetism* see **TUTTLE** (#3615).

LOVEJOY, Isaac see **MONROE COUNTY MEDICAL SOCIETY** (#2481).

2287. LOVERING, Anna Temple, d. 1914.
Hints in domestic practice and home nursing . . . Boston and Providence: Otis Clapp & Son, 1896.
    170 p. ; 19 cm.

The author was an 1889 graduate of the homeopathic Boston University School of Medicine. According to the A.M.A.'s *Directory of deceased American physicians, 1804-1929* (1:950), Anna Temple Lovering died on 22 February 1914.
    Part IV. of Lovering's domestic medicine (p. [129]-165) contains advertisements for homeopathic medicines, medicine chests, food supplements and equipment manufactured or sold by Otis Clapp & Sons. The firm was founded by Otis Clapp (1806-1886), who beginning in the 1840s made his mark as the publisher of homeopathic texts, the annual *Boston directory,* and Swedenborgian treatises. "He left publishing for politics eventually, serving as a representative in the [Massachusetts] state legislature in 1851 and 1854. Surprisingly, at his death . . . he

was the senior member of Otis Clapp & Son, homeopathic pharmacists" (John Tebbel, *A history of book publishing in the United States*, New York: R.R. Bowker, 1972, 1:446). The firm continued as a minor publisher of homeopathic texts into the first decade of the 20th century.

2288. LOVERING, Anna Temple, d. 1914.
Home treatment and care of the sick, including chapters on approaching maturity, marriage and maternity . . . Boston and Providence: Otis Clapp & Son, 1901.
[2], viii, 9-376 p. ; 20 cm.

2289. LOWE, Abraham Thompson, 1796-1888.
Fragments of physiology; or, essays on life, health, hygiene, disease, and cure of disease . . . Boston: Henry A. Young & Company, 1877.
vii, [7], [11]-220 p. ; 18 cm.

2290. LOWE, Edna Eugenia.
Health rules and danger signals . . . Chicago: Published by The Platform, the Lyceum and Chautauqua magazine, 1916.
154 p. : ill., port. ; 20 cm.

Lowe was a popular lecturer on the "health gospel" to Chautauqua audiences. In the publisher's foreword, Fred High writes that Lowe was a graduate of Northwestern University School of Oratory: "She was Director of Physical Education and Instructor in the department of Oratory in Carleton College, at Northfield, Minnesota, for five years. At present she is a member of the faculty of the Highland Park College, Des Moines, Iowa, where she is Director of Physical Education for Women and Instructor in the Oratory School" (p. 7). The seven plates are included in the pagination of the text. Frontispiece portrait of Lowe.

LOWELL YOUNG MEN'S TEMPERANCE SOCIETY see BARTLETT (#226).

2291. LOWRY, Edith Belle, 1878-1945.
Confidences. Talks with a young girl concerning herself . . . Chicago: Forbes & Company, 1914.
94, [2] p. ; 17 cm.

Brief chapters on reproduction and on hygiene written for girls ten to fourteen years of age. *Confidences* was issued five times between 1910 and 1919. The author was a 1905 graduate of the Jefferson Park Hospital Training School for Nurses (Chicago), and a 1907 graduate of the Bennett College of Medicine & Surgery (Chicago). In addition to her medical practice in Chicago, Lowry was acting chief of the Bureau of Hospitals for city's Dept. of Health during the First World War, and was in charge of field investigations on child hygiene in several southern states for the U.S.P.H.S. (1920-21). Lowry was the author of numerous popular books on women's health, child care, sex education, etc.

2292. LOWRY, Edith Belle, 1878-1945.
Herself. Talks with women concerning themselves . . . Chicago: Forbes & Company, [c1911].
221, [3] p. : ill. ; 19 cm.

First edition. Twenty-five chapters on the female reproductive system and its disorders, sex and sex behavior, the consequences of sexual immorality, sexually transmitted diseases, etc. In the chapter entitled "Women in business," Lowry expresses her admiration for women who choose a career rather than a loveless marriage "in which she must slave from morn till eve and then receive as recompense curses and fault-finding" (p. 189). In "Nervousness—a lack of control," she blames the nervous state often attributed to women on a lack of purposeful work, not on a condition inherent to the female nervous system.

2293. LOWRY, Edith Belle, 1878-1945.
Himself. Talks with men concerning themselves. By E.B. Lowry, M.D. . . . and Richard J. Lambert, M.D. Chicago: Forbes & Company, 1914.
[3]-216 p. : ill. ; 19 cm.

Himself was first published in 1912, and contains twenty-two chapters on the anatomy and physiology of the male generative organs, sexually transmitted diseases, masturbation, sex disorders, sexual behavior and ethics, prostitution, eugenics (including "sterilization of the unfit"), the necessity for contraception in family planning, etc. Lowry's co-author, Richard Jay Lambert (b. 1874), was also her husband. They married in 1911. Like his wife, Lambert was a 1907 graduate of the Bennett College of Medicine & Surgery. He later authored *Sex and mar-*

riage (1932), *Sex facts for women* (1936) and *Sex facts for men* (1936). Softcover.

2294. LOWRY, Edith Belle, 1878-1945.
Himself. Talks with men concerning themselves. By E.B. Lowry, M.D. . . . and Richard J. Lambert, M.D. Chicago: Forbes & Company, 1915.
[3]-216 p. : ill. ; 19 cm.

Himself was issued five times between 1912 and 1919.

2295. LOWRY, Edith Belle, 1878-1945.
The home nurse . . . Chicago: Forbes and Company, 1914.
224 p. : ill. ; 20 cm.

LOZIER, Clemence S. see HOLBROOK (#1690).

DR. LUCAS PRIVATE DISPENSARY see WILLIAMS (#3802).

2296. LUKENS, I.
The sick man's guide, or family director; compiled from the best botanic publications, with directions for using Dr. Samuel Thomsons medicines, bath, &c., for Thomsonians and all others who wish to prevent or cure their own diseases . . . Bridgetown, N.J.: G.S. Harris, printer, 1845.
189, [1] p. ; 16 cm.

"The compiler has collected many of the most valuable receipts by the most distinguished botanic writers and practitioners, such as Thomson, Howard, Curtis and others, that have been amply tested for a number of years in all forms of disease and have been found powerful and effective means in preventing and curing disease, in a convenient form, and at a price within the reach of every man" (p. [3]).

2297. LUSK, Graham, 1866-1932.
The fundamental basis of nutrition . . . New Haven: Yale University Press; London: Humphrey Milford, Oxford University Press, MDCCCCXIV [1914].
[8], 62 p. ; 18 cm.

"This lecture is published in its present form that educated people may be able to obtain a better understanding of the principles of nutrition than is to be derived from current popular writings" (*Pref.*). Trained as a physiologist, Lusk was the "founder of the science of nutrition in the United States" (*Dict. Amer. med. biog.,* 1:465). At the time this work was published, Lusk was on the faculty of the Cornell University Medical College, and had been appointed director of the Russell Sage Institute of Pathology. He was the author of *The elements of the science of nutrition,* which appeared in four editions between 1906 and 1928.

LUST, Benedict see JUST (#2056), KUHNE (#2167).

2298. LUST, Louise.
The practical naturopathic-vegetarian cook book. Cooked and uncooked foods . . . New York and "Yungborn." Butler, New Jersey: Published by Benedict Lust, 1907.
72 p., [1] leaf of plates : port. ; 20 cm.

Beneath her frontispiece portrait, Louise Lust, N.D., is described as director of Ladies' Department at the Naturopathic Institute and Health Home on East 59th St. in New York, and at Yungborn, the naturopathic sanitarium established by her husband Benedict Lust in Butler, N.J. She was also "instructor in practical naturopathy" at the American School of Naturopathy, another entity in Benedict Lust's naturopathic enterprise. Cover-title: *Naturopathic cook book.*

2299. LUTES, Della Thompson, 1872-1942.
Truth for girls . . . Cooperstown, N.Y.: The Arthur H. Crist Co., [191-?].
[12] leaves ; 15 cm.

In a pamphlet intended for pubescent females, Lutes explains "the meaning of menstruation" and puberty; the evil habit of masturbation; the difficulties young girls face in maintaining chastity ("Stay away—as you value your life—from cheap theatres, moving picture shows, public dance halls, railroad stations, postoffice lobbies," leaf [10] verso), and the consequences that follow if they allow themselves to be led astray. Series: *American motherhood* [leaflets]; *no. 44.*

A native of Jackson, Michigan, Lutes became a teacher in Jackson County at age sixteen, and later in the Detroit city schools. In 1907 she moved to Cooperstown, N.Y., where she edited the periodical *American motherhood* (1895-1919). Lutes was the author of numerous books, pamphlets, articles and newspaper columns on home economics and, cookery, as well as rural life.

2300. LYDSTON, George Frank, 1858-1923.
What to tell our boys. A practical guide to sex instruction prepared for aid of parents and teachers . . . [Chicago?: s.n., c1914].
[3]-304, [2] p., [16] p. of plates : ill. ; 20 cm.

A re-issue of *Sex hygiene for the male and what to say to the boy* published in 1912 (#2301) with additional plates. "Sex secrecy, false modesty and moral cowardice have been the upas tree under which quackery, ignorance, vice and venereal diseases have flourished and destroyed youth. A new era in the education of youth is dawning. It is an era of progress, and it is the duty of parent, teacher, sociologist and physician alike to help the good work along" (p. 10). In his first hundred pages, Lydston describes a hygienic regimen (exercise, diet, etc.) that strengthens and disciplines the adolescent male physically and morally. With chapter XI, he begins discussion of sexual hygiene: the dangers to moral propriety (suggestive literature, the theater, dance halls); the anatomy and physiology of the reproductive system; diseases and disorders of the male genitalia; and venereal diseases.

Lydston was an 1879 graduate of the Bellevue Hospital Medical School. In 1881 he moved to Chicago and joined the faculty of the newly formed College of Physicians and Surgeons (est. 1882), where for twenty years he lectured on the diseases and surgery urogenital system. "During his Chicago career he built up a large and select clientele and was said to have had the most lucrative practice of his time. From the time of his graduation Lydston was a prolific writer upon widely varying topics . . . For many years he was professor of criminal anthropology at the Kent College of Law. Many of his later contributions to medical literature were devoted to the possibilities of securing rejuvenation by gland transplantation. Lydston was a man of aggressive personality and was frequently involved in controversy . . . For years

he acted the part of the bad boy of the medical profession, hitting left and right among the organized fraternity. He was in perpetual feud with the American Medical Association" (*Dict. Amer. biog.*, 6:513-14).

2301. LYDSTON, George Frank, 1858-1923.
Sex hygiene for the male and what to say to the boy . . . Chicago, Illinois: The Hamming Publishing Co., [c1912].
[3]-304, [2] p., [7] leaves of plates : ill. ; 23 cm.

First edition. Re-issued in 1914 under the title: *What to tell our boys* (#2300).

2302. LYMAN, Henry Munson, 1835-1904.
The new American family physician. A reliable family physician giving in detail the cause, symptoms, treatment and history of all diseases of the human body and plain instructions for the care of the sick. With full directions for treating emergency cases. A description of the structure and functions of the human body, hygiene and rules of health. Written by Henry M. Lyman . . . Christian Fenger . . . H. Webster Jones . . . W.T. Belfield . . . Revised edition. Chicago: The Reilly & Britton Co., 1910.
[4], ix-xii, xv-xx, [6], [21]-940, [6], 941-973, [6], 974-1157 p., [18] p. of plates + [2] paper manikins : ill. ; 24 cm.

Originally published under the title *The practical home physician* in 1883, this manual of domestic medicine was also issued as: *The book of health* (1898), *The new American family physician* (1899); *The new century family physician* (1901); *The practical home doctor* (1907); and *The 20th century family physician* (1907). Most of these title variants appeared in multiple issues. The authors were all Chicago based allopaths, and all (except Jones) on the faculty of Rush Medical College. Henry Munson Lyman (1835-1904) was an 1861 graduate of the College of Physicians & Surgeons (New York); Christian Fenger (1840-1902), was a medical graduate of the University of Copenhagen; Henry Webster Jones was an obstetrician who appears neither in *Polk* nor the AMA directories; William Thomas Belfield (1856-1929), was an 1877 graduate of Rush Medical College who specialized in in urogenital diseases and surgery.

2303. LYMAN, Henry Munson, 1835-1904.
The practical home doctor. 20th century
household medical guide. A popular guide
for the household management of disease.
Giving the history, cause, means of prevention and symptoms of all diseases of men,
women and children and the most approved methods of treatment, with plain
instructions for the care of the sick. Full
and accurate directions for treating
wounds injuries, poisoning, etc. Also giving a concise account of the structure and
functions of the human body, hygiene and
rules of health. Written by Henry M. Lyman
. . . Christian Fenger . . . H. Webster Jones
. . . W.T. Belfield . . . Revised and enlarged
edition . . . Chicago: American Publishing
Company, [c1907].
    [4], ix-xii, xv-xx, [6], [21]-940, [6], 941-
973, [6], 974-1157 p., [18] p. of plates +
frontis. + [2] paper manikins : ill., port. ;
24 cm.

    Cover-title: *The practical family doctor.* Fron-
tispiece portrait of Christian Fenger.

2304. LYMAN, Henry Munson, 1835-1904.
The practical home physician. A popular
guide for the household management of
disease, giving the history, cause, means of
prevention and symptoms of all diseases of
men, women and children and the most
approved methods of treatment, with plain
instructions for the care of the sick. Full
and accurate directions for treating
wounds, injuries, poisoning, &c. Also giving a concise account of the structure and
functions of the human body, hygiene and
rules of health. Written by Henry M. Lyman
. . . Christian Fenger . . . H. Webster Jones
. . . W.T. Belfield . . . Albany, N.Y.: Selleck,
Ross & Co.; Chicago, Ill.: Western Publishing House, 1886.
    [2], ix-xii, xv-xx, [21]-1141 p., 20, [1]
leaves of plates + [2] paper manikins : col.
ill. ; 25 cm.

    Atwater copy lacking pages 1139-41.

2305. LYMAN, Henry Munson, 1835-1904.
The practical home physician. A popular
guide for the household management of
disease, giving the history, cause, means of

prevention and symptoms of all diseases of
men, women and children and the most
approved methods of treatment, with plain
instructions for the care of the sick. Full
and accurate directions for treating
wounds, injuries, poisoning, &c. Also giving a concise account of the structure and
functions of the human body, hygiene and
rules of health. Written by Henry M. Lyman
. . . Christian Fenger . . . H. Webster Jones
. . . W.T. Belfield . . . Albany, N.Y.: Selleck,
Ross & Co.; Chicago, Ill.: Western Publishing House, 1887.
    [2], ix-xii, xv-xx, [21]-1141 p., 18 leaves
of plates + [2] paper manikins : col. ill. ; 25
cm.

2306. LYMAN, Henry Munson, 1835-1904.
The practical home physician. A popular
guide for the household management of
disease giving the history, cause, means of
prevention and symptoms of all diseases of
men, women and children and the most
approved methods of treatment, with plain
instructions for the care of the sick. Full
and accurate directions for treating
wounds, injuries, poisoning, etc. Also giving a concise account of the structure and
functions of the human body, hygiene and
rules of health. Written by Henry M. Lyman
. . . Christian Fenger . . . H. Webster Jones
. . . W.T. Belfield . . . Revised and enlarged
edition . . . Chicago: Star Publishing Company, 1912.
    [12], ix-xii, xv-xx, [6], [21]-940, [6], 941-
973, [6], 974-1157 p., [2] paper manikins,
[4] leaves of plates, [16] p. of plates (labelled A-N), [18] p. of plates : ill. (some
col.) ; 25 cm. + atlas (28, [4] p. ; 17 cm.)

    The accompanying pamphlet (*Descriptive
plates illustrating the female reproductive organs*) is
issued in a brown paper envelope laid-in. This
slender, unbound atlas depicting the female
pelvis, organs of generation and gravid uterus
is nearly identical to the *Plates illustrating Maternity* inserted into the pocket on the back
board of Prudence Saur's *Maternity* (#3106,
3107), and the *Plates illustrating Tokology* inserted
into the back pocket of Alice Stockham's *Tokology* (#3367-3374)—also published in Chicago.
Though intended to accompany Lyman's *Practical home physician,* the *Descriptive plates* appar-

ently was not issued with the text, and had to be acquired for an additional $2.00, as evidenced by the priced manila envelope in which the pamphlet was issued. "Prevention of conception" is discussed briefly on the final unnumbered page of the *Descriptive plates*.

2307. LYMAN, Joseph Bardwell, 1829-1872.
The philosophy of house-keeping: a scientific and practical manual for the preparation of all kinds of food, the making up of all articles of dress, the preservation of health, and the intelligent and skilful performance of every household office. By Joseph B. Lyman . . . and Laura E. Lyman . . . Hartford: Goodwin and Betts, 1867.
xiv, 15-560 p. : ill. ; 20 cm.

First edition. A native of Chester, Mass., Lyman an 1850 Yale graduate. He taught school for several years in Connecticut and Mississippi before moving to Nashville (1853), and then to New Orleans (1855), where he both taught and studied law. Lyman practiced law for five years in New Orleans after receiving his law degree from the University of Louisiana in 1856. In 1858 he married Laura Elizabeth Baker (1831-1912), with whom he collaborated in writing *The philosophy of house-keeping*. At the outbreak of the Civil War, Lyman enlisted in the First Louisiana Cavalry. He was taken prisoner in September, 1863, but was released weeks later after taking an oath of allegiance to the United States. After a brief period in Massachusetts, Lyman moved to New York City where he began his career as an agricultural journalist. In 1869 he was made agriculture editor of the mass-circulation *New York Weekly Tribune*. A 14th edition of *The philosophy of housekeeping* was published at Hartford in 1869. The book's final appearance appears to be the 1882 Philadelphia edition entitled: *How to live: or the philosophy of housekeeping*.

LYMAN, Laura Elizabeth see **LYMAN** (#2307).

2308. LYTTON, Edward Bulwer Lytton, baron, 1803-1873.
Confessions and observations of a water patient: in a letter to the editors of the New Monthly Magazine. By Sir E. Lytton Bulwer, Bart. Baltimore: George H. Hickman, [1845].
20 p. ; 24 cm.

A successful novelist, playwright, editor and parliamentarian, Lytton is perhaps remembered today, if he is remembered at all, as the author of *The last days of Pompei* (1834), a work that was translated into an unfortunate American television "mini-series" in the late 1980s. Lytton's many literary and political activities, his tempestuous relationships with his mother and wife, etc., left him in a state of nervous fatigue that by 1844 rendered him "thoroughly shattered." Physicians and drugs had proved useless in restoring Lytton to a normal physical and emotional condition: "I was then under the advice of one of the first physicians of our age. I had consulted half the faculty. I had every reason to be grateful for the attention, and to be confident in the skill of those whose prescriptions had . . . flattered my hopes and enriched the chemist. But the truth must be spoken—far from being better, I was sinking fast" (p. 5). It was about this time that Lytton came across the hydropathic writings of R.T. Claridge (#646-648) and James Wilson (#3829, 3830). He visited the latter's water-cure at Malvern in Worcestershire, among others, and within months was fully restored in mind and body. "We ransack the ends of the earth for drugs and minerals," he writes, "we extract our potions from the deadliest poisons—but around us and about us, Nature, the great mother, proffers the Hygeian font, unsealed and accessible to all. Wherever the stream glides pure, wherever the spring sparkles fresh, there, for the vast proportion of the maladies which Art produces, Nature yields the benignant healing" (p. 12). Lytton's enthusiastic encomium of hydropathy was often included in other works on the water-cure. In printed wrapper. Running-title: *Bulwer on water cure*.

LYTTON, Edward Bulwer Lytton see also **BALBIRNIE** (#196-200) **BULWER AND FORBES** (#521); **FAIRCHILD** (#1117).

Subject Index to Volume 1

ALCHOHOL DRINKING
(see **TEMPERANCE**)

ANATOMY
(see **PHYSIOLOGY**)

BALNEOLOGY AND HYDROTHERAPY
(see also **HYDROPATHY**)

1803	Buchan (#472)
1808	Buchan (#476)
1828	Buchan (#488)
1831	Bell (#291)
1832	Buchan (#489)
1834	Buchan (#490)
1838	Ford (#1260)
1839	Buchan (#491)
1840	Gallup (#1325)
1842	Buchan (#493)
1843	Buchan (#494)
1844	Allen (#69)
1846	Buchan (#495)
1848	Allen (#70)
1850	Bell, J. (#289)
1850	Emmons (#1067)
1854	Fonda (#1207)
1855	Bell, J. (#290)
1858	Allen (#68)
1859	Bell, J. (#293)
186?	Buchan (#496, 497)
1860	Fonda (#1208)
1860	Jackson, R. (#1961)
1871	Cabell (#546)
1874	Dawson (#916)
1876	Fonda (#1209)
1885	Clifton Springs Sanitarium (#671)
1886	Bailey (#189)
1892	Irwin (#1899)
1896?	Clifton Springs Sanitarium (#669)
1897	*Baths and bathing* (#233)
1898	Bilz (#325)
1901	Kellogg (#2104)
1902	Clifton Springs Sanitarium (#670)
1902	Kellogg (#2105)
1903	Kellogg (#2106)
1917	Kuhne (#2167)

BOTANIC MEDICINE
(Thomsonian)

1832	Howard (#1803, 1811)

1833	Day (#919, 920)
1833	Howard (#1812)
1835	*Botanic sentinel* (#380)
1836	Bowker (#384)
1836	Curtis (#852)
1836	Emmons, S.B. (#1068)
1836	Gardner, M. (#1331)
1836	Hersey (#1637)
1836	Howard (#1813)
1837	*Botanico-medical recorder* (#381)
1837	Brown, J.A. (#430)
1837	Curtis (#854)
1837	Fonerden (#1210)
1841	Curtis (#855)
1841	Johnston (#2037)
1842	Chambers (#581)
1842	Emmons, S.B. (#1069)
1843	*Boston true Thomsonian* (#375)
1843	Comfort (#749)
1844	Abbott (#3)
1844	Colby (#705)
1845	Colby (#706, 707)
1845	Comfort (#751)
1845	Lukens (#2296)
1846	Coffin (#698)
1848	Colby (#708)
1849	Coffin (#699, 700)
1849	Hooker (#1778)
1850	Comfort (#750)
1850	Hooker (#1776, 1777)
1850	Howard (#1814)
1852	Howard (#1817)
1855	Curtis (#853)
1856	Howard (#1815)
1858	Howard (#1816)
1859	Howard (#1806)
1860	Howard (#1807)
1861	Hersey (#1638)
1861	Howard (#1804, 1808)
1863	Howard (#1809)
1865	Howard (#1810)
1874	Danforth & Bristol (#893)
1879	Colby (#703)
1879	Howard (#1805)
1880	Colby (#704)
1905	Clymer (#679)

BOTANIC MEDICINE
(non-Thomsonian)

1799	Culpeper (#842)

1805	Culpeper (#839)
1812	Carpenter (#565)
1816	Buchan (#498)
1816	Ewell (#1100)
1818	Hand (#1555)
1820	Hand (#1556)
1824	Culpeper (#838, 840)
1825	Buchan (#486)
1826	Culpeper (#843)
1828	Cobb (#684)
1831	Cobb (#687)
1833	Brooks (#421)
1833	Cooper, J.W. (#786)
1833	Gunn (#1459)
1834	Gunn (#1460, 1461)
1835	Gunn (#1462)
1836	Bowker (#384)
1836	Ewell (#1101)
1837	Carter (#571)
1837	Cobb (#686)
1838	Goodlett (#1380)
1838	Gunn (#1463, 1464)
1839	Gunn (#1465-1467)
1840	Cooper, J.W. (#787)
1842	Gunn (#1468)
1844	Gunn (#1469)
1845	*Every man's doctor* (#1099)
1845	Good (#1377)
1846	Cobb (#685)
1847	Ewell (#1102)
1847	Frisbee (#1315)
1847	Hand (#1557)
1848	Ewell (#1103)
1848	Hard (#1559)
1851	Downing (#980)
1852	Helme (#1615)
1852	Hillman (#1649)
1855	Hutchings (#1856, 1857)
1858	King (#2135)
1859	*The Family doctor* (#1121)
1863	Gunn (#1470)
1864	Gunn (#1471)
1864	King (#2136)
1866	Gunn (#1472)
1867	Brown, O.P. (#432)
1869	Jefferson (#2000)
1870	Briante (#405)
1870	*Every man his own doctor* (#1097)
1872	Brown, O.P. (#433)
1874	King (#2137)
1875	Brown, O.P. (#434)
1878	Gunn (#1475)
1882	Lighthall (#2251)
1887	Gunn (#1473)
1897	Fernie (#1155)
1901	Gunn (#1474)
1908	Harding (#1560)

CHILD CARE, HYGIENE AND DISEASES
(see also **INFANT CARE**)

1808	Jennings (#2007)
1811	Burns (#536)
1817	Ewell (#1109)
1820	Clarke (#658)
1826	Ewell (#1108)
1833	Bradford (#389)
1836	Alcott (#52, 53)
1837	Child (#634)
1838	Alcott (#54)
1841	Curtis (#855)
1842	Alcott (#38)
1843	Bakewell (#195)
1844	Barwell (#230)
1844	Bright (#412)
1844	Hancorn (#1554)
1845	Alcott (#39)
1846	Alcott (#55)
1848	Hard (#1559)
1848	Alcott (#56)
1850	Bright (#413)
1853	Bull (#520)
1854	Laurie (#2212)
1855	Alcott (#56.1)
1863	Hopkinson (#1785)
1866	Buchanan (#504)
1871	Chavasse (#621)
1872	Chavasse (#623)
1872	Fonssagrives (#1211)
1872	Jackson, J.C. (#1954)
1873	Chavasse (#614)
1874	Chavasse (#624)
1876	Boston. Board of Health (#373)
1878	Hanaford (#1550)
1879	Chavasse (#615)
1880	Chavasse (#616)
1880	Duncan (#999)
1881	Chavasse (#625)
1881	Ellis (#1053)
1881	Hale (#1484)
1881	Hedges (#1609)
1883	Jackson, J.C. (#1955)
1887	Chavasse (#627)
1888	Chavasse (#628)
1888	Hayes, H.E. (#1593)
1889	Chavasse (#629)
1889	Herrick (#1635)
1890	Bell, R. (#294)
1890	Keating (#2070)
1892	Chavasse (#626)
1895	Holt (#1743)
1896	Griffith (#1426)
1898	Griffith (#1427)
1900	Hogan (#1662)
1900	Holt (#1744)

COOKERY
(see also **DIET AND NUTRITION**)

DIET AND NUTRITION
(see also **HYGIENE, VEGETARIANISM**)

DRESS REFORM
(see **WOMEN'S HEALTH**)

DYSPEPSIA

ECLECTIC MEDICINE
(including **REFORMED PRACTICE** and **PHYSIO-MEDICALISM**)

ELECTROTHERAPY

EUGENICS
(see also **MARRIAGE**, **SEX EDUCATION**)

EXERCISE & EXERCISE THERAPY
(see also **HYGIENE**)

EYE DISEASES & HYGIENE

FIRST AID AND EMERGENCIES
(see also **MEDICINE**)

GYNECOLOGY
(see **WOMEN'S DISEASES**)

HEALTH RESORTS AND CLIMATE

ARKANSAS

COLORADO

EUROPE

FLORIDA

ILLINOIS

KENTUCKY

MASSACHUSETTS

MICHIGAN

MINNESOTA

1871   Bill (#323)

NEW HAMPSHIRE

187?   Birch Dale Mineral Spring Waters
          (#326)

NEW YORK

1841   De Veaux (#936)
1844   Allen (#69)
1848   Allen (#70)
1850   Emmons (#1067)
1854   Fonda (#1207)
1856   Clifton Springs Water-Cure (#672)
1858   Allen (#68)
1860   Fonda (#1208)
1874   Dawson (#916)
1876   Congress & Empire Spring Co. (#762)
1876   Fonda (#1209)
1878   Invalids & Tourists' Hotel (#1896)
188-?   John H. Gardner & Sons (#1332)
1880   Crystal Springs Hotel (#835)
1880   Elmira Water Cure (#1064)
1882   Elmira Water Cure (#1063)
1884   Hathorn Spring Water (#1583)
1884   Jackson, J.C. (#1935)
1885   Clifton Springs Sanitarium (#671)
1886   Bailey (#189)
1888   John H. Gardner & Sons (#1333)
1892   Foster (#1265)
1892   Irwin (#1899)
1896?   Clifton Springs Sanitarium (#669)
1902?   Clifton Springs Sanitarium (#670)
1905?   Jackson Health Resort (#1963)
1908   Jackson Health Resort (#1962)
1911   Glen Springs (#1371)

NORTH CAROLINA

1876   Gleitsmann (#1368)

PENNSYLVANIA

1872   Gettysburg Spring Co. (#1348)

UNITED STATES

1842   Drake (#982)
1855   Bell (#290)
1885   Billings, Clapp & Co. (#324)
1889   James (#1972)

VERMONT

1840   Gallup (#1325)

1851   Granger (#1400)
1872   Clarke, A. (#657)

VIRGINIA

1808   Caldwell (#552)
1853   Burke (#529)
1871   Cabell (#546)
188-?   Granville Sanitarium (#1404)

HERBAL MEDICINE
(see **BOTANIC MEDICINE**)

HOME ECONOMICS

1819   *Family receipt book* (#1127)
1831   Barnum (#224)
1832   Goodrich (#1381)
1833   Goodrich (#1382)
1836   Farrar (#1133)
1838   Farrar (#1134)
1839   Hale (#1485)
1841   Hale (#1486)
1843   Beecher (#283)
1846   Cooley (#783)
1847   Beecher (#284)
1849   Beecher (#285)
1853   *Arts revealed* (#161)
1855   *Arts revealed* (#162)
1857   Ellett (#1048)
1858   *Arts revealed* (#163)
1867   Lyman (#2307)
1869   Beecher (#274)
1872   Ellett (#1049)
1873   Beecher (#277, 282)
1873   Ellett (#1050)
1880   Farrar (#1135)
1880   Fowler, C.H. (#1270)
1881   Carlin (#563)
1889   Logan (#2272)
1905   Jefferis (#1991)

HOMEOPATHY

1834   *Allgemeine verstandliche Belehrung* (#71)
1837   Hering (#1633)
1839   Curie (#851)
1842   Dunsford (#1016)
1842   Everest (#1090)
1842   Holmes (#1741)
1843   Epps (#1072)
1843   Laurie (#2205)
1844   Hering (#1634)
1845   Jahr (#1971)
1846   Hempel (#1616)
1848   Hering (#1625)
1849   Chepmell (#631)
1849   Epps (#1073)

HYDROPATHY

HYDROTHERAPY
(see **BALNEOLOGY, HYDROPATHY**)

HYGIENE
(see also **CHILD CARE, HYGIENE, JUVENILE LITERATURE, WOMEN'S HEALTH**)

JUVENILE LITERATURE
(see **HYGIENE—JUVENILE LITERATURE,**
PHYSIOLOGY—JUVENILE LITERATURE)

MARRIAGE
(i.e., duties & responsibilities;
see also **SEX EDUCATION**)

1808	Jennings (#2007)
1837	Alcott (#57)
1840	Alcott (#58)
1842	Becklard (#263)
1842	Fowler, L.N. (#1276)
1843	Becklard (#265, 269)
1844	Becklard (#264, 269.1)
1844	Chabot (#580)
1845	Becklard (#266, 267)
1846	Becklard (#270)
1848	Fowler, O.S. (#1279, 1291)
1853	Josselyn (#2050)
1856	Alcott (#40-42)
1857	Alcott (#43)
1858	Foote (#1213)
1859	Alcott (#44)
1859	Foote (#1214)
1859	Fowler, O.S. (#1285)
1860	Alcott (#36)
1860	Ellis (#1056, 1059)
1860	Foote (#1215, 1216)
1866	Alcott (#45)
1866	Ellis (#1057)
1866	Foote (#1217)
1867	Fancher (#1128)
1867	Foote (#1218)
1868	Fowler, O.S. (#1292)
1869	Cowan (#808)
1869	Groves (#1439)
1870	Cowan (#809)
1870	Foote (#1219)
1870	Fowler, O.S. (#1300)
1870	Gardner (#1327, 1328)
1870	Jackson (#1903, 1904)
1871	Cooke (#779)
1871	Cowan (#810)
1871	Foote (#1220)
1872	Foote (#1221)
1872	Groves (#1440)
1872	Holland (#1696)
1873	Chavasse (#622)
1873	Duffey (#996)
1873	Foote (#1222)
1874	Cooke (#780)
1874	Davis, A.J. (#900)
1874	Foote (#1223, 1235)
1875	Faulkner (#1137)
1875	Fowler, O.S. (#1280)
1876	Duffey (#994)
1877	Cooke (#781)
1877	Foote (#1224)
1878	Cowan (#811)

188-?	Cowan (#812-815)
1880	Fowler, O.S. (#1299)
1881	Cooke (#782)
1882	Guernsey, H.N. (#1448)
1882	Holbrook (#1688)
1884	Foote (#1225)
1885	Duffey (#995)
1885	Foote (#1236)
1885	Greene, C.A. (#1410, 1411)
1885	Landis (#2194)
1886	Kinget (#2140)
1887	Jones, J.H. (#2041)
1887	Longshore-Potts (#2279)
1888	Foote (#1226)
1888	Lind (#2257)
1889	Bate (#232)
1889	Cowan (#818, 819)
1889	Foote (#1227)
189-?	Conger (#760)
189-?	Cowan (#816, 817)
189-?	Fowler, O.S. (#1281)
1890	Foote (#1228)
1890	Hoff (#1658)
1890	Kellogg (#2087)
1891	Beaumont (#261)
1891	Bigelow (#318)
1891	Capp (#556)
1891	Cowan (#820)
1891	Foote (#1229, 1237)
1892	Foote (#1230)
1893	Capp (#557)
1893	Hoff (#1659)
1894	Foote (#1231)
1894	Jefferis (#1993, 1995)
1895	Jefferis (#1994)
1896	Foote (#1232)
1896	Kellogg (#2088)
1897	Cowan (#821)
1897	Jefferis (#1996)
1899	Foote (#1233, 1238)
1899	*Confidential med. talks to women* (#759.1)
1900	Foote (#1239)
1901	Foote (#1246)
1901	Galbraith (#1319)
1902	Carpenter, E. (#566)
1902	Drake (#987)
1902	Foote (#1234)
1903	Cowan (#822)
1903	Foote (#1240)
1904	Beall (#253)
1904	Foote (#1247)
1904	Jefferis (#1997, 1998)
1904	Kellogg (#2089)
1906	Dressler (#991)
1907	Faulkner (#1138)
1908	Hoffmann (#1661)

MASTURBATION
(see also **SEX DISORDERS, SEX EDUCATION**)

MEDICINE
(i.e., home diagnosis & treatment of disease;
excludes **BOTANIC MEDICINE, ECLECTIC
MEDICINE, HOMEOPATHY, HYDROPATHY**)

MENTAL HEALING & SPIRITUALISM

MIDWIFERY
(i.e., labor, delivery, puerperium)

MUSEUMS, ANATOMICAL

NOSTRUMS see PROPRIETARY MEDICINES

OBSTETRICS see MIDWIFERY

ORAL HYGIENE

PEDIATRICS
see INFANT CARE and CHILD CARE

PHYSICAL FITNESS see EXERCISE & EXERCISE THERAPY

PHYSIOLOGY
(see also PHYSIOLOGY—JUVENILE LITERATURE)

PHYSIOLOGY—JUVENILE LITERATURE

PLANTS, MEDICINAL see **BOTANIC MEDICINE**

PREGNANCY
(i.e., hygiene, management, diseases)

PROPRIETARY MEDICINES & THERAPIES
(advertisements & promotional literature)

REFORMED PRACTICE see **ECLECTIC MEDICINE**

SEX DISORDERS
(e.g., impotence, spermatorrhoea;
see also **MASTURBATION**; **SEX EDUCATION**)

SEX EDUCATION
(i.e., reproduction, sexual anatomy, physiology
and hygiene; see also **CONTRACEPTION; SEX
ETHICS, SEXUALLY TRANSMITTED DISEASES**)

SEXUAL ETHICS AND BEHAVIOR
(see also **MARRIAGE**)

SEXUALLY TRANSMITTED DISEASES

SPERMATORRHEA
(see **SEX DISORDERS**)

TEMPERANCE
(see also **PHYSIOLOGY—JUVENILE
LITERATURE**)

THOMSONIAN MEDICINE see BOTANIC MEDICINE.

TOBACCO
(see also **HYGIENE—JUVENILE LITERATURE**)

TUBERCULOSIS, PULMONARY

VEGETARIANISM

WATER-CURE see **HYDROPATHY**.

WOMEN'S DISEASES
(see also **MEDICINE**)